DMU 0510304 01 1

AF846212

WITHDRAWN
DE MONTFORT UNIVERSITY
LIBRARY

Studies in Surface Science and Catalysis 137

INTRODUCTION TO ZEOLITE SCIENCE AND PRACTICE

2nd COMPLETELY REVISED AND EXPANDED EDITION

Studies in Surface Science and Catalysis

Advisory Editors: B. Delmon and J.T. Yates

Vol. 137

INTRODUCTION TO ZEOLITE SCIENCE AND PRACTICE

2nd COMPLETELY REVISED AND EXPANDED EDITION

Edited by

H. van Bekkum
Laboratory of Organic Chemistry and Catalysis
Delft University of Technology, 2628 BL Delft, The Netherlands

E.M. Flanigen
UOP Research and Molecular Sieve Technology,
Tarrytown Technical Center, Tarrytown, NY 10591, USA

P.A. Jacobs
Katholieke Universiteit, Centrum voor Oppervlaktechemie en Katalyse
Heverlee, (Leuven), Belgium

J.C. Jansen
Laboratory of Organic Chemistry and Catalysis
Delft University of Technology, 2628 BL Delft, The Netherlands

2001
ELSEVIER
Amsterdam - London - New York - Oxford - Paris - Shannon - Tokyo

ELSEVIER SCIENCE B.V.
Sara Burgerhartstraat 25
P.O. Box 211, 1000 AE Amsterdam, The Netherlands

© 2001 Elsevier Science B.V. All rights reserved.

This work is protected under copyright by Elsevier Science, and the following terms and conditions apply to its use:

Photocopying
Single photocopies of single chapters may be made for personal use as allowed by national copyright laws. Permission of the Publisher and payment of a fee is required for all other photocopying, including multiple or systematic copying, copying for advertising or promotional purposes, resale, and all forms of document delivery. Special rates are available for educational institutions that wish to make photocopies for non-profit educational classroom use.

Permissions may be sought directly from Elsevier Science Global Rights Department, PO Box 800, Oxford OX5 1DX, UK; phone: (+44) 1865 843830, fax: (+44) 1865 853333, e-mail: permissions@elsevier.co.uk. You may also contact Global Rights directly through Elsevier's home page (http://www.elsevier.nl), by selecting 'Obtaining Permissions'.

In the USA, users may clear permissions and make payments through the Copyright Clearance Center, Inc., 222 Rosewood Drive, Danvers, MA 01923, USA; phone: (+1) (978) 7508400, fax: (+1) (978) 7504744, and in the UK through the Copyright Licensing Agency Rapid Clearance Service (CLARCS), 90 Tottenham Court Road, London W1P 0LP, UK; phone: (+44) 207 631 5555; fax: (+44) 207 631 5500. Other countries may have a local reprographic rights agency for payments.

Derivative Works
Tables of contents may be reproduced for internal circulation, but permission of Elsevier Science is required for external resale or distribution of such material.
Permission of the Publisher is required for all other derivative works, including compilations and translations.

Electronic Storage or Usage
Permission of the Publisher is required to store or use electronically any material contained in this work, including any chapter or part of a chapter.

Except as outlined above, no part of this work may be reproduced, stored in a retrieval system or transmitted in any form or by any means, electronic, mechanical, photocopying, recording or otherwise, without prior written permission of the Publisher.
Address permissions requests to: Elsevier Science Global Rights Department, at the mail, fax and e-mail addresses noted above.

Notice
No responsibility is assumed by the Publisher for any injury and/or damage to persons or property as a matter of products liability, negligence or otherwise, or from any use or operation of any methods, products, instructions or ideas contained in the material herein. Because of rapid advances in the medical sciences, in particular, independent verification of diagnoses and drug dosages should be made.

First edition 2001

Library of Congress Cataloging in Publication Data
A catalog record from the Library of Congress has been applied for.

DE MONTFORT UNIVERSITY LIBRARY:
B/code
Fund: 15 Date: 2/6/04
Sequence:
Class: 549.68
Suffix: INT

ISBN: 0-444-82421-9

The paper used in this publication meets the requirements of ANSI/NISO Z39.48-1992 (Permanence of Paper).
Printed in The Netherlands.

Dedication

Dedicated to Richard M. Barrer (left) and Donald W. Breck (right),

two great scientists and pioneers in the field of zeolites and related materials.

Preface 1st Edition

Zeolite science and technology are still in the stage of rapid growth. This is testified to

- the exponential increase of papers and patent applications in the field, reflecting the work of many zeolite scientists, both industrial as well as academic;

- the increasing number of industrial zeolite-based processes and their present rapid expansion into organic chemicals manufacture, following important applications of zeolites in hydrocarbon processing;

- the recent progress in matters of zeolite accessibility range, matrix behaviour, lattice components and satellite structures. In fact, the zeolite horizon seems even wider now than some ten years ago;

- the recognition that zeolites, by virtue of their stability and regenerability, are perfectly suited for incorporation into a new generation of processes in which clean technology and environmental friendliness are major issues.

Some sixteen years have passed since the monograph "Zeolite Molecular Sieves; Structure, Chemistry and Use" by the late Donald W. Breck was published. The wealth of information and insights offered by this book, together with its high quality, guarantee its further intensive use for many years to come. In addition, the student or professional working in the field of zeolites needs efficient access to the latest achievements and developments. Some excellent books which deal with parts of the zeolite field are those by R.M. Barrer, P.A. Jacobs and J.A. Martens, R. Szostak, and A. Dyer.

The present book was conceived during the Summer School on Zeolites, held in 1989 at Zeist, The Netherlands, on the occasion of the 8th International Conference on Zeolites which was attended by some ninety - mainly young - persons from many countries. This Pre-Conference School was held in an informal atmosphere and generated much enthusiasm among teachers and participants. The programme of lectures delivered by experts in the zeolite field seemed perfectly suited to serve as a basis for an introductory book on zeolites and related materials, covering theory and practice.

All authors expanded their Summer School lectures into a review of the sector or technique dealt with, starting at an elementary level but giving access to the latest results and insights and referring to recent publications and reviews.

The Editors are convinced that the present text will not only be useful to students and workers entering the field. More experienced workers in the field will find a substantial coverage of the zeolite spectrum, including the latest views on zeolite structure, characterization and applications. It is particularly the combination of science *and* technology which will make the book a source of useful information for those chemists and chemical engineers with an interest in zeolites.

We would like to express our gratitude to the authors who, despite their heavy work-loads, have been so co-operative. In particular we would like to acknowledge Mrs. Mieke van der Kooij-van Leeuwen for typing several chapters of this book into such a pleasing format. She proved to be a perfect coordinator between authors, the editors and publishers in the preparation of the book and in organizing the Pre-Conference School.

Herman (H.) van Bekkum

Edith (E.M.) Flanigen

Koos (J.C.) Jansen

Delft/Tarrytown/Delft, September 1990

Preface 2nd Edition

More than ten years passed since the appearance of the first Edition of "Introduction to Zeolite Science and Practice" which book contained the expanded lectures delivered at the Summer School on Zeolites, preceding the 8th International Zeolite Conference (IZC).

Over 1600 copies of the 1st Edition found their way to an owner and judging by citation numbers the book is still lively in use.

In view of the substantial progress made in the last decade in the fields of zeolites and related materials it was decided to go for an extended 2nd Edition. Unfortunately - as often is the case - this process took more time than expected by the Editors.

In the mean time some new texts on zeolites were issued. Thus Weitkamp and Puppe edited the book Catalysis and Zeolites. Fundamentals and Applications (1999). Also in the 5 volumes of the Handbook of Heterogeneous Catalysis (Eds. G. Ertl, H. Knözinger, J. Weitkamp, 1997) much information on zeolites, their synthesis and modification, and particularly their use as catalysts can be found.

Nevertheless, the combination of data, discussion and deduction provided by the present book is a unique coverage of the field in the opinion of the Editors and utmost useful especially for people entering - at university or in industry - the fascinating world of zeolites and related materials.

In the present Edition the number of chapters rose from 16 to 22. The contributions can be divided into three categories:

- updated chapters by the original author(s). We mention for instance the chapter of Steve Wilson on phosphate-based molecular sieves, of which the number of pages doubled and the number of references multiplied by ten. Also the chapter by Henk van Koningsveld was completely transformed and is now fully devoted to zeolite sub-structures and how to build zeolite models, an important activity for someone entering the field.
- updated chapters by an expanded or new team of authors. We're for instance most happy that Douglas Ruthven was willing to joint Martin Post in writing the chapter on "Diffusion in zeolite molecular sieves". The not so easy task of setting the scene - performed in the 1st Edition by Leo Moscou, Chairman of the 8th IZC, has been taken over by the American-Dutch duo Bonnie Marcus/Theo Maesen, both experts in the zeolite markets. One of the authors of the chapter on "Techniques of zeolite characterization", Jan Willem Roelofsen, a fine colleague, passed away in 1995, and Andy Jentys and Johannes Lercher succeeded in compiling a fine new chapter on this topic.
- completely new chapters. First we feel privileged that Lynn McCusker and Christian Baerlocher contributed a new chapter on "Zeolite Structures". The Zürich couple is serving the zeolite community by providing splendid and up-to-date Internet pages containing zeolite structures on behalf of the International Zeolite Association (IZA). Another vacancy in the 1st Edition was a chapter on adsorption. The Editors are happy that Lovat Rees and Dongmin Shen were willing to give their expert-view on "Adsorption of gases in zeolite molecular sieves".

The ever expanding use of zeolites in organic chemical conversions induced us to split the 1st Edition chapter of Wolfgang Hölderich and Herman van Bekkum into a chapter on Bronsted and Lewis catalysis - by the authors mentioned - and a new chapter on oxidation catalysis. Two specialists, Giusseppe Bellussi and Marcello Rigutto, wrote a fine chapter on redox molecular sieves.

This 2nd Edition also contains new chapters on "Zeolite-based supramolecuar assemblies" (by Dirk De Vos and Pierre Jacobs, experts in this area) and on "The use of bulky probe molecules" (by Paul Kunkeler, Roger Downing and one of the Editors).

Finally, the super large pore zeolites and the fast growing area of ordered mesoporous materials are dealt with by Eelco Vogt, Charlie Kresge and Jim Vartuli. The latter two authors belong to the discoverers of the M41S family of mesoporous materials.

The Editors express their sincere gratitude to all authors for providing so much zeolite science and practice. We thank Roger Downing for much help in finalizing one of the chapters.

As was the case with the 1st Edition Mrs. Mieke van der Kooij-van Leeuwen was the indispensable and efficient coordinator between authors and Editors. Her efforts are acknowledged highly.

We also thank Wim Jongeleen for the cover illustration.

Finally we thank Mrs. Huub Manten and Kostas Marinakis of Elsevier Science for their confidence and their help in preparing this book.

Herman (H.) van Bekkum
Edith (E.M.) Flanigen
Pierre (P.A.) Jacobs
Koos (J.C.) Jansen

Delft/White Plains/Leuven/Delft - April 2001

List of contributors

Ch. Baerlocher
Laboratorium fur Kristallographie, ETH, Zurich, Switzerland

H. van Bekkum
Laboratory of Applied Organic Chemistry and Catalysis, Delft University of Technology, Julianalaan 136, 2628 BL, Delft, The Netherlands

G.Bellussi
Eniricherche S.p.A., Via, F. Maritano 26, 20097 San Donato, Milano, Italy

R.S.Downing
Avantium Technologies B.V., Zekeringstraat 29, 1014 BV, Amsterdam

G. Engelhardt
Institute of Chemical Technology I, University of Stuttgart, D-70550 Stuttgart, Germany

E.M.Flanigen
UOP Research and Development White Plains, NY 10603-3136, U.S.A.

B. van de Graaf
Laboratory of Applied Organic Chemistry and Catalysis, Delft University of Technology, Julianalaan 136, 2628 BL, Delft, The Netherlands

W.F.Holderich
University of Technology RWTH Aachen, Department of Chemical Technology and Heterogeneous Catalysis, Worringerweg 1, 52074 Aachen, Germany

K.Y. Jacobs
Department of Interphase Chemistry, Center for Surface Chemistry and Catalysis, K.U.Leuven, Kasteelpark Arenberg, 23, B-3001 Leuven, Belgium

P.A. Jacobs
Centrum voor Oppervlaktechemie en Katalyse, K.U.Leuven, Kard.Mercierlaan 92, B3001, Heverlee, Belgium

J.C.Jansen
Laboratory of Applied Organic Chemistry and Catalysis, Delft University of Technology, Julianalaan 136, 2628 BL, Delft, The Netherlands

A.Jentys
Technische Universitat Munchen, Lehrstuhl fur Technische Chemie II, Lichtenbergstrasse 4, D-85748 Garching, Germany

H.G.Karge
Fritz-Haber-Institut der Max-Planck-Gesellschaft, Faradayweg 4-6, 1000 Berlin 33 West, Germany

H. van Koningsveld
Laboratory of Applied Organic Chemistry and Catalysis, Delft University of Technology, Julianalaan 136, 2628 BL, Delft, The Netherlands

H.W.Kouwenhoven
Technisch Chemisches Laboratorium, ETH-Zentrum, 8092 Zuerich, Switzerland

Ch.T. Kresge, The Dow Chemical Company, Corporate R&D, 1776 Building, Midland, Michigan 48674, U.S.A.

B. de Kroes
Akzo Nobel Chemicals, Research Center amsterdam, PO Box37650, 1030 Amsterdam, The Netherlands

P.J.Kunkeler
Akzo PQ Silica, Ankerkade 111, 6222 NL Maastricht, The Netherlands

J.A.Lercher
Technische Universitat Munchen, Lehrstuhl fur Technische Chemie II, Lichtenbergstrasse 4, D-85748 Garching, Germany

B.Marcus
Zeolyst International, Conshohocken, PA, U.S.A.

Th.L.M. Maesen
Chevron Research and Technology Co.
100 Chevron Way 10-1512
Richmond, CA 94802, U.S.A.

J.A.Martens
Centrum voor Oppervlaktechemie en Katalyse, K.U.Leuven, Kard.Mercierlaan 92, B3001, Heverlee, Belgium

I.E.Maxwell
Avantium Technologies B.V., Zekeringstraat 29, 1014 BV, Amsterdam

L.B.McCusker
Laboratorium fur Kristallographie, ETH, Zurich, Switzerland

M.F.M.Post
Shell International Oil Products B.V., Shell Research and Technology Center, Amsterdam, Badhuisweg 3, 1031 CM, Amsterdam, The Netherlands

L.V.C.Rees
Department of chemistry, University of Edinburgh, West Mains Road, Edinburgh EH9 3JJ, Schotland, UK

D.M.Ruthven
Department of Chemical Engineering, University of Maine, Orono, ME 04469-5737, U.S.A.

M.S.Rigutto
Shell Global Solutions International b.v., P.O.Box 38000, 1030 BN Amsterdam, The Netherlands

R.A. van Santen
Schuit Institute of Catalysis, eindhoven, University of Technology Eindhoven, P.O. Box 513, 5600 MB Eindhoven, The Netherlands

R.A.Schoonheydt
Department of Interphase Chemistry, Center for Surface Chemistry and Catalysis, K.U.Leuven, Kasteelpark arenberg, 23, B-3001 Leuven, Belgium

D.Shen
The BOC Gases Technical Center, 100 Mountain Avenue, Murray Hill, NJ 07974, U.S.A.

B.Smit
Department of Chemical Engineering, University of Amsterdam, Nieuwe Achtergracht 166, 1018 WV, Amsterdam, The Netherlands

W.H.J.Stork
Shell Research and Technology Center, Amsterdam, Badhuisweg 3, 1031 CM,

R.Szostak
Department of Chemistry, Clark Atlanta university, 223 James P. Brawley Dr. SW, Atlanta, GA 30314 U.S.A.

R.P.Townsend
Chemistry Department, University of Southampton, Highfield, Southampton SO17 1BJ, UK.

J.C.Vartuli
Mobil Technology Company, Paulsboro Technical Center, Paulsboro, New Jersey 08066, U.S.A.

E.T.C.Vogt
Akzo Nobel Chemicals, Research Center amsterdam, PO Box37650, 1030 Amsterdam, The Netherlands

D. E. de Vos
Centrum voor Oppervlaktechemie en Katalyse, K.U.Leuven, Kard.Mercierlaan 92, B3001, Heverlee, Belgium

S.T.Wilson, UOP research and Development, UOP LLC, 25E. Algonquin Rd., Des Plaines, Il, USA

Contents

Dedication v
Preface vii
List of contributors xi

Chapter 1. The zeolite scene **1**
Th.L.M. Maesen and B. Marcus

Introduction 1
Major uses for synthetic zeolites 2
Markets 3
Research 5
Zeolite scene 7
References 8

Chapter 2. Zeolites and molecular sieves. An historical perspective **11**
E.M.Flanigen

Review of molecular sieves and zeolite fundamentals 11
Early history 15
Industrial history 16
History of molecular sieve materials 17
History of synthesis 27
The future 30
History of international conferences and organizations 31
Historical epilogue 31
References 32

Chapter 3. Zeolite structures **37**
L.B.McCusker and C. Baerlocher

Introduction 37
Zeolite framework types 38
Zeolite structures 54
Powder diffraction 58
Conclusions 65
References 65

Chapter 4. How to build zeolites **69**
H. van Koningsveld

Introduction 69
Building units 70
Zeolites containing zigzag chains 79
Zeolites containing saw chains 100
Zeolites containing crankshaft chains 111
Building schemes for the fibrous zeolites 127

Building schemes for the ABC family of zeolites and some related zeolites 129
The framework structure of disordered zeolites 138
References 171

Chapter 5. The preparation of oxide molecular sieves
A. Synthesis of zeolites **175**
J.C.Jansen

Introduction 175
Preparation 181
Zeolite product versus synthesis mixture 188
The low temperature reaction mixture 190
The temperature raise of the reaction mixture 196
The high temperature reaction process 198
Isolation of the zeolite product 204
Reaction parameters 205
All silica molecular sieves 211
Clathrasils 212
Examples of sytematic research 215
Literature sources pertaining zeolite preparation aspects 224

B. Phosphate-based molecular sieves **229**
St.T.Wilson

Introduction 229
Structural features 232
Synthesis 235
Conclusions 253
References 254

Chapter 6. Secondary synthesis methods **261**
R.Szostak

Introduction 261
Proton exchange: removal of aluminum from the framework 263
Cracking catalyst development: zeolite Y 265
Developing an acidic Y zeolite 266
Improving thermal stability 267
Ultrastable Y 267
Characteristics of ultrastabilized Y 270
Effect of sodium content 272
Other secondary synthesis methods 273
Summary of physical changes from various secondary synthesis methods 275
Mechanism of secondary synthesis 276
Mesopores 276
Framework and non-framework Al species 277
Details of key secondary synthesis methods 278
Catalytic properties and secondary synthesis methods 280

And than there is reality 280
Modification of high silica zeolites 281
Application of secondary methods to metallosilicates 283
Alumination 284
Insertion of other elements into silicate frameworks 286
References 288

Chapter 7. Clays: from two to three dimensions **299**
R.A.Schoonheydt and K.Y.Jacobs

Introduction 299
Smectites 301
Properties of smectites 303
Pillared clays 319
Conclusion 337
References 338

Chapter 8. Techniques of zeolite characterization **345**
A.Jentys and J.A.Lercher

Introduction 345
Lattice and pore structure 347
HTEM 350
Vibrational spectroscopy 351
Determination of sorption isotherms 353
Assessment of the heat of sorption 357
^{129}Xe NMR 358
Model reactions 359
Overall chemical composition 362
Determination of the framework Si/Al ratio 364
Determination of the zeolite acid-base functionality 369
Concluding remarks 382
References 383

Chapter 9. Solid state NMR spectroscopy applied to zeolites **387**
G.Engelhardt

Introduction 387
Basic principles and methods of high resolution solid-state NMR 388
Multinuclear high-resolution solid-state NMR of zeolites 392
NMR of aluminophosphate molecular sieves 414
References 415

Chapter 10. Introduction to zeolite theory and modelling **419**
R.A. van Santen, B. van de Graaf, and B. Smit

Introduction 419
Quantum-chemistry of zeolite acidity 420
Molecular dynamics 434
Monte Carlo simulations 448

Concluding remarks 460
References 461

Chapter 11. Ion exchange in zeolites **467**
R.P. Townsend and E.N. Coker

Introduction 469
A picture of the ion-exchange system 471
Equilibrium aspects of ion exchange in zeolites 476
Kinetics of ion exchange 492
Applications of zeolites in ion exchange processes 503
Ion exchange as a method of modification 510
Recent trends in ion exchange 512
Concluding remarks 518
References 519

Chapter 12. Diffusion in zeolite molecular sieves **525**
D.M. Ruthven and M.F.M. Post

Introduction 527
General principles 528
Theoretical prediction of intracrystalline diffusivities 532
Measurement of intracrystalline diffusion 532
Review of experimental diffusivity data for selected systems 552
Diffusion limitations in zeolite catalysis 566
Conclusions 569
Bibliography 570
References 572

Chapter 13. Adsorption of gases in zeolite molecular sieves **579**
L.V.C.Rees and D. Shen

Introduction 579
Thermodynamics of gas adsorption 580
Energetics of adsorption 586
Adsorption of mixtures 588
Experimental methods 591
Results and discussion 595
References 629

Chapter 14. Introduction to acid catalysis with zeolites in hydrocarbon Reactions **633**
J.A.Martens and P.A.Jacobs

Introduction 633
Alkylcarbenium and alkylcarbonium ions 634
Carbocations in zeolites 638
Reaction mechanisms of acid catalyzed conversions 644
Shape-selective catalysis in acid zeolites 663
References 669

Chapter 15. Preparation of zeolite catalysts **673**
H.W.Kouwenhoven and B. de Kroes

Introduction 673
Catalyst preparation procedure 674
Fluidized catalytic cracking (FCC) catalysts 691
Conclusion 703
References 703

Chapter 16. Coke formation on zeolites **707**
H.G.Karge

Introduction 707
Methods of coke investigation and results 709
IR spectroscopic investigations of coke formation 713
NMR spectroscopy 718
ESR spectroscopy 720
UV-vis spectroscopy 723
Photonelectron spectroscopy 726
Extraction of coked zeolite catalysts 727
Adsorption measurements 730
Particular phenomena related to coke formation 731
Effect of acidity 733
Mechanism and kinetics 734
Localization of coke deposition 736
Summary and concluding remarks 740
References 742

Chapter 17. Hydrocarbon processing with zeolites **747**
I.E.Maxwell and W.H.J.Stork

Introduction 747
Oil conversion processes 752
Basics of oil refining 752
Paraffin hydroisomerization 756
Skeletal isomerization of olefins 763
Olefin/paraffin alkylation 765
Gasoline-upgrading processes 765
Middle distillate upgrading processes 768
Catalytic dewaxing 769
Catalytic cracking 775
Hydrocracking 785
Residue hydroprocessing 797
Gas-conversion processes 801
Basics of gas conversion 801
Conversion of Methanol to Gasoline 802
Methanol conversion to lower olefins 805
Conversion of synthesis gas to gasoline 807
LPG conversion to aromatics 808
Olefin oligomerization 809

Conclusions and future developments 811
References 813

Chapter 18. Zeolites and related materials in organic syntheses. Bronsted and Lewis catalysis **821**
W.F.Holderich and H. van Bekkum

Introduction 821
General aspects 822
Non-catalytic use of zeolites in organic synthesis 829
Zeolite-catalyzed isomerizations 834
Electrophilic substitution of arenes 849
Zeolite catalytzed cyclizations 868
Nucleophilic substitution and addition 879
MPV reduction and oppenauer oxidation 890
Zeolite-catalyzed two- and multistep syntheses 892
Final remarks 896
References 897

Chapter 19. Metal ions associated to molecular sieve frameworks as catalytic sites for selective oxidation reactions **911**
G.Bellussi and M.S.Rigutto

Introduction 911
Titanium-containing silicalites 914
Vanadium-containing silicalites 934
Other transition metal-containing silicalites 943
Transition metal-containing aluminophosphates 943
Conclusive remarks 946
References 947

Chapter 20. Zeolite-based supramolecular assemblies **957**
D.E. de Vos and P.A.Jacobs

Supramolecular assembly 958
Catalysis with ship-in-the-bottle systems 966
Interaction of ship-in-thebottle systems with light 970
Cluster formation in zeolite environments 975
Polymers in zeolites 978
References 979

Chapter 21. The use of bulky molecules as probes for investigating the contributions of the external and internal pore-wall activities of zeolite catalysts **987**
P.J.Kunkeler, R.S.Downing and H. van Bekkum

Introduction 987
Selection of a probe molecule 992
Use of probe molecules for physico-chemical investigation 992
Probe molecules as substrates 993

Concluding remarks 1000
References 1000

Chapter 22. Beyond twelve membered rings **1003**
E.T.C.Vogt, C.T. Kresge and J.C. Vartuli

History 1005
Preparation 1009
Analysis 1009
Computer modeling 1013
Catalytic tests 1013
Applications 1018
Conclusion 1022
References 1022

Appendix 1029

Keyword index 1041

Studies in Surface Science and Catalysis 137
H. van Bekkum, E.M. Flanigen, P.A. Jacobs and J.C. Jansen (Editors)
© 2001 Elsevier Science B.V. All rights reserved.

Chapter 1

The Zeolite Scene – An Overview

Theo Maesen, Bonnie Marcus,
Zeolyst International, Conshohocken, PA, U.S.A.

Introduction:

Since the first edition of "Introduction to Zeolite Science and Practice" was published in 1991 (1), there has been continued growth in the area of zeolite science. The 12th International Zeolite Conference held in 1998 had 774 attendees from 37 countries; there were 566 papers and recent progress reports submitted. The increasing number of people attending conferences and studying zeolites is just one indication that these materials are finding more and more commercial uses in a variety of diverse fields.

The topics necessary to develop an understanding of zeolite science and practice are covered in depth in subsequent chapters of this publication. This chapter is to help familiarize the reader with the broad range of current applications for synthetic zeolites and zeolite-like materials and to look at areas where they may find commercial applicability in the future. It will also briefly cover areas that are of current academic interest.

A zeolite is a crystalline aluminosilicate with a three-dimensional framework structure that forms uniformly sized pores of molecular dimensions. As the pores preferentially adsorb molecules that fit snugly inside the pores and exclude molecules that are too large, they act as sieves on a molecular scale. Thus, zeolites are a subset of molecular sieves. They consist of robust, crystalline silica (SiO_2) frameworks. At some places in the framework Al^{3+} has replaced Si^{4+} and the framework carries a negative charge. Loosely held cations that sit within the cavities preserve the electroneutrality of the zeolite. Some of those cations are amenable to cation exchange and are able to reversibly interact with polar molecules. These properties have contributed significantly to the commercial success of zeolites (1, 2).

The use of molecular sieves with three-dimensional framework structures is well entrenched in areas as diverse as laundry detergents, oil refining and petrochemical industries, adsorbents, gas separations, agriculture and horticulture(2). Ever since the successful introduction of aluminosilicate molecular sieves (zeolites) in the late nineteen fifties (3), zeolites and the more recently discovered (silico)aluminophosphate molecular sieves have continuously improved current application areas and generated new ones (4-6).

Major Uses for Synthetic Zeolites:

When a zeolite framework contains an equal number of aluminum and silicon atoms, each oxygen atom is linked to one aluminum and one silicon atom and the cavities contain the maximum density of exchangeable cations. Synthetic zeolites with such a maximum cation exchange capacity are of interest as ion exchangers and adsorbents. In detergents, the largest ion exchange market for zeolites, the cation exchange capacity determines how well the zeolite can replace the ("hard") calcium and magnesium cations in the wash water with ("soft") sodium cations. This impedes the precipitation of calcium or magnesium surfactant salts, which result in a dull or unclean look. As an adsorbent, maximum non-framework cation density increases the extent to which the sieves are able to hold on to polar adsorbates.

In catalytic applications, it is desirable to have a more siliceous framework with cationic protons residing at well-separated exchange sites. The high silica content of the framework makes it resistant to the high temperatures that occur during the catalytic and regeneration cycles. A high dispersion of acidic protons assures that each proton has the maximum acid strength (7). A great deal of proprietary industrial research is done to try to modify these acidic sites and to tailor them for specific applications.

In addition, there is an on-going search for new molecular sieve structures because, a small change in the molecular dimension of the regular array of channels and cavities can determine its success or failure in an adsorption or catalytic conversion application. The molecular structure of the zeolite can lead to shape selective conversions by imposing steric constraints on the behavior of the adsorbed molecules (8) and by enhancing the formation of molecules with a shape commensurate with that of the pores (9).

Reflecting the importance of the optimum pore size and shape to adsorption and catalytic applications, the number of commercially synthesized molecular sieve structures (6) has more than doubled in the last decade. Assisted by the application of increased computing power to structure resolution (10-12), the number of known zeolite and silicoaluminophosphate structures is on a steady increase (13). The Structure Commission of the International Zeolite Association has compiled the majority of the known zeolite and other molecular sieve structures and has assigned official three-letter codes for the known structures (14). Currently this database contains some 130 different structure-types. Separately, a more inclusive compilation of molecular sieve data is available that cross-references structures with multiple unofficial names (15).

Most zeolites are synthesized by dissolving a source of alumina and a source of silica in a strongly basic aqueous solution. Ultimately, the solubility, the silica-to-alumina ratio, the nature of the cation and the synthesis temperature of the resultant gel determine what structure is formed (16). The aluminophosphate molecular sieves are formed by dissolving a source of alumina and a source of phosphate in an acidic aqueous solution. An amine or quaternary ammonium salt, may be used as a structure directing aid. Most of the current panoply of molecular sieve structures was obtained by screening a large range of organic cations (14, 15). However, it should be possible to further expand the field of synthesis done without the organic cations that are currently so popular, since molecular sieves with quite complex structures have been found in mineral deposits (17, 18).

Throughout the years that zeolites have been used commercially, the health aspects of these materials have been extensively studied. To date, the commercial materials have shown no adverse health affects. However, zeolite minerals with fibrous crystal morphologies have been found to be extremely powerful carcinogens (19). These fibrous zeolite minerals seem to require the assistance of a transition metal to reach their full lethal potential (20). There are no commercial synthetic materials that have a fibrous morphology.

Markets:

Much of the study of basic zeolite science was done on natural zeolites (3), but the need to assure reproducibility and a steady supply, has led to a commercial business with many manufacturers supplying synthetic zeolites (2). In 2000, there are approximately 130 molecular sieve structures registered with the Structure Commission of the International Zeolite Association (14); of those, only about 16 are of commercial interest and produced synthetically, *viz.* FAU, LTA, GIS, MFI, MOR, BEA, FER, LTL, MTT, TON, AEL, MWW, CHA, AFY, EDI, MER (2, 6, 21). In addition, one microporous crystalline silicotitanate (TAM-5) has been commercialized (6, 22, 23).

China and Cuba consume the largest quantity of natural zeolites; they are reported to use some 3 million tons per annum primarily to enhance the strength of cement (2). In the U.S.A., Western and Eastern Europe and Japan, natural zeolite consumption is 0.2 million tons per annum (2) - only 15% of the total annual zeolite consumption of 1.4 -1.6 million tons (2, 24) (Figure 1). In these regions zeolite minerals (particularly HEU-type zeolites (2)) are commodities. They serve as a nutrient release agent in agriculture and horticulture, as an odor control agent in animal husbandry, as pet litter and as a soil conditioner for golf course amendment and as ion exchangers to remove radioactive isotopes from the ground water and reactor effluents (2, 6).

In the U.S.A., Western and Eastern Europe and Japan, the largest tonnage of commercial zeolite sold is of the LTA-type (4A, NaA), primarily due to its use as a replacement for sodium tripolyphosphate (STPP) as a water softener in laundry detergents. (Figure 1, 2) (1, 2, 24). In the early 1980's, it was found that high concentrations of phosphate compounds in lakes, streams and rivers was causing eutrophication leading to large growths of algae on the surface of the water. The pressure to replace STPP by NaA zeolite was environmentally

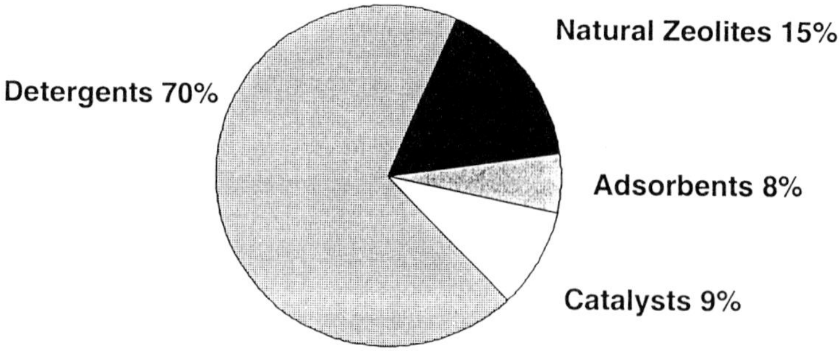

Figure 1: Estimated annual zeolite consumption (wt-% of total)[2] by the major individual applications.

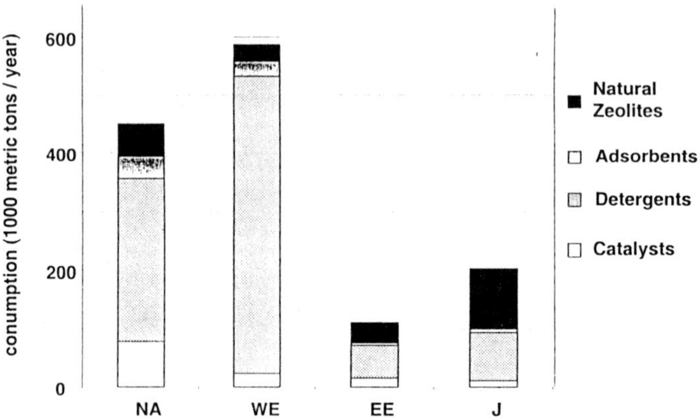

Figure 2: Regional estimated annual zeolite consumption (1000 metric tons / year). NA North America, WE Western Europe, EE Eastern Europe, J Japan.

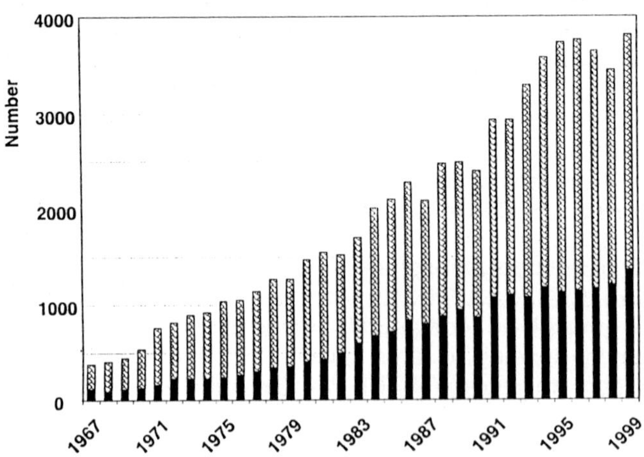

Figure 3: Number of publications on molecular sieves as patents (■), and in journals (▧).

driven, but due to their superior absorbency, they are preferred in concentrated detergents even where there are no phosphate restrictions.

Although they are not the largest volume use, the highest market value for synthetic zeolites is in catalysts. The synthetic zeolite that is the least expensive, FAU-type zeolite, accounts for more than 95% of the catalysis market (1, 2). Oil refineries consume large quantities of the FAU-type zeolites to manufacture gasoline from crude oil in the FCC process. Some refineries use additives such as MFI-type or less frequently BEA-type zeolites to the FCC process to optimize yield (1, 2). Due to this use, MFI-type zeolites are the second-most-used zeolite catalyst (1, 2, 4).

FAU-type zeolites are more expensive than LTA-type zeolites because they require considerable modification after synthesis before they can be added to the FCC process (1). Modified MFI-type zeolites are more expensive than modified FAU-type zeolites, because the latter can be synthesized at atmospheric pressure, whereas the former requires higher temperatures, and therefore, an autoclave or pressure cooker. When an organic structure directing material is used, the synthesis prices go up even further and the volume consumption tends to go down. Nonetheless there appears to be a small but flourishing market for high-end specialty zeolites that can be made this way.

The increase in the number and availability of hydrophobic zeolites such as silicalite, hydrophobic Y and beta have opened a market in adsorption of organic materials in applications such as automobile exhaust cleanup, volatile organic compound (VOC) abatement and specialty gas cleanup. This is an evolving area that will continue to grow as environmental restrictions are tightened.

Research:

In 1756, Axel F. Cronstedt christened hydrated aluminosilicate minerals “zeolites” (from the Greek zein (to boil) and lithos (stone)), because they formed a frothy mass when heated in the blowpipe (25). For the next 200 years, research efforts were sparse in part due to the limited availability of material (3). In the 1930’s Richard Barrer started systematic studies into zeolite synthesis under high pressure and temperature (3). Barrer’s results provided the impetus for research at Union Carbide that culminated in the late 1940’s in a route to synthetic zeolites at less extreme conditions (3, 16), and in the commercial success of the synthetic FAU-type zeolite in oil refining in the late 1950’s (3, 6). Research activities escalated when Mobil introduced organic cations into zeolite syntheses in the 1960’s, and started to discover the catalytic attractiveness of their MFI-type zeolites in the 1970’s (3). In the 1980’s, research activities were again escalated by the discovery of aluminophosphate molecular sieves and their derivatives, silico- metallo- metallo-silico- aluminophosphates by Union Carbide (3) and of titanosilicate molecular sieves by Eniricerche (26). By the 1990’s, research activity reached its current level of ~3600 publications a year (Figure 3), and the variety of actual and considered areas utilizing molecular sieves had grown phenomenally (2, 4, 5, 27). By that time even Mobil (28) and Toyota's major discovery of aluminosilicates with uniform mesopores did not incrementally increase the total number of yearly publications by more than 10%.

With the advent of uniformly mesoporous aluminosilicates that have many of the properties of amorphous materials, molecular sieve chemistry has gone full-circle and is now embracing the study of quasi-amorphous materials that zeolites have so successfully replaced in many markets. In principle, these aluminosilicates extend the size of molecules that can be shape-selectively processed into the 2-20 nm (i.e. mesopore) range. However, they lack short-range order and therefore also lack much of the desired catalytic and ion exchange properties of fully crystalline molecular sieves.

The search for stable materials that combine zeolite-like properties with larger pores continues. So far the (silico-)aluminophosphates that expand the pore size beyond 0.74 nm and towards the mesoporous domain (2-20 nm) have lacked thermal stability (29). Some new synthetic zeolites accessible through pores as large as 1 nm (UTD-1 (DON-type), SSZ-31) are thermally stable (14, 30, 31) and look promising. An alternative approach to increasing the accessibility of zeolite type acid sites is to maximize the site density exposed at the outside crystal surface. This can be accomplished by synthesizing zeolites that grow as very thin sheets, and subsequently keeping them separated by applying techniques such as pillaring (32, 33) and delamination (33, 34) borrowed from layered aluminosilicate molecular sieves (clays).

In addition to increasing the access to the zeolite-like acid sites, there is a search for molecular sieves that can do reactions other than acid-catalyzed ones. The discovery of titanosilicate molecular sieves (26) brought shape-selective oxidation reactions into the fold. The success of these sieves increased the efforts to incorporate more elements of the periodic table into molecular sieve structures (35-40). Another approach to producing molecular sieves with new catalytic functionalities is to encapsulate organometallic complexes (41) or organic functional groups (42) inside inorganic molecular sieves. This has led to the expansion of zeolite use into shape selective redox and base-catalyzed reactions (41, 42). It will be interesting to see how these partially organic molecular sieves perform when compared to the recently discovered molecular sieves with completely metal organic frameworks (43-46).

There is a significant research effort to introduce molecular sieves in new and novel areas such as sensors, membranes, optoelectronics , electrorheological fluids and functional nanomaterial fabrication (47-50). Generally, these applications require a very high degree of control over the morphology of individual crystals. Much progress in controlling the crystal growth and morphology has been made and is still being made, and this area of research is still growing rapidly. So far –to the authors' knowledge - these applications either cater to a tiny niche market or have not yet reached a commercial stage.

The formidable increase in computing power during the years that zeolites have been available has contributed to the rapid evolution of molecular modeling. Modeling becomes especially powerful if it is combined with the *in situ* study of chemical reactions (5154). Such a combination has recently shown how relatively mildly acidic zeolites (not stronger than an 80% sulfuric acid solution (55) are able to catalyze hydrocarbon conversions in much the same way as liquid superacids (significantly stronger than a 100% sulfuric acid solution) (51, 56-58). In both cases the stability of the protonated species appears to be more important than the ease with which the acid releases its proton.

Zeolite Science:
Though the number of patents and publications has remained relatively steady, the types of molecular sieves continue to proliferate and find their way in increasingly diverse applications. Reflecting this steady incline, there has been an increase in the number of scientists affiliated with zeolite science. In the last several years, independent zeolite associations have been started in many countries in Europe and the Far East. These groups have held their own meetings and symposia, and much collaborative work is being done among the universities in these areas. In addition, there are now workshops and sessions in the major catalysis, adsorption and environmental conferences devoted to zeolite science. In the face of the growth in the number of materials, in the number of applications and in the number of people studying zeolites, we can expect to see micropores playing an ever more important role in our society's technology, from housing construction to advanced, 21st century electronics.

It is safe to say that the future of zeolite science and technology is bright.

References

1 L. Moscou in "Introduction to Zeolite Science and Practice" (H. van Bekkum, E.M. Flanigen, J.C. Jansen, eds.), Stud.Surf.Sci.Catal. Vol.58, Elsevier, Amsterdam, 1991, pp.1-12.

2 M. Smart, T. Esker, A. Leder, K. Sakota in "The Chemical Economics Handbook", SRI international, pp.599.1000 A-599.1002 U.

3 E.M. Flanigen in "Introduction to Zeolite Science and Practice" (H. van Bekkum, E.M. Flanigen, J.C. Jansen, eds.), Stud.Surf.Sci.Catal. Vol.58, Elsevier, Amsterdam, 1991, pp.13-34.

4 K. Tanabe, W.F. Hölderich, Appl.Catal. 181 (1999) 399-434.

5 B.K. Marcus, W.E. Cormier, Chem.Eng.Prog. 95(6) (1999) 47-53.

6 J.D. Sherman, Proc.Natl.Acad.Sci. USA 96 (1999) 3471-3478.

7 G. Kühl in *Catalysis and Zeolites* (J. Weitkamp, L. Puppe, eds.), Springer, Berlin, 1999, pp. 81-197.

8 J. Weitkamp, S. Ernst, L. Puppe in *Catalysis and Zeolites* (J. Weitkamp, L. Puppe, eds.), Springer, Berlin, 1999, pp. 327-370.

9 M. Schenk, B. Smit, T.J.H. Vlugt, T.L.M..Maesen Angew. Chem. Int. Ed. 40(4) (2001) 736-739.

10 M.M. Treacy, M.J. Deem, J.M. Newsam, DIFFaX, Version 1.901 (1995).

11 R.W. Grosse-Kunstleve, L.B. McCusker, Ch. Baerlocher, Appl. Crystallogr. 30 (1997) 985-995.

12 M. Falcioni, M.W. Deem, J.Chem.Phys. 110 (1999) 1754-1766.

13 W.M. Meier, C. Baerlocher in Molecular Sieves (H.G. Karge, J. Weitkamp Eds.),Vol.2 (Structures and Structure Determination), Springer, Berlin, 1999, pp. 141-161.

14 a) Meier, W.M., Olson, D.H., and Baerlocher, Ch., "Atlas of Zeolite Structure Types", 4th, Elsevier, London, 1996; b) http://www.iza-structure.org/databases/ by Ch. Baerlocher, L.B. McCusker.

15 R. Szostak "Handbook of Molecular Sieves", 2nd Ed., Van Nostrand, New York, 1997.

16 D.W. Breck "Zeolite Molecular Sieves", Krieger, Malabar (Florida) 1984.

17 E. Galli, G. Vezzalini, S. Quartieri, A. Alberti, M. Franzini, Zeolites 19 (1997) 318-322.

18 R. Szostak in "Molecular Sieves, Science and Technology" Vol. 1 ("Synthesis"), H.G. Karge, J. Weitkamp eds., Springer, Berlin, 1998, pp. 157-185.

19 B. Baris, A.U. Demir, V. Shehu, Y. Karakoca, G. Kisacik, Y.I. Baris, J.Eniviron.Pathol., Toxicol. Oncol. 15(2-4) (1996) 183-189.

20 B.D. Hogg, P.K. Dutta, J.F. Long, A. Vaidyalingam in "Proceedings of the 12th international Zeolite Conference" (M.M.J. Treacy, B.K. Marcus, M.E. Bisher, J.B. Higgins eds.), Volume 4, Materials Research Society, Warrendal (PA), 1998, pp.2927-2933.

21 P.T. Barger, S.T. Wilson in Proceedings of the 12th international zeolite conference (M.M.J. Treacy. B.K. Marcus, M.E. Bisher, J.B. Higgins Eds.),Vol.1, Materials Research Society, Warrendale (PA), 1998, pp.567-590.

22 R.G. Anthony, D. Huckman, M. Gu, I. Latheef, C.V. Phillip, J.E. Miller, J.L. Krumhansl, D.E. Trudell, J.D. Sherman, D.J. Fennelly, T.J. Dangieri, 211th ACS National Meeting, New Orleans, LA, March 24-28 (1996) Book of Abstracts.

23 I.M. Latheef, M.E. Huckman, R.G. Anthony, Ind.Eng.Chem.Res. 39(5) (2000) 1356-1363.

24 A. Pfenninger in "Molecular Sieves" (H.G. Karge, J. Weitkamp Eds.),Vol.2 (Structures and Structure Determination), Springer, Berlin, 1999, pp. 164-199.

25 A.F. Cronstedt, Akad.Handl.Stockholm, 18 (1756) 120-130.

26 G. Perego, R. Millini, G. Bellussi in Molecular Sieves (H.G. Karge, J. Weitkamp Eds.),Vol.1 (Synthesis), Springer, Berlin, 1998, pp.188-228.

27 P. Espeel, R. Parton, H. Toufar, J. Martens, W. Hölderich, P.A. Jacobs in *Catalysis and Zeolites* (J. Weitkamp, L. Puppe, eds.), Springer, Berlin, 1999, pp. 377-436.

28 J.C. Vartuli, W.J. Roth, J.S. Beck, S.B. McCullen, C.T. Kresge in Molecular Sieves (H.G. Karge, J. Weitkamp Eds.),Vol.1 (Synthesis), Springer, Berlin, 1998, pp.97-119.

29 J.A. Martens, P.A. Jacobs in *Catalysis and Zeolites* (J. Weitkamp, L. Puppe, eds.), Springer, Berlin, 1999, pp. 53-80.

30 C.C. Freyhardt, M. Tsapatsis, R.F. Lobo, K.J. Balkus, M.E. Davis Nature(London) 381(6580) (1996) 295-298.

31 R.F. Lobo, M. Tsapatsis, C.C. Freyhardt, I.Chan, C.-Y. Cong, S.I. Zones, M.E. Davis J.Am.Chem.Soc. 119 (1997) 3732-3744.
32 W.J. Roth, C.T. Kresge, J.C. Vartuli, M.E. Leoniowicz, A.S. Fung, S.B. McCullen in "Catalysis by Microporous Materials" (H.K. Beyer, H.G. Karge, I. Kiricsi, J.B. Nagy, eds.), Stud.Surf.Sci.Catal. Vol.94, Elsevier, Amsterdam, 1994, pp.301-308.
33 A. Corma, U. Diaz, M.E. Domine, V. Fornes, Angew.Chem.Int.Ed. 39(8) (2000) 1499-1501.
34 A. Corma, V. Fornes, S.B. Pergher, Th.L.M. Maesen, J.G. Buglass Nature(London) 396 (1998) 353-356.
35 J.L. Guth in "Zeolite Microporous Solids: Synth., Struct., React." (E.G. Derouane, et al., eds.), Nato ASI Ser., Ser. C, Vol.352, 1992, pp.49-78.
36 T.E. Gier, X. Bu, P. Feng, G.D. Stucky, Nature(London) 395 (1998) 154-157.
37 S.A. Schunk, F. Schüth in "Molecular Sieves" (H.G. Karge, J. Weitkamp Eds.),Vol.1 (Synthesis), Springer, Berlin, 1998, pp. 229-263.
38 P.S. Halasyamani, S.M. Walker, D.O'Hare, J.Am.Chem.Soc. 121 (1999) 7415-7416.
39 H. Li, A. Laine, M. O'Keeffe, O.M. Yaghi, Science(Washington, D.C.) 283 (1999) 1145-1147.
40 J. Rocha, M.W. Anderson Eur.J.Inorg.Chem. (5) (2000) 801-818.
41 R.F. Parton, I.F.J. Vankelecom, M.J.A. Casselman, C.P. Bezoukhanova, J.B. Uytterhoeven, P.A. Jacobs, Nature(London) 370 (1994) 541-544.
42 C. Jones, T. Katsuyuki, M.E. Davis Nature(London) 393 (1998) 52-54.
43 D. Venkataraman, S. Lee, J. Zhang, J.S. Moore Natrue(London) 371 (1994) 591-593.
44 Y. Aoyama Top.Curr.Chem. 198 (Design of Organic Solids) (1998) 131-161.
45 H. Li, M. Eddaoudi, M. O'Keeffe, O.M. Yaghi, Nature(London) 402 (1999) 276-279.
46 B. Chen, M. Eddaoudi, S.T. Hyde, M. O'Keeffe, O.M. Yaghi, Science 291 (2001) 1021-1023.
47 D. Zhao, Y. Peidong, Q. Huo, B.F. Chmelka, G.D. Stucky Curr.Opin.Solid State Mater.Sci. 3 (1998) 111-121.
48 J. Caro, P. Kolsch, M. Noack, R. Schafer, I. Voigt, G. Fisher, P. Puhlfurss, H. Richter Chem.Tech. (Leipzig) 52 (2000) 23-27.
49 M. Tsapatsis, G. Xomeritakis, H. Hillhouse, S. Nair, V. Nikolakis, G. Bonilla, Z. Lai, CATTECH 3 (2000) 148-163.
50 K. Jansen, T. Maschmeyer, Ind.Catal.News (11) (1999) 2-10.
51 V.B. Kazansky in "Acidity and Basicity of Solids" (J. Fraissard, L. Petrakis eds.), Nato ASI Ser., Ser. C, Vol.444, 1994, pp.335-352.
52 J.B. Nicholas, T. Xu, J.F. Haw, Top.Catal. 6 (1998) 141-149.
53 H. Koller, G. Engelhardt, R.A. van Santen, Top.Catal. 9 (1999) 163-180.
54 H. Knözinger, S. Huber, J.Chem.Soc., Faraday Trans. 94 (1998) 2047-2059.
55 J.F. Haw, J.B. Nicholas, T. Xu, L.W. Beck, D.B. Ferguson, Acc.Chem.Res. 29 (1996) 259-267.
56 J.F. Haw, T.Xu, J.B. Nicholas, P.W. Goguen, Nature(London) 389 (1997) 832-835.
57 S.R. Blaszkowski, R.A. van Santen, Top.Catal. 4 (1997) 145-156.
58 V.B. Kazansky, Top.Catal. 11/12 (2000) 55-60.

Studies in Surface Science and Catalysis 137
H. van Bekkum, E.M. Flanigen, P.A. Jacobs and J.C. Jansen (Editors)
© 2001 Elsevier Science B.V. All rights reserved.

Chapter 2

Zeolites and Molecular Sieves: An Historical Perspective

Edith M. Flanigen, UOP Research and Development (retired)

White Plains, NY 10603-3136, U.S.A.

The history of zeolites and molecular sieves is reviewed from the discovery of the first zeolite mineral in 1756 through the explosion in new molecular sieve structures and compositions in the 1980's and 1990's.

R. M. Barrer's early pioneering work in adsorption and synthesis began the era of synthetic zeolites. The discovery of the commercially significant synthetic zeolites A, X and Y by R. M. Milton and D. W. Breck in the late 1940's to early 1950's led to their introduction by Union Carbide Corporation as a new class of industrial adsorbents in 1954, and in 1959 as hydrocarbon conversion catalysts. Today synthetic zeolites are used widely throughout the petroleum refining and chemical process industries as selective adsorbents, catalysts and ion exchangers, and represent an estimated nearly two billion dollar industry.

The last five decades have seen a chronological progression in molecular sieve materials from the aluminosilicate zeolites to the microporous silica polymorphs, the microporous aluminophosphate-based polymorphs, the metallosilicate and metallophosphate compositions, and to the octahedral-tetrahedral frameworks and mesoporous molecular sieves.

1. REVIEW OF MOLECULAR SIEVE AND ZEOLITE FUNDAMENTALS

Before detailing the history, a brief review of the fundamentals of zeolites and molecular sieves will be given, especially for those readers who are new to the field.

1.1. Molecular Sieves

Molecular sieves are porous solids with pores of the size of molecular dimensions, 0.3-2.0 nm in diameter. Examples include zeolites, carbons, glasses and oxides. Some are crystalline with a uniform pore size delineated by their crystal structure, e.g., zeolites. Others are amorphous, e.g., carbon molecular sieves. Most current commercial molecular sieves are zeolites. The pore size distribution for representative commercial adsorbent types (Fig. 1) contrasts the singular pore size of zeolites 5A and X or Y, with the broader distribution and larger mean pore diameter of the amorphous activated carbon and alumina adsorbents. The carbon molecular sieve has a mean pore diameter intermediate between that of 5A and X or Y zeolite but has a significantly broadened pore size distribution.

1.2. Zeolites

Zeolites are crystalline aluminosilicates of group IA and group IIA elements such as sodium, potassium, magnesium, and calcium (ref. 1). Chemically, they are represented by the empirical formula:

$$M_{2/n}O \bullet Al_2O_3 \bullet ySiO_2 \bullet wH_2O$$

where y is 2 to 10, n is the cation valence, and w represents the water contained in the voids of the zeolite. Structurally, zeolites are complex, crystalline inorganic polymers based on an infinitely extending three-dimensional, four-connected framework of AlO_4 and SiO_4 tetrahedra linked to each other by the sharing of oxygen ions. Each AlO_4 tetrahedron in the framework bears

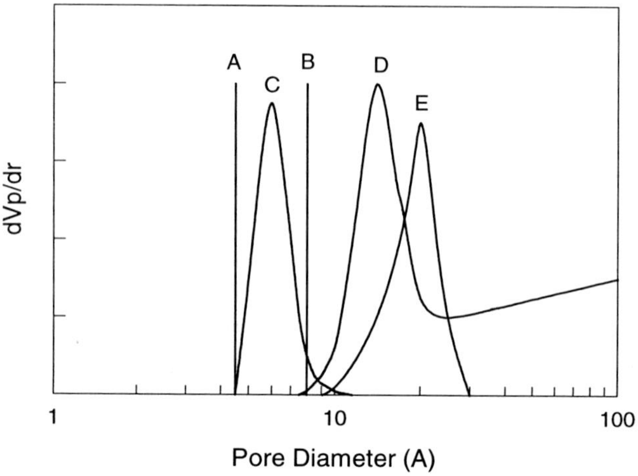

Figure 1. Pore size distribution of representative adsorbents. A) 5A (CaA) zeolite; B) X or Y zeolite; C) carbon molecular sieve; D) activated carbon; E) porous alumina

a net negative charge which is balanced by an extra-framework cation. The framework structure contains channels or interconnected voids that are occupied by the cations and water molecules. The cations are mobile and ordinarily undergo ion exchange. The water may be removed reversibly, generally by the application of heat, which leaves intact a crystalline host structure permeated by the micropores and voids which may amount to 50% of the crystals by volume.

The structural formula of a zeolite is based on the crystallographic unit cell, the smallest unit of structure, represented by:

$$M_{x/n}\ [(AlO_2)_x(SiO_2)_y] \bullet wH_2O$$

where n is the valence of cation M, w is the number of water molecules per unit cell, x and y are the total number of tetrahedra per unit cell, and y/x usually has values of 1-5. In the case of the high silica zeolites y/x is 10 to 100. There are two types of structures: one provides an internal pore system comprised of interconnected cage-like voids; the second provides a system of uniform channels which, in some instances, are one-dimensional channel systems. The preferred type has two- or three-dimensional channels to provide rapid intracrystalline diffusion in

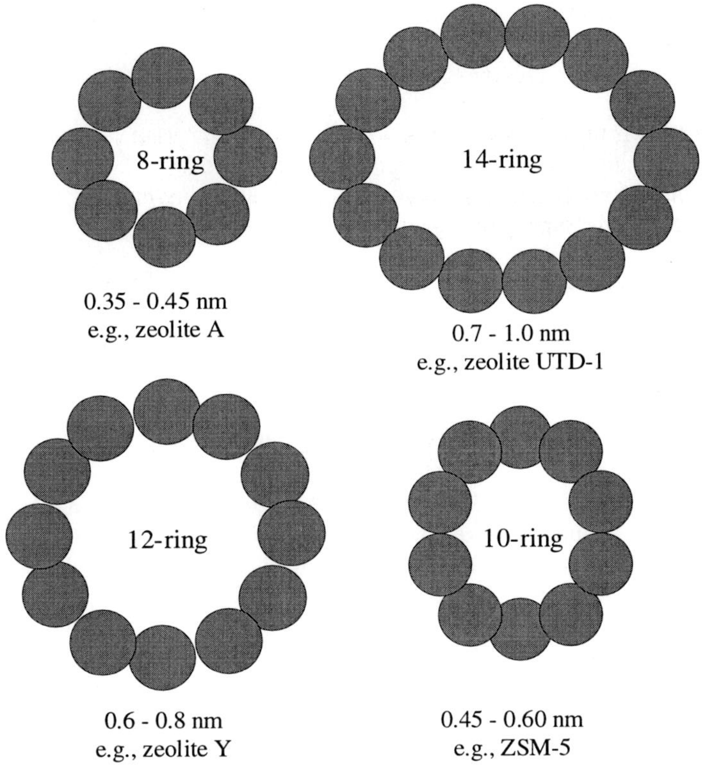

Figure 2. Typical zeolite pore sizes illustrated with oxygen packing models

adsorption and catalytic applications.

In most zeolite structures the primary structural units, the AlO_4 or SiO_4 tetrahedra, are assembled into secondary building units which may be simple polyhedra such as cubes, hexagonal prisms, or cubo-octahedra. The final framework structure consists of assemblages of the secondary units. (See Cpts. 3 and 4 in this volume.)

More than 70 novel, distinct framework structures of zeolites are known. They exhibit pore sizes from 0.3~1.0 nm, and pore volumes from about 0.10 to 0.35 cc/g. Typical zeolite pore sizes using oxygen-packing models are shown in Figure 2. They include small pore zeolites with eight-ring pores with free diameters of 0.30-0.45 nm, e.g., zeolite A; medium pore zeolites formed by a ten ring, 0.45-0.60 nm in free diameter, e.g., ZSM-5; large pore zeolites with 12-ring pores, 0.6-0.8 nm, e.g., zeolites X and Y; and extra-large pore zeolites with fourteen-ring pores, e.g., UTD-1. The molecular sieve effect is illustrated in Figure 3 for calcium A zeolite with an eight-ring pore of oxygens. Normal octane readily accesses the internal void through the pore whereas iso-octane is larger than the pore and is totally excluded.

The zeolite framework should be viewed as somewhat flexible, with the size and shape

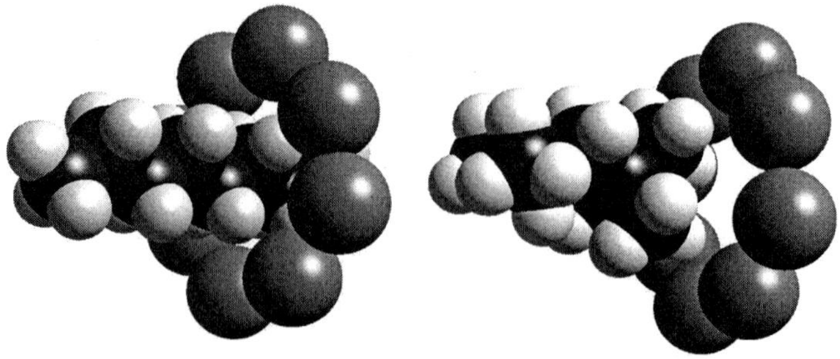

Figure 3. Illustration of the molecular sieve effect. The straight chain molecule normal octane (left) passes through the eight-ring aperture of 5A (CaA) zeolite; the branched molecule iso-octane (right) cannot.

of the framework and pore responding to changes in temperature and guest species. For example, ZSM-5 with sorbed neopentane has a 0.62 nm near-circular pore, but with substituted aromatics as the guest species the pore assumes an elliptical shape, 0.45 to 0.70 nm in diameter.

Some of the more important zeolite types, most of which have been used in commercial applications, include the zeolite minerals mordenite, chabazite, erionite and clinoptilolite, and the synthetic zeolites type A, X, Y, L, "Zeolon" mordenite, ZSM-5, beta, MCM-22, and zeolites F and W.

1.3 Nomenclature

There is no systematic nomenclature developed for molecular sieve materials. The discoverer of the synthetic species based on a characteristic x-ray powder diffraction pattern and chemical composition assigns trivial symbols. The early synthetic materials discovered by Milton, Breck and coworkers at Union Carbide used the Arabic alphabet, e.g., zeolites A, B, X, Y, L. The use of the Greek alphabet was initiated by Mobil and Union Carbide with the zeolites alpha, beta, and omega. Many of the synthetic zeolites which have the structural topology of mineral zeolite species were assigned the name of the mineral, for example, synthetic mordenite, chabazite, erionite and offretite. The molecular sieve literature is replete with acronyms: ZSM-5, -11, ZK-4 (Mobil); EU-1, FU-1, NU-1 (ICI); LZ-210 and AlPO, SAPO, MeAPO, etc. (Union Carbide, UOP); and ECR-1 (Exxon). The one publication on nomenclature by IUPAC in 1979 (ref. 2) is limited to the then known zeolite-type materials.

The *IZA Atlas of Zeolite Structure Types* (1st Ed. 1978; 2nd Ed. 1987; 3rd Ed. 1992; 4th Ed. 1996) published by the IZA Structure Commission assigns a three letter code to be used for a known framework topology irrespective of composition. Illustrative codes are LTA for Linde zeolite A, FAU for molecular sieves with a faujasite topology, e.g., zeolites X and Y, MOR for the mordenite topology, MFI for the ZSM-5 and silicalite topologies, and AFI for the aluminophosphate $AlPO_4$-5 topology. The acceptance of a newly determined structure of a zeolite or molecular sieve for inclusion in the official *Atlas* is reviewed and must be approved by the IZA Structure Commission. The IZA Structure Commission was given the authority in 1986 at the 7th

International Zeolite Conference in Tokyo, to approve and/or assign the three-letter structure code for new framework topologies.

The definition and usage of the term “zeolite” has evolved and changed especially over the last decade, to include non-aluminosilicate compositions and structures. The 4th Edition of the *Atlas* uses the term “zeolite and zeolite-like materials” and defines the structures included in the *Atlas* as tetrahedral oxide structures with a framework density less than about 20 T-atoms per 1000 A^3 irrespective of framework composition. Similarly the term zeolite has been broadened in the mineralogy literature to include tetrahedral framework compositions with T-elements other than Al and Si but where classical zeolite properties are exhibited, e.g., structures containing open cavities in the form of channels and cages, and reversible hydration-dehydration characteristics (ref. 3).

2. EARLY HISTORY

The history of zeolites began in 1756 when the Swedish mineralogist Cronstedt discovered the first zeolite mineral, stilbite (ref. 4). He recognized zeolites as a new class of minerals consisting of hydrated aluminosilicates of the alkali and alkaline earths. Because the crystals exhibited intumescence when heated in a blowpipe flame, Cronstedt called the mineral a "zeolite" derived from two Greek words, "zeo" and "lithos" meaning "to boil" and "a stone". In 1777 Fontana described the phenomenon of adsorption on charcoal (ref. 5). In 1840 Damour observed that crystals of zeolites could be reversibly dehydrated with no apparent change in their transparency or morphology (ref. 6). Schafhautle reported the hydrothermal synthesis of quartz in 1845 by heating a "gel" silica with water in an autoclave (ref. 7). Way and Thompson (1850) clarified the nature of ion exchange in soils (ref. 8). Eichhorn in 1858 showed the reversibility of ion exchange on zeolite minerals (ref. 9). St. Claire Deville reported the first hydrothermal synthesis of a zeolite, levynite, in 1862 (ref. 10). In 1896 Friedel developed the idea that the structure of dehydrated zeolites consists of open spongy frameworks after observing that various liquids such as alcohol, benzene, and chloroform were occluded by dehydrated zeolites (ref. 11). Grandjean in 1909 observed that dehydrated chabazite adsorbs ammonia, air, hydrogen and other molecules (ref. 12), and in 1925 Weigel and Steinhoff reported the first molecular sieve effect (ref. 13). They noted that dehydrated chabazite crystals rapidly adsorbed water, methyl alcohol, ethyl alcohol and formic acid but essentially excluded acetone, ether or benzene. In 1927 Leonard described the first use of x-ray diffraction for identification in mineral synthesis (ref. 13). The first structures of zeolites were determined in 1930 by Taylor and Pauling (ref. 15,16). In 1932 McBain established the term “molecular sieve" to define porous solid materials that act as sieves on a molecular scale (ref. 17).

Thus, by the mid-1930's the literature described the ion exchange, adsorption, molecular sieving and structural properties of zeolite minerals as well as a number of reported syntheses of zeolites. The latter early synthetic work remains unsubstantiated because of incomplete characterization and the difficulty of experimental reproducibility.

Barrer began his pioneering work in zeolite adsorption and synthesis in the mid-1930's to 1940's. He presented the first classification of the then known zeolites based on molecular size considerations in 1945 (ref. 18) and in 1948 reported the first definitive synthesis of zeolites including the synthetic analogue of the zeolite mineral mordenite (ref. 19).

3. INDUSTRIAL HISTORY

3.1. Synthetic Zeolites

Barrer's work in the mid to late 1940's inspired Milton of the Linde Division of Union Carbide Corporation to initiate studies in zeolite synthesis in search of new approaches for separation and purification of air. Between 1949 and 1954 R. M. Milton and co-worker D. W. Breck discovered a number of commercially significant zeolites, types A, X and Y. In 1954 Union Carbide commercialized synthetic zeolites as a new class of industrial materials for separation and purification. The earliest applications were the drying of refrigerant gas and natural gas. In 1955 T. B. Reed and D. W. Breck reported the structure of the synthetic zeolite A. In 1959 Union Carbide marketed the "ISOSIV" process for normal-isoparaffin separation, representing the first major bulk separation process using true molecular sieving selectivity. Also in 1959 a zeolite Y-based catalyst was marketed by Carbide as an isomerization catalyst (ref. 20).

In 1962 Mobil Oil introduced the use of synthetic zeolite X as a cracking catalyst. In 1969 Grace described the first modification chemistry based on steaming zeolite Y to form an "ultrastable" Y. In 1967-1969 Mobil Oil reported the synthesis of the high silica zeolites beta and ZSM-5. In 1974 Henkel introduced zeolite A in detergents as a replacement for the environmentally suspect phosphates. By 1977 industry-wide 22,000 tons of zeolite Y were in use in catalytic cracking. In 1977 Union Carbide introduced zeolites for ion-exchange separations.

Since their introduction as a new class of industrial materials in 1954, the annual market for synthetic zeolites and molecular sieves has grown immensely, to $1.6-1.7 billion worldwide in 1998 (ref. 21). The major application areas are as adsorbents, catalysts, and ion exchange materials. The largest single market by volume is the detergent application where zeolite A (and recently Type P) functions as an ion exchanger. In 1998, 2.3 billion pounds were consumed in that application. Although the second largest volume use is as catalysts, this is the largest value market for zeolites, about 55% of the total. Fluid catalytic cracking (FCC) catalysts, containing primarily silica-enriched forms of zeolite Y, represent more than 95% of total zeolite catalyst consumption, with smaller volumes used in hydrocracking and chemical and petrochemical synthesis. Catalyst consumption in 1998 is estimated at 360 million pounds (ref. 21).

Adsorption applications are varied, and include: drying and purification of natural gas, petrochemical streams, e.g., ethylene and propylene, refrigerants, and insulated windows; bulk separations, e.g., xylenes and normal paraffins; and in air separation to produce oxygen by pressure swing adsorption (PSA) or vacuum pressure swing adsorption (VPSA) processes. Adsorbent consumption in 1998 is estimated at 200 million pounds (ref. 21).

3.2 Natural Zeolites

For 200 years following their discovery by Cronstedt, zeolite minerals (or natural zeolites) were considered to occur typically as minor constituents in vugs or cavities in basaltic and volcanic rock. Such occurrences precluded their being obtained in mineable quantities for commercial use. From the late 50's to 1962 major geologic discoveries revealed the widespread occurrence of a number of natural zeolites in sedimentary deposits throughout the Western United States. The discoveries resulted from the use of x-ray diffraction to examine very fine-grained (1-5 um) sedimentary rock. Some zeolites occur in large near mono-mineralic deposits suitable for mining. Those that have been commercialized for adsorbent applications include chabazite, erionite, mordenite and clinoptilolite (ref. 22).

Japan is the largest user of natural zeolites (see Cpt. 1 in this volume by Maesen and Marcus). Mordenite and clinoptilolite are used in small volume in adsorbent applications including air separation and in drying and purification (ref. 23). Natural zeolites have also found use in bulk applications as fillers in paper, in pozzolanic cements and concrete, in fertilizer and soil conditioners and as dietary supplements in animal husbandry. World production of natural zeolites is estimated at 7.2 billion pounds in 1998, with the bulk applications representing most of that consumption (ref. 21).

The price of zeolites varies considerably depending on the application. The price of catalysts vary from about $1/pound for FCC to tens of dollars/pound for specialty catalysts; adsorbents from about $1.50 to $2/pound, up to tens of dollars/pound for specialty adsorbents; and about $0.30/pound for detergents. Natural zeolites in bulk applications sell for $0.02-0.10/pound, and in industrial adsorbent applications for $1-1.50/pound (ref. 21).

4. HISTORY OF MOLECULAR SIEVE MATERIALS

The theme and research on molecular sieve materials over the last fifty year period has been a quest for new structures and compositions. The major discoveries and advances in molecular sieve materials during that period are summarized in Table 1.

Table 1.
Evolution of Molecular Sieve Materials

Time of Initial Discovery	Composition
Late 40's to Early 50's	Low Si/Al Ratio Zeolites
Mid 50's to late 60's	High Si/Al Ratio Zeolites
Early 70's	SiO_2 Molecular Sieves
Late 70's	$AlPO_4$ Molecular Sieves
Late 70's to Early 80's	SAPO and MeAPO Molecular Sieves
Late 70's	Metallo-silicates, -aluminosilicates
Early to Mid 80's	$AlPO_4$-based Molecular Sieves
Early to Mid 90's	Metallophosphates Mesoporous Molecular Sieves Octahedral-tetrahedral Frameworks

The history of commercially significant molecular sieve materials from 1954 to 1979 was reviewed by this author in 1980 (ref. 24). Highlights from that review and the subsequent history are presented here. The reader is referred to Cpt. 3 in this volume by Baerlocher and McCusker for the structures of the materials.

4.1. Aluminosilicate Zeolites and Silica Molecular Sieves

The evolution of aluminosilicate zeolites is summarized in Table 2 based on increasing framework Si/Al composition. The four somewhat arbitrary categories are: 1) "low", 2) "intermediate", 3) "high" silica zeolites, and 4) "silica" molecular sieves.

Table 2.
The evolution of molecular sieve materials

Composition and examples
"Low" Si/Al Zeolites (1 to 1.5): A, X
"Intermediate" Si/Al Zeolites (~2 to 5):
A: Natural Zeolites: erionite, clinoptilolite, mordenite
B. Synthetic Zeolites: Y, L. large pore mordenite, omega
"High" Si/Al Zeolites (~10 to 100):
A. By thermochemical framework modification:
Highly siliceous variants of Y, mordenite, erionite
B. By direct synthesis: ZSM-5, beta
Silica Molecular Sieves: silicalite

The transition in properties accompanying the increase in the framework Si/Al is summarized in Table 3. The latter are generalized and should only be viewed as trends. The thermal stability increases from about 700°C in the low silica zeolites to 1300°C in the silica molecular sieves. The surface selectivity, which is highly hydrophilic in the low silica zeolites, is hydrophobic in the high silica zeolites and the silica molecular sieves. The acidity tends to increase in strength with increasing Si/Al ratio. As the Si/Al ratio increases, the cation concentration and ion exchange capacity (proportional to aluminum content) decreases. The structures of the low silica zeolites are predominantly formed with 4-, 6-, and 8-rings of tetrahedra. In the intermediate silica zeolites we see the onset of 5-rings in mordenite and omega zeolite. In the high silica zeolite structures and the silica molecular sieves we find a predominance of 5-rings of tetrahedra.

Table 3.
The transition in properties as framework Si/Al increases from 1 to ∞

Property	Transition
Stability	From ≤ 700°C to ~ 1300°C
Surface selectivity	From hydrophilic to hydrophobic
"Acidity"	Increasing strength, decreasing number
Cation concentration	Decreasing
Structure	From 4, 6, and 8-rings to 5-rings

The low silica zeolites represented by zeolites A and X are aluminum-saturated, have the highest cation concentration, and give optimum adsorption properties in terms of capacity, pore size and three-dimensional channel systems. They represent highly heterogeneous surfaces with a strongly hydrophilic surface selectivity. The intermediate Si/Al zeolites (Si/Al of 2-5) consist of the natural zeolites erionite, chabazite, clinoptilolite and mordenite, and the synthetic zeolites Y, mordenite, omega and L. These materials are still hydrophilic in this Si/Al range.

The high silica zeolites with Si/Al of 10-100 can be generated by either thermochemical framework modification of hydrophilic zeolites or by direct synthesis. In the modification route stabilized, siliceous variants of Y, mordenite, erionite, and over a half-dozen other zeolites have been prepared by steaming and acid extraction. These materials are reported to be hydrophobic and organophilic and represent a range of pore sizes from 0.4-0.8 nm. A very large number of high silica zeolites prepared by direct synthesis have now been reported, including beta, ZSM-5, -11, -12, -21, -34, NU-1 and FU-1, and ferrisilicate and borosilicate analogs of the aluminosilicate structures. Typical of the reported silica molecular sieves are silicalite, fluoride silicalite, silicalite-2 and TEA-silicate. ZSM-5 and silicalite have achieved commercial significance.

The difference in surface selectivity between the hydrophobic silica molecular sieve, silicalite, and the highly hydrophilic zeolite NaX can be observed by comparing their equilibrium adsorption isotherms for water, oxygen and n-hexane (Fig. 4). The hydrophilic NaX pore fills at

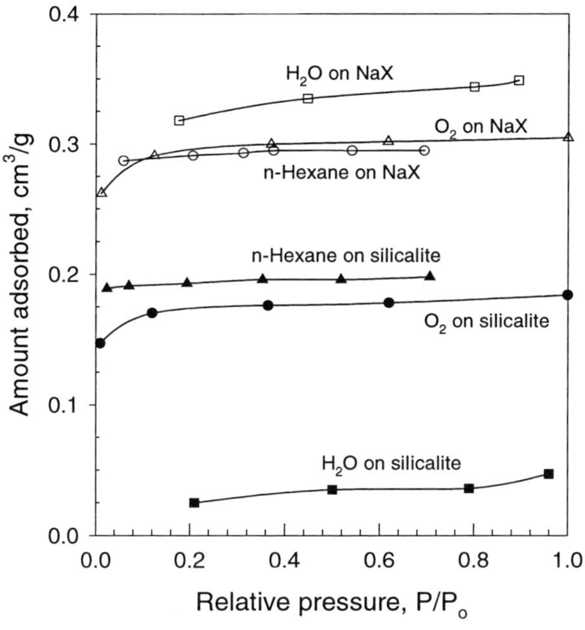

Figure 4. Comparison of adsorption equilibrium isotherms for water, oxygen, and n-hexane on NaX zeolite and silicalite. Water and n-hexane at 25C and oxygen at -183C.

low partial pressures with all three adsorbates giving a typical Langmuir-type isotherm shape. On silicalite, n-hexane and oxygen show a similar typical molecular sieve pore filling at low partial pressures, but illustrative of the hydrophobic characteristic of silicalite, only a small amount of water (~0.05 cc/g) is adsorbed even at a relative partial pressure near 1. It should be noted that the pore volume of silicalite is substantially smaller than that of NaX.

In summary, if we compare the properties of the low and intermediate zeolites with those of the high silica zeolites and silica molecular sieves, we find that their resulting properties allow the low and intermediate zeolites to remove water from organics and to carry out separations and catalysis on dry streams. In contrast, the hydrophobic high silica zeolites and silica molecular sieves can remove and recover organics from water streams and carry out separations and catalysis in the presence of water.

4.2. The 80's and the 90's

Overall the 80's and 90's can be described as a period of explosion in the discovery of new compositions and structures of molecular sieves. This can perhaps be seen most vividly by comparing the numbers of structure types contained in the various editions of the *Atlas of Zeolite Structure Types*. One hundred twenty-five structure type codes have been assigned through early 2000. The 1978 1st Edition contained 38 structure types, the 1987 2nd Edition 64, the 1992 3rd Edition 85, and the last 1996 4th Edition 98. Thus 87 new structure types have been discovered since 1978. However, the reader should be cautioned that a significant number of the structure types included in the *Atlas* are not truly microporous or molecular sieve materials, i.e., they are not stable to removal of the as-synthesized guest species, typically water or organic templates, and therefore cannot reversibly adsorb molecules or carry out catalytic reactions. Unfortunately, the *Atlas* does not give any information on the stability of the structures described.

4.2.1. The 80's

In the 1980's there was extensive work carried out on the synthesis and applications of ZSM-5 and a proliferating number of other members of the high silica zeolite family. In 1982 microporous crystalline aluminophosphate molecular sieves were described by Wilson et al. (ref. 25) at Union Carbide, and additional members of the aluminophosphate-based molecular sieve family, e.g., SAPO, MeAPO, MeAPSO, ElAPO and ElAPSO, subsequently disclosed by 1986 (ref. 26). Considerable effort in synthesizing metallosilicate molecular sieves was reported where the metals iron, gallium, titanium, germanium and others were incorporated during synthesis into silica or aluminosilicate frameworks, typically with the ZSM-5 (MFI) topology (ref. 27). Additional crystalline microporous silica molecular sieves and related clathrasil structures were reported.

The 80's saw major developments in secondary synthesis and modification chemistry of zeolites. Silicon-enriched frameworks of over a dozen zeolites were described using methods of: thermochemical modification (prolonged steaming); mild aqueous ammonium fluorosilicate chemistry; and by high temperature treatment with silicon tetrachloride and low temperature treatment with fluorine gas. Similarly, framework metal substitution using mild aqueous ammonium fluorometallate chemistry was reported to incorporate iron, titanium, chromium and tin into zeolite frameworks by secondary synthesis techniques. A review of modification and secondary synthesis chemistry is given in Cpt. 6 in this volume by Szostak.

4.2.1.1. Aluminophosphate-Based Molecular Sieves

In 1982 a major discovery of a new class of aluminophosphate molecular sieves was reported by Wilson et al. (ref. 25). By 1986 some thirteen elements were reported to be incorporated into the aluminophosphate frameworks, Li, Be, B, Mg, Si, Ti, Mn, Fe, Co, Zn, Ga,

Ge and As (ref. 26). These new generations of molecular sieve materials designated $AlPO_4$-based molecular sieves comprise more than two dozen structures and two hundred compositions.

Table 4.
Typical structures in $AlPO_4$-based molecular sieves

Species	Structure Type	Pore size, nm	Sat'n H_2O Pore vol. cc/g	Species	Structure Type	Pore size, nm	Sat'n H_2O Pore vol. cc/g
Very Large Pore				**Small Pore**			
VPI-5	Novel	1.25	0.35	14	Novel	0.4	0.19
8	Novel	0.9	0.24	17	ERI	0.43	0.28
				18	Novel	0.43	0.35
				26	Novel	0.43	0.23
Large Pore				33	Novel	0.4	0.23
5	Novel	0.8	0.31	34,44,47	CHA	0.43	0.3
36	Novel	0.8	0.31	35	LEV	0.43	0.3
37	FAU	0.8	0.35	39	Novel	0.4	0.23
40	Novel	0.7	0.33	42	LTA	0.43	0.3
46	Novel	0.7	0.28	43	GIS	0.43	0.3
				52	Novel	0.43	0.3
				56	Novel	0.43	0.3
				Very Small Pore			
Intermediate				16	Novel	0.3	0.3
11	Novel	0.6	0.16	20	SOD	0.3	0.24
31	Novel	0.65	0.17	25	Novel	0.3	0.17
41	Novel	0.6	0.22	28	Novel	0.3	0.21

The over two dozen structures of $AlPO_4$-based molecular sieves reported to date include zeolite topological analogues and a large number of novel structures. The major structures are shown in Table 4. They include fifteen novel structures as well as seven structures with framework topologies related to those found in the zeolites, CHA (-34, -44, -47), ERI (-17), GIS (-43), LEV (-35), LTA (-42), FAU (-37), and SOD (-20). Also shown is the pore size and saturation water pore volume for each structure type. The structures include very large pore (1.25 nm), large pore (0.7-0.8 nm), intermediate pore (0.6 nm), small pore (0.4 nm), and very small pore (0.3 nm) materials. Saturation water pore volumes vary from 0.16 to 0.35 cc/g comparable to the pore volume range observed in zeolites. The novel structures which have been determined include types 5, 11, 14, 16, 22, 33, 39, 46, 50, 52 and 56. The $AlPO_4$-based structures have been reviewed by Bennett et al. (ref. 28) and are described in this volume in Cpt 3 by Baerlocher and McCusker.

4.2.1.1.1. $AlPO_4$

The new family of aluminophosphate materials ($AlPO_4$-n) includes the first very large pore material, VPI-5 reported by Davis et al. (ref. 29). The VPI-5 structure is outlined by a unidimensional channel of an 18-membered ring with a free pore diameter of 1.25 nm. There is one 12-membered ring structure, $AlPO_4$-5, with a pore size of 0.8 nm; several medium pore structures outlined by 10-membered rings or elliptical 12-rings with pore diameters of 0.6-0.65 nm, for example, $AlPO_4$-11, -31 and -41; and small pore size materials such as $AlPO_4$-17 (ERI), with 8-membered ring pores and pore sizes of 0.4-0.43 nm in diameter.

The product composition expressed in terms of oxide ratios is:

$$xR \cdot Al_2O_3 \cdot 1.0{\pm}0.2\ P_2O_5 \cdot yH_2O$$

where R is an amine or quaternary ammonium ion. The $AlPO_4$ molecular sieve as synthesized must be calcined at 400 to 600°C to remove the R and water yielding a microporous aluminophosphate molecular sieve.

The characteristics of aluminophosphate molecular sieves include a univariant framework composition with Al/P = 1, a high degree of structural diversity, a wide range of pore sizes and volumes exceeding the pore sizes known previously in zeolite molecular sieves with the VPI-5 18-membered ring material. They are neutral frameworks and therefore have nil ion-exchange capacity. Their surface selectivity is mildly hydrophilic. They exhibit excellent thermal and hydrothermal stability, up to 1000°C (thermal) and 600°C (steam).

4.2.1.1.2. Silicoaluminophosphates (SAPO)

The next family of new molecular sieves are the silicoaluminophosphates (SAPO). Sixteen microporous structures have been reported to date, eight of which were never before observed in zeolites. The SAPO family includes a silicon analogue of the 18-ring VPI-5, Si-VPI-5 (ref. 29), a number of large pore 12-ring structures including the important SAPO-37 (FAU), medium pore structures with pore sizes from 0.6-0.65 nm, and small pore structures with pore sizes of 0.4-0.43 nm including SAPO-34 (CHA). The SAPO's exhibit both structural and compositional diversity.

The SAPO anhydrous composition is $0\text{-}0.3R(Si_xAl_yP_z)O_2$ where x, y and z are the mole fraction of the respective framework elements. The mole fraction of silicon, x, typically varies from 0.02 to 0.20 depending on synthesis conditions and structure type. Martens et al. have reported compositions with the SAPO-5 structure with x up to 0.8 (ref. 30). Van Nordstrand et al. have reported the synthesis of a pure silica analogue of the SAPO-5 structure, SSZ-24 (ref. 31).

The introduction of silicon into hypothetical phosphorus sites produces negatively charged frameworks with cation exchange properties and weak to mild acidic catalytic properties. Again, as in the case of the aluminophosphate molecular sieves they exhibit excellent thermal and hydrothermal stability.

4.2.1.1.3. Metal Aluminophosphates (MeAPO)

In the metal aluminophosphate (MeAPO) family the framework composition contains metal, aluminum and phosphorus (Ref. 26). The metal (Me) species include the divalent forms of Co, Fe, Mg, Mn and Zn, and trivalent Fe. As in the case of SAPO, the MeAPO's exhibit both structural diversity and even more extensive compositional variation. Seventeen microporous structures have been reported, eleven of these never before observed in zeolites. Structure types crystallized in the MeAPO family include framework topologies related to the zeolites, e.g., -34 (CHA) and -35 (LEV), and to the $AlPO_4$'s, e.g., - 5 and -11, as well as novel structures, e.g., -36 (0.8 nm pore) and -39 (0.4 nm pore). The MeAPO's represent the first demonstrated incorporation of divalent elements into microporous frameworks.

The spectrum of adsorption pore sizes and pore volumes and the hydrophilic surface selectivity of the MeAPO's are similar to those described for the SAPO's. The observed catalytic properties vary from weakly to strongly acidic and are both metal and structure dependent. The thermal and hydrothermal stability of the MeAPO materials is somewhat less than that of the $AlPO_4$ and SAPO molecular sieves.

The MeAPO molecular sieves exhibit a wide range of compositions within the general formula 0 - $0.3R(Me_xAl_yP_z)O_2$. The value of x, the mole fraction of Me, typically varies from 0.01 to 0.25. Using the same mechanistic concepts described for SAPO, the MeAPO's can be considered as hypothetical $AlPO_4$ frameworks that have undergone substitution. In the MeAPO's the metal appears to substitute exclusively for aluminum resulting in a negative (Me^{2+}), or neutral (Me^{3+}) framework charge. Like SAPO, the negatively charged MeAPO frameworks possess ion-exchange properties and Bronsted acid sites.

4.2.1.1.4. Other $AlPO_4$-Based Compositions

The MeAPSO family further extends the structural diversity and compositional variation found in the SAPO and MeAPO molecular sieves. These quaternary frameworks have Me, Al, P, and Si as framework species (Ref. 26). The MeAPSO structure types include framework

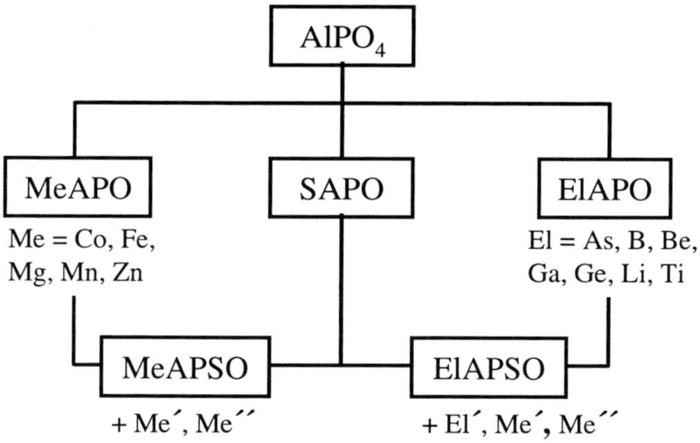

Figure 5. Schematic representation of the relationships in the aluminophosphate-based molecular sieves.

topologies observed in the binary $AlPO_4$ and ternary (SAPO and MeAPO) compositional systems and the novel structure -46 with a 0.7 nm pore. The structure of -46 has been determined (ref. 32).

Quinary and senary framework compositions have been synthesized containing aluminum, phosphorus and silicon, with additional combinations of divalent (Me) metals. In the ElAPO and ElAPSO compositions the additional elements Li, Be, B, Ga, Ge, As, and Ti have been incorporated into the $AlPO_4$ framework (ref. 26). Figure 5 shows the compositional relationships among the aluminophosphate-based families of molecular sieves.

4.2.1.2. Metallosilicate Molecular Sieves

A large number of metallosilicate molecular sieves have been reported particularly in the patent literature. Those claimed include silicates containing incorporated tetrahedral iron, boron, chromium, arsenic, gallium and germanium. Most of the work has been reported with structures of the MFI type. Others include metallosilicate analogues of ZSM-11, -12, Theta-1, ZSM-34 and beta. In only a limited number of the reported metal incorporations has sufficient characterization been presented to establish proof of incorporation. To date, only B, Ga, Fe and Ti have been sufficiently characterized to confirm structural incorporation. The metallosilicate molecular sieves are reviewed in detail by Szostak (ref. 27). The titanium-silicalite composition, TS-1, has achieved commercialization in selective oxidation processes, and iron-silicalite in ethylbenzene synthesis.

4.2.1.3. Other Framework Compositions

Crystalline microporous frameworks have been reported with compositions of: beryllosilicate, lovdarite (ref. 33); beryllophosphate (ref. 34); aluminoborate (ref. 35); aluminoarsenate (ref. 36); galloarsenate (ref. 37); gallophosphate (ref. 38); antimonosilicate (ref. 39); and germanosilicate (ref. 40).

Harvey et al. (ref. 34) reported the synthesis of alkali beryllophosphate molecular sieves with the RHO, GIS, EDI, and ANA structure topologies, and a novel structure, BPH. Simultaneously, the first beryllophosphate mineral species were reported, tiptopite, with the cancrinite (CAN) topology by Peacor et al. (ref. 41), and pahasapaite, with the RHO topology by Rouse et al. (ref. 42).

In the late 1980's Bedard et al. reported the discovery of microporous metal sulfides, based on germanium (IV) and Sn (IV) sulfide frameworks (ref. 43). The microporous sulfides are synthesized hydrothermally in the presence of alkylammonium templating agents. The GeS_4-based compositions include one or more of framework incorporated metals: Mn, Fe, Co, Ni, Cu, Zn, Cd, and Ga. Over a dozen novel structures were reported which have no analogues in the microporous oxides. Ozin et al. have extended this work to a large number of microporous sulfides and selenides (ref. 44). It should be noted that the microporous sulfides and selenides are prone to structure collapse upon calcination to remove the template species.

4.2.2. The 90's

The explosion in the discovery of new compositions and structures observed in the 80's continued through the 90's.

Some three dozen or more novel tetrahedral structures were synthesized in the 90's, based on aluminosilicate, silica, metallosilicate, and metallophosphate frameworks. Two are especially noteworthy. The gallophosphate cloverite has the first 20-ring pore, 0.4 x 1.32 nm in diameter, and the lowest observed framework density (number of T-atoms per 1000 A^3), 11.1 (ref. 45). The cloverite structure contains the D-4R and alpha cages reminiscent of the aluminum-rich zeolite Type A (LTA), combined with the rpa cage found in the aluminophosphate structures. It is an interrupted framework structure and thus has somewhat limited thermal stability. The siliceous zeolite UTD-1 contains a 14-ring pore 0.75 x 1.0 nm in diameter, and is the first aluminosilicate with a pore size larger than a 12-ring (ref. 46). CIT-5, a second 14-R structure with a pure silica composition and a 0.8 nm pore, was reported by Wagner et al. (ref. 50).

Gier et al. reported zinc and beryllium phosphates and arsenates with the X (FAU), ABW and SOD structures reminiscent of the early aluminum-rich synthetic zeolite chemistry. The synthesis of $ZnPO_4$-X (FAU) is especially spectacular. Crystallization occurs almost instantaneously at 0°C (ref. 47). Concurrent with ease of synthesis, the structure is thermally unstable.

Table 5 lists some of the major new structures reported in the 90's. Interestingly, there were no new aluminum-rich synthetic zeolites reported in either the 80's or the 90's. The new aluminosilicate structures were all high silica or pure silica in composition.

Table 5.
Major new synthetic structures of the 90's

Species	Structure Type	Pore size, nm	Ring Size	Ref.
MCM-22, 49	MWW	0.6	10	48,49
UTD-1	DON	1.0	14	46
CIT-5	CFI	0.8	14	50
EMC-2	EMT	0.7	12	51
Cloverite	-CLO	1.3	20	45

Not to be outdone by humankind there were a number of new zeolite minerals discovered in nature during the 90's. The zeolite mineral boggsite has a novel framework topology with 3-dimensional pores combining 10R's and 12R's, and has not yet been reproduced synthetically (ref. 52). Tschernichite is an aluminum-rich mineral analogue of the synthetic zeolite beta (ref. 53). Gottardiite is a new mineral analogue of synthetic zeolite Nu-87 (ref. 54). The zeolite mineral terranovaite has a novel structure with pentasil chains and a two-dimensional 10R channel (ref. 55). Mutinaite is a high silica zeolite mineral analogue of ZSM-5 with the highest silica content of all known zeolite minerals (Si/Al = 7.7) (ref. 56). The structure of the zeolite mineral perlialite was reported (ref. 57) to be the same as that of the synthetic zeolite L (LTL), some 35 years after the synthesis of zeolite L.

Tschortnerite surely is the most remarkable novel zeolite mineral discovered (ref. 58). Its unique framework topology contains five different cages, D-6R's, D-8R's, sodalite cages, truncated cubo-octahedra, and a unique 96-membered cage. Cu-containing clusters are encapsulated within the truncated cubo-octahedra. The pore structure is a 3-dimensional 8-R, and the framework density of 12.2 is among the lowest known for zeolites. The framework is alumina-rich with an Si/Al = 1, unusual for zeolite minerals.

4.2.2.1. New Directions in the 90's

Two major new classes of molecular sieve type materials were reported in the 90's, microporous frameworks based on mixed octahedral-tetrahedral frameworks in contrast to the previously described tetrahedral frameworks, and mesoporous molecular sieves with pore sizes ranging from about 2 nm to greater than 10 nm.

4.2.2.1.1. Octahedral-Tetrahedral Frameworks

The microporous materials described heretofore were all based on tetrahedral frameworks. Microporous titanosilicate materials with mixed octahedral-tetrahedral frameworks were reported in the 90's. The framework linkage is through TiO_6 octahedra and SiO_4 tetrahedra. Chapman and Roe described the titanosilicate GTS-1, a structural analogue of the mineral pharmacosiderite, with a 3-dimensional channel system and 8-R pores (ref. 59). Kuznicki and co-workers reported the synthesis of the titanosilicates ETS-4 and ETS-10 (ref. 60). Their respective pore sizes are 0.4

nm and 0.8 nm. ETS-4 is the synthetic analogue of the rare titanosilicate mineral zorite. The novel structure ETS-10 contains a 3-dimensional 12-R pore system and shows a high degree of disorder (ref. 61). ETS-10 has achieved commercial status in adsorption applications.

4.2.2.1.2. Mesoporous Molecular Sieves

A major advance in molecular sieve materials was reported in 1992 by researchers at Mobil. Kresge et al. and Beck et al. describe a new family of mesoporous silicate and aluminosilicate materials, designated M41S (ref. 62). The members of the family include: MCM-41, with a unidimensional hexagonal arrangement of uniform open channels 0.2-10 nm in diameter: a cubic structure, MCM-48, with a 3-dimensional channel system with pore sizes ~0.3 nm; and a number of lamellar structures. The order in the structure is derived from the channel arrangement. The silica or aluminosilicate wall outlining the channel is disordered and exhibits properties much like amorphous silica or silica-alumina.

Within the same time frame and independently, Inagaki and co-workers reported a mesoporous material designated FSM-16, prepared by hydrothermal treatment of the layered sodium silicate kanemite, $NaHSi_2O_5.3H_2O$ (ref. 63). Chen et al. substantiate that FSM-16 and MCM-41 bear a strong resemblance to each other, both with narrow mesopore distributions and similar physicochemical properties, but with FSM-16 having higher thermal and hydrothermal stability due to the higher degree of condensation in the silicate walls (ref. 64).

Both mesoporous materials are synthesized hydrothermally with a surfactant liquid crystal as the template (see section 5 below). They exhibit very high surface areas and pore volumes, of the order of 1000 m^2/g and 1.5 cc/g, respectively.

Since this initial work there has been a plethora of literature on mesoporous molecular sieves. In addition to the silica and aluminosilicate frameworks similar mesoporous structures of metal oxides now include the oxides of Fe, Ti, V, Sb, Zr, Mn, W, and others. Templates have been expanded to include nonionic, neutral surfactants, and block copolymers. Pore sizes have broadened to the macroscopic size length, in excess of 40 nm in diameter. A detailed review of the mesoporous molecular sieves is given in ref. 65. To date no commercial applications of the mesoporous molecular sieves have emerged.

In a tour-de-force of detective work, Di Renzo et al. uncovered an obscure US Patent filed in 1969 and issued in 1971 to Chiola et al., describing a low density silica. Reproduction of that patent resulted in a product having all of the properties of MCM-41 (ref. 66).

4.3. Characterization Techniques

The basic characterization techniques for zeolites and molecular sieves developed in the 50's and 60's, particularly by the Union Carbide workers, were x-ray powder diffraction, gravimetric adsorption measurements on multi-tube McBain-Bakr balances using different molecular-sized adsorbates, and various tests to screen applications. The characteristic x-ray powder diffraction pattern served as a structural identifier, and adsorption measurements gave the pore size and pore volume. A number of detailed structures were solved by the then known x-ray structure determination techniques. All of these basic characterization techniques are of course still in use today.

Application of state-of-the-art sophisticated characterization and structural techniques to molecular sieve materials in the 1980's and 1990's resulted in a major advance in our understanding of the structure and chemistry of molecular sieve materials. These include: the systematic development of hypothetical frameworks (ref. 67); the application of computational chemistry, computer modeling and ab initio calculations; the application of solid state NMR and high resolution electron microscopy techniques; ^{129}Xe NMR characterization; and the use of high energy radiation (synchrotron and intense pulsed neutron) and ab initio structural techniques to solve zeolite structures.

4.4. Applications

Applications of zeolites and molecular sieves in the 80's and 90's showed a growth in petroleum refining applications with emphasis on resid cracking and octane enhancement. ZSM-5 was commercialized as an octane enhancement additive in fluid catalytic cracking where Si-enriched Y zeolites serve as the major catalytic component in high-octane FCC catalysts. The use of zeolite catalysts in the production of organic (fine) chemicals appeared as a major new direction. Zeolites in detergents as a replacement for phosphates became the single largest volume use for synthetic zeolites worldwide (see Maesen and Marcus in this volume, Cpt. 1). Zeolite ion exchange products, both synthetic and natural, were used extensively in nuclear waste cleanup after the Three Mile Island and Chernobyl nuclear accidents. New applications emerged for zeolite powders in two potentially major areas, odor removal and as plastic additives.

In adsorption and separation applications the 80's and 90's saw a major growth in the use of pressure swing adsorption for the production of oxygen, nitrogen and hydrogen. Processes for the purification of gasoline oxygenate additives were introduced. Recent environmentally driven applications have arisen using the hydrophobic molecular sieves, highly siliceous Y zeolite and silicalite, for the removal and recovery of volatile organic compounds (VOC) that offer promise for significant market growth.

An exciting new scientific direction emerged in the 80's and 90's for exploring molecular sieves as advanced solid state materials. In a 1989 review, Ozin et al. (ref. 68) speculate "that zeolites (molecular sieves) as microporous molecular electronic materials with nanometer dimension window, channel and cavity architecture represent a 'new frontier' of solid state chemistry with great opportunities for innovative research and development". The applications described or envisioned include: molecular electronics; "quantum" dots/chains; zeolite electrodes; batteries; non-linear optical materials; and chemical sensors.

Zeolites have also been used as raw materials for ceramic compositions relevant to the electronic industry. Bedard et al. reported the high temperature processing of Mg-exchanged zeolite B (P) to form cordierite ceramic compositions (ref. 69).

5. HISTORY OF SYNTHESIS

A brief description of the history of the synthetic methods developed for the molecular sieves previously described is given here. The reader is referred to Chapter 5 in this volume by Jansen and Wilson for a detailed discussion of synthesis.

The method developed by Milton in the late 1940's involves the hydrothermal crystallization of reactive alkali metal aluminosilicate gels at low temperatures and pressures. The synthesis mechanism is generally described as involving solution mediated crystallization of the amorphous gel, involving concepts of cation templating. In the early alkali aluminosilicate synthesis of the low silica zeolites, it has been proposed that the hydrated alkali cation "templates" or stabilizes the formation of the zeolite structural sub-units. A schematic for the early zeolite synthesis is shown in Figure 6. Alkali hydroxide, reactive forms of alumina and silica, and H_2O are combined to form a gel. Crystallization of the gel to zeolite occurs at a temperature near 100°C.

The addition of quaternary ammonium cations to alkali aluminosilicate gels occurred in

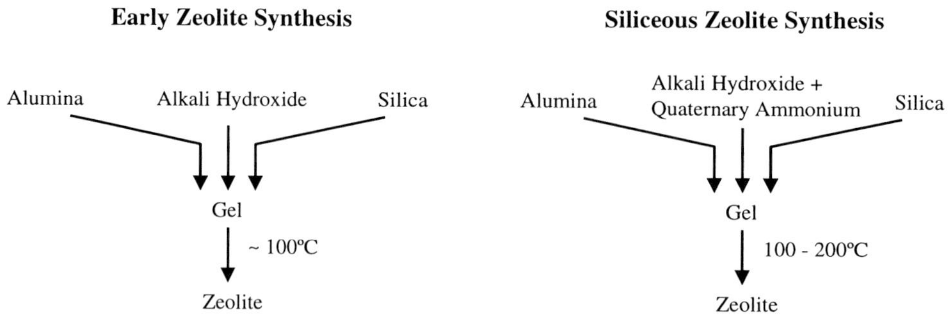

Figure 6. Schematic representation of synthesis methods for zeolite synthesis

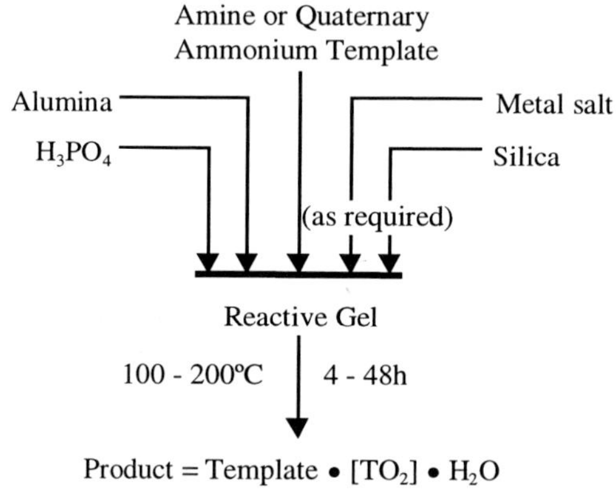

Figure 7. Schematic representation of the synthesis method for $AlPO_4$-based molecular sieves. The template and H_2O are removed by calcination at 400-600°C.

the early 1960's, first to produce intermediate silica zeolites, e.g., omega and N-A or ZK-4, and subsequently led to the discovery of the high silica zeolites and silica molecular sieves. A schematic for the synthesis of the siliceous zeolites is shown in Figure 6. The synthesis of high silica zeolites and silica molecular sieves involves synthetic chemistry similar to the initial low silica zeolite method with two important differences: the addition of the quaternary ammonium cation, and a crystallization temperature higher than 100°C, typically 125 to 200°C. The pH in both the low silica and the siliceous synthesis is basic with pH's in the region of 10 to 14.

A schematic of the synthesis method developed in the late 70's for the $AlPO_4$-based molecular sieves is shown in Figure 7. A reactive source of alumina is combined with phosphoric acid and an amine or quaternary ammonium template added to form a reactive gel. Silica or a metal salt is added optionally. The reactive gel is heated to 100 to 200°C for 4 to 48 hours to crystallize the molecular sieve product.

There are important differences in the $AlPO_4$-based synthesis compared to the aluminosilicate and silica systems: the amine or quaternary ammonium ion is frequently the only template species (no alkali metal); and the initial pH in the gel is typically mildly acidic to mildly basic.

The synthesis scheme for the mesoporous molecular sieves is shown in Figure 8. The structure directing species or template is typically a surfactant molecule.

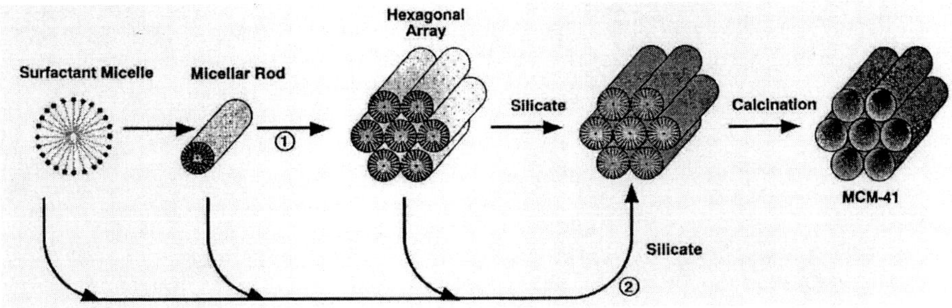

Figure 8. Schematic representation of the synthesis of MCM-41. Route 1 is a liquid crystal initiated mechanism; route 2 is a silicate anion initiated mechanism. Adapted from ref. 62

Table 6.
History of cation templates

Composition	Cation Types	Examples[a]	Structures
Low Si/Al	Alkali	Na^+ K^+	A, X, Y L
High silica zeolites Silica molecular sieves Metallosilicates	Alkali + Quaternary Ammonium	$Na^+ + Pr_4N^+$	ZSM-5 Silicalite TS-1 (Ti-silicalite)
Aluminophosphate-based Molecular sieves	Quaternary Ammonium or Amines	Pr_4N^+ $n\text{-}Pr_2NH$	$AlPO_4$-5 VPI-5
Mesoporous molecular Sieves	Alkali + surfactant molecules	Na^+ + $CTMA^+$	MCM-41, FSM-16

a. Pr_4N^+, tetrapropylammonium; $n\text{-}Pr_2NH$, di-n-propylamine; $CTMA^+$, cetyltrimethylammonium.

Flanigen and Patton first reported the use of fluoride as mineralizer in molecular sieve and zeolite synthesis, in the synthesis of F-silicalite (ref. 70). Subsequently, Guth and Kessler greatly expanded and developed this synthetic technique. The fluoride synthesis can be carried out in acidic pH's, unlike the basic pH's resulting when OH^- is used as the mineralizing species. Typically the fluoride route leads to larger, more perfect and often single crystals of silicate- and phosphate-based compositions, and in some cases novel structures. Cloverite was discovered using the fluoride synthesis technique. The fluoride ion seems to also play a templating role. A review of the fluoride synthesis methodology is contained in reference 71.

The history of cation templates used in molecular sieve synthesis is summarized in Table 6. Over the period of nearly fifty years of molecular sieve synthesis a number of concepts have been developed by various workers in cation-structure specificity. These are variously described as structure-directing, the templating effect, clathration phenomena, steriospecific replication, and nucleation effects. See Chapter 5 in this volume for a detailed discussion.

6. THE FUTURE

6.1. Materials

As noted in Sections 4.2.1. and 4.2.2. the 80's and 90's have seen an explosive and accelerating increase in discovery of new compositions and structural topologies. Based on the very high activity in this area in the last two decades, we can expect a continuation of the proliferation of new molecular sieve compositions and structures. Further advances can also be expected in novel compositions derived from modification and secondary synthesis chemistry. When we consider the very large number of structures and compositions now reported in the molecular sieve area and compare that with the number of commercial molecular sieves, what is the probability of future commercialization of a new material? There are many factors affecting achievement of commercial status: unique and advantageous properties of the material; the market need; the market size; the costs of development and marketing; and the cost and degree of difficulty in manufacturing. As a result it is likely, based on historical experience, that no more than a few of the prolific number of new molecular sieve materials of the 80's and 90's will achieve commercial status in the new millennium.

6.2. Applications

Molecular sieve adsorbents will continue to be used in the now-practiced separation and purification applications throughout the chemical process industry. New directions in the 90's include environmental and biopharmaceutical applications which have only recently received attention. Future trends in catalysis beyond the 90's include: a continuing accelerated discovery of new catalytic materials; an expanded use in petroleum refining particularly in the area of high octane gasoline and in the development of reformulated gasoline; commercial development in conversion of alternate resources to motor fuels and base chemicals; and as routes to organic chemical intermediate or end products.

The large application of zeolites as ion exchangers in detergents has leveled off in demand in North America, Western Europe and Japan in the late 90's, but should continue to grow during the 2000's, particularly in Asia, Australia and Latin America (ref. 21). The other applications of zeolites as ion exchangers in the nuclear industry, in radioactive waste storage and cleanup, and in metals removal and recovery, will probably remain a relatively small fraction of the worldwide market for molecular sieve materials.

Among the new application areas that could become large volume applications are the use of molecular sieves as functional powders, in odor removal, and as plastic additives. The use of zeolites in solid state applications is highly speculative. If ever practically realized that application would most probably represent a relatively small volume of the total zeolite consumption.

7. HISTORY OF INTERNATIONAL CONFERENCES AND ORGANIZATIONS

In 1957 the first informal molecular sieve conference was held at Pennsylvania State University in the U.S.A. In 1967 the first of a series of international molecular sieve conferences chaired by Professor R.M. Barrer was held in London. Subsequently, international meetings have been held every several years - in 1970 at Worcester, 1973 at Zurich, 1977 at Chicago, 1980 at Naples, 1983 in Reno, 1986 in Tokyo, 1989 in Amsterdam, 1992 in Montreal, 1994 in Garmisch-Partenkirchen, 1996 in Seoul, and 1998 in Baltimore.

An international molecular sieve organization was first formed in 1970 in conjunction with the Worcester Conference, and called the International Molecular Sieve Conference (IMSC). Its responsibility was to continue the organizational implementation of future international molecular sieve conferences on a regular basis. In 1977 at the Chicago meeting the name of the organization was changed to the International Zeolite Association (IZA) and its scope and purpose expanded to "promote and encourage all aspects of the science and technology of zeolitic materials", as well as organizing "International Zeolite Conferences" on a regular basis. The term zeolite in the new organization "is to be understood in its broadest sense to include both natural and synthetic zeolites as well as molecular sieves and other materials having related properties and/or structures" (ref. 72). International Zeolite Association regional affiliates have been established, and include: the British Zeolite Association (BZA), in 1980; the Japan Association of Zeolites (JAZ), in 1986; and regional zeolite associations in France, Italy, Hungary and Germany in the late 80's and early 90's. In 1995, a Federation of European Zeolite Associations (FEZA) was formed and presently includes the British, Bulgarian, Dutch, French, German, Hungarian, Italian, Romanian, Spanish, Polish and Georgian Zeolite Associations. FEZA sponsors several workshops in Europe each year, covering various aspects of zeolite science, and held its first International FEZA Conference in Hungary in 1999. The Northeast Corridor Zeolite Association (NECZA) was established in the United States in 1997 and holds an annual Symposium.

In addition to these organizations the IZA has several established Commissions. The first is the Structure Commission formed in 1977, which has published four editions of the *Atlas of Zeolite Structure Types* (1978, 1987, 1992, 1996). It should be noted that an up-to-date version of the *Atlas* is maintained on the World Wide Web at the Structure Commission's Web site (ref. 73). Subsequently commissions were established in the area of catalysis (1988) and synthesis (1992). An adsorption and ion exchange commission has been proposed. In 1988 the Consortium for Theoretical Frameworks was organized by J. V. Smith at the University of Chicago.

8. HISTORICAL EPILOGUE

Key factors in the growth of molecular sieve science and technology include: the pioneering work of Barrer: the key discoveries of Milton and Breck and associates at Union Carbide: the rapid commercialization of the new synthetic zeolites and their applications by

Union Carbide (1949-1954); the major development at Union Carbide in adsorption process design and engineering technology; major discoveries in hydrocarbon conversion catalysts at Union Carbide, Exxon, Mobil Oil, Shell and other industrial laboratories; the discovery and commercialization of sedimentary zeolite mineral deposits in the United States in the 60's; and last, but not least, the dedication and contribution of so many high quality scientists and engineers. It is estimated that by the end of the twentieth century there were over ten to twenty thousand such scientists and engineers in industry and academia dedicating a significant portion of their work to zeolite and molecular sieve science and technology.

REFERENCES

1. D.W. Breck, Zeolite Molecular Sieves, Structure, Chemistry and Use, John Wiley & Sons, New York, 1974; reprinted by Krieger, Malabar, Florida, 1984.
2. R.M. Barrer, Pure and Appl. Chem., 51 (1979) 1091-1100.
3. D.S. Coombs, A. Alberti, T. Armbruster, G. Artioli, C. Colella, E. Galli, J.D. Grice, F. Liebau, J.A. Mandarino, H. Minato, E.H. Nickel, E. Passaglia, D.R. Peacor, S. Quartieri,R. Rinaldi, M. Ross, R.A. Sheppard, E. Tillmanns, G. Vezzalini, Can. Mineral. 35 (1997) 1571-1606.
4. A.F. Cronstedt, Akad. Handl. Stockholm, 18 (1756) 120-130.
5. F. Fontana, Memorie Mat. Fis. Soc. Ital. Sci., 1 (1777) 679.
6. A. Damour, Ann. Mines, 17 (1840) 191.
7. G.W. Morey and E. Ingerson, Econ. Geol., 32 (1937) 607.
8. H.S. Thompson, J. Roy. Agr. Soc. Engl., 11 (1850) 68; J.T. Way, ibid. p. 313.
9. H. Eichhorn, Poggendorf Ann. Phys. Chem., 105 (1858) 126.
10. H. de St. Claire-Deville, Compt. Rend., 54 (1862) 324.
11. G. Friedel, Bull. Soc. Franc. Mineral. Cristallogr., 19 (1896) 94-118.
12. F. Grandjean, Compt. Rend., 149 (1909) 866-68.
13. O. Weigel and E. Steinhoff, Z. Kristallogr., 61 (1925) 125-54.
14. R.J. Leonard, Econ. Geol., 22 (1927) 18-43.
15. W.H. Taylor, Z. Kristallogr., 74 (1930) 1.
16. L. Pauling, Proc. Nat. Acad. Sci., 16 (1930) 453; Z. Kristallogr., 74 (1930) 213.
17. J.W. McBain, The Sorption of Gases and Vapors by Solids, Rutledge and Sons, London, 1932, Chap. 5.
18. R.M. Barrer, J. Soc. Chem. Ind., 64 (1945) 130.
19. R.M. Barrer, J. Chem. Soc., (1948) 2158.
20. R.M. Milton, in M.L. Occelli and H.E. Robson (Eds.), Zeolite Synthesis, ACS Sympos. Ser. 398, American Chemical Society, Washington, D.C., 1989, pp. 1-10.
21. M. Smart, T. Esker, A. Leder and K. Sakota, Zeolites Marketing Research Report in Chemical Economics Handbook, SRI International, August 1999, pp. 599.1000-1002.
22. F.A. Mumpton, in D. Olson and A. Bisio (Eds.), Proc. 6th Intl. Zeolite Conf., Reno, USA, July 10-15, 1983, Butterworths, Guilford, Surrey, UK, 1984, pp. 68-82.
23. K. Torii, in L.B. Sand and F.A. Mumpton (Eds.), Natural Zeolites, Occurrence, Properties, Use, Pergamon Press, New York, 1978, pp. 441-50.
24. E.M. Flanigen, in L.V.C. Rees (Ed.), Proc. 5th Intl. Conf. On Zeolites, Naples, Italy, June 2-6, Heyden, London, 1980, pp. 760-80.

25. S.T. Wilson, B.M. Lok, C.A. Messina, T.R. Cannan and E.M. Flanigen, J. Am.Chem. Soc., 104 (1982) 1146-47.
26. E.M. Flanigen, B.M. Lok, R.L. Patton and S.T. Wilson, in Y. Murakami, A. Ijima and J.W. Ward (Eds.), New Developments in Zeolite Science and Technology, Proc. 7th Intl. Zeolite Conf., Tokyo, Aug. 17-22, 1986, Kodansha Ltd., Tokyo and Elsevier, Amsterdam, 1986, pp. 103-12.
27. R. Szostak, Molecular Sieves, Principles of Synthesis and Identification, 2nd Ed., Blacklie Academic & Professional, London, 1998, pp. 208-244.
28. J.M. Bennett, W.J. Dytrych, J.J. Pluth, J.W. Richardson and J.V. Smith, Zeolites, 6 (1986) 349-60.
29. M.E. Davis, C. Saldarriaga, C. Montes, J. Garces and C. Crowder, Nature (London), 331 (1988) 698-9.
30. J.A. Martens, M. Mertens, P.J. Grobet and P.A. Jacobs, in P.J. Grobet, W.J. Mortier, E.F. Vansant and G. Schulz-Ekloff (Eds.), Innovation Zeolite Mater. Sci., Stud. Surf. Sci. Catal. 37, Elsevier, Amsterdam, 1988, pp. 97-105.
31. R.A. Van Nordstrand, D.S. Santilli and S.I. Zones, in W.H. Flank and T.E.Whyte, Jr. (Eds.), Perspect. Mol. Sieve Sci., ACS Symp. Ser. 368, 1988, pp. 236-45.
32. J.M. Bennett and B.K. Marcus, in P.J. Grobet, W.J. Mortier, E.F. Vansant and G. Schulz-Ekloff (Eds.), Innovation Zeolite Mater. Sci., Stud. Surf. Sci. Catal. 37, Elsevier Amsterdam, 1988, pp. 269-79.
33. S. Ueda, M. Koizumi, Ch. Baerlocher, L.B. McCusker and W.M. Meier, Preprints of Poster Papers 7th IZC (1986), p.23.
34. G. Harvey and W.M. Meier, in P.A. Jacobs and R.A. van Santen (Eds.), Zeolites: Facts, Figures, Future, Stud. Surf. Sci. Catal. 49A, Elsevier, Amsterdam, 1989, pp. 411-20.
35. J. Wang, S. Feng and R. Xu, in P.A. Jacobs and R.A. van Santen (Eds.), Zeolites: Facts, Figures, Future, Stud. Surf. Sci. Catal. 49A, Elsevier, Amsterdam, 1989, pp. 143-50.
36. G. Yang, L. Li, J. Chen and R. Xu, J. Chem. Soc. Chem. Commun., (1989) 948.
37. J. Chen and R. Xu, J. Solid State Chem., 80 (1989) 149-51.
38. J.A. Martens and P.A. Jacobs, in J. Weitkamp and L. Puppe (Eds), Catalysis and Zeolites, Fundamentals and Applications, Springer-Verlag, Berlin, 1999, pp. 53-80.
39. Y. Yamagishi, S. Namba and T. Vashima, in P.A. Jacobs and R.A. van Santen (Eds.), Zeolites: Facts, Figures, Future, Stud. Surf. Sci. Catal. 49A, Elsevier, Amsterdam, 1989, pp. 459-67.
40. Z. Gabelica and J.L. Guth, in P.A. Jacobs and R.A. van Santen (Eds.), Zeolites, Facts, Figures, Future, Stud. Surf. Sci. Catal. 49A, Elsevier, Amsterdam, 1989, pp. 421-30.
41. D.R. Peacor, R.C. Rouse and J.-H. Ahn, Am. Mineral., 72 (1987) 816-20.
42. R.C. Rouse, O.R. Peacor and S. Merlino, Am. Mineral., 74 (1989) 1195-202.
43. R.L. Bedard, S.T. Wilson, L.D. Vail, J.M. Bennett and E.M. Flanigen, in P.A. Jacobs and R.A. van Santen (Eds.), Zeolites: Facts, Figures, Future, Stud. Surf. Sci. Catal. 49A, Elsevier, Amsterdam, 1989, pp. 375-87.
44. R.W.J. Scott, M. MacLachlan, G.A. Ozin, Curr. Opin. Solid State Mater. Sci., 4 (1999) 113-121; C.L. Bowes and G.A. Ozin, Adv. Mat. (Weinheim), 8 (1996) 13-28.
45. M. Estermann, L.B. McCusker, Ch. Baerlocher, A. Merrouche and H. Kessler, Nature (London), 352 (1991) 320.
46. C.C. Freyhardt, M. Tsapatsis, R. F. Lobo, K.J. Balkus Jr., and M.E. Davis, Nature

(London), 381 (1996) 295; R.F. Lobo, M. Tsapatsis, C.C. Freyhardt, S. Khodabandeh, P. Wagner, C.-Y. Chen, K.J. Balkus Jr., S.I. Zones and M.E. Davis, J. Am. Chem. Soc., 19 (1997) 8474.
47. T.E. Gier and G.D. Stucky, Nature (London), 349 (1991) 508.
48. L. Puppe and J. Weisser, U.S. Patent No. 4 439 409 (1984); M.E. Leonowicz, J.A. Lawton, S.L. Lawton and M.K. Rubin, Science, 264 (1994) 1910.
49. S.L. Lawton, A.S. Fung, G.J. Kennedy, L.B. Alemany, C.D. Chang, G.H. Hatzikos, D.N. Lissy, M.K. Rubin, H.-K.C. Timken, S. Steuernagel and D.E. Woessner, J. Phys. Chem., 100 (1996) 3788-3798.
50. P. Wagner, M. Yoshikawa, M. Lovalla, K. Tsuji, M. Tsapatsis and M.E. Davis, J. Chem. Soc. Chem. Comm., 22 (1997) 2179.
51. F. Delprato, L. Delmotte, J.L. Guth and L. Huve, Zeolites, 10 (1990) 546.
52. J. J. Pluth and J.V. Smith, Am. Mineral., 75 (1990) 501.
53. R.C. Boggs, D.G. Howard, J. V. Smith and G.L. Klein, Am. Mineral., 78 (1993) 822.
54. A. Alberti, G. Vezzalini, E. Galli and S. Quartieri, Eur. J. Mineral., 8 (1996) 69-75.
55. E. Galli, S. Quartieri, G. Vezzalini, A. Alberti and M. Franzini, Am. Mineral., 82 (1997) 423.
56. G. Vezzalini, S. Quartieri, E. Galli, A. Alberti, G. Cruciani, A. Kvick, Zeolites, 19 (1997) 323.
57. Y.P. Menshikov, Zap. Vses. Mineral. O-va, 113 (1984) 607; G. Artioli and A. Kvick, Eur. J. Mineral., 2 (1990) 749.
58. H. Effenberger, G. Giester, W. Krause and H.-J. Bernhardt, Am. Mineral., 83 (1998) 607.
59. D. M. Chapman and A.L. Roe, Zeolites, 10 (1990) 730.
60. S. M. Kuznicki, K.A. Trush, F.M. Allen, S.M. Levine, M.M. Hamil, D.T. Hayhurst, M. Mansom, in M.L. Ocelli and H.E. Robson (Eds.), Synthesis of Microporous Materials, Van Nostrand Reinhold, New York, 1992, pp. 427-453 (Molecular Sieves, Vol. 1).
61. M.W. Anderson, O. Terasaki, T. Oshuna, A. Philippou, S.P. MacKay, A. Ferreira, J. Rocha and S. Lidin, Nature (London), 367 (1994) 347.
62. C.T. Kresge, M.E. Leonowicz, W.J. Roth, J.C. Vartuli and J.S. Beck, Nature (London), 359 (1992) 710-712; J.S. Beck, J.C. Vartuli, W.J. Roth, M.E. Leonowicz, C.T. Kresge, K.D. Schmitt, C.T.W. Chu, D.H. Olson, E.W. Sheppard, S.B. McCullen, J.B. Higgins, and J.L. Schlenker, J. Am. Chem. Soc., 114 (1992) 10835.
63. S. Inagaki, Y. Fukashima and K. Kuroda, J. Chem. Soc. Chem. Commun., (1993) 680; S. Inagaki, Y. Fukashima and K. Kuroda, in J. Weitkamp, H.G. Karge, H. Pfeifer and W. Holderich (Eds.), Zeolites and Related Microporous Materials: State of the Art 1994, Stud. Surf. Sci. Catal. 84, Elsevier, Amsterdam, 1994, pp. 125-132; S. Inagaki, Y. Fukashima, A. Akada, T. Kurauchi, K. Kuroda and C. Kato, in R. von Ballmoos, J.B. Higgins and M.M.J. Treacy (Eds.), Proceedings from the Ninth International Zeolite Conference, Montreal 1992, Butterworth-Heinemann, London, 1993, pp. 305-311; T. Yanagisawa, T. Shimizu, K. Kuroda and C. Kato, Bull. Chem. Soc. Jpn., 63 (1990) 998-992.
64. C.-Y. Chen, S.-Q. Xiao, and M.E. Davis, Microporous Materials, 4 (1995) 1-20.
65. J.S. Beck and J.C. Vartuli, Curr. Opin. Solid State Mater. Sci., 1 (1996) 4; D. Zhao, P. Yang, Q. Huo, G.D. Stucky and B.F. Chmelka, ibid., 3 (1998) 111.
66. F. Di Renzo, H. Cambon and R. Dutarte, Microporous Materials, 10 (1997) 283-286; V.

Chiola, J.E. Ritsko and C.D. Vanderpool, U.S. Patent No. 3 556 725 (1971), assigned to Sylvania Electric Products, Inc.

67. J.V. Smith, in P.A. Jacobs and R.A. van Santen (Eds.), Zeolites: Facts, Figures, Future, Stud. Surf. Sci. Catal. 49A, Elsevier, Amsterdam, 1989, pp. 29-47.
68. G.A. Ozin, A. Kuperman and A. Stein, Angew. Chem. Int. Ed. Engl., 28 (1989) 359-76.
69. R.L. Bedard and E.M. Flanigen, in R.L. Bedard, T. Bein, M.E. Davis, J. Garcia, V.A. Maroni and G.D. Stucky (Eds.), Synthesis/Characterization and Novel Applications of Molecular Sieve Materials, Materials Research Society, Pittsburgh, Mat. Res. Soc. Symp. Proc. Vol. 233, 1991, pp. 219-224.
70. E. M. Flanigen and R. L. Patton, U.S. Patent No. 4 073 865 (1978).
71. H. Kessler, J. Patarin and C. Schott-Darie, in J.C. Jansen, M. Stocker, H.G. Karge and J. Weitkamp (Eds.), in Advanced Zeolite Science and Applications, Stud. Surf. Sci. Catal. 85, Elsevier, Amsterdam, 1994, pp. 75-113.
72. International Zeolite Association, Newsletter 1, Aug. 22, 1977.
73. http://www.iza-structure.org/

Studies in Surface Science and Catalysis 137
H. van Bekkum, E.M. Flanigen, P.A. Jacobs and J.C. Jansen (Editors)
© 2001 Elsevier Science B.V. All rights reserved.

Chapter 3

Zeolite Structures

Lynne B. McCusker and Christian Baerlocher

Laboratorium für Kristallographie, ETH, Zürich, Switzerland

1. INTRODUCTION

The fascinating and wonderfully exploitable properties of zeolitic materials, such as their ion-exchange properties, their sorption capacity, their shape selectivity, their catalytic activity or their role as hosts in nanocomposite materials, are essentially determined by their structures. For example, sorption characteristics depend upon the size of the pore openings and the void volume; ion-exchange selectivity upon the number and nature of the cation sites and their accessibility; catalytic behavior upon the pore openings, the dimensionality of the channel system, the cation sites, and the space available for reaction intermediates; and host applications on the size and spacing of the cages. Consequently, structural analysis is a fundamental aspect of zeolite science.

Information on the framework type alone can elucidate many of the observed properties of a zeolite. The framework type, which just describes the connectivity (topology) of the framework tetrahedral atoms in the highest possible symmetry without reference to chemical composition, defines the size and shape of the pore openings, the dimensionality of the channel system, the volume and arrangement of the cages and the types of cation sites available. Nonetheless, the chemical composition of the framework, the nature of the species within the channels, and the type of post-synthesis modification also play a very important role in determining the specific properties of a particular zeolitic material. For example, an aluminosilicate framework has a negative charge whereas an aluminophosphate is neutral, a large cation can block or reduce the effective size of a pore opening, a small cation might distort a pore opening, or a sorbed species can influence the catalytic, optical, magnetic or electronic properties of a zeolite. Precise structural details, such as the nature of the distortion of a framework from ideal symmetry or the exact location of non-framework species, are often needed to fully understand the properties of a specific zeolite.

Unfortunately, most synthetic zeolitic materials are polycrystalline. That is, single crystals of a size suitable for the application of traditional crystallographic methods of structure analysis (i.e. ~ 50-100 μm on an edge) are rare. However, zeolites are crystalline with well-defined periodicity even if the crystallites are small, so structural elucidation using powder diffraction data is possible, though not quite so straightforward. Usually a number of analytical techniques are combined to probe the structure of a zeolite. These include sorption experiments (pore size and accessibility), solid state NMR (short range order), electron microscopy (symmetry, faulting), and powder diffraction.

The following sections of this chapter will cover (1) descriptions of selected zeolite framework types, (2) a discussion of some aspects of real zeolite structures, and (3) a summary of the information that can be extracted from a powder diffraction pattern.

2. ZEOLITE FRAMEWORK TYPES

Because zeolite scientists recognized very early that zeolite framework structures are fundamental to the understanding of zeolite chemistry [1], classification of zeolitic materials by framework type, first proposed by Meier and Olson in 1970 [2], has gained wide acceptance in the zeolite community. A framework *type*, as opposed to a framework *structure*, simply describes the connectivity of the tetrahedrally coordinated atoms (T-atoms) of the framework in the highest possible symmetry. The framework composition, the observed symmetry, and the actual unit cell dimensions are not considered. In this way, many different materials can be classified under one designation. For example, amicite, garronite, gismondine, gobbinsite, Na-P1, Na-P2 and SAPO-43 all have the gismondine framework type (**GIS**). A three letter code (e.g. **GIS**) is assigned to confirmed framework types by the Structure Commission of the International Zeolite Association according to rules set up by an IUPAC Commission on Zeolite Nomenclature [3,4]. The codes are normally derived from the name of the zeolite or "type material", e.g. **FAU** from the mineral **fau**jasite, **LTA** from **L**inde **T**ype **A**, and **MFI** from ZSM-5 (Zeolite Socony **M**obil - **fi**ve). Information pertinent to these framework types is published in the *Atlas of Zeolite Framework Types* [5] and on the internet at http://www.iza-structure.org/databases/. As new codes are approved, they are announced on the IZA Structure Commission's WWW pages (http://www. iza-structure.org/) and included in the internet version of the *Atlas*. As of January 2001, 133 zeolite framework types had been confirmed by the

Structure Commission. In this chapter, all references to materials whose framework types are known will be accompanied by the appropriate three letter code in boldface type.

2.1. Characteristics of zeolite framework types

The feature that is common to zeolite or zeolite-like materials is that they all have a 3-dimensional, 4-connected framework structure constructed from corner-sharing TO_4 tetrahedra, where T is any tetrahedrally coordinated cation. This framework structure is relatively open and characterized by the presence of channels and cavities. A description of a zeolite structure almost always begins with a description of the framework type in terms of the size of the pore openings and the dimensionality of the channel system.

Pore openings are characterized by the size of the ring that defines the pore, designated an *n*-ring, where *n* is the number of T-atoms in the ring. An 8-ring is considered to be a small pore opening, a 10-ring a medium one, and a 12-ring a large one, with free diameters (calculated using an oxygen radius of 1.35 Å) of approximately 4.1, 5.5 and 7.4 Å, respectively. Of course, rings can be distorted considerably so these numbers should only be used as a rough guide.

A number of structural features (cages, channels, chains, sheets) are

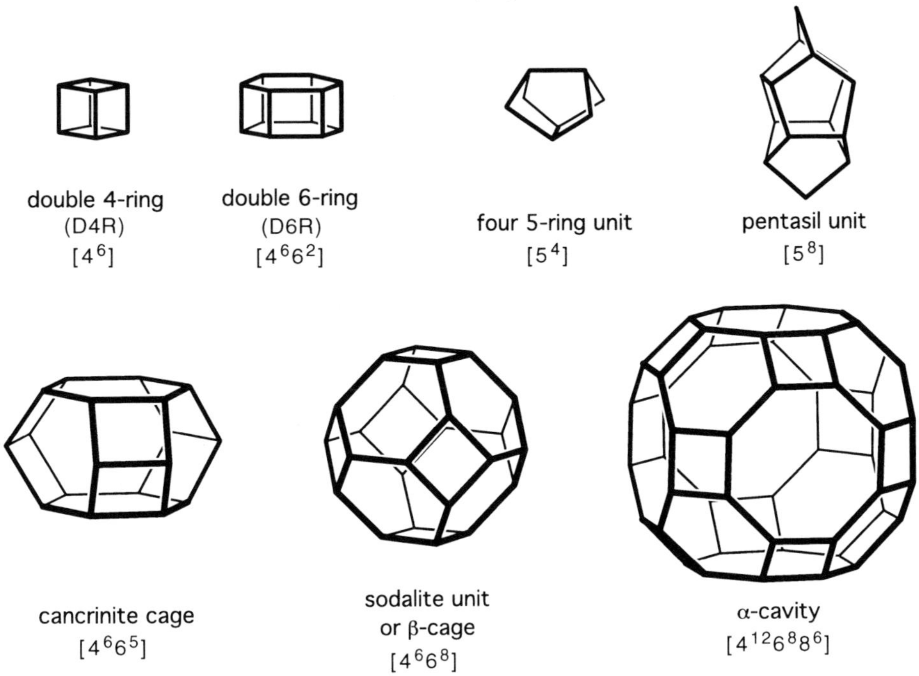

Figure 1. Some subunits and cages that recur in several framework types.

common to several different zeolite framework types, so designations such as α-cavity and β-cage, pentasil unit, crankshaft and double crankshaft chain, and 4.8^2 sheet or net have crept into common usage. To help the reader, some of these subunits are shown in Figures 1, 2 and 3. In these drawings, oxygen bridges have been omitted for clarity.

Since cages are sometimes described in terms of the *n*-rings defining their faces, these designations are also given in Figure 1. For example, a truncated octahedron (sodalite unit), whose surface is defined by six 4-rings and eight 6-rings, would be designated a $[4^6 6^8]$ cage. The three double chains in Figure 2 also occur as single chains in many zeolites, but these are so common that they are seldom discussed as a characteristic feature of a structure. However they can be put to good use in model building (see Chapter 4). The Narsarsukite chain is found more often in $AlPO_4$-structures than in silicates, whereas the pentasil chain of edge-sharing $[5^8]$ cages is characteristic of a family of high silica zeolites (**MFI**, **MEL**).

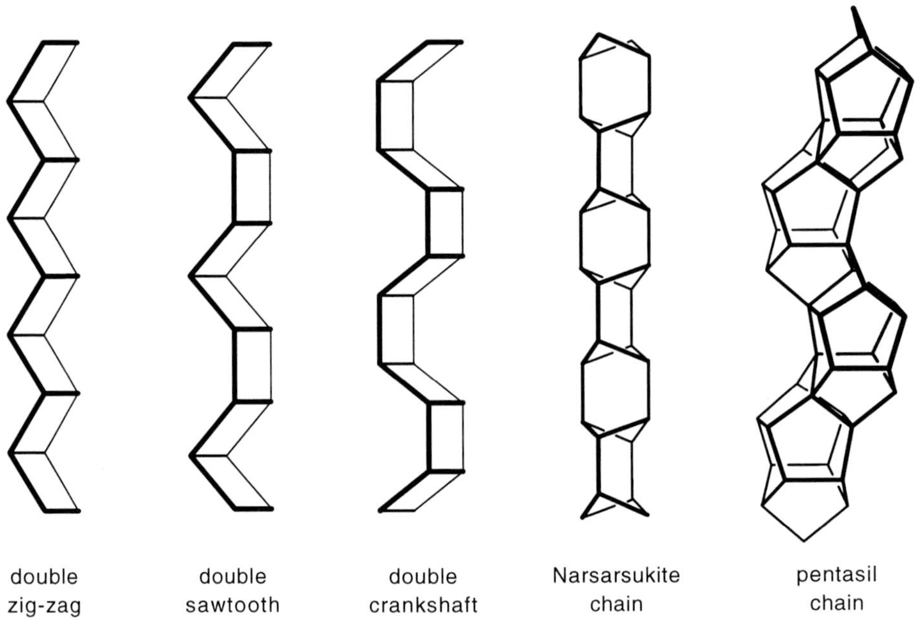

Figure 2. Some chains that recur in several framework types.

The channel walls of zeolites with 1-dimensional pores are often composed entirely of 6-rings. The two possible orientations of the 6-rings in such "6-ring wraps" are shown in Figure 3 for the 12-ring channels in **AFI** and **CAN**.

A nomenclature similar to that used for cages has also been developed to describe 2-dimensional, 3-connected sheets or nets. In this case, the sizes of the three *n*-rings associated with each node are used for the designation. In the net shown in Figure 3, for example, each node is associated with two 8-rings and one 4-ring and is therefore called a 4.8^2 net. To complete the 3-dimensional description, the orientation of the fourth connection can also be given as U or D (i.e. pointing **U**p or **D**own from the sheet). The example given in Figure 3 describes the 4.8^2 sheet found in the **GIS** framework type, where the connections from half of each 8-ring point up and the other half point down. The 8-rings containing the letters correspond to the central 8-ring (front layer) in the **GIS** framework type shown in the next section in Figure 10. Another example of a framework type that can be described in terms of a 4.8^2 sheet is that of **ABW**, which has an UUDUDDDUD orientation of tetrahedra around the 8-rings.

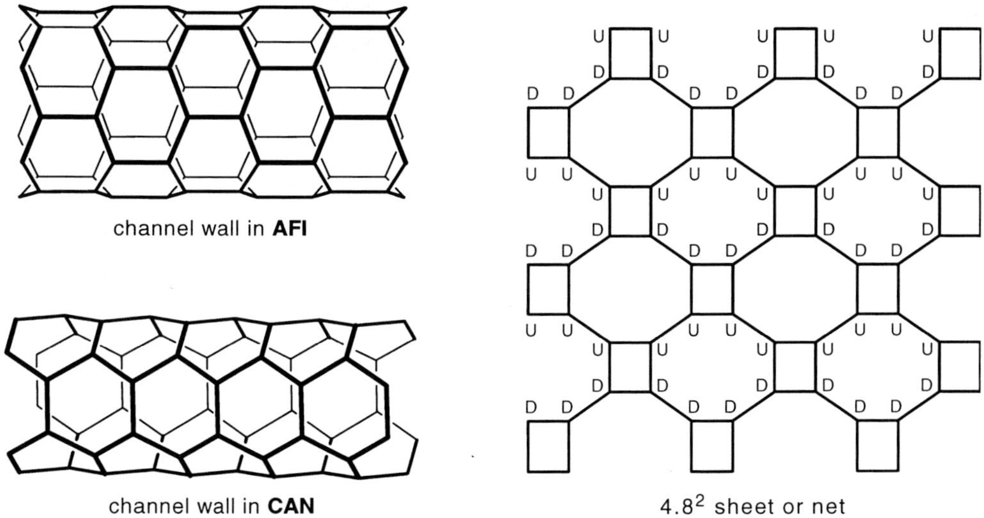

Figure 3. Two types of channel walls composed of 6-rings (left), and the **GIS** 4.8^2 sheet (right).

Some frameworks consist only of cages with a maximum ring size of six and have no channels (e.g. the pure-silica clathrasils), but the majority have at least 8-ring channels. These channels can intersect to form 2- and 3-dimensional channel systems, and this can be a critical feature for catalytic or sorption applications. For example, a 1-dimensional channel is much more easily blocked by the formation of coke deposits than is a higher dimensional one where "detours" are possible.

The stacking sequence of layers, cages or rings in zeolite frameworks is often described using the "ABC-system". This crystal chemistry terminology, which is usually used to describe the stacking of layers of closest packed spheres (atoms) in metals or oxides, has been adapted to describe stackings in certain types of zeolite structures. For example, fifteen of the 133 zeolite framework types can be described in terms of stackings of hexagonal arrays of 6-rings (Figure 4), and are know as the ABC-6 family of zeolite frameworks (see **SOD** and **CHA** in the following section). Other stackings described using the ABC terminology involve sheets of sodalite cages (see **FAU** and **EMT** in the following section). This concept of stacking sequences is not only an elegant way of describing a family of frameworks, but also appears to reflect the way nature builds real materials with such frameworks (see section 3.3).

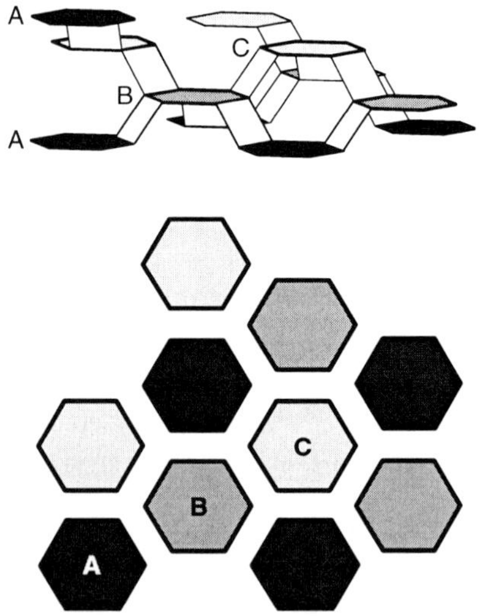

Figure 4. ABC-stacking of hexagonal arrays of 6-rings viewed from the side (top) and in projection (bottom).

Zeolite frameworks can be classified according to various schemes (e.g. by pore opening, by structural subunit, by channel system, by framework density, by loop configurations, and/or by coordination sequences). Most of these features are defined in the introductory pages of the *Atlas of Zeolite Framework Types* and then given for each framework type. It is perhaps worth noting that the set of coordination sequences and loop configurations for each of the T-atoms in a given framework type is unique, so this is a good way of determining whether or not the framework of a new zeolite is novel.

2.2. Selected zeolite framework types

Although there are 133 confirmed zeolite framework types, only a few of them describe zeolites or zeolite-like materials that are actually used in industrial applications. Fifteen have been selected for a more detailed description here. Some have been chosen because of their industrial relevance, and some because they illustrate specific structural features. They are presented in order of the historical development of zeolite synthesis from aluminosilicates to high silica zeolites to aluminophosphate molecular sieves to gallophosphates to zincosilicates. No ranking is implied. In the drawings of the frameworks

(Figures 5-19), the nodes represent T-atoms and the lines oxygen bridges. For clarity, rings with fewer than eight T-atoms have been made opaque.

The selected aluminosilicates are sodalite (**SOD**), zeolite A (**LTA**), faujasite (**FAU**), EMC-2 (**EMT**), chabasite (**CHA**) and gismondine (**GIS**). With the exception of **EMT**, all of these framework types have also been synthesized as aluminophosphates. The high silica zeolites, with a Si/Al ratio of at least 5, are ZSM-5 (**MFI**), ZSM-11 (**MEL**), mordenite (**MOR**), MCM-22 (**MWW**) and zeolite beta (***BEA**). The common feature of these framework types is the presence of 5-rings. To complete the spectrum, two aluminophosphates, $AlPO_4$-5 (**AFI**) and VPI-5 (**VFI**), one gallophosphate, cloverite (**-CLO**), and one zincosilicate, VPI-9 (**VNI**) will be discussed.

2.2.1. **SOD** (type material: sodalite)

In the strictest sense of the word, sodalite is not a zeolite, since it has only 6-ring windows and thus has only very limited sorption capacity. It has a framework density of 17.2 T-atoms per 1000Å^3. Nonetheless, it is an important material for creating simple periodic arrays of clusters, and is one of the most seriously investigated hosts for advanced materials [6]. Furthermore, it has much in common with some of the zeolites used in industrial applications. The **SOD** framework type (Figure 5) is best described as a body-centered cubic arrangement of β or sodalite cages (see Figure 1) joined through single 4- and 6-rings. It is also a member of the ABC-6 family of zeolites [7], and can be viewed as an ABCABC stacking of hexagonal arrays of single 6-rings in the [111] direction (the body diagonal of the cubic unit cell).

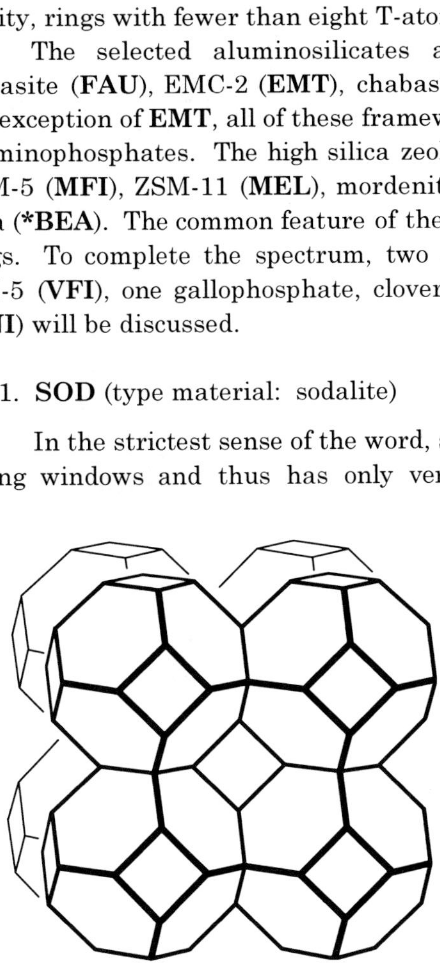

Figure 5. The **SOD** framework type

2.2.2. **LTA** (type material: Linde Type A)

The **LTA** framework type (Figure 6) is related to **SOD**, but in this case, the sodalite cages, in a primitive cubic arrangement, are joined via double 4-rings rather than single ones. This creates an α-cage (see Figure 1) instead of a β-cage in the center of the unit cell, and a 3-dimensional, 8-ring channel system. Alternatively, the framework can be described as a primitive cubic arrangement of α-cages joined through single 8-rings (producing a sodalite cage in the center). This is one of the more open zeolite framework types with a framework density of only 12.9 T-atoms per 1000Å³. Zeolite A is used as a desiccant both in the laboratory and between the panes of glass in double-glazed windows, and as an ion-exchanger in laundry detergents.

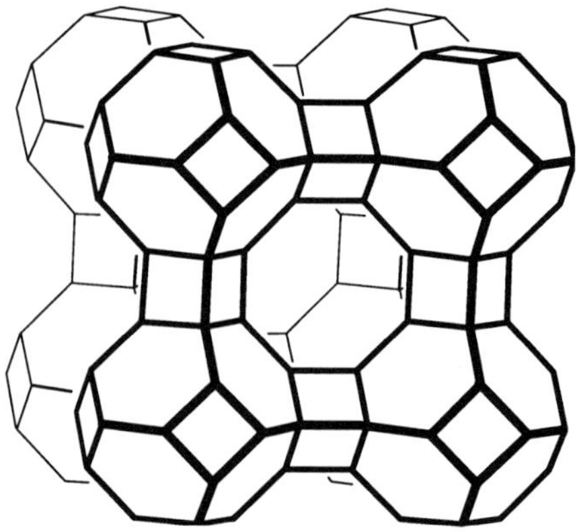

Figure 6. The **LTA** framework type

2.2.3. **FAU** (type material: faujasite)

There are also sodalite cages in the **FAU** framework type (Figure 7). In this case, they are arranged in the same way as the carbon atoms in diamond, and are joined to one another via double 6-rings. This creates the so-called supercage with four, tetrahedrally-oriented, 12-ring pore openings, and a 3-

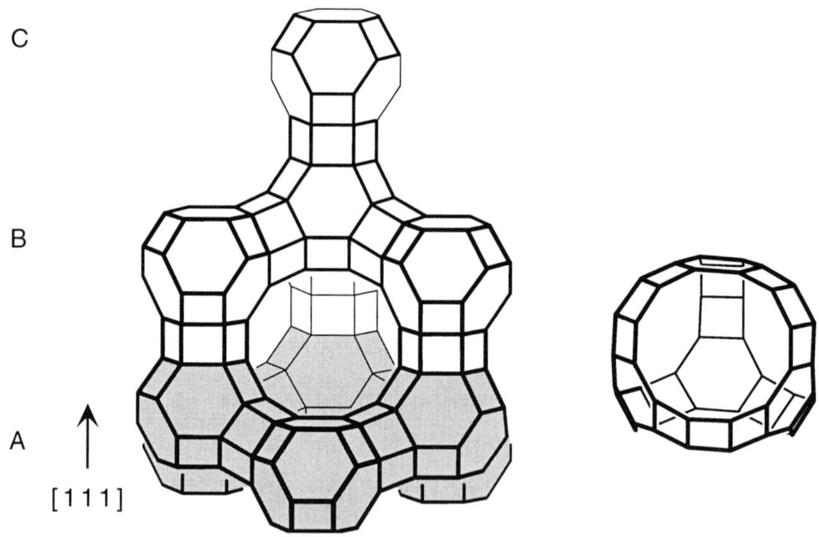

Figure 7. The **FAU** framework type and its supercage. The three different layers of sodalite cages are indicated with the letters A, B and C. Layer A is highlighted in gray.

dimensional channel system. The framework density, at 12.7 T-atoms/1000 Å^3, is even lower than that of **LTA**. There is a center of inversion in each of the double 6-rings, so the puckered layers of sodalite cages are related to one another by inversion. The framework type can also be described as an ABCABC stacking of such layers. The combination of large void volume (ca 50%), 12-ring pore openings and 3-dimensional channel system makes the thermally stable silicate materials with the **FAU** framework type ideal for many catalytic applications.

2.2.4. **EMT** (type material: EMC-2)

In the same way that lonsdaleite is a hexagonal analog of diamond (or wurtzite one of zinc blende), the **EMT** framework type (Figure 8) is the simplest

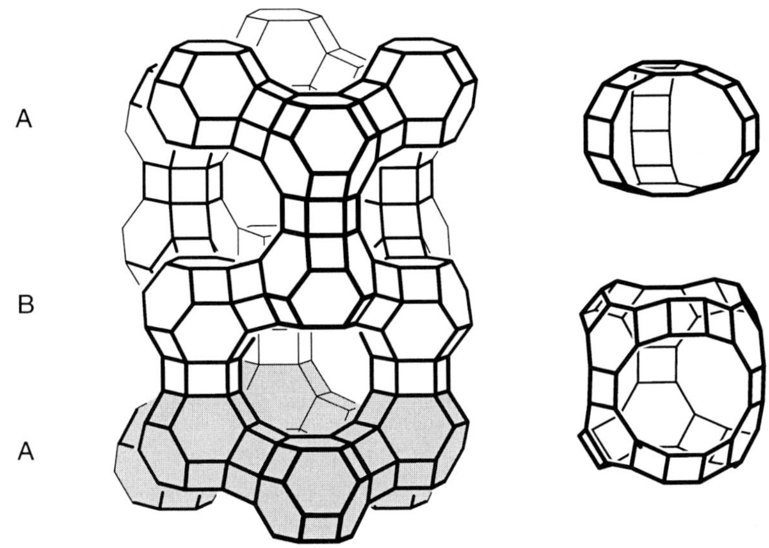

Figure 8. The **EMT** framework type showing the medium and larger cages separately. The two different layers of sodalite cages are indicated with the letters A and B. Layer A is highlighted in gray.

hexagonal analog of **FAU**. In **EMT**, the puckered sodalite cage layers are stacked in an ABAB sequence and the layers are related to one another by a mirror plane. This arrangement of sodalite cages creates a medium cage with three 12-ring pore openings and a larger cage with five. As in **FAU**, the resulting channel system is 3-dimensional with 12-ring pores, but the nature of the channel system and of the larger cages in the **EMT** framework type is significantly different. As might be expected, this framework type is also well-suited for catalytic applications.

2.2.5. **CHA** (type material: chabasite)

The **CHA** framework type (Figure 9) is another member of the ABC-6 family of zeolite frameworks. While **SOD** can be described in terms of an ABC stacking of hexagonal arrays of single 6-rings, **CHA** has an ABC stacking of double 6-ring arrays (or an AABBCC stacking of single 6-ring arrays). This

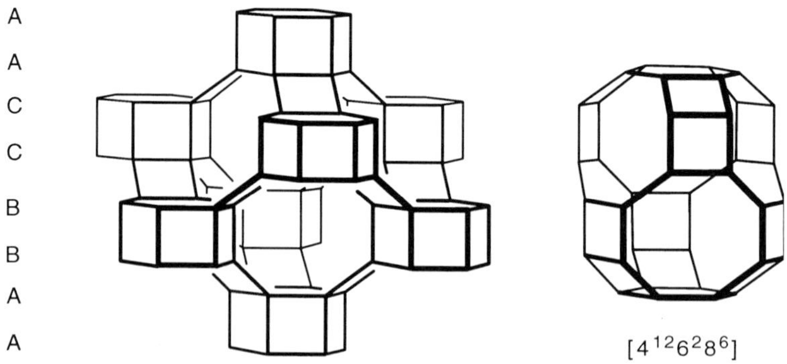

Figure 9. The **CHA** framework type (AABBCC 6-ring stacking indicated) and its cage.

stacking produces an elongated cage with six 8-ring pores and a 3-dimensional channel system. Unlike the previous examples, the channels in **CHA** are not straight. The silicoaluminophosphate with this framework type is used in the conversion of methanol to olefins and in the aldol condensation of aldehydes.

2.2.6. **GIS** (type material: gismondine)

The **GIS** framework type (Figure 10) can be described as a stacking of 2-dimensional arrays of double crankshaft chains (Figure 2). There are 8-ring channels running parallel to x and y, displaced with respect to one another along z. They intersect to form a 3-dimensional channel system. The double crankshaft chains are very flexible, and so is the **GIS** framework. Materials with this framework type have symmetries varying from monoclinic (e.g. gismondine) to orthorhombic (e.g. gobbinsite) to tetragonal (e.g. garronite) and the lattice parameters

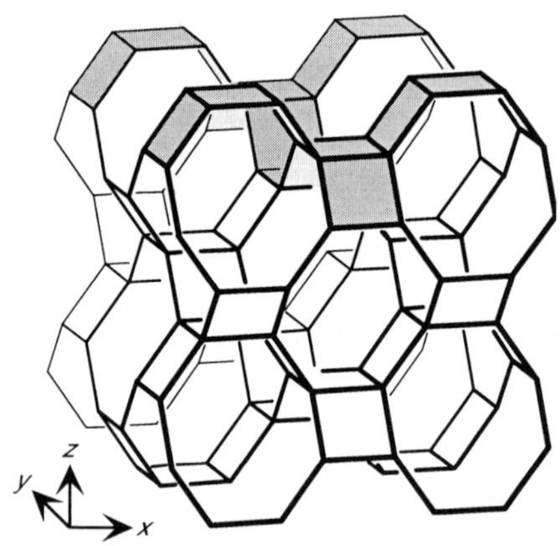

Figure 10. The **GIS** framework type with a double crankshaft layer highlighted.

can differ by as much as 6%. The framework type can also be described in terms of 4.8^2 nets stacked along the *x* or *y* direction (see section 2.1). The "maximum aluminum P" zeolite (or MAP for short), which is used as an ion-exchanger in laundry detergents, has this framework type.

2.2.7. **MFI** (type material: ZSM-5)

The framework type of the high silica zeolite ZSM-5 (Figure 11) can be described in terms of $[5^4]$ units, but it is easier to use pentasil units (Figure 1). These $[5^8]$ units are linked to form pentasil chains (Figure 2), and mirror images of these chains are connected via oxygen bridges to form corrugated sheets with 10-ring holes (e.g. the gray sheet perpendicular to *x* in Figure 11). Each sheet is linked by oxygen bridges to the next to form the 3-dimensional structure.

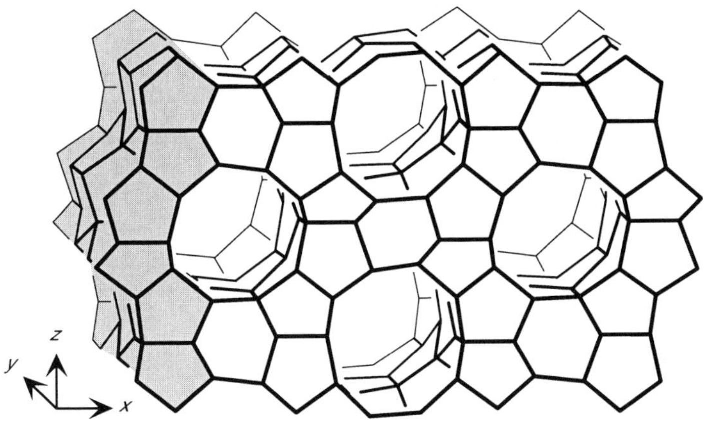

Figure 11. The **MFI** framework types with pentasil chains running parallel to *z*. One corrugated sheet perpendicular to *x* has been highlighted in gray. Adjacent sheets are related to one another by inversion centers (in the 6- and 10-rings).

Adjacent sheets are related to one another by an inversion center. This produces straight 10-ring channels parallel to the corrugations (along *y*), and sinusoidal 10-ring channels perpendicular to the sheets (along *x*). The latter channels link the straight channels to one another to form a 3-dimensional 10-ring channel system. Because the pore openings are 10-rings rather than 12-rings, the shape selectivity for sorption and catalysis is distinctly different from that of **FAU** or **EMT**, and this fact is exploited in catalysis applications. ZSM-5 has found many applications in refinery and petrochemical processes. With 12 T-atoms in the asymmetric unit, **MFI** is one of the most complex zeolite framework types known.

2.2.8. **MEL** (type material: ZSM-11)

In the **MEL** framework type (Figure 12), the corrugated sheets of pentasil chains that are found in **MFI** are also present (one is highlighted in gray). However, in **MEL**, adjacent sheets are related to one another by a mirror plane rather than by a center of inversion. This produces straight 10-ring channels

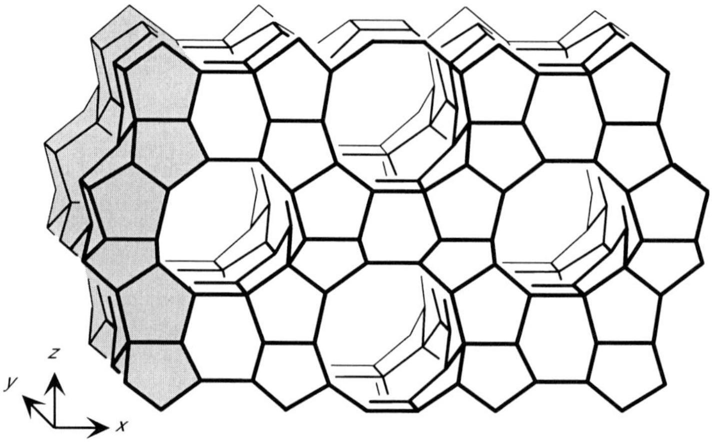

Figure 12. The **MEL** framework type with pentasil chains running parallel to z. One corrugated sheet perpendicular to x has been highlighted in gray. Adjacent sheets are related to one another by mirror planes (running through the 6- and 10-rings).

along both x and y. Because these channels are displaced from one another in z, a 3-dimensional channel system is formed. As might be expected, intergrowths of the **MEL** and **MFI** framework types can and do occur (see section 3.3).

2.2.9. **MOR** (type material: mordenite)

In the **MOR** framework type (Figure 13), units of four 5-rings [5^4] (Figure 1) are joined to one another via common edges to form chains. Mirror images of these chains are connected via oxygen bridges to form corrugated sheets (lying horizontal in Figure 13). These sheets, displaced by half a translation in c, are then connected to one another to form oval 12- and 8-rings along the corrugations. The lining of the 12-ring channels contains 8-rings, but the 8-ring openings of adjacent 12-ring channels are displaced with respect to one another, so only very limited access from one channel to the next is possible. Consequently, the channel system is effectively one dimensional.

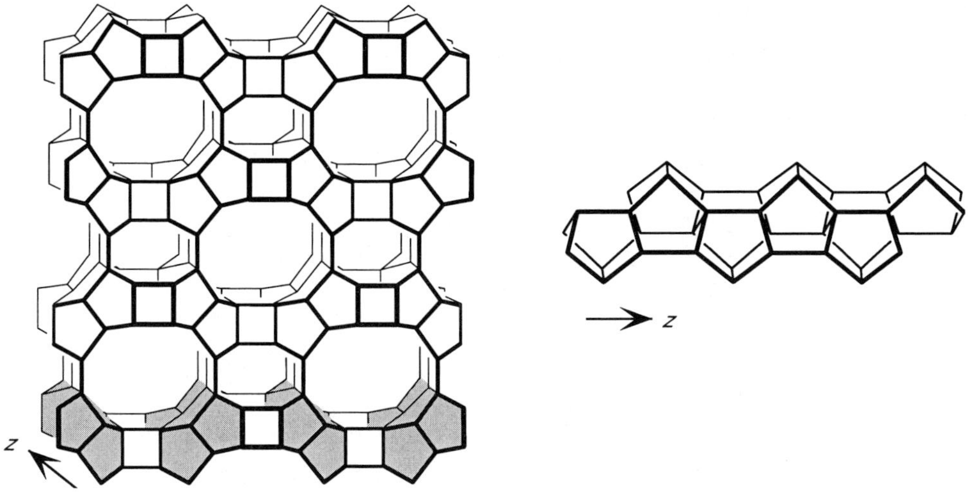

Figure 13. The **MOR** framework type (left) and the chain composed of edge-sharing [5^4] units (right). The chains in the first layer (related by mirror planes) are highlighted in gray.

2.2.10. **MWW** (type material: MCM-22)

The high-silica zeolite MCM-22 has a rather unusual framework structure (Figure 14). It can be viewed as a stacking of double layers joined by single oxygen

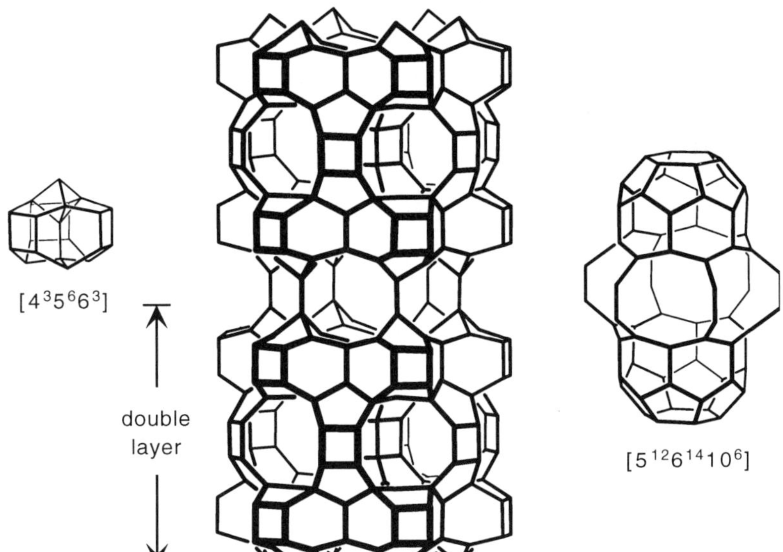

Figure 14. The **MWW** framework type showing the double layer, the small [$4^35^66^3$] cage, and the side pockets at the channel intersection between double layers.

bridges. The single layers consist of [$4^3 5^6 6^3$] cages sharing 4-ring faces, and are joined to a second layer via double 6-rings. The two layers of the double layer are mirror images of one another. This produces two non-intersecting, 2-dimensional, 10-ring, channel systems. One of these lies within the double layer, and the second between the double layers. The latter also has two side pockets (12-ring access) at each channel intersection (see Figure 14, right). While the [$4^3 5^6 6^3$] cage with a T-atom inside the cage may appear a little unusual, the geometry is quite reasonable.

2.2.11. ***BEA** (type material: zeolite beta)

Zeolite beta is disordered in the *c*-direction. That is, well-defined layers are stacked in a more or less random fashion. Since no ordered material has yet been produced, the three letter code is preceded by an asterisk to indicate that the framework type (Figure 15) described in the *Atlas* is an idealized end member of a series. [5^4] units are joined to one another via 4-rings to form layers with saddle-

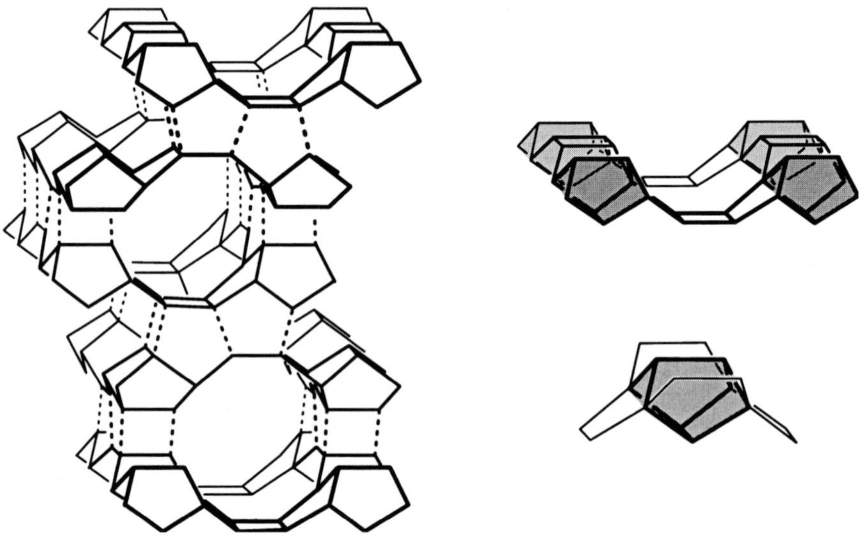

Figure 15. The idealized ***BEA** framework type with all layers related to one another via a counterclockwise rotation (connections between layers shown as dotted lines). The well-defined layer, and its building unit are shown separately.

shaped 12-rings. Adjacent layers are related to one another by a rotation of 90°. The disorder arises because this rotation can be in either a clockwise or a counterclockwise sense. If the counterclockwise or clockwise rotation were maintained throughout the crystal, the structure would be ordered and chiral.

However, all zeolite beta materials reported to date show extreme disorder. Interestingly enough, whatever the stacking sequence, a 3-dimensional 12-ring channel system results, so for catalytic applications, the stacking sequence is not important (unless, of course, the chirality of the channel system were to be exploited in some way).

2.2.12. **AFI** (type material: $AlPO_4$-5)

As for all $AlPO_4$-based molecular sieves, the framework of $AlPO_4$-5 (Figure 16) contains only even numbered rings, since Al and P alternate throughout the framework. In the **AFI** framework type, 6-rings are connected to three neighboring 6-rings via oxygen bridges to form 4-rings between the 6-rings and a hexagonal array of 12-rings. The tetrahedra are oriented in a strictly alternating fashion, so that every other one points up to the next layer while the others point down to the previous one. Mirror images of these layers are stacked on top of one another to form a 1-dimensional 12-ring channel system. Unlike the aluminosilicate molecular sieves, which tend to favor double crankshaft chains for connecting 4-rings in adjacent layers (e.g. tetrahedra oriented in an UUDD fashion), the aluminophosphates seem to prefer the Narsarsukite chain (Figure 2), in which diagonally related corners of the 4-rings form the bonds to the next layer (e.g. UDUD connections). The 12-ring channel in **AFI** is lined with 6-rings (Figure 3).

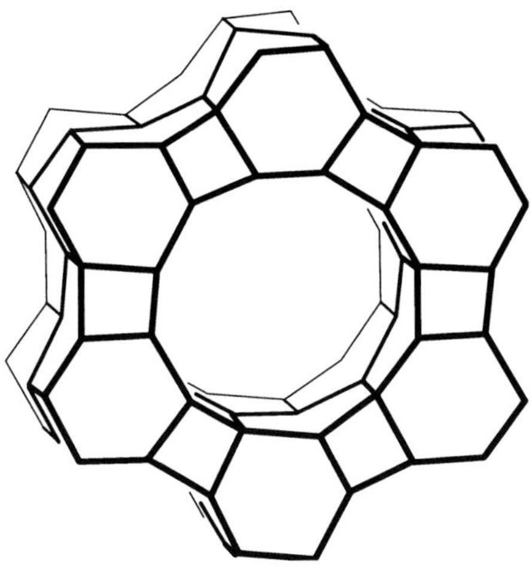

Figure 16. The **AFI** framework type.

2.2.13. **VFI** (type material: VPI-5)

The framework of the aluminophosphate VPI-5 (Figure 17) is closely related to that of $AlPO_4$-5. Instead of being linked via 4-rings, the 6-rings in the **VFI** framework type are linked via two 4-rings sharing a common edge (fused 4-rings). This produces an 18-ring in place of the 12-ring found in **AFI**. The tetrahedra are oriented in the same manner, and layers are stacked similarly. The 18-ring channel, with a free diameter of ca 12 Å, is also lined with 6-rings.

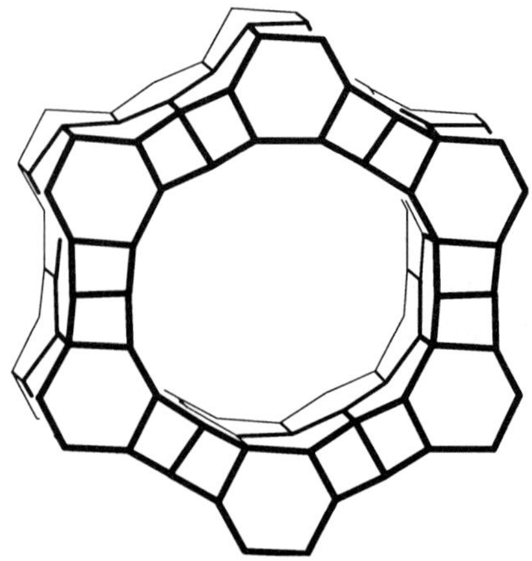

Figure 17. The **VFI** framework type.

One feature of the **VFI** framework type worth noting is the unusual conformation of the fused 4-rings. The geometry is highly strained if the T-atoms are assumed to be tetrahedral. The Al atom on the edge shared by the two 4-rings relieves this unfavorable situation by coordinating to two water molecules in addition to the four framework oxygens, and assumes an octahedral geometry [8]. Upon dehydration, these water molecules are lost, and VPI-5 transforms very easily into the related molecular sieve $AlPO_4$-8 with 14-rings and fewer fused 4-rings [9,10]. Under carefully controlled conditions, VPI-5 can be dehydrated and retain its framework type (albeit with considerable reduction in symmetry [11]).

2.2.14. **-CLO** (type material: cloverite)

Of the gallophosphate molecular sieves synthesized, probably the most exciting material from a structural point of view is cloverite. The **-CLO** framework type (Figure 18) consists of a primitive cubic array of α-cages joined to one another via two $[4^{8}6^{8}8^{2}]$ or *rpa* cages to produce an enormous cage with a body diagonal of ca 30 Å in the center of the cube. However, not all of the T-atoms in the framework are 4-connected. One eighth of the Ga and one eighth of the P form

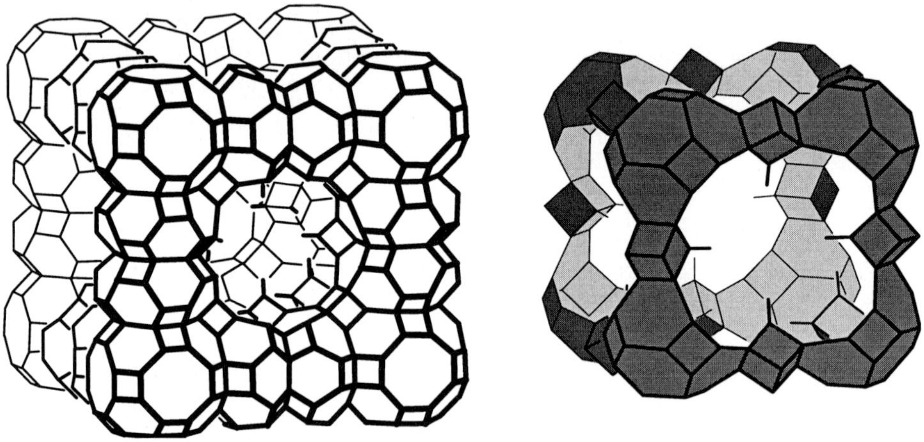

Figure 18. The **-CLO** framework type (left) and the large central cage (right).

only three bonds to framework oxygens. The fourth bond is to a terminal OH-group. That is, the framework is interrupted. The fact that not all T-atoms are 4-connected is indicated by a dash "-" in front of the three letter code. The terminal OH-groups protrude into the pore openings, and produce an unusual pore shape reminiscent of a 4-leafed cloverleaf (hence the name cloverite). The ring is composed of 20 T-atoms and 24 oxygens. There are two non-intersecting, 3-dimensional channel systems: one 20-ring (with cloverleaf-shaped pores) and one 8-ring (passing through the α-cages). The cage in the center is by far the largest yet observed, and the framework density (11.1 T-atoms per 1000 $Å^3$) the lowest. A further interesting aspect of the structure is that it can be constructed entirely from double 4-rings. In the structure of the as synthesized material, there is a fluoride ion in each double 4-ring, and this may be suggestive of a synthesis mechanism, since several other gallophosphate materials synthesized in the presence of HF have also been found to contain this unit (e.g. gallophosphate-**LTA**).

2.2.15. **VNI** (type material: VPI-9)

The framework of the zincosilicate VPI-9 (Figure 19) is fairly complicated, but can be described in terms of two types of layers (A and B) joined via isolated tetrahedra (drawn in gray in Figure 19, left). In layer A, pairs of $[5^3 3^1]$ polyhedra sharing common 3-rings are joined to adjacent pairs via edges to form sinusoidal

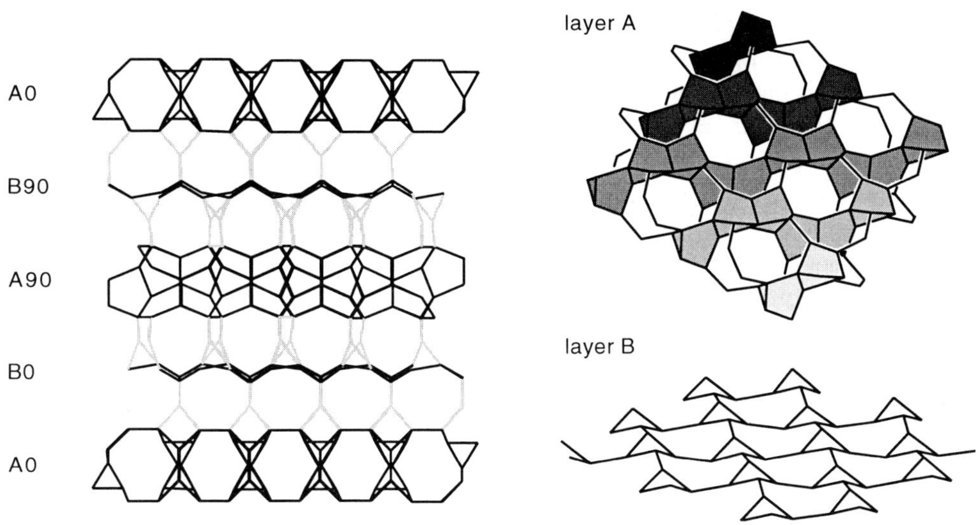

Figure 19. Projection of the **VNI** framework type in the [110] direction indicating the positions and rotations of layers A and B (left) with the two layers shown separately (right). The isolated tetrahedra linking the two types of layers are drawn in gray.

chains. These chains are connected via oxygen bridges to neighboring chains, displaced by half a period, to form 8-rings. These 8-rings are oval with a free diameter of only 2.3 x 4.3 Å, so the channels perpendicular to this layer are effectively blocked. Layer B is a simple 4.8^2 net with a UUDDUUDD arrangement of the tetrahedra in the 8-rings. The stacking of these layers (Figure 19 on the left) produces a 2-dimensional 8-ring channel system. Layer A at $z = 1/2$ is rotated 90° with respect to those at $z = 0$ and $z = 1$. Similarly, layer B at $z \approx 3/4$ is rotated 90° with respect to that at $z \approx 1/4$. The characteristic feature of the zincosilicates seems to be the presence of 3-rings, which are otherwise rare.

3. ZEOLITE STRUCTURES

The framework types discussed in the last section describe only the connectivities of the frameworks. While these characterize the basic framework structure in terms of approximate pore opening, cage arrangement and channel system, and facilitate comparison of related materials, they do not describe real materials. That is, the influence of framework composition, extra-framework cations, organic species, sorbed molecules, or structural defects, is not considered. These aspects are addressed in the following sections.

3.1. Framework composition

Many of the interesting properties of zeolites are based on the fact that the framework is anionic and the balancing cations exchangeable. A pure silica (SiO_2) framework is neutral, but if some of the tetravalent Si are replaced by trivalent Al to produce an aluminosilicate, the framework becomes negative and counterions such as Na^+ are needed to balance its charge. The neutral aluminophosphate or gallophosphate frameworks can be made anionic in a similar manner by inserting other elements into some of the T-sites. Even a small amount of a transition metal ion in the framework can make the material useful for catalysis applications. Many elements have now been incorporated into zeolite framework structures. What was originally the realm of aluminosilicates has expanded to include a significant portion of the periodic table. In some cases, only a few percent of the element is incorporated, while in others it is a major constituent.

The framework composition also affects the stability of a material. For example, a high silica zeolite usually has a higher thermal stability than does the corresponding aluminosilicate, an aluminosilicate tends to be more stable than an aluminophosphate, and a gallophosphate is generally more sensitive to moisture than is an aluminophosphate.

As has been indicated in the discussion of framework types, the chemical composition of a framework is sometimes reflected indirectly in certain features of the framework type. For example, double crankshaft chains are prevalent in aluminosilicates, 5-rings in high silica zeolites, Narsarsukite chains in aluminophosphates, and 3-rings in zinco- and beryllosilicates. Materials with strictly alternating T-atoms, such as Al and Si in aluminosilicates with a Si/Al ratio of 1, Al and P in aluminophosphates or Ga and P in gallophosphates, also require that only even numbered rings be present.

If there are two or more types of T-atoms and these are ordered (i.e. not randomly distributed over all T-sites), the ideal symmetry of the framework type is likely to be reduced. For example, Al and Si alternate in the framework structure of zeolite A (**LTA**). To illustrate the effect of this ordering on the symmetry, the **LTA** framework type with all nodes identical and with alternating nodes marked are shown in Figures 20a and b, respectively. The lattice constant (repeat distance) a and one of the mirror planes for the former is shown in Figure

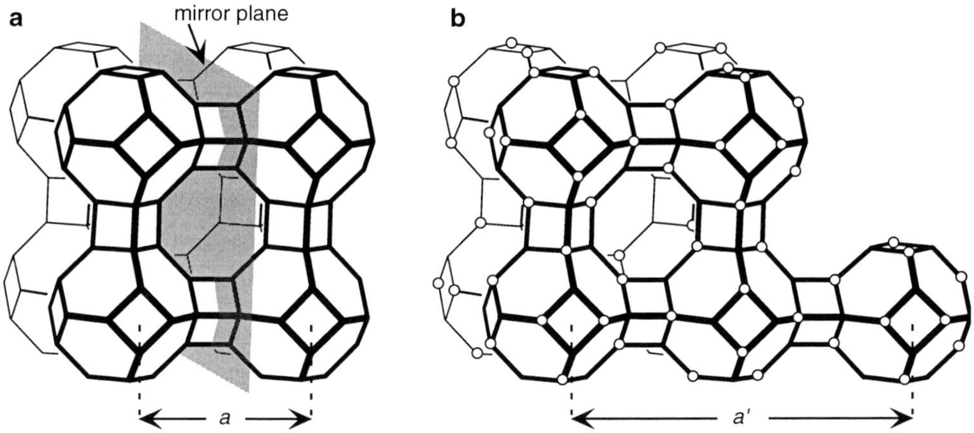

Figure 20. The **LTA** framework type (a) with all nodes identical, and (b) with alternating nodes marked. In (a) the repeat distance a and one of the mirror planes is indicated. In (b) the mirror plane shown in (a) is lost and the repeat distance is doubled in all directions ($a' = 2a$). For simplicity, the necessary doubling of the unit cell in (b) is shown in only one direction.

20a. In Figure 20b, the symmetry reduction dictated by the ordering of Si and Al is readily apparent. Two obvious effects of the alternation are that (1) the mirror planes between sodalite cages are gone, and (2) the unit cell has to be doubled along each of the axes. Similar effects are observed in other materials in which the T-atoms are ordered.

3.2. Extra-framework species

The channels and cages of a zeolite framework are usually filled with extra-framework species such as exchangeable cations, which balance the negative charge of the framework, removable water molecules, and/or organic species. These may come from the synthesis mixture or they may be the result of a post-synthesis treatment. Whatever their origin, it is often of interest to know where they are located.

Modern crystallographic techniques generally allow such information to be extracted from diffraction data, but there are some limitations that should be appreciated. The primary problem is the fact that extra-framework species do not generally follow the high symmetry of the framework, so they are what is called "disordered". For example, the Na^+ ion in an 8-ring of zeolite A is located off-center where it can approach three framework oxygens (Figure 21), but because there is a 4-fold axis running through the center of the 8-ring, there are four equivalent positions for the Na^+ ion. However, there is only room for one Na^+ ion per 8-ring. This Na^+ ion may hop between the four equivalent positions (dynamic disorder) or it may be stationary but occupy different positions in different 8-rings (static disorder). Conventional X-ray analysis cannot distinguish between these two possibilities, but whichever is the case, an electron density map generated from the diffraction data will show 1/4 of a Na^+ ion (e.g. 10/4 electrons) at each equivalent position rather than one ion (10 electrons) at a single position. This means that the peaks in the electron density map are weak, and that chemical sense (e.g. known chemical composition, feasible coordination numbers, sensible interatomic distances and angles, no fractional atoms possible, etc.) must be used to interpret them. In the case above, the interpretation is relatively simple, but for more complex molecules, the interpretation of the electron density map becomes more difficult and ambiguous.

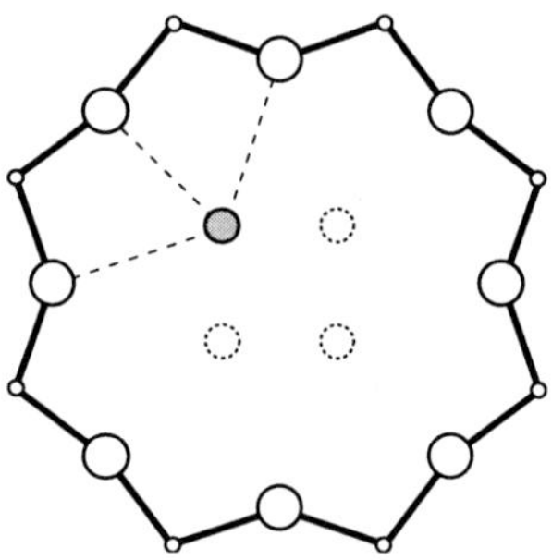

Figure 21. The 8-ring in zeolite A (**LTA**) showing the Na^+ ion position (full gray circle) and the 3 unoccupied symmetry equivalent positions (dotted circles).

Nonetheless, very useful information regarding the location of extra-framework species can be gleaned from a diffraction experiment. Examples include the location of the 18-crown-6 molecule required for the synthesis of pure EMC-2 (**EMT**) [12], the location of sorbed *m*- and *p*-xylene in Ba-exchanged zeolite X (**FAU**) at different loadings [13], and the location of naphthalene sorbed into ZSM-5 (**MFI**) [14].

3.3. Stacking faults

Closely related zeolite framework structures often form under very similar conditions, and this can lead to the formation of stacking faults or intergrowth structures. For example, both ZSM-5 (**MFI**) and ZSM-11 (**MEL**) contain pentasil sheets. The only difference between the two is the linkage between adjacent sheets (they are related by a center of inversion in **MFI** and by a mirror plane in **MEL**, see sections 2.2.7 and 2.2.8), and it is not uncommon for an occasional stacking fault to occur [15]. If substantial domains of two framework types are formed and these domains share a common face, the material is referred to as an intergrowth.

One of the first zeolite intergrowths to be examined was that of the natural zeolites offretite (**OFF**) and erionite (**ERI**), which are members of the ABC-6 family of structures (see section 2.1) with AABAAB and AABAAC 6-ring stacking sequences, respectively [16]. In this case, the stacking is critical, because a single stacking fault in offretite (i.e. a C instead of a B) blocks the 12-ring channel. As might be imagined, the ABC-6 family of zeolites is quite prone to stacking "mistakes". If a stacking fault occurs regularly (e.g. every second layer or every third layer), a new framework type with a new repeat period is formed. The two structures are then called the end members of an intergrowth series. For example, as the number of C layers increases in the **OFF/ERI** case, the end member **OFF** becomes the end member **ERI**:

pure **OFF**	→	faulted **OFF**	→	intergrowth	→	faulted **ERI**	→	pure **ERI**
AABAAB		a few AAC		domains of both **OFF** and **ERI**		a few AABAAB		AABAAC

increasing number of C layers →

High resolution electron microscopy is the technique of choice for the investigation of such structural defects. The high-resolution images of a faulted material will show the local stacking sequences and domain sizes quite clearly. As an example, a micrograph of a **FAU/EMT** intergrowth is shown in Figure 22. As was mentioned in sections 2.2.3 and 2.2.4, both zeolite structures can be built by stacking layers of sodalite cages in an ABCABC (for **FAU**) or ABAB (for **EMT**) sequence [17,18]. Both stacking sequences can be seen in Figure 22. In this case, the stacking faults do not block the channels, but the local environments are slightly different, so some of the properties of the intergrowth materials can differ from those of the pure end members.

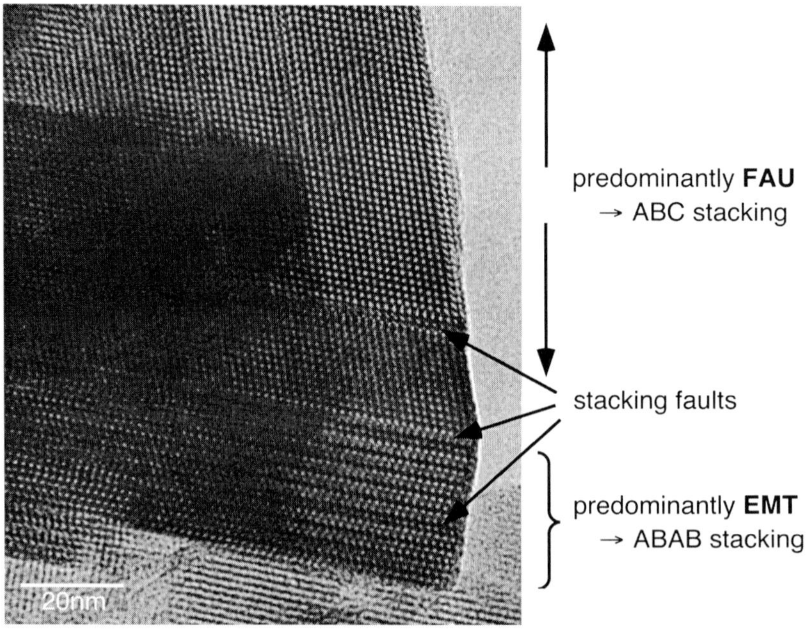

Figure 22. A high-resolution electron microscopy image of ECR-35 [16], an intergrowth of synthetic faujasite (**FAU**) and EMC-2 (**EMT**). Photograph kindly provided by M.M.J. Treacy.

Many such systems have been studied using electron microscopy techniques. Examples include studies of faulting in the zeolites beta (***BEA**) [19], ferrierite (**FER)** [20], and NU-86 [21]. For further examples and experimental details, the reader is referred to the review by Terasaki et al. [22].

4. POWDER DIFFRACTION

Since zeolite structural information is very often derived from laboratory X-ray powder diffraction data, it is perhaps appropriate to review a few aspects of the technique. The following sections will cover (1) a summary of the information contained in a powder diffraction pattern, (2) some of the common pitfalls in the collection of laboratory data, and (3) some common applications of the technique in everyday zeolite science. Additional information can be found in the paper by Baerlocher and McCusker [23].

4.1. Information in a powder diffraction pattern

A powder diffraction pattern has several features that can be of interest to a zeolite scientist: the positions of the peaks, the relative intensities of the

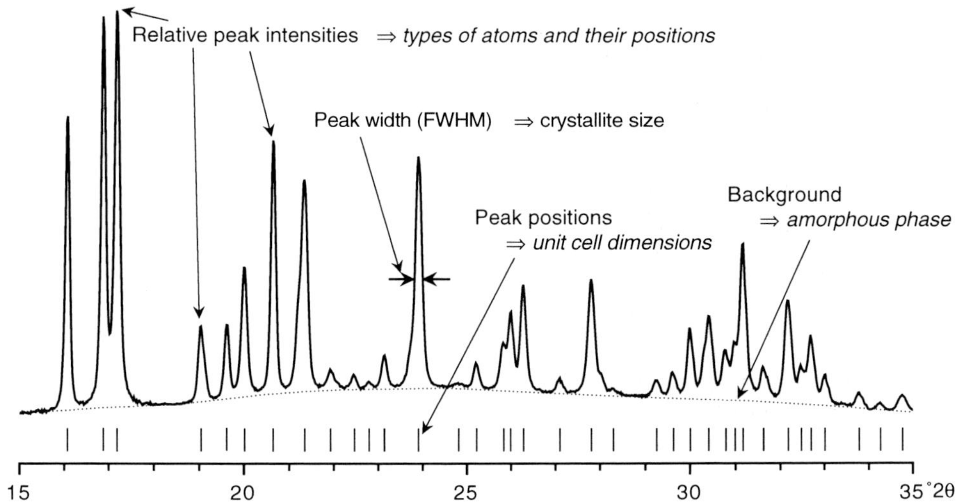

Figure 23. The relevant features of a powder diffraction pattern and their origin.

peaks, the width of the peaks, and the background (Figure 23). Each of these features can be interpreted relatively easily to yield useful information.

The *peak positions* in a powder pattern (usually measured in °2θ) are determined only by the geometry of the unit cell. Each peak represents at least one reflection (and often several that happen to have similar 2θ values). The 2θ value is related to the *d*-spacing of the corresponding reflection (spacing of the diffracting planes) by the Bragg equation

$$\lambda = 2d_{hkl} \sin\theta_{hkl} \quad (1)$$

where λ is the X-ray wavelength and *hkl* are the Miller indices of the reflection. The relationship between the *d*-values, the Miller indices *hkl* and the unit cell parameters *a*, *b*, *c*, α, β and γ is best seen in terms of the "reciprocal lattice parameters", a^*, b^*, c^*, α^*, β^* and γ^* (which can be derived from the unit cell parameters in a straightforward manner):

$$\frac{1}{d_{hkl}^2} = h^2a^{*2} + k^2b^{*2} + l^2c^{*2} + 2hka^*b^* \cos\gamma^* + 2klb^*c^* \cos\alpha^* + 2lhc^*a^* \cos\beta^* \quad (2)$$

The unit cell is a parallelepiped, which describes the 3-dimensional repeat unit of a crystal structure. To determine its size and shape, *hkl* values have to be assigned to each of the reflections (called "indexing the reflections"). For non-cubic systems, this is not a trivial problem, but there are now a number of reliable autoindexing programs available that can take a list of 2θ values (positions of peaks in the measured diffraction pattern), assign *hkl* indices, and determine the unit cell parameters [24].

The *relative intensities* of the peaks in a powder diffraction pattern are determined by the type and position of the various atoms within the unit cell. The intensity I_{hkl} for each reflection is proportional to the square of the structure factor F_{hkl},, and the structure factor can be calculated using the summation

$$F_{hkl} = \sum_{n} f_n e^{-B_n} e^{-2\pi i(hx_n + hy_n + lz_n)} \tag{3}$$

over all n atoms in the unit cell, where f_n is the atomic scattering factor (dependent upon the number of electrons), B_n is the atomic displacement factor, and $x_n y_n z_n$ are the coordinates of atom n. It is important to note in equation (3) that the intensity of each reflection is dependent upon the positions of *all* atoms in the unit cell. Changing just one atom (e.g. by ion exchange) will have an effect on the intensities of all reflections (to different degrees and in different directions).

The *widths* of the peaks in the pattern give an indication of the crystalline quality of the sample. They are dependent upon the intrinsic instrumental peak width, the crystallite size (or more precisely, the size of the diffracting domains), and the amount of stress or strain in the material. As the domains of coherent scattering decrease below ca 1000 Å, the peaks will broaden noticably. In a first approximation, the Scherrer equation

$$L = K\lambda / \beta_{cs} \cos\theta_{hkl} \tag{4}$$

can be used to estimate (± ca 10%) average crystallite size, if measurable broadening of the peaks (relative to a standard material) is observed. The full width at half maximum (FWHM) of a crystallite-size broadened peak is corrected for the instrumental contribution (either by simple subtraction of the instrumental FWHM or, more correctly, by deconvolution) to yield the quantity β_{cs} (°2θ converted to radians). L is the average dimension of the crystallite normal to the reflection planes, K is a constant close to 1 and λ is the wavelength. If the crystallites are very thin plates, reflections from planes perpendicular to the short dimension may be broader than those from planes in other directions. If, for example, the short dimension is assumed to be parallel to the *c*-axis, then the *00l* reflections will be broader than the *hk0* reflections. This is known as anisotropic line broadening. Similarly, stress and/or strain in the material will cause the peaks to broaden. Much more sophisticated line profile analysis is required to obtain more quantitative results or to separate crystallite size from stress/strain line broadening [25], but the qualitative information alone can be very useful.

Finally, the *background* in a powder pattern indicates whether or not an amorphous material is present in the sample. A high background relative to the peaks is usually caused either by a large amount of amorphous material (e.g. unreacted gel) or by X-ray fluorescence. The latter is observed, for example, if an Fe-containing sample is irradiated with CuKα radiation. It can be avoided by changing the wavelength.

4.2. Data collection

The reliability of the information extracted from a powder diffraction pattern is strongly dependent upon the quality of the data. The instrument calibration, sample preparation, instrumental setup and data collection parameters all affect that quality, so perhaps a few words about some of the more common problems in data collection are appropriate. Much more detailed information can be found in the book entitled *Modern Powder Diffraction* edited by Bish and Post [26], the review by Langford and Louer [27], or standard text books on powder diffraction techniques. Either reflection (e.g. Bragg-Brentano) or transmission (e.g. Guinier or Stoe) geometry can be used effectively, but each has its idiosyncrasies.

If the data are to be used to determine the unit cell parameters, accurate 2θ values are essential. To ensure this, the 2θ scale of the diffractometer must be carefully calibrated with a standard material whose peak positions are well established (NIST Si Standard 640b is usually used). Ideally the standard should be mixed with the sample under investigation so the calibration is done under conditions identical to those used for the data collection itself (internal standard). If this is not possible for some reason, calibration can be done separately (external standard), but then care must be taken that the positioning of the sample is reproducible. Displacement of the sample from the center of the 2θ (detector) circle, whether a flat plate or capillary sample is used, is the most common source of error in 2θ values.

The sharper the lines, the better the determination of the peak positions, so adjustment of the instrument to yield the narrowest peaks possible is recommended. Most laboratory diffractometers can be adjusted to give a FWHM of less than 0.10°2θ for the 111 reflection of the Si standard (28.44°2θ with CuKα radiation). The use of monochromatic radiation (e.g. $CuK\alpha_1$ rather than $CuK\alpha_{1,2}$) will also enhance the resolution in 2θ and make the indexing process easier.

If the relative intensities of the peaks are of interest (e.g. to identify a phase, or to determine whether or not a post-synthesis treatment has changed the structure - see next section), several factors need to be considered. Sample preparation is critical. The sample should (1) contain enough crystallites for all crystal orientations to be equally represented (rotation of the sample helps ensure this condition), (2) have no large crystals among the smaller ones (grinding and/or sieving the sample is recommended), (3) be "infinitely thick" for reflection geometry (i.e. the X-rays should not penetrate to the other side of the sample), (4) have negligible X-ray absorption for transmission geometry (capillary diameters of up to a 0.5mm are acceptable for most zeolites if CuKα radiation is used), and (5) be prepared in such a way that no crystallite orientation is preferred over another (preferred orientation is a particular problem for needle or platelet morphologies in a flat sample holder, but is generally less severe for capillary samples). If these conditions are not fulfilled, the relative intensities of the peaks in the measured diffraction pattern will be distorted from their correct values and interpretation of the data will be correspondingly more difficult.

For Bragg-Brentano geometries, it is important that the X-ray beam stays on the sample (i.e. not hit the sample holder) at all 2θ angles to ensure that a constant sample volume is irradiated. This means that at low angles a very small slit is required. A larger slit can be used at higher angles, as long as the parts of the pattern measured with different slit sizes are correctly scaled to one another. Too large a slit is often used, and this causes the intensities of the low angle peaks to appear to be too weak. Such problems can be avoided if a "variable divergence slit" system is used. Then the slits are opened continuously as a function of 2θ, so the peaks at higher angles will appear to have more intensity, because the volume of sample being measured is being increased continuously. For comparison with a conventional (constant slit) pattern, the data have to be normalized appropriately.

4.3. Common applications

Powder diffraction techniques are used on a routine basis by many zeolite scientists. Probably the most common application is the use of a powder diffraction pattern as a "fingerprint" in the identification of synthesis products. Ideally, a laboratory should have a set of "standard" zeolite patterns measured on the in house instrument for direct comparison. For laboratories without such a set of patterns or for those whose set is incomplete, the Structure Commission of the IZA has published a book entitled *Collection of Simulated XRD Powder Patterns for Zeolites* [28], which contains at least one representative powder diffraction pattern for each known framework type. Data for all framework types, including newly approved ones, are also available on the WWW (http://www.iza-structure/databases/).

Even a cursory examination of the patterns in the *Collection* shows that different materials with the same framework type can have markedly different diffraction patterns. To illustrate this point, the patterns for the natural zeolite gismondine and the synthetic zeolite Na-P1, both of which have the **GIS** framework type (Figure 10, section 2.2.6), are shown in Figure 24. The monoclinic gismondine has many more lines than does the tetragonal Na-P1. The pattern for Na-P1 is deceptively simple because the metric of the unit cell is cubic rather than tetragonal (i.e. $a = b = c$ although the tetragonal symmetry only requires that $a = b$).

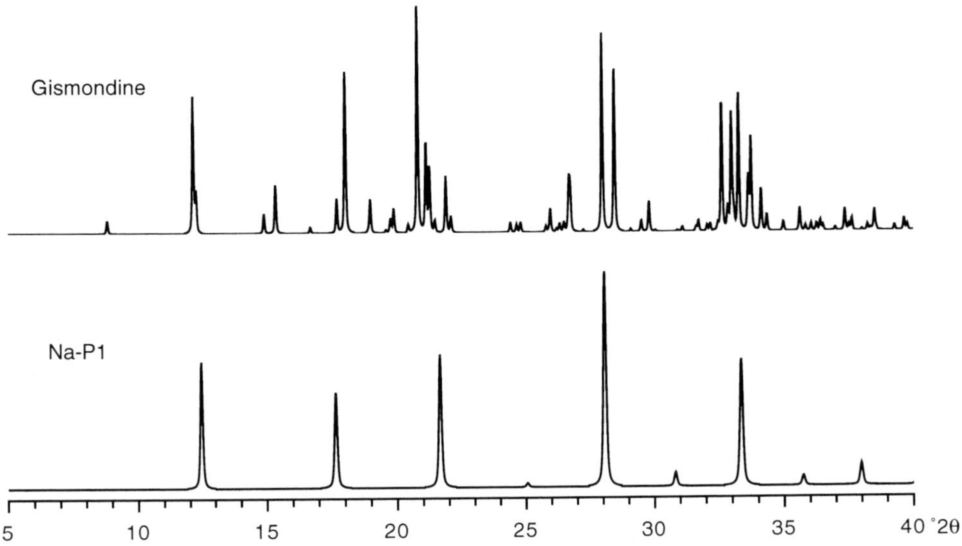

Figure 24. The powder diffraction patterns of two zeolites with the **GIS** framework type.

The identification of an unknown phase can sometimes be facilitated if the lattice parameters are determined. In the **GIS** example described above, the similarity of the diffraction patterns may not be recognized immediately, but the similarity of the lattice parameters is readily apparent:

Sample	**Space Group**	***a*** **[Å]**	***b*** **[Å]**	***c*** **[Å]**	α [°]	β [°]	γ [°]
Gismondine	$P2_1/a$	9.843	10.023	10.616	90	90	92.417
Na-P1	$I\bar{4}$	10.043	10.043	10.043	90	90	90

Of course, lattice parameters are also used to study the effects of post-synthesis treatment (e.g. ion exchange, calcination, dealumination, sorption, etc.), to estimate Si/Al ratios in well-calibrated systems such as faujasite (see Chapter 8), to monitor a phase transition as a function of temperature, or to begin the structural characterization of a new material. Indexing a pattern can also serve to establish that a phase is pure. If all lines can be indexed on a single unit cell, only one crystalline phase is present. If there are unindexed lines, either the indexing is incorrect or a crystalline impurity is present.

A simple method for the evaluation of whether or not a post-synthesis treatment has induced structural change is to look at the effect on the powder diffraction pattern. Changes in the relative intensities of the peaks indicate that a structural modification has occurred, changes in the positions of the lines indicate that the unit cell has deformed in some way, and broader (or narrower) lines indicate that the crystallinity has deteriorated (or improved). In general, non-framework species have a pronounced effect on the low angle region of the pattern. For example, a calcined material will tend to have higher relative intensities in this region than the corresponding as synthesized or loaded sample. The high angle region is usually less sensitive to the presence or absence of electron density in the channels and cages and more sensitive to distortions of the framework. More detail can be extracted from the powder pattern if a full Rietveld (whole-profile) structure refinement is performed [26,29], but even the simple qualitative evaluation of the pattern described above can be extremely informative. Similarly, the presence of an amorphous phase can be established very easily simply by looking at the diffraction pattern.

The determination of the structure of a zeolite with a new framework type remains a challenge to the powder method. Nonetheless, significant advances have been made in this area, and an increasing proportion of the new framework types are solved this way. While the techniques are rather sophisticated and beyond the scope of this chapter, it is perhaps important to know that methods of structure determination from powder diffraction data do exist and that all is not lost if single crystals of a new material cannot be synthesized. It may still be possible to solve the structure from the powder data [30].

Faulted materials (section 3.3) can also be recognized from their unusual powder diffraction patterns. The patterns are quite complicated, often with broadened lines for certain classes of reflections, and, until quite recently, could not be simulated easily. The calculation requires a knowledge of the structures of the end members, an estimate of the planar fault probability and an accurate description of the faulting involved. Given this information though, the program DIFFaX [31] can be used to simulate a powder pattern, which can then be compared with the observed one. In this way, the degree of faulting can be estimated. It should be noted that the powder pattern of an intergrowth of two phases will be quite different from that of a physical mixture of those two phases. The latter is simply a summation of the powder patterns of the constituent phases in proportion to the amount present, while the former involves a much more complicated calculation.

5. CONCLUSIONS

The key to understanding the properties of zeolites and zeolite-like materials lies in their structures. In this chapter, we have tried to convey the beauty and diversity of zeolite framework structures, to introduce some of the jargon used to describe them, and to outline some of the techniques used to characterize them.

The main feature of a zeolite structure is its framework type, which describes the arrangement of the cages, the dimensionality of the channel system and the approximate size of the pore openings. A few framework types, selected for their industrial relevance and/or to illustrate some of the more common structural nomenclature, have been presented. However, there are many more, and for more information about a specific framework type, the reader is referred to the relevant references in the *Atlas* and the *Collection*. To fully understand the properties of a real zeolitic material though, not only the framework type, but also the composition and true geometry of the framework, the location and nature of the extra-framework species, and the number and type of defects must be investigated.

Since most zeolites are only available as polycrystalline phases, powder diffraction is an essential structural characterization technique. A powder diffraction pattern can be used to identify a material, to determine unit cell parameters, to estimate the quality of a sample, to monitor phase transitions, to evaluate whether or not a post-synthesis treatment has induced structural changes, to establish whether or not impurities (amorphous or crystalline) are present, or to recognize the presence of a significant level of faulting.

6. ACKNOWLEDGMENTS

We thank Dr. M.M.J. Treacy for providing us with the micrograph in Figure 22. This work was supported in part by the Swiss National Science Foundation.

REFERENCES

1. K.F. Fischer and W.M. Meier, *Fortschr. Miner.*, 1965, **42**, 50.
2. W.M. Meier and D.H. Olson, *Adv. Chem. Ser.*, 1971, **101**, 155.

3. R.M. Barrer, *Pure and Appl. Chem.*, 1979, **51**, 1091.
4. L.B. McCusker, F. Liebau and G. Engelhardt, *Pure and Applied Chemistry*, 2001, in press
5. Ch. Baerlocher, W.M. Meier and D.H. Olson, "Atlas of Zeolite Framework Types", 5th edn., Elsevier, London, 2001 (formerly "Atlas of Zeolite Structure Types).
6. R. Jelinek, A. Stein and G.A. Ozin, *J. Am. Chem. Soc.*, 1993, **115**, 2390.
7. J.V. Smith and J.M. Bennett, *Am. Mineral.*, 1981, **66**, 777.
8. L.B. McCusker, Ch. Baerlocher, E. Jahn and M. Bülow, *Zeolites*, 1991, **11**, 308.
9. R.M. Dessau, J.L. Schlenker and J.B. Higgins, *Zeolites*, 1990, **10**, 522.
10. J.W. Richardson, Jr. and E.T.C. Vogt, *Zeolties*, 1992, **12**, 13.
11. J. de Oñate Martinez, L.B. McCusker and Ch. Baerlocher, *Micropor. Mesopor. Mater.*, 2000, **34**, 99.
12. Ch. Baerlocher, L.B. McCusker and R. Chiappetta, *Micropor. Mater.*, 1994, **2**, 269.
13. C. Mellot, D. Espinat, B. Rebours, Ch. Baerlocher and P. Fischer, *Catal. Lett.*, 1994, **27**, 159.
14. H. van Koningsveld and J.C. Jansen, *Micropor. Mater.*, 1996, **6**, 159-167.
15. G. Perego, M. Cesari and G. Allegra, *J. Appl. Crystallogr.*, 1987, **20**, 356.
16. J.M. Bennett and J.A. Gard, *Nature (London)*, 1967, **214**, 1005.
17. M.W. Anderson, K.S. Pachis, F. Prébin, S.W. Carr, O. Terasaki, T. Ohsuna and V. Alfredsson, *J. Chem. Soc., Chem. Commun.*, 1991, 1660.
18. M.M.J. Treacy, D.E.W. Vaughan, K.G. Strohmaier and J.M. Newsam, *Proc. R. Soc. London A*, 1996, **452**, 813.
19. J.M. Newsam, M.M.J. Treacy, W.T. Koetsier and C.B. de Gruyter, *Proc. R. Soc. London, Ser. A*, 1988, **420**, 375.
20. R. Gramlich-Meier, W.M. Meier and B.K. Smith, *Z. Kristallogr.*, 1984, **169**, 201.
21. M.D. Shannon in "Proc. 9th Int. Zeolite Conf., Montreal, 1992," eds. R. von Ballmoos, J.B. Higgins and M.M.J. Treacy, Butterworth-Heinemann, Boston, MA, 1993, p. 389.
22. O. Terasaki, T. Ohsuna, V. Alfredsson, J.-O. Bovin, D. Watanabe and K. Tsuno, *Ultramicroscopy*, 1991, **39**, 238.
23. Ch. Baerlocher and L.B. McCusker, *Stud. Surf. Sci. Catal.*, 1994, **85**, 391.
24. D. Louër, *NIST Spec. Publ.*, 1992, **846**, 92.
25. R. Delhez, T.H. de Keijser, J.I. Langford, D. Louër, E.J. Mittemeijer and E.J. Sonneveld in 'The Rietveld Method," ed. R.A. Young, Oxford University Press, Oxford, UK, 1993, p. 132.
26. "Modern Powder Diffraction", eds. D.L. Bish & J.E. Post, *Reviews in Mineralogy*, 1989, **20**, 369 pp.
27. J.I. Langford and D. Louër, *Rep. Prog. Phys.*, 1996, **59**, 131.

28. M.M.J. Treacy and J.B. Higgins, "Collection of Simulated XRD Powder Patterns for Zeolites", 4th edn., Elsevier, London, 2001.
29. R.A. Young, ed., "The Rietveld Method", Oxford University Press, Oxford, UK, 1993.
30. W.I.F. David, K. Shankland, Ch. Baerlocher and LB. McCusker, eds., "Structure Determination from Powder Diffraction Data", Oxford University Press, Oxford, in press.
31. M.M.J. Treacy, J.M. Newsam and M.W. Deem, *Proc. R. Soc. London, Ser. A*, 1991, **433**, 499.

Studies in Surface Science and Catalysis 137
H. van Bekkum, E.M. Flanigen, P.A. Jacobs and J.C. Jansen (Editors)
© 2001 Elsevier Science B.V. All rights reserved.

Chapter 4

How to build zeolites

H. van Koningsveld

Laboratory of Organic Chemistry and Catalysis, Delft University of Technology, Julianalaan 136, 2628 BL Delft, The Netherlands

This chapter gives a mainly pictorial description of several ways to build the (disordered) frameworks of 73 zeolite structure types. Infinite building units, like chains and layers, and finite building units, like single or double 6-rings, are used to build the zeolites. 6-Ring layers are frequently curled up to form tubular 6-ring walls. The (multiple) zigzag-, saw- and crankshaft chains are very probably not essential entities during the crystal growth of these zeolites. However, 6-rings, whether single, double or fused to (tubular) 6-ring sheets, may play an important role.

1. INTRODUCTION

The "Atlas of Zeolite Structure Types" [1] contains 98 topologically distinct tetrahedral TO_4 frameworks, where T may be Si, Al, P, Ga, B, Be etc.. The compiled structure types, characterized by mnemonic codes consisting of three capital letters, do not depend on composition, distribution of the various T atoms, cell dimensions or symmetry. Their frameworks exhibit such a diversity of four-connected three-dimensional nets, that finite and infinite component units were introduced to describe their topologies.

Finite units were introduced by Meier [2,3] and Smith [4]. The secondary building units (SBUs) of Meier, e.g., 4-, 5- or 6-rings, are invariably non-chiral. This means that only one kind of SBU rather than enantiomeric pairs is needed to assemble the three-dimensional framework. The assemblage of the structure does not necessarily involve crystallographic symmetry operations.

The finite structural subunits (SSUs) developed by Smith are often of greater complexity (f.e., polyhedral cages). The SSUs represent a structural feature. They are not, however, SBUs in the sense just mentioned because very often the framework cannot be constructed from SSUs alone. Frequently, SSUs need to share corners, edges or faces to complete the structure.

The SBUs, as such, are not meant to describe precursors from which the zeolite grows. On the other hand, inspection of the systematics in existing structure types may give clues for choosing targets for synthesis because equal segments in different frameworks, like (some of) the polyhedral cages, may play a role during crystal growth. Unfortunately, probably because of this potential ambiguity, the SBUs have not been listed in the latest issue of the Atlas [1].

Infinite units, f.e., chains and layers, were extensively discussed by several authors [4-8]. The 5-ring zeolites were described in terms of component chains [9] as well as in terms of component layers [10].

The infinite or finite building units are far from unique. However, they are common to several zeolite structure types and allow an easy description of the structures.

Section 2 of this contribution contains a general description of some (in)finite structural building units and summarizes the topological symmetry and unit cell data of zeolites which can be built with these (in)finite building units.

Subsequent Sections describe in more detail how to assemble 43 structure types using infinite building units constructed from different linkages of infinite zigzag chains (Section 3), saw chains (Section 4), crankshaft chains (Section 5) or fibrous chains (Section 6). The first paragraph of each Section summarizes the infinite building units used to assemble the structures. The second paragraph gives the building schemes and the perspective views of the structures. A third paragraph presents an alternative description of those structures which can (for the greater part) be built using (cylindrical) 6-ring sheets.

Section 7 describes how to build the 14 hexagonal ABC structure types by different stacking sequences of parallel single or double 6-rings, and how to assemble 7 additional structure types by linking non-parallel 6-rings.

Finally, Section 8 gives the construction schemes for the (disordered) frameworks of the intermediate structures and the pure end-members in the FAU/EMT faujasite zeolites, in the MFI/MEL pentasil and ITE/RTH zeolites, in the beta (*BEA) family of zeolites, in the SSZ-33 (CON) zeolites, in the lovdarite (LOV/VSV/RSN) and montesommaite (MON; and related WEI) zeolites and in the decasil (RTE) zeolites.

At the end of the paper, a Register of the described structure type codes is added.

All drawings are prepared using the ORTEP program of Carroll K. Johnson [11], to whom I am very much indebted. The topological symmetry and unit cell data of the zeolites are obtained from the web-site: http://www.iza-sc.ethz.ch, which site is maintained by Christian Baerlocher and Lynne B. McCusker. More detailed information on the described zeolite-types (f.e., type of T atoms involved, actual symmetry of the type material and isotypic frameworks) can be obtained by consulting the 'Atlas of Zeolite Structure Types' [1], the references therein and the web-site just mentioned. I thank Hermann Gies, Rich Kirchner and Mike Treacy, co-members of the "Disordered Structures Committee" (a sub-committee of the "Structure Commission" of the International Zeolite Association), for their input in the final version of Section 8.

This contribution does not describe several zeolites which can be assembled from (fused) single or double 4-rings; clathrasils, where the windows are too small to let pass encaged species or their decomposition products; and zeolites with interrupted frameworks.

It is planned to include more structure types in the near future.

2. BUILDING UNITS

2.1. Chains

Frameworks of many zeolites can readily be constructed from four (infinite) simple chains shown in Figure 1. Three of these chains, with identity periods of ~ m*2.5 Å, are referred to as zigzag (z), saw (s) and crankshaft (c) chains with m = 2, 3 and 4, respectively.

The number of T atoms in the independent repeat unit along the chain axis equals m. The presence of z, s, or c chains in a certain direction is very often reflected by the unit cell dimension in that direction.

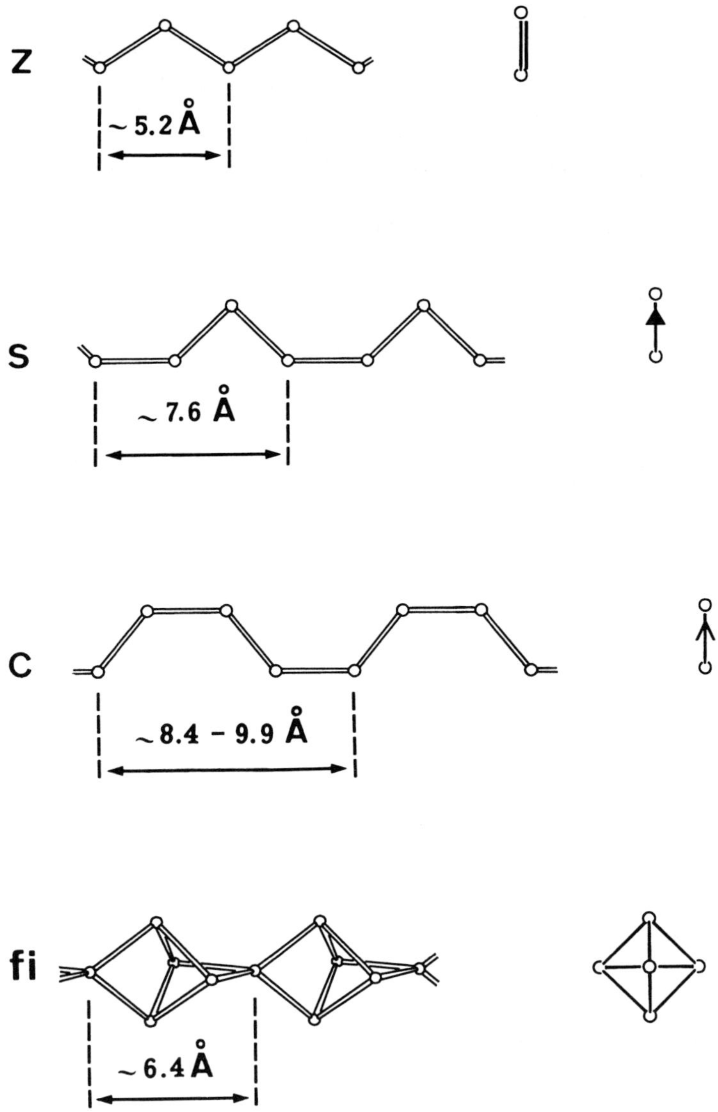

Figure 1. Four simple chains. Open circles denote T atoms at the intersection of lines. O atoms (not drawn) are about midway between T atoms. The identity period along the chain axis is indicated. The chain is depicted, followed by the symbolic representation of the projection along the chain axis. From top to bottom: zigzag (z), saw (s), crankshaft (c) and fibrous (fi) chain.

Tables 1 through 4 (p. 15, p. 34, p. 46 and p. 59, respectively) summarize the space group, cell dimensions, number of T atoms in the independent repeat unit (n) and the total number of T atoms in the unit cell (Z) of 43 framework topologies with maximum topological symmetry. Table 1 shows that 14 out of the 19 listed zeolites have crystallographic repeat distances (unit cell axes) that are indicative for the existence of z chains. The remaining 5 frameworks can be described using z chains with multiple identity periods. Table 2 gives the 9 frameworks with repeat distances which strongly imply the presence of saw chains. Two of the 9 frameworks (JBW and DAC) can also be assembled using zigzag chains. Table 3 presents the 15 zeolite frameworks which can be constructed using (multiple) crankshaft chains. One of the 15 frameworks (ATT) can as well be built using saw chains. The value of the crankshaft chain period in the zeolite lattice strongly depends on how the chain is connected to its neighbouring chains. Multiple crankshaft chains show considerable contraction of the chain period.

The fourth (infinite) simple chain, referred to as the fibrous (fi) chain, has an identity period of ~ 6.4 Å and can be obtained by linking 'capped' 4-rings (See Figure 1). The linkage of capped 4-rings can be more precisely described using the directionality of 4-rings. 4-Rings can be connected in different ways depending upon which oxygen atoms are involved in linking the 4-rings into ladders. The bridging oxygen atom can be above or below the 4-ring plane. The adjacent T atoms are said to have upwards (U) or downwards (D) directionality (See Figures 6,9 and 12). The fi chain can be constructed by connecting parallel UDUD-rings through single T atoms (Figure 1). The topological symmetry and unit cell data of the fibrous zeolites are summarized in Table 4.

2.2. 6-Ring sheets and single and double 6-rings

Many of the zeolites which can be assembled from linking chains can also be produced by connecting 6-ring sheets. The sheets consist of fused 6-rings which form, depending on the corrugation of the sheet, (distorted) zigzag- and crankshaft chains. Two examples are shown in Figure 2a. The 6-ring sheet can curl up to tubular pores where zigzag chains or crankshaft chains (Figure 2b) run parallel to the pore direction.

Several other zeolite frameworks can be constructed by connecting single or double 6-rings of T atoms in various ways. In a first group of zeolites the 6-rings are parallel to the hexagonal *ab* plane (translation period a ~ 13 Å) and are centered at either (0,0), (1/3,2/3) or (2/3,1/3) in the *ab* plane. The 6-rings can be connected to other 6-rings above and below the *ab* plane that also have their centres at (0,0,z), (1/3,2/3,z) or (2/3,1/3,z). These positions are usually denoted as the A, B and C positions [12]. The structures are described in terms of their repeating stacking sequences along the hexagonal c axis. The position of the first 6-ring in a sequence is, by convention, an A position. The distance between two successive 6-rings in a stack is about 2.55 Å. Figure 3 shows the stacking sequences for the ABABAB... (short: AB) and the ABCABCABC... (short: ABC) structures. The number of 6-rings per unit cell is equal to the number of stacks in the sequence: the AB structure has 2*6 = 12 T atoms per unit cell (Z = 12) and for the ABC structure Z = 18 (See Figure 3). Table 5 (page 134) summarizes 14 zeolite frameworks belonging to this so-called ABC family of zeolites. It can be anticipated that very frequently faults in the stacking sequences occur leading to intermediate, disordered frameworks (see next paragraph). The 14 listed zeolites exhibit, in addition to periodicity in the *ab* plane, translation symmetry along **c** and are examples of end-members in the ABC family of zeolties. Three of the frameworks (CAN, OFF and GME) can as well be assembled from linking infinite chains.

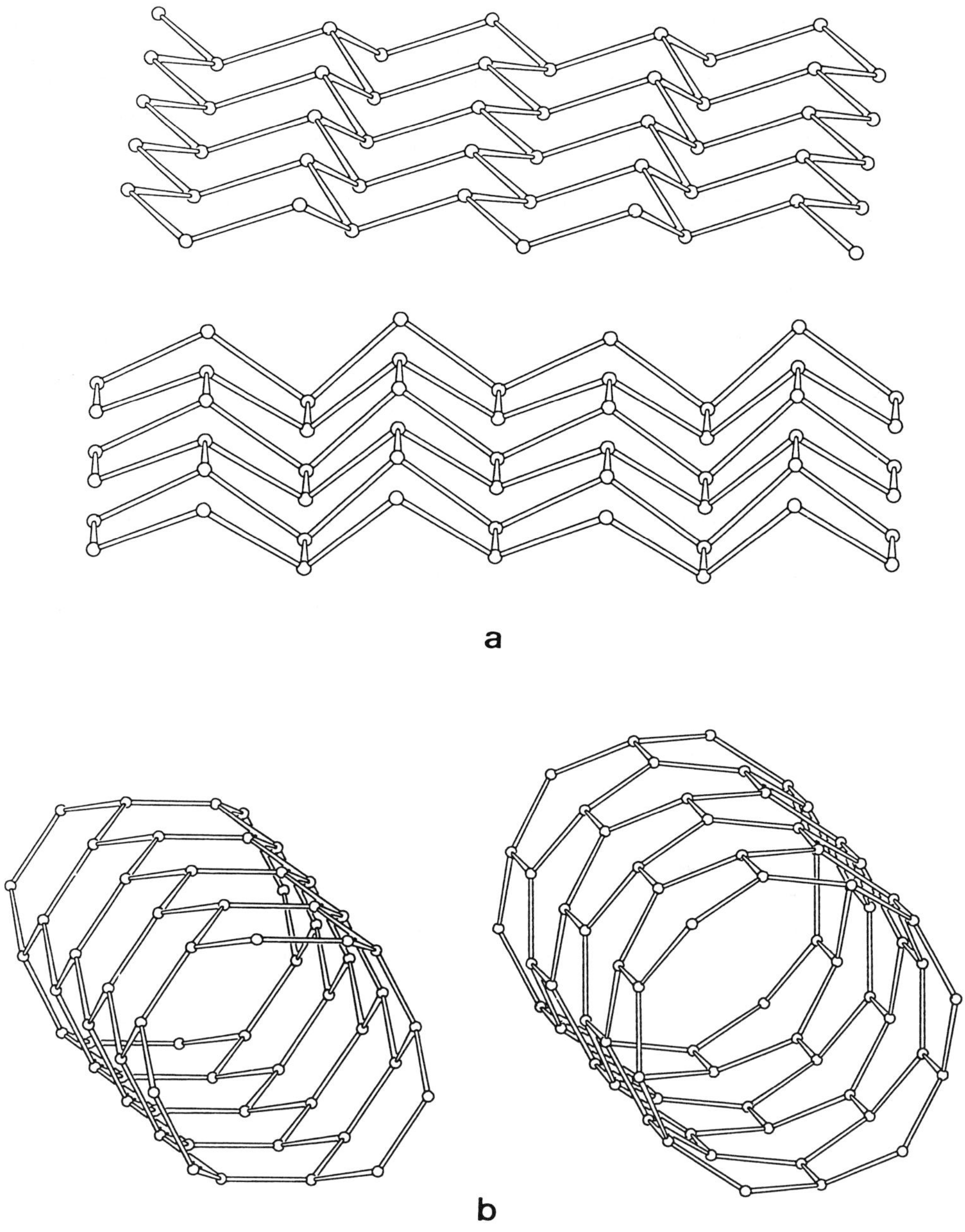

Figure 2. Planar and tubular 6-ring sheets.
(a): sheets with fused 6-ring chairs and 6-ring boats.
(b): tubes with zigzag or crankshaft chains parallel to the pore direction.

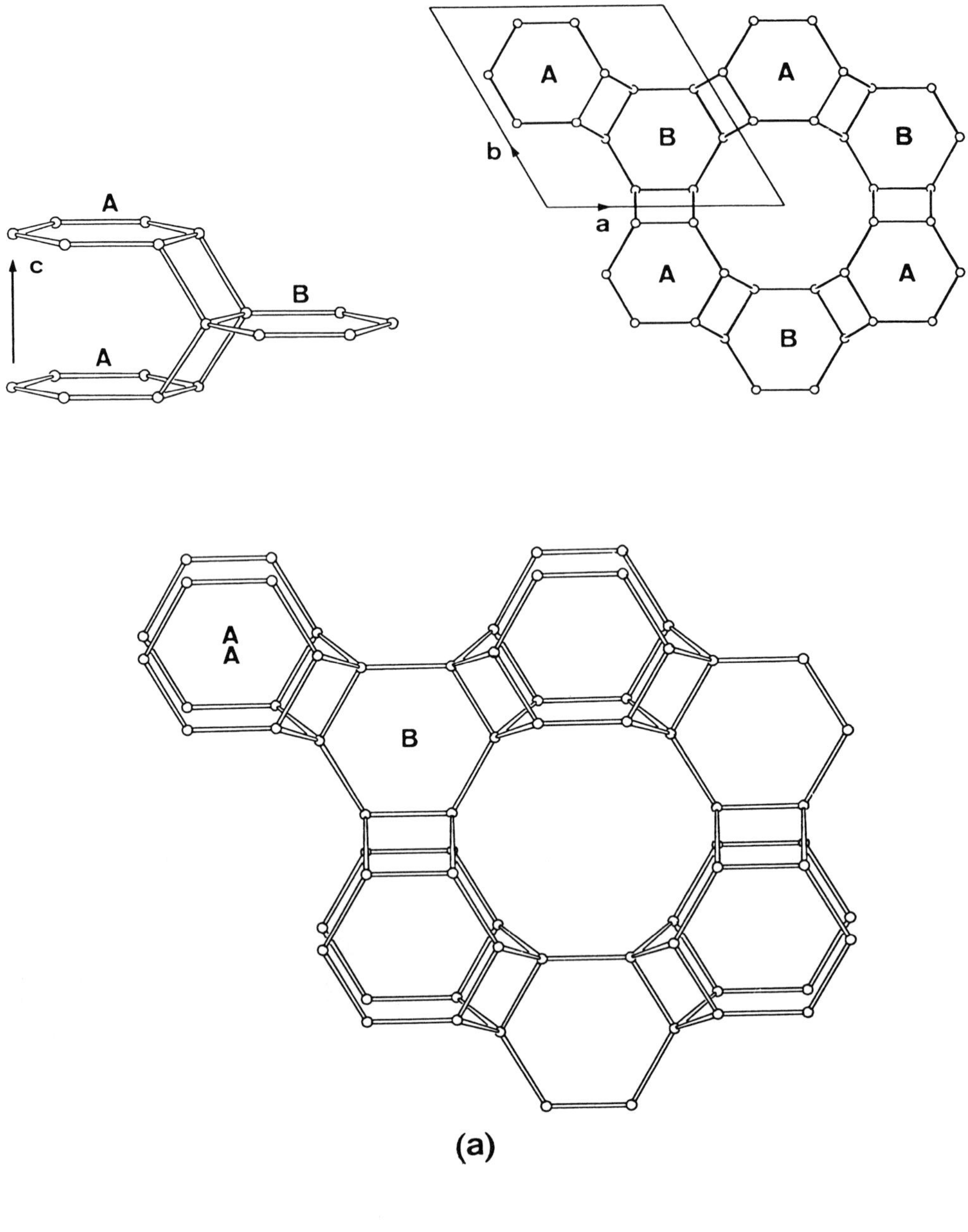

Figure 3. Stacking sequences of 6-rings in hexagonal zeolites.
(a): The AB stacking sequence is shown along **c** (top left), in parallel projection down **c** (top right) and in perspective view down **c**.

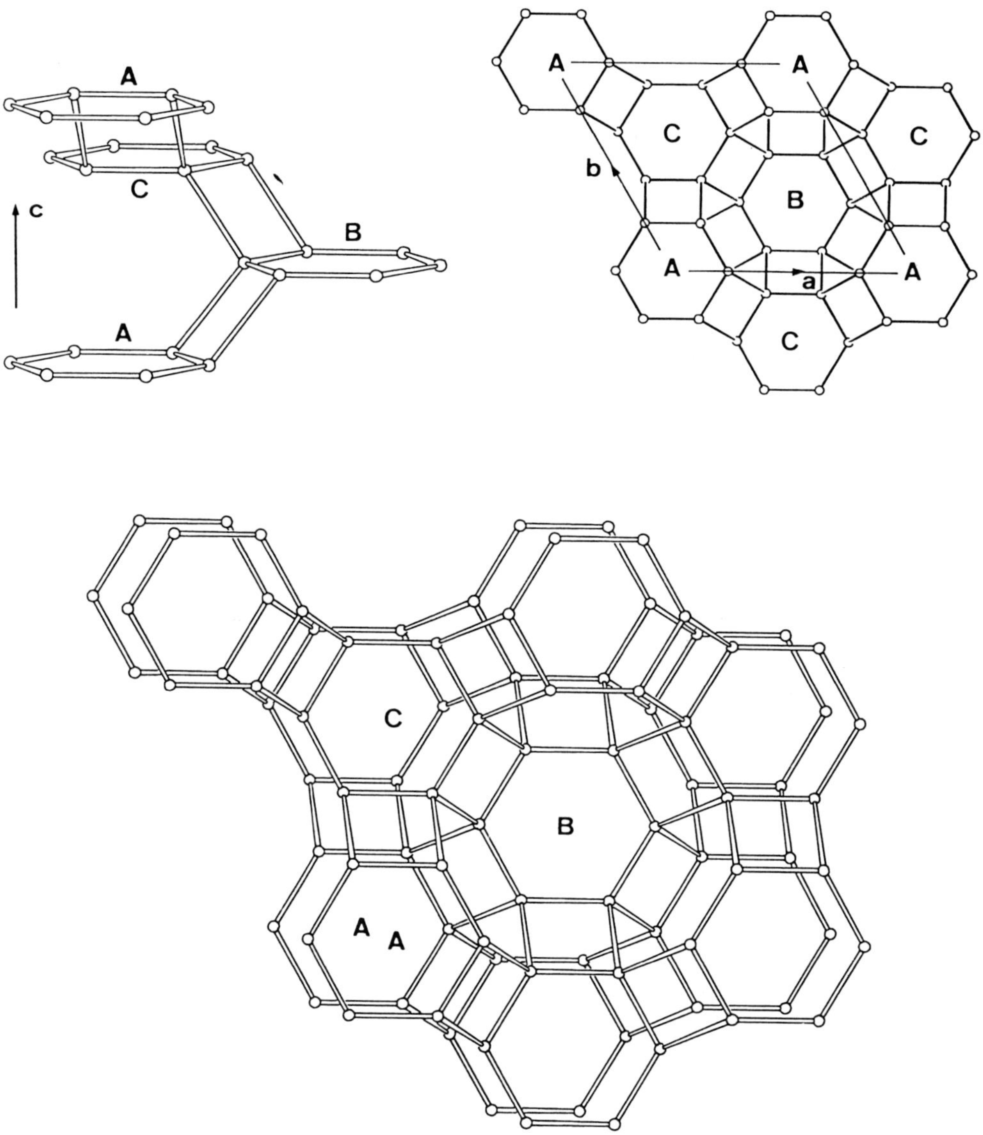

(b)

Figure 3. Cont'd.
(b): The ABC stacking sequence is shown along **c** (top left), in parallel projection down **c** (top right) and in perspective view down **c**.

In a second group of zeolites not all single or double 6-rings are parallel to each other. The structure type codes and crystallographic data of the zeolites in this group are listed in Table 6 (page 135).

An extensive alternative description of how to build the ABC family of zeolites appeared some time ago [8] and will not be given in this paper. In that description the hexagonal zeolites are assembled by connecting 6-ring sheets. The 6-rings in these (corrugated) sheets are connected through tilted 4-rings and not by fusion of the 6-rings.

2.3. Layers and chains in disordered frameworks

Many zeolites easily form twinned or intergrown structures. All these intermediate, disordered structures can be built by applying different symmetry operations to a basic building unit.

For example, zeolites which can form an infinite number of intermediate structures are zeolites with the cubic FAU or the hexagonal EMT structure type [13-16]. The basic building

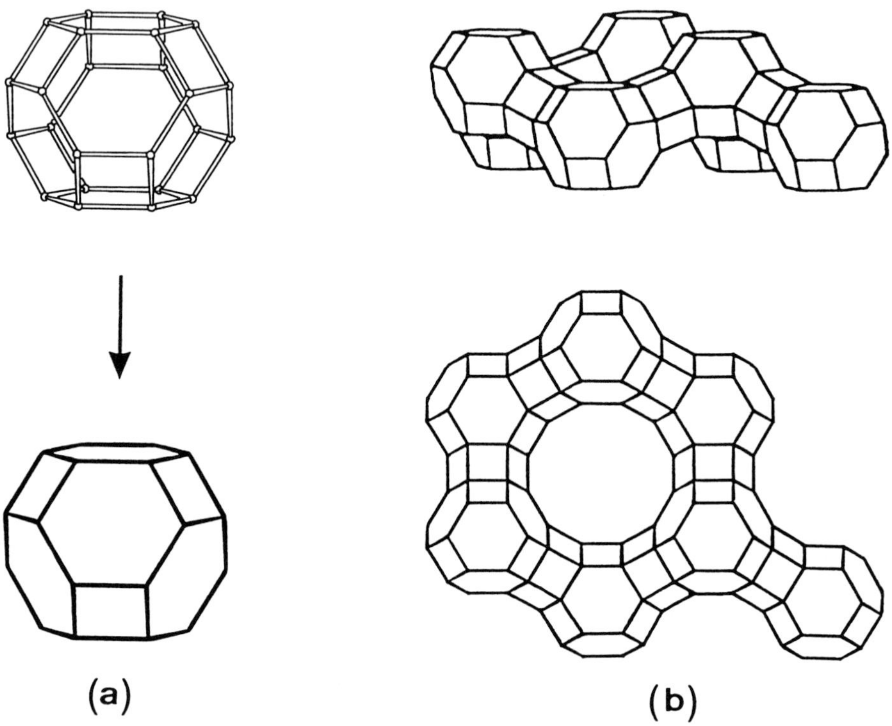

Figure 4. The sodalite or β cage.
(a): The (usual) transparent and polyhedral representations of the sodalite cage.
(b): Hexagonal layer, formed by linking sodalite cages through 6-rings, viewed from aside and from the top.

unit is a hexagonal layer that is comprised of rings of six so-called β or sodalite cages. Each sodalite cage consists of four tetrahedrally connected single 6-rings. Figure 4(a) shows the usual perspective drawing of the T atoms in a sodalite cage together with its polyhedral representation. The sodalite cages are linked through double 6-rings to form the basic hexagonal layer, referred to as faujasite layer and illustrated in Figure 4(b). FAU, EMT and all intermediate structures can be considered as stacks of these faujasite layers which are related by inversion or reflection.

A second example of zeolites which can easily form an infinite number of intermediate structures are the zeolites ZSM-5 and ZSM-11, with structure type MFI and MEL, respectively. The basic building unit in all these structures, the so-called pentasil layer, can be constructed in the following way. A building unit of 12 T atoms (Figure 5(a)) forms two types of chains along Z by applying a two-fold screw axis parallel to Z. Left- and right-handed chains are generated (Figure 5(b)). The chains, related by a mirror plane perpendicular to Y, are connected along Y to give the characteristic YZ pentasil layer shown in Figure 5(c). MFI, MEL and all intermediate structures consist of an array of parallel pentasil layers which are connected along X and which are, as in the previous example for FAU/EMT, related by inversion or reflection.

Another example, described in the same paragraph, regards the ITE/RTH zeolites [17-18]. The basic layer, depicted in Figure 28(a), is composed of chains consisting of building units of 16 T atoms. ITE, RTH and all intermediate structures are stacks of these basic layers which are, as in the faujasite (FAU/EMT) and pentasil (MFI/MEL) zeolites, related by inversion or reflection.

A third example of a family of zeolites which contains an enormous number of intermediate, disordered structures is the zeolite beta (*BEA) family [19-21]. The basic building unit is a tetragonal layer composed of T16 units (see Figure 30). All *BEA materials are stacks of these beta layers with lateral inter-layer shifts of ± ⅓ **a** or ± ⅓ **b.**

Another group of intermediate structures belong to the SSZ-33 (CON) family of zeolites [22-24]. All disordered SSZ-33 framework structures can be obtained by connecting layers depicted in Figure 33 with a lateral inter-layer shift of ± ⅓ **c**.

In the lovdarite- (LOV/RSN/VSV; [25-27]) and montesommaite- (MON; [28]) family of zeolites the end-members and intermediate structures can be constructed from the basic layers shown in Figures 36 and 39. In all structures neighbouring layers are related by reflection accompanied by a lateral inter-layer shift of zero or ± ½ **b**.

A final example of disordered framework structures belonging to the same family is the decasil (RTE) family of zeolites [29-30]. All decasil zeolites can be obtained by connecting chains (see Figure 43) in various ways.

In general, without specific steering (organic) template molecules, the distribution of the stacking modes between the layers or chains in the array can take any form. The nature of the cage- and/or (intersecting) channel-system(s) can be strongly dependent on the actual distribution of the connection modes in the array. Pure end-members are obtained when the connection modes induce three dimensional translation symmetry in the array. The asterisk in the code *BEA denotes that no pure end-members have yet been obtained. Crystal data of some simple end-members are listed in Table 7 (page 165).

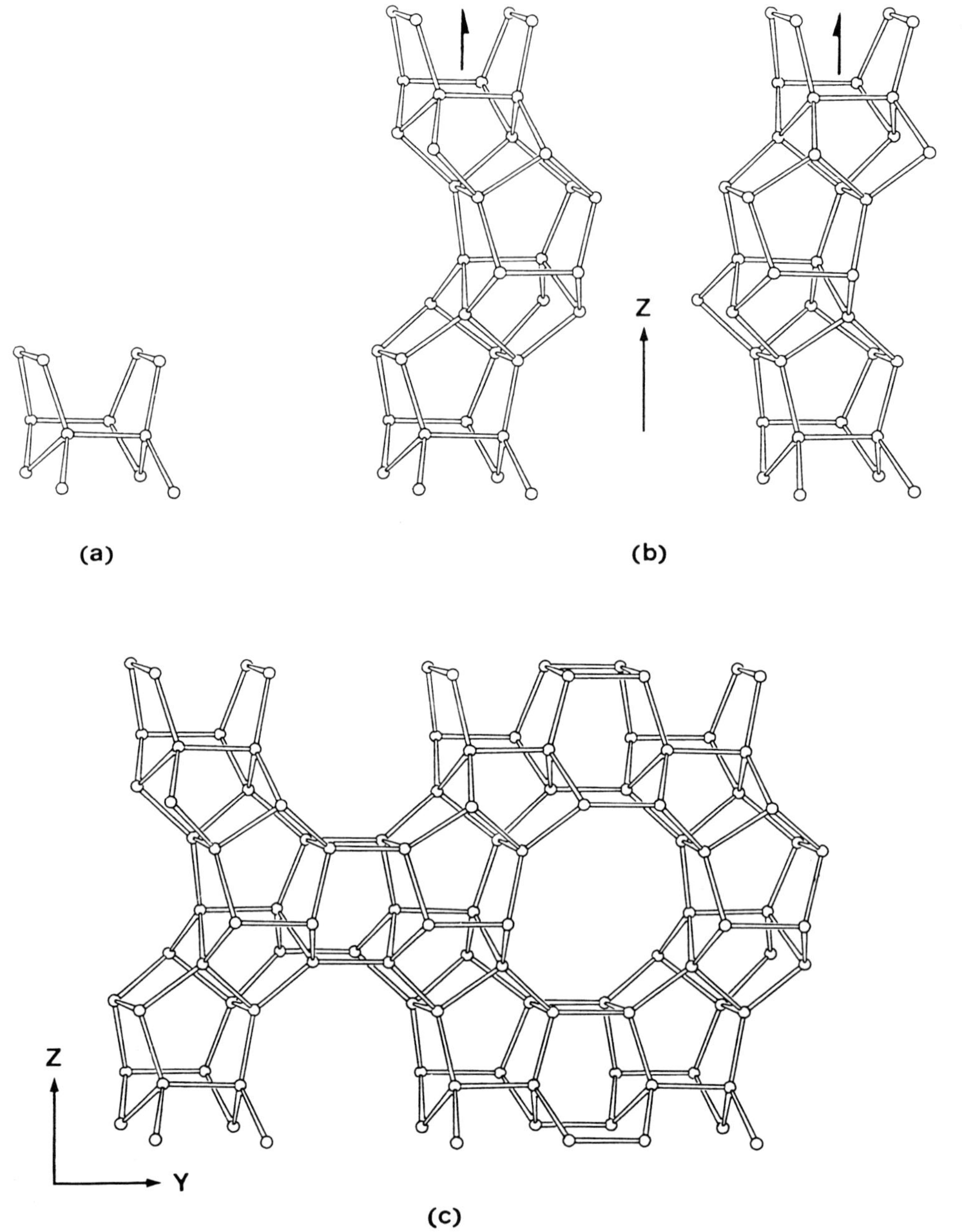

Figure 5. The T12-building unit (a) forms left- and right handed chains along Z (b). These chains, related by a mirror plane, generate the characteristic pentasil layer (c) which extends infinitely along Y and Z by applying translation symmetry.

3. ZEOLITES CONTAINING ZIGZAG CHAINS

3.1. Infinite buiding units from zigzag chains

Infinite building units (IBUs), from which the zeolites listed in Table 1 (page 83) can be assembled, are presented in Figure 6. The IBUs can be obtained from different linkages of z chains and are arranged in Figure 6 in order of increasing number of component z chains. Only parts of the 'infinite' IBUs are drawn. The drawings extend, in most cases, over more than the identity period along the chain axis. The independent unit is the smallest unit from which the IBU can be obtained by translation along the chain axis. The drawings of the infinite single and double zigzag chains (z1 and z2; Fig. 6a,b) cover about three unit cell edges and give three repeat units along the chain axis. In the single zigzag chain the independent unit consists of 2 T atoms (n=2) and in the double zigzag chain of 4 T atoms (a 4-ring ; (n=4)). The double chain can as well be described as an infinite ladder of 4-rings. For clarity reasons, starting with the triple zigzag chain given in Fig. 6c, only two repeat units are drawn in Fig. 6(c-l). The repeat unit shown in Fig. 6d, f.e., is a strongly deformed 6-ring (n=6). The IBU given in Fig. 6f very much resembles the JBW structure (see Figure 7).

Figure 6. Infinite building units from (multiple) zigzag chains. The building units are drawn in a perspective view with their symbolic representation of the projection along the chain axis underneath. The number of T atoms in the repeat unit along the chain axis equals n.

(a) z1: Single zigzag chain, n = 2.

(b) z2: Double zigzag chain (a 4-ring ladder), n = 4 (4-ring SBU).

(c) z3-1: Triple zigzag chain of the first kind, n = 6 (5-1 SBU).

(d) z3-2: Triple zigzag chain of the second kind forming a cylindric 6-ring wall, n = 6 (strongly deformed 6-ring).

(e) z4: Four zigzag chains form the z4 unit, n = 8.

(f) z6-1: Six zigzag chains of the first kind, n = 12.

(g) z6-2: Six zigzag chains of the second kind, n = 12. Can also be obtained by connecting two z3-2 units (or single 6-rings).

(h) z6-3: Six zigzag chains of the third kind. n = 12. Single 6-rings are connected differently from z6-2.

(i) z7: Seven zigzag chains form the z7 unit, n = 14. Can be obtained by connecting a z3-1 and a z4 unit.

(j) z8-1: Eight zigzag chains of the first kind, n = 16.

(k) z8-2: Eight zigzag chains of the second kind, n = 16.

(l) z-4ri: The additional 4-ring 'handle' leads to a doubling of the zigzag chain period. The repetition unit now contains 2*4 + 4 = 12 T atoms, n = 12.

(m): The drawings represent half of the independent repeat unit along the chain axis and the value of n given is in fact 1/2*n.

1: The additional 4-ring connects 6 zigzag chains consisting of 3 T atoms, n = 6*3 + 4 = 22. The total chain period consists of 2*3 = 6 T atoms and the chain length becomes ~3*5.2 Å.

2: The additional 4-ring connects 6 zigzag chains consisting of 4 T atoms, n = 6*4 + 4 = 28. The total chain period consists of 2*4 = 8 T atoms and the chain length becomes ~4*5.2 Å.

3: The additional 4-ring connects 6 zigzag chains consisting of 5 T atoms, n = 6*5 + 4 = 34. The total chain period consists of 2*5 = 10 T atoms and the chain length becomes ~5*5.2 Å.

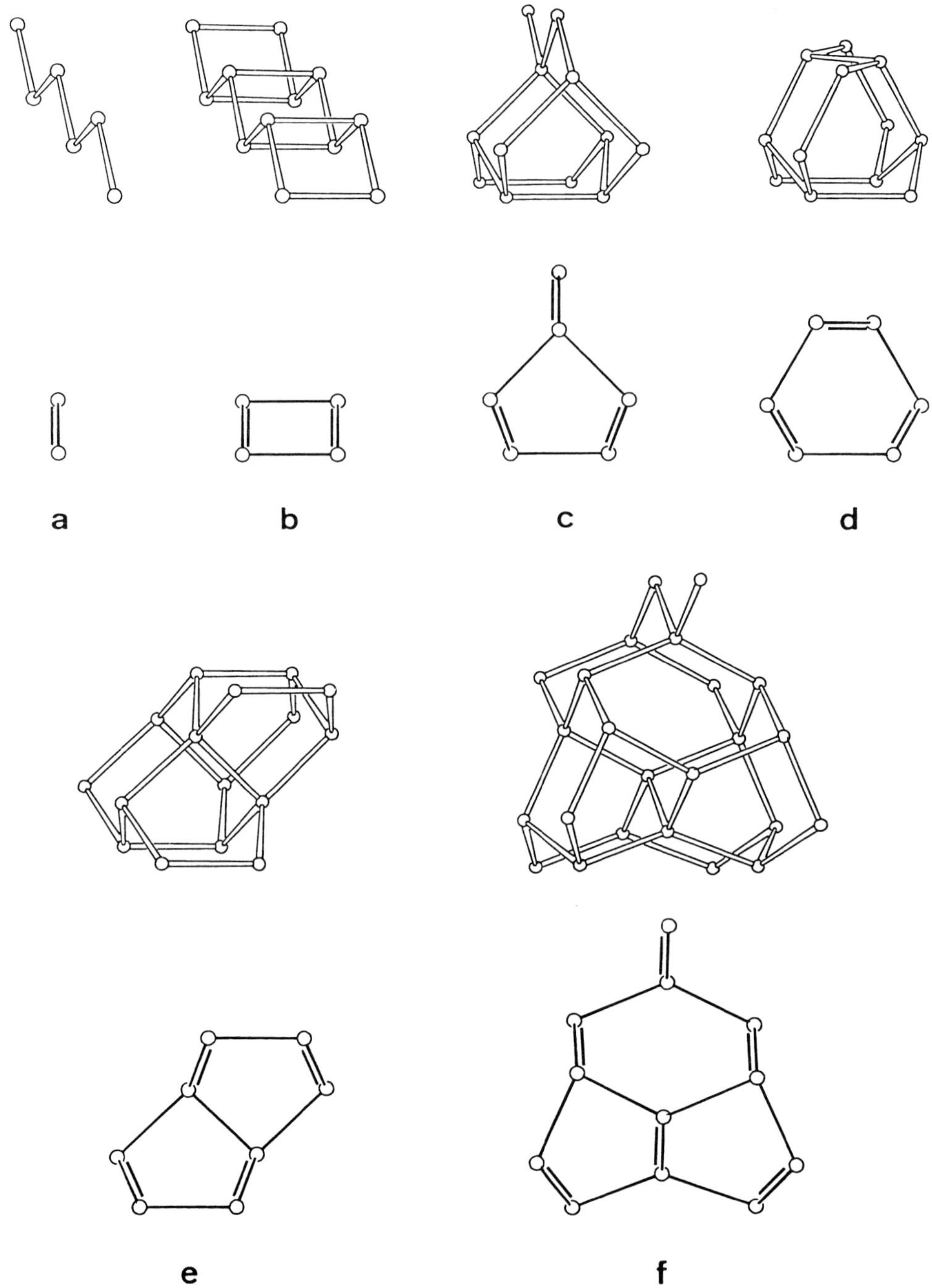

Figure 6. Legend: see previous page.

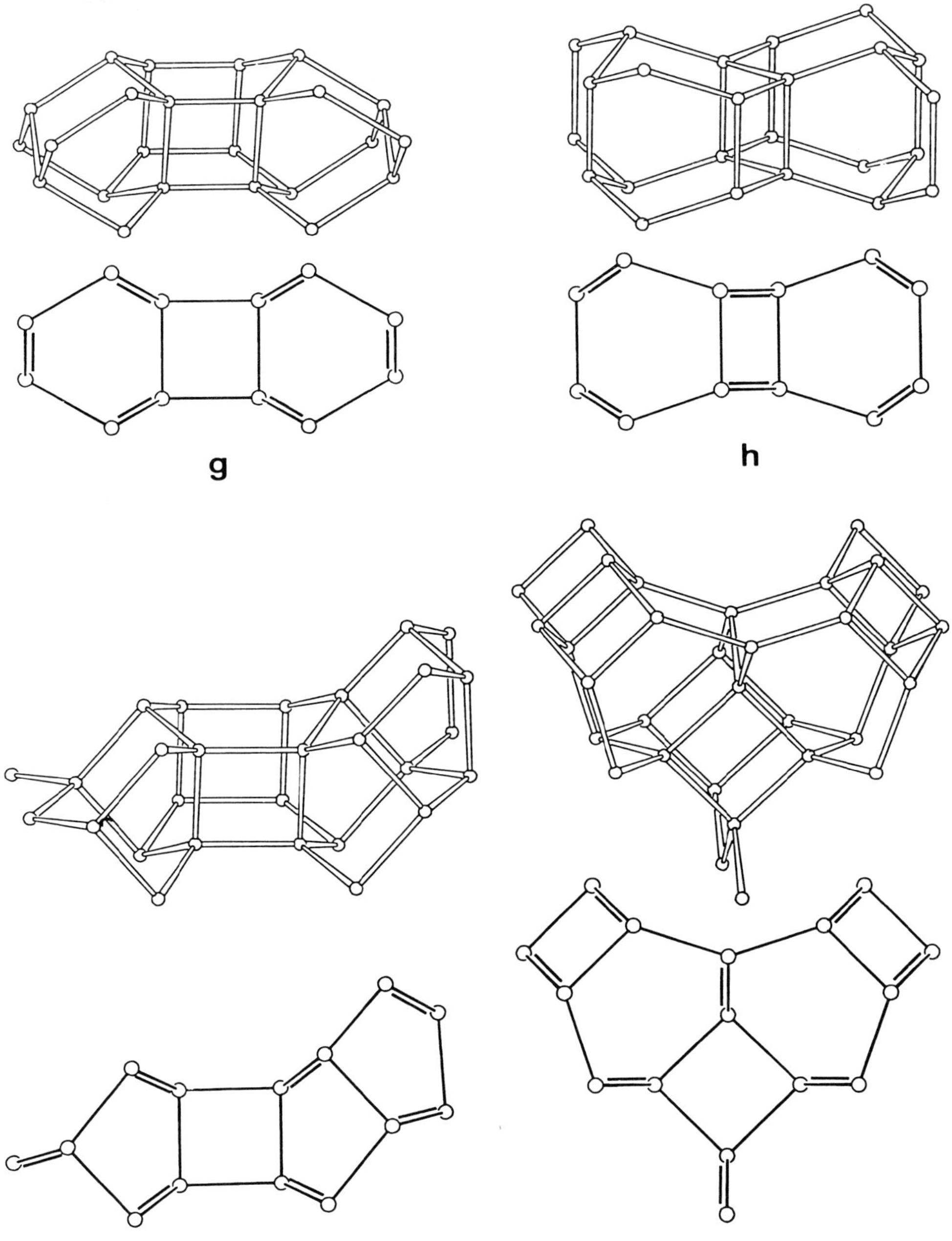

Figure 6. Cont'd.

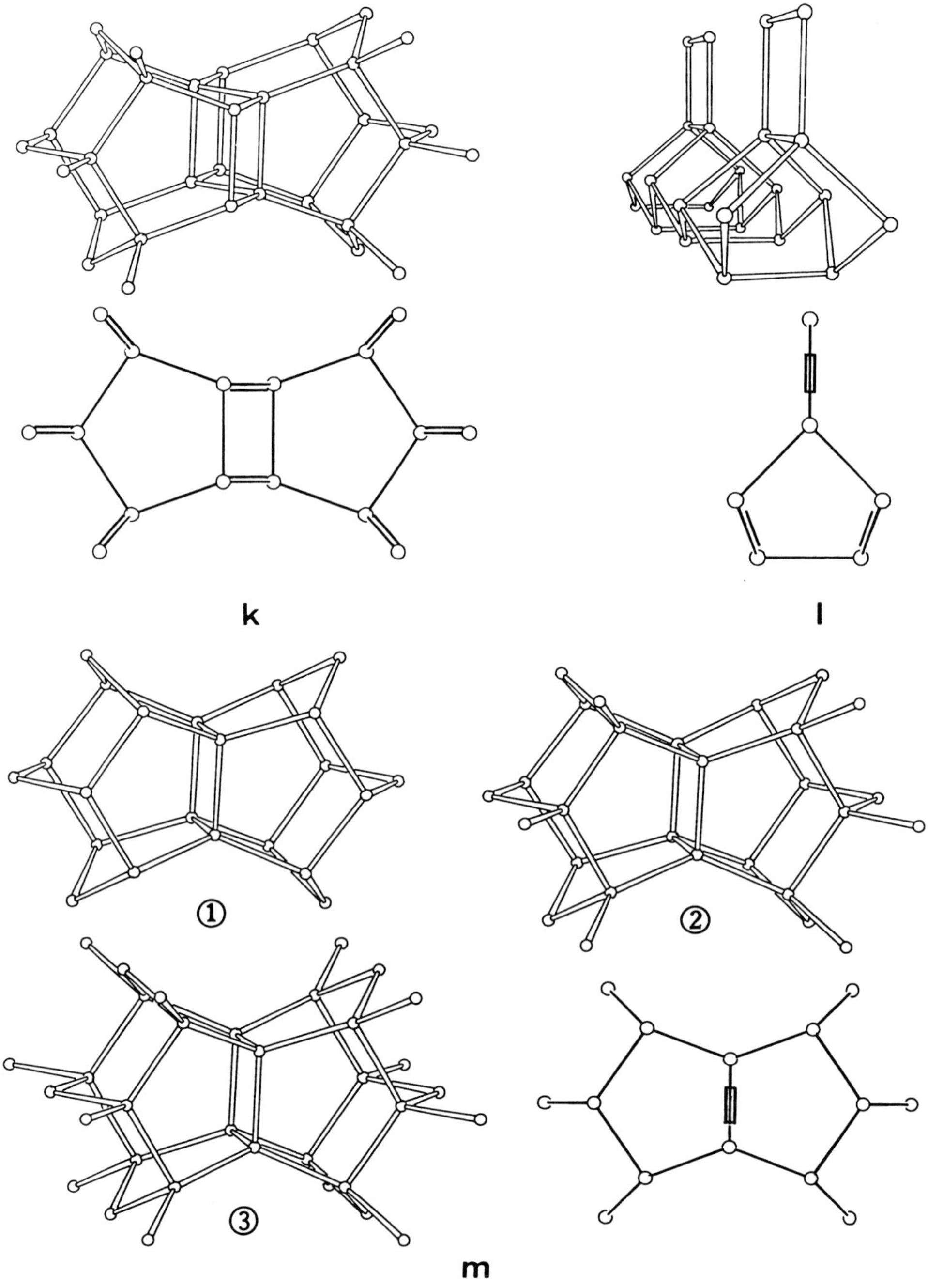

Figure 6. Final page.

Table 1
Zeolites containing (modified) zigzag chains. The topological unit cell and symmetry is given.
The chain axis is italicized

Code	space group	Cell dimensions (Å) a	b	c	β,γ (°)	n*	Z**
ABW	Imma	9.873	_5.254_	8.770		4	8
ATN	I4/mmm	13.071	13.071	_5.256_		4	16
ATO#	$R\bar{3}m$	20.914	20.914	_5.061_	120.0	12	36
ATS#	Cmcm	13.164	21.584	_5.257_		12	24
BIK	Cmcm	7.540	16.222	_5.256_		6	12
CAN#	$P6_3/mmc$	12.494	12.494	_5.254_	120.0	4	12
CAS	Cmcm	_5.256_	14.132	17.227		6	24
CFI	Imma	13.961	_5.264_	25.967		16	32
JBW#&	Pmma	_5.256_	7.450	8.159		4+2	6
MTT	Pmmn	_5.256_	22.031	11.384		12	24
MTW	C2/m	25.552	_5.256_	12.117	109.31	14	28
OSI	I4/mmm	18.506	18.506	_5.268_		16	32
TON	Cmcm	14.105	17.842	_5.256_		6	24
VET	$P\bar{4}$	13.048	13.048	_4.948_		8+1	17
DAC&	C2/m	18.574	7.542	_2*5.188_	108.92	8+4	24
EPI	C2/m	9.127	17.480	_2*5.188_	124.94	8+4	24
NON	Fmmm	22.862	_3*5.219_	13.935		18+4	88
EUO	Cmma	13.901	22.861	_4*5.145_		24+4	112
NES	Fmmm	_5*5.212_	13.884	22.860		30+4	136

* n = number of T atoms in the independent repeat unit along the chain axis (see Figure 6).
** Z = number of T atoms in the unit cell (see Figure 7).
\# Zeolite can also be constructed from single 6-rings.
& Zeolite can also be constructed from saw chains.

The IBUs in Fig. 6(l-m) contain, in addition to (deformed) z chains, a 4-ring. From these IBUs the 5 structure types listed at the bottom of Table 1 can be built. The drawings of the three IBUs shown in Fig. 6m extend only over half of the identity period. The number of T atoms in the independent unit (n) is given in Table 1.

3.2. Building schemes for zeolites containing (multiple) zigzag chains

Figure 7 presents the building schemes of the 19 zeolites listed in Table 1 using the IBUs given in Figure 6. Perspective views of the zeolites are accompanied by the parallel projection along the chain axis. The zeolites are, as are the IBUs in Figure 6, arranged in order of increasing complexity of the IBUs used.

In JBW a single z chain (2 T atoms/repeat unit; n = 2) is connected to a double z chain (n = 4) and the total number of T atoms per unit cell becomes Z = 4+2 = 6. As can clearly be seen from the perspective drawing of the structure in Figure 7, JBW can also be assembled by connecting single 6-rings. The JBW structure very much resembles the IBU shown in Fig. 6f.

The next three zeolites, ABW, ATN and CAN, contain 2 (Z = 8), 4 (Z = 16) and 3 (Z = 12) double z chain units (z2; Fig. 6b) per unit cell, respectively. CAN might as well be constructed by connecting single 6-rings.

BIK, TON and CAS can be built using a buiding unit composed of three zigzag chains: the z3-1 unit from Fig. 6c consisting of 6 T atoms per repeat unit. BIK has 2*6 = 12 T atoms per unit cell (Z = 12) and TON and CAS both have Z = 4*6 = 24. In VET a single T atom connects two z4 IBUs each containing 8 T atoms per repeat unit (z4; Fig. 6e) and Z = 17.

In MTT two z6-1 units (Fig. 6f) are connected: Z = 2*12 = 24. Three z6-2 units (Fig. 6g) form the unit cell of ATO (Z = 3*12 = 36). ATO can as well be constructed using six z3-2 units (Fig. 6d). The z3-2 unit can also be constructed using solely single 6-rings.

The same type of schemes show that in ATS, MTW, OSI and CFI the unit cell contains two z6-3 units (Z = 24; Fig. 6h), two z7 units (Z = 28; Fig. 6i), two z8-1 units (Z = 32; Fig. 6j) and two z8-2 units (Z = 32; Fig. 6k), respectively. The reader can verify that the unit cell of the ATS structure type can also be built by connecting six z2 units (Fig. 6b) or four single 6-rings. The IBU given to build MTW (Fig. 6i) can also be obtained by connecting the IBUs shown in Fig. 6c and Fig. 6e. An alternative (finite) building unit for OSI constitutes of four fused 6-rings drawn with bold lines in Figure 7.

The final five zeolites contain, in addition to zigzag chains, a 4-ring. In DAC and EPI this 4-ring 'handle' leads to a doubling of the chain axis period. The unit cell of DAC and EPI differ by a different connection of two z-4ri units (Fig. 6l; Compare also Fig. 8, Fig. 10 and Fig. 11). Both zeolites have Z = 2*12 = 24.

NON, EUO and NES show (F) centered layers perpendicular to the *b*, *c* and *a* axis, respectively. The thickness of this layer is 1.5 , 2 or 3 times the zigzag repeat and contains 3, 4 or 6 T atoms, respectively. The F centered layers can easily be constructed from the building units given in Figure 6m by translations only. Neighboring layers are related by a glide plane with an axial glide parallel to **c**, **a** or **b**, respectively. Linkage of layers produce perfect z chains in mirror planes at x = 0 and 1/2. In the z chains at x = +1/4 or - 1/4, serious torsional strain is accumulated: the angle between succeeding planes formed by the plane segments of the 3, 4 or 6 T atoms in NON, EUO and NES, respectively, is about 90°. In the drawing of the parallel projection one such strain-full z chain is indicated by an arrow. For clarity reasons, the perspective view only gives three building units. The number of T atoms per unit cell is 4*22 = 88, 4*28 = 112 and 4*34 = 136 in NON , EUO and NES, respectively.

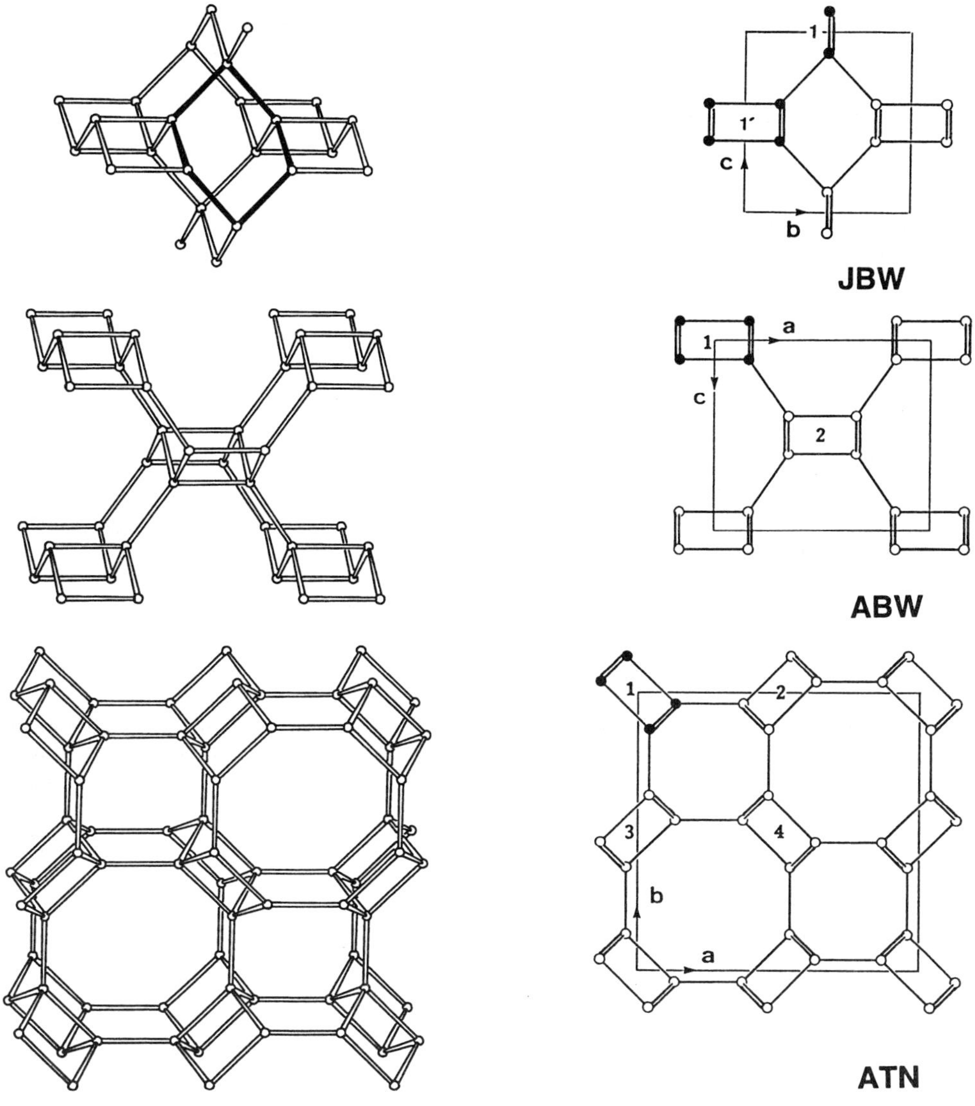

Figure 7. Perspective view and building scheme for 19 zeolites containing zigzag chains. The unit cell is outlined and the number of independent IBUs per unit cell is given. One of the IBUs is indicated by black dots. On each page the structure type code is given followed by the type of IBU used and the reference to Figure 6 in parentheses. One 6-ring is drawn with bold lines in those zeolites which can also be assembled from single 6-rings.
JBW (z1, Fig. 6a and z2, Fig. 6b); ABW and ATN (z2, Fig. 6b).

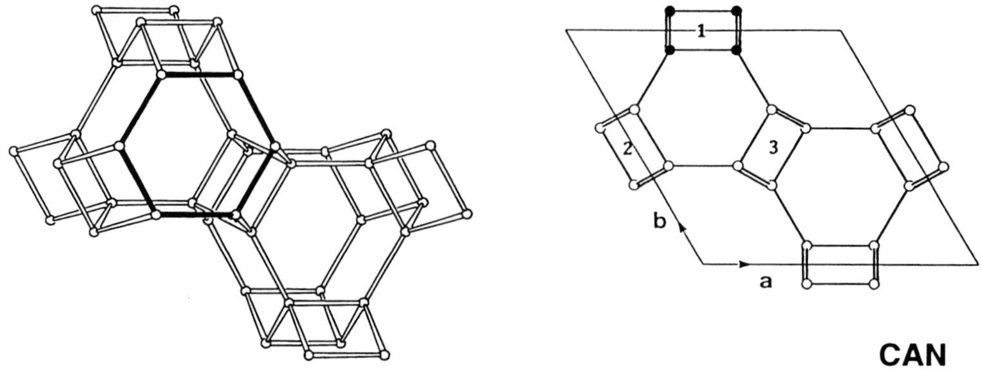

CAN

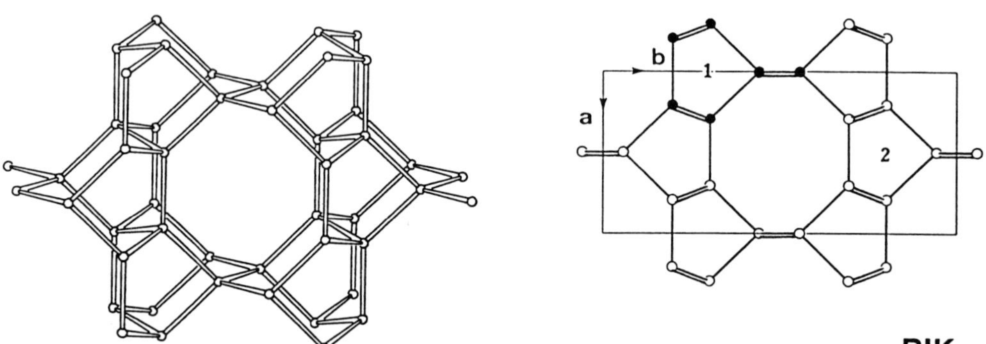

BIK

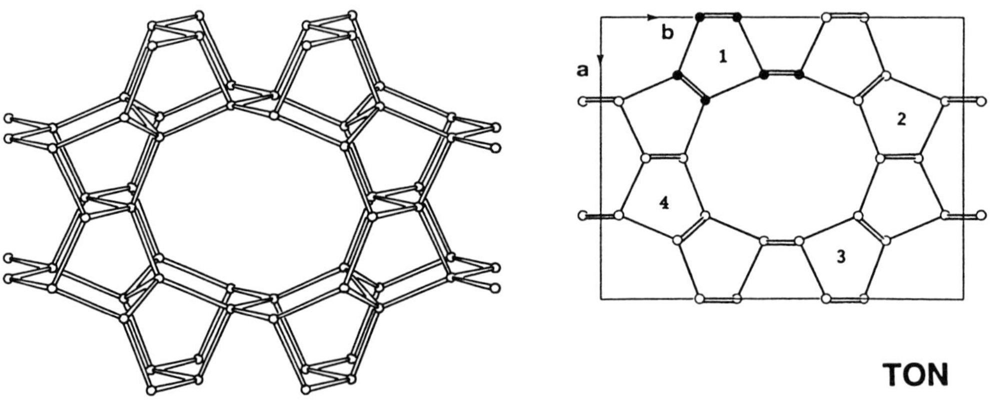

TON

Figure 7. Cont'd. CAN (z2, Fig. 6b); BIK and TON (z3-1, Fig. 6c).

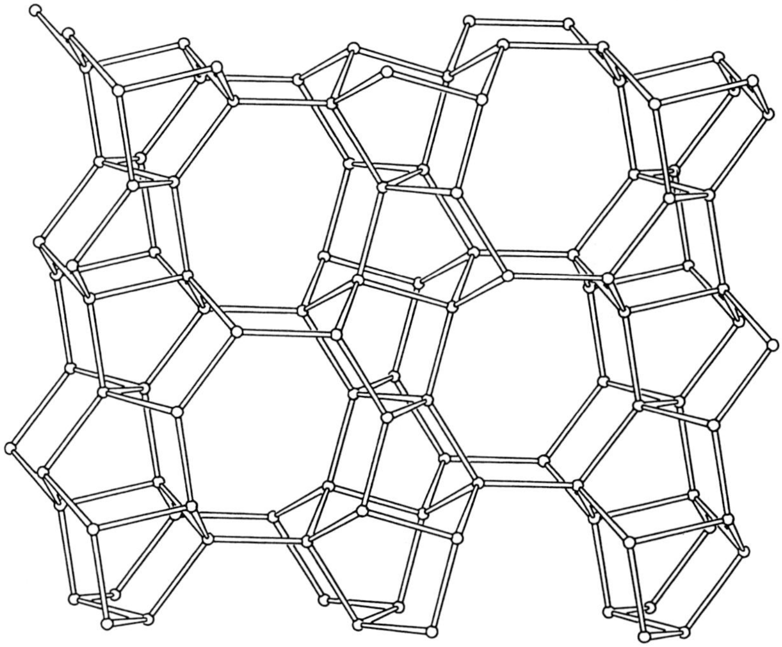

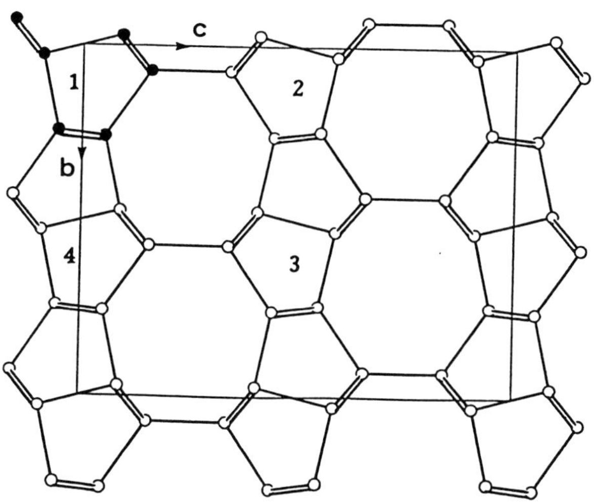

CAS

Figure 7. Cont'd. CAS (z3-1, Fig. 6c).

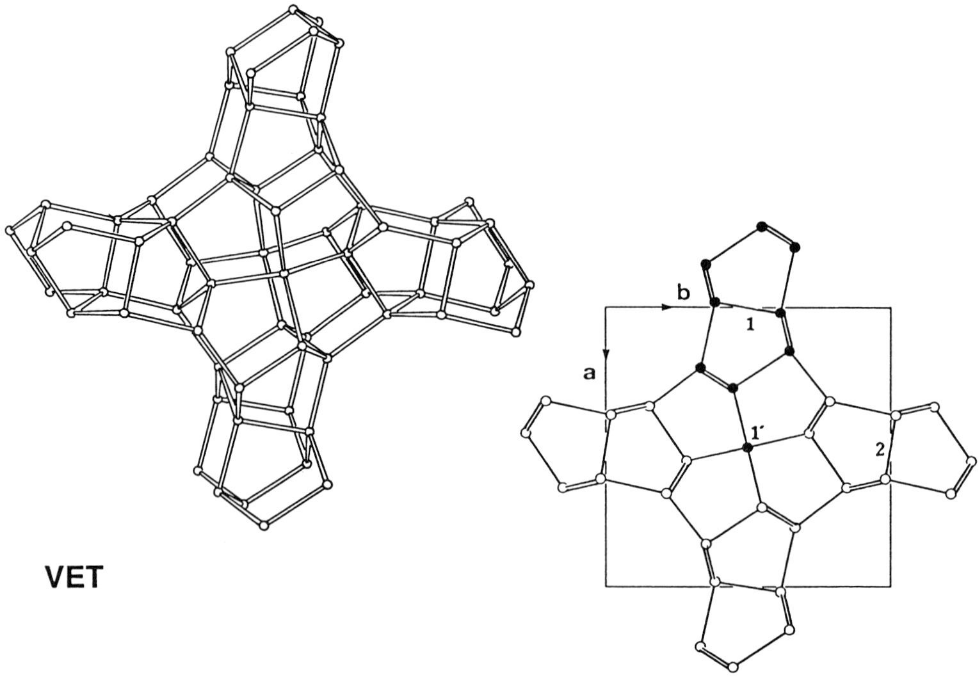

VET

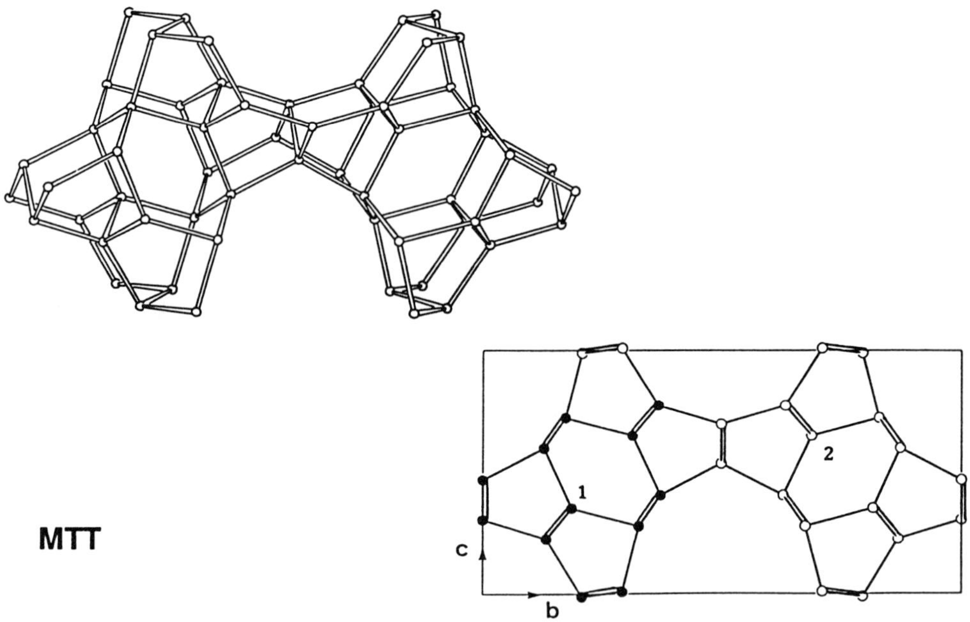

MTT

Figure 7. Cont'd. VET (z4, Fig. 6e); MTT z6-1, Fig. 6f).

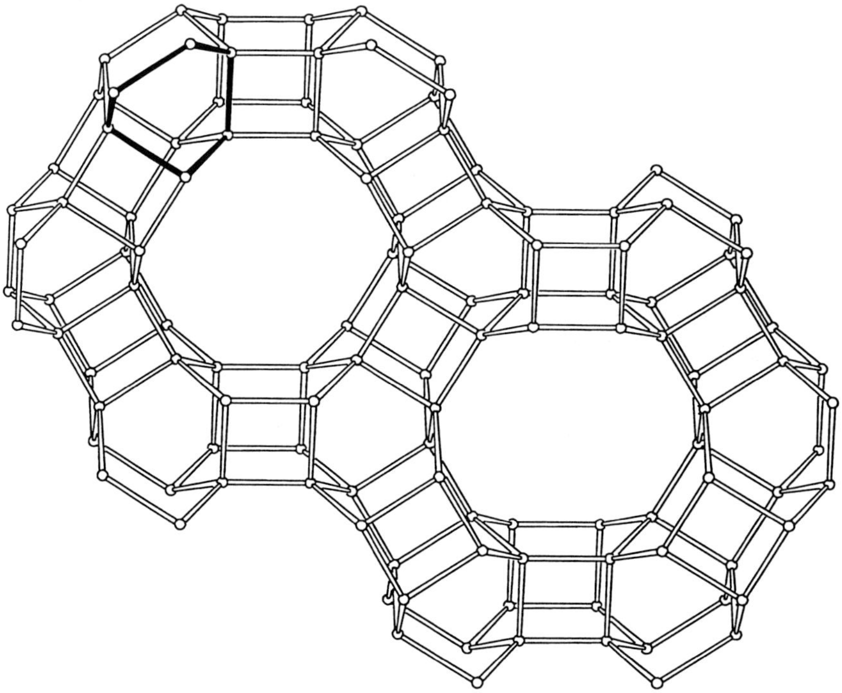

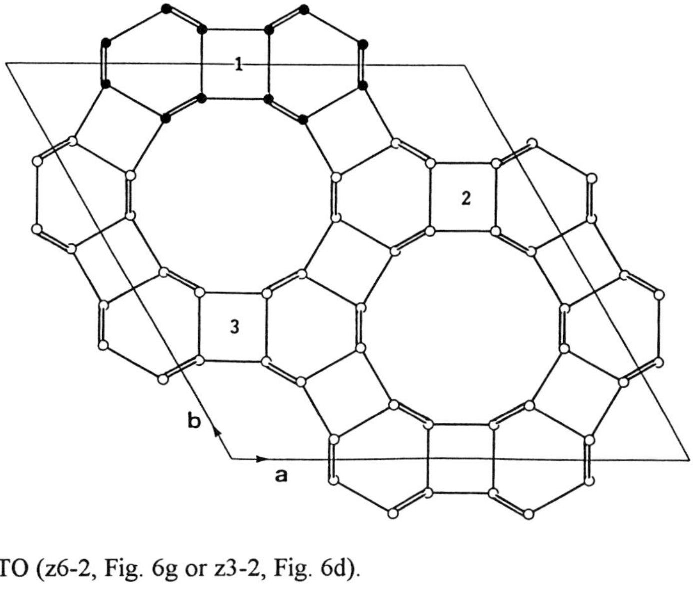

ATO

Figure 7. Cont'd. ATO (z6-2, Fig. 6g or z3-2, Fig. 6d).

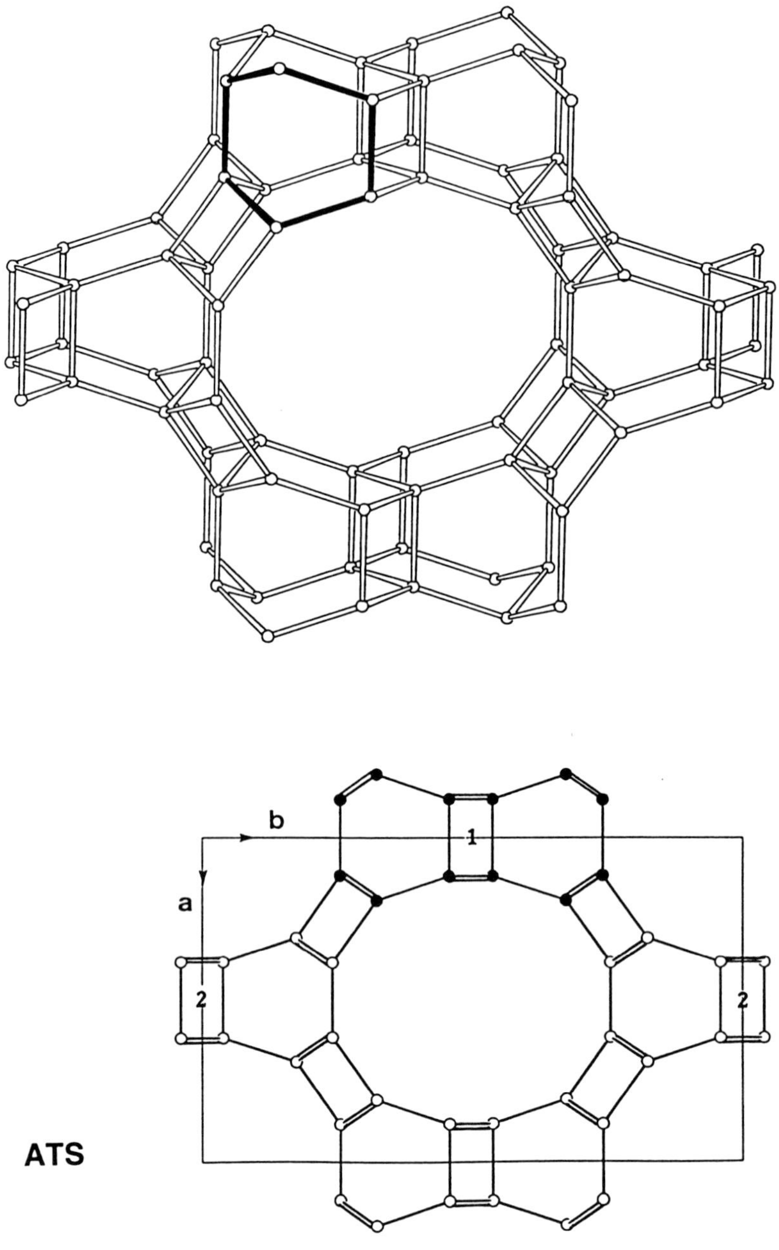

Figure 7. Cont'd. ATS (z6-3, Fig. 6h or z2, Fig. 6b).

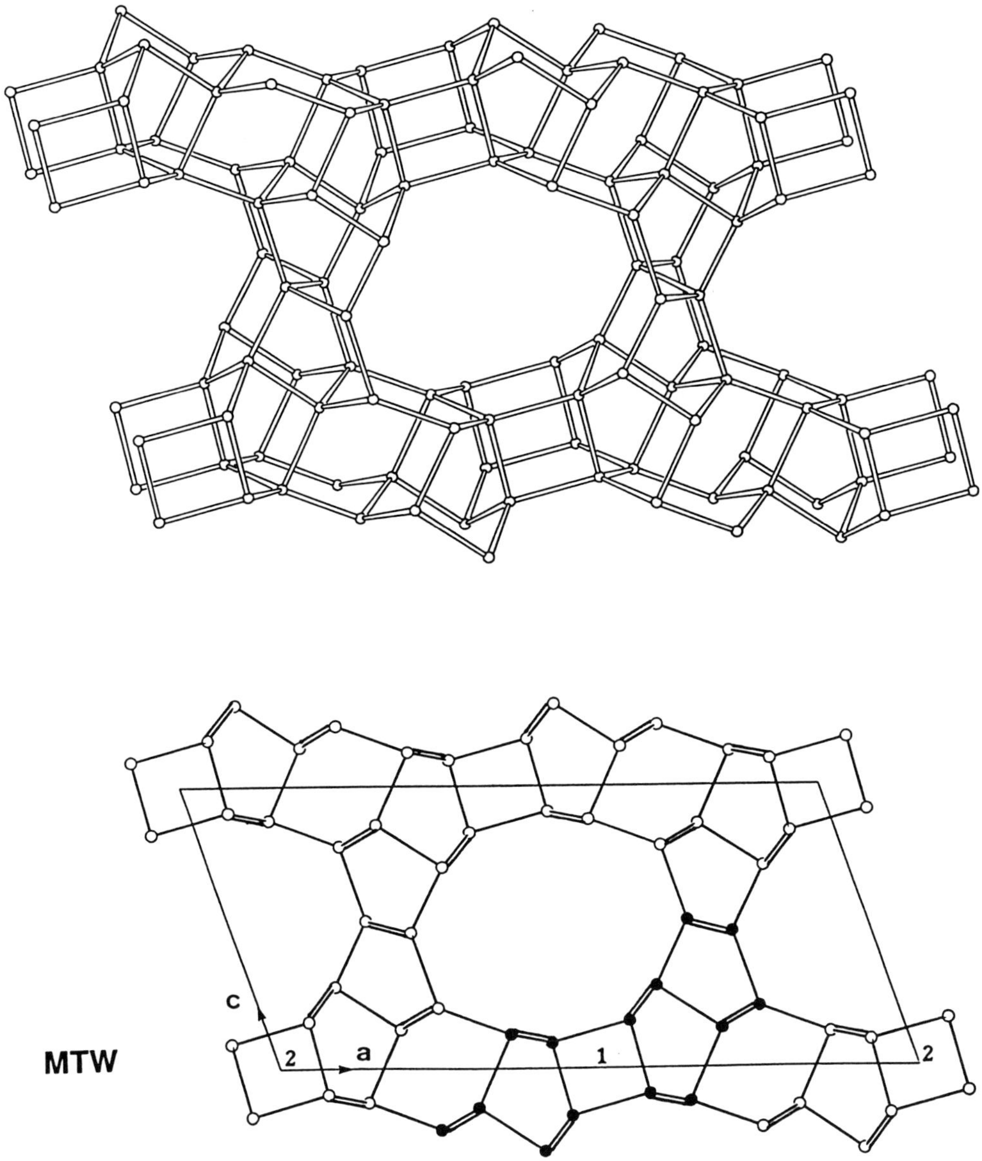

MTW

Figure 7. Cont'd. MTW (z7, Fig. 6i or z3-1, Fig. 6c and z4, Fig. 6e).

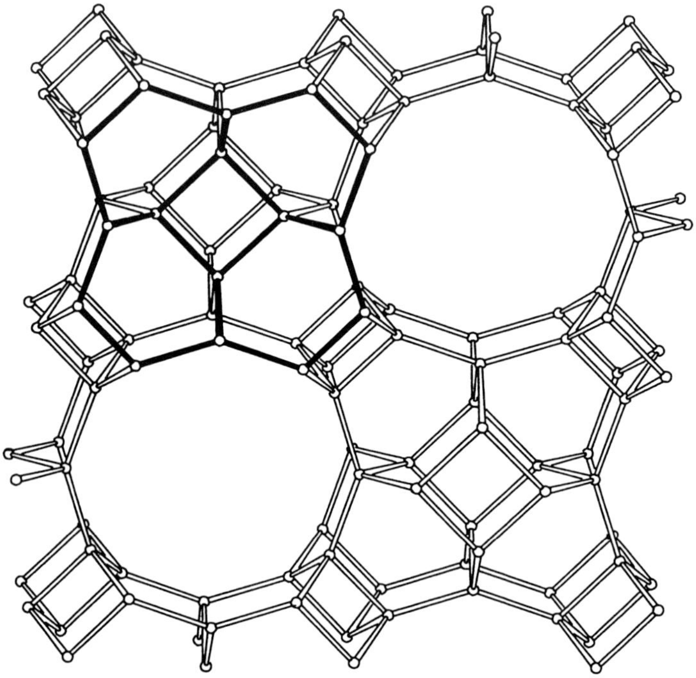

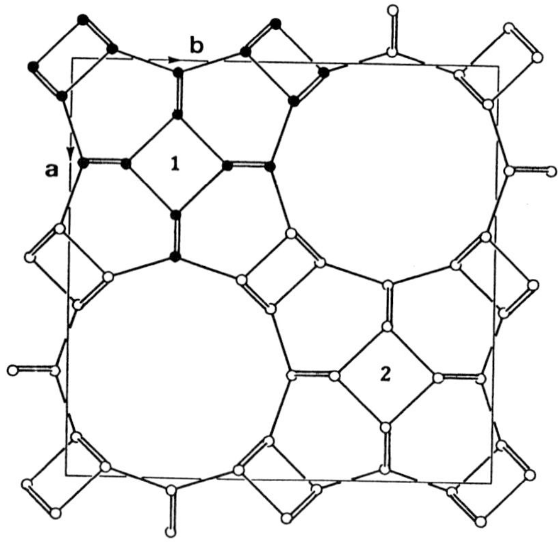

OSI

Figure 7. Cont'd. OSI (z8-1, Fig. 6j).

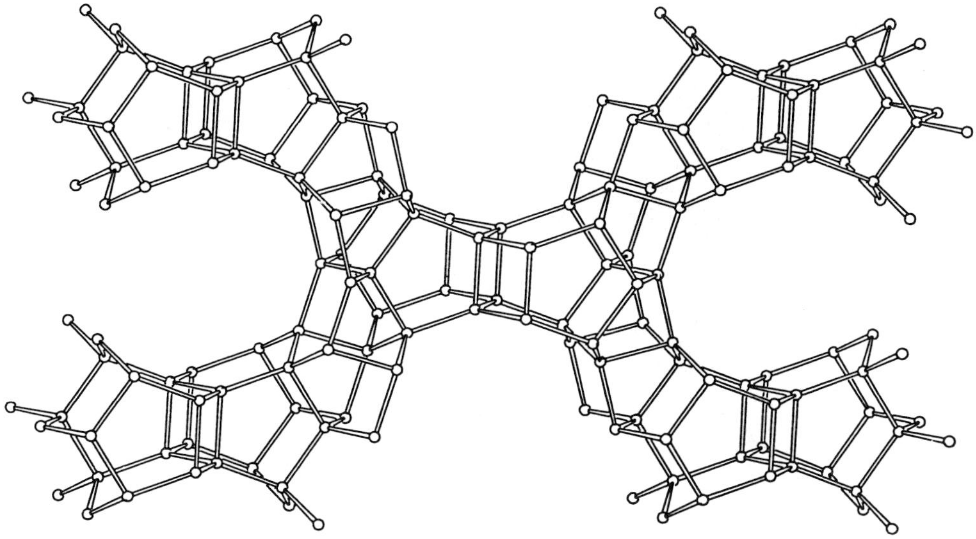

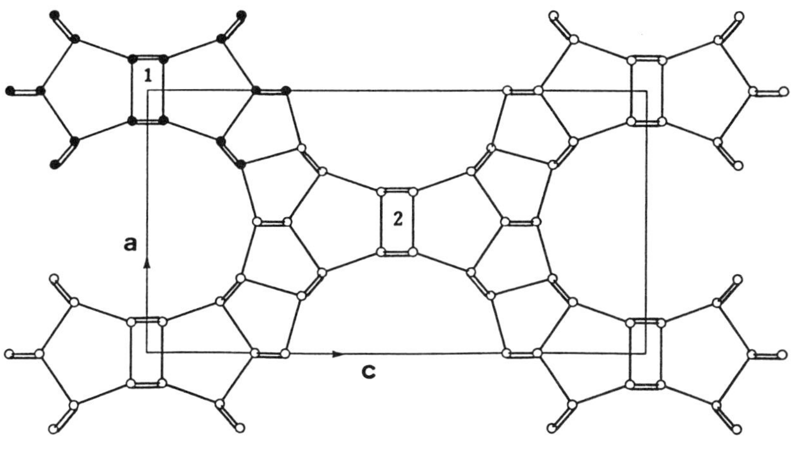

CFI

Figure 7. Cont'd. CFI (z8-2, Fig. 6k).

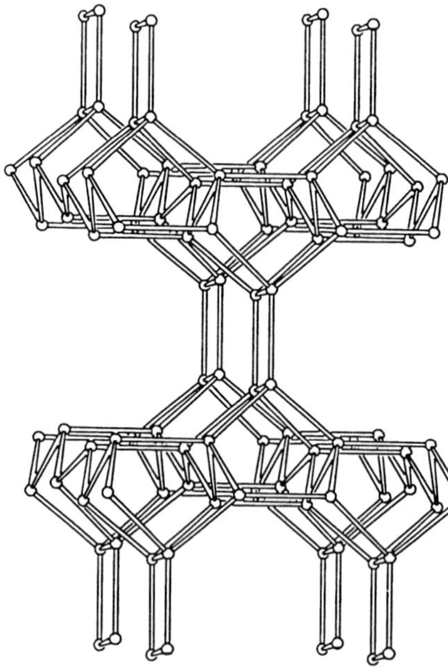

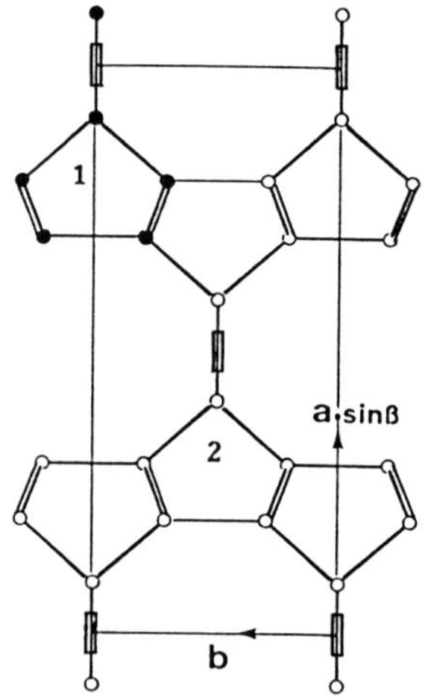

DAC

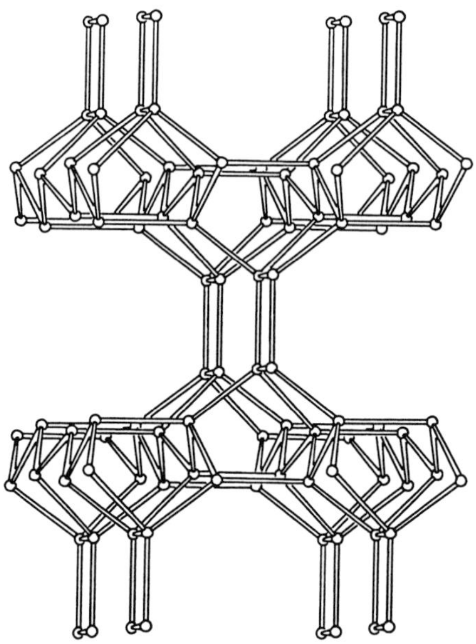

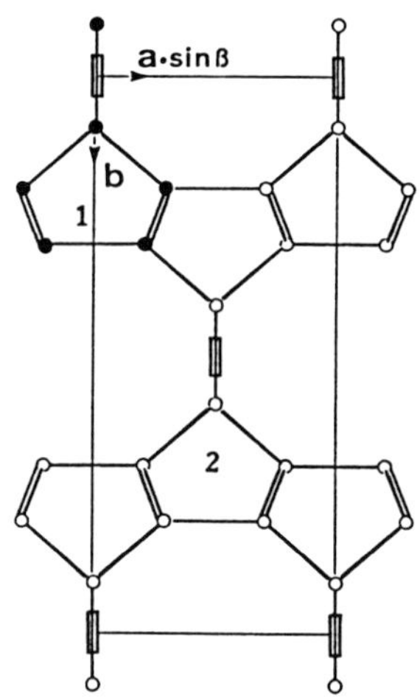

EPI

Figure 7. Cont'd. DAC and EPI (z-4ri, Fig. 6l).

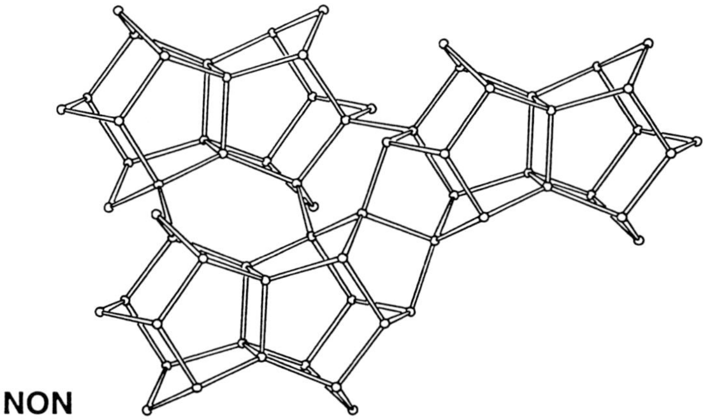

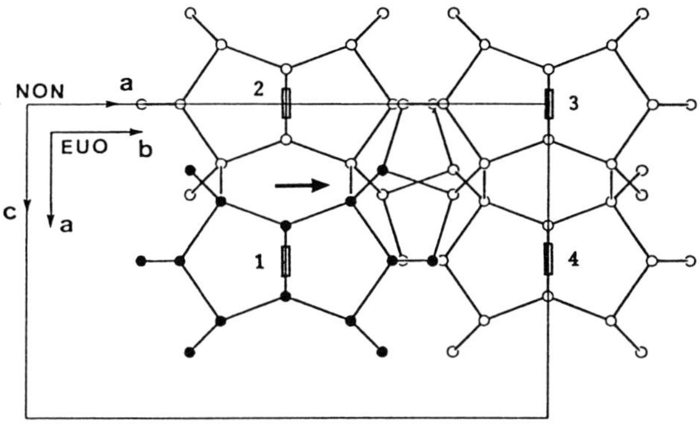

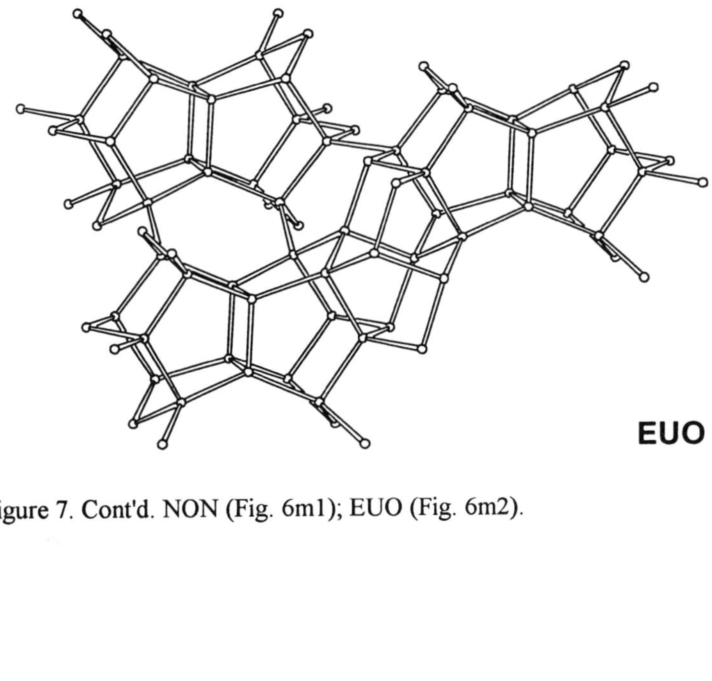

Figure 7. Cont'd. NON (Fig. 6m1); EUO (Fig. 6m2).

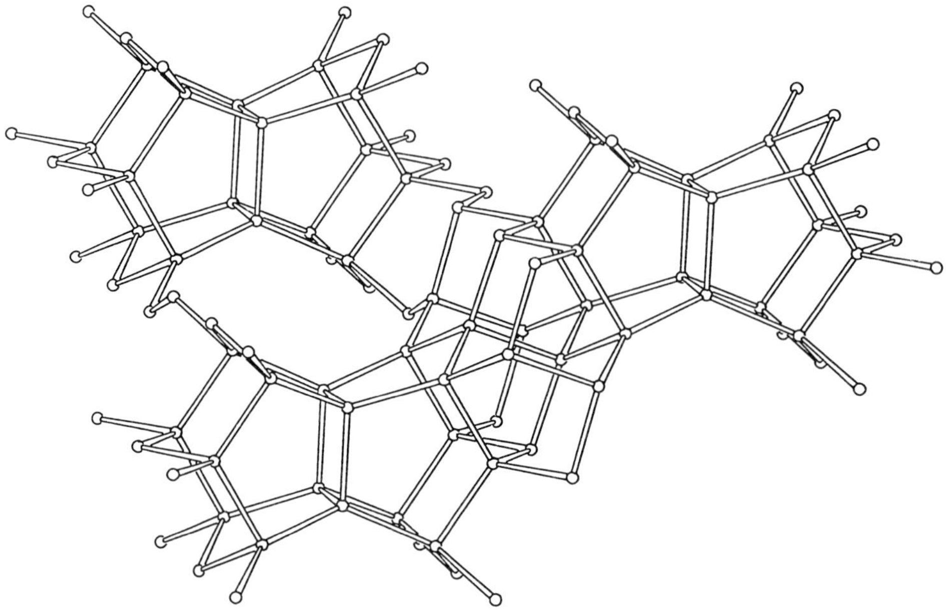

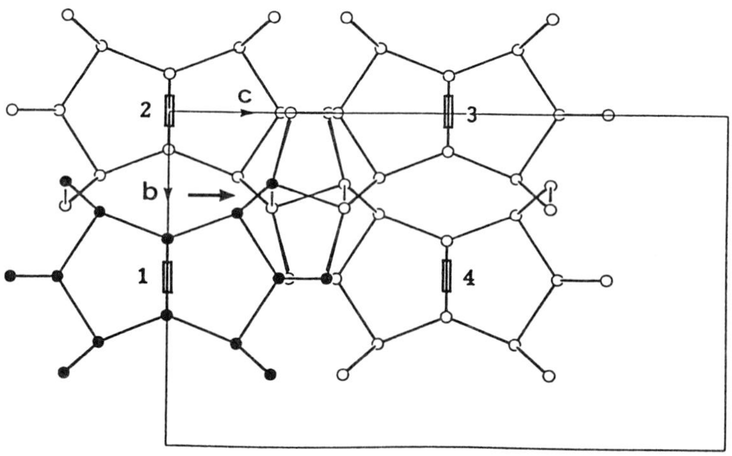

Figure 7. Final page. NES (Fig. 6m3).

3.3. Alternative descriptions using (cylindrical) 6-ring sheets

Figure 8 shows the (cylindrical) 6-ring sheets observed in 12 of the 19 structure types described in Section 3.2. Three types of 'cylindrical' (or tubular) sheets of fused 6-rings are found: the 6-ring walls form either 6-, 8- or 12-ring pores. In all three cases zigzag chains in the fused 6-ring wall run parallel to the channel axis. The independent repeat unit along the channel axis equals a ring consisting of 6, 8 or 12 T atoms, respectively. The (circular) crankshaft chains in the pore wall are perpendicular to the pore direction. The 6-ring pore tube equals the z3-2 building unit in Fig. 6d. The IBU in Fig. 6e (used to make the VET structure) is obtained when the T atoms of the 8-ring tube in Figure 8, indicated by arrows, are connected.

The unit cell of ATN is very easily constructed by linking two 8-ring tubes ($Z = 2*8 = 16$). ATO and CAN can both be built from 12-ring tubes. The unit cells of ATO and CAN contain three ($Z = 3*12 = 36$) and one ($Z = 12$) 12-ring(s), respectively.

The, sometimes strongly corrugated, 6-ring sheets are shown in the same order and in about the same orientation as they appear in the structures in Figure 7, except for the 6-ring sheet in ATO which is depicted after a rotation of about 30° around the view axis. Zigzag chains, formed by the fused 6-rings, run parallel to a channel axis in all structures. The crankshaft chains in the fused 6-ring sheets are now strongly distorted and hardly recognizable except for the 6-ring sheet in DAC and EPI where the crankshaft chains, perpendicular to the zigzag chains, are nearly undistorted.

In ABW 6-ring sheets are directly connected to form double z chains. In JBW the same type of 6-ring sheets as in ABW have single z chains in common.

In BIK, TON and CAS differently corrugated 6-ring sheets are connected through single z chains. Two neighbouring 6-ring sheets in ATO and MTW are connected in a rather simple way whereas the connection to the other neighbouring layer is very complex.

In CFI 6-ring sheets are directly connected to form double z chains. Two additional single z chains are needed to complete the unit cell content. In DAC and EPI the same type of 6-ring layer is connected to neighbouring layers by 4-rings in two different ways.

In DAC and EPI the 4-ring 'handles' are in a rectangular and oblique configuration, respectively. These configurations are illustrated in Figure 8, where, in addition to the 6-ring sheets, 2 T atoms of 4-ring handles have been drawn.

A majority of the structures contain (tubular) 6-ring sheets consisting of zigzag chains of fused 6-rings. It therefore looks like zigzag chains are a product of fusion of 6-rings and are not essential entities during crystal growth. In contrast, 6-rings, whether single or fused to (tubular) 6-ring sheets, may play an important role.

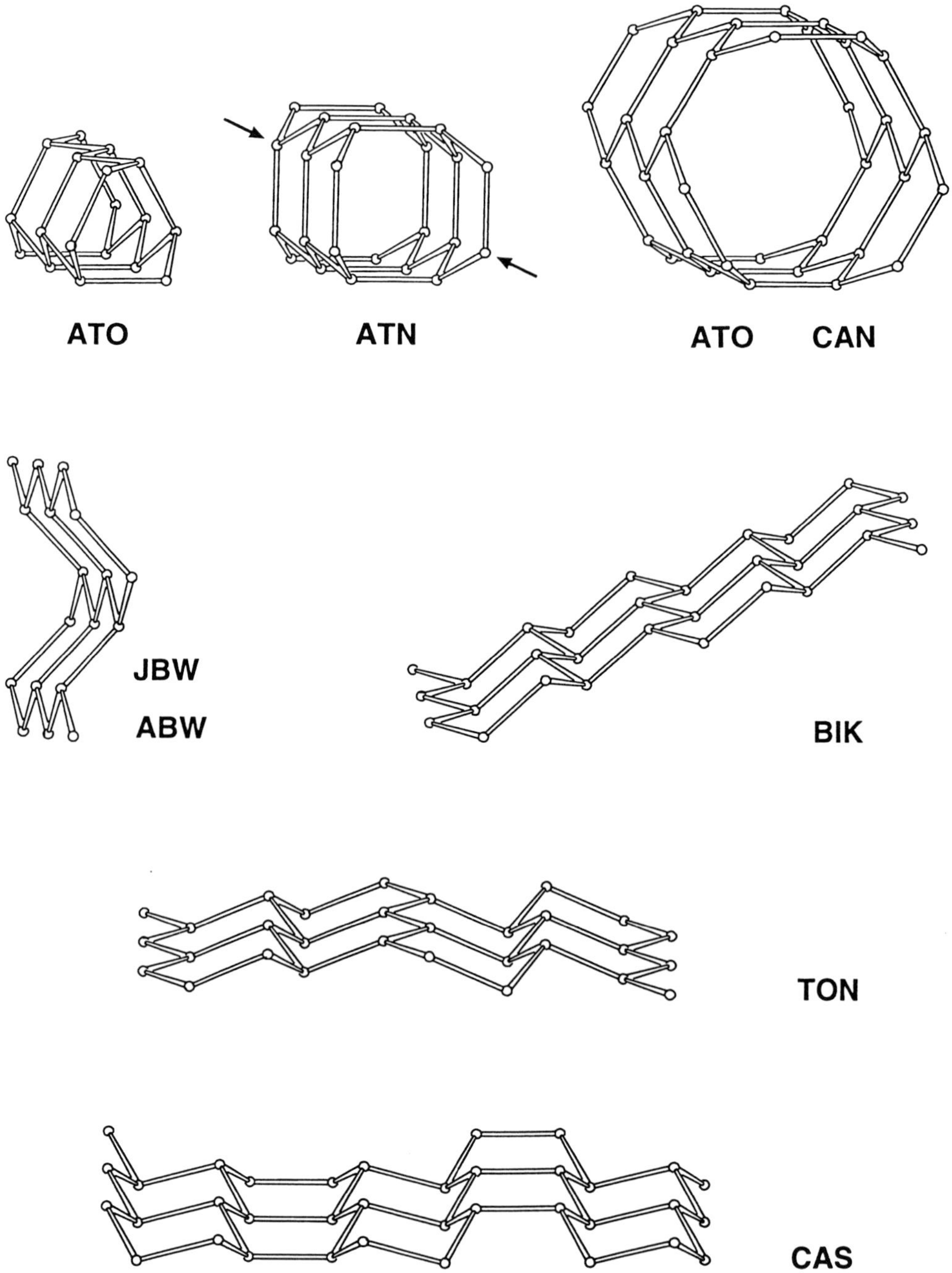

Figure 8. Tubular 6-ring walls and planar 6-ring sheets observed in zeolites that can be built from zigzag chains.

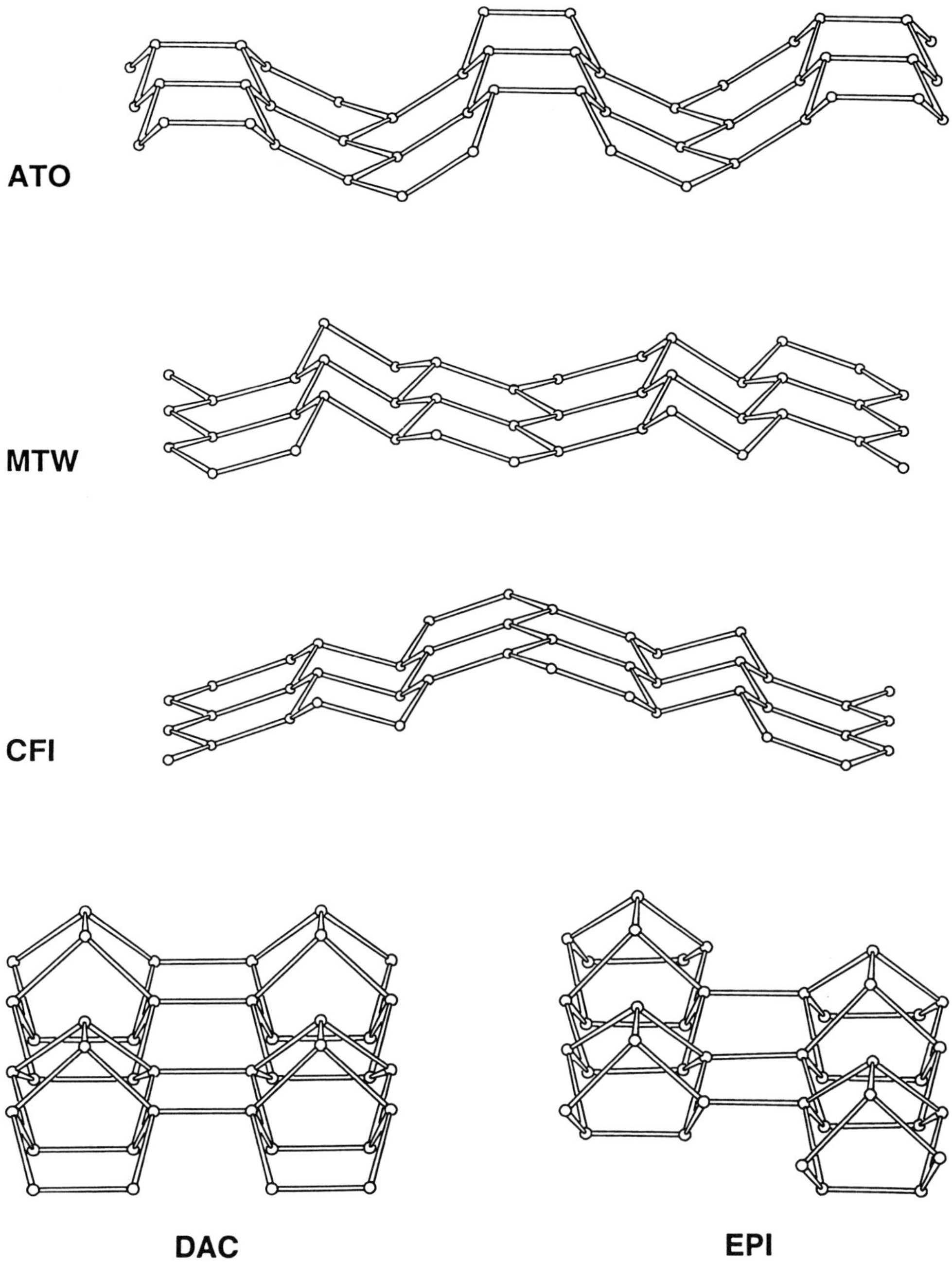

Figure 8. Final page. In DAC and EPI two T atoms of the 4-ring 'handles' have been added.

4. ZEOLITES CONTAINING SAW CHAINS

4.1. Infinite building units from saw chains

Parts of 'infinite' IBUs, from which the zeolites listed in Table 2 (page 102) can be assembled, are presented in Figure 9. The IBUs can be obtained from different linkages of saw (or s) chains and are arranged in Figure 9 in order of increasing number of component s chains. In the single saw chain (s1; Fig. 9a) the independent repeat unit consists of 3 T atoms (n = 3) and the double saw chain (s2; Fig. 9b) has n = 6. The double chain can as well be described as an infinite buckled ladder of 4-rings. A third saw chain can approach the s2 chain from 'above' or from 'below'. The triple saw chain s3-1 (Fig. 9c), with n = 9, is obtained when the three component saw chains are 'in phase'. In the triple saw chain s3-2 (Fig. 9d), also with n = 9, one saw chain is shifted with respect to the other two over half the chain period.

Finally, the s4 building unit (Fig. 9e), with n = 12, is obtained by connecting one more s chain to the s3-2 unit. The number of T atoms in the independent repeat unit (n) is added in Table 2.

4.2. Building schemes for zeolites containing (multiple) saw chains

Figure 10 gives the building schemes of the 9 zeolites listed in Table 2 using the IBUs presented in Figure 9. Perspective views of the zeolites are accompanied by the parallel projection along the chain axis. The zeolites are arranged, as are the IBUs in Figure 9, in order of increasing complexity of the IBUs used.

In JBW the unit cell contains two single s chains (s1; Fig. 9a) and the total number of T atoms per unit cell becomes Z = 2*3 = 6.

The next four zeolites, ATT, OFF, LTL and MAZ, can be constructed by linking two (Z = 12), three (Z = 18), six (Z = 36) and six (Z = 36) double s chain units (s2; Fig. 9b) per unit cell, respectively. JBW, ATT, OFF and LTL can as well be built using 6-rings only. This is illustrated in Figure 10 where one 6-ring in each structure is drawn with bold lines.

Figure 9. Infinite building units from (multiple) saw chains. The building units are drawn in a perspective view with their symbolic representation of the projection along the chain axis underneath. The arrow points towards the teeth of the saw. The number of T atoms in the independent repeat unit along the chain axis equals n. An open arrow denotes an s chain shifted over half the chain period with respect to a 'closed arrow' s chain.
(a) s1: Single saw chain, n = 3.
(b) s2: Double saw chain (a buckled 4-ring ladder), n = 6.
(c) s3-1: Triple saw chain of the first kind; the three saw chains are 'in phase', n = 9.
(d) s3-2: Triple saw chain of the second kind; one s chain is 'out of phase', n = 9.
(e) s4: Four saw chains form the s4 unit. Can be obtained by connecting one extra saw chain to the s3-2 unit.

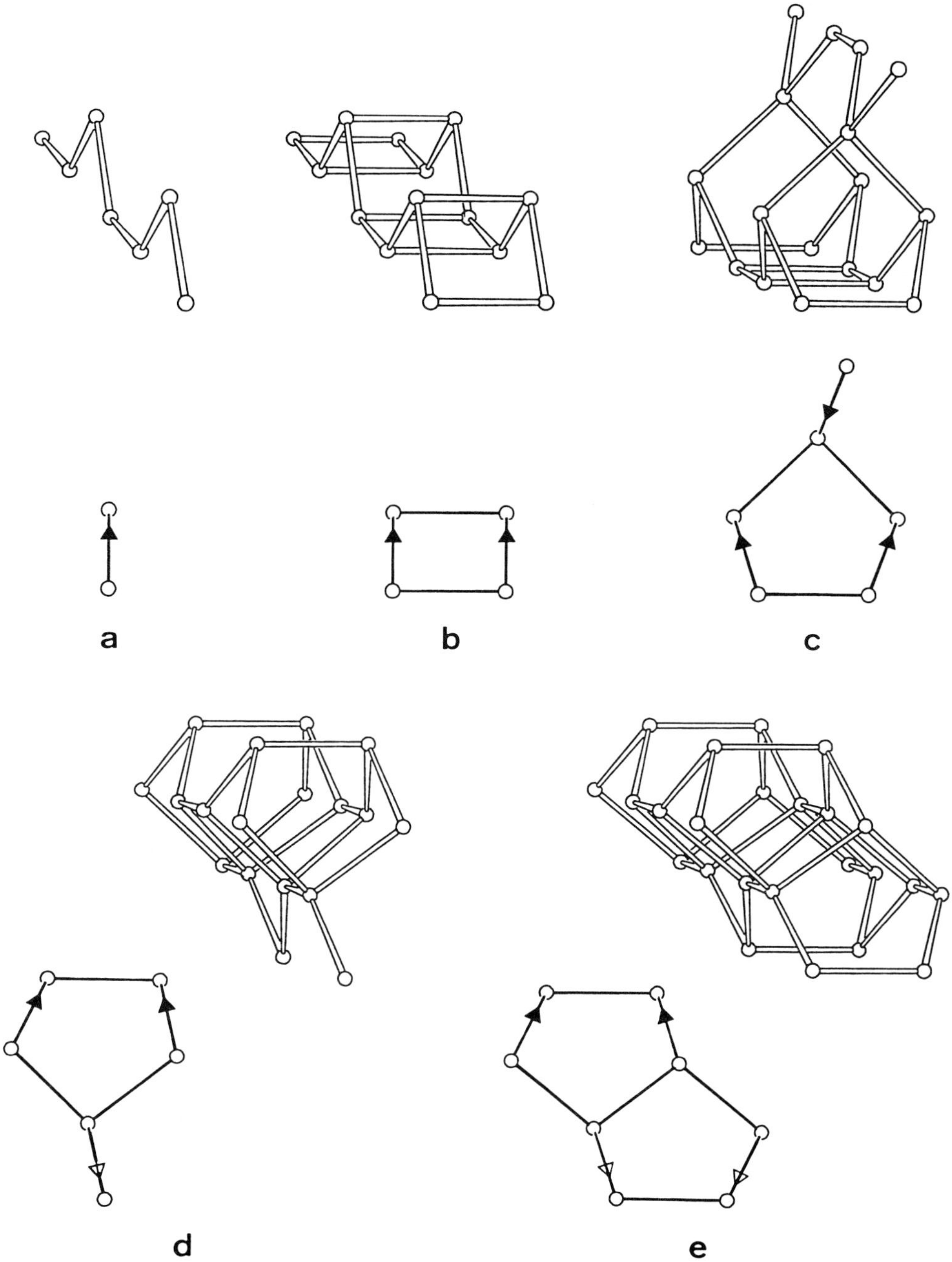

Figure 9. Legend: see previous page.

Table 2
Zeolites containing (modified) saw chains. The topological unit cell and symmetry is given. The chain axis is italicized

Code	space group	Cell dimensions (Å) a	b	c	β,γ (°)	n*	Z**
ATT#!	Pmma	9.980	*7.514*	9.369		6	12
DAC$	C2/m	18.574	*7.542*	10.377	108.92	12	24
FER	Immm	19.018	14.303	*7.541*		9	36
JBW#$	Pmma	5.256	*7.450*	8.159		3	6
LTL#	P6/mmm	18.126	18.126	*7.567*	120.0	6	36
MAZ	$P6_3/mmc$	18.102	18.102	*7.618*	120.0	6	36
MFS	Imm2	*7.543*	14.388	19.016		9	36
MOR	Cmcm	18.256	20.534	*7.542*		12	48
OFF#	$P\bar{6}m2$	13.063	13.063	*7.565*		6	18

* n = number of T atoms in the independent repeat unit along the chain axis (see Figure 9).
** Z = number of T atoms in the unit cell (see Figure 10).
# Zeolite can also be constructed from single 6-rings.
! Zeolite can also be constructed from crankshaft chains.
$ Zeolite can also be constructed from zigzag chains.

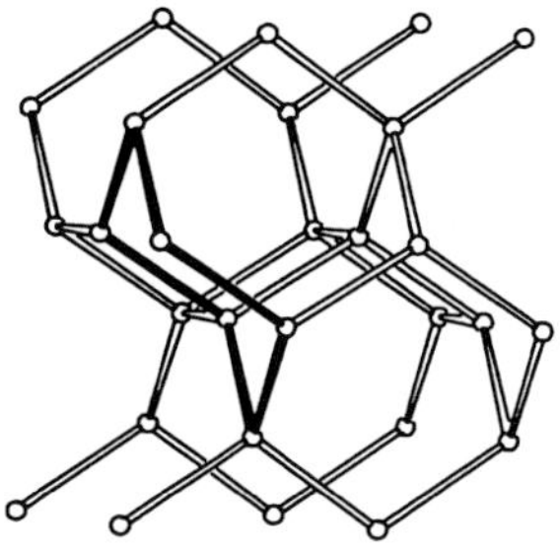

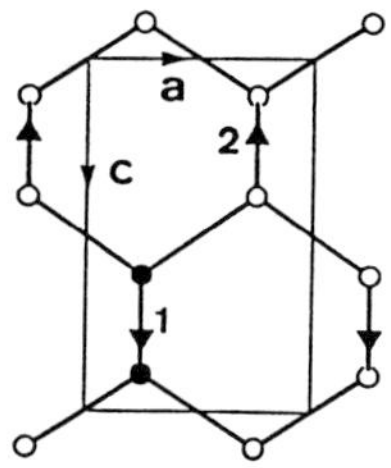

JBW

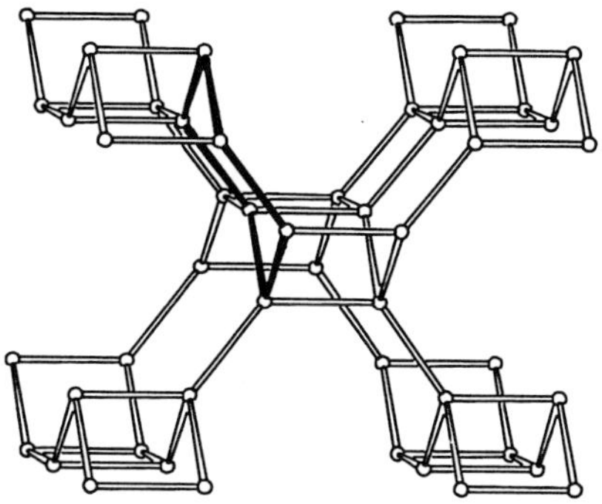

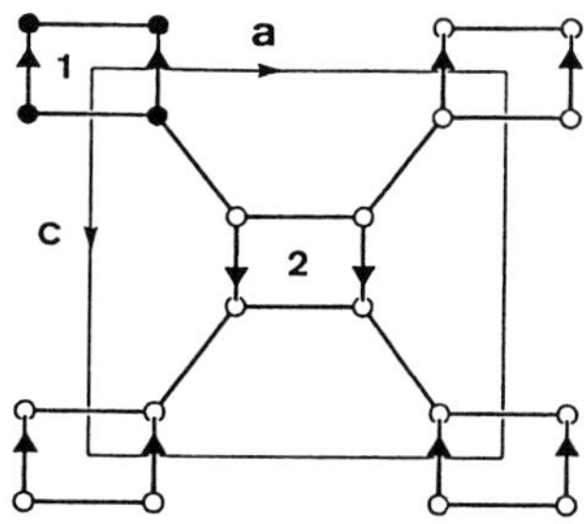

ATT

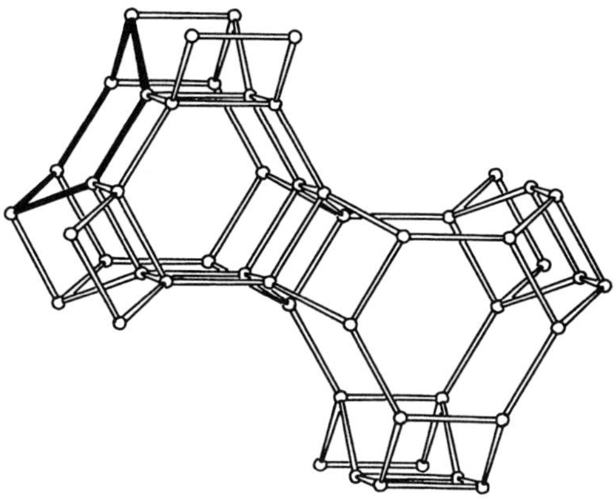

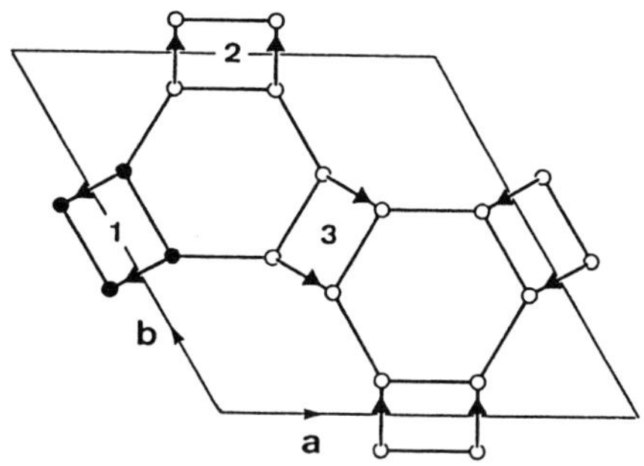

OFF

Figure 10. Perspective view and building scheme for 9 zeolites containing saw chains. The unit cell is outlined and the number of independent IBUs per unit cell is given. One of the IBUs is indicated by black dots. On each page the structure type code is given followed by the type of IBU used and the reference to Figure 9 in parentheses. One 6-ring is drawn with bold lines in those zeolites which can also be assembled from single 6-rings.
JBW (s1, Fig. 9a); ATT and OFF (s2, Fig. 9b).

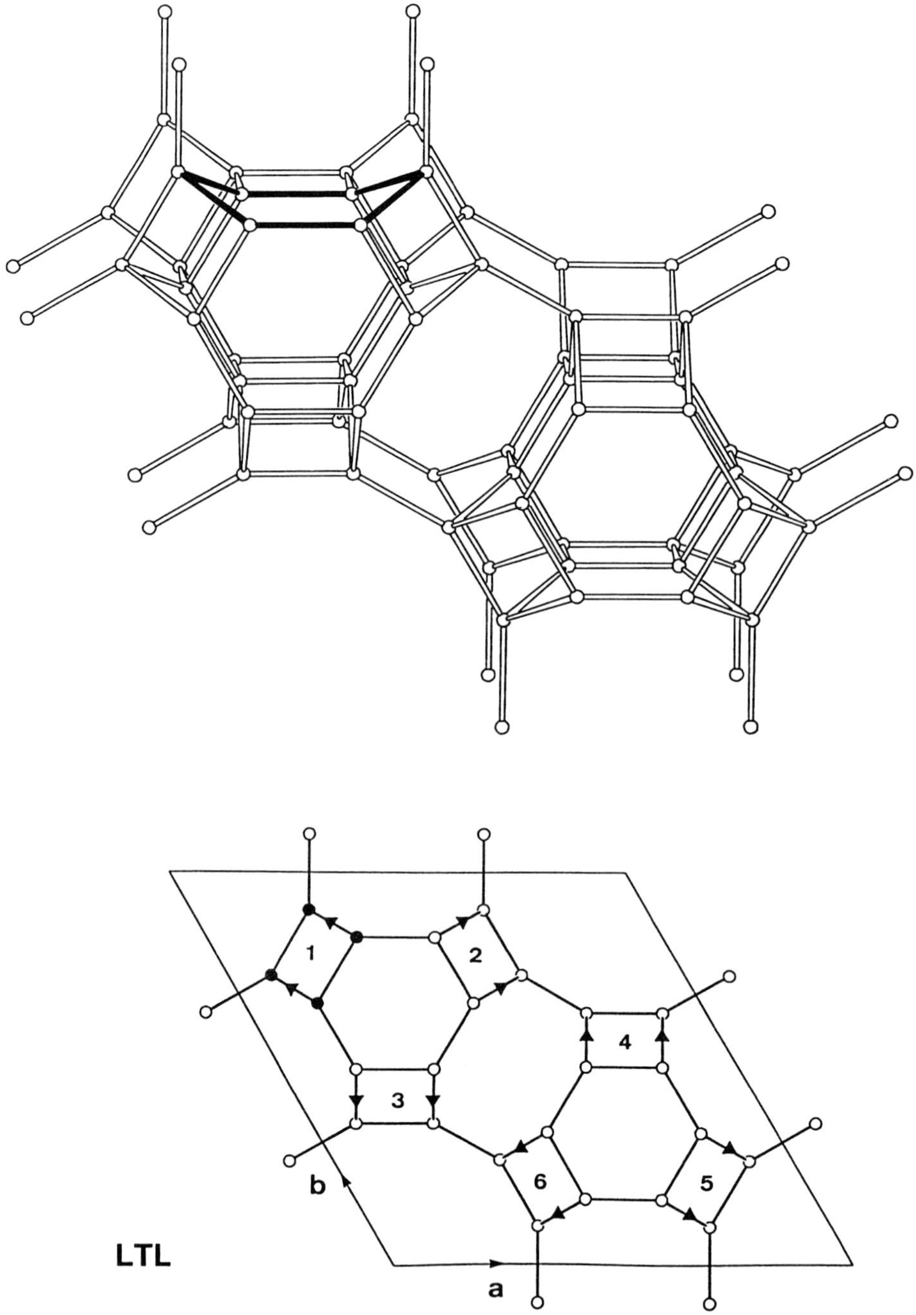

Figure 10. Cont'd. LTL (s2, Fig. 9b).

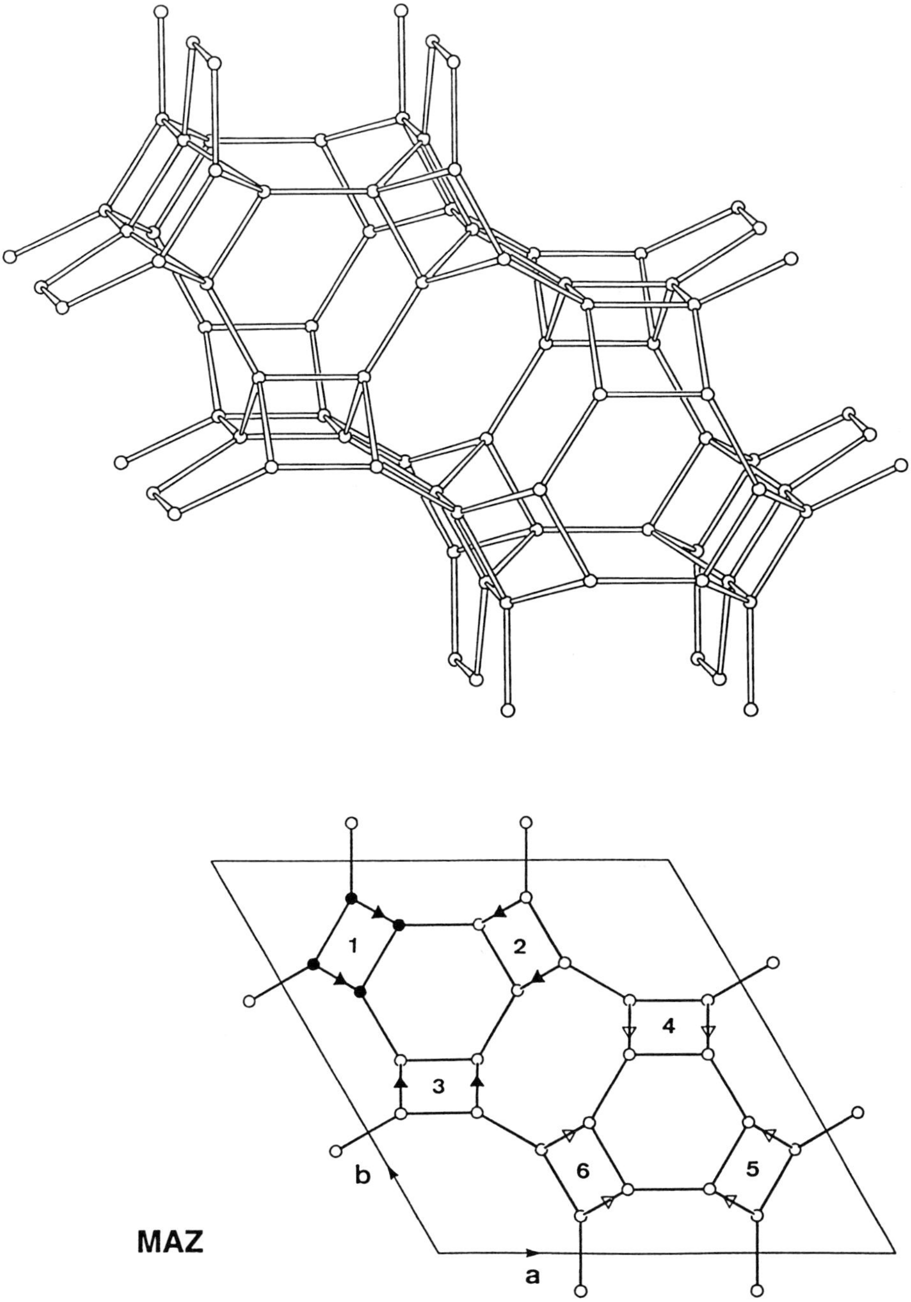

Figure 10. Cont'd. MAZ (s2, Fig. 9b).

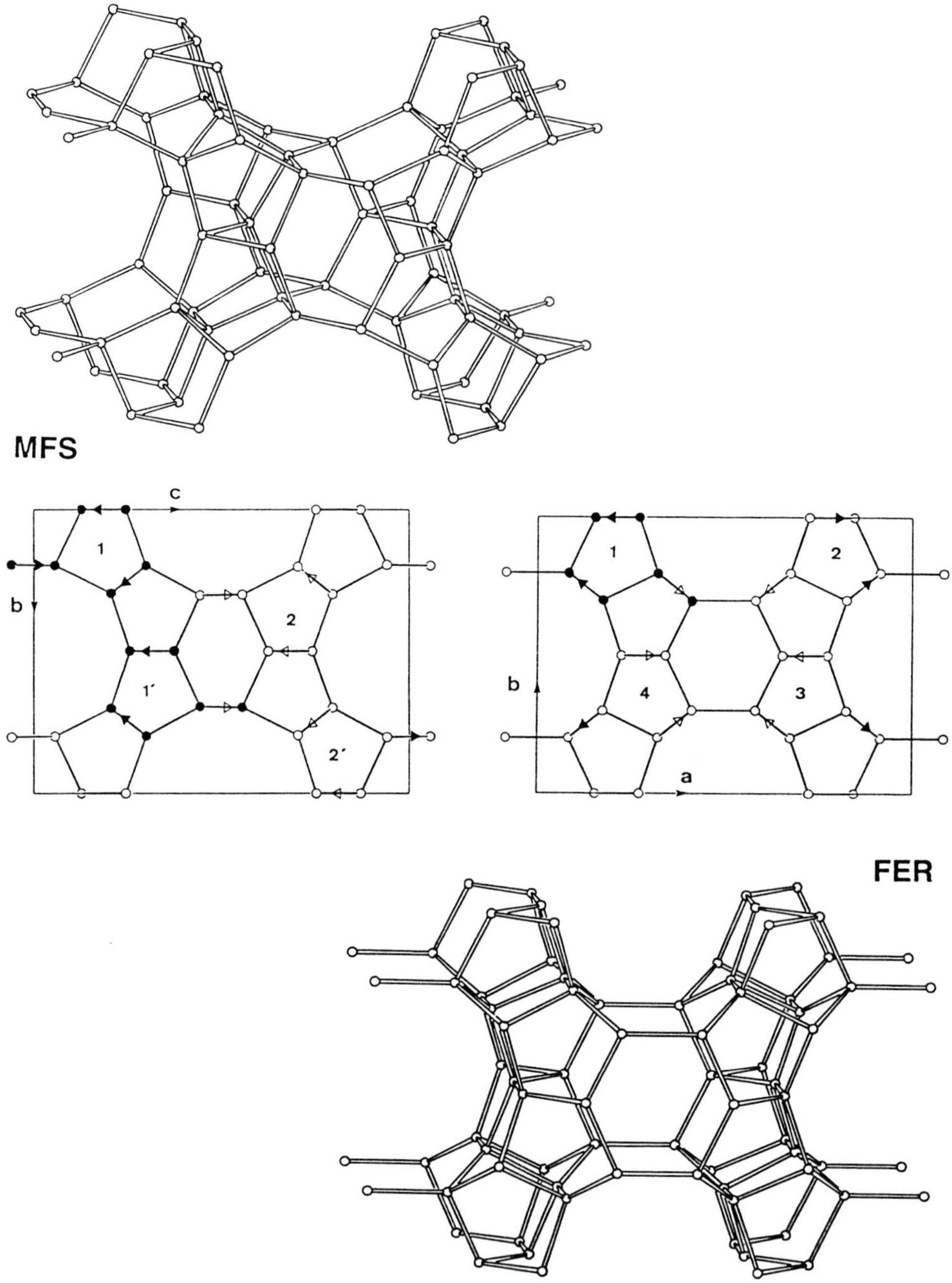

Figure 10. Cont'd. MFS (s3-1, Fig. 9c and s3-2, Fig. 9d); FER (s3-2, Fig. 9d).

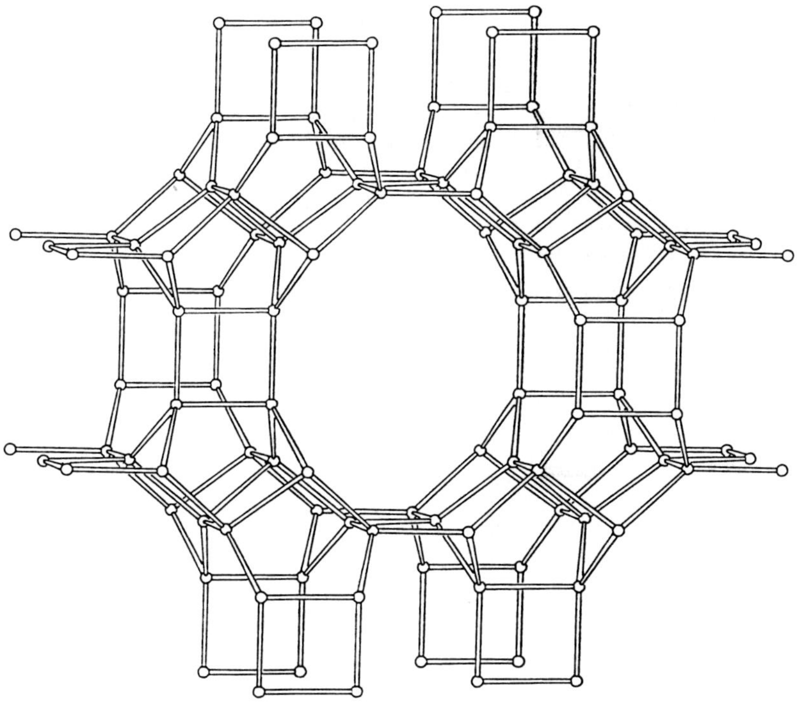

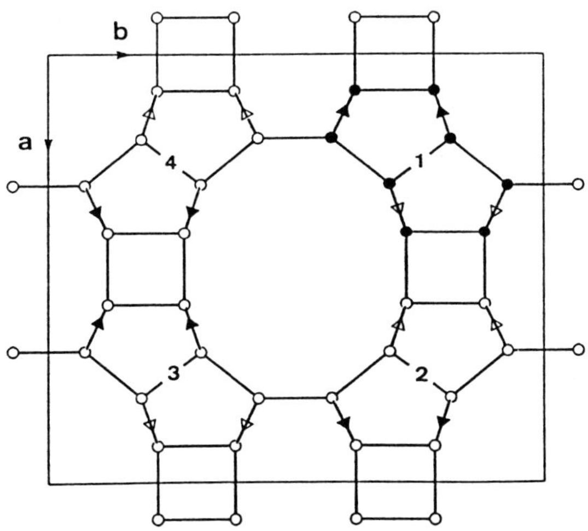

MOR

Figure 10. Cont'd. MOR (s4, Fig. 9e).

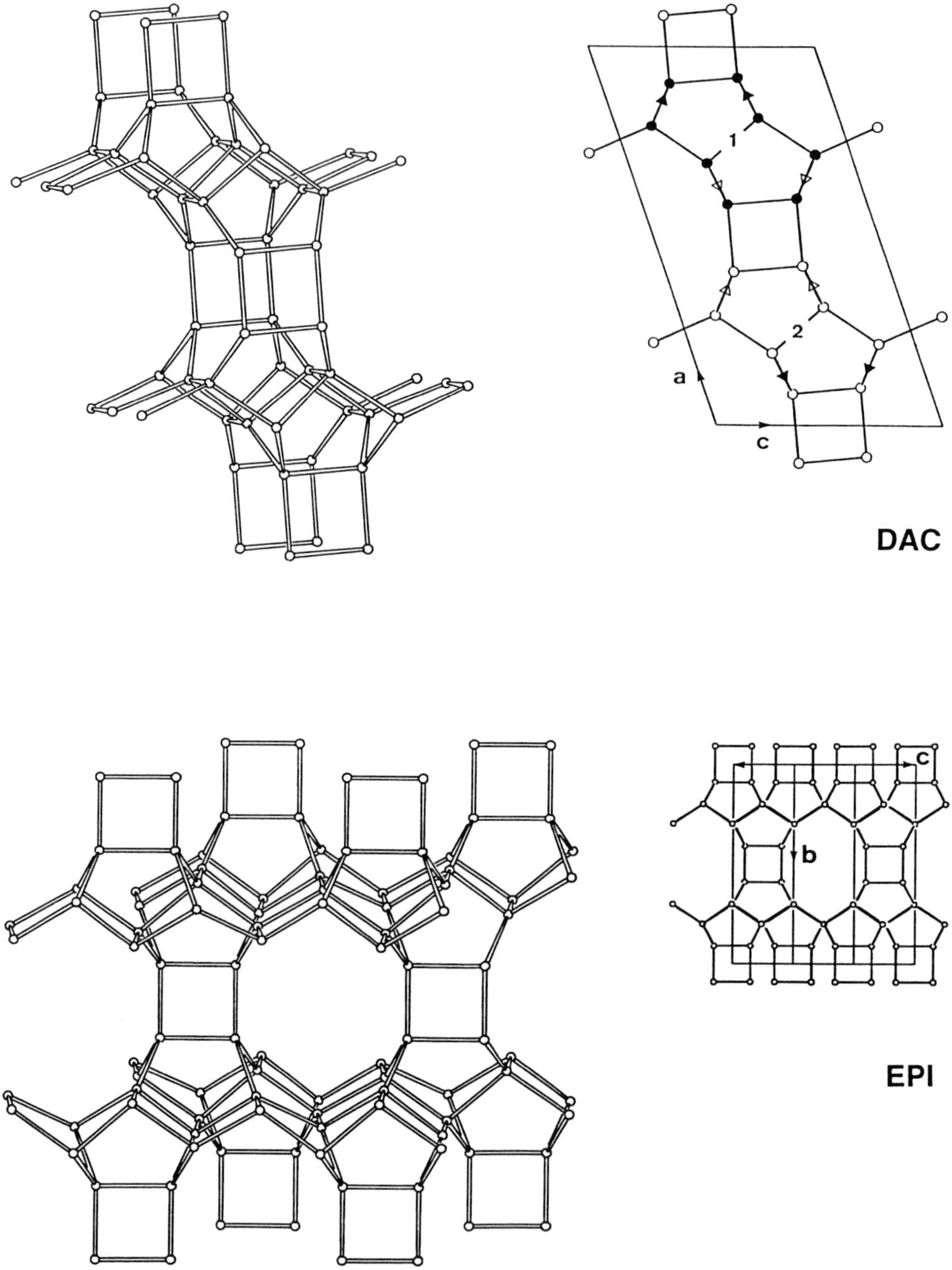

Figure 10. Final page. DAC (s4, Fig. 9e) and EPI (distorted saw).

In MFS there are two z3-1 units (Fig. 9c) and two z3-2 units (Fig. 9d) in the unit cell and Z = 2*9 + 2*9 = 36, whereas FER contains four z3-2 units (Z = 36).

MOR and DAC can be built using the building unit composed of four saw chains: the s4 unit from Figure 9e with 12 T atoms per repeat unit. In MOR and DAC there are four (Z = 48) and two (Z = 24) of these z4 units per unit cell, respectively.

The EPI structure has been added in Figure 10 for comparison with DAC. In DAC as well as in EPI saw chains can be recognized. However, the saw chains in EPI, which run nearly perpendicular to the plane of the paper, are strongly distorted: successive teeth of a saw chain point into different directions. The distance between two neighboring teeth in the saw is $a\sin\beta$ = 7.48 Å, which reflects the repeat distance in a saw chain. The same type of 6-ring layers (built from zigzag chains) run parallel to the *ac* plane in EPI and parallel to the *bc* plane in DAC (see also Sections 3.3 and 4.3).

4.3. Alternative descriptions using (cylindrical) 6-ring sheets

None of the 9 structure types listed in Table 2 exhibits a cylindric pore wall of fused 6-rings. The zeolites JBW, ATT, OFF and LTL can be built using single 6-rings. The other structure types (MAZ, MFS, FER, MOR and DAC) all contain a great number of 5-rings.

An alternative description of JBW using 6-ring sheets is already given in Paragraph 3.3. Figure 11 presents the 6-ring sheets in FER, MOR, and DAC (and EPI). FER is the only structure type in which strongly corrugated 6-ring sheets are connected through single 6-rings. The 6-ring sheet in FER has been rotated 90° about the view axis in Figure 10 for the drawing in Figure 11. Three additional T atoms from single 6-ring 'handles' have also been drawn.

MOR and DAC (and EPI) are examples of structures where the 6-ring sheets are connected through 4-rings. The alternative description of DAC, using 6-ring sheets, has already been illustrated in Figure 8. For an easy comparison with the MOR structure, the 6-ring sheet in DAC has been added here again, together with the 6-ring sheet in EPI, in about the same orientation as in Figure 10. Pairs of T atoms belonging to 4-ring 'handles' are drawn as well. [See also Figure 7 for a view on DAC and EPI along **c**].

In the 6-ring sheets crankshaft chains run parallel to a channel axis. The zigzag chains in the 6-ring sheet are now, except in the sheet of DAC and EPI, strongly distorted. Saw chains are generated when the atoms of the 'handle' are linked to the 6-ring sheets.

The repeat distances in the saw and crankshaft chains become equal. One can choose the crankshaft chains to build the 6-ring sheets and the crankshaft chains can be chosen in several ways. The 6-ring sheet can as well be built from zigzag chains. It is therefore tempting to suggest that saw- (and crankshaft- and zigzag-) chains are generated in the framework by the growth proces and are not essential entities during crystal growth whereas 6-rings may play an important role.

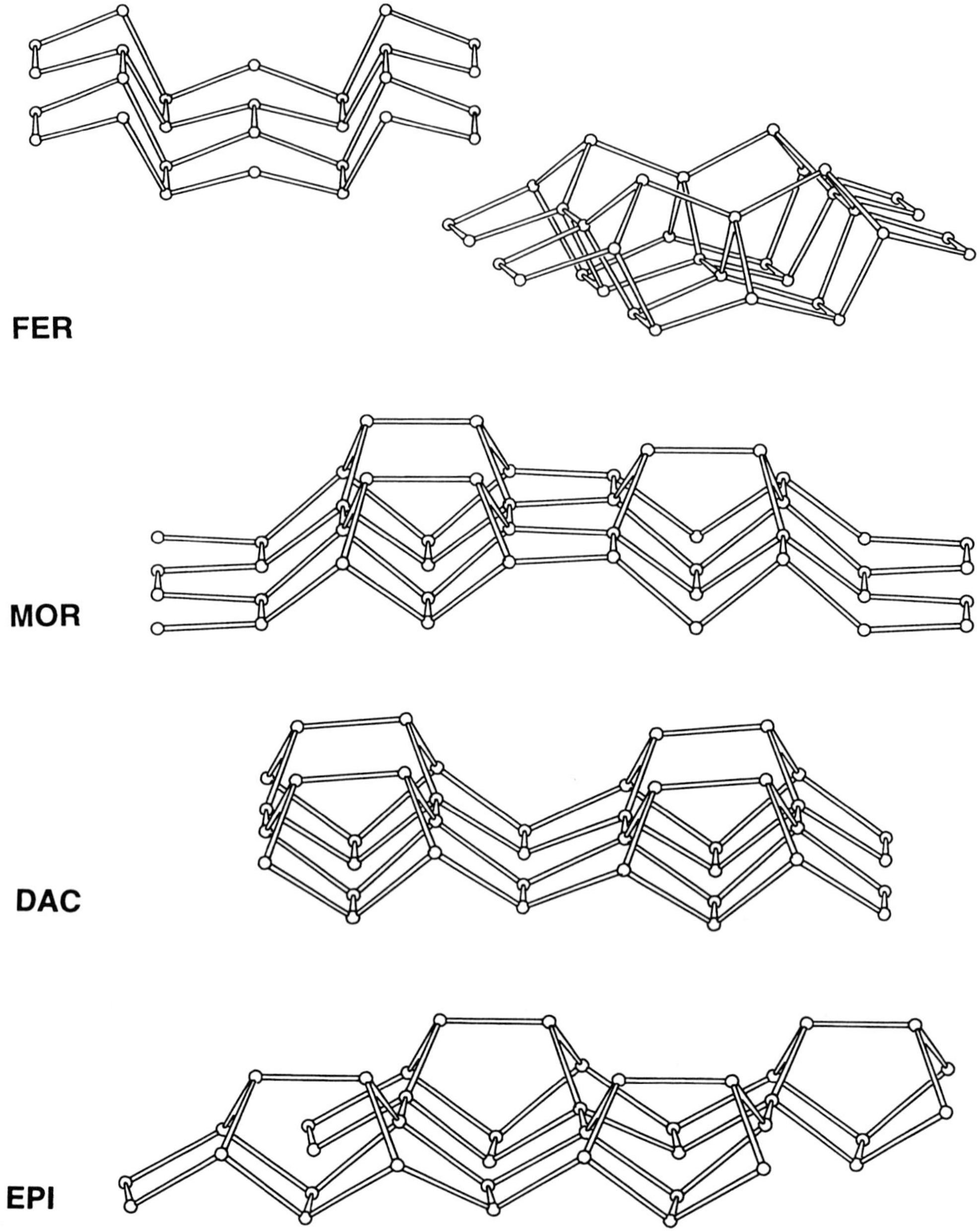

Figure 11. 6-Ring sheets in FER, MOR, DAC and EPI, zeolites which can be built from saw chains. For clarity reasons, the 6-ring sheet in FER is drawn without and including additional T atoms from 6-ring 'handles'. The drawings of the 6-ring sheets in MOR, DAC and EPI include pairs of T atoms belonging to 4-ring 'handles'.

5. ZEOLITES CONTAINING CRANKSHAFT CHAINS

5.1. Infinite building units from crankshaft chains

Infinite building units, from which the zeolites listed in Table 3 (page 114) can be assembled, are presented in Figure 12. The IBUs can be obtained from different linkages of crankshaft (c) chains and are arranged in Figure 12 in order of increasing number of component c chains. The drawings extend over nearly two identity periods along the chain axis.

The single crankshaft chain (c1; Fig. 12a) has an independent unit consisting of four T atoms (n = 4). The first type of a double crankshaft chain, c2-1 (n = 8; Fig. 12b), is obtained when two component c chains are 'in phase'. In c2-1 the 4-rings in the buckled ladder are connected through two T-T connections. A 'pseudo' double crankshaft chain (c2-2) is obtained when the 4-rings in the c2-1 ladder are connected through only one T-T connection (n = 8; Fig. 12c). In the third type of a double crankshaft chain, c2-3 (n = 8; Fig. 12d), one c chain is shifted with respect to the other over half the chain period. The directionality of the T atoms in the 4-rings in the buckled ladders of c2-1, c2-2 and c2-3 is different. The IBU c3-1 (n = 12; Fig. 12e) is generated when one more c chain is added to the c2-3 unit. One gets a tubular 6-ring pore of fused 6-rings, c3-2 (n = 12), when three c chains are connected as shown in Figure 12f.

Figure 12(g-j) shows IBUs constructed by linking four (n = 16) or five (n = 20) crankshaft chains. It is also possible to built tubular pores of fused 6-rings enclosing an 8- or 10-ring window from four or five single c chains, respectively. These latter IBUs will be discussed in Section 5.3 (See Figure 14).

Figure 12. Infinite building units from (multiple) crankshaft chains. The building units are drawn in a perspective view with their symbolic representation of the projection along the chain axis underneath. The number of T atoms in the independent repeat unit along the chain axis equals n.
(a) c1: Single crankshaft chain, n = 4.
(b) c2-1: Double crankshaft chain of the first kind (a buckled 4-ring ladder), n = 8.
(c) c2-2: 'Pseudo' double crankshaft chain: the 4-rings in the c2-1 ladder are connected through only one T-T connection. The crossed T atom indicates the linkage to the 4-ring below.
(d) c2-3: Double crankshaft chain of the third kind (the 4-ring pore wall consists of 6-rings only), n = 8.
(e) c3-1: Triple crankshaft chain of the first kind, n = 12.
(f) c3-2: Triple crankshaft chain of the second kind (the 6-ring pore wall is constituted solely from 6-rings), n = 12.
(g) c4-1: Four crankshaft chains of the first kind (8-ring pore; one crankshaft chain is 'out of phase'. If all four crankshaft chains are 'in phase' the 8-ring pore consists of 6-rings only (See Figure 14), n = 16.
(h) c4-2: Four crankshaft chains of the second kind (c3-2 plus c1), n = 16.
(i) c5-1: Five crankshaft chains of the first kind, n = 20.
(j) c5-2: Five crankshaft chains of the second kind, n = 20.

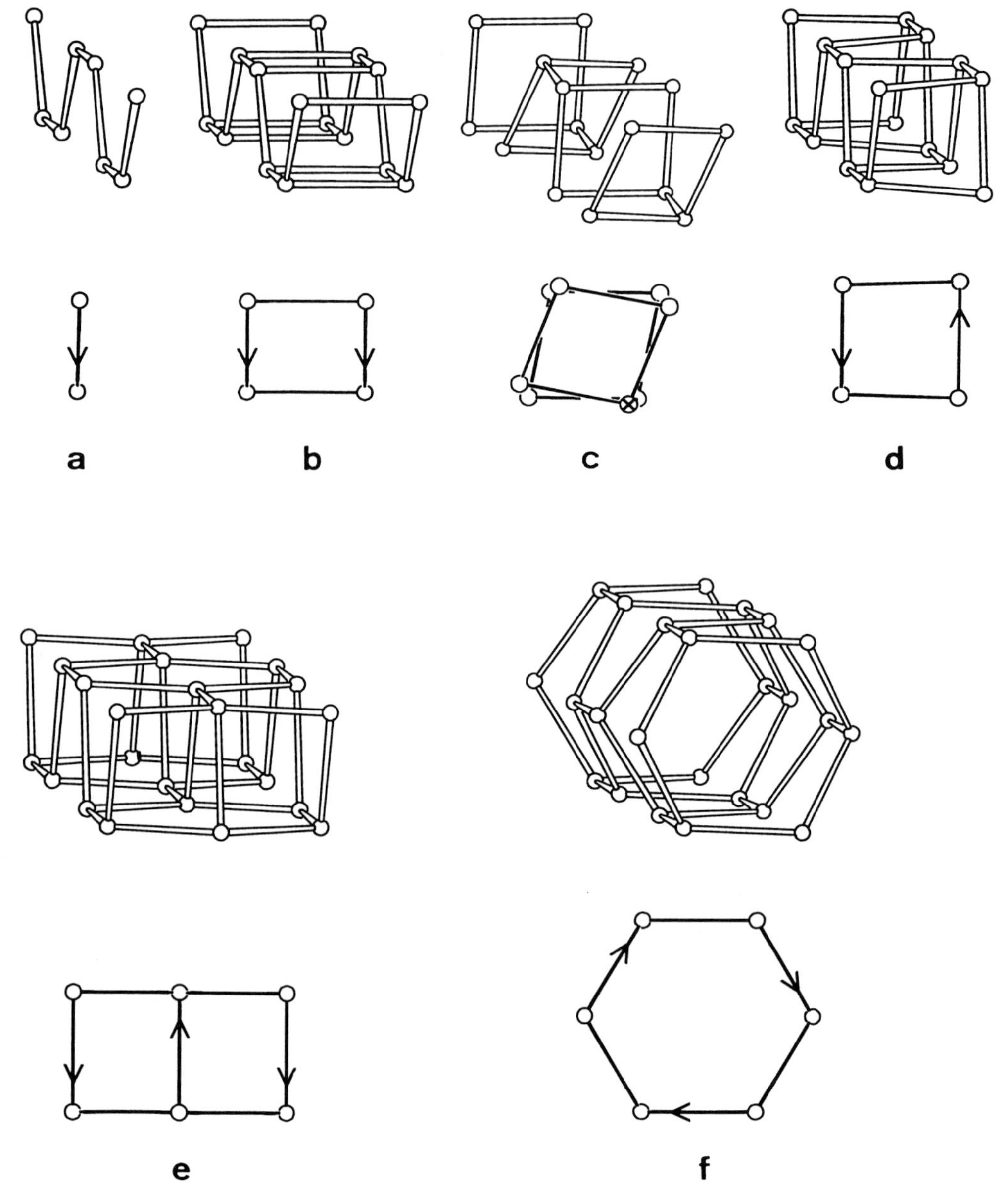

Figure 12. Legend: see previous page.

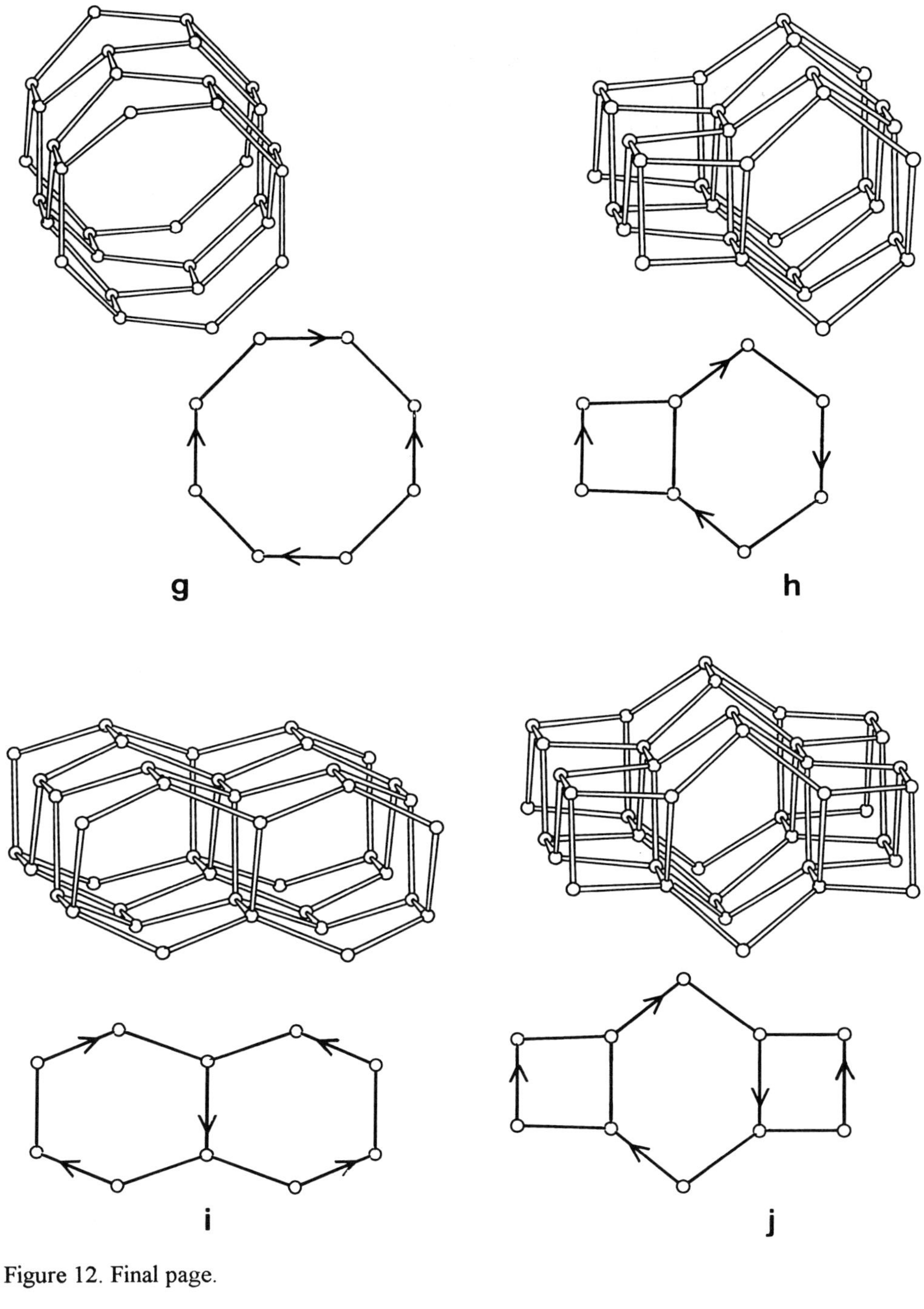

Figure 12. Final page.

Table 3
Zeolites containing (modified) crankshaft chains. The topological unit cell and symmetry is given. The chain axis is italicized

		Cell dimensions (Å)					
Code	space group	a	b	c	β,γ (°)	n*	Z**
APC	Cmca	*8.992*	19.356	10.392		8	32
ATT#&	Pmma	*9.980*	7.514	9.369		8;4	12
GIS	$I4_1$/amd	*9.801*	9.801	10.158		8	16
GME#	$P6_3$/mmc	13.672	13.672	*9.850*	120.0	8	24
MER	I4/mmm	14.012	14.012	*9.954*		8	32
PHI	Cmcm	*9.890*	14.064	14.046		8	32
YUG	C2/m	*10.248*!	13.788	6.782	111.55	8	16
AEL	Imma	*8.312*	18.729	13.392		8;4	40
AET	Cmcm	32.829	14.380	*8.374*		16;20	72
AFI#	P6/mcc	13.827	13.827	*8.580*	120.0	8	24
AFO	Cmcm	9.764	25.614	*8.327*		20	40
AHT	Cmcm	15.794	9.206	*8.589*		12	24
APD	Cmca	*8.724*	20.060	10.166		8	32
ATV#	Cmma	*8.579*	15.311	9.661		12	24
VFI	$P6_3$/mcm	18.284	18.284	*8.589*	120.0	12	36

* n = number of T atoms in the independent repeat unit along the chain axis (see Figure 12).
** Z = number of T atoms in the unit cell (see Figure 13).
# Zeolite can also be constructed from single 6-rings.
& Zeolite can also be constructed from saw chains.
! Cell length is about two times zigzag chain length.

5.2. Building schemes for zeolites containing (multiple) crankshaft chains

Figure 13 shows the building schemes of the 15 zeolites listed in Table 3 using the IBUs given in Figure 12. Perspective views of the zeolites are accompanied by the parallel projection along the chain axis. The zeolites are, as are the IBUs in Figure 12, arranged in order of increasing complexity of the IBUs used.

In ATT a single c chain (c1; Fig. 12a), containing 4 T atoms per repeat unit (n = 4), is connected to a double c2-1 chain (n = 8; Fig. 12b). The total number of T atoms per unit cell amounts to Z = 8+4 = 12. ATT can also be assembled by connecting single 6-rings, as can be seen from the perspective drawing in Figure 13. ATT very much resembles a 'doubled' JBW structure (See Figure 7). The next 5 zeolites, GIS, GME, APC, MER and PHI contain two (Z = 16), three (Z = 24), four (Z = 32), four (Z = 32) and four (Z = 32) double c chain units (c2-1; Fig. 12b) per unit cell, respectively. GME can as well be constructed by linking single 6-rings. The unit cell of YUG consists of two 'pseudo' double crankshaft chains (Z = 16). The linkage of the c2-2 chains (Fig. 12c) along c produces (deformed) single zigzag chains parallel to the c2-2 chains with a doubling of the zigzag chain axis period to 10.25 Å.

ATV and AEL can be built using the c1 chain and the c2-3 building unit (n = 8; Fig. 12d). In ATV there are two c1 and two c2-3 units per unit cell (Z = 2*4 + 2*8 = 24). The unit cell in AEL contains two c1 and four c2-3 units and Z = 2*4 + 4*8 = 40. AEL can also be made by connecting two c5-1 units (Fig. 12i). In APD and AFI four and three c2-3 units are linked leading to Z = 32 and Z = 24, respectively. ATV and AFI can as well be assembled from linking four single 6-rings or two tubular c3-2 units (Fig. 12f) and APD can also be produced by connecting two c4-1 units (Fig. 12g).

AHT and VFI can be built using two (Z = 2*12 = 24) and three c3-1 units (Z = 36; Fig. 12e), respectively.

When two c4-2 units (n = 16; Fig. 12h) and two c5-2 units (n = 20; Fig. 12j) are connected the AET structure (Z = 72) is made. AFO can be obtained by linking two c5-2 units (Z = 40).

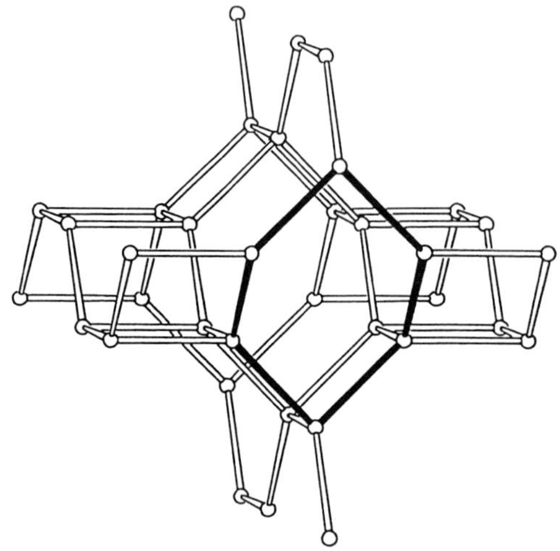

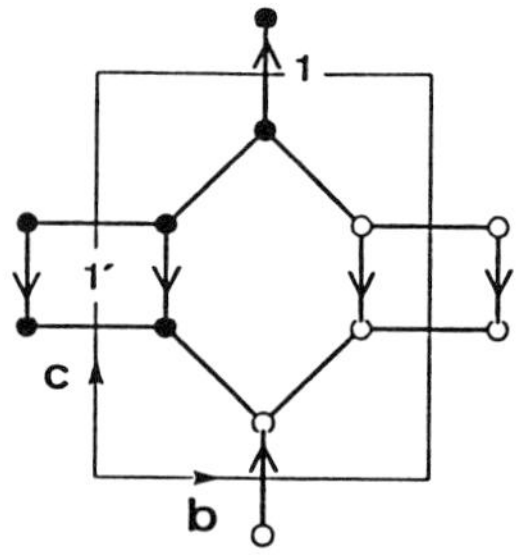

ATT

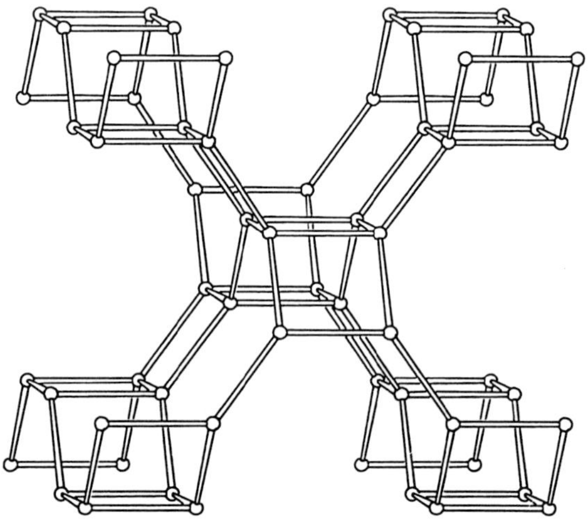

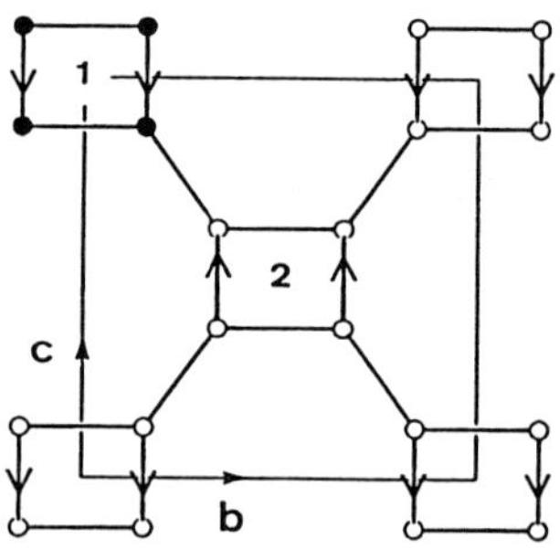

GIS

Figure 13. Perspective view and building scheme for 15 zeolites containing crankshaft chains. The unit cell is outlined and the number of independent IBUs per unit cell is given. One of the IBUs is indicated by black dots. On each page the structure type code is given followed by the type of IBU used and the reference to Figure 12 in parentheses. One 6-ring is drawn with bold lines in those zeolites which can also be assembled from single 6-rings.
ATT (c1, Fig. 12a and c2-1, Fig. 12b); GIS (c2-1, Fig. 12b).

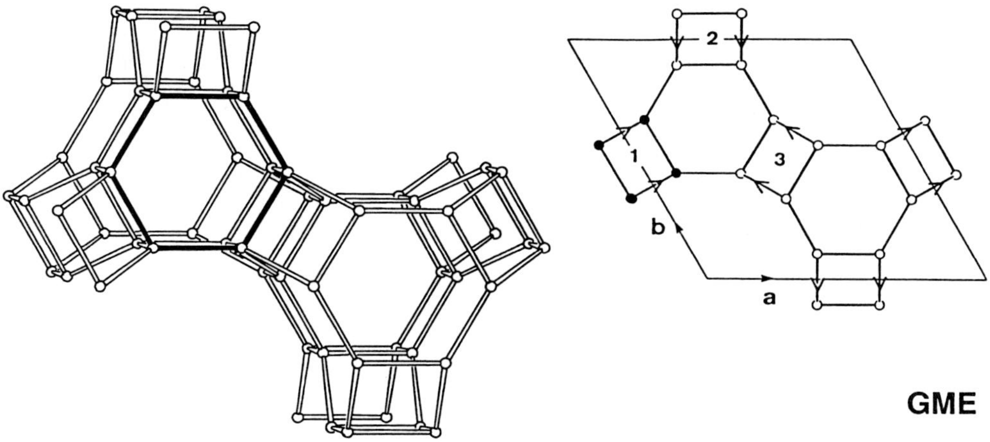

GME

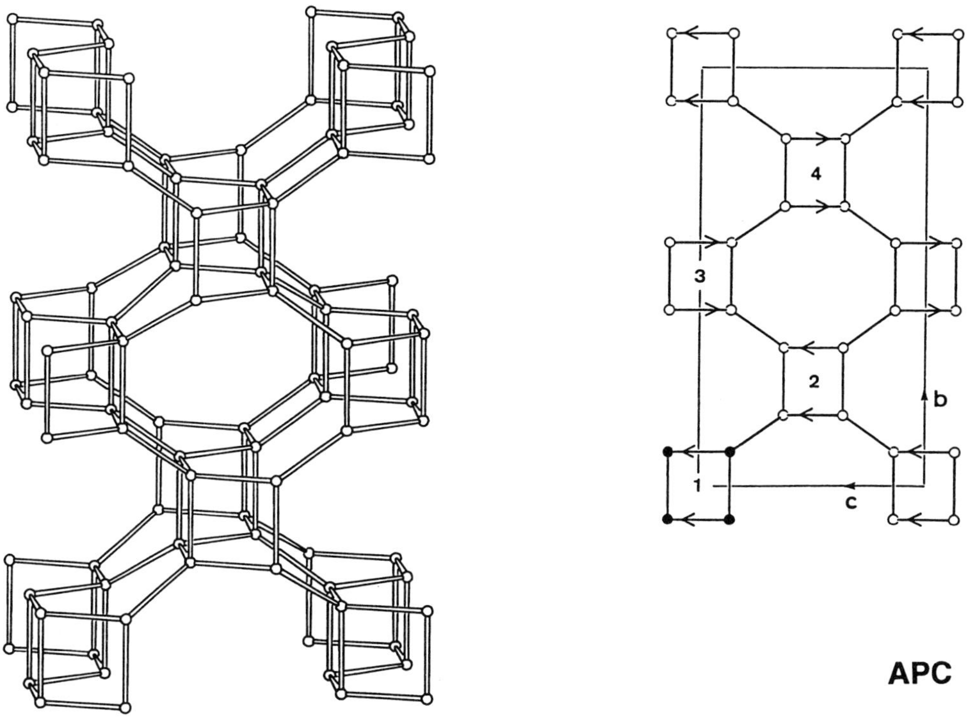

APC

Figure 13. Cont'd. GME and APC (c2-1, Fig. 12b).
and next page: MER and PHI (c2-1, Fig. 12b); YUG (c2-2, Fig. 12c).

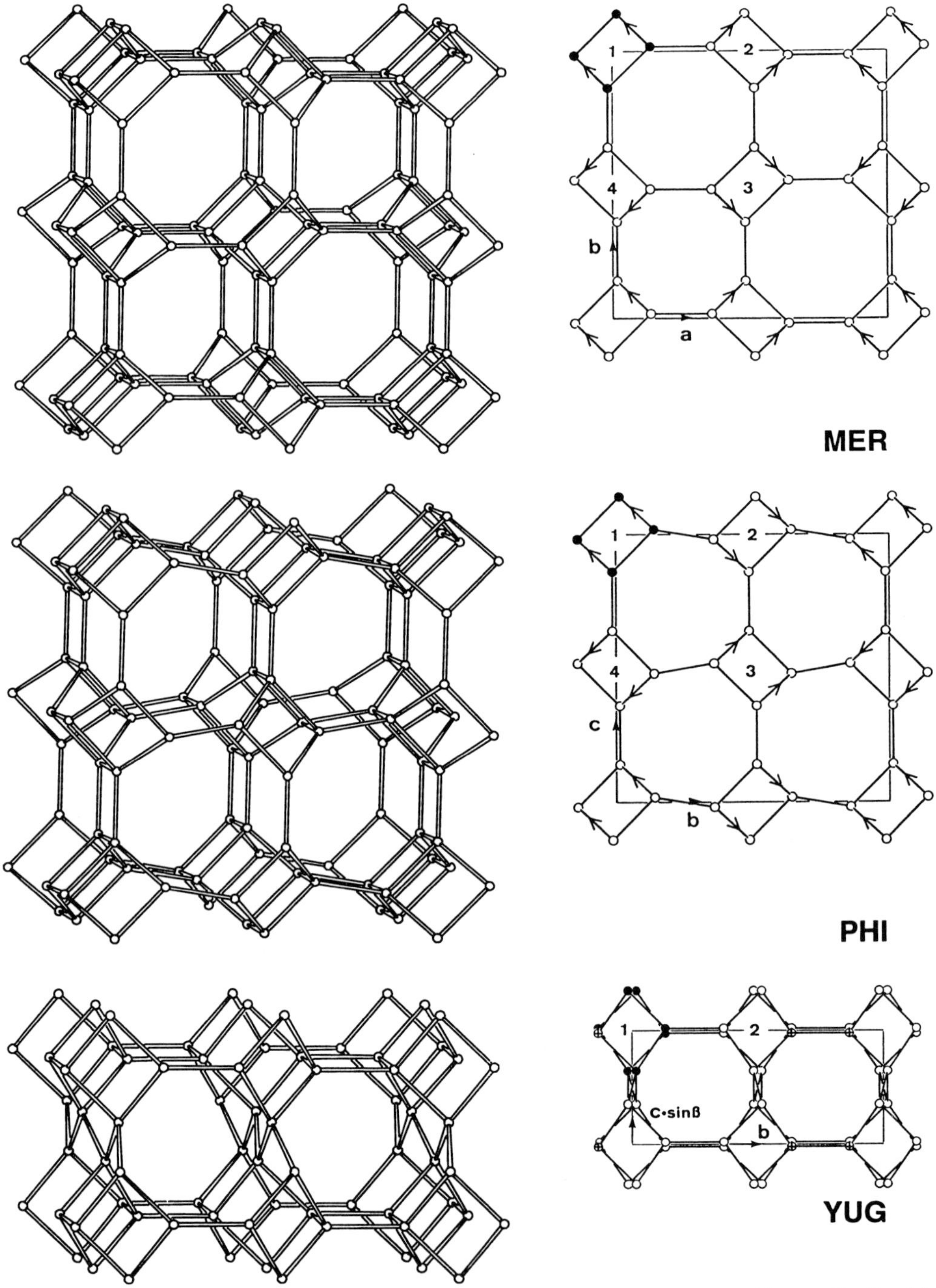

Figure 13. Cont'd.

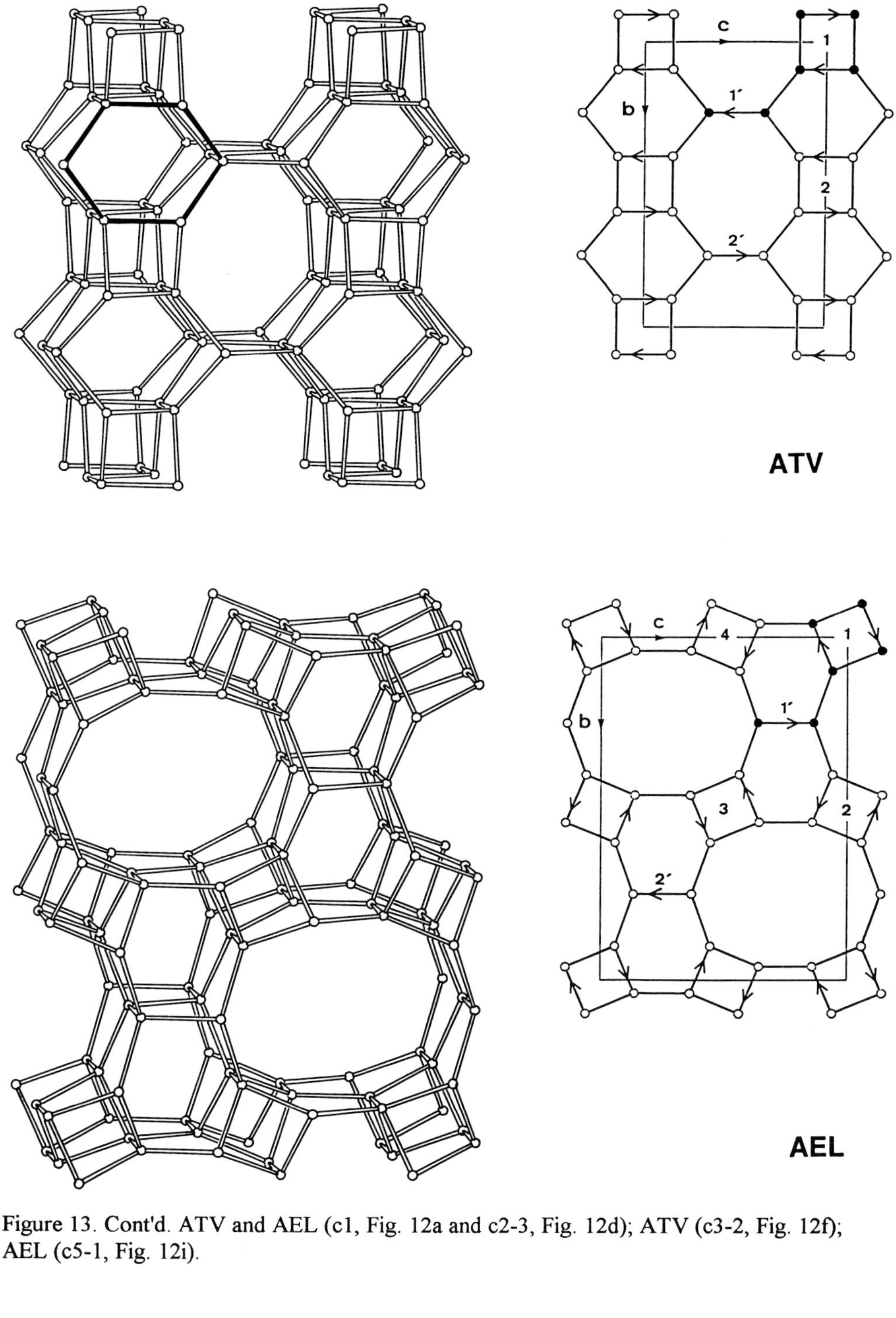

Figure 13. Cont'd. ATV and AEL (c1, Fig. 12a and c2-3, Fig. 12d); ATV (c3-2, Fig. 12f); AEL (c5-1, Fig. 12i).

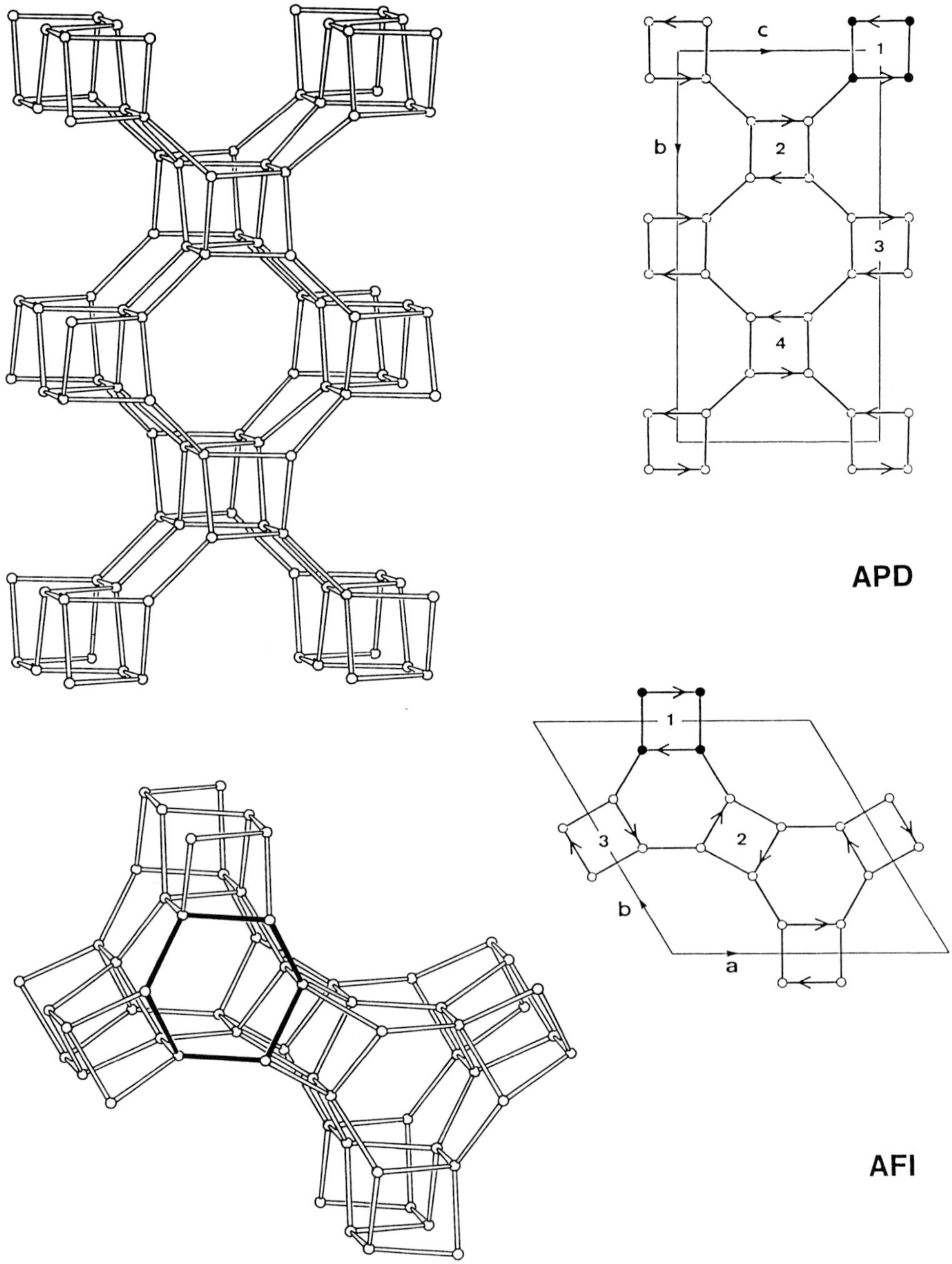

Figure 13. Cont'd. APD and AFI (c2-3, Fig. 12d); APD (c4-1, Fig. 12g); AFI (c3-2, Fig. 12f).

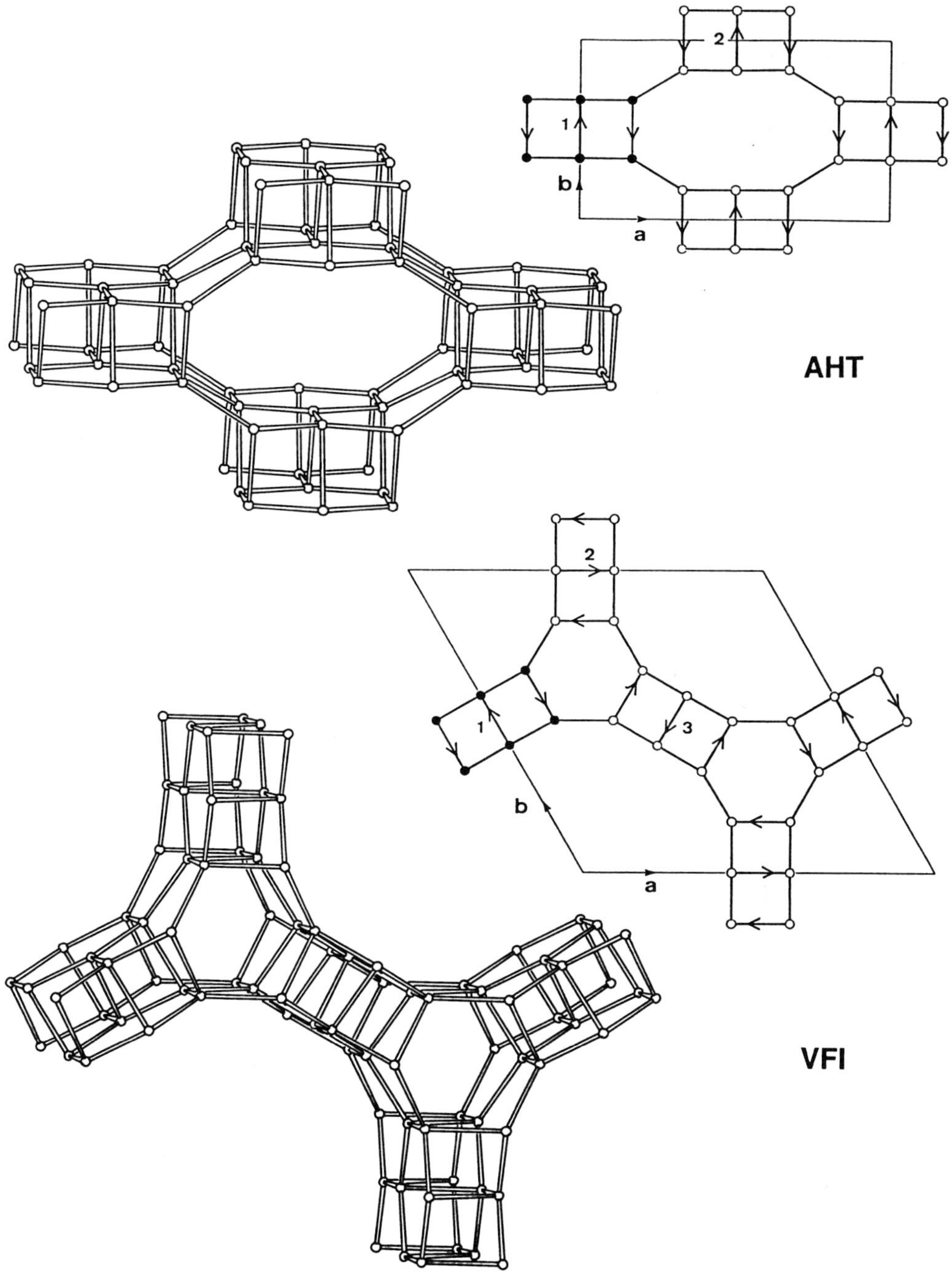

Figure 13. Cont'd. AHT and VFI (c3-1, Fig. 12e).

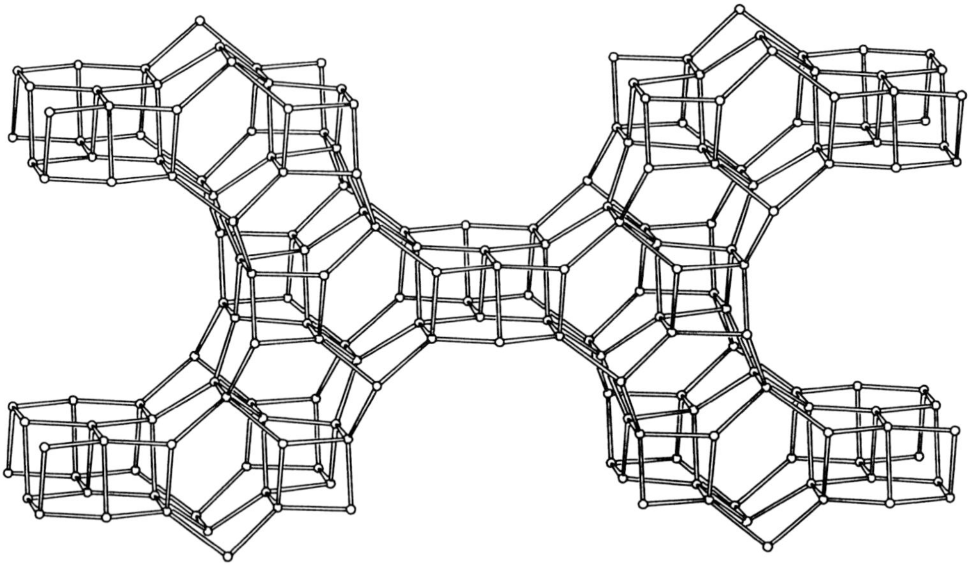

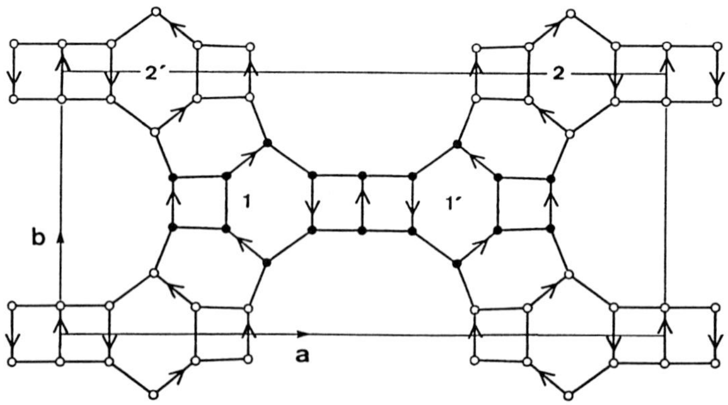

Figure 13. Cont'd. AET (c4-2, Fig. 12h and c5-2, Fig. 12j).

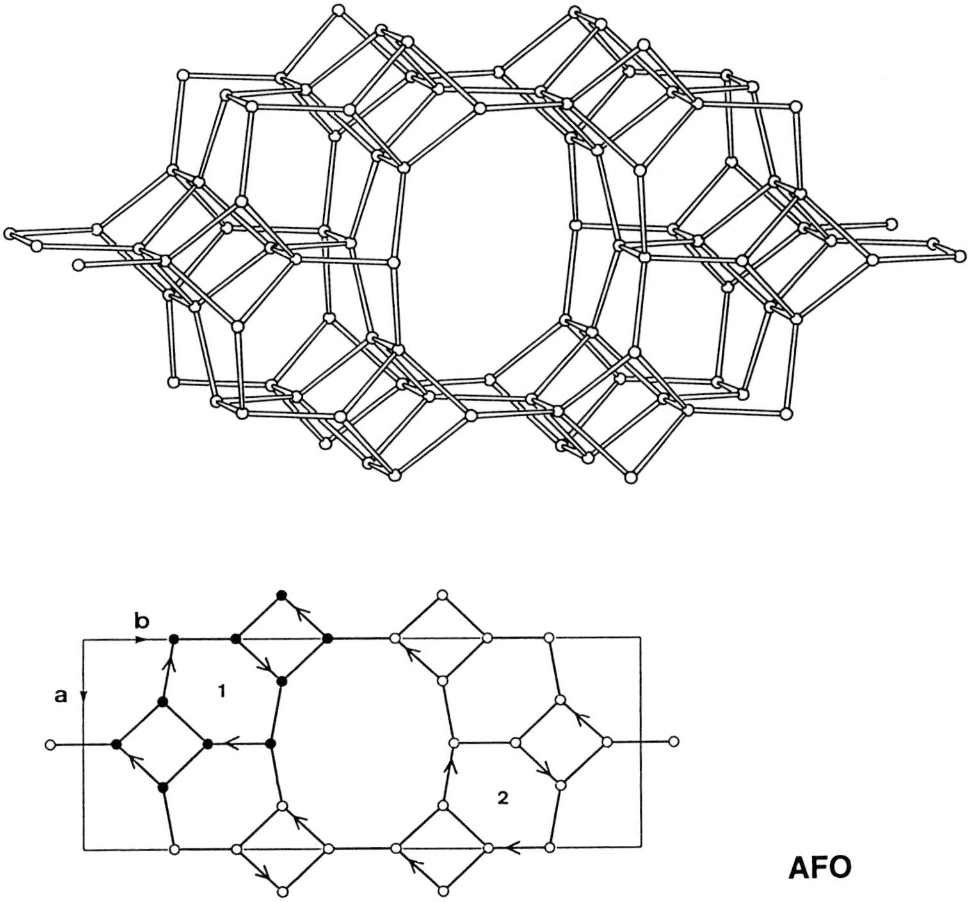

AFO

Figure 13. Final page. AFO (c5-2, Fig. 12j).

5.3. Alternative descriptions using (cylindrical) 6-ring sheets

Figure 14 depicts the tubular pores observed in 7 of the 15 structure types described in Section 5.2. The pore walls consist of fused 6-rings which form zigzag and crankshaft chains. Six types of tubes of fused 6-rings are found: the 6-ring walls form 6- (See Fig. 12f), 8-, 10-, 12-, 14- or 18-ring windows. 16-Ring windows are not observed (up till now)! The directions of the chains in the tubes are interchanged compared to the directions found in the 6-ring tubes described in Section 3.3. In all cases the crankshaft chains in the fused 6-ring wall run parallel to the pore direction and the repeat unit along the chanel axis consists of 12, 16, 20, 24, 28 or 36 T atoms, respectively. The (circular) zigzag chains in the pore wall are now perpendicular to the pore direction.

The 6-ring pore tube equals the c3-2 building unit in Figure 12f from which ATV and AFI can be built by direct connection of the c3-2 cylinders. AFI can also be obtained from 12-ring tubes: the structure of AFI equals the 12-ring tube and there is one such tube per unit cell (Z = 24). ATV can also be made by using 8-ring tubes. However, the 8-ring tubes are now partly fused. The cylindric pores have one c1 chain in common. The number of T atoms in the independent repeat unit has to be corrected accordingly to n = 16-4 = 12. Two of these units form ATV (Z = 24). ATV can as well be assembled by connecting 6-ring sheets perpendicular to **c**. Figure 15 depicts a 6-ring sheet in ATV after 90° rotation about **a** with respect to the orientation in Figure 13.

AEL and AHT can be assembled from linking 10-ring tubes. In AEL there are two tubes per unit cell (Z = 40); in AHT, in addition to one 10-ring cylinder, one c1 chain is needed to complete the unit cell content (Z = 20+4 = 24). AHT can also be built by connecting 6-ring sheets shown in Figure 15.

The construction of AFO is rather complicated: the 10-ring tubes are partly fused and single T atoms are needed to complete the structure. The reader is invited to check that the number of T atoms in the unit cell sums up to 40: two 10-ring tubes have one c1 chain in common; there are two independent 10-ring tubes per unit cell; there are four additional 'pairs' of T atoms.

The assemblage of AET using 14-ring pores is also difficult. In addition to two independent 14-ring tubes, four c1 chains are needed to complete the unit cell content (Z = 2*28 + 4*4 = 72). AET can also be made by linking 6-ring sheets depicted in Figure 15. Two additional c2-2 units per unit cell are needed to complete the structure.

The unit cell of VFI contains one 18-ring tube (Z = 36).

The crankshaft chains (and the zigzag chains) seem to be merely a result of fusion of 6-rings rather than being independent entities. Their role in crystal growth may therefore be of minor importance.

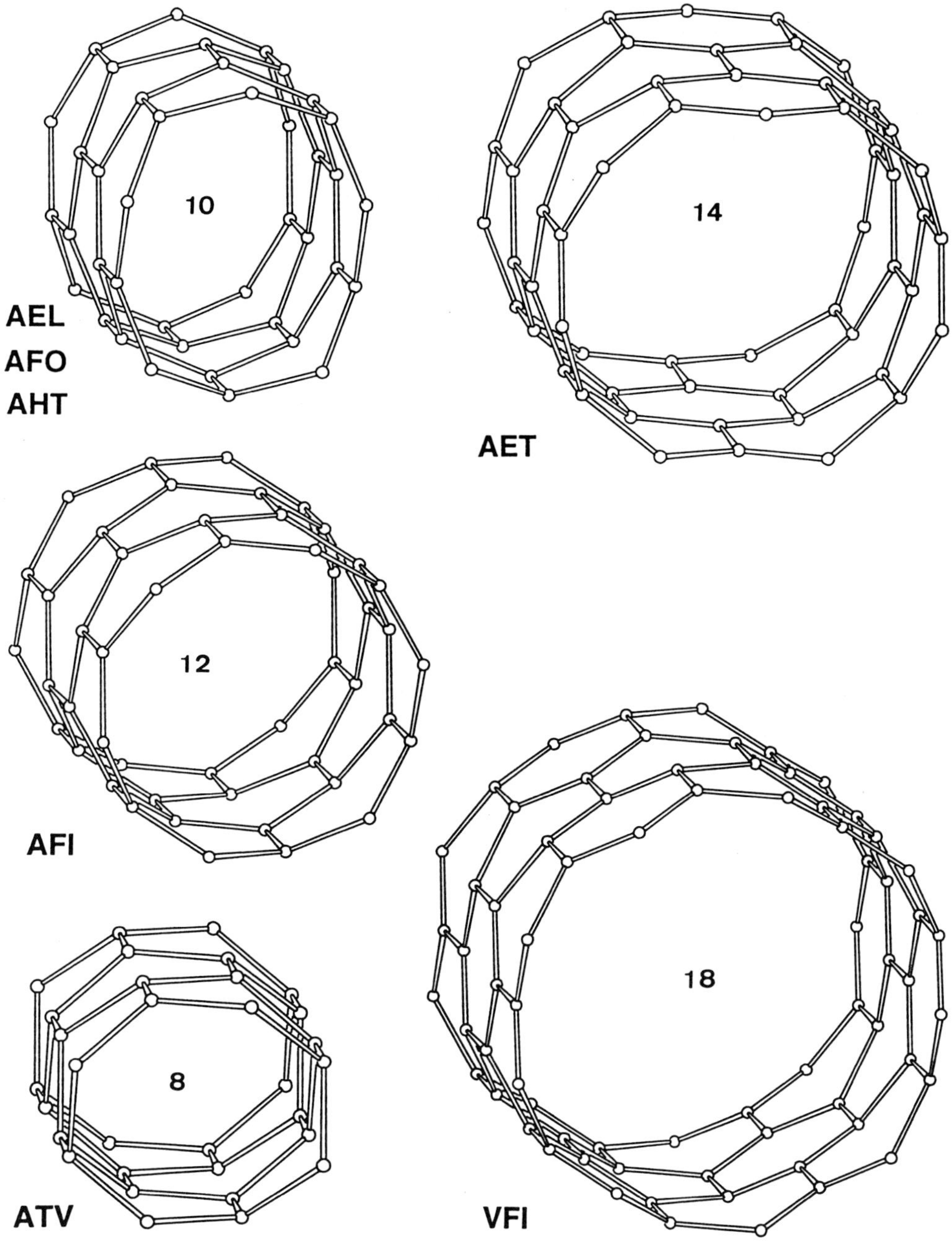

Figure 14. 6-Ring pore walls in zeolites which can be built from crankshaft chains. The number of T atoms in the pore window is given.

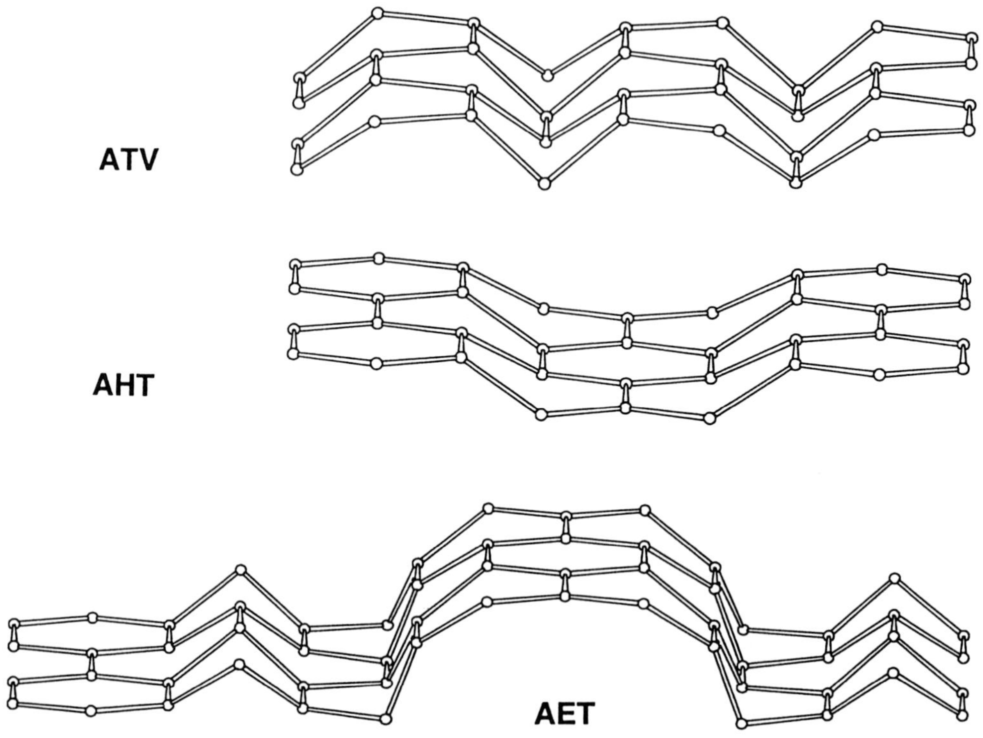

Figure 15. 6-Ring sheets in ATV, AHT and AET.

6. BUILDING SCHEMES FOR THE FIBROUS ZEOLITES

Figure 16 depicts the fibrous (fi) chain from Figure 1 in a different orientation. The linkages to neighbouring fi chains occur at levels A and B. The repeat unit along the chain axis is a capped 4-ring and the number of T atoms in the repeat unit along the chain is 5, as is illustrated in the down chain projection. Figure 17 shows the perspective views and building schemes of the fibrous zeolites in Table 4. The unit cell of EDI, NAT and THO contains one (Z = 5), four (Z = 20) and two (Z = 10) fi chains, respectively. The zeolites differ by different linkages of the fi chains.

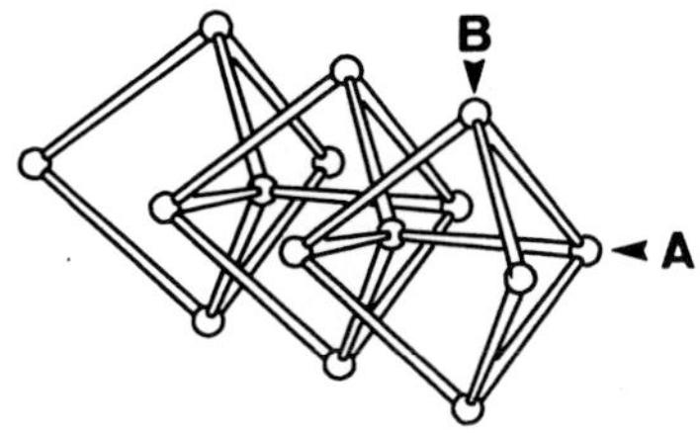

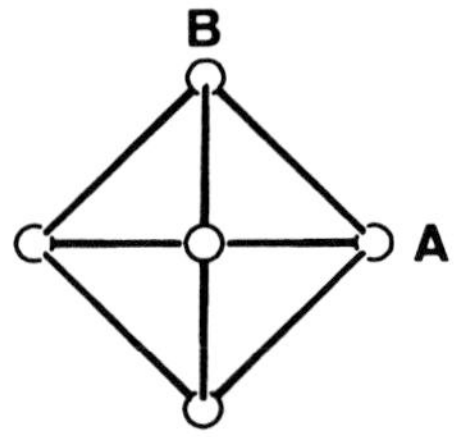

Figure 16. The fibrous (fi) chain as infinite building unit in perspective view and in down chain projection. Possible linkages to neighbouring fi chains are at levels A and B.

Table 4
Topological symmetry and unit cell of the fibrous zeolites. The fibrous chain axis is italicized

		Cell dimensions (Å)				
Code	space group	a	b	c	n*	Z**
EDI	$P\bar{4}m2$	6.926	6.926	*6.410*	5	5
NAT	$I4_1/amd$	13.850	13.850	*6.420*	5	20
THO	Pmma	14.000	7.000	*6.482*	5	10

* n = number of T atoms in the independent repeat unit along the chain axis (see Figure 16).
** Z = number of T atoms in the unit cell (see Figure 17).

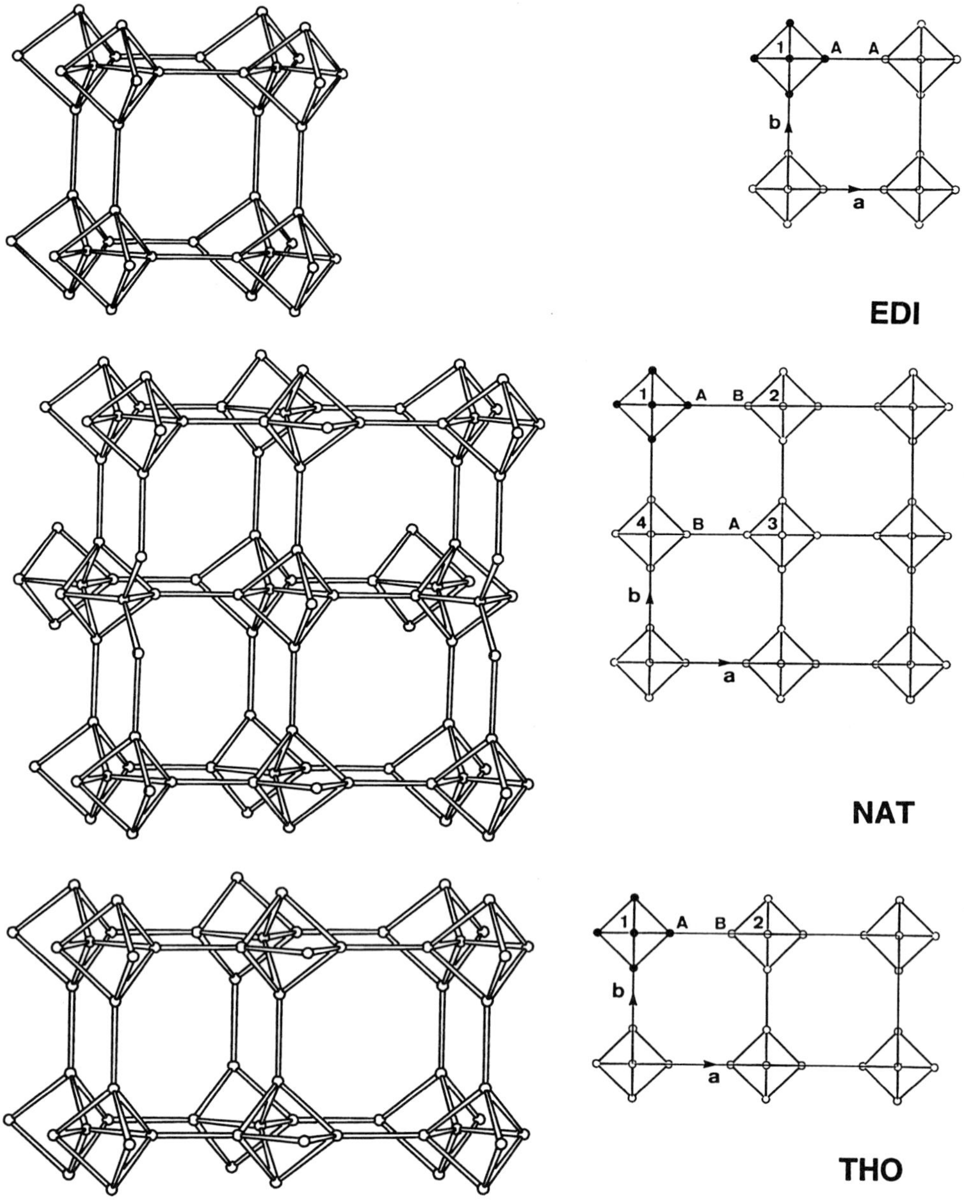

Figure 17. Perspective view and building scheme for the fibrous zeolites: EDI, NAT and THO. The unit cell is outlined and the number of independent IBUs per unit cell is given. One of the IBUs is indicated by black dots.

7. BUILDING SCHEMES FOR THE ABC FAMILY OF ZEOLITES AND SOME RELATED ZEOLITES

Figure 18 gives the stacking sequences in the 14 hexagonal structure types belonging to the ABC family of zeolites in the same order as listed in Table 5 (page 134). The stacking sequences cover one repeat unit along the (hexagonal) *c* axis. The content of only one unit cell is drawn. All structures can easily be extended by simple translations as illustrated in Figure 3. The CAN structure type is identical to the first example discussed in Section 2.2 and shown in Figure 3(a). The second example (Figure 3(b)) depicts the SOD structure in hexagonal description. Separate drawings of parallel projections of the frameworks along the hexagonal axis are not given because all these projections are equal to one of the examples shown in Figure 3.

Figure 19(a) presents the AABBCC framework of CHAB with an additional double 6-ring (D6R) at a second B position related to the first B position by a translation over one cell edge along **a**. (Compare Figures 3 and 18). The four D6R's form the $(\bar{1}1\bar{1})$ layer in CHAB. This layer, in which tilted D6R's are related by translations in two directions, is shown again in side view in Figure 19(b). In CHAB the layers are connected after a translation along $[\bar{1}1\bar{1}]$ as shown in Figure 20(a).

The same layers are observed in the zeolite with structure type AEI. In this zeolite neighbouring layers are related by a mirror plane (Figure 20(b); [31-32]).

Finally, the KFI structure type has a basic layer in which each D6R is related to all connected neighbouring D6R's by mirror symmetry as illustrated in Figure 20(c) [32]. A 18-hedral cavity, more commonly referred to as α cage, is obtained when the D6R's in KFI are replaced by single 6-rings (Figure 20(d)). The eight 6-rings of the α cage form the unit cell content of the zeolite RHO with Z = 48 (Figure 21(a)). In the frameworks of KFI and RHO α cages are linked together through 6-rings and 8-rings to form D6R's and double 8-rings, respectively. In AWW half α cages, consisting of four 6-rings, are connected through single 8-rings (Figure 21(b); [33]). When complete α cages are connected through 4-rings the sodalite cage, or β cage, is formed (See Figure 4(a)). The sodalite cage consists of four 6-rings and equals the unit cell content (Z = 24) of LTA (Figure 21(c)). In LTA sodalite cages are linked to form double 4-rings. In FAU and EMT sodalite cages are connected through D6R's in different ways (See next Section). Crystal data are summarized in Table 6 (page 135).

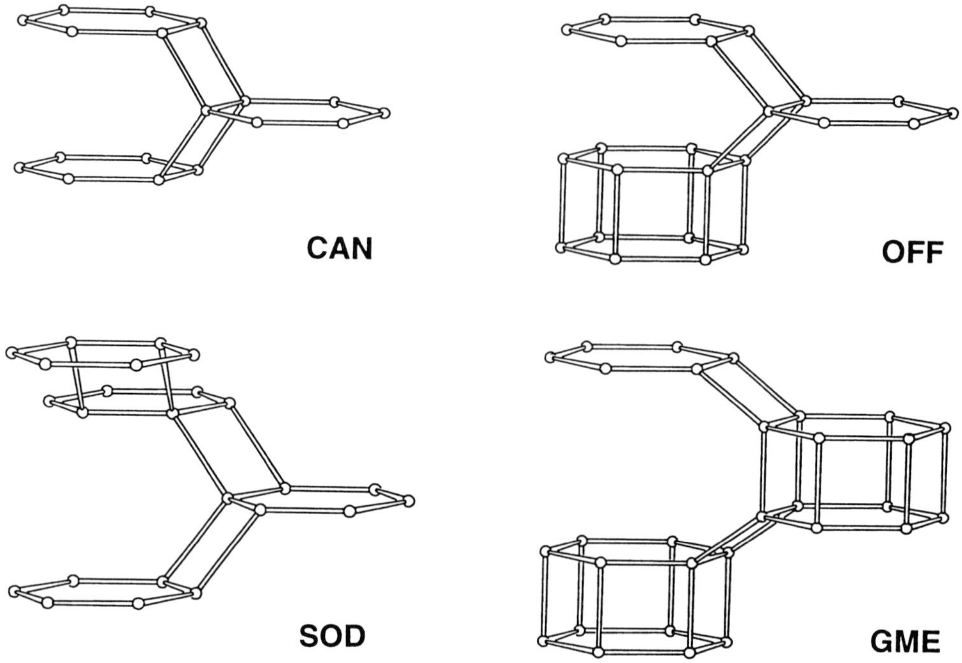

Figure 18. Building schemes for the ABC family of zeolites. The content of one unit cell is shown. An extended structure can be constructed by pure translations of 6-rings (Compare Figure 3).
On each page the structure type code is given followed by the stacking sequence. The stacking along the hexagonal *c* axis is completed to one repeat unit by the 6-ring denoted as: (A).
CAN: AB(A); OFF: AAB(A); SOD: ABC(A); GME: AABB(A).

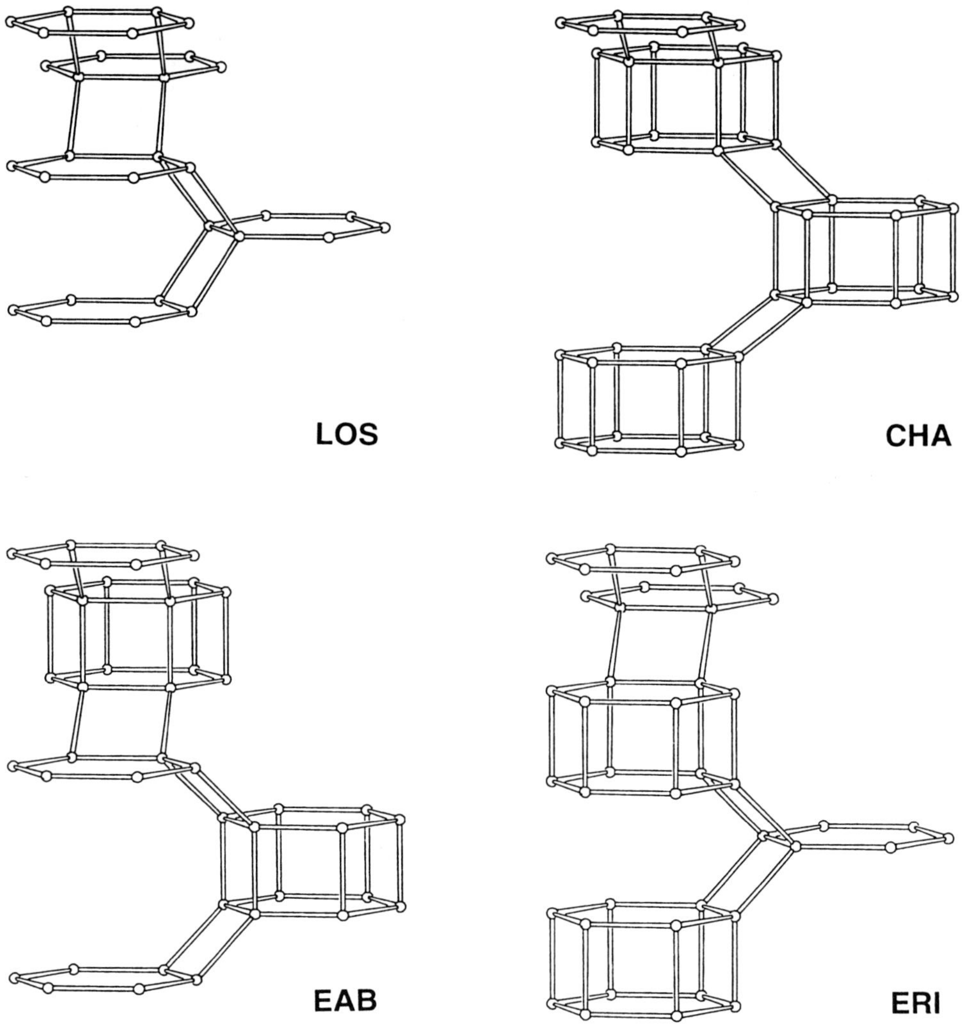

Figure 18. Cont'd. LOS: ABAC(A); CHA: AABBCC(A); EAB: ABBACC(A);
ERI: AABAAC(A);
and next page: LIO: ABABAC(A); AFX: AABBCCBB(A); AFG: ABABACAC(A);
LEV: AABCCABBC(A).

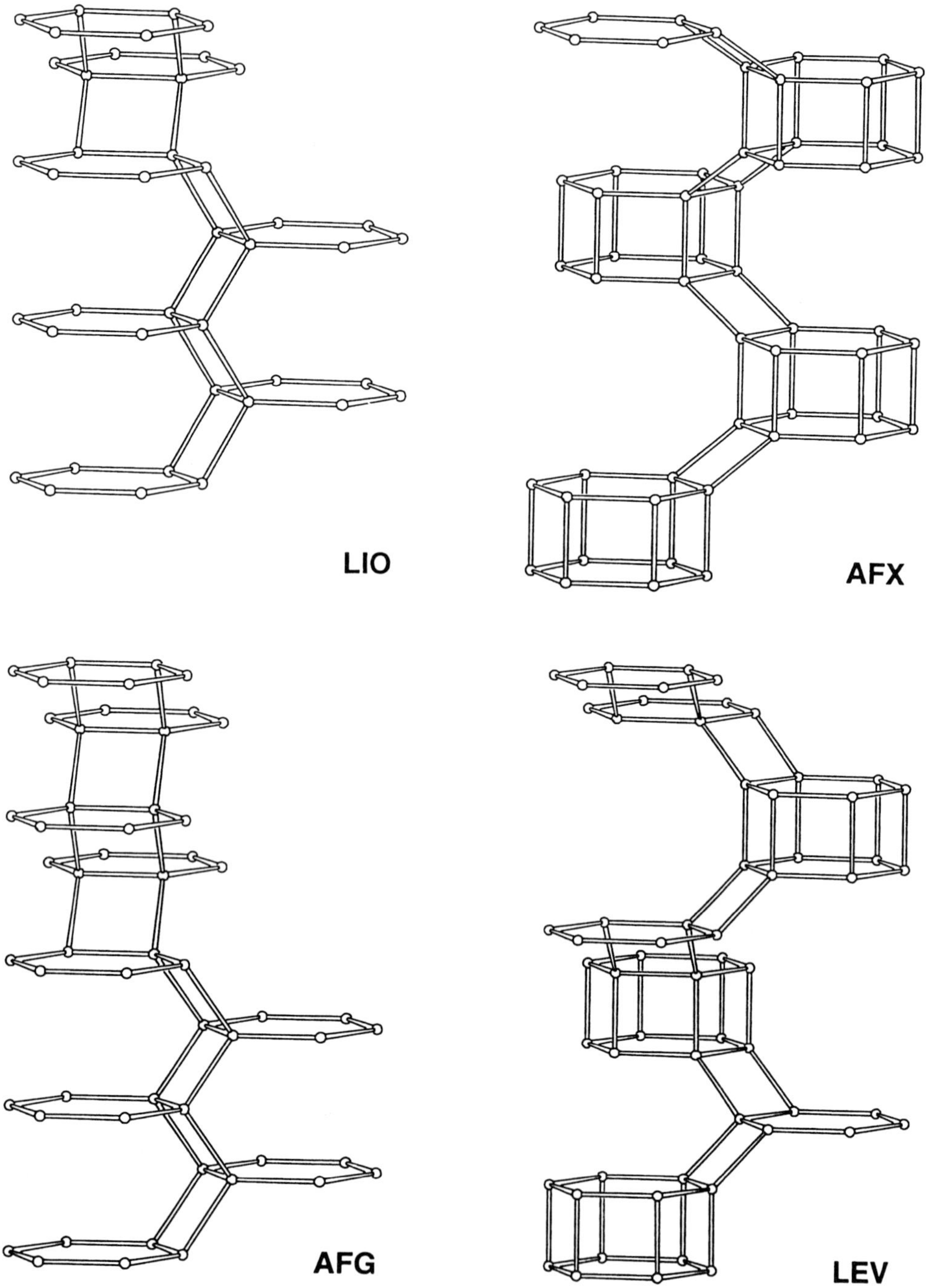

Figure 18. Cont'd. Legend: see previous page.

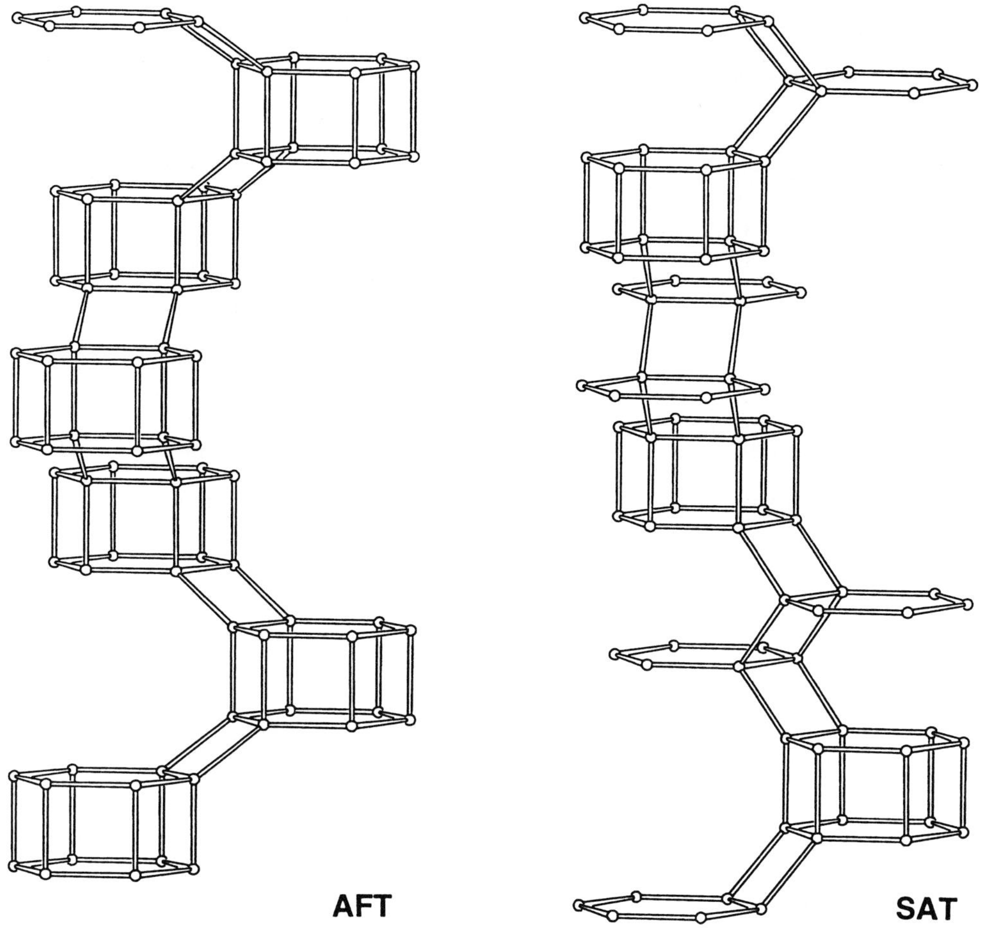

Figure 18. Final page. AFT: AABBCCAACCBB(A); SAT: ABBCBCCACAAB(A).

Table 5
Hexagonal zeolites belonging to the ABC family. The topological unit cell and symmetry is given. The zeolites are arranged in increasing complexity of the stacking sequence

		Cell dimensions (Å)			
Code	stacking sequence	a	c!	space group	Z*
CAN&	AB	12.494	5.254	$P6_3/mmc$	12
OFF+	AAB	13.063	7.565	$P\bar{6}m2$	18
SOD**	ABC	12.580	7.710	$R\bar{3}m$	18
GME$#	AABB	13.672	9.850	$P6_3/mmc$	24
LOS	ABAC	12.607	10.340	$P6_3/mmc$	24
CHA#	AABBCC	13.675	14.767	$R\bar{3}m$	36
EAB	ABBACC	13.178	15.005	$P6_3/mmc$	36
ERI	AABAAC	13.054	15.175	$P6_3/mmc$	36
LIO	ABABAC	12.281	15.632	$P\bar{6}m2$	36
AFX#	AABBCCBB	13.674	19.695	$P6_3/mmc$	48
AFG	ABABACAC	12.548	20.789	$P6_3/mmc$	48
LEV	AABCCABBC	13.168	22.578	$R\bar{3}m$	54
AFT#	AABBCCAACCBB	13.691	29.449	$P6_3/mmc$	72
SAT	ABBCBCCACAAB	12.871	30.577	$R\bar{3}m$	72

! c ~ (number of stacks)*2.55 Å.
* Z = number of T atoms per unit cell = (number of stacks)*6.
** Hexagonal setting; cell calculated from cubic cell.
& Zeolite can also be constructed from zigzag chains.
\+ Zeolite can also be constructed from saw chains.
$ Zeolite can also be constructed from crankshaft chains.
\# Zeolite can also be constructed from double 6-rings.

Table 6
Zeolites in which the single or double 6-rings are inclined with respect to each other. The topological unit cell and symmetry is given

Code	space group	Cell dimensions (Å) a	b	c	γ (°)	n*	Z**
AEI	Cmcm	13.677	12.607	18.497		8	48
AWW	P4/nmm	13.634	13.634	7.627		4	24
EMT	$P6_3/mmc$	17.215	17.215	28.082	120.0	16	96
FAU	$Fd\bar{3}m$	24.345	24.345	24.345		32	192
KFI	$Im\bar{3}m$	18.578	18.578	18.578		16	96
LTA	Pm3m	11.919	11.919	11.919		4	24
RHO	$Im\bar{3}m$	14.919	14.919	14.919		8	48

* n = number of 6-rings per unit cell.
** Z = number of T atoms per unit cell.

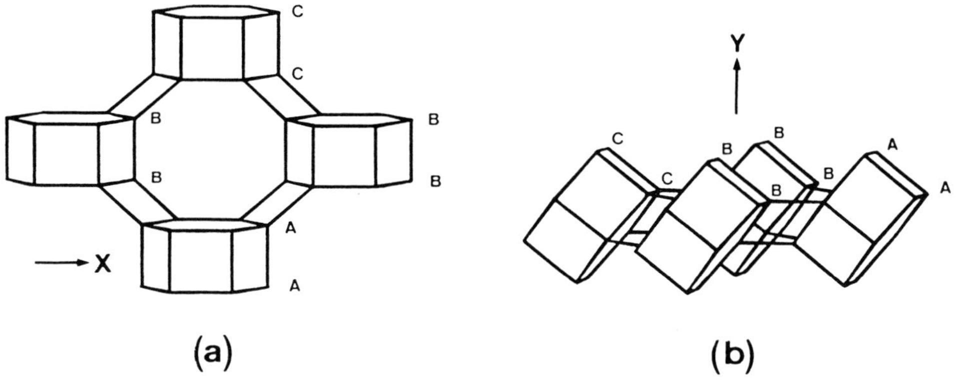

Figure 19. The double 6-ring layer in CHA.
(a): Unit cell content of CHA extended with one double 6-ring.
(b): Side view of the layer with tilted parallel double 6-rings. The orientation is obtained after rotation of the layer in (a) by ~30° about the X-axis followed by a rotation of ~80° about the Y-axis.

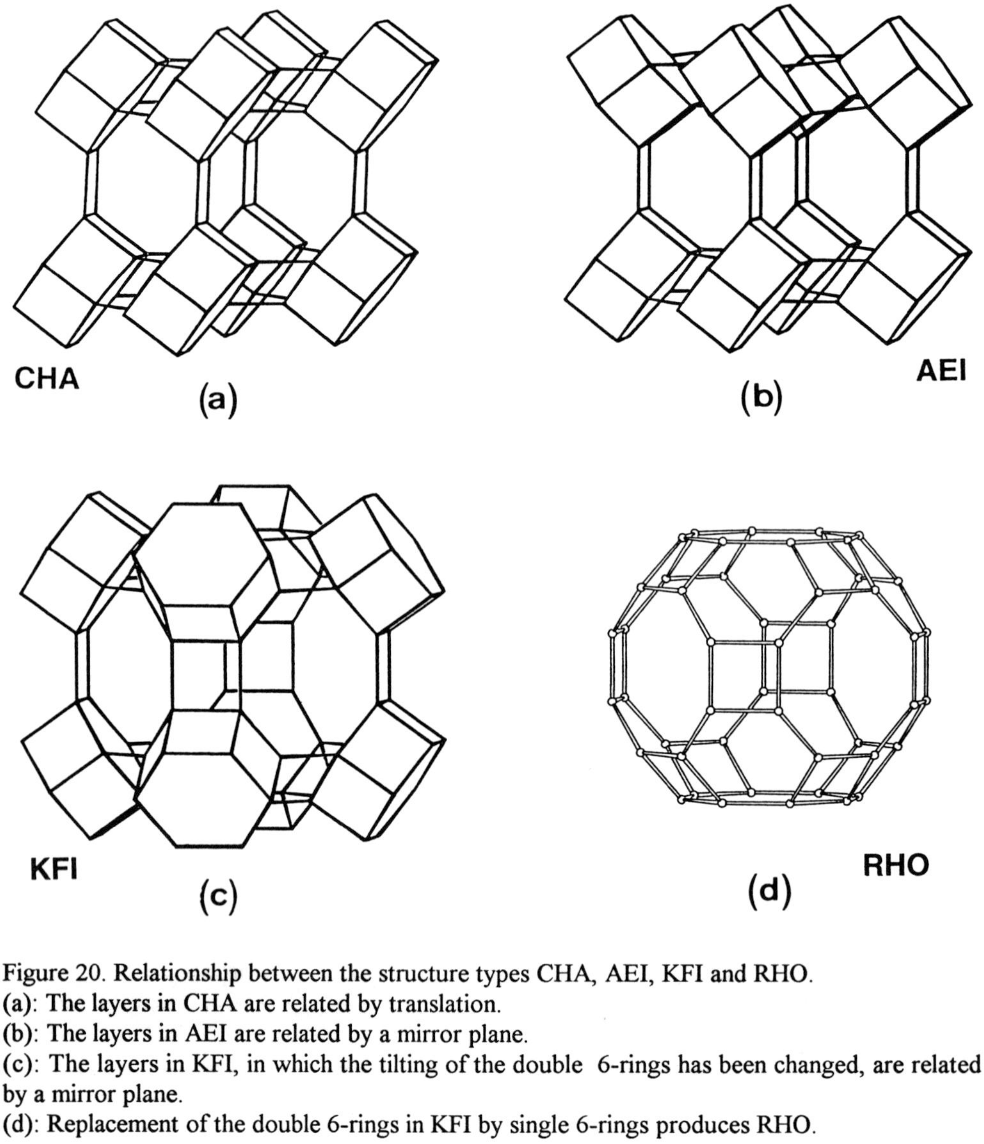

Figure 20. Relationship between the structure types CHA, AEI, KFI and RHO.
(a): The layers in CHA are related by translation.
(b): The layers in AEI are related by a mirror plane.
(c): The layers in KFI, in which the tilting of the double 6-rings has been changed, are related by a mirror plane.
(d): Replacement of the double 6-rings in KFI by single 6-rings produces RHO.

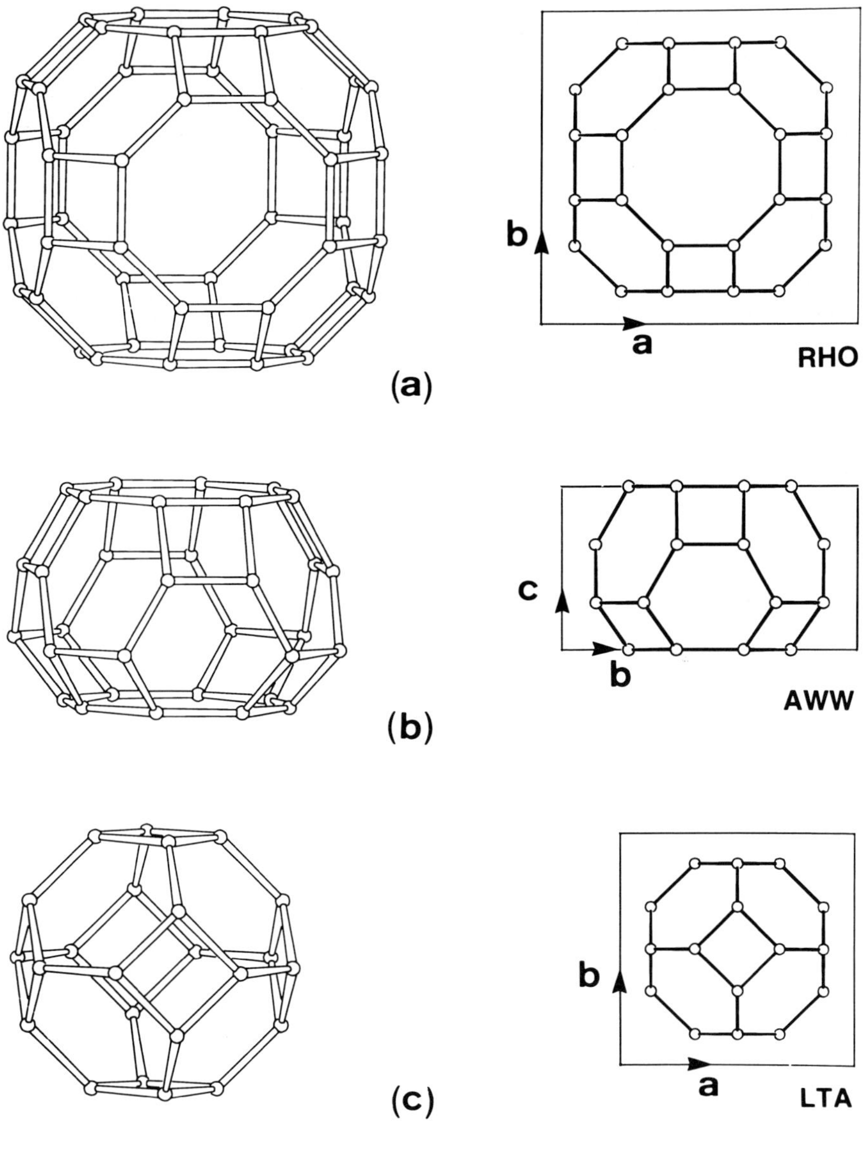

Figure 21. Cages in RHO, AWW and LTA.
(a): In RHO the α cages are connected through 8-rings to form double 8-rings.
(b): In AWW half α cages are connected along **c** through single 8-rings.
(c): In LTA sodalite cages (or β cages) are connected through 4-rings to form double 4-rings.

8. THE FRAMEWORK STRUCTURE OF DISORDERED ZEOLITES

8.1 The FAU/EMT family of zeolites

The characteristic basic sodalite cage layer, referred to as faujasite layer (See Section 2.3), is shown in Figure 22. The layers, depicted in side view in Figures 22(b) and 22(c), are identical and related by a 60° rotation about the plane normal or by a mirror plane perpendicular to the plane normal. FAU and EMT and all intermediate structures can be considered as stacks of these faujasite layers. Crystal data are given in Table 7 (page 164).

The FAU framework can be built by recurrently stacking of layers in the same orientation and shifted with respect to each other over ⅓(-**a** + **b**). The lateral translation of the fourth layer amounts to (-**a** + **b**) which is equivalent to a zero shift with respect to the first layer. This type of connecting the layers introduces inversion centres in the D6R's formed by adjoining the layers (Figure 23(a)).

The EMT framework is produced when the next layer is rotated over 60° about the layer

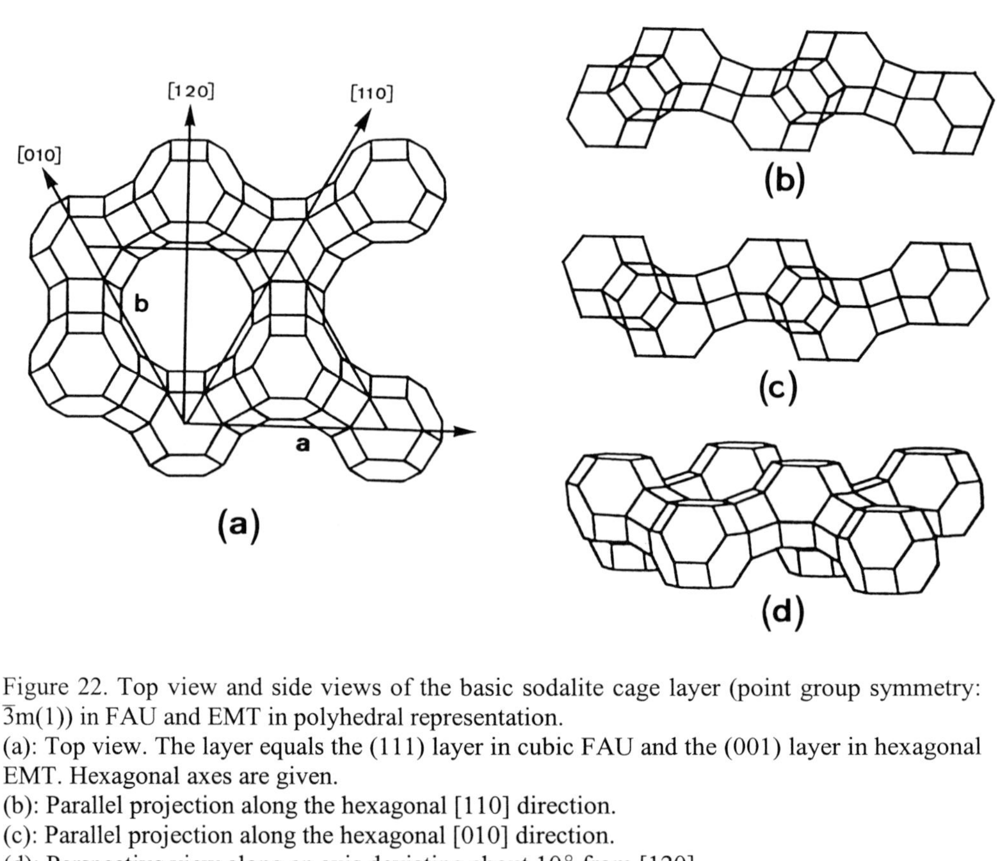

Figure 22. Top view and side views of the basic sodalite cage layer (point group symmetry: $\bar{3}m(1)$) in FAU and EMT in polyhedral representation.
(a): Top view. The layer equals the (111) layer in cubic FAU and the (001) layer in hexagonal EMT. Hexagonal axes are given.
(b): Parallel projection along the hexagonal [110] direction.
(c): Parallel projection along the hexagonal [010] direction.
(d): Perspective view along an axis deviating about 10° from [120].

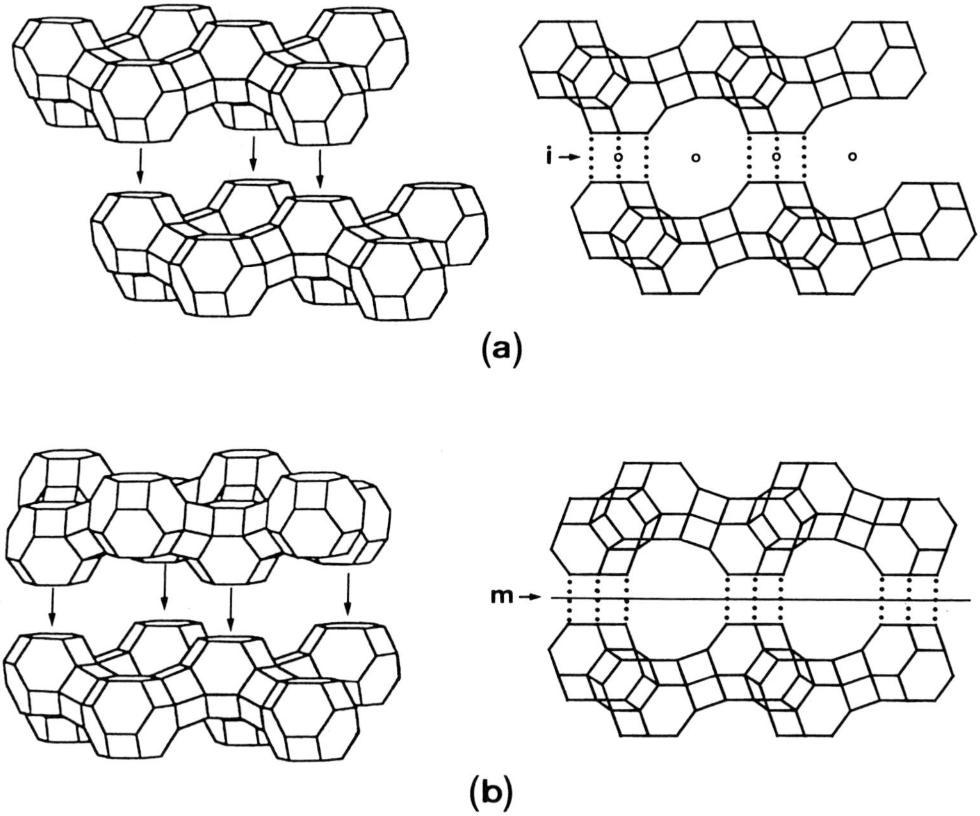

Figure 23. Perspective view (~ along [120]) and parallel view (along [110]) of the faujasite layer illustrating the two possible ways in which successive layers can be connected along [001].

(a): The top layer is shifted over ⅓(-**a** + **b**) before connecting it to the bottom layer. Inversion centres i (o) are induced between the layers.

(b): The top layer is rotated over 60° about the layer normal before connecting it to the bottom layer. Mirror planes m (|) are induced passing through the double 6-rings connecting the layers.

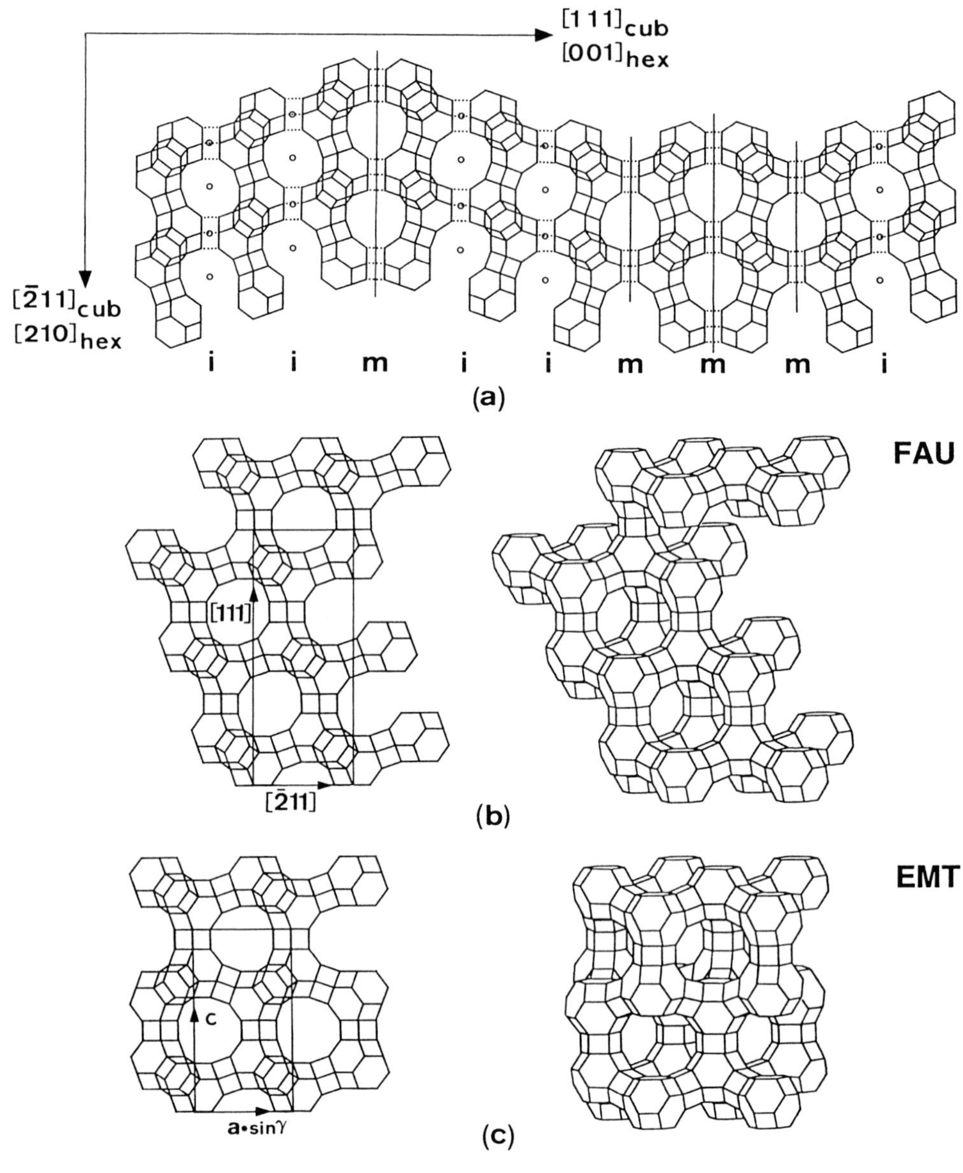

Figure 24. Structures in the FAU/EMT family of zeolites.
(a): Example of an intermediate structure.
(b): Pure FAU is obtained when neighbouring faujasite layers are exclusively related by inversion.
(c): Pure EMT is obtained when neighbouring layers are exclusively related by reflection.

normal, followed by the same shift as in FAU, before connecting it to the previous layer. This stacking introduces mirror planes between the layers. The mirror planes pass through the D6R's formed when the layers are linked (Figure 23(b)).

In general, without steering (organic) template molecules, the distribution of the two symmetry operations between the faujasite layers can take any form. An example of an intermediate FAU/EMT zeolite is illustrated in parallel projection in Figure 24(a).

The cell dimensions along the hexagonal [001] axis is not defined because there is no translation symmetry along [001]. The pure end-members are obtained when the stack exhibits only inversion (in FAU; Figure 24(b)) or reflection (in EMT; Figure 24(c)) between the faujasite layers. The distribution of the two symmetry elements between the layers may be influenced by (organic) template molecules (f.e., crown ethers [13,14]). Unique zeolites are synthesized when growth conditions reproducibly lead to the same distribution of the symmetry elements. The geometry of the cage system is strongly dependent on the actual stacking sequence.

8.2 The MFI/MEL- and ITE/RTH-family of zeolites

The characteristic pentasil layer, discussed in Section 2.3, is shown again in Figure 25. All structures, belonging to the MFI/MEL family of zeolites, consist of an array of parallel YZ-pentasil layers extending along X. The neighbouring pentasil layers can be connected in two different ways. Firstly, successive pentasil layers can be connected after a rotation of 180° about X (or Y) with respect to each other. The resulting connectivity exhibits inversion symmetry between the layers as illustrated in Figure 26(a). Secondly, successive pentasil

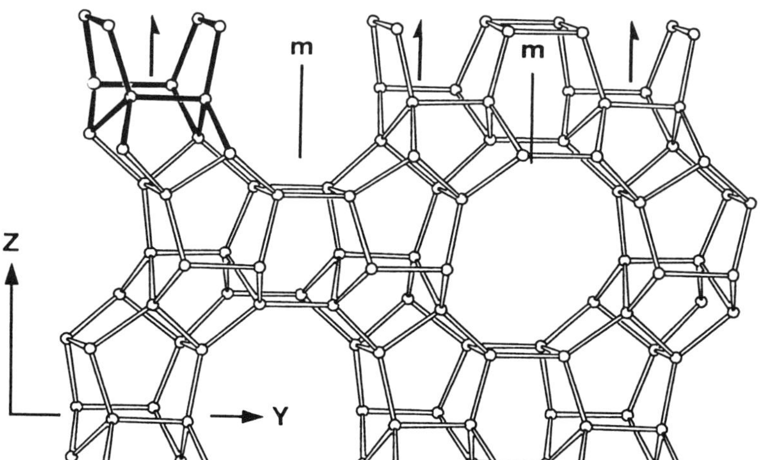

Figure 25. The basic building unit of the pentasil family of zeolites, the pentasil layer, is composed of T12-units (in bold) forming left- and right-handed chains along Z. The chains, related by mirror planes perpendicular to Y, are connected along Y to give the characteristic YZ pentasil layer.

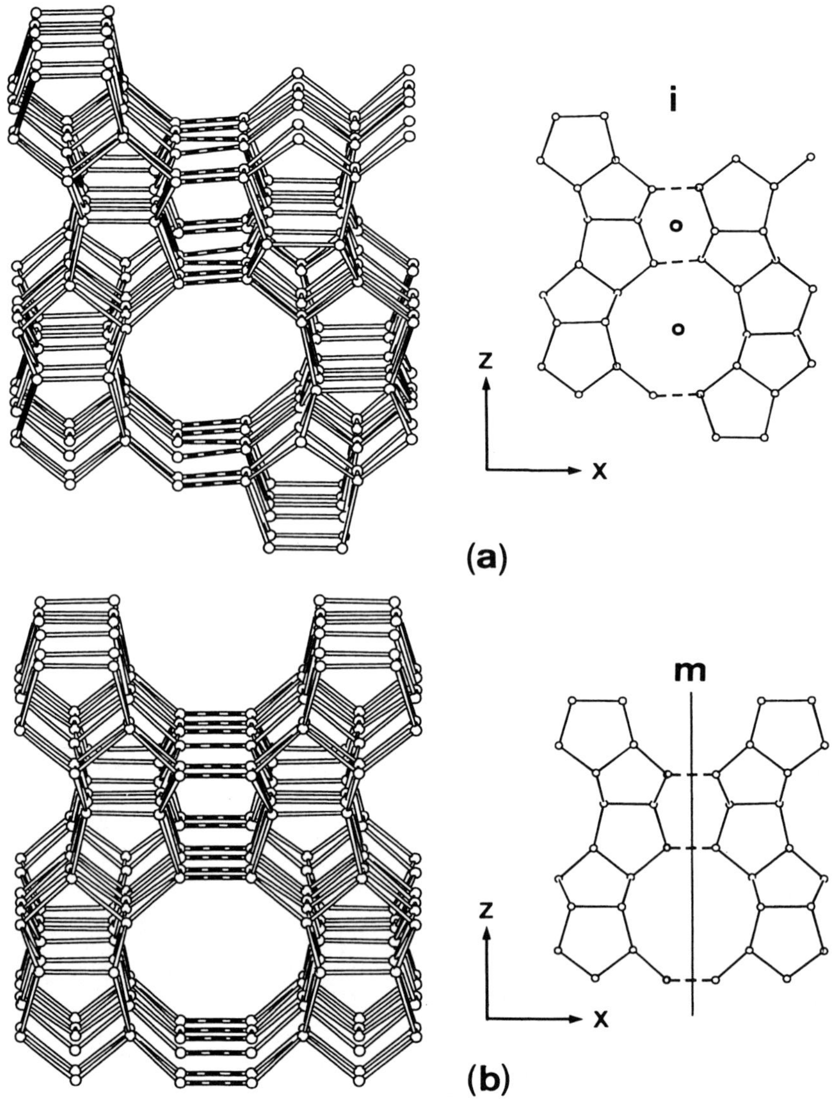

Figure 26. Perspective and parallel view along Y illustrating the two possible ways in which successive pentasil layers perpendicular to [100] can be connected. Inter-layer connections are striped.

(a): The second pentasil layer is rotated over 180° about X. There are inversion centres i (o) between the layers.

(b): The second pentasil layer is rotated over 180° about Z. There is a mirror plane m (|) between the layers.

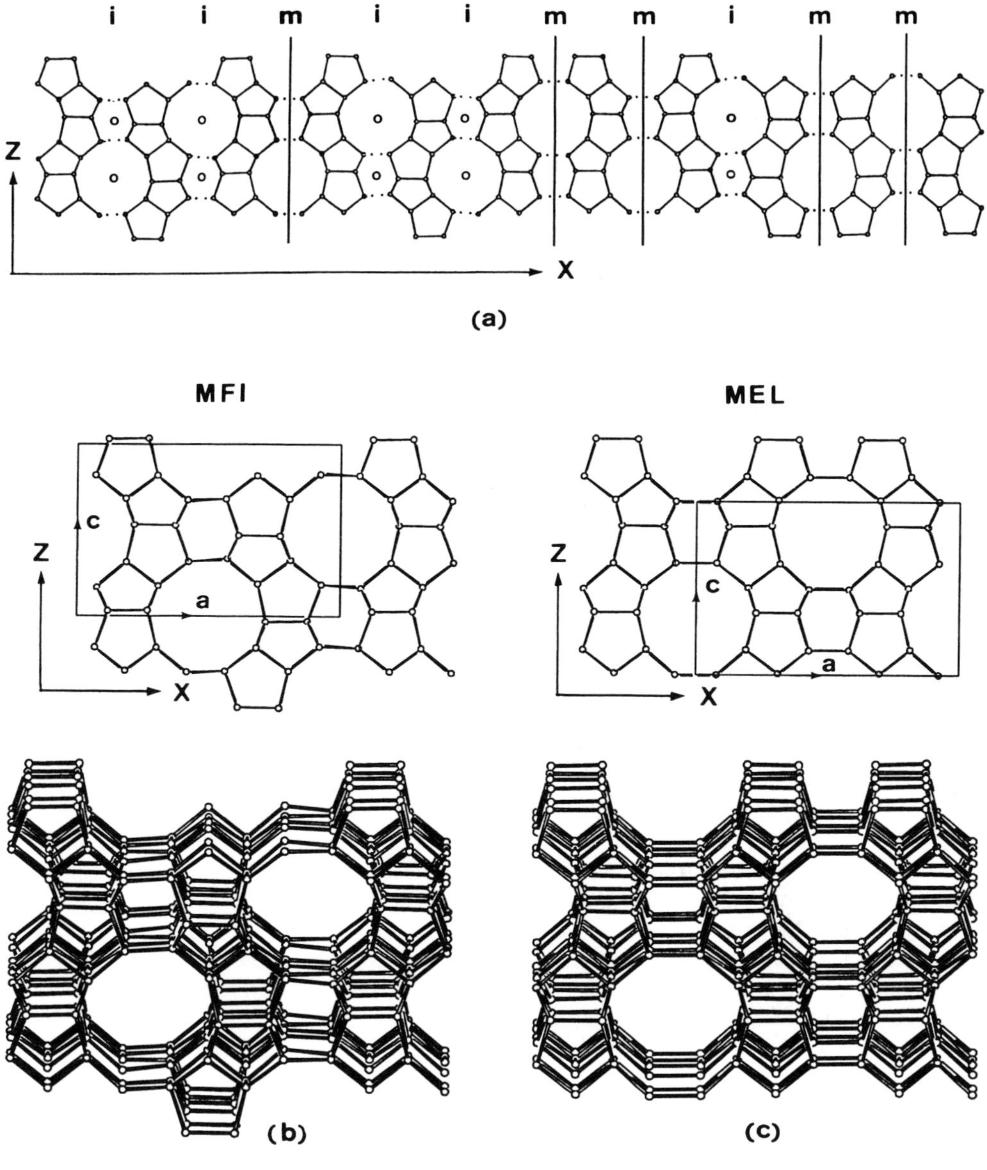

Figure 27. Structures in the MFI/MEL family of zeolites.
(a): Example of an intermediate structure.
(b): Pure MFI is obtained when neighbouring pentasil layers are exclusively related by inversion.
(c): Pure MEL is obtained when neighbouring pentasil layers are exclusively related by reflection.

layers can be connected after a rotation with respect to each other of 180° about Z. The connectivity now shows mirror symmetry between successive layers (Figure 26(b)).

An example of the structure of an intermediate MFI/MEL zeolite is illustrated in parallel projection in Figure 27(a). The cell dimension along X is not defined because there is no translation symmetry along X. The pure end-members, MFI and MEL, exclusively have inversion centres (in MFI) or mirror planes (in MEL) between successive pentasil layers. In MFI and MEL there is, in addition to the existing translation symmetry along Y and Z, translation symmetry along X and a unit cell can be indicated (Figure 27(b,c)).

The distribution of i and m along [100], D(x;i,m), can probably be influenced during the growth of the zeolite by using different (organic) template ions. The end-members can be obtained by using, f.e., tetrapropylammonium- and N,N-diethyl-3,5-dimethylpiperidinium-ions in the synthesis of MFI and MEL, respectively. Other unique zeolites within the MFI/MEL series are synthesized when growth conditions reproducibly lead to specific values of D(x;i,m). The nature of the intersecting channel system in the (intermediate) structures, which determine to a great extend the physical properties, is strongly dependent on the actual value of D(x;i,m).

The same type of disorder is observed in the ITE/RTH family of zeolites. T16-units, consisting of three 4-rings and four 5-rings (see Figure 28(a)), are connected to chains after pure translations along X have been applied. This linkage of T16-units in ITE/RTH is different from the linkage of T12-units in MFI/MEL. In MFI/MEL neighbouring T12-units are related by a two-fold screw axis in stead of by a pure translation. Chains are connected along Y after a shift of ½ **a**, a rotation of 180° about Y or after a mirror operation in a plane perpendicular to Y has been applied to the subsequent chain with respect to the previous one. The basic layer, point group c1m1, is shown in Figure 28(a). Neighbouring basic layers can be connected along Z in two different ways as illustrated in Figure 28(c). Firstly, layers can be recurrently stacked in the same orientation along Z. The resulting connectivity exhibits inversion symmetry between the layers. Secondly, successive basic layers can be connected after a rotation of 180° about Z (or Y) with respect to each other. The connectivity now shows mirror symmetry between neighbouring layers.

An intermediate structure in the ITE/RTH family of zeolites is shown in Figure 29(a). The resemblance of the connection modes in intermediate FAU/EMT- and MFI/MEL-structures with those in the ITE/RTH intermediates is striking.

Two simple end-members, ITE and RTH, are shown in Figure 29(b,c). Both structures have been observed as pure single crystal material. The three dimensional periodic end-member structures comply to a specific sequence of symmetry elements between the basic layers. RTH and ITE exclusively exhibit inversion centres (RTH) and mirror planes (ITE) between successive basic layers.

Crystal data are given in Table 7 (page 164).

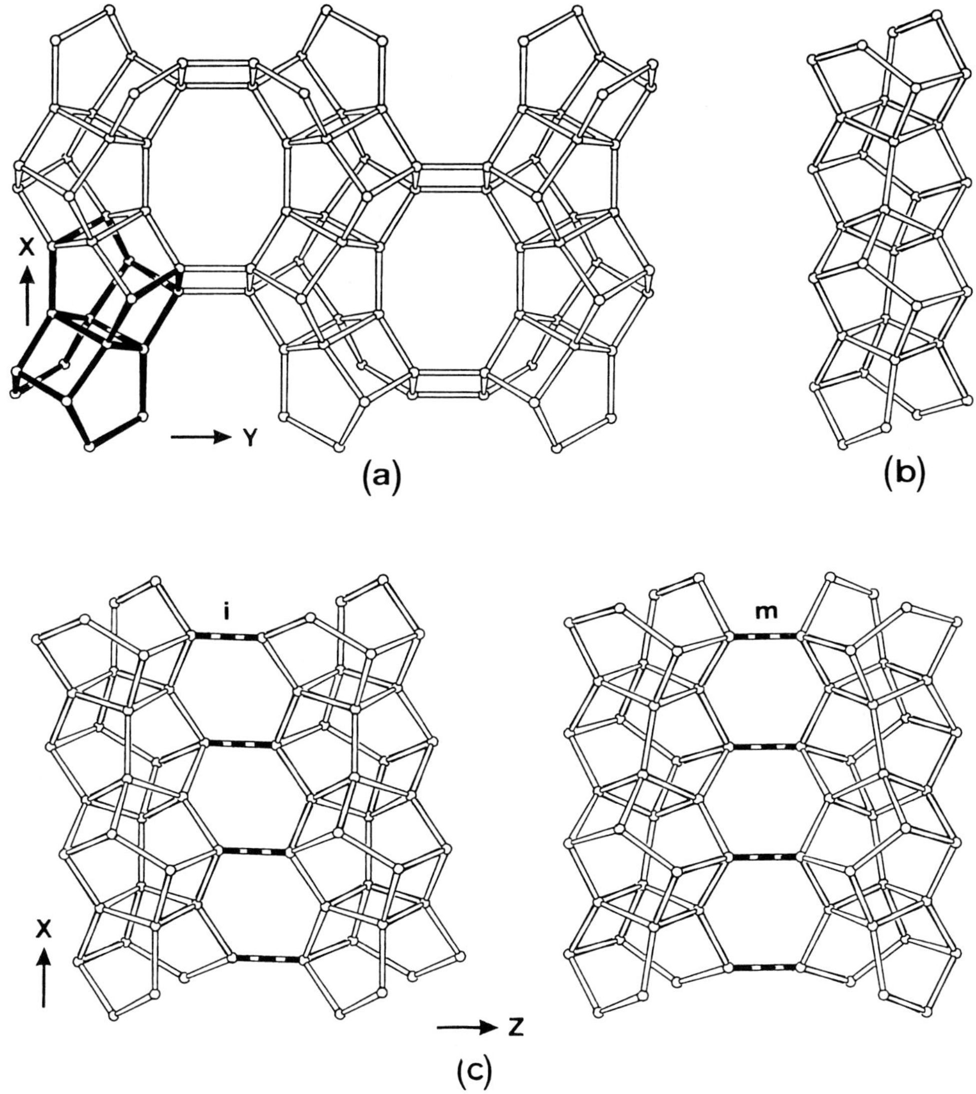

Figure 28. The basic layer in the ITE/RTH family of zeolites.
(a): Top view of the basic XY plane, layer symmetry c1m1, composed of T16 units (in bold) forming chains along X. Chains, related by mirror planes perpendicular to Y, are connected along Y.
(b): Parallel view down Y.
(c): Parallel view down Y illustrating two ways to connect basic layers along Z: the second layer is connected in the same orientation (left) or after a rotation of 180° about Z (right). Between the layers, inversion centres (i) and mirror planes (m) are induced.

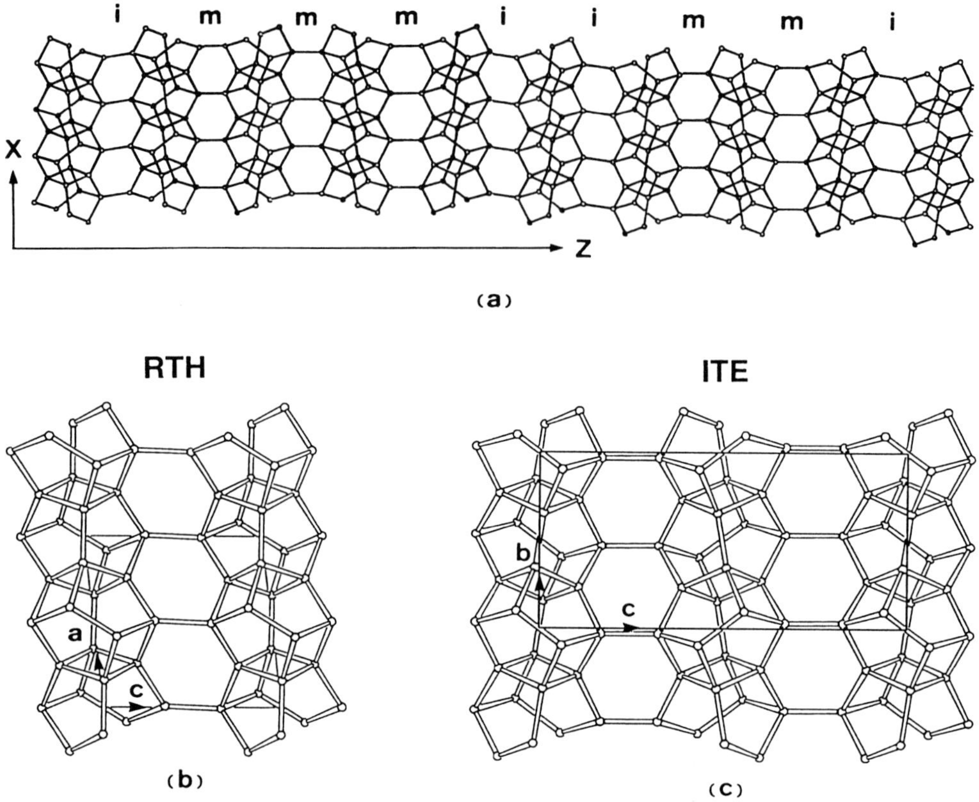

Figure 29. Structures in the ITE/RTH family of zeolites.
(a): Example of an intermediate structure.
(b): Pure RTH is obtained when neighboring basic layers are related by inversion.
(c): Pure ITE is obtained when neighboring basic layers are related by reflection.

8.3 The beta family of zeolites

The characteristic basic layer of the beta family of zeolites is shown in Figure 30(a). This tetragonal beta layer is composed of T16-units (in bold) related by pure translations along **a** and **b**. The layers, depicted in Figures 30(b) and 30(c), are identical and related by a 90° rotation about the plane normal or by a mirror plane perpendicular to the plane normal. All end-members and all intermediate beta structures can be considered as stacks of basic beta layers. Subsequent layers are connected to previous ones after a mirror operation on the previous layer (or after 90° rotation of the layer) accompanied by a lateral shift of zero, $+^1/_3$**a**, $-^1/_3$**a**, $+^1/_3$**b** or $-^1/_3$**b**. The connectivity of neighbouring layers with a lateral shift of zero has not yet been observed. The lateral shift alternates strictly between **a** and **b**. The three possibilities to connect neighbouring layers along [001] is illustrated in Figure 31. The distribution of the lateral shifts between the layers stacked along [001] defines the 3-dimensional structure.

All three-dimensional periodic beta structures comply to a specific sequence of the lateral shifts along [001]. The five simplest end-members are shown in Figure 32 and their topological symmetry and unit cell data are given in Table 7 (page 164). None of the end-members has yet been observed as pure single crystal material. The sequences of the lateral inter-layer shifts in the simplest end-members, leading to additional translation symmetry along [001], are given in Table 8 (page 165).

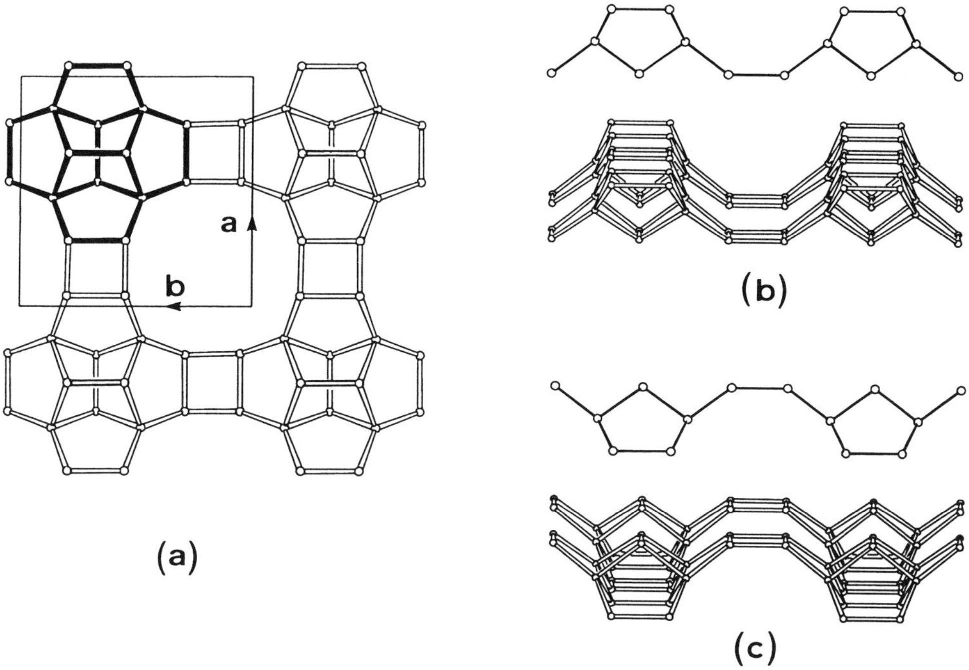

Figure 30. Top view and side views of the basic beta layer in the zeolite *BEA family.
(a): Top view. Basic (001) layer composed of T16-units (in bold). Tetragonal axes are given.
(b): Parallel projection (top) and perspective view (bottom) along [100].
(c): As (b) along [010].

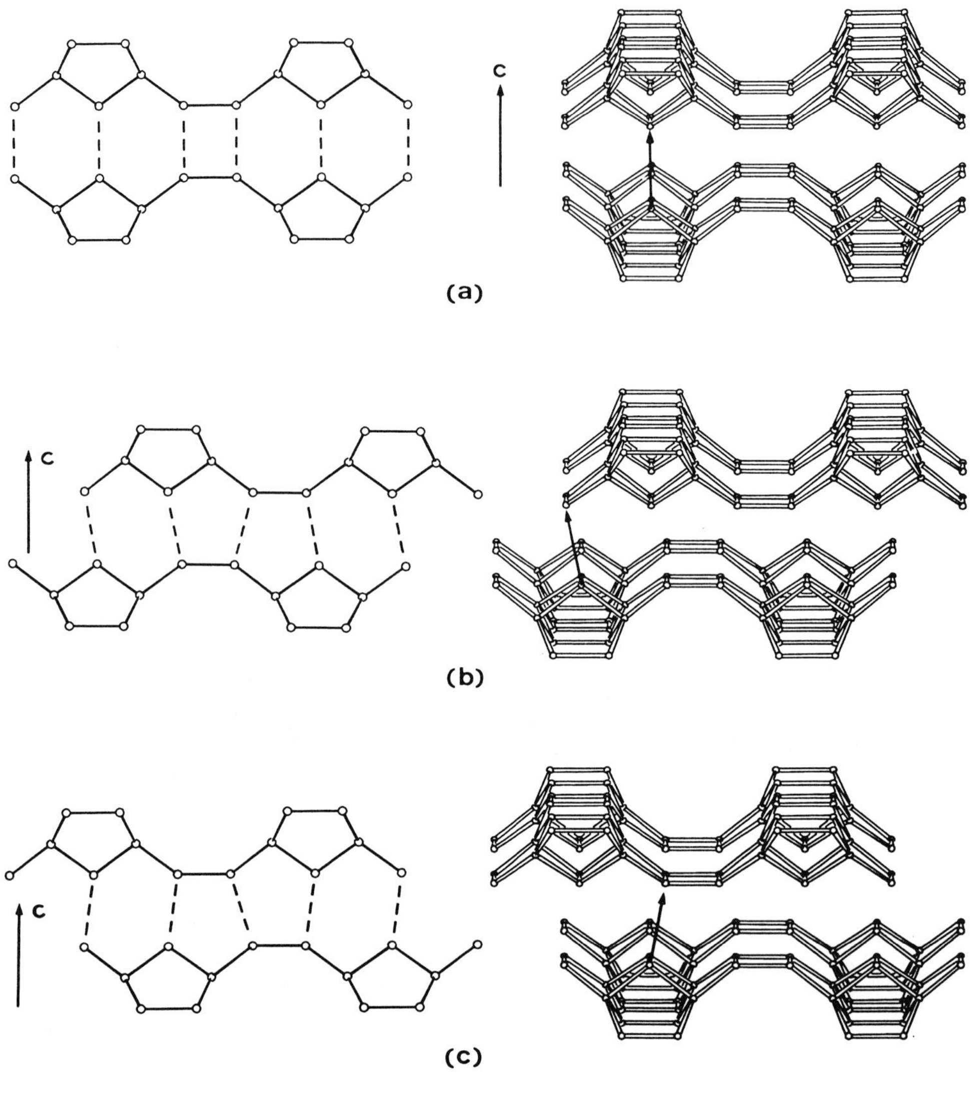

Figure 31. Parallel and perspective side views of the three possible ways in which beta layers, related by mirror planes, can be connected. The arrow joins a pair of T atoms to be connected.
(a): The lateral shift between layers is zero. This connection mode has not yet been observed.
(b): The lateral shift is +1/3 (along **a** or **b**).
(c): The lateral shift is -1/3 (along **a** or **b**).

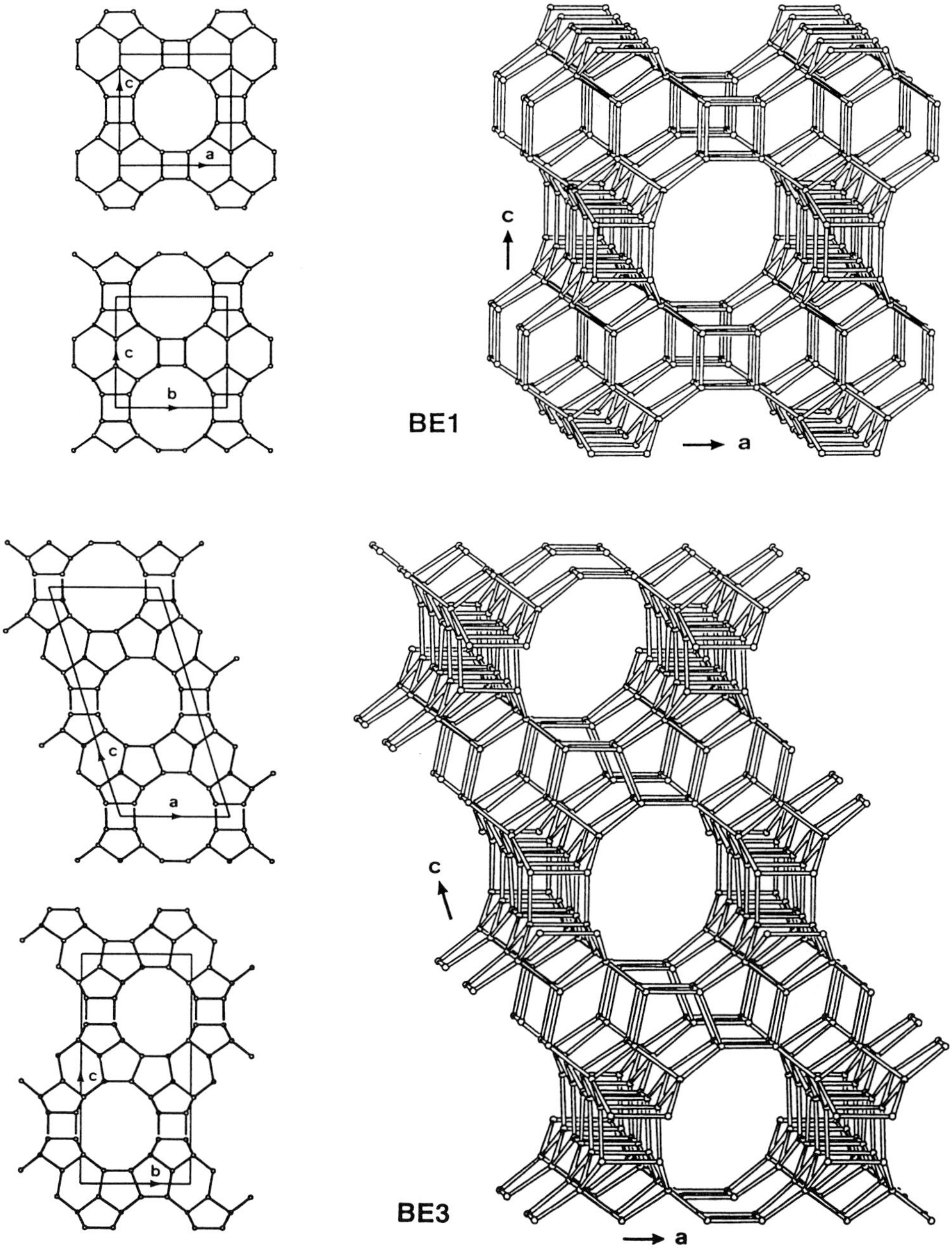

Figure 32. The simplest ordered end-members of the beta family. None of them has yet been observed as pure single crystal material.

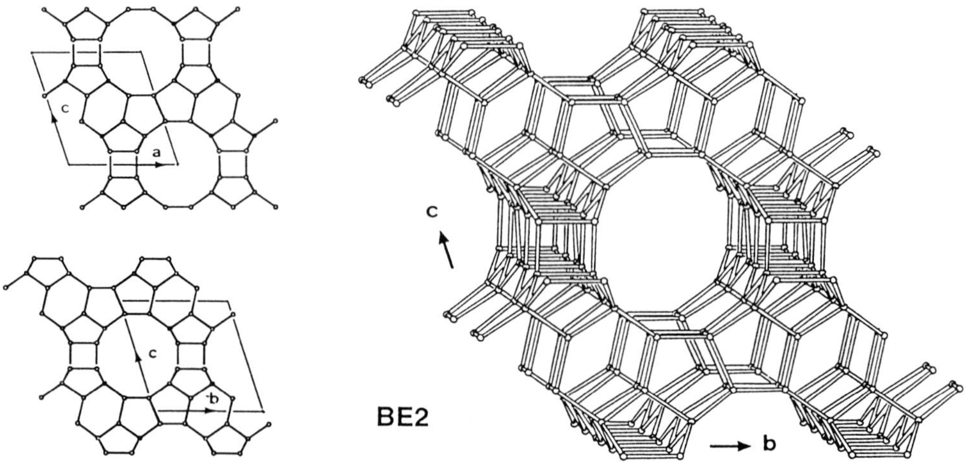

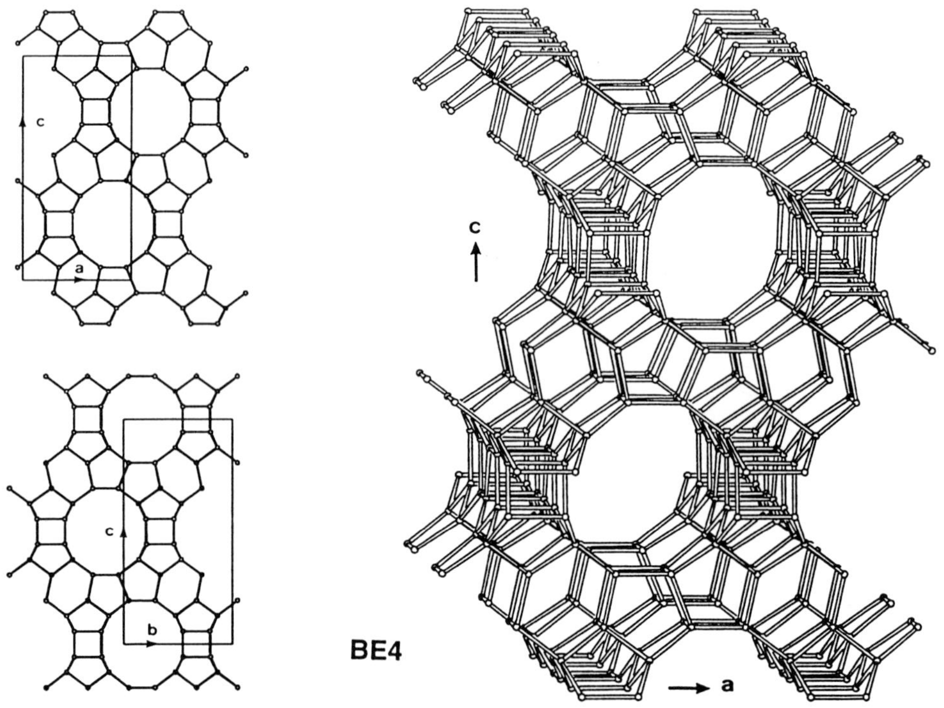

Figure 32. Final page. BE4 corresponds to the structure with type code *BEA. BE5 is the enantiomorph of BE4 and has not been drawn separately.

8.4 The SSZ-33 family of zeolites

The basic layer of all structures belonging to the SSZ-33 family of zeolites is composed of parallel chains (Figure 33(a)). The chains are formed by connecting T14-units (in bold) related by pure translations along **c**. In the (puckered) basic layer, two neighbouring chains are related by a 180° rotation about [010] (or [001]) as illustrated in Figures 33(b) and 33(c). The two layers, shown in perspective side view in Figure 33(c), are identical and related by a 180° rotation about [010] (or [001]), by a mirror operation perpendicular to the plane normal or by a shift of the origin over ½**b**. The basic layer, formed by connecting chains along [010], is puckered as depicted in Figure 33(d). All end-members and all intermediate SSZ-33 structures can be considered as stacks of these basic layers along [100].

Neighbouring layers, related by a mirror plane, can be connected along [100] through oxygen bridges after a lateral shift of zero, $+^{1}/_{3}\mathbf{c}$ or $-^{1}/_{3}\mathbf{c}$ has been applied. The connection mode where the lateral shift is zero, has not yet been observed. The three possible connectivities of neighbouring layers are shown in Figure 34. The analogy with the (two-dimensional) connection modes in the beta family is evident. The 3-dimensional structure is defined when the sequence of the lateral shifts between the layers is known.

The three simplest end-members, exhibiting additional translation symmetry along [100], are shown in Figure 35 and their topological symmetry and unit cell data are added in Table 7 (page 164). There is no difference in the projections of the structure of the end-members along [001]. End-member CO2 represents the framework structure with type code CON. The other end-members have not yet been observed as pure single crystal material. The sequences of the lateral inter-layer shifts in the simplest end-members are added in Table 8 (page 165).

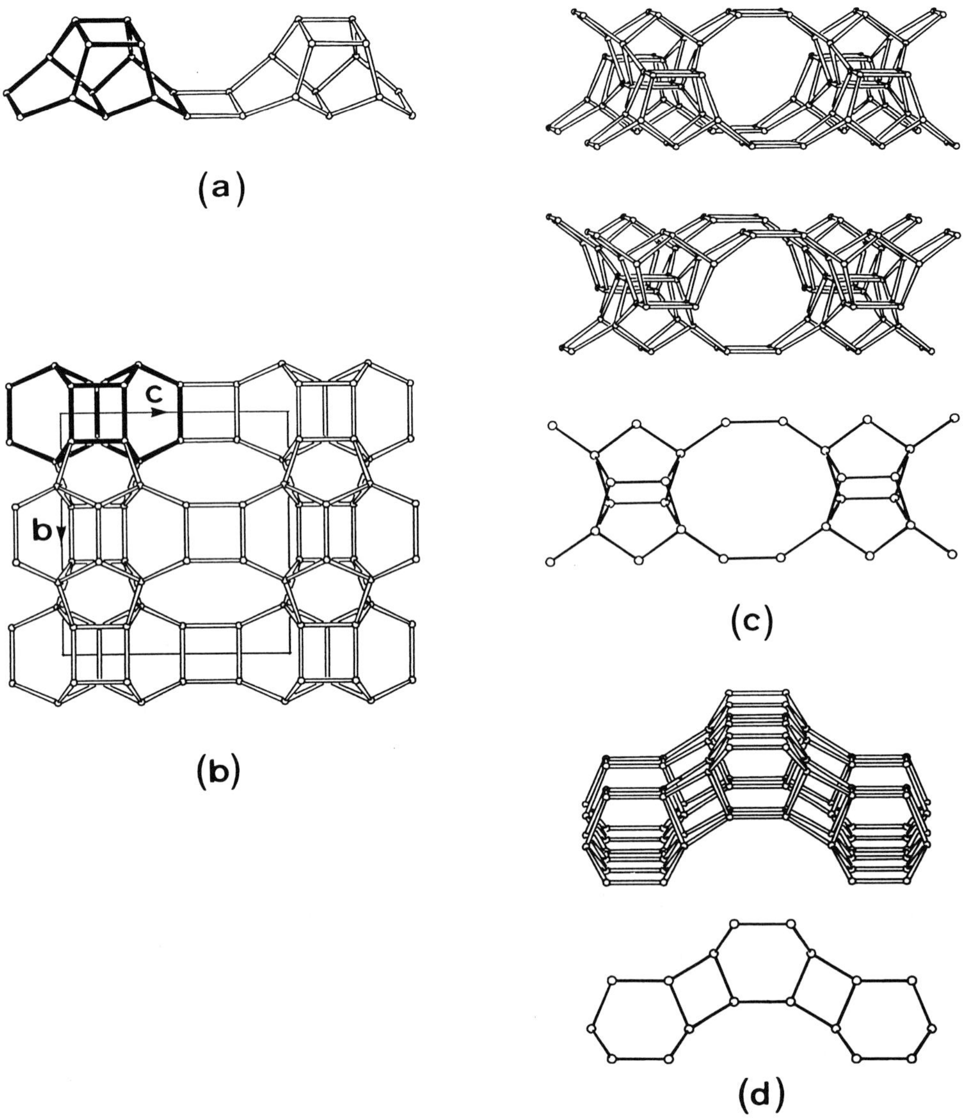

Figure 33. Top view and perspective side views of the (chain and) basic layer, point group symmetry (2)mm, in the SSZ-33 family of zeolites.
(a): The chain composed of T14-units (in bold).
(b): Top view of the basic (100) layer.
(c): Perspective view (top) and parallel projection (bottom) along [010].
(d): As (c) along [001].

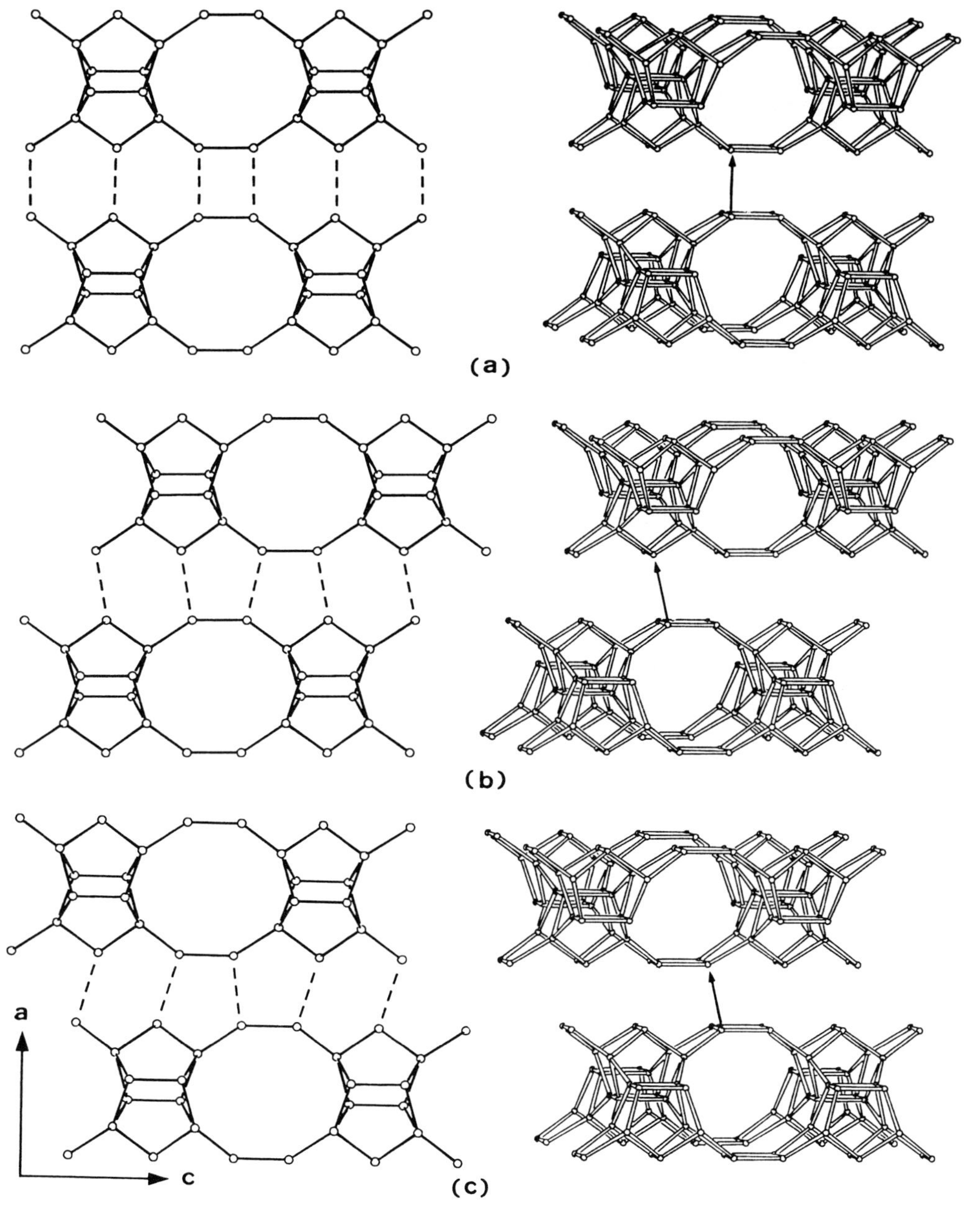

Figure 34. Side views of the three possible ways in which SSZ-33 layers, related by mirror planes, can be connected. The arrow joins a pair of T atoms to be connected. The lateral shift is (a): zero, (b): +⅓**c** and (c): -⅓**c**. Mode (a) has not yet been observed.

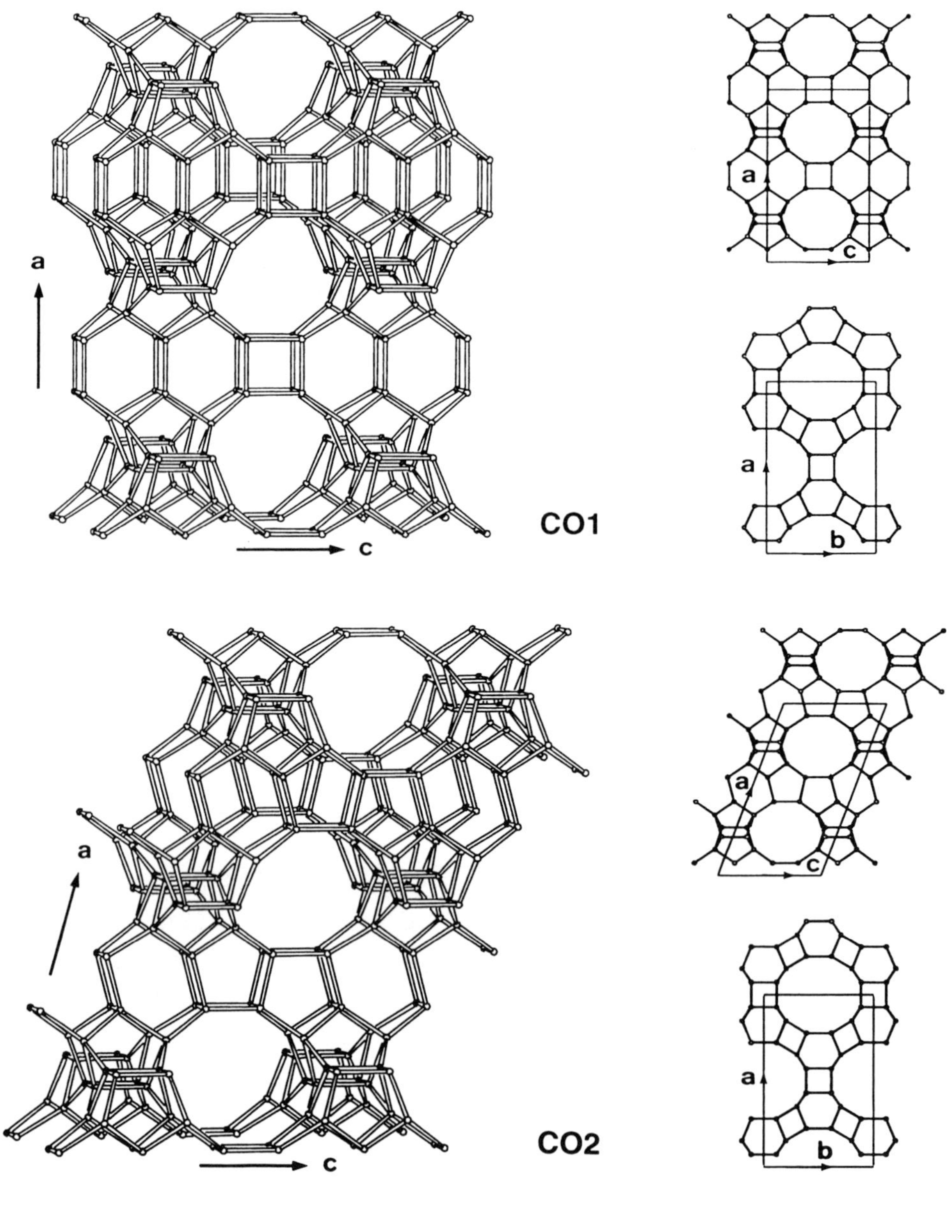

Figure 35. The simplest ordered end-members of the SSZ-33 family. CO2 corresponds to the end-member with structure type code CON.

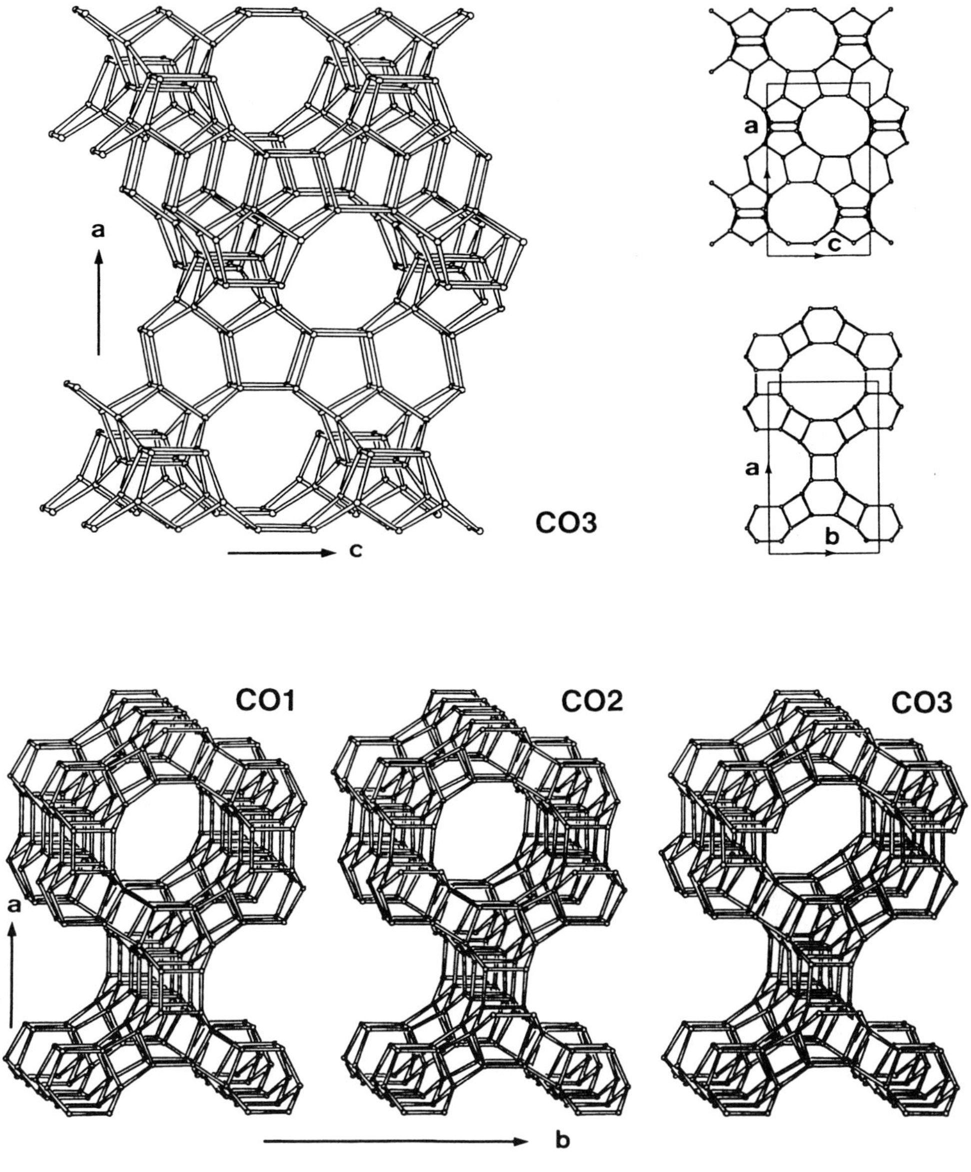

Figure 35. Final page. CO1 and CO3 have not been observed yet as pure single crystal material. The lateral shifts are given in Table 8.

8.5 The lovdarite- and montesommaite-zeolites and the WEI structure type

The characteristic layer in all structures belonging to the lovdarite family of zeolites consists of two 4-ring layers which are connected through single T atoms as illustrated in Figure 36(a). The 4-rings have UDUD directionality (see Section 2.1). This basic layer can be constructed by connecting T9-units (in bold) related by pure translations along **a** and **b** (see Figure 36(b)). The layers, depicted in Figure 36(a) and 36(c), are identical and related by a 90° rotation about the plane normal or by a mirror operation perpendicular to the plane normal.

In all end-members and intermediate structures subsequent layers are connected to previous ones after a mirror operation on the previous layer (or after a 90° rotation of the layer) accompanied by a lateral inter-layer shift of zero, (±)½**a** or (±)½**b**. The connection modes of neighbouring layers along [001] is illustrated in Figure 37.

All three-dimensional periodic lovdarite structures comply to a specific sequence of the lateral shifts along [001]. The three simplest end-members are shown in Figure 38. The topological symmetry and unit cell data of the three structure types are added in Table 7 (page 164). The sequences of the lateral inter-layer shifts are given in Table 8 (page 165). All three end-members have been observed as pure single crystal material.

When the 4-ring layers are connected directly and not through single T atoms, the montesommaite family of zeolites is constructed. The basic layer consists of 4-rings (in bold) related by pure translations along **a** and **b** (see Figure 39(a)). The layers, depicted in Figure 39(b) and 39(c), are identical and related by a 90° rotation about the plane normal or by a mirror operation perpendicular to the plane normal. The analogy with the lovdarite family is evident: subsequent layers are connected to previous ones after a mirror operation on the previous layer followed by lateral inter-layer shifts of 0, (±)½**a** or (±)½**b** leading to the same type of connection modes as in the lovdarite zeolites (compare Figures 37 and 40).

The end-member, with structure type MON (see Figure 41), can be built by recurrently stacking of layers related by mirror planes (or by a 90° rotation about the layer normal), followed by a lateral shift of ½**a** or ½**b**. This type of connecting the layers introduces inversion centres between the layers. The topological symmetry and unit cell data of MON are given in Table 7 (page 164) and the sequence of the lateral shifts are added in Table 8 (page 165). Except in parts of the APD framework, the connection mode with lateral shift zero, has not yet been observed.

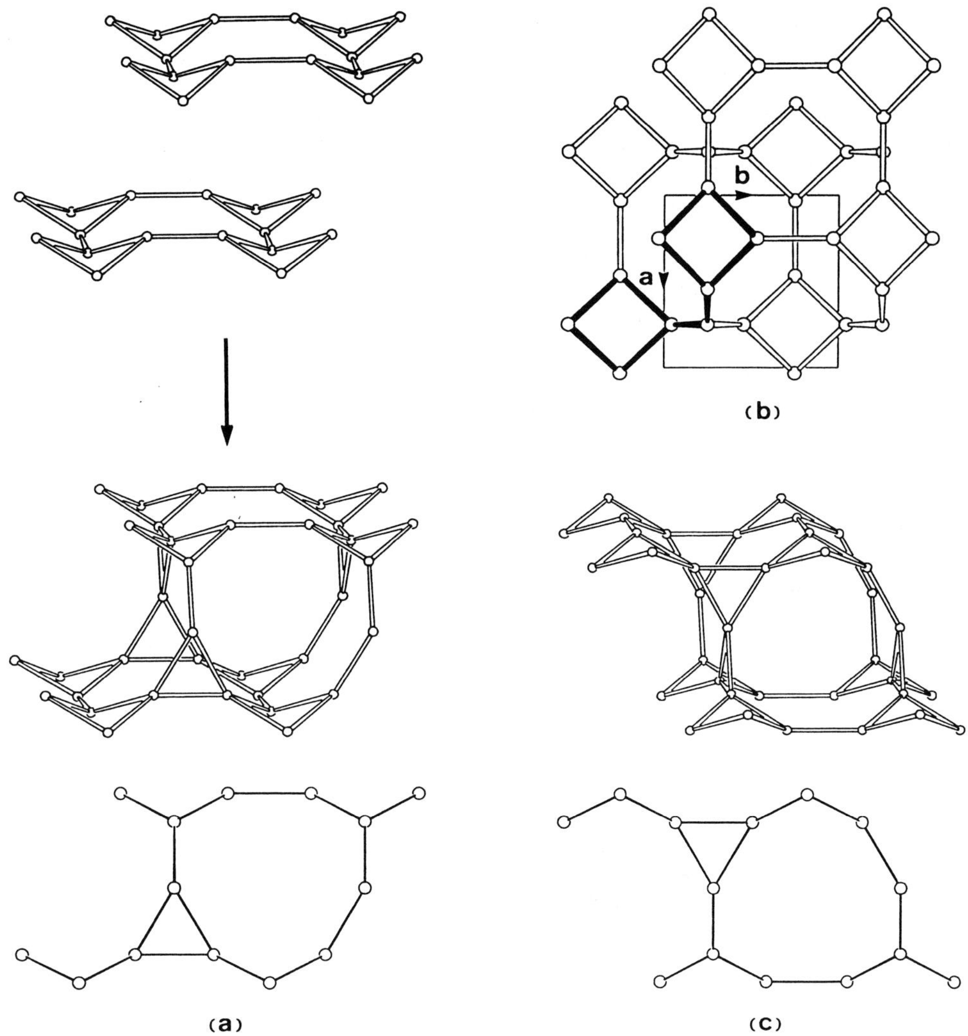

Figure 36. Basic (001) layer, point group symmetry mm(2), in the lovdarite zeolites.
(a): Construction, perspective view and parallel projection of the basic layer down [100].
(b): Top view of the basic layer composed of T9-units (in bold). The T9-unit consists of two 4-rings with UDUD directionality connected through a single T atom.
(c): As (a) along [010].

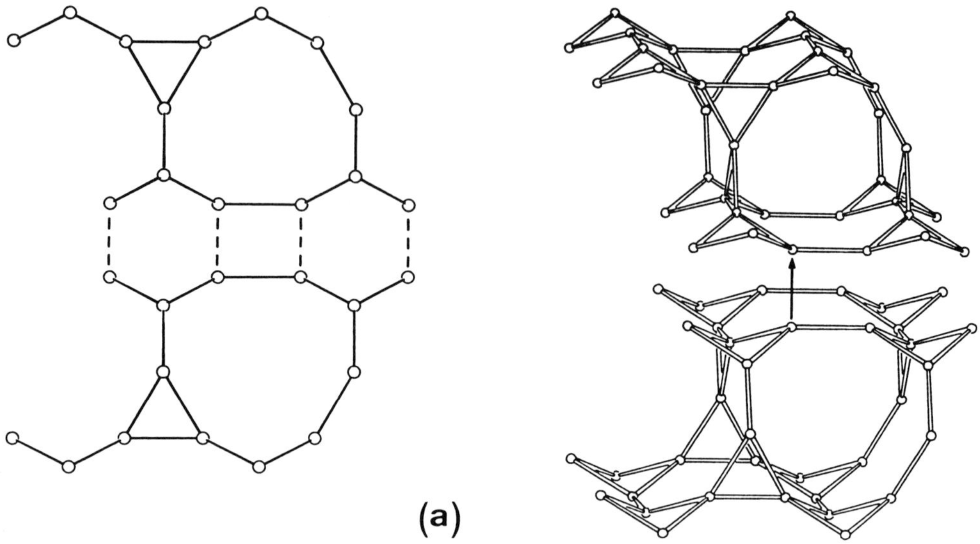

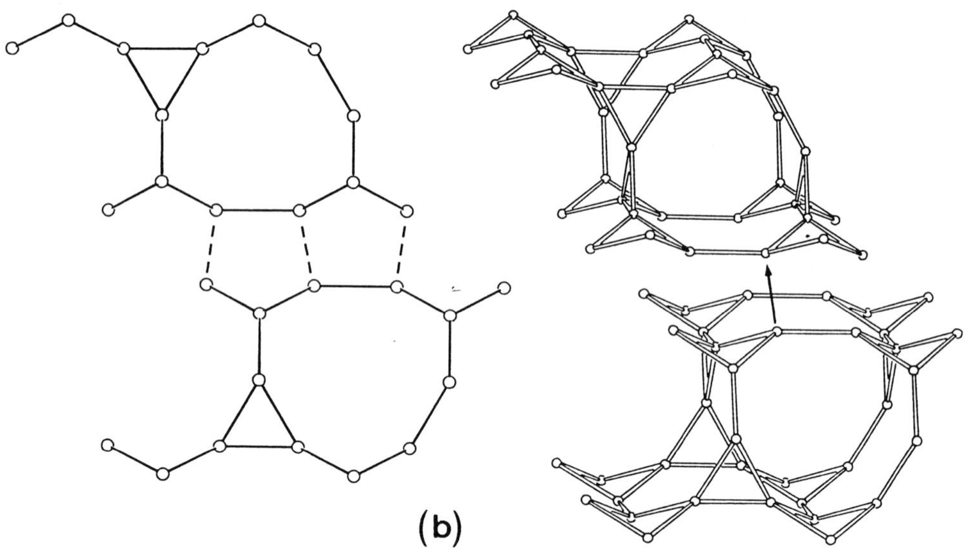

Figure 37. Parallel and perspective views of two ways in which basic lovdarite layers, related by mirror planes, can be connected.
(a): The lateral shift between layers is zero.
(b): The lateral shift is (±) ½ (along **a** or **b**).

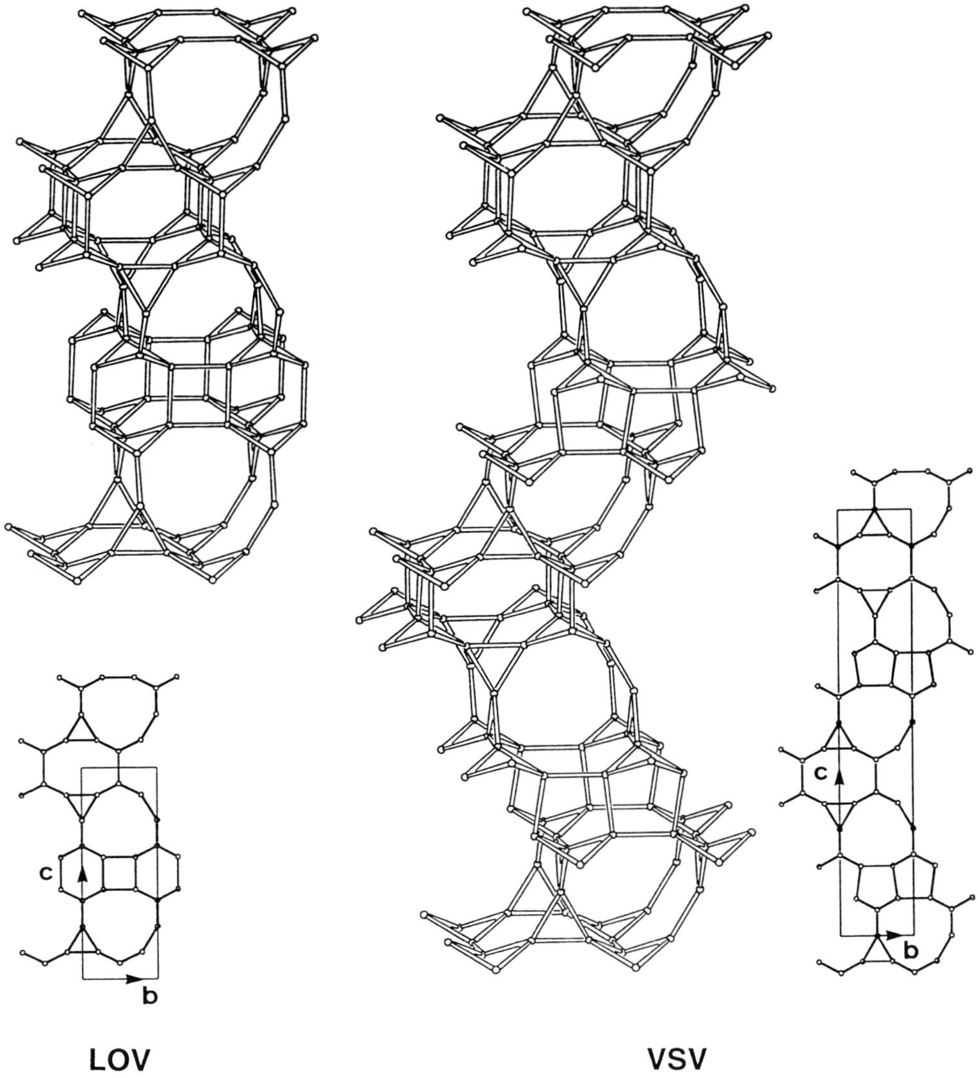

Figure 38. Three simplest ordered end-members of the lovdarite family. All three structures have been observed as pure single crystal material. The lateral shifts are added in Table 8 (page 165). In LOV and VSV the *ac* projection is equal to their *bc* projection.

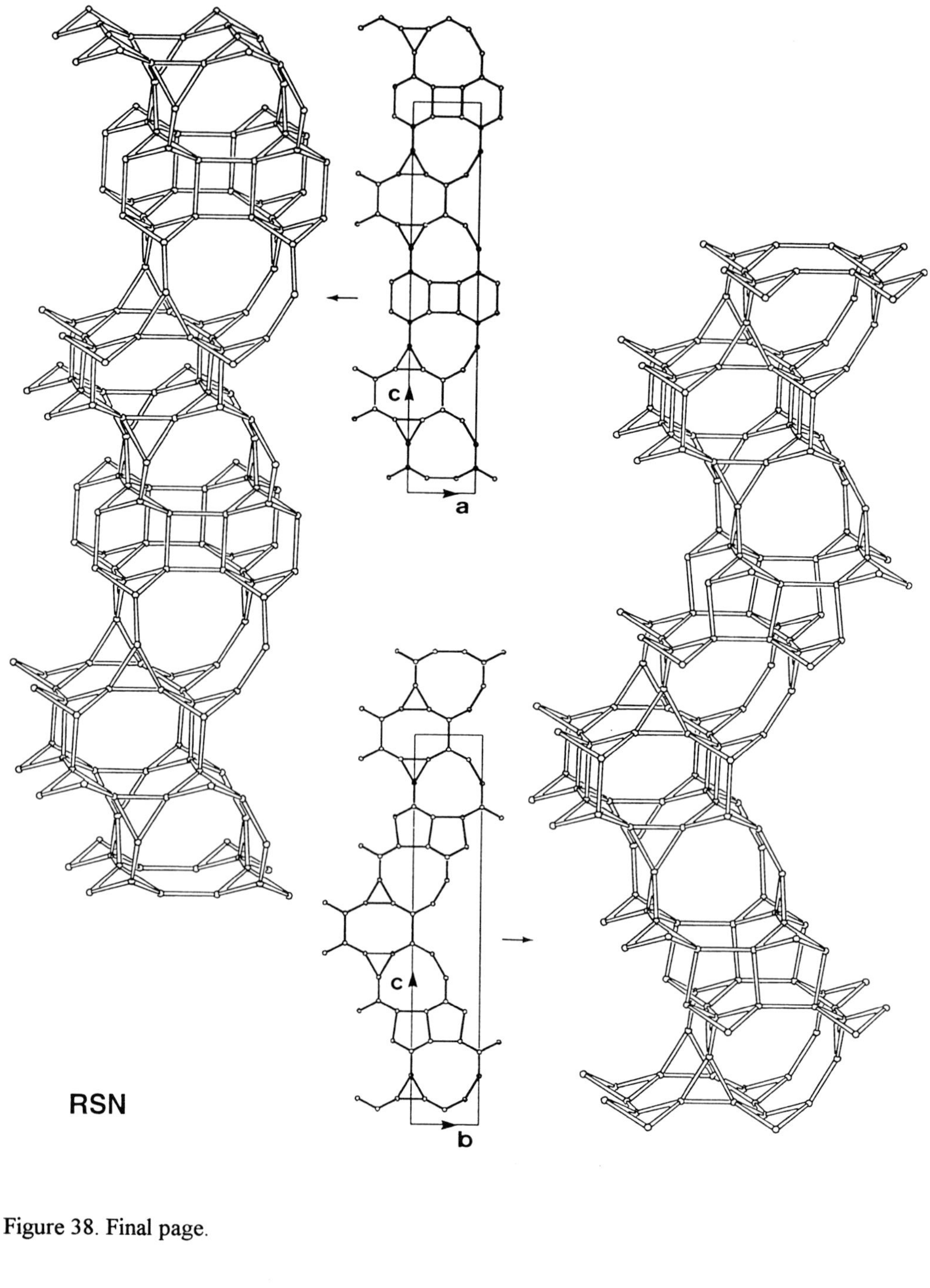

RSN

Figure 38. Final page.

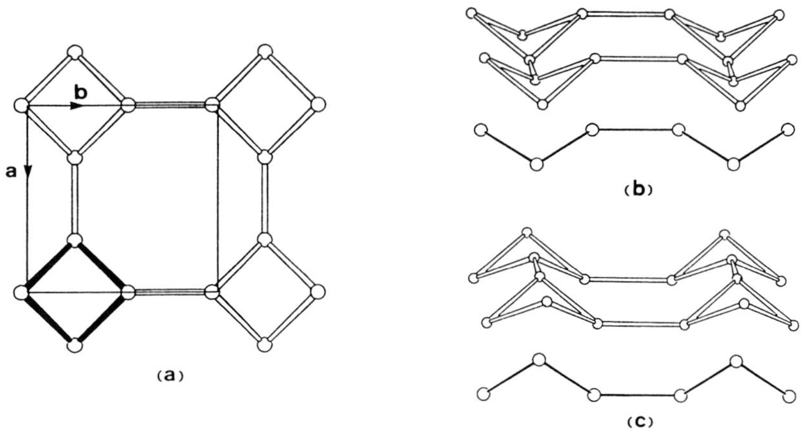

Figure 39. Basic layer, point group symmetry ($\overline{4}$)m2, in the montesommaite zeolites.
(a): Top view of the basic (001) layer composed of T4-rings (in bold) with UDUD directionality.
(b): Perspective view (top) and parallel projection (bottom) along [100].
(c): As (b), along [010].

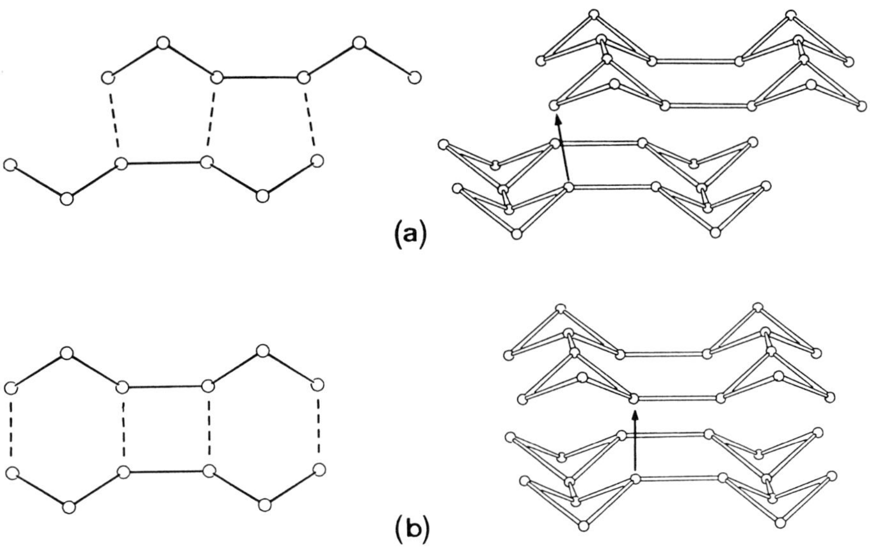

Figure 40. Parallel and perspective views of the two possible ways in which basic montesommatie layers, related by mirror planes, can be connected.
(a): The lateral shift between successive layers is (±) ½ (along **a** or **b**).
(b): The lateral shift is zero. This connection mode has not yet been observed.

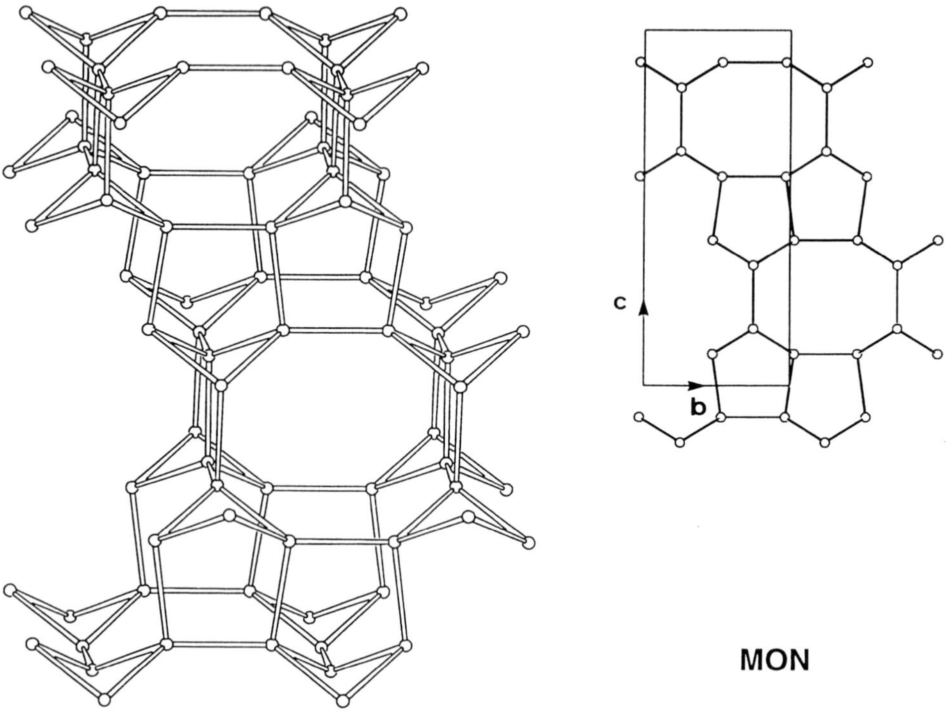

Figure 41. The end-member with structure type code MON has been observed as pure single crystal material. The lateral inter-layer shifts in MON are summarized in Table 8 (page 165).

The 4-rings in the layer shown in Figure 42(a) exhibit a directionality (UUDD) which is different from the 4-rings in the basic layers in lovdarite and montesommaite. In the basic layer in the zeolite with structure type WEI [34], a single T atom connects two 4-rings within the plane as illustrated in Figure 42(b). The basic layer can be constructed by connecting T10-units (consisting of two 4-rings + 2 single T atoms) after pure translations along **b** and **c**. The WEI structure, shown in Figure 42(d), is produced when identical neighbouring layers are connected after a lateral shift of (±)½**b** has been applied. The connection mode with lateral shift (±)½**c** is unlikely. Crystal data of and lateral shifts in WEI are summarized in Tables 7 and 8 on page 164 and 165, respectively.

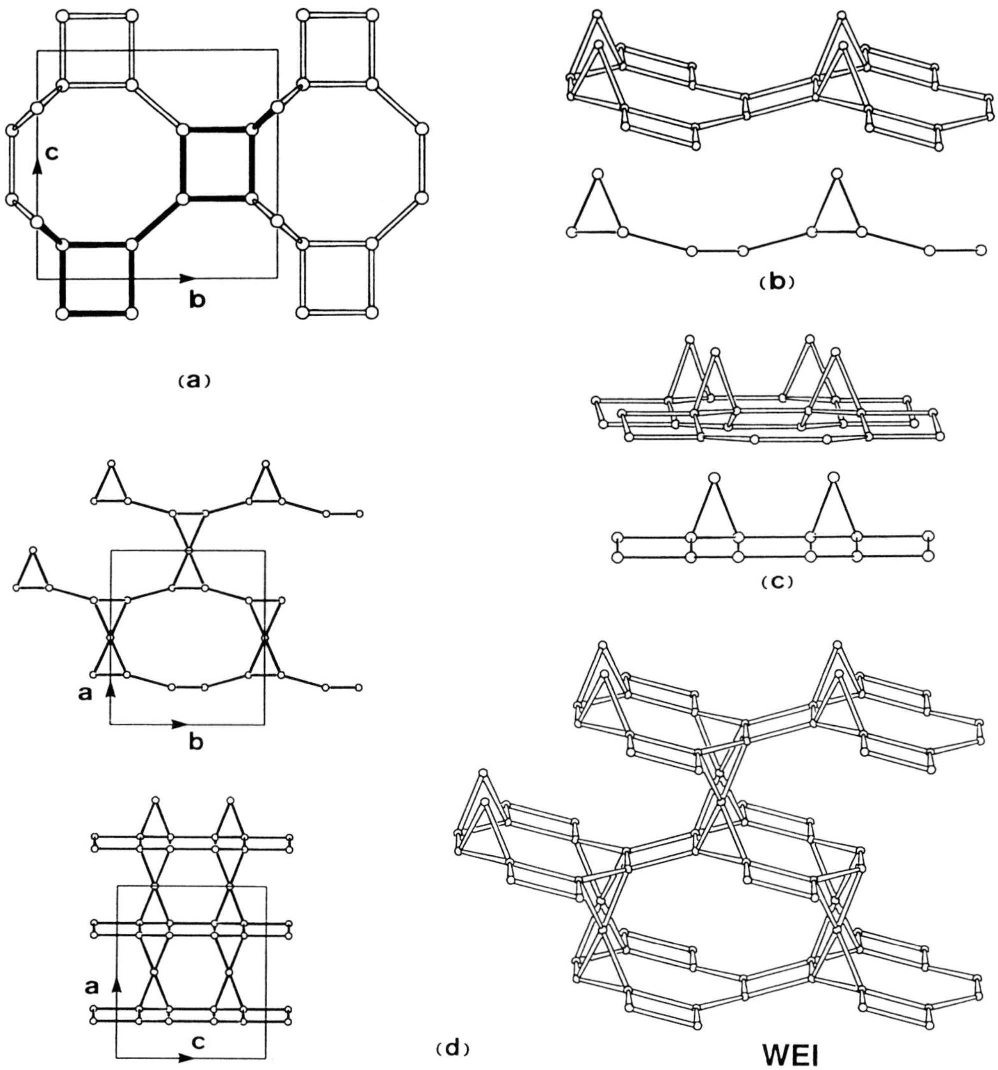

Figure 42. Basic (100) layer in WEI.
(a): Top view of the layer composed of T10-units (in bold).
(b): Perspective view (top) and parallel projection (bottom) along [001].
(c): As (b), down [010].
(d): Perspective view along [001] and parallel projections of the WEI structure illustrating the connection mode with lateral shift of ½ **b** (and zero along **c**).

Table 7
Crystal data of some end-members in zeolite families with disordered frameworks

		Cell dimensions (Å)					
Code	space group	a	b	c	β (°)	n*	Z#
FAU&	$R\bar{3}m$	17.215	17.215	42.167	120.0!	24**	144
EMT	$P6_3/mmc$	17.215	17.215	28.082	120.0!	16**	96
MFI	Pnma	20.090	19.738	13.142		12	96
MEL	$I\bar{4}m2$	20.270	20.270	13.459		12	96
ITE	Cmcm	20.753	9.804	20.009		16	64
RTH	C2/m	9.762	20.530	9.996	96.90	16	32
BE1	$P4_2/mmc$	12.81	12.81	13.00		16	32
BE2$	$P\bar{1}$	12.69	12.69	14.42	89.98!	16	32
BE3	P2/c	12.47	12.47	27.61	107.52	16	64
BE4@	$P4_122$	12.632	12.632	26.186		16	64
BE5	$P4_322$	12.632	12.632	26.186		16	64
CO1	Pmmm	not DLS optimized yet				14	56
CO2%	C2/m	22.684	13.373	12.553	69.48	14	56
CO3?	Pncm	21.08	13.26	12.33		14	56
LOV	$P4_2/mmc$	7.163	7.163	20.876		9	18
VSV	$I4_1/amd$	7.156	7.156	41.842		9	36
RSN	A2/m	7.158	7.155	41.826	90.00	9	36
MON	$I4_1/amd$	7.135	7.135	17.809		4	16
WEI‡	Cccm	11.803	10.278	9.989		10	20
RT1##	C2/m	14.098	13.670	7.431	102.42	12	24
RT2	$P2_1/m$	9.7	19.6	7.4	98.9	12	24
RT3	P4/nmm	19.4	19.4	7.4		12	48

* n = number of T atoms in the independent repeat unit.
\# Z = number of T atoms per unit cell.
& Rhombohedral cell in hexagonal setting.

Table 7 - Continued

** n = number of 6-rings per unit cell.
$ Topological symmetry of BE2 has been transformed from C2/c to $P\bar{1}$ with $\alpha = \beta =$ 107.0°.
! Value of angle γ.
@ BE4 corresponds to the structure with type code *BEA. BE4 and BE5 are enantiomorphs.
% CO2 corresponds to the structure with type code CON.
? For comparison reasons the standard space group Pmna has been transformed to Pncm.
‡ The length of half of the diagonals in the *bc* plane is 7.166 Å.
RT1 corresponds to the structure with type code RTE.

Table 8
Lateral shifts in some simple end-members of the *BEA, the SSZ-33, the lovdarite and the montesommaite family of zeolites. The lateral shifts are between successive (mirrored) basic layers stacked along the direction in which disorder occurs

end-member	lateral shift in fractions of the periods in the basic layer				
BE1	(0,0);	(0,0);	(0,0);...		
BE2*	(0,-⅓);	(0,-⅓);	(0,-⅓);...		
BE3*	(0,-⅓);	(-⅓,0);	(0,+⅓);	(-⅓,0);	(0,-⅓);.........
BE4	(-⅓,0);	(0,-⅓);	(+⅓,0);	(0,+⅓);	(-⅓,0);. ......
BE5	(+⅓,0);	(0,-⅓);	(-⅓,0);	(0,+⅓);	(+⅓,0);........
CO1	(0,0);	(0,0);	(0,0);...		
CO2*	(0,+⅓);	(0,+⅓);	(0,+⅓);...		
CO3*	(0,+⅓);	(0,-⅓);	(0,+⅓);...		
LOV	(0,0);	(0,0);	(0,0);...		
VSV	(0,½);	(½,0);	(0,½);	(½,0);	(0,½);..........
RSN	(0,½);	(0,0);	(0,½);	(0,0);	(0,½);..........
MON	(0,½);	(½,0);	(0,½);	(½,0);	(0,½);..........
WEI	(½,0);	(½,0);...			

* The same centrosymmetric structure is obtained by reversing the sign of all lateral shifts.

8.6 The decasil family of zeolites

The basic unit in the decasil family of framework structures is the decasil chain, Figure 43(a), formed by connecting T12-units (in bold) related by pure translations along the chain axis. As orientation sensitive indicator one of the 4-rings in the T12-unit is shaded. In the description of the decasil chain we did not use a T12-unit composed of two 6-rings connected through (oxygen) bridges T2-T3 and T6-T7 because in that case the shaded 4-ring is formed by connecting T12 units and the use of the shaded 4-ring as an orientation sensitive indicator is less clear. The numbered T atoms are used in describing the connection modes. Figure 43(b) shows a perspective and parallel view down the chain axis.

Neighbouring chains can be connected through oxygen bridges in several ways. Eight possibilities are depicted in Figure 44. Chains can be connected after pure translations. The resulting connection modes are numbered as 1, 4, 5 and 7. The connection modes 2 and 3 are generated when the translation is accompanied by a 180° rotation about the chain axis. The connection modes 6 and 8 are obtained in case the translation is followed by a +90° and -90° rotation about the chain axis, respectively. Connection modes 2 and 3, 5 and 7, and 6 and 8 are pairwise identical. The modes in each pair are related by a 180° rotation about an axis parallel or (except for pair 5 and 7) perpendicular to the connecting T-T bridges. The connection modes 5 and 7 are also related by a mirror operation parallel to the plane of the connected chains.

Only connection modes 5,6,7 and 8 have been observed up till now. Once the distribution of the connection modes in two dimensions is known the 3-dimensional framework is defined.

Three of the simplest ordered end-members of the decasil family are given in Figure 45. The topological symmetry and unit cell data are added in Table 7 (page 164). Only end-member RT1 has been observed as pure single crystal material. RT1 represents the framework structure with type code RTE. The parallel projections on the right-hand side in Figure 45 represent layers from which the RTn framework can be built: RT1 contains exclusively layers of type 1 (or of type 2); RT2 can be constructed from layers of type 1 and 2 or exclusively from type 3 layers; and RT3 is composed of type 3 layers related by a 180° rotation about [001]. The connection modes are given in Table 9.

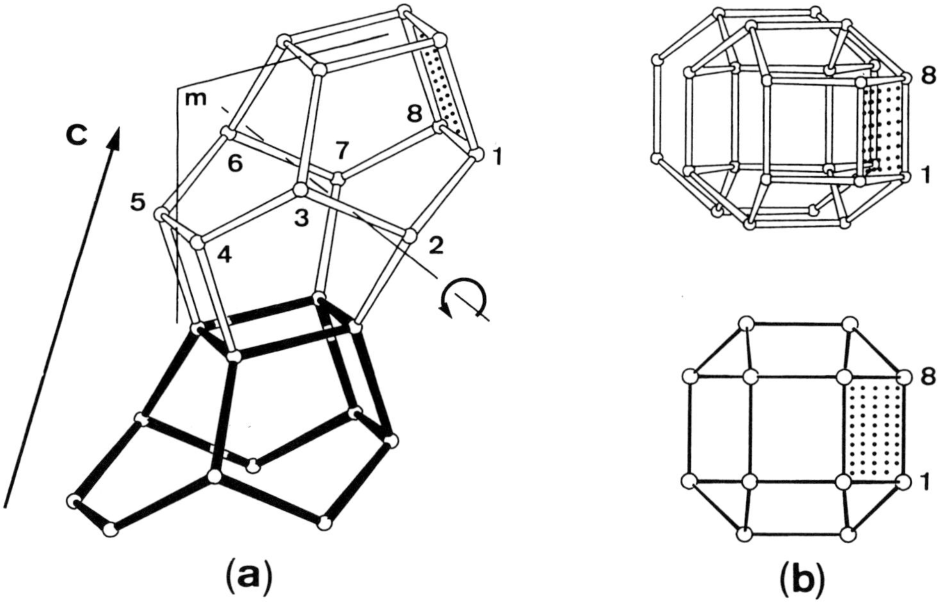

Figure 43. The basic chain in the decasil family of zeolites.
(a): The decasil chain composed of T12-units (in bold). The rod symmetry (2/m) is indicated. The numbered atoms are used in describing the connection modes.
(b): Perspective view (top) and parallel projection (bottom) along the chain axis. As orientation sensitive indicator one of the 4-rings is shaded.

Table 9
Connection modes between decasil chains stacked along **a** and **b**

end-member	sequence of the connection modes: (along a,......; along b,......;)
RT1*	(5,5,......;7,7,......)
RT2	(7,7,......;8,6,8,....)
RT3	(8,6,8,....;8,6,8,....)

* This is the end-member with structure type code RTE; the sequence of the connection modes is along (-**a** + **b**) and (**a** + **b**), respectively.

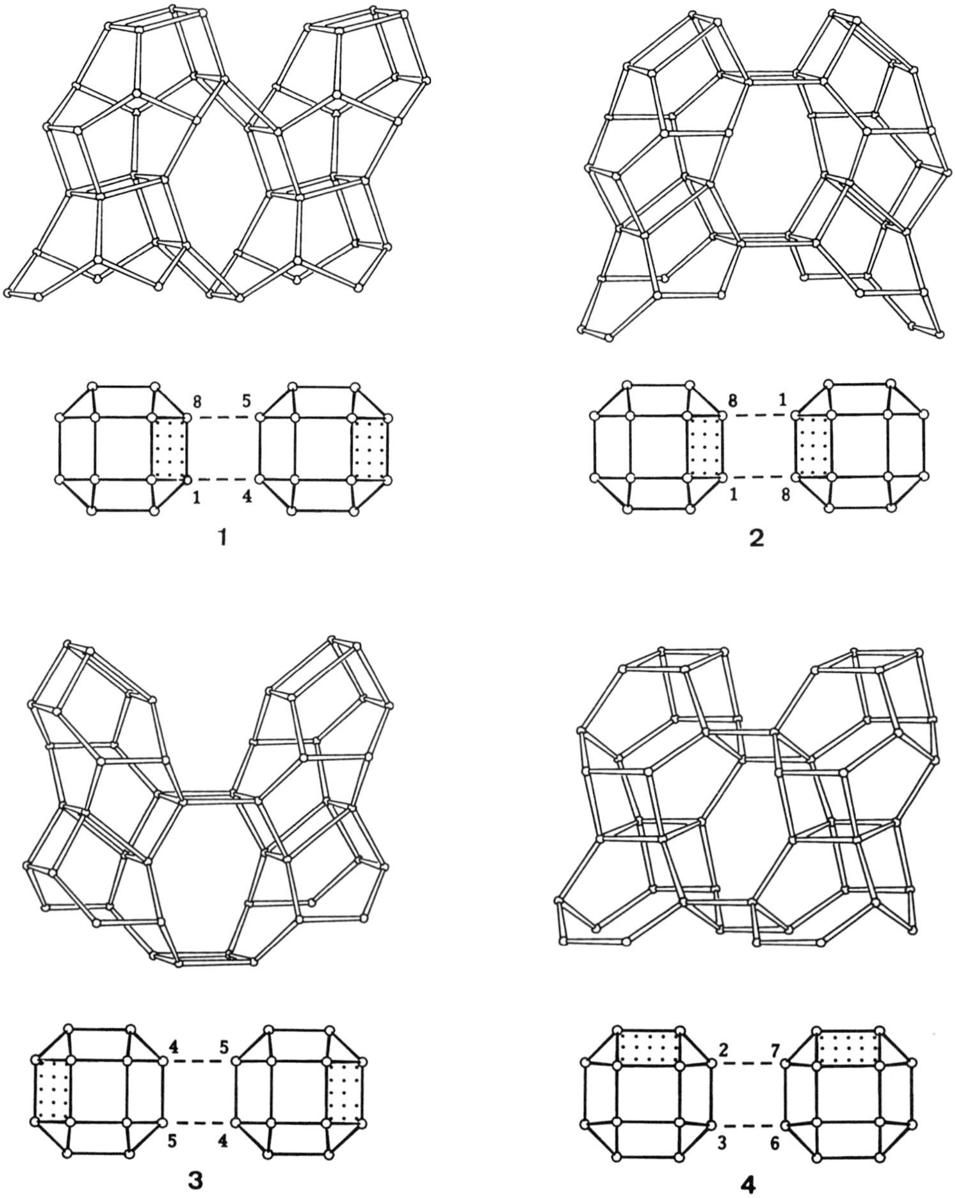

Figure 44. Connection modes in the decasil zeolites. Perspective drawings (top) are perpendicular to the chain axis. Parallel projections (bottom) are along the chain axis.

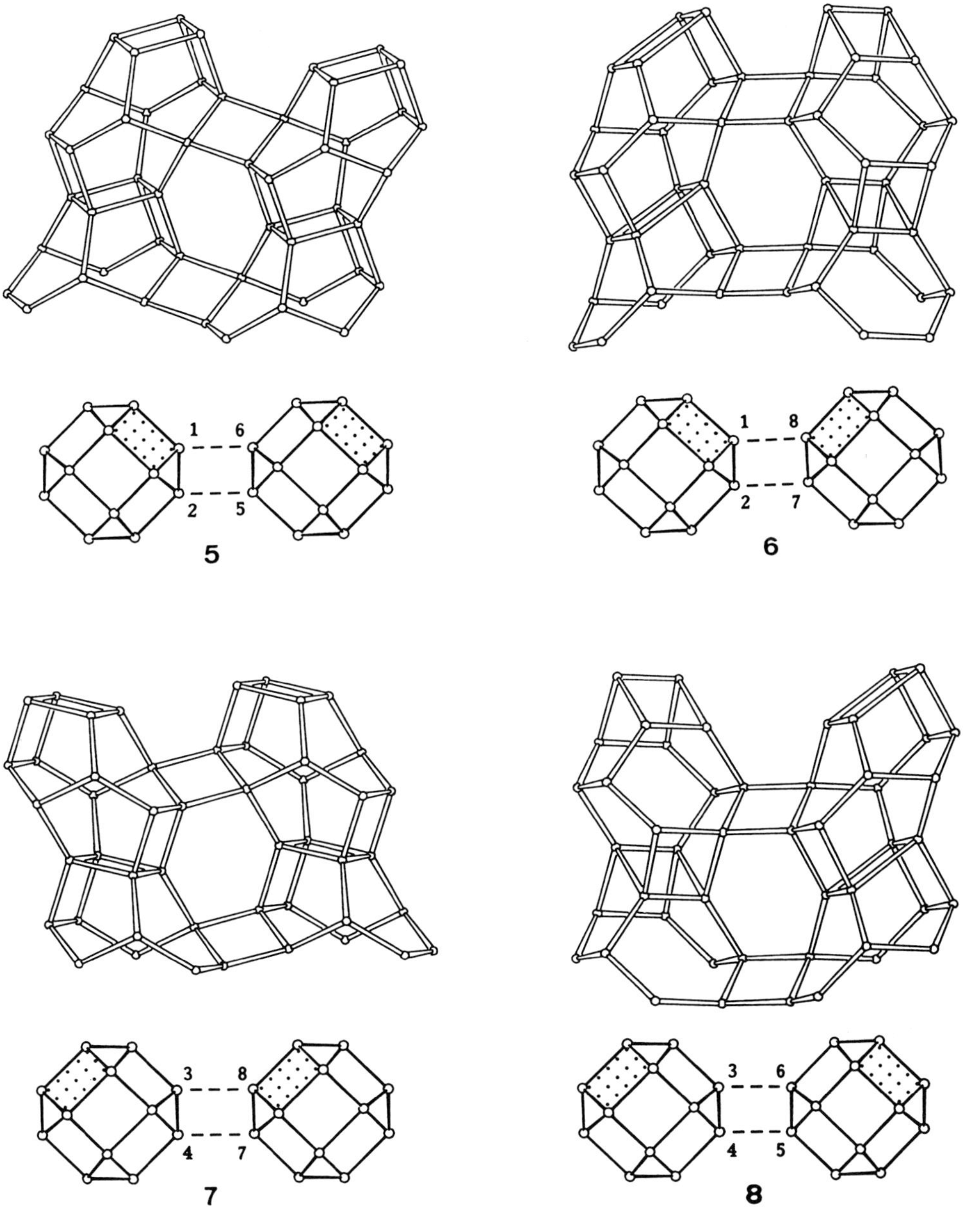

Figure 44. Final page. The T-atom numbering refers to the numbering in Figure 34. Only connection modes 5,6,7 and 8 have been observed up till now.

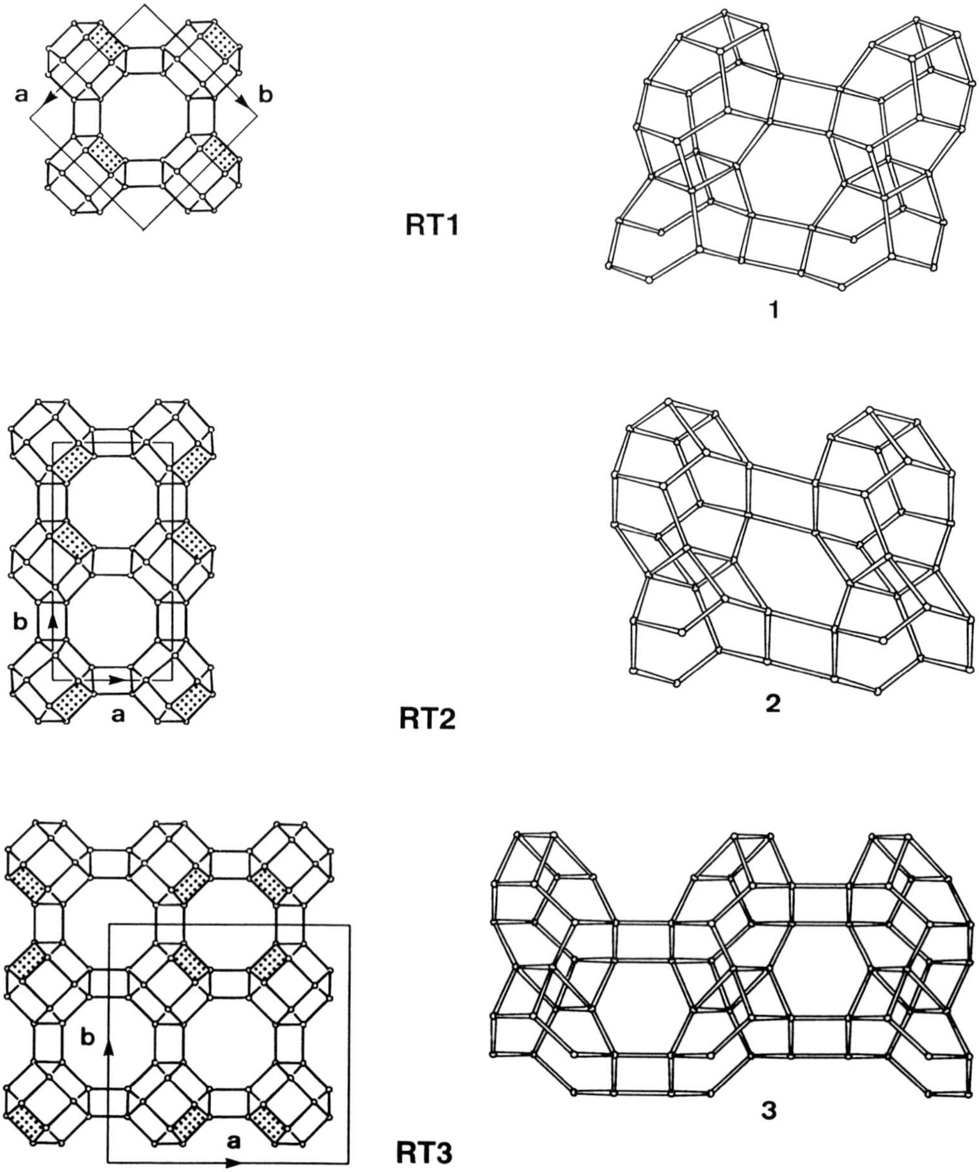

Figure 45. Three end-members of the decasil family. RT1 represents the structure with type code RTE. The parallel projections 1,2 and 3 are along [110], [010] and [010] in RT1, RT2 and RT3, respectively. The connection modes are summarized in Table 9.

REFERENCES

1. W.M. Meier, D.H. Olson and Ch. Baerlocher. Atlas of Zeolite Structure Types, 4th rev. ed., Elsevier, London, 1996.
2. W.M. Meier, in Molecular Sieves, Soc. of Chem. and Ind., London, 1968, pp. 10-27.
3. W.M. Meier and D.H. Olson, in Atlas of Zeolite Structure Types, 2nd rev. ed., Butterworths, London, 1987, p. 5.
4. J.V. Smith, Chem. Rev., 88 (1988) 149.
5. F. Liebau, in Structural Chemistry of Silicates, Springer-Verlag, Berlin, Heidelberg, 1985.
6. G. Gottardi and E. Galli, in Natural Zeolites, Volume 18, Minerals and Rocks, Springer-Verlag, Berlin, Heidelberg, New York, Tokyo, 1987.
7. W.M. Meier, in Natural Zeolites, Occurrence, Properties and Use (Eds. L.B. Sand and F.A. Mumpton) Pergamon Press, Oxford, 1978, pp. 99-103.
8. H. van Koningsveld, in Introduction to Zeolite Science and Practice (Eds. H. van Bekkum, E.M. Flanigen and J.C. Jansen) Elsevier, Amsterdam; Stud. Surf. Sci. Catal., 58 (1991) 35.
9. H. van Koningsveld, Zeolites, 12 (1992) 114.
10. D.E. Akporiaye, Z. Kristallogr., 188 (1989) 103.
11. C.K. Johnson. ORTEP. Report ORNL-3794, revised June 1970. Oak Ridge National Laboratory, Tennessee, USA.
12. G.W. Noble, P.A. Wright and Å. Kvick, J. Chem. Soc. Dalton Trans., (1997), in press.
13. J.M. Newsam, M.M.J. Treacy, D.E.W. Vaughan, K.G. Strohmaier and W.J. Mortier, J. Chem. Soc., Chem. Commun., (1989) 493.
14. S.L. Burkett and M.E. Davis, Microporous Mater., 1 (1993) 265.
15. Ch. Baerlocher, L.B. McCusker and R. Chiapetta, Microporous Mater., 2 (1994) 269.
16. M.M.J. Treacy, D.E.W Vaughan, K.G. Strohmaier and J.M. Newsam, Proc. R. Soc. Lond. A, 452 (1996) 813.
17. M.E. Camblor, A. Corma, P. Lightfoot, L.A. Villaescusa and P.A. Wright, Angew. Chem. Int. Ed. Engl., 36 (1997) 2659.
18. S. Vortmann, B. Marler, H. Gies and P. Daniels, Microporous Mater., 4 (1995) 111.
19. M.M.J. Treacy and J.M. Newsam, Nature, 332 (1988) 249.
20. J.M. Newsam, M.M.J. Treacy, W.T. Koetsier and C.B. de Gruyter, Proc. R. Soc. Lond. A, 420 (1988) 375.
21. J.B. Higgins, R.B. La Pierre, J.L. Schlenker, A.C. Rohrman, J.D. Wood, G.T. Kerr and W.J. Rohrbaugh, Zeolites, 8 (1988) 446.
22. R.F. Lobo, M. Pan, I. Chan, H.X. Li R.C. Medrud, S.I. Zones, P.A. Crozier and M.E. Davis, Science, 262 (1993) 1543.
23. R.F. Lobo, M. Pan, I. Chan, R.C. Medrud, S.I. Zones, P.A. Crozier and M.E. Davis, J. Phys. Chem., 98 (1994) 12040.
24. R.F. Lobo and M.E. Davis, J. Am. Chem. Soc., 117 (1995) 3764.
25. S. Merlino, Acta Cryst., A37 (suppl.) (1981) C189.
26. C. Röhrig and H. Gies, Angew. Chem. Int. Ed. Engl., 34 (1995) 63.
27. J. Annen, M.E. Davis, J.B. Higgins and J.L. Schlenker, J. Chem. Soc., Chem. Commun., (1991) 1175.
28. R.C. Rouse, P.J. Dunn, J.D. Grice, J.L. Schlenker and J.B. Higgins, Am. Mineral., 75 (1990) 1415.

29. B. Marler, A. Grünewald-Lüke and H. Gies, Zeolites, 15 (1995) 388.
30. A. Grünewald-Lüke and H. Gies, Microporous Mater., 3 (1994) 159.
31. A. Simmen, L.B. McCusker. Ch. Baerlocher and W.M. Meier, Zeolites, 11 (1991) 654.
32. K.P. Lillerud and D. Akporiaye, in Zeolites and Related Microporous Materials (Eds. J. Weitkamp, H.G. Karge, H. Pfeifer and W. Hoelderich) Elsevier, Amsterdam; Stud. Surf. Sci. Catal., 84 (1994) 543.
33. J.W. Richardson, Jr., J.J. Pluth and J.V. Smith, Naturwissenschaften, 76 (1989) 467.
34. F. Walter, Eur. J. Mineral., 4 (1992) 1275.

REGISTER

Code	Section	Table	Figure
ABW	3.2;3.3	1	7,8
AEI	7	6	20
AEL	5.2;5.3	3	13,14
AET	5.2;5.3	3	13,14,15
AWW	7	6	21
AFG	7	5	18
AFI	5.2;5.3	3	13,14
AFO	5.2;5.3	3	13,14
AFT	7	5	18
AFX	7	5	18
AHT	5.2;5.3	3	13,14,15
APC	5.2	3	13
APD	5.2	3	13
ATN	3.2;3.3	1	7,8
ATO	3.2;3.3	1	7,8
ATS	3.2	1	7
ATT	2.1;4.2;4.3;5.2	2,3	10,13
ATV	5.2,5.3	3	13,14,15
*BEA	1;2.3;8.3	7,8	30,32
BIK	3.2;3.3	1	7,8
CAN	2.2;3.2;3.3;7	1,5	3,7,8,18
CAS	3.2;3.3	1	7,8
CFI	3.2;3.3	1	7,8
CHA	7	5	18,19,20
CON	1;2.3;8.4	7,8	35
DAC	2.1;3.2;3.3;4.2 4.3	1,2	7,8,10,11
EAB	7	5	18
EDI	6	4	17
EMT	1;2.3;7;8.1	6,7	22,24
EPI	3.2;3.3;4.2;4.3	1	7,8,10,11
ERI	7	5	18
EUO	3.2	1	7
FAU	1;2.3;7;8.1	6,7	22,24
FER	4.2;4.3	2	10,11
GIS	5.2	3	13
GME	2.2;5.2	3,5	13,18
ITE	1;2.3;8.2	7	29
JBW	2.1;3.1;3.2;3.3 4.2;4.3;5.2	1,2	7,8,10
KFI	7	6	20
LEV	7	5	18
LIO	7	5	18
LOS	7	5	18
LOV	1;2.3;8.5	7,8	38
LTA	7	6	21
LTL	4.2;4.3	2	10
MAZ	4.2;4.3	2	10
MEL	1;2.3;8.2	7	27
MER	5.2	3	13
MFI	1;2.3;8.2	7	27
MFS	4.2;4.3	2	10
MON	1;2.3;8.5	7,8	41
MOR	4.2;4.3	2	10,11
MTT	3.2	1	7
MTW	3.2;3.3	1	7,8
NAT	6	4	17
NES	3.2	1	7
NON	3.2	1	7
OFF	2.2;4.2;4.3	2,5	10,18
OSI	3.2	1	7
PHI	5.2	3	13
RHO	7	6	20,21
RSN	1;2.3;8.5	7,8	38
RTE	1;2.3;8.6	7,9	45
RTH	1;2.3;8.2	7	29
SAT	7	5	18
SOD	7	5	3,18
THO	6	4	17
TON	3.2;3.3	1	7,8
VET	3.2;3.3	1	7
VFI	5.2;5.3	3	13,14
VSV	1;2.3;8.5	7,8	38
WEI	1;2.3;8.5	7,8	42
YUG	5.2	3	13

Studies in Surface Science and Catalysis 137
H. van Bekkum, E.M. Flanigen, P.A. Jacobs and J.C. Jansen (Editors)
© 2001 Elsevier Science B.V. All rights reserved.

Chapter 5A

THE PREPARATION OF OXIDE MOLECULAR SIEVES

A. Synthesis of zeolites J.C. Jansen

B. Synthesis of $AlPO_4$ -based molecular sieves S.T. Wilson

A. Synthesis of zeolites

J.C. Jansen
Delft University of Technology,
Laboratory of Applied Organic Chemistry and Catalysis,
Julianalaan 136, 2628 BL Delft, The Netherlands

I. INTRODUCTION

a. General

Nature provided mankind with zeolites (ref. 1). Massive zeolite deposits have been discovered at many places in the world (ref. 2). The occurrence of natural zeolites can be assigned to certain geological environments or hydrological systems (refs. 3,4).
Natural zeolites generally form by reaction of mineralizing aqueous solutions with solid aluminosilicates. The main synthesis parameters are:
(i) the composition of the host rock and interstitial solutions; pH ~ 10
(ii) the time; thousands of years
(iii) the temperature; often $< 100^0$ C
The first systematic studies on zeolite syntheses could thus be guided by the geological and mineralogical findings of the natural species (ref. 5). From 1946 on many additional zeolite types without a natural counterpart have been synthesized (ref. 6). The evolution in the preparation of one of the most studied zeolites is illustrated in Figure 1 by the number of papers and patents on the material denoted as ZSM-5 (ref. 7 and Section XII of this chapter).

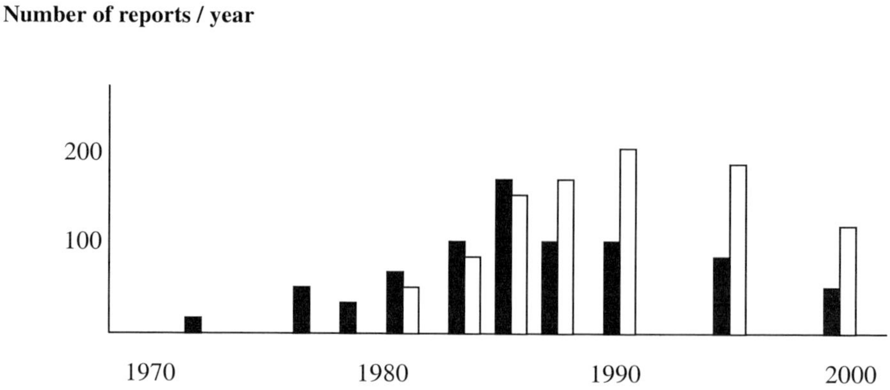

Fig. 1. The annual number of papers (□) and patents (■) on the synthesis of zeolite ZSM-5 since the first publication in 1972.

Throughout the last six decades molecular sieves were mainly prepared by reactive crystallization or precipitation from an aqueous mixture of reagents at $6 < pH < 14$ and temperatures between 100-200 °C. As shown in Scheme 1 a relatively large effort is needed in the optimization of the synthesis mixture,

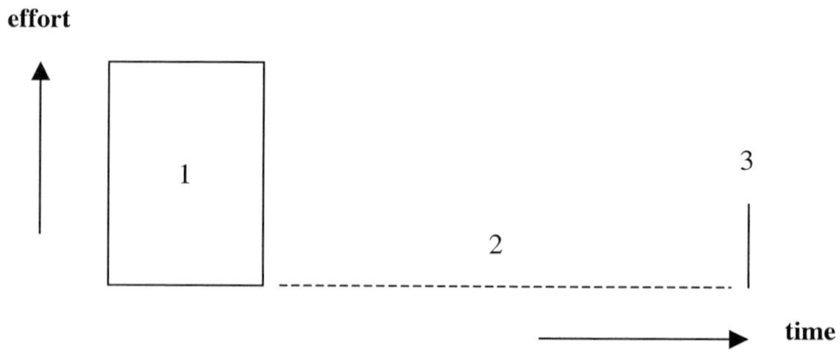

Scheme 1. The effort in the optimization of the synthesis mixture (1), the reaction (2) and the isolation (3) versus time.

whereafter the hydrothermal reaction process (2) runs autoclaved for a few minutes, days or weeks without intervention. Isolation (3) of the crystalline material is a simple final step in the synthesis procedure.

The zeolite synthesis field is not only extended and refined by useful data from modern zeolite characterization and application techniques but also by the interfacing activities with physical, chemical and mathematical science, see Scheme 2.

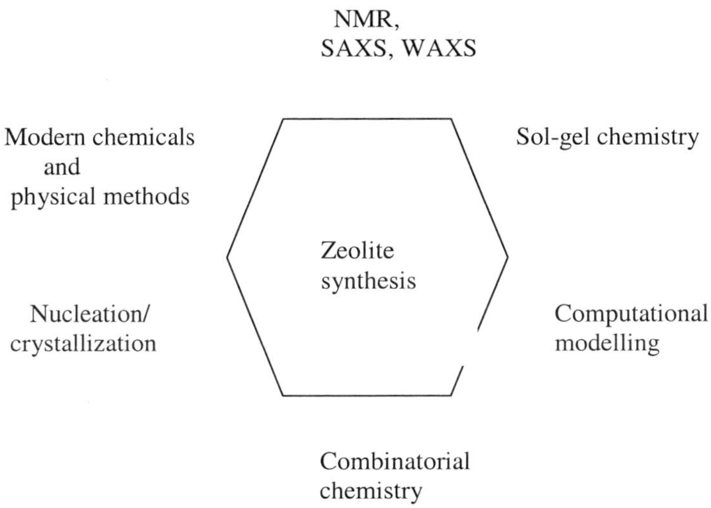

Scheme 2. Areas of chemical, physical and mathematical science interfacing the zeolite preparation field.

Studies in the sol-gel chemistry and NMR analysis area have contributed substantially to the knowledge of the hydrothermal reaction process. A new approach, combinatorial chemistry is a recently developed technique to screen and optimize with a high throughput a zeolite synthesis procedure (ref.12).

Papers and reviews regarding subjects within the different areas which are mentioned in Scheme 2 and which are of interest for zeolite synthesis are given in Table 1.

Besides the annually new zeolite preparations the extensive exploratory efforts of "zeolite scientists" in the last four decades has resulted in the synthesis of porous materials like the $AlPO_4$ -group (part B of this chapter), the metal-sulfides (ref. 30) and the clathrasils (ref. 31). Accordingly, zeolite synthesis appears to remain a promising area for future research. The crystallinity of different synthesis products is well illustrated in Plate 1. The morphology and forms of the crystals give a first indication of the type of zeolite present and the purity of the product.

Table 1. Examples of subjects from areas of physical, chemical and mathematical science which delivered contributions to the knowledge of the zeolite synthesis process together with references.

Area	Subject	Reference
Sol-gel chemistry	Hydrolysis and condensation of silicates	8, 9
	The sol-gel process	10
NMR	Structure of (alumino)silicate-clusters in solution	11a
SAXS, WAXS	Determination of nano-sized particles	11b
Combinatorial chemistry	Efficient routes to synthesise zeolite materials	12
Computational modelling	Lattice energy calculation	13
	Local interactions, template-framework	14
Modern chemical and physical methods	Alkoxides as reagents	15
	Fluorides as reagents and mineralizing agents	16
	Gravity - reduced	17
	- elevated	18
	CVD (chemical vapour deposition)	19
	Microwave	20
Nucleation/ crystallization theory- practice	Mathematical analyses of zeolite crystallization. A review.	21
	Are the general laws of crystal growth applicable to zeolite synthesis	22
Zeolite Characterization	ZSM-5/-11 intergrowth	23
Application	Catalysis	
	- The catalytic site activity	24-26
	- The catalytic properties and the crystal size	27

b. *This chapter*

In this part of the chapter the preparation of two subgroups of the micro-porous tectosilicates (see Chapter 3) i.e. the aluminosilicates and silicates, both including the clathrasils, will be presented. The division between aluminosilicates and silicates is often discussed on Al-poor rather than Al-free level (ref. 30). The aluminosilicates, starting from Si/Al ratio 1 up to e.g. Si/Al ratio of 10000, do reveal the presence of Al in synthesis, in characterization as well as in application, see Fig. 2 (ref. 31).

The Al-poor zeolites show no, at least no detectible, Al-dependent behaviour and are therefore, together with the Al-free materials, denoted as silicates.

The presence of aluminum, the guest-host interaction and the nucleation and crystallization all contribute to the synthesis events which are chronologically described in Sections II to VII of this chapter.

Section VIII is focussed on the reaction parameters.

In Sections IX and X the silicates and clathrasils are presented.

Examples of research syntheses performed with certain procedures and/or mixture compositions are listed in Section XI.

Sections XII and XIII contain literature sources on zeolite preparations and the references, respectively.

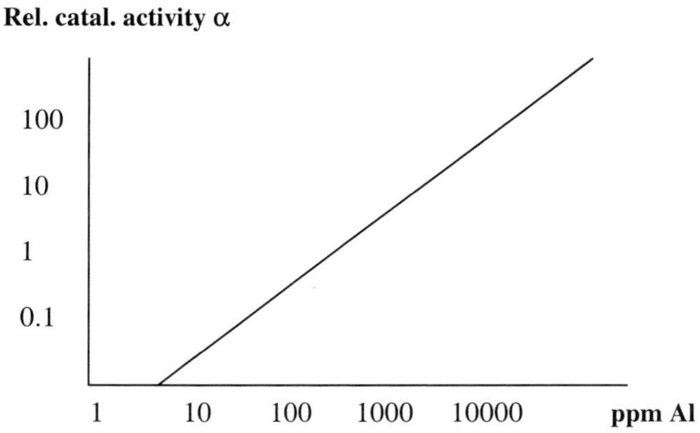

Fig. 2. The relative catalytic activity of H-ZSM-5 versus Al content on ppm scale (ref. 31).

Plate 1. The crystalline nature of zeolites. a) Single crystals of zeolite A and b) and c) of analcime and of natrolite, respectively. d) A batch of zeolite L, e) typical needle aggregates of zeolite

mordenite and f) of Nu-10.

II. PREPARATION

a. Reactants

The chemical sources which are in principle needed for zeolite syntheses are given in Table 2.

Table 2. Chemical sources and their function in zeolite synthesis.

Sources	Function(s)
SiO_2	Primary building unit(s) of the framework
AlO_2^-	Origin of framework charge
OH^-	Mineralizer, guest molecule
Alkali cation, template	Counterion of AlO_2^-, guest molecule
Water	Solvent, guest molecule

Within each type of source a variety of chemicals (ref. 32), has been used as the differences in physical nature and chemical impurities strongly influence the zeolite synthesis kinetics (ref. 34) and sometimes the properties as catalysts (refs. 24-27).

Data on the specifications of regularly used chemical sources are given in the following survey.

- SiO_2 -sources

Recent synthesis papers of the Proceedings of the International Zeolite Conferences (refs. 34-37) and of other zeolite conferences (refs. 38-40) reveal that for laboratory scale particular Si-sources are often used, see Table 3a.

Depending upon the particular synthesis a certain Si-source might favour a specific crystallization. For instance, the Aerosil 200 product can be readily dissolved compared to the Optipur and Gold label material because of the difference in particle size, see Figure 3. As the rate of dissolution can influence the rate of nucleation and crystallization (ref. 41) the product formation can be affected. At the same time the Al and other potentially Si replacing impurities are more than 10000 times higher in the Aerosil 200 product compared to the Optipur and Gold label materials.

The influence of impurities can change the crystal form (ref. 42) and the chemical properties (refs. 24-26).

Table 3. Specifications and the suppliers of recent, regular used sources, and high-pure, Si-, and Al-sources.

Si-source (a) Al-source (b)	Specifications Phys.	 Chem. impurities (ppm)	Reference manufacturer
(a)			
Silicon compounds			
$Si(OCH_3)_4$			
Tetramethyl orthosilicate (TMeOS)	liquid	Na,Ca < .5	Merck
$Si(OC_2H_5)_4$			
Tetraethyl orthosilicate (TEOS)	liquid	Al,Pt < .2	
$Na_2SiO_3 .9H_2O$		Al < 200 Fe < 120 Ti < 60	"N" Philidelphia Quartz Co.
Water glass		heavy metals < 50	
Colloidal silica			
Ludox-AS-40 sol		Al < 500	DuPont de Nemours
SiO_2 40 wt %		Zr	
NH_4^+ (counterion)		Fe < 50	
Ludox-HS-40		Ti	
SiO_2 40 wt %		B < 10	
Na^+ (counterion)			
Fumed silica		Al < 10	
Aerosil-200	Dp ~ 12 nm	Fe < .6	Degussa
CAB-O-SIL M-5		Ti < 10	BDH
* Silica			
Optipur	Dp ~ 200 mm	Al < .001	Merck
Gold label	Dp ~ 800 mm	Fe < 0.01	Aldrich

Table 3, continued

(b)

- $NaAlO_2$			Riedel de Hahn
Na_2O 54%		Fe < 4	Carlo Erba
Sodium aluminate			BDH Ltd.
- AlOOH			
Pseudo-boehmite	Dp ~ mm		
Al_2O_3 70%		Fe < 4	Vista
H_2O 30%		Ti < 40	
Catapal-B			
- $Al(OH)_3$			
Gibbsite		Fe < 3	Merck
- $Al(NO_3)_3.9H_2O$			
- Al_2O_3	Dp ~ nm		
* Aluminiumoxide (Ultrex)		Fe < 0.01	Baker

Therefore, a careful choice of the reactants is needed. The high grade Si- alkoxides of which even double alkoxides like -Si-O-Al- are available (ref. 43) do not have the above discussed disadvantages, except for the rate of hydrolysis of the alkoxide groups.

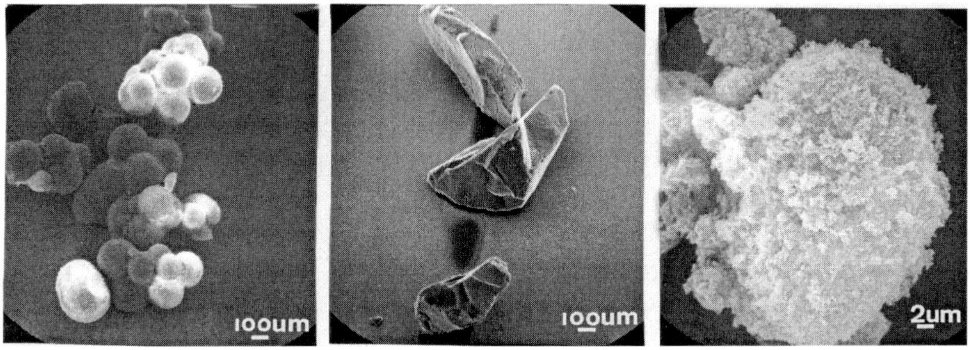

Fig. 3. SEM photographs of (a) Optipur, (b) Gold label, and (c) Aerosil 200.

- AlO_2^- sources

Often used Al-sources, collected from the same references as given for the Si-sources, are listed in Table 3b together with the main chemical impurity. Though the very pure Al_2O_3 product consists of small particles it is not easily dissolved.

- Alkali cation/template

The inorganic cations in the zeolite synthesis are mainly alkaline or ammonium ions. The organic cations/templates used may be divided in charged and neutral molecules containing functional atoms or groups. The large number of organic molecules used in zeolite synthesis is extensively listed in several publications (refs. 44, 45) together with the specific zeolite product formed. To illustrate in general the variation in organic template molecules some of the more common templates are listed in Table 4.

- OH^-

Most zeolite syntheses are performed under basic conditions using OH^- as a mineralizing agent. A second agent is F^- (refs. 16 and 46) of which the different nature compared to OH^- will be discussed in the section on reaction parameters.

Both anions are the counterion of the inorganic or organic cations used for the syntheses. Depending upon the quality of the mineralizing agent impurities such as Al^{3+} and Fe^{3+} are present at ppm scale.

- The overall reactant mixture

In general the chemical behaviour of impurities like Fe^{3+} and Ti^{4+} are of minor importance compared to Al^{3+} in high Si/Al zeolites in the heterogeneous catalysis when based on Bronsted activity. However, in the case of an all silica zeolite, or modified zeolites like B-ZSM-5, Fe-ZSM-5 and Ga-ZSM-5 traces of Al^{3+} from reactants as given in Table 3 may play an unexpected dominant role in the Bronsted activity (refs. 24-26). Extensive information on this point is given in Chapter 6 on the modification of zeolites. Another example of the influence of impurities from reactants is K^+. The crystallization time can be retarded by factors when K^+ is present in the synthesis of e.g. zeolite Na-A or Na-ZSM-5 (refs. 47, 48).

Impurities like trivalent metal ions sometimes change physical conditions in the reaction mixture indicated by the crystal form or morphology (ref. 42).

Table 4. Type of organic templates, functional atoms/groups and references.

Organic template	Functional atom/group	Ref.	Organic template	Functional atom/group	Ref
amine	$-N<$	49	penta-erythritol	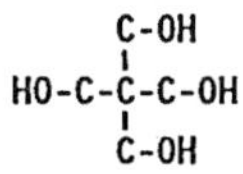	65
	 n=4,5	50	amine + alcohol	$>N-(C_n-OH)_x$ n=2,3 x=1–3	66
		51	ammonium + alcohol	$-N^+-C_n-OH$ n=2	67
di-amine	$>N-C_n-N<$ 3≤n≤9	52	acetal	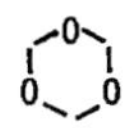	68
	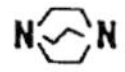	53	amine + ether	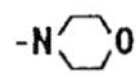	69
ammonium	$-N^+-$	54	N-oxide + ammonium	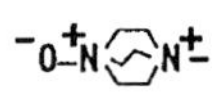	70
	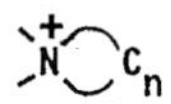 n=4,5	55	phosphonium	$-P^+-$	71
	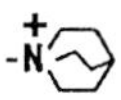	56		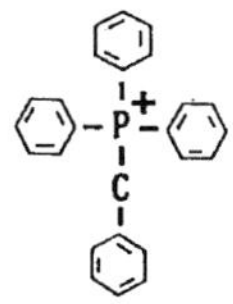	72
	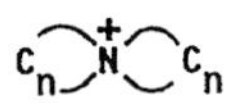 n=4,5	57	di-phosphonium	$-P^+-C_6-P^+-$	73
di-ammonium	$-N^+-C_n-N^+-$ 3≤n≤9	58			
	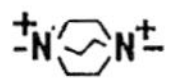	59			
tri-ammonium	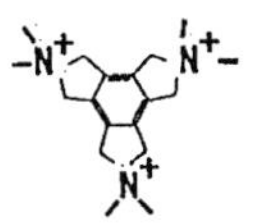	60			
amine + ammonium	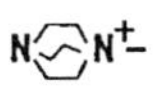	61			
alcohol	C_n-OH n=1–6	62			
di-ol	$HO-C_n-OH$ n=2–6	63			
tri-ol	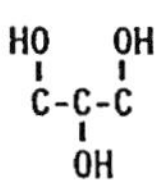	64			

b. The reaction vessel and hydrothermal conditions

Depending upon the reaction temperature chosen, mainly between 60 °C and 300 °C, reaction vessels can vary as shown in Table 5. Various Teflon inserted autoclaves, see Fig. 4a, together with relatively low priced (stainless steel) and high priced (reenforced polyether-ether-ketone) autoclaves are shown in Fig. 4b and 4c.

To follow the course of the events taking place in the synthesis mixture, a lookthrough autoclave, see Fig. 4d, can be used. It was concluded, using such experimental conditions, that nucleation and crystallization of zeolite ZSM-5 occurred *on* and *in*, respectively, gelspheres of about 2 mm which were not present at the starting clear solution (ref. 42).

Another possibility to monitor synthesis events *in situ* is the application of IR internal reflectance using a crystal (TlBr) embedded in an autoclave, as shown in Fig. 5 (refs. 74, 75).

Table 5. Regular used lab-scale reaction vessels, the typical impurities and temperature range.

Reaction vessel	Volume	Impurity	Temperature
Plastic bottle	< 1 l	Zn	< 100 °C
Stainless steel	< 5 l	Fe ,Cr	< 200 °C
Stainless steel + teflon lining	< 2 l	nuclei of preceeding synthesis	< 200 °C
Quartz autoclave	< 5 ml	Si	< 200 °C

The autoclaves must be filled between 30 and 70 vol % in the case of an aqueous reaction mixture between 100 and 200 °C to maintain a liquid phase (ref. 76).

Cleaning of the reaction vessels can be considered in some cases, e.g. the teflon-lined autoclaves. As memory effects caused by nuclei of preceeding syntheses in cavities of the teflon wall can be encountered in subsequent experiments it is important to clean the vessel with either HF and water at room temperature or NaOH and water at the reaction temperature.

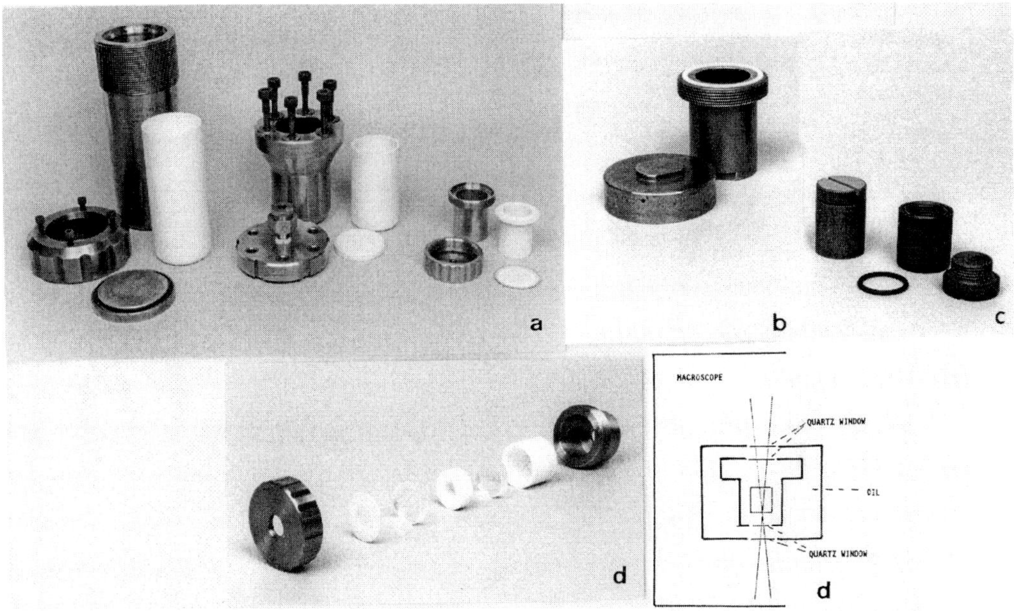

Fig. 4. Different autoclaves for laboratory use. a) Teflon lined autoclaves up to 1000 ml, b) stainless steel autoclave of 25 ml, c) "Arlon" (polyether ether ketone [PEEK]) reenforced with carbon fiber or glass fiber) autoclaves and d) stainless steel look through autoclave with quartz windows and Teflon inserts in exploded view together with a schematic drawing of the experimental set up.

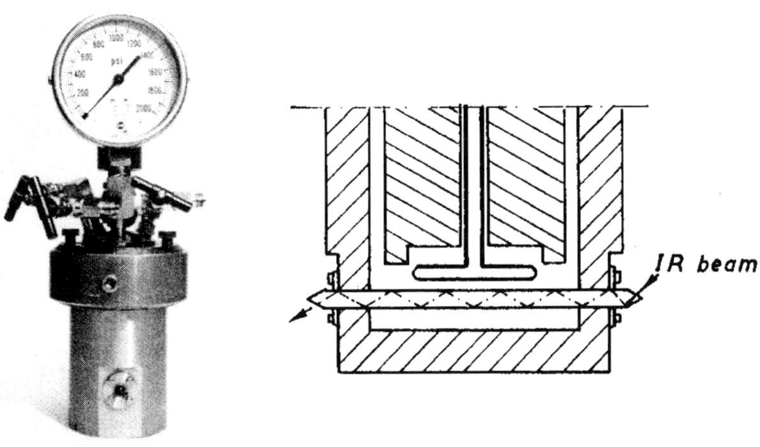

Fig. 5. Parr mini autoclave. IR internal reflection via a crystal embedded in the autoclave makes monitoring of zeolite synthesis events possible (ref. 74).

III. ZEOLITE PRODUCT VERSUS THE SYNTHESIS MIXTURE

a. *Two synthesis examples*

The fine-tuning and differences in the preparation of each zeolite type is too complex to be discussed in this introduction on the synthesis of zeolites. In specific parts of the chapter, however, is chosen for a more detailed presentation of the synthesis of two substantial different zeolite types, i.e. zeolite Na-A and zeolite TPA-ZSM-5. These two zeolites present roughly all the groups in which zeolite types are divided (ref. 77). The synthesis mixtures and chemical and physical properties of both zeolites are given in Table 6.

Table 6. The synthesis mixtures, physical and chemical properties of zeolites Na-A and TPA-ZSM-5 (refs. 78-80).

Na-A		TPA-ZSM-5
	- An example of synthesis mixtures - (molar oxide ratio)	
1	SiO_2	1
.5	Al_2O_3	< .14
1	Na_2O	.16
	TPA_2O	.3
17	H_2O	49
< 100	T (°C)	> 150
	- Physical and chemical properties -	
3D, cages connected via windows	pore arrangements	2D, intersecting channels
1.28	density (g/cm^3)	1.77
.37	pore volume (cm^3/g)	.18
Na^+, H_2O	lattice stabilization	TPA^+
1	Si/Al	>12
low	Bronsted activity	high
hydrophylic	affinity	hydrophobic

b. *Zeolite product versus synthesis mixture*

The most simple zeolite product composition can be given by the overall Si/Al ratio and the cation type/content.

More often the unit cell composition of the zeolite crystal is expressed, e.g.

Na-A: Na_{12} [$Al_{12}Si_{12}O_{48}$].27 H_2O

TPA-ZSM-5: 4 TPA[$Al_nSi_{96-n}O_{192}$]H_2O
* n < 4

* At higher loadings than 4 Al/uc, TPA^+ is replaced by the smaller cation Na^+ (ref. 81).

The zeolite reaction mixture is often formulated in the molar oxide ratio of the reactants, e.g. SiO_2 :Al_2O_3 :Na_2O:(TPA_2O):H_2O. The ratios of H_2O/SiO_2, OH^- /SiO_2 , SiO_2 /Al_2O_3 and (Na_2O/TPA_2O) then give an impression of the concentration, solubility and the expected zeolite types, respectively (ref. 82). Correlation between the synthesis mixture and the product can be obtained from ternary composition diagrams (see Fig. 6a,b) (refs. 83-86), or from graphs of crystallization fields of zeolite types as a function of reactant ratios, see Fig. 6c and Section XI.b.3.

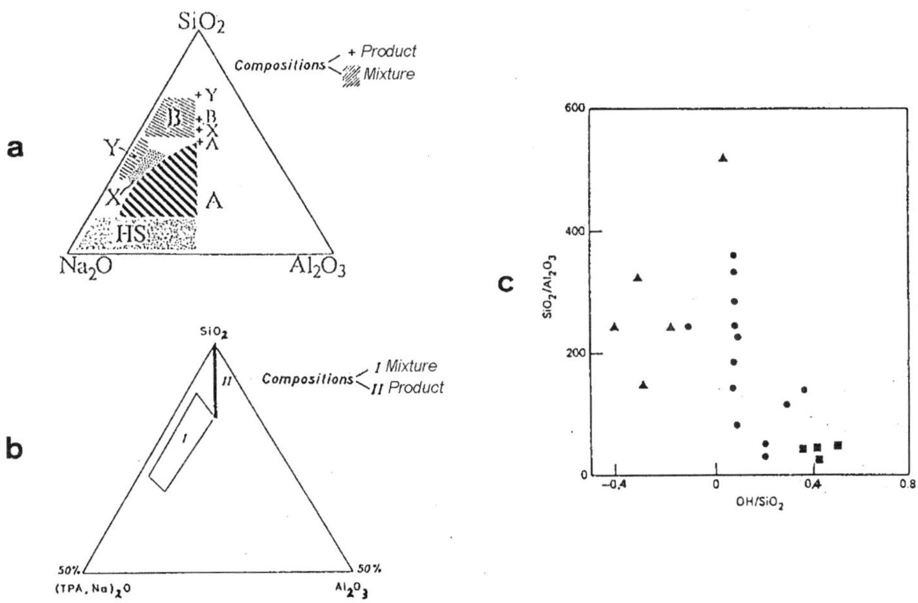

Fig. 6. Zeolite product versus the synthesis mixture. a) and b), ternary composition diagrams with an inorganic and organic cation/template, respectively. c) Crystallization fields, indicating (●) ZSM-5, (■) ZSM-35, and (▲) ZSM-39 (ref. 87).

The product fields at certain P,T, depicted in Fig. 6, are obtained from experimental data which are not always expected from a thermodynamic point of view. As the inevitably heterogeneous synthesis mixture contains micro-domains with different reactant ratios, kinetic parameters might induce other product phases than those derived from the ternary synthesis composition diagram.

Because particularly the nucleation is kinetically determined it is thus of interest to understand the different factors, e.g. type of Si-source, cation, Al-source, additives and physical parameters, influencing the kinetic stage of nucleation. The influence of these factors can be recognized in the subsequent events of the zeolite preparation which are given in Table 7 and discussed hereafter in detail.

Table 7. The subsequent events present in the course of the zeolite preparation.

Temperature	Subsequent events
Low (< 60 ^{o}C)	Reactant solutions Reactant mixture - gel formation
Low T high (< 60°C T < 200 ^{0}C)	Gel rearrangement Dissolution of gel Dissociation of silicate
High (< 200 ^{0}C)	Pre-nucleation phase Nucleation Crystallization
Low (< 60 ^{0}C)	Isolation

IV. THE LOW TEMPERATURE REACTION MIXTURE

a. *Introduction*

The reaction mixture events occurring at low temperature (< 60 ^{o}C) will be discussed for two reasons.

i) Reaction mixtures are prepared at low temperature. Chemical and physical changes should take place through adequate aging. For example, zeolite A mixtures might still comprise silicate and aluminate phases which are not mixed on a molecular level, after poor aging,. This is observed, in contrast to heating in a hot air oven, when the mixture is fast heated with microwaves to the synthesis temperature. The product distribution then shows pure zeolite, silicate, aluminate and an amorphous aluminosilicate (ref.88).

ii) Substantial knowledge about the zeolite reaction mixture at low temperature has been obtained using characterization methods such as the molybdate method (ref. 89a), the paper chromatography method (ref. 89b), the trimethylsilylation method (ref. 90), IR- and laser-Raman spectroscopy (ref. 91), single crystal structure analysis (refs. 92, 93) and the NMR technique (ref. 94).

Mostly, starting reaction mixtures typically consist of a gel phase and a liquid phase which means that nucleation is initiated at high temperature on/in the residual gel phase, though there are a few exceptions (refs. 91, 95). The (alumino)silicate gel phase consists of either a homogeneous dispersed phase of branched chains of sol particles, see Fig. 7a, or a more separated solid phase of an ordered aggregate of sol particles (like opals), see Fig.7b. The mechanisms and kinetics of formation have been extensively investigated and reviewed (refs. 96a and b, 97).

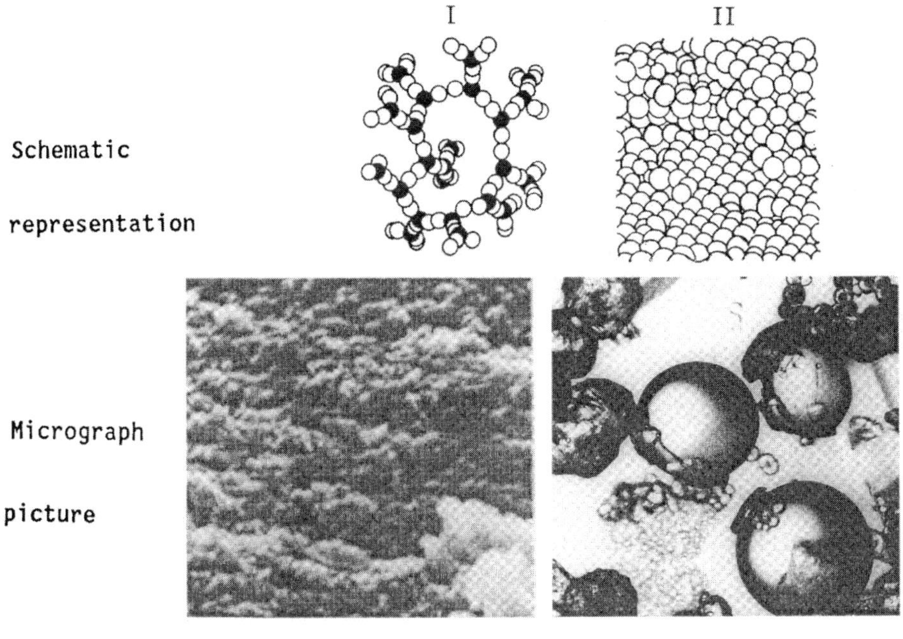

Fig. 7. Alkaline gel forms. Schematic representation and micrograph picture of I) a dispersed low density gel (ref. 96b) of branched chains of sol particles and II) a separated high density gel form resulting in spheres consisting of an ordered aggregate of sol particles (like opals).

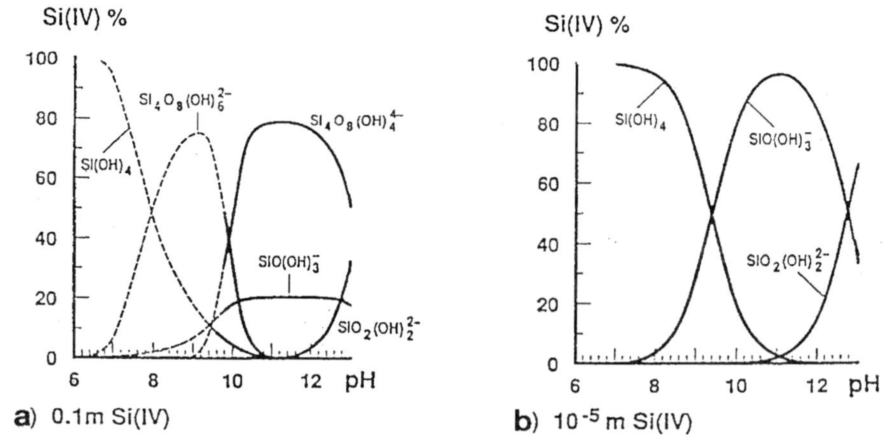

Fig. 8. Silicate distribution versus pH at a) high and b) low silicate concentration (ref. 98).

The pH of the liquid phase, in the case of OH^- as the mineralizing agent, lies generally between 8-12. As depicted in Fig. 8 the most abundant form(s) of Si-species at relatively high pH are the monomeric ions, whereas at lower pH value monomeric neutral Si-species can be formed, when the Si-concentration is low.

At high concentration, however, cyclic tetramers are the most abundant species (ref. 98).

b. *Hydrolysis and condensation of silicate*

Monomers and oligomers in solution are in equilibrium with the gel phase. At this ambient stage of the reaction mixture monomeric silica species can be released from the gel via hydrolysis reactions and are present in solution as e.g. $Si(OH)_3O^-$ and $Si(OH)_2O_2^{2-}$. The dissolution of the gel is promoted by the OH -coordination of silicon above four, thus weakening the other siloxane bonds to the gel network. This nucleophilic mechanism may occur via an S_{n2}-type transition state as shown in Scheme 3a (ref. 9).

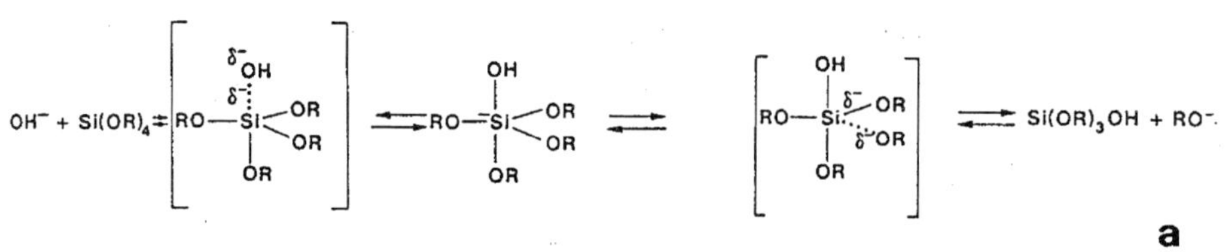

a

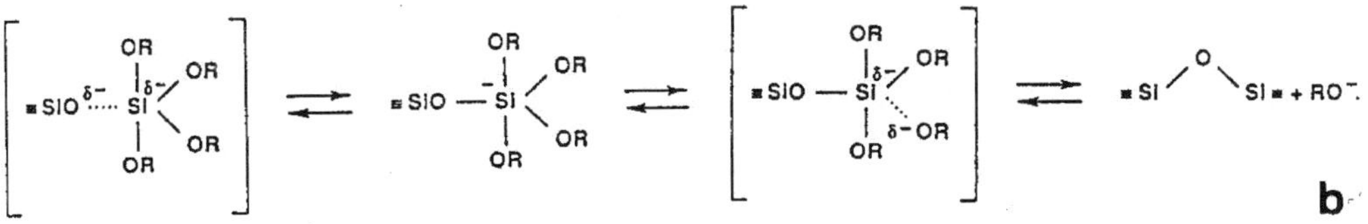

Scheme 3. a) Hydrolysis and b) condensation mechanism of silicate species at room temperature.

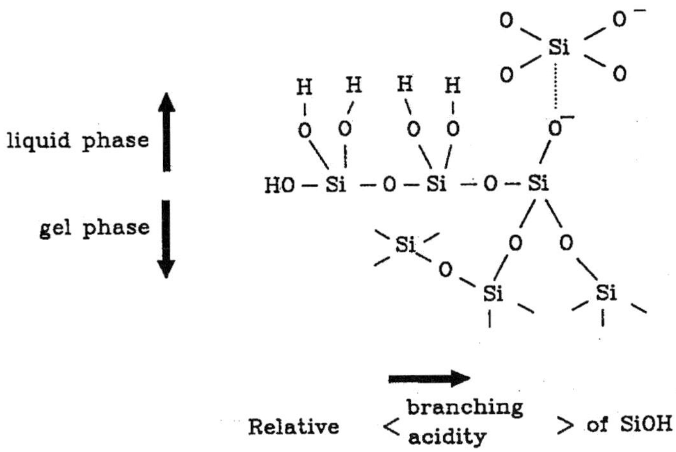

Scheme 3c. Growth site in the gel phase for monomers from solution.

The mechanism of the condensation reactions in aqueous systems at high pH involves the attack of a nucleophilic deprotonated silanol group on a monomeric neutral species as represented in Scheme 3b (ref. 9).

The acidity of the silanol group depends on the number and type of substituents on the silicon-atom. The more silicon substituents are present, the more acidic the OH-groups of the central silicon atom. As shown in Scheme 3c, at high pH the most favourable polymerization is the reaction between large most highly branched species and the monomer silica species.

At more neutral pH, hydrolysis and condensation of clusters, containing specific bonding configurations, see Fig. 9, indicate that inversion in the pentacoordinate state of Si, illustrated in Scheme 3, is not essential (ref. 99).

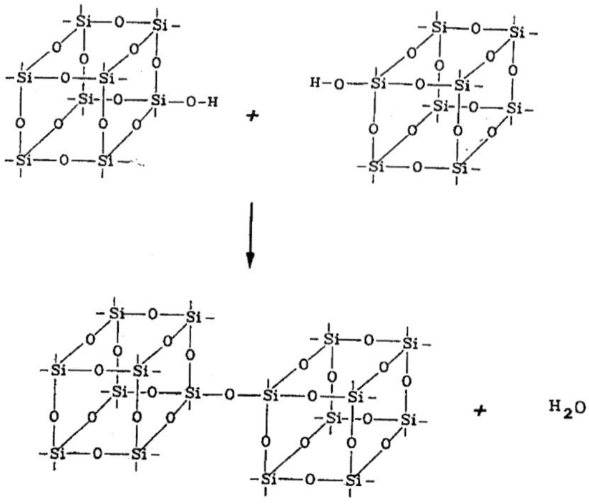

Fig. 9. Condensation of octamers, A) with retention of the configuration.

The pentacoordinate silicon intermediate state is, however, confirmed crystallographically (ref. 100).

Condensation of the monomers leads, as the pH of the zeolite synthesis mixture is above the isoelectric point of silica (ref. 101) to ramified clusters. Such clusters can be reorganized into fewer larger particles with a corresponding reduction in surface energy, according to the Ostwald ripening principle. The structural evolution of a growing cluster is schematically given in Fig. 10.

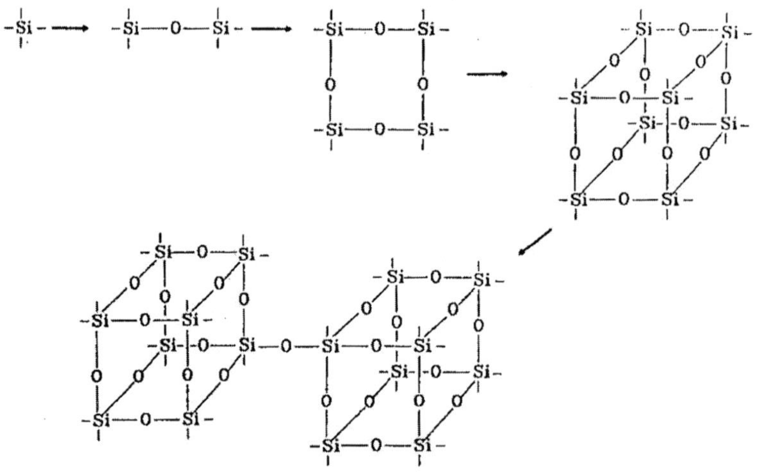

Fig. 10. Structural evolution of silicate clusters.

c. *Evidence for silicate clusters*

In the course of the gel dissolution the monomers form dimers, according to ^{29}Si-NMR studies (ref. 94), via condensation reactions whereafter trimers and tetramers, cyclic trimers and tetramers and higher order rings are observed as condensation products, see Fig. 11.

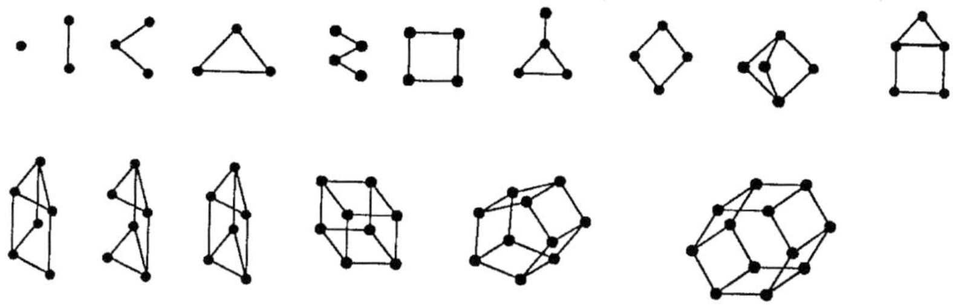

Fig. 11. Numerous oligomers characterized in solution at low temperature by ^{29}Si-NMR (ref. 101).

Evidence for the existence of e.g. double four rings resulted from the single crystal structure analysis of so-called pseudo-A, a material, *not a zeolite*, crystallized at ambient temperature from a mixture of SiO_2 , TBAOH and H_2O, see Fig. 12 (ref. 93).

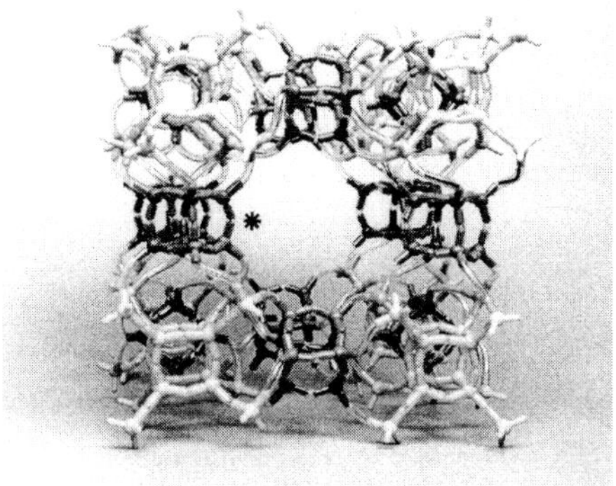

Fig. 12. Model of a part of the framework of pseudo-A; the double four ring units are indicated by asterisks (*).

The silicate species identified in the liquid phase by a.o. NMR, SAXS (ref. 102) and IR, (ref. 103) are products in a simple reaction mixture of SiO_2 , NaOH and H_2O at room temperature.

The interaction of alkali-ions in such systems is not clear. It is often suggested (ref. 104) that the ordered hydration sphere of a.o. Na^+ stabilizes silicate species. Recent NMR results indicate that interactions between cations and silicate species (ref. 105) do occur.

An organic cation/template added as ingredient(s) to the simple reaction mixture shows in typical experiments according to NMR measurements interaction with the gel and silicate species, respectively (refs. 106-108). However, the highly complicated set of interactions and fast changing equilibria, due to the increased number of type of species after addition of template and/or Al has not been unravelled yet.

V. THE TEMPERATURE RAISE OF THE REACTION MIXTURE

Temperature raise, from < 60 °C up to < 200 °C, can be performed in several ways as shown in Fig. 13 for one type of autoclave and reaction mixture. The different heating rates are achieved in static systems.

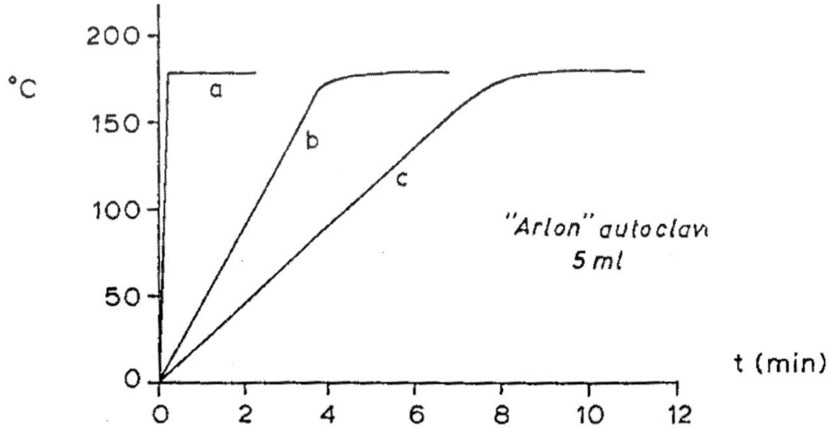

Figure 13, different heating rates for one type of autoclave achieved by (a) microwave, (b) hot sand bath and (c) hot air oven.

The size of the autoclave, the viscosity of the reaction mixture and the way of agitating e.g. static, tumbling or turbo-stirring are factors modulating the temperature raise of the reaction mixture.

During the temperature raise of the reaction mixture from ambient to reaction conditions primary events are:

- Accelerated dissolution of the gel into monomeric silicate species.
- Dissociation of silicate oligomers in solution and increase of monomers as measured by NMR up to ~ 100 °C (refs. 109-112). As shown in Fig. 14 a model study with NMR on trimethylsilylated silicate confirms (ref. 109) a shift of the silicate anion equilibrium from relatively high-molecular, mainly double four rings, to low-molecular weight, monomers and dimers.

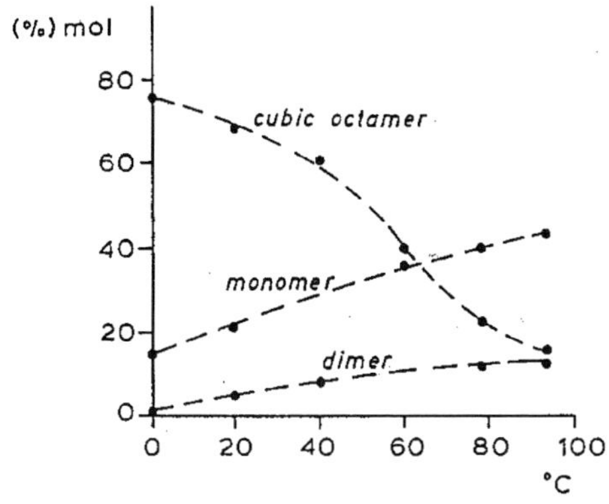

Fig. 14. Main changes in composition (% mol) of trimethylsilylated silicate solution versus temperature.

- Higher concentration and mobility of monomeric silicate- and eventually aluminate species.
- Association of primary building units.
- Possible nucleation and crystallization of unwanted (metastable) phases.

Some secondary events are:
- The start of the degradation of quaternary ammonium ions, which can be substantial in a ZSM-5 synthesis (ref. 42) as depicted in Fig. 15.
- Start of the drop in pH caused by the Hoffman degradation.

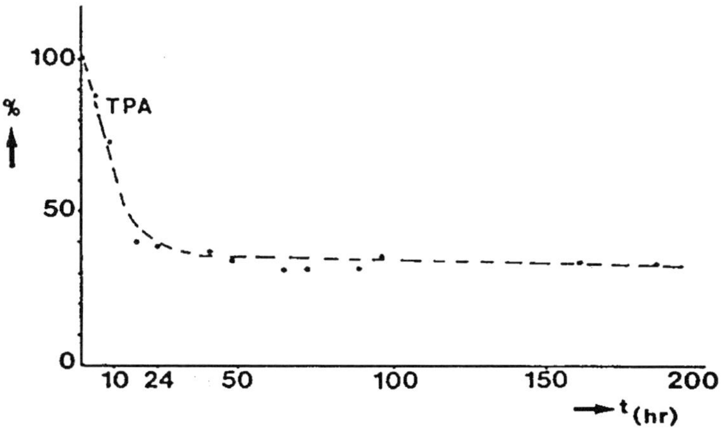

Fig. 15. Degradation of tetrapropylammonium versus time.

VI. THE HIGH TEMPERATURE REACTION PROCESS

a. *Introduction*

The main event occurring in the synthesis mixture at the reaction temperature is the formation of zeolites from amorphous material. Chemical reaction processes accelerated by the high temperature lead to:
i) further reorganization of the low temperature synthesis mixture;
ii) whereafter primary (homogeneous or heterogeneous) and secondary (seed crystals (ref. 113)) nucleation;
iii) and finally, precipitation (based on reactions) as a form of crystallization.

b. *Nucleation*

At the high temperature of the reaction mixture the zeolite crystallization is expected after an induction period in which the nucleation occurs. During the induction period the gel and species in solution (aforementioned in the low temperature section) rearrange from a continuous changing phase of monomers and clusters, e.g. polysilicates and aluminosilicates. These clusters form and disappear through inhomogeneities in the synthesis mixture via condensation and hydrolysis processes. The continuous dissolution of the gel phase increases, however, the amount of clusters and the possibility of further association of the clusters and cations. In the course of this process particles become stable. Nuclei of certain dimensions, e.g. 1.0 nm for zeolite Na-A (ref. 114) and 2.0 nm for zeolite ZSM-5 (ref. 115), are formed and crystallization starts.

c. *Crystallization*

The lines along which ideas on zeolite crystal formation are developed, either based on bulk and macroscopic observations or on molecular mechanistic scale are described in this paragraph. Four cases of nucleation and crystallization are schematically presented in Table 8.
(a) Zeolite crystallizations, which might occur in clear synthesis solutions, or, more often, in heterogeneous reaction mixtures where
(b) highly dispersed or
(c) dense gel forms are present, see also Fig. 7.
In some occasions (d) metastable solid phases undergo transformation during synthesis. Homogeneous nucleation whereafter crystallization has been observed in (a) clear solution experiments (refs. 91, 92).

Table 8. Four cases of crystal growth environment and schematic representation of nucleation and crystallization.

Crystal growth environment	Nucleation (•) Crystallization (⟶)
(a) Clear solution	
(b) Dispersed low density gel	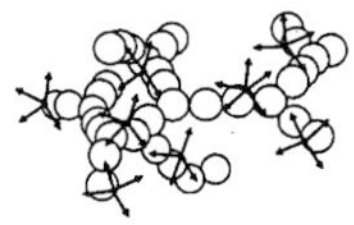
(c) Separated high density gel	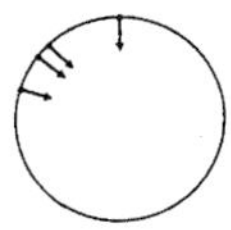
(d) Solid phase	

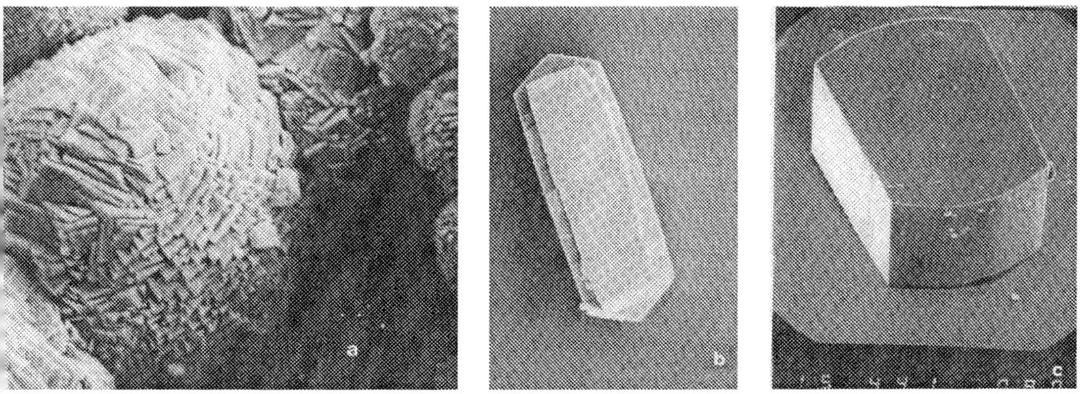

Fig. 16. a) Powder and b) a twinned elongated prismatic crystal of ZSM-5 from a dispersed gel phase and c) a cubic form of ZSM-5 from a dense gel phase.

Nucleation (heterogeneous) occurs at the liquid-gel interface in the dispersed gel-solution mixtures (b) (ref. 108). The forms of the crystallization products in the case of a dispersed gel phase are shown for ZSM-5 in Fig. 16a,b.
Similarly to the clear synthesis solutions, the driving force for crystallization is equal in all directions as shown in Table 8a,b. In the case of a dense gel phase present in the synthesis mixture, see Table 8c, crystallization proceeds into the gel (ref. 42) as shown schematically in Fig. 17.
Deviating crystal forms compared to crystal forms from dispersed gel systems are then observed, as shown in Fig. 16c.

Generally, the typical form and morphology of a zeolite crystal reveals not only information on the type of the zeolite formed but also on the crystal growth history, as shown above.

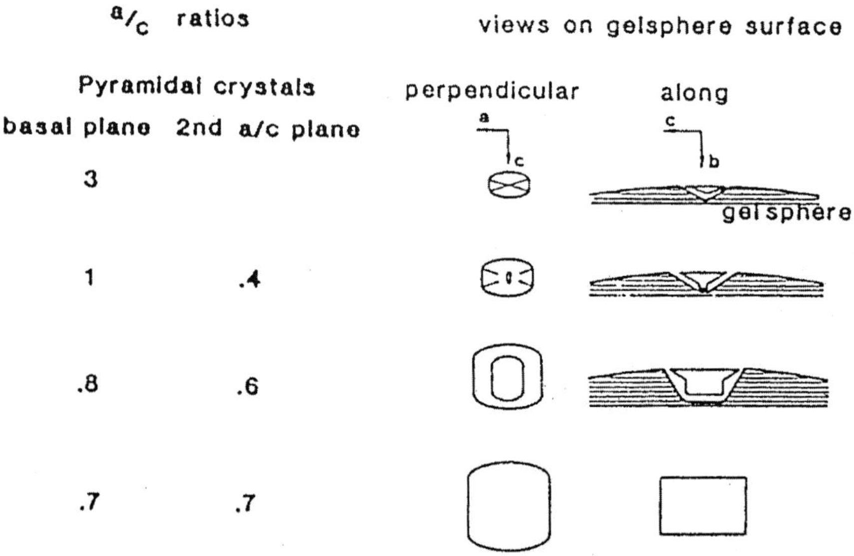

Fig. 17. Average a/c ratios of developing crystals and schematic drawing of growth process in the gel spheres.

As a liquid phase is continuously present between the dissolving dense gel phase and the growing crystal, the crystallization is, however, still solvent mediated.

When a metastable solid phase, e.g. a zeolite, is present in the synthesis mixture, a transformation into a more stable phase is possible, according to the Ostwald rule of successive transformations (ref. 116).

The nucleation and crystallization of the new phase, illustrated in Table 8d, occurs in the supersaturated solution generated by the dissolution of the former phase (ref. 117).

In the last three cases of Table 8 dynamic equilibria between successive steps of dissolution, ion transportation and precipitation, can be recognized (ref. 118). Especially, the precipitation/crystallization step, i.e. the type of crystal building units and the way of crystal growth on molecular level, has been subject to many studies.

d. *Crystal building units*

At least three types of crystal building units have been suggested which are described below.

d.1. The primary building unit

Arguments that primary building units, i.e. tetrahedral monomeric species, can be involved in the crystallization are:
i) The general view from crystal growth theories that crystals are formed via primary building units (ref. 119);
ii) The general view in sol/gel chemistry (refs. 8, 10) that the most favoured condensation reaction occurs between a monomeric and polymeric species. In terms of the zeolite crystallization: between a primary building unit and a crystal surface; see Section IVb;
iii) At raising temperatures (up till 100 °C) the concentration of monomers increases (ref. 109) at the expense of clusters. Though *in situ* measurements (till 200 °C) are not actually performed, the above experimental results might indicate that at reaction temperatures mainly monomers are present;
iv) Studies on the crystallization of zeolite have shown that the growth of a zeolite occurs by a surface reaction of monomeric anions (ref. 120).

d.2. A typical cluster as building unit

As shown in Chapter 4 of this book secondary building units (SBU's) are relatively low (F 16-Si-tetrahedra) polymer units. SBU's were introduced several decades ago (ref. 121) and used since to present structural (ref. 6) and further physical features of the zeolites. At the same time SBU's acting as non-chiral independent units can generate a certain zeolite structure.

It is, however, though the SBU's show sometimes a superficial resemblance to silicate anions, not likely that SBU's are the building blocks of the growing crystal (ref. 122). On the other hand, the building of the porous and different zeolite frameworks with monomers condensating in the right topology seems less favourable compared to a typical cluster building unit (ref. 123). From this point of view suggestions are raised about a typical or common cluster building unit for all zeolite structures.

d.3. The cation templating theory

Organic as well as inorganic cations show structure directing, i.e. water-ordering, properties. Typical examples are given in a review of single crystal structure analysis of organic water clathrated cations (ref. 124 a). The water molecules comprising a tetrahedral network in the first layer around the cation might be partly replaced by silicate and aluminate anionic tetrahedra.

The clathrated cations might serve this way as crystal building units.
An example of such a templating/clathrating role is the formation of sodalite with tetramethylammonium (TMA) cations (ref. 124 b).

d.4. Particular building blocks for one type of zeolite
Recently such blocks have been identified in the formation of MFI frameworks indicating that building units can be present for a specific type of zeolite. (ref. 125)

The high temperature events, discussed above, are summarized in Scheme 4.

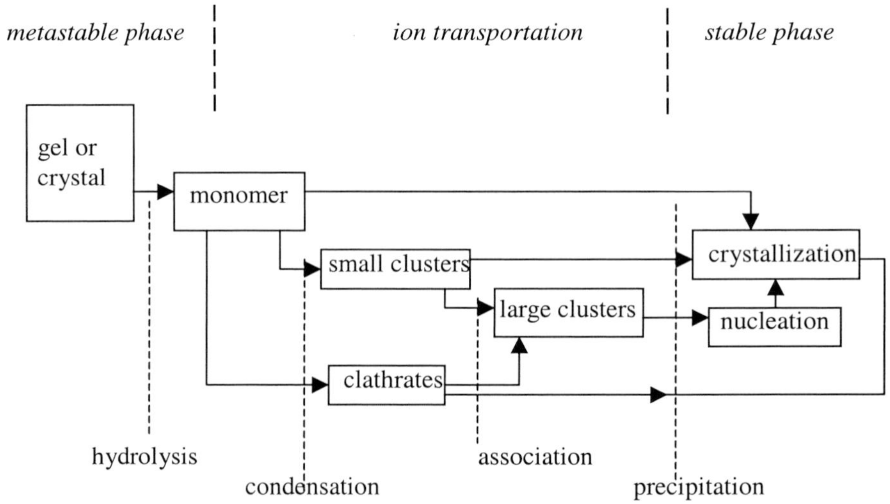

Scheme 4. Representation of successive steps in the evolution of the reactive crystallization of zeolite

Nucleation and crystallization events are generally illustrated by characteristic S-shaped crystallization curves (ref. 126). The yield (wt % of crystalline material), often determined by indirect methods, plotted against time gives an impression of the nucleation and crystallization time and certain reaction temperatures. More accurate information on the crystallization kinetics can be provided when, based on crystal size and size distribution, the linear crystal growth rate and the rate of nucleation can be determined. Of the studies (ref. 127) on zeolite crystallization, one contribution (ref. 128) reporting on a method to collect kinetic data is briefly described here.

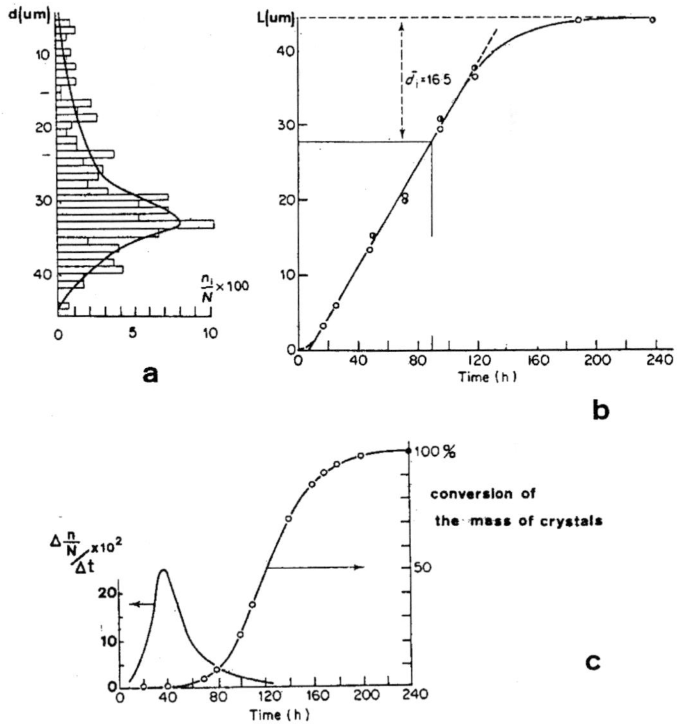

Fig. 19. a) Histogram of the crystal size distribution in the final product, b) diameter of the largest crystals of different unfinished crystallization runs versus time, resulting in the crystal growth rate graph and c) (i) the nucleation rate (number of crystals of each unfinished crystallization run versus time) together with (ii) the yield curve.

A number of identical synthesis experiments, but differing in total synthesis time, were performed. The average diameter of the largest crystals which could be collected from the various products was measured. In the case of zeolite Na-X it was found that in a plot of crystal size versus time the linear crystal growth rate (.5 L/ t) was constant, irrespective of the crystal size, even until near exhaustion of the crystal building units, see Fig. 19b.

The nucleation time can be determined now for any crystal in the final product of this Na-X crystallization, knowing the growth rate. For instance, a crystal of 16.5 mm nucleated at t ~ 90 h. Together with the particle size distribution curve, Fig. 19a, the rate of nucleation was found, see Fig. 19c.

The nucleation rate curve and the yield curve calculated from both the growth rate and particle size distribution curve, indicate that as soon as the crystallization starts the chemical nutrients are consumed for crystal growth.

The formation of fresh nuclei is from then on largely suppressed. In conclusion, it can be said that zeolite synthesis, resulting in a good crystalline product can deliver accurate information on nucleation and crystallization kinetics.

f. *Energy of activation*

Though zeolitic material can be prepared at low temperature (20-60 °C) most nucleation and crystallization processes are performed at temperatures between 60 and 250 °C. The choice of the reaction temperature is governed by the energy of activation required for the zeolite crystallization.

Table 9 shows the energy of activation (E_a) as a function of the Si/Al ratio.

Table 9. E_a 's of different zeolite framework types and Si/Al ratios.

Guest molecule	Framework	Si/Al	E_a (kcal/mol)	Ref. 129
Na^+ ; H_2O	Y	1.5	11.8	a
		1.8	12.3	
		2.2	14.1	
		2.5	15.6	
TPA^+	MFI	30	7	b
TPA^+	MFI	∞	11	
Na^+; H_2O	MFI	80	18	c

It appears that the E_a 's are not related to diffusion of crystal building units in solution (E_a (diff.) < 5 kcal mol) but to condensation reactions between the crystal surface and crystal building unit. As shown in Table 9 the E_a of Na-Y changes as a function of the Si/Al ratio which indicates that the more silicious the zeolite, the larger the E_a . Generally, this trend is also observed between different zeolites, although the contribution to E_a of cations and templates, as shown for ZSM-5, can be substantial.

VII. ISOLATION OF THE ZEOLITE PRODUCT

Products of zeolite preparations can be composed of either one pure zeolitic phase, a mixture of zeolitic phases or a mixture of a zeolitic phase and e.g. quartz, cristobalite or gel phase. Mostly the product is isolated by decantation/centrifugation or filtration.

If the product consists of crystals with a uniform crystal form which is recognized as characteristic for the expected product, the zeolite can be separated by decanting the mother liquor followed by washing with water.

If there is, however, e.g. some gel phase present, this may be either co-precipitated as a separate phase or adsorbed on the crystals. Careful dissolution of the gel phase with e.g. a dilute basic OH^- solution at slightly elevated temperature is strongly advisable prior to the isolation of the zeolite. Especially in the case of adsorbed gel on the crystal surface elemental analysis (AAS, ICP or EMPA) is required to control the Si/Al ratio of the crystals before and after the washing procedure (ref. 130). The final step in the zeolite preparation is the drying or calcination procedure after which the zeolite void volume is free for different modification and/or application.

VIII. REACTION PARAMETERS

a. *Introduction*

The type of reactants, the way the reactant mixture is made, the pH and the temperature typically affect the crystallization kinetics and product formation.
Furthermore the pre-treatment of the reaction mixture, the addition of crystal growth inhibitors, the reaction mixture temperature trajectory and the use of seeds have an influence on the zeolite preparation.

Some aspects of the type of the above mentioned factors are discussed in the following paragraphs. Illustrations are mainly given on the zeolite A and ZSM-5 formation.

b. *The Si-source*

As mentioned in Section II of this chapter the different types of the Si-sources contain impurities which may affect zeolite crystallization. Another parameter, the specific surface area of these sources, can result in different nucleation and crystallization times as shown for zeolite A in Fig. 20a (ref. 47). The shorter induction and crystallization times lead to more and smaller crystals, see Fig. 20b.

a

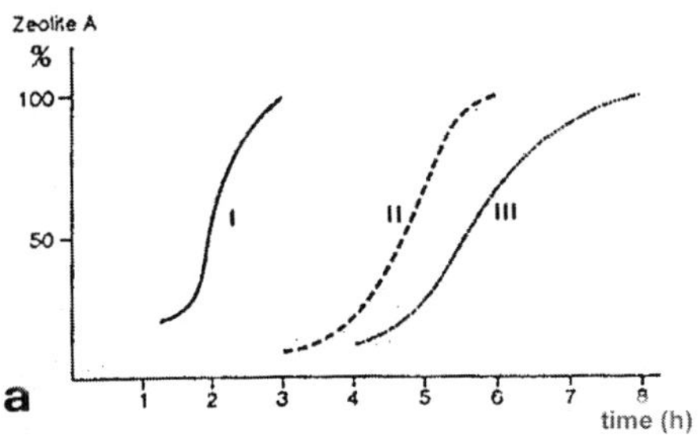

b

Silica source	Crystals rel. number	size
I	48	.7
II	30	2.6
III	15	4.8

Fig. 20. a) The yield of zeolite A versus time of different silica sources. b) The specific surface areas of the silica sources (I > II > III) result in different amounts and sizes of crystals in μm.

b.The type of template

Many types of template are regularly used (see e.g. Section II of this chapter). The surprising performance of certain templates on stabilizing the type of zeolite framework formed is illustrated in Table 10. One type of template can be used to crystallize various zeolites whereas the same type of zeolite may be crystallized while using different templates.

Table 10. Single and mixture of templates/cations in the preparation of different zeolite types.

Single template	Zeolite	Ref.	Mixture of template/cation	Zeolite	Ref.
TMA^+	Sodalite	131	TMA^+, Na^+	A, X, Y	135
	Gismondine	132		L (+ K)	136
				Sodalite, P, S and R	136
				ZSM-6 and ZSM-47	137
				Omega	138
TPA^+	ZSM-5	133	TEA, EDA, Ethanolamine, Propanolamine, Ethanol, Glycerol, Morpholine, Hexanediol, TPA^+ (each with Na^+)	ZSM-5	139
Na^+	ZSM-5	134			

The role of the single template/cation in stabilizing subunits of different zeolite types is not unravelled yet.

A common factor, however, appears to be the size of a certain free void diameter in the structures of sodalite and gismondine, 6.8 A and 7.0 A, respectively, and the diameter of . 6.7 A of the template TMA , see Fig. 21.

Fig. 21. Models of a) the sodalite and b) the gismondine void and the void filler/template/cation TMA^+.

Although TPA^+ and Na^+ are rather different templates/cations a common factor might be the stabilization of voids (either intersection of channels or channel windows), see Fig. 22.

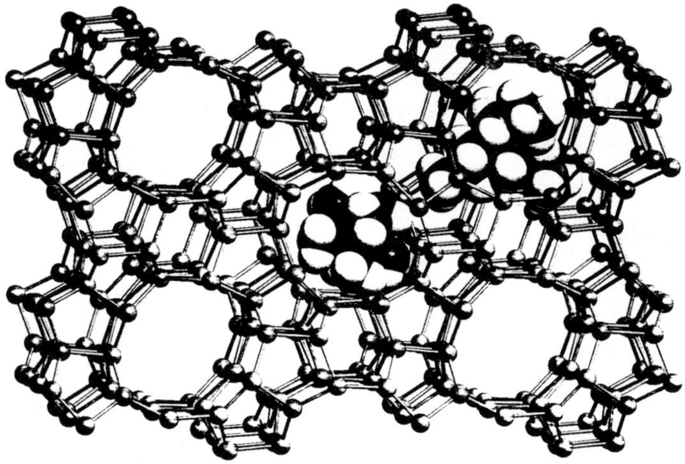

Fig. 22. View along straight channels of wire model of ZSM-5 with either TPA^+ (*) or hydrated Na^+ (O) on intersections of channels and channel windows, respectively. (courtesy St. Bromley)

Charged templates compensate negative framework charges, due to isomorphous substitution of Si^{4+} by Al^{3+}. A range of Si/Al ratios is possible, see Scheme 5. If, however, the number of charged templates required for charge compensation cannot be accommodated for dimensional reasons the zeolite combines charged templates with e.g. Na^+. This way, still various Si/Al ratios for one zeolite type are possible as shown in Scheme 5 for zeolite ZSM-5.

Sodalite can be prepared with two different Si/Al ratios.

ZSM-5	Si/Al	Sodalite	Si/Al
TPA^+	23 - <10000	TMA^+	5
TPA^+/Na^+	23 - 11		
Na^+	11	Na^+	1

Scheme 5. Different Si/Al ratios for ZSM-5 and sodalite.

d.1 *The reactant mixture*

The way reactant mixtures are made, e.g. the addition sequence of the reactants, the stirring and gel aging can result in method-dependent factors influencing nucleation. As shown in Fig. 23 crystals of zeolite A started growing in a zeolite X synthesis mixture whereafter zeolite X crystals started growing on and over the zeolite A crystal (refs. 140 a and b).

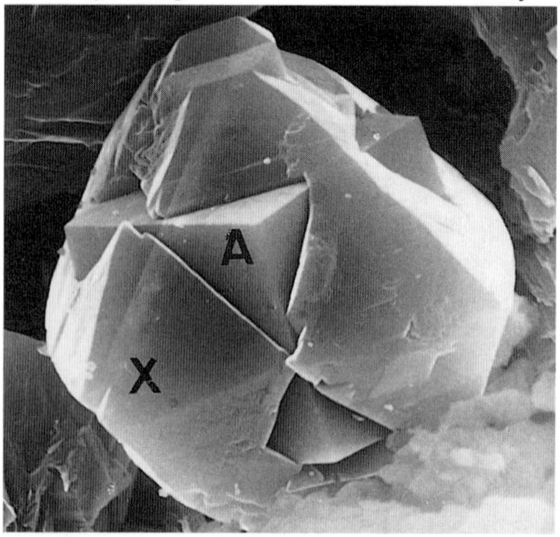

Fig. 23. Overgrowth of zeolite X onto zeolite A.

Though the thermodynamic variables were correctly chosen to prepare zeolite X, synthesis mixtures of zeolite A and X, given below, do have comparable elements and apparently local kinetic factors initiated the synthesis of zeolite A.

	$Na_2SiO_3.9H_2O$	$NaAlO_2$	Triethanolamine	H_2O	Ref. (141)
zeolite A:	.4	.1	.7	28	(molar)
zeolite X:	.4	.05	.7	28	

d.2 *nano particles from clear solutions*

The synthesis of so-called nano particles of zeolite (short diffusion path) has gained interest in the last decennium. Though high concentrations and temperature induce high nucleation rates, thus many nuclei, it is also possible to use diluted solutions and relatively low temperatures, typically below 100 °C, in order to synthesize nano-particles (ref. 142 a,b,c).

d.3 *in-situ crystallization of zeolite on supports*

Another recent development is the nucleation of zeolite from/on supports of metal or ceramics. Crystals, ranging in size from a few tens of nanometers parallel to the support surface up to a few tens of µm perpendicular to the support surface can be grown in a controlled way (142 d). For the important field of zeolites grown to a closed layer onto a porous support so as to give zeolitic membranes the reader is referred to the recent reviews (142 e-i).

e. The pH

e.1. *Introduction*

The pH and the solubility of reactants in the synthesis mixture are governed by the presence of OH^- or F^-.

An advantage of F^- compared to OH^- is the higher solubility of e.g. Fe^{3+} and Ti^{4+} and the condensation capability for e.g. Ge^{4+}. A too high concentration of F^-, however, prevents the polycondensation mechanisms. A compromise between solubility of certain elements and inhibition of zeolite framework formation leads to F^- synthesis mixtures which are less supersaturated than OH^- media. Hence, zeolite types are obtained this way as well, (ref. 16 a and b).

e.2. OH^-

Raising pH of synthesis mixtures using OH^-, mainly influences the crystallization of a certain zeolite in a positive way within the synthesis field. As depicted in Fig. 24a and b for zeolite A and zeolite ZSM-5, respectively, increasing the pH shows an increase in the crystallization rate. The OH^- is a strong mineralizing agent for bringing reactants into solution.
The higher the pH and thus the concentration of dissolved reactants the more the rate of crystal growth of zeolites is enhanced (refs. 47, 143).

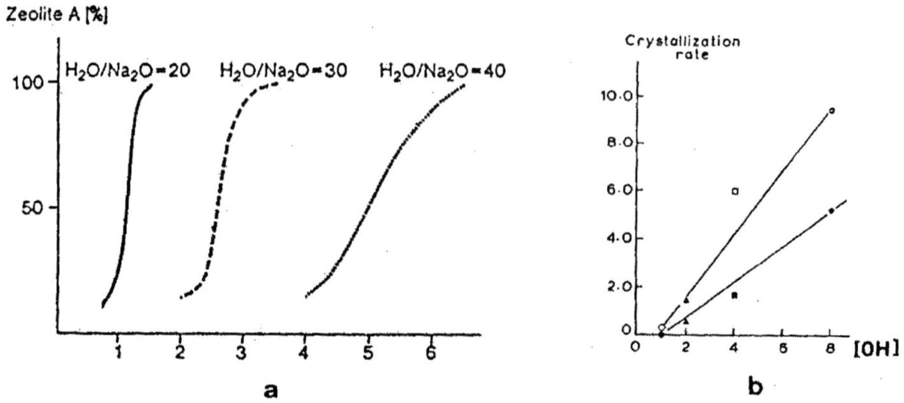

Fig. 24. The influence of alkalinity on a) zeolite A and b) ZSM-5 crystallization.

e.3. F^-

After the first publications on synthesis with F (ref. 46) extensive studies have been undertaken to investigate the effect of F^- and possibilities in zeolite synthesis (refs. 16, 46). Replacing OH by F^- the pH values of the synthesis mixture lies generally between 3 and 10. A typical synthesis formulation is given in Section XII of this chapter.
Zeolites obtained so far by this route are silica-rich materials and include:

ZSM-5
Ferrierite
Theta-1
ZSM-23
and Beta

f. *The temperature*

It has been shown for many zeolites that raising synthesis temperatures within a certain zeolite synthesis field increases the crystal growth rate (refs. 47, 144, 145). As shown in Fig. 25a and b for zeolite A and zeolite ZSM-5, increasing temperature influences the crystal growth rate whereas in the case of zeolite A the crystal size does not change significantly compared to substantial variations in the ZSM-5 product.

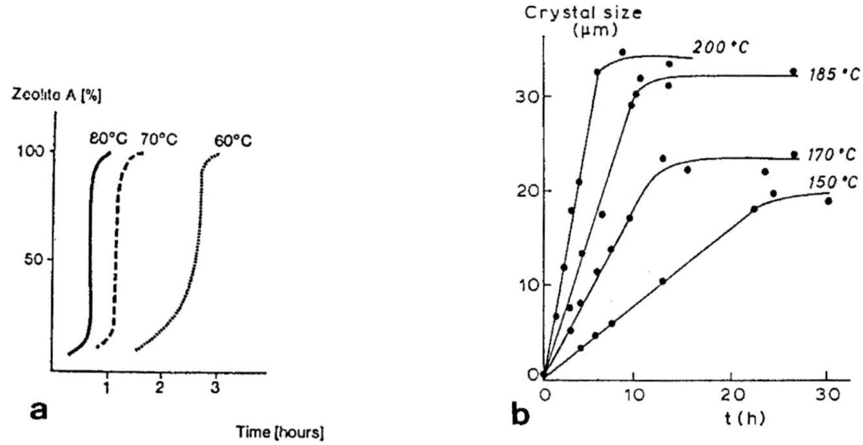

Fig. 25. Influence of temperature on the crystallization of a) zeolite A and b) zeolite ZSM-5.

IX. ALL SILICA MOLECULAR SIEVES

a. *Introduction*

Two preparation routes can be followed to obtain all silica molecular sieves:
i) A direct synthesis to crystallize molecular sieves with an SiO_2 composition and well known zeolite topologies.
ii) A secondary synthesis. After the direct synthesis of a zeolite a de-alumination procedure, e.g. acid oxalic treatment, steaming (ref. 146), ammonium silicon hexafluoride (ref. 147) or silicon tetrachloride (ref. 148) can lead to an all silica molecular sieve.

b. *Synthesis*

Though the neutral all silica molecular sieves do formally not need to be stabilized with cations the silica structures usually contains the cations used in the synthesis. For example, tetrapropylammonium for silicalite-1, tetrabutylammonium for silicalite-2, and tetraethylammonium for silica-ZSM-12.
Recently, amines, di-amines (ref. 149) and poly-amines (ref. 150) have been used as templates.

Table 11 contains a list of all silica molecular sieves with two examples of synthesis recipes and references.

Table 11. All silica zeolites, recipes and references.

Product	Ref.
Silica-ZSM-48 Recipe: 14.6 g of triethylenetetramine is dissolved in 18 ml H_2O whereafter the solution is stirred into dry 1.2 g SiO_2 . The smooth dispersion is then autoclaved between 120-180 °C for 28-105 days, respectively.	150
Silica-ferrierite Recipe: 0.75 g $Si(OCH_3)_4$ is added to a solution of 1.2 g ethylenediamine (EDA) in 10 ml H_2O. After adding 2 ml 1 M aqueous boric acid the solution is sealed in a silica tube and heated at . 170 °C for 56 days.	151
Silicalite-1 (Silica-ZSM-5)	152
Silicalite-2 (Silica-ZSM-11)	153
Silica-ZSM-22	149

c. *Remark*

The main property of the silica molecular sieves is the strong hydrophobic character of the pores. The preferential uptake of e.g. traces of organic compounds (ref. 152) from water, which is not accommodated (ref. 154), in silicalite-1 is a good example.

X. CLATHRASILS or actually silicates and zeolite molecular sieves?

a. *Introduction*

The name "clathrasil" has been introduced for a subclass of porous tectosilicates different from zeolites. The windows of the framework, connecting the cages, are too small to let guest species, stabilized during the synthesis, pass.
This characteristic of a clathrate together with the all silica composition is considered as specific for the members of the clathrasils (ref. 29).

There are, however, exceptions. The recently synthesized decadodecasil-3R (DD-3R) (ref. 155) contains windows of eight-rings of oxygen, indicating that diffusion of small molecules through the porous structure is possible after calcination. This structure can therefore be considered to form an interface between the clathrasils and the silica molecular sieves.

A modified type of DD-3R denoted as Sigma-1 (ref. 156) can, however, be seen as a link between clathrasils and zeolites, because some Si-framework sites are isomorphously substituted by Al. Finally, there is a tectosilicate,

Sigma-2 (ref. 157), of which the structure contains two different polyhedra, see Fig. 26, with a free diameter of .75 nm. Sigma-2 has been prepared in the silicalite as well as in the zeolite form and can thus be considered as an intermediate between clathrasils, zeolites and silicates.

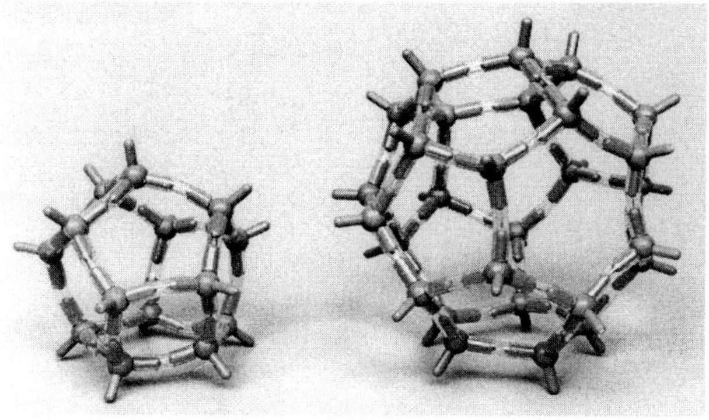

Fig. 26. The nonahedral and eikosahedral cages of Sigma-2.

b. *Experimental*

The clathrasils can be synthesized generally from .5 M silica, prepared by hydrolyzing an alkoxysilane, e.g. $Si(OCH_3)_4$, in solutions containing an amine as guest template molecule. The syntheses are mainly carried out between 160-240 ^0C. The clathrasils, together with detailed synthesis data and products expected, are given in Table 12.

Table 12. Clathrasils with references to synthesis prescriptions and two examples of synthesis recipes.

Product	Ref.
Melanophlogite	158
Dodecasil 3C	159
Dodecasil 1H	160
Silica-sodalite	161
Sigma-1	156
DD-3R (silica)	155
Recipe: 0.75 g $Si(OCH_3)_4$ is added to a solution of 1.2 g ethylenediamine (EDA) in 10 ml H_2O. After adding 350 mg 1- adamantaneamine the solution is sealed in a silica tube and heated at 170 °C for . 70 days.	

Table 12, continued,

Sigma-2 (silica and aluminosilicate)		157
Synthesis example: The molar oxide ratio of the synthesis system is:		
Na_2O	3	
AN	20 (1-adamantanamine)	
Al_2O_3	(0.6) (Al-wire)	
SiO_2	60 (colloidal silica)	
H_2O	2400	
The system was crystallized at 180 °C and continuously stirred for a few days.		

c. *Remark*

The templating role of some of the guest molecules is illustrated in Fig. 26. Polyhedra of different clathrasils are filled with a guest molecule. As only 4-, 5- or 6-ring faces are present in most of the polyhedra it looks like the crystal building units have formed around the guest molecule as this molecule is too large to pass through one of the rings. The clathrate formation might therefore be obtained and based on single building units in solution and/or at the growing crystal surface (ref. 162).

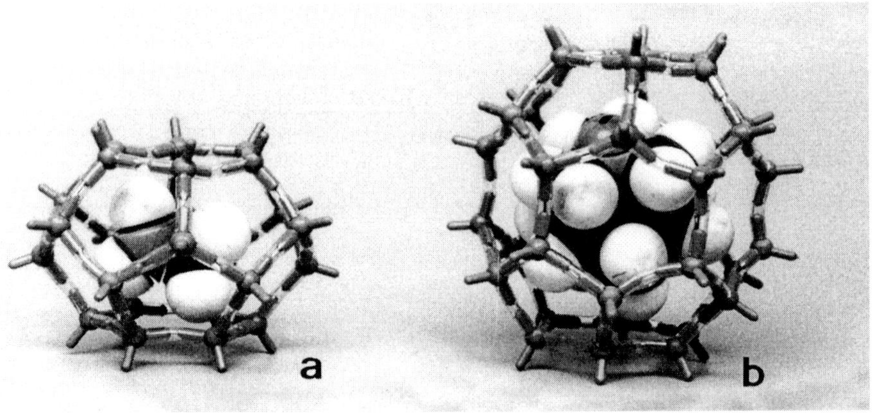

Fig. 27. Orientation of various guest molecules in clathrasils (ref. 163). a) H_3CNH_2 in [$5^{12}6^2$] and b) adamantylamine in [$5^{12}6^8$] of melanophlogite and dodecasil 1H, respectively.

XI. EXAMPLES OF SYSTEMATIC RESEARCH in the field of molecular sieves preparation to reach various objectives

a. *Introduction*

A main thrust of research is:
- to synthesize new molecular sieves
- further optimization of recipes
- to gain knowledge on the essential functions of reactants, e.g. structure directing role of cation/template
- to prepare relative large single crystals for fundamental studies

The list can be longer, however, the examples given below illustrate generally the purpose and variety in the research of molecular sieves preparation.

b. *Research examples*

Objective	Parameter(s)
1) Preparation of zeolites	Non-aqueous solvents
2) Preparation of zeolites	F^- as mineralizing agent
3) Investigation of crystallization fields with pyrrolidine as template	Na_2O-Al_2O_3 –SiO_2 –H_2O system was varied
4a) Investigation of template-zeolite interaction	Systematic variation of template
4b) Directing role of template in the crystallization	Use of bis-quaternary ammonium compounds
5) Large single crystals	Knowledge on nucleation/ crystallization
6) Morphology and form of zeolite products	Change of [SiO_2], template, cation or additives

b.1. *The use of non-aqueous solvents*

In contrast to the rich crop of zeolite types synthesized in aqueous systems the results in non-aqueous solvents are poor (refs. 164, 165). Solvents used, of which the choice was a.o. based on boiling point (100-200 °C) and relative permittivity (10-45) (water: 78), are given in Table 13.

Table 13. Zeolite products formed.

Solvents	Na^+	K^+	Li^+	Ca^{++}
Glycol	HS	-	-	-
Glycerol	HS	-	-	-
DMSO	HS	-	-	-
Sulfolane	HS	-	-	-
C_6 C_7 alcohol	HS	-	-	-
Ethanol	HS	-	-	-

HS: Hydroxysodalite.

Generally mixtures within the following molar oxyde ratio were used:

MeO	1-20
Al_2O_3	1
SiO_2	1-100
Solvent	5-350
and MeO/SiO_2	0.1-10

As shown in the Table zeolite products could only be obtained in the case of Na . The use of other inorganic and organic cations was not successful.

As the boiling point is a less critical factor than high relative permittivity (reduces the Coulomb force between ions and polar compounds thus enhancing dissolution) other non-aqueous solvents for zeolite crystallization which might be subject to zeolite synthesis tests are given below.

Non-aqueous solvent	E_r
formic acid	57
formamide	84
hydrogen peroxide	93
hydrocyanic acid	95

b.2. *The use of F^- (ref. 166)*

The compositional ratios of the reaction mixtures used were:

	Al or B/Si	F/Si	Templ./Si	H_2O/Si
in molar ratio	0-0.5	0.05-6	0.05-6	4-500

with the pH of the mixtures between 1.5-10. The reaction mixtures were heated at 60-250° C and autoclaved for a few hours to a few months. After isolation the products were washed with water and dried.

A typical example of a "F " synthesis of ZSM-5 is given below:

- Reaction mixture composition:

36 g	Ammonium aluminiumsilicate (Si/Al ~ 7; NH_4 /Al ~ 1)	
18.5 g	NH_4F	pH = 7
33.2 g	TPABr	temp. = 172 °C
180 g	H_2O	time = 11 days

- Product

Unit cell composition:

1.8 NH_4^+ + 4.1 TPA^+ [$Al_{2.9}$ $Si_{93.1}$ O_{192}]

Crystal size

30 x 12 μm

Advantages and differences using F^- instead of OH^- as concluded so far:

- Low pH compared to OH^-
- Incorporation in the framework of elements sparingly soluble in alkaline medium, e.g. Fe^{+++}
- Synthesis without alkaline cations
- New possibility to directly incorporate cations as NH_4^+ and divalent cations such as Co^{++} as well
- Good stability of usual templates such as TAA in this medium
- Highly crystalline materials

b.3. *Pyrrolidine as template (ref. 87)*

The crystallization of zeolites in the system Na_2O-Al_2O_3 –SiO_2 –H_2O + pyrrolidine as a template was studied. The reaction mixture compositions used are given in Table 14 in molar oxyde ratio.

Table 14. Reaction mixture ranges in molar oxide ratios.

Na_2O	0.05-0.5	
Al_2O_3	0.002-0.05	
SiO_2	1	~. 0.7 pyrrolidine
H_2SO_4	0-0.4	
H_2O	20-80	

Two procedures were used:

I. To a stirred aluminium sulfate solution, calculated amounts of sodium silicate, sulfuric acid and pyrrolidine were added dropwise.
II. Calculated amounts of aluminium nitrate, colloidal silica and pyrrolidine were added to a stirred sodium hydroxide solution.

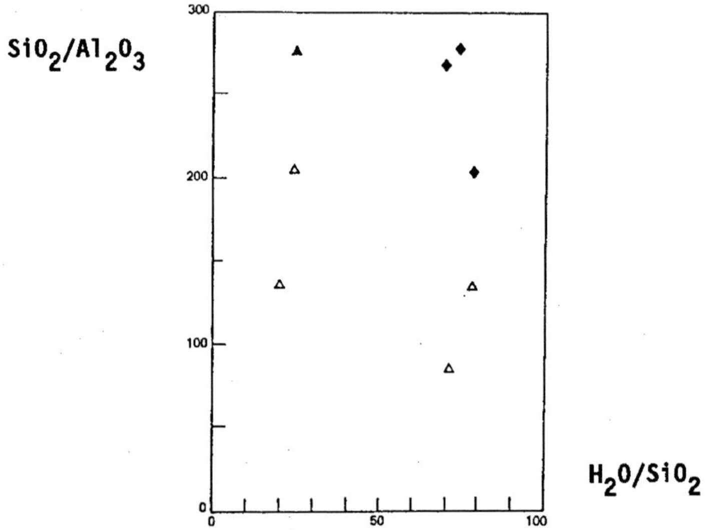

Fig. 28. Crystallization fields of product SiO_2/Al_2O_3 versus H_2O/SiO_2 of ZSM-39 (▲), ZSM-48 (◆) and KZ-1 (△).

The follow-up of both procedures was to autoclave the reaction mixture for 7-40 h at 423-435 K with stirring. After isolation the product was washed with water and dried.

The results of the experiments are given in Fig. 6 for procedure A whereas the results of the experiments with procedure B are given in Fig. 28.

The main conclusion of the study is that pure ZSM-5, ZSM-35, ZSM-39, ZSM-48 and KZ-1 can be crystallized with pyrrolidine in the aforementioned synthesis system. No common factor, based on the use of pyrrolidine, could be recognized in the various zeolite products.

b.4.a. *The use of bis-quaternary ammonium compounds in molecular sieves synthesis (ref. 58)*

The objective in this study was the systematic variation of template in the synthesis. An example of the synthesis is given below together with the products formed, see Table 15.

The general formula of this bis-quaternary template (T) is:

$$T = [(CH_3)_3N(CH_2)_xN(CH_3)_3]^{2+}$$

Table 15. Synthesis mixtures and product formation with bis-quat as template.

Synthesis conditions molar oxide ratio		x	Product formation
SiO_2	60		
Al_2O_3	1		Zeolite phases
Na_2O	10	3	ZSM-39
TBr_2	10	4	ZSM-12
H_2O	3000	5,6	EU-1
		7,8	ZSM-23
			Silica phases
		3	EU-4
(without Al_2O_3)			
		4,9	EU-2

The reaction conditions were 180 °C, three days and crash-cooling after the synthesis was terminated.

b.4.b. *Another example (ref. 167)*

Systematic variation of the chain length of the template (T) given below in the general formula

$$T = H_2N\text{-}(CH_2)x\text{–}N\text{-}H_2$$

resulted in the products, given in Table 16.

Table 16. Zeolite formations, obtained with α ,ω -diamines

x	Zeolite phase
2-5	Ferrierite
	ZSM-5
5-6	ZSM-5
7-10	ZSM-11

The full synthesis description is given in ref. 167.

b.5. The synthesis of relatively large single crystals of molecular sieves

b.5.1. *Introduction*

Pertaining to e.g. the viscosity of the synthesis mixture several systems, clear solution, diluted gel and dense gel phase have been investigated.

b.5.2. Crystallization of ZSM-22 from a clear solution (ref. 149)

In a typical experiment tetramethoxysilane was hydrolyzed in 3 M aqueous diethylamine (DEA) according to the following reactions:

$$Si(OCH_3)_4 + 2\,H_2O \xrightarrow[\substack{180\,^{\circ}C \\ 100\ days}]{} SiO_2 + 4\,CH_3OH \xrightarrow{DEA} 2\,SiO_2\,(C_2H_5)_2NH$$

Single crystals of silica-ZSM-22 of 45 x 100 x 225 μm were isolated and used for structure determination.

b.5.3. *Synthesis of elongated prismatic ZSM-5 crystals*

The objective of this study was to obtain large single crystals of ZSM-5.
Systems using Na^+-TPA^+, Li^+-TPA^+ and NH_4^+-TPA^+ were investigated applying a reaction mixture given in molar oxide ratio for e.g. NH_4^+ -TPA^+:

TPA_2O	4	
$(NH_4)_2O$	123	T = 453 K
Al_2O_3	1	t = 7 days
SiO_2	59	
H_2O	2280	

Products

Alkaline-free, homogeneous elongated prismatic single crystals of ZSM-5 of 350 μm in length at maximum (ref. 168).

b.5.4. *Synthesis of cubic shaped single crystals of ZSM-5 (ref. 169)*

The synthesis of this type of crystals developed recently (ref. 169) was subject of a study on the crystal growth history (ref. 42) of this type of crystals. The objective was: to pinpoint the driving forces which change the ZSM-5 crystal form from elongated prismatic into cubic. The crystal growth history study revealed that the cubic crystal growth occurred in a dense gel phase.

Perfect single crystals up to 500 μm of zeolite (ZSM-5) and all silica (silicalite-1) molecular sieve type, see Fig. 16c, could be obtained using the following molar oxyde ratio:

	ZSM-5	Silicalite-1
SiO_2	12	12
Al_2O_3	1	
Na_2O	44	44
TPA_2O	44	44
H_2O	2000	2000

After 5 days at 180 °C crystals could be isolated and selected from the product.

b.5.5. *Synthesis of single crystals of zeolite A and X (ref. 142)*

Single crystals of zeolite A and X up to 100-500 mm in size could be obtained using the following procedures.

Procedure for zeolite A:
Solution I: 100 g $Na_2SiO_3 .9H_2O$ in 350 ml H_2O + 50 ml TEA
Solution II: 80 g $NaAlO_2$ in 350 ml H_2O + 50 ml TEA
Both solutions are filtered with milipore filters, whereafter solution II is added to solution I with stirring. The crystallization is performed at 75-85 °C for 2-3 weeks, without stirring.

Procedure for zeolite X:
Identical to the procedure for zeolite A, only 40 g of $NaAlO_2$ is used in solution II now. The crystallization time is 3-5 weeks.

Remark

Careful filtering of the starting solutions substantially reduces the amount of heterogeneous nuclei such as dust and foreign particles in the starting chemicals. The lower the number of nuclei, the larger the crystals.

b.6. *Morphology and form of mordenite and ZSM-5*

The morphology and/or form of zeolite crystals appear generally to be influenced by:
- [SiO_2]
- Guest molecule type
- Cation (ref. 171)
- Crystal growth inhibitors

A frequently observed crystal form of mordenite is the needle form (with pore channel system parallel to needle direction), see Fig. 29a.

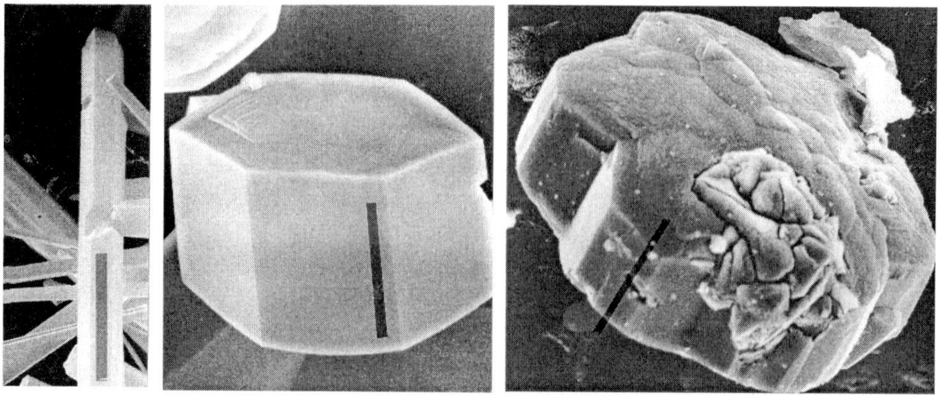

(a) (b) (c)

Fig. 29. Different forms of mordenite. The needle form a), the intermediate forms b) and c) and the disk form d). The pore direction is indicated by a bar (ref. 85).

As shown in Fig. 29b, c and d, completely different crystal forms of mordenite can be prepared. According to the synthesis system used (ref. 85) the main influence in the shape of the crystals seems to be the [SiO_2]. The higher the [SiO_2], i.e. the more the crystallization occurs in a dense gel, the more the elongated form is reduced and changed into a disk form.
The increase in pore entries and decrease in pore length going from needle to disk form is evident and may be of interest in catalysis (ref. 170).

The elongated prismatic form is the most frequently found crystal form of ZSM-5. Changing the [SiO_2] can change the crystal form as shown in Fig. 30a and b for relatively low and high [SiO_2] concentrations, respectively.

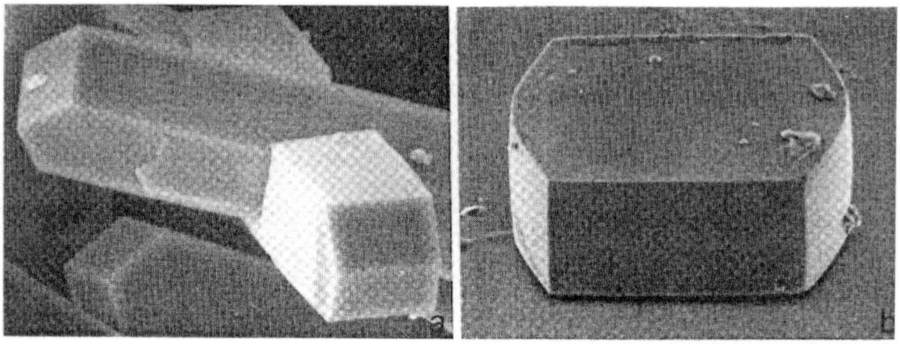

Fig. 30. The elongated prismatic crystal form (a) and the cubic crystal form (b) of zeolite ZSM-5.

Changing the template type, i.e. replacing TPA for the divalent bi-quaternary ammonium ion, hexapropyl-1,6-hexanediammonium, resulted in different crystal
forms for low as high [SiO_2] as well, see Fig. 31a and b.

Fig. 31. The modification of the crystal form at low (a) and at high (b) [SiO_2] of zeolite ZSM-5 prepared with biquat as template.

In the case of an additive (inhibitor) like boric acid an enrichment of crystal faces in the c-direction was observed as shown in Fig. 32.

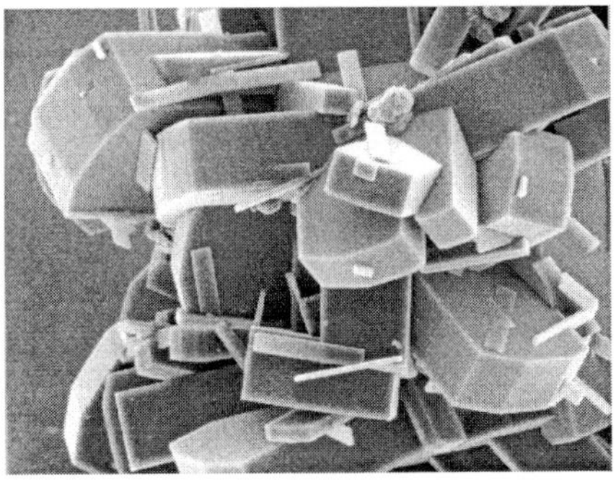

Fig. 32. Additional crystal face {001} compared to regular elongated prismatic form of zeolite ZSM-5.

XII. LITERATURE SOURCES PERTAINING ZEOLITE PREPARATION ASPECTS

Though most of the literature sources are given in Section XIII, a more extended list of sources is given below for reasons of clarity and ease.

- *Atlas of Verified Zeolite Synthesis*

Microporous and Mesoporous Materials, vol.22 no 4-6 (1998), updated in 2001.

- *Chemical Abstracts*

A literature search in the Chemical Abstracts (CA) can be successful when Controlled Vocabulary Index Terms (CVIT's) are used. As CVIT's after 1976 are not only assigned to words in the title and the abstract, but also throughout the text of the paper (open literature or patent) the search will be thoroughly. The choice of CVIT's must be correct. In the case the word "synthesis" is used instead of "preparation" the main part of the search "hits" will pertain to reactions with the help of zeolites whereas the preparation of zeolites is then difficult to extract.

- *Proceedings of International Zeolite Conferences (IZC)*

1. "Molecular Sieves", Soc. Chem. Ind., London, 1968; Proceedings of the 1st IZC, London, U.K., 1967.
2. "Molecular Sieves I and II", Adv. Chem. Ser., 101 and 102, ACS, Washington, D.C., 1971; Proceedings of the 2nd IZC, Worcester, Mass., U.S.A., 1970.
3. "Molecular Sieves", Adv. Chem. Ser., 121, ACS, Washington, D.C., 1973; W.M. Meier and J.B. Uytterhoeven, Eds., Proceedings of the 3rd IZC, Zürich, Switzerland, 1973.
4. "Molecular Sieves-II", ACS Symp. Ser., 40, ACS, Washington, D.C., 1977; J.R. Katzer, Ed., Proceedings of thre 4th IZC, Chicago, Ill., U.S.A., 1977.
5. "Proceedings of the 5th International Conference on Zeolites", Heyden, London, Philadelphia, Rheine, 1980; L.V.C. Rees, Ed., Proceedings of the 5th IZC, Naples, Italy, 1980.
6. "Proceedings of the 6th International Conference on Zeolites", Butterworths, Guildford, 1984; D. Olson and A. Bisio, Eds., Proceedings of the 6th IZC, Reno, Nev., U.S.A., 1983.
7. New Developments in Zeolites Science and Technology", Kodansha, Tokyo, Elsevier, Amsterdam, Oxford, New York, Tokyo, 1986, Stud. Surf. Sci. Catal., 28; Y. Murakami, A. Iijima and J.W. Ward, Eds., Proceedings of the 7th IZC, Tokyo, Japan, 1986.
8. "Zeolites: Facts, Figures, Future", Elsevier, Amsterdam, Oxford, New York, Tokyo, 1989, Stud. Surf. Sci. Catal., 49; P.A. Jacobs and R.A. van Santen, Eds., Proceedings of the 8th IZC, Amsterdam, Netherlands, 1989.
9. Proceedings from the 9th IZC, Butterworth-Heinemann, Boston, London, 1992, R. von Ballmoos, J.B.Higgins and M.M.J.Treacy, Montreal, Canada,1992
10 "Zeolites and Related Microporous Materials", State of the Art 1994, Elsevier, Amsterdam, Oxford, New York, Tokyo, 1994, Stud.Surf.Sci.Catal. 84; J.Weitkamp, H.G.Karge, H.Pfeifer and W.Holderich, Eds. Proceedings of the 10th IZC, Garm isch Partenkirchen, Germany, 1994.

11 “Progress in Zeolite and Microporous Materials”, Elsevier, Amsterdam, Oxford, New York, Tokyo, 1996, Hakze Chon, Son-ki Ihm and Young Sun Uh, Eds. Proceedings of the 11th IZC, Seoul, Korea, 1996.
12 ”Proceedings of the 12th IZC”, MRS, Warrendale, Pennsylvania, M.M.J.Treacy, B.K.Marcus, M.E.Bisher and J.B.Higgins, Eds. Baltimore, Maryland, U.S.A., 1998

- *Synthesis part on International Conferences*

"Zeolites, Synthesis, Structure, Technology and Application", Elsevier, Amsterdam, Oxford, New York, Tokyo, 1985, Stud. Surf. Sci. Catal., 24; B. Drzaj, S. Hocevar and S. Pejovnik, Eds.
"Innovation in Zeolite Materials Science", Elsevier, Amsterdam, Oxford, New York, Tokyo, 1988, Stud. Surf. Sci. Catal., 37; P.J. Grobet, W.J. Mortier, E.F. Vansant and G. Schulz-Ekloff, Eds.
"Zeolite Synthesis", ACS Symp. Ser., 398, ACS, Washington, D.C., 1989; M.L. Occelli and H.E. Robson, Eds.

- *Journals*

Zeolites, L.V.C. Rees and R. von Ballmoos, Eds., Publishers, Butterworth, Heinemann, Stoneham, MA, U.S.A., until 1993.
Microporous Materials, J.Weitkamp, Ed., P.A.Jacobs Consulting Ed., Elsevier, Amsterdam-London-New York-Tokyo., until 1997.
Microporous and Mesoporous Materials, J.Weitkamp Ed., S.L.Suib, R.W.Thompson, and K.Kuroda, Regional Eds., Elsevier, Amsterdam-London-New York-Tokyo. , as from 1998.

- Books

"Zeolite Molecular Sieves", Structure, Chemistry and Use, John Wiley & Sons, New York, London, Sydney, Toronto, 1974; D.W. Breck.
"Hydrothermal Chemistry of Zeolites", Academic Press, London, New York, 1982; R.M. Barrer FRS.
"Synthesis of High-Silica Aluminosilicate Zeolites", Elsevier, Amsterdam, Oxford, New York, Tokyo, 1987, Stud. Surf. Sci. Catal., 33; P.A. Jacobs and J.A. Martens, Eds.
"Molecular Sieves, Principles of Synthesis and Identification", Van Norstrand Reinhold, New York, 1989; R. Szostak.
"An Introduction to Zeolite Molecular Sieves", John Wiley and Sons, Chichester, 1988, A. Dyer.
“ Molecular Sieves”, Vol. I, ‘Synthesis’,, Springer, Berlin, Heidelberg, New york, Tokyo, 1998, J.Weitkamp and H.G.Karge, Eds.
Chapter I of J.L.Guth and H. Kessler in “Catalysis and Zeolites”, Fundamentals and Applications, J.Weitkamp and L.Puppe(Eds.), Springer, Berlin (1999)

ACKNOWLEDGMENT. I like to thank Dr. H. W. Kouwenhoven for reading the manuscript.

XIII. REFERENCES

1 G. Gottardi and E. Galli, Minerals and Rocks, Natural Zeolites, Springer-Verlag, Berlin, 1985.
2 L.B. Sand and F.A. Mumpton, Natural Zeolites, Occurrence, Properties and Use, Pergamon Press, Oxford, 1978.
3 R.L. Hay, Geologic Occurrence of Zeolites, in: L.B. Sand and F.A. Mumpton (Eds.), Natural Zeolites, Pergamon, Oxford, 1978, pp. 135-143.
4 a) A. Iijima, Geology of Natural Zeolites and Zeolitic Rocks, in: L. Rees (Ed.), Proc. 5th Int. Conf. on Zeolites, Naples, Italy, June 2-6, 1980, Heyden, London, 1980, pp. 103-118., b) R.W. Tschernich, Zeolites of the World, Geoscience Press, Inc. Phoenix, Arizona,USA, 1992.
5 R.M. Barrer, Synthesis of Molecular Sieve Zeolites, in: Molecular Sieves, London, England, Soc. Chem. Ind., London, 1968, pp. 39-46.
6 W.M. Meier and D.H. Olson, Atlas of Zeolite Structure Types, 2nd edn., Butterworths, London, 1987.
7 C.A.G. Konings, Delft University of Technology, Central Library. The Library search was performed using Controlled Vocabulary Index Terms of the Chemical Abstracts.
8 C.J. Brinker, D.E. Clark and D.R. Ulrich (Eds.), Symp. Proc. Mat. Res. Soc., Vol. 32, Better Ceramics through Chemistry, Albuquerque, U.S.A., February, 1984, Elsevier, New York, 1984.
9 C.J. Brinker, J. Non-Crystalline Solids, 100, 1988, pp. 31-50.
10 C.J. Brinker, D.E. Clark and D.R. Ulrich (Eds.), Symp. Proc. Mat. Res. Soc., Vol. 73, Better Ceramics through Chemistry II, Palo Alto, U.S.A., April 15-19, 1986, Mat. Res. Soc., Pittsburgh, 1986.
11a) S.D. Kinrade and T.W. Swaddle, Inorg. Chem., 27, 1988, 4253-4259. b) A.V. McCormick, A.T. Bell and C.J. Radke, J. Phys. Chem., 93, 1989, pp. 1741-1744.
12 P.P.Pescarmona, J.C. van der Waal, I.Maxwell and Th. Maschmeyer, Angew. Chem. Int. Ed. 2001, pp.740-743
13 R.A. van Santen, G. Ooms, C.J.J. den Ouden, B.W. van Beest and M.F.M. Post, Computational Studies of Zeolite Framework Stability, in: M.L. Occelli and H.E. Robson (Eds.), Zeolite Synthesis, ACS Symp. Ser. 398, ACS, Washington, DC, 1989, pp. 617-633.
14 A.G. Pelmenshschikov, G.M. Zhidomirov and K.I. Zamaraev, Quantum- chemical Interpretation of Intrazeolite Chemistry Phenomena, in: P.A. Jacobs and R.A. van Santen (Eds.), Stud. Surf. Sci. Catal., 49B, Elsevier, Amsterdam, 1989, pp. 741-752.
15 D.C. Bradley, Chem. Rev., 89, 1989, pp. 1317-1322.
16 a) J.L. Guth, H. Kessler, J.M. Higel, J.M. Lamblin, J. Patarin, A. Seive, J.M. Chezeau and R. Wey, Zeolite Synthesis in the Presence of Fluoride Ions, in: M.L. Occelli and H.E. Robson (Eds.), Zeolite Synthesis, ACS Symp. Ser., 398, ACS, Washington, DC, 1989, pp. 176-195. b) P.A.Barrett, E.T.Boix, M.A.Camblor, A.Corma, M.J. Diaz-Cabanas, S.Valencia and L.A. Villaescusa, Proc. 12th IZC, M.M.J.Treacy,B.K.Marcus,M.E.Bisher and J.B.higgins, (Eds.) (1998), Baltimore U.S.A., MRS, Warrendale, Pennsylvania.

17 L.B. Sand, A. Sacco, Jr., R.W. Thomson and A.G. Dixon, Zeolites, 7, 1989, pp. 387-392.
18 D.T. Hayhurst, P.J. Melling, Wha Jung Kim and W. Bibbey, Effect of Gravity on Silicalite Crystallization, in: M.L. Occelli and H.E. Robson (Eds.), Zeolite Synthesis, ACS Symp. Ser., 398, ACS, Washington, DC, 1989, pp. 233-243.
19 M. Niwa and Y. Murakami, J. Phys. Chem. Solids, 50, 1989, 487-496.
20 Chu Pochen, F.G. Dwyer and V.J. Clarke, Crystallization Method Employing Microwave Radiation, E.P. 0358827.
21 R.W. Thomson and A. Dyer, Zeolites, 5, 1985, pp. 202-210.
22 F. Di Renzo, F. Fajula, F. Figueras, S. Nicolas and T. des Courieres, Are the General Laws of Crystal Growth Applicable to Zeolite Synthesis?, in: P.A. Jacobs and R.A. van Santen (Eds.), Stud. Surf. Sci. Catal., 49A, Elsevier, Amsterdam, 1989, pp. 119-132.
23 L.Y. Hou and L.B. Sand, Determinations of Boundary Conditions of Crystallization of ZSM-5/ZSM-11 in one System, in: D. Olson and A. Bisio (Eds.), Proc. 6th Int. Conf. on Zeolites, Reno, U.S.A., July 10-15, 1983, Butterworths, London, 1984, pp. 887-893.
24E.G. Derouane, L. Baltusis, R.M. Dessau and K.D. Schmitt, Quantitation and Modification of Catalytic Sites in ZSM-5, in: B. Imelik (Ed.), Catalysis by Acids and Bases, Elsevier, Amsterdam, 1985, pp. 135-146.
25C.T-W. Chu, G.H. Kuehl, R.M. Lago and C.D. Chang, J. of Catal., 93, 1985, pp. 451-458.
26M.F.M. Post, T. Huizinga, C.A. Emeis, J.M. Nanne and W.H.J. Stork, An Infrared and Catalytic Study of Isomorphous Substitution in Pentasil Zeolites, in: H.G. Karge and J. Weitkamp (Eds.), Stud. Surf. Sci. Catal., 46, Proc. of an Int. Symp., Sept. 4-8, 1988, W

Studies in Surface Science and Catalysis 137
H. van Bekkum, E.M. Flanigen, P.A. Jacobs and J.C. Jansen (Editors)
© 2001 Elsevier Science B.V. All rights reserved.

Chapter 5B

Phosphate-based molecular sieves: novel synthetic approaches to new structures and compositions

Dr. Stephen T. Wilson, UOP Research and Development, UOP LLC, 25 E. Algonquin Rd., Des Plaines, IL, US

The first aluminophosphate molecular sieve, $AlPO_4$-5, was synthesized in 1978 using Pr_4NOH as template and synthesis conditions significantly different from those of zeolites. Using readily accessible organic amines and quaternary ammonium hydroxides as structure directing agents, over 20 structures, large, medium and small pore, were discovered within a few years, many of them unobtainable in zeolites. Framework elements such as Si, Mg, Co, Mn, and Ti were incorporated into the $AlPO_4$ lattice, producing Bronsted acidity or redox activity.

By 2000 the number of structures rose to over 50 as new phosphate-based compositions and novel synthesis media were explored. Remarkable structures with very large pore sizes, such as VPI-5 and cloverite, received a lot of attention and aroused some controversies. The number of elements that can be incorporated into phosphate-based molecular sieves is currently greater than 15.

Although novel templates were initially the primary source of this structural diversity, other synthesis modifications quickly led to new families of materials. The recent preparation and structural analysis of chain and sheet-like aluminophosphate structures has led to speculation that these may be similar to intermediates in the formation of $AlPO_4$-based molecular sieves and thus may provide mechanistic clues.

Most of the catalytic interest has centered on SAPO's which have weak to moderate Bronsted acidity and two of these have been commercialized, SAPO-11 in lube oil dewaxing by Chevron and SAPO-34 in methanol-to-olefins conversion by UOP/Norsk Hydro. Spurred by the success of TS-1 in oxidation catalysis, there is renewed interest in Ti, Co, V, Mn and Cr substituted $AlPO_4$-based molecular sieves for selective partial oxidation.

1. INTRODUCTION

1.1. Scope

The aim of this chapter is to update an earlier review of $AlPO_4$-based molecular sieve synthesis [1], with an emphasis on expansions in the original synthesis methods that have led to novel structures or compositions and to greater mechanistic insight. Earlier reviews offer additional useful background. In 1986 and 1988 Flanigen and coworkers [2, 3] first surveyed the structural and compositional diversity and introduced physicochemical principles to rationalize elemental site preferences and bonding patterns. In 1998 Szostak in Molecular Sieves: Principles of Synthesis and Identification devoted a chapter to aluminophosphate molecular sieves [4]. Most recently, Martens and Jacobs in their reviews of crystalline microporous phosphates updated the list of structure-types and further elaborated the physicochemical properties, particularly the bonding patterns [5, 6]. Kessler has also included a section on phosphate-based materials in his review of the synthesis of molecular sieves [7]. Cheetham has reviewed the broader field of open-framework inorganic materials, emphasizing the dramatic expansion during the last decade. The scope of microporous materials is no longer limited to aluminosilicate zeolites, Al phosphates and related systems, but now embraces frameworks based upon as many as 25 elements of the periodic table, and includes non-oxide phases found in the realm of sulfides, chlorides, and nitrides, as well as in coordination compounds [8].

1.2. Nomenclature

The nomenclature used by Union Carbide workers for the first $AlPO_4$-based molecular sieves assigned an acronym that reflected the framework composition ($AlPO_4$, SAPO, . . .) and a number that identified the structure-type (e.g. 5, 11). The naming conventions, as in zeolites, have gotten more complicated as more research groups prepare novel materials, and multiple names frequently exist for the same structure type (e.g. VPI-5, AlPO-54, $AlPO_4$-H1). The original $AlPO_4$ designation has often been shortened to AlPO in the literature for greater consistency with SAPO, MAPO, etc. Structures with similar numerical designations may be quite different, in fact. The use of IZA structure codes (where they have been assigned) is often necessary to differentiate such structure-types; for example AlPO-5 (AFI), VPI-5 (VFI), and DAF-5 (CHA) are all $AlPO_4$-based molecular sieves, but they differ in structure.

1.3. Compositions

The $AlPO_4$ [9, 10], SAPO [11, 12] and various substituted $AlPO_4$-based molecular sieves that followed the original discovery demonstrated the remarkable ability of the $AlPO_4$ composition to incorporate other elements into either Al or P lattice sites. The most common compositional families and their acronyms are shown in Table 1.

Additional acronyms were created to encompass multiple families of materials, such as MeAPO or MeAPSO (where Me is one of the divalent cations in Table 2). The terms ELAPO and ELAPSO were coined to reflect a broad range of compositions where the element "EL" is defined based on crystal chemical principles and represents one or more elements capable of forming a tetrahedral framework oxide unit.

In 1986 Flanigen et al initially enumerated 13 elements (Table 2) which can be incorporated alone or in combination into the $AlPO_4$ framework. Over the intervening years four additional elements have been added to the list (V, Cr, Ni, Cu), although the evidence is less persuasive, the number of structure-types is more limited, and the levels of incorporation are significantly lower.

Table 1
Original families of $AlPO_4$-based compositions

Acronym	US Patent	Year
$AlPO_4$ (or AlPO)	4310440	1982
SAPO	4440871	1984
TAPO (or TiAPO)	4500651	1985
FAPO (or FeAPO)	4554143	1985
CoAPO	4567029	1986
MAPO (or MgAPO)	4567029	1986
MnAPO	4567029	1986
ZAPO (or ZnAPO)	4567029	1986
FeAPSO	4683217	1987
MnAPSO	4686092	1987
TiAPSO	4684617	1987
CoAPSO	4744970	1988
MAPSO (or MgAPSO)	4758419	1988
ZAPSO (or ZnAPSO)	4935216	1990

The ability of some of these elements to assume tetrahedral coordination in an oxide lattice stretches the accepted crystal chemical concepts regarding radius ratio and T-O distance. Since the divalent elements like Mn^{2+}, Fe^{2+}, and Zn^{2+} are significantly larger than Al^{3+}, they might not be expected to fit into an Al^{3+} vacancy. The successful siting of such elements was attributed to the greater spatial accommodation of the microporous structures that allows larger cations to "bulge" into the open channels. For the second and third row transition elements the tendency toward higher coordination numbers and larger size should preclude them from occupying tetrahedral sites in the $AlPO_4$ lattice.

Table 2
Abbreviated Periodic Table showing elements incorporated into $AlPO_4$-based molecular sieves and their typical oxidation states.

+1	+2		+4	+4 +5	+3	+2 +3	+2 +3	+2 +3	+2	+2	+2	+3	+4	+5
Li	Be											B		
Na	Mg											Al	Si	P
K	Ca	Sc	Ti	V	Cr	Mn	Fe	Co	Ni	Cu	Zn	Ga	Ge	As

Shaded = good evidence; Underlined = weak evidence

Based in part on the $AlPO_4$ model, compositional families without Al or P have also been discovered. Binary phosphate frameworks include $GaPO_4$, $M(BePO_4)$, and $M(ZnPO_4)$ where the M are non-framework species, either alkali or organic cation, balancing framework charge. The divalent elements that have been most successful at forming microporous phosphates are the same elements that were seen to substitute for Al in the initial $AlPO_4$-based compositions. The synthesis of assorted gallophosphates, in particular, has benefited from the expansion in synthetic methods, such as the use of fluoride-containing media or alcoholic media. (see below)

The compositional complexity of the $AlPO_4$-based molecular sieves required a flexible method of expressing framework compositions. This led to the normalized TO_2 formula, $(El_xAl_yP_z)O_2$, where the relative concentration of the framework elements El, Al, and P is expressed as a mole fraction, $x + y + z = 1.0$. For example a typical framework composition of SAPO-34 would be $(Si_{0.08}Al_{0.50}P_{0.42})O_2$.

2. STRUCTURAL FEATURES

2.1. Structure-types

In the first review of aluminophosphate structures with Al/P =1, Bennett et al described ten novel structures [13]. There are currently over 50 phosphate-containing molecular sieve structures [14] recognized by the IZA Structure Commission and many of these have no zeolite analogs. The examples in Table 3 illustrate the compositional diversity observed for each structure-type so far. Since the focus of this chapter is on synthesis, most of the references in this table were chosen because they include significant synthetic details. In addition to these structures, there are numerous other tetrahedral nets containing significant phosphate concentration which do not yet have IZA codes

Table 3 (pt 1)
Compositional Variants of Phosphate-based Structures in IZA Atlas

IZA	Example	Ring T-atoms	AlPO	SAPO	MeAPO	Other PO_4	Zeolite	Stable
ABW	ZnPO4	8				[15]	x	N
ACO	ACP-1	8			[16]			N
AEI	AlPO-18	8		[17]	[18]			Y
AEL	AlPO-11	10	[9, 10]	[11, 12]	[19]			Y
AET	AlPO-8	14	[9, 10]	[20]				Y
AFN	AlPO-14	8	[9, 21]			[22]		Y
AFI	AlPO-5	12	[9, 10]	[11, 12]	[19]		x	Y
AFO	AlPO-41	10	[23, 24]	[12, 25]	[26]			Y
AFR	SAPO-40	12	[12, 27]	[28]	[27]			Y
AFS	MAPSO-46	12		[29]	[30, 31]			Y(dry)
AFT	AlPO-52	8	[32]					Y
AFX	SAPO-56	8		[33, 34]	[35]		x	Y
AFY	CoAPO-50	12			[30, 36]			N
AHT	AlPO-H2	10	[37, 38]					N
ANA	BePO4-P	6	[9]		[16]	[39, 40]	x	N
APC	AlPO-C	8	[37, 41]					Y
APD	AlPO-D	8	[37, 41]					Y
AST	AlPO-16	6	[9, 10]	[12]	[19]		x	Y
ATN	MAPO-39	8	[42]	[43]	[19, 44]			Y

Table 3 (pt 2)
Compositional Variants of Phosphate-based Structures in IZA Atlas

IZA	Example	Ring T-atoms	AlPO	SAPO	MeAPO	Other PO_4	Zeolite	Stable
ATO	SAPO-31	12	[9, 10]	[12, 45]	[46]			Y
ATS	MAPO-36	12	[47]	[33]	[19]			Y
ATT	AlPO-33	8	[48]					Y
ATV	AlPO-25	8	[9, 10]					Y
AWO	AlPO-21	8	[9, 10]		[49]			N
AWW	AlPO-22	8	[9, 10]					Y
BPH	BePO4-H	12				[39, 50]	x	N
CAN	Tiptopite	12				[51]	x	N
CGF	CoGaPO-5	8				[52]		N
CGS	CoGaPO-6	8				[53]		N
CHA	SAPO-34	8	[54, 55]	[12]	[19]	[55]	x	Y
CZP	Chiral $ZnPO_4$	12				[56]		Y
DFO	DAF-1	12			[57]			Y(dry)
DFT	DAF-2	8				[58]		N
EDI	BePO4-E	8				[39]	x	N
ERI	AlPO-17	8	[9, 10]	[12]	[19]		x	Y
FAU	SAPO-37	12		[12, 59]	[60]	[61]	x	Y(dry)
GIS	SAPO-43	8	[62]	[12]	[30, 63]	[64]	x	N
LAU	CoGaPhosphate					[65]	x	N
LEV	MAPO-35	8	[66]	[12]	[19]		x	Y
LOS	Li-BePO4	8				[67]	x	N
LTA	SAPO-42	8	[68, 69]	[12, 70]	[68]	[71]	x	Y
MER	ACP-MER2	8			[16]		x	ND
OSI	UIO-6	12	[72]					Y
PHI	ACP-2	8			[73]		x	ND
RHO	Pahasapaite	8				[39, 74]	x	ND
SAO	STA-1	12			[35]			N
SAS	STA-6	8			[75]			Y
SAT	STA-2	8			[76]			Y
SAV	STA-7	8			[77]			Y
SBE	UCSB-8	12			[73]			N
SBS	UCSB-6	12			[73]	[73]		N
SBT	UCSB-10	12			[73]	[73]		N
SOD	AlPO-20	6	[9, 10]	[12]	[19]	[78]	x	Y
THO	ACP-THO1	8			[16]	[16]	x	ND
VFI	VPI-5	18	[79, 80]	[81, 82]	[83]			Y
WEI	Weinebeneite	10				[84]		ND
ZON	ZAPO-M1	8	[85]		[86]			Y
-CLO	Cloverite	20				[87]		Y

x = zeolite structural analog exists, N = no, Y = yes, ND = not reported, Y(dry) = thermally stable when dry

Some of these structure-types have been observed in very limited compositions, due in part to insufficient synthetic effort, but in some cases the limitations probably reflect strong compositional preferences. With the dramatic increase in the number of new structures described as "potentially" microporous, the stability of such structures to removal of occluded species (by thermal or other means) has become an issue. Many but not all of the structures in Table 3 are stable to removal of the occluded organic or water. Where stability data has been reported, this is indicated in Table 3, but the author recognizes that stability is not easily defined for some of these structures and is certainly dependent on composition. There are at least three examples of structures that are thermally stable to template removal, but lose structure on rehydration at room temperature.

2.2. Bonding concepts

Based on considerable compositional and structural data and the limitations imposed by hydrothermal synthesis conditions, Flanigen et al [2] proposed a number of guiding structural principles or crystal chemical rules (Table 4). These rules have remained valid for all the stable, calcined structures describable as tetrahedral nets. Based on these concepts Flanigen et al proposed rules for framework cation siting of metal and Si in AlPO frameworks: 1) Me incorporation into a hypothetical Al site, 2) Si into a hypothetical P site, and 3) 2 Si for Al+P.

Table 4
Bonding concepts in $AlPO_4$-based molecular sieves [3]

Observed	Not likely
Al-O-P	P-O-P
Si-O-Si	P-O-Si
Si-O-Al	Al-O-Al
Me-O-P	Me-O-Al
Me-O-P-O-Me	Me-O-Me
Electroneutral and negatively charged frameworks	Positively charged frameworks

Some of the as-synthesized $AlPO_4$ structures do contain additional Al-OH bonds (e.g. AlPO-17, AlPO-18, AlPO-14, and AlPO-EN3) that expand the coordination sphere of the Al to 5 or 6. These oxo-species are thought to be bridging hydroxyls, producing a negative framework charge to balance the positively charged, occluded templates. The hydroxo-species are lost along with the template on calcination, resulting in a tetrahedral network. Other structures containing higher concentrations of such bonds may undergo a more disruptive solid state transformation upon thermal removal of the template (e.g. the conversion of AlPO-21 to AlPO-25).

Martens and Jacobs [5] further elaborated the types of isomorphic substitution according to various substitution mechanisms (SM). In particular, monovalent (SM Ia), divalent (SM Ib), and trivalent (SM Ic) elements substitute for Al, and thus bond through oxygen to P. Tetravalent (SM IIa) and pentavalent (SM IIb) elements substitute for P. This behavior avoids positively charged frameworks or too high negative charge density. These authors also recognized that the avoidance of Si-O-P bonds precludes the simple "homogeneous" substitution of 2 Si for Al+P (SM III ho), so a third type of "heterogeneous" substitution mechanism (SM III he) was needed that reflected the presence of SiO_2 patches with only Si-O-Si and Si-O-Al bonds. All the framework charge associated with this patch is concentrated at the boundary. The presence of additional Al within the SiO_2 patch produces a zeolitic region.

3. SYNTHESIS

3.1. General features

In 1961 F. D'Yvoire reported the synthesis of various hydrates of $AlPO_4$ (designated H1, H2, H3, H4) under relatively acidic conditions and in the absence of any organic species [37]. The structures of three of these hydrates {H2 (AHT), H3 (APC), H4} and two dehydrated conversion products of H3 {C (APC), D (APD)} have been determined. H1 has been claimed to be isostructural with VPI-5, but there is some disagreement in the literature. Structure-types APC and APD can be considered microporous since they can adsorb H_2O [41], but H2 and H4 do not survive dehydration, converting to dense phases.

The first $AlPO_4$-based molecular sieves prepared using organic structure-directing agents were the binary oxide $AlPO_4$ (AlPO) compositions. These were prepared from gel compositions [9, 10] containing equimolar amounts of Al and P, typically R • Al_2O_3 • P_2O_5 • 40 H_2O where R is an organic amine or quaternary ammonium hydroxide. Many of the early structures were prepared by simply varying the template or structure-directing agent R. Other synthesis parameters like template concentration, crystallization time and temperature also influence structure formation. One example of this phenomenon is the ability of di-n-propylamine (nPr_2NH) to template AlPO-11, -31, -39, -41, or VPI-5 (Table 5). With one exception (VPI-5) these structures contain significant amounts of occluded template. The template can be removed by calcination with retention of structure. The VPI-5 typically does not contain occluded template when analyzed, but this may be the result of extensive washing during work-up coupled with weak adsorption of the template. In general, crystallization occurs at 100-200°C, in the absence of alkali elements, and typically in the pH range 3-10.

Table 5
Products from base gel x n-Pr_2NH •: Al_2O_3 • P_2O_5 • 40 H_2O: Additive.

Structure	x	Time	Temp	Additives	Ref.
AlPO-11	1.0	24	200		[9]
AlPO-31	1.0	46	200		[9]
AlPO-39	2.0	168	150	1.0 CH_3COOH	[42]
AlPO-41	2.0	22	200		[23]
VPI-5	1.0	22	142		[79]

The $AlPO_4$ products generally exhibit strict alternation of Al and P in the calcined, dehydrated form. Most of the early syntheses utilized a reaction mixture composition that reflected this structural preference of the product, i.e. equimolar Al and P. Orthophosphoric acid is almost always the source of P, and pseudoboehmite alumina or aluminum alkoxide are the most common Al reagents. The organic template was typically present in excess of the stoichiometric requirement and served a dual role, providing structure-direction and pH control. It has been found advantageous to adjust the initial pH to a value greater than 3 to avoid formation of dense phases such as $AlPO_4$-tridymite or -quartz. This adjustment has most often been accomplished by addition of template. Elevated pH (> 10) tends to either prevent crystallization or solubilize product, reducing yield.

By far the most common templating agents are the amines and quaternary ammonium hydroxides. To date >125 different amines and >35 quats have been successfully utilized. These are predominantly commercially available compounds. Some custom synthesis templates, rigid polycyclic organic quats that have been so successful in the synthesis of novel high-silica structures, have also been tested in AlPO and SAPO synthesis, but only previously known structures such as AlPO-5 and SAPO-34 were reported [88]. The more complex SAPO and MeAPO compositions were initially prepared using reaction mixtures of composition: wR • xAl_2O_3 • P_2O_5 • 40 H_2O • z [other oxides]. In the preparation of SAPO molecular sieves almost all reactive forms of silica have been used successfully – fumed, colloidal, precipitated, alkoxide. The other oxide-forming elements have usually been added as water-soluble salts, preferably acetates or nitrates. Although the organic template [R] is thought to exert the strongest structure direction, the addition of other framework elements can also influence structure direction. This is particularly true of divalent substitution where framework charge is introduced. For example, the structures like CoAPO-50 and UCSB-6 only exist as highly charged frameworks.

3.2. Templating

For many of the early structures the structure-directing role of the template appeared to be dominated by stereospecific space-filling and stoichiometry between the template and the framework, and influenced to a lesser extent by framework charge compensation. This was particularly true of the charge neutral $AlPO_4$ compositions. The increasing numbers of structure determinations for as-synthesized $AlPO_4$ materials, however, have indicated the presence in the lattice of bonded OH-species producing a temporary negative charge balancing the occluded cationic template, whether quaternary ammonium or protonated amine. Calcination removes both cationic template and extraframework hydroxide. For many of the more recent, highly charged frameworks with high levels of divalent cation substitution (e.g. UCSB-6, -8, and -10), charge compensation by the template has reasserted its importance. Hydrogen-bonding between the protonated amine groups of the template and lattice oxygen is also increasingly significant.

Since 1991 thirty-two phosphate-containing structures have been recognized by the IZA structure commission. This includes fifteen $AlPO_4$ structures that pre-existed that date, but were granted IZA codes after 1991 (AEI, AEN, AET, AFN, AFO, AFR, AFT, AHT, ATN, ATO, ATS, ATV, AWO, AWW, VFI). Of the remaining structures (Table 6) the majority were templated with diamines or diquats. Of these structures, seven are stable to conventional removal of the templating agent via calcination. The lower stability of the remaining 10 structures seems to correlate with a higher concentration of divalent cation in the lattice, rather than lower framework density.

As in zeolite synthesis, structure-template specificity for the $AlPO_4$-based structures varies remarkably. Some structures readily form with many different templates, e.g. AFI readily crystallizes with > 85 different templates, the smallest being isopropylamine (3 carbons) and the largest, hexabutyl-1,6-hexanediammonium (30 carbons). This relative lack of specificity is similar to that of the zeolitic MTW structure which also possesses a one-dimensional channel structure bounded by a 12-ring. Greater than 49 templating agents have been reported for MTW. The CHA structure forms with > 30 templates, ranging in size from isopropylamine to N,N,N-trimethyl-1-adamantanammonium (13 carbons), and AEL forms with >20 templates, the largest being 1,2-bis(4-pyridyl)ethane. In contrast, CoAPO-50 (AFY) and SAPO-40 (AFR) have only been prepared with nPr_2NH and Pr_4NOH, respectively. It is interesting to note that many of the structures with the largest cages, such as FAU, -CLO, or SBS, were not templated by a particularly large template, but rather actually contain multiple organic cations in each cage. For example, the large cage in cloverite (-CLO) is thought to contain 8 quinuclidinium cations.

Table 6
Phosphate-containing structure-types given IZA codes since 1991.

Species	IZA	Template	Stable[a]	% divalent cation[b]	FD[c]
SAPO-56	AFX	N,N,N',N'-tetramethyl-hexanediamine	Yes	0	14.3
UiO-6	OSI	Tetraethylammonium	Yes	0	18.7
STA-1	SAO	Heptane-1,7-bis-quinuclidinium	Yes	9	13.9
STA-6	SAS	1,4,8,11-tetramethyl-1,4,8,11-tetraazacyclotetradecane	Yes	9	15.0
STA-2	SAT	Butane-1,4-bis-quinuclidinium	Yes	8	16.6
STA-7	SAV	1,4,8,11-tetramethyl-1,4,8,11-tetraazacyclotetradecane	Yes	10	14.5
UiO-7	ZON	Tetramethylammonium, F	Yes	0	17.0
Chiral-ZnPO	CZP	Na	---	50	16.7
Cloverite	-CLO	Quinuclidine, F	Partial	0	11.1
ACP-1	ACO	Ethylenediamine	No	45	16.0
CoGaPO-5	CGF	DABCO	No	22	17.4
CoGaPO-6	CGS	1,2-diaminocyclohexane	No	25	15.5
DAF-1	DFO	Decamethonium	No	13	14.1
DAF-3	DFT	Ethylenediamine	No	50	16.5
UCSB-8	SBE	1,9-diaminononane	No	25	13.7
UCSB-6	SBS	1,7-diaminoheptane	No	25	13.7
UCSB-10	SBT	4,7,10-trioxa-1,13-tridecanediamine	No	25	13.7

[a] Stable to template removal; [b] % divalent cation in framework sites; [c] Framework density ($TO_2/1000A^3$). See Table 3 for references.

There are a growing number of examples of what appears to be cooperative structure-direction in the crystallization of new structures (Table 7). SAPO-37 (FAU) is prepared using a mixture of TPAOH and TMAOH. Solid state NMR, chemical analysis, and thermal analysis support the presence of TMA in the sodalite cages, and TPA in the supercages. The amounts of TPAOH and TMAOH must be carefully balanced to avoid formation of SAPO-5 and SAPO-20 impurities. In SAPO-LTA, the TMA is again present in the sodalite cages, diethanolamine occupies the α-cage and 8R channel system, and fluoride is found in the center of all D4R units. In these two examples the organics occupy unique locations in the structure [70]. In the synthesis of AlPO-52, both Pr_3N and TEAOH appear to be necessary, but only one of the organics (TEAOH) is occluded in the structure. The large-pore gallium oxyfluorophosphate GaPO-TREN, prepared by Weigel et al., using a combination of two amines, has two parallel channel systems delineated by 12- and 8-membered rings. The organic amines are ordered in the channels, TREN in the 12-ring channel and pyridine in the 8-ring one. The authors propose that one of the amines acts as a structure-directing agent and the other as a space-filling agent. Of these six examples all but the GAPO-TREN are reasonably stable to removal of the occluded organic.

Table 7
Examples of cooperative structure-directing effects

Structure	Structure directing agents	Proposed location	Ref
SAPO-37	TPAOH	Supercage	[89]
	TMAOH	Sodalite cage	
AlPO-52	TEAOH	Large cage	[32]
	Pr_3N	Not occluded	
SAPO-LTA	TMAOH	Sodalite cage	[70]
	Diethanolamine	Supercage	
	F	D4R	
GaPO-TREN	TREN	Large channel	[90]
	Pyridine	Small channel	
	F		
SAPO-40	TPAOH	Large channel	[28]
	TMAOH	Not occluded	
	Alkali	Not occluded	
UiO-6	TEAOH	Large channel	[72]
	F	Occluded	
	Alkali	Occluded	

TPAOH = tetrapropylammonium hydroxide, TMAOH = tetramethylammonium hydroxide, TEAOH = tetraethylammonium hydroxide, TREN = tris(2-aminoethyl)methane

The participation of alkali cations in the synthesis of $AlPO_4$-based structures is relatively rare. The earliest reports of SAPO-40 synthesis indicated that trace amounts of Na^+ in the TPAOH were present during synthesis and apparently beneficial [12]. A later study found that the trace amounts of alkali seemed to suppress SAPO-5 nucleation [91]. Further synthesis optimization eliminated the need for trace alkali and found that, although the TMA cation was not required, it was still desirable in order to suppress AFI nucleation [92]. The 12-ring structure UiO-6 also appears to benefit from the presence of both K^+ and F^- during synthesis and both species are occluded, although no structural details have been reported [72]. Alkali cations have been most successful in templating highly charged frameworks such as MZnPO4 (see below).

The synergistic use of two or more templating agents to produce novel structures has barely been exploited, but the number of possible combinations is daunting and the success rate is likely to be low.

As an alternative to the amines and quaternary ammonium cations, Balkus and coworkers have explored the use of positively charged, stable metal-ligand complexes as cationic templates. Cobalticinium ion, Cp_2Co^+, was found to be a template for the AFI and AST structures [93]. Cloverite and UTD-10 were prepared using the metal-ligand complex $[Co(NMe_3)_2sar]^{5+}$ (where $(NMe_3)_2sar$ = 1,8- bis(trimethylammonio)-3,6,10,13,16,19-hexaazabicyclo[6.6.6]- icosane) [94]. In this complex the Co^{3+} is completely encapsulated by the bicyclic ligand.

Alkali cation/crown ether complexes have been successful as templating agents in the synthesis of the zeolite compositions EMT, FAU, and KFI. In aqueous gallophosphate media, the K^+/18-crown-6 complex was not sufficiently stable to play a templating role. However, in alcoholic media (MeOH) and in the presence of KF, this complex yielded $GaPO_4$-LTA. In general, the N-containing aza-crown ethers have been far more successful as templates in $AlPO_4$-based or gallophosphate synthesis [95]. MgAPO and CoAPO compositions with AEI, LTA, SAS or SAV structure-types were prepared with aza-crown ethers. Structural analysis revealed that the crown was occluded, but no cations remained within the macrocycles, suggesting that these organics were simply acting as cyclic amines [96].

Although novel templating agents continue to hold promise for preparation of novel structures, other synthesis parameters have also been exploited with significant success.

3.3. Synthesis with water as solvent

Most of the early synthesis efforts employed H_2O as the only solvent, and the amount of H_2O present in these gels was usually 10 H_2O/TO_2. Over time the range of H_2O concentrations effective in crystallization has been greatly expanded. For example, AlPO-5 has been successfully synthesized using the gel composition $Et_3N \bullet Al_2O_3 \bullet P_2O_5 \bullet xH_2O$, where H_2O/TO_2 varies from 10 to 2.5 [97]. When the water content was low (2.5 H_2O/TO_2), the conversion of reactants to AlPO-5 was nearly stoichiometric and the synthesized products had higher crystallinity and the ideal Al/P =1 ratio. SEM also reveals that there is a preference for the formation of thin hexagonal platelets in AlPO-5 samples synthesized from starting molar gel compositions with $H_2O/TO_2 = 2.5$ to 5. The "dry gel" method has also been employed to prepare AlPO-5, AlPO-11 and the corresponding SAPO forms [98]. In this method the minimum amount of water necessary to support crystallization was found to be approximately 1.1-1.4 H_2O/Al_2O_3. At the other extreme, 50 H_2O/TO_2 were used to prepare AlPO-5 using tropine as template [99] and SAPO-5 crystals have been grown from solutions with >180 H_2O/TO_2. (see below)

The synthesis of aluminophosphates from clear solutions was first reported by Lesch and Patton [100] using aqueous solutions containing Al, P, template, and optionally one or more additional elements (e.g. Si, Mg, Co). Relatively high template, phosphate, and water concentrations (Table 8) characterized these solutions and product yields were low. Although a variety of previously known structures were observed, this method also produced the first preparation of AlPO-34, a CHA structure without Si or metal present.

Later, clear solution synthesis of AlPO-11 was demonstrated using fluoride media [101], e.g. $2.4\ nPr_2NH \bullet Al_2O_3 \bullet 1.26\ P_2O_5 \bullet 1.5\ HF \bullet 70\ H_2O$. Crystallization initially produced low yields of very thin, 1 mm size needles (2h). After 18 h, the yield increased 50% to 75% at the expense of the crystal size, with smaller 5-10 μm crystals resulting. The authors proposed that a secondary nucleation occurs, with the redistribution of nutrients between the larger, initially-formed crystals and the smaller growing crystals. The solution phase plays a critical role in AlPO crystallization, while both transient amorphous and crystalline phases appear to act as a source of Al and P nutrients. In both these approaches to solution synthesis, Al_2O_3 is the limiting reactant, and there is an attempt to balance the acidic (H_3PO_4, HF) and basic constituents (amine or quaternary ammonium hydroxide) in order to avoid overly acidic or basic conditions. Most syntheses of $AlPO_4$-based molecular sieves still occur in the pH range 3-10.

Table 8
Synthesis of $AlPO_4$-based molecular sieves from solution [100]

Template, w	Oxide, x	P_2O_5, y	H_2O, z	Seed	Temp °C	Time h	Products
8.0 TEAOH		4.0	149	AlPO-5	100	192	AlPO-34, tr 5
8.0 TEAOH		4.0	149	None	100	474	AlPO-34, tr 18
10.0 TPAOH		5.0	444	None	150	24	AlPO-5
8.0 QUIN		4.0	180	None	150	189	AlPO-17
10.0 TEAOH		5.0	153	None	150	216	AlPO-18
10.0 TEAOH	0.4 MgO	5.0	179	None	100	17	MAPO-34
10.0 TEAOH	0.8 SiO2	5.0	179	None	100	169	SAPO-34

w Template • x oxide • Al_2O_3 • y P_2O_5 • z H_2O. Reactant concentrations normalized to 1.0 Al_2O_3. QUIN = quinuclidine, TEAOH = tetraethylammonium hydroxide, TPAOH = tetrapropylammonium hydroxide, tr = trace

3.4. Large crystals

Solution or near-solution synthesis has been employed to make large crystals. A recent review of the synthesis of large, single molecular sieve crystals includes examples of phosphate-based structures [102]. These have been prepared in a variety of reaction media including predominantly organic or aqueous solutions, sometimes including weak acids or fluoride ions.

Reaction media similar to that used for solution synthesis, but with even higher water concentrations, have been used by Finger et al to grow large crystals of AlPO-5 and SAPO-5 from solution or near solution with Al_2O_3 as the limiting reagent [103]. For example, the reaction mixture 3.4 Et_3N • Al_2O_3 • 2.4 P_2O_5 • 0.25 SiO_2 : 1300 H_2O tends to give large crystals up to 0.5 mm in length [104]. Using similar methods somewhat smaller crystals of SAPO-31 were also prepared [105].

Using the gel composition 1.55 Et_3N • Al_2O_3 • 1.03 P_2O_5 • n SiO_2 • 750 H_2O, Demuth et al studied the effects of Si source and concentration on hydrothermal SAPO-5 synthesis [106]. They found that large crystals of hexagonal rod-like morphology (av. size $\leq 220 \times 50$ μm) and high optical quality were produced in high yields only if an aluminum oxide hydrate sol or a pseudohydrargillite was used as aluminum source. It was possible to control the morphology and the crystal size of SAPO-5 by the silica source and amount used. Further, it was found that the type of silica in the synthesis gel drastically influences the rate of crystal growth parallel to the crystallographic c-axis. The amount of pyrogenic silica used controls the morphology of the crystals, while the reactivity of the silica determines the crystal size obtained. Using powder x-ray diffraction and TGA-DTA it was demonstrated that Si predominantly substitutes for P sites if a Si molar fraction [Si/(Al + P + Si)] of <0.06 is used in the synthesis gel.

Schunk and coworkers found a compositional gradient for silicon as T-atom substituent in large optically clear crystals with the AFI structure using the electron microprobe [107]. The silicon concentration in the center of the crystals is lower by a factor of two to three compared to the other part of the crystals. This corresponds to the findings of their kinetic studies and previous findings concerning the correlation between silicon content and crystal size. During the early stages of AFI crystal growth, 'pencil type' crystals form that are depleted in Si. These later develop into hexagonal prisms, undergoing a flattening of the tips during crystal growth, as larger amounts of Si are incorporated. Increasing the initial silicon content of the gel accelerates the morphological change and inhibits lengthwise growth of the crystallites.

A different method was given by Kuperman and coworkers for growing uniformly large and morphologically well defined single crystals of aluminophosphate molecular sieves with dimensions of 0.4-5.0 mm [108]. Their method employs a nonaqueous solvent (usually pyridine), a mineralizer (HF), an additional organic templating species and only reagent amounts of water. Both large-pore and small-pore materials of various framework compositions and structure type were claimed. Crystallization from alcoholic media also tends to produce larger crystals.

The synthesis of VPI-5 (VFI) was optimized for the production of large single crystals with dimensions up to 400 x 35 x 35 μm [109]. All syntheses produced aggregates of needle-like crystallites. In most cases, these were highly intergrown, but under the optimized conditions, individual hexagonal needles radiating from a central point were formed. Key synthesis parameters were found to be the use of polyphosphoric acid (longer chain length preferred), the H_2O content of the synthesis mixture, the ultrasonic treatment of the gel, the selection of the organic additive and the reaction temperature. With a gel composition of $Al_2O_3 \bullet P_2O_5 \bullet 95.4$ $H_2O \bullet$ dibutylamine and a reaction temperature of 123°C, large single crystals could be grown.

Many of the novel binary and ternary phosphate structures discovered by the UCSB group were grown as relatively large single crystals from alcohol-rich media (see below).

3.5. Compositional gradients

As noted above for SAPO-5, both long- and short-range compositional gradients have been observed in a variety of $AlPO_4$-based molecular sieves. Pellet et al accomplished this by growing crystals containing Al and P as part of the composition then, by a change in crystallization conditions or media composition, growing the same crystal structure epitaxially on these seed crystals, but with a different elemental composition [110]. Examples included SAPO-11 on AlPO-11, SAPO-31 on AlPO-31, CoAPSO-47 on SAPO-47. By manipulating the synthesis media, compositional gradients were induced even in small crystals.

Compositional gradients can also result during a "typical" crystallization where all the ingredients are present at the onset of crystallization. The Si gradients observed in the large crystals of SAPO-5 cited above reflect a time dependency in the availability and reactivity of Si during the crystallization. Another case of compositional gradients was reported in a comparison of CoAPO- and CoAPSO-34 (CHA) [111]. The presence of Co in the framework was established by a number of experimental techniques. Using electron microprobe analysis, Co was found to be homogeneously distributed in the CoAPO, while in the CoAPSO it had higher concentration near the center of the crystal that progressively decreased toward the edges as the Si content increased. These two examples reflect long-range compositional heterogeneity, primarily from center-to-edge of a crystal. These results are consistent with the idea that divalent framework species like Co^{2+} or Mg^{2+} are more readily incorporated than Si^{4+} into the growing framework. The rate of incorporation may reflect the relatively low solubility of silica species at the typical synthesis pH values observed in SAPO synthesis.

Shorter range compositional gradients were invoked by Martens et al to explain the catalytic activity of SAPO-5, SAPO-11, and SAPO-37 samples for the conversion of decane [112, 113]. By elemental analysis, global framework Si T-atom fractions ranging from 1-42% were determined. Solid state ^{29}Si-MASNMR detected non-uniform distributions of Si, Al, and P, which was attributed to the presence of both zeolitic aluminosilicate crystal domains (SA) and silicoaluminophosphate (SAPO) crystal domains. The acid sites located in the SA and SAPO domains and at the domain boundaries all contribute to the catalytic activity, but the contribution of each type is strongly dependent on the SAPO-n structure type (see below).

3.6. Mixed solvents – alcohol/water

The use of alcohol and alcohol/water mixtures as solvent has been exploited to produce novel structures and to isolate possible synthesis intermediates. The synthesis of molecular sieves from non-aqueous solvents has been recently reviewed [114].

Table 9
SAPO structures formed in n-hexanol-H_2O media [115]

Species	IZA code	Template
MCM-1	APC	TPAOH
MCM-2	CHA	TEAOH
MCM-6	AFI	
MCM-7	AEL	nPr_2NH
MCM-8	SOD	TMAOH
MCM-9	VFI + AEL	nPr_2NH
MCM-37	AET	nBu_2NH

In one of the earliest uses of alcohol solvents for the synthesis of phosphate molecular sieves, Derouane and von Ballmoos in various Mobil patent applications [115] employed biphasic media – n-hexanol/water – to synthesize SAPO molecular sieves, which they designated MCM-1 through 10 (Table 9). They essentially prepared an aqueous AlPO gel (Al_2O_3 + H_3PO_4 + organic + H_2O) which they combined with an n-hexanol solution of tetraethyl orthosilicate (TEOS). Since n-hexanol is not very soluble in water at ambient conditions, slow hydrolysis of the TEOS presumably occurs and silica species are partitioned between the two phases. It is difficult to predict the partitioning of the other reactants in such a complex mixture, particularly at elevated temperatures.

More recently Franco and coworkers studied hydrothermal crystallization of SAPO-5 in a biphasic n-hexanol-H_2O system in the presence of surfactants [116]. The chemical composition of a typical synthesis gel was 1.5 Et_3N • Al_2O_3 • 1.0 P_2O_5 • 0.5 SiO_2 • 40 H_2O • 4.4 n-hexanol • x surfactant, using the cationic surfactant hexadecyltrimethylammonium chloride. At the highest cationic surfactant concentration (above the critical micelle concentration), no phase separation was observed and the product was large spherical aggregates (20-30 μm). In the absence of n-hexanol, hexagonal plates were the dominant morphology, and the influence of the surfactant was quite small. The Si content of the crystals was not significantly affected by surfactant and n-hexanol. ^{29}Si-MASNMR measurements showed modification of the mechanism of Si incorporation, with the surfactant favoring smaller pure Si domains and larger aluminosilicate domains. Neutral or anionic surfactants were ineffective in promoting these changes.

R. Xu and his coworkers have prepared reaction media using alcohols that are highly miscible with water, like ethylene glycol (EG) and propylene glycol [117]. The alumina is supplied as aluminum isopropoxide and the only water present in these gels comes from the phosphoric acid [118]. Thirty-two different amines were examined using the following reaction mixture composition: 0.5-20 amine • Al_2O_3 • 1.2 - 3.0 P_2O_5 • 40 EG • xH_2O. Crystallization was generally slower than in aqueous media, typically 5-30 days at 180°C, and single crystals were more common. Two examples of previously known phases prepared under such conditions are:

AlPO-5: 5.9 Et_3N • Al_2O_3 • 1.8 P_2O_5 • 45 propylene glycol • x H_2O
AlPO-11:5.0 nPr_2NH • Al_2O_3 • 1.8 P_2O_5 • 60 ethylene glycol • x H_2O

Other structures prepared were AlPO-41, AlPO-21, AlPO-20, MgAPO-43, and CoAPO-36.

The addition of divalent cations like cobalt, magnesium and zinc to such alcohol-rich aluminophosphate gel compositions was utilized by Feng and coworkers to generate many structures with relatively high framework charge, designated ACP-xxxx [15]. These include synthetic zeolite analogs, some zeolite analogs previously known in mineral form only, and some zeolite-like structures only previously known theoretically. A typical gel reported for preparing the thomsonite structure-type ACP-THO1 has the approximate composition: 9.8 R • 3.8 CoO • Al_2O_3 • 5.2 P_2O_5 • 17.9 H_2O • 128 EG. In general, these reaction mixtures tend to have Al_2O_3 as the limiting reagent, and quite high levels of template, phosphate, and metal. Several of the structure-types (Table 10) were prepared with 2 or more templates.

Table 10
Selected cobalt aluminophosphate zeotypes prepared in alcoholic media [16].

IZA	Formula	Co	Al	P	Template (R)
ANA	$Cs_2AlCo_2P_3O_{12}$	.333	.167	.500	Cs
CHA	$RCoAl_2P_3O_{12}$	.167	.333	.500	1,6-diaminohexane
GIS	$RCoAlP_2O_8$	.250	.250	.500	Dimethylamine
MER	$RAlCo_3P_4O_{16}$	.375	.125	.500	Ethylenediamine
PHI	$R(NH_4)AlCo_3P_4O_{16}$	.375	.125	.500	2,2-dimethyl-1,3-diaminopropane
SOD	$RAlCo_2P_3O_{12}$	.333	.167	.500	Piperazine
THO	$RAlCo_4P_5O_{20}$	.400	.100	.500	N-methyl-ethylenediamine
ACO	$RAl_{0.88}Co_{7.12}P_8O_{32}$	.445	.055	.500	Ethylenediamine

In these cases, the concentration of the divalent T-atom in the product appears to be controlled by varying the charge density and geometry of the structure-directing, protonated amine molecules. The most effective templates appear to be the diamines, which would have the highest charge density (protonated) and the best hydrogen-bonding ability. Many of these compositions have divalent framework cation contents (and framework charges) higher than any of the CoAPO species reported by the Union Carbide group. The zeolite structure-types here are comparable to low Si/Al framework topologies, exhibiting high framework charge density. There is extensive hydrogen-bonding between the templates and framework oxygen, helping to stabilize these open frameworks. The attempted removal of the template by calcination causes structure collapse.

In a second paper by the same authors [73], novel zeolite-type structures with large cages interconnected by multidimensional 12-ring channels were synthesized using larger α,ω-diamines (Table 11). More than a dozen large-pore materials were created in three different topologies with aluminum (or gallium), cobalt (or manganese, magnesium, or zinc), and phosphorus at the tetrahedral coordination sites. As an example, the Co aluminophosphate form of UCSB-6 was prepared from the gel composition: 2.7 1,7-diaminoheptane • 2.0 CoO • Al_2O_3 • 3.8 P_2O_5 • 35

EG • 112 H_2O, digested at 180C for 4 days. Tetragonal UCSB-8 has an unusually large cage built from 64 tetrahedral atoms and connected by an orthogonal channel system with 12-ring apertures in two dimensions and 8-ring apertures in the third. Rhombohedral UCSB-10 and hexagonal UCSB-6 are structurally related to faujasite and its hexagonal polymorph, EMT, respectively, and have large cages connected by 12-ring channels in all three dimensions. In each of these structures, there is extensive hydrogen bonding between amines and framework, contributing to their stability. Unfortunately, none of these structures is thermally stable on removal of the occluded organic.

Table 11
Novel large-pore phosphate structures [73]

IZA	Species	Formula	Co(Zn)	Al(Ga)	P	Template
SBS	UCSB-6	$Co_{24}Ga_{24}P_{48}O_{192}$	.25	.25	.500	1,7-diaminoheptane
SBE	UCSB-8	$Co_{32}Al_{32}P_{64}O_{256}$	.25	.25	.500	1,9-diaminononane
SBT	UCSB-10	$Zn_{36}Ga_{36}P_{72}O_{288}$	.25	.25	.500	4,9-dioxa-1,12-diamino-dodecane

3.7. Two-dimensional structures and interrupted frameworks in alcohol-rich media

Numerous layered or sheet-like aluminophosphate (and gallophosphate) structures have also been obtained in alcoholic media, particularly with diamines and medium size primary amines, some of which do not normally template 3D fully connected AlPO nets in purely aqueous systems. A few examples are shown in Table 12, but a more complete listing is to be found in two recent references [119, 120]. The layer structures consist of macroanionic sheets with well-defined pores and possessing P/Al >1. The two major classes of sheets have stoichiometries that can be expressed as $[Al_2P_3O_{12}]^{3-}$ and $[Al_3P_4O_{16}]^{3-}$. Within these sheets the Al and P are strictly alternating and each Al is either 4- or 5-coordinate and typically bonds to 4 phosphate groups. The phosphate groups, on the other hand, may bond to 2, 3, or 4 Al within the sheet, leaving 0, 1 or 2 pendant hydroxyl or oxo groups to participate in hydrogen bonding with the template. The layers are separated by occluded organic amines, which are protonated and extensively involved in hydrogen bonding. Interestingly, despite the preponderance of alcohol in the solvent mix, there have been few reports of occluded alcohol in these structures. So far, these layered structures have been almost exclusively reported with Al or Ga present. There has been at least one account of a layered structure with Si or divalent metal present, based on elemental analysis and MASNMR, but structural details have not been revealed [121]. It is interesting to note that some of the diamine templates are the same ones that produce 3D nets when metals such as Co or Zn are present during synthesis.

Table 12
Two-dimensional AlPO sheet structures.

Layer Composition	Interlayer Template	Ring sizes in layer	4-connected nets from same template	Ref
$[Al_2P_3O_{12}]$	2-$BuNH_2$	4,6	AFN, CHA, AFI	[122]
$[Al_2P_3O_{12}]$	Pyridine	4,8	CHA	[122]
$[Al_2P_3O_{12}]$	4-Mepyridine	4,6	AFI	[119]
$[Al_2P_3O_{12}]$	Cyclohexylamine	4,6	AFI, CHA, LEV, ERI	[123]
$[Al_2P_3O_{12}]$	Cyclohexylamine	4,6,8		[123]
$[Al_3P_4O_{16}]$	Ethylenediamine	4,6,8	AlPO-12, AFN, DFT	[124]
$[Al_3P_4O_{16}]$	n-butylamine	4,6,12		[125]
$[Al_3P_4O_{16}]$	1,2-diaminopropane	4,6		[126]
$[Al_3P_4O_{16}]$	$[C_3H_5N_2]$	4,D6R,8		[127]
3D interrupted				
$[Al_4P_5O_{20}H]$	Ethylenediamine	4,D6R,8,12		[128]
$[Al_5P_6O_{24}H]$	Triethylamine	4,6,8,10,20	AFI, CHA	[129, 130]

Under some conditions, novel interrupted structures with 3D connectivity were observed, such as the extra large ring aluminophosphate JDF-20, $[Al_5P_6O_{24}H] \cdot 2H_2O$. This was prepared by heating the gel composition 4.7 $Et_3N \cdot Al_2O_3 \cdot 1.8\ P_2O_5 \cdot 18$ triethylene glycol $\cdot\ xH_2O$ for 5 days at 180°C. JDF-20 has a one dimensional channel, bounded by a 20 T-atom elliptical aperture, 6.2 x 14.5A. This structure is not stable to removal of the template, but converts to AlPO-5 at >300°C in air or hydrothermally at >200°C [130].

Using a fixed gel composition of 5 $Et_3N \bullet Al_2O_3 \bullet 1.8\ P_2O_5 \bullet$ x alcohol $\bullet\ yH_2O$, Ruren Xu and coworkers examined synthesis in 20 alcohols of varying polarity. The fully connected structure AlPO-5 was observed in the solvents of highest polarity, with interrupted structures such as JDF-20 and AlPO-CI observed at progressively lower polarity [117]. The interrupted and layer structures typically have much more organic template occluded.

In a more detailed study (Table 13) of the effects of controlled amounts of water in alcoholic media, Oliver and coworkers used the reaction mixture 5.9 $Et_3N \bullet 0.9\ Al_2O_3 \bullet 1.8\ P_2O_5 \bullet xH_2O \bullet$ 14 tetraethylene glycol to produce a progression of structures of increasing level of condensation [131]. The authors found that under intermediate conditions the chain structure $[AlP_2O_8H_2]^-[Et_3NH]^+$ and the JDF-20 phase co-crystallize. Since these structures possess similar building units, the authors suggested that together they provide new insight into the mode of formation of aluminophosphate molecular sieves and allow one to postulate reaction pathways for the transformation between the two phases.

The H_2O content of the reaction mixture appears to play a crucial role in determining the nature of the final products. The hypothetical hydrolysis of polymeric $[AlP_2O_8H_2]^-$ into various related chain and sheet structures followed by their ordered condensation, permits the formation of the building blocks of a wide range of $AlPO_4$-n structures, such as VPI-5, $AlPO_4$-5, $AlPO_4$-11, tinsleyite and many others, as well as dense aluminophosphate phases, such as $AlPO_4$-cristobalite. The authors propose that crystallization proceeds through chains to sheets to 3D lattice structures, with hydrogen bonding helping to orient the cationic templates and the macroanions. Condensation occurs by means of progressive hydrolysis and extrusion of phosphate and organic into solution, as the ratio P/Al changes from 2.0 in the chain structure, to 1.20-1.33 in the sheets, to 1.00 in the 3D lattices. Solution transport involves phosphorus-rich, anionic aluminophosphate species. A subsequent review [132] goes into greater detail.

Table 13.
Effect of H_2O concentration on low-dimensional structure formation in alcoholic synthesis of AlPO-5 [131]

Composition	Structure	X (moles H_2O)
$[AlP_2O_8H_2]^-$ $[Et_3NH]^+$	Chain polymer	3
$[Al_5P_6O_{24}H]^{2-}$ $2[Et_3NH]^+$	JDF-20	10
$[Al_3P_4O_{16}]^{3-}$ $3[Et_3NH]^+$	Macroanionic sheet	13
$[Al_{12}P_{12}O_{48}]$ $1.5[Et_3N]$	Some AlPO-5	43
$[Al_{12}P_{12}O_{48}]$ $1.5[Et_3N]$	AlPO-5	93, no alcohol

3.8. Mixed solvents - HF/water

Following the initial report of the synthesis of silicalite in fluoride media by Flanigen and Patton [133], J.L. Guth and coworkers began their extensive exploration of the synthesis of zeolites in the presence of fluoride ion [134, 135]. In 1989 they extended their methods to the crystallization of phosphate-based molecular sieves [136], where three effects were identified for the fluoride: 1) a mineralizing effect without structure directing, 2) structure-directing without incorporation, and 3) structure-directing with incorporation [137].

The initial preparation of several novel, phosphate-based compositions supported a structure-directing role, with the incorporation of fluoride in small cages such as D4R or as species bridging 2 Al in 4-rings (Table 14). The use of F, in particular, expanded the number of gallophosphate structures [138]. The novel $GaPO_4$ structure cloverite (-CLO) was the first molecular sieve with a pore opening comprising 20 tetrahedrally coordinated atoms. This structure was obtained in the presence of quinuclidine as a template and is an interrupted framework containing D4R units with encapsulated fluoride [139]. The structure name derives from the shape of this opening, which suggests a 4-leaf clover, and is due to the presence of terminal OH groups on the D4R units. The supercage formed at the intersection of the 3-dimensional channel system has a body diagonal of 30Å. Extensive characterization including thermal analysis coupled with mass spectroscopy has revealed significant thermal stability to removal of much of the occluded organic and structural fluoride [140].

In the presence of fluoride, $GaPO_4$-LTA crystallized from a mixture containing di-n-propylamine as template, whereas in the absence of fluoride, the more dense species GaPO4-a was formed [141]. The $GaPO_4$-LTA is stable to at least 500°C in a dry atmosphere and shows significant n-hexane adsorption capacity (14,8% at $p/p_0 = 0.5$). The F anions found in the D4Rs of the LTA and -CLO structures presumably play a co-structuring role in addition to their mineralizing effect. In contrast to the tendency of F to bond to one Si at a time in pure silica frameworks [142], the fluoride tends to form bridging bonds with two or more Al or Ga in the phosphate structures. The fluoride incorporated in the oxide lattice can be viewed as a source of negative charge balancing the positively charged template species, in much the same way as the OH^- species present in as-synthesized AlPO-17. There appear to be different compositional constraints governing the formation of phosphate-based LTA and -CLO, since LTA can be formed in a great variety of compositions, but -CLO is limited to a $GaPO_4$ composition, so far.

Table 14
Selected phosphate-based structures facilitated by fluoride

IZA	Type	Composition	Template (R)	Location of F	Ref.
LTA	AlPO		TMAOH, diethanolamine	Inside D4R	[68]
	AlPO	$[AlPO_4]_{96} \bullet 8RF_2 \bullet 64\ H_2O$	Kryptofix 222		[143]
	CoAPO		TMAOH, diethanolamine		[70]
	SAPO		TMAOH, diethanolamine		[70]
	GaPO4	$[GaPO_4]_{12} \bullet 3RF$	nPr_2NH		[141]
	GaPO4		K^+, 18-crown-6		[95]
CLO	GaPO4	$Ga_{768}P_{768}O_{2976}(OH)_{192}F_{192}] \bullet$ 192(RH)	Quinuclidine	Inside D4R	[139]
CHA	AlPO	$[Al_6P_6O_{24}F]\ [C_4H_{10}NO]$	Morpholine	Bridging two Al in 4R	[144]
GIS	AlPO	$[Al_2P_2O_8F]\ [NH_2Me_2]$	Me_2NH	Bridging two Al in 4R	[145]
-----	GaPO4	ULM-16 $[Ga_{32}(PO_4)_{32}F_{16}]\ [C_6H_{14}N]_{12}$	Cyclohexylamine	Bridging two Ga	[146]

Ferey and his coworkers have recently exploited gallophosphate synthesis in the presence of fluoride using predominately diamine templates [147]. The resulting structures frequently contain GaO_4F and GaO_4F_2 polyhedra and tend to be thermally unstable with respect to removal of F and organic. One exception is the unusually stable structure ULM-16 (Table 14) which is made with cyclohexylamine and contains a 1-D channel delimited by a 16-membered ring [146]. Water and amine are lost progressively between 40 and 400°C with retention of structure. The powder XRD pattern of the residue at 400°C is very similar to that obtained at room temperature, and is retained up to 800°C. At this temperature, fluorine is lost with destruction of the structure. So far, only the AlPO-CHA structure seems to survive the complete removal of the structural F.

3.9. Structure-Composition Relationships

The possibility that certain structure-types might form in a limited compositional range is still likely, but there are a number of instances where apparent compositional restrictions have been overcome with changes in template or solvent. Table 15 shows some examples of synthesis modifications that led to new compositions previously thought unlikely. For example, the original template reported for AlPO-18 was TEAOH, but in the presence of Si or Co, SAPO-34 or CoAPO-34 were more likely to form with this template, particularly when higher concentrations of those elements were present. The recent discovery of iPr_2EtN as template yielded a robust synthesis of SAPO-18 and MeAPO-18, resistant to formation of CHA. In contrast, the tendency of TEAOH to template AlPO-18 was an impediment to AlPO-34 synthesis until fluoride was added to the synthesis media.

Table 15
Synthesis modifications leading to additional compositions (ref. in Table 3)

IZA	Initial Composition	Synthesis modification	Later Compositions
AET	AlPO-18	Novel template, iPr_2EtN	SAPO-18; CoAPO-18
CHA	SAPO-34 MeAPO-34	Fluoride, solution methods	AlPO-34
ATS	MgAPO-36	Seeding, gel composition Temperature, aging	SAPO-36 AlPO-36
ATN	MgAPO-39	Additional acids, anions	AlPO-39; SAPO-39
AFR	SAPO-40	Gel composition	AlPO-40; CoAPO-40
LTA	SAPO-42	Fluoride, dual template	AlPO-LTA; CoAPO-LTA
AFS	MgAPSO-46	Gel composition	SAPO-46; MgAPO-46

Of all the elements incorporated in $AlPO_4$-based structures, Si shows the greatest compositional range, and this range has been progressively expanded over time for certain well-studied structure-types. The substitution pattern for Si in SAPO molecular sieves is known to be sensitive to many synthesis parameters, such as structure-type, total Si content, organic template, amine/Al_2O_3 ratio, P_2O_5/Al_2O_3 ratio, pH, crystallization time and temperature. At low framework concentrations, the Si atoms are individually incorporated into P sites and the number of Bronsted acid sites tracks Si content. At higher concentrations, the number of acid sites lags behind Si content as Si islands are produced, leading to <1 acid site per Si [148]. The further incorporation of Al into these Si islands produces regions of zeolitic composition, and presumably more zeolitic acidity. Several groups have used ^{29}Si-MASNMR to try to quantify SAPO and zeolitic environments in selected SAPO's (Table 16).

Of the SAPO structures that have been studied most thoroughly, four have zeolite structure analogs (FAU, CHA, ERI, AFI) and two do not (AEL, AEI). SAPO-34 and SAPO-37 show fairly high levels of Si incorporation by means of direct Si for P substitution (SAPO domain). SAPO-34 (or 44) can be synthesized with a great variety of templates and the substitution pattern appears to be dependent on the template used. SAPO-5, -11, and -18 show mixed substitution even at relatively low levels of Si incorporation.

Deconvoluted ^{29}Si MASNMR spectra were used by Martens et al to calculate the relative contributions of SAPO and zeolitic (SA) domains to the compositions SAPO-5, SAPO-11, and SAPO-37. SAPO-11 shows evidence of SA domains even at the lowest levels of Si incorporation. Low levels of Si incorporation have never been achieved with SAPO-37.

X-ray diffraction and ^{27}Al and ^{29}Si MASNMR were used by Vomscheid and coworkers to study the interaction of the templates (morpholine or Et_4NOH) with the SAPO-34 framework (CHA) and to characterize the distribution of Si in the lattice [149]. They found that all samples showed pore filling by the two templates, equivalent to two morpholine or one TEAOH per CHA cage. At the higher Si levels, the ^{29}Si MASNMR was consistent with isolated Si in SAPO domains and Si islands, rather than the more zeolitic Si/Al regions. Synthesis efforts by Ashtekar [150] using TEAOH achieved even higher levels of Si incorporation and the solid-state MASNMR results show significant evidence of Si in zeolitic environments.

Table 16
Observed level and type of Si substitution in selected SAPO structures based on MASNMR.

Species	Frwrk mole fraction			Template	Si domains[a]		Ref.
	Si	Al	P		SAPO	SA	
SAPO-5	0.01	0.50	0.49	Cyclohexylamine	100	0	[151, 152, 153]
	0.15	0.44	0.41		90	10	
	0.28	0.39	0.33		75	25	
SAPO-11	0.04	0.52	0.44	n-Pr_2NH	94	6	[5, 112]
	0.21	0.43	0.36		72	28	
	0.42	0.31	0.27		54	46	
SAPO-37	0.16	0.49	0.36	Pr_4N^+,Me_4N^+	92	8	[5, 112]
	0.23	0.43	0.34		79	21	
	0.29	ND	ND		69	31	
SAPO-34	0.03	0.50	0.47	TEAOH	100	0	[149, 150, 154]
	0.10	0.49	0.41		<100	>0	
	0.16	0.48	0.36		58	42	
	0.19	0.48	0.33		48	52	
SAPO-34	0.10	0.50	0.41	Morpholine	100	0	[149, 155]
	0.18	0.49	0.33		<100	ND	
	0.33	0.40	0.27		<100	ND	
SAPO-44	0.12	0.50	0.38	Cyclohexylamine	81	19	[150, 156]
	0.21	0.45	0.34		61	39	
	0.30	0.42	0.28		31	69	
SAPO-18	0.03	ND	ND	iPr_2EtN	70	30	[157, 158]
	0.09	ND	ND		48	52	
	0.13	0.46	0.31		ND	ND	

[a] %Si assigned to SAPO or silica-alumina (incl. Si islands) domains. ND = not reported

The CHA composition SAPO-44, prepared with cyclohexylamine, was also synthesized with varying amounts of Si and characterized by XRD, SEM, FTIR spectroscopy, and thermal analyses [150]. The results of these studies along with the solid-state MASNMR show that for low levels of Si incorporation, the major part substitutes for isolated P atoms in the structure. There appears to be significant amounts of Si in Si patches and zeolitic Si/Al environments at the higher levels of incorporation. It is interesting to note that the cyclohexylamine can produce both SAPO-5 and SAPO-44 depending on synthesis conditions. The crystallization of SAPO-44 appears to be favored over SAPO-5 by a higher template concentration, different Al_2O_3 source, and a longer crystallization time.

For a structure like AFI with >85 known templates, the template effect on Si incorporation has barely been explored. Some of the synthesis methods discussed earlier, such as alcohol media or F-mediated synthesis, that have been so successful in incorporating high concentrations of divalent cations into lattice positions, have not yet been successful in SAPO synthesis.

3.10. Mesoporous aluminophosphates

Mesoporous aluminophosphates have been synthesized using many of the same strategies successful with the mesoporous aluminosilicates (Table 17). The earliest work employed long chain amines in aqueous and non-aqueous solvents and tended to give mesolamellar structures that were unstable to removal of the templating agent [159, 160, 161]. With the use of cationic surfactants such as hexadecyltrimethylammonium chloride (CTAC) or bromide (CTAB) for structure direction and TMAOH to control pH, several different groups have reported the preparation of thermally stable structures (Table 18).

In the work of Khimyak [162] the synthesis of stable hexagonal AlPO structures rather than lamellar structures depends on the complex interplay of P/Al ratio, pH, water content, and surfactant concentration. The lamellar products tend to be phosphate-rich with Al/P ratios of 0.67-0.78 while the more stable hexagonal products exhibit Al/P ratios 0.73-1.05. There is compositional overlap between these two types of structures, and the synthesis pH lies in the range of 6-8.

Table 17
Synthesis conditions for selected mesoporous aluminophosphate structures

Species	Surfactant[a]	Al_2O_3	P_2O_5	SiO_2	TMAOH	H_2O	Time h	Temp C	Ref
	1.0 dodecylamine	0.80	1.0			60	24	100	[161]
L1	0.27 CTAC	0.33	1.0		2.40	112		130	[162]
L3	0.25 CTAC	0.50	1.0		3.07	163		150	[162]
Hex-1	0.58 CTAC	0.82	1.0		1.93	246	72	25	[162]
Hex-2	0.25 CTAC	0.50	1.0		2.57	158	72	25	[162]
UHM-1	0.50 CTAC	0.50	1.0		2.60	430	72	25	[163]
UHM-3	0.50 CTAC	1.24	1.0	0.21	2.60	430	72	25	[163]
S-III	0.44 CTAB	0.38	1.0	0.03	0.50	85	48	110	[164]
C_{16}-AlPO	1.00 CTAC	1.00	1.0		2.00	65	120	130	[165]
C_{22}-AlPO	1.00 C_{22}TMA	1.00	1.0		2.00	65	4	70	[165]

[a]CTAC = $C_{16}H_{33}N(CH_3)_3Cl$, CTAB = $C_{16}H_{33}N(CH_3)_3Br$, C_{22}TMA = $C_{22}H_{45}N(CH_3)_3Cl$

Chakraborty has reported the preparation of mesoporous SAPO compositions from P_2O_5-rich mixtures using CTAB as template [164]. The XRD patterns of the calcined samples indicate a more crystalline nature than typically seen with MCM-41. TEM studies reveal faceted particles with a hexagonal external morphology. Solid state MAS NMR is consistent with Al, P, and Si environments similar to those in microporous SAPO structures, i.e. tetrahedral Al and P, and tetrahedral Q4 Si in Si(xAl) environments where x = 0-4.

The mesostructured organo-aluminophosphates C_{16}-AlPO and C_{22}-AlPO prepared by Kimura illustrate the effects of using surfactants with different alkyl chain length (C16 and C22). Calcination yielded thermally stable mesoporous aluminophosphates with surface areas >700 m^2/g and with pore diameters in the range 1.8-3.9 nm. The hexagonal mesostructured materials before calcination had less condensed frameworks as determined by ^{31}P MAS NMR, which caused certain shrinkage during calcination for the formation of the porous structures. Before calcination the ^{27}Al MASNMR of both materials showed significant amounts of six-coordinate Al, but after calcination, only four-coordinate Al was observed. H_2O and benzene adsorption data indicated that these aluminophosphate materials had comparatively hydrophilic mesopores in contrast to mesoporous silicas. The Al/P ratios are the highest of any of the thermally stable mesostructures.

Table 18
Characterization of calcined mesoporous aluminophosphates

Species	Structure	Surfactant	Al/P	Si/P	d(100) nm	BET m^2/g	Calc °C
UHM-1	Hex	CTAC	> 1.0	---	3.6	772	500
UHM-3	Hex	CTAC	> 1.0	0.20	3.6	928	500
S-I	Hex	CTAB	0.8	0.01	2.8	920	600
S-II	Hex	CTAB	1.0	0.02	2.8	940	600
S-III	Hex	CTAB	1.1	0.03	2.8	980	600
Hex-1	Hex	CTAC	1.1		1.9	760	500
Hex-2	Hex	CTAC	0.9		2.0	890	500
C_{16}-AlPO	Hex	CTAC	1.49		3.2	980	600
C_{22}-AlPO	Hex	C_{22}TMA	1.52		4.3	760	600

Ref in Table 17.

3.11. Beryllium and Zinc Phosphates/Arsenates

Of all the divalent cations incorporated into the $AlPO_4$-based structures, Be and Zn have shown the greatest tendency to form microporous binary phosphates (Al-free). Harvey and Meier prepared a range of microporous beryllophosphates containing extra-framework alkali metal cations and water [39]. The structures were designated R, G, E, P, and H (Table 19). The frameworks all have strictly alternating Be and P, resulting in a rather high net negative charge, which is balanced by alkali cations. These cations appear to play a structure-directing role, analogous to their role in the synthesis of low Si/Al zeolites. Beryllophosphate-H requires the presence of both Na^+ and K^+, and each cation is found in a specific site in the structure.

In the preferred synthesis method, equimolar beryllium sulfate and phosphoric acid in water are combined with a solution of the alkali metal chloride, to give a clear solution of pH 1. Aqueous TEAOH is then added to adjust the pH to 6, forming a gel, which is crystallized at 100-200°C. The gel pH appears to be critical, since dense phases form at pH $\leq$ 4, and no crystals form at pH $\geq$ 7. The use of alkali hydroxide for pH control also yields most of the products, but the TEAOH method is preferable and more reproducible. No TEAOH is present in any of the products. On heating these structures, most of the water is lost at 200-300C. At temperatures above 300°C, all but the beryllophosphate-P decompose or densify. The thermal stability is considerably lower than that of aluminosilicate or aluminophosphate molecular sieves, which the authors attribute to the greater ionic character of the Be-O bonds, compared to T-O bonds in other molecular sieve materials.

Stucky and coworkers [166, 167, 168] have prepared new families of hydrated zincophosphates/arsenates and beryllophosphates/arsenates with many structures that are also related to the low Si/Al zeolitic aluminosilicates (Table 19). These materials were prepared from gels over a wider range of pH and at lower temperatures than are currently possible for aluminophosphates. This behavior was attributed to the greater solubility of the framework elements. Analogs of SOD, RHO, ABW and FAU are easily prepared at pH values 2-12 and temperatures of 4-100°C. Growth of crystals adequate for x-ray structural analysis seems to be easier than is the case for aluminosilicate/phosphate materials. The presence of the alkali cations tends to stabilize these highly charged frameworks.

Table 19
Selected Be and Zn phosphate structures with extraframework alkali cations

Structure	Name	Framework	Extra-framework species	Ref.
ABW		$[ZnPO_4]_4$	$(NH_4)_4$	[169]
ANA	Beryllophosphate-P	$[BePO_4]_{24}$	Cs_{16} , H_8	[39]
BPH	Beryllophosphate-H	$[BePO_4]_{14}$	Na_7 K_7 , 20 H_2O	[39]
CAN	Tiptopite	$[BePO_4]_6$	$K_2(Li_{2.9}Na_{1.7}Ca_{0.7})(OH)_{2.1}$ • $3H_2O$	[51]
CZP	Chiral zincophosphate	$[ZnPO_4]_{12}$	Na_{12}, 12 H_2O	[56]
EDI	Beryllophosphate-E	$[BePO_4]_{10}$	K_{10}, 10 H_2O	[39]
EDI	ZP-4	$[ZnPO_4]_{20}$	K_{20}, 16 H_2O	[170]
FAU	ZnPO-X	$[ZnPO_4]_{96}$	Na_{67} $(Me_4N)_{12}$ Zn_8 , 192 H_2O	[61]
FAU	BePO-X	$[BePO_4]_{96}$	Na_{96}, 192 H_2O	[61]
GIS	Beryllophosphate-G	$[BePO_4]_8$	Na_8, 10 H_2O	[39]
LOS	BePO-LOS	$[BePO_4]_6$	Li_8, HPO_4, H_2O	[67]
MER	BeP-MER1	$[BePO_4]_{32}$	$(NH_4)_{32}$, 4 H_2O	[171]
RHO	Beryllophosphate-R	$[BePO_4]_{24}$	Li_{24}, 40 H_2O	[39]
SOD		$[ZnPO_4]_6$	Na_6, 8 H_2O	[78]

The faujasite analogs containing Zn/P (ZnPO-X) or Be/P (BePO-X) as the only framework tetrahedral atoms are the most remarkable [61]. These phases were prepared at near ambient temperature. The compounds have strictly alternating (Zn/Be):P tetrahedra, as confirmed by ^{31}P-NMR data. BePO-X is marginally stable to calcination to 350°C, while ZnPO-X irreversibly loses its structure upon H_2O loss at 250°C. Both BePO-X and ZnPO-X display typical zeolitic ion-exchange chemistry, however.

In addition to the essentially "inorganic" zincophosphate structures shown in Table 19, several have been prepared with occluded organic. A series of structures were prepared with diazabicyclo[2.2.2]octane (DABCO), each consisting of a three-dimensional network of ZnO_4 and PO_4 tetrahedra, linked through oxygen vertexes. The first two, ZnPO/dab-A and –B, contain one-dimensional eight-ring channels, occupied by cationic, diprotonated DABCO and water. Unlike the SOD and FAU structures, some of the zinc and phosphorus polyhedra contain pendant H_2O and OH groups making these interrupted frameworks [172]. ZnPO/dab-C contains two distinct 1-dimensional 8-ring channels, occupied by extraframework, protonated DABCO. ZnPO/dab-D is a new framework type containing bridged 3-rings of tetrahedra. The DABCO molecule coordinates to the framework in this phase via Zn-N links. The guest-framework and framework-framework hydrogen-bonds are a key structural feature of these new phases. Numerous Zn/P structures with Zn>P or P>Zn have also been prepared by Harrison and coworkers . These tend to be interrupted frameworks with evidence of extensive hydrogen-bonding and low thermal stability.

Two Zn phosphate materials with Zn>P (ZnPO-W and ZnPO-TMA), both containing infinite Zn-O-Zn chains as part of the three-dimensional framework, were prepared by hydrothermal methods and characterized by single-crystal x-ray diffraction [173]. ZnPO-W has one 8R channel along the a-axis with H_2O molecules in the channel. ZnPO-TMA has two different 8R channels also along the a-axis with trimethylammonium cations located within the channels.

Table 20
Selected ZnPO structures with occluded organic

Name	Framework	Extra-framework species	Ref.
ZnPO/dab-A	$[Zn_2(HPO_4)_3]$	(H_2dabco)	[172]
ZnPO/dab-B	$[Zn_4(PO_4)_2(HPO_4)_2(H_2O)_2]$	(H_2dabco) , H_2O	[172]
ZnPO/dab-C	$[Zn_5(PO_4)_2(HPO_4)_4]$	(H_2dabco) , H_2O	[174]
ZnPO/dab-D	$[Zn_3(PO_4)_4(HPO_4)_2]$	Hdabco	[174]
ZnPO-TMA	$[Zn_4(H_2O)(PO_4)_3]$	$NHMe_3$	[173]
UiO-21	$[ZnPO_4]_{18}$ • 3Zn	1-(2-aminoethyl)piperazine	[175]
DAF-3	$[ZnPO_4]_{16}$	ethylenediamine	[176]

UiO-21 is a CHA structure analog with additional Zn cations at the center of each D6R. Each CHA cage contains two doubly protonated template molecules, which, along with the Zn^{2+} in the D6R, balance the high framework charge. In this structure the organic was not found to be hydrogen bonding with lattice oxygen. DAF-3 contains a 3-dimensional network of channels delineated by 8-rings and containing occluded ethylenediamine.

Two approaches to the synthesis of $NaZnPO_4 \bullet H_2O$, an open-framework, chiral, hydrated Na zincophosphate phase (CZP), have been reported [56]. The ordered, alternating, vertex-sharing ZnO_4 and PO_4 groups form a new framework topology, which includes novel 4-ring squares and edge-sharing helices of 4-rings. The disordered extra-framework species (Na^+, H_2O) are located in pear-shaped cavities (12-ring diameter), interconnected through 8-rings and 6-rings.

The hydrothermal syntheses and some properties of a new family of microporous zincophosphate/arsenate materials, denoted $M_3Zn_4O(XO_4)_3 \bullet nH_2O$ (M = Li, Na, K, Rb, Cs, ...; X = P, As; n = 4-6), were reported by Stucky and coworkers [177]. All these materials are based on a flexible, anionic $Zn_4O(XO_4)_3$ network built up from vertex-linked tetrahedral ZnO_4 and XO_4 units. Four ZnO_4 units share a vertex, resulting in novel OZn_4 centers. This network contains only pairs of 3-rings (a spiro-5 unit) and 8-rings, and no 4-rings or 6-rings: it encloses roughly spherical cavities connected by a three-dimensional network of 8-ring channels propagating in the orthogonal [100], [010], and [001] directions (cubic unit-cell axes). The use of mixed four-coordinate anion types (XO_4^{3-}, O^{2-}) results in a number of interesting structural and stability properties. Extraframework cations and H_2O molecules occupy these cavities and channels. These phases display typical zeolitic dehydration/rehydration and ion-exchange reactions with thermal stabilities up to 600°C, the highest known for this type of open-framework Zn(P/As)O material.

Bedard has reported the hydrothermal synthesis of framework substituted zincophosphates [178]. These molecular sieves have a 3-dimensional microporous framework structure of ZnO_2, PO_2, and MO_2 tetrahedral units and an intracrystalline pore system. M can be a variety of elements with valences 2, 3, or 4, such as Si, Mg, Cu, Ga, Al, among others.

4. CONCLUSIONS

The study of phosphate-based molecular sieves continues to flourish. Novel structures are being discovered at an accelerating pace, but the discovery of novel structures that are stable to removal of occluded species remains a daunting challenge. The AlPO and SAPO compositions are still the most stable, probably because they are the most covalent in their bonding. Novel templates are still responsible for most of the new structures, but novel media such as alcohol-rich solvents and aqueous fluoride have also aided in the quest. Physico-chemical characterization of known structures is enhancing our appreciation of the similarities and differences between $AlPO_4$-based molecular sieves and zeolites.

REFERENCES

[1] Wilson, S. T. Stud. Surf. Sci. Catal. (1991), 58 (Introd. Zeolite Sci. Pract.), 137-51.
[2] Flanigen, E. M.; Lok, B. M.; Patton, R. L.; Wilson, S. T. Stud. Surf. Sci. Catal. (1986), 28 (New Dev. Zeolite Sci. Technol), 103-12. Ibid., Pure Appl. Chem (1986), 58, 1351-8.
[3] Flanigen, E. M.; Patton, R. L.; Wilson, S. T. Stud. Surf. Sci. Catal. (1988), 37 (Innovation Zeolite Mater. Sci), 13-27.
[4] Szostak, R. Molecular Sieves: Principles of Synthesis and Identification, 2nd Edition. Blackie Academic and Professional: London. 1998, pp. 251-274.
[5] Martens, J. A.; Jacobs, P. A. Stud. Surf. Sci. Catal. (1994), 85 (Advanced Zeolite Science and Applications), 653-685.
[6] Martens, J. A. and Jacobs, P. A. Catal. Zeolites (1999), Weitkamp, J. and Puppe, L. (eds.), Springer, pp. 53-80.
[7] Kessler, H. Compr. Supramol. Chem. (1996), pp 425-464.
[8] Cheetham, A.K.; Ferey, G.; Loiseau, T. Angew. Chem. Int. Ed. (1999), 38(22), 3268-3292.
[9] Wilson, S. T.; Lok, B. M.; Flanigen, E. M. US Patent No. 4 310 440 (1982).
[10] Wilson, S. T.; Lok, B. M.; Flanigen, E. M. J. Amer. Chem. Soc. (1982), 104, 1146.
[11] Lok, B. M.; Flanigen, E. M.; Patton, R. L.; Messina, C. A.; Gajek, R. J. Amer. Chem. Soc. (1984), 106, 6092.
[12] Lok, B. M.; Messina, C. A.; Patton, R. L.; Gajek, R. T.; Cannan, T. R.; Flanigen, E. M. US Patent No. 4 440 871 (1984).
[13] Bennett, J. M.; Dytrych, W. J.; Pluth, J. J.; Richardson, J. W. Jr.; Smith, J. V. Zeolites (1986), 6, 349-60.
[14] Meier, W. M.; Olson, D. H.; Baerlocher, Ch. Atlas of Zeolite Structure Types, Fourth Rev. Ed., Elsevier, 1996; IZA Structure Commission website: www.iza-structure.org/databases/
[15] Bu, X.; Feng, P.; Gier, T. E.; Stucky, G. D. Zeolites (1997), 19(2/3), 200-208.
[16] Feng, P.; Bu, X.; Stucky, G. D. Nature (London) (1997), 388(6644), 735-741.
[17] Chen, J.; Thomas, J. M.; Wright, P. A. 1994, Catal. Lett. (1994), 28(2-4), 241-8.
[18] Chen, J.; Thomas, J. M. J. Chem. Soc. 1994, Chem. Commun. (1994), (5), 603-604.
[19] Wilson, S. T.; Flanigen, E. M. US Patent No. 4 567 029 (1986).
[20] Maistriau, L.; Gabelica, Z.; Derouane, E. G.; Vogt, E. T. C.; Van Oene, J. Zeolites (1991), 11, 583-92.
[21] Zibrowius, B.; Lohse, U.; Szulzewsky, K.; Fichtner-Schmittler, H.; Pritzkow, W.; Richter-Mendau, J. Stud. Surf. Sci. Catal. (1991), 65(Catal. Adsorpt. Zeolites), 549-62.
[22] Parise, J. B. Acta Crystallogr., Sect. C: Cryst. Struct. Commun. (1986), C42, 670-3.
[23] Lesch, D. A.; Wilson, S. T. EP 254 075 B1 (1988).
[24] Clark, H. W.; Rievert, W. J.; Olken, M. M. Microporous Mater. (1996), 6(3), 115-124.
[25] Prakash, A. M.; Chilukuri, S. V. V.; Bagwe, R. P.; Ashtekar, S.; Chakrabarty, D. K. Microporous Mater. (1996), *6*(2), 89-97.
[26] Sinha, A. K.; Satyanarayana, C. V. V.; Srinivas, D.; Sivasanker, S.; Ratnasamy, P.

Microporous Mesoporous Mater. (2000), 35-36, 471-481.
[27] Lourenco, J. P.; Ribeiro, M. F.; Ribeiro, F. R.; Rocha, J.; Onida, B.; Garrone, E.; Gabelica, Z. Zeolites (1997), 18(5/6), 398-407.
[28] Lourenco, J. P.; Ribeiro, M. F.; Ribeiro, F. R.; Rocha, J.; Gabelica, Z.; Dumont, N.; Derouane, E. G. Stud. Surf. Sci. Catal. (1994), 84(Zeolites and Related Microporous Materials, Pt. A), 867-7.
[29] Prakash, A. M.; Ashtekar, S.; Chakrabarty, D. K.; Chilukuri, S. V. V. J. Chem. Soc., Faraday Trans. (1995), 91(6), 1045-50.
[30] Wilson, S. T.; Mercer, W. C. EP 293 920 (1988).
[31] Akolekar, D. B.; Kaliaguine, S. J. Chem. Soc., Faraday Trans. (1993), 89, 4141-7.
[32] Wilson, S. T.; King, L. EP 293 938 (1988).
[33] Wilson, S. T. US Patent No. 5 370 851 (1994).
[34] Wilson, S.T.; Broach, R.L.; Blackwell, C.S.; Bateman, C.; McGuire, N. K.; Kirchner, R.M. Microporous Mesoporous Mater. (1999), 28(1), 125.
[35] Noble, G. W.; Wright, P. A.; Lightfoot, P.; Morris, R. E.; Hudson, K. J.; Kvick, A.; Graafsma, H. Angew. Chem. Int. Ed. Engl. (1997), 36(1/2), 81-83.
[36] Akolekar, D. B. Zeolites (1995), 15(7), 583-9.
[37] D'Yvoire, F. Bull. Soc. Chim. France (1961), 1762-1776.
[38] Li, H. X.; Davis, M. E. J. Chem. Soc., Faraday Trans. (1993), 89, 951-6.
[39] Harvey, G.; Meier, W. M. Stud. Surf. Sci. Catal. (1989), 49A, 411-420.
[40] Coker, E. N.; Rees, L. V. C. J. Chem. Soc., Faraday Trans. (1992), 88, 263-72.
[41] Keller, E. B.; Meier, W. M.; Kirchner, R. M. Solid State Ionics (1990), 43, 93-102
[42] Wilson, S. T.; King, L. US Patent No. 4 663 139 (1987).
[43] Sinha, A. K.; Hegde, S. G.; Jacob, N. E.; Sivasanker, S. Zeolites (1997), 18(5/6), 350-355.
[44] Akolekar, D. B.; Kaliaguine, S. K. Zeolites (1994), 14(8), 620-4.
[45] Zubowa, H. L.; Alsdorf, E.; Fricke, R.; Neissendorfer, F.; Richter-Mendau, J.; Schreier, E.; Zeigan, D.; Zibrowius, B. J. Chem. Soc., Faraday Trans. (1990), 86, 2307-12.
[46] Zubowa, H. L.; Richter, M.; Roost, U.; Parlitz, B.; Fricke, R. Catal. Lett. (1993), 19, 67-79.
[47] Zahedi-Niaki, M. H.; Joshi, P. N.; Kaliaguine, S. Chem. Commun. (Cambridge) (1996), (11), 1373-1374.
[48] Patton, R. L.; Gajek, R. T. US Patent No. 4 473 663 (1984).
[49] Chen, J.; Sankar, G.; Thomas, J. M.; Xu, R.; Greaves, G. N.; Waller, D. Chem. Mater. (1992), 4, 1373-80.
[50] Harvey, G.; Baerlocher, C.; Wroblewski, T. Z. Kristallogr. (1992), 201(1-2), 113-23.
[51] Peacor, D. R.; Rouse, R. C.; Ahn, J. H. Am. Mineral (1987), 72, 816-20.
[52] Chippindale, A. M.; Cowley, A. R. Zeolites (1997), 18(2/3), 176-181.
[53] Chippindale, A. M.; Cowley, A. R. Microporous Mesoporous Mater. (1998), 21(4-6), 271-279.
[54] Lesch, D. A.; Patton, R. L.; Woodward, N. A. US Patent No. 4 938 937 (1990).
[55] Schott-Darie, C.; Kessler, H.; Benazzi, E. Stud. Surf. Sci. Catal. (1994), 83 (Zeolites and Microporous Crystals), 3-10.
[56] Harrison, W. T. A.; Gier, T. E.; Stucky, G. D.; Broach, R. W.; Bedard, R. L. Chem. Mater. (1996), 8(1), 145-51.
[57] Wright, P. A.; Sayag, C.; Rey, F.; Lewis, D. W.; Gale, J. D.; Natarajan, S.; Thomas, J. M. J.

Chem. Soc., Faraday Trans. (1995), 91(19), 3537-47.
[58] Chen, J.; Jones, R. H.; Thomas, J. M.; Natarajan, S.; Hursthouse, M. B. Angew. Chem. Int. Ed. Engl. (1994), 33(6), 639.
[59] Maistriau, L.; Dumont, N.; Nagy, J. B.; Gabelica, Z.; Derouane, E. G. Zeolites (1990), 10, 243-50.
[60] Sierra, L.; Patarin, J.; Guth, J. L. Microporous Mater. (1997), 11(1-2), 19-35.
[61] Harrison, W. T. A.; Gier, T. E.; Moran, K. L.; Nicol, J. M.; Eckert, H.; Stucky, G. D. Chem. Mater. (1991), 3, 27-9.
[62] Paillaud, J.-L.; Marler, B.; Kessler, H. Chem. Commun. (1996), (11), 1293-1294.
[63] Akolekar, D. B.; Kaliaguine, S. 1994, Microporous Mater. (1994), 2(2), 137-44.
[64] Cowley, A. R.; Chippindale, A. M. Chem. Commun. (1996), (5), 673-4.
[65] Bond, A. D.; Chippindale, A. M.; Cowley, A. R.; Readman, J. E.; Powell, A. V. Zeolites (1997), 19(5/6), 326-333.
[66] Zhu, G.; Xiao, F.-S.; Qiu, S.; Hun, P.-C.; Ruren Xu; Ma, S.; Terasaki, O. Microporous Mater. (1997), 11(5-6), 269-273.
[67] Harrison, W. T. A.; Gier, T. E.; Stucky, G. D. Zeolites (1993), 13, 242-8.
[68] Sierra, L.; Deroche, C.; Gies, H.; Guth, J. L. 1994, Microporous Mater. (1994), 3(1-2), 29-38.
[69] Schreyeck, L.; D'agosto, F.; Stumbe, J.; Caullet, P.; Mougenel, J. C. Chem. Commun. (1997), (13), 1241-1242.
[70] Sierra, L.; Patarin, J.; Deroche, C.; Gies, H.; Guth, J. L. Stud. Surf. Sci. Catal. (1994), 84C, 2237-4.
[71] Merrouche, A.; Patarin, J.; Soulard, M.; Kessler, H.; Anglerot, D. Synth. Microporous Mater. (1992), 1, 384-99.
[72] Akporiaye, D. E.; Fjellvag, H.; Halvorsen, E. N.; Haug, T.; Karlsson, A.; Lillerud, K. P. Chem. Commun. (1996), (13), 1553-1554.
[73] Bu, X.; Feng, P.; Stucky, G. D. Science (Washington, D. C.) (1997), 278(5346), 2080-2085.
[74] Rouse, R. C.; Peacor, D. R.; Merlino, S. Am. Mineral. (1989), 74, 1195-202.
[75] Patinec, V.; Wright, P. A.; Lightfoot, P.; Aitken, R. A.; Cox, P. A. J. Chem. Soc., Dalton Trans. (1999), (22), 3909-3911.
[76] Noble, G. W.; Wright, P. A.; Kvick, A. J. Chem. Soc., Dalton Trans. (1997), (23), 4485-4490.
[77] Wright, P. A.; Maple, M. J.; Slawin, A. M. Z.; Patinec, V.; Aitken, R. A.; Welsh, S.; Cox, P. A. J. Chem. Soc., Dalton Trans. (2000), (8), 1243-1248.
[78] Nenoff, T. M.; Harrison, W. T. A.; Gier, T. E.; Stucky, G. D. J. Amer. Chem. Soc. (1991), 113, 378-9.
[79] Li, H.-X.; Davis, M. E. Catal. Today (1994), 19, 61-106.
[80] Bedard, R. L.; Patton, R. L.; Wilson, S. T. US Patent No. 5 013 535 (1991).
[81] Martens, J. A.; Balakrishnan, I.; Grobet P. J.; Jacobs, P. A. Stud. Surf. Sci. Catal. (1991), 69(Catalysis, Reaction Kinetics, and Inorganic Reaction Mechanisms), 135-43.
[82] Kraushaar-Czarnetzki, B.; Dogterom, R. J.; Stork, W. H. J.; Emeis, K. A.; van Braam Houckgeest, J. P. J. Catal (1993), 141, 140-7.
[83] Joly, J. F.; Cauffriez, H.; Guth, J. L. EP 496 647 (1992)
[84] Walter, F. Eur. J. Mineral (1992), 4, 1275-83.

[85] Akporiaye, D. E.; Fjellvag, H.; Halvorsen, E. N.; Hustveit, J.; Karlsson, A.; Lillerud, K. P. J. Phys. Chem. (1996), 100(41), 16641-16646.
[86] Marler, B.; Patarin, J.; Sierra, L. Microporous Mater. (1995), 5(3), 151-9.
[87] Merrouche, A.; Patarin, J.; Kessler, H.; Soulard, M.; Delmotte, L.; Guth, J. L.; Joly, J. F. Zeolites (1992), 12, 226-32.
[88] Nakagawa, Y. US Patent No. 5 501 848 (1996).
[89] Briend, M.; Lamy, A.; Peltre, M. J.; Man, P. P.; Barthomeuf, D. Zeolites (1993), 13, 201-11.
[90] Weigel, S. J.; Weston, S. C.; Cheetham, A. K.; Stucky, G. D. Chem. Mater. (1997), 9(6), 1293-1295.
[91] Dumont, N.; Gabelica, Z.; Derouane, E. G.; Di Renzo, F. Microporous Mater. (1994), 3(1-2), 71-84.
[92] Lourenco, J. P.; Ribeiro, M. F.; Ribeiro, F. R.; Rocha, J.; Onida, B.; Garrone, E.; Gabelica, Z. Zeolites (1997), 18(5/6), 398-407.
[93] Balkus, K. J. Jr.; Gabrielov, A. G.; Sandler, N. Mater. Res. Soc. Symp. Proc. (1995), 368(Synthesis and Properties of Advanced Catalytic Materials), 369-75.
[94] Balkus, K. J. Jr.; Kim, S. J.; Sargeson, A. M. Mater. Res. Soc. Symp. Proc. (1997), 454(Advanced Catalytic Materials - 1996), 217-224.
[95] Reinert, P.; Schott-Darie, C.; Patarin, J. Microporous Mater. (1997), 9(3,4), 107-115.
[96] Maple, M. J.; Philp, E. F.; Slawin, A. M. Z.; Lightfoot, P.; Cox, P. A.; Wright, P. A. J. Mater. Chem. (2001), 11(1), 98-104.
[97] Wan, Y.; Williams, C. D.; Duke, C. V. A.; Cox, J. J. J. Mater. Chem. (2000), 10(12), 2857-2862.
[98] Bandyopadhyay, M.; Bandyopadhyay, R.; Kubota, Y.; Sugi, Y. Chem. Lett. (2000), (9), 1024-1025.
[99] Zhu, G.; Xiao, F.-S.; Qiu, S.; Hun, P.-C.; Ruren Xu; Ma, S.; Terasaki, O. Microporous Mater. (1997), 11(5-6), 269-273.
[100] Lesch, D. A.; Patton, R. L. EP 293 919 (1988).
[101] Szostak, R.; Duncan, B.; Aiello, R.; Nastro, A.; Vinje, K.; Lillerud, K. Synth. Microporous Mater. (1992), Issue 1, 240-7. Editors: Occelli, M. L.; Robson, H. E. Van Nostrand Reinhold, New York, N. Y.
[102] Qiu, S.; Pang, W.; Xu, R. Stud. Surf. Sci. Catal. (1997), 105A(Progress in Zeolite and Microporous Materials, pt. A), 301-308.
[103] Finger, G.; Richter-Mendau, J.; Buelow, M.; Kornatowski, J. Zeolites (1991), 11, 443-8.
[104] Finger, G.; Jahn, E.; Zeigan, D.; Zibrowius, B.; Szulzewsky, K.; Richter-Mendau, J.; Buelow, M. Bull. Soc. Chim. Belg (1989), 98, 291-5.
[105] Baur, W. H.; Joswig, W.; Kassner, D.; Kornatowski, J.; Finger, G. Acta Crystallogr., Sect. B: Struct. Sci (1994), B50(3), 290-294.
[106] Demuth, D.; Stucky, G. D.; Unger, K. K.; Schuth, F. Microporous Mater. (1995), 3(4-5), 473-87.
[107] Schunk, S. A.; Demuth, D. G.; Schulz-Dobrick, B.; Unger, K. K.; Schueth, F. Microporous Mater. (1996), 6(5-6), 273-285.
[108] Kuperman, A.; Nadimi, S.; Oliver, S.; Ozin, G. A.; Garces, J. M.; Olken, M. M. Nature (1993), 365(6443), 239-42.

[109] De Onate Martinez, J.; Falamaki, C.; Baerlocher, C.; McCusker, L. B. Microporous Mesoporous Mater. (1999), *28*(2), 261-269.
[110] Pellet, R. J.; Coughlin, P. K.; Springer, A. R.; Gajek, R. T. US Patent No. 4 861 739 (1989).
[111] Ashtekar, S.; Chilukuri, S. V. V.; Prakash, A. M.; Harendranath, C. S.; Chakrabarty, D. K. J. Phys. Chem. (1995), 99(18), 6937-43.
[112] Martens, J. A.; Grobet, P. J.; Jacobs, P. A. J. Catal (1990), *126*, 299-305.
[113] Mertens, M.; Martens, J. A.; Grobet, P. J.; Jacobs, P. A. NATO ASI Ser. B (1990), *221*, 1-52.
[114] Morris, R. E.; Weigel, S. J. Chem. Soc. Rev. (1997), 26(4), 309-318.
[115] Derouane, E. G.; Von Ballmoos, R. N. EP 146385, 146386, 146387, 146388, 146389, 146390 (1985).
[116] Franco, M. J.; Mifsud, A.; Perez-Pariente, J. Zeolites (1995), 15(2), 117-23.
[117] Xu, R.; Huo, Q.; Pang, W. Proc. Int. Zeolite Conf., 9th (1993), Meeting Date 1992, Volume 1, 271-8. Editors: Von Ballmoos, R.; Higgins, J. B.; Treacy, M. M. J. Butterworth-Heinemann, Boston, Mass.
[118] Huo, Q.; Xu, R. J. Chem. Soc., Chem. Commun. (1990), 783-4.
[119] Yu, J.; Sugiyama, K.; Hiraga, K.; Togashi, N.; Terasaki, O.; Tanaka, Y.; Nakata, S.; Qiu, S.; Xu, R. Chem. Mater. (1998), 10(11), 3636-3642.
[120] Li, J.; Yu, J.; Yan, W.; Xu, Y.; Xu, W.; Qiu, S.; Xu, R. Chem. Mater. (1999), *11*(9), 2600-2606.
[121] Kraushaar-Czarnetzki, B.; Stork, W. H. J.; Dogterom, R. J. Inorg. Chem. (1993), 32, 5029-33.
[122] Chippindale, A. M.; Powell, A. V.; Bull, L. M.; Jones, R. H.; Cheetham, A. K.; Thomas, J. M.; Xu, R. J. Solid State Chem. (1992), 96(1), 199-210.
[123] Oliver, S.; Kuperman, A.; Lough, A.; Ozin, G. A. Chem. Commun. (1996), (15), 1761-1762.
[124] Jones, R. H.; Thomas, J. M.; Xu, R.; Huo, Q.; Cheetham, A. K.; Powell, A. V. J. Chem. Soc., Chem. Commun (1991), 1266-8.
[125] Chippindale, A. M.; Cowley, A. R.; Huo, Q.; Jones, R. H.; Law, A. D.; Thomas, J. M.; Xu, R. J. Chem. Soc., Dalton Trans. (1997), (15), 2639-2643.
[126] Williams, I. D.; Gao, Q.; Chen, J.; Ngai, L.-Y.; Lin, Z.; Xu, R. Chem. Commun. (1996), (15), 1781-1782.
[127] Yu, J.; Terasaki, O.; Williams, I. D.; Quiv, S.; Xu, R. Supramol. Sci. (1998), 5(3-4), 297-302.
[128] Yu, J.; Sugiyama, K.; Zheng, S.; Qiu, S.; Chen, J.; Xu, R.; Sakamoto, Y.; Terasaki, O.; Hiraga, K.; Light, M.; Hursthouse, M. B.; Thomas, J. M. Chem. Mater. (1998), 10(5), 1208-1211.
[129] Huo, Q.; Xu, R.; Li, S.; Xu, Y.; Ma, Z.; Yue, Y.; Li, L. Proc. Int. Zeolite Conf., 9th (1993), Meeting Date 1992, Volume 1, 279-86.
[130] Jones, R. H.; Thomas, J. M.; Chen, J.; Xu, R.; Huo, Q.; Li, S.; Ma, Z.; Chippindale, A. M. J. Solid State Chem. (1993), 102, 204-208.
[131] Oliver, S.; Kuperman, A.; Lough, A.; Ozin, G. A.; Garces, J. M.; Olken, M. M.; Rudolf, P. Stud. Surf. Sci. Catal. (1994), 84(Zeolites and Related Microporous Materials, Pt. A), 219-2.

[132] Oliver, S.; Kuperman, A.; Ozin, G. A. Angew. Chem., Int. Ed. (1998), 37(1/2), 46-62.
[133] Flanigen, E.M., Patton, R.L., US Patent No. 5 073 865 (1978).
[134] Guth, J. L.; Kessler, H.; Wey, R. Stud. Surf. Sci. Catal. (1986), 28(New Dev. Zeolite Sci. Technol.), 121-8.
[135] Guth, J. L.; Kessler, H.; Higel, J. M.; Lamblin, J. M.; Patarin, J.; Seive, A.; Chezeau, J. M.; Wey, R. ACS Symp. Ser. (1989), 398(Zeolite Synth.), 176-95.
[136] Kessler, H. Mater. Res. Soc. Symp. Proc (1991), 233, 47-55.
[137] Guth, J. P., Kessler, H., Caullet, P., Hazm, J., Merrouche, A., and Patarin, J., Proc. Int. Zeolite Conf., 9th (1993), Meeting Date 1992, Volume 1, 215-22. Editor(s): Von Ballmoos, Ronald; Higgins, John B.; Treacy, Michael M. J. Publisher: Butterworth-Heinemann, Boston, Mass.
[138] Kessler, H.; Patarin, J.; Schott-Darie, C. Stud. Surf. Sci. Catal. (1994), 85(Advanced Zeolite Science and Applications), 75-113.
[139] Estermann, M.; McCusker, L. B.; Baerlocher, C.; Merrouche, A.; Kessler, H. Nature (1991), 352(6333), 320-3.
[140] Bedard, R. L.; Bowes, C. L.; Coombs, N.; Holmes, A. J.; Jiang, T.; Kirkby, S. J.; Macdonald, P. M.; Malek, A. M.; Ozin, G. A.; et al. J. Am. Chem. Soc (1993), 115, 2300-13.
[141] Simmen, A.; Patarin, J.; Baerlocher, Ch. Proc. Int. Zeolite Conf., 9th (1993), Meeting Date 1992, Volume 1, 433-40. Editor(s): Von Ballmoos, Ronald; Higgins, John B.; Treacy, Michael M. J. Publisher: Butterworth-Heinemann, Boston, Mass.
[142] Koller, H.; Wolker, A.; Valencia, S.; Villaescusa, L. A.; Diaz-Cabanas, M. J.; Camblor, M. A. Proc. Int. Zeolite Conf., 12th (1999); pp 2951-2954. Editors: Treacy, M. M. J.; Marcus, B. K.; Bisher, M. E.; Higgins, J. B. Publisher: Materials Research Society, Warrendale, PA.
[143] Schreyeck, L.; Stumbe, J.; Caullet, P.; Mougenel, J.-C.; Marler, B. Microporous Mesoporous Mater. (1998), 22(1-3), 87-106.
[144] Harding, M. M.; Kariuki, B. M. Acta Crystallogr, Sect. C: Cryst. Struct. Commun (1994), C50(6), 852-4.
[145] Paillaud, J.-L.; Marler, B.; Kessler, H. Chem. Commun. (1996), (11), 1293-1294.
[146] Loiseau, T.; Ferey, G. J. Mater. Chem. (1996), 6(6), 1073-1074.
[147] Ferey, G. C. R. Acad. Sci., Ser. IIc: Chim. (1998), 1(1), 1-13.
[148] Barthomeuf, D. Zeolites (1994), 14(6), 394-401.
[149] Vomscheid, R.; Briend, M.; Peltre, M. J.; Man, P. P.; Barthomeuf, D. J. Phys. Chem. (1994), 98(38), 9614-18.
[150] Ashtekar, S.; Chilukuri, S. V. V.; Chakrabarty, D. K. J. Phys. Chem. (1994), 98, 4878.
[151] Martens, J. A.; Mertens, M.; Grobet, P. J.; Jacobs, P. A. Stud. Surf. Sci. Catal. (1988), 37(Innovation Zeolite Mater. Sci), 97-105.
[152] Mertens, M.; Martens, J. A.; Grobet, P. J.; Jacobs, P. A. NATO ASI Ser., Ser. B (1990), 221(Guidelines for Mastering Prop. Mol. Sieves), 1-52.
[153] Martens, J. A.; Grobet, P. J.; Jacobs, P. A. J. Catal (1990), 126, 299-305.
[154] Wilson, S. and Barger, P. Microporous Mesoporous Mater. (1999); 29(1-2), 117-126.
[155] Prakash, A. M.; Unnikrishnan, S. J. Chem. Soc., Faraday Trans. (1994), 90(15), 2291-6.
[156] Liu, Z.; Guo, X. PCT Patent No. WO. 9919254 (1999).
[157] Chen, J.; Wright, P. A.; Thomas, J. M.; Natarajan, S.; Marchese, L.; Bradley, S. M.; Sankar, G.; Catlow, C. R. A.; Gai-Boyes, P. L. J. Phys. Chem. (1994), 98(40), 10216-24.

[158] Unpublished work.
[159] Oliver, S.; Kuperman, A.; Coombs, N.; Lough, A.; Ozin, G. A. Nature (1995), 378(6552), 47-50.
[160] Gao, Q.; Che, J.; Xu, R.; Yue, Y. Chem. Mater. (1997), 9(2), 457-462.
[161] Sayari, A.; Moudrakovski, I.; Reddy, J. S.; Ratcliffe, C. I.; Ripmeester, J. A.; Preston, K. F. Chem. Mater. (1996), 8(8), 2080-2088.
[162] Khimyak, Y. Z.; Klinowski, J. Phys. Chem. Chem. Phys. (2000), 2(22), 5275-5285.
[163] Zhao, D.; Luan, Z.; Kevan, L. Chem. Commun. (1997), (11), 1009-1010.
[164] Chakraborty, B.; Pulikottil, A. C.; Das, S.; Viswanathan, B. Chem. Commun. (1997), (10), 911-912.
[165] Kimura, T.; Sugahara, Y.; Kuroda, K. Chem. Commun. (1998), (5), 559-560; ibid., Microporous Mesoporous Mater. (1998), 22(1-3), 115-126.
[166] Gier, T. E.; Stucky, G. D. Nature (1991), 349, 508-10.
[167] Nenoff, T. M.; Harrison, W. T. A.; Gier, T. E.; Stucky, G. D. J. Am. Chem. Soc. (1991), 113, 378-9.
[168] Harrison, W. T. A.; Gier, T. E.; Stucky, G. D.; Broach, R. W.; Bedard, R. L. Chem. Mater. (1996), 8(1), 145-51.
[169] Bu, X.; Feng, P.; Gier, T. E.; Stucky, G. D. Zeolites (1997), 19(2/3), 200-208.
[170] Broach, R. W.; Bedard, R. L.; Song, S. G.; Pluth, J. J.; Bram, A.; Riekel, C.; Weber, H.-P. Chem. Mater. (1999), 11(8), 2076-2080.
[171] Bu, X.; Gier, T. E.; Stucky, G. D. Microporous Mesoporous Mater. (1998), 26(1-3), 61-66.
[172] Harrison, W. T. A.; Martin, T. E.; Gier, T. E.; Stucky, G. D. J. Mater. Chem. (1992), 2, 175-81.
[173] Bu, X.; Feng, P.; Stucky, G. D. J. Solid State Chem. (1996), 125(2), 243-248.
[174] Harrison, W. T. A.; Nenoff, T. M.; Eddy, M. M.; Martin, T. E.; Stucky, G. D. J. Mater. Chem. (1992), 2, 1127-34.
[175] Kongshaug, K. O.; Fjellvag, H.; Petter Lillerud, K. Microporous Mesoporous Mater. (2000), 39(1-2), 341-350.
[176] Jones, R. H.; Chen, J.; Sankar, G.; Thomas, J. M. Stud. Surf. Sci. Catal. (1994), 84C, 2229-36.
[177] Harrison, W. T. A.; Broach, R. W.; Bedard, R. L.; Gier, T. E.; Bu, X.; Stucky, G. D. Chem. Mater. (1996), 8(3), 691-700.
[178] Bedard, R. L. US Patent No. 5 302 362 (1994).

Studies in Surface Science and Catalysis 137
H. van Bekkum, E.M. Flanigen, P.A. Jacobs and J.C. Jansen (Editors)
© 2001 Elsevier Science B.V. All rights reserved.

Chapter 6

Secondary Synthesis Methods

R. Szostak

Department of Chemistry, Clark Atlanta University, 223 James P. Brawley Dr. SW, Atlanta, GA 30314 USA

1. SUMMARY

Secondary or post-synthesis treatments have been utilized to produce many of the desired catalytic properties that could not be achieved by direct synthesis. These include increased catalytic acid activity, improved thermal and hydrothermal stability, and extra porosity. Historically, research in this area has been driven by the demand for improved petroleum cracking catalysts. Treatments used include: acid leaching, high temperature steam, and chemical treatments with chelates or fluoride compounds as well as combinations of these treatments. These processes are described along with the effects these methods have on the structural, physical and catalytic properties of the material.

2. INTRODUCTION

Post-synthesis modification of microporous materials encompasses a variety of techniques to further control the catalytic activity and/or the shape selectivity of a specific microporous structure. For the zeolite materials, three property changes have been noted: improved acid activity, increased thermal stability, and porosity change. The extent to which the properties are changed is dependent on the zeolite treated as well as the method of treatment. After over 40 years of study into the post-synthesis modification of zeolites, the following conclusions can be drawn: 1) Each post-synthesis treatment method modifies the properties of the zeolite in slightly different ways. 2) Acid solutions can selectively dealuminate a zeolite structure and wash the aluminum out of the crystal. 3) Hydrothermal dealumination removes the aluminum from the framework but does not remove it from the crystal. 4) Heterogeneity in the concentration of framework and non-framework aluminum is dependent on the modification method used.

Amorphous aluminum can collect on the surface of the crystal (from hydrothermal treatment). The surface can be depleted of aluminum centered acid sites (from fluorosilicate treatment). 5) A secondary pore system (mesopores) can sometimes be generated. 6) Hydroxyl nests can be annealed. Silicon, aluminum and other metal ions can be introduced into the framework at these nest sites improving stability as well as altering the catalytic properties.

For each method, a number of characterization techniques have been applied in order to quantitatively identify the changes brought about within the material upon treatment. Adsorption and x-ray powder diffraction were the first tools used by researchers to explore the impact of the treatment of these structures on their porosity, site heterogeneity and structural stability. Infrared spectroscopy helped develop an understanding the nature of the Bronsted acid sites and of the hydroxyl nests within the structure. Nuclear Magnetic Resonance (NMR) has proved to be critical in understanding of the location and state of the aluminum. Electron microscopy and surface sensitive techniques such as ESCA, AUGER, and SIMS, have provided insight into surface concentration of aluminum. Of all the different techniques utilized, examination of the resulting catalytic properties remains to be the most powerful tool to confirm the effect the modification method has on the material.

Improved catalytic performance has been the critical driving force for the high degree of activity in the area of zeolite synthesis as well as in post-synthesis modification. This is best illustrated in Table 1 where the relative activities for n-hexane cracking are compared for an amorphous silica-alumina catalyst, and synthetic faujasites (FAU) that have been modified through various methods.[1]

The history of the development of post-synthesis modification methods is short, with the key developments occurring between the late 1960s to the early 1980s. The understanding of post-synthesis modification of microporous materials parallels the improvements in characterization techniques and continues today. The application of these methods will continue as long as new microporous materials are discovered.

Table 1: Relative activities for n-hexane cracking, the role of modification on improving catalytic cracking performance of the faujasite (FAU) family of zeolites.[1,2,3]

Catalysts	**K/K_{std}**
Amorphous silica-alumina	1
Sodium zeolite X(FAU)	1.2
Rare earth hydrogen zeolite Y(FAU)	460
Rare earth hydrogen zeolite X(FAU)	7,800
Hydrogen zeolite Y(FAU)	30,000
Steamed-USY*(FAU)	23,000
Steamed, chemically dealuminated-USY*(FAU)	870,000

*Ultra stabilized Y(FAU)

3. PROTON EXCHANGE: REMOVAL OF ALUMINUM FROM THE FRAMEWORK

Very early studies had shown that not all of the zeolite materials could be directly ion exchanged with protons due to loss in crystallinity as the pH of the solution was lowered.[4] The very aluminum rich zeolite A(LTA) lost significant crystal structure in acidic solution. This characteristic, however, was not observed for the higher silica containing materials such as clinoptilolite(HEU). L.B. Sand in his 1961 paper noted "preliminary results show that this material is a unique cation exchanger, operating over the entire pH range...without change in the crystal structure from the parent material.[5]"

Though structural degradation like that observed for the zeolite A(LTA) was not observed, other significant changes were found to have occurred through these acid treatments. Barrer and Makki illustrated this in 1964.[6] They observed that progressive acid treatment of this zeolite could produce a fully decationated structure. Barrer proposed that the first step was the replacement of the metal cation with H_3O^+ followed by $[Al^{+3} + H_3O^+]$ removal leaving what they referred to as a hydroxyl next. The evidence for this was the change in adsorption properties of the treated crystals. Energetic homogeneity was shown by the adsorption of krypton and carbon dioxide. The development of such nests has been well established in the hydrogarnets and confirmed through neutron diffraction studies.[7] A depiction of the process of dealumination and hydroxyl nest formation is shown in Figure 1.

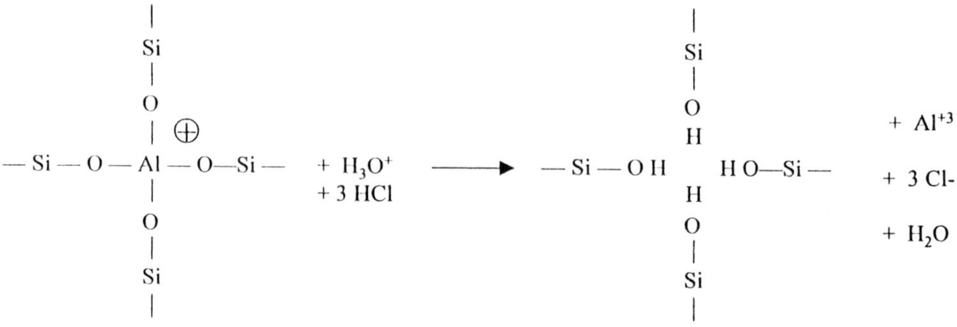

Figure 1: The process of acid mediated dealumination is proposed to take place through the extraction of aluminum from the framework and the generation of a hydroxyl nest.

The selective removal of aluminum from a silicate matrix is not surprising. Neither is the almost complete dealumination of a zeolite. This is because, 1) hydrated alumina becomes appreciably soluble at a pH of around 4 but in the presence of hydrated silica the solubility of alumina is only slightly suppressed. This would indicate that a zeolite should be susceptible to acid attack and subsequent aluminum removal. And 2) in most of the zeolites, the aluminum is essentially surface aluminum occupying sites that are either on the surface of the crystal or the surface of the pore. This is unlike many clays which are acid leaching resistant because much of the aluminum is located between layers of silicate and not in contact with the aqueous acid phase. The difference between the location of aluminum in the zeolite and the clay is shown in Figure 2.

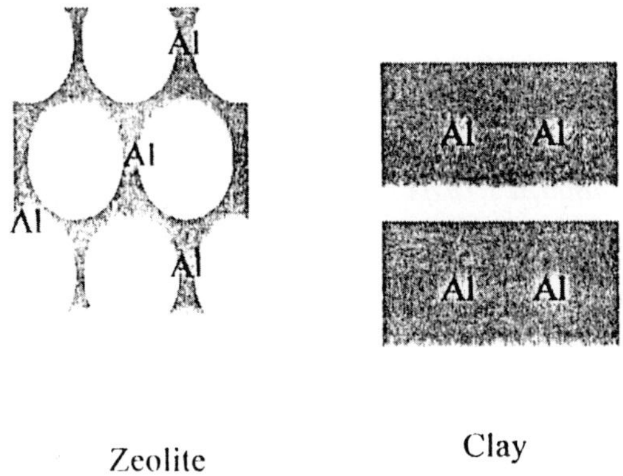

Figure 2: The difference between the location of aluminum in the zeolite structure and that of a clay. In the zeolite all the aluminum is accessible to the surface (pore surface or exterior of the crystal). In the clay, much of the aluminum is shielded from the surface by a layer of silica.

Mordenite (MOR), erionite (ERI) and zeolite L (LTL), all silica-rich zeolites, could also be dealuminated by direct leaching with strong mineral acids. SiO_2/Al_2O_3 ratios of 600 for acid dealuminated mordenite (MOR) have been reported. The dealumination of mordenite was of special interest to the early researchers due to 1) its availability, 2) its potential application as a catalyst.[8]

4. CRACKING CATALYST DEVELOPMENT: ZEOLITE Y(FAU)

The key needs for a successful cracking catalyst are strong acid activity and high thermal and hydrothermal stability. It is well known that the higher the SiO_2/Al_2O_3 ratio the more stable the zeolite structure. Unfortunately few zeolites can be prepared with the desired SiO_2/Al_2O_3 through primary synthesis alone. The large pore zeolite Y(FAU) contains the intrinsic properties of high surface area and three dimensional pore structure. The SiO_2/Al_2O_3 of this zeolite is limited to a maximum value around six by primary synthesis routes. Three secondary synthesis methods have been used to produce the desired properties: 1) hydrothermal treatment, 2) chemical treatment, and 3) a combination of hydrothermal and chemical treatment.[9] The details of these methods are summarized in Table 2.

Table 2: Preparation methods of aluminum-deficient zeolites.[9]

Methods to prepare dealuminated Y (FAU) zeolites
a. Hydrothermal treatment of NH_4 Y (FAU)
(Formation of ultrastable Y (FAU))
b. Chemical treatment
i. Reaction with chelating agents (EDTA, ACAC)
ii. Reaction with $(NH_4)_2SiF_6$ in solution
iii. Reaction with $SiCl_4$ (or other halide) vapor
iv. Reaction with F_2 gas
c. Hydrothermal and chemical treatment of ultrastable Y(FAU) with:
i. acids (HCl, HNO_3)
ii. bases (NaOH)
iii. salts (KF)
iv. chelating agents (EDTA, ACAC)

5. DEVELOPING AN ACIDIC Y ZEOLITE

As illustrated in Table 1, the acid activity for the hydrogen form of zeolite Y represented a substantial improvement over that of the amorphous silica-alumina catalyst. The acid form of zeolite Y(FAU) can be prepared by one of two methods: direct ion exchange with dilute mineral acids or by preparing the ammonium form of the zeolite then calcining it to release the ammonia. The acid treatment, as described in the previous section was found to do more to the structure than replace the metal cations with protons. Acid treatment of zeolite Y(FAU) as a method of proton ion exchange also proved difficult and very sensitive to the pH of the solution. The critical pH value for zeolite Y(FAU) which resulted in framework aluminum extraction followed by structural collapse was found to be between 2.5 and 3. Thus the need for careful control of the pH made direct acid exchange an undesirable process.[10] Ammonium ion exchange followed by dry calcination at temperatures between 300-500C would also produce the proton form of Y (H-Y(FAU)) and rapidly became the method of choice for obtaining an acid Y(FAU) catalyst.[11-13]

One would reason that a completely hydrogen ion exchanged zeolite Y (FAU) prepared via the NH_4^+ route would exhibit the greatest acid activity. This proved to be incorrect. It was observed that when the calcination temperature exceeded 500C, loss of the aluminum ions from the framework and hence decrease in the proton ion exchange capacity occurred.[14,15] The term 'decationization' was coined and defined as the loss in ion exchange capacity for at least 10% of the aluminum ions in the zeolite. This was later found to produce improved catalytic properties.

6. IMPROVING THERMAL STABILITY

To be useful as a cracking catalyst, the material must be able to withstand temperatures in the range of 700-950C and a steam environment. Though H-Y(FAU) represented a major advance, thermal stability was still limited to 500C. It was discovered that even at room temperatures the presence of moisture in an H-Y(FAU) catalyst would destroy the structure.[16] Stability of the zeolite structure, both thermal and hydrothermal, could be achieved through ion exchange with multivalent cations.[17] Stabilization of hydrogen Y(FAU), for example could be achieved with the addition of only 3.7 La^{+3} cations per unit cell.[18]

7. ULTRASTABLE Y (FAU)

Simultaneous with the development of stabilization methods using rare earth ions such as lanthanum as described above, researchers investigated other methods to achieve thermal stability. McDaniel and Maher in 1967 first reported the preparation of a thermally stable zeolite Y(FAU).[19-22] They called the highly stable form of Y(FAU) that they prepared, "ultrastable Y". This material was prepared from the ammonium Y(FAU) and calcining the material above 500C. This modified Y(FAU) exhibited properties that were similar to the decatonated zeolites described previously, i.e. the ultrastabilized Y(FAU) exhibited reduced ion-exchange capacity and a smaller unit cell than the parent material.[23] Both features are indicative of a removal of aluminum from framework sites.

The importance of framework aluminum removal was not immediately recognized and much discussion pursued concerning the role residual sodium ions might have in structural stability of ultrastabilized Y(FAU) as well as the basis for wide variations observed in the activity of the H-Y(FAU) catalysts.[24-26] The role of sodium in secondary synthesis is described in more detail later in this paper. Further research identified the critical role of residual steam as the key to enhancing stability in the process of ammonium ion removal to form H-Y(FAU).[27-34] Kerr and Ward both pointed out that the shape of the container used to calcine NH_4-Y(FAU) was critical in producing stability in the resulting zeolite.[27,35] Samples that were prepared using a shallow bed produced H-Y(FAU) from NH_4-Y(FAU), while samples that were prepared in deep beds with longer contact time with the steam being generated by the moist sample, produce the ultrastabilized product. Treating ammonium-Y(FAU) with steam at temperatures around

500C produced similar results to those of the deep-bed calcined materials. The production of the same ultrastabilized product using in-situ reactor activation methods was also pointed out.[35-37] These studies helped clarify the basis for vastly differing activities observed for samples of H-Y(FAU) prepared from NH_4-Y(FAU). Figure 3 shows the dependence of the degree of dealumination on the partial pressure of steam in a shallow bed treatment of zeolite Y(FAU) at 550C.[38] The critical difference in stability between the stabilized and non-stabilized Y(FAU) zeolites is shown in Table 3 where the effects on the properties of the zeolite, such as unit cell volume and thermal stability differ significantly.

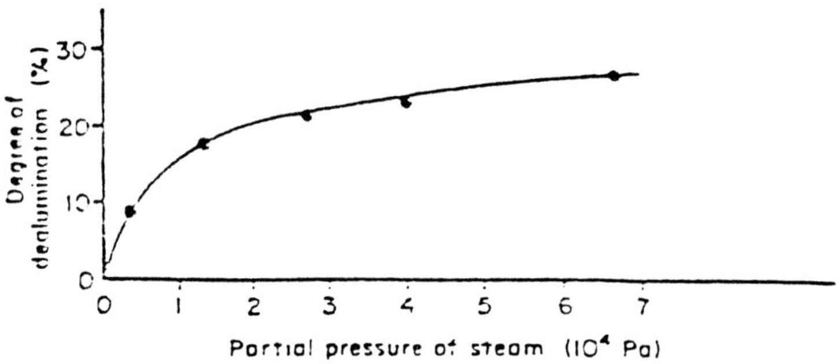

Figure 3: Dependence of the degree of dealumination on the partial pressure of steam in the shallow bed treatment of zeolite Y(FAU) at 550C.

Table 3
Composition and thermal stability of zeolite Y(FAU) one which is stabilized by steam treatment and not stabilized.

	Stabilized	*Not Stabilized*
SiO_2/Al_2O_3 of starting zeolite Y	5.2	5.2
Chemical analysis		
Na_2O	0.10%	0.10%
Al_2O_3	21.10%	21.20%
SiO_2	78.77%	78.77%
SiO_2/Al_2O_3	6.35	6.35
Unit Cell Å	24.34	24.69
Unit cell of starting zeolite Y,Å	24.65	24.65
Surface area after two hours at:		
700F	837(m^2/g)	1008(m^2/g)
1500	851	254
1550	793	132
1650	842	18
1700	743	15
1725	678	
1800	542	

Though the ultrastabilized Y(FAU) still contained all of its aluminum, the non-framework aluminum could be washed out of the structure after stabilization by washing with dilute acid.[39] In fact very highly silica zeolite Y(FAU) maintaining structural integrity could be obtained through the process of hydrothermal treatment followed by acid washing.[40-42] This is shown in a series of experiments summarized in Table 4. With steam treatment only, the bulk SiO_2/Al_2O_3 ratio remains constant with the only change to indicate framework aluminum removal being the change in the unit cell volume. Unit-cell volume decreases with increasing temperature (severity) of steaming. Crystallinity after treatment was not reported in this study; however, the indication is that, depending on the treatment employed, only 20% of the sample remains crystalline under the more severe treatments. The bulk SiO_2/Al_2O_3 ratio was altered through subsequent leaching of the sample with acid as can be seen from the data presented. Changes are observed in the amount of residual sodium as well as the bulk aluminum content, with little observed change in the unit cell volume, indicating that the role of the acid leaching is mainly to remove the spurious aluminum species not contained within the framework.

Table 4: Treatment of zeolite Y(FAU) under various conditions and its effects on SiO_2/Al_2O_3 ratio and on the unit cell dimensions.[43]

Samples	*Treatment*	*Na%*	*SiO_2/Al_2O_3 (a)*	*A_o/Å (b)*
NH_4Y(FAU)	NaY exchanged in NH_4Cl	2.2	4.8	24.70
Deal 6-2	NH_4Ysteamed 600C in 100% H_2O, exchanged with NH_4Cl, steamed 820C in 60% H_2O	0.27	4.8	24.35
Deal 6-3	Same but final treatment 890C in 100% H_2O	0.27	4.8	24.32
Deal 6-4	Deal 6-3 steamed 920C in 100% steam	0.27	4.8	24.22
Deal 9-2	Same as Deal 6-2 but final treatment 890C, 100% H_2O	0.27	4.8	24.19
Deal 9-2ext	Deal 9-2 treated with HCl (1M) at 90C	0.07	48	24.20
Deal 10 ext	Same as Deal 9-2, ext but steaming and extraction repeated two times	0.07	98	24.18
Deal 16 ext	Same as Deal 9-2 ext	0.04	156	24.20
HY	NaY exchanged in NH_4Cl (eight times) outgassed at 300C in vacuum	1	4.8	

(a) from chemical analysis
(b) unit cell constant

8. CHARACTERISTICS OF ULTRASTABILIZED Y(FAU)

Crystals of ultrastabilized Y(FAU) contains a complex mixture of materials: 1) the intact faujasite structure dominating the crystal (as seen by the maintenance of x-ray crystallinity), 2) hydroxyl nests, 3) non-framework aluminum species, 4) amorphous silica-alumina debris and 5) a secondary mesoporous system.

The X-ray diffraction evidence of a shrinking unit cell upon steam or hydrothermal treatment has been used as an indication of the removal of aluminum ions from the framework since the Si-O bonds are shorter than the Al-O bonds. A decrease in the unit cell dimensions of the order of 1-1.5% relative to the parent material has been observed. The exact amount of such decrease depends on 1) sodium content, 2) degree of aluminum extraction and 3) the calcination conditions. The range of changes observed in the unit cell of type Y(FAU) with treatment starting with differing SiO_2/Al_2O_3 parent materials is shown in Figure 4. The SiO_2/Al_2O_3 of the ordinate is the framework ratio not representative of the bulk SiO_2/Al_2O_3.

Breck and Flanigen were the first to show that the number of Al ions in the structure of Y(FAU) can be determined from the unit-cell constants.[44] Kerr has summarized the relationship in the form:[45]

$$N(Al) = m(a_o - x)$$

Where N(Al) equals the framework aluminum ions per unit cell and a_o is the unit cell parameter in Å. The values for slope and intercept have been determined by various groups and their results are provided in Table 5.

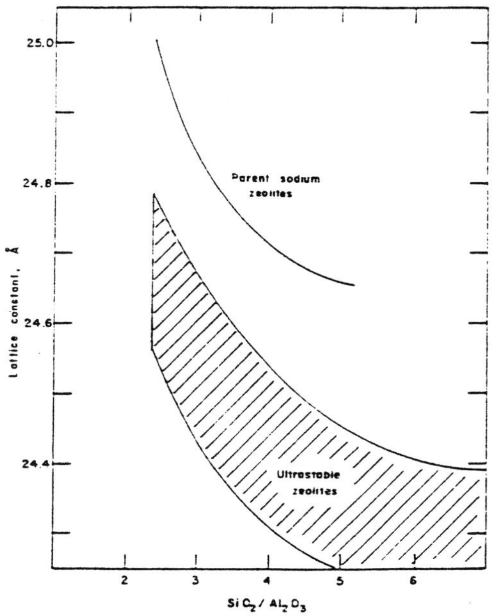

Figure 4: Unit cell contraction in ultrastable zeolites.

Table 5: Values for the slope and intercept used to determine the amount of framework aluminum in zeolite type Y(FAU) based on the unit cell parameter.

	Slope, m	*Intercept, X*
Breck and Flanigen[44]	115.2	24.191
Kerr [44]		
Dempsey, et al.[46]	112.1	24.222
Sohn et al.[47]	107.1	24.238

In using the relationship between unit cell size and framework aluminum content it is critical for reproducibility to have all samples equilibrated at constant humidity before the measurement of the cell size.[44,48] The variations reported by other groups can generally be attributed to the presence of defect 'holes' in the structure, which are dependent on the method used to dealuminate the zeolite. The amount of healing or replacement of aluminum with silicon in the structure is method dependent. Since the aluminum-depleted defect site is proposed to be larger than that of the healed (silicon-substituted) site it is difficult to directly compare framework aluminum contents of the different materials on the basis of unit cell size alone.[49] The presence of such method dependent defects has been identified through a number of other techniques.[50,51] Care must therefore be taken in using unit cell data to exclusively quantitatively determine the amount of framework aluminum. Still, determination of the unit cell size has been an important technique utilized at the refineries.[52] In catalytic cracking, the size of the unit cell has been shown to correlate well with the RON (research octane number) and MON (motor octane number), the ratio of olefin/paraffin and the yields of several of the fractions including C3 gas, gasoline, light cycle oil, bottoms, and coke. The smaller unit cell volume correlates with the lower production of coke. In the refinery the unit cell size is used to predict RON and the dry gas yield.

The unit cell size, however, is only one step in the characterizing of the type Y(FAU) cracking catalysts.[53,54] In addition to acid strength and site density which can be changed by controlling the modification procedure; levels and types of the non-framework component as well as mesoporosity development within the crystal also alter the performance of the cracking catalyst. As discussed above, these cannot be ascertained from the unit cell measurements. The variability in sites based on the type of modification employed is proposed to change the cracking intermediates from the carbocations to radical cation type species.

9. EFFECT OF SODIUM CONTENT

Figure 5 shows the important role of sodium in the contraction of the unit cell of zeolite Y(FAU).[122] The presence of sodium has a two-fold effect on stability. High temperature stability is limited by its presence; a long recognized fact in the catalyst art.[55] The presence of greater than 25% sodium prevents unit cell contraction during calcination of the partially ammonium exchanged zeolite, as the presence of sodium inhibits aluminum extraction. Generally speaking, the unit cell contraction is influenced by sodium content, but it is not a direct result of sodium removal. Little contraction occurs with ion exchange alone; in fact, a slight increase in the unit cell is observed when the sodium ions are replaced by ammonium ions.

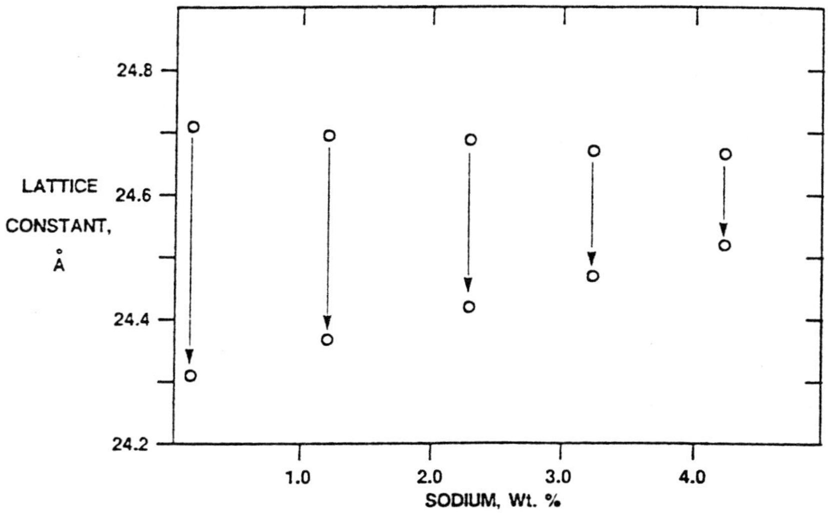

Figure 5: Unit cell contraction of ammonium-exchanged zeolite Y(FAU) during stabilization as a function of sodium content.

Extraction of aluminum from the framework occurs when the partially exchanged form undergoes an intermediate calcination and not during the exhaustive exchange process. Low sodium levels are of practical importance as they aid in enhancing vanadium tolerance in the cracking catalyst.

10. OTHER SECONDARY SYNTHESIS METHODS

Several groups have explored other methods of producing thermally and hydrothermally stable cracking catalysts. These include the use of EDTA, $SiCl_4$ vapor, and $(NH_4)_2SiF_6$. Each of these methods modify the zeolite in slightly different ways allowing for the ability to better custom tailor the zeolite to the catalytic application.[38,51,53,56,57] 27Al NMR, for example, has been used to compare the effect these various treatments have on the nature of the non-framework aluminum generated. The results are shown in Figure 6. Dealumination with EDTA, which extracts the aluminum from the structure, shows little difference in the 27Al NMR spectrum from that of the parent material. The differences in the peak positions for the non-framework aluminum in ultrastabilized Y(FAU) and Y(FAU) treated with $(NH_4)_2SiF_6$ in the figure are attributed to the presence of fluoride bound to the non-framework aluminum in the silicon fluoride treated sample. Chemical analysis substantiates the presence of residual fluoride.[63]

NMR has been applied to try to quantify the degree of mesoporosity in the zeolite.[58-60] The 'Swiss cheese' model of this dealuminated zeolite is linked to the severity of the high temperature, hydrothermal treatment. Mesopore volume is not as evident in type Y(FAU) samples dealuminated via the $SiCl_4$ or $(NH_4)_2SiF_6$ route.[61] A comparison was made of the difference in the extent of mesopores in three dealuminated samples of

Y(FAU) prepared by three different methods: steam treatment (high level of mesopores), $SiCl_4$ treatment (moderate level of mesopores), and treatment with $(NH_4)_2SiF_6$ (some mesoporosity).[97,98] However, subsequent reports on thoroughly washed and well characterized samples of the $(NH_4)_2SiF_6$ treated Y(FAU) have indicated no mesoporosity can be observed after such treatment.[62] Treatment with EDTA is also shown to induce a mesopore system.[63]

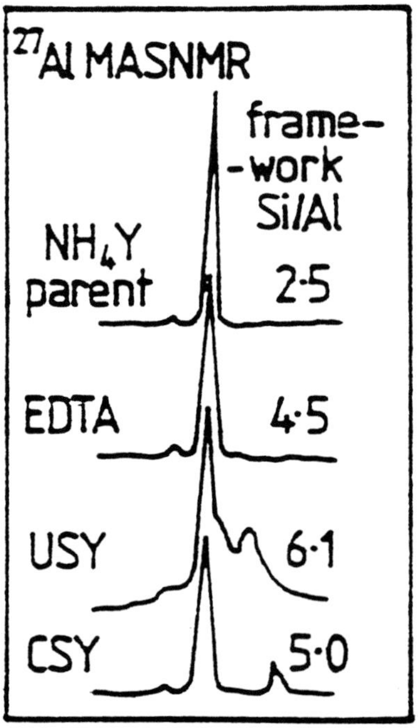

Figure 6: 27 Al NMR of type Y(FAU) (NH4Y parent), EDTA treated Y(EDTA), steam stabilized Y (USY) and (NH4)2SiF6 treated Y (CSY). Framework Si/Al provided on the right side of the figure.[63]

11. SUMMARY OF PHYSICAL CHANGES FROM VARIOUS SECONDARY SYNTHESIS METHODS

EDTA method

The EDTA method utilizes the acid leaching properties of the acid form of EDTA. Kerr first reported this method in 1968.[29,64] The resulting material differs from ultrastabilized Y(FAU) because of inhomogeneity within the crystal. The acid form of EDTA preferentially attacks the outer surface of the crystal producing a large number of silanol nests.[65-68]

SiCl4 treatment

Beyer first reported silicon tetrachloride vapor at elevated temperatures in 1980.[69,70] The advantage of this method is that the reagent that removes the aluminum from the framework also supplies the needed silicon to fill the vacancies.[71] Like the EDTA method, however, heterogeneity within the crystal is noted. Aluminum chloride migrates to the surface of the crystal. Upon contact with water, the aluminum chloride on the surface converts to intractable aluminum oxide.[68 ,72-74] Moreover, residual aluminum chloride can remain trapped within the sample. Methods have been used to improve the removal of aluminum from the sample. Lithium exchange, for example, helps the aluminum chloride volatilize more readily.[75]

Ammonium fluorosilicate method

Refluxing the zeolite with ammonium fluorosilicate was first explored by Skeels and Breck and reported in 1984.[76,77] Like the EDTA treatment, this method is considered to be a very mild method for removing aluminum. Because of the presence of additional silicon in the reagent, defect-free, silicon rich samples can be obtained. The critical difference between this method and ultrastabilization techniques described earlier is that materials treated with ammonium fluorosilicate do not contain mesopores.[78]

Other fluoride treatments

Fluorine treatment is sensitive to the type of zeolite as well as the conditions of the treatment. Both aluminates and silicates react with fluorine under extremely mild conditions. In a zeolite dealumination will occur with the subsequent formation of AlFx(OH)y species, depending on the amount of fluoride and water present when exposed to a fluorine containing speices. F2,[79],[80] HF,[81-84], boron trifluoride,[85] aluminum fluoride or silicon fluoride containing species,[86] have all been reported to structurally modify zeolites. In the reaction with F_2, not all zeolites treated show an increase in thermal stability with treatment. An important feature of this method of modification is the persistence of fluorine residue after treatment that may not be easily extracted.

12. MECHANISM OF SECONDARY SYNTHESIS

Kerr proposed that the extra-framework aluminum was generated through a multi-step process. The first step was the removal of aluminum from the framework (dealumination) producing an extra-framework aluminum species, $Al(OH)_3$ and a hydroxyl nest. This aluminum hydroxide species could then be dehydrated producing $Al(OH)_2^+$, $Al(OH)^{+2}$, and Al^{+3} species which could act to balance the remaining framework aluminum charge. It was proposed that these species would provide a similar framework stabilizing effect as that which has been observed for rare earth exchanged Y(FAU). The framework stabilizing feature of the non-framework aluminum was questioned when it was shown that aluminum ion exchange did not exhibit similar stabilizing effect and when the non-framework aluminum was removed from the ultrastabilized Y(FAU), the structure maintained stability.[28,35,37,87,88] The population of aluminum in framework sites within the parent Y(FAU) also dictated its hydrothermal stability and the amount of mesopores formed after secondary synthesis.[89]

A number of proposals for the fate of the hydroxyl nest remaining in the aluminum deficient structure have been entertained. These included formation of new Si-O-Si bonds though geometrical reorientation,[90] condensation reactions with nearby silica in the zeolite,[26] or migration of extraneous non-framework silicon atoms into the site.[91]

A more detailed study looking at the kinetics of dealumination using energy dispersive x-ray diffraction has shown that calcination of NH_4-Y(FAU) results in two effects which occur at slightly different rates: a decrease in the unit cell size and a decrease in the crystallinity.[92] The reduction in the unit cell size is proposed to be due to the generation of separated aluminum vacancies in the structure while the loss in crystallinity observed during this process is due to amorphization from the collapse of areas of the zeolite structure when several adjacent aluminum vacancies are formed. This amorphization is proposed to give rise to the mesopore structure in the material.

13. MESOPORES

Mesopores can be observed in the textural changes between an unsteamed and steamed type Y(FAU) zeolite under transmission electron microscopic investigation.[93-95] High resolution images of the steamed sample show many areas of low contrast which are attributed to the presence of amorphous domains or holes and macrocavities within the zeolite crystal. Adsorption studies also indicate the presence of a mesoporous network in the materials.[96-100] The development of the mesopores is attributed to the extraction of neighboring aluminum ions from the framework which destabilizes the region causing localized loss in structure and generation of amorphous silica-alumina debris.

14. FRAMEWORK AND NON-FRAMEWORK ALUMINUM SPECIES

A significant number of studies have been performed on the dealuminated Y(FAU) materials to establish that the primary catalytic activity is based on the protons associated with the tetrahedral aluminum sites.[101] The intrinsic activity of the acid site is dependent on the number of sites within the structure up to certain aluminum content after which the strength of the individual acid sites remains constant.[9,102-112] An example of the change in catalytic activity as a function of the amount of framework aluminum per unit cell is shown in Figure 7.[113]

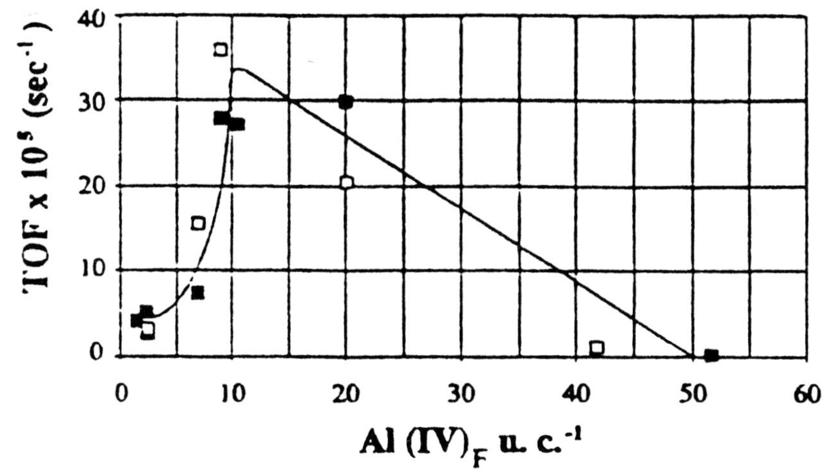

Figure 7: Evolution of the TOF in hydroisomerization of heptane (full squares) and decane (open squares) as a function of framework aluminum content.

A number of techniques have been used to establish the different types of aluminum in the dealuminated samples[114-117] Three types of aluminum have been identified: 1) framework aluminum (Bronsted acidity), 2) octahedral non-framework aluminum, and 3) NMR invisible aluminum.[118-120] This third type of aluminum is proposed to be due to non-framework aluminum on the surface of the zeolite that is present in several coordinations.[113]

The initial increase observed in the activity can be accounted for by the increase in site acidity as the number of aluminum ions per unit cell are decreased. After the maxium is reached, a gradual decrease in activity is observed. It has been postulated that this is due to the removal of extra-framework aluminum species that have been removed during acid treatment that is necessary to obtain these low Al contents,[121] or perhaps to the low turnover frequency of hidden sites.[54]

15. DETAILS OF KEY SECONDARY SYNTHESIS METHODS

Steam stabilization methods (Methods A and B)[122]

Method A:

This procedure illustrates the stabilization of the structure following a simple ammonium exchange while the zeolite still contains 10-25% of its original alkali-metal cations.

1. A Y(FAU)-type zeolite is ion exchanged with an ammonium salt solution to reduce the level of sodium to about 10-25% of its original value. The remaining sodium is increasingly difficult to remove by conventional exchange because of its location in the small sodalite cages of the zeolite.

2. The partially exchanged zeolite is washed free of excess salt and stabilized by heating in a static or steam-containing atmosphere to a temperature between 600C and 825C, so that the unit cell shrinks by about 1% or about 0.2 to 0.3Å.

3. The thermal stability of the stabilized structure can be increased still further by removing the major portion of the remaining sodium which has now become readily exchangeable.

Method B: Complete alkali removal.

1. A Y(FAU)-type zeolite is ion exchanged with an ammonium salt solution to reduce the level of the sodium to about 10 to 25% of its original value.

2. The partially exchanged zeolite is washed free of excess salts and heated to a temperature between 200 and 600C so that the sodium ions have been redistributed but the structure has not undergone significant shrinkage.

3. The major portion of the remaining sodium is removed by again ion exchanging with an ammonium salt solution. The zeolite at this stage is in an extremely precarious metastable state and can be easily made unstable by a variety of conditions not yet fully understood.

4. The low-sodium, metastable zeolite is rapidly heated in a static or steam-containing atmosphere to a temperature between 600 and 800C to minimize the time during which the hydrogen form might be present. This stabilizing calcination causes a shrinkage of approximately 1 to 1.5% in the unit cell.

Steam Stabilization/Acid Extraction Method

This method is based on the steam treatments described above with the exception that the extra-lattice aluminum is removed via an acid leaching step.

1. NH_4Y(FAU) is prepared from NaY(FAU) then calcined at 540C for two hours in steam.

2. The material is ammonium ion exchanged and further calcined at 815C for three hours in steam.

3. The resulting material is treated for two hours with HCl(1N).

EDTA Method [29, 64, 122]

This illustrates the preparation of an ultrastable structure in which framework aluminum is extracted from a Y(FAU) type zeolite with a solution of the acid form of EDTA.

1. Approximately 0.25 to 0.50 mole of H_4EDTA per equivalent of zeolite cation is slowly added to a slurry of the zeolite in water under refluxing conditions, the complete addition requiring at least 18 hours.

2. The aluminum deficient zeolite is heated with a purge of inert gas to 800C resulting in a contraction of the lattice of about 1%.

$SiCl_4$ Treatment [70,123]

Dealumination takes place in an upflow ractor with a bed depth of about 50 mm and diameter of 30 mm.

1. Hydrated NaY(FAU)(or HY(FAU)) powder was placed in the reactor and the temperature increased 10C/min to 347K and held constant for two hours and purged with nitrogen.

2. The reactor was cooled to 247C, still under nitrogen. The N_2 stream was then diverted through a bubbler containing $SiCl_4$ until a constant temperature was obtained (the reaction being exothermic). Further heating of 10C/min was undertaken until the final exposure temperature was reached. It was then held at that temperature for 40 minutes after which pure N_2 was reintroduced, the tube purged of $SiCl_4$ and the reactor cooled to room temperature.

Ammonium Fluorosilicate Method [124,125]

1. $NaNH_4Y$(FAU) zeolite with 80% of the total Na^+ exchanged by NH_4^+ was introduced in a 0.4M solution of $(NH_4)_2SiF_6$ at 70C, the zeolite to $(NH_4)_2SiF_6$ ratio being 3.3 g/g.

2. After the zeolite was well dispersed, the temperature of the bath was increased to 95C and kept at this value during three hours with agitation. Then the product was filtered and thoroughly washed with boiling water. The sample was dried and exchanged twice at 80C with an ammonium solution and calcined at 550C. Further dealumination by this procedure required intermediate calcination. It must be noted that dealumination further than 25 Al/u.c. provokes an appreciable loss in crystallinity. The unit cell size generally range from 24.60Å to 24.35Å and the material has been reported to be more stable than other dealuminated Y(FAU) preparations.[126]

16. CATALYTIC PROPERTIES AND SECONDARY SYNTHESIS METHODS

Though a variety of different methods have been developed to accomplish the same goal, that of stabilizing the zeolite Y(FAU) structure with the removal of framework aluminum, it appears that each modifies a different aspect of the material. For example, the higher silica containing materials prepared from treatment with ammonium fluorosilicate makes 2-5% more gasoline, 0.2 to 0.4% more light cycle oil and 1 to 2% less coke than catalysts made with the steam treated Y(FAU). In the steam-stabilized Y(FAU) more coke is formed which is attributable to amorphous debris in the pores which is decreased with the fluorosilicate treated materials.[126] The SiCl4 treatment of type Y(FAU) produces differing product distributions in the conversion of n-decane when compared to the steamed zeolite.[127] It is evident from the catalytic results in both model cracking reactions, as well as in actual use of these catalysts under conditions similar to the industrial cat crackers, that the differences observed between modification techniques determined through physical characterization methods translate to observed changes under catalytic operating conditions. The complex subtleties of these differences, are still not completely understood.

17. AND THEN THERE IS REALITY

It is important to note that any ‘initial’ material used as an FCC catalyst immediately begins a course of change upon entering the cracking unit. “Equilibrium” catalysts in the unit differ from the “initial” materials so carefully characterized using the techniques described in this chapter. The “equilibrium” catalyst is a mixture of many different modified zeolites. As the catalysts in the unit are exposed to very high temperatures in the regenerator, dealumination of the framework continues. Such exposures to the regenerator temperatures are for differing lengths of time, ranging from a few hours to a few months. The acid activities of the "equilibrium" catalyst may therefore span an order of magnitude, from extremely high activity found in recently added fresh catalyst to the other extreme, near zero activity for a catalyst that has been circulating for several months. These "equilibrium" catalysts differ from unit to unit, from feed type to feed type, not only from catalyst to catalyst. In the optimization of cracking catalysts, a compromise must be reached between high thermal stability and enough acid sites to maintain activity. Selectivity towards gasoline and related products of high commercial importance is the ultimate goal, regardless of the method.

18. MODIFICATION OF HIGH SILICA ZEOLITES

The application of post-synthesis modifications to already thermally stable zeolites containing a high proportion of silica has been explored in detail over the last 20 years. Increased hydrophobicity, making the material more organophilic and further diluting the number of highly acidic Bronsted sites within the structure has been the basis for much of the work in this area. Enhancement in activity with steam treatment was observed by Mirandatos and Barthomeuf for mordenite(MOR).[128,129] A number of groups have reported similar behavior in the hydrothermal treatment of ZSM-5(MFI).[130-133] It has been proposed that this enhanced activity may be a result of interplay between active sites rather than merely a decrease in the concentration of framework aluminum.[134] The enhancement in activity for the higher silica zeolites only occurs when mild steaming conditions are employed. Under severe steaming conditions, reduced activity is found in direct proportion to the loss of aluminum from the framework sites. The enhancements observed when the zeolite is mildly steamed are not reaction specific, i.e., paraffin (n-butane, n-hexane) cracking, toluene disproportionation and paraffin isomerization reactions all reflect such enhancement in activity of zeolite ZSM-5(MFI).[130,131,135,136]

Lago et al. provided a detailed examination of ZSM-5(MFI) correlating the degree of enhancement with the SiO_2/Al_2O_3 ratio as well as with the water vapor pressure. These are shown in Figures 8 and 9. The lower the water vapor pressure, the greater the enhancement in activity, with pressures less than 100 torr providing the greatest enhancement in cracking activity towards n-hexane. Such low vapor pressures are indeed easily obtained under normal calcination conditions where high environmental humidity may be encountered or localized steaming environments generated (i.e. decomposition of the organic template to produce steam)) suggesting a need for care in template removal. A wide range of cracking activities are encountered with the mild steam enhancement of ZSM-5(MFI) A maximum between activity and the number of aluminum ions per unit cell was determined to be two which is consistent with the proposed site pairs for aluminum in the framework. An examination of the cracked product distribution for butane cracked over mildly steamed ZSM-5(MFI) shows little difference in selectivity based on the mixture of products from that observed for the parent material. This suggests a similar Bronsted activity in both the parent and steam enhanced material.

On the other hand, a detailed NMR study of these materials does not appear in agreement with the proposed aluminum site pairing model proposed by Lago.[137] No signal in the NMR was observed which indicates the presence of Si(2Al) which makes up 5% of the aluminum atoms in the material.

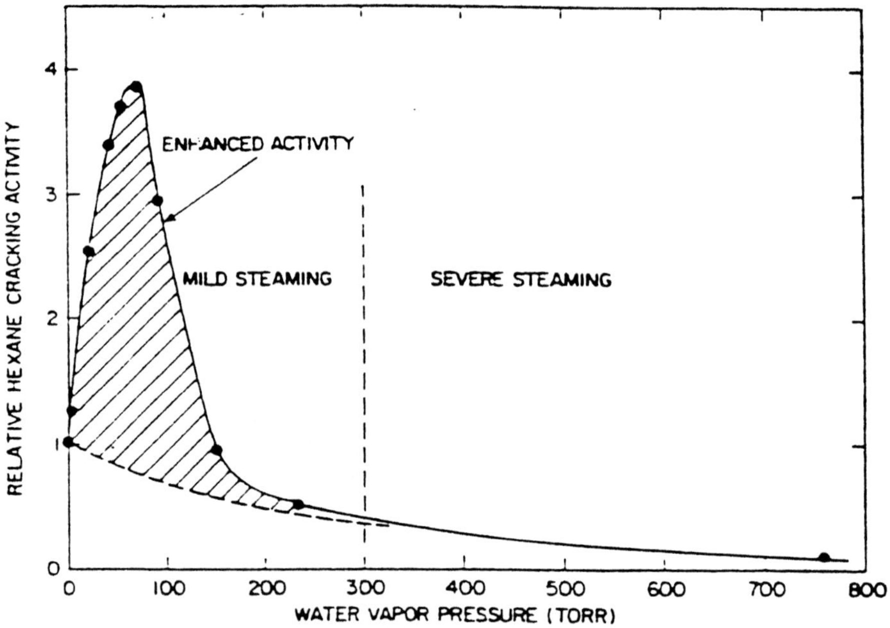

Figure 8: The effect of steaming severity (pressure) on the hexane cracking activity of HZSM-5(MFI).

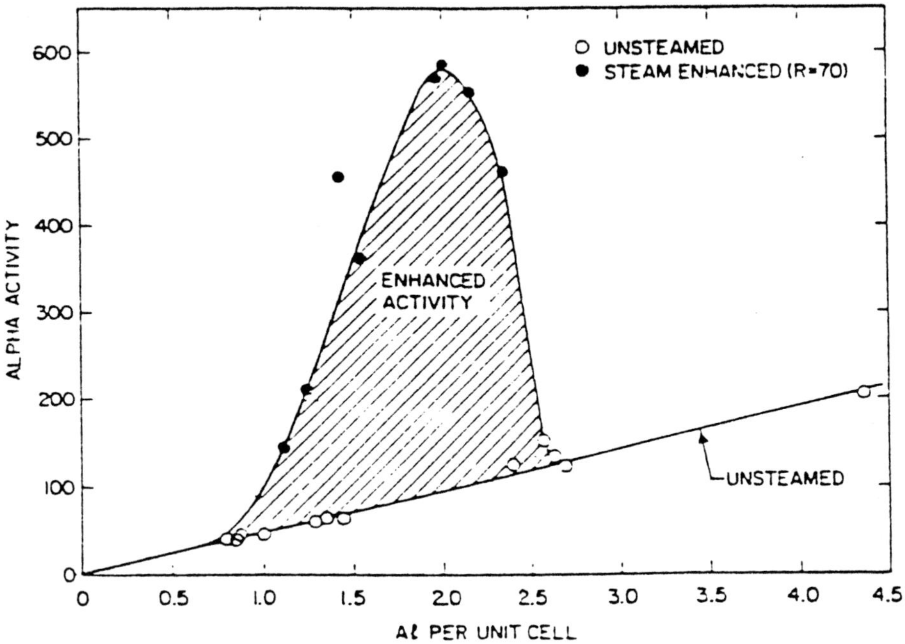

Figure 9: Dependence of alpha activity on the tetrahedral Al content of unsteamed and steam enhanced HZSM-5(MFI).

Though numerous types of species have been proposed for the non-framework aluminum in dealuminated zeolites, including cationic, anionic as well as neutral species as described previously, agglomerate formation of the aluminum species appears to be low in these mildly steamed materials. Rehydration results in the detection of AlOOH and $Al(H_2O)_6^{+3}$ species. Brunner[137] as well as Ione and coworkers[138] indicate that these non-framework species are composed of non-charged, non-hydroxylated and non-stable small complexes and propose, that as electron pair acceptors, they behave as Lewis sites. Due to the low concentration of such species, direct evidence of their presence is difficult to obtain.[139]

19. APPLICATION OF SECONDARY METHODS TO METALLOSILICATES

As described above, hydrothermal modification of high silica zeolites, depending on conditions of steaming, results in an initial increase in activity followed by a decay in that activity as increasing amounts of aluminum are removed from the framework. Similarly, demetallation results when a metallosilicate is contacted with steam. Significant differences arise when a potentially catalytically active metal component, originally in the framework is dislodged. Liberation from the framework can increase catalytic metal activity. Two metallosilicate systems have been examined in detail, the gallosilicate and the iron silicate analogs of zeolite ZSM-5(MFI).

Physical characterization of the materials shows a decline in acidity based on ammonia adsorption with time of steam treatment as well as a decrease in ion exchange capacity.[140] In the case of the iron silicate, a significant change in color is observed with steaming, shown to be due to the formation of non-framework octahedral Fe-O-Fe species.[141] The iron system lends itself to analysis with a wide variety of techniques including Mossbauer, which substantiates the increasing presence of octahedral non-framework iron in the material with severity (time) of steaming. HREM images of the steamed samples show the generation of particles in the range of 15Å to 150Å depending on the conditions of treatment, duration and temperature.[142] Steaming temperatures around 700C produce very large agglomerates of iron oxide while milder temperatures (550C) induced smaller, more highly dispersed particles to form.

The catalytic activity of these materials was increased with the increase strongly dependent on the reaction employed. In the cracking of butane, a distinct enhancement in catalytic activity was seen for the modified gallosilicate. In addition, the product distributions were significantly altered due to the presence of non-framework gallium(oxide) with the predominant formation of olefins. The modified iron silicate catalyst also resulted in a change in product selectivity when used in the cracking of butane. Methane formation was enhanced due to the presence of the non-framework iron (oxide). A comparison between the cracked product distributions for the modified alumino-, gallo-, and iron silicates is shown in Figure 9. Comparisons with free metal oxide show that the production of specific cracked products directly resulted from the presence of non-framework metal oxide component.

20. ALUMINATION: INSERTION OF ALUMINUM INTO SILICATE FRAMEWORKS

Several authors have reported the generation, or regeneration of acid sites within a zeolite structure by secondary synthesis methods.[143-148] Aluminum has been inserted into the framework sites either from addition of an external source of aluminum or through reinsertion of non-framework or detrital aluminum. The insertion of aluminum was confirmed with the observed increase in Bronsted acidity. The reaction of aluminum halides appears to be non-specific as deposition of extra-lattice aluminum is observed. Steaming samples containing binder was also shown to increase the aluminum content of the microporous component due to migration of aluminum from the binder to the silicate.[149,150] Ammonia desorption studies show a distinct similarity between zeolites synthesized with aluminum in the framework and zeolites which have been treated with $AlCl_3$. 29Si NMR studies of the aluminated materials show the presence of a signal at –106 ppm, attributed to the presence of Si-(1Al) in the material.[151] Development of new hydroxyl bands upon treatment with $AlCl_3$ also suggests that such insertion is possible. The reaction with $AlCl_3$, however, does not occur under mild conditions, however realumination with $NaAlO_2$ and NaOH does.[152] The alkaline conditions appear to facilitate the exchange.[153] High temperatures are required. Mechanistically, it has been proposed that such insertions or exchanges between elements are a result of filling of vacancies left in the framework by reaction with the silanol groups at the defect sites. These silanol defects (hydroxyl nests) have been shown to be stable to temperatures up to 400°C.[154]

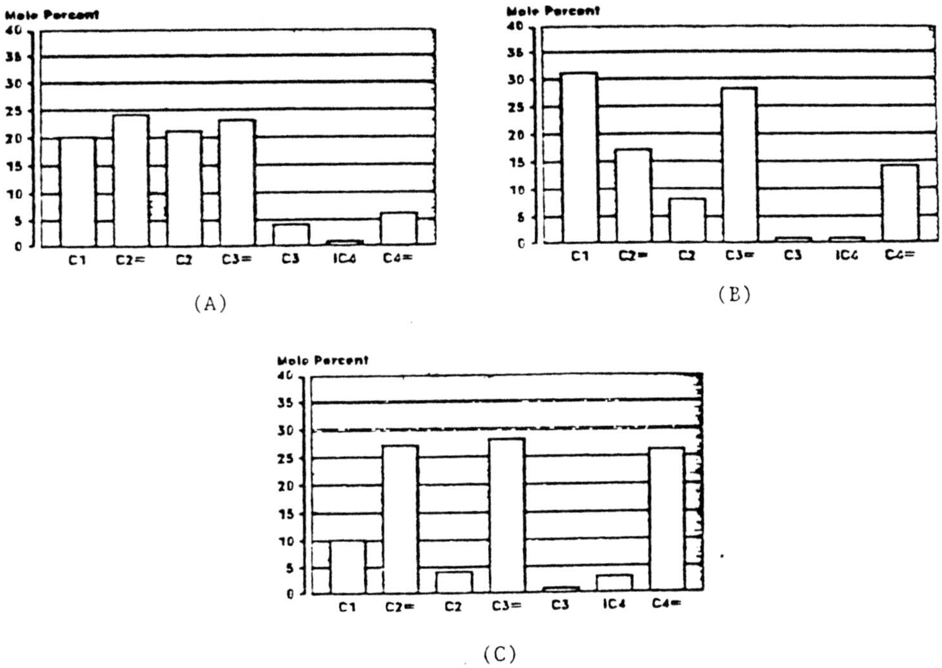

Figure 10: A comparison of the distribution of products from the cracking of butane over modified metal silicate molecular sieves with the ZSM-5(MFI) structure. (a) HZSM-5(MFI), (b) HFeZSM-5(MFI) steamed one hour, (c) HGaZSM-5(MFI) steamed one hour.

Exchanges between aluminum and other trivalent elements such as boron have also been noted.[155] The method of preparing aluminosilicates from the borosilicates through secondary methods has found substantial use. There has been a great increase in the number of novel high-silica borosilicates with structures that are not found in the aluminosilicates. These borosilicates can be transformed into potentially useful active aluminosilicates catalysts through aluminum insertion. These include SSZ-24(AFT) and UTD-1.

21. INSERTION OF OTHER ELEMENTS INTO SILCIATE FRAMEWORKS

Gallium has been shown to substitute for aluminum in the large pore zeolites under mild conditions.[156] The ammonium fluorogallate (from gallium nitrate plus ammonium fluoride) in solution phase results in the following replacement reaction:

$$GaF_x^{3-x} + [AlO_4]^- \Rightarrow [GaO4]^- + AlF_x^{3-x}$$

The gallium appears to substitute for framework aluminum in zeolite Y(FAU) and for non-framework aluminum in ultrastabilized Y(FAU).[157] The gallium containing compounds were shown to be catalytically active with an increase in isomerization, dehydrogenation and oligomerization. This might be valuable for enhanced octane gasoline.[158]

22. HYDROTHERMAL TREATMENT AND THE SILICOALUMINO-PHOSPHATES

A large number of new molecular sieves with framework compositions other than silica/alumina have been prepared since the first announcement of the synthesis of the aluminophosphate family of molecular sieves by Wilson in 1982. Recently some of these materials have been exploited for industrial catalyst applications. UOP in a joint venture with Norsk Hydro developed a methanol-to-olefin process that utilizes SAPO-34(CHA) as the catalyst. The UOP/HYDRO MTO process represents an environmentally green process that uses wet methanol, which can be converted selectively to high quality ethylene and propylene.[159] Because the MTO process produces one mole of water per mole of methanol used and regeneration conditions used to remove coke deposits are rigorous, the desirable catalyst would be one that is both thermally and hydrothermally stable. SAPO-34(CHA) has been found to be stable in both high temperature and high steam conditions.[160,161] Studies have shown that better than 80% crystallinity is maintained after treatment for 50 hours in 100% steam at temperatures of 775C. This is shown in Figure 11.

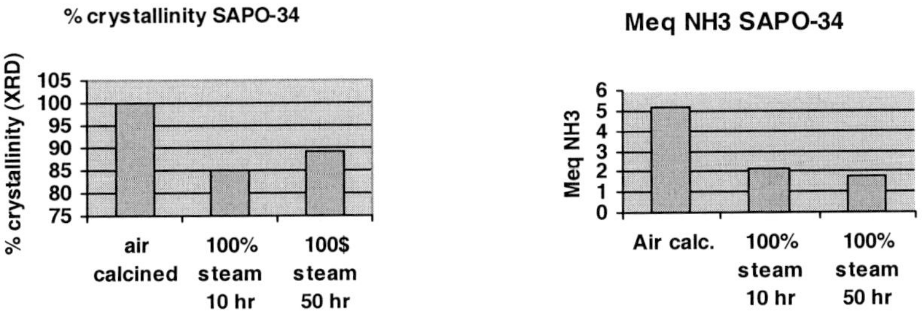

Figure 11: Comparison of the change in crystallinity and acid site number for SAPO-34(CHA) after calcination and steam treatments.

Though the number of acid sites do decline with time under these severe conditions, as measured by ammonia desorption experiments, the catalytic activity was not adversely affected. In fact mild steaming was shown to improve light olefin selectivity as well as lessen the amount of propane produced. Researchers at Exxon modified the SAPO-34(CHA) MTO catalyst to improve ethylene yield using an acid extraction method instead of steam.[162] The exact nature of these new modified materials remains to be seen and should prove to be a rich area of further research.

ACKNOWLEDGEMENTS

This paper is dedicated to Gary Skeels (the late) who, during his career, has contributed significantly to the advances in this area and, over the years, has provided me with lively discussions on this topic.

REFERENCES

---

[1] Haynes, H. W., *Catal. Rev.-Sci Eng.*, **17**(2), 273 (1978)

[2] Myers, C.G., B.W. Rope, W E. Garwood, U.S. Patent 3, 84,572 (1968)

[3] Kerr, G. T., J. N. Miale, R. J. Mikovsky, U.S. Patent 3, 493,519 (1970)

[4] Kunin, R. Elements of Ion Exchange, Reinhold Publ. New York, 25 (1960)

[5] Keough, A. H., L. B. Sand, *J. Amer. Chem. Soc.*, **83**, 3536 (1961)

[6] Barrer, R. M., M.B. Makki, *Can. J. Chem.*, **42**, 1481 (1964)

[7] Barrer, R. M., B. Coughlan, in Molecular Sieves, Society of Chemical Industry, (London) Monograph, 141 (1968)

[8] Kranich, W.L., Y.H. Ma, L. B. Sand, A. H. Weiss, I. Zwiebel, Adv. Chem. Ser., **101**, 502 (1971)

[9] Scherzer, J., in Catalytic Materials, American Chemical Society, Washington, DC, 157 (1984)

[10] Rabo, J.A., P. E. Pickert, J. E. Boyle, U.S. Patent 3,130,006 (1964)

[11] Uytterhoeven, J.B., L. G. Christner, W. K. Hall, *J. Phys. Chem.*, **69**, 2117 (1965)

[12] Ward, J.W., *J. Catal.*, **9**, 225 (1967)

[13] Stevenson, R.L., *J. Catal.*, **21**, 113 (1971)

[14] Hickson, D.A., S. M. Csicsery, *J. Catal.*, **10**, 27 (1968)

[15] Ward, J.W., R. C. Hansford, *J. Catal.*, **13**, 364 (1969)

[16] Eberly, P.E., S. M. Laurent, H. E. Robson, U.S. Patent 3,506,400 (1970)

[17] Bremer, H., W. Morke, R. Schodel, F. Vogt, Adv. Chem. Ser., **121**, 249 (1973)

[18] Ballivet, D., P. Pichat, D. Barthomeuf, Adv. Chem. Series, **121**, 469 (1973)

[19] Maher, P.K., C. V. McDaniel, U. S. Patent 3, 293, 192 (1966)

[20] Maher, P.K., C. V. McDaniel, U.S. Patent 3,402,996 (1968)

[21] McDaniel, C.V., P. K. Maher, U.S. Patent 3,449,070 (1969)

[22] McDaniel, C.V., P.K. Maher, in Molecular Sieves, Soc.Chem. Ind. (London), Monograph, 186, (1968)

[23] Siantar, D.P.,W.S. Millman, J.J. Fripiat, *Zeolites*, **15**, 556 (1995)

[24] Ambs, W.J., W. H. Flank, *J. Catal.*, **14**, 118 (1969)

[25] Ambs, W.J., W. H. Flank, *J. Catal.,* **18**, 238 (1970)

[26] Kerr, G.T., D. H. Olson, E. Dempsey, *J. Catal.*, **18,** 236 (1970)

[27] Ward, J. W., *J. Catal.*, **27**, 157 (1972)

[28] Kerr, G.T., *J. Phys. Chem.*, **71**, 4155 (1967)

[29] Kerr, G.T., G. F. Shipman, *J. Phys. Chem.*, **72**, 3071 (1968)

[30] Hansford, R.C., U. S. Patent, 3,354,077 (1967)

[31] Peri, J.B., Catalysis, vol. 1, J. W. Hightower, ed., Elsevier, New York, 329 (1973)

[32] Scherzer, J.,J. L. Bass, *J. Catal.*, **28**, 101 (1973)

[33] Eberly, P.E., S. M. Laurent, H. E. Robson, U.S. Patent 3,506,400 (1970)

[34] Kerr, G.T., J.N.Miale, R. J. Mikovsky, U.S.Patent 3,493,519 (1970)

[35] Kerr, G.T., *J. Catal.*, **15**, 200 (1969)

[36] Kerr, G.T., Adv. Chem. Ser., American Chemical Society, Washington, DC, **121**, 219 (1973)

[37] Jacobs, P., J.B. Uytterhoeven, *J. Catal.*, **22**, 193 (1971)

[38] Engelhardt, G., U. Lohse, V. Patzelova, M. Magi, E. Lippmaa, *Zeolites*, **3**, 239 (1983)

[39] Scherzer, J., *J. Catal.* **54**, 285 (1978)

[40] Lohse, U., E. Alsdorf, H.Z. Stach, *Anorg. Allg. Chem.*, **447**,64 (1978)

[41] Bosacek, V., Patzelova, V., Z. Tvaruzkova, Dd. Freude, U. Lohse, W. Schirmer, H. Stach, H. Thamm, *J. Catal.*, **61**, 435 (1980)

[42] Lynch, J., F. Raatz, P. Dufrensne, *Zeolites*, **7**, 333 (1987)

[43] Mauge, F., A. Auroux, J. C. Courcelle, Ph.Engelhard, P. Gallezot, J. Grosmangin, in Catalysis by Acids and Bases, B. Imelik, et al., eds., Elsevier, Amsterdam, 91 (1985)

[44] Breck, D. W., E. M. Flanigen, in Molecular Sieves, Soc. Of Chem. Ind. London, 47 (1968)

[45] Kerr, G. T., *Zeolites*, **9**, 350 (1989)

[46] Dempsey, E., *J. Catal.*, **33,** 497 (1974)

[47] Sohn, J. R., S. J. Decanio, J. H. Lunsford, D. J. O'Donnell, *Zeolites*, **6**, 225 (1986)

[48] Ritter, R.E., J.E. Creighton, T.G. Roberie, J.S. Chin, C.C.Wear, NPRA Annual Meeting Los Angeles, March 23 paper AM 86-45

[49] Zi, G., Yi, T., *Zeolites*, **8**, 232 (1988)

[50] B. Nagy, J., E. G. Derouane, in Perspectives in Molecular Sieve Science, Flank and Whyte, eds., ACS Symp. Series, 2 (1989)

[51] Hanke, W., K. Moller, *Zeolites*, **4,**244 (1984)

[52] Pine, L.A., P.J. Maher, W.A. Wachter, *J. Catal.,* **85**, 466 (1984)

[53] Corma, A., V. Fornes, A. Martinez, A.V. Orchilles, in Perspectives in Molecular Sieve Science, Flank and Whyte, eds, ACS Symp. Series, 542 (1988)

[54] Corma, A., V. Fornes, J. Perez-Pariente, E. Sastre, J.A. Martens, P.A. Jacobs, in Perspectives in Molecular Sieve Science, Flank and Whyte eds, ACS Symp. Series, 555, (1988)

[55] Bond, G. R. U.S. Pat 2,617,712 (1949)

[56] Engelhardt, G., U. Lohse, V. Patzelova, M. Magi, E. Lippmaa, *Zeolites*, **3**, 233 (1983)

[57] Corma, A., V. Fornes, F. V. Melo, J. Herrero, *Zeolites*, **7**, 559 (1987)

[58] Klinowski, J., *J. Chem. Soc., Chem. Commun.*, 570 (1981)

[59] Klinowski, J., J. M. Thomas, C. A. Fyfe, G.C. Gobbi, *Nature,* **296**, 533 (1982)

[60] Klinowski, J., J. M. Thomas, M. W. Anderson, C.A. Fyfe, G.C Gobbi, *Zeolites*, **3**, 5 (1983)

[61] Zukal, A., V. Patzelova, U. Lohse, *Zeolites*, **6**, 133 (1986)

[62] Skeels, G.W., E. M. Flanigen, in Zeolites: Facts, Figures, Future, Jacobs and van Santen, eds., Elsevier, Amsterdam, 331 (1989)

[63] Akporiaye, D., A. P. Chapple, D. M. Clark, J. Dwyer, I.S. Elliott, D. J. Rawlence, in New Developments in Zeolite Science and Technology, Murakami et al., eds., Elsevier, Amsterdam, 351 (1986)

[64] Kerr, G.T., *J. Phys. Chem.*, **72**, 2594 (1968)

[65] Sulikowski, B., J. Phys. Chem., 97,1420 (1993)

[66] Shyu, J.Z., E.T. Skopinski, J.G. Goodwin, A. Sayari, *Appl. Surf. Sci.*, **21**, 297 (1985)

[67] Dwyer, J., F.R. Fitch, G. Qin, J. C. Vickerman, *J. Phys. Chem.*, **86**, 4574 (1982)

[68] Dwyer, J., F.R. Fitch, F. Machado, *J. Chem. Soc. Chem. Commun.*, 422 (1981)

[69] Beyer, H.K. I. Belenkaya in Catalysis by Zeolites, B. Imelik, C. Naccache, Y. Ben Taarit et al., eds *Studies in Surface Science and Catalysis*, **5**, Elsevier, Amsterdam 203 (1980)

[70] Beyer, H.K., I. Belenkaya, F. Hange *J. Chem. Soc. Faraday Trans.*, **81**, 2889 (1985)

[71] Freude, D., J. Klinowski, H. Hamdan, *Chem. Phys. Lett.*, **149,** 355 (1988)

[72] Andera, V., Kubelkova, L., J. Novakova, *Zeolites,* **5**, 67 (1985)

[73] Corma, A., V. Fornes, O. Pallota, *J. Chem. Soc.,, Chem. Commun*, 333 (1986)

[74] Kubelkova, L., L. Dudikova, Z. Bastl, *J. Chem. Soc. Faraday Trans.,* **83**, 511 (1987)

[75] Klinowski, J., J. M. Thomas, M. W. Anderson, *Zeolites*, **3**, 5 (1983)

[76] Skeels, G. W., D. W. Breck, in Proceedings of the 6th International Zeolite Conference, D. H. Olson, A. Bisio, eds., Butterworth, Guildford, UK, 87 (1984)

[77] Garralong, G., V. Fornes, A. Corma, *Zeolites*, **8**, 268 (1988)

[78] Skeels, G.W., E. M. Flanigen, in Zeolites: Facts, Figures, Future, P.A. Jacobs, R. A. van Santen, eds., *Studies in Surface Science and Catalysis*, **49**, Elsevier, Amsterdam, 331 (19889)

[79] Lok, B.M. F. P. Gortsema, C.A. Messina, H. Rastelli, T.P.J. Izod, in Intrazeolite Chemistry, Stucky and Dwyer, eds., *ACS Symposium Series*, **218**, 41 (1983)

[80] Lok, B.M., T.P.J. Izod, *Zeolites*, **2**, 6 (1982)

[81] Shell, Netherlands Appl. 6,626,352 (1968)

[82] Voorhies, A., U.S. Patent 3,630,965 (1971)
[83] Takase, S., T. Shioiri, M. Ushio, Ger Pat. 2,219,736 (1972)

[84] Szostak, R., A.H. Weiss, Y. Yang, 17th ACS MidAtlantaic Regional Meeting April 6-8 (1983)

[85] Hervert, G.L., U.S. Patent 3,467,728 (1969)

[86] Fishel, N.A., U.S. Patent 3,413,370 (1968)

[87] Wang, K.M., J. H. Lunsford, *J. Catal.*, **24**, 262 (1972)

[88] Breck, D.W., G. W. Skeels, Molecular Sieves—II, Am. Chem. Soc., Washington, DC, 271 (1977)

[89] Yoshida, A., K. Inoue, Y. Adachi, *Zeolites,* **11**, 223 (1991)

[90] Kerr, G.T., *J. Phys. Chem.*, **73,** 2780 (1969)

[91] Maher, P.K., F. D. Hunter, J. Scherzer, *Adv. Chem. Ser.*, **101**, 266 (1971)
[92] Steel, A.T., E. Dooryhee, *Zeolites*, **13**, 336 (1993)

[93] Mauge, F., A. Auroux, J.C. Courcelle, Ph. Engelhard, P. Gallezot, J. Grosmangin, in Catalysis by Acids and Bases, B. Imelik, et. al , eds, Elsevier, Amsterdam, 91 (1985)

[94] Lynch, J., F. Raatz, P. Dufresne, *Zeolites*, **7**, 333 (1987)

[95] Patzelova, V., N.I. Jaeger, *Zeolites*, **7**, 240 (1987)

[96] Lohse, U., Z. *Anorg allg. Chem.*, **460**,179 (1980)

[97] Corma, A., V. Fornes, A. Martinez, F. Melo, O. Pallota, in Innovation in Zeolite Material Science, P.J. Grobet, et al., eds., Elsevier, Amsterdam, 495 (1988)

[98] Corma, A., V. Fornes, F. V. Melo, J. Herrero, *Zeolites, 7,* 559 (1987)

[99] Fernandez, C., J.C. Vedrine, J. Grosmangin, G. Szabo, *Zeolites*, **6**, 484 (1986)

[100] Fernandez, C., A. Auroux, J.C. Vedrine, J. Grosmangin, G. Szabo, in New Developments in Zeolite Science and Technology, Murakami et al., eds., Elsevier, Amsterdam, 345 (1986)

[101] DeCanio, S.J., J. Rack Sohn, P.O. Fritz, J.H. Lunsford, *J. Catal.,* **101**,132 (1986)

[102] Shertukde, P.V., Hall, W.K., Dereppe, J.-M, Marcelin, G., *J. Catal.*, **139**, 468 (1993)

[103] Dwyer, J., P. J. O'Malley, in Keynotes in Energy-Related Catalysis, Studies in Surface Science and Catalysis; Elsevier, Amsterdam, **35**, 5 (1988)

[104] Beaumont, R., D. Barthomeuf, *J. Catal.*, **27**, 45 (1972)

[105] Lunsford, J.H., *J. Phys. Chem.*, **72**, 4163 (1968)

[106] Mirodatos, C., D. Barthomeuf, *J. C. S. Chem. Commun.*, 39 (1981)

[107] Dwyer, J., F. R. Fitch, G. Qin, J. C. Vickerman, *J. Phys. Chem.*, **86**, 4574 (1982)

[108] Barthomeuf, D., *Mat. Chem. Phys.*, **17**, 49 (1987)

[109] Kasansky, V.B., in Structure and Reactivity of Modified Zeolites, *Studies in Surface Science and Catalysis*, Elsevier, Amsterdam, **18**, 61 (1984)

[110] Mortier, W.J., *J. Catal.*, **55**, 138 (1978)

[111] Jacobs, P.A., *Catal. Rev. Sci. Eng.*, **24**, 415 (1982)

[112] Dwyer, J., F.R. Fitch, E.E. Nkang, *J. Phys. Chem.*, **87**, 5402 (1983)

[113] Remy, M.J., D. Stanica, P.A. Jacobs, *J. Phys. Chem.*, **100**, 12440 (1996)

[114] Karge, H.G., V. Dondur, J. Weitkamp, *J. Phys. Chem.*, **95**, 283 (1991)

[115] Miessner, H., H.Kosslick, U. Lohse, B. Parlitz, Vu-A. Tuan, *J. Phys. Chem.*, **97**, 9741 (1993)

[116] Barthomeuf, D., *Zeolites*, **10**, 131 (1990)

[117] Janin, A., M. Maache, J.C. Lavalley, J.F.Joly, F. Raatz, N. Szydlowski, *Zeolites*, **11**,391 (1991)

[118] Rocha, J., J. Klinowski, *J. Chem. Soc. Chem. Commun.*, 1121 (1991)

[119] Wolf, I., D. Freude, *Microporous Materials*, **5**, 69 (1995)

[120] Wu, X., L.G. Butler, *Microporous Materials*, **4**, 265 (1995)

[121] Wang, Q.L., G. Giannetto, M.Guisnet, *J. Catal.*, **130**, 471 (1991)

[122] McDaniel, C. V., P.K. Maher, in Zeolite Chemistry and Catalysis, J.A. Rabo, ed., American Chemical Society, Washington, DC, 285 (1974)

[123] Bayer, H.K., I. Belenkaya, in Catalysis by Zeolites, B. Imelik, et. Al., eds., Elsevier, Amsterdam, 203 (1980)

[124] Skeels, G. W., Breck, D.W, in Proceedings of the Sixth International Zeolite Conference, D. H. Olson, A. Bisio, eds., Butterworths, 87 (1984)

[125] Garralon, G., V. Fornes, A. Corma, *Zeolites*, **8**, 268 (1988)

[126] Rabo, J.A., R.J.Pellet, J.S. Magee, B.R. Mitchell, J.W. Moore, W.S. Letzsch, L.L. Upson, J.E. Magnusson, presented at the NPRA Annual Meeting; Los Angeles; March 23, paper AM-86-27

[127] Jacobs, P.A., J.A. Martens, H.K. Beyer, in Catalysis by Acids and Bases, Imelik, et al., eds., Elsevier, Amsterdam, 399 (1985)

[128] Mirandatos, C., D. Barthomeuf, *J. Chem. Soc., Chem. Commun.*, 39 (1981)

[129] Mirandatos, C., B.H. Ha, K. Otsuka, D. Barthomeuf, in Proceedings of the 5th International Conference on Zeolites, Rees, ed., Heyden, London, 382 (1980)

[130] Lago, R. M., W.O. Haag, R. J.Mikovski, D. H. Olson, S.D. Hellring, K. D. Schmidt, G. T. Kerr, in New Developments in Zeolite Science and Technology, Murakami, et al., Elsevier, Amsterdam, 677 (1986)

[131] Haag, W.O., R. M. Lago, Eur. Pat. 34444 (1981)

[132] Sendoda, Y., Y. Ono, Zeolites, 8, 101 (1988)

[133] Szostak, R., R. Kuvadia, T.L. Thomas, Report, Multi-Client Meeting, March1988, Atlanta, GA

[134] Barthomeuf, D., *J. Phys. Chem.*, **83**, 249 (1979)

[135] Ashton, A.G., S. Batmanian, D.M. Clark, J. Dwyer, F.R. Fitch, A. Hinchcliffe, F. J. Machado, in Catalysis by Acids and Bases, Imelik, et al., eds., Elsevier, Amsterdam, 101 (1985)

[136] Ashton, A.G., J. Dwyer, I.S. Elliott, F. R. Fitch, M. Greenwood, G. Qin, J. Speakman, in Proceedings of the Sixth International Conference on Zeolites, Olson and Bisio, eds., Butterworths, 704 (1984)

[137] Brunner, E., H. Ernst, D. Freude, M. Hunger, C.B. Krause, D. Prager, W. Reschetilowski, K. H. Schwieger, *Zeolites*, **9**, 282 (1989)

[138] Ione, K.G., V. G. Stepanov, G.V. Echevskii, A.A. Shubin, E.A. Paukshtis, *Zeolites*, **4**, 114 (1984)

[139] Hong, Y., J.J. Fripiat, *Microporous Materials*, **4**, 323 (1995)

[140] Szostak, R., V. Nair, D.K. Simmons, T.L. Thomas, R. Kuvadia, B. Duncan, D.C. Shieh, in Innovations in Zeolite Materials Science, Grobet et al., eds., Elsevier, Amsterdam, 403 (1988)

[141] Nair, V., PhD Thesis, School of Chemical Engineering, Georgia Institute of Technology, Atlanta, GA, USA (1987)

[142] Csencsits, R., PhD Thesis, 1988, Materials and Chemical Sciences Division, Univ. of California, Berkeley, USA

[143] Dessau, T.M., G.T. Kerr, *Zeolites*, **4**,315 (1984)

[144] Chang, C.D., C.T.W. Chu, J.N.Miale, R.F.Bridger, R.B.Calvert, *J. Amer. Chem.Soc.*, **106,**8143 (1984)

[145] Anderson, M.W., J. Klinowski, L. Xinsheng, *J. Chem.Soc. Chem. Commun.* 1596 (1984)

[146] Yashima, T., K. Yamagishi, S. Nakata, S. Asaoka, in Innovation in Zeolite Material Science, P.J. Grobet, et al., eds., Elsevier, Amsterdam, 175 (1988)

[147] Jacobs, P.A. M. Tielen, J. Ben Nagy, G. Debras, E.G. Derouane, Z. Gabelica, in Proceedings of the Sixth International Conference on Zeolites, Olson and Bisio, eds., Butterworths, 783 (1984)

[148] Wu, P., T. Komatsu, T. Yashima, *J. Phys. Chem.*, **99**, 10923 (1995)

[149] Shahabi, D.S., W.E. Garwood, P.Chu, et al., *J.Catal.,* **93**, 471 (1985)

[150] Chang, C.D., S.D. Hellring, J.N.Miale, K.D. Schmitt, *J. Chem.Soc. Faraday Trans.*, **81**, 2215 (1985)

[151] Anderson, M.W., J. Klinowski, X. Liu, *J. Chem. Soc., Chem Commun.*, 1596 (1984)

[152] Zhang, Z., X.Liu, Y. Xu, R.Xu, *Zeolites*, **11**, 232 (1991)

[153] Barrie, P.J., L.F. Gladden, J. Klinowski, *J. Chem. Soc., Chem.Commun.*, 592 (1991)

[154] Ray, G.J., A.G. Nerheim, J.A. Donohue, J.A. Hriljac, A.K. Cheetham, *Colloids and Surfaces*, **63,** 77 (1992)

[155] Derouane, E.G., L. Baltusis, R.M.Dessau, K.D. Schmitt, in Catalysis by Acids and Bases, Imelik et al., eds., Elsevier, Amsterdam, 135 (1985)

[156] Dwyer, J., K. Karim, *J. Chem.Soc., Chem. Commun.*, **14**,905 (1991)

[157] Karim, K., J. Dwyer, D.J. Rawlence, M. Tariq, A. Nabhan, *J. Mater. Chem.*, **2**,1161 (1992)

[158] Karim, K., J. Dwyer, D.J. Rawlence, *Applied Catalysis*, A, **109**, 235 (1994)

[159] UOP Information Sheet, UOP 2699B-1 1197032D Olefins and Derivatives, UOP/HYDRO MTO Process, Methanol to Olefins Conversion ; **1999**; S. Wilson, P. Barger, *Microporous Mesoporous Mater.* **29**, 117 (1999)

[160] Barger, P.T.; Lesch, D.A.; *Arabian J. Sci. Eng.*; **21,** 263 (1996)

[161] Watanabe, Y.; Koiwai, A.; Takeuchi, H.; Hyodo, S.; Noda, S.; *J. Catal.*; **143**, 4230 (1993)

[162] Janssen, M.J.G.; U.S. Patent 5,475,182 (1995)

Studies in Surface Science and Catalysis 137
H. van Bekkum, E.M. Flanigen, P.A. Jacobs and J.C. Jansen (Editors)
© 2001 Elsevier Science B.V. All rights reserved.

Chapter 7

CLAYS: FROM TWO TO THREE DIMENSIONS.

R. A. Schoonheydt and K.Y. Jacobs

Department of Interphase Chemistry, Center for Surface Chemistry and Catalysis, K.U.Leuven, Kasteelpark Arenberg, 23, B-3001 Leuven, Belgium

SUMMARY

In the first part the structures and structural chemistry of smectites are described. Charge density, cation exchange capacity, surface area, swelling and pore volumes are defined. The crystal chemistry is discussed with emphasis on the vibrations of the structural hydroxyls and the ^{29}Si MAS-NMR data. The surface chemistry is discussed in relation to the acidic and redox properties of smectitic surfaces.
In the second part a pillared clay is defined and the different pillaring agents are summarized. The Al-pillared clays are discussed in depth as prototypes of clays with oxidic pillars.

1. INTRODUCTION

The first known traces of the use of clays by mankind are 25,000 years old. Clay figures, pottery and ceramics were made by the primitive people in Europe and Asia (1). Today clay is an important industrial raw material. Hundreds of million tons are worked up in ceramics, in the paper industry, in the iron ore and metal industries, in oil drilling and in pet feeding. Clays are used as adsorbents, decoloration agents, ion exchangers, supports and as catalysts (2, 3). In agriculture the clay-organic and clay-oxidic complexes determine the structure, texture, water retention and the inorganic and organic fertilizer composition of the soils.
Clay minerals are phyllosilicates: layered or two-dimensional silicates. The basic building blocks are the $Si(O,OH)_4$ tetrahedra and the $M(O,OH)_6$ octahedra with M = Al^{3+}, Mg^{2+} or $Fe^{2+,3+}$ (see Figure 1). The condensation of a monolayer of tetrahedra with a layer of octahedra gives the 1:1 or TO minerals, known as kaolinites with M = Al^{3+}, and serpentines when M = Mg^{2+}. They are also respectively called dioctahedral and trioctahedral 1:1 minerals.

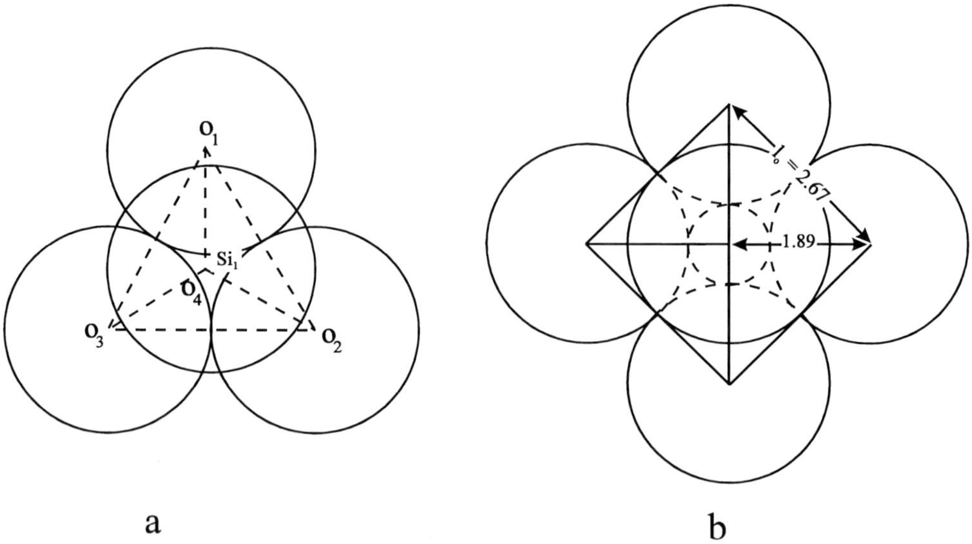

Figure 1: (a) The SiO_4 tetrahedron. O_1, O_2 and O_3 are the basal oxygens; O_4 is the apical oxygen. (b) The AlO_6 octahedron. Distances are in Angstroms. (Nemecz, 14)

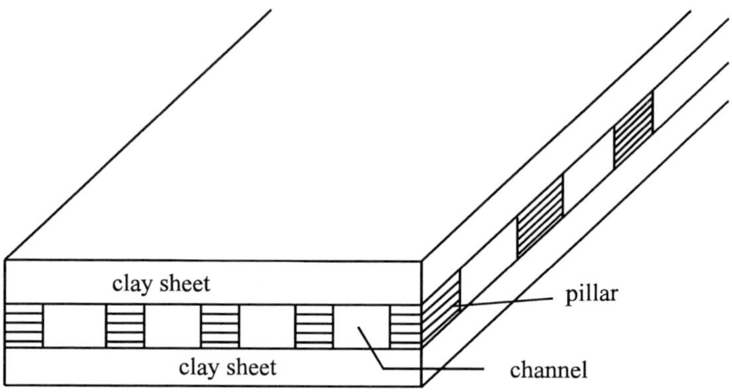

Figure 2: Schematic drawing of a pillared clay.

When a monolayer of octahedra is sandwiched between two monolayers of Si-tetrahedra 2:1 type clay minerals are obtained. They are dioctahedral in the case of Al-octahedra and trioctahedral for Mg-octahedra. In the former case two out of three octahedral sites are occupied, in the latter case all three are occupied. The transformation of the layered 2:1 clay minerals into three-dimensional structures with molecular sieve properties is our present subject. The aim is to obtain a rigid two-dimensional channel network (Figure 2). The size of the channels is preferably in the microporous and mesoporous range. This two-dimensional interconnected network of channels, if it can be realized, is unique with respect to zeolites and microporous aluminium phosphates.

2. SMECTITES

Table 1 gives the classification of the 2:1 clay minerals and Figure 3 shows the ideal structure. These structures are electroneutral. This can be easily seen by assigning ionic charges to the atoms (Si^{4+}, Al^{3+}, O^{2-}, H^{+}, Mg^{2+}). These electroneutral structures are found in nature as pyrophyllite (dioctahedral) and talc (trioctahedral).

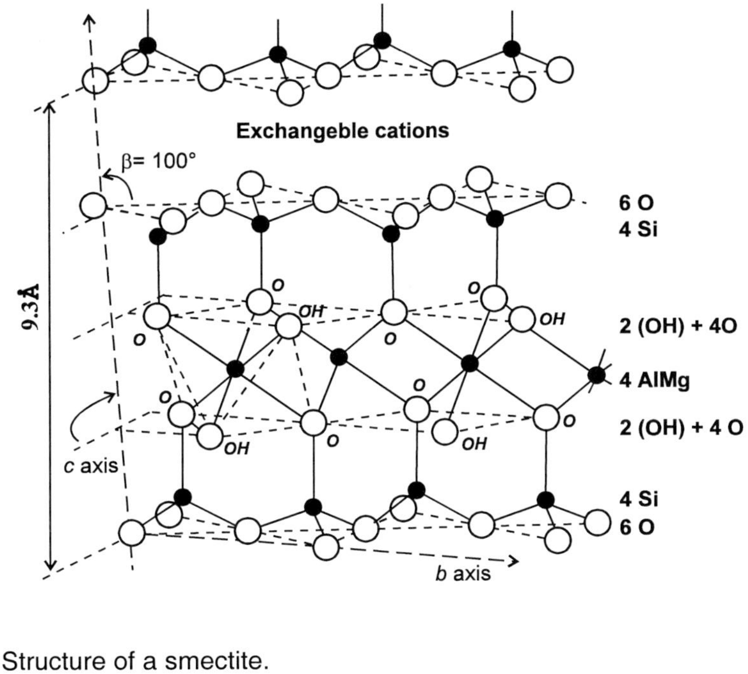

Figure 3: Structure of a smectite.

TABLE 1: Classification of 2:1 clay minerals.

Charge/ formula	group	subgroup	name
0	pyrophyllite	dioctahedral	pyrophyllite
	talc	trioctahedral	talc
0.25 - 0.6	smectite	dioctahedral	montmorillonite
			beidellite
		trioctahedral	saponite
			hectorite
0.6 - 0.9	vermiculite	dioctahedral	vermiculite
		trioctahedral	vermiculite
1	mica	dioctahedral	muscovite
		trioctahedral	biotite

The other minerals of Table 1 are obtained by isomorphous substitution of Si^{4+} in the tetrahedral layer or Al^{3+} and Mg^{2+} in the octahedral layer. When the incoming cation has a lower valency than the outgoing cation, the lattice is negatively charged and electrical neutrality is maintained by the exchangeable cations. The general formula representation is

$$(Si_{8-x}M_x)^{IV}(Al_{4-y}M'_y)^{VI}OH)_4O_{20}M"_{(x+y)/n}mH_2O$$

for dioctahedral clays or

$$(Si_{8-x}M_x)^{IV}(Mg_{6-y}M'_y)^{VI}(OH)_4O_{20}M"_{(x+y)/n}\ mH_2O$$

for trioctahedral clays. The superscripts IV and VI refer to the tetrahedral and octahedral layers respectively. M and M' have one unit of charge less than the cation they replace and M" is the exchangeable cation with valency n.

As shown in Table 1, three groups are distinguished depending on the degree of isomorphous substitution. For smectites x + y = 0.25 - 0.6, corresponding to a cation exchange capacity of 0.64 - 1.50 meq/g. For micas the sum x + y is at least one and vermiculites take an intermediate position.

Isomorphous substitution is mainly localized in the octahedral layer. One obtains the montmorillonites and the hectorites for dioctahedral and trioctahedral smectites respectively. If, however, isomorphous substitution is localized in the tetrahedral layer the minerals are beidellite (dioctahedral) and saponite (trioctahedral). If the octahedral layer is composed of Fe^{3+} instead of Al^{3+} one has nontronites.

3. PROPERTIES OF SMECTITES

Natural materials like smectites invariably contain impurities. The most common crystalline impurities are quartz, micas, calcite and feldspars. They can be recognized by X-ray diffraction (XRD) and removed by fractionation. When a clay material is centrifuged so as to keep the < 2 μm fraction in suspension, the precipitate mainly contains the impurities, while the clay remains in suspension. Amorphous oxidic impurities of Fe and Al are often adsorbed on the clay particles and difficult to remove. Treatment procedures with dithionite have been developed to remove them; however, not without problems (6). These tedious, but substantive procedures, together with the need for 'custom-made' materials for specific applications has given a renewed interest in the synthesis of clays. Though this is not the subject of this article, examples of hectorite-, saponite- and beidellite-synthesis are found in the literature (7-13).
In any case, in the following pure smectite clay minerals are supposed and their most important properties are described in general terms. A non-exhaustive list of books with more detailed information is given in the list of references. (14-19).

3.1. XRD patterns and unit cell parameters

In layer silicates the average Si-O distance is 0.162 nm and the average Al^{VI}-O distance is 0.192 nm. The O-O distance in the ideal SiO_4 tetrahedron is 0.267 nm. These tetrahedra are joined together in a two-dimensional array by sharing three oxygens (Figure 4). The basal oxygens are coplanar with the layer plane and common to two adjacent tetrahedra. The apical oxygen is free. The chemical formula of the layer is $(Si_2O_5^{2-})_n$. The Si atoms at the centre of gravity of each tetrahedron form a regular hexagonal array with a = 0.530 nm and b = 0.916 nm.

The O-O distance in the ideal AlO_6 octahedron is 0.267 nm. The two-dimensional condensation of these octahedra gives a gibbsite sheet or dioctahedral layer with a = 0.462 nm and b = 0.801 nm (Figure 4). The slight differences between the a and b dimensions of the Si- and Al-sheets indicate that (1) condensation of these sheets to form clay structures is possible; (2) the sheets are distorted in the actual structures to match their dimensions. These distortions are usually represented by a rotation of the Si-tetrahedra such that the Si-O-Si angle lies between 180° (hexagonal symmetry) and 120° (ditrigonal symmetry) (Figure 4).

The extent of the distortion depends on the isomorphic substitution (degree, type of ion in tetrahedral or in octahedral position). As a result, the values of the a and b parameters deviate from the ideal values and the dependence of the b parameter on the chemical composition as been expressed as (20):

$$b = (0.8944 + 0.0096\ Mg^{2+} + 0.0096\ Fe^{3+} + 0.0037\ Al^{VI}) + 0.0012\ \text{nm} \qquad (1)$$

Mg^{2+}, Fe^{3+} and Al^{VI} are the number of Mg, Fe^{3+} and octahedral Al^{3+} per half unit cell.

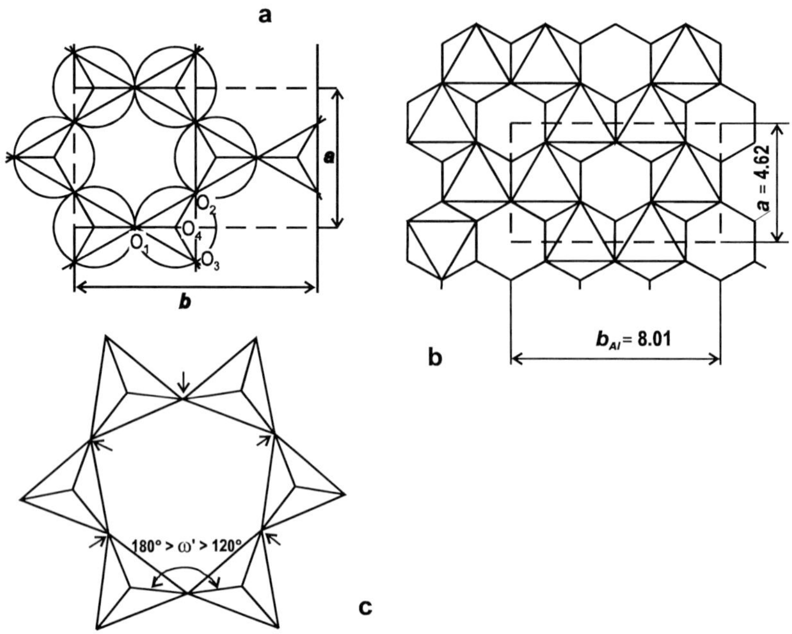

Figure 4: (a) The $(Si_2O_5^{2-})_n$ sheet; (b) the dioctahedral Al sheet; (c) the tetrahedral array of ditrigonal symmetry. Distances are given in Ångström (Nemecz, 14).

The c parameter is variable and depends on the size of the exchangeable cation and the amount of water in the interlamellar space. The distance between two sheets is usually taken as 0.96 nm. It increases to 1.25 nm, to 1.50 - 1.55 and to 1.80 - 1.90 nm with respectively a monolayer, a bilayer and a triple layer of water molecules in the interlamellar space (Figure 5).

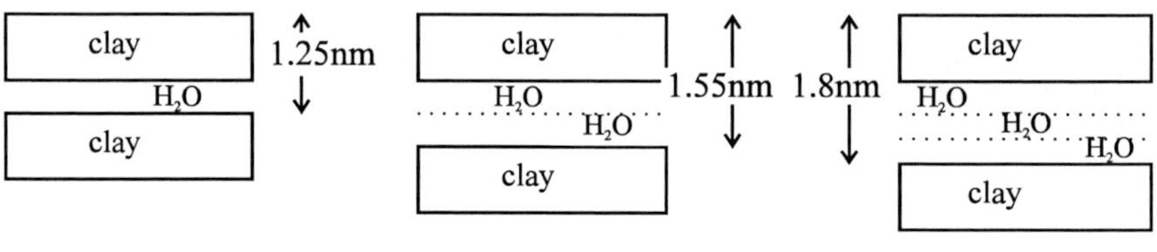

Figure 5: One-, two- and three-layers hydrate of smectites.

A typical XRD pattern of an air-dry smectite is shown in Figure 6 together with the hkl indices. The powder patterns are composed of two sets of lines: 00l lines or basal reflections, which are sensitive to the state of the minerals (hydration, exchangeable cation....); and hk0 + hkl reflections, which are diagnostic of the mineral.

Thus, for dioctahedral smectites the 060 line is at 0.149-0.152 nm and for trioctahedral smectites at 0.153-0.154 nm. It is related to the b parameter ($b = 6.d_{060}$) and therefore to the chemical composition of the octahedral layer. More detailed identification is based on chemical composition and the Hoffmann-Klemen effect. The unit cell is monoclinic and ideal unit cell formulae are:

Montmorillonite:

$(Si_8)(Al_{3.15}Mg_{0.85})O_{20}(OH)_4X_{0.85}nH_2O$

Beidellite:

$(Si_{7.15}Al_{0.85})(Al_4)O_{20}(OH)_4X_{0.85}nH_2O$

Saponite:

$(Si_{8-y}Al_y)(Mg_{6-z}R_z^{3+})O_{20}(OH)_4X_{y-z}nH_2O$

Hectorite:

$(Si_8)(Mg_{6-z}Li_z)O_{20}(OH)_4X_znH_2O$

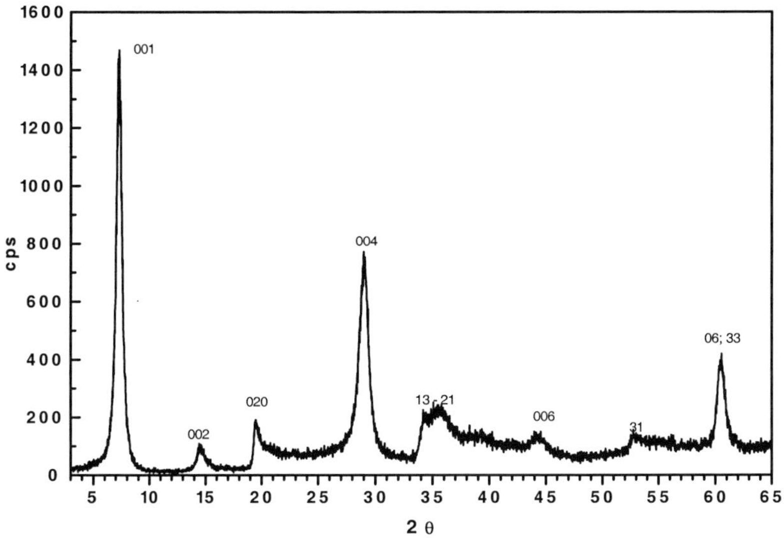

Figure 6: X-ray diffraction pattern of Na-Saponite.

Beidellite usually does not contain Mg, while hectorite has almost no Al.
The Hoffmann-Klemen effect is the irreversible migration of Li^+ upon heating from ion exchange positions into the octahedral layers via the ditrigonal holes of the tetrahedral layers. There are two conditions to be fulfilled: the octahedral layer must

have vacant sites and must carry a negative charge. Only montmorillonites fulfill these conditions. The result is a decrease in the cation exchange capacity (CEC) and the absence of swelling in water or in polyalcohols. The Hoffmann-Klemen effect is not restricted to Li^+ but to any exchangeable cation small enough to migrate through the ditrigonal holes. Examples are transition metal ions such as Cu^{2+} and Ni^{2+} (21).

3.2. Charge density and cation exchange capacity

The structural formulae give the amount of negative charges per unit cell generated by isomorphous substitution. They are compensated by an equal number of positive charges of the exchangeable cations. This is the so-called pH-independent part of the CEC. At the edges of the crystals the sheet structure is broken and the crystal is terminated by OH-groups. At high pH they deprotonate, the edges become negatively charged and have a cation exchange capacity. At low pH they adsorb protons and become positively charged. As a consequence there is a pH-dependent part in the experimentally determined CEC. Maes et al. (22) proposed the following equations for the pH dependence of the CEC in the pH range 3.9 - 5.8 for clays saturated with monovalent and divalent cations respectively:

$$\text{CEC (meq/100g)} = 79.9 + 5.04\ \text{pH} \quad (2)$$

$$\text{CEC (meq/100g)} = 96.1 + 3.93\ \text{pH} \quad (3)$$

In Figure 7 the crystal terminating groups of dioctahedral smectites without and with isomorphous substitution are shown. For a pH of 6.5 White and Zelazny calculated a negative charge density of 5.32 μmole/m^2 (23). If we assume clay particles in the form of a square of 2 μm and 1 nm thick, a CEC of 1 mmole/g and a density of 2.3 g/cm^3, this corresponds to 4.63 μmole/g or 0.46% of the CEC. For particles of 100 nm one obtains 92.6 μmole/g or 9.26% of the CEC. Thus, the contribution of the edge sites to the total CEC is not negligible and will depend on the experimental conditions: crystal size and shape, type of exchangeable cation and pH.

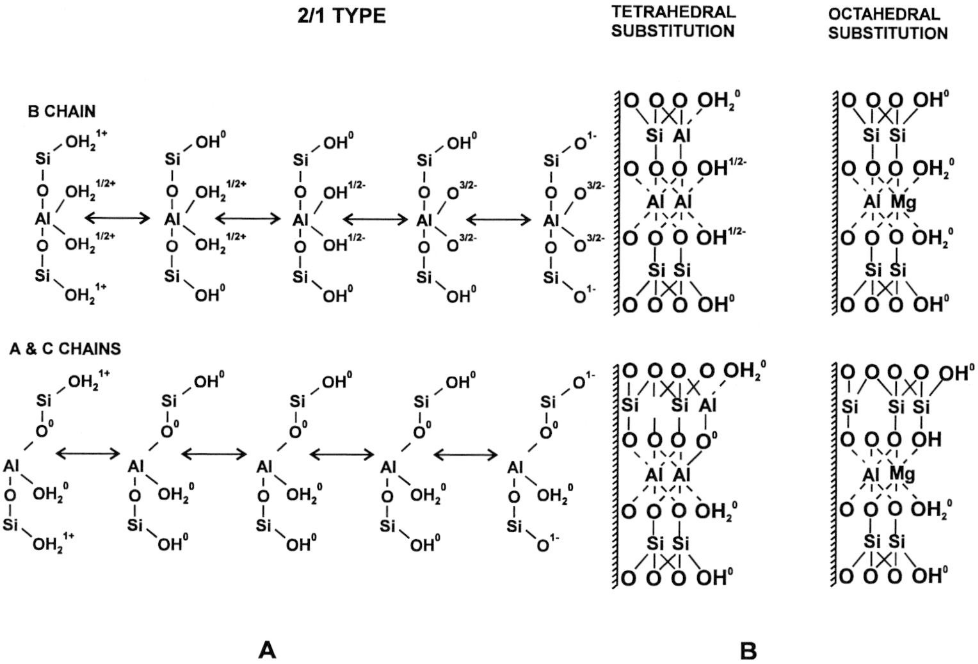

Figure 7: (a) Effect of pH on the edge charges of smectite layers without isomorphous substitution. The pH increases from left to right. (b) Effect of octahedral and tetrahedral substitution on the edge charges at pH = 6.5. (White and Zelazny, 23)

The determination of the charge density σ, defined as CEC/S with S the surface area in m^2/g and CEC in meq/g or, better, in C/g (1 meq = 96.485 C) is subject to the same problems as the determination of the CEC. The determination of the surface area presents particular problems (see next section) and it is better to calculate σ from the structural formula:

$$\sigma\,(C/m^2) = \frac{\text{CEC (meq/g) e FW}}{1000 * 2ab} \tag{4}$$

e is the elementary charge equal to $1.6 * 10^{-19}$ C; FW is the formula weight of the unit cell and 2ab is the surface area of the unit cell in m^2. If the pH-dependent part of the CEC is known and subtracted, one obtains the charge density, due to isomorphous substitution.

Lagaly and Weiss (24) pioneered the alkylammonium method to determine this layer charge density. The method consists in the ion exchange of alkylammonium ions with varying chain lengths. When the average distance between the negative lattice charges is larger than the length of the alkylammonium molecules, the latter form a monolayer with d_{001} = 1.36 nm. When the average distance between the layer charges is smaller than the length of the molecules, a bilayer is adsorbed with d_{001} = 1.77 nm. The transition between both situations occurs at a critical chain length, which depends on the charge density. If all the particles of the clay have the same charge density, the transition is sharp. In reality, this is not the case (25-27). The charge density varies from one particle to another and the experimentally determined value is an average one. There are problems involved with the application of the alkylammonium method, mainly related to the estimate of the amount of chains protruding out of the clay sheets. This has made it difficult to establish relationships between charge density and particle size.

3.3 Surface area swelling and sorption

There are a number of ways to calculate the surface area of smectite, which is commonly expressed in m^2/g. An exact calculation is possible, if the density and the size of a single sheet are known:

$$S = \frac{S_l \, 10^{-6}}{V_l * \rho} \tag{5}$$

where S_l (m^2) represents the sum of the planar surface area or the interlamellar surface area and the surface area of the edges of one sheet; V_l is the volume of one sheet in cm^3 and ρ the density (2.5 g/cm^3). Hence , for 2 μm particles the surface area is 800 m^2/g and for 50 nm particles it is 832 m^2/g, the difference being entirely caused by the difference in edge surface (Figure 8).

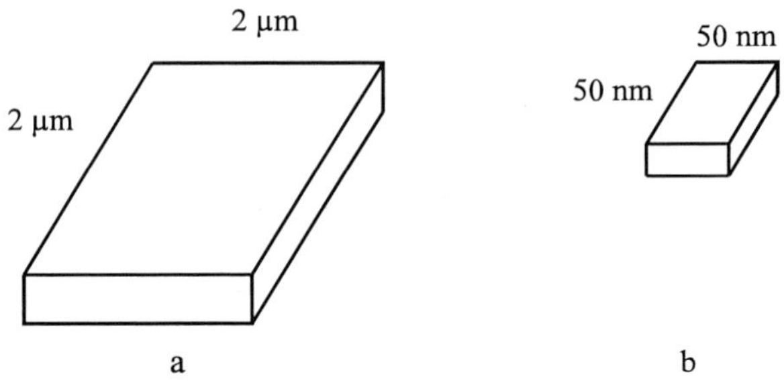

Figure 8: Total surface areas of individual, regularly shaped clay-particles: (a) 8.008 10^{-12} m^2/particle; 2.5 10^{14} particles/cm^3. (b) 5.2 10^{-15} m^2/particle; 4 10^{17} particles/cm^3.

A second way to calculate the surface area is by substitution of the unit cell dimensions (a,b) in the following formula:

$$S = \frac{N_A * 2\,a\,b}{FW} \tag{6}$$

FW is the formula weight of the unit cell in g, N_A is Avogadro's number and 2ab is the surface area of a single sheet in m^2.

Finally, if one has only experimental values of density and the interlayer distance d_{001}, the surface area can be obtained as:

$$S = \frac{2\ 10^3}{\rho\ d_{001}} \tag{7}$$

where ρ is given in g/cm^3 and d_{001} in nm. In equations (6) and (7) the edge surface is not taken into account.

These surface areas are never measured experimentally. There are several reasons for the discrepancy between theory and practice. In the first place, smectites are characterized by a distribution of particle sizes. Secondly, they do not occur as elementary sheets, but the sheets are stacked together into aggregates. Ideally, three types of stacking can be distinguished: face-to-face, edge-to-edge and face-to-edge. An air-dry powder is a collection of such aggregates of different sizes and shapes, organized into a microfabric. The result is a powder with micropores and mesopores. The surface area and the pore size distribution can only be investigated experimentally e.g. by analysis of the N_2 adsorption and desorption isotherms. Typical results are summarized in Table 2.

Table 2: Surface areas (m^2/g) and pore volumes (μl/g) of montmorillonites (28)

smectite	cation	drying	S_{BET}	S_{mi}	S_{me}	V_t	V_{mi}	V_{me}
MO	Na^+	Lyo	124	69	54	116	40	76
		Sed	115	65	53	120	43	77
	Cs^+	Lyo	157	107	45	120	49	71
		Sed	151	95	47	118	59	59
	Ca^{2+}	Lyo	104	47	48	81	26	55
		Sed	107	76	27	86	39	47
OT	Na^+	Lyo	144	82	57	142	54	88
WB			44	32	10	37	18	19

MO = Moosburg montmorillonite, 0T = 0tay montmorillonite and WB = Wyoming bentonite; Lyo = freeze dried; Sed = dried by sedimentation. S_{BET}, S_{mi} and S_{me} are respectively the total surface area, the microporous surface area and the mesoporous surface area. V_t, V_{mi} and V_{me} are respectively the total pore volume, the microporous volume and the mesoporous volume.

The data of Table 2 are obtained on the size fraction < 0.5 μm. When electron micrographs are obtained on the same size fraction the average sizes of Table 3 are obtained (28).

Table 3: Average particle sizes (nm) of the < 0.5 μm fraction of Na^+-saturated smectites.

Otay	20.0 x 20.0
Moosburg	73.3 x 73.3
Camp Berteau	103.2 x 76.4
Wyoming bentonite	129.0 x 96.6
Greek white	213.9 x 160.0

The comparison of the data of both tables shows that clays with larger particles tend to have smaller surface areas. This is expected when only the "external" surface area of the aggregates is measured i.e. N_2 molecules do not penetrate in the interlamellar space between the individual clay sheets of the aggregates. It is also shown that the measured surface areas are modulated by the drying procedure and, especially, the type of exchangeable cation. This indicates that the number of elementary clay sheets per aggregate and the agglomeration of the aggregates in a powder are cation and drying procedure dependent.
The measurement of the total surface area (interlamellar surface area included) requires a probe molecule of known size, which penetrates in the interlamellar space to form either monolayers or bilayers. Polar molecules are preferred because of strong ion-dipole interactions with the exchangeable cations. Ethyleneglycol, glycerol, ethyleneglycol monoethylether (EGME), methylene blue, cetyl pyridinium bromide (PB) and water have been proposed (29). The main problem with these measurements is that the packing density of these molecules is dependent on the charge density of the clays. Consequently, the surface area covered by one molecule varies from clay to clay.
In liquid water smectites swell. Water is adsorbed in the interlamellar space, the distance between the elementary sheets in the aggregate increases to such an extent that the aggregate disintegrates into elementary sheets. Ultimately, the latter are randomly distributed in the suspension, which is transformed into a gel. Swelling is influenced by the charge density of the clay, the type of exchangeable cation and the particle size. In general the size of the aggregates in aqueous suspension increases with the size of the exchangeable cation ($Li^+ < Na^+ < K^+ < Rb^+ < Cs^+$), with its charge ($Ca^{2+} > Na^+$) and, for one type of exchangeable cation, with the charge density of the clay.
Clay dispersions are not in thermodynamic equilibrium. They tend to minimize their interfacial energy by aggregation. The dispersion ages. This process is also called flocculation or coagulation. This is a slow process because - mainly - two opposing forces cooperate: electrical double layer repulsion and London-Vander Waals attraction. They result in a secondary energy minimum, when the mutual interaction energy of two clay sheets is plotted versus distance (Figure 9). The primary energy

minimum is that of the two collapsed clay sheets i.e. the distance between the sheets is one or two layers of water molecules. This can only be achieved by surmounting the energy barrier between the primary and secondary energy minimum. This theory, called the Derjaguin, Landau, Verwey, Overbeek (DLVO) theory, is now recognized as too simplistic (30, 31) but it is sufficient for a qualitative discussion such as the present one.

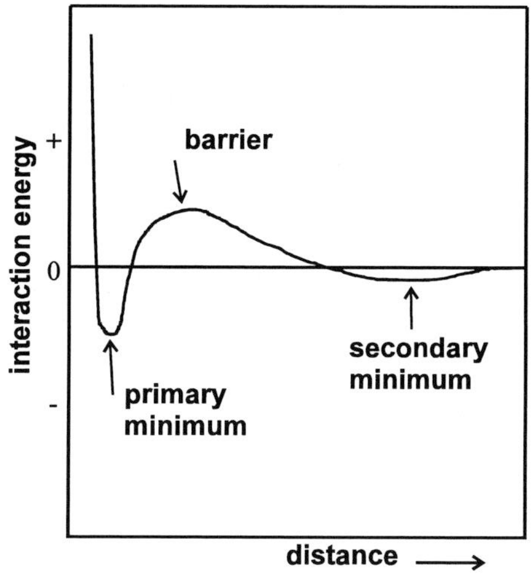

Figure 9: The dependence of the interaction energy of two clay sheets in aqueous suspension on the distance between the sheets.

Because of its obvious industrial and economic importance the clay-water system has attracted considerable scientific attention. When the properties of the water phase in the double layer next to the surface are measured spectroscopically, no significant difference with the properties of bulk water is observed, except for the 1 - 3 water layers, which are closest to the surface (32). Moreover, all the measurements indicate that something like a single clay sheet does not exist in aqueous suspension. Usually one finds aggregates or tactoids. Fripiat et al. (33) found that the average size of the tactoids was independent of the concentration. It looks then as if the clay particles in suspension are the same as those of the air-dry powder. What changes is the distance between the elementary sheets in the aggregate, due to the adsorption of water. The clay particles swell, but do not desintegrate completely. Somehow they "remember" their original composition.

3.4. Crystal chemistry

Our knowledge of the crystal chemistry of smectites is largely based on results of infrared spectroscopy (IR) and magic angle spinning nuclear magnetic resonance (MAS-NMR).

IR spectroscopy allows to distinguish dioctahedral from trioctahedral minerals. Thus, the stretching vibration of the structural OH groups of dioctahedral smectites occurs around 3630 cm^{-1}, while for trioctahedral smectites it appears at 3676 cm^{-1}. Furthermore, the principal Si-O stretching band as well as the OH bending band arise at distinctly lower frequencies for trioctahedral structures (Table 3). A fourth difference is seen on the low frequency side: only one single band near 460 - 450 cm^{-1} is observed for Si-O-Mg, compared to two or three strong bands in this area for dioctahedral minerals (34).

Table 3: Typical IR vibrations of clay minerals (34, 35).

	OH stretch		SiO stetch				OH bend			
dioctahedral										
Pyrophylliet	3674	3645	1120	1068	1050		853	835	814	
							AlAl-	AlFe	AlMg	Fe_2^{3+}
Otay Montm	3622		1088	1029			915	890	840	
Camp Bert.	3627			1038			918	890	847	
Beidellite	3632			1038			922	890		
Nontronite	3556		1085		1004					815
trioctahedral										Mg_3
talc	3676	3660		1044	1018	688				669
Hectorite	3678		1074		1010	678				655
Saponite	3720 * - 3677			1054	1002	690				654
Stevensite	3676			1053	1013	680				657

Band positions of OH stretching bands are given for dehydrated samples, *: band only appears after dehydration of the sample in KBr at 150°C.

A further differentiation among trioctahedral smectites can be made after dehydrating a K- or Li- exchanged sample, or even a simple KBr-prepared disc. Since their hydroxyls vibrate in the ditrigonal hole of the tetrahedral layer, they can interact with penetrating water molecules and cations. The removal of water causes a repulsion between these cations and the H of the hydroxyls. An OH stretching band appears at 3720 cm^{-1} if, as for saponite, a highly localized negative charge exits on the Si-O surface (34, 36).

The bending hydroxyl modes are most sensitive to the chemical composition of the octahedral layer of dioctahedral structures (35-37). Typical bending frequencies, ranging from 920 cm^{-1} for a AlAl -OH grouping to 655 cm^{-1} for a trioctahedral MgMgMg-OH entity can be seen in Table 3.

The coordination of the cations in the octahedral and tetrahedral layers is conveniently studied by MAS-NMR. The SiO_4 tetrahedra share three oxygens of the basal plane and form the so-called Q^3 structures. The fourth oxygen is shared with the octahedral layer. The ^{29}Si resonances are therefore determined in the first place by the chemical composition of the tetrahedral layers i.e. the isomorphous substitution of Si by Al, giving Q^3(OAl), Q^3(1Al), Q^3(2Al) and Q^3(3Al) with between brackets the number of nearest neighboring Al in the tetrahedral layer. The typical resonance positions are given in Table 4.

Table 4: ^{29}Si and ^{27}Al MAS-NMR chemical shifts for smectites (38-43; 46-48).

clay	Q^3(OAl)	Q^3(1Al)	Q^3(2Al)	Q^3(3Al)	Al^{IV}	Al^{VI}
dioctahedral						
Wyoming M	-93.5				-	3.2
Cheto M	-94.1					
Otay M	-93.4					
Polkville M	-93.3					
Texas M	-93.6					
Tatatilla M					68.8	3.2
Beidellite	-92.6	-87.7	-83.2		70.0	4.1
synth beid.	-92.7	-88.4	-82.3		69.9	3.9
trioctahedral						
hectorite	-94.2					
laponite	-94.4					
saponite	-95.6	-90.7	-84.8		65.7	-
synth. sapon.	-93.1	-88.5	-83.2		65.7	7.54

M = montmorillonite

As for zeolites, the ^{29}Si chemical shift increases (= deshielding) with increasing amounts of neighbouring Al^{3+}. The distribution of Al^{3+} in the tetrahedral layer is based on Loewenstein's rule and on the principle of the maximum dispersion of charge i.e. Al^{3+} is homogeneously distributed over the tetrahedral layer (43, 44). The chemical shift of the trioctahedral smectites is typically 2 - 3 ppm more negative than that of the dioctahedral ones. This is due to the difference in electronegativity of the cations in the octahedral layer. Indeed, in trioctahedral smectites the apical oxygen of the Si tetrahedron is connected to 3 Mg^{2+}, while it is bonded to 2 Al^{3+} and one vacancy in the case of dioctahedral smectites. In the first case the total electronegativity of the cations is 3 x 1.31 = 3.93 and 2 x 1.61 = 3.32 in the second case. It follows also that the atom-for-atom substitution of an element with lower electronegativity in the octahedral layer results in deshielding and a less negative ^{29}Si chemical shift (42). Layer charge arises from the isomorphous substitution of one element by a less electronegative element either in the tetrahedral or in the octahedral layer. The effect is deshielding at the Si position and a less negative chemical shift. Figure 10 shows that a linear relation exists between the ^{29}Si chemical shift and the total layer charge.

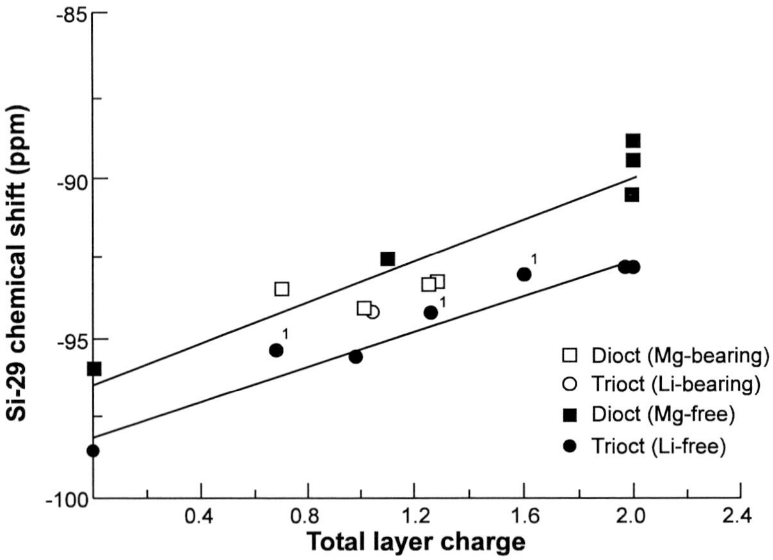

Figure 10: Relation between the ^{29}Si chemical shift and the total layer charge of clays (42)

Deshielding also means less electron density at the Si nucleus or weaker Si-O bonds. It is therefore not surprising to see a linear correlation between the Si(2p) binding energies and the ^{29}Si chemical shift, as shown in Figure 11 for smectites and micas (45).

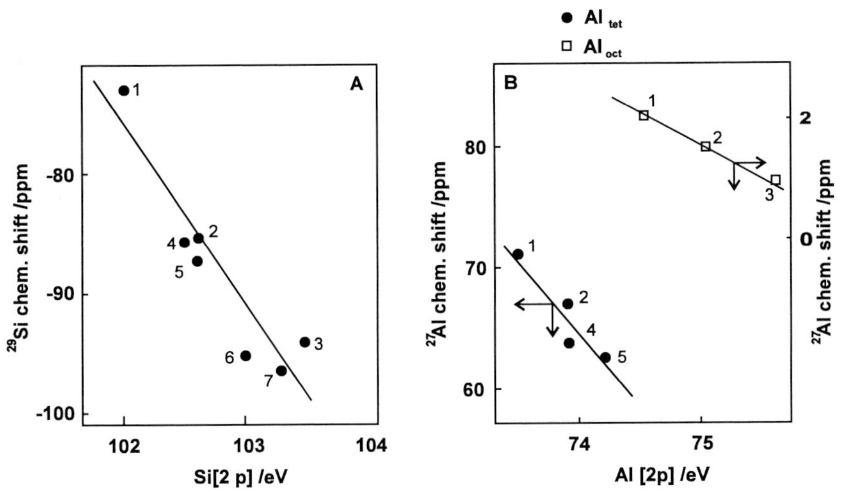

Figure 11: Relation between the ^{29}Si and ^{27}Al chemical shifts and the binding energies of Si and Al respectively (45)

The spectra of the quadrupolar nucleus ^{27}Al are more difficult to obtain, because of the combined effect of chemical shift anisotropy, bulk magnetic susceptibility and second order quadrupolar interaction (49). It is advisable to obtain spectra at the highest magnetic field possible in combination with a high spinning frequency. The effect of the magnetic field is shown in Figure 12. At low field the spectra are dominated by second order quadrupolar broadening and at high field by magnetic susceptibility anisotropy (due to paramagnetic ions in the lattice or present as impurities). One remarks the considerable line sharpening at high fields albeit with the production of more spinning side bands. However, the latter hardly interfere with true ^{27}Al resonance lines at high spinning frequencies, and line positions of 3 - 10 ppm and 65 - 75 ppm are obtained for respectively octahedral and tetrahedral Al^{3+} (Table 4). The same trends, mentioned for the chemical shift of ^{29}Si, hold for that of tetrahedrally coordinated Al^{IV}. Its value increases with increasing Al^{IV} for Si substitution and with increasing total layer charge, while a better shielding is observed for trioctahedral structures (48).

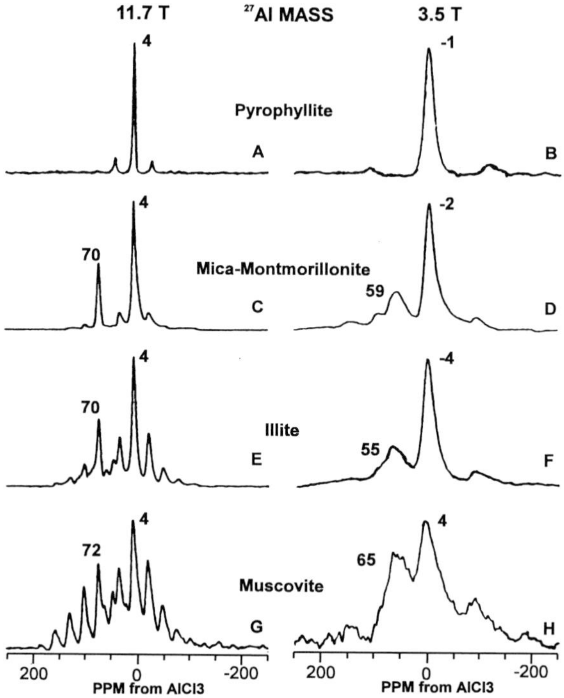

Figure 12: ^{27}Al MAS-NMR spectra of clays (40).

From ^{29}Si MAS-NMR spectra, it is possible to determine the Si/ Al^{VI} ratio, ^{27}Al MAS-NMR spectra yield the Al^{IV}/Al^{VI} ratio. The combination with chemical analysis data, permits the writing of the structural formula. Finally, ^{27}Al MAS-NMR is useful for quantifying the Al^{IV}/Al^{VI} ratio (48).
Both IR and MASNMR spectroscopy are valuable tools to study the reactions, which occur upon heating smectites. Thus, IR spectra show that the removal of the structural hydroxyls occurs in the range 500 - 700°C for montmorillonites and 700 - 900°C for hectorite (50). Lattice breakdown occurs after dehydroxylation, which is also evidenced by the disappearance of octahedral Al from the ^{27}Al spectrum. Presumably Al acquires a distorted fivefold coordination; the resonance line broadens and is lost in the background (50). Finally, new dense crystalline phases appear above 900°C.

3.5 Surface chemistry

Because of their small particle size and their occurrence as aggregates smectites have an external surface area, which is not negligible in surface characterization studies. In the case of dioctahedral clays Brönsted and Lewis acid sites are present on the edges of the crystals. In Figure 7 the Brönsted acid sites are shown. A Lewis site is an exposed or three-coordinated Al^{3+}, substituting for Si^{4+} in the tetrahedral layer. The acid strength of the Brönsted sites is determined either by Hammett indicators (H_o) in an aprotic solvent or by the butylamine titration method of Benesi (49). Typical data are summarized in Table 5.

Table 5: Acid strength and acidity of smectites (49).

smectite	H_0
Na^+-montmorillonite	+ 1.5 - - 3.0
NH_4^+-montmorillonite	+ 1.5 - - 3.0
H^+-montmorillonite	- 5.6 - - 8.2
acid activated clay	< - 8.2

clay	surface area (m^2/g)	butylamine titre (mmol/g):		
		+3.3 - +1.5	*+1.5 - -3.0*	*<-3.0*
Na^+-M	41	0.03	0.01	0.00
H^+-M	52	0.10	0.00	0.55

M = montmorillonite

The 40 μmol/g of acid sites on Na^+-M (Table 5) are not directly comparable to the theoretical number of acid sites on the edges because the particle size and the particle size distribution of Na^+-M are unknown. Also, the number of acid sites on H^+-montmorillonite, titrated with n-butylamine, is smaller than the CEC (0.98 meq/g). This means that in benzene, the solvent used for the titration, the interlamellar space is not totally available for reaction. This is a general problem for all the titration methods and not only for smectite clays. Another possibility is that the lattice of the protonated montmorillonite is partially destroyed.

The acidity of smectites is strongly influenced by the residual water content and by the type of exchangeable cation. In a pioneering study Mortland et al. (51, 52) showed that the residual water in the coordination sphere of the exchangeable cations in the interlamellar space is acidic enough to protonate NH_3 with formation of NH_4^+. The higher the polarizing power of the cation the larger the amount of NH_4^+. The reaction can easily be followed by IR spectroscopy. Quantitative data are shown in Table 6.

Table 6: Amount of NH_4^+(mmol/g) in montmorillonite

Cation	low water	high water
Li	0.23	0.17
Na	0.16	0.10
K	0.10	0.11
Ca	0.80	0.16
Mg	1.01	0.74
Al	1.01	1.00

Smectites are also electron acceptors. The electron-accepting or oxidizing sites are located at the edges and in the structure. The edge sites are supposed to be trigonal Al^{3+}, acting as Lewis acids. Fe^{3+} in the lattice sites are the structural oxidizing sites (52). A nice test of the presence of these sites is the benzidine colouration test. When benzidine is adsorbed on the smectite, an electron is abstracted by the clay (Figure 13) and the blue monovalent benzidine cation is formed. The intensity of the colour is dependent on the amount of oxidizing sites. When the clay surface is acidic, the blue monovalent cation is transformed to the yellow divalent radical cation.

Redox properties may also be induced by the exchangeable cations, such as Cu^{2+}, Ag^+, Fe^{3+} and Ru^{3+} (53-55). Arene molecules, adsorbed in the coordination sphere of these cations in the interlamellar space undergo radicalar polymerization at low water contents.

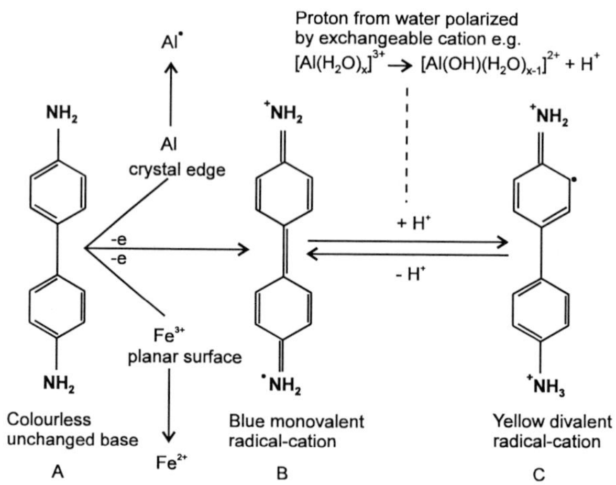

Figure 13: The colour reaction of benzidine on clays (Theng, 16).

4. PILLARED CLAYS

Pillaring is the process by which a layered compound is transformed into a thermally stable micro- and/or mesoporous material with retention of the layer structure. The material obtained is a pillared compound or pillared layered solid. A pillared derivative is distinguished from an ordinary intercalate by virtue of its intracrystalline porosity made possible by the lateral separation of the intercalated guest.
Pillared clays (PILC's) form a special class of pillared layered solids (PLS). The goal of the pillaring process is to introduce micro- and meso-porosity in the clay minerals. This is achieved by a combination of a smectite with a relatively low negative charge density and a pillaring agent prop with a large positive charge. Then, under the condition that the pillaring agents are homogeneously distributed over the surface e.g. by an ion exchange reaction, a two-dimensional channel network is created (Figure 2). Other requirements are that the clay sheet is rigid i.e. does not bend; that the adsorption of props on the external surface is negligible and that all the clay sheets are pillared (Figure 14).

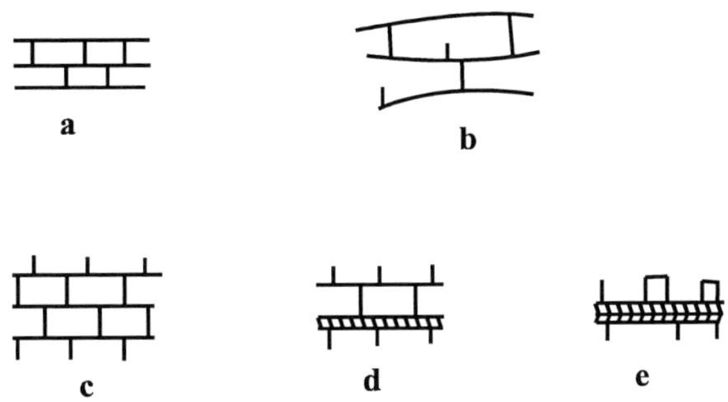

Figure 14: Schematic drawings of possible PILC's: (a) ideal PILC; (b) PILC with bent sheets; (c) ideal PILC with pillars on the external surface; (d) incompletely pillared clay; (e) clay mineral with microporosity created by the pillars on the external surface

4.1 Pillaring agents

In principle, every molecule or particle, able to penetrate into the interlamellar space is a possible pillaring agent. However, to create micro- or meso-porosity, molecules or particles are needed which are larger than the N_2 molecule (0.315 - 0.353 nm).

4.1.1 Organic pillars

Alkylammonium ions (mono-alkyl to tetra-alkyl) with varying alkyl chain length and bicyclic amine cations were the first pillaring agents studied. The use of such "organoclays" as sorbents has been reviewed by R.M. Barrer, the pioneer of the field

(56, 57). These organoclays have an interlamellar volume, which depends on the charge density of the clay and the chain length of the molecules. They adsorb organic solvents such as alkanols to the extent that swelling occurs. This is the basis for their industrial use as gelling agents, thickeners and fillers. The adsorption of mono-alkylammonium ions of varying chain length can be used to estimate the charge density of a particular smectite (24, 58). The organoclays obviously lack, among other things, thermal stability to be used successfully in gas-phase catalysis.

4.1.2 Metal trischelates

M(2,2'-bipyridine)$_3^{n+}$ and M(o-phenanthroline)$_3^{n+}$ (M = transition metal ion) are extremely selectively adsorbed on smectite surfaces to the extent that the ion exchange is followed by an intercalation i.e. the adsorption of the cation and its associated anion in the interlamellar space (59). The size of the complexes is about 1.0 nm. In principle they are pillaring agents which can induce microporosity. On one hand, these complexes lack thermal stability; on the other hand, they are rather unreactive. The clay surface can be decorated with one of the two enantiomers of these complexes, giving chiral clay minerals, which are successfully applied as packing material in a column for separation of optical isomers (60, 61).

4.1.3 Polyoxocations

Polyoxocations are by far the most intensively studied pillaring agents. The cations can be prepared in solution or in situ i.e. in the interlamellar space of the clay. The first method is the preferred one, because it allows a better control of the type of polyoxocation. Al, Zr, Fe, Cr, Ti, Be oxocations have been exchanged on smectites. Several review papers have been published on the subject (62-66). Clay minerals with multimetallic pillars have been prepared too. There are three types: those with mixed metal oxocations; those with two (or more) different monometallic pillars and clays with monometallic pillars on which a second cation is grafted (67-71).
Al-PILC's are the most intensively studied. They will be used as model systems to describe the pillaring process and the properties of the pillared clays in subsequent paragraphs.

4.1.4 Organometallic pillars

This involves the adsorption of organometal complex precursors and their controlled hydrolysis in the interlamellar space. An example is Si(acac)$_3$ (acac = acetylacetonate), but also polynuclear organometallics can be envisaged (72, 73). Intercalated organometal complexes may have interesting catalytic properties as such. Thus, cationic Rh-phosphine complexes, intercalated in the interlamellar space, are hydrogenation and hydroformylation catalysts. Although their activity is not always comparable to that of the homogeneous catalysts, they show some unusual selectivity, depending on the residual water content of the clay (acidity), the spacing between the sheets and the swelling in a particular solvent (49).

The surface of acid-activated clays can be derivatized with molecules containing groups, which bind catalytically active transition metals. This group of catalysts has been extensively studied by Choudary in catalysis of organic reactions (74).

4.1.5 Metal cluster cations

Complexes such as $(Mo_8Cl_8)^{4+}$, $(Nb_6Cl_{12})^{n+}$ and $(Ta_6Cl_{12})^{n+}$ are ion exchanged on the clay. They contain an additional amount of water molecules which partially hydrolyze on the surface. Upon heating, an oxidation reaction converts the Cl-complexes to oxides, which function as pillars (75). The reactions can be written as:

$$Nb_6Cl_{12}(H_2O)_6^{(n-1)+} + H_2O \longrightarrow Nb_6Cl_{12}(OH)(H_2O)_5 + H_3O^+ \qquad (8)$$

$$Nb_6Cl_{12}(OH)(H_2O)_5 + 9\ H_2O \longrightarrow 3\ Nb_2O_3 + H^+ + 12\ HCl + 8\ H_2 \qquad (9)$$

When heated in air, the reaction becomes:

$$Nb_6Cl_{12}(OH)(H_2O)^{(n-1)+} \xrightarrow{O_2} 3\ Nb_2O_5 + x\ HCl + y\ Cl_2 + z\ H_2O \qquad (10)$$

4.1.6 Metal oxides

This involves the direct intercalation of oxide sols of, for instance, SiO_2, Al_2O_3 and silica-aluminas. The intercalation of the tubular imogolite, $Si_2Al_4O_6(OH)_8$ has been successfully achieved (77). The point is to create stable nanometer or subnanometer particles with a positive charge on their surface and to exchange them on the clay surface. An exciting new development has been produced in the group of Pinnavaia with the templated synthesis of intragallery mesoporous silica with cetyltrimethylammonium cations, decylamine and tetraethylorthosilicate (76).

4.2 Synthesis of Pillared Clay Minerals

It follows from the definition of a pillared clay mineral that pillaring with polyoxocations is the most popular route. This is essentially an ion exchange reaction. The micropore volume will be maximized by a combination of a polyoxocation with a high charge and a smectite with a low charge density. To fix the ideas consider a clay with a negative charge density of 1 meq/g and an interlamellar surface of 700 m^2/g. Then, the average surface area per charge is 1.16 nm^2. When this clay is pillared with a cation of charge +7 and a size of 1 nm and the pillaring agent is regularly distributed in a planar array, the average distance between the pillaring agents is 5 nm and the microporous volume is 0.596 cm^3/g. This is an ideal case and the number must be considered as an upper limit.
Polyoxocations are obtained from partially hydrolyzed solutions of polyvalent cations. The chemistry is well documented for Al^{3+}, much less for other cations. Therefore, Al^{3+} will be used as an example.

4.2.1 Solution chemistry of Al

Al^{3+} easily hydrolyzes in aqueous solution (78-85). The pK values of some of the elementary steps are given below. Keep in mind that these pK values are dependent on the ionic strength, the Al concentration and the temperature.

$$Al(H_2O)_6^{3+} + H_2O \underset{pK_1 = 4.99 - 5.3}{\longleftrightarrow} [Al(OH)(H_2O)_5]^{2+} + H_3O^+ \quad (11)$$

$$Al(H_2O)_6^{3+} + 2\,H_2O \underset{pK_2 = 8.71 - 10.91}{\longleftrightarrow} [Al(OH)_2(H_2O)_4]^{+} + 2\,H_3O^+ \quad (12)$$

$$Al(H_2O)_4(OH)_2^{+} \underset{pK = 6.7}{\longleftrightarrow} Al(OH)_4^{-} + 2\,H_3O^+ \quad (13)$$

The Keggin ion $[Al_{13}O_4(OH)_{24}(H_2O)_{12}]^{7+}$, Al_{13}^{7+}, is stable in a broad range of conditions: $[Al^{3+}] < 0.01M$, $[OH^-]/Al^{3+} < 2.4$, $pH < 5$. It consists of a central tetrahedral AlO_4 unit with every O connected to 3 Al $(OH)_4(H_2O)_2$ octahedra (Figure 15).

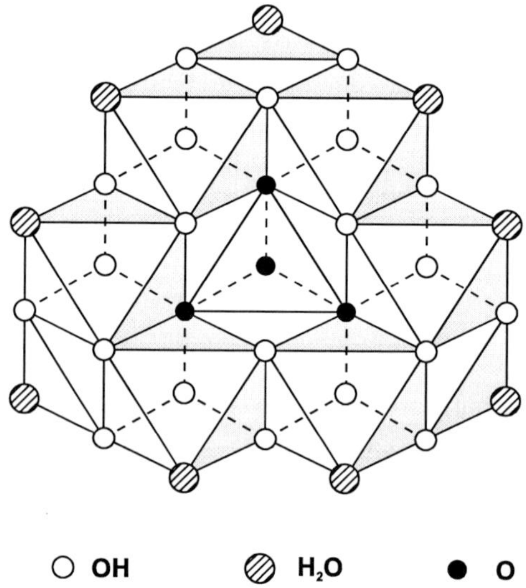

Figure 15: The Al Keggin ion.

Partial neutralization occurs by deprotonation of the coordinated water molecules (86-87):

$$[Al_{13}O_4(OH)_{24}(H_2O)_{12}]^{7+} + 2\,H_2O \underset{pK_1 = 12.48}{\longleftrightarrow} [Al_{13}O_4(OH)_{26}(H_2O)_{10}]^{5+} + 2\,H_3O^+ \quad (14)$$

$$[Al_{13}O_4(OH)_{26}(H_2O)_{10}]^{5+} + 2\ H_2O \underset{pK_2\ =\ 12.86}{\longleftrightarrow} [Al_{13}O_4(OH)_{28}(H_2O)_8]^{3+} + 2\ H_3O^+ \quad (15)$$

$$[Al_{13}O_4(OH)_{28}(H_2O)_8]^{3+} + 2\ H_2O \underset{pK_3\ =\ 13.13}{\longleftrightarrow} [Al_{13}O_4(OH_{30}(H_2O)_6]^{+} + 2\ H_3O^+ \quad (16)$$

These pK values are very close, indicating that the coordinated water molecules have similar properties. As the charge of the cluster decreases, the tendency to form aggregates increases. Al_{13}^{7+} and associated forms can be distinguished on the basis of their characteristic AlOH vibrations, summarized in Table 7 (87). Al_{13}^{7+} is also easily detected by ^{27}Al NMR by the line of its central tetrahedral Al^{3+}, at 62.5 ppm, the octahedral Al^{3+} are NMR silent in solution, but appear in the solid state with a complicated line shape, due to quadrupolar coupling (Figure 16).

Table 7: AlOH vibrations of oligomeric Al

Species	δ(OH), cm^{-1}	γ(OH), cm^{-1}
Al_{13}^{7+}	980	780
associated Al_{13}	960-980	720-730
polymeric Al_{13}	880-960	-
pseudoboehmite	1070-1080	740-760

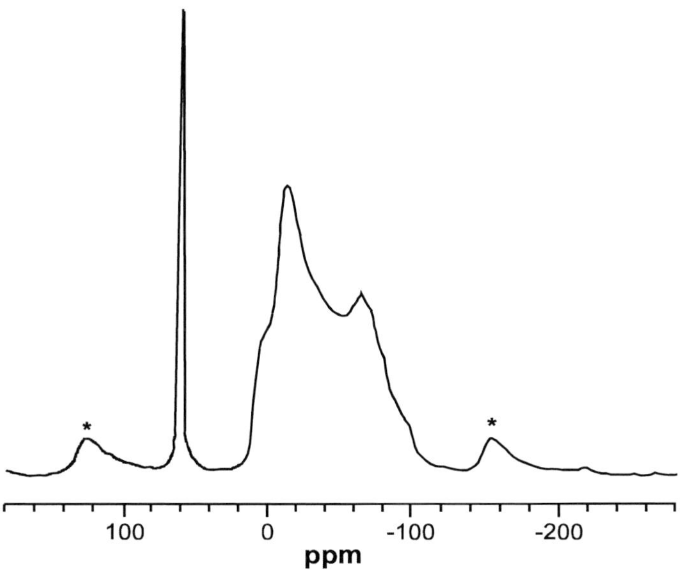

Figure 16: The ^{27}Al MAS-NMR spectrum of the Al_{13}^{7+}-sulfate salt. Spinning side bands are indicated by an asterisk.

4.2.2 Ion Exchange of Al_{13}^{7+}

Under optimized conditions typical partially neutralized solutions of Al^{3+} contain 90-95% of Al^{3+} in the form of Al_{13}^{7+}. The main ion exchange reaction with a smectite clay in the Na^+ form is:

$$7\,\overline{Na^+} + Al_{13}^{7+} \longrightarrow \overline{Al_{13}^{7+}} + 7\,Na^+ \qquad (17)$$

with the bar denoting "on the surface". Two regions of uptake can be distinguished in the isotherms (Figure 17): a quantitative selective uptake of Al_{13}^{7+} up to 2 - 2.2 mmol Al^{3+}/g and an additional continuous uptake at higher Al concentrations in solution (88-89).

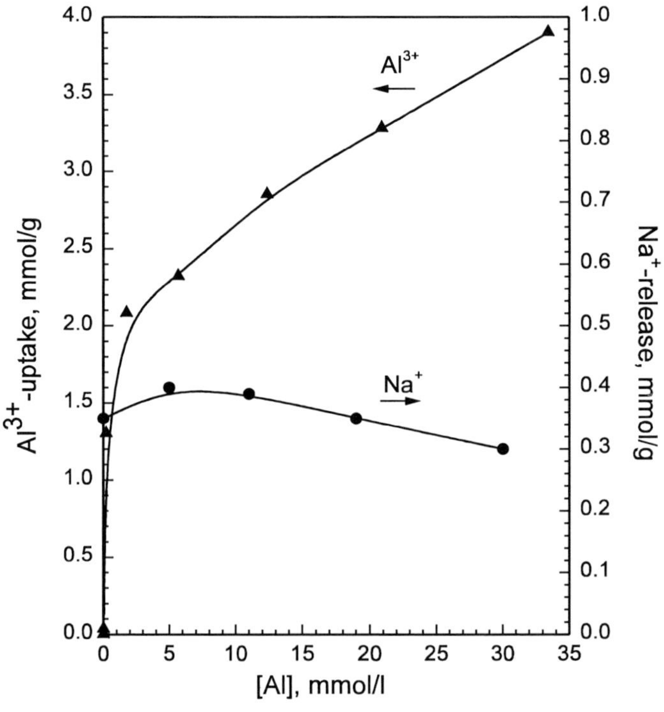

Figure 17: Uptake of Al and release of Na upon ion exchange of Al_{13}^{7+} with saponite.

Na^+ is released in the first region of the isotherm and thus the second regime of Al uptake is precipitation of Al_{13} on the surface.

In the beginning of the exchange the amount of Na^+ released is slightly more than required by the stoechiometry of eq. 17. This is indicative for a hydrolysis of Al_{13}^{7+} upon contact with the clay mineral. The reaction can be represented as:

$$[Al_{13}O_4(OH)_{24}(H_2O)_{12}]^{7+} \longleftrightarrow [Al_{13}O_4(OH)_{24+x}(H_2O)_{12-x}]^{(7-x)+} + x\,H^+ \qquad (18)$$

$$x\,H^+ + \overline{Na^+} \longleftrightarrow x\,Na^+ + \overline{(H^+)_x} \qquad (19)$$

Because of the ion exchange character of the adsorption of Al_{13}^{7+} the amount adsorbed is proportional to the CEC (90). The upper limit corresponds to a monolayer of Al_{13}^{7+} in the interlamellar space. This has been established by the group of Lambert (91) as 1 Al_{13}^{7+} per 6 $O_{20}(OH)_4$ formula units or 1 negative charge per 6 unit cells.

4.2.3 Washing and Calcination

After ion exchange a material is obtained with d_{001} = 1.25 nm or with a distribution of d_{001} spacings in the range 1.25 - 1.90 nm. Washing leads to a uniform spacing of 1.8-1.9 nm and a better ordering of the individual clay sheets into aggregates. The latter is evidenced by the sharpening of the d_{001} line after washing. In ideal cases higher order reflections are visible. What happens chemically is unknown. However, as the washing water has typically a pH = 6-7 additional hydrolysis of the adsorbed Al_{13} clusters is expected.

The chemistry of the calcination step can be written as

$$2\,[Al_{13}O_4(OH)_{24}(H_2O)_{12}]^{7+} \longrightarrow 13\,Al_2O_3 + 14\,H^+ + 41\,H_2O \qquad (20)$$

in the case of total transformation or as

$$2\,[Al_{13}O_4(OH)_{24}(H_2O)_{12}]^{7+} \xrightarrow{O_2} 2\,Al_{13}O_{11}(OH)_{17} + 32.5\,H_2O + 14\,H^+ \qquad (21)$$

in the case of a minimal transformation to a neutral cluster. The exact chemistry is unknown. The fact, however, that the spacing before and after calcination is constant within 0.1 nm is an indication that the pillars do not agglomerate.

Reactions (20) and (21) lead to neutral clusters. This means that the interaction pillar-clay mineral is purely physical (van der Waals). This idea has been challenged in the case of clay minerals with tetrahedral isomorphous substitution (beidellite, saponite) and of fluor-hectorite and Cheto-type montmorillonite (90-93). For these clay minerals there is ^{29}Si and ^{27}Al NMR evidence that chemical anchoring between the pillar and the clay mineral surface occurs with inversion of a Si or Al tetrahedron of the tetrahedral layer of the clay mineral. This is schematically shown in Figure 18. It is interesting to note that a clay mineral structure with inverted tetrahedra is the old (and abandoned) Edelman-Favajee model of a clay mineral (16).

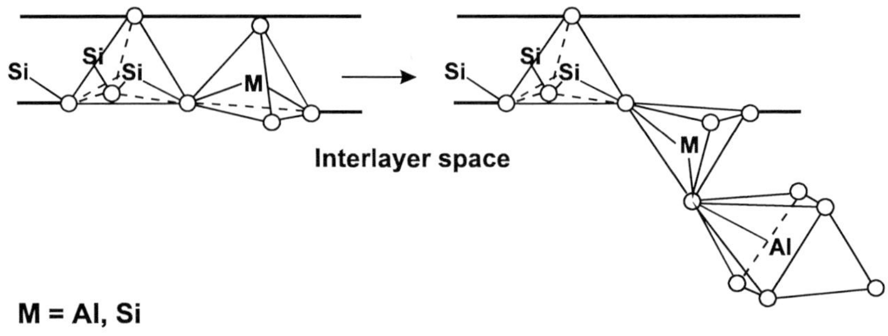

Figure 18: Schematic representation of the chemical anchoring between the pillar and the clay layer.

4.3. The Pore Structure of Pillared Clays

A pillared clay mineral has the following characteristics: (1) the layers are propped apart vertically and do not collapse upon removal of solvent; (2) the minimum increase of the basal spacing is the diameter of the N_2 molecule; (3) the pillaring agent has molecular dimensions and is laterally spaced in the interlamellar space on a molecular length scale; (4) the interlamellar space is porous and accessible to molecules at least as large as N_2. The d_{001} spacing gives only information about the average distance between the clay sheets but not about the distribution of the pillars in the interlamellar space and the resulting pore structure. There are two ways to study the pore structure: direct observation with high resolution transmission electron microscopy (HRTEM) or by atomic force microscopy (AFM) and indirectly, via the analysis of N_2 adsorption-desorption isotherms.

Electron microscopy studies of pillared clay minerals are scarce (94, 95). They confirm the 1.8-1.9 nm spacing and are indicative of the heterogeneous nature of the materials in that areas where face-to-face, face-to-edge and edge-to-edge aggregates co-exist. The latter become more important as the particle size decreases. These studies also confirm that the materials deteriorate slowly upon aging: the 1.8-1.9 nm spacing disappears and Al_2O_3 particles are formed. This was also observed by other groups (96). With AFM it was impossible to resolve the individual pillars, but the surfaces of the pillared clay mineral aggregates were found to be relatively free of Al_2O_3 debris (97-101).

Indirect information on the distribution of the pillars in the interlamellar space is obtained by the analysis of N_2 adsorption/desorption isotherms. However, their analysis is not straightforward, as both the Langmuir equation and the BET theory fail (102). Three more refined analyses have been proposed in the literature. Gil et al. (103) carefully analyzed the isotherm in the very small p/p_0 regime with the Horvath-Kawazoe theory. Maes et al. (104) and Zhu and Vansant (105, 106) proposed a logarithmic plot of the isotherm in the same pressure regime. They could distinguish five different regions corresponding to pores which allow adsorption of one to five layers of N_2 molecules. From this the surface area and pore volumes of the different pore sizes could be derived. To avoid the very low pressure regime Remy et al. (102) analyzed the "classical" isotherms ($0.01 < p/p_0 < 1$) with two assumptions: (1) adsorption at the external surface is governed by multilayer adsorption in agreement with the BET theory; (2) the micropore volume is completely filled in the pressure domain in which the BET theory is applied ($0.05 < p/p_0 < 0.25$). Representative data are shown in Table 8. It shows that the BET equation gives systematically an under-estimation of the surface area; the Langmuir formula an over-estimation. Apparently similar samples can have quite different surface areas and pore volumes. Reproducibility of the preparation of pillared clay minerals is indeed one of the major drawbacks of these materials.

Table 8: Surface and pore volumes of pillared clay minerals.

	Surface area, m^2g^{-1}			Pore volume cm^3g^{-1}		
	BET	Langmuir	New method	micropore	mesopore	total
			Model analysis of Remy et al. (102)			
Al-PM	253	-	325	0.0912		
Al-PB	187	-	245	0.0687		
			Method of Gil et al. (103)			
Al-PM-1	-	289	-	0.107	0.054	0.161
Al-PM-3	-	291	-	0.109	0.051	0.160
			Method of Zhu et al. (105)			
Al-PM	342	403	409	0.1505		
Ti-PM	282	-	254	0.1392		
Zr-PM	227	286	251	0.0978		

PM: pillared montmorillonite; PB: pillared beidellite

According to Zhu and Vansant (105, 106) the most abundant pore size is that with diameter in the range 0.73 - 1.09 nm (Figure 19). One might expect that the micropore volume decreases with increasing number density of Al_{13} pillars. This was shown to be the case for CO_2 adsorption at room temperature but not for N_2 adsorption at 77K. (Table 9) (107).

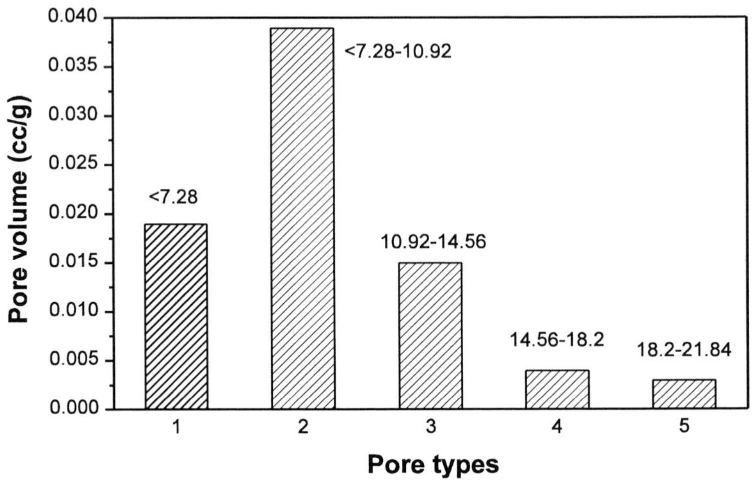

Figure19: Pore size distribution of an Al-PILC, as analysed by the method of Zhu et al. (106). The pore size corresponding to the five pore types are indicated in Å.

Table 9: Comparson of micropores volumes of pillared saponites.

layer charge	$V(CO_2)$, cm^3/g	$V(N_2)$, cm^3/g
0.8	0.148	0.088
1.0	0.133	0.124
1.2	0.0499	0.141
1.4	0.085	0.124

Several reasons may be invoked to explain the difference between CO_2 adsorption and N_2 adsorption: some CO_2 may chemisorb on the alumina pillars; both molecules may probe different surfaces because of differences in kinetic diameters. In any case the pore volumes are much smaller than one can expect theoretically (see introduction). This points towards non-uniform distribution of pillars and heterogeneity of the material, as evidenced by HRTEM.

4.4 Thermal Stability and Pore Size Control

From the very beginning of pillared clay mineral research there has been a drive to obtain materials with improved thermal stability, larger surface areas and larger pore volumes. The thermal stability is determined by the stability of the pillar and of the clay mineral. The latter is turned in its acidic form (see eq. 18), which is known to be thermally less stable. Thus, the thermally stable pillared clay minerals must have excellent crystallinity and if, in addition, they have isomorphous substitution in the tetrahedral layer, the interaction pillar-clay mineral is maximized, eventually via chemical anchorage, and so is the thermal stability. The most successful clay mineral in this regard is rectorite, a regular mica-smectite interstratification compound, the structure of which is shown in Figure 20. The exceptional thermal stability of pillared rectorite is ascribed to the layer rigidity that the mica layers impart on the pillared structure (99, 108). Retention of the d_{001} spacing has been reported up to 1073 K under steaming conditions. According to Urabe et al. (109) pillared synthetic mica-montmorillonite has a similar thermal stability.

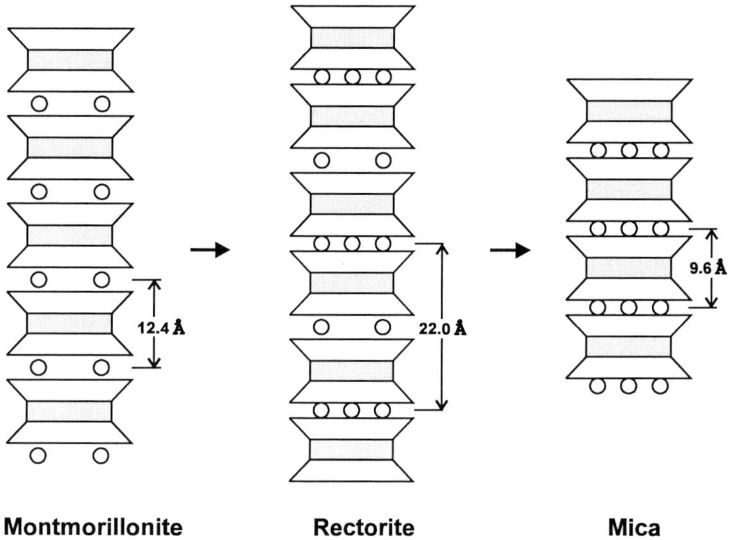

Figure 20: Schematic representation of rectorite as a regularly interstratified mica-montmorillonite. The indicated interlayer spacings correspond to a 1 layer hydrate.

Pore size control is closely related to thermal stability and to the distribution of the pillars in the interlamellar space. In the case of a non-uniform distribution some pillars may be so close that upon calcination they merge into one pillar. In the case of a uniform distribution the average distance between the pillars is equal for all pillars and the distance may be too large for the pillars to aggregate in a relatively short time span. Several methods for improving the homogeneity of the distribution of the pillars have been proposed in the literature: (1) competitive ion exchange $Al_{13} \leftrightarrow NH_4^+$ (110); (2) CEC control by using synthetic clay minerals with different charge density (90); (3) pre-adsorption of alkylammonium cations (111); (4) co-adsorption of poly(ethyleneoxide) (112) or poly(vinylalcohol) (113).

4.5 Mesoporous Materials

The following strategies are reported in the literature to obtain mesoporous materials out of a layered silicate: (1) hydrothermal pillaring with rare earth-Al pillar moieties (114); (2) adsorption of colloidal sols (118); (3) gallery templated polymerization of e.g. $Si(OEt)_4$ (76); (4) topochemical transformation of layered silicates in the presence of surfactants (115-117).

In the first method commercial, partially neutralized Al^{3+} solutions are modified by addition of La^{3+} or Ce^{3+} and the pillaring process of a smectite clay mineral is performed under reflux or under hydrothermal conditions. A material is obtained which is, at least partially, expanded to a d_{001} spacing of 2.6 nm. The structure and composition of the pillaring oligomer is unknown nor is it known to what extent the La^{3+} or Ce^{3+} cations are incorporated into the pillar. In view of the pH conditions of the pillaring process only traces are expected to be incorporated in the pillar; some may be ion exchanged or adsorbed on the pillar. Anyhow, the data of Table 10 clearly show the increase in surface area and pore volume with respect to the regularly pillared materials.

Table 10: Mesoporous pillared layered silicate.

Material	d_{001}, nm	Surface area *BET, m^2g^{-1}*	Pore volume, cm^3g^{-1}: *total*	*micro*
La, Al-PM	2.6	493	0.28	0.15
Ce, Al-PB	2.57	431	0.23	0.00
La, Al-PB-AD	2.54	426	0.24	0.00
Ti, Si-PB-AD	-	403	0.50	
Ti, Si-PB-SCD	-	501	4.8	
Zr, Si-PB-AD	-	354	0.54	
Zr, Si-PB-SCD	-	521	5.0	
Ti, Si-PB-AD	-	125	0.03	
Ti, Si-PB-SCD	-	467	1.5	
Imogolite-PM	-	480	0.20	
PCH	-	470-750		
FSM-16	3.7-5	900-1200		

PM: pillared montmorillonite, PB: pillared beidellite, AD: air dried, SCD: super critically dried

When sols of TiO_2-SiO_2, SiO_2-Al_2O_3 or ZrO_2-SiO_2 are used in combination with supercritical drying, pillared smectites are obtained which are almost X-ray amorphous. The materials can be characterized as expanded clay structures with a 5 - 10 -fold increase of pore volume, mainly due to the effect of supercritical drying

(Table 10). When imogolite, a tubular aluminosilicate, is used as pillaring agent, two types of micropores are formed: intratubular and intertubular (Figure 21) with a very small mesoporosity (66).

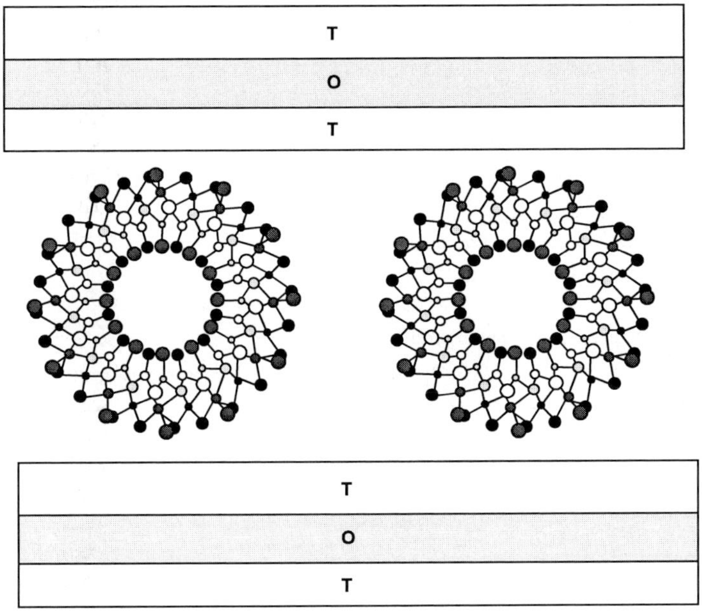

Figure 21: Cross sectional view of Na^+ montmorillonite, intercalated with a monolayer of imogolite ($SiAl_2O_3(OH)_4$).

True mesoporous materials can be obtained via two routes. The first one, proposed by Pinnavaia's group, consists of an expansion of the individual clay mineral sheets in aqueous suspension with a cationic surfactant in conjunction with a neutral co-surfactant, followed by interlamellar hydrolysis and condensation polymerization of neutral inorganic precursors e.g. $Si(OEt)_4$ (76). The surfactants form intragallery micelles around which the controlled hydrolysis and condensation of $Si(OEt)_4$ takes place (Figure 22 a). By calcination the surfactants are removed and SiO_2 pillars are formed. A mesoporous material is obtained with pores of 2.0 - 2.5 nm and a surface area as large as that of typical zeolites.
Fukushima's group modifies layered silicates such as kanemite with cationic surfactants. Under controlled pH conditions the surfactant forms micelles and the silicates undergoes a topotactic transformation by folding around the micelles (Figure 22 b). After calcination, to remove the surfactant, a mesoporous material is obtained.

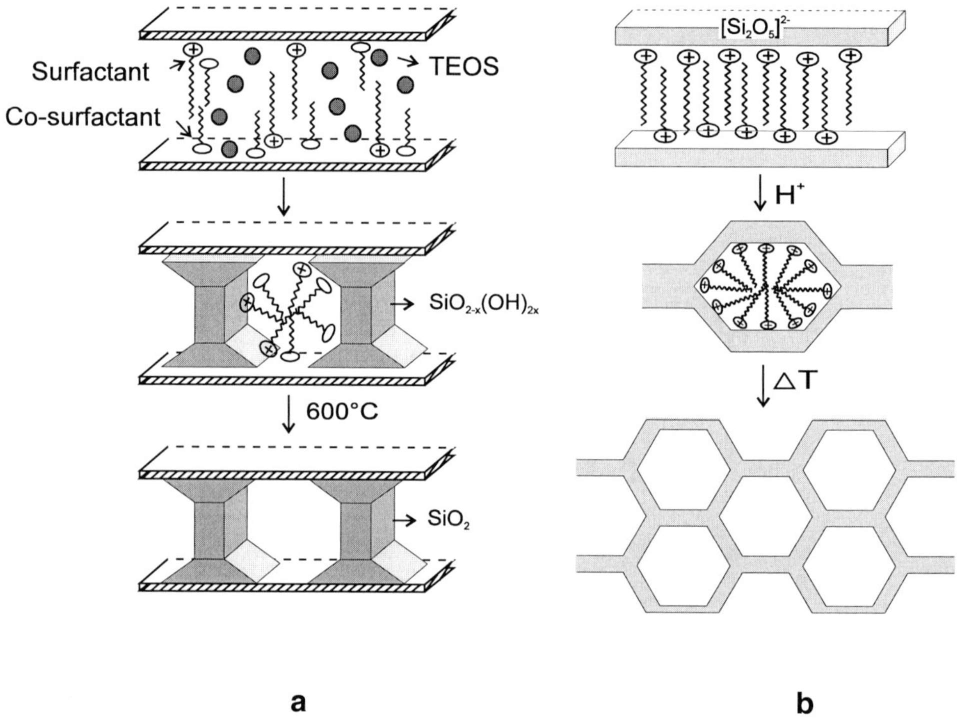

Figure 22: Formation of mesoporous materials: a) by synthesis of structures within the interlayer of clays, using intercalated surfactants as templates; b) by micell-formation of surfactants, intercalated in a layered silicate.

4.6 Generalization of the Pillaring Process

The principles underlying the Al pillaring of smectites have been extended in two directions: pillaring of non-smectitic clay minerals and pillaring with other elements than Al. There are now literature data on Al pillaring of acid activated clays (119-121), taeniolite (122-124), vermiculite (125) and talc (126, 127). In the latter two cases the clay minerals have to be attacked severely e.g. with Na_2SiF_6 to make them swellable before pillaring. Typical data are given in Table 11. The textural properties of these materials are very similar to those of Al pillared smectites. In the case of acid activated clays mesoporosity is significant and a broad range of pore sizes is measured, extending into the mesoporous range.

Table 11: Properties of modified pillared clay minerals:

clay mineral	pillaring agent	d_{001}, nm	$V_{micropore}$, $cm^3\ g^{-1}$	V_{pore}, $cm^3\ g^{-1}$	surface area $cm^2\ g^{-1}$
acid activated montmorillonite	Al		0.09	0.29	257
taeniolite*	Al	1.84	0.133	-	344
talc	Al	1.87			320

* taeniolite: $Li(Mg_2Li)Si_4O_{10}(OH)_2$

Variations in the pillaring process may be sought in the use of other pillaring agents, the best known being Zr, but transition metal ions such as Cr, Fe and Ti have been popular too. In the case of Zr the pillaring species is considered to be the tetranuclear complex $[Zr_4(OH)_8(H_20)_{16}Cl_x]^{(8-x)+}$ (128). Depending on the hydrolysis conditions 2 or 3 of these sheets link together into stacks, resulting in broad micopore size distributions. Hydrolysis of $TiCl_4$ with HCl gives the octamer $[(TiO)_8(OH)_{12}]^{4+}$ which polymerizes into large units with non-uniform size. Partially pillared products are obtained, often accompanied by the formation of an anatase phase, deposited on the external surface of clay particles (129). Fe-polymers formed by hydrolysis have a low positive charge and a high molecular weight. The interlayer region is stuffed with pillars, or intercalation is limited to the outer rims of the clay particles, due to slow diffusion, or hydroxy-oxidic polymers of Fe may simply precipitate. The obtained product resembles more a mixed oxide than a pillared clay. Pre-adsorption of butylammonium cations is twofold: (1) they induce face-to-face aggregation of the clay platelets; (2) they are only partially exchanged by the cationic intercalating agents. Thus after calcination microporosity is left. This is illustrated in Table 12 (129-134). Also included in table 12 are data of pillaring of laponite, where the benificial role of ethylenediamine is the face-to-face aggregation. It is also shown that the use of $[Fe_3(Ac)_7(OH)_2(H_2O)]^+$ (Ac = acetate), which can be better formulated as $[Fe_3(Ac)_6(HAc)(H_2O)_2]^+$, allows a more controlled hydrolysis and a better pillaring (129-134).

Table:12: Properties of pillared clays (129-134).

Clay mineral/ pillaring agent	BET surface area ($m^2 g^{-1}$)		micropore volume cc g^{-1}		d_{001} nm
	amine	*no amine*	*amine*	*no amine*	
Montmorillonite					
Al	-	341	-	0.132	1.86
Ti	313	282	0.147	0.154	2.3 - 2.7
Zr	284	271	0.104	0.094	1.96 + >
Fe	233	134	0.121	0.037	1.21 + >
Fe_3	-	200 - 272	-	0.082-0.148	1.56 - 1.78
Cr	-	241 - 286	-	0.015 - 0.09	2.66
Laponite					
Al	354	286	0.329	0.199	-

4.7 Properties and Use of Pillared Clay Minerals.

The acidity of pillared clay minerals is the major property studied in view of maximizing catalytic activity in acid catalysis. The subject is discussed in detail by Butruille and Pinnavaia (63). There are three sources of acidity in pillared clay minerals: (1) protons generated by the pillaring process, which migrate into the clay structure (Brönsted); (2) the edges of the clay mineral particles (Brönsted and Lewis); (3) the pillar itself (Lewis and Brönsted). Both Brönsted and Lewis acidity have indeed been found with the pyridine adsorption/desorption technique. The Lewis acidity is primarily determined by the pillar for Al-pillared materials; Brönsted acidity by the clay mineral particles (135, 136). In the case of Zr- and Ti-pillared materials Lewis acidity is fully controlled by the pillars; Brönsted acidity by the clay mineral. To increase the Brönsted acidity, smectites with tetrahedral substitution are preferred, besides rectorite, acid activated clays and synthetic mica-type montmorillonite.
One can question the relevance of the pyridine technique because it supposes that all sites react with one pyridine molecule (some sites may be hidden). A potentiometric titration gave three groups of sites in the pK ranges 5-6, 6-7 and 8-10 (124). However, the method did not allow measurement of sites with pK < 4 or > 10. Direct observation of acidic OH group by FTIR is also not evident and hampered by the intense OH stretching vibrations of the parent clay. Acidic OH groups have been reported in the literature for beidellite at 3440 cm^{-1}, bentonite at 3700 cm^{-1} and saponite at 3594 cm^{-1} (135, 137). These O-H stretching vibrations have the characteristic that they disappear upon adsorption of pyridine and reappear upon

desorption. However, the wide frequency range is uncommon and cannot be explained at this moment.
The acidity can also be probed by acid catalyzed reactions such as the hydroconversion n-alkanes. Figure 23 shows the typical data for hydroconversion of heptane (137-139). The general sequence is zeolite β, zeolite Y > Al-pillared saponite > SAPO-11 > strongly dealuminated zeolites > Al-pillared montmorillonite. The position of Al-pillared saponites is the same as that of Al-pillared beidellite. Thus the general rule is that clay minerals with tetrahedral substitution are superior to Al-pillared clay minerals with isomorphous substitution in the octahedral layer (137-139).

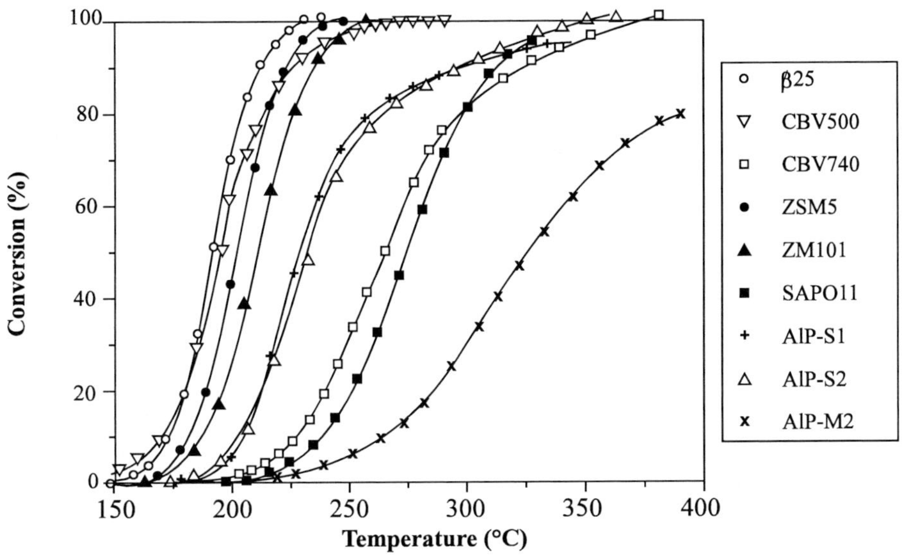

Figure 23: The hydroconversion of heptane over Pt/ Al-pillared clay minerals as compared to some Pt/ zeolites. The legend refers to the following structures: β/25: zeolite beta (P.Q., Si/Al=12.5), CBV500: zeolite Y (P.Q., Si/Al=2.6), CBV740: dealuminated zeolite Y (P.Q., Si/Al= 21), ZSM-5: CBV8020 (Si/Al= 40), ZM101: dealuminated mordenite, AlP-S1: Al-pillared saponite (Ballerat), AlP-S2: Al-pillared saponite (Esmectica de Yunclillos), AlP-M2: Al-pillared montmorillonite (Serrata Blanca) (Poncelet, 137-139).

The typical microporous volume and pore sizes make pillared clay minerals good candidates for sorption and molecular sieving.
Recently the potential of pillared clay minerals for gas separation has been studied. Two routes have been explored to improve the molecular sieving effect: increase of pillar density and modification with cations. Both have a positive effect. Thus, a high

charge density clay mineral such as Arizona montmorillonite has better adsorption properties than Wyoming montmorillonite (Table13). Improvements in adsorption and N_2-O_2 separation can also be seen with cation modified pillared clay minerals. This is due to the quadrupolar interaction between N_2 molecules and the positive charge on the cations. Thus N_2:O_2 ratios of 3 to 4 have been obtained (140-142).

Table13: Use of Al-pillared clay minerals for the separation of gases.

	Adsorption, mmol g^{-1}		
Clay mineral	N_2	O_2	CH_4
Arizona Montmorillonite	0.037	0.032	0.083
Li-modified Arizona Montm.	0.076	0.024	-
Wyoming Montmorillonite	0.032	0.029	0.047
Montmorillonite*	0.02	0.01	-
Sr-modified Montm*	0.04	0.01	-

All clay minerals are Al-pillared; adsorption at 1 atm and 297K.
*: adsorption at 0.45 atm and 273K

Pillared clays have intermediate hydrophilic-hydrophobic character (143, 144) and thus some possibility in removal of organic pollutants (145, 146). This is illustrated in Figure 24 by the adsorption of chlorinated hydrocarbons. In all cases, the adsorption capacity of the mixed Fe-Zr pillared montmorillonite is a factor 4 higher than that of the parent clay. They also have some residual ion exchange capacity. The latter can be almost fully restored to the CEC of the original clay mineral by an ammonia treatment followed by ion exchange of the formed NH_4^+ by the desired cation. Thus, heavy metal cations can be removed from polluted waters (147).

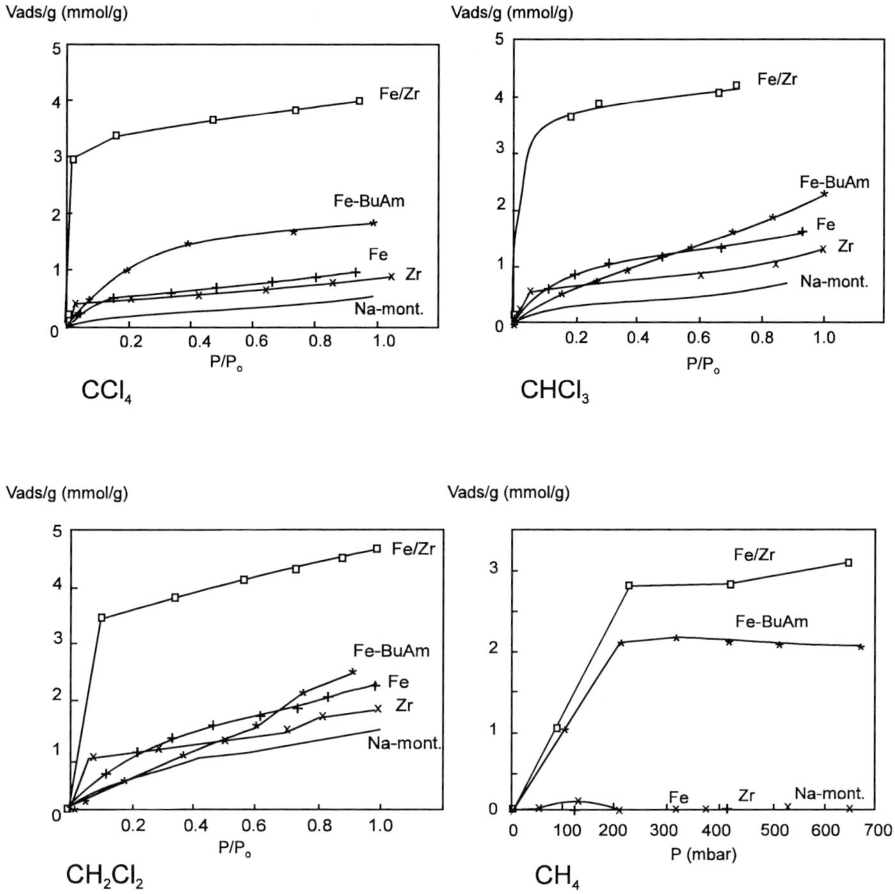

Figure 24: Adsorption isotherms of some (chlorinated) hydrocarbons on different pillared clays compared to the adsorption on Na-montmorillonite (129).

CONCLUSION

Pillared clay minerals, PILC's, were originally devised as materials with pore sizes complementary to those of zeolites. Progress in the field of zeolite synthesis and in the preparation of PILC's has been such that this complementarity in pore size does not exist anymore. PILC's should now be considered as a set of materials in their own right with potential applications in heterogeneous catalysis, selective adsorption and environmental chemistry. The main drawback has been the reproducibility of preparation and the lenghty, time-consuming synthesis procedure. In the case of Al-pillaring this has been solved to a large extent. Swelling clay minerals with excellent crystallinity and isomorphous substitution in the tetrahedral layer together with the Al-

Keggin ion are the preferred combination. The conditions for pillaring (to reduce the workload to a minimum) are now established (148). For all other types of pillars the chemistry of the cations (especially transition metal ions) in the partially neutralized aqueous solutions is too complicated or simply not known to be able to perform a reproducible pillaring process.

In any case, due to pillaring the interlamellar space of the clay minerals becomes available also at high temperatures. The clay mineral surface essentially behaves as a Brönsted acid. However, the pillar itself has also an active surface, being of the Lewis type in the case of Al-pillars. It is this combination of properties of the clay surface and properties of the pillars that offers a large potential for exploring the catalytic and adsorptive properties of these materials.

REFERENCES

1 B. Hulthén, in N.A. Shaikh and N.-G. Wik (Eds.), Clay Minerals - Modern Society, Nordic Society for Clay Research, Uppsala, 1986, pp. 99-109.

2 I.E. Odom, in L. Fowden, R.M. Barrer and P.B. Tinker (Eds.), Clay Minerals: their Structure, Behavior and Use, The Royal Society, London, 1984, pp. 171-189.

3 W.B. Jepson, in L. Fowden, R.M. Barrer and P.B. Tinker (Eds.), Clay Minerals: their Structure, Behavior and Use, The Royal Society, London, 1984, pp. 191-212.

4 D.W. Breck, Zeolite Molecular Sieves: Structure, Chemistry and Use, J. Wiley, New York, 1974.

5 M.E. Davis, C. Saldarriaga, C. Montes, J. Garces and C. Crowder, Zeolites, 8 (1988) 362-367.

6 M.S. Stul and L. Van Leemput, Clay Min., 17 (1982) 209-215.

7 R. M. Barrer and L. W. R. Dicks, J. Chem. Soc. (A), (1967) 1523-1529.

8 K.A. Carrado, P. Thyagarajan, R. W. Winans and R. E. Botto, Inorg. Chem., 30 (1991) 794-799.

9 A. Decarreau, Bull. Mineral., 103 (1980) 579-590.

10 J. T. Kloprogge, J. Breukelaar, J. W. Geus, B. H. Jansen, Clays and Clay Minerals, 42 (1994) 18-22.

11 V.C. Farmer, W.J. Hardy, F. Elsass and M. Robert, Clays and Clay Minerals, 42 (1994) 180-186.

12 J. T. Kloprogge, A.M.J. Van der Eerden, B. H. Jansen, J. W. Geus and R.D. Schuiling, Clays and Clay Minerals, 41 (1993) 423-430.

13 Yanagisawa, T. Kusunose, K. Ioku, N. Yamasaki, P.B. Malla and S. Komarneni, J. Materials Science Letters, 15 (1996) 861-863.

14 E. Nemecz, Clay Minerals, Akadémiai Kiado, Budapest, 1981.

15 R.E. Grim, Clay Mineralogy, McGraw Hill, New York, 1968.

16 B.K.G. Theng, The Chemistry of Clay-Organic Reactions, Adam Hilger, London, 1974.

17 H. Van Olphen and J.J. Fripiat, Data Handbook for Clay Minerals and other Non-Metallic Materials, Pergamon, Oxford, 1979.
18 L. Fowden, R.M. Barrer and P.B. Pinker, Clay Minerals: their Structure, Behavior and Use, The Royal Society, London, 1984.
19 A.C.D. Newman (Ed.), Chemistry of Clays and Clay Minerals, Longman Scientific and Technical, Harlow, 1987.
20 see ref. 14, pp. 36-38.
21 F. Velghe, R.A. Schoonheydt and J. B. Uytterhoeven, Clays and Clay Min., 25 (1977) 375-380.
22 A. Maes, P. Peigneur and A. Cremers, in S.W. Bailey (Ed.), Proc. Int. Clay Conf., Mexico City, July 16-23, 1975, Applied Publishing, Wilmette, 1976, pp. 319-329.
23 G.N. White and L.W. Zelazny, Clays and Clay Min., 36 (1988) 141-146.
24 G. Lagaly and A. Weiss, in S.W. Bailey (Ed.), Proc. Int. Clay Conf., Mexico City, july 16-23, 1975, Applied Publishing, Wilmette, 1976, pp. 157-172.
25 M.S. Stul and W.J. Mortier, Clays and Clay Min., 22 (1974) 391-396.
26 A. Maes, M.S. Stul and A. Cremers, Clays and Clay Min., 27 (1979) 387-392.
27 D.A. Laird, A.D. Scott and T.E. Fenton, Clays and Clay Min., 37 (1989) 41-46.
28 M.S. Stul and L. Van Leemput, Surface Technology, 16 (1982) 89-100; 101-112.
29 see ref. 19, pp. 245-247.
30 see ref. 19, pp. 203-224.
31 R. Kjellander, S. Marcelja and J.P. Quirk, J. Colloid and Interface Sci., 126 (1988) 194-211.
32 G. Sposito and R. Prost, Chem. Rev., 82 (1982) 553-573.
33 J.J. Fripiat, J. Cases, M. Francois and M. Letellier, J. Colloid and Interface Sci., 89 (1982) 378-400.
34 J.D. Russell and A.R. Fraser in M.J. Wilson (Ed.), Clay Mineralogy: Spectroscopic and Chemical Determinative Methods, Chapmann & Hall (1994) pp 11-64.
35 V.C. Farmer in V.C. Farmer (Ed.), The Infrared Spectra of Minerals, Mineralogical Society, London, 1974, p. 331.
36 H. Suquet, R. Prost and H. Pezerat, Clay Min., 17 (1982) 231- 241.
37 G. Sposito, R. Prost and J.-P. Gaultier, Clays and Clay Min., 31 (1983) 9-16.
38 J.G. Thompson, Clay Min., 19 (1984) 229-236.
39 M. Lipsicas, R.H. Raythatha, T.J. Pinnavaia, I.D. Johnson, R.F. Giese, Jr., P.M. Costanzo and J.-L. Robert, Nature, 309 (1984) 604-607.
40 R.A. Kinsey, R.J. Kirkpatrick, J. Hower, K.A. Smith and E. Oldfield, Am. Min., 70 (1985) 537-548.
41 S. Kormarneni, C.A. Fyfe, G.J. Kennedy and H. Strobl, J. Am. Ceram. Soc., 69 (1986) C45-C47.
42 C.A. Weiss, Jr., S.P. Altaner and R.J. Kirkpatrick, Am. Miner., 72 (1987) 935-942.
43 J. Sanz and J.M. Serratosa, J. Am. Chem. Soc., 106 (1984) 4790-4793.
44 C.P. Herrero, J. Sanz and J.M. Serratosa, Solid State Comm., 53 (1985) 151-154.

45 A.R. Gonzales - Elipe, J.P. Espinos, G. Munuera, J. Sanz J.M. Serratosa, J. Phys. Chem., 92 (1988) 3471-3476.
46 J.T. Kloprogge, Ad. M.J. van der Eerden, J. Ben H. Jansen and J.W. Geus, Geologie en Mijnbouw 69 (1990) 351-357.
47 Nadeau, P.H., V.C. Farmer, W.J. McHardy and D.C. Bain, Am. Mineral. 70 (1985) 1004-1010
48 D. W. Woesnner, Amer. Mineral. 74 (1989) 203-215.
49 J.P. Rupert, W.T. Granquist and T.J. Pinnavaia, in A.C.D. Newman (Ed.), The Chemistry of Clays and Clay Min., Mineralogical Society, London, 1987, pp. 275-318.
50 D. Tilak, B. Tennakoon, J.M. Thomas, W. Jones, T.A. Carpenter and S. Ramdas, J. Chem. Soc. Faraday Trans. I, 82 (1986) 545-562.
51 M.M. Mortland, J.J. Fripiat, J. Chaussidon and J. Uytterhoeven, J. Phys. Chem., 67 (1963) 248-258.
52 see ref. 16, pp. 261-269.
53 M.M. Mortland and H.E. Doner, USP 3,78,457.
54 Y. Soma, M. Soma and I. Harada, J. Phys. Chem., 88 (1984) 3034-3038.
55 Y. Soma, M. Soma and I. Harada, J. Contaminant Hydrology, 1 (1986) 95-106.
56 R.M. Barrer, in L. Fowden, R.M. Barrer and P.B. Pinker (Eds.), Clay Minerals: their Structure, Behavior and Use, The Royal Society, London, 1984, pp. 113-132.
57 R.M. Barrer, Clays and Clay Min., 37 (1989) 385 - 395.
58 G. Lagaly, in L. Fowden, R.M. Barrer and P.B. Pinker (Eds.), Clay Minerals: their Structure, Behavior and Use, The Royal Society, London, 1984, pp. 95-112.
59 R.A. Schoonheydt, J. Pelgrims, Y. Heroes and J.B. Uytterhoeven, Clay Min., 13 (1978) 435-438.
60 A. Yamagishi and R. Ohnishi, Inorg. Chem., 21 (1982) 4233-4236.
61 A. Yamagishi, J. Chem. Soc. Dalton Trans., (1983) 679-681.
62 R. Burch (Ed.), Pillared Clays, Elsevier, Amsterdam, Catalysis Today, 2 (1988) 185 - 367.
63 J.R. Butruille and T.J. Pinnavaia in G. Alberti and T. Bein (Eds.), Comprehensive Supramolecular Chemistry, Vol 7: Solid-state Supramolecular Chemistry: Two- and Three-dimensional Inorganic Networks, Pergamon Press, Oxford, UK, 1996, pp. 219-250.
64 R. Szostak and C. Ingram in H.K. Beyer, H.G. Karge, I. Kirisci and J.B. Nagy (Eds.), Catalysis by Microporous Materials, Stud. Surf. Sci. Catal.,Vol 94, Elsevier Science, 1995, pp. 13-38.
65 T.J. Pinnavaia and H. Kim in E.G. Derouane, F. Lemos, C. Naccache and F. Ramao-Ribeiro (Eds.), Zeolite Microporous Solids: Synthesis, Structure and Reactivity, NATO ASI Series C, vol. 352, Kluwer Academic, Dordrecht, 1992, pp. 79-90.
66 T.J. Pinnavaia, T Kwon and S.K. Yun in E.G. Derouane, F. Lemos, C. Naccache and F. Ramao-Ribeiro (Eds.), Zeolite Microporous Solids: Synthesis, Structure and Reactivity, NATO ASI Series C, vol. 352, Kluwer Academic, Dordrecht, 1992, pp. 91-104.
67 J. Shabtai and J. Fijal, USP 4 579 832 (1986).
68 J. Sterte and J. Shabtai, Clays and Clay Min. 35 (1987) 429 - 439.

69 D.E.W. Vaughan, USP 4 666 877 (1987).
70 J.R. McCauley, WO 88/06488
71 W.Y. Lee, R.H. Raythatha and B.J. Tatarchuk, J. Catalysis, 115 (1989) 159 - 179.
72 C.G. Manos, Jr., M.M. Mortland and T.J. Pinnavaia, Clays and Clay Min.,32 (1984) 93-98.
73 R.M. Lewis, K.C. Ott and R.A. Van Santen, USP 4,510,257 (1985).
74 B.M. Choudary, K. Rair Kumar, Z. Jamil and G. Thyagarajan, J. Chem. Soc. Chem. Comm., (1985) 931-932.
75 S.P. Christiano, J. Wang and T.J. Pinnavaia, Inorg. Chem., 24 (1985) 1222-1227.
76 A. Galarneau, A. Barodawalla and T.J. Pinnavaia, Nature, 374 (1995) 529-531.
77 I.D. Johnson, T.A. Werpy and T.J. Pinnavaia, J. Am. Chem. Soc., 110 (1988) 8545-8547.
78 V. Seefeld, R. Bertram, P. Starke and W. Gessner, Silikattechnik, 39 (1988) 239-241.
79 V. Seefeld, R. Bertram, H. Görz, W. Gessner and S. Schönherr, Z. Anorg. Allg. Chemie, 603 (1991) 129-135.
80 V. Seefeld, R. Bertram, D. Müller and W. Gessner, Silikattechnik, 42 (1991) 305-308.
81 J.T. Kloprogge, D. Seykens, J.B.H. Jansen and J.W. Geus, J Non-Crystalline Solids, 142 (1992) 94-102.
82 J.T. Kloprogge, D. Seykens, J.W. Geus and J.B.H. Jansen, J Non-Crystalline Solids, 142 (1992) 87-93.
83 J.T. Kloprogge, D. Seykens, J.B.H. Jansen and J.W. Geus, J Non-Crystalline Solids, 152 (1993) 207-211.
84 J.T. Kloprogge, D. Seykens, J.W. Geus and J.B.H. Jansen, J Non-Crystalline Solids, 160 (1993) 144-151.
85 G. Furrer, C. Ludwig and P.W. Schindler, J. Colloid and Interface Sci, 149 (1992) 56-67.
86 W.O. Parker,jr and I.Kiricsi, Applied Catalysis A: General, 121 (1995) L7-L11.
87 J. Fijal and Z. Klapyta in J. Konta (Ed.), Proc. 11th Conf. Clay Min. & Petrol., Univ. Karlova, Praha, 1993, 99-109.
88 R.A. Schoonheydt, H. Leeman, A. Scorpion, I. Lenotte and P. Grobet, Clays and Clay Min., 42 (1994) 518-525.
89 R.A. Schoonheydt, J. Van den Eynde, H. Tubbax, H. Leeman, M. Stuykens, I. Lenotte and W.E.E. Stone, Clays and Clay Min., 41 (1993), 598-607.
90 L. Bergaoui, J.F. Lambert, R. Franck, H. Suquet and J.L. Robert, J. Chem. Soc. Faraday Trans., 91 (1995) 2229-2239.
91 D. Plee, F. Borg, L. Gatineau and J.J. Fripiat, J. Am. Chem. Soc., 107 (1985) 2362-2369.
92 T.J. Pinnavaia, S.D. Landau, M.-S. Tzou, I.D. Johnson and M. Lipsicas, J. Am. Chem. Soc., 107 (1985) 7222-7224.
93 K. Bukka, J.D. Miller and J. Shabtai, Clays and Clay Min.,40 (1992) 92-102.
94 J. Fijal, Z. Klapyta, K. Dyrek, S. Witkowki and G. Djega-Mariadassou, Proc. 7th Euroclay Conf. Dresden, 1991, 347-352.

95 S. Witkowski, K. Dyrek, Z. Sojka, G. Djega-Mariadassou, J. Fijal and Z. Klapyta, Clay Minerals, 29 (1994) 743-749.
96 S. Chevalier, R. Franck, J.F. Lambert, D. Barthomeuf and H. Suquet, Clay Minerals, 27 (1992) 245-248.
97 M.L. Occelli, J.V. Senders and J. Lynch, J. Catalysis, 107 (1987) 557-565.
98 M.L. Occelli, B. Drake and S.A.C. Gould, J. Catalysis 142 (1993) 337-348.
99 M.L. Occelli,.S.A.C. Gould and B. Drake, Microporous Materials, 2 (1994) 205-215.
100 H. Hartman, G. Sposito, A. Yang, S. Manne, S.A.C. Gould and P.K. Hansma, Clays and Clay Min., 38 (1990) 337-342.
101 M.L. Occelli, J.V. Senders and J. Lynch, J. Catalysis, 107 (1987) 557-565.
102 M.J. Remy, A.C. Vieira Coelho and G. Poncelet, Microporous Materials, 7 (1996), 287-297.
103 A. Gil, L. Guia, P. Grange and M. Montes, J. Phys. Chem., 99 (1995) 301-312.
104 N. Maes, H.Y. Zhu and E.F. Vansant, J. Porous Materials, 2 (1995) 97-105.
105 H.Y. Zhu and E.F. Vansant, J. Porous Materials, 2 (1995) 107-113.
106 H.Y. Zhu, N. Maes and E.F. Vansant in J. Rouquerol, F. Rodriquez-Reinoso, K.S.W. Sing and K.K. Unger (Eds.), Characterization of Porous Solids III, Stud. Surf. Sci. Catal., Vol 87, Elsevier Science, 1994, 457-466.
107 L. Bergaoui, J.F. Lambert, M.A. Vincente-Rodriguez, L.J. Michot and Villiéras, Langmuir, 11 (1995) 2849-2852.
108 W.L. Ijodo, T. Lee and T.J. Pinnavaia, Adv. Materials, 8 (1996) 79-83.
109 K. Urabe, N. Kouno, H. Sakurai and Y. Izumi, Adv. Materials, 3 (1991) 558-561.
110 F. Figuéras, Z. Klapita, P. Massiani, Z. Mountassu, D. Tichit, F. Fajula, C. Gueguen, J. Bousquet and A. Auroux, Clays and Clay Min., 38 (1990),257-264.
111 I. Heylen, C. Vanhoof and E.F. Vansant, Microporous Materials, 5 (1995) 53-60.
112 E. Montarges, L.J. Michot, F. Lhote, T. Fabien and F. Villieras, Clays and Clay Min., 43 (1995) 413-426.
113 J. Guan and T.J. Pinnavaia, Materials Sci. Forum, 152-153 (1994) 109-113.
114 E. Booij J.T. Kloprogge, J.A.R. van Veen, Appl. Clay Sci., 11 (1996) 155-162.
115 S. Inagaki, Y. Fukushima and K. Kuroda in J. Weitkamp, H.G. Karge, H. Pfeifer, W. Hölderich (eds.), Stud. Surf. Sci. Catal., vol. 84A, Elsevier, Amsterdam, 1994, pp 125-132.
116 S. Inagaki, Y. Fukushima, A. Okada, T. Kurauchi, K. Kuroda and C. Kato in R. von Ballmoos, J.B. Higgins and M.M.J. Treacy (eds.), Proc. 9th International Zeolite Conference, Butterworth Heinemann, Boston, 1993, pp. 305-311.
117 S. Inagaki, Y. Yamada and Y. Fukushima in H. Chon, S.-K. Ihm and Y.S. Uh, (eds.), Progress in Zeolite and Microporous Materials, Stud. Surf. Sci. Catal. vol 105A, 1997, 109-116
118 M.L. Occelli, P.A. Peaden, G.P. Ritz, P.S. Iyer and M. Yokoyama, Microporous Materials 1 (1993), 99-113.
119 R. Mokaya and W. Jones, J. Porous Materials, 1 (1997) 97-110.
120 R. Mokaya and W. Jones, J. Catalysis, 153 (1995) 76-85.
121 J. Bovey and W. Jones, J. Mater Chem, 5 (1995) 2027-2035.
122 H. Sakurai, K. Urabe and Y. Izumi, Bull. Chem. Soc. Jpn, 64 (1991) 227-235.
123 H. Sakurai, K. Urabe and Y. Izumi, J. Chem. Commun., (1988) 1519-1520.

124 T. J. Bandosz, J. Jagiello and J.A. Schwarz, J. Phys. Chem., 99 (1995) 13522-13527.
125 J.B. d'Espinose de la Caillerie and J.J. Fripiat, Clays and Clay Min., 39 (1991) 270-280.
126 K. Urabe, I. Kenmoku and Y. Izumi, J. Phys. Chem. Solids, 57 (1996) 1037-1041.
127 K. Urabe, I. Kenmoku, K. Kawabe and Y. Izumi, J. Chem. Soc., Chem. Comm., (1991) 867-869.
128 P. Cool and E.F. Vansant, Microporous Materials, 6 (1996) 27-36.
129 N. Maes, I. Heylen, P. Cool and E.F. Vansant, Applied Clay Science, 12 (1997) 43-60.
130 N. Maes and E.F. Vansant, Microporous Materials, 4 (1995) 43-51.
131 M.A. Martin-Luengo, H. Martins-Carvalho, J. Ladriere and P. Grange, Clay Minerals, 24 (1989) 495-504.
132 T. Mishra, K.M. Parida and S.B. Rao, J. Colloid and Interface Sci., 183 (1996) 176-183.
133 C. Volzone, Clays and Clay Min., 43 (1995) 377-382.
134 M. Sychev, N. Kostoglod, E.M. van Oers, V.H.J. de Beer R.A. van Santen, J. Kornatowski and M. Rozwadowski in H.K. Beyer, H.G. Karge, I. Kirisci and J.B. Nagy (Eds.), Catalysis by Microporous Materials, Stud. Surf. Sci. Catal.,Vol 94, Elsevier Science, 1995, pp. 39-46.
135 S. Chevalier, R. Franck, H. Suquet, J.F. Lambert and D. Barthomeuf, J. Chem. Soc. Faraday Trans., 90 (1994) 667-674.
136 S.A. Bagshaw and R.P. Cooney, Chem. Mater., 5 (1993) 1101-1109.
137 S. Moreno, R. Sun Lou and G. Poncelet, J.Catalysis, 162 (1996) 198-208.
138 R. Molina, S. Moreno, A. Vieira-Coelho, J.A. Martens, P.A. Jacobs and G. Poncelet, J. Catalysis, 148 (1994) 304-314.
139 S. Moreno, R. Sun Lou and G. Poncelet, J. Phys. Chem B, 101 (1997) 1569-1578.
140 L.S. Cheng and R.T. Yang, Ind. Eng. Chem. Res. 34 (1995) 2021-2028.
141 L.S. Cheng and R.T. Yang, Microporous Materials, 8 (1997) 177-186.
142 A. Molinard and E.F. Vansant, Adsorption, 1 (1995) 49-59.
143 P.B. Malla and S. Komarneni, Clays and Clay Min., 38 (1990) 363-372.
144 S. Yamanaki, P.B. Malla and S. Komarneni, J. Colloid and Interface Sci., 134 (1991) 51-58.
145 H.Y. Zhu, Z.H. Hu, N. Maes and E.F. Vansant in E.F. Vansant (ed.), Separation Technology, Elsevier science, 1995, pp.715-726.
146 T.G. Danis, T.A. Albanis, D.E. Petrakis and P.J. Pomonis in A. Copin, G. Houins, L. Pussemier and J.F. Salembier (Eds.), Proc. 5th Int. Symp Environmental Behaviour of Pesticides and Regulatory Aspects II, 1994, pp. 148-151.
147 L Bergaoui, J.F. Lambert, H. Suquet and M. Che, J. Chim. Phys., 92 (1995) 1486-1505.
148 J. Miehé, R. Le Dred, J. Baron and D. Saehr in M.L. Ocelli and H. Kessler, Synthesis of Porous Materials, Marcel Dekker Inc., 1997, pp. 523-526.

Studies in Surface Science and Catalysis 137
H. van Bekkum, E.M. Flanigen, P.A. Jacobs and J.C. Jansen (Editors)
© 2001 Elsevier Science B.V. All rights reserved.

Chapter 8

Techniques of zeolite characterization

A. Jentys and J. A. Lercher

Technische Universität München, Lehrstuhl für Technische Chemie II,
Lichtenbergstr. 4, D-85748 Garching, Germany.

8.1 INTRODUCTION

Information on the structural, chemical and catalytic characteristics of zeolites is essential for deriving relations between their chemical and physicochemical properties on the one side and the sorptive and catalytic properties on the other. Such relations are of high importance, as they allow the rational development of sorbents, catalysts and advanced structural materials.

Individual analysis techniques typically probe only a particular aspect of the material and, consequently, a combination of methods is necessary to give a balanced description of the frequently complex zeolite. This places a high experimental demand upon a chemist or physicist who seeks to use and improve a microporous material for a dedicated purpose. Of the three main uses of zeolites, as cation-exchangers, sorbents/hosts and catalysts, only characterization with respect to the latter two applications are being dealt with in depth in this chapter.

In general, the characterization of a zeolite has to provide information about (i) its structure and morphology, (ii) its chemical composition, (iii) its ability to sorb and retain molecules and (iv) its ability to chemically convert these molecules. A major difficulty arises from the fact that the latter two are surface chemical properties. The micropore surface of a zeolite is, however, a periodic three-dimensional intrinsic property of the material and, thus, cannot be straightforwardly probed with surface sensitive techniques. Therefore, many of the approaches chosen rely on bulk characterization techniques and focus on the description of the interaction of the zeolite with atomic or molecular probes. Typical analytical approaches and the probed chemical or structural information are summarized in Table 8.1. Excellent reviews on these techniques exist for the individual methods [1-7] that go, in scope and depth, far beyond the current overview.

Here, we focus on an outline of a series of characterization methods that should provide a guideline for characterizing molecular sieves. Various techniques will be briefly described and their advantages and limitations will be discussed using specific examples whenever possible. At the end of the chapter we will outline a minimum set of experiments, which should be performed after activation of the materials and reported in publications for each of the main uses of molecular sieves.

Table 8.1: Compilation of characterization techniques

Technique	Synonym	Structure	Pore size	Chemical composition	Functional groups
Electron microscopy	HRTEM SEM EDX	•	•	•	
Nuclear magnetic resonance (Magic angle spinning)	MAS-NMR		•	•	•
Sorption of probe molecules			•		•
Model reactions			•	•	•
Thermogravimetry, Differential scanning calorimetry,	TGA/DSC		•	•	•
Temperature programmed desorption	TPD			•	•
Vibrational spectroscopy		•	•	•	•
X-ray absorption spectroscopy	XAS	•		•	
X-Ray diffraction	XRD	•		•	
X-ray fluorescence spectroscopy	XRF			•	
X-ray photoelectron spectroscopy	XPS			•	•

8.2. LATTICE AND PORE STRUCTURE, PORE SIZE DISTRIBUTION AND STRUCTURAL INTEGRITY

In principle, three regions of interfaces in a molecular sieve can be identified, i.e., the outer surface terminating an individual crystal, the pore entrance (pore mouth) and the inner surface of the pore walls. While the first two originate from the termination of the three-dimensional lattice, the latter surface is part of the crystalline structure and, therefore, it is possible to describe the pore size, geometry and the idealized surface corrugation by characterizing the bulk structure of a zeolite. In order to arrive at a more realistic description of the pores and their surface chemistry, it is, however, necessary to use atomic or molecular probes and to determine their sorption and interaction in the pores. These interactions and their relation to the equilibrium pressure can be monitored by various techniques, which will also be outlined in this chapter.

8.2.1 X-Ray diffraction (XRD)

For determining the long-range order and the phase purity of zeolites, X-ray diffraction is the most frequently used technique. For most of the applications, the investigated materials are polycrystalline and, therefore, only powder X-ray diffraction techniques will be covered here. For samples available as larger crystals or in a crystal morphology that favors a preferred orientation, it has to be assured that all individual crystallites are randomly oriented when analyzed, which can be achieved by, e.g., careful sample grinding.

In polycrystalline materials each reflection observed at a given angle 2Θ in a powder X-ray diffraction pattern, measured at the wavelength λ, is related to lattice planes with a distance d and an orientation indicated by the Millers indices hkl as described by the Bragg's law:

$$n\lambda = 2\,d_{(hkl)}\sin\Theta$$

For crystals with cubic symmetry (e.g., faujasite) the size of the unit cell a_0 can be determined from the angular positions of the reflections.

$$a_0 = \frac{\lambda\sqrt{h^2+k^2+l^2}}{2\sin\Theta}$$

The unique structure of the different zeolites, in terms of atom positions and unit cell dimensions, is reflected in characteristic positions and relative intensities of the peaks observed. The XRD patterns of known structure types of zeolites and other microporous molecular sieves are compiled in the "Collection of Simulated XRD Patterns for Zeolites" [8] as well as in the online data basis of the IZA [9].

In the simplest case, the structure of a particular zeolite sample can be identified by comparing its experimentally measured powder diffractogram with reference diffractograms. Also combinations of zeolite structures can be identified in this way. Before applying qualitative measurements, it is important to compare the concentration and location of cations at the exchange positions in the samples and the reference, as differences may lead to additional reflections and/or different intensities of diffraction peaks. Also changes in the size of the unit cell resulting from differences in the framework composition (see section 8.4.1 for determination of the Si/Al ratio by XRD) and the presence of sorbed molecules inside the pores (e.g., template molecules) influence the positions and the intensity of the XRD peaks [4, 10].

The integrated intensity of selected reflections can be used to determine the crystallinity of the sample by normalizing their intensities to those of a standard sample. The crystallinity is defined according to the ASTM Standards on Catalysis [11] as:

$$\text{X-ray crystallinity} = \frac{\text{Intensity of peak}_{(\text{hkl})}\ \text{sample}}{\text{Intensity of peak}_{(\text{hkl})}\ \text{standard}} \times 100\%$$

The full width at half maximum B of diffraction reflections can be used to estimate the average crystallite size t in [nm] according to [12]:

$$t = \frac{0.9 \times \lambda}{B \cos\Theta}$$

The structure of novel microporous molecular sieves can be deduced and refined from high quality X-ray powder diffraction data, typically collected using X-ray sources of high brilliance such as synchrotrons, by applying the Rietveld analysis method [13, 14]. This method is based on the full profile fitting of the powder diffractogram, which can be considered as a superposition of individual, partially overlapping *hkl*-reflection profiles. Starting from a theoretical model of the structure, the diffraction pattern is calculated and compared to the data observed. This method follows a least square refinement in order to minimize the difference between the calculated and observed diffractograms. The quality of the refinement is given in values called *R-factors*. Here, the R-factor R_{WP} (WP=weighted profile) is given as an example:

$$R_{WP}^2 = \frac{\sum \left[\left| I_{i(obs)} \right| - \left| I_{i(calc)} \right| \right]^2}{\sum \left| I_{i(obs)} \right|^2}$$

where $I_{i(obs)}$ and $I_{i(calc)}$ are the observed and calculated intensities at a defined point i. As the system has a high degree of freedom with respect to the parameters that can be varied during the fitting process (e.g., atomic coordinates, unit cell parameters, parameters of the profile function, thermal coefficients, preferred orientation), a reasonably good starting model of the zeolite structure is needed. An example of the structure refinement for zeolite NW-3 is shown in Fig. 8.1 (adapted from ref. [15]).

X-Ray diffraction utilizes the periodicity of the crystal lattice and, thus, can be used to deduce information about crystal structure, crystallite size, location and concentration of exchanged cations and crystallinity of a zeolite sample. It is, however, conceptually not possible to obtain information on non-periodic properties of the crystals such as lattice defects, stacking faults and hydroxyl groups resulting from an incomplete condensation of Si-O-Si linkages during the synthesis. In addition, crystalline phases in concentrations below ~ 5% and amorphous phases cannot be detected [16].

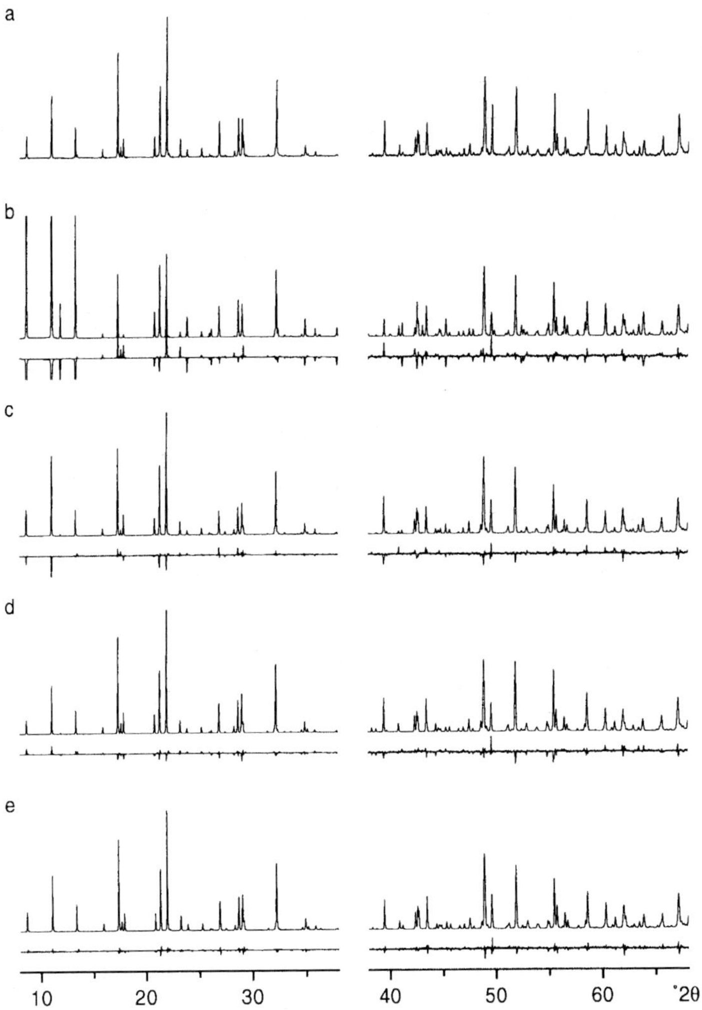

Figure 8.1: Example for Rietveld refinement of zeolite NW-3: (a) data observed, (b) initial model with framework atoms, (c) C and N positions added, (d) H positions simulated (e) final model, all parameters refined. (The scale for the second half of the profiles has been increased by a factor of 5). After ref [15].

8.2.2 (High resolution) Transmission electron microscopy

In TEM, a high intensity primary electron beam (typically 100-200 keV) passes through the sample and the transmitted and diffracted electrons are collected [17]. Therefore, the sample has to be placed in vacuum (< 10^{-8} torr) and needs to be very thin (10–500 Å). A two-dimensional projection of the sample is generated from the attenuation of the beam that depends on the density and thickness of the sample. If an objective aperture is used that allows only the transmitted electrons to contribute, but excludes all diffracted electrons, the so-called bright field image is generated. For low magnification views this imaging mode is preferred. If only diffracted electrons are collected, which are slightly off axis from the transmitted beam, while the undiffracted central beam is excluded by the aperture, a dark field image is obtained. For supported catalysts, dark field images show an enhanced contrast for the metal particles. The high resolution transmission electron microscopy (HRTEM) technique is performed with axial illumination using an objective aperture, which is larger and, therefore, allows to combine several diffracted beams with the axial transmitted beam to form the image. The size of the aperture determines which of the diffracted beams contribute to the image, since spatial frequencies below the resolution limit only contribute to the image noise and, thus, reduce the contrast. Images obtained under these conditions can be directly related to the atomic structure of the material. However, careful image processing is often required to enhance the contrast.

While transmission electron microscopy provides high-resolution information of the sample, the three-dimensional information is lost due to the projection into a two-dimensional image. However, the recent development of electron tomography, which reconstructs a three-dimensional picture from a series of two-dimensional TEM images collected at various tilt angles of the sample, allows the generation of a three dimensional picture of a zeolite particle (3D-TEM) [18]. Fig. 8.2 shows electron micrographs of a steamed and acid-leached crystal of faujasite. A conventional TEM image (Fig. 8.2 a) displays a projection of the crystal on a plane. The mesopores can be deduced from the scattering contrast variation but cannot be viewed separately. In Fig. 8.2 b a thin slice from a reconstructed 3D-image is shown. Now the mesopores can be viewed with a clarity that allows detailed studies, e.g., of the mechanism of formation.

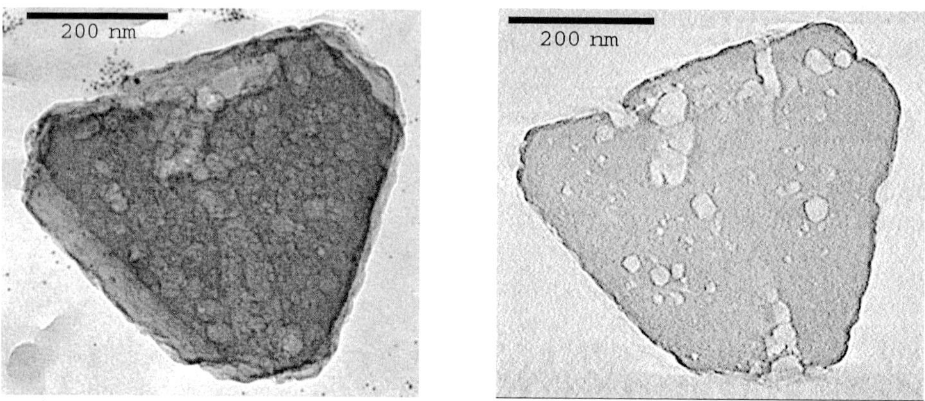

a) b)

Figure 8.2: 2D-TEM image of a steamed and acid-leached faujasite crystal taken at a magnification of 20 000 (a) and a thin (1.25 nm) slice through the 3D-TEM reconstruction of this crystal (b). Small gold particles (5 nm), used as markers, are observed in (a). (Courtesy K. P. de Jong, Utrecht University, The Netherlands)

8.2.3 Vibrational spectroscopy of zeolite framework vibrations

The vibration frequencies of the zeolite lattice, which result from stretching and bending modes of the T-O units, are observed in the range between 200 and 1500 cm^{-1} [19, 20]. In principle IR and Raman spectroscopy can be used to monitor these vibrations. Different selection rules apply for each method and, therefore, the transition probabilities of the vibration modes observed are different, which leads to conceptual differences in the spectra with respect to the intensity of the bands. As the underlying principle of the lattice vibrations, however, is the same, these techniques will be discussed together in this section. The main advantage of Raman spectroscopy over IR-Spectroscopy for studying zeolite lattice vibrations is that zeolites and other oxides such as silica, alumina and titania show strong IR absorption in the spectral region below 1200 cm^{-1}. Additionally, due to the excitation with lasers having typically a wavelength between 488 and 785 nm in Raman spectroscopy, the experimental setup and the reactions conditions can be varied over a wide range [21]. Disadvantages of Raman spectroscopy are that on some samples a strong fluorescence is observed, which can be overcome by varying the wavelength of the laser used for excitation and the generally lower sensitivity compared to IR-spectroscopy.

Another technique to study the bending modes of the T-O units and of adsorbed molecules is inelastic neutron scattering (INS) [22]. With the present neutron spectrometers a spectral range of 1 to 4000 cm^{-1} can be covered. Compared to IR and Raman spectroscopy, neutrons can penetrate optically opaque materials and, due to the large incoherent cross section of protons, INS is very sensitive to hydrogen.

In general, the lattice modes can be divided into structure sensitive and structure insensitive vibrations (summarized in Table 8.2). The position of the strong vibration around 500 cm^{-1}, resulting from mixed stretching and bending modes, strongly depends on the T-O-T angles and, therefore, was used to characterize the aluminum content of the lattice. For Al-free faujasite, two sharp bands at 488 and 510 cm^{-1}, corresponding to T-O-T angles of 141 and 147°, respectively, were observed in the Raman spectra [23]. These bands were significantly broadened and gradually shifted to 516 and 467 cm^{-1} with decreasing Si/Al ratio due to the different length of the Al-O (1.73 Å) and Si-O (1.62 Å) bonds. For ZSM-5 zeolites, a correlation between the position of the lattice vibrations observed at 544 cm^{-1} (ZSM-5 lattice deformation) and 790 cm^{-1} (T-O-T stretching vibration) and the Al content was reported [24]. (see Fig. 8.3).

Table 8.2: Structure sensitive and insensitive lattice vibrations of zeolites

Structure insensitive vibrations	**wavenumber [cm^{-1}]**
Asymmetric stretching vibrations	1200 - 1000
Symmetric stretching vibrations	850 - 700
Bending vibrations	600 - 400

Structure sensitive vibrations	**wavenumber [cm^{-1}]**
Asymmetric stretching vibrations	1050 – 1150
Symmetric stretching vibrations	750 - 820
Double ring vibrations	500 - 650
Pore opening vibrations	300 - 420

At frequencies below 300 cm^{-1} stretching vibrations of the charge balancing cations relative to the zeolite lattice are observed. The frequencies of these vibrations were shown to depend on the charge, mass and location of the cations in the zeolite lattice. For example for Na^+ in faujasite bands at 160, 106, 188 and 90 cm^{-1} were observed, which were assigned to cations in S_I, $S_{I'}$, S_{II} and S_{III} positions, respectively [25].

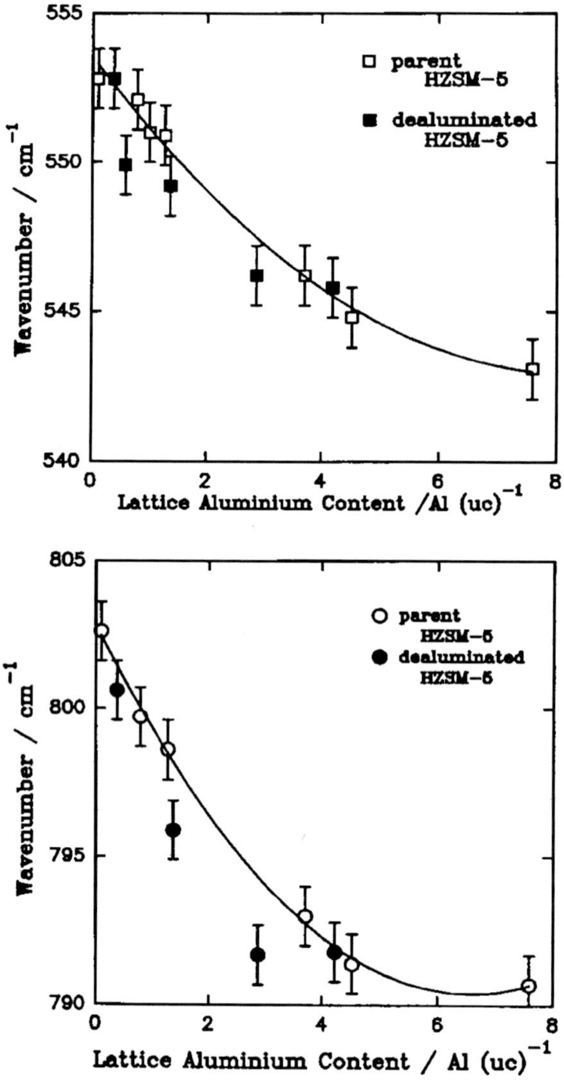

Figure 8.3: Frequencies of the framework vibrations of ZSM-5 as function of the Al content.

8.2.4 Determination of sorption isotherms

According to the IUPAC classification, the diameter of micropores ranges from 0.3 to 2 nm, of mesopores from 2 to 50 nm and of macropores from 50 nm up to 100 µm [26]. Using small noble gases (Ar) or small molecules (N_2), the total pore volume and the distribution between macro-, meso- and micro-pores can be determined. With this method it is, however, impossible to accurately differentiate between the size of micropores, which is usually achieved by measuring the sorption isotherms of molecules with increasing minimum kinetic diameter.

8.2.4.1 Sorption of noble gases and small molecules

The sorption isotherms of relatively inert gases such as N_2 and Ar, typically measured at the condensation temperature of the corresponding gas, reflect the size and shape of the pores [27]. These sorption isotherms can be classified into six principal types [28] as shown in Fig. 8.4. The characteristics and examples of the different types of isotherms are summarized in Table 8.3.

Table 8.3: Type and characteristics of sorption isotherms

Type of isotherm	Characteristics	Example
I	Adsorption in micropores	Benzene on zeolites
II	Multilayer adsorption on flat surfaces	N_2 on SiO_2, Al_2O_3
III	Weak gas-solid interactions	H_2O on noble metals
IV	Multilayer adsorption and pore condensation	N_2 on MCM-41 type materials
V	Weak gas-solid interactions	H_2O on noble metals
VI	Non porous adsorbent	Kr on graphite

From type II and type IV isotherms, the specific surface area of the adsorbents can be determined by applying the BET equation [29].

$$\frac{p}{(p_0 - p)V} = \frac{1}{cV_{mon}} + \frac{(c-1)\,p}{cV_{mon}\,p_0}$$

where V is the volume of gas adsorbed, V_{mon} the volume of gas adsorbed in a monolayer, p_0 the vapor pressure of the gas, c is proportional to $exp(-(\Delta H_d - H_{vap})/RT)$, ΔH_d is the enthalpy of adsorption in the first layer and H_{vap} the heat of vaporization. The BET surface area is determined from the correlation of p/p_0 *vs.* $p/V(p_0-p)$ using the pressure range $0.05 < p/p_0 < 0.3$.

This method can only be applied to materials having meso- and macropores, as the concept of multilayer adsorption is not applicable, when the size of the molecules adsorbed and of the pores probed are in the same range. In microporous materials the adsorbed molecules fill the pores completely and, because of the limited space multilayer adsorption is suppressed. While it is, therefore, fundamentally meaningless, the BET surface area is frequently reported for zeolites (and other microporous materials) and can be used as a purely empirical value to compare the quality and porosity of materials of the same kind.

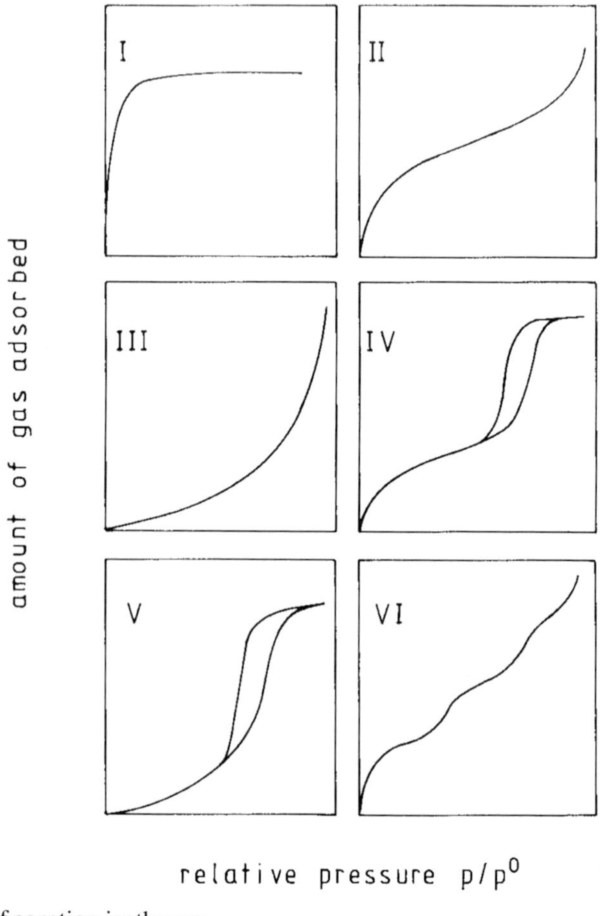

Figure 8.4: Types of sorption isotherms

The *t*-plot method can be applied for a quantitative determination of the micropore volume [30]. For a large number of non-porous solids the shape of the N_2 sorption isotherm can be represented with a single curve, by plotting the ratio V_a/A (V_a being the volume of N_2 adsorbed, A the surface area of the sample) as a function of the relative pressure (p/p_0). The resulting, so-called *t*-curve, which is in principle a BET type II isotherm, reflects the statistical thickness of the adsorbed N_2 multilayer and can be calculated according to:

$$t[nm] = 0.1\sqrt{\frac{13.99}{-\log\left(\frac{p}{p_0}\right) + 0.034}}$$

In the range of partial pressures from 0.1 to 0.8, this equation gives a good description of the course of the common *t*-curve.

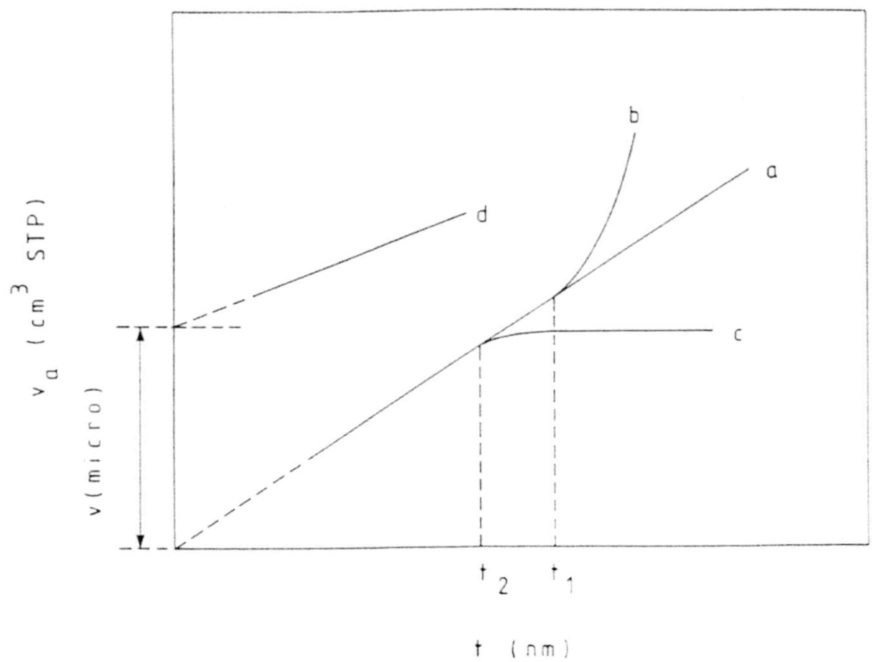

Figure 8.5: Different courses of the *t*-plot

Deviations of the *t*-plot from the ideal behavior allow the deduction of the nature of pores and to determine the micropore volume (see Fig. 8.5). Curve *a* in Fig. 8.5 is linear and extrapolates through the origin, which indicates an ideal *t*-material and, therefore, *A(t)* can be calculated from the tangent. Curve *b* shows a strong upward deviation from linearity at a certain pressure p/p_o, i.e., more N_2 is sorbed than might be expected, which indicates that starting from t_1 capillary condensation takes place in addition to adsorption. In contrast, if some narrow pores are filled by multilayer adsorption (curve *c*), further adsorption does not occur in this part, as that surface has become unavailable. At this point the *t*-plot begins to deviate downwards from the straight line (t_2 in Fig. 8.5). This situation is, for instance, encountered in the presence of slit-shaped pores.

The presence of micropores is indicated by a positive intercept at the y-axis as shown in line *d*. A straight line is found and the surface area *A(t)* can be calculated from its tangent. The positive intercept is caused by a relatively large nitrogen uptake at a very low *t*-values. Such a high N_2 uptake at very low pressure is due to the stronger adsorption in micropores and, hence, allows one to calculate the micropore volume V_{micro} from the intercept.

8.2.4.2 Sorption of molecules with varying kinetic diameter

The most direct way of assessing the distribution of the micropore size is to compare the sorption isotherms of molecules with different minimum kinetic diameters. For this purpose, the uptake of a series of sorbates with increasing minimal kinetic diameter on a solid is explored. The drop in the adsorbed amount with increasing size of the sorbate defines the minimum pore diameter of the tested solid. Usually, molecules ranging from N_2 to substituted aromatic compounds such as trimethylbenzenes are used as probes. The minimum kinetic diameter of various probe molecules and the diameter of the zeolite pores are compared in Fig. 8.6 [31]. In the presence of distinct functional groups, the use of *in situ* spectroscopy can serve to evaluate the fraction of sites located inside and outside the particles of the microporous material. The natural limit of pore sizes to be probed depends on the size and the vapor pressure of the sorbate.

Note that the interactions of molecules adsorbed in micropores can be very strong and, therefore, relaxations of the lattice and, consequently, changes in the effective pore diameter of the material can be observed, which also results in a strong temperature dependence of the accessibility [32].

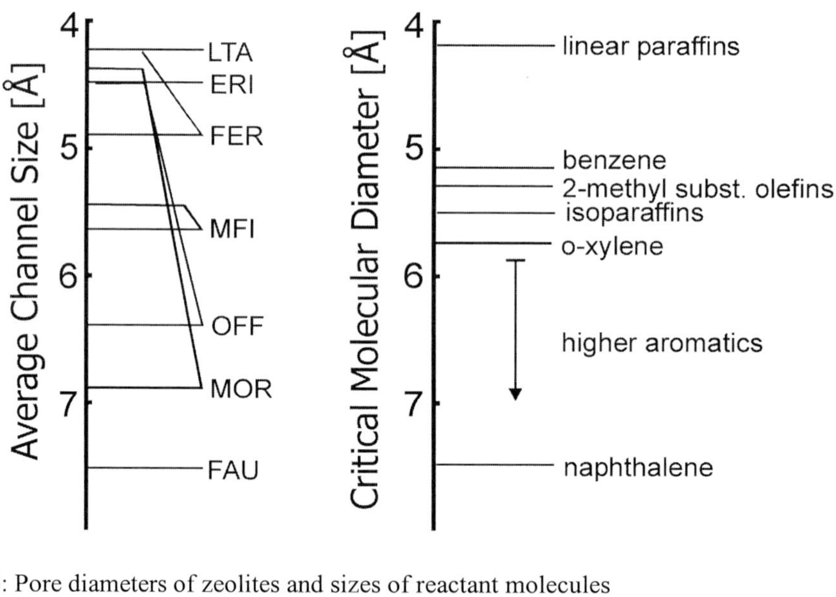

Figure 8.6: Pore diameters of zeolites and sizes of reactant molecules

8.2.5 Assessment of the heat of sorption of alkanes and enthalpy-entropy relationships

The effective pore size of molecules in zeolites can be assessed through the relation between the adsorption enthalpy and entropy of adsorbed alkanes shown by Eder *et al.* [33] using the adsorption of n-alkanes on a series of molecular sieves.

The initial adsorption enthalpy was measured directly by calorimetry, while the adsorption entropy was calculated from the adsorption constant measured independently by gravimetry. Fig. 8.7 shows the linear relation between the adsorption enthalpy and entropy. This indicates a compensation effect, i.e., for a given zeolite the stronger bound molecule also loses more entropy compared to the gas phase. The slope for the various molecular sieves is, however, markedly different. For the zeolite with the smaller pore size a higher entropy loss, i.e., the lower entropy of sorption for a given sorption enthalpy, is observed. This clearly emphasizes that the confinements of the micropore exert a larger influence on the entropic term of the sorption than on the enthalpic term, which results from the restricted spatial orientation of the n-alkanes in the micropores.

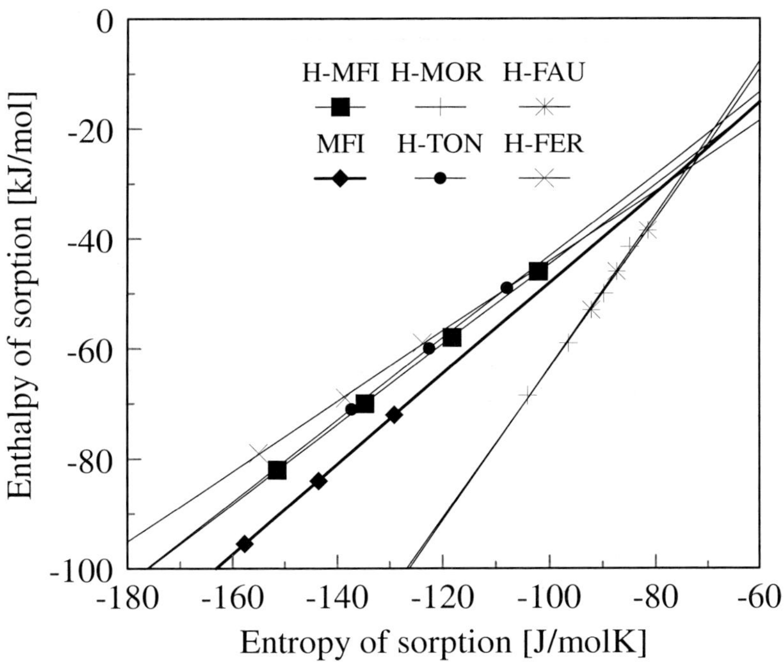

Figure 8.7: Sorption enthalpy and entropy for a series of n-alkanes on several zeolites

8.2.6 ^{129}Xe Nuclear magnetic resonance spectroscopy

The size distribution of micro- and mesopores in molecular sieves can indirectly been assessed from the shift in ^{129}Xe NMR of sorbed Xe [34]. Due to the size of Xe only pores larger than 4.36 Å can be probed. The shift of ^{129}Xe adsorbed in a porous material can be expressed as:

$$\delta = \delta_0 - \delta_e - \delta_m - \left(\delta_{(Xe-Z)} \cdot \rho_Z\right) - \left(\delta_{(Xe-Xe)} \cdot \rho_{Xe}\right)$$

The chemical shift in the ^{129}Xe NMR (δ) relative to the reference (δ_0), consists of contributions from the electric field (δ_e), the magnetic field (δ_m), the interaction of Xe with the zeolite ($\delta_{(Xe-Z)}\rho_Z$) and ($\delta_{(Xe-Xe)}\rho_{Xe}$) the interaction between the Xe atoms. ρ_Z is a characteristic parameter reflecting the zeolite structure and ρ_{Xe} is the concentration of Xe atoms in the pores.

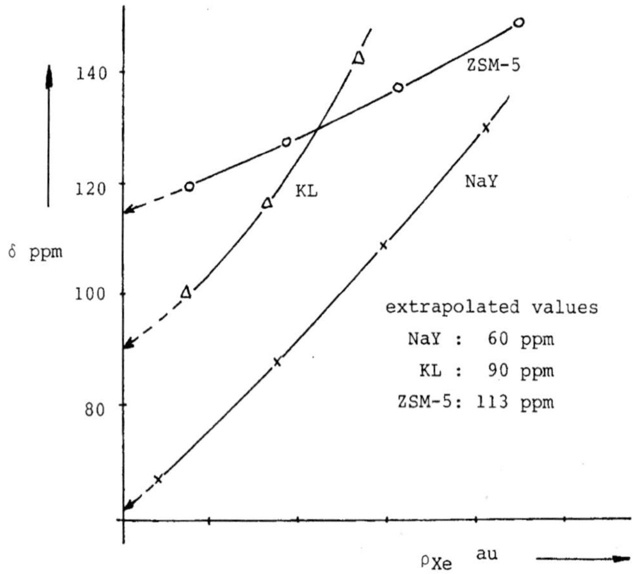

Figure 8.8: ^{129}Xe chemical shift as function of the ρ_{Xe} for three types of zeolites

At infinitely low Xe partial pressure the chemical shift is determined by the electric field term (δ_e) and the interactions between Xe and the zeolite ($\delta_{(Xe-Z)}$), which are increasing with decreasing channel or cavity size of the zeolite. Therefore, the extrapolation of the chemical shift to zero Xe concentration (see Fig. 8.8) allows to estimate the pore size of the zeolite. Note, that the chemical shift in the ^{129}Xe NMR increases with decreasing mean free path for the Xe atoms. The slope of the curves is an indication of the void space accessible to Xe, while the relative intensity of the signal can be related to the crystallinity of the material.

8.2.7 Model reactions

In addition to the spectroscopic characterization, model reactions are an effective tool to obtain information on the size of the pore system and the accessibility of reactant molecules to the active sites in the catalysts. The main advantage is that the information is obtained under conditions that are typical for the application of the materials in the catalytic reactions. In the following three typical model reactions to evaluate the void space of the zeolite will be described.

8.2.7.1 Constraint index (CI)

The Constraint Index is determined from the relative rates of n-hexane and 3-methylpentane cracking [35] and is defined as the selectivity observed during the cracking of a 1:1 mixture:

$$CI = \frac{k_{\text{n-hexane cracking}}}{k_{\text{3-methylpentane cracking}}}$$

The values of the Constraint Index assigned to different classes of pore diameters are summarized in Table 8.4, examples observed on various zeolites are compiled in Table 8.5.

For amorphous silica-alumina a CI of 0.6 was reported, which is in the same range as that of Y type zeolites, while for zeolites with smaller pores a significant increase was found. This method can be used for zeolites with medium to small pore sizes, while it is not sensitive enough to classify zeolites with pores consisting of 12 or more oxygen atoms.

Table 8.4: Classification of pore size by the Constraint Index

CI	Pore size classification
CI < 1	large pores
1 < CI < 12	intermediate pores
CI > 12	small pores

Table 8.5: Constraint index and pore sizes of zeolites

Molecular sieve	Constraint index	Number of oxygen atoms in the pore aperture
Erionite	38.0	8
ZSM-11	8.7	10
ZSM-5	8.3	10
Offretite	3.7	12 & 8
ZSM-12	2.0	12
Silica-Alumina	0.6	-
Mordenite	0.5	12 & 8
ReY	0.4	12

8.2.7.2 Refined Constraint Index (RCI)

The conversion of n-decane over a Pt loaded zeolite (1 wt%) in its protonic form (bifunctional catalyst) leads to the Refined Constraint Index [36]. Note that the Pt loading has to be sufficiently high to ensure that dehydrogenation/hydrogenation on the metal is not the rate-limiting step.

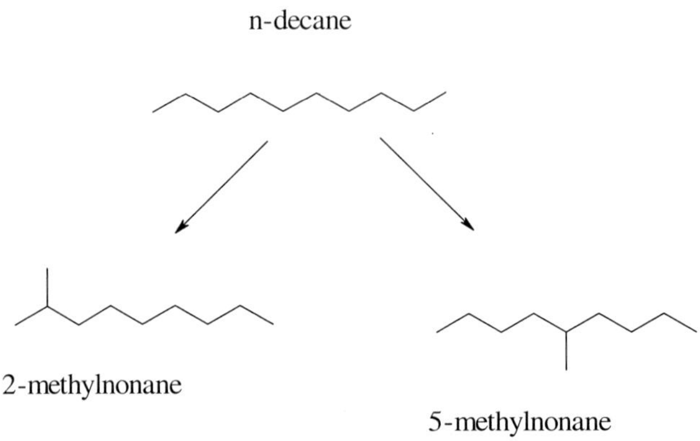

The Refined Constraint Index is calculated from the ratio of 2-methylnonane and 5-methylnonane formed using reaction conditions under which 5 % isomerization of n-decane is obtained.

$$RCI = \frac{c_{2\text{-methylnonane}}}{c_{5\text{-methylnonane}}}$$

The space requirement for the formation of 5-methylnonane is larger than that for 2-methylnonane and, therefore, the RCI will increase with decreasing pore diameter. Due to the size of the reactant and probe molecules, this method is especially well suited for the characterization of zeolites with 10 MR pore systems.

8.2.6.3 Spaciousness index (SI)

The Spaciousness Index is determined from the ratio between iso-butane and n-butane formed during hydrocracking of n-butylcyclohexane [37].

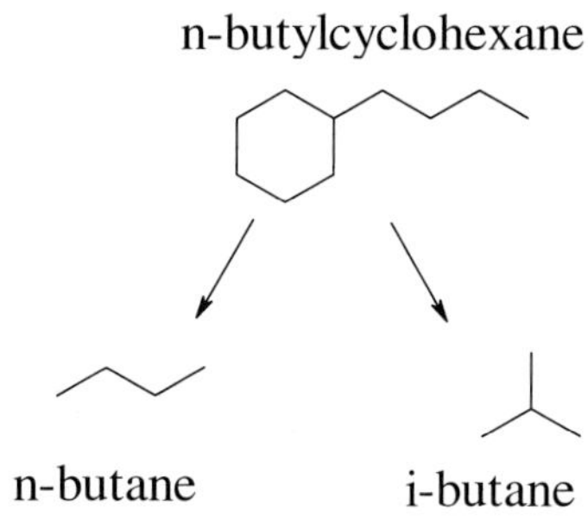

iso-Butane is formed *via* a smaller transition state than n-butane. Therefore, the SI decreases with increasing pore size.

$$SI = \frac{c_{\text{iso-butane}}}{c_{\text{n-butane}}}$$

This reaction is especially sensitive for the characterization of 12 MR pore systems and, thus, complements the description of zeolite pore systems with the Constraint Index and Refined Constraint Index.

8.3 OVERALL CHEMICAL COMPOSITION

The determination of the chemical composition of a molecular sieve is a must for all studies applied. In many cases this is performed now by external services and only for special applications dedicated methods are employed. Thus, the descriptions are held very brief when standard analytical methods are employed and only the restrictions are especially highlighted. The methods for determining the chemical composition can be divided into two groups.

8.3.1 Complete dissolution of the material and analysis of the liquid

The sample is dissolved in an appropriate solvent such as an aqueous solution of HF and the solution is subsequently analyzed by analytical methods such as AAS or ICP. These methods require the complete dissolution of the solid sample and depend on an accurate chemical analysis of the solution.

8.3.2 Physical analysis methods

8.3.2.1 X-ray fluorescence spectroscopy (XRF)

The sample is exposed to high-energy X-ray radiation, which leads to the excitation of electrons from a core orbital of the element being probed. The main relaxation mechanism is that electrons from an energetically higher orbital fill the vacancy, which results in the generation of radiation (X-ray fluorescence) or in the ejection of an electron (from an other level) by the so called Auger effect (see Fig 8.9). However, as the experiments are carried out under atmospheric pressure, only the X-ray radiation can be detected.

The energy and intensity of the X-ray fluorescence radiation is characteristic for the type and concentration of the emitting atoms and, thus, can be used for qualitative and quantitative chemical analysis. The samples are typically prepared as a solid mixture with $LiBO_4$. For calibration a sample with known composition is used.

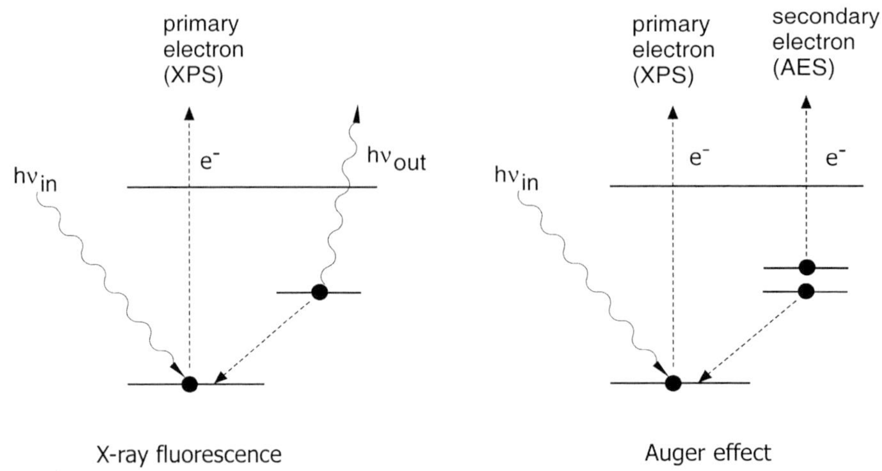

Figure 8.9: Schematic representation of the relaxation processes

8.3.2.2 X-ray photoelectron spectroscopy (XPS)

In X-ray photoelectron spectroscopy, the sample is exposed under ultra high vacuum conditions to a monochromatic X-ray radiation, typically produced by MgKα (1253.6 eV) or AlKα (1486.3 eV) radiation and the energy spectrum of the ejected primary electrons is measured (see Fig. 8.9) [16]. The binding energy of electrons is calculated according to:

$$E_{kin} = h\nu - E_b - \Phi$$

where E_{kin} is the kinetic energy of the photoelectrons, $h\nu$ the energy of the primary X-ray beam, E_b the binding energy of an electron and Φ the work function. The binding energies are characteristic for each chemical element and, therefore, XPS can be used to analyze the composition of the sample. The probing depth of XPS varies between 1.5 and 6 nm according to the mean free path of the photoelectrons and, thus depends on the element being probed [7].

The energy levels of the core electrons depend on the chemical state and environment of the atoms. Thus, the binding energies are not only element-specific, but also contain information on the (partial) charge of the elements. For example, it was shown that the Si_{2p}, Al_{2p} and O_{1s} binding energies of sodium exchanged zeolites including X, Y, mordenite and ZSM-5 increase with the Si/Al ratio [7].

Also, the variation of the composition as a function of the sample depth and the formation of surface layers can be studied by XPS. These methods rely on the energy dependence of the mean free path length of the photoelectron [38].

8.3.2.3 Energy dispersive analysis of X-ray fluorescence (EDAX, EDX)

Scanning electron microscopes focus the electron beam on a small spot on the surface to produce an image of the sample. In addition, the X-ray fluorescence and the Auger electrons emitted by the electron beam can be analyzed (see Fig 8.10).

As the electron beam scans the surface a spatially resolved elemental analysis can be obtained. If the Auger electrons are detected (energy range ~100 eV) the chemical information is surface sensitive, while the detection of the X-ray fluorescence, having typically a wavelength where the mean free path-length is in the order of the particle size of the zeolite crystals, the bulk composition of the zeolite is observed.

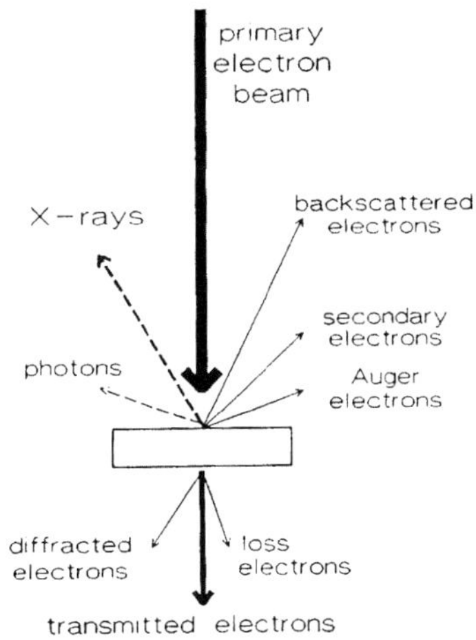

Figure 8.10: Interaction of the primary electron beam with the sample

8.4 DETERMINATION OF THE FRAMEWORK Si/Al RATIO

With the physicochemical methods described in the previous section, it is not possible to analyze the location of the aluminum atoms in the zeolite framework. This is, however, highly important as only aluminum atoms, which are tetrahedrally coordinated and part of the zeolite lattice generate Brønsted type acid sites. On the other hand aluminum oxide clusters inside the zeolites pores may act as Lewis type acid sites and/or may enhance the strength of Brönsted type acid sites. Therefore, it is essential to determine the fraction of framework and non-framework aluminum atoms.

8.4.1 Lattice constant determined by X-Ray diffraction

In general, the size of the unit cell increases with decreasing Si/Al ratio, as the length of a Al-O bond (1.73 Å) is larger than that of a Si-O bond (1.62 Å). For zeolites with cubic symmetry, i.e., with equivalent 100, 010 and 001 planes, the lattice parameter a_0 (see 8.2.1) is directly proportional to the Si/Al ratio.

For X and Y type zeolites relationships between the Si/Al ratio and the unit cell size were reported by Breck and Flanigen [39], Fichtner-Schmittler *et al.* [40] and Sohn *et al.* [41]:

$$N_{Al} = 101.202\left(a_0/æ - 24.2115\right)$$

where N_{Al} is the concentration of aluminum atoms in the lattice and a_0 the unit cell size. The correction factor *æ*, introduced by Jorik [42], is 1 for as synthesized faujasite and 0.9957 and 0.9966 for sodium X and Y type zeolites, respectively. The method is especially popular for determining the Si/Al ratio in fully formulated zeolites after frequent regenerations and steaming. With non-cubic zeolites the method is more complex to apply and may give misleading results, if aluminum is preferentially inserted or removed from particular lattice positions.

8.4.2 ^{29}Si MAS-NMR

By applying the Magic Angle Spinning (MAS) technique, high resolution NMR spectra of solids are obtained, which can be used to characterize the chemical and structural environment of the lattice atoms in zeolites [6, 43-45]. A more detailed discussion of this method is given in Chapter 9, thus, only the determination of the Si/Al ratio of zeolites is described here.

In the ^{29}Si MAS-NMR spectra of zeolites up to 5 well separated peaks are observed, which can be assigned to Si atoms connected via oxygen atoms to other framework atoms such as Si or Al (i.e., Si(nAl) n = 0…4). For each Al atom in the direct neighborhood to a Si atom a downfield shift of 5 ppm is observed (see Fig. 8.11).

According to Löwenstein's rule, each Al atom is surrounded by 4 Si atoms, while the Si atoms can be surrounded by up to four Al atoms. Therefore, the Si/Al ratio of the framework can be calculated directly from the ^{29}Si MAS-NMR spectra using the relative intensities of the Si(nAl) peaks ($I_{Si\text{-}nAl}$):

$$Si/Al = \frac{\sum_{n=0}^{4} I_{Si(nAl)}}{\sum_{n=0}^{4} 0.25n \cdot I_{Si(nAl)}}$$

This method is independent of the zeolite structure and the type of charge balancing cations and includes only Al atoms that are part of the zeolite lattice. Therefore, NMR yields the framework Si/Al ratio, whereas chemical analysis provides the total sample composition including also non-framework Al species.

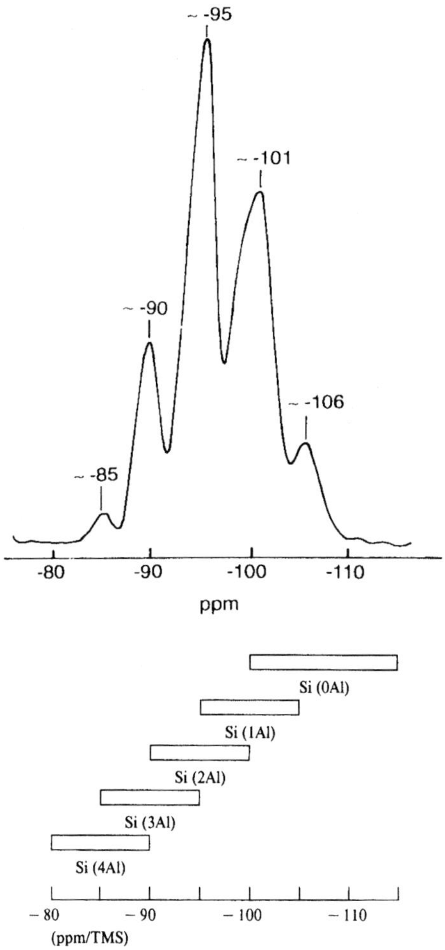

Figure 8.11: ^{29}Si MAS NMR spectrum of Na Y zeolite (Si/Al ratio=2.6).

8.4.3 Catalytic cracking of light alkanes

Protolytic cracking of small alkanes has been shown to reflect the concentration of Brønsted acid sites of molecular sieves. For H/ZSM-5 (Si/Al=15 to 100 000) the rate constant for hexane cracking was found to be proportional to the aluminum content [46] (see Fig. 8.12).

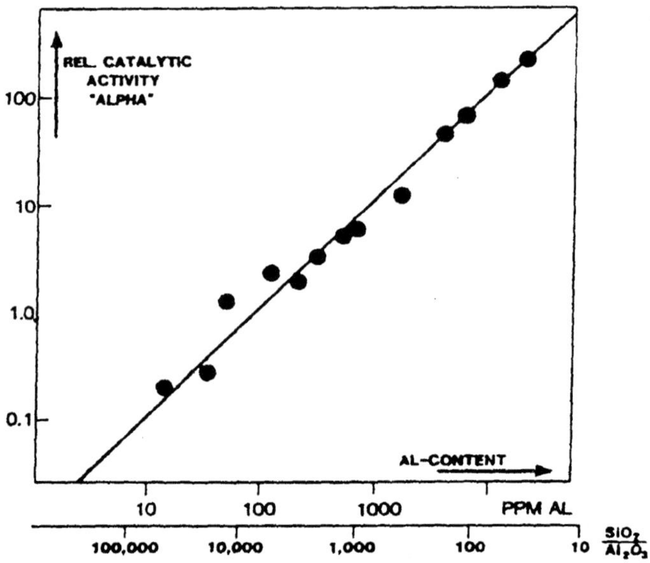

Figure 8.12: Hexane cracking activity vs. Si/Al ratio for H/ZSM-5

These data indicate that all Brønsted sites are isolated and ideally diluted and, therefore, have the same activity. The homogeneity of the sites was verified by partially blocking the Brønsted acid sites by ion exchange with Cs^+, which has been found to selectively poison the strongest acid sites first. For fresh H/ZSM-5 a linear poisoning curve was found (see Fig. 8.13 a), which indicates that all sites are of equal activity. In contrast, after mild steaming of the same zeolite, the activity for n-hexane cracking was found to increase (see Fig. 8.13 b).

This suggests the presence of Al pairs, in which one Al^{3+} is removed from the framework and forms a Lewis type acid site that withdraws electron density from the neighboring Brønsted type acid site and, thus, enhances its acidity [47]. These sites are present in a minor concentration (~4% of all Al-sites present in H/ZSM-5 with Si/Al = 35) and result in inhomogeneity of the Brønsted type acid sites as shown by the non linear dependence of the activity on the Cs^+ exchange degree (see Fig. 8.13.b).

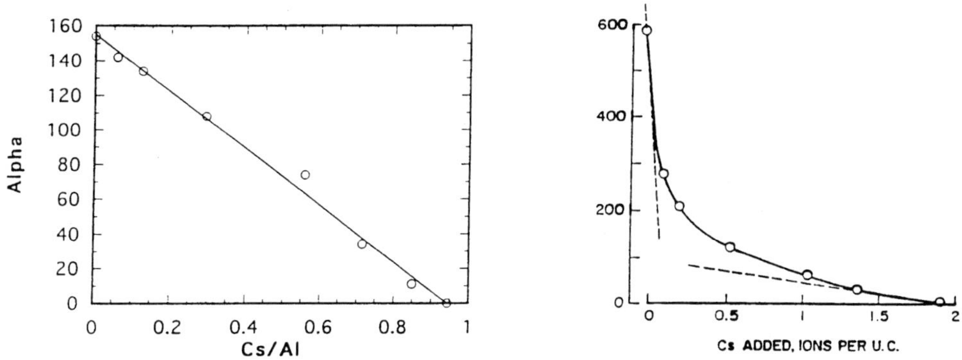

Figure 8.13: Effect of Cs^+ exchange on the n-hexane cracking activity for H/ZSM-5 (Si/Al = 35) (a) as prepared and (b) mildly steamed ratio for H/ZSM-5

Another commonly used test reaction to characterize zeolite acidity is cracking of n-butane [48]. The rate of the reaction k_A is determined by:

$$k_A = \frac{F \ln(1-c)}{W}$$

with c the mole fraction of n-butane converted, F the flow rate in (cm^3/min) and W the sample weight (g). For a series of zeolites the rate of n-butane cracking is shown in Fig. 8.14.

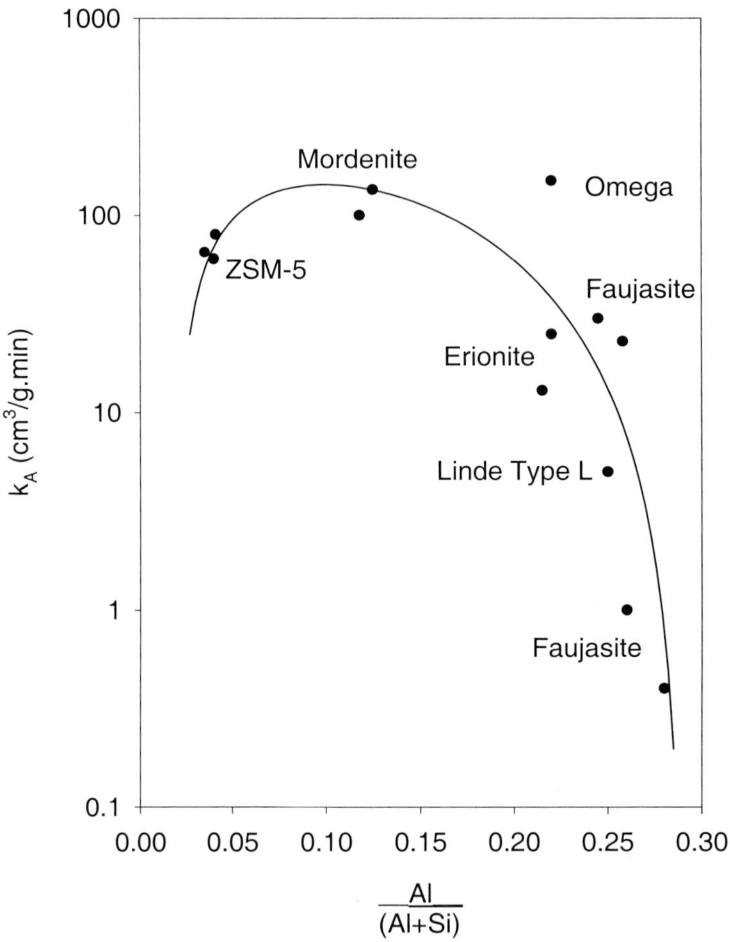

Figure 8.14: Activity as function of the Si/Al ratio for n-butane cracking. (After ref [48]).

With increasing Si/Al ratio the strength of the acid sites increases and, therefore, a sharp increase in k_A is observed. At some point the decrease in the concentration of acids sites overcompensates the activity increase and starting from this Si/Al ratio a decreasing activity is observed.

A good agreement between the results from n-butane cracking and n-hexane cracking exists. Therefore, both test reactions can be applied to materials with a broad range of acidity.

8.5 DETERMINATION OF THE ZEOLITE ACID-BASE FUNCTIONALITY

8.5.1. Titration methods

In principle the acid strength of a zeolite can be determined by the reaction with a series of aniline bases (Hammett indicators), which change color during the interaction with an acid of a certain strength. The applicability of this method, however, is strongly limited as the indicator molecules often cannot enter the pores of medium pore zeolites and thus react only with sites on the external surface and in the pore mouth region. Thus, their use is only recommended for special cases and great caution should be used in the interpretation of the results [49].

8.5.2 Determination of functional groups by vibrational spectroscopy

Generally, all forms of IR-spectroscopy (*i.e.*, transmission absorption, specular reflection, diffuse reflectance, emission, acoustic methods, etc.) can be applied [3]. For most practical experimental reasons, however, the use of the transmission-absorption technique is the best suited, which is mostly related to the design of the cells being used [50]. Typical designs for cells can be found, e.g., in refs. [50-52]. As the catalysts usually have to be pressed into self-supporting discs, it is always necessary to assure that this procedure does not influence catalyst properties. Note that the development of IR-microscopy allows to reduce the analytical volume to an extent that it is possible to study the transport and surface chemistry in single zeolite crystals [53, 54].

Two experimental requirements have to be fulfilled to enhance the value of the experiments. First, the sorbate has to be introduced to the catalyst sample in sufficiently small quantities to achieve a gradual increase of the coverage with the sorbate. If the strength of the acid sites is studied, it is essential that the adsorption/desorption equilibrium is established at every point of the experiment. Otherwise a parallel irreversible sorption of the sorbate on the sites will occur, which would falsely suggest a rather homogeneous acid site distribution even for cases where acid sites of very different strengths exist [55]. A good example for this phenomenon is the sorption of ammonia at ambient temperature. The strongly basic ammonia will sorb on all (Brønsted) acidic hydroxyl groups irrespective of their acid strength at any reasonable experimental partial pressure. If the admission of doses of sorbate does not hinder subsequent adsorption, the method, however, could be potentially used to titrate the acid sites from the outside to the inside of a catalyst particle given a homogeneous concentration of the base molecule exists in the reactor.

The second important experimental detail is that a good mixing of the sorbate within the cell is required and that the change in partial pressure of the sorbate is rapid with respect to the uptake by the catalyst. In this way, the time dependence of the uptake can be used to estimate the diffusion coefficient at a given equilibrium pressure and temperature. Such experiments have been reported for vacuum [56], carrier gas [57, 58] and by IR-microscopy [50, 59] providing unique information on molecular aspects of the transport in catalyst pores.

8.5.2.1 Surface hydroxyl groups

IR-spectroscopy is the most direct method to study type, density and acidity of hydroxyl groups in zeolites. The frequency of the stretching vibration of a hydroxyl group (ν_{OH}) is a function of the force constant between the O and H atoms and, thus, reflects the type and, at least in a homologous series, the acid strength of the hydroxyl group. In such a series of hydroxyl groups a weaker O-H bond, i.e., a more acidic hydroxyl group, is reflected by a lower stretching frequency (see Fig. 8.15). It has to be emphasized, however, that the position of the band cannot be used as a direct indication for the acid strength of a hydroxyl group, as there is no fundamental correlation between the curvature of the OH-potential well and the heterolytic O-H dissociation energy.

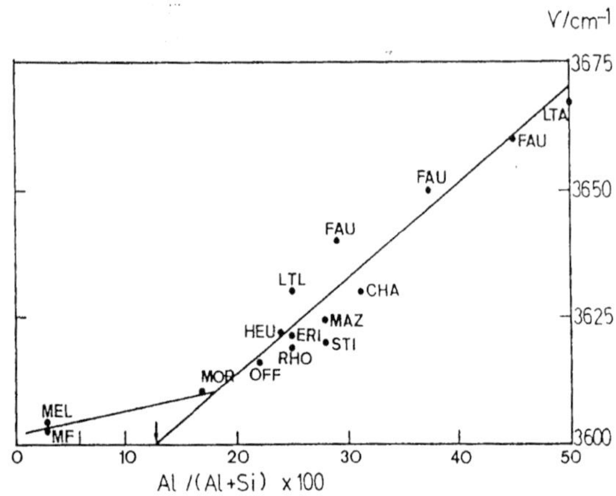

Figure 8.15: Frequency of the hydroxyl vibration in H zeolites as function of the Si/Al ratio

A less subtle and more robust application results from the fact that the wavenumber of an O-H stretching vibration is strongly influenced by (i) the type of coordination of the oxygen (terminal versus bridging oxygen), (ii) the structural environment of the hydroxyl group, i.e., the type of zeolite lattice, and (iii) the perturbation of the OH group through the surroundings by lattice or extra-lattice oxygen. A comparison of the OH stretching vibrations typically observed on different types of zeolites and a schematic depiction of the corresponding hydroxyl groups are shown in Fig. 8.16. Bridging hydroxyl groups (SiOHAl groups, Brønsted type acid sites) are observed on most types of zeolites between 3650 and 3550 cm^{-1}. Additionally, two types of silanol groups (Si-OH groups) are usually present on zeolites: (i) OH groups on the outer surface terminating the zeolite crystal and (ii) OH groups on structural defects resulting from an incomplete condensation or from the removal of lattice atoms. The frequency of the OH stretching vibrations of silanol groups on the external surface is in the range of 3740–3745 cm^{-1}, while that of silanol groups related to defects depends on their local environment. The presence of hydrogen bonding interactions shifts the frequency to ~ 3500 cm^{-1}, while isolated silanol groups are only slightly shifted to lower frequencies, i.e., 3735-3715 cm^{-1}, compared to the silanol groups on the external surface. Silanol groups on the external surface are always observed in the IR-spectra of zeolites. Their relative intensity depends on the primary size of the zeolite particles, i.e., samples with a large particle size show a relatively low intensity of the Si-OH groups on the external surface and *vice versa*.

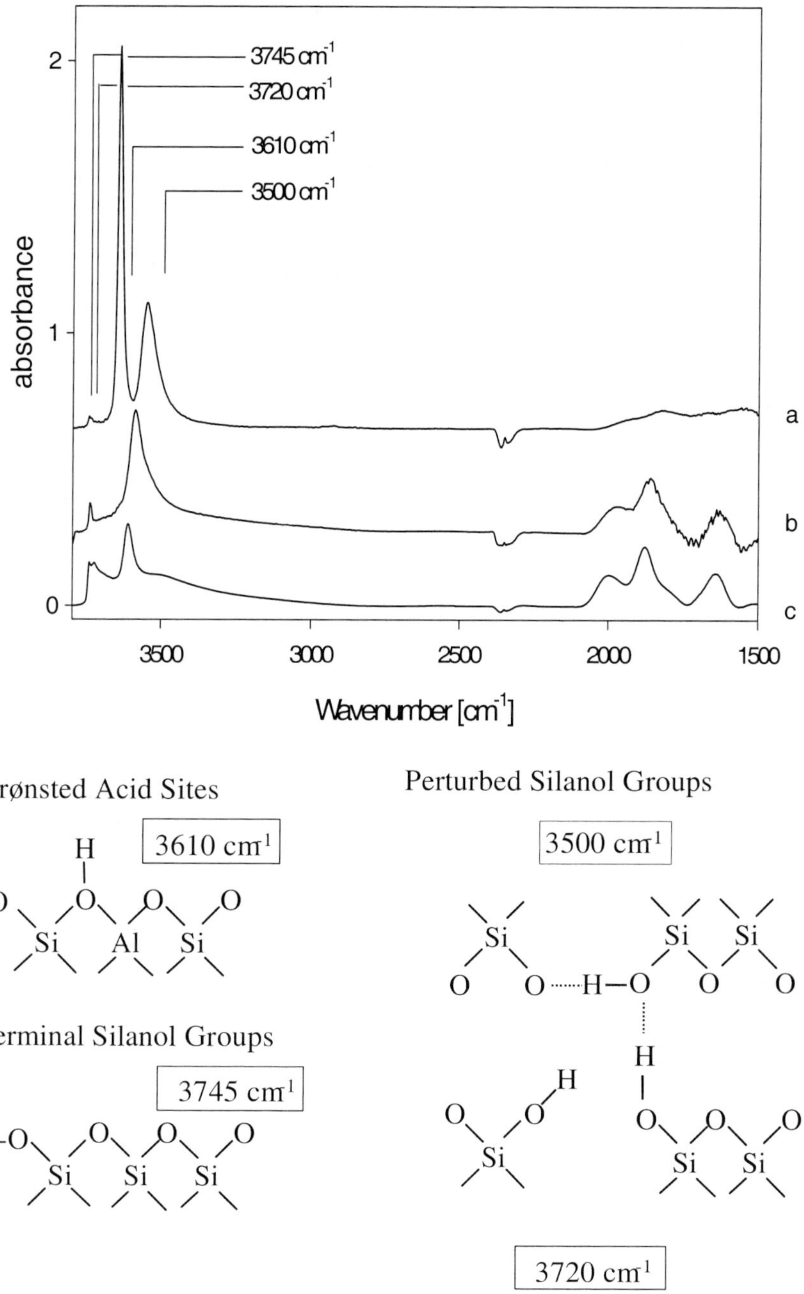

Figure 8.16: Comparison of the OH stretching vibrations observed on (a) Faujasite, (b) Ferrierite and (c) ZSM-5 and the assignment of the hydroxyl groups observed on H/ZSM-5.

A comparison between IR-spectra of zeolites allows an estimation of the Brønsted acid site density, the primary particle size and the presence of defects. The IR spectra of two ZSM-5 samples are compared in Fig. 8.17.

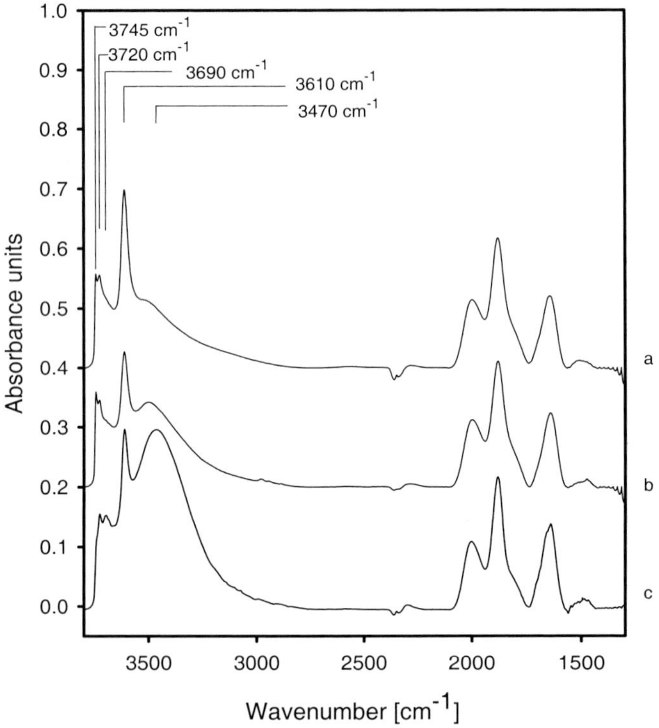

Figure 8.17: Comparison of IR-spectra of H/ZSM-5 samples

Spectra *a* and *b* show the same H/ZSM-5 sample with a ratio Si/Al=40 after having applied different calcination conditions. Spectrum *c* shows a different H/ZSM-5 sample having a Si/Al ratio of 27 (all determined from chemical analysis). The milder calcination procedure used for the sample shown in Fig. 8.17 *a* compared to that shown in Fig. 8.17 *b* led to a higher concentration of Brønsted acid sites and a lower concentration of structural defects as indicated by the relative intensities of the bands at 3610 cm^{-1} and at 3720 and 3470 cm^{-1}, respectively. The IR-spectrum shown in Fig. 8.17 *c* clearly indicates that although more Al is present in this sample (Si/Al=27), less is incorporated into the framework of the zeolite, which is in line with the significantly higher concentration of lattice defects. The similar intensities of the silanol groups on the external surface (3745 cm^{-1}) indicate that both samples have a comparable primary particle size.

In order to compare the IR-spectra from different zeolite samples, the spectra have to be normalized to the amount of material irradiated by the IR-beam (sample thickness). For zeolites with identical structures, or more precisely with the same density of T-atoms, the intensity of the Si-O vibration overtones, three bands observed between 2074 and 1560 cm^{-1}, can be used (see Figs. 8.16 and 8.17) as a parameter linearly related to the sample thickness. With MCM-41 type materials it has been shown [60] that the integral intensity of the structural vibrations I_{LV} shows a linear correlation to the amount of sample in the IR-beam:

$$\frac{w}{r^2 \cdot \pi} = k \cdot I_{LV}$$

This formula describes the relation between the sample weight w, normalized to the size of the wafer ($A=r^2\pi$) and the integral intensity of the lattice vibrations I_{LV}. The use of the normalized weight makes the factor k independent from the wafer size used. Two series of experiments with different sample diameters measured on different IR-spectrometers are compared in Fig. 8.18. Note that the value of k depends on the density of T-atoms of the zeolite and, therefore, it has to be determined for each type of molecular sieve individually.

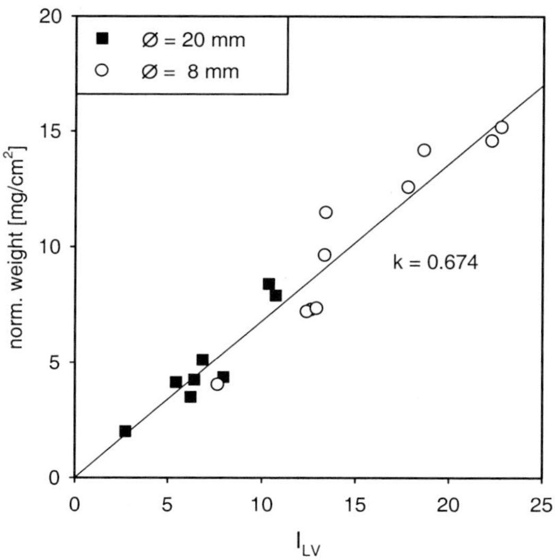

Figure 8.18: Correlation between the intensity of the structural vibrations of MCM-41 and the sample weight

8.5.2.2 Probing the acid character using IR spectroscopy of adsorbed probe molecules

As the type of probe molecule chosen might critically influence the characteristics of the probed solid and thus also might affect the structure-activity-relationships derived, the choice of the appropriate probe molecule is crucial. Therefore, whenever possible it is advantageous to use the reactants as probe molecules. In the following practical criteria to select probe molecules will be discussed, which combine the earlier suggestions of Paukshtis *et al.* [61], Knözinger [62] and Lercher *et al.*[3]. An overview of the potential application of frequently used probe molecules is compiled in Table 8.6

Table 8.6: Frequently used probe molecules

Probe molecule	Type of sites probed	Comment
Ammonia	Type and concentration of acidic sites	Small size, can access almost all sites in microporous oxides
Aliphatic amines	Type and concentration of acidic sites	Intramolecular reactions possible
Pyridine	Type and concentration of acidic sites	Can only enter into 10 MR channels and above. Molar absorption coefficients available
Substituted pyridine	Type and concentration of acidic sites accessibility of sites	Sorption on sites close to the pore openings
Benzene, Toluene, Xylene	Acid strength, accessibilities of sites	Possible reactions (isomerization) on strongly acid sites, slow uptake for bulky xylene isomers on 10 MR zeolites
Alkanes	Acid strength, accessibility of sites	Good choice for probing sites for hydrocarbon conversion
Pyrrole	Type and concentration of basic sites	Intermolecular reactions possible
CO_2	Type and concentration of basic sites	Formation of carbonates possible
Nitriles	Type and concentration of acidic sites	
CO	Metal cations, Type and concentration of acidic sites	Low temperature necessary for adsorption on zeolites
NO, NO_2	Metal cations, Type and concentration of acidic sites	Nitrate formation on the surface

1. Criterion: The probe molecule should have dominating base and rather weak acidic properties.

Upon sorption, the probe molecule should primarily interact with its base electron pair donor function, while the acidic electron pair acceptor function should hardly interact with the solid. Molecules, which fulfill these requirements ideally, are strong Lewis type bases such as ammonia and most amines including pyridine. Taking ammonia as an example, it was argued that the primary interaction on acidic surfaces takes place between the electron lone pair at the nitrogen atom and the Brønsted or Lewis type acid sites on the surface. Theoretical calculations for ammonia sorption on protonic zeolites [63], however, indicate that the interaction between the hydrogens of ammonia and the oxygens of the oxidic lattice contribute markedly to the stabilization of the formed sorption complex. Thus, the acid strength of the hydroxyl group is not the only parameter determining the strength of interaction in the sorption complex. This should be taken as an indication that more complex molecules that are frequently used as probes for acidity might have even more pronounced simultaneous acid and base interactions, which makes it difficult to determine the acid properties of the solid unambiguously.

2. Criterion: The IR-spectrum of the sorbed probe molecule should allow the distinction between sorption on Brønsted and Lewis type acid sites.

Significant differences in the IR-spectrum of Brønsted and Lewis bound molecule, e.g., for ammonia and pyridine [64, 65], should exist. Alternatively, changes in the IR-spectrum of the solid upon sorption of the probe molecule characteristic for protonation or coordinative bonding can be evaluated. The latter applies to hydroxyl groups and their disappearance or the shift in the absorption maxima of their IR-bands upon sorption of the probe molecule.

In Fig. 8.19 the spectra of an activated H/ZSM-5 zeolite is subtracted from the spectra obtained after equilibration of the sample with benzene and pyridine.

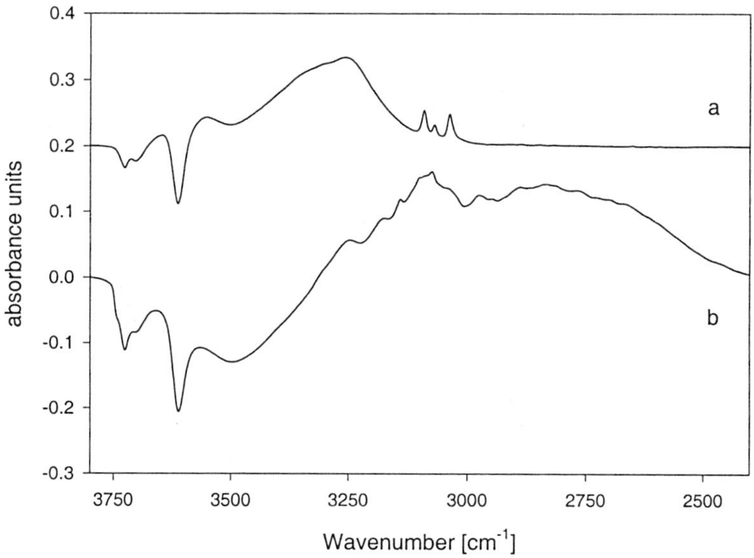

Figure 8.19: Proton transfer *vs.* hydrogen bonding: Sorption of (a) benzene and (b) pyridine on H/ZSM-5

In these spectra bands, which increase in intensity or result from the probe molecule are pointing upwards, while bands which decrease in intensity during the adsorption of the probe molecules are pointing downwards. Clearly, it can be seen that during the adsorption of the probe molecules the intensity of the unperturbed SiAlOH groups at 3610 cm^{-1} (Brønsted acid sites) decreases, as a negative band is observed. The reason for this, however, is different for the two probe molecules discussed. For benzene a hydrogen bonding interaction is being established shifting the OH stretching vibration to a lower frequency. Therefore, in parallel to the decrease of the band at 3610 cm^{-1} a new band around 3300 cm^{-1} appears in the IR-spectra. In contrast, during adsorption of pyridine a pyridinium ion is formed. Therefore, the bridging hydroxyl group is not present anymore, which is reflected by its disappearance in the

IR-spectra. Additionally pyridine is adsorbed on non-protonic hydroxyl groups (e.g., terminal hydroxyl groups), which results in a perturbation of the hydroxyl group and is manifested in a broad band with a maximum at approximately 2900 cm^{-1}.

3. Criterion: The probe molecule should allow to differentiate between acid sites of the same type, but of different strength.

The IR-spectra of the molecule sorbed on sites with different acid strength should show detectable differences. Depending on the electron donor strength of the probe molecule and on the acidity of the hydroxyl group a hydrogen bonding interaction or proton transfer to the probe molecule occurs. Therefore, a strong base such as pyridine is better suited to differentiate between subtle differences in acid strength as long as it is not protonated, while a weak base, e.g., benzene [66, 67], will be more suitable to differentiate between sites with large differences in strength. However, when interpreting the interaction between the acid site and the base probe molecule not only their acid-base strengths, but also the polarizability of the orbitals [68] is involved in the acid-base bonding and has to be taken into account.

A general treatment of that concept was given by Pearson [69] and early applied by Burwell *et al.* [70] for a qualitative description of sorbate-sorbent interactions. In general, the concept predicts that interactions between acid-base sites of comparable polarizability are the most favorable ones, meaning that hard-hard and soft-soft interactions are preferred to hard-soft interactions.

4. Criterion: The size of the probe molecule should be comparable to the size of the reactant to probe the concentration of acid sites relevant for a particular reaction.

The smallest probe molecules that are frequently used are NH_3 and CO, with minimum kinetic diameters of 0.165 nm and 0.073 nm, respectively. Both molecules will probe the maximum concentration of acid sites that might be involved in a catalyzed reaction. As the size of the molecule increases, fewer acid sites can be reached and, thus, steric constraints around the acid sites can be probed by selecting a series of probe molecules with increasing size or steric constraints around the electron pair donor (base) function (see section 8.5.2.3).

In all cases, the quantitative determination of the concentration of sorbed molecules is crucial for the appropriate interpretation of the IR-spectra. The adsorption stoichiometry of probe molecules per sorption site can be determined (i) spectroscopically from the disappearance or the perturbation of the functional groups of the solid acid (e.g., hydroxyl groups) and the increase of characteristic bands of the sorbed molecules or (ii) by parallel gravimetric or volumetric measurements. The known stoichiometry is mandatory to avoid misinterpretation due to chemical molecule-molecule interactions, leading to the formation of larger clusters interacting with one particular site and also due to purely spectroscopic phenomena such as dipole-dipole coupling.

8.5.2.3 Probing the accessibility of hydroxyl groups and metal cations in zeolites

Adsorption of molecules with different kinetic diameter leads to a direct assessment of the pore diameter and as a consequence on the accessibility of the sites in these pores. The most sensitive probe molecule are alkanes, which are sorbed at the Brønsted type acid sites *via* dipole induced hydrogen bonding [33]. To visualize the accessibility of molecules inside the pores various commercial computer programs exist [71, 72].

While this is a feasible approach to determine constraints around Lewis acid sites, care must be taken with Brønsted acid sites. It could be shown that the protons of hydroxyl groups can be attracted by base molecule from sites, which are sterically well shielded. This imposes a special problem for the determination of the accessibility of sites in zeolites, where interactions of strong bases with hydroxyl groups in virtually inaccessible locations were observed. The classic example is the IR-spectra of mordenite after adsorption of NH_3, pyridine and benzene as shown in Fig. 8.20.

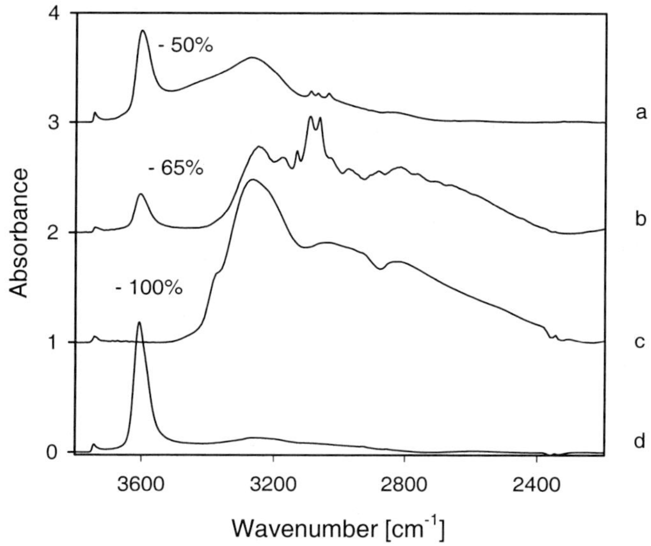

Figure 8.20: Adsorption of (a) benzene, (b) pyridine and (c) NH_3 on (d) H/MOR

Particularly interesting in the structure of mordenite is that perpendicular to the uni-dimensional main channels with a diameter of 6.5 x 7.0 Å small cavities (side pockets) with a diameter of 2.6 x 5.7 Å exist. The IR-spectra clearly indicate that NH_3 interacts with Brønsted type acid sites in the main as well as in the side channels. The bands observed around 3380 and 3280 cm^{-1} are the antisymmetric and symmetric N-H stretching vibrations of the NH_4^+ group, respectively [64]. Pyridine and benzene, due to their larger size, can only adsorb on hydroxyl groups in the main channel of mordenite. For pyridine and benzene C-H stretching vibrations of the aromatic C-atoms were observed above 3000 cm^{-1} together with a perturbed hydroxyl group, as described in section 8.5.2.2. Although benzene and pyridine have a similar kinetic diameter, pyridine interacts with a larger number of acid sites compared to benzene as the strength of interaction is significantly higher and, thus, pyridine also reaches protons from hydroxyl groups in the side pockets.

8.5.2.4 Probing the concentration of acidic hydroxyl groups in zeolites

The molar extinction coefficients for pyridine adsorbed on Brønsted and Lewis type acid sites, 1.67 and 2.22 cm/μmol, respectively, were derived by Emeis [73] using a series of zeolites with different acid site concentration and strength. The concentration of Brønsted and Lewis type acid sites can be determined directly from the integral intensity of the corresponding bands in the IR-spectra after adsorption of pyridine:

$$c_{(Lewis)} = \frac{1.42 \cdot \Delta I_L \cdot r^2}{w} \quad \left[\frac{mmol}{g}\right]$$

$$c_{(Brønsted)} = \frac{1.88 \cdot \Delta I_B \cdot r^2}{w} \quad \left[\frac{mmol}{g}\right]$$

$c_{(Lewis)}$ is the concentration of Lewis type acid sites, $c_{(Brønsted)}$ is the concentration of Brønsted type acid sites, ΔI_L and ΔI_B are the integral intensities of the IR-bands at 1455 cm^{-1} (Lewis type acid sites and weakly acidic silanol groups) and at 1545 cm^{-1} (Brønsted type acid sites) [74], r is the diameter and w is the weight of the wafer. A crucial point during the determination of the site concentration with this method is to ensure the complete coverage of all sites and the absence of physisorbed pyridine, which is typically achieved by outgassing the sample in vacuum at 150 °C after the adsorption of pyridine.

8.5.3 Temperature programmed desorption to assess zeolites acidity

The principle of temperature programmed desorption is to monitor the changes in the surface coverage (i.e., the rate of desorption) of probe molecules pre-adsorbed on the zeolite as a function of the temperature by following the partial pressure of the probe molecules in the gas phase over the sample with, e.g., a mass spectrometer. The changes in the partial pressure are directly proportional to the desorption rate under the assumptions: (i) that the rate of removal of the molecules from the gas phase over the sample is significantly higher than the desorption rate and (ii) that re-adsorption of the probe molecules does not occur [75]. Note, that both criteria are not (ideally) fulfilled for most zeolite samples.

If molecules are chosen that preferentially interact with a certain type of surface site, the strength of the interaction is reflected in the temperature profile of the desorption rate. In principle, an increasing activation energy for the desorption process, e.g, resulting from the interaction of the probe molecules with sites of increasing acid strength, causes the maximum of the desorption rate to appear at higher temperatures. Additionally, the integral area under the peaks is proportional to the number of probe molecules adsorbed and after a calibration, e.g., by using an external standard, the concentration of the sorption sites can be directly determined from the integral area of the desorption peaks.

The comparison of the NH_3-TPD on mordenite, ZSM-5, SAPO-11 and AlPO-11, shown in Fig 8.21, indicates that each sorption site of different acidity is reflected by a maximum in the desorption rate. On all materials weakly acidic sites are present, which result in the maximum at lower temperature (~ 373 K). On mordenite, ZSM-5 and SAPO-11 additionally strongly acidic sites are present, the difference in the position of the second desorption maximum indicates that the SiOHAl groups on ZSM-5 and mordenite are more strongly acidic than those on SAPO-11.

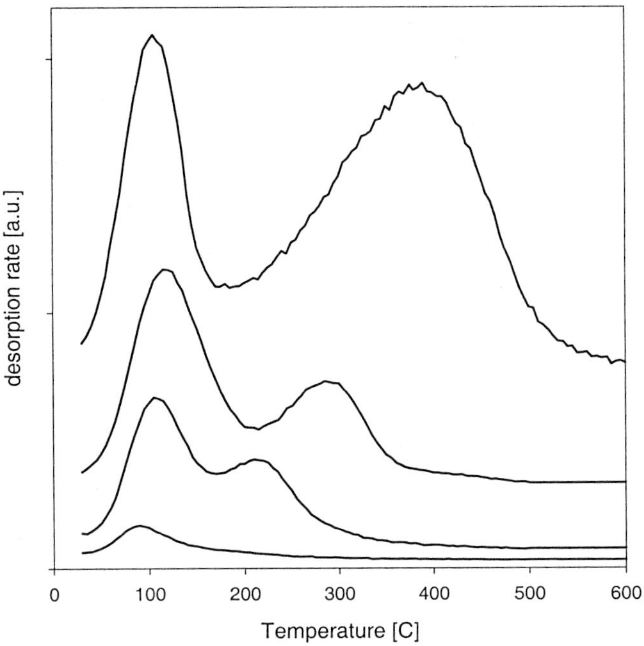

Figure 8.21: Comparison of the NH_4-TPD on (a) Mordenite, (b) ZSM-5, (c) SAPO-11 and (d) AlPO-11

While the example discussed reflects the trends in the acidity for the four samples, significant limitations exist for NH_3-TPD. First, it can not differentiate between Brønsted and Lewis acid sites and second, the strong interaction of ammonia with the acid sites might result in a partial re-adsorption of NH_3, which can cause a shift of the desorption maxima of up to 150 K, depending on the experimental conditions and the particle size of the sample. Therefore, it might be preferable to use reactive amines such as isopropylamine as probe molecules [76].

$$HRNH_2 + H\text{-}O\text{-}Zeol \rightarrow HRNH_3^+ \ldots {}^-O\text{-}Zeol \rightarrow R + NH_3 + H\text{-}O\text{-}Zeol$$

During TPD the desorption of propene and ammonia results from the decomposition of the amine occurring only at Brønsted type acid sites, while the desorption of isopropylamine indicates the presence of aprotic Lewis type acid sites.

An alternative approach is to follow the desorption of NH_3 after ion exchanging the zeolite in a 1 molar NH_4OH solution for 24 h at room temperature [77]. During the liquid phase ion exchange, the protons are replaced by NH_4^+ ions. These act as the charge balancing cations and, therefore, are only bound to the Brønsted type acid sites, while on Lewis type acid sites ammonia is not bound. Thus, during TPD only ammonia desorbing from Brønsted type acid sites is observed.

8.5.4 Characterization acidity by XPS of sorbed probe molecules

The XPS binding energies of chemisorbed nitrogen containing basic molecules such as ammonia and pyridine are characteristic for the acid strength of the sorption sites. This method consists of a deconvolution of the N 1s level and on the determination of the relative intensities of the peak components [2, 7].

For pyridine, the N 1s peak in the XPS consists typically of three contributions. The peak with the highest N 1s binding energy, observed at 402.4 and 401.8 eV for faujasite [78] and ZSM-5 [79], respectively, was assigned to pyridine adsorbed on Brønsted type acid sites. The peak at binding energies of 400.4 and 398.7 eV (faujasite and ZSM-5) was assigned to pyridine adsorbed on Lewis type acid sites. On both samples a third peak at 400.0 eV was observed, which was assigned to pyridine adsorbed on weak Brønsted type acid sites.

8.5.5 Calorimetry/gravimetry of probe molecule sorption

By combining differential scanning calorimetry (DSC) and thermogravimetry (TGA) the heat of adsorption can be directly correlated to the molar uptake of the sorbates. Typically, the differential heat of adsorption, defined as the heat evolved during the adsorption of a small quantity of gas at constant temperature [1], is determined as a function of the surface coverage. The heat of adsorption of a basic molecule on an acid site is claimed to be characteristic for the strength of the site. To obtain an acid strength distribution from the heat of adsorption as function of the coverage the sample temperature has to be sufficiently high to assure that the probe molecules reach the sorption equilibrium on the sites being probed. Therefore, the temperature for the TGA/DSC experiments should be above the temperature of the TPD maxima of the probe molecule on the zeolite sample.

The dependence of the heat of adsorption on the coverage provides detailed information on the interaction of the probe molecules with zeolites (see Fig. 8.22).

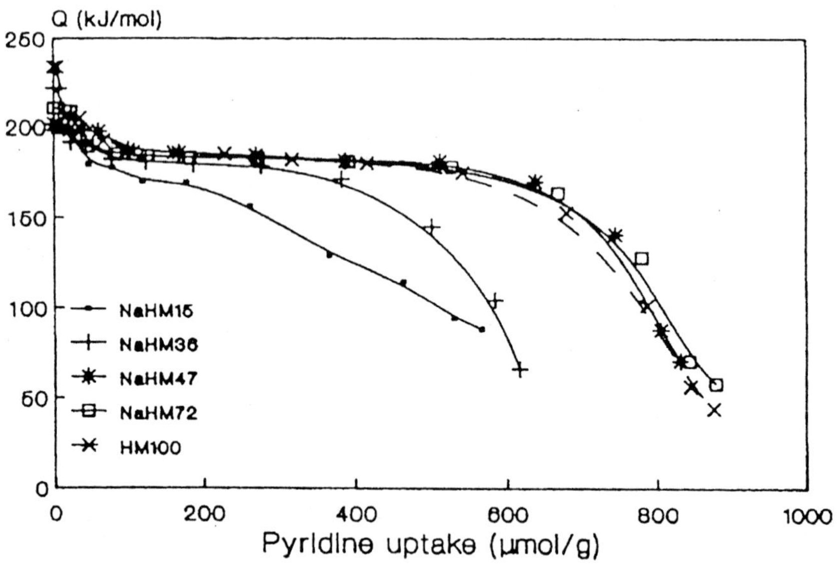

Figure 8.22: Differential heat of adsorption of pyridine on NaH/mordenites

As an example pyridine adsorption on NaH/mordenites is used. The differential heat of adsorption shows three regions [80]. The sharp decrease of Q_{diff} at low coverage indicates the presence of a small concentration of very strong Lewis type acid sites. The plateau of constant heat of adsorption results from the adsorption of pyridine on the Brønsted type acid sites, which decreases sharply after all Brønsted type acid sites are covered. For the adsorption of pyridine on Na^+ cations only (sample HM-15) the heat of adsorption is lower and decreases continuously with the coverage. An increasing heat of adsorption as function of the coverage would indicate the presence of strong intermolecular sorbate-sorbate interactions.

8.5.6 X-ray absorption spectroscopy

X-ray absorption spectra are typically measured on high intensity X-ray radiation sources such as synchrotrons. The XANES (X-Ray Absorption Near Edge Structure) region provides information on the electronic structure and the chemical environment, while the EXAFS (Extended X-Ray Absorption Fine Structure) region is characteristic for the structural properties of the atoms in the sample [81]. The advantage of this method over others such as NMR, XPS and XRD is that the experiments can be carried out under reaction conditions and provide element specific information.

X-ray absorption spectroscopy was frequently used over the last years to describe the location of metal cations in zeolites and the framework substitution [82, 83]. More recent interest has focused on the use of low energy X-rays to directly study the framework atoms of zeolites. The X-ray absorption edges of Si and Al are observed at 1839 and 1559 eV (K edges resulting from the 1s → 3p transition), the so-called soft X-Ray region. In this region, the background absorption of the atmosphere is significantly higher than in the energy region used to characterize transition metals. Therefore, specially designed experimental stations and experiential setups in order to generate and detect low energy X-ray photons and to minimize the absorption from the atmosphere have to be used.

The Al-K edge was shown to be sensitive to the type of cation exchanged and to the acid strength of a zeolite [84]. The withdrawal of electron density from the aluminum is reflected in the intensity of the peak above the X-ray absorption edge (the so called "white line"). With decreasing electron density on the aluminum atoms the intensity of the while line increases, which is an indication for an increasing acid strength of the zeolites.

The XANES at the Al-K edge can also be applied to determine the aluminum coordination in zeolites. Using the XANES of reference materials with Al in tetrahedral and octahedral coordinations, a curve fitting procedure was introduced using a set of Gaussian and arctangent functions, which can differentiate between framework and non-framework Al atoms [85].

The analysis of the EXAFS of the Al and Si atoms in the lattice results in Al-O [84] and Si-O [86] distances, which can be related to subtle changes in the framework geometry. It has to be mentioned, however, that X-ray absorption spectra result from all absorber atoms in the sample and, therefore, a volume average of the structural details is obtained.

8.5.7 On the characterization of zeolites basicity

Similar to the acid properties, the basic properties of zeolites can be probed by NMR [43], XPS [87], titration with acids [88] and by ad- and desorption of probe molecules followed with spectroscopic methods [3], temperature programmed desorption and calorimetry combined with thermogravimetry [89]. In principal, characterization of basicity follows two approaches, one, which characterizes the charges on the oxygen in the zeolite lattice (e.g., XPS) and the other, which characterizes the strength and concentration of basic sites *via* interaction with probe molecules. However, a significantly smaller number of probe molecules is available for the characterization of basic sites compared to acid sites, which leaves the choice of the probe molecule the main limitation for these experiments [90]. Many probe molecules will either adsorb/react also on acid sites or show strong intermolecular interactions, which might lead to a difficult interpretation of the results.

CO_2 is commonly used for the characterization of basic sites, but it can interact with several types of sites (cations, hydroxyl groups and framework oxygen atoms) and may give rise to surface reactions [89, 91]. Pyrrole, which is an amphoteric molecule, interacts with the framework oxygen atoms and from the shift in the N-H stretching vibration the basicity of the materials can be determined [92]. Also the N 1s binding energy (measured by XPS) can be used to study the basic properties, as shown on a series of alkali containing faujasites [93], where a linear correlation between the N 1s binding energy and the HN stretching frequency as well as the heat of adsorption was observed. Other molecules such as organic acids, i.e., benzoic acid [94], acetic acid [94] and trichloroacetic acid [95] have been also applied for the characterization of basic sites.

8.6 Concluding remarks

The techniques described here allow the characterization of the complex pore structure and the acid/base related properties of molecular sieves, while redox and host/guest properties have only partially been covered. New physicochemical methods increase in importance, such as X-ray absorption spectroscopy and positron emission imaging [96], while others such as small angle scattering [97-99], which is important to understand the particle formation, were not discussed in the present chapter.

However, as emphasized in the introduction always several methods are needed to describe the properties of a molecular sieve. This is not only important for studies describing the synthesis of new materials, but it is equally important for studies where a particular application of a molecular sieve is explored. For these cases, i.e., when sorptive, catalytic or ion exchange properties are investigated, the experimental part should not only contain a detailed description of the preparation procedures, but should also give a minimum set of characterization data in order to be able to repeat the experiments and/or to compare them with results of own materials.

These techniques should include:

(i) Chemical composition

(ii) N_2 or Ar sorption isotherms measured at the condensation temperature of the gas

(iii) X-Ray diffraction data

(iv) Sorption of probe molecules monitored by semiquantitative methods, e.g., gravimetry, TPD or molecular spectroscopy to the assess the general acid/base properties

(v) Pyridine adsorption (IR-spectroscopy), TPD of the samples after liquid phase NH_4OH exchange or ^{29}Si and ^{27}Al MAS-NMR to characterize Brønsted and Lewis type acid sites.

REFERENCES

1 N. Cardone-Martinez and J. A. Dumesic, Adv. Catal., 149 (1992) 38.

2 M. Stöcker, Microp. Mater., 6 (1996) 235.

3 J. A. Lercher, C. Gründling and G. Eder-Mirth, Catal. Today, 27 (1996) 353.

4 G. Perego, Catal. Today, 41 (1998) 251.

5 R. F. Howe, Stud. Surf. Sci. Catal., 102 (1996) 97.

6 J. Klinowski, Anal. Chim. Acta, 283 (1993) 929.

7 S. Kaliaguine, Stud. Surf. Sci. Catal., 102 (1996) 191.

8 M. M. J. Treacy, J.B. Higgins & R. von Ballmoos, *in* "Collection of Simulated XRD Powder Patterns for Zeolites"

9 http://www.iza-structure.org/databases/

10 R. Millini, E. P. Massara, G. Perego and G. Bellussi, J. Catal., 137 (1992) 479.

11 ASTM Standards on Catalysis D3906-85a (1988).

12 P. Scherrer, Göttinger Nachrichten, 2 (1918) 89.

13 H. M. Rietveld, J. Appl. Cryst., 2 (1969) 65.

14 R. A. Young, in "The Rietveld Methode", Oxford University Press, Oxford (1993).

15 Ch. Baerlocher and L. B. McCusker, Stud. Surf. Sci. Catal., 85 (1994) 391.

16 J. W. Niemantsverdriet in „Spectroscopy in Catalystis", VCH Weilheim (1995).

17 A. K. Datye in "Handbook of Heterogeneous Catalysis", (G. Ertl, H. Knözinger and J. Weitkamp eds.), VCH Verlag, Weinheim, p. 493 (1997).

18 A. J. Koster, U. Ziese, A. J. Verkleij, A. H. Janssen, and K. P. de Jong J. Phys. Chem. B., 104 (2000) 9368.

19 E. M. Flanigen, in "Zeolite Chemistry and Catalysis" (J. Rabo ed.), ACS Monograph Ser., 171 (1976) 93.

20 E. M. Flanigen, H. Khatami and H. A. Szymanski, Adv. Chem. Ser., 10 (1971) 201.

21 H. Knözinger and G. Mestl, Topics in Catalysis, 8 (1999) 45.

22 M. J. Wax, R. R. Cavanagh, J. J. Rush, G. D. Stucky L. Abrams and D. R. Corbin, J. Phys. Chem., 90 (1986) 532.

23 P. K. Dutta and J. Twu, J. Phys. Chem., 95 (1991) 2498.

24 S. M. Campbell, D. M. Bibby, J. M. Coddington, R. F. Howe and R. H. Meinhold, J. Catal., 161 (1996) 338.

25 H. Esemann and H. Forster, Z. Physik. Chem., 93 (1989) 1409.

26 IUPAC Manual of Symbols and Terminology, Appendix 2, Part1, Colloid and Surface Chemistry, Pure Appl. Chem., 31 (1972) 578.

27 J. A. Lercher in "Catalysis - an Integrated Approach", (R. A. van Santen, P. van Leeuwen, J. A. Moulijn and B. A. Averill, eds.) Elsevier, Amsterdam, p. 543 (1999).

28 S. Brunauer, L. S. Deming, W. S. Deming and E. Teller, J. Am. Chem. Soc., 62 (1940) 1723.

29 S. Brunauer, P. H. Emmet and E. Teller, J. Am. Chem. Soc., 60 (1938) 1553.

30 B. C. Lippens, B. G. Linsen and J. H. de Boer, J. Catal., 3 (1964) 32.

31 D. W. Breck, in "Zeolite Molecular Sieves", John Wiley & Sons (1974).

32 E. N. Coker, D. P. Roelofsen, R. M. Barrer, J. C. Jansen and H. van Bekkum, Microp. and Mesop. Mater., 22 (1998) 261.

33 F. Eder and J.A. Lercher, J. Phys. Chem., 101 (1997) 1273.

34 J. Fraissard and T. Ito, Zeolites, 8 (1988) 351.

35 V. J. Frilette, W. O. Haag and R. M. Lago, J. Catal., 67 (1981) 218.

36 J. A. Martens and P. A. Jacobs, Zeolites, 6 (1986) 334.

37 S. Ernst, R. Kumar, M. Neuber, and J. Weikamp, in „Characterization of Porous Solids", (K. K. Unger ed.), Elsevier, Amsterdam (1988).

38 Shimada, H., Matsubayashi, N., Imamura, M., Sato, T., and Nishijima, A., Catal. Lett., 39 (1996) 125.

39 D. W. Breck and E. M. Flanigen, in "Molecular Sieves", Soc. Chem. Ind., 47 (1968).

40 H. Fichtner-Schmittler, U. Lohse, G. Engelhardt and V. Patzelova, Cryst. Res. Tech., 19 (1984) 1.

41 J. R. Sohn, S. J. DeCanio, J. H. Lunsford and D. J. O'Donnel, Zeolites, 6 (1986) 225.

42 V. Jorik, Zeolites, 13 (1993) 187.

43 G. Engelhardt, Stud. Surf. Sci. Catal., 52 (1991) 151.

44 G. Engelhardt in "Handbook of Heterogeneous Catalysis", (G. Ertl, H. Knözinger and J. Weitkamp eds.), VCH Verlag, Weinheim, p. 525 (1997).

45 M. Stöcker, Stud. Surf. Sci. Catal., 102 (1996) 141.

46 D. H. Olson, W. O. Haag and R. M. Lago, J. Catal., 61 (1989) 390.

47 R. M. Lago, W. O. Haag, R. J. Mikovsky, D. H. Olson, S. D. Hellring, K. D. Schmitt and G. T. Kerr, Proc. of the 7th Int. Zeol. Conf. Amsterdam, p. 677 (1986).

48 H. Rastelli, B. M. Lok J. A. Duisman, D. E. Earls and J. T. Mullhaupt, Can. J. Chem. Eng., 60 (1982) 44.

49 M. Deeba and W. K. Hall, J. Catal., 60 (1979) 417.

50 J. A. Lercher, V. Veefkind and K. Fajerwerg, Vibr. Spectroscopy, 19 (1999) 107.

51 J. B. Peri, in "Catalysis-Science and Technology", (J. R. Anderson and M. Boudart eds.), Vol. 5, Springer, Berlin, p. 172 (1984).

52 W. C. Hecker and A. T. Bell, J. Catal., 71 (1981) 216.

53 M. Nowotny, H. Kessler and J. A. Lercher, Zeolites, 11 (1991) 454.

54 G. Müller, T. F. Narbeshuber, G. Mirth and J. A. Lercher, J. Phys. Chem., 94 (1994) 7436.

55 D. J. Parillo, R. J. Gorte and W. E. Farneth, J. Am. Chem. Soc., 115 (1993) 12441.

56 G. Mirth and J. A. Lercher, J. Phys. Chem., 95 (1991) 3736.

57 G. Mirth, J. Cejka and J. A. Lercher, J. Catal., 139 (1993) 24.

58 W. Niessen and H. G. Karge, Microp. Mater., 1 (1993) 1.

59 M. Hermann, W. Niessen and H.G. Karge, Stud. Surf. Sci. Catal., 94 (1995) 131.

60 A. Jentys, K. Kleestorfer and H. Vinek, Microp. and Mesop. Mater., 27 (1999) 321.

61 E. A. Paukshtis and E. N. Yurchenko, Russ. Chem. Rev., 52 (1983) 242.

62 H. Knözinger, in "Elementary Reaction Steps in Heterogeneous Catalysis", (R. W. Joyner and R. A. van Santen eds.) Kluwer Academic Publishers, Dordrecht, p. 267 (1993).

63 E. H. Teunissen, R. A. van Santen and A. P. J. Jansen, J. Phys. Chem., 97 (1993) 203.

64 L.H. Little, in "Infrared Spectra of Adsorbed Species", Academic Press, London (1966)

65 E. P. Parry, J. Catal., 2 (1963) 371.

66 P. A. Jacobs, Catal. Rev.-Sci. Eng., 24 (1982) 415.

67 P. A. Jacobs, J. A. Martens, J. Weitkamp and H. K. Beyer, Far. Disc. Chem. Soc., 72 (1981) 351.

68 R. G. Parr and R. G. Pearson, J. Am. Chem. Soc., 105 (1983) 7512.

69 R. G. Pearson, J. Am. Chem. Soc., 85 (1963) 3533.

70 R. L. Burwell Jr., J. F. Read, K. C. Taylor and G. L. Haller, Z. Phys. Chem., 64 (1969) 18.

71 Cerius^2, Molecular Simulations Inc.

72 T. F. Nagy, S. D. Mahanti and J. L. Dye, Zeolites, 19 (1997) 57.

73 C. A. Emeis, J. Catal., 141 (1993) 374.

74 J. W. Ward, J. Catal., 9 (1967) 225.

75 R. P. H. Gasser in „An introduction to chemisorption and catalysis by metals", Clarendon Press, Oxford (1985).

76 R. J. Gorte, Catal. Lett., 62 (1999) 1.

77 M. Stockenhuber and J. A. Lercher, Microp. Mater., 3 (1995) 4547.

78 C. Defosse and P. Canesson, J. Chem. Soc., Faraday Trans., 1 (1976) 2565.

79 R. Borade, A, Sayari, A. Adnot and S. Kaliaguine, J. Phys. Chem., 94 (1990) 5989.

80 A. Auroux and J. Dakta, Appl. Catal. A, 165 (1997) 473.

81 D. C. Koningsberger and R. Prins, eds., Principles, Applications, Techniques of EXAFS, SEXAFS and XANES, Chemical Analysis Vol. 92, Wiley, New York (1988).

82 Z. C Zhang; G. D. Lei and W. M. H. Sachtler, in "Series on Synchrotron Radiation Techniques and Applications", (Y. Iwasawa ed.), World Scientific, Singapore ,Vol. 2, p. 137 (1996).

83 K. J. Chao, A. C. Wei, H. C. Wu and J. F. Lee, Catal. Today, 49 (1999) 277.

84 D. C. Koningsberger and J. J. Miller, Catal. Lett., 29 (1994) 77.

85 K. Shimizu, Y. Kato, T. Yoshida, H. Yoshida, A. Satsuma and T. Hattori, J. Chem. Soc., Chem. Commun., 1681 (1999).

86 M. Fröba, J. Wong, P. Behrens, P. Sieger, M. Rowen T, Tanaka, Z. Rek and J. Felschke, Physica B, 208 & 209 (1995) 65.

87 J. F. Tempere, D. Delafosse and J. P. Contour, Chem. Phys. Lett., 33 (1975) 95.

88 J. I. Take, N. Kikichi and Y. Yoneda, J. Catal., 21 (1971) 164.

89 A. Auroux and A. Gervasini, J. Phys. Chem., 94 (1990) 6371.

90 J. C. Lavalley, Catal. Today, 27 (1996) 377.

91 J. C. Lavalley, Trends Phys. Chem., 2 (1991) 305.

92 P. O. Scokart and P. G. Rouxhet, J. Chem. Soc., Farady Trans., 76 (1980) 1476.

93 M. M. Huang, S. Kaliaguine, M. Muscas and A. Auroux, J. Catal., 157 (1995) 266.

94 E. A. Paukshtis, N. S. Kotsarenko and L. G. Karakchiev, React. Kin. Katal. Lett., 12 (1979) 315.

95 W. Przystajko, R. Fiedorow, and I. G. Dalla Lana, Zeolites, 7 (1987) 477.

96 B. G. Anderson, R. A. van Santen and L. J. van IJzendoorn, Appl. Catal. A, 160 (1997) 125.

97 M. Linden, J. Blanchard, S. Schacht, S. A. Schunk and F. Schüth, Chem. Mater., 11 (1999) 3002.

98 P. E. A. de Moor, T. P. M. Beelen, B. U. Komanschek and R. A. van Santen, Microp. and Mesop. Mater., 21 (1998) 263.

99 P-P. E. A. de Moor, T. P. M. Beelen, R. A. Van Santen, L. W. Beck and M. E. Davis, J. Phys. Chem. B, 104 (2000) 7600.

Studies in Surface Science and Catalysis 137
H. van Bekkum, E.M. Flanigen, P.A. Jacobs and J.C. Jansen (Editors)
© 2001 Elsevier Science B.V. All rights reserved.

Chapter 9

SOLID STATE NMR SPECTROSCOPY APPLIED TO ZEOLITES

G. Engelhardt
Institute of Chemical Technology I, University of Stuttgart, D-70550 Stuttgart, Germany

1. INTRODUCTION

Since the phenomenon of nuclear magnetic resonance (NMR) was discovered in 1946 by Bloch and Purcell, NMR spectroscopy has emerged as one of the most powerful tools in investigating structure and dynamics of molecular systems in the various branches of chemistry. NMR spectroscopy relies on the fact that transitions can be induced between magnetic spin energy levels of certain atomic nuclei in a magnetic field. The main application of NMR in chemistry is based on the early observation that the transition frequency measured in the NMR spectrum of an atomic nucleus in a particular chemical or structural environment is a very sensitive probe of that environment. Provided that the distinct "chemically shifted" transitions of nuclei of the same element residing in different surroundings can be resolved in the NMR spectrum, unique information on structure, conformation and dynamics of the system under study can be obtained. It is this distinction of individual resonance lines or specific line patterns for resonance atoms in chemically or structurally distinct locations which defines a "high-resolution" NMR spectrum.

Until the late 1970's, the application of high-resolution NMR spectroscopy in chemistry was more or less restricted to liquids since the straightforward application of the conventional NMR techniques to solids yields in general spectra with very broad and almost featureless resonance lines, masking any differences in the transition frequencies of resonance atoms in distinct structural environments. However, the development of novel sophisticated NMR techniques in the early 1970's opened the possibility to measure NMR spectra of solids comparable in spectral resolution to those of liquids. Subsequently, high-resolution solid-state NMR spectroscopy has developed rapidly into an effective and widely used means of elucidating subtle details of the structural properties of a wide range of different kinds of solid materials (for a comprehensive survey see [1]). Since high-resolution solid-state NMR spectroscopy gives, in general, information on the local order of the structure and can be applied to crystalline, microcrystalline and amorphous samples, it is a valuable complement to X-ray and other diffraction techniques which probe the long-range order of crystalline materials.

The first application of high-resolution solid-state NMR to zeolites was presented in 1979 at a conference [2], and a series of pioneering papers were published in 1981 (see, e.g., [3-6]). In this early work mainly ^{29}Si NMR but also ^{27}Al NMR [7] were used to study the aluminosilicate framework of various zeolites. Subsequently the interest in the application of multinuclear solid-state NMR to zeolites and other silicates and aluminosilicates as well as aluminophosphate molecular sieves has been growing very rapidly and several thousand papers on the subject have been published during the last years. The progress in the field has been summarised in several reviews (see, for example, [8-10]) and a book [11].

In this article a brief introduction will be given into the basic principles and experimental techniques of high-resolution solid-state NMR, followed by a concise survey on the main information retrievable from the NMR spectra and its application to structural studies of zeolites and related systems. Emphasis is laid on the framework structure of aluminosilicate zeolites and their isomorphously substituted variants, but non-framework constituents are also briefly considered. Selected literature is cited in each chapter but the number of references is restricted and I wish to apologise to all authors who made significant contributions to the field but are not cited in the present article.

2. BASIC PRINCIPLES AND METHODS OF HIGH-RESOLUTION SOLID-STATE NMR

2.1 High-resolution NMR: Liquids versus solids

High-resolution NMR spectra of liquid samples exhibit, in general, sharp lines characterised by distinct chemical shifts (line positions) and intensities (integrated peak areas). Both parameters can easily be extracted from the spectra and are closely related to the chemical structure of the investigated sample. The chemical shift reflects very sensitively the immediate structural surroundings of the resonance nucleus, while the intensity is directly proportional to the concentration of the different nuclei in the sample.

For solids the situation is different in that considerable line broadenings arise in the NMR spectra owing to specific interactions of the nuclear spins tightly bound in the rigid lattice of the solid sample. If the conventional NMR technique is applied to solids, the line broadening prevents, in general, the resolution of different peaks for structurally distinct resonance atoms, or even renders the NMR spectra unobservable.

The critical differences between NMR of liquids and solids arise from the directional dependence of the various spin interactions. The fixed orientations of nuclear environments in the solid lattice affect the shape and position of the NMR lines measured for different orientations of the sample with respect to the direction of the external magnetic field, B_o. In highly ordered single crystals, were only a small number of symmetry-related orientations of nuclei is present, narrow lines are observed, the positions of which change with the orientation of the crystal in the external field. Microcrystalline or amorphous powders, however, are characterised by a random distribution of different orientations and the observed NMR pattern is a broad superposition of individual resonances of all orientations. In liquids, the different orientations are averaged by the fast and essentially isotropic motions of the molecules which reduce the spin interactions to a discrete average value or to zero.

2.2 Nuclear spin interactions and line broadening in solids

There are three main interactions which may contribute to line broadenings in solid-state NMR spectra:

- The *dipolar coupling* arises from the direct dipole-dipole interactions between the magnetic moments of the observed nucleus and those of neighbouring nuclei. The dipolar interaction depends on the magnitude of the magnetic moments of the interacting nuclei, falls off very rapidly with the internuclear distance ($1/r^3$), and is independent of the applied magnetic field B_o. Dipolar coupling with protons, such as ^1H-(^1H), ^{13}C-(^1H) or ^{29}Si-(^1H), is the dominant line-broadening interaction in ^1H-NMR spectra and in NMR spectra of other spin-1/2 nuclei in proton-containing solids and can range up to about 100 kHz. The homonuclear dipolar interaction of nuclei with low natural abundance (e.g., ^{13}C or ^{29}Si) or low chemical concentration in the sample can generally be neglected owing to the large internuclear distances.

- The *chemical shift anisotropy* (CSA) follows from the spatial dependence of the nuclear shielding which is determined by the symmetry of the electronic charge distribution around the resonance nucleus. The CSA interaction increases linearly with the strength of the external magnetic field B_0. Typical line broadenings from CSA range up to about 100 ppm but may be considerably larger in special cases.

- The *quadrupolar interaction* is observed only for nuclei with a nuclear spin quantum number $I > 1/2$ (e.g. ^{11}B, ^{17}O, ^{23}Na, ^{27}Al). Nuclei with $I > 1/2$ possess a nuclear electric quadrupole moment , eQ, which interacts with the electric field gradient (EFG), eq, produced at the nuclear site by a non-spherical charge distribution around the nucleus. The quadrupolar interaction is usually characterised by the quadrupole coupling constant, $QCC = e^2qQ/h$, and the asymmetry parameter of the EFG, $\eta_Q = (eq_{xx} - eq_{yy})/eq_{zz}$. QCC can range up to several MHz and, owing to the directional dependence of the quadrupolar interaction, strong line broadenings may appear in the spectra of polycrystalline solids. The quadrupolar powder patterns of nuclei with non-integer spin I are mainly affected by the quadrupolar interaction to second order which decreases with increasing B_0.

Although the line broadenings resulting from the above interactions obscure the "high-resolution" characteristics of the solid-state NMR spectra of powders, the interactions itself are, in principle, also a potential source of information on structure and bonding. Dipolar interactions yield internuclear distances and molecular geometries, chemical shift anisotropies give information on coordination symmetry, and from quadrupolar interactions information on the electric field gradient, i.e. the symmetry of the charge distribution around the nucleus may be derived. It is, however, often difficult to separate the different interactions and to extract the pertinent interaction parameters from the spectra.

2.3 Experimental techniques for line narrowing and sensitivity enhancement

High-resolution NMR spectra of solids with discrete resonance lines for structurally distinct nuclei can only be obtained when the line broadening interactions considered above are removed or at least substantially reduced. Line narrowing in solids is achieved, in particular, by three different experimental techniques:

- *Dipolar Decoupling* (DD) removes the heteronuclear dipolar interaction by irradiating the resonance frequency of the nucleus giving rise to the dipolar broadening (e.g. ^{1}H) while observing the nucleus under study (e.g. ^{13}C, ^{29}Si). Dipolar decoupling is mainly used to remove line broadening effects of dipolar interactions with protons in the NMR spectra of other nuclei.

- *Multiple-Pulse Sequences* (MPS) are used to remove homonuclear dipolar interactions by irradiation of precisely defined sequences of short, intense radio-frequency pulses which average the dipolar interaction by reorienting the nuclear spins. Multiple-pulse sequences are preferably applied to remove ^{1}H-^{1}H dipolar interactions in ^{1}H NMR spectra.

- *Magic-Angle Spinning* (MAS). Line broadenings due to dipolar interactions and chemical shift anisotropy can be removed by fast rotation of the sample about an axis inclined at the angle $\theta = 54°44'$ to the direction of the external magnetic field. Both the dipolar coupling and the chemical shift anisotropy contain an angular dependence of the form $(3\cos^2\theta - 1)$. Thus, if θ is chosen to be 54°44', the "magic angle", this term becomes zero and the dipolar interactions vanish, the chemical shift anisotropy is averaged to its isotropic value, and the quadrupolar broadening is reduced. The rotation frequency in MAS experiments should be of the order of the line width (in Hz) of the static spectrum. Spinning at lower rates yields a spectrum where the central peak is flanked by a series of "spinning side bands" separated by the spinning frequency. Although rotation frequencies of several kHz (typically 4-6 kHz with rotors of 7 mm diameter, 8-12 kHz with 4 mm rotors, and up to 25 kHz with 2.7 mm rotors) can be achieved, complex side band patterns appear in the MAS spectra of systems characterised by large dipolar or quadrupolar interactions, or chemical shift anisotropies.

Quadrupolar line broadenings can also be reduced but not fully removed by MAS. There are, however, special techniques to obtain spectra free of quadrupolar broadening. These are, in particular, Double Rotation (DOR) [12], Dynamic Angle Spinning (DAS) [13] , and Multiple-Quantum Magic-Angle Spinning (MQMAS) [14]. In addition, Satellite Transition Spectroscopy (SATRAS) [15] is used by which the spinning side bands of the satellite transitions are evaluated to derive the quadrupole parameters.

Another important technique in high-resolution solid-state NMR which improves the sensitivity (i.e. the signal-to-noise ratio) of the spectra of nuclei with low natural abundance (e.g. ^{13}C or ^{29}Si) but does not affect the resolution of the spectrum is

- *Cross-Polarisation* (CP). In this technique an indirect excitation of dilute spins S (which are observed) by polarisation transfer from abundant spins I (mostly ^{1}H) is employed. The CP experiment consists of three steps: (i) excitation of the I spins by a 90° pulse, (ii) polarisation transfer from I to S spins by simultaneous irradiation of the rf fields B_{1I} and B_{1S} of the I and S spins according to the Hartmann-Hahn-condition $\gamma_I B_{1I} = \gamma_S B_{1S}$ (where γ_I and γ_S are the magnetogyric ratios of the nuclei I and S), and (iii) acquisition of the FID of the S spins with continued irradiation of the ^{1}H field for heteronuclear dipolar decoupling (see above). The gain in sensitivity is given by the ratio γ_I / γ_S and amounts to a maximum factor of about four for ^{1}H-^{13}C and about five for ^{1}H-^{29}Si CP experiments. In addition, shorter pulse repetition times can be applied in CP experiments than in measurements with direct excitation of the S spins, because the optimum pulse delay in CP depends on the ^{1}H relaxation time $T_1(^{1}H)$ which is usually much shorter than T_1 of the S spins ^{13}C or ^{29}Si. The CP experiment may thus be repeated with much shorter intervals, giving a further increase in the signal-to-noise ratio of the S nucleus spectrum in a given period of time. The CP efficiency depends on the strength of the I-S dipolar interaction, i.e. on the distance between the I and S nuclei. The CP technique may thus also be used to detect close coordinations between protons and, e.g., Si or C atoms.

To achieve optimum line narrowing and sufficient signal-to-noise ratio in the solid-state NMR spectra, the experimental procedures considered above may be used in combination, e.g. CP-MAS, DD-MAS or CRAMPS (Combined Rotation And Multiple Pulse Spectroscopy). However, in zeolites and other inorganic systems often no or only few hydrogen atoms are present. In these cases the CP technique cannot be used or is not very effective and dipolar proton decoupling is not always necessary. Thus, application of MAS is sufficient to remove the chemical shift anisotropy and any small dipolar interaction, and to narrow quadrupolar broadened lines, if present. Provided that the chemical shift anisotropies are not too large, it is often advantageous to carry out these experiments at the highest possible magnetic field strength since sensitivity and chemical shift dispersion is maximised at high field, whereas the dipolar interaction is not affected and the line broadening due to (second order) quadrupolar interactions is minimised. Magnetic field strengths up to 17.6 T are available in state-of-the-art NMR instruments, but 9.4 T is in general sufficient for most applications in the zeolite field.

2.4 Two-dimensional (2D) solid-state NMR

The application of 2D NMR techniques to liquids is well established and widely used, e.g., to determine connectivities between atoms within molecular structures. In spite of many restrictions, 2D NMR experiments can be applied to solids as well and provide valuable information on the three-dimensional connectivities of the atoms within the framework of zeolites and related systems. In particular, ^{29}Si MAS 2D COSY (Correlation Spectroscopy) and INADEQUATE (Incredible Natural Abundance Double Quantum Transfer Experiment) pulse sequences have been applied to high silica zeolites [9]. In these experiments the scalar coupling interactions in the ^{29}Si-O-^{29}Si fragments are used to establish the connectivity of the silicate framework. In the resulting ^{29}Si 2D NMR spectra both frequency axes represent ^{29}Si chemical shifts and the coupled ^{29}Si nuclei are characterised by cross-peaks on either side of the diagonal of the plot. In addition, 2D TEDOR (Transferred-Echo Double Resonance) and REDOR (Rotational Echo Double Resonance) experiments has been used to determine the connectivities between quadrupolar and spin-1/2 nuclei (e.g. ^{27}Al and ^{29}Si) [16], and ^{1}H-^{29}Si 2D HETCOR (Heteronuclear Correlation) is applied to study SiOH groups of surface species of silicas and zeolites [17].

2.5 Important NMR parameters and related structural information

Most of the structural information that can be derived from a high-resolution NMR spectrum of solids relies on the following parameters:

- the number of signals in the spectrum,
- the relative signal intensities, I,
- the isotropic chemical shift, δ_{iso},
- the quadrupole coupling constant, QCC, and the asymmetry parameter of the EFG, η_Q (for nuclei with I > 1/2).

Number of signals and signal intensities, I. Provided that well resolved signals appear in the NMR spectrum (or may be separated by spectra simulation and line decomposition), the number of signals in the spectrum gives immediately the number of different structural environments of the observed nucleus present in the sample. The relative (integrated) signal intensities correspond to the relative occupancies of the different environments. It must be noted, however, that for quadrupolar nuclei the line intensities obtained experimentally from MAS NMR spectra are not a priori proportional to the site populations but must often be corrected for possible contributions from satellite transitions and/or spinning side bands [18].

Isotropic chemical shift, δ_{iso}. For spin-1/2 nuclei, δ_{iso} is the position of the line in the MAS NMR spectrum relative to that of a standard sample and is simply measured from the location of the signal maximum on a calibrated δ-scale (in ppm). For quadrupolar nuclei, the situation is again more complicated because the line position (given by the center of gravity of the quadrupolar powder pattern in the MAS spectrum or by the location of the sharp line in the DOR spectrum) is shifted by second-order quadrupolar interactions and must, therefore, be corrected for the quadrupolar shift contributions. δ_{iso} provides valuable information on the chemical environment of the resonance atom, such as number and type of neighbouring atoms, coordination number, and bonding geometry.

Quadrupole coupling constants, QCC, and *asymmetry parameters of the EFG, η_Q,* ($0 \leq \eta_Q \leq 1$ with $\eta = 0$ for axial symmetric EFG) cannot directly be obtained from the NMR spectra but have to be calculated from distinct features of the quadrupolar powder pattern (total line width, peak separation, shoulders) [19] or determined by computer simulations of the line shape. While the first method works well for isolated powder patterns, the latter is also applicable to overlapping or fully superimposed patterns. Good estimates of QCC can further be obtained from MAS or DOR experiments at different B_o field strenghts. Since the quadrupolar interaction depends on the charge distribution around the nucleus under observation, valuable information on the local electronic environment of that nucleus can be derived from the quadrupole interaction parameters, particularly on distortions of the coordination symmetry.

3. MULTINUCLEAR HIGH-RESOLUTION SOLID-STATE NMR OF ZEOLITES

3.1 Which nuclei are candidates?

In principle, each of the three basic atomic constituents of the aluminosilicate framework of zeolites - silicon, aluminium and oxygen - are amenable to NMR measurements by its naturally occurring isotopes ^{29}Si, ^{27}Al, and ^{17}O, respectively. However, the ^{17}O isotope has a very low natural abundance of only 0.037% and a nuclear quadrupole moment giving rise to line broadening, which together make the application of ^{17}O NMR difficult. Isotopic ^{17}O enrichment is possible but hardly practicable for routine use. Therefore, ^{17}O NMR has been used only sparingly so far, although some interesting and promising ^{17}O NMR studies of of ^{17}O enriched silicates and zeolites haven been published [20]. In contrast to ^{17}O, the ^{27}Al isotope is 100% abundant but has also a nuclear quadrupole moment and may thus exhibit often quadrupolar broadened lines. The isotope ^{29}Si has a natural abundance of 4.7% and a nuclear spin of I = 1/2, i.e. it has no quadrupole moment and thus gives rise to narrow resonance lines. Owing to their important role as framework constituents, their favourable properties in NMR experiments, and especially due to the high sensitivity of their NMR spectra against structural effects (see below), ^{29}Si and ^{27}Al nuclei play the dominant role in NMR studies of the framework structure of zeolites and other silicates and aluminosilicates.

Several other elements replacing silicon or aluminium in tetrahedral sites of the aluminosilicate framework of zeolites, such as phosphorous, boron, gallium, germanium, beryllium, vanadium, and others, can also be profitably studied by solid-state NMR of their respective NMR-active nuclides, for example ^{31}P, ^{11}B, ^{69,71}Ga, ^{73}Ge, ^{9}Be, ^{51}V. However, except ^{31}P which has I = 1/2 and 100% natural abundance, most of the other nuclei possess a quadrupole moment and are often characterised by low relative receptivities in the NMR experiment. ^{31}P NMR (together with ^{27}Al NMR) has been proved to be a useful method for structural studies of aluminophosphate molecular sieves and related systems. ^{11}B NMR, which has a reasonably high sensitivity, has been successfully applied to study incorporation of boron in the framework of high-silica zeolites. Only few in-depth studies of the other nuclei has been performed so far.

Most elements acting as charge-balancing cations in zeolites have also isotopes suitable for NMR experiments, but besides ^{7}Li, ^{23}Na, ^{133}Cs, ^{139}La, and ^{205}Tl only a few others have been studied so far. Organic species present as templates or sorbates in the zeolite cavities can be advantageously investigated by ^{13}C MAS NMR.

Finally, high-resolution solid-state ^{1}H NMR plays an important role in investigating bridging Si(OH)Al and terminal SiOH or AlOH groups, adsorbed H_2O, and organic molecules in zeolites.

In summary, the potential of solid-state NMR in zeolite science relies on the application of a wide range of nuclei and it is significant that an environmental probe exists for almost every atomic constituent which may be present in the zeolite framework or in non-framework species enclathrated in the zeolite cavities. Table 1 summarises the NMR properties of selected nuclei which has been used in solid-state NMR of zeolites and related materials.

Table 1
NMR properties of selected nuclei applied in NMR of zeolites

Isotope	Spin	NMR frequency (MHz)[a]	Natural abundance (%)	Relative receptivity[b]
^{1}H	1/2	400.00	99.985	1.000
^{7}Li	3/2	155.45	92.58	0.272
^{11}B	3/2	128.34	80.42	0.133
^{13}C	1/2	100.58	1.108	$1.76 \cdot 10^{-4}$
^{17}O	5/2	54.23	0.037	$1.08 \cdot 10^{-5}$
^{23}Na	3/2	105.80	100.00	$9.27 \cdot 10^{-2}$
^{27}Al	5/2	104.23	100.00	0.207
^{29}Si	1/2	79.46	4.70	$3.69 \cdot 10^{-4}$
^{31}P	1/2	161.92	100.00	$6.63 \cdot 10^{-2}$
^{51}V	7/2	105.15	99.76	0.38
^{133}Cs	7/2	52.46	100.00	$4.82 \cdot 10^{-2}$
^{205}Tl	1/2	230.83	70.50	0.140

[a] at $B_o = 9.4$ T
[b] product of natural abundance (%) and NMR sensitivity $[\gamma^3 I(I+1)]$

3.2 Silicon-29 NMR

Line broadening in solid-state ^{29}Si NMR spectra of microcrystalline zeolite powders is mainly due to the chemical shift anisotropy of the ^{29}Si nucleus. In addition, heteronuclear dipolar interactions between ^{29}Si and ^{1}H or ^{27}Al, and possibly other NMR-active nuclei present in the zeolite structure may contribute to the line width. The application of the MAS technique is therefore essential and mostly sufficient in obtaining highly resolved ^{29}Si NMR spectra of zeolites. The simultaneous application of high-power dipolar ^{1}H decoupling is sometimes helpful in removing residual dipolar interactions of ^{29}Si spins with nearby ^{1}H nuclei. The spectra obtained under these conditions show, in general, distinct signals for structurally different Si sites in the zeolite structure.

The most direct, and rather fundamental, information which follows immediately from the number of distinct lines in the spectrum is the number of structurally inequivalent Si sites present in the sample. Moreover, from the normalised peak intensities the relative proportions of Si atoms in the various sites can be determined. Information on the local environment of the SiO_4 tetrahedra forming the zeolite framework can be derived from the chemical shift data.

3.2.1 Structure correlation of ^{29}Si chemical shifts

The ^{29}Si chemical shifts of silicates and aluminosilicates depend sensitively on the number and type of T-atoms (T = Si, Al or other tetrahedral framework atoms) connected with a given SiO_4 tetrahedron. Characteristic high-field shifts are observed with increasing number of SiOT bridges formed by the given SiO_4 tetrahedron (degree of SiO_4 polymerisation), and typical low-field shifts of the resonances follow from the replacement of Si by Al in the second coordination sphere of the central Si atom with a given number of SiOT bridges (degree of tetrahedral Al substitution). In zeolites and other aluminosilicates with a three-dimensional framework structure, all silicon atoms are connected via oxygen bridges with four other T-atoms, i.e. there are five different structural units of the type $Si(OSi)_{4-n}(OAl)_n$ with n = 0 - 4, conventionally designated "Si(nAl)". In a first approximation, neglecting the presence of crystallographically inequivalent Si(nAl) sites, the ^{29}Si NMR spectrum of a zeolite may thus consist of one to five peaks corresponding to the five possible Si(nAl) environments in the zeolite framework. With increasing number n of Al atoms, the peaks are systematically shifted to low field, whereby each Al substitution results in a shift contribution of about 5 ppm. As shown in Figure 1, such shift differences can readily be resolved in the ^{29}Si MAS NMR spectra of microcrystalline zeolites, and well separated peaks of the whole series of Si(nAl) units appear in the spectrum of zeolite ZK-5 with Si/Al = 2.2. In addition to the chemical environment, the ^{29}Si chemical shift is affected by the bonding geometry around the Si atom under study, i.e. by SiO bond lengths and SiOT bond angles (see below). Therefore, chemically equivalent but crystallographically inequivalent Si atoms may exhibit different chemical shifts.

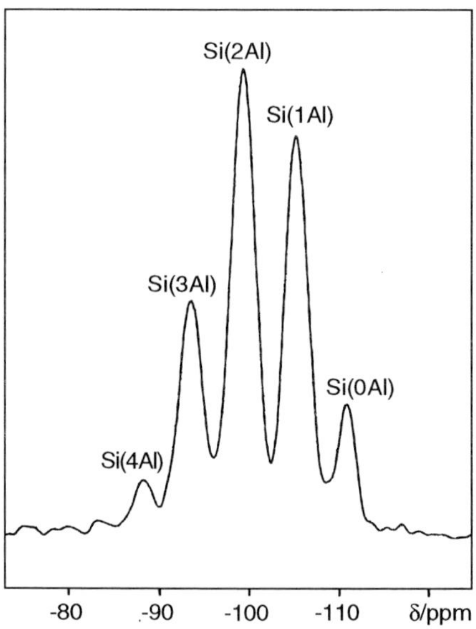

Fig.1. ^{29}Si MAS NMR spectrum of zeolite ZK-5 (Si/Al = 2.2)

^{29}Si chemical shift data of selected zeolites are given in Table 2, and typical shift ranges for the five Si(nAl) environments established from a large body of shift data measured in various types of zeolites and other framework silicates are shown in Figure 2. Although the different shift ranges partly overlap, the assignment of well separated peaks to the corresponding Si(nAl) environments of a particular zeolite is, in general, feasible. The largest shift range observed for a certain Si(nAl) unit is that of Si(4Al) in aluminosilicate sodalites. Depending on the type of salts enclathrated in the sodalite cages, the ^{29}Si chemical shift changes by more than 20 ppm [21] (see also Section 3.2.2).

Table 2
^{29}Si chemical shifts (in ppm versus TMS) of selected zeolites [a]

Zeolite	Si/Al	Si(4Al)	Si(3Al)	Si(2Al)	Si(1Al)	Si(0Al)
LiCl-sodalite	1.0	-76.4				
NaK-X	1.0	-84.6				
Na-A	1.0	-89.6				
K-sodalite[b]	1.0	-96.7				
ZK-4	1.4	-89.1	-93.9	-99.5	-106.1	-110.7
NaK-P1	1.9	-87.6	-91.9	-97.3	-102.4	-107.0
Na-Y	2.5	-83.8	-89.2	-94.5	-100.0	-105.5
Offretite[c]	2.9	-	-93.5 -97.5	-97.5 -101.9	-101.9 -106.9	-106.9 -112.5
Omega[c]	3.1	 -89.1	-89.1 -93.7	-93.7 -98.8	-98.8 -107.0	-103.4 -112.0
TMA-sodalite	4.7			-104.6	-110.5	-116.2
Na-mordenite	5.0			-100.1	-105.7	-112.1

[a] from [11]
[b] dehydrated
[c] two crystallographically inequivalent Si sites (see text)

3.2.2 Framework structure studies

Local environment of SiO_4 tetrahedra. As has been shown above, distinct signals appear in the ^{29}Si NMR spectra of zeolites for the different Si(nAl) environments forming the tetrahedral zeolite framework. The relative signal intensities are directly related to the relative concentrations of the various Si(nAl) units present in the zeolite structure. Figure 3 shows the ^{29}Si NMR spectra of a series of zeolites X and Y with different Si/Al ratios [4]. With increasing Si/Al ratio, the decrease of the signal intensities of the aluminium-rich Si(nAl) units and a corresponding increase of the aluminium-poor Si(nAl) peaks is clearly visible. Consequently, from a careful analysis of the chemical shifts and line intensities, the specific

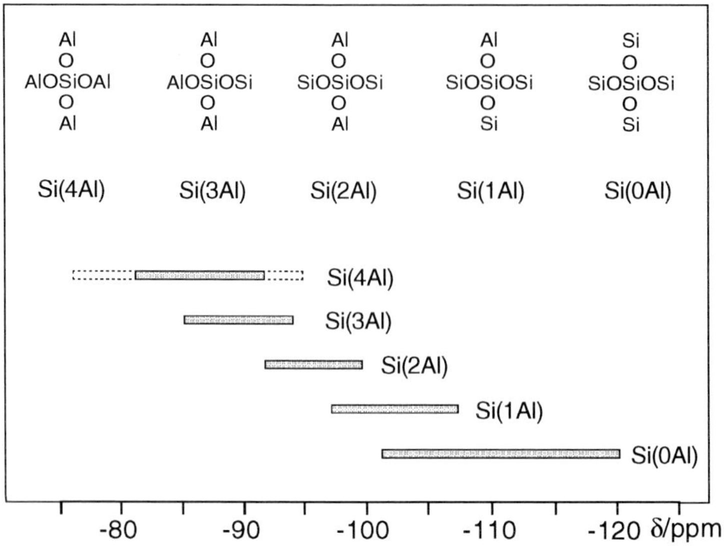

Fig.2. Ranges of [29]Si chemical shifts of Si(nAl) units in zeolites. The dotted lines for Si(4Al) units designate the shift range for sodalites (see text).

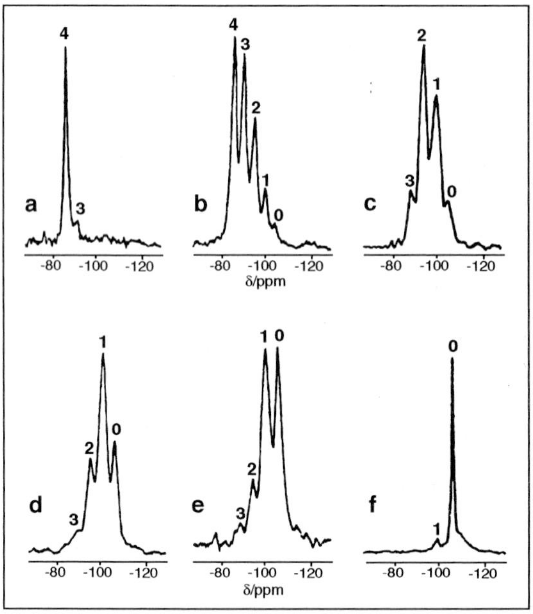

Fig.3. ^{29}Si MAS NMR spectra of zeolites NaX and NaY with different Si/Al ratios: a - Si/Al = 1.02, b - 1.41, c - 2.45, d - 3.77, e - 4.88, f - ca. 50. Peak assignments are given by the number n of the respective Si(nAl) units

types and relative populations of the distinct Si(nAl) units present in a zeolite can, in principle, be determined. Moreover, if other tetrahedral atoms replacing aluminium and/or silicon are present in the framework, characteristic shifts of the Si(nT) lines may be observed. A pertinent example are the gallosilicate zeolites in which the Si(nGa) peaks are shifted to low field by 1.5 to 2 ppm per Ga substitution in comparison to their Si(nAl) counterparts [22].

Silicon atoms bearing OH groups, e.g. $Si(OH)(OSi)_{3-n}(OAl)_n$, located in framework defects or in amorphous parts of the zeolite material exhibit peaks characteristically shifted to low field from the corresponding Si(nAl) units and may overlap with the Si[(n+1)Al] signals. The SiOH peaks can, however be clearly identified by their intensity enhancement in the CP spectra [23].

Si/Al ratio and Si,Al ordering. Provided that the ^{29}Si NMR spectrum of a zeolite is correctly interpreted in terms of the Si(nAl) units and no AlOAl linkages are present (i.e. Loewenstein's rule applies), the quantitative ratio of tetrahedral Si and Al atoms in the zeolite framework can directly be calculated from the peak intensities according to the equation

$$Si/Al = \Sigma I_n / \Sigma (0.25 \cdot n \cdot I_n) \qquad (1)$$

where I_n are the intensities of the Si(nAl) peaks and summation is from n = 0 to n = 4 [4]. Equation (1) is independent of the specific structure of the zeolite and includes only that aluminium which is substitutionally incorporated into the tetrahedral framework, but excludes any non-framework aluminium frequently present, for example, in thermally or chemically treated zeolites (see below).

Information on Si, Al ordering in the tetrahedral T-sites of the zeolite framework can be derived by comparison of the relative Si(nAl) populations obtained from NMR peak intensities with model generated populations. The latter were derived from ordering schemes constructed by distributing the respective number of Si and Al atoms on the T-sites of a certain cutout (repeating unit) of the framework structure. This procedure has successfully been applied to study the Si, Al ordering in faujasite-type zeolites of different Si/Al ratios [4, 24] and in several other zeolite types (see e.g. [25]). Figure 4 shows the ordering schemes proposed for a series of zeolites X and Y from the NMR spectra presented in Figure 3.

Crystallographically non-equivalent Si sites. As mentioned above, crystallographically inequivalent Si sites in chemically equivalent environments may have different chemical shifts which leads to splittings of the NMR signal of a certain Si(nAl) unit. Although such splittings often complicate the spectra interpretation, they may provide valuable information on the presence of crystallographically inequivalent sites and their populations in the zeolite framework. To avoid complications due to superposition of the spectral patterns for the different Si(nAl) units, it is advantageous to investigate highly siliceous zeolites containing only Si(0Al) environments. Provided that the framework geometry is retained in the high-silica form of the zeolite, the spectrum is greatly simplified and any multiple line pattern indicates the presence of crystallographically non-equivalent Si sites in the framework.

One of the most striking examples is represented in Figure 5 by the essentially aluminium-free variant of the zeolite ZSM-5 (silicalite) which shows a temperature dependent monoclinic - orthorhombic phase transition. The ^{29}Si MAS NMR spectrum at 295 K exhibits 20 well resolved lines (two with three-fold intensity) representing the 24 crystallographically inequivalent Si sites of the monoclinic form, whereas at 393 K only 10 lines appear in the spectrum (two with double intensity) which correspond to the 12 inequivalent Si sites in the orthorhombic form [26]. Moreover, dramatic and fully reversible changes in the spectra are observed upon loading the ZSM-5 zeolite with organic sorbates, such as p-xylene,

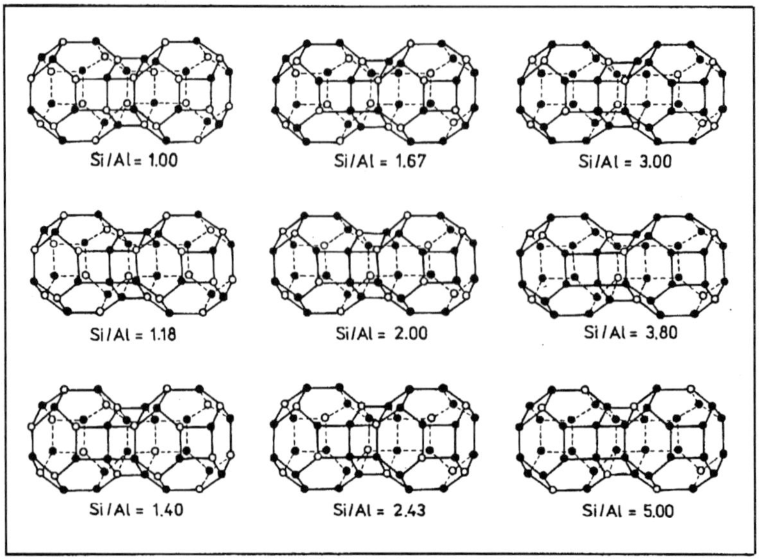

Fig.4. Si,Al ordering schemes of zeolites X and Y as derived from ^{29}Si NMR. Black dots designate Si atoms, white dots Al atoms

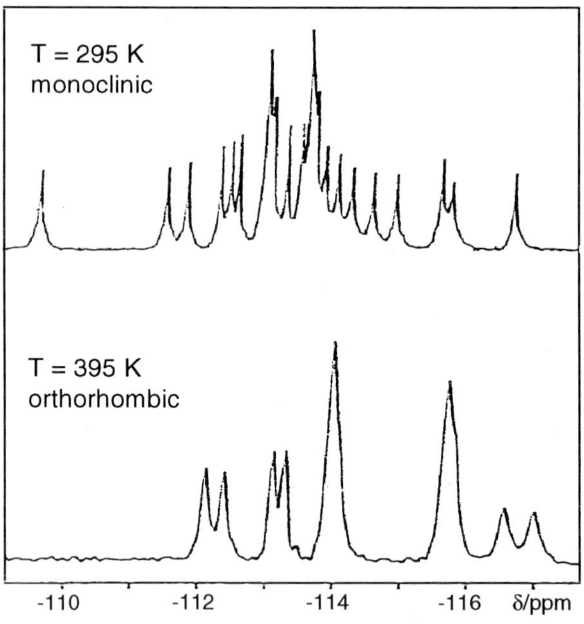

Fig.5. ^{29}Si MAS NMR spectra of high-silica zeolite ZSM-5 [9]

acetylacetone, pyridine [26], n-decane, cyclohexane, or benzene [11], which, together with X-ray powder diffraction studies, provide detailed information on the sorbate-framework interactions in these materials.

The ^{29}Si NMR spectra of the high-silica zeolites are also very helpful in the correct assignment of the spectra of the corresponding variants with low Si/Al ratio. Figure 6 shows the ^{29}Si NMR spectra of dealuminated mordenite, offretite, and zeolite omega, and of the corresponding low Si/Al parent materials [9]. Three lines with the intensity ratio 2:1:3 are observed in the spectrum of high-silica mordenite, corresponding to the four inequivalent Si positions in the population ratio $2Si_1$: $1Si_4$: ($2Si_2$+$1Si_3$), whereas two lines with the intensity ratio of 2:1 appear for the two inequivalent Si sites with the population ratio $2Si_1$:$1Si_2$ in the spectra of offretite and zeolite omega. Depending on the shift differences of the crystallographically inequivalent Si sites in the high-silica form, line broadening (mordenite), deceptively simple spectra (offretite), or additional signals (omega) result from the superposition of the Si(nAl) sub-spectra in the aluminium-containing parent zeolites. Clearly, the observed peak overlap must be taken into account when the Si/Al ratios were calculated from those and similar spectra (see e.g. [11, 27]).

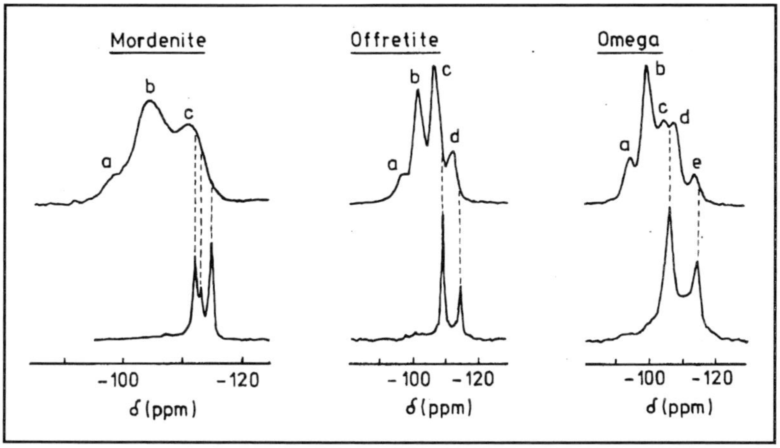

Fig.6. ^{29}Si MAS NMR spectra of mordenite (Si/Al = 5), offretite (5.1), and zeolite omega (4.2), and their high-silica variants (lower spectra). Peak assignments (upper spectra): mordenite a - Si(2Al), b - Si(1Al), c - Si(0Al); offretite a - Si_1(2Al), b - Si_1(1Al) + Si_2(2Al), c - Si_1(0Al) + Si_2(1Al), d - Si_2(0Al); zeolite omega a - Si_1(2Al), b - Si_1(1Al), c - Si_1(0Al) + Si_2(2Al), d - Si_2(1Al), e - Si_2(0Al) [9].

Correlations between SiOT bond angles and ^{29}Si chemical shifts. The differences in the ^{29}Si chemical shifts of Si atoms located in the same chemical environment are mainly due to changes in the bonding geometry of the respective Si(nAl) structural unit. It has been shown by empirical correlations [21,28,29], and rationalised by theoretical considerations [28,30,31], that the ^{29}Si chemical shift δ of a certain Si(nAl) unit is linearly correlated with the average value of the four SiOT bond angles α at the central silicon atom. Quantitative relationships between δ and α, secα, sin(α/2) and cosα/(cosα - 1) have been established by linear regression analysis of a large number of experimental data. The correlation works with these different functions of α since they are all approximately linear in the range of α involved. The correlation between δ and cosα/(cosα - 1) follows from theoretical considerations [30]. Linear regression of the data of 52 different Si(nAl) units in zeolites and silica polymorphs has led to the following relationship [31] with a linear correlation coefficient of 0.90:

$$\delta = -223.9 \cos\alpha/(\cos\alpha - 1) + 5n - 7.2 \qquad (2)$$

where n is the number of Al atoms in the Si(nAl) unit under study, and δ is given in ppm from tetramethylsilan, and α in degrees. Although eq. (2) can be applied to any Si(nAl) environment, similar relationships with correlation coefficients better than 0.96 have been derived for Si(0Al) [29a,30] and Si(4Al) units [21,29b] in zeolites. The quantitative relationships between δ and α can be used to estimate mean SiOT bond angles from ^{29}Si chemical shifts or, vice versa, if the bond angles are known from X-ray studies, to calculate shift data which may be helpful in the correct assignment of the spectra.

The quantitative evaluation of the various relationships between ^{29}Si chemical shifts and SiOT bond angles shows that a chemical shift change of about 0.6 ppm is to be expected for 1° change in the bond angle, with high-field shifts for larger angles. Since for highly crystalline materials a peak-to-peak resolution of 0.1 ppm can be achieved in the ^{29}Si NMR spectra (see, e.g., Fig.5), differences in the bond angles of about 0.2° may be detected. This extremely high sensitivity of the chemical shifts against small atomic dislocations in the zeolite framework explains the dramatic changes of the ^{29}Si NMR spectra of zeolite ZSM-5 upon sorbate loading or temperature variation considered above and shown in Figure 5. The relationship between SiOT bond angles and ^{29}Si chemical shifts offers also an explanation of the large shift differences observed for the Si(4Al) environments in sodalites with different enclathrated salts: The shift range between -76.4 ppm and -96.7 ppm correlates with SiOAl bond angles between 125° and 159° [21].

3.2.3 Modification of the zeolite structure

The catalytic, sorptive and ion-exchange properties, and especially the chemical and thermal stability of zeolites depend strongly on the content of tetrahedral aluminium in the framework. It is, therefore, of particular interest to modify the Si/Al ratio of the zeolite while retaining the topology and crystallinity of the framework structure. Dealumination of the zeolite framework can be achieved e.g. by hydrothermal treatment of the ammonium-exchanged form of the zeolite ("ultrastabilisation"), or by chemical treatment of the zeolite with suitable reagents in solution (e.g. acids, chelating agents, salts) or in the vapour phase at high temperatures (e.g. $SiCl_4$). A review of various techniques of dealumination is given in [32] and [11].

The quantitative determination of the extent of framework dealumination by chemical analysis has proven to be difficult since, after being removed from the tetrahedral framework, the aluminium may retain in the zeolite pores as charged or neutral Al-oxo-hydroxo complexes. This non-framework aluminium can, in principle, be extracted from the zeolite by acid leaching, but its total removal cannot always be achieved and the extraction of further framework aluminium cannot be ruled out.

Application of ^{29}Si NMR has been shown to be especially useful in the characterisation of dealuminated zeolites, since it provides direct information on the composition and Si,Al distribution of the tetrahedral framework, independently of the presence of non-framework Al species. Thus, the process of framework dealumination can be monitored in detail by the ^{29}Si NMR spectra. A great number of papers devoted to this topic have appeared in the literature, the early work is summarised in Chapter V.3.4 of [11].

As an example, Figure 7 shows the ^{29}Si NMR spectra of a series of hydrothermally dealuminated Y zeolites, measured without (upper spectra) and with cross-polarisation (lower spectra) [23a]. Using the signal intensities of the spectra registered without CP, the Si/Al ratio

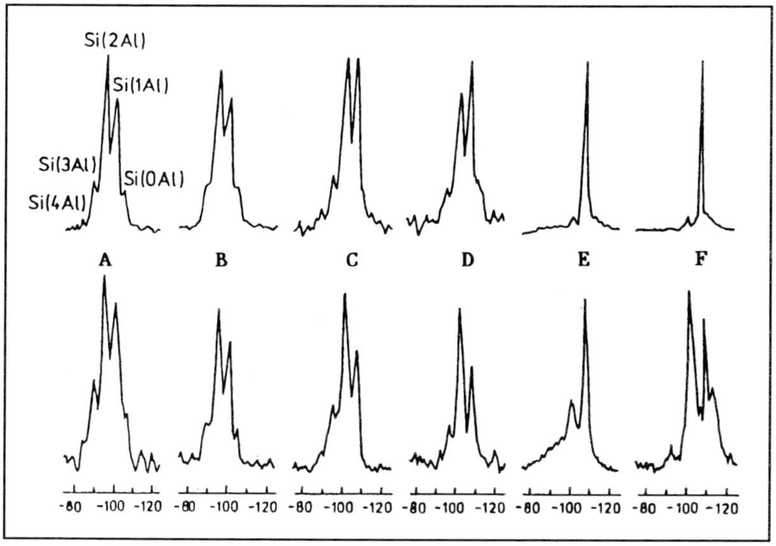

Fig.7. ^{29}Si MAS NMR spectra of dealuminated zeolites Y measured without (upper spectra) and with cross-polarization (lower spectra). A - parent NaY (Si/Al = 2.5); B - NH_4Y; C - sample B after hydrothermal treatment; D - acid leached sample C; E - sample D after NH_4 - exchange and hydrothermal treatment; F - acid leached sample E [23a].

of the zeolite framework can be calculated by eq. (1). The Si/Al ratio amounts to 2.5 for both the parent NaY (sample A) and ammonium-exchanged $NaNH_4Y$ zeolites (sample B), but increases to 4.9 after hydrothermal treatment of sample B for 3h at 540°C (sample C), indicating framework dealumination with the concomitant formation of non-framework Al species. Acid treatment of sample C with 0.1M HCl at 100°C results in a further increase of the Si/Al ratio to 6.6 in sample D which shows that not only the non-framework, but also some framework Al is extracted. After exchange of the residual Na^+ cations in sample D against NH_4^+ and another hydrothermal treatment for 3h at 815°C, the most part of the tetrahedral aluminium is removed from the framework, yielding a Si/Al ratio of more than 50 in sample E. Finally, sample F is obtained by acid leaching of sample E. The enhanced intensity of the signal at about -100 ppm in the CP spectra of samples C - F clearly indicate the formation of $(SiO)_3SiOH$ groups in the zeolite framework, resulting from incomplete reoccupation of the Al vacancies by Si and an increasing number of SiOH groups at the surface of the secondary mesopores created by Si migration into the vacancies of the framework [33]. The coincidence of the peaks of Si(1Al) and $(SiO)_3SiOH$ environments at about -100 ppm introduces some error in the determination of the Si/Al ratio from the non-CP spectra, since the intensity contribution of the latter unit must be ascribed to the Si(0Al) rather than to the Si(1Al) silicon atoms. Therefore, the Si/Al ratio determined by ^{29}Si NMR for zeolite samples containing $(SiO)_3SiOH$ groups will be a minimum value. It should be noted, that the CP spectra are generally not reliable to quantitative evaluation since the intensity enhancement factors by cross-polarisation are not known. Nevertheless, CP technique is a very useful method for the detection of SiOH groups in zeolites. Besides the $(SiO)_3SiOH$ environments, even geminal OH groups in $(SiO)_2Si(OH)_2$ structural groups may be detected by the CP signal at -90 ppm, clearly visible, e.g., in the CP spectrum of sample F. Moreover, the signal at -110 ppm in this spectrum can be attributed to Si(0Al) units close to surface OH groups in non-framework silica, formed by partial lattice destruction during the acid treatment.

Another important process applied in the preparation of a zeolite catalyst is decationation, i.e. the removal of the cations compensating the negative charge of the tetrahedral framework aluminium and their replacement by protons. Decationation is usually achieved by treatment of the zeolite with aqueous acids or by deammoniation of the NH_4-exchanged zeolite at higher temperatures. ^{29}Si NMR has shown that both reactions are usually accompanied by a certain degree of framework dealumination, even when the decationation is performed under very mild conditions, e.g. by shallow-bed calcination of $NaNH_4Y$ at 150-300°C under vacuum [34]. Figure 8 shows the ^{29}Si and ^{27}Al MAS NMR spectra of $NaNH_4Y$ zeolite and its protonated form obtained by shallow bed calcination at 300°C. The dealumination is evidenced by the changes in the Si(nAl) peak intensities and also from the appearance of a signal for octahedral aluminium in the ^{27}Al NMR spectrum (see Section 3.3.2). Framework dealumination may also occur during the use of the zeolite in a catalytic reaction and may be a possible reason for catalyst deactivation. ^{29}Si NMR of freshly prepared, activated, and used zeolite catalysts is an efficient method to monitor such dealumination processes.

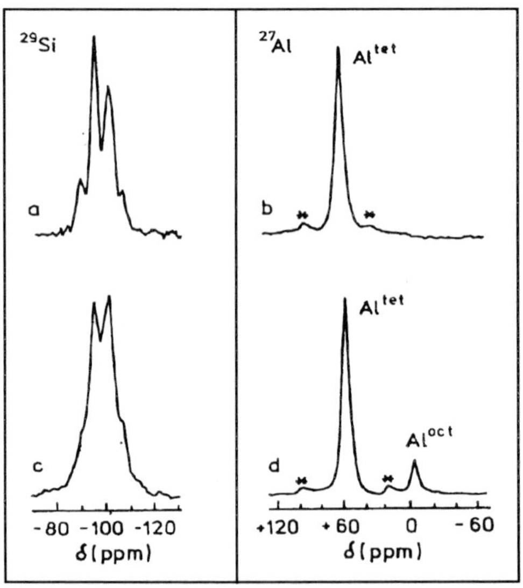

Fig.8. ^{29}Si and ^{27}Al MAS NMR spectra of zeolite $NaNH_4Y$ (a, b) and zeolite NaHY (c, d) prepared from NH_4Y by shallow bed calcination at 300°C [34].

Following the procedure outlined in Section 3.2.2, the ordering of Si and Al atoms in the tetrahedral framework of dealuminated Y zeolites has been studied and compared with the parent zeolite [35] or with samples of similar Si/Al ratio prepared by direct synthesis [36]. The ordering patterns which comply with the intensity distribution of the Si(nAl) peaks in the corresponding NMR spectra indicate that, in addition to the replacement of Si for Al, a mutual site exchange of an adjacent pair of Si and Al atoms occurs during the dealumination process. This site exchange involves a rearrangement of the tetrahedral framework in the course of the dealumination reaction which is in line with the assumption of local recrystallisation processes during the reinsertion of silicon atoms into the Al vacancies of the framework.

As discussed above, the content of framework aluminium of a zeolite can be decreased by hydrothermal treatment of its ammonium-exchanged form, but it is also possible to reverse the dealumination process by reinsertion of the non-framework aluminium into tetrahedral framework positions. This Al reinsertion has been achieved by a simple treatment of a dealuminated Y-zeolite (Si/Al = 4.26) with a 0.25M aqueous KOH solution at 80°C for 24 h [37]. The realumination is clearly visible from the ^{29}Si NMR spectra of the zeolite before and after the treatment: In the realuminated product the intensity of the Si(0Al) signal is greatly reduced, while the intensities of the Si(2Al), Si(3Al) and Si(4Al) signals are correspondingly increased, indicating that a considerable amount of aluminium has re-entered the zeolite framework. The spectrum of the realuminated product yields a Si/Al ratio of 2.56, which is very close to that of the parent Y zeolite used in the dealumination reaction, i.e. the non-framework Al has completely reinserted into the framework. Interestingly, the intensity distribution of the signals in the ^{29}Si NMR spectrum of the realuminated product is different from that of the parent Y zeolite, which means that the Si,Al ordering in the framework must also be different. Dealumination of Y zeolites by H_4EDTA or $SiCl_4$ treatment removes the most part of the non-framework aluminium from the sample. However, ^{29}Si NMR shows clearly, that these samples can be realuminated by treatment with aqueous $NaAlO_2$/NaOH [38] solution.

3.2.4 Zeolite synthesis

The process of zeolite synthesis from reaction mixtures containing silicate and aluminate anions, alkali metal cations, and possibly organic template molecules usually involve the formation of an intermediate aluminosilicate gel from which, after an induction period, the zeolite crystallises. The synthesis mixture thus forms a complex heterogeneous system composed of the amorphous gel, the supernatant solution and the emergent zeolite crystals. ^{29}Si NMR can be applied to both liquid and solid phases occurring in the course of zeolite synthesis and has provided valuable information on the silicate and aluminosilicate anion distribution in the starting solutions, the process of gel formation from the liquid phase, and the transformation of the gel into the crystalline zeolite.

Liquid state ^{29}Si NMR permits a detailed characterisation of the type and quantitative distribution of the $Si(O^-)_{4-n}(OSi)_n$ structural units (conventionally designated Q^n) in the silicate solutions, from which a mean degree of silicate polymerisation may be inferred. In addition, a variety of distinct silicate anions formed by various Q^n groups has been identified in the solutions, and dissolved aluminosilicate anions have been detected in Al-containing silicate solutions (for details see Chapter III of [11] and [39]). It is, however, still only hypothesised that some of the species detected by NMR function in fact as precursor units of the zeolite structure under crystallisation conditions. The distribution of the silicate species in the solutions depends on the silica concentration, the cation-to-silicon ratio and the pH-value.

Figure 9 shows the ^{29}Si NMR spectra of two sodium silicate solutions used as starting solutions in zeolite NaA syntheses following two different routes, A and B [40]. Solution A with c_{SiO_2} = 1M and Na/Si = 1.9 contains mainly low molecular weight anion species consisting of Q^0, Q^1 and Q^2 units, whereas in solution B with c_{SiO_2} = 1.65M and Na/Si = 0.6 polymer silicate anions containing branching (Q^3) and cross-linking groups (Q^4) predominate.

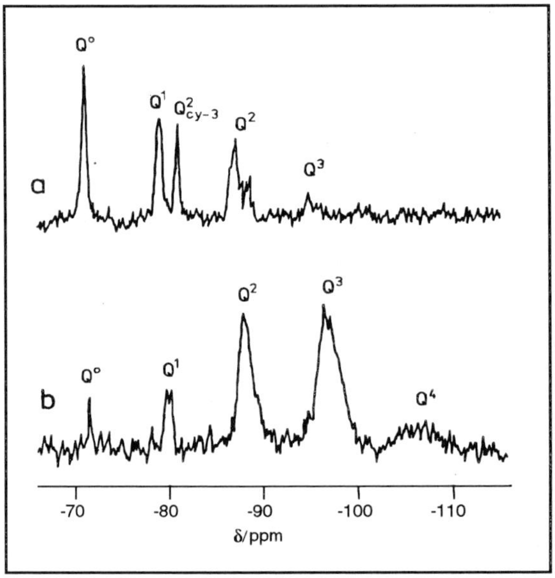

Fig.9. ^{29}Si NMR spectra of silicate solutions used in zeolite NaA synthesis.
a - solution A, c_{SiO_2} = 1M, Na/Si = 1.9; b - solution B, c_{SiO_2} = 1.65M, Na/Si = 0.6.

The intermediate aluminosilicate gels formed from the liquid reaction mixtures can be studied by solid-state ^{29}Si NMR. Owing to the amorphous structure of the gels, broad lines appear in the spectra which, however, provide information on the distribution of the various Si(nAl) units present in the highly disordered three-dimensional framework of the gel. As an example the above mentioned synthesis of zeolite NaA will be considered [40,41]. After mixing the sodium silicate solution A with sodium aluminate solution a gel is formed which gives a broad NMR signal at -85 ppm, indicative of Si(4Al) units in amorphous or highly disordered environments. When, however, the silicate solution B is mixed with aluminate solution (yielding the same overall composition of the reaction mixture as for solution A), the initial gel shows an even broader NMR peak at -93 ppm, typical of a silica-rich amorphous aluminosilicate composed mainly of Si(2Al) and Si(3Al) units. In agreement with these findings, chemical analysis yields a Si/Al ratio of 1.0 for the gel from solution A, but of 2.4 for that from solution B.

The crystallisation of the zeolite from the initial gel can be followed by the ^{29}Si NMR spectra of the solid products separated from the reaction mixture after increasing periods of time. A gradual narrowing and shifting of the NMR peak to higher field is observed for the products withdrawn from the reaction mixture A after heating to 80°C for 4 and 6 hours, and after 17 h at 80°C a narrow line at -89.4 ppm is measured, which is characteristic of highly crystalline zeolite NaA (see Table 2). In contrast, with increasing reaction time, the NMR line of the initial gel obtained from reaction mixture B shifts to low field, indicating transformation of the Si(3Al) and Si(2Al) units into Si(4Al) by incorporation of aluminium from the solution in the gel. After about 1 h at 80°C the spectrum is very similar to that of the initial gel of mixture A, and the subsequent zeolite crystallisation occurs in the same way as described above. The ^{29}Si NMR spectrum of the final sample (after 22 h at 80°C) is again typical of highly crystalline zeolite NaA. Thus, reaction mixtures with the same overall composition, but prepared from sodium silicate solutions of different degree of silicate condensation give rise to different intermediate aluminosilicate gels. However, crystalline zeolite NaA is formed from both types of intermediates but after different total times of crystallisation.

Similar ^{29}Si NMR investigations as discussed above have been carried out for the synthesis of zeolite NaA from metakaolinite [42] and several other zeolite types, such as zeolite Y [43,44], mordenite [44], hydroxosodalite [45], and ZSM-5 [44,46]. The mechanism of the structure directing role of tetrapropylammonium (TPA) ions in the synthesis of pure-silica ZSM-5 has been studied by ^{1}H-^{29}Si CP MAS NMR between the protons of TPA and the silicon atoms of the zeolite framework precursors [47]. The results indicate that short-range intermolecular interactions between the TPA ions and the gel structure are established prior to development of long-range order indicative of the ZSM-5 structure.

3.3 Aluminium-27 NMR

3.3.1 Quadrupole interaction, ^{27}Al chemical shifts, and local environments of Al sites

Unlike ^{29}Si which has a nuclear spin I = 1/2, ^{27}Al has I = 5/2 and therefore a nuclear quadrupole moment (eQ) which interacts with an electric field gradient (eq) produced by a non-spherically symmetric charge distribution around the ^{27}Al nucleus. These quadrupole interactions may give rise to strong line broadening and shifts of the line position (center of gravity) in the spectrum to high field from the isotropic chemical shift. Both quadrupolar line broadening and line shifts render the interpretation of the ^{27}Al NMR spectra with respect to chemical shifts and quantitative evaluation more difficult. On the other hand, valuable information on the symmetry of the charge distribution at the Al nucleus may be obtained from the quadrupole interaction described in terms of the quadrupole coupling constant $QCC = e^2qQ/h$ and asymmetry parameter of the electric field gradient η_Q.

In powder samples, the only transition usually observed is the central +1/2, -1/2 transition which is not affected by quadrupolar interactions to first order but by the much smaller second order interaction. The latter decreases with increasing strength of the external magnetic field B_o and can be reduced (but not fully averaged) by magic angle spinning. Thus application of both high B_o fields and MAS may result in considerable line narrowing in the ^{27}Al NMR spectra of solid powders.

The ^{27}Al NMR spectra of zeolites are, in general, much simpler than their ^{29}Si NMR counterparts since according to Loewenstein's rule (which forbids AlOAl pairings) only a single tetrahedral Al environment, namely $Al(OSi)_4$, exists in the zeolite framework. The AlO_4 units in hydrated zeolites show only small deviations from tetrahedral symmetry and thus only weak quadrupole interactions (QCC of about 1-2 MHz). Consequently, a single narrow line is usually observed in the ^{27}Al MAS NMR spectra of hydrated zeolites (see e.g. Fig.8b) with a typical chemical shift of about 60 ppm (from aqueous $Al(NO_3)_3$ solution). The isotropic ^{27}Al chemical shifts of tetrahedral framework aluminium in hydrated zeolites of different structure type cover the relatively small range from about 55 to 68 ppm, and no definite relationships could be found between the shifts and the Si/Al ratio or Si,Al ordering of the zeolite framework. However, similarly to ^{29}Si NMR, a linear relationship between ^{27}Al chemical shifts and mean AlOSi bond angles has been established with shift data carefully corrected for quadrupolar shift contributions [48].

In dehydrated zeolites the symmetry of the AlO_4 tetrahedra is heavily distorted which causes large quadrupole interactions (QCC of about 5-8 MHz) and thus much broader ^{27}Al resonances. Even larger quadrupole couplings (QCC of about 10-16 MHz) have been observed in dehydrated H-zeolites, and the resulting extremely broad resonances could only be measured at static samples (i.e. without MAS) applying static quadrupole-echo techniques [49,50].

Non-framework aluminium in zeolites which has typically an octahedral AlO_6 coordination gives rise to signals at about 0 ppm, i.e. well separated from the tetrahedral Al sites of the framework (see e.g. Fig.8d). However, if the non-framework Al exists as polymeric aluminium oxides or oxide hydrates in the zeolite cavities, strong quadrupolar line broadening may be observed owing to heavy distortions of the octahedral symmetry of the AlO_6 sites.

3.3.2 Framework and non-framework aluminium - quantitative aspects

The high inherent sensitivity of ^{27}Al NMR (^{27}Al has a natural abundance of 100% and generally short relaxation times) permits the detection of very small contents of tetrahedral Al in the framework of high-silica zeolites which are not seen in the ^{29}Si NMR spectra. For example, traces of tetrahedral framework Al has been detected by a peak at about 55 ppm in the ^{27}Al NMR spectrum of silicalite, prepared without any intentional addition of aluminium in the synthesis mixture [51].

An important application of ^{27}Al NMR to zeolites is the detection and characterisation of non-framework aluminium species formed, e.g., by various thermal or hydrothermal treatments of zeolites as usually applied in dealumination processes or the preparation of the acidic H-forms of zeolites. As mentioned above (and shown e.g. in Fig.8d), well separated signals at about 60 ppm and about 0 ppm appear in the ^{27}Al NMR spectra for four-coordinated framework Al and six-coordinated non-framework Al, respectively. Provided that all aluminium is visible in the spectrum, i.e. no signal intensity is lost as a result of very strong quadrupolar line broadening, the relative proportions of framework and non-framework Al in the zeolite can be directly determined from the intensities of the signals at about 60 ppm and 0 ppm, I_{60} and I_0. Moreover, if the total Si/Al ratio, $(Si/Al)_{tot}$, of the sample is known from chemical analysis, the Si/Al ratio of the tetrahedral framework, $(Si/Al)_{fr}$, can be determined according to

$$(Si/Al)_{fr} = (Si/Al)_{tot} (I_{60} + I_0)/I_{60} \qquad (3)$$

However, it has been demonstrated in several papers that the quantitative results derived from ^{27}Al NMR does not always agree with those obtained from ^{29}Si NMR, and that non-framework Al is often underestimated by the former method. Although several methods have been proposed which may be helpful to render all aluminium "NMR visible" (e.g., application of very strong B_o fields and high spinning rates, rehydration of the sample or its impregnation with acetylacetone), care must always be exercised to ensure that all aluminium is detected in the ^{27}Al NMR spectrum. As mentioned above, static quadrupole-echo measurements are helpful in detecting all framework Al in dehydrated H-zeolites.

Further complications may arise in the application of ^{27}Al NMR if broad signals of Al sites subject to different quadrupolar interactions overlap in the spectrum. The separation of such resonances can be achieved by the two-dimensional quadrupole nutation MAS NMR technique [52] which allows the resolution of ^{27}Al sites with different quadrupole coupling constants in the F_1 dimension, while in the F_2 dimension the ordinary spectrum is presented. ^{27}Al nutation MAS NMR has been applied to separate the strongly overlapping lines observed in the ^{27}Al MAS NMR spectra of hydrothermally dealuminated Y zeolites [53]. Figure 10 shows the MAS NMR spectra and the component lines resolved by ^{27}Al nutation NMR for two differently treated samples. From the intensities of the distinct line shapes and the contents of framework and non-framework aluminium derived from ^{29}Si NMR, it has been concluded that the ^{27}Al NMR spectra are superpositions of three lines which correspond to framework (Al^F), non-framework octahedral (Al^{NFO}) and non-framework tetrahedral (Al^{NFT}) aluminium. By this procedure, it was possible to account for all the aluminium in the samples, although the nature of the newly discovered non-framework tetrahedral aluminium is not fully understood. The presence of Al^{NFT} species in dealuminated Y zeolites has also been proved by ^{27}Al DOR NMR studies [54].

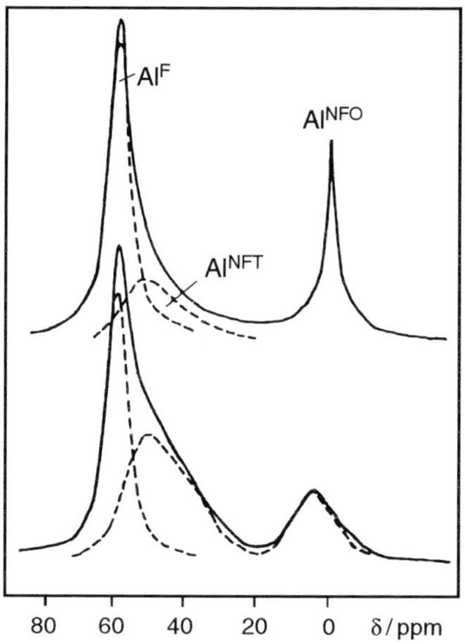

Fig. 10. ^{27}Al MAS NMR spectra of dealuminated Y zeolites containing non-framework aluminium. The dotted component lines were derived from 2D nutation MAS experiments.

3.4 NMR of other framework atoms

3.4.1 Oxygen-17 NMR

Since oxygen is a main constituent of the zeolite framework, the possibilities of carrying out detailed ^{17}O NMR studies in the solid state are particular attractive. However, as mentioned in Section 3.1, the ^{17}O isotope has a very low natural abundance and a quadrupole moment which renders the observation of ^{17}O NMR spectra very difficult without (expensive!) ^{17}O enrichment of the samples. Therefore, ^{17}O NMR studies of zeolites have been performed throughout at isotopically ^{17}O enriched samples. The main features of the ^{17}O NMR of spectra of zeolites were demonstrated for NaA and NaY, the latter being studied with different degree of dealumination [20a]. In addition, zeolite ZSM-5 [20b], some gallosilicate zeolites, and aluminophosphate molecular sieves [20c] were investigated by ^{17}O NMR.

As a typical example, Figure 11 shows the static and MAS ^{17}O NMR spectra of zeolite NaY (Si/Al = 2.74). The static spectrum indicates the presence of two overlapping second-order quadrupolar broadened lines which are attributed to the two types of chemically inequivalent oxygen species, SiOSi and SiOAl, in the NaY framework. The line shapes of the two spectral components can be simulated using the quadrupole coupling data of zeolite NaA for SiOAl and of highly dealuminated NaY for SiOSi, and their relative intensities can be calculated from the Si/Al ratio. Addition of the corresponding component spectra results in the simulated spectrum shown in Figure 11b. The MAS NMR spectra are markedly narrowed but the fine-structure of the lineshapes is less pronounced.

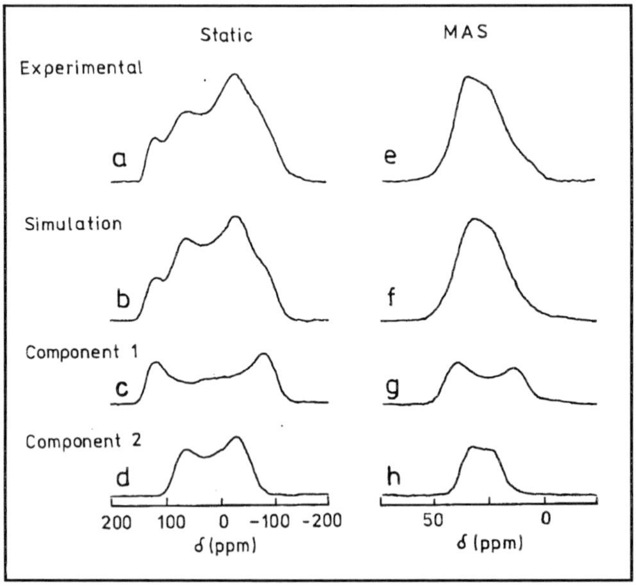

Fig. 11. ^{17}O NMR spectra of zeolite NaY. a - static spectrum; b - simulation of spectrum (a) using component 1 (c) for SiOSi and component 2 (d) for SiOAl; e - MAS spectrum; f - simulation of spectrum (e) using components g and h [20a].

The ^{17}O NMR studies of zeolites, gallosilicate zeolites and aluminophosphates may be summarised as follows:

- The spectra consist of characteristic resonances from chemically distinct oxygen species, i.e. SiOSi, SiOAl, SiOGa and AlOP which all exhibit a second order quadrupole broadened line shape. Crystallographically non-equivalent oxygen sites present in zeolites A and Y cannot be resolved.

- The presence of two types of SiOT fragments in a zeolite structure, e.g. SiOSi and SiOAl in Y zeolites, or SiOSi and SiOGa in gallosilicate zeolites, leads to a superposition of resonances of either ^{17}O species with intensities according to their relative populations in the zeolite.
- The oxygen species of the various TOT fragments are characterised by QCC in the range of 3.1 to 5.7 MHz, asymmetry parameters η_Q up to 0.3, and isotropic chemical shifts δ_{iso} between 28 and 67 ppm (versus $H_2{}^{17}O$). The observed order of QCC is AlOP > SiOSi > SiOGa > SiOAl, and that of δ_{iso} is AlOP > SiOSi > AlOSi > GaOSi.

3.4.2 Boron-11 NMR

The NMR-related properties of the ^{11}B nucleus are in many respects similar to those of ^{27}Al: ^{11}B has a nuclear spin of 3/2 and therefore a quadrupole moment, the sensitivity of ^{11}B NMR is reasonably high, and the relaxation times are sufficiently short for spectra with good signal-to-noise ratio to be readily registered within relatively short measuring times.

In boron-containing zeolites, boron may occur as tetrahedral BO_4 or trigonal BO_3. At high magnetic field and under MAS conditions tetrahedral boron gives narrow lines, whereas trigonal boron exhibits a clear quadrupolar pattern owing to its larger quadrupolar interaction. The ^{11}B chemical shift range is relatively small and the signals of both coordinations may overlap. There is, however, usually sufficient resolution to differentiate clearly the two species and to determine their relative amounts present in the sample. Tetrahedral and trigonal boron has, e.g., been detected during the dehydration of H-boralite, the boron containing variant of zeolite ZSM-5 [55,56]. A sharp line of tetrahedral BO_4 appears in the hydrated sample, while in the dehydrated material a broad doublet of trigonal BO_3 predominates which is superimposed by the line of residual BO_4 species. Rehydration leads to reconstitution of the original spectrum showing the single BO_4 line.

^{11}B NMR may also be used to prove the incorporation of tetrahedral boron into the zeolite framework and to discriminate against non-framework tetrahedral boron. $B(OSi)_4$ environments in the zeolite framework are characterised by ^{11}B chemical shifts between -1 to -4 ppm (from $BF_3{\cdot}Et_2O$), while slightly but definitely different shifts in the range of +1 to +2 ppm are observed for other BO_4 coordinations, e.g. in borates. Similar to ^{27}Al NMR, the high sensitivity of ^{11}B NMR permits the detection of very small amounts of tetrahedral BO_4 in different environments [57].

3.5 Hydrogen-1 NMR

While the proton-proton dipolar interactions in dehydrated zeolites are comparatively small owing to the large proton-proton distances, ^1H NMR of zeolites is complicated by heteronuclear dipolar interaction of the protons with the quadrupolar ^{27}Al nuclei, and the generally narrow range of proton chemical shifts. Nevertheless, ^1H MAS NMR at high magnetic fields and fast spinning rates of carefully dehydrated samples provides sufficient spectral resolution to identify different proton sites in zeolites, and to characterise, quantitatively, their distribution in the zeolite structure. Numerous papers have appeared on ^1H NMR of zeolites and the recent developments were surveyed in several reviews (see, e.g., [8,58-60]). In the following only a short summary of some general aspects will be given.

Several distinct types of protons can be identified in dehydrated zeolites by their chemical shifts: terminal SiOH (1.3 - 2.2 ppm), bridging hydroxyl groups SiO(H)Al (3.6 - 5.6 ppm), and AlOH at non-framework aluminium (2.6 - 3.6 ppm). In HY zeolites the resonance of bridging hydroxyl groups (Brönsted sites) is splitted into two components for OH-groups pointing into the supercage (at about 3.9 ppm) and into the sodalite cages (at 4.9 ppm). Further resonances may appear from ammonium ions (6.5 - 7.5 ppm) and OH-groups connected with metal cations (-0.5 to 0.5 ppm), e.g., CaOH or MgOH in CaNaY or MgNaY zeolites. Characteristic low-field shifts are observed for bridging or metal OH-groups involved in hydrogen bonds. The resonance line of water in hydrated zeolites appears typically between 3.5 and 5 ppm. However, in hydrated H-zeolites this line may be shifted to lower field due to rapid exchange of the acidic protons with hydroxonium ions which resonate at about 13 ppm.

The ^1H chemical shifts has been related to the acidity of the bridging and other OH-groups in zeolites, and the acid strength of the protons has been, at least qualitatively, characterised by the chemical shift differences of the hydroxyl protons in different zeolite types and differently treated samples [49,58,61]. In addition, the changes in the proton chemical shifts due to interactions with adsorbed probe molecules, such as ammonia, pyridine, carbon monoxide, acetonitrile, ethylene, and others, have been used to characterise acidity and accessibility of OH-groups in zeolites [59,60,62,63].

^1H NMR has also been applied to study the proton distribution in hydrothermally treated zeolites Y and ZSM-5 [64,65], and quantitative estimates of the concentrations of terminal, bridging, and extra-framework-Al hydroxyls and their changes by different treatment conditions have been derived from the spectra. As an example, Figure 12 displays the 500 MHz ^1H MAS NMR spectra of a series of dealuminated H-ZSM-5 zeolites with framework

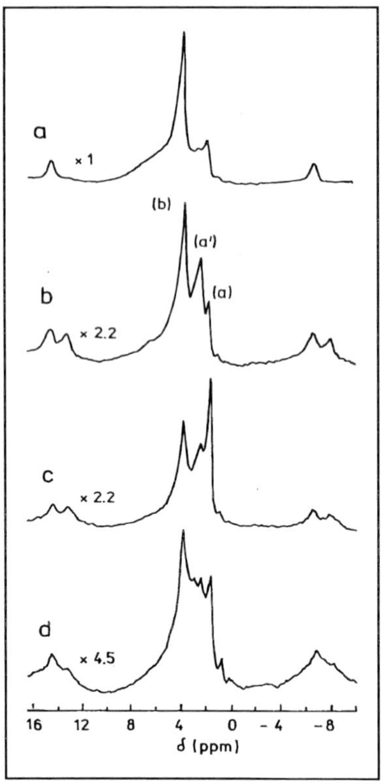

Fig.12. ^1H MAS NMR spectra (500 MHz) of dehydrated H-ZSM-5 zeolites prepared from NH_4-ZSM-5 by hydrothermal treatment. Line assignment: (a) terminal SiOH, (a') AlOH at non-framework Al, (b) bridging hydroxyl groups SiOHAl. a - 500°C, 2 h, shallow bed, Si/Al = 22; b - 500°C, 1 h, steam; Si/Al = 47; c - 500°C, 3 h, steam, acid leached, Si/Al = 62; d - 500°C, 3 h, steam , Si/Al = 110. The weak lines at about -7 and 14 ppm are spinning side bands. The spectra were measured in absolute intensity mode. The expansion of the vertical scale is indicated.

Si/Al ratios of 20 to 110. The spectra exhibit three lines which were attributed to terminal SiOH (at 1.8 ppm), non-framework AlOH (at 2.5 ppm) and bridging Si(OH)Al (at 4.0 ppm). The absolute intensity of the line at 4 ppm decreases with increasing Si/Al ratio of the framework and correlates linearly with the number of framework Al atoms per unit cell. This correlation shows that the intensity of the 4 ppm line is a direct measure of the number of strong acid sites (Brönsted sites) in the zeolite. Further information on the formation of lattice defects and partial amorphisation of the ZSM-5 material, and on the degree of hydroxylation of the non-framework aluminium in the course of the hydrothermal treatment can be derived from the intensity variations of the lines of AlOH and SiOH groups.

3.6 NMR of non-framework cations in zeolites

In addition to the nature of the tetrahedral framework, the catalytic, adsorptive and ion-exchange properties of zeolites depend strongly on the kind, position and mobility of the cations balancing the negative charge brought about by the AlO_4-tetrahedra in the zeolite framework. Since the NMR characteristics of the cation resonances are related to the intracrystalline environment and mobility of the cations, NMR can, in principle, provide information on the coordination, site occupancy and dynamics of cations in zeolites. Out of the great variety of cations which may be present in the zeolite cavities, only a small selection have been studied by NMR so far, these are, in particular, ^{23}Na, ^{7}Li, ^{139}La, ^{133}Cs, and ^{205}Tl. Except ^{205}Tl (which has I = 1/2), all of these are quadrupolar nuclei.

3.6.1 Sodium-23 NMR

Solid state ^{23}Na NMR offers considerable potential for determining the location of sodium cations present in the zeolite cavities. However, complex spectral patterns are generally observed due to superposition of the quadrupolar line shapes of different sodium sites, and an array of NMR techniques has to be applied for line separation. These include fast MAS, DOR, 2D nutation MAS, experiments at different magnetic field strengths, and, most important, careful computer simulations of the individual resonances in the spectra [66]. In addition, theoretical calculations of the EFG at the ^{23}Na nucleus based on crystal structure data or sensible structure models [67] are very helpful for spectra interpretation.

Detailed ^{23}Na NMR investigations on sodium siting in dehydrated zeolites NaX and NaY with Si/Al ratios between 1.1 and 4.0 have been performed applying all the methods mentioned above [68]. The typical cation sites in faujasite-type zeolites are displayed in Figure 13a. As an example, Figure 13b shows the ^{23}Na MAS NMR spectrum (measured at 105.8 MHz) of dehydrated zeolite NaY and its simulation by three components. The two broad quadrupolar patterns are attributed to sodium at sites SII (QCC = 3.9 MHz, δ_{iso} = -12 ppm) and SI' (4.8 MHz, -4 ppm), located above the centers of the six-ring windows in the supercages and ß-cages, respectively, and the narrow line at low field corresponds to sodium at SI (ca. 1 MHz, -12 ppm) in the hexagonal prism. In NaX zeolites (Si/Al < 1.3) up to five distinct resonances could be identified which are assigned to SI, SI', SII and two slightly different SIII' sites (2.5 and 1.5 MHz, -14 and -28 ppm, respectively) located in the supercage close to the fused four-rings (see Fig.13a). The experimentally determined QCC-values agree well with those calculated from crystal structure data by a simple point charge model. From the intensities of the simulated component lines, the quantitative populations of the various sites are obtained which agree well with those derived from X-ray diffraction studies. It should be noted that the simulations and assignments of the MAS NMR spectra are greatly supported by the results obtained from measurements at different B_0 fields (7.0 to 17.6 T), DOR and 2D nutation experiments, and the above mentioned field gradient calculations [68].

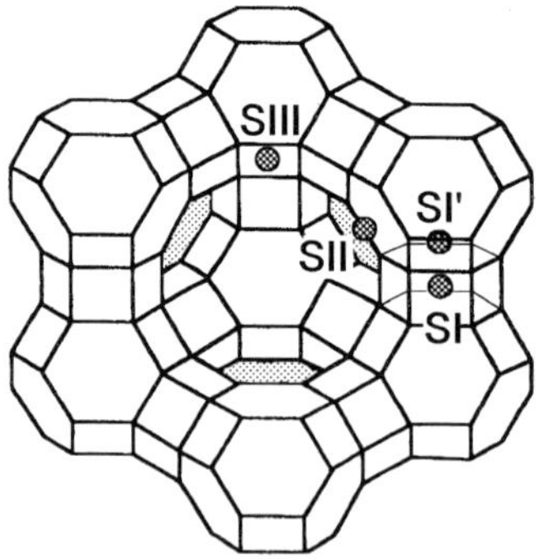

Fig.13a. Cation sites in faujasite-type zeolites

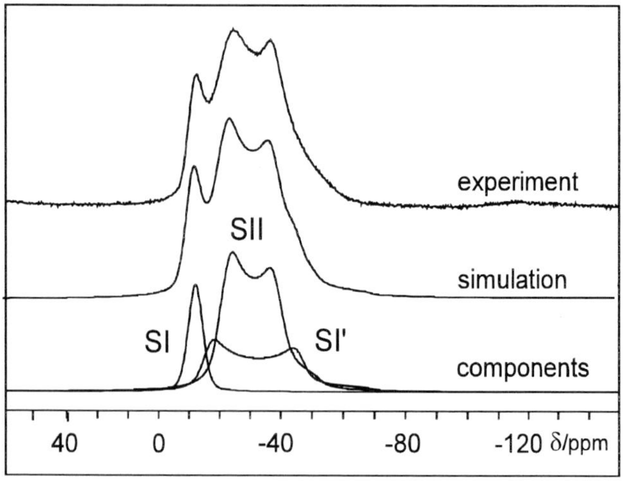

Fig.13b. Experimental and simulated ^{23}Na MAS NMR spectra (105.8 MHz) of dehydrated zeolite NaY (Si/Al = 2.4)

^{23}Na NMR has also been applied to study the cation distribution in various cation-exchanged zeolite, e.g., NaCsY [69], NaLaY [70] and NaLiX [71]. For these materials, complementary information on the siting of the cations other than Na^+ has been derived from ^{133}Cs, ^{139}La and ^{7}Li NMR studies, respectively. Furthermore, the location and migration of sodium cations in NaY zeolites loaded with organic sorbates [71] or metal carbonyls [72] were studied by ^{23}Na MAS and DOR NMR.

It is obvious from the foregoing that multinuclear solid-state NMR has great potential in investigating cation distributions in zeolites which, however, has certainly not yet fully exploited. Once the general procedure for measuring the spectra and their evaluation has been elaborated for a certain cation and zeolite type, NMR is comparatively fast and quantitatively reliable. It is, therefore a valuable complement to diffraction techniques the application of which to analyse cation sitings in microcrystalline zeolite powders is often time-consuming and cumbersome.

3.7 NMR of organic templates and adsorbed molecules in zeolites

Studies on the structure, position and interactions of organic species present in the zeolite cavities are of great interest for a deeper understanding of adsorption and catalytic processes in zeolites, and, if structure-directing agents ("templates") are used in synthesis, of the mechanisms of zeolite formation. High-resolution solid-state ^{13}C and ^{1}H MAS NMR has great potential in such investigations. The isotropic ^{13}C chemical shift is highly sensitive to the environment of the carbon nucleus, and the ^{13}C NMR spectra often display narrow and well resolved resonances of the distinct carbon atoms of the organic guest species in the zeolite.

The high resolution of the ^{13}C NMR spectra of organic templates in zeolites is demonstrated in Figure 14 by the ^{13}C NMR spectrum of the tetrapropylammonium ion (TPA^+) enclathrated in the zeolite ZSM-5 structure in the course of synthesis [11]. For comparison, the spectrum of an aqueous solution of TPAOH is also shown. The general agreement of the two spectra reveals that the occluded TPA ions are chemically intact in the ZSM-5 channels. There is, however, a clear splitting of the methyl resonances which arises from the different surroundings of the propyl chains extending in the two different channels of the ZSM-5 structure. The ^{13}C NMR spectra of a variety of other organic templates present in as-synthesised zeolites reveal that most of the templates remain intact within the zeolite channels, although characteristic shift effects indicate some interaction of the functional group of the organic species with the zeolite framework. For tetramethylammonium ions (TMA), these shift effects are characteristic of the size of the cavity in which the TMA ion is trapped [73]. Thus, the site of TMA ions in zeolites containing cages of different dimensions may be determined, and the size of the ion trapping cage may be estimated from the ^{13}C chemical shifts, even when the detailed structure of the zeolite is unknown.

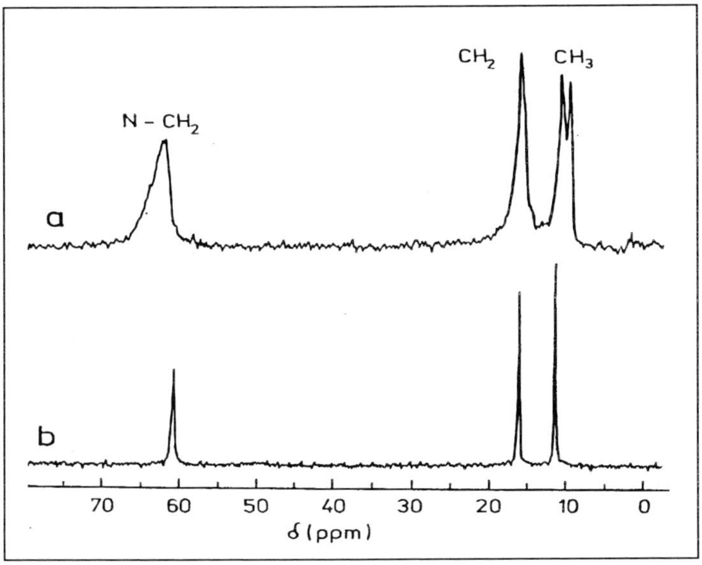

Fig. 14. ^{13}C MAS NMR spectra of (a) Na,TPA-ZSM-5 zeolite and (b) TPAOH in aqueous solution (TPA = tetrapropylammonium)

Adsorbed organic molecules in zeolites can successfully be studied by ^{1}H and ^{13}C MAS NMR, and for nitrogen-containing compounds also by ^{15}N NMR. Numerous papers devoted to this subject have appeared in the literature, and reviews of the field are presented in [8,11]. Owing to the large range of ^{13}C chemical shifts in organic molecules, well resolved resonances can generally be observed in the ^{13}C MAS NMR spectra of adsorbed molecules, and even their chemical transformation in the zeolite can be followed in detail (see, for example [74]). For ^{1}H NMR the situation is more difficult due to the small differences of the ^{1}H chemical shifts and line broadenings induced by restrictions of the molecular mobility and by local magnetic fields. However, very high resolution can be achieved in the ^{1}H NMR spectra of adsorbed molecules using high external magnetic fields and MAS with sealed samples of very low concentrations of paramagnetic impurities. As an example, four very narrow lines were observed in the ^{1}H MAS NMR spectrum of butene-1 adsorbed on ultra-pure NaX and NaY zeolites, and even line splittings due to J couplings between protons and between protons and ^{13}C atoms could be detected. [75]. The high resolution permits the application of two-dimensional experiments originally designed for liquids, e.g., COSY, NOESY, or ^{1}H-^{13}C HETCOR, which facilitate line assignments and may provide further improvements of the resolution in the two-dimensional spectrum.

3.8 In-situ NMR studies of catalytic reactions in zeolites

NMR has also been proven to be an excellent tool to study catalytic reactions in zeolites. In particular, ^{13}C and ^{1}H MAS NMR are applied to follow the chemical transformations of reacting molecules by temperature- and time-dependent changes in the spectra. In-situ studies of catalytic reactions require MAS experiments at variable temperature under controlled atmosphere at the sample which is still a technical challenge. However, considerable progress has been achieved in recent years with two different approaches. In the first approach, which models batch reactor conditions, sealed glass ampoules containing the zeolite catalyst and the organic adsorbate are used which fit precisely in the MAS rotors. The NMR spectra are measured in-situ while the reaction is in progress at a suitably high temperature. Alternatively, the experiment may be divided into reaction periods at higher temperatures and measuring periods at lower temperature. A wide range of reactions has been studied by this technique, for details see, e.g., [76] and the literature cited therein. The second approach applies continuous-flow conditions which simulate to some extent the situation in a catalytic flow-reactor. In this method, a continuous stream of a carrier gas and the gaseous educts flows through the zeolite catalyst during the NMR experiment. The gas is introduced through a narrow tube placed in an axial hole of the cylindrical catalyst bed within the MAS rotor and leaves the rotor through a hole in the rotor cap [77]. In a similar probe design, the reactant gas is led to the top of the catalyst bed in the rotor and passes the catalyst aided by the pressure differential [78]. It is obvious that NMR studies of catalytic reactions using reagent flow better approximate the actual conditions in a catalytic flow-reactor than those carried out in sealed ampoules. However, the former technique is still in its infancy and only few in-depth studies have been published so far.

4. NMR OF ALUMINOPHOSPHATE MOLECULAR SIEVES

Although the aluminophosphate molecular sieves, $AlPO_4$-n (n denotes a specific structure type), and their silicon or metal atoms containing variants, SAPO-n and MAPO-n, cannot be considered as belonging to the zeolite family, they have zeolite-like three-dimensional frameworks consisting of alternating AlO_4 and PO_4 tetrahedra in the $AlPO_4$-n series, and of AlO_4, PO_4, and SiO_4 or MeO_4 in the SAPO-n and MAPO-n materials.

Aluminophosphate molecular sieves are attractive materials for solid-state NMR studies since they contain two different kinds of 100% abundant nuclei, ^{31}P and ^{27}Al. The ^{31}P MAS NMR spectra generally show symmetrical lines in the shift range of -14 to -35 ppm (from 85% H_3PO_4), consistent with tetrahedral PO_4 units in the framework of these materials [79,80]. The ^{27}Al NMR spectra are more complex and may exhibit broad and asymmetrical line shapes due to quadrupolar interactions of the ^{27}Al nuclei. The latter arises from significant distortion of the AlO_4 tetrahedra caused by interaction with entrapped organic template molecules and/or H_2O in the cavities of the $AlPO_4$ structure. In hydrated samples, some of the framework Al atoms can become hydrated, forming octahedral $Al(OP)_4(OH_2)_2$ environments. The ^{27}Al chemical shifts attributed to tetrahedrally coordinated Al in $Al(OP)_4$ units usually cover the range from about 29 to 46 ppm, whereas resonances of octahedrally coordinated Al appear in the shift range of -7 to -23 ppm [79].

Multiple ^{31}P and ^{27}Al resonances appear for $AlPO_4$ structures characterised by crystallographically inequivalent T sites in the framework. The hydrated 18-ring aluminophosphate molecular sieve VPI-5 is a case in point: Three lines of equal intensity were observed in both the ^{31}P (at -23.7, -27.5, and -33.2 ppm) and the ^{27}Al spectrum (at 43.6, 41.6, and -10.4ppm), in full agreement with the three crystallographically distinct P and Al atoms in this structure. The ^{27}Al resonance at high field (-10.4 ppm) corresponds to doubly hydrated, six-fold coordinated aluminium, while all the other resonances are typical of tetrahedral P and Al atoms (see above) [81].

The ^{27}Al and ^{31}P NMR spectra of SAPO's have been found to be comparable to those of the $AlPO_4$ materials. The ^{29}Si NMR spectra usually show single lines with chemical shifts between -89 an -92 ppm, consistent with the presence of one $Si(OAl)_4$ environment which implies that silicon substitutes for phosphorous in the aluminophosphate framework [80]. However, SAPO crystals often contain silicon-rich aluminosilicate domains and phosphorous-rich aluminophosphate domains which causes broadening or even characteristic splittings in the ^{29}Si NMR spectra due to the presence of different Si(nAl) environments. As an example, five well separated lines of different intensities have been observed in the ^{29}Si MAS NMR spectrum of a SAPO-44 sample, suggesting the existence of aluminosilicate domains containing the five Si(nAl) surroundings with n = 0-4 [82]. Furthermore, a doublet splitting of the ^{29}Si resonances due to two crystallographically distinct Si sites has been detected in dehydrated SAPO-17 [83].

REFERENCES

1 C.A.Fyfe, Solid State NMR for Chemists, CFC Press, Guelph, 1983.
2 G.Engelhardt, D.Kunath, M.Mägi, A.Samoson, M.Tarmak, and E.Lippmaa, Workshop on Adsorption of Hydrocarbons in Zeolites, Berlin-Adlershof, 1979
3 E.Lippmaa, M.Mägi, A.Samoson, M.Tarmak, and G.Engelhardt, J.Amer.Chem.Soc. 103 (1981) 4992
4 G.Engelhardt, U.Lohse, E.Lippmaa, M.Tarmak, and M.Mägi, Z.anorg.allg.Chem. 482 (1981) 49
5 S.Ramdas, J.M.Thomas, J.Klinowski, C.A.Fyfe, and J.S.Hartman, Nature 292 (1981) 228
6 J.Klinowski, J.M.Thomas, C.A.Fyfe, and J.S.Hartman, J.Phys.Chem. 85 (1981) 2590
7 D.Freude and H.J.Behrens, Cryst.Res.Technol. 16 (1981) 1236
8 H.Pfeifer and H.Ernst, Ann.Rep.NMR Spectr. 28 (1994) 91
9 C.A.Fyfe, Y.Feng, H.Grondey, G.T.Kokotailo, and H.Gies, Chem.Rev. 91 (1991) 1525
10 M.Stöcker, Stud.Surf.Sci.Catal. 85 (1994) 429; dto. 102 (1996) 141
11 G.Engelhardt and D.Michel, High-Resolution Solid-State NMR of Silicates and Zeolites, Wiley, Chichester 1987
12 A.Samoson, E.Lippmaa, and A.Pines, Mol.Phys. 65 (1989) 1013)
13 K.T.Mueller, B.Q.Sun, G.C.Chingas, J.W.Zwanziger, T.Terao, and A.Pines, J.Magn.Reson. 86 (1990) 470)
14 A.Medek, J.S.Harwood, and L.Frydman, J.Am.Chem.Soc. 117 (1995) 12779
15 C.Jäger, G.Kunath, P.Losso, and G.Scheler, Solid State Nucl.Magn.Reson. 1 (1992) 261

16 C.A.Fyfe, K.C.Wong-Moon, Y.Huang, H.Grondey, and K.T.Mueller, J.Phys.Chem. 99 (1995) 8707
17 A.J.Vega, J.Am.Chem.Soc. 110 (1988) 1049
18 D.Massiot, C.Bessada, J.P.Coutures, and F.Taulelle, J.Magn.Reson, 90 (1990) 231.
19 G.Engelhardt and H.Koller, Magn.Reson.Chem.29 (1991) 941
20 (a) H.K.C.Timken, G.L.Turner, J.P.Gilson, L.B.Welsh, and E.Oldfield, J.Amer.Chem.Soc. 108 (1986) 7231; (b) S.Yang, K.D.Park, and E.Oldfield, J.Am.Chem.Soc. 111 (1989) 7278; (c) H.K.C.Timken, N.Janes, G.L.Turner, S.L.Lambert, L.B.Welsh, and E.Oldfield, J.Amer.Chem.Soc. 108 (1986) 7236; (d) T.H.Walter, G.L.Turner, and E.Oldfield, J.Magn.Reson. 76 (1988) 106
21 G.Engelhardt, S.Luger, C.Buhl, and J.Felsche, Zeolites 9 (1989) 182
22 D.E.W.Vaughan, M.T.Melchior, and J.J.Jacobsen, ACS Symp.Ser. 218 (1983) 231; J.M.Thomas, J.Klinowski, S.Ramdas, M.W.Anderson, and C.A.Fyfe, ACS Symp.Ser. 218 (1983)159; S.Hayashi, K.Suzuki, S.Shin, K.Hayamizu, and O.Yamamoto, Bull.Chem.Soc.Jap. 58 (1985) 52
23 (a) G.Engelhardt, U.Lohse, A.Samoson, M.Mägi, M.Tarmak, and E.Lippmaa, Zeolites 2 (1982) 59; (b) E.F.Rakiewicz, K.T.Mueller, T.P.Jarvie, K,J.Sutovich, T.G.Roberie, and A.W.Peters, Microporous Mater. 7 (1996) 81
24 J.Klinowski, S.Ramdas, J.M.Thomas, C.A.Fyfe, and J.S.Hartman, J.Chem.Soc.Farad.Trans.2 78 (1982) 1025
25 T.Takaishi, M.Kato, and K.Itabashi, Zeolites 15 (1995) 21
26 C.A.Fyfe, H.Strobl, G.T.Kokotailo, G.J.Kennedy, and G.E.Barlow, J.Amer.Chem.Soc. 110 (1988) 3373
27 R.H.Jarman, A.J.Jakobsen, and M.T.Melchior, J.Phys.Chem. 88 (1984) 5748
28 G.Engelhardt, Stud.Surf.Sci.Catal. 52 (1989) 151
29 (a) C.A.Fyfe, Y.Feng, and H.Grondey, Microporous Mater. 1 (1993) 393; (b) J.M.Thomas, J.Kennedy, S.Ramdas, B.K.Hunter, and T.B.Tennakoon, Chem.Phys.Lett. 102 (1983) 158; (c) J.M.Newsam, J.Phys.Chem. 91 (1987) 1259
30 G.Engelhardt and R.Radeglia, Chem.Phys.Lett. 108 (1984) 271;
31 R.Radeglia and G.Engelhardt, Chem.Phys.Lett. 114 (1985) 28
32 J.Scherzer, ACS Symp.Ser. 248 (1984) 157
33 A.Zukal, V.Patzelova, and U.Lohse, Zeolites 6 (1986) 133
34 G.Engelhardt, U.Lohse, M.Mägi, and E.Lippmaa, Stud.Surf.Sci.Catal. 18 (1984) 23
35 G.Engelhardt, U.Lohse, V.Patzelova, M.Mägi, and E.Lippmaa, Zeolites 3 (1983) 329
36 J.Dwyer, K.Karim, W.J.Smith, N.E.Thompson, R.K.Harris, and D.C.Apperly, J.Phys.Chem. 95 (1991) 8826
37 H.Hamdan, B.Sulikowski, and J.Klinowski, J.Phys.Chem. 93 (1989) 350
38 Z.Zhang, X.Liu, Y.Xu, and R.Xu, Zeolites 11 (1991) 232
39 R.F.Morlock, A.T.Bell, A.K.Chakraborty, and C.J.Radke, J.Phys.Chem. 95 (1991) 4501
40 G.Engelhardt, B.Fahlke, M.Mägi, and E.Lippmaa, Zeolites 5 (1985) 49
41 G.Engelhardt, B.Fahlke, M.Mägi, and E.Lippmaa, Zeolites 3 (1983) 292
42 J.Rocha, J.Klinowski, and J.M.Adams, J.Chem.Soc.Farad.Trans. 87 (1991) 3091
43 N.Dewaele, P.Bodart, Z.Gabelica, and J.B. Nagy, Acta Chim.Acad.Sci.Hung. 119 (1985) 233
44 P.Bodart, J.B.Nagy, Z.Gabelica, and E.G.Derouane, J.Chim.Phys.Phys.Chim.Biol. 83 (1986) 777

45 S.Hayashi, K.Suzuki, S.Shin, K.Hayamizu, and O.Yamamoto, Chem.Phys.Lett. 110 (1984) 54
46 K.F.M.G.J.Scholle, W.S.Veeman, P.Frenken, and G.P.M.van der Velden, Appl.Catal. 17 (1985) 233
47 S.Burkett and M.E.Davis, J.Phys.Chem. 98 (1994) 4647).
48 E.Lippmaa, A.Samoson and M.Mägi, J.Amer.Chem.Soc. 108 (1986) 1730
49 D.Freude, H.Ernst, and I.Wolf, Solid State Nucl.Magn.Reson. 3 (1994) 271
50 M.Hunger, T.Horvath, G.Engelhardt, and H.G.Karge, Stud.Surf.Sci.Catal. 94 (1995) 756
51 C.A.Fyfe, G.C.Gobbi, J.Klinowski, J.M.Thomas and S.Ramdas, Nature 296 (1982) 530
52 A.Samoson and E.Lippmaa, Chem.Phys.Lett. 100 (1983) 205
53 A.Samoson, E.Lippmaa, G.Engelhardt, U.Lohse and H.G.Jerschkewitz, Chem.Phys.Lett. 134 (1987) 589
54 G.J.Ray and A.Samoson, Zeolites 13 (1993) 410).
55 K.F.M.G.J.Scholle and W.S.Veeman, Zeolites 5 (1985) 118
56 R.de Ruiter, A.P.M.G.Kentgens, J.Grootendorst, J.C.Jansen, and H. van Bekkum, Zeolites 13 (1993) 128
57 J.-Ch.Buhl, G.Engelhardt, and J.Felsche, Zeolites 9 (1989) 40
58 H.Pfeifer, NMR - Basic Principles and Progress, 31 (1994) 31
59 M.Hunger, Solid State Nucl.Magn.Reson. 6 (1996) 1
60 E.Brunner, J.Mol.Struc. 355 (1995) 61
61 L.Heeribout, P.Batamack, C.Doremiex-Morin, R.Vincent, and J.Fraissard, Colloids.Surf. A115 (1996) 229
62 J.L.White, L.W.Beck; and J.F.Haw, J.Am.Chem.Soc. 114 (1992) 6182
63 F.Yin, A.L.Blumenfeld, V.Gruver, and J.J.Fripiat, J.Phys.Chem.B 101 (1997) 1824
64 D.Freude, M.Hunger and H.Pfeifer, Z.Phys.Chem. NF, 152 (1987) 171
65 G.Engelhardt, H.-G.Jerschkewitz, U.Lohse, P.Sarv, A.Samoson, and E.Lippmaa, Zeolites 7 (1987) 289
66 G.Engelhardt, M.Hunger, H.Koller, and J.Weitkamp, Stud.Surf.Sci.Catal. 84A (1994) 421
67 H.Koller, G.Engelhardt, A.P.M.Kentgens, and J.Sauer, J.Phys.Chem. 98 (1994) 1544
68 M.Feuerstein, M.Hunger, G.Engelhardt, and J.P.Amoureux, Solid State Nucl.Magn.Reson. 7 (1996) 95
69 H.Koller, B.Burger, A.Schneider, G.Engelhardt, and J.Weitkamp, Micropor.Mater. 5 (1995) 219
70 M.Hunger, G.Engelhardt, and J.Weitkamp, Micropor.Mater.3 (1995) 497
71 M.Feuerstein, PhD Thesis, University of Stuttgart
72 R.Jelinek, S.Özkar, H.O.Pastore, A.Malek, and G.Ozin, J.Am.Chem.Soc. 115 (1193) 563; H.Koller, A.R.Overweg, R.A. van Santen, and J.W.de Haan, J.Phys.Chem. B 101 (1997) 1754
73 R.H.Jarman and M.T.Melchior, J.Chem.Soc.Chem.Commun. 1984, 414; S.Hayashi, K.Suzuki, S.Shin, K.Hayamizu and O.Yamamoto, Chem.Phys.Lett. 113 (1985) 368
74 T.Xu, E.J.Munson, and J.F.Haw, J.Am.Chem.Soc. 116 (1994) 1962
75 U.Schwerk and D.Michel, Z.Phys.Chem. 189 (1995) 29

76 J.F.Haw, NMR Techniques in Catalysis, (A.T.Bell and A.Pines, Eds.) Marcel Dekker, Inc., New York, 1994, p. 139; I.I.Ivanova and E.G.Derouane, Stud.Surf.Si.Catal. 85 (1994) 357; J.F.Haw, J.B.Nicolas, T.Xu, L.W.Beck, D.B.Ferguson, Acc.Chem.Res. 29 (1996) 259; H.Ernst, D.Freude, T.Mildner, and I.Wolf, Solid State Nucl.Magn.Reson. 6 (1996) 1477; M.Hunger and D.Horvath, J.Chem.Soc., Chem.Commun. 1995, 1423
78 P.Goguen and J.F.Haw, J.Catal. 161 (1996) 870
79 C.S.Blackwell and R.L Patton, J.Phys.Chem. 88 (1984) 6135
80 C.S.Blackwell and R.L.Patton, J.Phys.Chem. 92 (1988) 3965
81 P.J.Grobet, A.Samoson, H.Gerts, J.A.Martens, and P.A.Jacobs, J.Phys.Chem. 95 (1991) 9620
82 A.M.Prakash, S.Unnikrishnan, and K.V.Rao, Apl.Catal. A110 (1994) 1
83 B.Zibrovius and U.Lohse, Solid State Nucl.Magn.Reson. 1 (1992) 137

Studies in Surface Science and Catalysis 137
H. van Bekkum, E.M. Flanigen, P.A. Jacobs and J.C. Jansen (Editors)
© 2001 Elsevier Science B.V. All rights reserved.

Chapter 10

Introduction to zeolite theory and modelling

R.A. van Santen,[a] B. van de Graaf,[b] and B. Smit[c]

[a]Schuit Institute of Catalysis, Eindhoven, University of Technology Eindhoven, P.O. Box 513, 5600 MB Eindhoven, The Netherlands

[b]Laboratory for Organic Chemistry and Catalysis, Delft University of Technology Julianalaan 136, 2628 BL Delft, The Netherlands

[c]Department of Chemical Engineering, Universiteit van Amsterdam Nieuwe Achtergracht 166, 1018 WV Amsterdam, The Netherlands

In this Chapter some of the recent advances in zeolite theory and modelling are present. In particular we discuss the current status of computational chemistry in Brønsted acid zeolite catalysis, Molecular Dynamics simulations of molecules adsorbed in zeolites, and novel Monte Carlo technique to simulate the adsorption of long chain hydrocarbons in zeolites.

1. Introduction

In the previous edition of these book various theoretical and computational approaches to study zeolites have been discussed [1]. The focus was on methods to compute zeolite structure and stability, computer graphics to visualise the complex zeolite structures, and Molecular Dynamics and Monte Carlo calculations to simulate the adsorption and diffusion of small molecules in these materials.

Although most of these topics are still of interest, many new developments have taken place since the appearance of the first edition. Novel algorithms, such as ab-initio molecular dynamics, embedding, or novel Monte Carlo techniques have been applied to zeolites and allow us to use computational for many more problems in zeolite science. Furthermore, the enormous increase in computer power allows us to study systems that are more realistic and closer to experiments.

A disadvantage of this development is that it is not possible to cover the entire field in a single chapter. The limited space forces us to focus on two aspects zeolite acidity and the adsorption and diffusion of molecules in the pores. The first topic deals with the chemistry of the protons in zeolites and in this part recent developments in quantum chemistry calculations are discussed. Since molecules can not be converted if they do not adsorb or diffuse in the pores of the zeolites, diffusion and adsorption are important aspects contributing to the activity of a zeolite. These properties have been studied using Molecular Dynamics or the Monte Carlo technique and in the second part of this chapter recent developments in this area are discussed.

2. Quantum-chemistry of zeolite acidity

2.1. Introduction

Current electronic structure calculations have reached a stage that they can be usefully applied to analyze the energetics of reacting systems representing the catalytically reacting site [2,3]. Here we will discuss the current status of this approach with respect to the protonation reaction in zeolites.

Using transition state reaction rate theory [4] rate constants for the elementary reaction steps can be computed. Quantum-chemical calculations enable not only the prediction of interaction energies in the ground state, but also the energies that correspond to the transition state. If in addition a normal mode analysis is done in reactant as well as transition state the activation entropy can be computed. Within the harmonic approximation the corresponding expression for the activation entropy is:

$$\Delta S^* = k_B \ln \frac{\prod_j^{N-1} \left[1 - \exp(-h\nu_j/k_B T)\right]^{-1}}{\prod_k^N \left[1 - \exp(-h\nu_k'/k_B T)\right]^{-1}}. \tag{1}$$

With ν_j and ν_k the normal mode frequencies of the complex in the transition state and ground state respectively. k_B is Boltzmanns constant and T the temperature.

The expression for the activation energy becomes:

$$E_{\text{act}} = \Delta E^* + \frac{1}{2} h \left(\sum_j^{N-1} \nu_j - \sum_k^N \nu_k' \right) \tag{2}$$

ΔE^* is the computed electronic barrier energy. The correction to ΔE^* is due to the zero point frequencies of the nuclei. Within the harmonic approximation the normal modes can currently be routinely computed by most of the available ab-initio quantum-chemical routines. To compute such properties is also very important in case one wishes to validify computational predictions. Measurements of rate constants of elementary reaction steps would be very useful, but are rarely available for zeolites. Spectroscopic data are widely available of adsorbates interacting with protons. This holds especially for Infrared and NMR data. It has been shown for instance that proton NMR chemical shifts can be used to probe the state of adsorbed CH_3OH [5]. Also Nuclear Quadruple Coupling constants have been computed [6] to probe the local coordination of H_2O. Of course the results depend sensitively on the accuracy or level of the calculations. For the computation of infrared spectra an additional complexity arises from the intrinsic anharmonicity of the OH modes. This is typically of the order of a few 100 cm^{-1}. In case one wishes to use the infrared spectrum as a probe for the adsorption state, this shift cannot be ignored. Methods [2,7] have been developed to not only solve the anharmonicity problem in one dimension, but in several dimensions. In essence one has to compute the potential energy surface with a certain dimensionality and than one solves the equations of motion of the atoms in this potential [7]. Such approaches have also been used to develop force fields or potentials so as to predict the structure and physical properties of zeolites. This has been reviewed elsewhere [8] and will not be the subject of this paper. However to study chemical reactivity one has to exclusively use quantum-chemical approaches.

An important issue is the use of clusters to model the local environment around the zeolitic proton. Until recently no quantum-chemical computer routines were available to solve the protonation problem in the three dimensional solid. The main technical limitation was the need to optimize the geometry of the lattice to identify ground states and transition states with the fully relaxed structures. Molecular mechanics studies based on classical potentials have provided ample proof of the flexibility of the zeolite lattice and local relaxation upon protonation [9]. Quantum-chemical cluster calculations have only given agreement with experiment when the geometry had been allowed to relax. In fact the now popular Density Functional Theory methods became only useful, once programs as DGAUSS, DMOL, DMON or AOF were provided with routines that enables geometry optimization. This situation has only quite recently changed for the solid state problem. The breakthrough has been provided for by the implementation of the Car-Parrinello method [10–13]. It is not our intention to provide an introduction to quantum-chemical methods, but here we will shortly explain the essential electronic structural principles that are basic to each technique. This will later help to appreciate and to compare the different techniques.

2.2. Computational methods

To study chemical reactivity quantitatively one has to use ab-initio quantum-chemical techniques. Nonetheless the use of semi-empirical methods can be quite useful. For instance to obtain a starting structure of a transition state to be completed by a first principle computational method, transition states obtained by the much less computational demanding semi-empirical methods can be quite useful. A technique that has been used in our work is MNDO [14,15]. In the transition state a computation of the normal mode frequencies should give one imaginary mode.

Most ab-initio programs have now the option to not only find the stationary point and local energy minima, but also to compute the saddle points that correspond to the transition states.

The electronic structure codes can be defined in two categories. One class is based on the use of the Hartree-Fock method [16] and subsequent improvements. Gaussian [17], Gamess [18] and Crystal [19,20] belong to this class, Gaussion and Gamess being codes suitable for the study of clusters. Crystal can be used to compute properties of the three dimensional solid.

Important issues here are choice of basis set and with which correction the Hartree-Fock method has to be used. Crystal does not yet have the option for geometry optimization. Recently Gaussian as well as Crystal have obtained options to include Density Functional Theory options. The essence of the Hartree-Fock method is that one-electron wave functions, corresponding to molecular orbitals, are computed using a potential where the electron-electron interaction is deduced from the electrostatic interaction with other electrons distributed in the other one-electron wave functions. It implies that correlation between the motion of the individual electron is ignored. It appears that to compute hydrogen bonds and protonation this is not an acceptable approximation and correlation corrections are important. A reasonable approach appears the use of the Møller-Plesset 2 (MP2) correction [21,22], implemented in most commercial packages. This is an important issue, because the correlation correction to the Hartree-Fock energy also contains the van der Waals interactions.

When a molecule adsorbs into a zeolite it not only experiences hydrogen bonding with the protons but also an important interaction with the highly polarizable oxygen atoms in the wall of the zeolite. Per oxygen atom this interaction is usually quite small, typically of the order of 10 kJ/mole. Therefore, it is called physical adsorption. However, a larger molecule has contact with many channel oxygen atoms, hence the total interaction can be quite large. It gives arise to the substantial adsorption energy of molecules adsorbed in zeolites.

This physical adsorption is mainly due to the dispersive van der Waals interactions and therefore very important to compute. Configuration interaction methods as the Møller-Plesset method are quite important for this purpose. Many technical aspects determine the quality of a calculation. Most important are choice of basis set and geometry optimization. Often geometrys are optimized in the Hartree-Fock approximation and energies are improved by using better approximations for chosen geometry's. For the subtle problem of protonation of polar molecules, this leads to erroneous results as we will discuss in the next section.

In the Density Functional Theory methods [11,12,23–31], to be discussed now, the van der Waals problem has in essence not been solved. Density Functional Theory has initially been invented to provide a shortcut to the computational demanding Hartree-Fock theory. In Local Density Theory, the initial version of Density Functional Theory, the electron exchange term in Hartree-Fock theory is replaced by a function that depends on the electron-density similarly as a free-electron gas. Surprisingly to compute structures this appears to be a very useful approach. Not however to compute bond energies. They appear often to be overestimated by a factor of two. A dramatic improvement is found when the local density approximation is corrected for by so called density gradient correction terms to correct for electron-correlations as well as electron density dependence.

Different versions of these gradient correction terms are available all having a particular advantage, dependent on systems to be studied. Cluster versions as well as solid-state versions of the DFT programs are available.

For the zeolite protonation DFT methods have demonstrated themselves to be very useful and to produce ground state interaction energies often correct with an accuracy of $\approx$5 kJ/mole. Transition state energies may be underestimated by $\approx$30 kJ/mole. To compute infrared frequencies the situation is not that satisfactory. To produce accurate values application of an adequate electron correlation density function is critical. The B3LYP functional appears to be best, but anharmonicity corrections have to be included. It appears that weakening of vibrational bonds by hydrogen bonding is significantly overestimated [32]. As mentioned earlier van der Waals interactions cannot be reliably computed, they appear to be underestimated ($\approx$10 kJ/mole). Whereas geometry optimization methods have been widely implemented for clusters, this is not the case for calculations of the three dimensional solid.

Car-Parrinello methods [10–13] intrinsically provide a solution for this, because their purpose is to study the molecular motion in the field of the computed electronic structure. Again it is essential to use a version that contains the gradient-corrected functional for the electronic correlation energy. Other issues affecting accuracy relate to the k-cut-off in the number of electronic plane-wave functions to be used in the Car-Parrinello method and the choice of pseudo potentials. So far in zeolites no transition states have been studied using Car-Parrinello techniques. They have great promise because application of

Car-Parrinello method do not require the cluster approximation and enable determination of optimum geometries.

2.3. The cluster approximation and embedding methods

To judge which aspects of the zeolite catalysis problem can be treated using clusters to model the Brønsted acidic site, one has to distinguish which kinetic parameters depend on structural differences between zeolites from the question whether the chemical bonding aspects of the interaction with a zeolitic proton can be usefully studied with the cluster approximation [33].

Properties that depend strongly on zeolite structure are heats of adsorption and diffusion rates. Clearly clusters cannot be used to study these aspects of the kinetics problem. Also zeolites may control the overall kinetics of a reaction by preventing access of molecules that are too large to enter a micropore or by preventing the formation of product molecules that are too large for the zeolite micropore. Product formation may be prevented if transition states are required that are too large for the micropore. These are extremely important effects. The use of zeolite Y as a cracking catalyst is based on its reduced coke formation. This is prevented because condensation of aromatic rings is suppressed by micropore size.

In the case of methanol conversion catalysis by the medium-pore size ZSM-5 leads to stable operation and formation of benzene and methylbenzenes. On wider pore zeolites methanol will also be converted to higher hydrocarbons, but zeolite operation is not stable due to formation of deactivating carboneous materials. The narrow pore chabasite type zeolite SAPO-34 [34,35] will convert methanol selectively to ethylene. So in considerations of the overall kinetics of zeolites one has to explicitly treat the structure dependent aspects, that are mainly determined by van der Waals interactions [36].

Clearly the prediction of the size of the transition state of a particular reaction implies that an estimate can be made of the minimal pore-size requirement for the reaction to proceed.

Important to the choice of a cluster is how well physical properties are properly computed. Initially it has been proposed that a charged cluster should be cut from the zeolite and be embedded in an electrostatic lattice, so as to compute the long range Madelung effects correctly. It is now clear that in such an approach chemical bonding features become completely dominated by edge effects and lead to physically unrealistic results.

There is now a general agreement that clusters should be chosen neutrally by terminating them with hydroxyl or hydride bonds. Studies of the deprotonation energy of such clusters show significant fluctuations in computed values [37]. Hence it is not obvious which cluster choice is best. From NMR [38] and IR [39] studies of the zeolitic proton it is well known that its chemical bond is affected by compositional changes in structure at positions that are next nearest neighbor of the tetrahedra lattice bridged by protonated oxygen atoms, that forms the Brønsted acidic site. This implies the use of clusters at least of a size of 26 tetrahedra. These effects are not electrostatic, but covalent and can be well understood using the Bond Order Conservation principle [40]. It is for instance reflected in the dependence of the intrinsic acidity of protons on the Al-concentration on zeolite Y. As follows from calorimetric ammonia adsorption measurements [41,42] the ammonia protonation energy drops as soon as a fraction of the protons reacted with ammonia. This is very similar to the allosteric effect that is found in enzymes with several sites of interaction. Clearly clusters containing only three to five tetrahedra, as often used to study transition states cannot describe these composition dependent effects. However termination of the cluster can be altered so as to influence the intrinsic acidity. In small clusters, it can be mimicked by assigning different bond lengths to terminal SiO or SiOH bonds and optimizing all other parameters [43]. For instance lengthening of terminating SiH bonds shortens neighboring SiH bonds and as a consequence weakens a neighboring proton-oxygen bond. Also the same reaction can be studied as a function of cluster size. A plot can be made of the activation energy of that reaction against deprotonation energy of the clusters [44,45] (see Figure 1). Then one can extrapolate the curve to that deprotonation- energy for which one desires to compute the activation energy that for instance can be estimated for a proton based on its infrared shift induced by a weakly adsorbing molecule [46]. One then uses the Brønsted-Polanyi principle that proposes the relation:

$$\Delta E_{\text{act}} = c\Delta E_{\text{reaction}}. \tag{3}$$

The change in activation energy is proportional to the change in reaction energy as long as there is no change in reaction mechanism.

We have found [33] that once neutral clusters are large with 10 or 12 tetrahedral sites, electrostatic embedding affects the interaction energy of ammonia with a zeolitic proton with less than 10%. Many physical properties of zeolitic protons are described with high accuracy by very small clusters.

2.4. General concepts and critical evaluation of computational results

Calculations on the Brønsted acidic site in a zeolite have been done with many different methods. A representative cluster is shown in Figure 2 [6]. This cluster has five T-sites and the Brønsted acidic site is the protonated bridging atom between Si and Al. The predicted geometry is from a DFT calculation. The calculation is obtained from a non geometry constrained relaxation.

The result is very similar to that obtained with a high quality HF/MP2 calculation. The H-Al distance can be compared with NMR estimates and the predicted OH stretch frequency, when corrected for anharmonicity will be within 50 cm^{-1} of that of the typical low alumina spectrum. Proper AlOH distances are obtained only when clusters have at least three T-sites. The deprotonation energy of such small clusters is usually a few hundred kJ/gat larger than experimental values.

The H-chemical shift as well as Al-NMR Quadrapole Coupling constants can be compared well with experimental values. One deduces a charge less than 0.1 e.u. for the proton, consistent with strong covalent bonding of the OH bond. Several analyses have been made of the dependence of the deprotonation energy on the location of the Si-OH-Al unit in the zeolite. One expects each non equivalent O positron to have a different deprotonation energy. The approaches to be taken can be quantum chemical or classical.

In quantum-chemical calculations one can constrain the positions of the atoms that terminate the cluster. These terminating positions can be chosen according to crystallographically determined positions or positions obtained from structural simulations of the siliceous or aluminum phosphate topologies. For the latter accurate procedures are

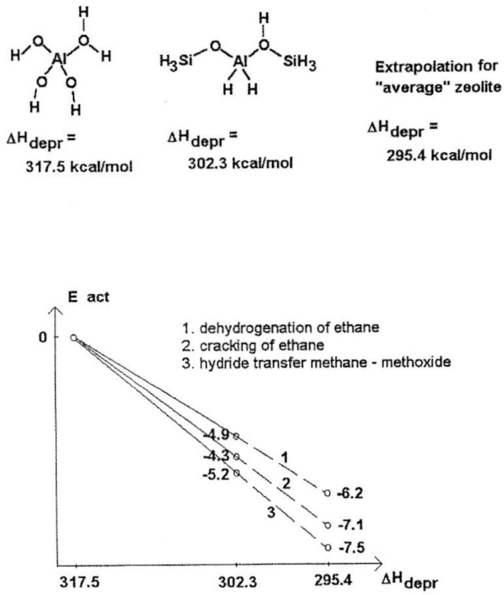

Figure 1. Brønsted-Polanyi dependence of the activation energy as a function of cluster size [44].

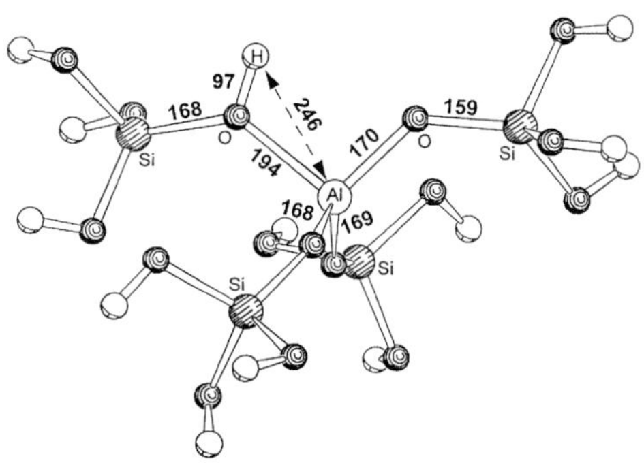

Figure 2. A cluster representing the Brønsted acidic site. Distances are in ppm [6].

available. Such cluster choices imply choices for particular sites. The implicit assumption underlying the constrained cluster model is that the changes in site structure due to the replacement of a Si-O-Si or Al-O-P unit by a Si-OH-Al or equivalent unit are local. This has indeed been found in molecular mechanics simulations of the extended lattice. These simulations employ classical potentials deduced from quantum-chemical cluster calculations. Due to the very shallow Si-O-Si bonding potential the increase in local volume that occurs upon substitution is accommodated by small changes of the SiOSi angles. A few T-sites away from the acidic sites the structure is the same as before substitution.

In such calculations a spread in deprotonation energies of 80 kJ/mole is found for low Al concentration systems [47,48]. In addition to molecular mechanics simulations of extended systems also recently quantum-mechanical Car-Parrinello calculations have been done for Offretite [49]. With in essence the same result. The largest difference in deprotonation energies on different Offretite oxygen atoms is now 40 kJ/mole. These differences are in quite good agreement with mobility NMR measurements [50]. The conclusion of both approaches is that the main parameter that determines the difference in deprotonation energy is not the long range electrostatic interaction, as might have been expected, but differences in local geometry. In structures with short Si-O-Si distances the constraint on the Al-OH-Si unit is largest resulting in slightly different Al-O-Si angles. This is the main reason for the lower acidity at sites that are most expandable. One can also use a combination of a quantum-chemical calculation and classical structure prediction, by embedding the cluster in a three dimensional lattice connected to the quantum-chemical part of the system by classical forces. In this way a smaller cluster can be used and its spatial relaxation is determined from the classical part of the calculation [48].

Interaction with a weakly adsorbing molecule as CO or NO shifts the OH stretch frequency downwards with $\approx$1100cm^{-1}. A strongly interacting molecule as acetonitrile shifts the OH frequency downwards by $\approx$100cm^{-1}. Calculations using different calculational approaches on clusters reproduce this bondweakening of the OH bond, but with different accuracy [51]. DFT seems to overestimate the bondweakening. It is important that anharmonic corrections to the frequencies are fully included. HF overestimates frequencies by 10%. HF-MP2 calculations are to be preferred. HF-MP2 and DFT calculations agree that acetonitrile is not protonated.

An important issue concerns the interpretation of the broadened downwards shifted intensity due to the weakened OH stretching frequency. Especially when strong bases as acetonitrile, H_2O or methanol interact.

The high intensity is due to the enhancement of the transition dipole moment. This is due to the polarization of the OH band in contact with the doubly occupied lone pair electrons of the adsorbing base. The resulting Pauli-repulsion is decreased by polarization of electrons away from the interaction region. In case of the interaction with acetonitrile the charge on the proton has increased to 0.3 e.u. [52]. Cluster calculations reproduce this quite well. One finds that the acidic nature of the zeolitic Brønsted OH bond relates to its easiness of polarization. The OH bridging a three valent and four valent lattice cation appears highly polarizable. The broadened infrared spectrum of the perturbed OH bond in contact with a strong base is found to be split in two or three peaks. Many theoretical studies have been devoted to the interpretation of this band structure.

For acetonitrile or acetone it has been proposed that the peaks describe a hydrogen bonded in equilibrium with a protonated species.

For methanol and H_2O the bands have also been assigned to the asymmetric and symmetric OH stretching modes of the protonated methoxonium or hydronium system [53]. Now according to the most accepted interpretation it is an A, B, C band system due to Fermi resonance of hydrogen bonded base, with overtones of in-plane and out-plane OH bending modes [3]. Cluster calculations have been very useful to analyze this. They also have shown the limitations of different calculational approaches.

It appears that DFT theory including gradient exchange-correlation correction terms, properly optimized, generally predicts that the hydrogen bonded species is the ground state. It is essential to analyze explicitly whether a stationary point is a local minimum or a transition state.

HF-MP2 theory is essential, with a good quality basis set to describe the interaction with the proton, if one wishes to use a HF based method. The structure has to be optimized at the HF/MP2 level. To compute frequencies full anharmonicity has to be included. For interacting acetonitrile the A, B, C band profile has been theoretically reproduced. Coupling between the OH modes and acetonitrile has been computed by solving the dynamics of the proton in the potential deduced from quantum-mechanical cluster calculations [7].

With increasing acidity of the cluster the intensity maximum shifts to lower frequencies.

Also frequency shifts of molecular modes of the base molecule can be used to analyze acidity [54]. Here a scaled HF approach suffices. These are sensitive to Lewis or Brønsted acidic interactions.

To summarize this discussion it may be useful to emphasize that most critical to the computation of proper OH frequency shifts is the right prediction of the structure of the interacting system. Small bond distance differences can have a large effect on the final results. As we will see it will also be strongly influenced by cluster choice. Because of the large heterolytic bond dissociation energy of the zeolitic OH bond ($\approx$1250 kJ/mol), protonation of a molecule, even a very basic one can only occur because of the presence of an additional stabilizing term:

$$[ZH \cdot B]_I \rightarrow [Z^- \cdot HB^+]_{II}. \tag{4}$$

Structure I represents the hydrogen bonded situation, structure II is the case where the basic molecule has become protonated.

The energy change between structure I and II can be considered to consist of three terms: The deprotonation energy of the zeolite, the protonation energy of the base (the sum of these two is usually endothermic) and the electrostatic stabilization of the Zwitterionic pair of structure II. It will appear that optimization of these backbinding interactions determines the structure of groundstates and especially transition states to a considerable extent. The importance of this zeolite wall effect can be deduced from cluster calculations on the protonation of NH_3. This problem has been studied for NH_3 interacting with a single T-site cluster as $[Al(OH)_2H_2]$ H [55–58] as well as large ring systems [33,58–62]. The problem has been studied with different qualities of basis sets, correlation energy as well as cluster relaxation.

For historic reasons it may be of interest to mention that Hall *et al.* [63] were the first to propose that decomposition of NH_4^+ to NH_3 would lead to the zeolite Brønsted acidic site we here consider.

One finds that NH_4^+ becomes only formed, when two or three of its protons are close to the negatively charge (Lewis basic) oxygen atoms around aluminum. When clusters increase in size, these oxygen atoms can become non-equivalent and hence the back binding NH$\cdots$O bonds become non-equivalent. This is observed in cluster calculations [33] as well as electrostatically embedded systems [62].

Electrostatic embedding calculations, especially of small clusters have to be considered with care. Sometimes full electrostatic potentials are used or negatively charged clusters and the embedding is used to correct for this deficiency. It appears that boundary effect corrections can be significant and can make such cluster results quite unreliable. When charges are used derived from quantum-chemical calculations and at least ten-T site clusters are used, the additional energy changes as the energy due to electrostatic embedding have so far been found to be at maximum 10% [33]. Detailed studies of cluster size are available. Brand *et al.* [60] used a constrained relaxation method and corrected for basis set deficiencies. Whereas the deprotonation energy only slowly converges with cluster size, it appears that the protonation energy of ammonia to ammonium is much less dependent on cluster size. This is a very general observation and derives from a compensation effect. The covalent backbinding interaction of the NH_4^+ protons changes in a direction opposite from that of the deprotonation energy.

The same has been found in a HF/MP2 study by Zygmunt *et al.* [64] on H_2O interacting with clustermodels of the protonic site in ZSM-5. The most important effect of cluster size is on the geometry of the adsorbed molecule. The non-equivalence of oxygen atoms around Al tends to make two- or three point adsorption more asymmetric. For H_2O and CH_3OH, Greatbanks *et al.* [65] find correct of 17 kJ/mol, 31 kJ/mol, to the heat of adsorption of H_2O or CH_3OH, respectively.

They studied the effect of electrostatics on clusters of 3 T sites. 3-21 G optimized geometries were used and correlation effects were computed in this geometry. Whereas this may cause an overestimate of the embedding effect, interestingly the total adsorption energies computed of 85 kJ/mol for H_2O and 110 kJ/mol for CH_3OH are very similar to values found by Shah *et al.* [13] using Car-Parrinello Plane Wave calculations, which necessarily uses DFT. For methanol this value is $\approx$30 kJ/mol higher than the interaction energy computed by Blaszkowski *et al.* [66] for methanol interacting with a 3 T acid site cluster by DFT. This amounts to a correction of 30% to the energy computed on the small cluster. For the heat of adsorption a correction of the same order of magnitude is necessary for calculations on small clusters. Compared to the single tetrahedron case and especially when clusters with tetrahedral rings are considered two new geometric options arise. Firstly for steric reasons to the backbinding molecule not always three oxygen atoms around Al may be available, but only two. As we will later see this geometry constraint may inhibit not only formation of particular adsorption modes but also transition states. Secondly in a ring also other atoms than the one connected to Al may interact with the protonated molecule. Such crossing-overs have been found for $NH4^+$ adsorbed to large ring system clusters by Sauer *et al.* [58]. Shah *et al.* [67] and Nusterer *et al.* [68,69], using Car-Parrinello calculations on extended systems representing chabasite and sodilite found the same for methanol. When the cross connecting oxygen atoms share also a lower valent lattice ion as Al, the oxygen atoms will also be basic and lead to favorable Zwitterionic interactions. There is ample crystallographic evidence that in many situations such situations arise [70]. However it is important that experimentally quite high loadings of molecules are used. In such a case a network of hydrogen bonded molecules will be present leading to additional stabilizing interactions. For instance cluster calculations [33,54] agree that a single H_2O molecule will only hydrogen bond to a zeolitic proton. For this strong spectroscopic evidence exists, not only from H-NMR [32] but also from the analysis of Infrared spectra [54]. However when two water molecules interact the di-molecular complex will protonate [32]. The additional H_2O molecule stabilizes the hydronium ion.

NH_4^+ is also stabilized by interaction with a second ammonia molecule. This also accompanied by a much weaker interaction with the zeolite lattice oxygen atoms and one point adsorption becomes possible [70,71].

For hydrogen-bonded adsorption two or three point adsorption can be important. This is schematically illustrated for hydrogen bonded methanol in Figure 3a, b and c. In this structure a Brønsted acidic proton interacts with the basic oxygen from methanol. The proton of methanol backbinds to one of the basic oxygen atoms around Al. Such bonding where a basic alcohol oxygen atom is hydrogen bonded to Brønsted acidic proton and the alcohol proton is hydrogen bonded to a Lewis basic surface atom has been first proposed by Knözinger *et al.* [72]. As mentioned earlier in Car-Parrinello Plane Wave calculations on three dimensional systems backbinding is also observed with more remote oxygen atoms further from the Al site.

This indicates that backbinding in ringsystems connected to the three-dimensional zeolitic oxygen atom is rather weak. The interaction is H(2) - O(2) (Figure 3b is overestimated because of the equivalence of the oxygen atoms around Al. The calculations Greatbanks *et al.* [65] are a strong indication of this. The relative small downwards shift of the CH_3OH O(3) - H(2) stretch frequency to 3540 cm-1 [73,74] provides experimental support. An analogous result has been reported for the weakening of H_2O stretching frequencies [75]. It is important to realize that DFT [66] as well as HF/MP2 [5] cluster calculations overestimate this bond weakening by more than 200 cm-1. The 3250 cm-1 belongs to the asymmetric stretching frequency of proton [3] in the dimethanol complex sketched in Figure 4. Quantum-chemistry has been extensively applied to study the problem [5,13,65–67] whether CH_3OH is adsorbed in the hydrogen bonded state or protonated as in the methoxonium ion (Figure 3c).

Analogous studies are available for the protonation problem of H_2O [27–30,64,65,70,76] acetone [77,78], acetonitrile [7,79], and dimethyl ether [80]. The earlier reported increases in computed adsorption energy due to embedding indicate that this mainly due to an enhanced H(1) - O(3) (Figure 3a) interaction. This implies an enhanced Brønsted acidity of the H(1) - O(1).

All cluster calculations so far indicate only hydrogen bonding with this strongly basic molecules. This agrees with the current interpretation of infrared spectra and H-NMR nuclear shifts. Only the more basic pyridine and ammonia are present as protonated species.

As demonstrated experimentally [78] there is a parallel between gasfase proton affinity and protonation energy of basic molecules in the zeolite. Pyridine and ammonia are molecules with the highest protonation energy.

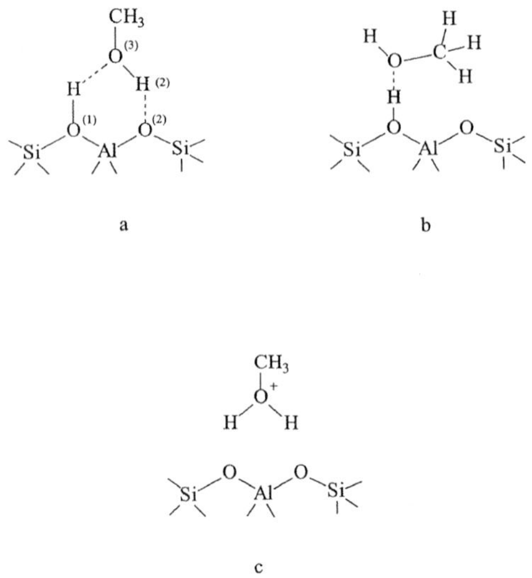

Figure 3. Adsorption modes of methanol (schematic): a end-on adsorbed methanol; b side-on adsorbed methanol; c protonated methanol

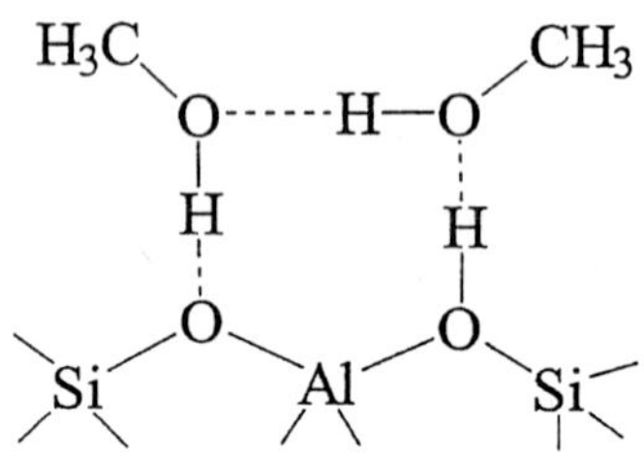

Figure 4. Di-methanol complex (schematic)

An important technical issue in the calculations is the calculation of local energy minimum versus local saddlepoints. The latter correspond to transition states. It has become clear that for strongly Brønsted basic molecules the difference in energy between the protonated and hydrogen bonded molecule is of the order of 10 kJ/mol. Initially with clusters erroneous results were reported because geometries were used as obtained with the Hartree-Fock method or an artificial geometry constraint had been applied to reduce computer cost.

It has become clear that optimized geometries within the HF/MP2 approximation and a high quality basis set are essential and no symmetry constraint should be applied. Cluster calculations then show as typical result for H_2O or CH_3OH the double- minimum potential [5,32] as shown in Figure 5. The methoxonium of hydronium ion is a transition state structure a top of the double-minimum and the hydrogen bonded situations are local minima. Of course the local minima should become different in energy for embedded situations. Such potential energy curves are not yet available. Computed H-NMR shifts [5] and QCC constants [6] correspond to a dynamically strongly perturbed H_2O or CH_3OH molecule that hops between the two minima in the double-minimum potential.

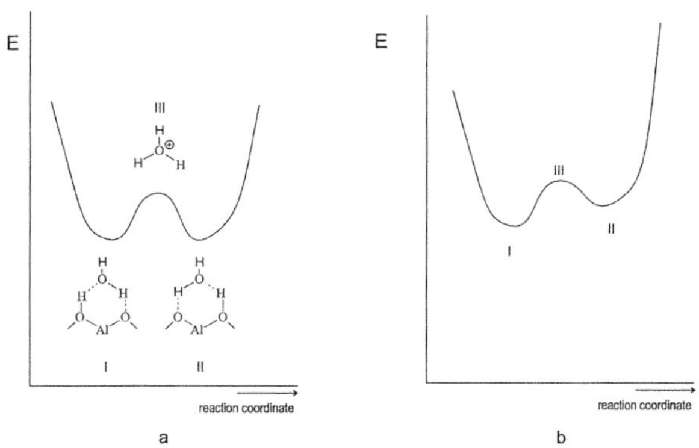

Figure 5. Potential energies of adsorbed methanol as a function of proton positions: a. surface atoms equivalent b. surface atoms non-equivalent

A recently published neutron scattering study [76] of water adsorbed in H-ZSM-5 shows very nice agreement between computed loss frequencies and the hydrogen bonded water model.

Car-Parrinello Plane Wave calculations [13,68] also indicate very small energy differences between the protonated and hydrogen bonded state. No transition state is yet available in such calculations.

In a combined experimental and theoretical study [81,82] of the methane-deuterium exchange reaction, the non-equivalence of the oxygen atoms around Al has been shown to be responsible for the difference in turnover frequency per Al in faujasite compared to ZSM-5 (MFI). Because of the larger unit cell of the MFI structure, the differences in deprotonation energy of non-equivalent O atoms is much less in the MFI structure than in the zeolite Y structure.

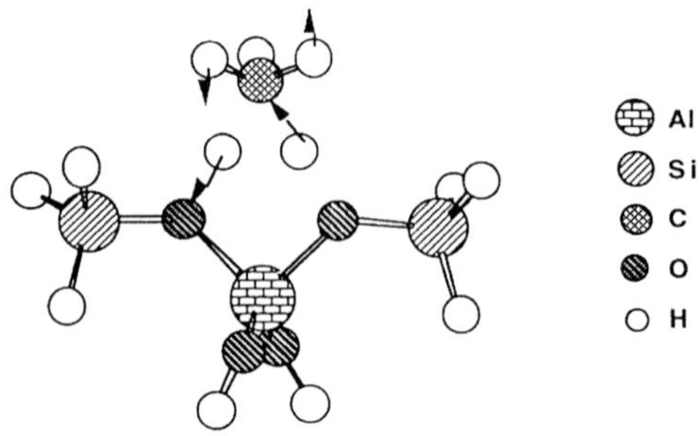

Figure 6. Transition state of hydrogen/deuterium exchange of methane [66]

The computed transition state for the deuterium exchange reaction (Figure 6), in which the carbon atom becomes five coordinated as in a carbonium-ion, with the computed reaction coordinate shows that a different oxygen atom accepts the proton from CH_4D^+ than the oxygen atom that donates deuterium.

Using the Brønsted-Polanyi relation, equation 3 and computed deprotonation energies [44,45] the average TurnOver Frequencies for zeolite Y and ZSM-5 have been computed. The result is a significantly larger TOF for ZSM-5. Important is now also the theoretical prediction that the carbonium ion CH_5^+ is not a stable local energy minimum intermediate, but a transition state with its energy maximized as a function of the reaction coordinate. The calculations have been confirmed by Blaszkowski *et al.* [83] and Evleth e.a. [84]. This is a very general result and consistent with the previous discussion.

Whereas in the case of CH_5^+ backbinding to the negatively charged lattice atoms is rather strong and bond distances are such that the interaction can be considered nearly covalent, usually transition states are rather ionic.

In the case of H/D exchange of methane the Brønsted-Polanyi relation between ΔE_{act} and the difference in deprotonation energies of the oxygen atoms that give H-D exchange is found to be [80]:

$$\Delta E_{\mathrm{act}} = 0.7 \Delta E_{\mathrm{prot}}. \tag{5}$$

However in case one deals with a ionic transition state this dependence has become much less typically

$$\Delta E_{\mathrm{act}} = 0.1 \Delta E_{\mathrm{prot}}. \tag{6}$$

If one now compares the difference in reactivity of different sites, the activation will be dominated by differences in deprotonation energy of the Brønsted acidic proton. The computed bond distances between transition state atom positions and lattice oxygen atoms are often a good discriminator between these situations.

In calculations one can also test the Brønsted-Polanyi relation on deprotonation energies by varying the position of the atoms that bound the cluster. Such an approach has been used for proton activation in alkanes in [83]. These transition states except for H/D exchange are usually found to be quite ionic.

Another important carbocation is the carbenium ion formed in superacids by protonation of an olefin. By contrast in solid acids carbenium ions are transition states. Senchenya and Kazansky [85,86] and Pelmenschikov *et al.* [87] were the first to propose that carbenium ion intermediates are not stable intermediates but instead are transition states. After protonation the protonated olefin becomes converted to a stable alkoxy species (s-complex) [84–86,88–90]. One has to distinguish between the hydrogen bonded ethylene (π-complex) and the alkoxy complex (s-complex). Cant and Hall [91] were the first to measure for ethylene adsorption the adsorption energy of the π-complex. They found a value of 36 kJ/mol for zeolite Y. The generation of alkoxy species has been demonstrated by NMR spectroscopy [92–94] as well as infrared spectroscopy [95].

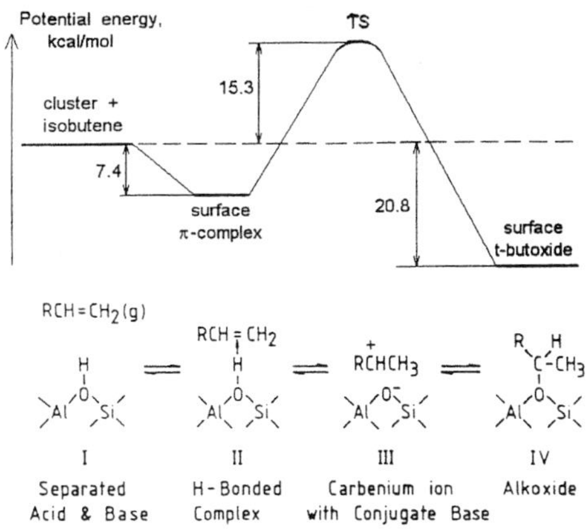

Figure 7. Reaction energy diagram of the protonation of isobutene [44] and the corresponding steps for a therminal alkene.

Figure 7 gives the computed energy reaction diagram representative of the energy changes in the different elementary reaction steps that leads to formation of the protonated alkene. The energies given have been computed for the protonation of isobutene on a three tetrahedral cluster. It is important to realize that the energy changes are only due to the interaction of the p electronic system of the molecule upon interaction with the zeolitic proton. Hence the free state of zero interaction energy is the state of isobutene adsorbed in the micropore of a zeolite. There the interaction energy is dominated by the

Van der Waals interaction and is strongly micropore dimension dependent [96]. If one wishes to compare the computed protonation energy (s-complex) with the measured heat change when a molecule adsorbs from the gasfase, one has to add to the computed protonation energy the heat of adsorption of isobutene in a siliceous zeolite channel. With respect to the gas phase one now realizes that the heat of protonation is different for different zeolites because of the differences in Van der Waals interaction of the adsorbed molecules with the zeolite micropore. The interaction with the zeolite proton is often nearly the same. In Figure 7 the different interaction stages are shown. As mentioned the reference state is isobutene adsorbed in zeolitic micropore. A first step is the formation of a hydrogen bond between proton and olefin to form a π-bonded complex. This is a non activated process and corresponds to a relatively small interaction energy.

Protonation of the olefin is an activated process. The OH bond has to be ruptured. The transition state (Figure 7) has the geometry of a carbenium ion. Protonation occurs to the primary carbon atom and most of the positive charge becomes located on the tertiary carbon atom that has become part of a planar geometry and can be considered sp2 hybridized. The positively charged side of the molecule is directed towards one of the negatively charged oxygen anions, which results in a strong electrostatic stabilization. The transition state for ethylene protonation is rather covalent and becomes more ionic for isobutene. The ground state that corresponds to the protonated ethylene molecule is the s-bounded complex, the alkoxy species. The difference in energy between this state and the transition state energy is of course dominated by the covalent O-C bond, where strength only slightly varies between the different oxygen atoms. The difference in relative stability of the carbenium ion transition states controls the activation energy differences. NMR experiments have shown that a stabilized carbenium-ionic state can be converted to stable carbenium ion. An example is the cyclopentenyl cation [94].

3. Molecular dynamics

3.1. Introduction

In recent years, molecular dynamics (MD) has been used in an increasing number of studies on zeolite systems. These studies include both the adsorption and diffusion of guest molecules and the structure and dynamics of zeolite lattices. Some of these studies will be discussed below to give the reader an idea of the scope and limitations of MD simulations but first a short introduction to this computational tool is given. For a more thorough discussion of MD the reader is referred to the books by Allen and Tildesley [97] or Frenkel and Smit [98].

3.2. MD as computational tool

In MD classical mechanics is used to study the evolution of a system in time. First the trajectory of the system is generated. In the often used Verlet algorithm this is done by a Taylor expansion (until third order) of the positions (r) for each of the particles (atoms) around time t:

$$r(t+\Delta t) = r(t) + \frac{\mathrm{d}r(t)}{\mathrm{d}t}\Delta t + \frac{1}{2}\frac{\mathrm{d}^2 r(t)}{\mathrm{d}t^2}\Delta t^2 + \frac{1}{6}\frac{\mathrm{d}^3 r(t)}{\mathrm{d}t^3}\Delta t^3, \quad (7)$$

$$r(t-\Delta t) = r(t) - \frac{\mathrm{d}r(t)}{\mathrm{d}t}\Delta t + \frac{1}{2}\frac{\mathrm{d}^2 r(t)}{\mathrm{d}t^2}\Delta t^2 - \frac{1}{6}\frac{\mathrm{d}^3 r(t)}{\mathrm{d}t^3}\Delta t^3. \quad (8)$$

Combination of these equations, in which Δt is the time step of the simulation, gives:

$$r(t+\Delta t) = 2r(t) - r(t-\Delta t) + \frac{\mathrm{d}^2 r(t)}{\mathrm{d}t^2}\Delta t^2. \tag{9}$$

which can be solved in combination with Newton's equation of motion:

$$m\frac{\mathrm{d}^2 r(t)}{\mathrm{d}t^2} = F_{\text{Newton}} = -\frac{\mathrm{d}\mathcal{U}_{\text{pot}}}{\mathrm{d}r}. \tag{10}$$

In the latter equation m is the mass of the particle (atom) and $\mathcal{U}_{\text{pot}}$ is the potential energy. The velocity of the particle follows from:

$$v(t) = \frac{\mathrm{d}r(t)}{\mathrm{d}t} = \frac{r(t+\Delta t) + r(t-\Delta t)}{2\Delta t}. \tag{11}$$

From the velocities the kinetic energy of the system can be calculated:

$$\mathcal{U}_{\text{kin}} = \frac{1}{2}\sum_{i=1}^{N} m_i v_i(t)^2. \tag{12}$$

In equation (12) N is the number of particles (atoms) in the system. The temperature of the system follows from:

$$T(t) = \frac{2}{3}\frac{\mathcal{U}_{\text{kin}}}{N k_B} \tag{13}$$

in which k_B is the Boltzmann constant.

Generally, MD simulations are carried out on rather small systems ($N < 10^4$). Surface or wall effects can be avoided and the link with real systems can be preserved by implementing periodic boundary conditions. In this technique the simulation box is replicated in space to form an infinite lattice. The trick is that, when during the simulation a molecule (atom) leaves the central box, one of its images will enter the box at the opposite side. The effect of periodic boundary conditions on the thermodynamic properties is generally small but certain properties cannot be studied in this way [97].

With simulations in the NVE ensemble (constant number of particles, constant volume, constant total energy) the instantaneous kinetic energy fluctuates around an average but the total energy $\mathcal{U}_{\text{tot}} = \mathcal{U}_{\text{kin}}(t) + \mathcal{U}_{\text{pot}}(t)$ should be constant (conserved) during the simulation. This is only achieved after the system has been equilibrated long enough to erase all start-up effects and when the time step t is sufficiently short (order of 10^{-15} s (fs)). During equilibration the temperature of the system can be set approximately by coupling it temporarily to a temperature bath. Such a coupling, in which the velocities and the positions of the particles are corrected when the temperature calculated with equation (13) deviates from the simulation temperature that is desired, is central in simulations in the NVT ensemble (constant number of particles, constant volume, constant temperature). The NVT algorithm is less expensive in computer time because it largely avoids the long equilibration period. It has been argued [1] that the NVT algorithm reflects physical reality better in simulations in which the zeolite lattice is kept fixed.

To simulate the trajectory one has to calculate in each time step (see equation (10)) the potential energy and its derivatives (the forces on the particles (atoms)). In most simulations on zeolite systems a force field has been used for this purpose. Each force field, and there are many, has to be validated against experimental data or against results from ab initio calculations. It is clear that a quantum chemical calculation of the potential energy and the forces would be preferable. However, at the moment the computational costs of such a simulation are prohibitive.

After the trajectory of the system has been generated one can find the thermodynamic properties $< A >$ by averaging over time:

$$< A >= \frac{1}{n} \sum_{i=1}^{n} A(t_i). \tag{14}$$

where n is the number of sampling points and $A(t)$ is the value of the particular property at a specific time. Examples are the temperature in an NVE simulation and the heat of adsorption for a guest molecule in a zeolite. When the simulation is performed with a rigid lattice the latter can be calculated from:

$$\Delta H_{\mathrm{ads}} = RT - < \Delta \mathcal{U}_{\mathrm{ads}} >, \tag{15}$$

$$< \Delta \mathcal{U}_{\mathrm{ads}} > = < \Delta \mathcal{U}_{\mathrm{nb}} > + < \Delta \mathcal{U}_{\mathrm{guest}} >, \tag{16}$$

in which $< \Delta \mathcal{U}_{\mathrm{nb}} >$ is the average value of the non-bonded interactions and $< \Delta \mathcal{U}_{\mathrm{guest}} >$ is the average strain in the guest molecule as a result of adsorption. The calculation of some other properties from the simulated trajectory will be discussed below.

In general, for good statistics the number of sampling points n should be large, which makes MD simulations expensive both in computer time and in storage requirements because the trajectory has to be saved for later analysis. Of course the possibilities in MD simulations depend strongly on the length of the time step, the number of the degrees of freedom and the available computer facilities. At the moment the practical limit seems to be a total simulation time of a few nanoseconds.

3.3. Adsorption and diffusion

Most MD studies in zeolite science were carried out on the diffusion of relatively small molecules in zeolite channels. The self-diffusion coefficients can be obtained from the simulated trajectories with the Einstein relationship:

$$D = \lim_{t \to \infty} \frac{< \Delta r(t)^2 >}{6t}, \tag{17}$$

in which $< \Delta r(t)^2 >$ is the averaged mean square displacement of the centre of mass of the molecules:

$$< \Delta r(t)^2 >=< [r(t_0 + t) - r(t_0)]^2 > . \tag{18}$$

The averaging is generally carried out over all molecules and over independent time origins t_0. The latter implies that different points along the trajectory are taken as origin for a part of the trajectory.

Of course, the limit $t \to \infty$ (equation (17)) cannot be reached but total simulation time should be sufficiently long to obtain accurate values for D; e.g. Bull *et al.* [99] estimated that a simulation time of around 200 ps is required for an accurate determination of D

for benzene in NaY at room temperature. In practice [100] this limits the application of MD in diffusion studies to molecules with $D > 10^{-10}$ m^2s^{-1}. These are rather small molecules; in the n-alkane series n-hexane is the largest member for which diffusion has been studied by MD [101]. For large molecules other modelling techniques have to be used. One possibility is to use transition state theory with a hopping-model in which molecules pass an energy barrier when they migrate from one adsorption site to the other. Mosell *et al.* [102] have used MD simulations to obtain transmission coefficients for cage-cage migration of xenon in zeolite NaY. These transmission coefficients are then used [103] as corrections in a simulation based on transition state theory. The results are in accordance with those of conventional MD simulations. Both the value of the diffusion coefficient and that of the activation energy are in excellent agreement. This validates the use of transition state theory in simulations of molecules with diffusion coefficients too small on the time scale of conventional MD simulations. An interesting approach is also that of Maginn *et al.* [104] who use concepts from Brownian motion theory and transition state theory in a study of long n-alkanes, up to C_{20}, in silicalite.

Most MD simulations on diffusion of molecules in zeolites have been carried out with a rigid framework and often also with a rigid molecule. The rigid framework approximation neglects the influence of the framework vibrations on the diffusion motion. It appears that for small molecules this does not make a significant difference in the calculated values of D and ΔH_{ads}. The framework vibrations, however, do influence the pair distribution functions [100,105]. The calculated diffusion coefficients generally agree well with experimentally obtained microscopic diffusion coefficients (see Figure 8)).

The purpose of most MD studies on diffusion in zeolites, however, was not the numerical value of the self-diffusion coefficients but to get microscopic picture of the diffusion mechanism. One of such studies [102,103] has been discussed above; a few more examples follow.

The diffusion of methane in silicalite-1 has been studied by several groups [100,106–111]. Figure 9 taken from the work of Goodbody *et al.* [108] shows the expected anisotropy of the diffusion and its dependence on loading.

Shen and Rees [112] studied the diffusion of carbon dioxide in the pure siliceous frameworks of theta-1 and silicalite-1. Their results (see Figure 10) show that the number of CO_2 molecules in the simulation box does not affect significantly the diffusion in the three-dimensional channel system of silicalite-1 but it does in the one-dimensional channel system theta-1. This is ascribed to single-file diffusion of CO_2 in theta-1. The activation energy for diffusion can be obtained from the temperature dependence of the diffusivity (see Figure 11). Again there is reasonable agreement between the experimental observations and the data obtained by simulation. Single-file diffusion can be inferred from MD simulations by an increase of the mean square displacement $< \Delta r(t)^2 >$ that is proportional to the square root of the time [113,114]. In the study of Shen and Rees [112], however, the total simulation time was too short for this observation.

Smirnov and Van de Graaf [111] studied the diffusion of methane in silicalite-1 and 2 with a CIEEM MD simulation. The CIEEM force field [115] has geometry dependent atomic charges and allows polarization to be taken into account in the MD simulation. It was found that the polarization of the methane molecules reaches a maximum at the edges of the channel intersections. Also there is a significant difference between silicalite-1

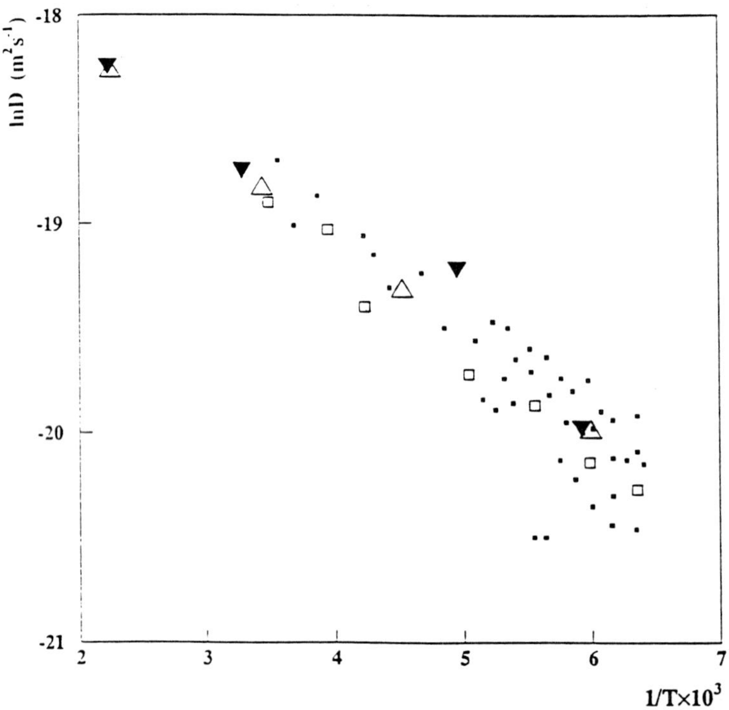

Figure 8. Log plot of diffusion coefficients for 12 molecules of methane per unit cell: computed from MD simulations, (open triangles) vibrating framework, (solid triangles) fixed framework; experimental values, (solid squares) ZSM-5 with various aluminium content and crystal size and (open squares) silicalite versus $1/T$. (reproduced with permission from ref. [99]).

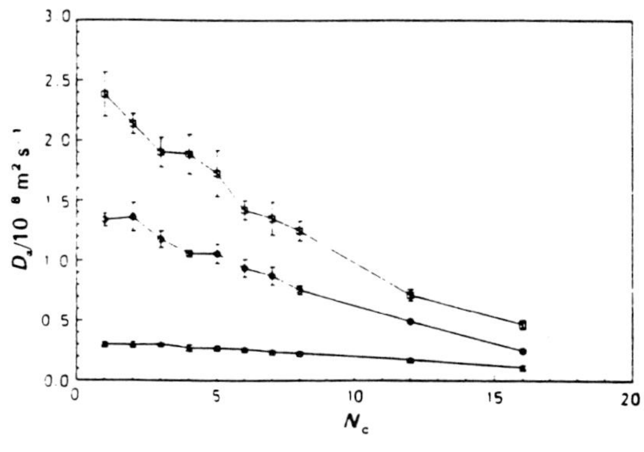

Figure 9. Diffusion tensor elements (in increasing order: c-axis, a-axis , b-axis) for methane in silicalite at 298 K as a function of loading. (reproduced with permission from ref. [108].

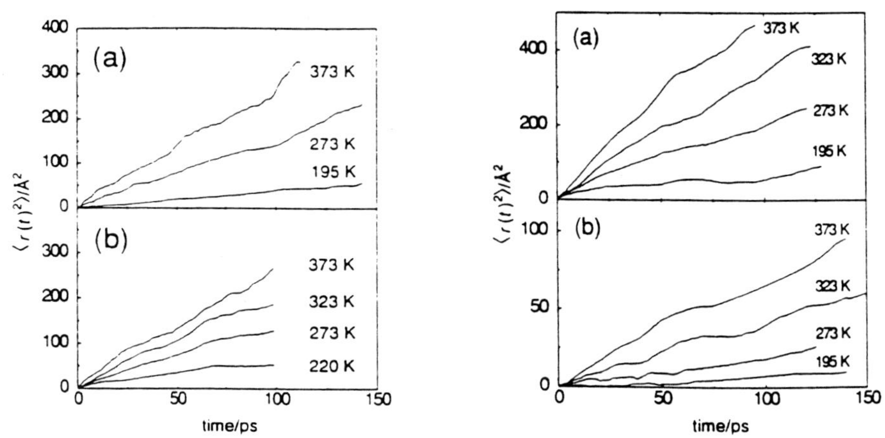

Figure 10. Mean-square displacements of carbon dioxide in silicalite-1 (on the left) and theta-1 (on the right): (a) one and (b) two molecules per simulation box. (reproduced with permission from ref. [112]).

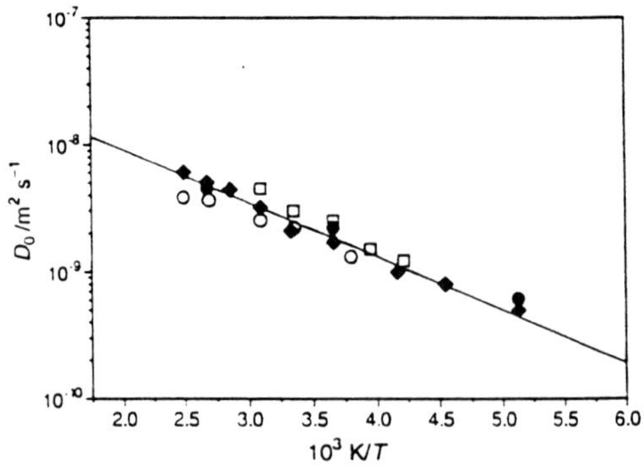

Figure 11. Temperature dependence of carbon dioxide diffusivity in silicalite: (open squares) frequence response measurements; (open circles) microscopic pulsed-field gradient NMR; (solid circles, solid squares) MD with one and two molecules per box, respectively. (reproduced with permission from ref. [112]).

and 2 with respect to the polarization of the methane molecules (see Figure 12). It was expected that this difference will be expressed in a difference in catalytic activity.

Yashonath and co-workers have performed extensive MD simulations to investigate the effect of the diameter of the adsorbed molecules on the diffusion coefficient [116–122]. Figure 13 shows the diffusion coefficient of Lennard-Jones molecules adsorbed in NaCaA as a function of the diameter of these adsorbed molecules. The simulations indicate that as the diameter of the molecules is increased the diffusion coefficient first decreases. However, if this diameter has approximately the same size as the window between the cages, then the diffusion coefficient increases and a maximum is observed. A further increase of the diameter of the adsorbed molecules gives again a decrease of the diffusion coefficient. Such a maximum or "levitation effect" has been observed in simulations of various other zeolites (Y [117,120], silicalite [118,122], and VPI-5 [122]) and is both observed in simulations using a rigid [117] and a flexible [121] zeolite framework. Experimental evidence of this effect has not yet been reported.

The diffusion coefficient in zeolite A is determined by the hopping of the adsorbed molecules from one cage to another through a window with a given diameter. If one inserts molecules with a larger diameter, the diffusion coefficient is expected to decrease since it will be more difficult for those bigger molecules to go through this window. Yashonath and co-workers discovered that indeed if the molecules are small compared to the window diameter the window is an energetic barrier. However, for molecules that have approximately the same diameter as the window this energy barrier disappears and hence the diffusion is larger that one would expect. Of course, if the molecules are too big the diffusion drops down to zero.

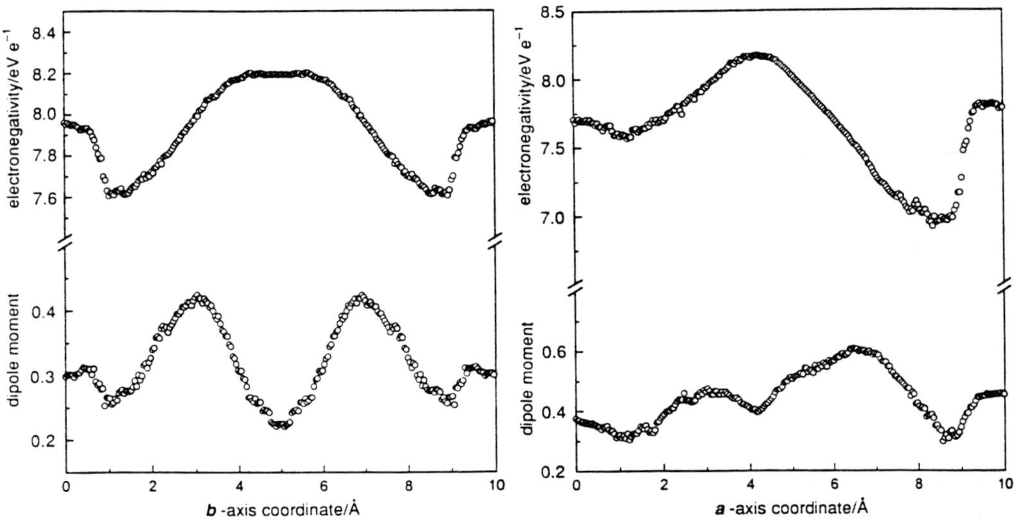

Figure 12. Dependence of the molecular electronegativity and the molecular dipole moment (D) of methane molecules on the position of the molecules in the channel between the intersections in MFI (left; straight channel) and MEL (right). Each point corresponds to the average over all segments. The minimum and maximum values of the horizontal axis correspond to the centres of the intersections. (reproduced with permission from ref. [111]).

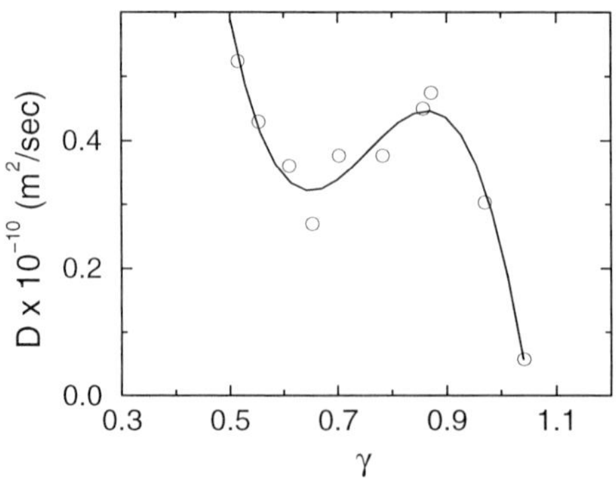

Figure 13. Diffusion coefficient of Lennard-Jones particles as a function of their reduced size in the zeolite NaCaA. The size of the molecule is written as $\gamma = 2^{1/6}\sigma_{ss}/\sigma_w$, where σ_{ss} is the diameter of the Lennard-Jones molecules and σ_w the diameter of the window. The factor $2^{1/6}$ is introduced to take into account the complete repulsive part of the potential. On the x-axis the Lennard-Jones diameters of the nobles gasses are indicated. The line is a guide to the eye. The simulation results are taken from ref.[119].

Another interesting phenomena that has been speculated upon theoretically (see references in [123]) is the so-called resonant diffusion. These theories predict the diffusion coefficient as a function of chain length to be periodic. This periodicity is related to a matching of the size of the molecule with the periodicity of the lattice. Runnebaum and Maginn [124] have performed extensive Molecular Dynamics simulations of the linear alkanes ranging form C_4 to C_{20} in silicalite and found evidence for this phenomena.

3.4. Vibrational characteristics of adsorbed molecules

When the molecules are treated as flexible in the MD simulation it is relatively simple to calculate the vibrational characteristics from the trajectory. The power spectrum is obtained by Fourier transform of the velocity auto-correlation function $< v(t)^2 >$. A power spectrum results from all motion because in its calculation no selection rules are applied. The IR spectrum can be obtained by Fourier transform of the dipole moment auto-correlation function $< \mu(t)2 >$.

An example of a calculation [111] is given in Figure 14. It shows the power spectrum and IR spectra computed for methane in silicalite-1. In the power spectrum the band at 100 cm^{-1} relates to translational and rotational degrees of freedom. The two IR spectra were computed with different force fields: one with fixed atomic charges (FCA) and one with geometry dependent charges (CIEEM force field). In the latter spectrum a low-intensity band is present at 2826 cm^{-1} that is absent in the FCA spectrum. The band relates to the normally forbidden symmetric stretch and is observed in the CIEEM spectrum as a result of polarization of the methane molecules by the zeolite lattice.

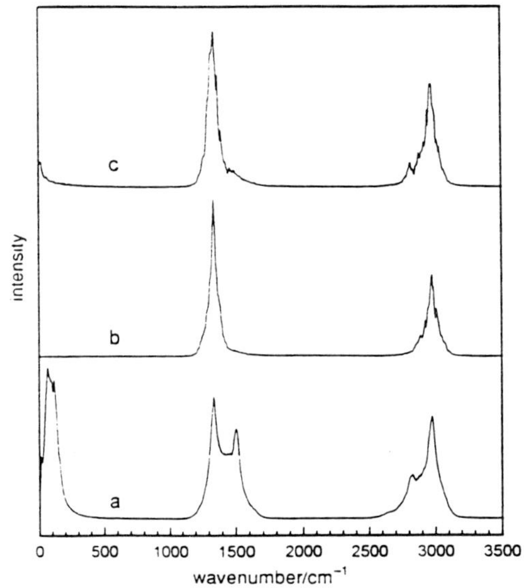

Figure 14. Calculated vibrational spectra of methane molecules adsorbed in MFI: power spectrum of the hydrogen atoms (a) and IR spectra from a MD run using a fixed charge approximation (b) and a CIEEM MD run (c). (reproduced with permission from ref. [111]).

The applications of Raman scattering in zeolite science and the calculation of Raman spectra from MD simulations for molecules adsorbed in zeolites have been reviewed recently by Brémard and Bougeard [125].

3.5. Zeolite structure and lattice dynamics

MD is also used increasingly in studies on the structure of zeolites and on lattice dynamics. Of course, such studies depend on an adequate force field being available. The advantage of MD studies over energy minimization studies is that MD is closer to physical reality: entropic effects are automatically taken into account and in the calculation of the lattice vibrations it is not necessary to invoke the harmonic approximation. The calculation of the lattice vibrations follow the same procedure as the calculation of the vibrational characteristics of adsorbed molecules. Again the results of a few studies will be discussed to illustrate the scope of MD simulations for these subjects.

Yamahara *et al.* [126] have studied the thermal behavior of silicalite-1 with MD. In agreement with the experiment they find that the monoclinic structure is the stable one at low temperature and that the orthorhombic structure is the stable one at high temperature (see Figure 15). They note that the observation of this particular phase transition in MD simulations depends critically on the balance of forces in the applied force field. Several published force fields give only one of the two structures.

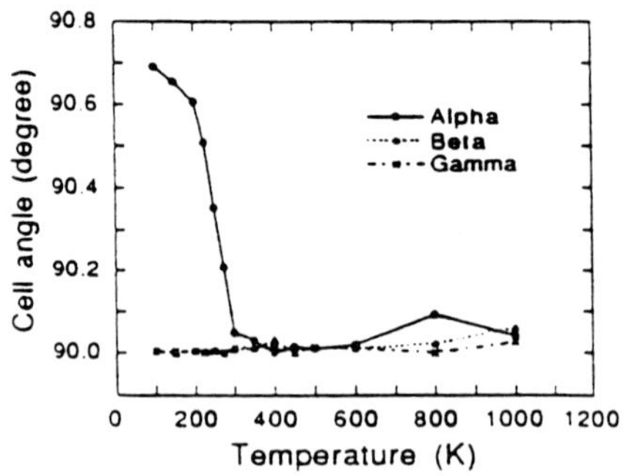

Figure 15. Temperature dependence of unit cell parameters (angles) of MFI obtained from MD runs starting with the orthorhombic structure. (reproduced with permission from ref. [126]).

MD simulation has also be applied to isomorphic substitution. Oumi *et al.* [127] have investigated the titanium substitution in TS-1. They studied the influence of this substitution on the lattice parameters and by comparing their results with experimental data (X-ray diffraction) they concluded that T8 was the most probable site for substitution. However, in a more recent study [111] it was shown that the experimental trends in the lattice parameters can be reproduced satisfactorily with titanium atoms placed at random

T positions (see Figure 16). Nevertheless, the approach by Oumi *et al.* is quite original as it does not a priori assumes thermodynamic equilibrium as has been done in all other theoretical studies on isomorphic substitution.

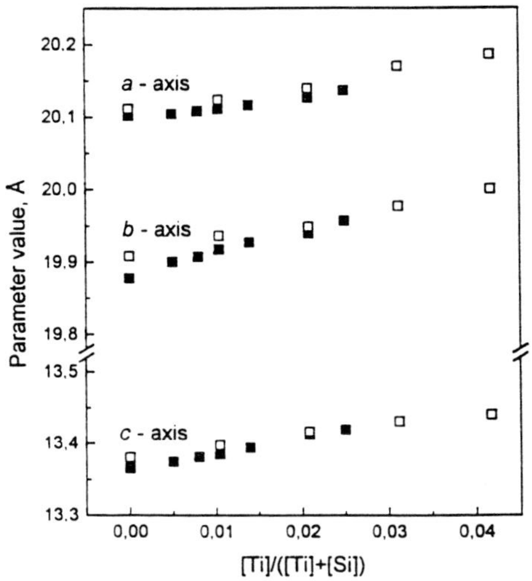

Figure 16. Dependence of the lattice parameters on the [Ti]/([Ti] + [Si]) ratio. Open squares denote calculated (MD) values, solid squares denote experimental values. (reproduced with permission from ref. [128]).

Smirnov and Bougeard [129] have studied the lattice dynamics of some siliceous zeolites with MD. Their results show that for zeolite A the presence of sodium or potassium ions leads to drastic changes in the distribution of the window diameter (see Figure 17). In view of these results one may well question simulations on the adsorption of larger molecules in which the zeolite lattice is kept fixed.

MD simulations can provide a microscopic explanation of features observed in experimental spectra. An example [128] is given in Figure 18 which provides an answer to the origin of the band at 960 cm^{-1} in the vibrational spectra of titanium substituted zeolites. From the power spectra calculated for the TiO_4 tetrahedra it is concluded that the band relates to the localized Si-O mode in the Si-O$\cdots$Ti bridges.

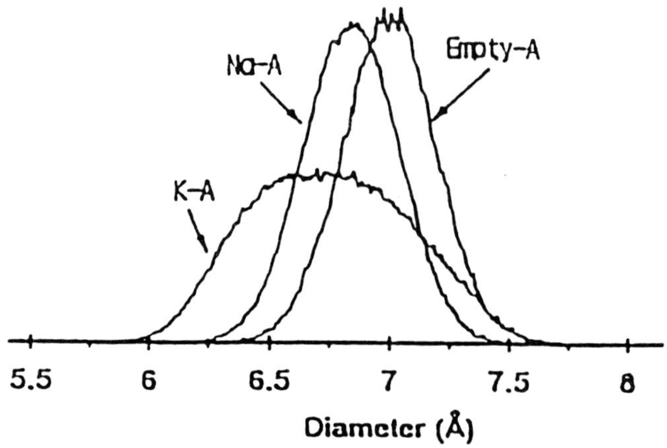

Figure 17. The distribution function of the window diameter of zeolite A without cations (empty-A), Na zeolite A (Na-A) and K zeolite A (K-A). (reproduced with permission from ref. [129]).

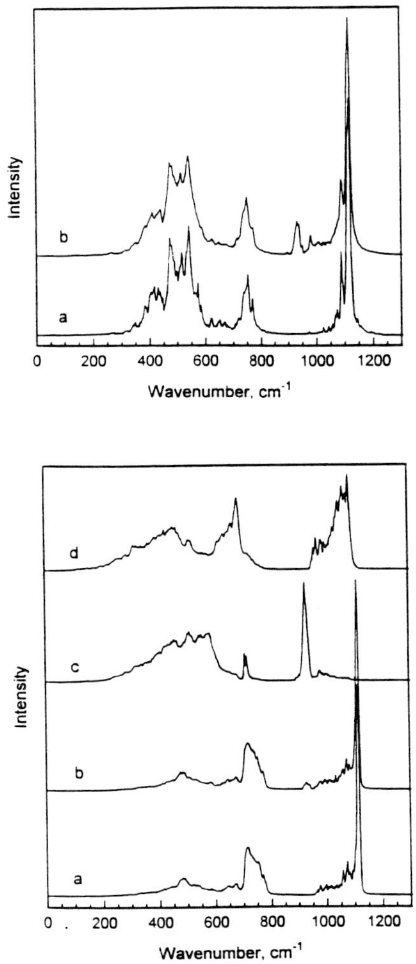

Figure 18. At the top, calculated IR spectra for silicalite (a) and Ti containing MFI (b); at the bottom, spectra of oxygen vibrations in TiO_4 tetrahedra: (a) SiO_4 in silicalite, (b) in SiO_4 and (c) in TiO_4 tetrahedra in the Ti-containing MFI, and (d) in TiO_4 tetrahedra in the model in which the same values of force constants for Ti-0 and Si-0 bonds are used. (reproduced with permission from ref. [128]).

4. Monte Carlo simulations

4.1. Introduction

In the previous edition of this book, the basic Molecular Dynamics and Monte Carlo techniques have been discussed [1]. At that time, standard simulation techniques were used for the first time to simulate the adsorption [130–132] and diffusion [133,134] in zeolites. As shown in the previous section, these simulations gave important insights in the behavior of small molecules (noble gasses, methane, or ethane) adsorbed in the pores of the zeolites. A review of these simulation results can also be found in ref. [135]. These simulations also made clear that simulating long chain molecules would require much more cpu time than one could afford. Moreover, one would have to wait many decades before computers would be sufficiently fast such that one would be able to perform such simulations. Next we discuss some of the recent attempts to develop schemes to make the simulations of such systems more efficient.

Before we start with the technical aspects of the Monte Carlo technique, let us discuss the previous statements of the limitations of conventional simulation techniques in more detail. Since the conventional simulation techniques such as Molecular Dynamics and the Monte Carlo technique are equally valid for simple and complex fluid, one may wonder whether a simulation of a complex system is simply a question of changing the force fields. In some cases it may indeed be as simple as that, in particular if the computers are sufficiently powerful to deal with these more complex systems in a reasonable amount of time. For some problems, such as the sorption of long-chain hydrocarbons, however, the increase in computer time can be prohibitively large; it may take many years of super-computer time before a calculations is finished.

Let us illustrate this with a specific example. Over the last decade many simulation studies on the sorption of molecules in zeolites have been published. A careful look at these studies reveals that most simulations concern the adsorption of noble gases or methane, only a few studies of ethane or propane have been published. In petro-chemical applications of zeolites, however, we are interested in the behavior of much longer alkanes such as octane and decane. The reason why only small molecules have been studied becomes clear from, for example, the work of June *et al.* [136] and Hernández and Catlow [101], in which Molecular Dynamics simulations were used to investigate the diffusion of butane and hexane in the zeolite silicalite. June *et al.* showed that the diffusion of butane from one channel of the zeolite into another channel is very slow compared to diffusion of bulk butane. As a consequence many hours of super-computer time were required to obtain reliable results. In addition, these results show that the diffusion coefficient decreases significantly with increasing chain length. Hence, extrapolation of these results suggests that many years of super-computer time would be required to obtain comparable results for the longer alkanes.

The above example illustrates one of the main limitations of Molecular Dynamics, in such a simulation, the approach is to mimic the behavior of the molecules as realisticly as possible. If successful, all properties will be like in nature, including the diffusion. If the molecules diffuse slowly this will be reflected in very long simulation times and in the case of long chain alkanes these simulation times turned out to be much longer than we can currently afford. In principle, one can circumvent this intrinsicly slow dynamics by using a Monte Carlo technique. In a Monte Carlo simulation one does not have to follow the "natural path" and one can, for example, perform a move in which it is attempted to displace a molecule to a random position in the zeolite. If such a move is accepted, it corresponds to a very large jump in phase space.

Again, utilization of these type of un-natural Monte Carlo moves turned out to be limited to small molecules. For example, Goodbody *et al.* [108] have used this Monte Carlo trick to determine the adsorption isotherms of methane in a zeolite. In such a simulation one can observe that out of the 1000 attempts to move a methane molecule to a random position in the zeolite 999 attempts will be rejected because the methane molecule overlaps with a zeolite atom. If we were to perform a similar move with an ethane molecule, we would need 1000 $\times$ 1000 attempts to have one that was successful. Clearly, this random insertion scheme will break down for any but the smallest alkanes.

The above example, the adsorption of chain molecules in the pores of a zeolite, is used to illustrate the problems that may occur if one uses a conventional simulation techniques to more complex systems. It is interesting to note that similar problems may occur in the simulation of phase equilibria of chain molecules, simulations of polymers, or liquid crystals. For many of these systems it is relatively straightforward to implement the force fields to simulate these systems, however, the required simulation times to determine reliable equilibrium properties may be prohibitively long. These simulation times may even be so extreme that it can not be expected that increasing computer power will be of any help. To be able to perform simulations on complex systems it is therefore important to develop novel algorithms that are orders of magnitude more efficient than the conventional algorithms. In this chapter such algorithms are discussed.

4.2. Simulation techniques

We start the discussion on simulation techniques with a short review on the conventional (Metropolis) Monte Carlo technique. For a more extensive discussion the reader is referred to the book by Allen and Tildesley [97] or the more recent book by Frenkel and Smit [98].

4.2.1. Basic Monte Carlo simulations

The prime purpose of the kind of Monte Carlo program that we shall be discussing is to compute equilibrium properties of classical many-body systems. The Monte Carlo method is a technique to generate a sequence of configurations, each of these configurations occurs with a certain probability and the Monte Carlo scheme should be developed in such a way that the desired probability distribution will be sampled. In the canonical ensemble (constant temperature, volume, and number of particles) we have to sample a Boltzmann distribution:

$$\mathcal{N}(n) \propto \exp[-\beta \mathcal{U}(n)] \tag{19}$$

which states that the probability of finding configuration n is proportional to the exponent of the energy of this configuration $\mathcal{U}(n)$ times the Boltzmann factor $\beta = 1/k_B T$. One way of generating these configurations is, the basic Monte Carlo algorithm, as introduced by Metropolis *et al.* [137], proposed:

1. Select a particle at random, and calculate its energy $\mathcal{U}(\mathbf{r}^N)$.

2. Give the particle a random displacement; $r' = r + \Delta$, and calculate its new energy $\mathcal{U}(\mathbf{r}'^N)$.

3. Accept the move from $\mathbf{r}^N$ to $\mathbf{r}'^N$ with probability

$$\mathrm{acc}(o \to n) = \min\left(1, \exp\{-\beta[\mathcal{U}(\mathbf{r}'^N) - \mathcal{U}(\mathbf{r}^N)]\}\right). \tag{20}$$

The above algorithms assume we have a simple atomic system. If we are simulating molecules rather than atoms we must also generate trial moves that change the molecular orientation. A possible approach is, for example, to consider a system consisting of N linear molecules. We specify the orientation of the ith molecule by a unit vector $\hat{\mathbf{u}}_i$. One possible procedure to change $\hat{\mathbf{u}}_i$ by a small, random amount is the following. First, we generate a unit vector $\hat{\mathbf{v}}$ with a random orientation. Next we multiply this random unit vector $\hat{\mathbf{v}}$ by a scale factor γ. The magnitude of γ determines the magnitude of the trial rotation. We now add $\gamma\hat{\mathbf{v}}$ to $\hat{\mathbf{u}}_i$. Let us denote the resulting sum vector by $\mathbf{t}$: $\mathbf{t} = \gamma\hat{\mathbf{v}} + \hat{\mathbf{u}}_i$. Note that $\mathbf{t}$ is not a unit vector. Finally, we normalize $\mathbf{t}$, and the result is our trial orientation vector $\hat{\mathbf{u}}_i'$. We still have to fix γ, which determines the acceptance probability for the orientational trial move. We have not yet indicated whether or not the translational and orientational trial moves should be performed simultaneously. Both procedures are acceptable. However, if rotation and translation correspond to separate moves, then the selection of the type of move should be probabilistic rather than deterministic.

Only slightly more complex is the case of a nonlinear rigid molecule. It is conventional to describe the orientation of nonlinear molecules in terms of the Eulerian angles (ϕ, θ, ψ). However, for most simulations, use of these angles is less convenient because all rotation operations should then be expressed in terms of trigonometric functions, and these are computationally expensive. It is usually better to express the orientation of such a molecule in terms of quaternion parameters (for a discussion of quaternions in the context of computer simulation, see [97] or [98]).

If the molecules under consideration are not rigid then we must also consider Monte Carlo trial moves that change the internal degrees of freedom of a molecule. In practice, it makes an important difference whether or not we have frozen out some of the internal degrees of freedom of a molecule by imposing rigid constraints on, say, bond lengths and possibly even some bond angles. If not, the situation is relatively simple: we can carry out normal trial moves on the Cartesian coordinates of the individual atoms in the molecule (in addition to center of mass moves). If some of the atoms are strongly bound, it is advisable to carry out small trial moves on those particles (no rule forbids the use of trial moves of different size for different atoms, as long as the moves for one particular atom are always sampled from the same distribution).

However, when the bonds between different atoms become very stiff, this procedure does not sample conformational changes of the molecule efficiently. In Molecular Dynamics simulations it is common practice to replace very stiff intramolecular interactions by rigid constraints. For Monte Carlo simulations this is also possible. In fact, elegant techniques have been developed for this purpose [138]. However, the corresponding MD techniques [139] are so much easier to use, in particular for large molecules, that we cannot recommend the use of the Monte Carlo technique for any but the smallest flexible molecules with internal constraints.

4.2.2. Monte Carlo simulations of chain molecules

It is possible to perform simulations beyond the Metropolis technique. These type of simulations have been developed by Siepmann and Frenkel [140] to make Monte Carlo moves of long chain molecules on a lattice possible. This so-called configurational-bias Monte Carlo (CBMC) technique for is based on the early work of Rosenbluth and Rosenbluth [141] and Harris and Rice [142]. This technique has since been extended to continuum models by Frenkel *et al.* [143] and de Pablo *et al.* [144].

4.2.3. Beyond Metropolis

The general idea of biased sampling is best explained by considering a simple example. Let us assume that we have developed a Monte Carlo scheme that allows us to generate trial configurations with a probability that depends on the potential energy of that configuration:

$$\alpha(o \to n) = f[\mathcal{U}(n)].$$

For the reverse move, we have

$$\alpha(n \to o) = f[\mathcal{U}(o)].$$

Suppose we want to sample the N,V,T ensemble, which implies that we have to generate configurations with a Boltzmann distribution (19): To proof that the correct distribution is sampled, we have to demonstrate that detailed balance is obeyed

$$K(o \to n) = K(n \to o), \tag{21}$$

where $K(o \to n)$ is the flow of configuration o to n. This flow is given by the product of the probability of being in configuration o, the probability of generating configuration n, and the probability of accepting this move,

$$K(o \to n) = \mathcal{N}(o) \times \alpha(o \to n) \times \mathrm{acc}(o \to n). \tag{22}$$

If we imposing detailed balance, we get as a condition for the acceptance rule,

$$\frac{\mathrm{acc}(o \to n)}{\mathrm{acc}(n \to o)} = \frac{f[\mathcal{U}(n)]}{f[\mathcal{U}(o)]} \exp\{-\beta[\mathcal{U}(n) - \mathcal{U}(o)]\}.$$

A possible acceptance rule that obeys this condition is

$$\mathrm{acc}(o \to n) = \min\left(1, \frac{f[\mathcal{U}(n)]}{f[\mathcal{U}(o)]} \exp\{-\beta[\mathcal{U}(n) - \mathcal{U}(o)]\}\right). \tag{23}$$

It is instructive to apply the above to the Metropolis scheme. For the Metropolis scheme, there is no difference in the probability to generate the old or new configuration:

$$\alpha(o \to n) = \alpha(n \to o) = \alpha.$$

If we substitute this equation in equation (23), we recover the Metropolis acceptance rule (20).

This derivation shows that we can introduce an arbitrary biasing function $f(\mathcal{U})$ in the sampling scheme and generate a Boltzmann distribution of configurations, provided that the acceptance rule is modified in such a way that the bias is removed from the sampling scheme.

We now illustrate the use of non-Metropolis sampling techniques to demonstrate how they can be used to enhance the efficiency of a simulation of a system containing chain molecules.

4.2.4. Configurational-Bias Monte Carlo

The starting point for the configurational-bias Monte Carlo technique is the scheme introduced by Rosenbluth and Rosenbluth in 1955 [140–144]. The Rosenbluth scheme itself also was designed as a method to sample polymer conformations. A drawback of the Rosenbluth scheme is, however, that it generates an unrepresentative sample of all polymer conformations; that is, the probability of generating a particular conformation using this scheme is *not* proportional to its Boltzmann weight. Rosenbluth and Rosenbluth corrected for this bias in the sampling of polymer conformations by introducing a conformation-dependent weight factor W. However, as was shown in detail by Batoulis and Kremer [145], this correction procedure, although correct in principle, in practice works only for relatively short chains.

The solution of this problem is to bias the Rosenbluth sampling in such a way that the correct (Boltzmann) distribution of chain conformations is recovered in a Monte Carlo sequence. In the configurational-bias scheme to be discussed next, the Rosenbluth weight is used to bias the *acceptance* of trial conformations generated by the Rosenbluth procedure. As we shall show, this guarantees that all chain conformations are generated *with the correct Boltzmann weight.*

The idea of the CBMC scheme is to grow a molecule atom by atom. During the growing, a set of trial orientations is generated and out of these trial orientations the optimum orientation is selected with the highest probability. The bias which is introduced by this growing scheme is subsequently removed by adjusting the acceptance rules. The configurational-bias Monte Carlo algorithm consists of the following steps:

1. Generate a trial conformation using the Rosenbluth scheme to grow the entire molecule, or part thereof, and compute its Rosenbluth weight $W(n)$.

2. "Retrace" the old conformation and determine its Rosenbluth factor, $W(o)$.

3. Accept the trial move with a probability

$$\text{acc}(o \to n) = \min[1, W(n)/W(o)]. \tag{24}$$

The generation of a trial conformation n of a chain consisting of ℓ atoms is generated using an algorithm based on the method of Rosenbluth and Rosenbluth. An important point that we have to consider is the way in which trial conformations of a chain molecule are generated. For example, one could generate trial segments with orientations distributed uniformly on a unit sphere. However, for many models of interest this procedure is not very efficient, in particular when there are strong intramolecular interactions (e.g., bending and torsion potentials). The efficiency of a configurational-bias Monte Carlo algorithm depends to a large extent on the method used of generating the trial orientations. For example, an isotropic distribution of trial directions is well suited for completely flexible chains. In contrast, for a stiff chain (e.g., liquid-crystal forming polymer), such a trial position will almost always be rejected because of the intramolecular interactions.

From the preceding discussion, it follows that the intramolecular interactions should be taken into account in generating the set of trial conformations Here, we consider the case of a flexible molecule *with* contributions to the internal energy due to bond bending and torsion. The fully flexible case then follows trivially. Consider a chain of ℓ linear segments, the potential energy of a given conformation $\mathcal{U}$ has two contributions:

1. The *bonded potential energy* $\mathcal{U}^{\text{bond}}$ is equal to the sum of the contributions of the individual joints. A joint between segments i and $i+1$ (say) has a potential energy u_i^{bond} that depends on the angle θ between the successive segments. For instance, $u_i^{\text{bond}}(\theta)$ could be of the form $u_i^{\text{bond}}(\theta)=k_\theta(\theta-\theta_0)^2$. For realistic models for polyatomic molecules, u_i^{bond} includes all local bonded potential energy changes due to the bending and torsion of the bond from atom $i-1$ to atom i.

2. The *external potential energy* $\mathcal{U}^{\text{ext}}$ accounts for all interactions with other molecules and for all the nonbonded intramolecular interactions. In addition, interactions with any external field that may be present are also included in $\mathcal{U}^{\text{ext}}$.

In what follows we shall denote a chain in the absence of the external interactions as the *ideal* chain. Note that this is a purely fictitious concept, as real chains always have nonbonded intramolecular interactions.

To perform a configurational-bias Monte Carlo move, we apply the following "recipe" to construct a conformation of a chain of ℓ segments. The construction of chain conformations proceeds segment by segment. Let us consider the addition of one such segment. To be specific, let us assume that we have already grown $i-1$ segments and are trying to add segment i. This is done in two steps. First we generate a trial conformation n, next we consider the old conformation o. A trial conformation is generated as follows:

1. Generate a fixed number (say k) trial segments. The orientations of the trial segments are distributed according to the Boltzmann weight associated with the bonded interactions of monomer i (u_i^{bond}). We denote this set of k different trial segments by

$$\{\mathbf{b}\}_k = \{\mathbf{b}_1, \cdots, \mathbf{b}_k\},$$

where the probability of generating a trial segment $\mathbf{b}$ is given by

$$p_i^{\text{bond}}(\mathbf{b})\mathrm{d}\mathbf{b} = \frac{\exp[-\beta u_i^{\text{bond}}(\mathbf{b})]\mathrm{d}\mathbf{b}}{\int \mathrm{d}\mathbf{b}\exp[-\beta u_i^{\text{bond}}(\mathbf{b})]} = C\exp[-\beta u_i^{\text{bond}}(\mathbf{b})]\mathrm{d}\mathbf{b}. \tag{25}$$

2. For all k trial segments, we compute the external Boltzmann factors $\exp[-\beta u_i^{\text{ext}}(\mathbf{b}_i)]$, and out of these, we select one, denoted by n, with a probability

$$p_i^{\text{ext}}(\mathbf{b}_n) = \frac{\exp[-\beta u_i^{\text{ext}}(\mathbf{b}_n)]}{w_i^{\text{ext}}(n)}, \tag{26}$$

where we have defined

$$w_i^{\text{ext}}(n) = \sum_{j=1}^{k}\exp[-\beta u_i^{\text{ext}}(\mathbf{b}_j)]. \tag{27}$$

3. The selected segment n becomes the ith segment of the trial conformation of the chain.

4. When the entire chain is grown, we calculate the Rosenbluth factor of the chain:

$$W^{\text{ext}}(n) = \prod_{i=1}^{\ell} w_i^{\text{ext}}(n), \tag{28}$$

where Rosenbluth factor of the first monomer is defined by

$$w_1^{\text{ext}}(n) = k \exp[-\beta u_1^{\text{ext}}(\mathbf{r}_1)], \tag{29}$$

where $\mathbf{r}_1$ is the position of the first monomer.

For the old configuration, a similar procedure to calculate its Rosenbluth factor is used.

1. One of the chains is selected at random. This chain is denoted o.

2. The external energy of the first monomer is calculated. This energy involves only the external interactions. The Rosenbluth weight of this first monomer is given by

$$w_1^{\text{ext}}(o) = k \exp[-\beta u_1^{\text{ext}}(o)]. \tag{30}$$

3. The Rosenbluth factors of the other $\ell - 1$ segments are calculated as follows. We consider the calculation of the Rosenbluth factor of segment i. We generate a set of $k-1$ orientations with a distribution prescribed by the bonded interactions (25). These orientations, together with the actual bond between segment -1 and i, form the set of k orientations $(\mathbf{b}_o, \mathbf{b}'^*)$. These orientations are used to calculate the external Rosenbluth factor:

$$w_i^{\text{ext}}(o) = \sum_{j=1}^{k} \exp[-\beta u_i^{\text{ext}}(\mathbf{b}_j)]. \tag{31}$$

4. For the entire chain the Rosenbluth factor of the old conformation is defined by

$$W^{\text{ext}}(o) = \prod_{i=1}^{\ell} w_i^{\text{ext}}(o). \tag{32}$$

After the new configuration has been generated and the Rosenbluth factor of the old configuration has been calculated the move is accepted with a probability

$$\text{acc}(o \to n) = \min[1, W^{\text{ext}}(n)/W^{\text{ext}}(o)]. \tag{33}$$

In [98] it is demonstrated that this algorithm samples the desired distribution.

4.3. Applications

Here the use of the CBMC technique is illustrated with some examples.

4.3.1. Simulating the energetics and siting

In the introduction we have used the adsorption of molecules as an example to illustrate the type of problem one can encounter in simulating systems that exhibit slow diffusion. Smit and Siepmann have used the CBMC technique to study the energetics and siting of alkanes in silicalite [146,147]. In these simulations alkane molecules are modelled with a united atom model, i.e., a CH_3 and a CH_2 group are considered as a single interaction centre and the zeolite is modelled as a rigid crystal. The zeolite-alkane interactions are assumed to be dominated by dispersive interaction with the oxygen atoms, which are described with a Lennard-Jones potential. A closely related technique was used by Maginn *et al.* [148]. Figure 19 shows that the simulations of Smit and Siepmann and Maginn *et al.* predict heats of adsorption of the longer chain alkanes in silicalite that are in good agreement with the experimental data and the simulations of Bigot and Peuch [149].

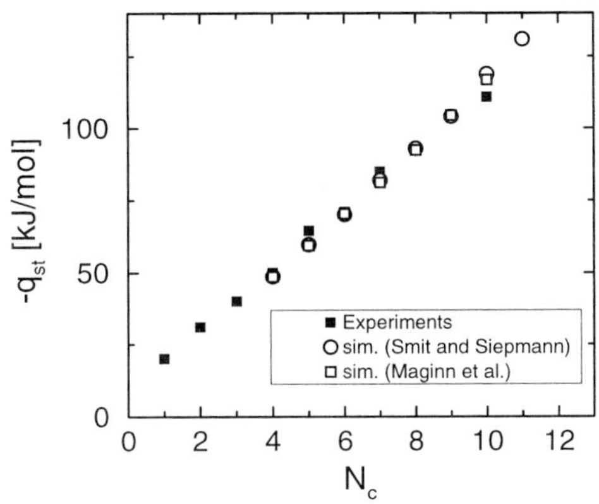

Figure 19. The heats of adsorption of the alkanes in silicalite as a function of the total number of carbon atoms. The closed symbols are experimental and the open symbols are simulation data from Maginn *et al.* [148] and Smit and Siepmann [147,146].

These simulations give some interesting insights in the distribution of the alkanes over the channels. In Fig. 20 the distribution of butane and dodecane is shown graphically. It is interesting to compare the probability distribution of butane with the one of dodecane. Whereas the plots of butane show an equal density of points in the straight and zig-zag channels, dodecane has a significantly lower probability to be in the zig-zag channels. In a sense, silicalite becomes more and more "unidimensional" with increasing length of the hydrocarbon.

Bates *et al.* [96] performed CBMC simulations to investigate the energetics of n-alkanes in various zeolites. In particular, Bates *et al.* . studied the heat of adsorption of n-alkanes as a function of the pore diameter. The results of these simulations (Figure 21) indicate an optimum sorption of n-alkanes in these all-silica zeolites occurs for zeolites with an average

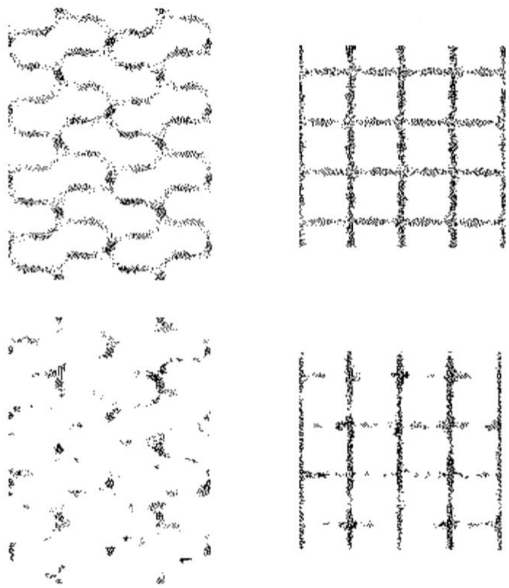

Figure 20. Distribution of butane (top) and dodecane (bottom) over the various channels of silicalite. At regular intervals during the simulations, a dot representing the position of the alkane molecule is plotted. The density of the dots is a measure of the probability of finding an alkane in a particular section of the zeolite. The left figures are projections on the side plane and the right figures projections on the top plane.

pore diameter between 4 and 5 Å. Details on how the zeolite influences the conformation of these hydrocarbons in various zeolites can be found in [150]. Other applications of the CBMC technique can be found in [151].

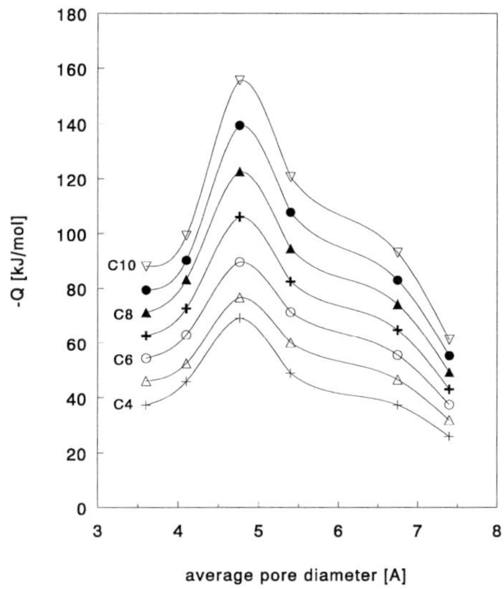

Figure 21. Heats of adsorption (Q) of n-alkanes in zeolites (RHO, LTA, FER, MFI, MOR, FAU) as a function of mean pore diameter (A spline is fitted to the data as a guide to the eye).

4.3.2. Simulating adsorption isotherms

Most Molecular Simulations are performed either in the micro-canonical ensemble or in the canonical ensemble. In the micro-canonical ensemble the thermodynamic variables that are kept constant are the temperature, volume, and total energy, whereas in the canonical ensemble the temperature, volume and number of particles are fixed [98,97]. To determine an adsorption isotherm we have to know the number of particles inside the zeolite as a function of the pressure and temperature of the reservoir which is in contact with the zeolite. A naive, but theoretically valid approach would be to use the Molecular Dynamics technique (micro-canonical ensemble) and simulate the experimental situation: an adsorbent in contact with a gas. Such a simulation is only possible for very simple systems. In real life experiments equilibration may take minutes or even several hours, depending on the type of gas molecules. These equilibration times would be reflected in a Molecular Dynamics simulation, the difference being that a minute of "experimental" time takes of the order of 10^9 seconds on a computer. In most cases we are not interested in the properties of the gas phase, yet a significant amount of cpu-time will be spent on the simulation of this phase. Furthermore, in such a simulation there is an interface between the gas phase and the zeolite. In this interfacial region the properties of the system are different from the bulk properties in which we are interested. Since in a simulation the systems are small compared to experiments —hence the interfacial region is a relatively large part of such a system— we have to simulate an unnecessary large system to minimize the influence of this interfacial region.

Most of the above-mentioned problems can be solved by a careful choice of ensembles. For adsorption studies a natural ensemble to use is the grand-canonical ensemble (or μ, V, T-ensemble). In this ensemble the temperature, volume, and chemical potential are fixed. In the experimental setup the adsorbed gas is in equilibrium with the gas in the reservoir. The equilibrium conditions are that the temperature and chemical potential of the gas inside and outside the zeolite must be equal. The gas that is in contact with the adsorbent can be considered as a reservoir that imposes a temperature and chemical potential on the adsorbed gas. We therefore have to know only the temperature and chemical potential of this reservoir to determine the equilibrium concentration inside the adsorbent. This is exactly what is mimicked in the grand-canonical ensemble; the temperature and chemical potential are imposed and the number of particles is allowed to fluctuate during the simulation. This makes these simulations different from the conventional ensembles where the number of molecules is fixed.

The grand-canonical Monte Carlo method works best if the acceptance of trial moves by which particles are added or removed is not too low. For atomic fluids this condition effectively limits the maximum density at which the method can be used to about twice the critical density. Special tricks are needed to extend the grand-canonical Monte Carlo method to somewhat higher densities [98]. The grand-canonical Monte Carlo technique is easily implemented for mixtures and inhomogeneous systems, such as fluids near interfaces. In fact, some of the most useful applications of the grand-canonical Monte Carlo method are precisely in these areas of research. Although the grand-canonical Monte Carlo method can be applied to simple models of non-spherical molecules, special techniques are required since the method converges very poorly for all but the smallest poly-atomic molecules. For chain molecules the grand-canonical ensemble can be combined with the CBMC technique. Techniqual details are discussed in ref. [152], here we discuss an application of this combination.

Adsorption isotherms are of practical importance since they give information on the number of molecules adsorbed in the pores of a zeolite as a function of the pressure of the reservoir. Adsorption isotherms are also of fundamental interest since they may signal phase transitions, such as capillary condensation or wetting, of the fluid inside the pores [153]. For example, if a system exhibits capillary condensation, one would measure a stepped adsorption isotherm with hysteresis. Steps or kinks without hysteresis are occasionally observed on flat substrates [154]. Since the pores of most zeolites are of molecular dimensions, adsorbed alkane molecules behave like a (pseudo) one-dimensional fluid. In a one-dimensional system phase transitions do not occur and therefore one would expect that for alkanes the adsorption isotherms are of type I, i.e., do not show kinks or steps. If steps occur they are usually attributed to capillary condensation in the exterior secondary pore system formed by the space between different crystals [155]. For silicalite, adsorption isotherms have been determined for various n-alkanes: for the short chain alkanes (methane—pentane) the isotherms are indeed of type I, also for decane a type I isotherm is observed. For hexane and heptane, however, a kink or step is observed.

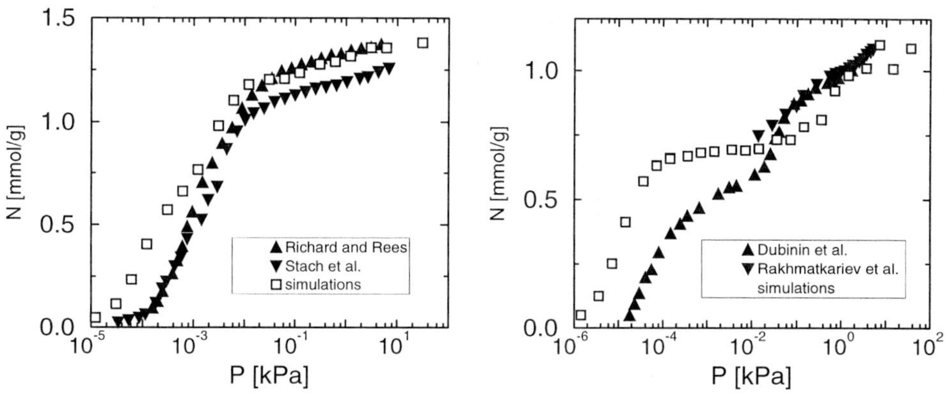

Figure 22. Adsorption isotherms of hexane (left), and heptane (right) in silicalite, the closed symbols are experimental data and the open symbols the results from simulations at $T = 298$ [K].

Adsorption isotherms are conveniently determined from simulations in the grand-canonical ensemble. In this ensemble the temperature and chemical potential are imposed, but the number of particles is allowed to fluctuate. Adsorption isotherms of alkanes in silicalite have been simulated by Smit and Maesen [156]. The simulated adsorption isotherms for hexane and heptane are shown in Figure 22. The agreement of the simulation results with the experimental data is good at high pressures, but at low pressures deviations exist which indicate that the zeolite-alkane model may be further improved. It is interesting to note that for heptane both the experiments and the simulations show a step at approximately half the loading. Also for hexane detailed inspection of the calculated adsorption isotherm shows a kink at this loading. Since the simulations are performed on a perfect single crystal, these deviations from the type I isotherm must be due to a transition of the fluid inside the pores and can not be attributed to the secondary pore system.

Smit and Maesen attribute this transition to a commensurate "freezing" in the channels of a zeolite. The length of a hexane molecule is of the order of the length of the period of the zig-zag channel. At low chemical potential, the hexane molecules move "freely" in these channels and the molecules will be part of their time in the intersections. If part of the intersection is occupied, other molecules can not reside in the straight channels at the same time. At high pressures, almost all hexane molecules are exactly fitting into the zig-zag channel, they do no longer move freely and keep their nose and tail out of the intersection. In such a configuration the entire straight channel can now be tightly packed with hexane molecules (see Figure 23). This may explain the plateau in the adsorption isotherm; in order to fill the entire zeolite structure neatly, the hexane molecules located in zig-zag channels have first to be "frozen" in these channels. This "freezing" of the positions of the hexane molecules implies a loss of entropy and will therefore only occur if the pressure (or chemical potential) is sufficiently high to compensate for this loss. Further experimental evidence for this unexpected behavior of hexane and heptane can been found in refs. [157–159].

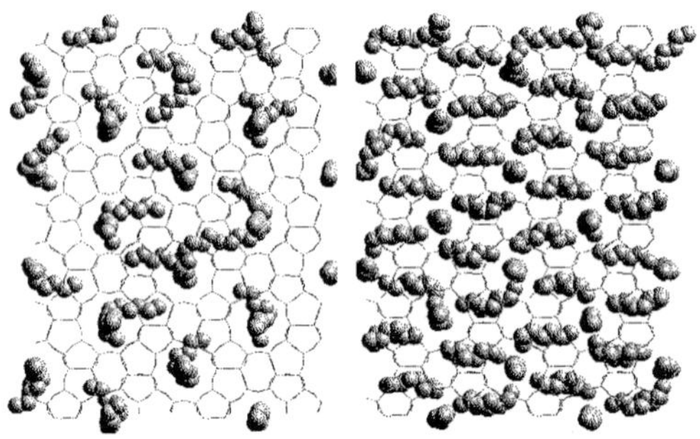

Figure 23. Hexane in silicalite: the left figure is at approximately half the maximum loading and the right figure is at almost maximum loading. The figures are projections, the zig-zag channels are in the plane of the figure and straight channels are perpendicular to the plane of the figure. The hexane molecules are represented by gray spheres and red/orange lines represent the zeolite framework.

5. Concluding remarks

Current Density Functional Theory based cluster calculations enable predictions of transition-states and their energies of quite complicated molecules.

This is a useful tool to probe acid catalyzed reaction-mechanism in zeolites. There is a need to extend this approach with embedding techniques. One expects this to become feasible in the near future.

In studies on absorption and diffusion Molecular Dynamics simulations are limited to rather small systems. However, MD simulations are valuable to test other computational approaches and is the method of choice to get a microscopic picture of diffusion mechanisms. One may expect that MD simulations will replace energy minimization more and more as the tool in studies on zeolite structures and dynamics. The increase in computer power will certainly enhance the number of studies in which MD is combined with ab-initio methods and embedding.

We have discussed novel algorithms to make Monte Carlo simulations of chain molecules adsorbed in the pores of zeolites more efficient. In particular, it is demonstrated that the configurational-bias Monte Carlo can be used to simulate the adsorption of long chain hydrocarbons in zeolites. For these type of applications, this technique is many orders of magnitude more efficient than the conventional Monte Carlo techniques or Molecular Dynamics. An important limitation of the Monte Carlo technique is that information on transport properties can not be obtained directly. However, in some cases one can get some information on diffusion if this is an activated process, where one needs to know the free energy barrier. This free energy barrier can be computed with Monte Carlo techniques. Some preliminary work in this direction can be found in [160].

REFERENCES

1. R.A. van Santen, D.P. de Bruyn, C.J.J. den Ouden, and B. Smit, in *Introduction to Zeolite Science and Practice*, Studies in Surface Science and Catalysis, edited by H. van Bekkum, E.M. Flanigen, and J.C. Jansen (Elsevier, Amsterdam, 1991).
2. J. Sauer, P. Ugliengo, E. Garrone, and V.R. Saunders, Chem. Rev. **94**, 2095 (1994).
3. R.A. van Santen and G.J. Kramer, Chem. Rev. **95**, 637 (1995).
4. R.A. van Santen and J.W. Niemantsverdriet, *Chemical Kinetics and Catalysis* (Plenum, New York, 1995).
5. F. Haasse and J. Sauer, J. Am. Chem. Soc. **117**, 3780 (1995).
6. H. Koller, E.L. Meijer, and R.A. van Santen, 1997.
7. E.L. Meijer, R.A. van Santen, and A.P.J. Jansen, J. Phys. Chem. **100**, 9282 (1996).
8. R.A. van Santen and A.J.M. de Man, in *Molecular Modeling*, edited by M.A.C. Nascimento (World Scientific, Singapore, 1994), pp. 1–43.
9. R.A. van Santen, A.J.M. de Man, W.P.J.H. Jacobs, E.H. Teunissen, and G.J. Kramer, Catal. Lett. **9**, 273 (1991).
10. R. Car and M. Parrinello, Phys. Rev. Lett. **55**, 2471 (1985).
11. M.C. Payne, M.P. Teter, D.C. Allan, T.A. Arias, and J.D. Joannapoulos, Rev. Mod. Phys. **64**, 1045 (1992).
12. L.J. Clarke, I. Stich, and M.C. Payne, Comp. Phys. Chem. **72**, 14 (1992).
13. R. Shah, J.D. Gale, and M.C. McPayne, J. Phys. Chem. **100**, 11688 (1996).
14. M.J.S. Dewar, E.G. Zoebisch, E.F. Healy, and J.J.P. Stewart, J. Am. Chem. Soc. **107**, 3902 (1985).
15. M.J.S. Dewar and E.G. Zoebisch, THEOCHEM **49**, 1 (1988).
16. W.J. Hehre, L. Radom, P.V.R. Schleyer, and Pople. J.A., *Ab-initio molecular orbital therory* (Wiley, New York, 1986).
17. Gaussian 94, Gaussian Inc. Pittsburg PA, 1993.
18. M.F. Guest and J. Kendrick, 1986, gamess Users manual, CCP1/86/1; SERC Daresbury Laboratory.
19. R. Doven and V.J. Saunders, 1992, cRYSTAL Q2 Users manual, Gruppo di Chimica Tecnica. Universita di Torino and Daresbury, SERC Laboratory.
20. C. Pisani, R. Dovesi, and C. Rotti, *Ab-initio treatment of cystalline systems* (Springer, Berlin, 1988).
21. A. Szabo and Oslund, *Modern Quantum Chemistry* (MacMillan, , 1982).

22. C. Møller and M.S. Plesset, Phys. Rev. **46**, 618 (1934).
23. DGauss, Unichem Chemistry Codes, Cray Research Inc. Eagan, MN.
24. DMOL User Guide, version 3.0.0, Molecular Simulations, Inc. San Diego, 1995.
25. Plane Wave User Guide, version 3.0.0, Molecular Simulations, Inc. San Diego, 1995.
26. G. te Velde, 1993, aDF User's Guide, Vrije Universiteit Amsterdam.
27. E.J. Baerends, D.E. Ellis, and P. Ros, Chem. Phys. **2**, 41 (1973).
28. E.J. Baerends and P. Ros, Chem. Phys. **2**, 52 (1973).
29. E.J. Baerends and P. Ros, Chem. Phys. **8**, 412 (1975).
30. T. Ziegler, Chem. Rev. **91**, 659 (1991).
31. P.E. Blöchl, Phys. Rev. B **50**, 17953 (1994).
32. M. Kossner and J. Sauer, J. Phys. Chem. **100**, 6199 (1996).
33. E.H. Teunissen, A.P.J. Jansen, R.A. van Santen, and R. Orlando, J. Phys. Chem. **101**, 5865 (1996).
34. S. Nawaz, S. Kolboe, K.-P. Lillerud, M. Stöcker, and H.M. Oren, Stud. Surf. Sci. Catal. **61**, 421 (1991).
35. C.T.-W. Chu and C.D. Chang, J. Catal. **86**, 297 (1984).
36. R.A. van Santen, J. Mol. Catal. A **107**, 5 (1996).
37. H.V. Brand, L.A. Curtiss, and L.E. Iton, J. Phys. Chem. **97**, 12773 (1993).
38. M. Hunger, M.W. Anderson, A. Ojo, and H. Pfeifer, Microporous Mat. **1**, 17 (1993).
39. D. Bartomeuf, *Catalysis 1987* (Elsevier, Amsterdam, 1988), p. 177.
40. R.A. van Santen, B.W.H. van Beest, and A.J.M. de Man, in *Guidelines for mastering the properties of molecular sieves*, Vol. 221 of *NATO ASI Series B*, edited by D. Barthomeuf, E.G. Derouane, and W. Hölderich (Plenum, New York, 1990), pp. 201–224.
41. N. Cardona-Martinez and J.A. Dumesic, Adv. Catal. **38**, 149 (1992).
42. U. Lohse and Jänchen, in *Elementary reaction steps in heterogeneous catalysis*, edited by R.W. Joyner and R.A. van Santen (Kluwer, Amsterdam, 1993), p. 113.
43. G.J. Kramer, A.J.M. de Man, and R.A. van Santen, J. Am. Chem. Soc. **113**, 6435 (1991).
44. V.B. Kazansky, M.V. Frash, and R.A. van Santen, Stud. Surf. Sci. and Catal. **105**, 2283 (1997).
45. V.B. Kazansky and I.N. Senchenya, J. Catal. **109**, 108 (1989).
46. E.A. Paukshtis and E.N. Yurchenko, Russ. Chem. Rev. **52**, 426 (1983).
47. G.J. Kramer and R.A. van Santen, J. Am. Chem. Soc. **115**, 2887 (1993).
48. J.-R. Hill and J. Sauer, J. Phys. Chem. **97**, 6579 (1993).
49. L. Campana, A. Selloni, J. Weber, A. Pasquarello, I. Papai, and A. Goursot, Chem. Phys. Lett. **226**, 245 (1994).
50. P. Sarv, T. Tuherm, E. Lippmaa, K. Keskinen, and A. Root, J. Phys. Chem. **99**, 13763 (1995).
51. K.M. Neyman, P. Strodel, S.P. Ruzankin, N. Schlensog, H. Knözinger, and W. Rösch, Catal. Lett. **31**, 273 (1995).
52. A.G. Pelmenschikov, R.A. van Santen, J. Jänchen, and E.L. Meijer, J. Phys. Chem. **97**, 11071 (1993).
53. J.D. Gale, C.R.A. Carlow, and J.R. Carruthers, Chem. Phys. Lett. **216**, 155 (1993).
54. A.G. Pelmenschikov and R.A. van Santen, J. Phys. Chem. **97**, 10678 (1993).

55. E.H. Teunissen, F.B. van Duyneveldt, and R.A. van Santen, J. Phys. Chem. **96**, 366 (1992).
56. E. Kassab, E. Gerti, and M. Allavena, J. Phys. Chem. **95**, 9425 (1991).
57. E. Kassab, J. Fongreet, M. Allavena, and E.M. Evleth, J. Phys. Chem. **97**, 9034 (1993).
58. J. Sauer and F. Haase, J. Phys. Chem. **98**, 3083 (1994).
59. A.T. Bell and J.N. Theodorou, J. Phys. Chem. **99**, 1505 (1995).
60. H.V. Brand, L.A. Curtiss, and L.E. Iton, J. Phys. Chem. **96**, 7725 (1992).
61. S.J. Cook, A.K. Chakraborty, A.T. Bell, and D.N. Theodorou, J. Phys. Chem. **97**, 6679 (1993).
62. S.P. Greatbanks, P. Sherwood, I.H. Hillier, R.J. Hall, N.A. Burton, and I.R. Gould, Chem. Phys. Lett. **234**, 367 (1995).
63. J.B. Uytterhoeven, L.G. Christner, and W.K. Hall, J. Phys. Chem. **69**, 2117 (1965).
64. S.A. Zygmunt, L.A. Curtiss, L.E. Iton, and M.K. Erhardt, J. Phys. Chem. **100**, 6663 (1996).
65. S.P. Greatbanks, I.H. Hillier, and P. Sherwood, J. Phys. Chem. **105**, 3770 (1996).
66. S.R. Blaszkowski and R.A. van Santen, J. Phys. Chem. **101**, 2292 (1997).
67. R. Shah, M.C. Payne, M.-H. Lee, and J.D. Gale, Science **271**, 1395 (1996).
68. E. Nusterer, P.E. Blöchl, and N. Schwarz, Ang. Chem. **108**, 187 (1996).
69. E. Nusterer, P.E. Blöchl, and N. Schwarz, Chem. Phys. Lett. **253**, 448 (1996).
70. L. Smith, A.K. Cheetham, R.E. Monis, J.M. Thomas, and P.A. Wright, J. Chem. Science **27**, 799 (1988).
71. E.H. Teunissen, R.A. van Santen, A.P.J. Jansen, and F.B. van Duyneveldt, J. Phys. Chem. **97**, 1993 (1993).
72. H. Knözinger, H. Bühl, and K. Kochloefl, J. Catal. **24**, 57 (1972).
73. G. Mirth and J. Lercher, in *Natural Gas Conversion*, edited by A. Holmen e.a. (Elsevier, Amsterdam, 1991), p. 437.
74. G. Mirth, J. Lercher, M.W. Anderson, and J. Klinowski, J. Chem. Soc. Faraday Trans. **86**, 3039 (1990).
75. F. Wakabayaski, J. Kondo, K. Domen, and C. Hirose, Catal. Lett. **21**, 257 (1993).
76. H. Jobic, A. Tuel, M. Krossner, and J. Sauer, J. Phys. Chem. **100**, 19545 (1996).
77. J. Florian and L Kubelkova, J. Phys. Chem. **98**, 8734 (1994).
78. D.J. Parillo, R.J. Gorte, and W.E. Farneth, J. Am. Chem. Soc. **115**, 12441 (1993).
79. L Kubelkova, J. Kortla, and J. Florian, J. Phys. Chem. **99**, 10285 (1995).
80. T. Fujino, M. Kashitani, J.N. Kondo, K. Domen, C. Hirose, F. Ishida, F. Goto, and F. Wakabayaski, J. Phys. Chem. **100**, 11649 (1996).
81. G.J. Kramer, R.A. van Santen, C.A. Emeis, and A.K Nowak, Nature **363**, 529 (1993).
82. G.J. Kramer and R.A. van Santen, J. Am. Chem. Soc. **117**, 1766 (1995).
83. S.R. Blaszkowski, M.C.A. Nascimento, and R.A. van Santen, J. Phys. Chem. **100**, 3463 (1996).
84. E.M. Evleth, E. Kassab, and L.R. Sierra, J. Phys. Chem. **98**, 1421 (1994).
85. V.B. Kazansky, Kinet. and Catal. **21**, 159 (1980).
86. I.N. Senchenya and V.B. Kazansky, Catal. Lett. **8**, 317 (1991).
87. A.G. Pelmenschikov, W.U. Zhanpeisov, E.A. Paukshtis, L.V. Malyssheva, G.M. Zhidomirov, and K.I. Zamaraev, Dokl. Akad. Nauk. SSSR **293**, 915 (1987).

88. P.M. Viruella, C.M. Zicovich-Wilson, and A. Corma, J. Phys. Chem. **97**, 13713 (1993).
89. A. Corma, P.M. Viruella, and C.M. Zicovich-Wilson, J. Phys. Org. Chem. **7**, 364 (1994).
90. V.B. Kazansky, Acc. Chem. Res. **24**, 379 (1991).
91. N.W. Cant and W.K. Hall, J. Catal. **52**, 161 (1972).
92. J.F. Haw, B.R. Richardson, I.S. Ohiro, W.O. Lazo, and A.J. Spled, J. Am. Chem. Soc. **111**, 2052 (1989).
93. J.F. Haw, J.B. Nicholas, T. Xu, L.W. Beck, and D.F. Ferguson, Acc. Chem. Res. **29**, 259 (1996).
94. T. Xu and J.F. Haw, J. Am. Chem. Soc. **116**, 7753 (1994).
95. I. R. Forester and R.F. Howe, J. Am. Chem. Soc. **109**, 5076 (1987).
96. S.P. Bates, W.J.M. van Well, R.A. van Santen, and B. Smit, J. Am. Chem. Soc. **118**, 6753 (1996).
97. M.P. Allen and D.J. Tildesley, *Computer Simulation of Liquids* (Clarendon Press, Oxford, 1987).
98. D. Frenkel and B. Smit, *Understanding Molecular Simulations: from Algorithms to Applications* (Academic Press, San Diego, 1996).
99. L.M. Bull, N.J. Henson, A.K. Cheetham, and S.J. Heyes, J. Phys. Chem. **97**, 11776 (1993).
100. P. Demontis and G.B. Suffritti, in *Proceedings of the 9th International Zeolite Conferenence*, Montreal, edited by R. von Ballmoos, J.B. Higgins, and M.M.J. Treacy (Butterworth-Heinemann, Boston, 1992), p. 345.
101. E. Hernández and C.R.A. Catlow, Proc. R. Soc. Lond. A **448**, 143 (1995).
102. T. Mosell, G. Schrimpf, C. Hahn, and J. Brickmann, J. Phys. Chem. **100**, 4571 (1996).
103. T. Mosell, G. Schrimpf, and J. Brickmann, J. Phys. Chem. **100**, 4582 (1996).
104. E.J. Maginn, A.T. Bell, and D.N. Theodorou, J. Phys. Chem. **100**, 7155 (1996).
105. G. Schrimpf and J. Brickmann, J. Comput.-Aided Mater. Des. **2**, 49 (1995).
106. R.L. June, A.T. Bell, and D.N. Theodorou, J. Phys. Chem. **94**, 8232 (1990).
107. J.R. Hufton, J. Phys. Chem. **95**, 8836 (1991).
108. S.J. Goodbody, K. Watanabe, D. MacGowan, J.P.R.B. Walton, and N. Quirke, J. Chem. Soc., Faraday Trans. **87**, 1951 (1991).
109. K.S. Smirnov, Chem. Phys. Lett. **229**, 250 (1994).
110. D. Dumont and D. Bougeard, Zeolites **15**, 650 (1995).
111. K.S. Smirnov and B. van de Graaf, J. Chem. Soc. Faraday Trans. **92**, 2475 (1996).
112. D. Shen and L.V. Rees, J. Chem. Soc. Faraday Trans. **92**, 487 (1996).
113. D. Keffer, A.V. McCormick, and H.T. Gavis, Mol. Phys. **87**, 367 (1996).
114. K. Hahn, J. Kaerger, and V. Kukla, Phys. Rev. Lett. **76**, 2762 (1996).
115. K.S. Smirnov and B. van de Graaf, J. Chem. Soc. Faraday Trans. **92**, 2469 (1996).
116. S. Yashonath and P. Santikary, J. Phys. Chem. **97**, 13778 (1993).
117. S. Yashonath and P. Santikary, J. Phys. Chem. **98**, 6368 (1994).
118. S. Yashonath and S. Bandyopadhyay, Chem. Phys. Lett. **228**, 284 (1994).
119. S. Yashonath and P. Santikary, J. Chem. Phys. **100**, 4013 (1994).
120. S. Bandyopadhyay and S. Yashonath, Chem. Phys. Lett. **223**, 363 (1994).
121. P. Santikary and S. Yashonath, J. Phys. Chem. **98**, 9252 (1994).
122. S. Bandyopadhyay and S. Yashonath, J. Phys. Chem. **99**, 4286 (1995).

123.J. Wei, Ind. Eng. Chem. Res. **33**, 2467 (1994).
124.R.C. Runnebaum and E.J. Maginn, J. Phys. Chem. B **101**, 6394 (1997).
125.D. Brémard and D. Bougeard, Adv. Mater. **1**, 10 (1995).
126.YK. Yamahara, K. Okazaki, and K. Kawamura, Catal. Today **23**, 397 (1995).
127.Y. Oumi, K. Matsuba, M. Kubo, T. Inui, and Miyamoto A., Microporous Mater. **4**, 53 (1995).
128.K.S. Smirnov and B. van de Graaf, Microporous Mater **7**, 133 (1996).
129.K.S. Smirnov and D. Bougeard, J. Mol. Struc. **348**, 155 (1995).
130.H.J.F. Stroud, E. Richards, P. Limcharoen, and N.G. Parsonage, J. Chem. Soc., Faraday Trans. I **72**, 942 (1976).
131.S. Yashonath, J.M. Thomas, A.K. Nowak, and A.K. Cheetham, Nature **331**, 601 (1988).
132.B. Smit and C.J.J. den Ouden, J. Phys. Chem. **92**, 7169 (1988).
133.S.D. Pickett, A.K. Nowak, J.M. Thomas, B.K. Peterson, J.F. Swift, A.K. Cheetham, C.J.J. den Ouden, B. Smit, and M. Post, J. Phys. Chem. **94**, 1233 (1990).
134.A.K. Nowak, C.J.J. den Ouden, S.D. Pickett, B. Smit, A.K. Cheetham, M.F.M. Post, and J.M. Thomas, J. Phys. Chem. **95**, 848 (1991).
135.C.R.A Catlow editor, *Modelling of Structure and Reactivity in Zeolites* (Academic Press, London, 1992).
136.R.L. June, A.T. Bell, and D.N. Theodorou, J. Phys. Chem. **96**, 1051 (1992).
137.N. Metropolis, A.W. Rosenbluth, M.N. Rosenbluth, A.N. Teller, and E. Teller, J. Chem. Phys. **21**, 1087 (1953).
138.M. Fixman, Proc. Nat. Acad. Sci. USA **71**, 3050 (1974).
139.G. Ciccotti and J.P. Ryckaert, Comp. Phys. Rep. **4**, 345 (1986).
140.J.I. Siepmann and D. Frenkel, Mol. Phys. **75**, 59 (1992).
141.M.N. Rosenbluth and A.W. Rosenbluth, J. Chem. Phys. **23**, 356 (1955).
142.J. Harris and S.A. Rice, J. Chem. Phys. **88**, 1298 (1988).
143.D. Frenkel, G.C.A.M. Mooij, and B. Smit, J. Phys.: Condens. Matter **4**, 3053 (1992).
144.J.J. de Pablo, M. Laso, and U.W. Suter, J. Chem. Phys. **96**, 2395 (1992).
145.J. Batoulis and K. Kremer, J. Phys. A: Math. Gen. **21**, 127 (1988).
146.B. Smit and J.I. Siepmann, J. Phys. Chem. **98**, 8442 (1994).
147.B. Smit and J.I. Siepmann, Science **264**, 1118 (1994).
148.E.J. Maginn, A.T. Bell, and D.N. Theodorou, J. Phys. Chem. **99**, 2057 (1995).
149.B. Bigot and V.-H. Peuch, J. Phys. Chem. **99**, 8206 (1995).
150.S.P. Bates, W.J.M. van Well, R.A. van Santen, and B. Smit, J. Phys. Chem. **100**, 17573 (1996).
151.S. Bandyopadhyay and S. Yashonath, J. Phys. Chem. B **101**, 5675 (1997).
152.B. Smit, Mol. Phys. **85**, 153 (1995).
153.R. Evans, in *Liquides aux Interfaces/Liquids at interfaces*, Les Houches, Session XLVIII, 1988, edited by J. Charvolin, J.F. Joanny, and J. Zinn-Justin (North Holland, Amsterdam, 1990), pp. 1–98.
154.S.J. Gregg and K.S.W. Sing, *Adsorption, Surface Area and Porosity* (Academic Press, London, 1982).
155.H. Stach, U. Lohse, H. Thamm, and W. Schirmer, Zeolites **6**, 74 (1986).
156.B. Smit and T.L.M. Maesen, Nature **374**, 42 (1995).

157.W.J.M. van Well, J.P. Wolthuizen, B. Smit, J.H.C. van Hooff, and R.A. van Santen, Angew. Chem. (Int. Ed.) **34**, 2543 (1995).
158.D.H. Olsen and P.T. Reischmann, Zeolites **17**, 434 (1996).
159.M.S. Sun, O. Talu, and D.B. Shah, J. Phys. Chem. **100**, 17276 (1996).
160.B. Smit, L.D.J.C. Loyens, and G.L.M.M. Verbist, Faraday Discuss. **106**, 93 (1997).

Studies in Surface Science and Catalysis 137
H. van Bekkum, E.M. Flanigen, P.A. Jacobs and J.C. Jansen (Editors)
© 2001 Elsevier Science B.V. All rights reserved.

Chapter 11

ION EXCHANGE IN ZEOLITES

Rodney P. Townsend[a*] and Eric N. Coker[b]

[a] Chemistry Department, University of Southampton, Highfield, Southampton SO17 1BJ, UK.

[b] BP, Chemicals Research & Engineering, Sunbury-on-Thames, Middlesex TW16 7LL, UK. Email: CokerEN@bp.com

SUMMARY

Aspects of ion exchange in zeolites are reviewed, with special reference to those properties of zeolites which give rise to characteristic, and sometimes unique, ion exchange behaviour. As well as discussing basic principles, some thermodynamic and kinetic aspects of the theory of ion exchange are covered with particular reference to their utility for predicting exchange behaviour in zeolites and for the use of ion exchange techniques in detergency building, radionuclide separation, waste water treatment and the preparation of zeolitic catalysts and sorbents. Recent developments on the use of zeolites in detergency are next reviewed in more detail, together with progress in the field of ion exchange using high silica zeolites or solid state techniques.

[*] Address for correspondence: Royal Society of Chemistry, Burlington House, Piccadilly, London W1V 0BN, UK. Email: townsendr@rsc.org

1. INTRODUCTION
2. A PICTURE OF THE ION-EXCHANGE SYSTEM
   2.1. Specific zeolite properties
      2.1.1. Ion sieving and volume exclusion
      2.1.2. Different exchange sites
      2.1.3. Framework flexibility and hysteresis
      2.1.4. Zeolite stability
3. EQUILIBRIUM ASPECTS OF ION EXCHANGE IN ZEOLITES
   3.1. Basic concepts
      3.1.1. The ion-exchange isotherm
      3.1.2. Selectivity and affinity
      3.1.3. The mass action quotient and the thermodynamic equilibrium constant
      3.1.4. The standard free energy of exchange
   3.2. Prediction of ion exchange equilibria
      3.2.1. Prediction procedures
      3.2.2. Results and limitations
   3.3. Concentration scales and ideal behaviour
4. THE KINETICS OF ION EXCHANGE
   4.1. Some basic concepts
   4.2. Hierarchical model of a zeolite particle or pellet
      4.2.1. Film mass transfer resistance
      4.2.2. Transport processes within the pellet or agglomerate
      4.2.3. Diffusion within the crystallite micropores
   4.3. Experimental tests of theoretical models
   4.4. Concluding comment
5. APPLICATIONS OF ZEOLITES IN ION EXCHANGE PROCESSES
   5.1. Detergent Building
      5.1.1. Criteria for an Effective Zeolite Builder
      5.1.2. Developments
   5.2. Radionuclide Separation
   5.3. Wastewater Treatment
6. ION EXCHANGE AS A METHOD OF MODIFICATION
   6.1. Preparation of Catalysts
   6.2. Preparation of Sorbents
7. RECENT TRENDS IN ION EXCHANGE
   7.1. Ion Exchange in High-Silica Zeolites
   7.2. Solid State Ion Exchange
   7.3. Case Study: Zeolite A *versus* MAP
8. CONCLUDING REMARKS
9. REFERENCES

1. INTRODUCTION

Ion exchange is an intrinsic property of most zeolites. As a consequence the phenomenon has either given rise to an admittedly few but nevertheless important number of direct applications, or the phenomenon is used indirectly, as a means of "tailoring" zeolite structure and hence properties when these materials are used in other ways, such as in catalysis or gas sorption.

There are some analogies which one can draw between zeolites on one hand and clay minerals on the other. Both classes of materials are mineral in origin, comprising similar elements (*viz* silicon, aluminium, alkali and alkaline earth metals with perhaps smaller quantities of other metals such as iron and titanium). Water is also an important component in both types of minerals. Both have an intrinsic ability to exchange cations, and this intrinsic ability arises as a consequence of isomorphous replacement. In the case of clay minerals this isomorphous replacement may be of trivalent cations by divalent or of a tetravalent cation by a trivalent one. In the case of zeolites, the isomorphous replacement is always of the tetravalent framework cation (*i.e.*, silicon) by a cation of lower charge (normally aluminium). As a consequence of this substitution a net negative charge arises on the framework of the zeolite which has to be neutralised by the presence of cations within the pores. These cations may be any of the metals or complexes of the same, or alkylammonium cations, and it is the great variety in nature and extent of exchange of different cationic species which may occur in zeolites which gives rise to the richness of the chemistry.

For aluminophosphate molecular sieves ($AlPO_4$'s) the origin of an ion exchange capability derives also from isomorphous exchange but the situation here is more subtle than for zeolites. In topological terms, both types of molecular sieves are described as 4:2 nets of T:O atoms ("T" here denoting a tetrahedrally co-ordinated atom within the framework such as Si, Al or P). Since the formal charge on the silicon ion within the framework is +4, for a pure silica zeolite (SiO_2), provided there are no defects present, the topology gives rise naturally to an overall neutral charge. Pure $AlPO_4$ molecular sieves are similar to pure silica in their overall electroneutrality, but for the aluminium phosphates this electroneutrality arises from the strict alternate ordering of the aluminium and phosphorus atoms within the framework (*i.e.*, every aluminium atom is linked through four oxygens to four phosphorus atoms and *vice versa* - aluminium or phosphorus atoms are never linked to their own kind *via* an oxygen bridge). Thus in contrast with pure silica zeolites, where the sum of the charges of the T atom (+4) and the four shared oxygens is zero at each T atom site, with the $AlPO_4$'s the charges at each T atom site sum to -1 and +1 for aluminium and phosphorus, because the formal charges for these discrete ions are +3 and +5, respectively. The strict alternate ordering for the aluminium and phosphorus atoms within the $AlPO_4$ framework is seen therefore to be a requirement for electroneutrality combined with framework stability (*viz*, the avoidance of two adjacent T atom sites having a unit charge of the same sign).

In principle therefore, for $AlPO_4$'s a net charge on the framework and a consequent ion exchange capability could arise through isomorphous replacement of either aluminium or phosphorus by T atoms having a different formal charge. Such isomorphous replacements do occur. If silicon is involved this gives rise to the so-called SAPO family of silicoaluminophosphates. With other heteroatoms in addition or not to silicon, materials belonging to the so-called MeAPSO and MeAPO families are formed. Consider as an example isomorphous replacement to give SAPO molecular sieves. Two possibilities could in principle arise: the replacement of either aluminium or phosphorus by silicon. If the former were to take place without a concomitant replacement of phosphorus by silicon, then a situation not seen with zeolites would arise in that the framework would become positively charged, thus creating an anionic exchange material. However, experimental studies have as yet only succeeded in producing negatively charged or neutral frameworks, the former arising where silicon replaces phosphorus only and the latter where silicon replaces aluminium and phosphorus in equal quantities.

Zeolites and microporous materials such as $AlPO_4$'s are analogous to clay minerals in another respect, in that an ion-exchange capacity can arise not from an intrinsic property which occurs as a consequence of isomorphous substitution in the framework, but rather from unsatisfied valencies occurring at the termination of the crystal edges and faces, or from faults within the structure. The most common origin of an ion-exchange capacity arising as a consequence of such faults, dislocations or edges comes from the presence of silanol groups or hydroxyls attached to edge aluminiums. In clay minerals up to 20% of the exchange capacity may arise from these sources; in the case of zeolites the extent of exchange capacity arising from such sources is normally small relative to the intrinsic one. In the case of high-silica zeolites, these secondary sources of ion exchange capacity plus non-homogeneous distribution of aluminium in the framework can lead to significant and unexpected effects [1].

In fact, isomorphous substitution into the framework itself can be regarded quite properly as a form of ion exchange; in recognition of this a complete statistical thermodynamic formulation for framework substitution has been drawn up [2]. The phenomenon has given rise to a whole area of zeolite chemistry involving the substitution of aluminium by other trivalent metals such as iron, and indeed it has been shown that aluminium can be re-inserted into the framework subsequent to its removal by an ion-exchange process [3]. Ion exchange may or may not be isomorphous in the case of zeolites. Usually, when one cation is exchanged for another within the zeolite no change in the overall structure of the zeolite occurs; however there are well documented cases where a phase transformation can occur, especially when monovalent cations are involved [4,5]. In this respect zeolites have a property which is in common with some clay minerals [6], but not with ion-exchange resins.

When zeolites display ion-exchange properties that are distinct from clay minerals and resins, these are normally in situations where the presence of micropores within the crystalline framework is manifested. This is particularly true with respect to exclusion effects: because the microporous channels in the zeolite are of comparable size to the typical cation size, cations may be excluded from all or part of the internal surface of the zeolite on the basis of their size. Alternatively, while being accepted, the size of the cation may be such that complete exchange is not possible without filling all the available space. These matters are discussed further below. In some instances, size or volume exclusion may exhibit a temperature dependency, where a critical temperature must be exceeded before a given cation is able to enter certain parts of the structure.

In addition to considering applications, any discussion on ion-exchange in zeolites must consider formulations, be they thermodynamic or kinetic in nature. Formulations for ion exchange which have been developed are not unique to zeolites and are usually applicable also to other classes of ion-exchangers, such as clay minerals or resins. Because different workers who are familiar with particular types of materials have separately developed these formulations, the subject as a whole is a complicated one [7-9] and it will only be discussed in broad outline within this chapter, with a view to emphasising the important principles. In addition, no attempt will be made to review systematically all the literature which has been published on ion exchange within zeolites. Rather, the aim will be to give the reader a broad up-to-date overview of the subject which will allow him/her to access the appropriate literature.

2. A PICTURE OF THE ION-EXCHANGE SYSTEM

Before considering zeolites in particular, it is important to understand some basic principles. One can regard ion exchange as a chemical reaction (a double decomposition) between two phases. In reality there may be any number of phases involved and any number of exchanging species, and there may be net movement of solvent(s) from one phase to another as the reaction proceeds. However, for the sake of simplicity we will restrict our considerations in this section mainly to just two reacting cations and two phases.

There are various matters which are worth emphasising at this point. First of all, and especially in the case of zeolites, the ion-exchange reaction is one which is best regarded as involving *bulk* phases. It is not essentially a surface reaction. Secondly, the reaction is stoichiometric, with electrical neutrality of each phase being maintained throughout the exchange process. For these reasons it is not helpful (as is often done) to refer to the process as one of adsorption, not just because adsorption is frequently non-stoichiometric in nature, but because the term itself implies that the reaction is one confined to surfaces or interfaces. In terms of the bulk phases, one usually refers to the exchanger (zeolite) and the external solution, but actually it is important to bear in mind that the external solution is *also* an exchanger phase. For the very simplest case one has therefore two exchanger phases, whether those two exchanger phases be a zeolite plus the external electrolyte solution, or a molten salt in equilibrium with a zeolite, or two zeolites together [10]. In addition to the two exchanger phases, a third phase also has to be taken into account, even for the simplest system, which is the vapour, as the solvent vapour pressure (normally water) is used as a thermodynamic reference point for all three phases. The solvent can have a profound effect on various properties of the exchange reaction including the rate of attainment and the actual position of the exchange equilibrium. Thus in solid state exchange reactions, (involving two zeolites), a small amount of water can greatly speed the reaction [11]. By contrast, if very dry ethanol is used in place of water as solvent the rate of reaction for (say) calcium in zeolite A is very greatly slowed [12].

Let us now "home-in" on the two exchanger phases which are present in the simplest system and compare their properties. Consider first the zeolite. At the start, and before the reaction occurs, one may regard the zeolite as a salt (*e.g.,* $NaAlSiO_4$) in which some water is dissolved. The other phase comprises mainly water in which is dissolved a salt (*e.g.,* KCl). After reaction has taken place we have a solution phase comprising water in which are now dissolved two salts (NaCl and KCl), while the zeolite is now a mixture of two salts. This salt mixture is usually a solution of one of the salts in the other, such as for example $K_x(AlSiO_4)_x{\bullet}yH_2O$ dissolved in $Na_{(1-x)}(AlSiO_4)_{(1-x)}{\bullet}zH_2O$, or the converse. Alternatively, phase separation may occur, with a new phase nucleating out within the old (*e.g.*, the Sr/Na-X system - [13]). (Of course, phase separation on exchange may also occur in the aqueous solution if the salt corresponding to the cation which is leaving the zeolite happens to be insoluble in water). Thus the reaction involves cations which are initially present in each separate phase moving between the two phases until an equilibrium composition within each phase is attained.

What about the anions? Here, a fundamental difference between the two phases may be observed. Whilst most of the cations within the zeolite channels are normally free to move, and of course are mobile in the external solution phase, the "anions" within the zeolite are not, since they comprise the anionic framework which constitutes the zeolitic phase. However, in solution, both the cations *and* anions are free to move between phases. This has two consequences:

(a) In the absence of salt imbibition [14], the "normality" (that is the number of moles of unit charge per given volume) of a given zeolite is virtually fixed. This is *not* true for the solution phase, where (within the limits of solubility of the salts in the solvent) the normality can be varied at will. In fact, the only simple way that one can alter the "salt" concentration in the zeolite phase is to change the silicon to aluminium ratio within the framework; in essence, this produces a different exchange system (*e.g.*, one involving zeolite X instead of Y).

(b) While the anions in aqueous solution can in principle move freely out of and into the zeolite, the anionic framework of the zeolite cannot do the converse with respect to the solution phase. Thus the anions in solution can only move into the zeolite by taking stoichiometric quantities of cations with them; otherwise the requirement that each phase would remain electrically neutral would be violated. If cations and anions move together into the zeolite, there is then a *net-shift* in salt concentration from the solution to the zeolite phase. This phenomenon is sometimes called salt imbibition [14]; in effect the ion-exchange capacity of the two phases has changed [15,16]. That of the solution has declined while that of the zeolite has increased concomitantly. Donnan exclusion [17] stops the transfer of salt from one phase to another at low salt concentrations in solution, but the phenomenon can occur either at high solution concentrations or when the second exchanger phase is a molten salt.

2.1. Specific zeolite properties

In drawing up this picture of the ion-exchange system as a whole, let us next concentrate on the properties of the zeolite phase itself. In particular, let us consider those characteristics of zeolites which are significant when it comes to applications. Zeolites are normally highly crystalline materials. This means that the anionic framework comprises an extended regular array of silicon, aluminium and oxygen ions which enclose microporous channels, these latter being of molecular dimensions. This makes zeolites significantly different from either a clay mineral (where most of the exchange occurs in inter-lamellar space), or a typical amorphous ion exchange resin. Since the micropores within the zeolite are of molecular dimensions, various consequences follow, which are described in the following sections.

2.1.1. Ion sieving and volume exclusion

The size of the pores may be such that some ions are too large to enter some of the cages and/or interconnecting channels within the zeolite structure. In effect, these ions are therefore "sieved-out" of part of the space which is available within the crystal. This phenomenon may lead to the complete exclusion of one sort of cation from a particular zeolite [18], or partial exchange only occurring [19], with a clear maximum level of exchange for the entering cation which is less than that one would expect on the basis of the framework Si/Al ratio.

Alternatively, channels within the zeolite structure may be large enough for the ions to diffuse through without severe restriction, but the *volume* of the individual cation may be such that the sum of the volumes of all the cations of a particular type required to neutralise the anionic framework is greater than the available space within the zeolite [18,20]. When this occurs, partial exchange with respect to the entering cation is again observed, not because of a sieving effect, but because of a volume exclusion effect.

2.1.2. Different exchange sites

The three dimensional extended array which constitutes the zeolite framework may have associated with it quite clearly defined exchange sites. Thus, per unit cell, one can commonly define particular sites within the zeolite which differ one from another in terms of the energies of interaction associated with them. A very good example of this is the faujasitic group of zeolites. In these materials a large number of different sites are identified, and associated with each type of site is a particular ion population and site energy [21].

It is important to recognise that it can be a difficult and slow process to remove or exchange cations from certain of these sites even though both the leaving and entering cations are able to move through the channels and cages in which these sites are found. As a consequence, the experimental conditions used in preparing a particular ion-exchanged form of a zeolite, whether in a mixed or homoionic form, can be absolutely critical.

To understand this, it is necessary to use a more rigorous formalism. The whole zeolite framework may be described as comprising a set of intimately mixed sub-lattices between which different cations can distribute differently. At equilibrium one would expect that *provided ion sieving and volume exclusion effects are absent* then the relative proportions of each particular cation type in each sub-lattice will be described by a Boltzmann distribution. The particular equilibrium distribution will depend on the cation type because of the manner in which ion-ion, ion-solvent and ion-framework energies of interaction are dependent of the nature of the cation. Thus, providing equilibrium can be attained without the experimental conditions becoming so stringent that damage to the zeolite occurs, different mixed or homoionic forms of exchanged zeolites may be prepared reproducibly. However, the kinetics of exchange may be so slow that equilibrium is difficult or impossible to attain under practical reaction conditions; under these circumstances the experimental conditions employed (*e.g.*, level of the hydration, time of heat treatment, temperature employed) can have a *critical* effect on the final population distributions. This has particular significance with respect to the preparation, activity and long term stability of catalysts such as "sodium free" zeolite Y for hydrocarbon cracking [22]. The importance of controlling carefully the experimental conditions used to prepare homoionic or mixed ion form zeolites may explain much of the diverse X-ray diffraction data observed by different workers and summarised by Mortier [23].

Given this situation, it is easy to appreciate that it is possible to have for example two samples of zeolite having exactly the same overall bulk composition yet have markedly different properties because of different non-equilibrium distributions of cations between the different sublattices within the material, these being simply a consequence of using different preparative procedures when synthesising the samples.

A further point to note is that even when the kinetics are fast and equilibrium can be attained readily within a short period, the presence of a set of sublattices can lead to complex non-ideal exchange behaviour *even when exchange onto each and every sub-lattice is ideal* [24]. This has practical consequences when trying to predict and model exchange behaviour for applications of industrial importance and is discussed further later on (Section 3.2 *et sequens* and Section 4.2.3).

2.1.3. Framework flexibility and hysteresis

All zeolite structures exhibit some degree of flexibility as evidenced by for instance the sorption of molecules whose maximum diameters exceed the nominal pore openings of the zeolite [25,26]. Distortion of the zeolite framework is necessary to allow passage of the sorbate into the channels. This flexibility allows different cation-exchanged forms of a given zeolite to adopt different symmetries and unit cell sizes, depending upon the rigidity of the framework, the sizes of the cations, the strength of the interaction of the cations with the framework and the degree of hydration. One particularly flexible structure is the gismondine-type zeolite P which is known to convert between three symmetries, depending on hydration [27] and cation content [28]. For example, zeolite P readily changes symmetry from pseudo-cubic to pseudo-tetragonal on replacing Na^+ by K^+ cations. The symmetry change is accompanied by a contraction of the unit cell, associated with the lower degree of hydration of the K-form. The Na^+/K^+ isotherm is reversible, but has an unusual contour, exhibiting increasing selectivity for K^+ as the fraction of K^+ in the solid phase increases. This phenomenon is due to the decline in the energy of interaction between Na^+ and the framework as the latter contracts and sheds water molecules and due to preferential clustering of K^+ in the solid phase (*vide infra*). Despite the differing unit cell sizes and symmetries of the pure Na- and K- forms of zeolite P, they are fully miscible, and can co-exist over the entire range of solid compositions. The implications of the flexibility are discussed further in Section 7.3.

In many cases the end-members of exchange (pure entering-ion form and pure leaving-ion form) have limited mutual miscibility. Isotherms describing such exchange reactions exhibit hysteresis where the forward and reverse exchange paths do not coincide. In the range of compositions where two solid phases co-exist the forward and reverse branches of the isotherm run approximately parallel to the solid phase composition axis (*i.e.*, the solid composition changes while the solution composition remains constant). Hysteretic isotherms have been reported for, among others, Na^+/K^+ and Na^+/Li^+ in zeolite F [29], Na^+/K^+ and K^+/Rb^+ in analcime [30,31] and Na^+/Ag^+ and Li^+/Ag^+ in cancrinite [31]. Extremely large hysteresis loops were reported for Na^+/K^+ and Na^+/NH_4^+ in a beryllophosphate analogue of zeolite P ($BePO_4$-P, see Figure 4) [32]. The contrast between the ion exchange behaviour of zeolite P, which does not display hysteresis, and $BePO_4$-P probably arises from the smaller unit cell of the latter and different flexibilities of the frameworks. In addition, the difference in formal valency of the T-atoms in $BePO_4$-P of three units (Be^{2+} *vs* P^{5+}) leads to a much higher ionic field strength when compared to aluminosilicate P.

Barrer and Klinowski adopted a statistical thermodynamic approach to describe the conditions under which hysteresis could be expected [2]. When the interactions between two or more entering ions on adjacent exchange sites is favourable (exothermic), the ions tend to cluster together, forming islands rich in that ion. If the clustering is sufficiently favoured, the zone rich in entering ion develops into a separate phase, and the sample thus contains two phases, one rich in leaving ions and one rich in entering ions. As the number of entering ions in the solid phase increases, so does the proportion of that phase in the mixture. The size of the hysteresis loop (or solubility gap) depends upon the difference in structure between the two end members of exchange and the stiffness of the framework. If the clustering is not so favoured as to lead to the development of separate phases, hysteresis does not occur but selectivity for the entering ion will increase with exchange, as seen for zeolite P.

According to the phase rule, the two co-existing solid phases should be in thermodynamic equilibrium with a solution of one distinct composition only [33]. Therefore the forward and reverse branches of the isotherm should coincide. However, the rearrangements of the framework from one symmetry to the other is delayed by interfacial and strain free-energies and the original framework persists until it becomes supersaturated to a certain level with respect to the entering ion. The same phenomenon occurs in both forward and reverse exchanges, leading to the opening of a hysteresis loop in the isotherm. Hysteretic isotherms display increasing selectivity for the entering ion as its concentration in the solid phase increases over the range in which two solid phases co-exist. As more entering ions exchange into the solid, the proportion of the sample possessing the symmetry of that end-member increases, as does the selectivity for the entering ion.

The non-coincidence of forward and reverse exchange paths in an isotherm may be caused, however, by factors other than limited miscibility of the exchange end-member, *i.e.,* when isotherms are measured without allowing sufficient time for attainment of equilibrium, or when irreversible changes occur to the sample during forward exchange. When any of these factors are operational, equilibrium ion exchange isotherms will not be obtained.

2.1.4. Zeolite stability

Finally, one should note the limits of stability of zeolites when ion-exchange processes are undertaken. Many commonly used zeolites, such as zeolites A and X, are hydrolytically unstable in solutions of even mildly acid *p*H, and removal of aluminium from the framework will occur readily. In high aluminium zeolites this can lead ultimately to breakdown of the crystalline structure. This is a phenomenon which has often been neglected in the past, has led to irreproducibility of data [22] and is of particular significance in detergency [34,35]. In addition it is worth noting the limits outside which precipitation of basic salts may occur. Zeolites exhibit an alkaline reaction to their environment and in the presence of many transition metal cations, basic salts may be precipitated if adequate care is not taken [36,37]. Also, one should be aware that many metal ions are speciated in solution [38]. If speciation occurs to a significant extent, then the exchange capacity observed will normally be higher than expected [39], and this behaviour has been observed in clay minerals not only with copper [39], but also with calcium and magnesium [40,41].

It is important to recognise the problems which may arise because of zeolite and/or cation hydrolysis and cation speciation. Ignorance of these phenomena and consequent failure to take proper experimental precautions can lead to meaningless data and flawed conclusions. This is because zeolite hydrolysis gives rise to a large number of "impurity species" which interfere with the behaviour of the simple binary cation exchange under study. Consider a very simple case: almost invariably the starting material used for exchange studies with zeolites has been the "homoionic" sodium form. However, simply measuring the *p*H of a suspension of this zeolite in water will show a marked alkaline reaction because of the equilibrium

$$Na^{+}_{(zeolite)} + 2H_2O = Na^{+}_{(soln)} + H_3O^{+}_{(zeolite)} + OH^{-}_{(soln)}.$$

The more aluminium there is in the framework, the more sodium ions there are initially present in the zeolite; consequently, this tendency to hydrolyse increases with declining silicon:aluminium ratio. This systematic trend has been observed experimentally [42] and recently has been studied extensively [43-47] for trace level exchange of calcium ions in zeolites X and Y [43,44,46] as well as for caesium exchange in synthetic mordenite [47], a higher silica material.

This hydrolysis process leads on to other secondary phenomena. Exposure of the alkaline solution to air leads to the dissolution of acidic gases present in the atmosphere, especially carbon dioxide and sulphur dioxide. These react with the hydroxide ions in solution, shifting the equilibrium and leading to further elution of sodium from the zeolite [45]. In addition, the presence of hydroxide ions encourages the dissolution and removal of T-atoms from the framework, which not only can result in partial destruction of the crystalline integrity of the zeolite but can also introduce aluminate and silicate species into solution. Although this process is slow under the conditions we consider here [45], nevertheless these species can polymerise and co-polymerise to form suspended colloidal particles which are difficult to filter out and which also lead to experimental error [47].

Further details on these phenomena are given elsewhere, including a discussion on experimental errors arising from inadequate evaluation of maximal exchange levels [48], impurities in salts [48], errors in inferring correctly the cation uptake into the zeolite through the use of difference measurements in the solution phase [48], interference from colloidal particulate species [47], hydronium ion exchange *per se* [45], as well as association [46] and hydrolysis [49] of metal cations in the presence of hydroxide, carbonate and bicarbonate anions. In addition, the interested reader is referred to a recent detailed discussion of the subject [48].

3. EQUILIBRIUM ASPECTS OF ION EXCHANGE IN ZEOLITES

Over 20 years ago, in his comprehensive book on ion exchange, Helfferich compared the current theoretical understanding of both the thermodynamic and kinetic aspects of ion exchange, noting how much better understood were the equilibrium aspects of ion exchange processes [49]. While Helfferich was primarily discussing ion exchange resins, for zeolites it is still true that equilibrium processes have been more comprehensively considered and better understood than kinetic ones. The equilibrium properties of many experimental systems have been studied, with the usual aim of obtaining various thermodynamic parameters from the experimental data. Such studies can often be an end in themselves, and part of the purpose of this section will be to emphasise what can and cannot be done with such measurements.

3.1. Basic concepts

In order to appreciate how the ion-exchange properties of zeolites may be important in applications such as detergency, pollution control or in the preparation of sorbents and catalysts, it is not necessary here to discuss in depth all the details of the thermodynamics, but it is important to clear some common misconceptions out of the way.

3.1.1. The ion-exchange isotherm

The usual aim of those who study the equilibrium aspects of ion-exchange in zeolites is to obtain a measure of the preference the zeolite displays for one ion over either another, or a group of other, ions. For this purpose, it is normal to construct isotherms. The experimental methods by which an isotherm may be constructed have been described elsewhere [50]. Essentially, one plots equilibrium compositions attained by the two phases when in intimate contact with each other as a function of differing relative concentrations of the exchanging cations. While the term "isotherm" indicates that the temperature must be kept constant, in addition both the zeolite and the solution phase are kept isonormal (see Section 2 above). The isotherm is then constructed by plotting at equilibrium the equivalent fraction $E_{(S)}$ of the incoming cation present within the solution phase against the equivalent fraction E of the same cation in the zeolite. The equivalent fraction may be defined as that proportion of the exchange capacity of the phase which is neutralised by the given cation A, and is expressed for the two phases as follows:

$$E_{A(S)} = z_A n_{A(S)} / (\Sigma_i z_i n_{i(S)})$$
$$E_A = z_A n_A / (\Sigma_i z_i n_i) \qquad (1)$$

Here z_A, z_i are the valencies of exchanging cations A and i and n_A, n_i the moles per unit volume of A and i within each phase. The subscript "(s)" indicates "solution phase". For reasons of convenience, to do with the evaluation of the thermodynamic parameters, it is convenient to plot the isotherm with the equivalent fraction of the metal ion in the solution phase $E_{A(S)}$ as the dependent variable; this custom is often not followed by researchers working on resins.

One may broadly classify isotherms into four kinds. Examples of isotherms of the first kind are shown in Figures 1 and 2 and represent relatively simple systems in which the exchanger is either selective for the incoming ion (Figure 1) or un-selective (Figure 2). For isotherms of the second kind (Figure 3) the plot is sigmoid, indicating a change in selectivity as a function of E_A. Isotherms of the third kind (Figure 4) are characterised by a plateau, and within this plateau region non-reversibility of the isotherm may be seen (a "hysteresis loop" (see Section 2.1.3.)). Finally, isotherms of the fourth kind are seen often with zeolites (Figure 5), in which a clear limit to exchange is observed which is lower than the theoretical exchange capacity of the material (see comments above in Section 2). This limit may be due either to ion sieving or volume exclusion (see Section 2.1.1).

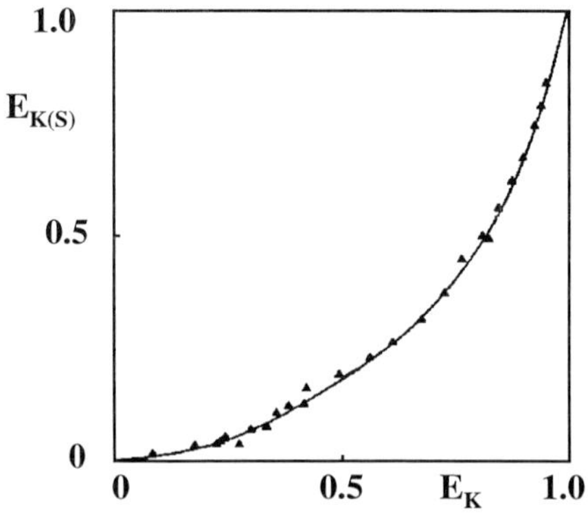

Figure 1. Isotherm type I: exchanger selective for ingoing ion. Na/K exchange in ZSM-5, 25°C, 0.05N. Reproduced with permission from reference 143.

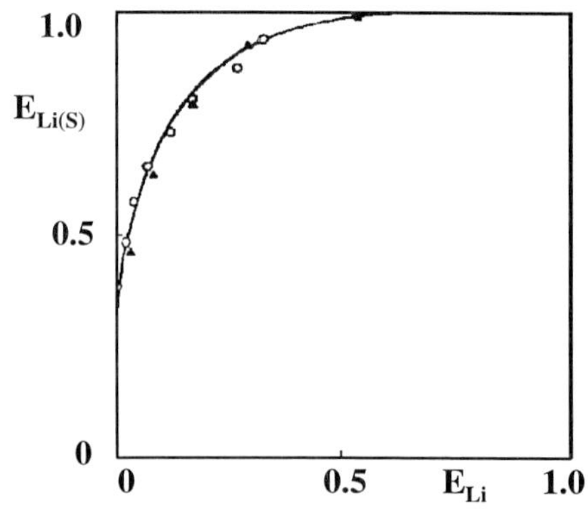

Figure 2. Isotherm type I: exchanger unselective for ingoing ion. Na/Li exchange in zeolite P, 25°C, 0.1N. Reversibility is shown by the coincidence of forward (○) and reverse (▲) exchange datum points. Reproduced with permission from reference 28.

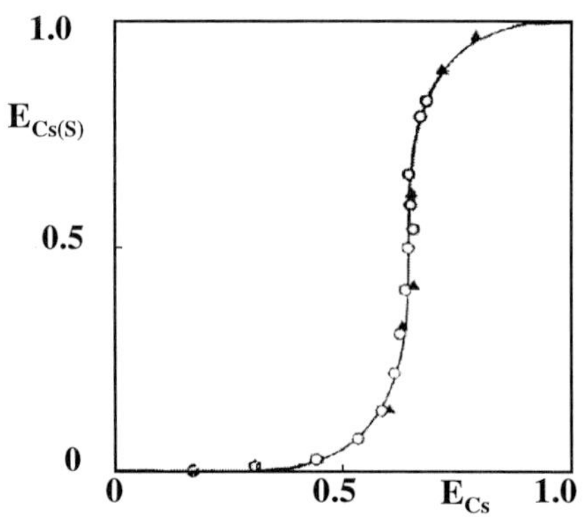

Figure 3. Isotherm type II: selectivity for ingoing ion changes. Na/Cs exchange in zeolite K-F, 25°C, 0.1N. Reversibility is shown by the coincidence of forward (○) and reverse (▲) exchange datum points. Reproduced with permission from reference 29.

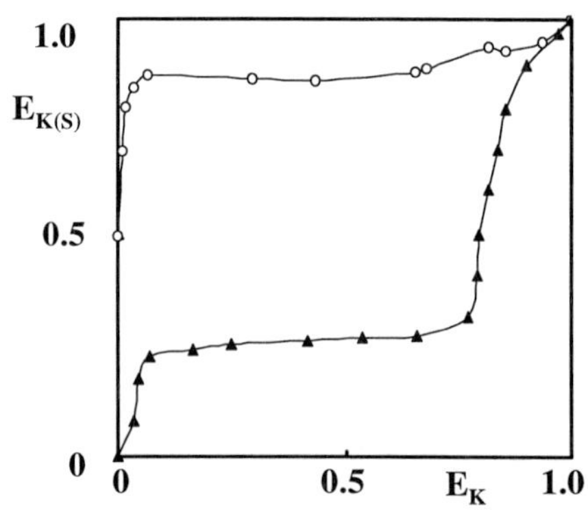

Figure 4. Isotherm type III: non-reversible exchange with hysteresis loop. Na/K exchange in a beryllophosphate with the structure of zeolite P, 25°C, 0.05N. Forward (○) and reverse (▲) exchange branches do not coincide. Reproduced with permission from reference 32.

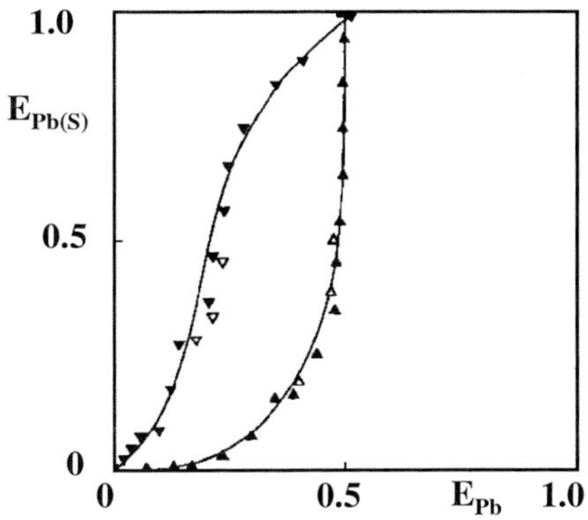

Figure 5. Isotherm type IV: limit to exchange is below theoretical capacity. 2Na/Pb (▲) and $2NH_4$/Pb (▼) exchange in mordenite, 25°C, 0.1N. Reversibility is shown by the coincidence of forward (filled) and reverse (open) datum points. Reproduced with permission from reference 130.

3.1.2. Selectivity and affinity

To avoid potential confusion, it is most important to distinguish between the concepts of preference, selectivity and affinity. The actual preference which an exchanger exhibits for one cation compared with another is an equilibrium property, but its value depends on many different conditions of the system as a whole (*viz,* the temperature, pressure, overall ionic strength of the solution, solution *p*H, composition of both solution and zeolite phases, solvent properties, *etc.*). Thus the magnitude of the preference which the zeolite may exhibit will vary with the conditions used in the experiment. When a particular and comprehensive set of experimental conditions is specified, then a unique value for the preference will be observed but *only if all these variables are strictly controlled and sufficient time has elapsed for the system to be at equilibrium.* When, and only when, these requirements are fulfilled, then the preference can be quantified by means of a *selectivity coefficient,* which is just a mass action quotient, commonly derived from an ion-exchange reaction equation of the form:

$$z_B A^{zA+}{}_{(S)} + z_A BL_{zB} = z_A B^{zB+}{}_{(S)} + z_B AL_{zA} \qquad (2)$$

where the subscript '(s)' refers to the solution phase, and L is that portion of zeolitic framework holding unit negative charge. There are many differently defined selectivity coefficients in use, according to the concentration scales one chooses to employ [48]; an example of one of the most commonly employed definitions is that given for $K_{m(E)}$ in equation 6 below (see Section 3.1.3).

Another function is sometimes used, which is not a selectivity coefficient and should not be confused with such. This is the *separation factor*, which may be expressed mathematically as a quotient of two ratios, with the two ratios expressing the relative quantities of the given cations found within the two phases at equilibrium:

$$\alpha = (E_A/E_{A(S)}) / (E_B/E_{B(S)}) = (E_A E_{B(S)} / E_B E_{A(S)}) \qquad (3)$$

It is evident from equations 2 and 3 that α is only a mass action quotient in the unusual case where all concentrations are defined in terms of equivalent fractions and where $z_A = z_B = 1$.

Nevertheless α is related in a relatively simple manner to the selectivity coefficient $K_{m(E)}$ [51] and can be evaluated from the relative values of two areas as shown in Figure 6. A few moment's inspection of Figures 1-5 will then show that α will normally vary as a function of exchanger phase composition. This point is most important: α and consequently $K_{m(E)}$ may vary markedly as E_A is increased (see Figure 3). Therefore, the selectivity coefficient, which provides a precise quantitative description of the preferences exhibited by the zeolite as the relative concentrations of exchanging ions are varied, is a directly measurable composition-dependent function, unlike the thermodynamic equilibrium constant (see Section 3.1.3).

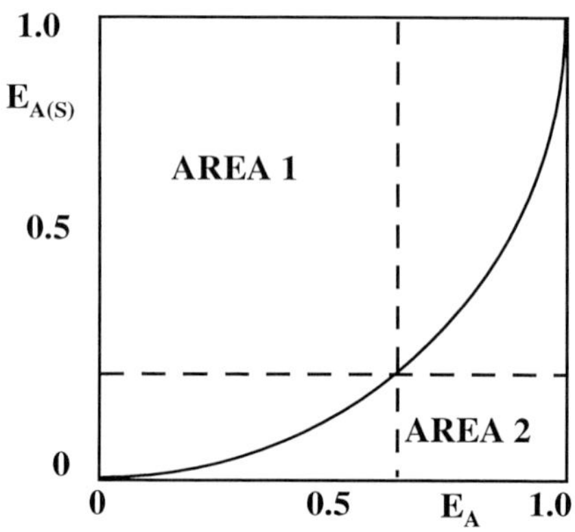

Figure 6. Graphical representation of the separation factor (equation 3), where α = (Area 1) / (Area 2).

It is precisely for this reason that the terms "preference" and "affinity" should also not be used interchangeably. Affinity is a term which has a particular thermodynamic meaning, in terms of the chemical potentials μ for each component in each phase within the system overall. The ion-exchange reaction given in equation 2 can be written more concisely as:

$$0 = z_A z_B \sum_{B \,(all\ phases)} (B/z_B) \quad (4)$$

from which the definition of affinity *A* follows as [52]:

$$A = - \sum_{B \,(all\ phases)} (1/z_B)\,(\mu_B - \mu_B^{eq}) \quad (5)$$

where μ^{eq} refers to the chemical potential of a particular component at equilibrium. The affinity is therefore seen as a sum of the *differences* in chemical potential for each separate component within the chemical system under the currently existing conditions on one hand, and what those chemical potentials would be at equilibrium on the other. Overall, in terms of magnitude the affinity *A* defines the *potential* for the ion-exchange reaction to proceed further, while in terms of sign it defines the *direction* of reaction. While A remains > 0 the extent of reaction as defined by the reaction equation 4 will increase; when A is < 0 the ion exchange reaction will move in the reverse direction to that defined [52].

Since *A* is defined in terms of chemical potentials its value depends not on ionic concentrations but on the *activities* of the exchanging components. This is in contrast with the definition of a selectivity coefficient. Consequently, the terms affinity, preference and selectivity should never be used interchangeably.

3.1.3. The mass action quotient and the thermodynamic equilibrium constant

From equation 2 a mass action quotient $K_{m(E)}$ may be defined as:

$$K_{m(E)} \quad = \quad (m_B^{zA}\ E_A^{zB}) / (m_A^{zB}\ E_B^{zA}) \tag{6}$$

where m_A, m_B are the molal concentrations (mol kg^{-1}) of ions A and B in solution.

We noted above (Section 3.1.2) that the mass action quotient is readily obtainable from experimental data as shown on the isotherm. The thermodynamic equilibrium constant is not so readily accessible from experimental data. It is defined also using the reaction equation 2 given above, as

$$K_a \; = \; (a_{B(S)}^{zA}\ a_A^{zB}) / (a_{A(S)}^{zB}\ a_B^{zA}) \tag{7}$$

Where $a_{(S)}$, a are respectively the activity of the cation in solution or crystal phase. One can see that the essential difference between the mass action quotient and the thermodynamic equilibrium constant is that while the former is defined in terms of concentrations the latter is defined in terms of activities; it is precisely because of this that the former is not constant with zeolite phase composition whilst the latter is.

Why is this? Imagine the zeolite initially to be wholly in the B cationic form. Then the B cations are interacting with the framework, water in the channels, and with each other. One can visualise the system as having a given amount of energy and that energy will vary directly with the amount of B zeolite which is present. One can therefore partition the energy of the system into a set of "packets", each packet comprising that amount of framework (with associated water) that is required to neutralise one B cation. If one therefore halved the amount of B zeolite then the energy of the system would also halve. Now imagine A cations entering the zeolite framework and replacing B cations, until the quantity of B zeolite is half what it was originally. However, one cannot assume that the energy of the system arising from B zeolite is now half what it was before, because new interactions are present which were not present when all the zeolite was in the B form. These new factors include interactions between A and B cations and the effect overall on the framework of A cations interacting with it, and the effect that A cations have on the total water content in the zeolite. The entire converse argument can be proposed in terms of A zeolite. Thus in effect, because of these new interactions, the behaviour of the system will not be predictable simply on the basis of the concentrations of A, B and water in the zeolite by extrapolating from the properties of the pure A and B homoionic forms.

Consequently, each of these concentrations has to be multiplied by a correction factor, which for each of these components is separately changing with the overall composition of the zeolite, as the relative strengths of all the different interactions with respect to that particular component change. When we multiply each component concentration by its appropriate correction factor, we then obtain *activities,* the magnitudes of which take into account these new interactions. Thus the thermodynamic equilibrium constant, defined in terms of activities, is independent of exchanger composition, in contrast to the mass action quotient. Since (see below) the standard free energy ΔG° is directly related to the thermodynamic equilibrium constant K_a, it follows that it is highly misleading to attempt to use standard free energy values to draw mechanistic conclusions about composition dependent selectivity trends in the real system [53,54].

3.1.4. The standard free energy of exchange

The standard free energy of exchange ΔG° is related to the thermodynamic equilibrium constant as follows

$$\Delta G^\circ = -(RT/z_A z_B).\ln K_a \qquad (8)$$

It is evident that since the thermodynamic equilibrium constant is defined in terms of activities, then before the standard free energy can be determined it is necessary to apply the appropriate correction factors to the concentration terms. These correction factors are called *activity coefficients.* Activity coefficient values have to be determined for each component in each phase in order that the thermodynamic equilibrium constant can be found. The means by which this is done is complicated and has been described in detail elsewhere [8,9,50,54]. What is important is to recognise that the magnitudes of the activity coefficients are determined with reference to appropriate *standard states* for components within the system. For the salts in solution the Henry law convention is normally applied and these reference states are usually the hypothetical ideal molal (mol kg^{-1}) or molar (mol dm^{-3}) solutions (*i.e.,* for the molal scale the reference state is a solution containing 1 mol kg^{-1} of the salt in water with the cations and anions from which the salt is constituted all still behaving as though the only forces acting on them are those present in the infinitely dilute solution). For dilute solutions only, molal ≈ molar and the corresponding activity coefficients γ and y are essentially equal in magnitude.

For each form of the zeolite (*i.e.,* the A form or the B form) the standard states are defined following the Raoult convention and are therefore respectively these homoionic forms of the zeolite each respectively immersed in infinitely dilute solutions of the same cation [15]. This ensures that the activity of the water within the zeolite is equal to the value found in the external solution (*i.e.,* the activity of the water in both phases is equal to that in the vapour phase and is unity in the standard state). In contrast to the dilute solution phase, the choice of concentration scale is *very* important here (see Section 3.3).

Following the determination of activities, the standard free energy can be determined. In terms of reaction equation 2 above, the standard free energy is defined in terms of *standard* chemical potentials μ° as:

$$\Delta G^\circ = \underset{(both\ phases)}{\Sigma_B} (1/z_B)\, \mu_B^\circ \qquad (9)$$

Thus the standard free energy change for the ion-exchange reaction, as defined by equation 2, is that free energy change which occurs when l/z_B moles of homoionic B zeolite in its standard state react with l/z_A moles of A ion in its standard state solution to give *entirely* l/z_B moles of B in its standard state solution plus l/z_A moles of A zeolite also in its standard state, all at the prescribed standard temperature and pressure.

3.2. Prediction of ion exchange equilibria

From the discussion above, it is evident that $\Delta G°$ is the overall free energy change which occurs when the system moves *wholly* from one set of idealised standard states to another. One may ask, therefore, what use is this function? Indeed, what use is thermodynamics? It is evident that the standard free energy, which is composition independent, cannot tell us about the magnitude of the selectivity as a function of composition, or the reasons why selectivity values may vary with composition. So what use are thermodynamic functions and the equations in understanding zeolite ion exchange processes? McGlashan elegantly and succinctly answers this question [55]: "What can ....an equation 'tell one' about one's system or process? Or, in other words, what can we learn from such an equation about the microscopic explanation of a macroscopic change? Nothing whatever. What then is the use of thermodynamic equations? They are useful because some quantities are easier to measure than others. If every equilibrium property of every kind of system had been measured with high accuracy then thermodynamics would be useless (though it would still be beautiful)"(*sic*).

This quotation encompasses what the purpose of ion exchange thermodynamics is for zeolite systems, or indeed for any system. Essentially its purpose it not to enable us to speculate about mechanism, but to enable us to predict the behaviour of the system over ranges of conditions (temperature, pressure, concentration) which may be difficult to measure directly because of experimental limitations. Having determined the thermodynamic equilibrium constant for one set of conditions by carrying out the appropriate activity corrections to the experimental isotherm, then one can in principle use these data to predict how the zeolite will behave under any other set of conditions. The development and testing of such predictive procedures has been a major aim of experimentalists over the last few years. Essentially, three things are required to enable one to predict ion exchange equilibria accurately. These are:

- an adequate thermodynamic formulation;
- very accurate experimental data for at least one set of conditions;
- accurate activity coefficient data for the dissolved salts in the external solutions for the range of conditions over which one wishes to predict behaviour.

The ability to predict exchange equilibria in this manner becomes increasingly important when one is considering the use of zeolites in ion exchange applications. For example, in detergency one deals frequently with a range of temperatures, a range of concentrations, and a multicomponent system involving sodium, calcium, magnesium and oxonium cations exchanging into the zeolite [22]. It is clearly impossible to measure the behaviour of a zeolite under every conceivable set of conditions. It is in such situations that the ability to predict exchange behaviour becomes important (see Section 7.3). This is obviously a subject of great complexity; consequently, only the basic principles are outlined below.

3.2.1. Prediction procedures

Numerous approaches have been developed by workers on such diverse materials as resins, clay minerals and zeolites in order to predict exchange behaviour for multicomponent systems from binary exchange data only. Many of these methods have been reviewed [7,56]. Approaches which use binary data in order to predict multicomponent equilibria can be broadly classified in two groups [56]. In the first group are methods where ternary or multicomponent activity coefficients for the exchanger phase are derived using as data experimentally determined activity coefficients for the conjugate binary systems in conjunction with a semi-empirical mathematical procedure; these derived data are then used to predict the ternary or multicomponent equilibrium compositions. The second type of approach is to predict ternary or multicomponent equilibrium compositions by graphical manipulation of conjugate binary exchange data. Both predictive methods have the merit that they require less experimental work than does the rigorous thermodynamic approach (*i.e.,* only the acquisition of accurate binary exchange data); however detailed studies on various zeolitic systems has shown that both these types of approaches are deficient compared to a rigorous thermodynamic one, for zeolites at least [56].

How then can thermodynamic data be used to predict exchange selectivities? To answer this question we need just to consider a little more carefully the relationship between the thermodynamic equilibrium constant and the mass action quotient (equations 6 and 7). The relationship between them is:

$$K_a = (m_B^{zA}/m_A^{zB})\,(E_A^{zB}/E_B^{zA})\,(\gamma_B^{zA}/\gamma_A^{zB})\,(g_A^{zB}/g_B^{zA}) = K_{m\,(E)}\,\Gamma\,\Phi \qquad (10)$$

where Γ (= $(\gamma_B^{zA}/\gamma_A^{zB})$) and Φ (= (g_A^{zB}/g_B^{zA})) are respectively overall activity corrections for the solution and crystal exchanger phases. It is common practice, and indeed convenient, to define another function K_E (commonly called the corrected selectivity quotient or, for historical reasons [57], the Kielland quotient), which is related to $K_{m(E)}$:

$$K_E = K_{m(E)}\,\Gamma \qquad (11)$$

Gaines and Thomas [15] demonstrated long ago that provided the variation of K_E with the crystal phase composition is known, it is a relatively easy matter to thence obtain the thermodynamic equilibrium constant:

$$\ln K_a = (z_B - z_A) + \int_0^1 \ln K_E\, dE_A + \nabla + \Psi \qquad (12)$$

Equation 12 is the form which one obtains when all factors are taken into account, including possible changes in water activity within the zeolite framework and salt imbibition. The two functions ∇ and Ψ respectively take these two phenomena into account (for further details see equation 24 in ref. [16]). One can rearrange the above equation readily to express it in terms of the corrected selectivity quotient at a particular composition P, to give:

$$ln\ K_{E(P)} = \int_0^1 ln\ K_E\ dE_A - ln\ \Phi_{(P)} + (z_B - z_A) + \nabla + \Psi \qquad (13)$$

Barrer and Klinowski [51] have shown that the water term ∇ is usually insignificant in magnitude for zeolitic ion exchange processes. Furthermore, Barrer and Walker [14] showed that salt imbibition occurred to a negligible extent when the external electrolyte concentration was low. Consequently the activity correction for the crystal phase Φ should be normally virtually independent of the strength of the external electrolyte concentration used [51].

It is this latter fact which is the key to enabling us to predict exchange equilibria in zeolites under constant temperature and pressure. If the conditions hold, then it is evident from equation 13 above that in the absence of salt imbibition, the whole of the RHS is constant as the total external electrolyte concentration (not composition) is changed, and therefore of course so is K_E. It is then evident from equation 11 that if K_E does not vary as the total external electrolyte concentration is changed, then the *only* way by which $K_{m(E)}$ can change is if Γ changes (*i.e.,* the activity coefficients for the salts in solution change).

We emphasised above how important it was to have accurate experimental data for one set of conditions (*i.e.,* solution concentration, temperature and pressure). It should also now be clear how important it is to be able to evaluate Γ accurately as we change the overall external electrolyte concentration in solution at a given temperature and pressure. Procedures to do this have been worked out, and involve measuring ratios of individual salt activity coefficients ($\gamma_{\pm MX}$) *within the mixed electrolyte solutions* ($\gamma_{\pm MX(NY)}$). The methods for doing this are complicated and are described elsewhere [54].

3.2.2. Results and limitations

That the prediction procedure outlined above works has been demonstrated not only for binary systems [53] but also on studies on some ternary exchange systems [58,59]. The first of these comprised a detailed examination of both the binary and ternary exchange behaviour of calcium, magnesium and sodium in zeolite A [58]; this work was prompted by the use of zeolite A as a detergent builder and the results are discussed further in Section 5.1.1. Suffice to say at this point that the procedures were successful in the accurate prediction of exchange compositions at different overall electrolyte concentrations. Success was also normally achieved for the Na/K/Cd-zeolite X exchange (Figure 7) although systematic errors were observed at higher external solution concentrations [59]. The fact that these errors were systematic suggested that the cause of the errors lay not so much with the quality of the experimental measurements which had been carried out on the Na/K/Cd-X system [59] but *either* because in this case the values of ∇ and/or Ψ were not negligible (see discussion above, following equation 13) *or* because of systematic errors in the magnitude of Γ (see equation 10 and discussion at the end of Section 3.2.1). This latter point is an important one. It was noted in the previous section that the degree to which one can accurately predict trends in equilibrium exchange compositions involving zeolites depends heavily on Γ, a function which

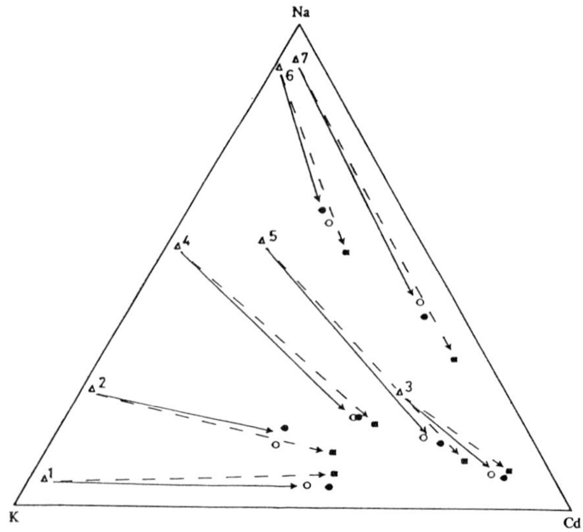

Figure 7. Experimental and predicted zeolite phase compositions for the Na/K/Cd-zeolite X system, 25°C, 0.4N. (△) solution composition at 0.1N; (○) predicted zeolite composition at 0.4N; (●) experimental zeolite composition at 0.4N; (■) zeolite composition at 0.1N. Reproduced with permission from reference 59.

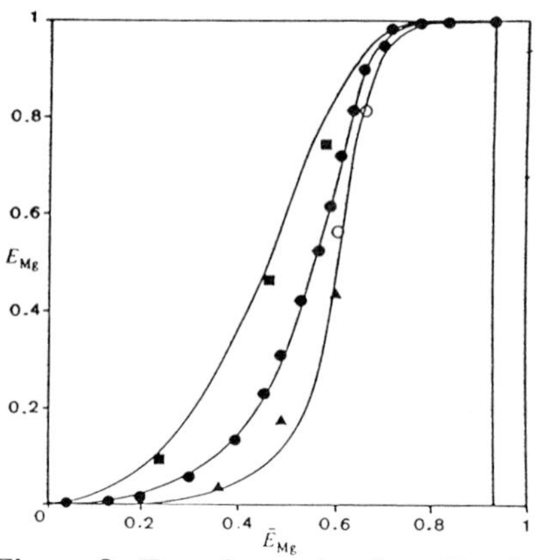

Figure 8. Experimental and predicted isotherm data for the 2Na/Mg-zeolite X (Si/Al = 1.27) system, 25°C. Experimental datum points: (■) 0.4N; (●) 0.1N; (▲) 0.025N; (○) 0.1N, reverse exchange. Solid line at 0.1N is best fit, otherwise predicted curves. Reproduced with permission from reference 16.

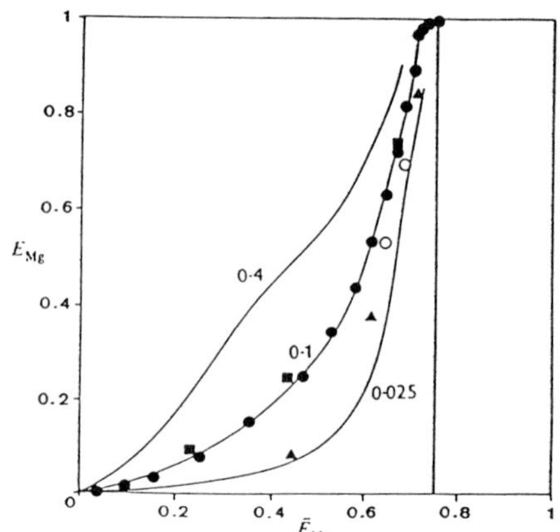

Figure 9. Experimental and predicted isotherm data for the 2Na/Mg-zeolite Y (Si/Al = 2.44) system, 25°C. Experimental datum points: (■) 0.4N; (●) 0.1N; (▲) 0.025N; (○) 0.1N, reverse exchange. Solid line at 0.1N is best fit, otherwise predicted curves. Reproduced with permission from reference 16.

is derived algebraically from ratios of the values of the mean molal stoichiometric activity coefficients $\gamma_{\pm MX}$ in solution. The quality of the $\gamma_{\pm MX}$ data used is clearly a key issue here; the many different sources where these data have been tabulated do not all agree in some cases. This matter is dealt with most usefully in a recent book on thermodynamics [60] by Fletcher (see in particular his appendices 2 and 3).

However, poor quality activity coefficient data cannot be the sole explanation for observed discrepancies. For example, in another study by Franklin and Townsend [16] on both binary and ternary exchange behaviour of sodium, ammonium and magnesium ions in various X and Y zeolites, at higher electrolyte concentrations increasingly serious deviations between prediction and observation were seen as the zeolite framework charge was lowered from X through to a high silica Y material, irrespective of whether the Na/Mg, Na/NH_4 or Na/NH_4/Mg systems were examined. This is shown in Figures 8 and 9, using the binary Na/Mg exchanges in X and high silica Y respectively as examples. Since the same $\gamma_{\pm MX}$ data sources were used for the studies on the two Y and one X zeolites, the increasing discrepancy between prediction and observation from X to high-silica Y had to be to do with the properties of the zeolites themselves and not from errors in the magnitude of Γ. Estimates of the magnitudes, and trends in magnitudes, of ∇ and Ψ were made [16] and it was concluded that the most likely cause of the discrepancies was a degree of anion imbibition as the ion pair $MgCl^+$ at higher electrolyte concentrations, a process which one would expect to occur more [14,49] in the lower charge, higher silica Y materials, as observed [16]. Consequently, the authors proposed that similar imbibition of ion pairs may take place with calcium, cadmium and copper in higher silica materials such as Y, resulting in similar discrepancies to those seen by them with magnesium [16]. Past work had shown that this might be expected to occur especially when chloride salts were used [38-41].

However, further investigations have shown that the situation is not that simple, while not excluding the suggestion that salt imbibition is the cause of the deviations seen with Mg/Na exchange in Y. Trace exchange of calcium into Na-X and Na-Y zeolites was studied, with particular care being taken to measure and take into account possible hydrolysis effects [43,44]. Apparent significant deviations from the law of mass action, as conventionally formulated for binary ion exchange reactions (equation 2), were observed with trace calcium exchange in dilute solution. These deviations could arise from a failure to take into account the influence of other components present in low quantities which affect the equilibrium and consequently the form of the ion exchange reaction and mass action quotient. They could not be explained purely in terms of hydrolysis effects [43]. Subsequent attempts at predicting the Ca/Na equilibrium in X and Y [44] showed good agreement at moderate or higher concentrations (note that this is the opposite to what was observed for magnesium [16]), but in this case predictions were poor for *dilute* solutions. This failure was again ascribed to the apparent deviation from the law of mass action described above, but the cause remained obscure [44]. In this case, because the solutions were dilute or very dilute, salt imbibition could be ruled out [14,44], and neither hydronium exchange or zeolite dissolution, or a combination of these, seemed to provide an adequate explanation. Further work [46] led to the suggestion that the presence of colloid-size particles arising from hydrolysis, which could not be separated from the solution phase before analysis, was the cause of this problem, but this was not proven. Thus the reasons for these prediction failures remain obscure, but probably involve combinations of all these factors to differing degrees, according to the solution pH and concentration employed. In many important applications, prediction of ion exchange equilibria has to be carried out in dilute solutions, so these issues are not esoteric ones. Rigour in theoretical treatments must be matched with similar rigour in experimental measurements (see also Section 2.1.4). Further details on this, and other predictive procedures which have been developed, are given elsewhere [48].

Lastly, we should note that all the foregoing has been concerned with prediction of exchange equilibria at constant temperature and pressure. Elementary thermodynamics [61] shows how quantities such as the standard free energy and equilibrium constant change with temperature and pressure:

$$R d\ln K_a = (\Delta H^o/T^2)dT - (\Delta V^o/T)dP \Rightarrow (\partial \ln K_a/\partial T)_p = \Delta H^o/RT^2;\ (\partial \ln K_a/\partial P)_T = -\Delta V^o/RT \quad (14)$$

As is usually the case with reactions in condensed phases, standard volume changes associated with exchange are very small. Therefore, unless very high pressures are employed pressure changes will have very little effect on the value of $\ln K_a$. Enthalpy changes associated with ion exchange reactions in zeolites are usually fairly small but are nevertheless measurable. Even when a ΔH^o value can be measured accurately however, it is often questionable to use equation 14 to predict $\ln K_a$ at other temperatures since the relative fractions of cationic charge ε_i associated with each of the i sublattices within the zeolite framework may not only vary with cationic composition but also with temperature [62]. When this happens the magnitude of ΔH^o may change markedly with temperature and then equation 14 (which holds if $(\partial \Delta H^o/\partial T) \rightarrow 0$) is not applicable.

3.3. Concentration scales and ideal behaviour

In Section 3.1 above, a brief comment was made on the importance of specifying carefully one's choice of concentration scale for defining ion exchange levels in the zeolite phase. Some discussion of this is important, even in an overview of ion exchange such as here, because thermodynamic quantities such as $\Delta G°$, definitions of ideal behaviour and the magnitudes of activity coefficients can all differ according to the concentration scale chosen. When comparing one's own experimental data with that in the literature, it is *essential* to ensure that one is comparing like with like, or that the appropriate data are being used for the prediction of ion exchange behaviour. As the subject can be very involved [8,63], only a brief outline of some important definitions and principles is given here.

Rational concentration scales have been adopted near universally to define ionic concentrations in ion exchange materials, whether these be resins, clay minerals or zeolites. The obvious choice would seem to be that of mole fraction. Since even a binary exchange process normally involves a solvent also, the minimum number of components is normally three (n_A moles of homoionic A-zeolite, n_B moles of homoionic B-zeolite and n_S moles of solvent). The definition of mole fraction x_A is then

$$x_A = n_A / (n_A + n_B + n_S) \quad (15)$$

The reference state for each component follows the Raoult law convention, *viz.* that

$$a_i = x_i f_i \tag{16}$$

with

$$\lim_{(x_i \to 1)} a_i = x_i \tag{17}$$

where f_i is a rational activity coefficient. From equations 16 and 17 it is obvious that for the zeolite to exhibit ideal behaviour f_i must be unity for all values of x_i.

However, x_i is rarely, if ever, employed, for the simple reason that for either x_A or x_B to equal unity, n_S must be zero, *i.e.,* the standard states with respect to exchanging ions A and B are the respective homoionic forms of the zeolite in the *dry state* [63]. Consequently, for ion exchanges in which the zeolite is immersed in an external electrolyte solution such as water, the standard free energy of exchange will include two terms extra to those associated with moving from the pure B to pure A form of the material (see equation 2 and Section 3.1.4), *viz.*, a reversible isothermal hydration of the B-form before any exchange takes place and then a reversible isothermal dehydration of the A form after the exchange process is complete [63].

To avoid this complication, it has long been the practice to use alternative rational concentration scales. Although these were introduced long ago, two of them are used extensively to this day. The first, the cation equivalent fraction, has already been introduced (Section 3.1.1):

$$E_A = z_A n_A / (z_A n_A + z_B n_B) \tag{1}$$

The second, the cation mole fraction, is defined as:

$$X_A = n_A / (n_A + n_B) \tag{18}$$

In both equations 1 and 18 a mole of homoionic exchanger is defined identically, following Vanselow [64], in terms of one mole of exchange cation (*e.g.,* A^{z_A+}) in association with z_A equivalents of negatively charged framework L^-, to give AL_{z_A}, BL_{z_B} moles *etc.* of exchanger. Note that although the definitions of cation equivalent fraction and cation mole fraction appear now to exclude the moles of solvent present (*cf.* equation 15), the solvent is now included within the definition of the reference states for either ion A or B, in that these reference states are now the respective homoionic forms of the zeolites *immersed in infinitely dilute solutions of the same ion* [15]. In effect, with these definitions, the zeolite is regarded as comprising a mixture of two *salt hydrates*, which may or may not be miscible (see also Section 2.1.3) [15]. Nevertheless, although the quantity of solvent in each homoionic form of the zeolite in its standard state is strictly defined, the solvent remains an independently variable component of the system.

In passing, note that another rational scale was introduced by Gapon [65]. This is also a cation mole fraction but here the mole was defined not as AL_{z_A}, BL_{z_B} but rather as LA_{1/z_A}, LB_{1/z_B}. Thus the mole of Gapon contains one "equivalent" of exchange capacity compared to z_A equivalents for the mole of Vanselow [63,64]. Numerically it is identical to the cation equivalent scale as given in equation 1, but despite this, if this concentration scale is used as a basis for a thermodynamic formulation [66], activity coefficients values are *not* identical to those obtained using the cation equivalent fraction scale of Gaines and Thomas [15,63]. The approach of Hogfeldt [66], based on the Gapon mole, has fallen into disuse.

It is however important to be aware of and understand the other two rational concentration scales (equations 1 and 18), because these were the basis for the development of two parallel and complementary thermodynamic formulations by Gaines and Thomas [15] and Bonner, Argersinger and Davidson [67], both of which remain in common use today. Using cation equivalent and cation mole fractions respectively, the corresponding selectivity coefficients are:

$$K_{m(E)} \quad = \quad (m_B^{zA}/m_A^{zB})\,(E_A^{zB}/E_B^{zA}) \tag{6}$$

and

$$K_{m(X)} \quad = \quad (m_B^{zA}/m_A^{zB})\,(X_A^{zB}/X_B^{zA}) \tag{19}$$

whence the thermodynamic equilibrium constant is also respectively:

$$K_a \quad = \quad K_{m(E)}\ \Gamma\ (g_A^{zB}/g_B^{zA}) \ = \ K_E\ \Phi \tag{20}$$

from equations 10, 11 and 12, or

$$K_a \quad = \quad K_{m(X)}\ \Gamma\ (f_A^{zB}/f_B^{zA}) \ = \ K_X\ \Xi \tag{21}$$

where f_A,f_B are rational activity coefficients corresponding to the cation mole fraction scale. If the same choice of concentration units is made in both cases for the external solution phase, then for a given overall solution concentration and composition Γ will have the same value in both equations 20 and 21. However, if $z_A \neq z_B$, $E_A \neq X_A$ and $E_B \neq X_B$ and consequently the non-ideality corrections Φ and Ξ will not have the same value.

It is essential therefore to note that even though for both these treatments the definition of the mole as well as the form of the exchange reaction are identical (equation 2), and consequently the values of thermodynamic quantities such as $\Delta G°$ are the same [63], the values of the activity coefficients g and f are *not* identical when $z_A \neq z_B$ and cannot be used interchangeably.

Similarly, the definitions of ideal behaviour differ also. This point is more subtle and may seem surprising, since equations 20 and 21, combined with exactly similar definitions for ideal behaviour, *viz.*

$$g_A = g_B = 1 \text{ for all } E_A,E_B,$$
$$f_A = f_B = 1 \text{ for all } X_A,X_B,$$

lead to

$$K_a = K_E = K_X \tag{22}$$

when the system is behaving ideally, since then for all compositions $\Phi = \Xi = 1$.

However, we noted earlier that the solvent (*e.g.,* water) remains an independently variable term. Consequently, it is necessary to define also the conditions for ideal behaviour to include the solvent as well as ions A and B. The Gibbs-Duhem equations corresponding to the cation equivalent fraction and cation mole fraction scales (as defined in equations 1 and 18) are respectively [15,63,67]

$$(E_A/z_A)\, d\ln g_A E_A + (E_B/z_B)\, d\ln g_B E_B + [n_S/(z_A n_A + z_B n_B)]\, d\ln a_S = 0 \qquad (23)$$

and

$$X_A\, d\ln f_A X_A + X_B\, d\ln f_B X_B + [n_S/(n_A + n_B)]\, d\ln a_S = 0 \qquad (24)$$

Applying now the conditions of ideal behaviour to equation 24, *i.e.,* that $f_A = f_B = 1$ for all X_A,X_B, we readily obtain further the condition that the solvent must also fulfil for ideal behaviour when the cation mole fraction scale is employed, *viz.*

$$[n_S/(n_A + n_B)]\, d\ln a_S = 0 \qquad (25)$$

which, since a_S must be unity in the two homoionic standard states, is simply that $a_S = 1$ for all X_A,X_B, as one would expect. Using the cation equivalent scale does not lead to such a simple result. Assuming, in analogy to equation 24, that the conditions for ideal behaviour are now that $g_A = g_B = 1$ for all E_A,E_B, then equation 23 becomes [63]

$$[n_S/(z_A n_A + z_B n_B)][d\ln a_S/dE_A] = (z_A - z_B)/z_A z_B = \text{constant} \qquad (26)$$

which is quite different for solvent ideal behaviour to that shown in equation 25 for the mole fraction scale. When $z_A = z_B$, this reduces to $a_S = 1$ for all E_A,E_B, which is identical to the result for the mole fraction scale. When $z_A \neq z_B$, one has the rather odd result that while at the extrema of the isotherm (*viz.*, $E_A = 0,1$) a_S by definition must be unity, for other compositions the value of $\ln a_S$ varies *linearly* with E_A, in accordance with equation 26. This implies inevitably that at some point there will be discontinuity in the plot of $\ln a_S$ against E_A, and in fact the position of the discontinuity is determined by the requirement that when $z_A \neq z_B$, the value of the water integral ∇ should equal $(z_A - z_B)$ [9]. Otherwise equation 12 leads to an apparently absurd result that $\ln K_a \neq \ln K_E$ when the system is behaving ideally [9,63].

The complexities which can arise because of the existence of different concentration scales have recently prompted Hall [68] to propose an alternative general formalism, which accommodates also swelling of the exchanger phase, salt imbibition and the influence of solvent activity on exchange behaviour, as well as discussing enthalpy and volume changes associated with the exchange reaction (see the discussion above, following equation 14). He also proposes a definition of ideal behaviour [68] which corresponds closely with Nernst-Planck equations describing ideal ion transport behaviour [69] (see also Section 4.3 below).

4. THE KINETICS OF ION EXCHANGE

If the formulation and consequent prediction of ion exchange equilibria have proved to be complicated, understanding and predicting the dynamics and kinetics of these systems has proved to be an even less tractable task. As a consequence, there have been relatively few systematic studies on exchange kinetics in zeolites in recent years, despite the obvious importance of this subject in areas such as catalyst preparation or detergency. A series of early studies on mainly natural materials by Barrer, Rees and co-workers [70-77] threw considerable light on one important aspect of the subject, *viz.*, intracrystalline diffusion. These seminal studies were reviewed in the first edition of this volume [78]; this time, a broader picture of ion transport phenomena will be presented, with a stronger emphasis on fundamental principles and their significance in applications. Experimental tests of theories, and attempts to predict ion exchange rates in zeolites, are also reviewed.

4.1. Some basic concepts

It is most important to recognise that the rate of a particular ion exchange reaction may be determined by one or more of several distinct mechanisms. It is commonly assumed that the rate-controlling step is one mediated by some diffusion process within the individual zeolite crystallites but this need not be the case. For example, within the zeolite crystallites themselves, *kinetic exchange* between cations diffusing through the micropores and cations bound to specific sites within the framework could be the rate determining step. For this to occur, the requirement is merely that the rate of exchange of ions at sites within the zeolite channels should be slower than is the rate at which these ions diffuse through the micropores. Indeed, a formulation including both exchange rate constants and intracrystalline self-diffusivities, that takes this possibility specifically into account, was published some time ago [79,80].

Having recognised that the rate of exchange at sites within the zeolite framework may be rate-determining, let us now concentrate on different aspects of mass transfer resistance. Note that even if the rate controlling transport process is diffusion, there is still the possibility that this rate-controlling diffusion process may not be within the crystallite micropores themselves but instead limited by transport through a near-static boundary layer interposed between the external solution and the crystallite surface. Such a process is often described as *film* or *boundary layer diffusion*. However, there are still further possibilities: if the individual crystallites have aggregated to form agglomerates, the rate-determining steps may include other possible diffusion processes within the macropores which then exist between individual crystallites within the crystal agglomerate. Alternatively diffusion through a quiescent boundary layer between the external electrolyte solution and the overall agglomerate of crystallites itself may be the predominant process.

The picture is still not complete. *Fluid mass transfer* processes, which should be distinguished sharply from the molecular transport mechanisms which characterise diffusion, may also contribute to the overall rate of ion exchange, depending on the hydrodynamic regime which predominates. An example of such a process is convective flow. *Momentum transfer* (shear) can also be indirectly important (*e.g.*, the effects that different impeller speeds may have on film mass transfer resistance when stirring a suspension of particles - see Section 4.2.1).

Many terms have been introduced here. Their relative significance can be appreciated by next looking more systematically at transport within an agglomerate of aggregated zeolite crystallites, or a pellet of zeolitic material, whichever is appropriate.

4.2. Hierarchical model of a zeolite particle or pellet

Agglomerates of zeolite crystallites were described above - both these and discrete crystallites are used in detergency applications. In many manufacturing applications, whether these are in catalysis, sorption or ion exchange, pelletised forms of the zeolites are often utilised, where the pellet may contain in addition to the zeolite other components such as amorphous aluminosilicate binder and perhaps a diluent (typically a clay mineral). For ion exchange applications it is most important to be aware of this, as these other components may also influence the overall ion exchange behaviour of the composite pellet, both kinetic and equilibrium aspects. However, whether one is considering an agglomerate of aggregated crystallites, or a composite pellet, a *hierarchical model* (Figure 10) provides a useful approach for distinguishing the nature of the different transport processes which may operate at different length scales, and for discerning their relative importance in a particular situation. This hierarchy, described in the following subsections, effectively operates at three levels or length scales.

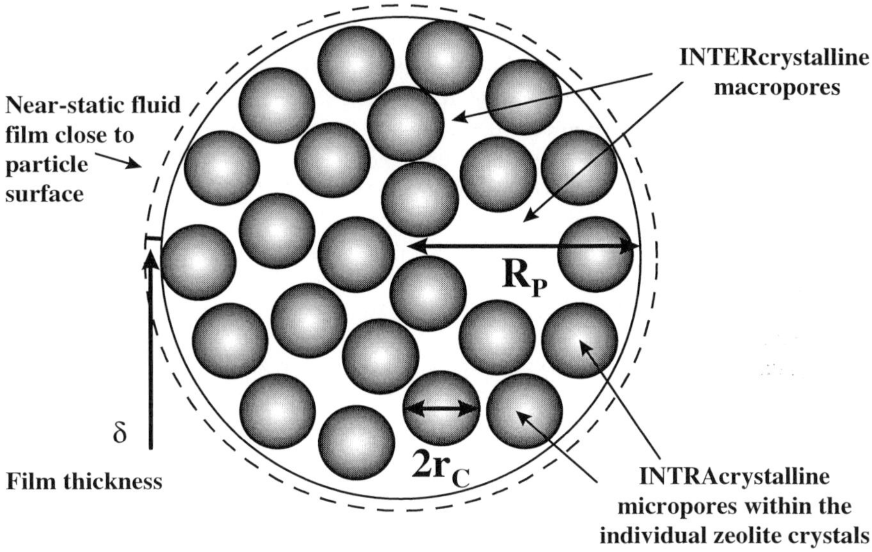

Figure 10. An idealised agglomerate of zeolite crystals where the various hierarchical levels of diffusion resistance are apparent (film diffusion, inter-crystalline transport and intra-crystalline diffusion; see text for more details). R_P - radius of agglomerate or pellet; r_c - radius of zeolite crystallite.

4.2.1. Film mass transfer resistance

The highest hierarchical level is concerned with the pellet or agglomerate *in toto.* The key question here is whether transport of ions through a near-static fluid film, which acts as a boundary layer between the external solution and the macroparticle itself, may affect or even determine the overall rate of exchange. That film mass transfer resistance may be rate controlling has been recognised for a long time [49,81]; the transport mechanism involved is normally one of diffusion, quantified by a *film diffusion coefficient* D_F. A useful dimensionless criterion [82], originally developed by Helfferich [49], enables one to predict broadly whether the exchange rate for a particular system is likely to be controlled by film diffusion, *viz.* :

$$\{(m_\Sigma D_F \delta / m_{\Sigma(S)} D_P R_P)(5 + 2\alpha)\} << 1 << \{(m_\Sigma D_F \delta / m_{\Sigma(S)} D_P R_P)(5 + 2\alpha)\} \quad (27)$$

(particle diffusion control) *(film diffusion control)*

Here $m_\Sigma, m_{\Sigma(S)}$ are the total ionic concentrations in the particle and solution respectively and D_P is an *intraparticulate diffusivity* (*i.e.,* within the pellet or agglomerate and *not* necessarily within individual crystallites - this would be an *intracrystalline diffusivity* (Section 4.2.3). The nature of D_P is discussed further in Section 4.2.2). δ and R_P are the film thickness and mean particle radius respectively (Figure 10). α is the separation factor, as given in equation 3.

Apart from the obvious point that the relative magnitudes of D_F and D_P will affect whether film diffusion or particle diffusion is rate controlling, relation 27 highlights several other dimensionless ratios of key significance. One would expect the magnitude of the electrochemical potential gradient at the solution/particle interface to be influential: this is a function of $m_\Sigma/m_{\Sigma(S)}$, which shows that the more dilute the external electrolyte solution is, the greater is the chance that film diffusion control will occur.

The magnitude of the ratio δ/R_P is of prime importance: two points should be noted here. First, film diffusion control will be favoured by the use of small pellets or agglomerates. Secondly, the hydrodynamic properties of the system come to the fore, because the thickness of the near-static boundary layer δ is affected by the velocity of the fluid (in this case electrolyte solution) passing parallel to the interface. Higher Reynolds numbers N_{Re} will favour smaller values of δ depending on the nature of the flow [83]; consequently film diffusion will be favoured either by the use of low impeller speeds in the case of a batch reactor, or by low throughput rates if an ion exchange column is employed. In addition, the mechanical integrity of the particle or pellet can also be very important here. Consider for example detergent particle agglomerates in a laundry powder. Typically these may consist of agglomerates of zeolite crystallites held together by adhesive, viscoelastic surfactant bridges. These are designed through a judicious combination of shear and dissolution processes to break up and disperse, leaving highly dispersed smaller particles or even individual zeolite crystallites which may find themselves in regions of low agitation and consequently in regions where film diffusion control predominates. In this situation, it is not the properties of the particular zeolite which may determine the rate of exchange, but rather the degree to which the machine agitates the wash liquor suspension.

Finally, we note that equation 27 shows that film-controlled diffusion is favoured by high values of the separation factor α [49].

4.2.2. Transport processes within the pellet or agglomerate

Moving now down to the second level within the hierarchical model, the focus is on whether and which transport processes may control, or contribute to, the overall ion exchange rate within the macropores which exist in the agglomerate or pellet *between* the individual zeolite crystallites, and how these processes may couple with diffusion within the zeolite micropores themselves. Thus one is considering particle-controlled diffusion which is both *inter*crystalline and *intra*crystalline in nature. Figure 10 is an idealised depiction of an irregularly shaped agglomerate. If one assumes that the agglomerate or pellet can be regarded as approximating to spherical symmetry with a mean pellet radius R_P, then the differential mass balance for one diffusing species i is given by [73]

$$(\partial X_{i,P} / \partial t) = (1/R^2).\partial/\partial R[R^2 D_{eff,P} (\partial X_{i,P} / \partial R)] \qquad (28)$$

where t is time, $X_{i,P}$ is a dimensionless concentration of component of species i within the pellet at R, $D_{eff,P}$ is an *effective intraparticulate diffusivity* describing overall homogeneous [84] transport behaviour of the species in the pellet and $0 < R < R_P$. X_i (the cation mole fraction, equation 18) has been chosen here to describe the dimensionless (rational) concentration but this is an arbitrary choice provided one is consistent throughout; one could equally well choose E_i or even x_i (see Section 3.3). Boundary conditions for integration of equation 28 include

$$(\partial X_{i,P} / \partial R) = 0 \quad \text{at} \quad R = 0 \quad \text{for} \quad t \geq 0 \qquad (29)$$

with initial conditions (one component i only) of

$$X_{i(S)} = 1, \quad X_{i(\sigma)} = X_{i,P} = 0 \quad \text{at} \quad t = 0 \qquad (30)$$

where $X_{i(S)}$, $X_{i(\sigma)}$ are respectively the dimensionless solution concentrations of species i in the bulk, or at R_P (*viz., still in solution* but right at the interface between the zeolite pellet and the external solution). The boundary condition in equation 29 merely describes the symmetry of the system using radial co-ordinates. The second is more subtle: note that $X_{i(\sigma)}$ is *in equilibrium* with the solid phase ($X_{i,P}$ at $R = R_P$) for $t \geq 0$. In addition, if film diffusion control is insignificant, $X_{i(S)} = X_{i(\sigma)} = 1$ for all $t > 0$ [84].

The key issue, however, is to understand which transport mechanisms contribute to the effective particulate diffusivity $D_{eff,P}$ (equation 28). In a pellet or agglomerate of zeolite, $D_{eff,P}$ will be made up of contributions from transport through the macropores between the individual crystallites and intracrystalline diffusion within the zeolite micropores (Figure 10). For a fluid, at least four mechanisms can contribute to transport in the macropore. These are *Knudsen diffusion* (a molecular transport process which can be important when the fluid is essentially gaseous in nature, where collisions between the diffusing species and the pore walls predominate), *surface diffusion* (diffusive motion of adsorbed species over a surface or interface), *bulk diffusion* (a molecular transport process mediated primarily by collisions between the diffusing species themselves) and *Poiseuille flow* (not diffusion but rather mass transfer by laminar flow down a pressure gradient in the pore). With liquids in macropores and in the ion exchange context, only bulk and surface diffusion can be expected to make significant contributions to the effective macropore diffusivity $D_{i,p}$. The porous structure of the pellet or agglomerate means that the liquid bulk diffusion path will be lengthened by the nature of the macropore structure. This is usually allowed for by applying a simple numerical correction factor called the tortuosity τ to the liquid bulk diffusivity $D_{i,l}$. If surface diffusion occurs, this will enhance the rate of transport over the liquid bulk value, which means that the effective macropore diffusivity becomes [84]

$$D_{i,p} = (D_{i,l}/\tau) + D_{i,\sigma} \tag{31}$$

where $D_{i,\sigma}$ is the surface diffusivity.

The overall effective intraparticulate diffusivity $D_{eff,P}$ will then be some function of both $D_{i,p}$ and a *micropore* or *intracrystalline diffusivity* $D_{i,c}$ (the nature of $D_{i,c}$ is discussed further in the next section). Robinson, Arnold and Byers [84] considered two such heterogeneous diffusion extrema, in which the macropore and micropore diffusion processes in the pellet occurred completely either in parallel or in series. Taking as an example the parallel case, the differential mass balance shown in equation 28 is modified to describe two fluxes operating together, one for the macropores and the other for the micropores in the pellet:

$$\xi(\partial X_{i,p}/\partial t) + (\partial X_{i,c}/\partial t) = (1/R^2).\partial/\partial R\{R^2\,[\xi D_{i,p}\,(\partial X_{i,p}/\partial R) + D_{i,c}\,(\partial X_{i,c}/\partial R)]\} \tag{32}$$

where ξ is a correction to the pellet concentration of species i which takes into account the proportion of binder in the pellet. This approach, as well as the heterogeneous model for fluxes operating in series, were tested on real systems comprising either individual crystals or pellets; the results are reviewed briefly in Section 4.3.

The discussion here has been in general terms. It is important to recognise specific properties of diffusion processes in the ion exchange context. This is done in the next section, which concentrates solely on intracrystalline diffusion in the zeolite micropores themselves.

4.2.3. Diffusion within the crystallite micropores

At the lowest level within this hierarchical model, and in the absence of rate limitation by the exchange process itself (Section 4.1) the focus turns now on to particle-controlled diffusion which is purely *intra*crystalline in nature, *viz.*, within the zeolite micropores themselves. Four properties of such a diffusion process are emphasised here:

(a) The limiting case: during the exchange process, in the limiting case when intracrystalline exchange is completely rate determining, radial concentration gradients with respect to each exchanging cation will exist within each individual crystallite, but concentration gradients will not be present in the pellet mesopores between the crystallites (Figure 10) [82]. Thus *in this limiting case* $(dc_i/dR_P) = 0$ but $(dc_i/dr_c) \neq 0$, and the rates at which the cations move to equalise concentrations throughout the zeolite crystallite domains will therefore be wholly determined by the magnitudes of time-dependent coupled chemical and electrical potential gradients that exist *within each individual crystallite*. This is a limiting case of the coupled macropore and micropore diffusion processes described above in

Section 4.2.2 (see equation 32) and one in which micropore diffusion overwhelmingly predominates as the rate-controlling process:

$$(\partial X_{i,c} / \partial t) = (1/r^2).\partial/\partial r[r^2 D_{i,c} (\partial X_{i,c} / \partial r)] \tag{33}$$

The boundary and initial conditions for integration are similar to those described above (spherical symmetry assumed, *cf.* equations 28, 29 and 30), *viz*:

$$(\partial X_{i,c} / \partial t) = 0 \text{ at } r = 0 \text{ for } t \geq 0 \tag{34}$$

and

$$X_{i(S)} = 1, \; X_{i(\sigma)} = X_{i,c} = 0 \text{ at } t = 0 \tag{35}$$

It is commonly assumed that equation 33 generally holds for ion exchange in zeolites. However, this assumption is often made without any validating checks being carried out and in some cases it is unjustified when dealing with pellets or larger agglomerates [84].

(b) The effect of concentration: for any systems which do not behave ideally, the values of diffusivities will vary with the magnitude of the local concentrations of the diffusing species themselves. This is especially important for the intracrystalline diffusion of cations in zeolites because these systems may behave in a strongly non-ideal manner. The dependency of diffusivities on concentration in non-ideal systems is a simple consequence of the fact that Fick's laws of diffusion are formulated in terms of *concentration gradients*:

$$J_i = - D_i \, grad \, c_i \tag{36}$$

and

$$(dJ_i / dt) = div \, (D_i \, grad \, c_i) \tag{37}$$

where J is the flux, D a diffusion coefficient and c the concentration of species i. However, in actuality, the determinant for diffusion is either a *chemical potential gradient* or (for ion exchange) an *electrochemical potential gradient* with respect to the diffusing species. Consequently, for systems which do not behave ideally (where $c_i \neq a_i$ for all c_i - see Section 3.3) the so-called Fickian "diffusion constant" is anything but constant as the concentration varies (for this reason it is much better to describe D_i as a diffusion coefficient or diffusivity). To a reasonable approximation [85] the dependence of diffusivity D_i on concentration is generally described by the Darken relation:

$$D_i = D^*_i \, (d \, ln \, a_i / d \, ln \, c_i) \tag{38}$$

where D^*_i refers to a reference state in which the system behaves ideally. For gas sorption, this can be a tracer diffusivity, measured in the limit as the adsorbate concentration $c_i \to 0$ in the zeolite (*i.e.,* Henry Law definition of ideal behaviour). For ion exchange, the Darken relation becomes

$$D_{AA} = D^*_{AA} \,(d\, ln\, a_A / d\, ln\, c_A) \tag{39}$$

where D_{AA} is an *intrinsic* or *self-diffusivity*. In essence, D_{AA} describes the rate at which A^{zA+} cations exchange with their own kind for all compositions where E_A is <1. D^*_{AA} is then the corresponding A,A self-diffusivity observed in the *homogeneous* A form of the zeolite, in equilibrium with an infinitely dilute external solution containing the same cation, which is of course the standard state for A-zeolite (Raoult law definition of ideal behaviour - Sections 3.1.4 and 3.3, with $a_A \rightarrow c_A$ as $E_A \rightarrow 1$). Consequently, equation 39 shows that for non-ideal systems it is only *in the limiting case* that $D_{AA} = D^*_{AA}$, since it is only in the limit, as ideal behaviour is approached, that $(d\, ln\, a_A / d\, ln\, c_A) = 1$. Otherwise (and this point needs emphasis!) $D_{AA} \neq D^*_{AA}$.

Values of D_{AA} and D^*_{AA} can be determined experimentally by carrying out radiotracer measurements of self exchange [74] on *equilibrated systems*, *i.e.,* on systems where there is no chemical potential gradient in A. For D_{AA}, these measurements would be carried out on a range of mixed A,B forms of the zeolite in equilibrium with external electrolyte solutions having corresponding equilibrium A,B compositions as defined by the isotherm (Figures 1 - 6). Similarly, D^*_{AA} can be determined to a good approximation using radiotracer measurements on the pure A-form of the zeolite immersed in a very dilute solution of an A salt.

There are therefore, even for a simple binary ion exchange process, *five* diffusivities to consider, *viz.*, the self-diffusivities D^*_{AA}, D^*_{BB} for the homoionic A,B forms of the zeolite, the concentration dependent intrinsic diffusivities D_{AA}, D_{BB}, and the *inter-diffusion coefficient* D_{AB}, which is discussed further below.

(c) The effect of temperature: the diffusion mechanism is likely to be a strongly activated process, not only because the diffusing species are of comparable sizes to the microporous channels in which they are moving, but because both they and the zeolite framework are electrically charged. Consequently, the dependency of the diffusion rate on temperature exhibits Arrhenius-type behaviour:

$$D_{AA} = D^o_{AA}\, exp\, (-E_a / RT) \tag{40}$$

where E_a is an energy of activation. Transition state theory [86] gives for the pre-exponent that

$$D^o_{AA} = (kT/h)\, \lambda^2\, exp\, (\Delta S^{\ddagger} / R) \tag{41}$$

where k, h and $\Delta S^{\ddagger}$ are respectively the Boltzmann constant, Planck constant and entropy of activation. λ is a *jump distance*, signifying the distance between successive energy minima as the ion diffuses through the crystal matrix. Note that equations 40 and 41 are in terms of self diffusivities; the inter-diffusion coefficient D_{AB} may not necessarily follow such a simple temperature dependent behaviour because of the way these diffusivities are related (see equations 43 and 44).

Intracrystalline diffusion in zeolites shares this Arrhenius-type behaviour in common with surface diffusion in particular (Section 4.2.2) although only partly for similar reasons.

(d) Different exchange sites and channel sizes: It was noted earlier that zeolites may be regarded as comprising regular sets of sublattices which are characterised by different exchange site energies (Sections 2.1.2 and 3.2.2). Add to this the different channel sizes which provide access to the exchange sites and it is evident that the medium within the zeolitic domain in which the ions diffuse is heterogeneous yet strongly periodic in energetic terms. These periodicities can be correlated with diffusion jump distances λ (see equation 41) and where exchange behaviour can be attributed fairly unambiguously to contributions from the different sublattices (such as in faujasitic zeolites) attempts have been made to rationalise observed exchange rate behaviour in terms of more than one distinct diffusion path operating in parallel [87]. Although such attempts may seem rather artificial they serve to emphasise an important point, *viz.*, that the existence of long range order with respect to both the micropores and different exchange sites in the crystallite means that zeolites are very different to, for example, a typical ion-exchange resin. This can add complexity to the kinetic behaviour of the system and consequently the prediction of exchange rates.

With these four properties (a) to (d) in mind, let us next try to picture the processes which are going on within the crystallite. Consider the simplest case of two types of cations exchanging within the channels or cages of the zeolite. Clearly, as a consequence of different ion sizes (and possibly also charges) the two types of cations will have different mobilities; in addition water will be present. If concentration gradients are present within the crystal, then the different types of cations will move in the directions of negative concentration gradients in order to equalise their concentrations throughout the system (this process is, of course, described by Fick's first law). However, the mobilities of the two types of cations are, as we have already noted, different. We are visualising a counter diffusion involving the two types of cations: since the mobilities are different, in the absence of other constraints, one would expect that the faster moving cation would tend to build its concentration faster than the slower. If this process were to continue unconstrained, charge separation would occur and an electrical potential gradient would build as the two types of ions move towards each other within the crystal. In reality charge separation does not happen to any significant degree, as the electrical potential gradient which forms as charge separation begins to occur operates in such a way as to slow the faster moving ion and, conversely, accelerate the slower moving ion. Thus it is important to consider not only the elimination of the concentration gradients by diffusion (as described by Fick's laws) but also the maintenance of an absence of an electrical potential gradient throughout the crystal. The necessity to consider also the electrical potential gradient means that the net flux of ion exchanging species is properly described by [82,88]

$$J_A = -D_{AB}\,[\mathit{grad}\ c_A + (z_A c_A F/RT)\ \mathit{grad}\ V] \qquad (42)$$

where F is the Faraday and V the electrical potential. This expression is called the Nernst-Planck equation [68,69,88]. It is one which holds under ideal conditions, and a set of Nernst-Planck expressions, one for each type of exchanging ion, have to be solved simultaneously. By combining this expression with corresponding irreversible thermodynamic formulations, Barrer and Rees [70] derived an appropriate general expression for the inter-diffusion coefficient D_{AB}, and then applied it to ion-exchange kinetics in zeolites. Their complete expression takes into account not only non-ideality of the exchanger, but couplings between ion-ion and ion-water fluxes. The complete expression is complicated [70]; however, the essence of the relationship between D_{AB}, $D^{\star}_{AA}$ and $D^{\star}_{BB}$ is seen if one assumes no coupling between fluxes other than that due to the electrical potential gradient [73]:

$$D_{AB} = D^{\star}_{AA}\, D^{\star}_{BB}[z_B^2 c_B(\partial ln\, a_A / \partial ln c_A) + z_A^2 c_A(\partial ln a_B / \partial ln c_B)]\, [z_A^2 c_A D^{\star}_{AA} + z_B^2 c_B D^{\star}_{BB}]^{-1} \quad (43)$$

Note that equation 43 incorporates the Darken relationship discussed above. Consequently, a further simplification can be made if the exchanger behaves ideally for all exchanger compositions. In this case (see the discussion after equation 39) the terms $(\partial ln\ a_i / \partial ln\ c_i) = 1$, so that

$$D_{AB} = D^{\star}_{AA}\, D^{\star}_{BB}[c_A z_A^2 + c_B z_B^2]\, [c_A z_A^2 D^{\star}_{AA} + c_B z_B^2 D^{\star}_{BB}]^{-1} \quad (44)$$

This equation was first derived by Helfferich and Plesset [89,90].

4.3. Experimental tests of theoretical models

To test the validity of models describing intracrystalline ion diffusion in zeolites, outlined in the preceding section, it is necessary to generate curves describing the fractional attainment of equilibrium versus time using a theoretical model and compare these with experimental data. These curves can be generated by solving equation 37 using the appropriate boundary conditions, with D_i being replaced by the interdiffusion coefficient D_{AB}:

$$(dJ_A / dt) \quad = \quad div\ (\ D_{AB}\ grad\ c_A) \quad (45)$$

or, in the absence of knowledge of how D_{AB} changes with concentration, by using the approximate form:

$$(dJ_A / dt) \quad = \quad \overline{D}_{AB}\ div\ (grad\ c_A) \quad (46)$$

where $\overline{D}_{AB}$ is an integral diffusivity. Brooke, Duffy and Rees examined the kinetics of exchange of calcium with strontium [73,74] and sodium with potassium [74-76] in chabazite using experimentally determined activity corrections derived from equilibrium isotherm data (Gaines and Thomas convention [15] - see equation 20 and Section 3.3) and they also determined experimentally the composition dependence of the self-diffusivities D_{NaNa} and D_{KK} in chabazite [75]. Consequently, they were able to examine and model rigorously the time dependent exchange rate behaviour of these systems using equation 45 rather than the approximate form given by equation 46. By employing numerical integration procedures [73] and assuming spherical symmetry (see also equation 33) to solve the diffusion equations, they were able to demonstrate only qualitative agreement between observed and predicted rates of exchange [73,74,76]. Activity corrections turned out to be *highly* significant for the Ca/Sr-chabazite system [73]; indeed, on applying the non-ideality corrections a discontinuity in the plot of $d(D_{CaSr})/dE_{Ca}$ was observed and overall D_{CaSr} varied by an order of magnitude. For the Na/K-chabazite system [75,76], non-ideality corrections had little effect on predicted rates but it proved necessary to incorporate the measured concentration dependence of the self-diffusivities D_{NaNa} and D_{KK} in order to obtain reasonable agreement between theory and experiment.

More recently, an experimental study was published which examined the relative importance of *all* the mass transfer processes which could contribute to overall exchange rate behaviour (see Sections 3 above). Exchange rates in chabazite crystal agglomerates ("Ionsiv IE-90", mean diameter ~ 24μm) were compared with the same zeolite crystallites pelletised using clay mineral binder ("Ionsiv IE-96", ranging in diameters from 48μm to 590μm) using similar experimental conditions [84]. Binary and multicomponent exchanges were studied, involving the ions Cs^+, Mg^{2+}, Ca^{2+} and Sr^{2+}, in a batch reactor of the Carberry type [91], with the envelope containing the zeolite sample under study attached to the impeller. This system ensured very effective flow around and close to the surfaces of the zeolite particles over the whole range of particle sizes employed. Controlled impeller speeds could be employed, ranging from 500 to 1000 min^{-1}, and under these conditions film mass transfer resistance was demonstrated to be insignificant [84].

Having set up the experimental conditions so that any artefactual influences from film diffusion control could be definitely excluded, Robinson, Arnold and Byers were in a position to assess the relative contributions which (macroporous) bulk liquid and surface diffusion on one hand (Section 4.2.2), and (microporous) intracrystalline diffusion on the other (Section 4.2.3), made to the overall rates observed, for both binary and multicomponent exchange processes. They were able to draw the following main conclusions [84]:

- In contrast to Ionsiv IE-90 crystal agglomerates, where a homogeneous model involving only intracrystalline diffusion described the behaviour very adequately (equation 33 with $D_{i,c} = 10^{-14} m^2 s^{-1}$), *macropore diffusion contributed significantly* to overall mass transfer resistance within the Ionsiv IE-96 chabazite pellets.
- Effective macropore diffusivities ($D_{eff,P}$) values were obtained for pelleted chabazites ($10^{-8} m^2 s^{-1}$) which were some two orders of magnitude higher than what one would expect for the bulk liquid diffusivity. This showed that *surface diffusion also contributed significantly* to the macroporous diffusion mechanism (see equation 31, $D_{i,\sigma} \sim 10^{-9} m^2 s^{-1}$).
- Both the surface diffusivity $D_{i,\sigma}$ and the intracrystalline diffusivity $D_{i,c}$ were *concentration dependent* and followed Darken's equation (see Section 4.2.3.b and equations 38 and 39).
- Generally, binary diffusivities could *not* be used to predict multicomponent diffusivities when broad simplifying assumptions were made using concentration independent effective diffusivities. This parallels to some extent the situation observed for equilibrium studies of ion exchange in zeolites, where binary activity coefficient data could not be used to predict multicomponent activity coefficients, although this does appear to be possible for clay minerals and resins [7,56].
- *Only the most rigorous heterogeneous model* (Section 4.2.2) *adequately predicted multicomponent exchange rates.* This model involved bulk liquid and surface diffusion operating in parallel, but these two macroporous diffusion mechanisms operated in series with the microporous intracrystalline diffusion process. Examples of the good fits obtained using this model are shown in Figure 11.

Although the exchange rate studies on chabazite described above employed a batch reactor, the results have implications which impact on the modelling of column exchange processes, since many studies concerned with the modelling and prediction of exchange behaviour in ion-exchange columns assume local equilibrium is attained at each stage in the process [92].

This is often a justifiable assumption; taking just one recent elegant example, significant success in predicting exchange behaviour has been demonstrated by coupling fundamental statistical thermodynamic theory with Monte Carlo techniques in order to predict ion exchange behaviour on resin columns [93]. However, the studies of Robinson, Arnold and Byers [84] demonstrate that macroporous diffusion is an important additional factor which should be taken into account *in addition* to microporous diffusion if one changes the product form from zeolite crystallites to pellets. Consequently, if for zeolite crystallite agglomerates our experimental studies of exchange rates indicate that we can rule out kinetic control when employing this product form in a column, we should *not* assume the same would be true for a column comprising pellets of the same zeolite.

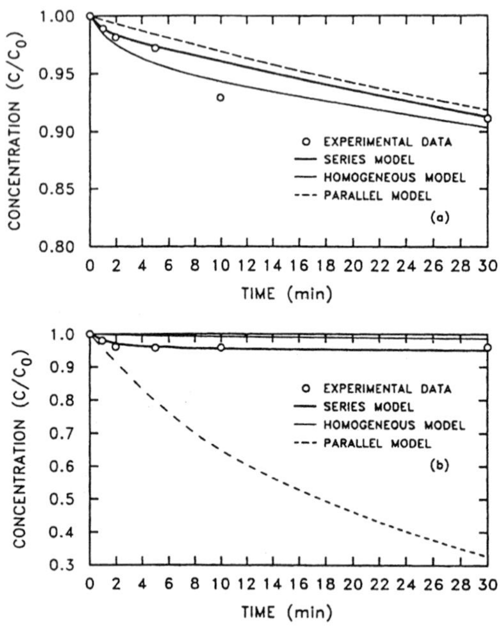

Figure 11. Experimental and predicted multi-component ion exchange kinetics in Ionsiv-96 (chabazite pellets), 24-25°C, 0.013N. Solution contains Sr^{2+}, Ca^{2+}, Mg^{2+}, Cs^{+} and Na^{+}; (a) Sr^{2+} exchange; (b) Mg^{2+} exchange. Reproduced with permission from reference 84.

4.4. Concluding comment

Bearing in mind that when studying the kinetics of ion exchange in zeolites one is dealing with non-steady state ion transport phenomena in a concentrated solid electrolyte which contains also water, it is hardly surprising that the relatively simplistic models described above often show their inadequacies when compared with experimental data. However, although the *a priori* prediction of rates of ion exchange is presently beyond our grasp, and "more fundamental studies are needed to aid in the development of models for zeolite ion-exchange systems" (*sic.*, [84]) a good general picture of the factors which influence exchange rates does now exist.

5. APPLICATIONS OF ZEOLITES IN ION EXCHANGE PROCESSES

There are several applications in which the ion exchange properties of zeolites are exploited directly. These include water softening (or "building") in detergents, wastewater treatment (including municipal, industrial, agricultural and radioactive) and animal food supplementation (*e.g.*, to regulate ammonia or ammonium levels in the stomach). Some of these direct applications of zeolitic ion exchange will be discussed below.

5.1. Detergent Building

Until concerns arose in the 1970's about its environmental impact, sodium tri-polyphosphate was the builder (or water softener) of choice for detergent formulations. Increased levels of phosphates in natural waters were leading to eutrophication, that is, the phosphate acted as a fertiliser and promoted the growth of algae, which eventually choked the waterways and depleted the amount of oxygen dissolved in the water, thus killing indigenous flora and fauna. The contribution of detergent-based phosphate was quite small, however, compared to that arising from agricultural and industrial sources.

Sodium tri-polyphosphate forms strong complexes with Ca^{2+} and Mg^{2+} ions in solution, preventing their precipitation with the surfactant or as carbonates (reducing deposition of solids onto clothing and avoiding loss of detergent surfactant). Zeolites possessing selectivity for calcium and/or magnesium over sodium are obvious candidates to replace phosphates; indeed, legislative and environmental pressures favour the use of zeolites over phosphates in detergents around the world [94,95]. The principal obstacles to ion exchange of Ca^{2+} and Mg^{2+} are the large hydration shells surrounding the ions. The hydrated cations are too large to enter the zeolite pores without shedding a number of their associated water molecules. Because of its higher charge density, the Mg^{2+} ion, bare radius 72 pm, has a larger and more tightly held hydration sphere than Ca^{2+}, bare radius 112 pm. In addition, the *rate* of dissociation of water molecules from the hydration shells is slower for Mg^{2+} than Ca^{2+}; consequently, the removal of Mg^{2+} from solution by a zeolite is typically less efficient than that of Ca^{2+}. While most zeolite types, both naturally occurring and synthetic, have been tested for their water softening capabilities, zeolite A has been mainly used (high calcium selectivity and kinetics), sometimes in conjunction with zeolite X (which removes magnesium more effectively than zeolite A). It is estimated [96] that approximately 800,000 tonnes of zeolite A are used currently in laundry detergents per annum. Recently, a P-type zeolite, possessing a Si:Al ratio of 1:1 (Maximum Aluminium P, or MAP) has been developed and introduced into detergent formulations [96]. Section 7.3 provides a detailed account of the properties of zeolites A and MAP in water softening applications.

Franklin and Townsend [97] re-examined binary exchange equilibria involving the ion pairs Na/Ca, Na/Mg and Ca/Mg in zeolite A. They then constructed the first complete ternary isotherm for this system (Figure 12). The binary exchange experiments clarified ambiguities previously seen regarding the maximum level of magnesium exchange which can occur in zeolite A and also demonstrated the huge preference that this material exhibits for calcium over magnesium when only these two ions are present [97]. They also attempted to reconcile discrepancies in earlier work which appeared to be in part due to experimental error, but were also probably due to hydrolysis occurring, with the intervention of oxonium ions (see also [34,35,98]). Subsequently, Franklin and Townsend used these experimental data [97] to predict successfully (Figure 13) both binary and ternary exchange equilibria for this system

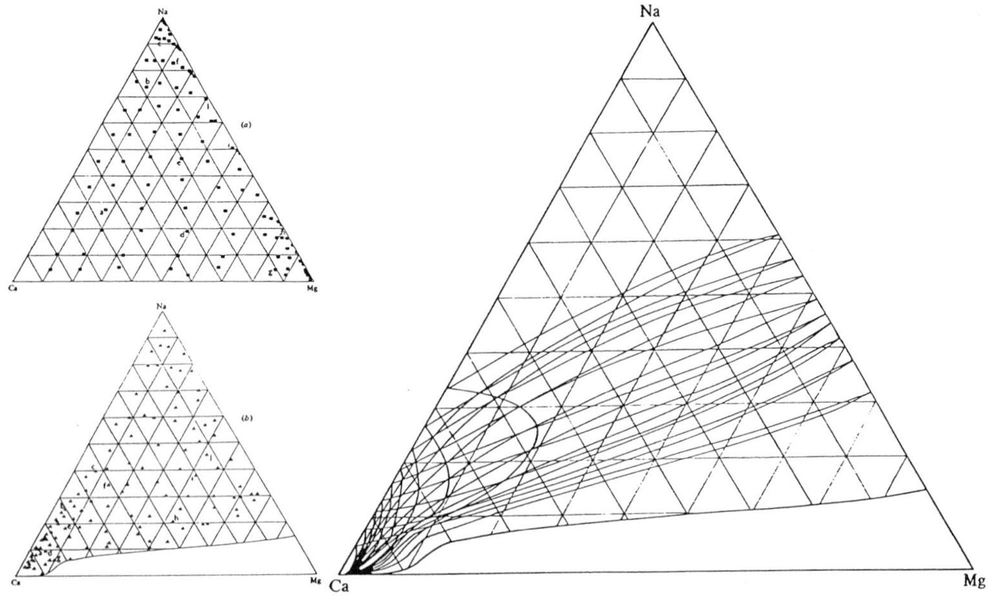

Figure 12. Ternary exchange isotherm for the 2Na/Ca/Mg-zeolite A system, 25°C, 0.1N. For ternary exchange reactions the isotherm comprises a surface as opposed to a curve which is seen for binary exchanges (*cf.* Figures 1-5). Top left: solution composition data; bottom left: corresponding solid phase data. In the ternary diagram on the right, the solution phase grid has been distorted such that all solution compositions fall on top of the corresponding solid phase equilibrium compositions (un-distorted grid). Reproduced with permission from reference 97.

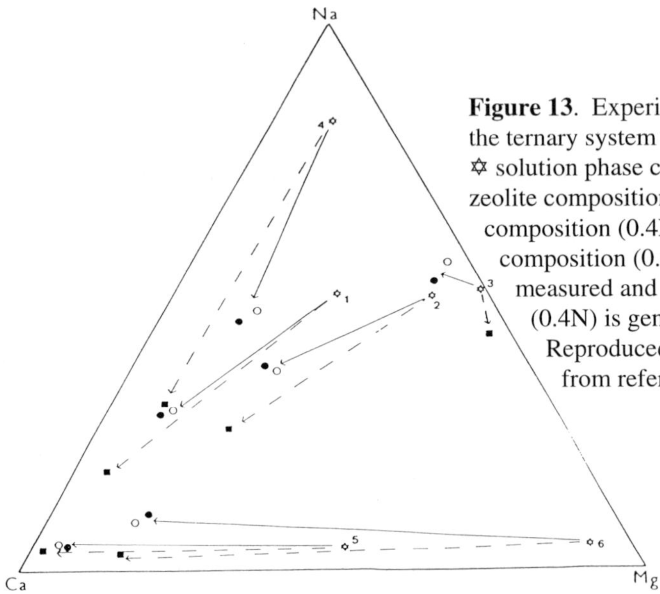

Figure 13. Experimental and predicted data for the ternary system 2Na/Ca/Mg-zeolite A, 25°C. ✡ solution phase composition; ■ measured zeolite composition (0.1N); ● measured zeolite composition (0.4N); ○ predicted zeolite composition (0.4N). Agreement between measured and predicted compositions (0.4N) is generally good. Reproduced with permission from reference 58.

over a range of concentrations [58]. At about the same time as this work [58,97] another study on the prediction of ternary compositions for the Na/Ca/Mg-A system was published [99]. These predictions were carried out by evaluating Wilson interaction parameters [7] for the binary conjugate systems, then using these to predict the ternary equilibrium compositions [99].

Detailed studies of the kinetics of ion exchange systems relevant to detergency have been relatively few. The only major one was by Drummond, DeJonge and Rees [98]. They used an earlier theoretical approach of Danes and Wolf [100], and demonstrated the inhibiting effect on the rate of calcium exchange which magnesium may exert (Figure 14) as well as the need to take into account sodium/oxonium exchange and possible hydrolysis of the zeolite [98].

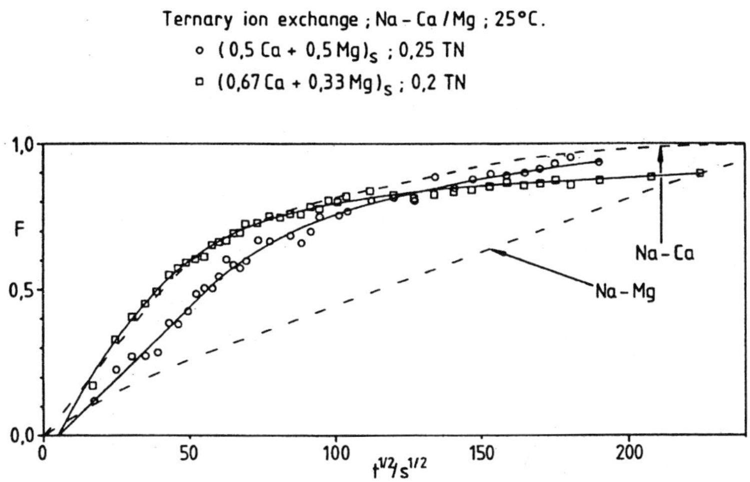

Figure 14. Ternary exchange kinetics for 2Na/Ca/Mg in zeolite A, 25°C (datum points and solid curves). The degree of exchange, F, is the total amount of Na replaced by *either* divalent cation. As the amount of Mg in solution increases, the overall exchange rate decreases. Binary data are included for comparison (broken curves). Reproduced with permission from reference 98.

In Section 4, it was noted that particle controlled diffusion is normally the significant mechanism in ion exchange kinetics involving zeolites, but that as one reduced the crystallite size, the consequent increase in rate which will occur will reach a limit as film-controlled diffusion takes over. In the case of self diffusion, there is a simple relationship between the rate of exchange and the characteristic dimension of the crystal (*viz.* the rate is inversely proportional to the square of the radius of the crystal). For exchange between two different types of ions the relationship is not so straightforward, but nevertheless decreasing crystallite size will markedly increase the rate. The significance of this is seen for a country such as Japan, where typical washing conditions place stringent demands on the zeolite compared to those required in either the USA or Europe [95]. Essentially, a typical Japanese wash involves short washing cycles and cold water washings (*e.g.*, 5°C). In order to boost performance, therefore, a major aim for the Japanese market is to produce crystallites of very small size, and one way in which this problem is being overcome is to prepare zeolite A from acid-treated clay precursors [95].

It has been emphasised that crystallite size can be important in another respect, namely, when one is designing a liquid detergent [101]. The concern here is not primarily the rate of exchange, but rather the rate of sedimentation of the solid zeolite builder within the liquid on standing. Unfortunately, the solution is not just to reduce the average crystallite size, because below a certain average crystallite size the rheological properties become unsuitable. Clearly, in this situation, it is important to add other ingredients to the liquid mixture which may modify the interaction between the liquid and the zeolite [101].

Ideally, a detergent builder should be able to remove calcium and magnesium in two ways. The first of these is the removal of the ions from solution, as discussed above; here zeolite A is effective for control of calcium levels [102]. With regard the second aspect, the situation is different. This is where calcium, in particular, is bound with soil on the fabric itself and is helping the soil to adhere to the surface of the textile. Zeolites are, as a rule, not effective in removing such deposits, and "co-builders" are commonly used to boost the zeolite performance. Co-builders are usually present in the form of polycarboxylates [94,103] or sequestrants [104]. Despite the necessity for co-builders, zeolite-containing detergent formulations have been claimed to be better than the traditional phosphate systems in overall performance because of certain other advantages, such as decreased overall wear-and-tear of the textiles and of the washing machine itself [94]. In addition, extensive testing has revealed no deleterious effects of the introduction of zeolite A into natural waterways. In fact, it has been shown that when zeolite A decomposes in natural waterways, the by-products form stable phosphate compounds, thus helping to remove phosphates from the environment [105].

5.1.1. Criteria for an Effective Zeolite Builder

Leaving aside obvious criteria, such as that the zeolite should be environmentally acceptable itself and of a reasonable cost, the most important requirement for a successful zeolite builder is its *capacity*. If this statement seems over-simplistic, consider further what is required in terms of capacity for a zeolite builder. Obviously the capacity should be the maximum possible. In terms of zeolites this means that the silicon to aluminium ratio of the framework should be as near as possible to unity. However, this in itself is *not* sufficient, because it is also required that the zeolite be selective for calcium and (if possible) magnesium. This is because, as explained above (Sections 2 and 3) the effective capacity is not simply the overall capacity of the exchanger as determined by its silicon to aluminium ratio, but rather is also a function of the selectivity of the material for calcium and/or magnesium (*i.e.,* the shape of the isotherm and whether partial exchange limitations are observed). Furthermore, when it is necessary to remove simultaneously both calcium and magnesium, the *relative* selectivity of the zeolite for these two ions is of importance, (*i.e.,* we have to concern ourselves with at least a ternary exchange equilibrium).

There is another factor which also has to be taken into account. This is the kinetic factor. It is not sufficient just that the zeolite has a high intrinsic exchange capacity and that it be selective for one or both of calcium and magnesium, but also the rate at which these ions are removed from solution must be sufficiently high to ensure their removal early on in the washing cycle. Thus, in conclusion, the "capacity" of a zeolite builder is a complicated function of the framework composition of the zeolite, the selectivity manifested for the hardness ions, the levels to which those ions can exchange into the material and the rate at which they are removed.

It is hardly surprising therefore, that a considerable amount of effort has gone into examining the equilibrium and kinetic aspects of calcium and magnesium exchange into sodium zeolites. Because zeolite A has been used traditionally as the detergent builder, most of this effort in the past has obviously concentrated on this material.

5.1.2. Developments

By far the most significant recent development in the field of zeolite builders has been the introduction of zeolite MAP, which exhibits superior performance to zeolite A. Section 7.3 will discuss this issue in depth.

Another alternative to zeolite A is the amorphous, or semi-crystalline precursor gel, which is formed during the crystallisation of zeolite A or X. Studies have shown that these materials may show better selectivity to Ca^{2+} and Mg^{2+} than their crystalline counterparts [106-108]. Kinetic studies have indicated that, although exchange may be slow, the precursor materials partially eliminate the difference in exchange rates for Ca^{2+} and Mg^{2+} of an order of magnitude seen in crystalline zeolites; Ca^{2+} and Mg^{2+} ions exchange into the precursors at roughly comparable rates [108,109].

In the search for builders effective for Mg^{2+}, a further improvement of Mg^{2+} exchange rates compared to those of Ca^{2+} may be realised by moving to porous materials possessing increased electric field gradients. While beryllium compounds cannot be used in any commercial product for toxicity reasons, a microporous beryllophosphate material with the structure of zeolite P serves as an example to demonstrate the possibilities of utilising such materials. The high electric field gradient (due to the difference in valency of the T-atoms of three units, as opposed to one unit in aluminosilicate zeolites) leads to an exchange rate *three times faster for Mg^{2+} than for Ca^{2+}* [110].

The reader should be reminded that not only zeolitic materials are candidates for detergent builder applications. Various silicates have also been tested. For instance, the crystalline layered silicate SKS-6 ($Na_2Si_2O_5.xH_2O$) is worthy of mention in this section. It has been proposed as an alternative detergent builder [111], and demonstrates promising performance for Mg^{2+}, while it is less efficient for Ca^{2+} [112]. SKS-6 acts predominantly as a precipitating builder, as it is quite soluble in water, while its calcium and particularly magnesium salts are less soluble. Solution state ^{29}Si NMR could not detect any interaction between soluble silicate species and Ca^{2+} or Mg^{2+}, suggesting that SKS-6 does not behave as a sequestering builder [112].

5.2. Radionuclide Separation

Zeolites, as well as other materials with ion exchange properties, are commonly used to remove certain radio-nuclei from low- and medium-level nuclear waste. Natural zeolites are of great interest due to their (sometimes) high abundance and low cost. A large number of publications describe the use of natural zeolites for the treatment of radioactive waste [113,114]. In this section, the basic principles are discussed with relation to some relatively recent literature.

The principal radioactive components of nuclear waste are typically $^{90}Sr^{2+}$ and $^{137}Cs^{+}$; removal of these nuclides from effluents which may contain significant quantities of competing ions, such as Ca^{2+}, Mg^{2+}, Na^{+} and a range of anions is of particular interest. The composition of waste to be treated varies considerably, depending upon its source. For instance, nuclear "pond water", which is the water in which spent fuel elements are stored for long periods to allow them to cool (both thermally and through radioactive decay), is typically alkaline, and very dilute in radio-nuclides; Table 1 lists a typical pond water composition. Wastes of higher concentration may contain, *e.g.*, 10^{-4} mol dm^{-3} Cs^{+} in a matrix of 4.5 mol dm^{-3} with respect to sodium [115], thus highly selective exchangers are needed. Much nuclear waste is acidic due to the method of recycling spent fuel rods from nuclear power stations, which is to dissolve them in concentrated HNO_3. Stability of the zeolites under these conditions is an issue, particularly with samples of low Si:Al ratio (*i.e.*, high exchange capacity). Clinoptilolite and chabazite are often used for radioactive waste treatment [116,117], as they are abundant natural minerals, with good ion exchange selectivities for the common radionuclides. Natural phillipsite has also found applications [118], while natural laumontite in its calcium form has a high selectivity for $^{90}Sr^{2+}$ [119]. The synthetic zeolites 4A and 13X have been demonstrated to be useful for the removal of $^{90}Sr^{2+}$ from simulated low-level waste [120], while a synthetic mordenite effectively removed $^{137}Cs^{+}$ from simulated and real wastes [121]. The commercial zeolites Ionsiv IE-96 and A-51 (from Union Carbide, now UOP) were used effectively for the decontamination of high-level wastewaters generated in the reactor during the Three Mile Island accident [114].

Table 1: Typical composition of nuclear pond water, *p*H 11.5 (only ions competing for exchange are listed) [114]

Ion	Concentration / μg dm^{-3}
Na^{+}	100,000
Ca^{2+}	1500
Mg^{2+}	700
Cs^{+}	17
Sr^{2+}	0.52

The speciation of many of the radionuclides in nuclear waste is complex and usually varies considerably with *p*H, hence the ion exchange properties of the zeolites may be difficult to predict. Indeed, in many instances, ion exchange is not the major mechanism for removal of radionuclides; the zeolite crystals may act as substrates for the electrostatic surface interaction of ionic species, or for the precipitation of neutral ones. Depending upon the *p*H, temperature, *etc.*, the radionuclides may exist in solution as cationic, neutral, or even anionic species [122]. Olguin *et al.* [122] reported the sorption of a number of nuclei which are products of the fission of ^{235}U and ^{239}Np on zeolites Y and natural erionite and on the clay mineral bentonite. Only in a few cases, and under certain *p*H conditions, was ion exchange itself believed to be involved in the removal of fission products by the zeolites. Considering the removal of uranium using clinoptilolite, ion exchange occurs only under certain conditions where UO_2^{2+} exists, while adsorption of uranium species onto the crystal surface at near-neutral *p*H is a major factor in the removal of uranium-containing species from solution [123].

Due to the complex composition of most nuclear waste streams, the effectiveness of a particular zeolite in treating nuclear waste may not be predictable from binary exchange data, or trials with simulated waste. Laboratory scale testing using actual nuclear waste is often the only way to obtain a reliable prediction of a zeolite's performance when applied.

In some cases, zeolites are used to concentrate certain nuclides for recycling, *i.e.*, once the desired nuclide has exchanged into the solid phase, the zeolite is treated with a concentrated salt solution to elute the radionuclide, and the zeolite is re-used (regenerative zeolite bed). In others, the radionuclides are removed from the environment and permanently stored in the zeolite (non-regenerative zeolite bed). In non-regenerative processes, once the radioactive species have been immobilised in, or on the zeolite, the possibility of transport of the contamination back into the environment has to be reduced. The contaminated zeolites are often solidified with concrete or glass, and when possible densified to sodalite, thus preventing the contamination from leaching out [124]. This is one area where zeolites offer a distinct advantage over ion exchange resins, since the latter are not very compatible with the concrete and glass matrices used to solidify and contain the ion-exchanger phase once radioactive. Another advantage of zeolites is their stability to radiation, for instance, it has been demonstrated the clinoptilolite may endure up to 10^8 Greys (100 MGy) of radiation without any noticeable change in structure or ion exchange properties [125].

In addition to their use in treating radioactive wastewaters, zeolites have been used to decontaminate animals, particularly sheep and reindeer, which have ingested radiation following nuclear disasters, such as that at Chernobyl. Addition of zeolites to contaminated soils has been shown to reduce uptake of radionuclides by plants, and hence, reduce the spreading of contamination through the food chain.

5.3. Wastewater Treatment

Zeolites are used commonly to remove ammonia and ammonium ions from municipal and agricultural wastewater [115]. Where deposits of natural zeolites, especially clinoptilolite, chabazite, mordenite and phillipsite, are abundant, they have been used extensively for this purpose. Clinoptilolite, in particular, shows good selectivity for NH_4^+, while a number of synthetic zeolites have also been tested [126-128], with zeolites F and W showing higher exchange capacities for ammonium than the natural materials. During the testing of zeolites for the treatment of nuclear waste (Section 5.2) it has been found that those which exhibit high selectivity for caesium were usually also highly selective for ammonium [126]. The build up of toxic levels of ammonia and ammonium in fish breeding tanks is avoided through the use of clinoptilolite in some fish farms [129].

In order to remove the ammonia and ammonium, the wastewater is commonly passed through a bed or column of zeolite; this is often more practicable with natural zeolites, as their physical form tends to be more granular, and thus more suitable for columnar use than synthetic samples. Once saturated with ammonium ions, the zeolites may be regenerated or used directly as fertiliser, where they slowly release ammonium ions into the soil. Regeneration involves elution of NH_4^+ from the zeolite by a competing ion. Treatment of the resulting regenerant solution then becomes an issue; one example being the conversion of NH_4^+ to NH_3 by raising the *p*H of the solution, counter-flowing air through a column to strip off the ammonia and subsequent conversion of this into nitrogen in the presence of chlorine [115,126]

Heavy metal cations, transition metal ions and various trace elements may be removed from wastewaters using zeolites [127,130], but these applications are far less common than the removal of ammonia and ammonium. Depending upon the conditions during treatment, removal of these other species may in fact occur through precipitation rather than ion exchange, as was discussed in Section 5.2.

6. ION EXCHANGE AS A METHOD OF MODIFICATION

The technique of ion exchange is frequently used to alter the properties of zeolites, *e.g.*, to prepare acidic or basic catalysts, to tailor the pore size for specific adsorption processes and the introduction of specific adsorption sites. A number of uses of ion exchange as a technique of modification will be outlined below.

6.1. Preparation of Catalysts

Besides obvious advantages such as shape selectivity and recyclability, environmental considerations are one of the major driving forces for the replacement of homogeneous catalysts by zeolites in industrial processes [131,132]. Homogeneous catalysts are often difficult to separate from product streams, non-reusable, and result in the generation of large volumes of waste materials.

The most common application of ion exchange here is the preparation of ammonium forms of zeolites *en route* to the generation of acidic catalysts. Upon calcination of the ammonium zeolite, ammonia is driven off, leaving behind the acidic form of the zeolite, where protons reside at the ion exchange sites. It is important to note, however, that highly aluminous zeolites, such as types A and X, cannot be transformed into acidic forms in this way; upon calcination of their ammonium-exchanged forms, the structures of zeolites A and X collapse [133-135]. Alternatively, acidic zeolites may be generated in some cases by direct ion exchange in an acidic solution.

In addition to the ubiquitous H-form, other cationic forms of zeolites may be employed as catalysts. Common examples are rare-earth (RE) exchanged zeolites, such as RE-Y, used in alkylation reactions (particularly for octane enhancement in the petroleum industry).

The discussion above has considered acidic catalysts; basic zeolite catalysts also find applications [131]. Alkali metal-containing zeolites, particularly caesium exchanged (or impregnated) ones, have been studied for their basic activity for reactions such as double-bond isomerisation [136], ethoxylation [131] and alcohol decomposition [137].

6.2. Preparation of Sorbents

The shelves of almost every chemical laboratory house bottles of Molecular Sieve 3A, 4A or 5A, used principally for solvent drying. These are all type A zeolite, possessing different charge-balancing cations which tune the effective pore size to approximately that given in its name, *i.e.*, 3A is potassium-exchanged A with a pore size of 0.33 nm, 4A is the sodium form, with a pore size of 0.41 nm and 5A is calcium-exchanged, with a nominal 0.44 nm pore opening. The strong influence which the exchangeable cations have on the effective pore size is due to the particular positions which the cations adopt within the structure [138].

Potassium ions, for instance, prefer positions close to the 8-membered ring pore opening, resulting in a restriction of the effective window diameter. Changes in effective pore size upon cation exchange may be enhanced also by framework distortions.

Calcium, sodium and lithium forms of zeolites A and X have been employed in the non-cryogenic separation of N_2 and O_2 from air [139]. In the case of lithium zeolites, striking dependencies of N_2 capacity and N_2/O_2 selectivity on the degree of Li^+-exchange have been observed. Interestingly, the capacities and selectivities of LiNaX do not vary with Li^+ content in a linear fashion, rather they remain effectively constant at low Li^+ exchange levels, then increase when a certain threshold Li^+ content is reached (Figure 15). It has been demonstrated that the increase in capacity and selectivity for N_2 occurs when Li^+ begins to occupy site III in the FAU structure. Further improvements have been made by replacing residual Na^+ with certain divalent cations, such as Ca^{2+}, Sr^{2+}, Mg^{2+}, Zn^{2+}, Mn^{2+} and Ni^{2+}.

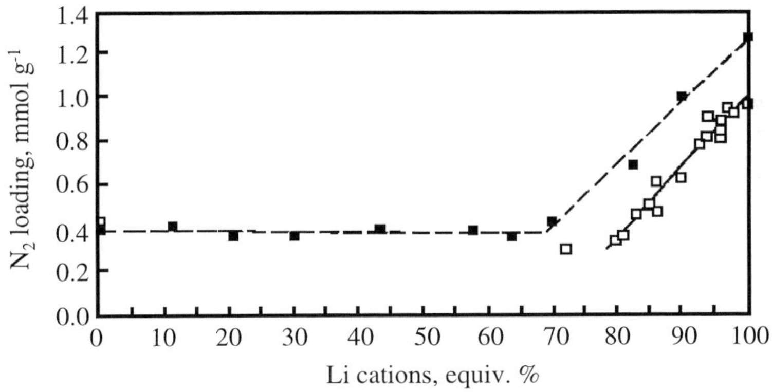

Figure 15. Nitrogen capacity of (Li,Na)-X zeolites as a function of Li^+-exchange level. □ : Si/Al = 1.25; ■ : Si/Al = 1.0 ("low-silica X"). Reproduced with permission from reference 139.

A further example of the utility of zeolite adsorbents is the strong adsorption of unsaturated hydrocarbons onto transition metal-, and especially silver-exchanged zeolites. The resulting complexes are stable, and the structures of a number of these materials have been determined by single-crystal X-ray techniques [140,141]. An example of the commercial utilisation of such adsorbents is in the storage of perishables. Ripening bananas and other fruit give off ethylene, which accelerates the ripening process; in order to increase shelf life, they may be stored in the presence of a silver-exchanged zeolite [142].

7. RECENT TRENDS IN ION EXCHANGE

In the following sections, a few areas which have developed significantly since the first edition of this book will be discussed. This account is far from exhaustive, but three main areas will be focused on: ion exchange studies of high silica zeolites, ion exchange in the solid state, and ion exchange properties of the new detergent zeolite MAP.

7.1. Ion Exchange in High-Silica Zeolites

For most applications of ion exchange, low silica zeolites such as zeolites A and X are used due to their high exchange capacities. Such zeolites possess high ion exchange site densities, relatively high electric field gradients and large numbers of water molecules. When one considers high-silica zeolites, however, the exchange site density, electric field gradient and degree of hydration are much lower. While the distance between exchange sites is usually of little importance for univalent ions, multivalent ions may be unable to satisfy multiple sites which are beyond a certain distance apart. Cation hydration effects become more evident in ion exchange involving high-silica zeolites. The presence, or absence, of water within the zeolite can have a profound effect upon the "bridging distance" of a multivalent cation, while the preference of cations of high charge density for the solution phase (where their need for hydration can be met) is evident. A further consideration when investigating ion exchange in high-silica zeolites is the contribution of defect sites (SiO^-) and framework-terminating SiOH groups to the ion exchange capacity. For low silica zeolites, this contribution is negligible compared to the capacity of the intrinsic ion exchange $(AlOSi)^-$ sites. A few studies have been made which demonstrate the interesting ion exchange properties of high-silica zeolites and these are summarised below.

Matthews and Rees reported isotherms for the exchange of Ca^{2+} , Sr^{2+}, Ba^{2+}, Fe^{2+} and La^{3+} as well as a number of univalent ions into Na-ZSM-5 with 2.4 Al atoms per unit cell [143]. While each univalent ion achieved 100% exchange, the multivalent cations achieved only limited maximal exchange. Ion sieving was not the reason, as the larger Cs^+ ion was able to replace all Na^+ ions. The degree of exchange attainable could be correlated to the degree of hydration of the ions; the higher the charge density of the ion, the larger its hydration sphere, and hence the greater the shielding of the charge. With highly shielded charges, only pairs or triplets of Al atoms sufficiently close together could be satisfied by the cation. Thus the observed maximal exchange cut-offs increased in the order $La^{3+} \approx Ca^{2+} < Sr^{2+} < Ba^{2+}$. For a given multivalent ion, increasing the temperature resulted in an increase in the maximal exchange limit, probably due to more facile dehydration of the cation. This is another example of how particular zeolite properties can give rise to specific, sometimes surprising, ion exchange behaviour (*cf.* Section 2.1).

McAleer, Rees and Nowak studied these ion exchange effects in greater detail, and modelled the distribution of aluminium atoms within the framework [144]. The exchanges of K^+, Ca^{2+}, Sr^{2+} and Ba^{2+} into three Na-ZSM-5 samples, with varying Si:Al ratios (corresponding to 1.1, 2.0 and 4.2 Al atoms per unit cell) were studied. As expected, K^+ was able to replace all sodium ions, while the divalent ions did not achieve 100% exchange under the conditions employed. For the most siliceous (and thus, least hydrated) sample, the maximal degree of exchange varied little with the cation, while the lesser siliceous samples showed a clear increase in ion exchange cut-off as the cation radius increased. These observations are consistent with the notion that the degree of hydration of a cation determines

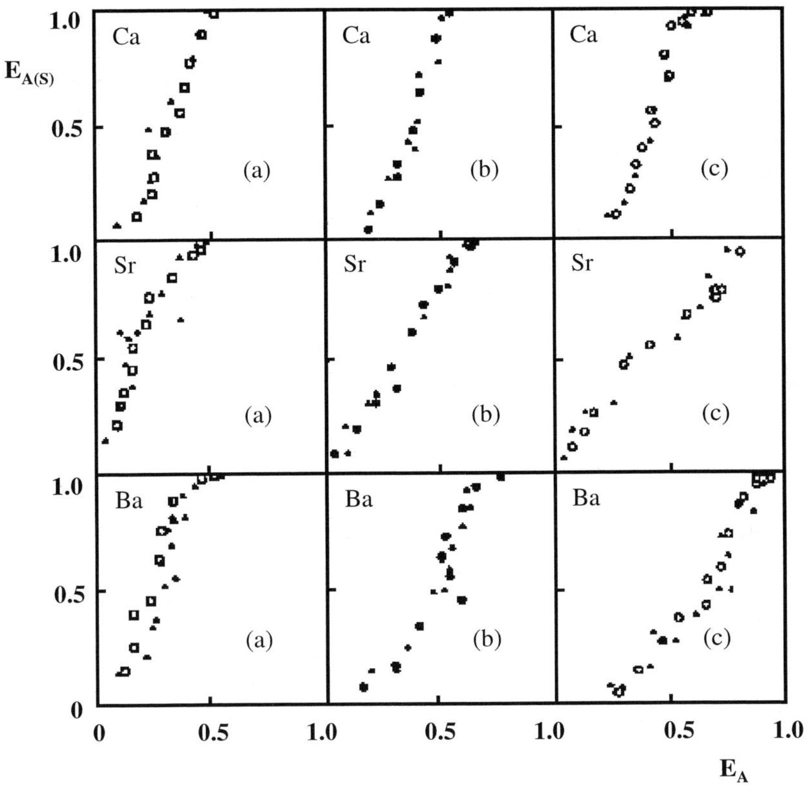

Figure 16. 2Na/divalent cation exchange in ZSM-5 with various Si/Al ratios, 65°C. The ingoing cation is indicated in the top left-hand corner of each isotherm. (a): 1.1 Al per u.c.; (b): 2.0 Al per u.c.; (c): 4.2 Al per u.c. Reproduced with permission from reference 144.

7.2. Solid State Ion Exchange

The first reports of solid state ion exchange in zeolites appeared in the early 1970's [148,149], but the phenomenon did not attract much attention until the mid 1980's. Solid state ion exchange is a useful route to the preparation of zeolite catalysts, and has recently been reviewed in detail [150]. As its name implies, solid state ion exchange involves the exchange of ions between two solids, with little or no water present. The two solids may be a zeolite and a metal salt, or two zeolite samples. Alkaline, alkaline-earth, rare-earth, transition and noble metal cations have all been successfully introduced into zeolite channels and cavities via solid state reactions. Only details essential to the understanding of the basis of solid state ion exchange will be given here; the interested reader should refer to the review [150] for further experimental procedures and results.

The solid state ion exchange reactions between stoichiometric quantities of a metal salt and a zeolite in its sodium-, or hydrogen-form are represented in equations 47 and 48 respectively.

$$MCl_n + nNaZ = xMZ_n + (1-x)MCl_n + (n-nx)NaZ + nxNaCl = MZ_n + nNaCl \qquad (47)$$

$$MCl_n + nHZ = xMZ_n + (1-x)MCl_n + (n-nx)HZ + nxHCl\uparrow = MZ_n + nHCl\uparrow \qquad (48)$$

Where M is an n-valent metal cation and Z is a mono-valent zeolite fragment (ion exchange site).

The usual method of preparing a catalyst via solid state ion exchange is to prepare an intimate mixture of the zeolite and metal salt containing the entering ion. The mixture may be prepared by either intense mechanical mixing (milling), or by suspending both components in a volatile solvent, mixing thoroughly and allowing the solvent to evaporate. This is followed by thermal treatment, typically at 300 to 400°C under high vacuum, or in a stream of inert gas. A few hours are usually required for maximum exchange.

Performing ion exchange reactions in the solid state may be advantageous to traditional solution mediated exchange in a number of cases. For instance, cation hydration effects, which are often restrictive in ion exchange reactions in solution are avoided in the solid state, thus allowing easier access of ions into the channels and cavities of zeolites. This advantage of solid state ion exchange is most clearly seen in cases where entering ions of high charge density, or relatively small-pore zeolites are concerned. The exchange reaction may be taken to completion in one step if a few simple conditions are met. Starting with a zeolite in its H^+ or NH_4^+ form, and given stoichiometric quantities of zeolite and metal salt, volatilisation of the leaving ion (H^+) as, say, HCl under the conditions of the exchange drives the exchange to completion (equation 48). In addition, sparingly soluble, or insoluble metal salts may be used in solid state ion exchange, while they obviously would be poor choices for traditional solution-mediated ion exchange. Solid state ion exchange has thus made available a wider range of metal-containing zeolites. From a practical point of view, performing an ion exchange reaction in the solid state avoids the handling and disposal of large quantities of electrolyte solution and cuts down on the costs of wasted chemicals, as well as the associated environmental impact.

The direct exchange of ions between two zeolites has been reported by Fyfe *et al.* [10]. For instance, on gently mixing Li-A and Na-A, the two phases could be distinguished by XRD and MAS NMR, however, after only 10 seconds of vigorous mixing, an equilibrium phase, containing only NaLi-A was obtained (Figure 17).

Although the presence of water vapour may be instrumental in certain solid state reactions [151], it has been shown that some reactions may proceed in the (near) total absence of water [152].

The thermodynamics, kinetics and mechanisms of solid state ion exchange are presently not well understood, although some detailed studies have recently appeared which try to address these points [153-155].

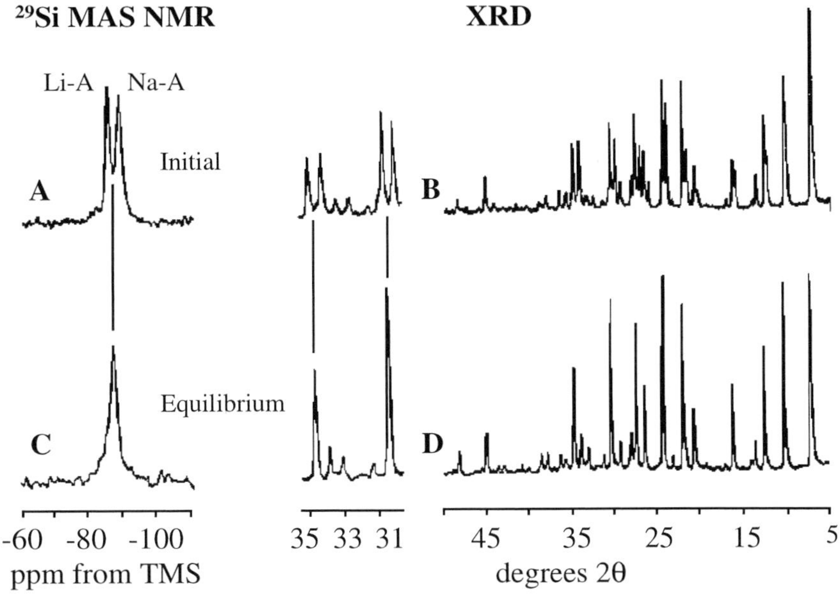

Figure 17. ^{29}Si MAS NMR spectra (A,C) and XRD powder patterns (B,D) of a mixture of LiA and NaA immediately after mixing (A,B) and at equilibrium (C,D). Reproduced with permission from reference 10.

7.3. Case Study: Zeolite A *versus* MAP

Recently, studies on the properties of a P-type zeolite (MAP), claimed to be more effective than A for use as a detergent builder, were published [96]. The development and commercialisation of this material exemplifies many of the issues touched on earlier in this chapter, including the importance and utility of prediction procedures (Section 3.2), the influence of framework flexibility on structural changes induced by exchange (Section 2.1.3) and how crystallite size, morphology and aggregation behaviour affect the exchange characteristics (Section 4.2 *et sequens*). Above all, one sees how the subtle interplay between the kinetic and equilibrium ion-exchange characteristics of the zeolite combine to produce a selectivity behaviour which makes zeolite MAP a better detergent builder than zeolite A [96].

MAP is a zeolite belonging to the P family, whose members are characterised by the gismondine (GIS) framework [156]; the idealised unit cell content of these materials is $Na_xAl_xSi_{16-x}O_{32}.yH_2O$ [157]. The Löwenstein rule [158], which precludes the existence of Al-O-Al bridges within a zeolite framework under normal conditions for energetic reasons [2], sets therefore for the P family a lower limit of eight framework silicons per unit cell, which corresponds to a 1:1 Si:Al framework ratio, and this composition is of course also the one with the maximum possible aluminium content. MAP corresponds to this composition, and is therefore an end-member of the zeolite P family in compositional terms. Under synthesis conditions commonly employed [157,159] it is normal to obtain more siliceous P materials than MAP.

Since the ion exchange capacity is set by the quantity of aluminiums substituted into the framework (Section 1), it also follows that MAP has the highest exchange capacity of any of the members of the GIS structure type family. It is therefore in an analogous position to that in which zeolite A is found within the family having the LTA structure type [156]. Zeolite MAP also shares, in common with A, a strictly alternating Si:Al framework ordering [96].

We noted earlier (Section 5.1) that kinetic behaviour during ion exchange is one key factor for assessing satisfactory performance of a candidate zeolite as a detergent builder. Bearing in mind that intracrystalline exchange proceeds by an activated diffusion process within zeolites (Section 4.2.3), a comparison of channel sizes alone [156] for the two zeolites would lead one to expect that provided average particle sizes for the two materials are similar, then in terms of kinetic performance at least A should be superior to MAP as a detergent builder since not only are the channel sizes in A somewhat larger than in MAP but the channel connectivity is higher. Indeed, this has been claimed [160]. However, this is to neglect the fact that MAP can be routinely prepared with average crystallite agglomerate sizes of $<1 \mu m$ [96], which tips the balance in favour of MAP, even in terms of kinetic behaviour. This demonstrates aptly the importance, emphasised earlier (Sections 4.1 to 4.3) of looking at *all* the transport processes within the zeolite agglomerate rather than just at intracrystalline exchange.

However, kinetic behaviour is only one term in the equation. The *preference* displayed by the zeolite for calcium (and magnesium) over sodium as a function of composition is another (Section 3.1.1), and it is here that another property of the P zeolites, alluded to by Meier and Olson [156], proves to be of considerable importance. This is the unusual flexibility of the GIS framework [96,157], which means that relatively small changes in the chemistry induced by ion exchange [27,96,159], or in the level of adsorbed species [96,159,161], can change the conformation of the framework, resulting in clearly discernible different phases. The subtleties of these changes, coupled with the difficulties of unambiguously assigning structures from X-ray studies for a variety of reasons [157] has led to considerable confusion in the literature. However, it now appears to be established that for the sodium-exchanged form at least three phases exist as a function of silicon content [157], *viz*:

Designation	NaP1 (or "P_c" [157])	NaP2	High-silica P
Si atoms per u.c.	8 - 10	10 - 12	12 - 13
Symmetry	Not yet established	Orthorhombic	Tetragonal

The situation concerning P1 is of particular interest. Symmetry considerations [157] would lead one to expect NaP1 to be monoclinic, as is gismondine ($Ca_4Al_8Si_8O_{32}$) itself [156] but the assignment of the correct symmetry, or indeed "symmetries" [*sic.*, 157] is still ambiguous, especially for materials with compositions identical, or closely identical, to MAP [157]. This is a testimony to the great flexibility of the framework, so that only small changes in chemical composition induce marked structural distortions of the framework.

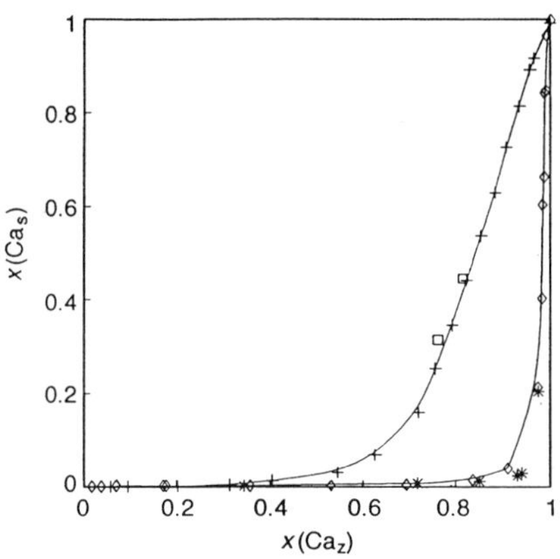

Figure 18. 2Na/Ca exchange in zeolite MAP (◊,✳) and in zeolite A (+,□), 25°C, 0.1N. (◊,+) Forward exchange; (✳,□) reverse exchange. The higher selectivity exhibited by zeolite MAP for Ca^{2+} at $x(Ca_z)$ > 0.3 is clearly seen. Reproduced with permission from reference 161.

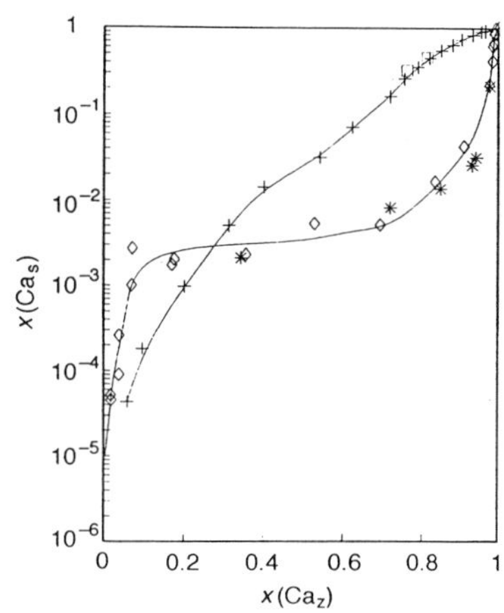

Figure 19. Semi-logarithmic plot of data from Figure 18. The higher selectivity of zeolite A for Ca^{2+} at $x(Ca_z)$ < 0.3 can be seen. Symbols as in Figure 18. Reproduced with permission from reference 161.

The importance of this flexibility phenomenon has been demonstrated recently in detail [161]. Ion exchange isotherms were constructed for the Na/Ca-MAP system and these were compared with ones previously published by Franklin and Townsend [58,97] on the Na/Ca-A system. MAP generally showed a much greater preference for calcium over sodium (Figure 18) but a semi-logarithmic representation showed that "at low calcium loadings, although in both zeolites the selectivity for calcium is high, zeolite A is more selective for Ca than is zeolite MAP" [*sic.*, 161]. In fact, a plateau region was observed on the isotherm for MAP

(Figure 19), characteristic of a composition region where two solid phases were co-existing (see the discussion on this phenomenon in Section 2.1.3). The overall much greater preference displayed by MAP for calcium over sodium is seen therefore to arise from the ready induction of a calcium-rich phase through co-operative binding of calcium ions within the zeolite [161]. As described in Section 2.1.3, the mechanism by which this clustering gives rise to new phases was described in statistical thermodynamic terms a while ago by Barrer and Klinowski [2]; what is important to note here is that the *marked flexibility* of the MAP framework allows this phase transformation to happen very easily without phase separation and consequent hysteretic behaviour. Thus the exchange of calcium for sodium occurs readily, rapidly and *reversibly*, with the preference for calcium actually *rising* as the calcium level in the material increases up to $E_{Ca} = 0.97$ [161]. It is this behaviour, in combination with the kinetic properties, which makes MAP much more attractive than A for application as a detergent builder, especially when superior bleach resistant and liquid holding properties are also taken into account [96].

Finally, we note that Adams *et al.* [161] applied the prediction methods of Franklin and Townsend [58] to their zeolite MAP data. They were able to predict successfully the selectivity behaviour of MAP over a range of solution concentrations once a limitation placed by Franklin and Townsend [58] on the iteration procedure had been removed. This required modification of the iteration procedure was a particular consequence of the unusual variation of $ln\ K_E$ with E_{Ca} which was observed for the Na/Ca-MAP system and was necessary to ensure that the global, rather than a local minimum, for the iteration criterion was attained [161]. The only case where predictions were not entirely satisfactory were for the most dilute solutions, an observation in agreement with those of Harjula *et al.* for other systems [46,48] (see also Sections 2.1.4 and 3.2.2), and one which is still not fully understood.

8. CONCLUDING REMARKS

Ion exchange studies do not figure prominently in current conference proceedings concerning zeolites and related materials. Nevertheless, the examples given above demonstrate clearly how the basic understanding of the theory of ion exchange in zeolites which has been developed over the years, coupled with careful experimental measurements, has borne fruit in major up-to-date applications of these materials in processes of high commercial importance. There is a continuing need for further good-quality basic research on this topic.

In the previous edition of this book [78], it was noted how new measurement techniques were pushing forward our understanding and capabilities in this area. A new and powerful tool which will significantly advance our understanding and predictive modelling of ion exchange processes in zeolites is computational modelling. Computational techniques have already been applied extensively to structural, catalytic and sorption studies in molecular sieves; the emphasis is now shifting towards ion exchange also. Examples of this are some recent studies which simulate successfully hydration, ion exchange and dehydration processes in clinoptilolite [162,163]. For both the equilibrium and kinetic aspects of ion exchange, computer simulation studies undoubtedly will be an exciting research area to watch in the future.

9. REFERENCES

1 G.P. Handreck and T.D. Smith, *J. Chem. Soc. Faraday Trans. 1*, **85** (1989) 645.
2 R.M. Barrer and J. Klinowski, *Phil. Trans. R. Soc. Lond.*, **285** (1977) 637.
3 H. Hamdan, B. Sulikowski and J. Klinowski, *J. Phys. Chem.* **93** (1989) 350.
4 H.S. Sherry, in E.M. Flannigen and L.B. Sand (eds), "Molecular Sieve Zeolites - I", *ACS Symp. Ser.* **101** (1971) 361.
5 R.M. Barrer, *Bull. Soc. Fr. Mineral. Cristallogr.*, **97** (1974) 89.
6 J.T. Iiyame, *Bull Soc. Fr. Mineral. Cristallogr.*, **87** (1964) 532.
7 P. Fletcher, K.R. Franklin and R.P. Townsend, *Phil. Trans. R. Soc. Lond.*, **A312** (1984) 141.
8 R.M. Barrer and R.P. Townsend, *J. Chem. Soc. Faraday Trans. 2*, **80** (1984) 629.
9 R.P. Townsend, *Pure & Appl. Chem.*, **58** (1986) 1359.
10 C.A. Fyfe, G.T. Kokotailo, J.D. Graham, C. Browning, G.C. Gobbi, M. Hyland, G.J. Kennedy and C.T. De Schutter, *J. Am. Chem. Soc.*, **108** (1986) 522.
11 G.T. Kokotailo, S.L. Lawton and S. Sawruk, in J.R. Katzer (ed.), "Molecular Sieves - II", ACS Symp. Ser., Vol **40**, ACS, Washington DC, (1977) 439.
12 A. Dyer and R.B. Gettins, *J. Inorg. Nucl. Chem.*, **32** (1970) 2401.
13 D.H. Olson and H.S. Sherry, *J. Phys. Chem.*, **71** (1968) 4095.
14 R.M. Barrer and A.J. Walker, *Trans. Faraday Soc.*, **60** (1964) 171.
15 G.L. Gaines and H.C. Thomas, *J. Chem. Phys.*, **21** (1953) 714.
16 K.R. Franklin and R.P. Townsend, *J. Chem. Soc. Faraday Trans. 1*, **84** (1988) 2755.
17 F. Helfferich, "Ion Exchange", McGraw-Hill, London, 1962, pp. 134-139.
18 R.M. Barrer, R. Papadopoulos and L.V.C. Rees, *J. Inorg. Nucl. Chem.*, **29** (1967) 2047.
19 P. Fletcher and R.P. Townsend, *Zeolites*, **3** (1983) 129.
20 B.K.G. Theng, *N.Z. J. Sci.*, **14** (1971) 1026.
21 R.M. Barrer, "Zeolites and Clay Minerals as Sorbents and Molecular Sieves", Academic Press, London, 1978, pp. 76-87.
22 K.R. Franklin, R.P. Townsend, S.J. Whelan and C.J. Adams, in Y. Murakami, A. Iijima and J.W. Ward (eds.), *Proc. 7th Int. Zeolite Conf. Tokyo*, Elsevier 1986, p. 289.
23 W.J. Mortier, "Compilation of Extra-Framework Sites in Zeolites", Butterworth-Heinemann, Guildford, 1982.
24 R.M. Barrer and J. Klinowski, *J. Chem. Soc., Faraday Trans. 1*, **68** (1972) 73.
25 H. van Koningsveld and J.C. Jansen, *Microp. Mat'ls*, **6** (1996) 159.
26 E.N. Coker, D.P. Roelofsen, R.M. Barrer, J.C. Jansen and H. van Bekkum, *Microp. Mesop. Materials* **22** (1998) 261.
27 A.M. Taylor and R. Roy, *J. Chem. Soc.* (**1965**) 4028.
28 R.M. Barrer and B.M. Munday, *J. Chem. Soc.* (A), (**1971**) 2909.
29 R.M. Barrer and B.M. Munday, *J. Chem. Soc.* (A), (**1971**) 2914.
30 R.M. Barrer and L. Hinds, *J. Chem. Soc.*, (**1953**) 1879.
31 R.M. Barrer and J.D. Falconer, *Proc. Roy. Soc.* A, **236** (1956) 227.
32 E.N. Coker and L.V.C. Rees, *J. Chem. Soc. Faraday Trans.*, **88** (1992) 263.
33 S.T. Bowden, "The Phase Rule and Phase Reactions" Macmillan, London, 1950.
34 T.E. Cook, W.A. Cilley, A.C. Savitsky and B.H. Wiers, *Environ. Sci. Technol.*, **16** (1982) 344.

35 H.E. Allen, S.H. Cho and T.A. Neubecker, *Water Res.*, **17** (1983) 1871.
36 R.M. Barrer and R.P. Townsend, *J. Chem. Soc., Faraday Trans. 1*, **72** (1976) 661.
37 P. Fletcher and R.P. Townsend, *J. Chromatogr.*, **238** (1982) 59.
38 P. Fletcher and R.P. Townsend, *J Chem. Soc. Faraday Trans. 1*, **81** (1985) 1731.
39 G. Sposito, K.M. Holtzclaw, C.T. Johnston and C.S. Lavesque-Madore, *Soil Sci. Soc. Am. J.*, **45** (1981) 1079.
40 G. Sposito, K.M. Holtzclaw, L. Charlet, C. Jouany and A.L. Page, *Soil Sci. Soc. Am. J.*, **47** (1983) 51.
41 G. Sposito. C. Jouany, K.M. Holtzclaw and C.S. Levesque, *Soil Sci. Soc. Am. J.*, **47** (1983) 1081.
42 R.P. Townsend, K.R. Franklin and J.F. O'Connor, *Adsorption Sci. Technol.*, **1** (1984) 269.
43 R. Harjula, J. Lehto, J.H. Pothuis, A. Dyer and R.P. Townsend, *Proc. Int. Conf. Ion Exchange, ICIE '91*, Kodansha Ltd Tokyo, p.157.
44 R Harjula, A. Dyer, S.D. Pearson and R.P. Townsend, *J. Chem. Soc. Faraday Trans.*, **88** (1992) 1591.
45 R. Harjula, J. Lehto, J.H. Pothuis, A. Dyer and R.P. Townsend, *J. Chem. Soc. Faraday Trans.*, **89** (1993) 971.
46 R. Harjula, A. Dyer and R.P. Townsend, *J. Chem. Soc. Faraday Trans.*, **89** (1993) 977.
47 R. Harjula, J. Lehto, J.H. Pothuis, A. Dyer and R.P. Townsend, *J. Chem. Soc. Faraday Trans.*, **89** (1993) 1877.
48 R.P. Townsend and R. Harjula, in H.G. Karge and J. Weitkamp (eds.) "Molecular Sieves - Science and Technology Volume 3: Post-Synthesis Modification", Springer-Verlag Heidelberg, in print.
49 F. Helfferich, "Ion Exchange", McGraw-Hill, London, 1962.
50 A. Dyer, H. Enamy and R.P. Townsend, *Sep. Sci. Technol.*, **16** (1981) 173.
51 R.M. Barrer and J. Klinowski, *J. Chem. Soc., Faraday Trans. 1*, **70** (1974) 2080.
52 M.L. McGlashan, "Chemical Thermodynamics", Academic Press, London, 1979, pp. 78-79; 106-107.
53 R.P. Townsend, P. Fletcher and M. Loizidou, in D. Olson and A. Bisio (eds.), *Proc. 6th Int. Zeolite Conf.*, Reno, 1983, p. 110.
54 P. Fletcher and R.P. Townsend, *J. Chem. Soc., Faraday Trans. 2*, **77** (1981) 2077.
55 M.L. McGlashan, "Chemical Thermodynamics", Academic Press, London, 1979, pp. 111-112.
56 K.R. Franklin and R.P. Townsend, *Zeolites*, **8** (1988) 367.
57 J. Kielland, *J. Soc. Chem. Ind.*, **54** (1935) 232T.
58 K.R. Franklin and R.P. Townsend, *J. Chem. Soc., Faraday Trans. 1*, **81** (1985) 3127.
59 K.R. Franklin and R.P. Townsend, *J. Chem. Soc., Faraday Trans. I*, **84** (1988) 687.
60 P. Fletcher, "Chemical Thermodynamics for Earth Scientists", Longman Scientific and Technical UK, 1993, pp. 430-444.
61 K. Denbigh, "The Principles of Chemical Equilibrium", Cambridge University Press 1968, p. 300.
62 R.M. Barrer, *Zeolites*, **4** (1984) 361.
63 R.M. Barrer and R.P. Townsend, *Zeolites,* **5** (1985) 287.

64 A.P. Vanselow, *J. Am. Chem. Soc.,* **54** (1932) 1307.
65 Y.N. Gapon, *J. Gen. Chem. USSR*, **3** (1933) 144.
66 E. Hogfeldt, *Arkiv for Kemi*, **5** (1953) 147.
67 O.D. Banner, W.J. Argersinger and A.W. Davidson, *J. Am. Chem. Soc.*, **74** (1952) 1044.
68 D.G. Hall, *J. Phys. Chem.*, **100** (1996) 15599.
69 W.E. Morf, "The Principles of Ion-Selective Electrodes and of Membrane Transport", in Studies in Analytical Chemistry Vol II, Elsevier, Amsterdam, 1981.
70 R.M. Barrer and L.V.C. Rees, *J. Phys. Chem. Solids*, **25** (1964) 1035.
71 R.M. Barrer, R.F. Bartholomew and L.V.C. Rees, *J. Phys. Chem. Solids*, **24** (1963) 51.
72 R.M. Barrer, R.F. Bartholomew and L.V.C. Rees, *J. Phys. Chem. Solids*, **24** (1963) 309.
73 N.M. Brooke and L.V.C. Rees, *Trans. Faraday Soc.*, **64** (1968) 3383.
74 N.M. Brooke and L.V.C. Rees, *Trans. Faraday Soc.*, **65** (1969) 2728.
75 S.C. Duffy and L.V.C. Rees, *J. Chem. Soc. Faraday Trans.*, **70** (1974) 777.
76 S.C. Duffy and L.V.C. Rees, *J. Chem. Soc. Faraday Trans.*, **71** (1975) 602.
77 S.C. Duffy and L.V.C. Rees, *J. Chromatogr.*, **102** (1974) 149.
78 R.P. Townsend, in H. van Bekkum, E.M. Flanigen and J.C. Jansen (eds.), "Introduction to Zeolite Science and Practice", Elsevier, Amsterdam, 1991, p.359.
79 L.M. Brown, H.S. Sherry and F.J. Krambeck, *J. Phys. Chem.*, **75** (1971) 3486.
80 L.M. Brown and H.S. Sherry, *J. Phys. Chem.*, **75** (1971) 3855.
81 G.E. Boyd, A.W. Adamson and L.S. Myer Jr., *J. Am. Chem. Soc.*, **69** (1947) 2836.
82 M. Streat and F.L.D. Cloete, in R.W. Rousseau (ed.), "Handbook of Separation Process Technology", Wiley-Interscience, New York, 1987, p. 697.
83 C.N. Satterfield and T.K. Sherwood, "The Role of Diffusion in Catalysis", Addison-Wesley, Massachusetts, 1963, p.31 *et sequens*.
84 S.M. Robinson, W.D. Arnold and C.H. Byers, *AIChE Journal*, **40** (1994) 2045.
85 R.M. Barrer, "Zeolites and Clay Minerals as Sorbents and Molecular Sieves", Academic Press, London, 1978, pp. 270-271.
86 H. Eyring and E.M. Eyring, "Modern Chemical Kinetics", Chapman Hall, London, 1963.
87 E. Hoinkis and H.W. Levi, *Z. Naturforsch.*, **A24** (1969) 1784.
88 M. Franzreb, W.H. Höll and H. Sontheimer, in M.J. Slater (ed.), "Ion Exchange Advances", Proceedings of IEX'92, Elsevier, London, 1992, p.167.
89 F. Helfferich and M.S. Plesset, *J. Chem. Phys.*, **28** (1958) 418.
90 M.S. Plesset, F. Helfferich and J.N. Franklin, *J. Chem. Phys.*, **29** (1958) 1064.
91 D.M. Ruthven, "Principles of Adsorption and Adsorption Processes", Wiley, New York, 1984.
92 H. Sherry, *Zeolites*, **13** (1993) 377.
93 F.F. de Kock and J.S.J. van Deventer, *Chem. Eng. Commun.*, **135** (1995) 21.
94 H. Upadek and P. Krings, in H.G. Karge and J. Weitkamp, (eds.) "Zeolites as Catalysts, Sorbents and Detergent Builders", Elsevier, London, 1989, p. 701.
95 I. Yamane and T. Nakazawa, in Y. Muakami, A. Iijima and J.W. Ward, (eds.) Proc. 7th Int. Zeolite Conf., Elsevier, 1986, p. 991.
96 C.J. Adams, A. Araya, S.W. Carr, A.P. Chapple, K.R. Franklin, P. Graham, A.R. Minihan, T.J. Osinga and J.A. Stuart *Stud. Surf. Sci. Catal.*, **105** (1997) 1667.

97 K.R. Franklin and R.P. Townsend, *J. Chem. Soc. Faraday Trans. I*, **81** (1985) 1071.
98 D. Drummond, A. DeJonge and L.V.C. Rees, *J. Phys. Chem.*, **87** (1983) 1967.
99 E. Costa, A. de Lucas, J. Zarca and F.J. Sanz, *Lat. Am. J. Chem. Eng. Appl. Chem.*, **17** (1987) 135.
100 F. Danes and F. Wolf, *Z. Phys. Chem. (Leipzig)*, **252** (1973) 15.
101 W. Leonhardt and B.M. Sax, in H.G. Karge and J. Weitkamp, (eds.) "Zeolites as Catalysts, Sorbents and Detergent Builders", Elsevier, London, 1989, p. 691.
102 C.P. Kurzendorfer, M. Liphard, W. von Rybinski and M.J. Schwuger, *Colloid and Polymer Sci.*, **265** (1987) 542.
103 M.J. Schwuger and M. Liphard, in "Zeolites as Catalysts, Sorbents and Detergent Builders", H.G. Karge and J. Weitkamp, (eds). Elsevier, London, 1989, p. 673.
104 T. Mukaiyama, H. Nishio and O. Okumura, in Y. Muakami, A. Iijima and J.W. Ward, (eds.) *Proc. 7th Int. Zeolite Conf.*, Elsevier, 1986, p. 1017.
105 P. Kuhm and W. Lortz, *Stud. Surf. Sci. Catal.*, **84** (1994) 2303.
106 Y. Tsuruta, T. Satoh, T. Yoshida, O. Okumura and S. Ueda, in Y. Muakami, A. Iijima and J.W. Ward, (eds.) *Proc. 7th Int. Zeolite Conf.*, Elsevier, 1986, p. 1001.
107 V.C. Mole and L.V.C. Rees, in P.A. Williams and M.J. Hudson, (eds.) "Recent Developments in Ion Exchange", Elsevier Applied Science, London, 1987, p. 264.
108 L.V.C. Rees in "Zeolites as Catalysts, Sorbents and Detergent Builders", H.G. Karge and J. Weitkamp, (eds). Elsevier, London, 1989, p. 661.
109 E.N. Coker, "Ion exchange studies of microporous solids", Ph.D. Thesis, University of London, 1990.
110 E.N. Coker and L.V.C. Rees, *J. Chem. Soc. Faraday Trans.* **88** (1992) 273.
111 H.-P. Rieck, US Patent 4,664,839 assigned to Hoechst, 12th May, 1987.
112 E.N. Coker and L.V.C. Rees, *J. Mater. Chem.* **3** (1993) 523.
113 L.B. Sand and F.A. Mumpton (eds.); "Natural Zeolites: Occurrence, Properties, Use", Pergamon Press, 1978, and references therein.
114 A. Dyer; "An Introduction to Zeolite Molecular Sieves", John Wiley and Sons, Bath, 1988.
115 B.W. Mercer and L.L. Ames, in L.B. Sand and F.A. Mumpton (eds.), "Natural Zeolites: Occurrence, Properties, Use", Pergamon Press, 1978, 451.
116 S.M. DePaoli and J.J. Perona; *AIChE J.*, **42** (1996) 3434.
117 R.T. Pabalan and F.P. Bertetti; *Mater. Res. Soc. Symp. Proc.*, **333**(Scientific Basis for Nuclear Waste Management XVII) (1994) 731.
118 P. Burn, D.K. Ploetz, A.K. Saha, D.C. Grant and M.C. Skriba; *AIChE Symp. Ser.* **83**(Recent Prog. Adsorpt. Ion Exch.) (1987) 259.
119 A. Dyer, A.S. Abdel Gawad, M. Mikhail, H. Enamy and M. Afshang; *J. Radioanal. Nucl. Chem.* **154** (1991) 265.
120 P.K. Sinha, K.B. Lal, P.K. Panicker and V. Krishnasamy; *Radiochim. Acta*, **73** (1996) 157.
121 P.K. Sinha, P.K. Panicker, R.V. Amalraj and V. Krishnasamy; *Waste Management*, **15** (1995) 149.
122 M.T. Olguin, M. Solache, J.L. Iturbe, P. Bosch and S. Bulbulian; *Sep. Sci. Technol.*, **31** (1996) 2021.

123 R.T. Pabalan, J.D. Prikryl, P.M. Muller and T.B. Dietrich; *Mater. Res. Soc. Symp. Proc.*, **294**(Scientific Basis for Nuclear Waste Management XVI) (1993) 777.
124 C. Pereira, M.A. Lewis and J.P. Ackerman; *Proc. Embedded Top. Meet. DOE Spent Nucl. Fuel Fissile Mater. Manag.* (1996) 129.
125 R. Fullerton; US Atomic Energy Comm., Document No. HW-69256, (1961) [from reference 115]
126 B.W. Mercer, L.L. Ames, C.J. Touhill, W.J. van Slyke and R.B. Dean; *J. Water Pollut. Contol Fed.*, **42**(part 2) (1970) R95.
127 J.D. Sherman, in C. Calmon and H. Gold, (eds.), "Ion Exchange for Pollution Control", Vol. II, CRC Press, Florida, 1979, p. 227.
128 R.W. Tschernich "Zeolites of The World", Geoscience Press, 1992.
129 K. Torii, in L.B. Sand and F.A. Mumpton (eds.), "Natural Zeolites: Occurrence, Properties, Use", Pergamon Press, 1978, 441.
130 M. Loizidou and R.P Townsend, *Zeolites*, **7** (1987) 153.
131 P.E. Dai, *Catal. Today*, **26** (1995) 3.
132 S.T. Sie, *Stud. Surf. Sci. Catal.*, **85** (1994) 587.
133 G.H. Kühl *J. Catal.*, **29** (1973) 270.
134 V. Dondur and V. Rakic *Thermochim. Acta.*, **93** (1985) 753.
135 P.J. Davis, H. van Bekkum and E.N. Coker, *J. Chem. Education*, **76** (1999) 469.
136 H. Tsuji, F. Yagi and H. Hattori, *Chem. Lett.*, (**1991**) 1881.
137 P.E. Hathaway and M.E. Davis, *J. Catal.*, **116** (1989) 263.
138 C.G. Coe, in E.F. Vansant and R. Dewolfs, (eds.) "Process Technology Proceedings, 8, Gas Separation Technology", Elsevier, Amsterdam, 1990, p. 149.
139 T.R. Gaffney, *Curr. Opinion Solid State Mat. Sci.*, **1** (1996) 69.
140 Y.H. Yeom, Y. Kim, S.H. Song and K. Seff, *J. Phys. Chem. B*, **101** (1997) 2138.
141 P.E. Riley and K. Seff, *Inorg. Chem.*, **14** (1975) 714.
142 T. Maruoka, S. Maeda and M. Matsushita, Japanese Patent 87-306758 (1987).
143 D.P. Matthews and L.V.C. Rees, in T.S.R. Prasado Rao (ed.), "Advances in Catalysis Science and Technology" Wiley Eastern, New Delhi, India, 1985, p. 493.
144 A.M. McAleer, L.V.C. Rees and A.K. Nowak, *Zeolites*, **11** (1991) 329.
145 T.C. Watling and L.V.C. Rees, *Zeolites*, **14** (1994) 687.
146 N.A. Briscoe, D.W. Johnson, M.D. Shannon, G.T. Kokotailo and L.B. McCusker, *Zeolites*, **8** (1988) 74.
147 T.C. Watling and L.V.C. Rees, *Zeolites*, **14** (1994) 693.
148 J.A. Rabo, M.L. Poutsma and G.W. Skeels, in J.W. Hightower, (ed.), *Proc. 5th Int'l Congr. Catal.*, North Holland Publishing Co., New York, 1973, p. 1353.
149 A. Clearfield, C.H. Saldarriaga and R.C. Buckley, in J.B. Uytterhoeven, (ed.), *Proc. 3rd Int'l Conf. on Mol. Sieves*; Recent Research Reports, University of Leuven Press, Leuven, Belgium, Paper No. 130, 1973, p. 241.
150 H.G. Karge, *Stud. Surf. Sci. Catal.*, **105** (1997) 1901.
151 J. Koy and H.G. Karge, in preparation.
152 H.G. Karge, V. Mavrodinova, Z. Zheng and H.K. Beyer, *NATO ASI Series B, Physics*, **221** (1990) 157.
153 H.G. Karge, *Stud. Surf. Sci. Catal.*, **75**(Discussion) (1993) 269.

154 M. Jiang and H.G. Karge, *J. Chem. Soc. Faraday Trans.*, **91** (1995) 1845.
155 M. Jiang, J. Koy and H.G. Karge, in M.L. Occelli and H. Kessler, (eds.) *Proc. 3rd Int'l Symp. Synth. of Zeolites, Expanded Layer Compounds and other Crystalline Microporous or Mesoporous Solids*, ACS Meeting, Anaheim, CA, USA, April 2-7, (1995) Marcel Dekker Inc. New York.
156 W.M. Meier and D.H. Olson, Atlas of Zeolite Structure Types, 3rd Edition, Butterworth, 1992.
157 S. Hansen, U. Håkansson, A.R. Landa-Canovas and L. Fälth, *Zeolites*, **13** (1993) 276.
158 W. Löwenstein, *Amer. Mineralog.*, **39** (1954) 92.
159 A.M. Taylor and R. Roy, *Amer. Mineralog.*, **49** (1964) 656.
160 E.v.R. Borgstedt, H.S. Sherry and J.P. Slobogin, *Stud. Surf. Sci. Catal.*, **105** (1997) 1659.
161 C.J. Adams, A. Araya, K.J. Cunningham, K.R. Franklin and I.F. White, *J. Chem. Soc., Faraday Trans.*, **93** (1997) 499.
162 A.J. Richards, P. Barnes, D. Collins, F. Christodoulos and S.M. Clark, *J. Mol. Graphics*, **13** (1995) 28-56.
163 D. O'Connor, "Modelling of Cation Exchange, (De)Hydration and Adsorption in Zeolites", PhD-thesis, University of London, 1996.

Studies in Surface Science and Catalysis 137
H. van Bekkum, J.C. Jansen, P.A. Jacobs and E.M. Flanigen
© 2001 Elsevier Science B.V. All rights reserved

Chapter 12

DIFFUSION IN ZEOLITE MOLECULAR SIEVES

DOUGLAS M. RUTHVEN
Department of Chemical Engineering
University of Maine
Orono, ME 04469-5737 (USA)

MARTIN F.M. POST
Shell International Oil Products B.V.
Shell Research and Technology Centre, Amsterdam
Badhuisweg 3, 1031 CM Amsterdam (The Netherlands)

ABSTRACT

Intracrystalline diffusion of adsorbed molecules in zeolite crystals is critically important in relation to the application of zeolites as adsorbents and shape selective catalysts. Over the past fifty years this subject has attracted considerable research, aimed both at developing improved experimental methods for measurement of intracrystalline diffusion and at increasing our basic understanding of the transport mechanism. The main focus of this chapter is on the experimental techniques and their advantages and limitations. This is followed by a review of some of the reported diffusivity values, highlighting the agreement and discrepancies between different measurement techniques and the observed trends of diffusivity and activation energy with molecular size/shape, zeolite composition and channel geometry. The final section focuses on the quantitative relationship between zeolite diffusion and catalyst activity.

CONTENTS

1. **INTRODUCTION**

2. **GENERAL PRINCIPLES**
   - **2.1 Definition of Diffusivity**
   - **2.2 Driving Force for Diffusion**
   - **2.3 Stefan-Maxwell Equations**
   - **2.4 Diffusion in a Multicomponent System**
   - **2.5 Single File Diffusion**
   - **2.6 Non-Isotropic Systems**

3. **THEORETICAL PREDICTION OF INTRACRYSTALLINE DIFFUSIVITIES**

4. **MEASUREMENT OF INTRACRYSTALLINE DIFFUSION**
   - **4.1 Microscopic Techniques**
   - **4.2 Macroscopic Techniques**
     - *Uptake Rate Measurements*
     - *Constant Molar Flow Method*
     - *Frequency Response*
     - *Chromatographic Measurements*
     - *ZLC Method*
     - *DAB*
     - *Membrane Measurements*
     - *Effectiveness Factor*

5. **REVIEW OF EXPERIMENTAL DIFFUSIVITY DATA FOR SELECTED SYSTEMS**
   - **5.1 Self-Diffusion**
   - **5.2 Transport Diffusion**
     - *Variation of Crystal Size*
     - *Concentration Dependence*
     - *Effect of Chain Length*
     - *Effect of Molecule Shape*
     - *Effect of Si/Al Ratio*
   - **5.3 Diffusion in a Binary Adsorbed Phase**
     - *Vapor Systems*
     - *Liquid Phase Systems*

6. **DIFFUSION LIMITATIONS IN ZEOLITE CRYSTALS**
   - *Determination of Diffusivities from Catalytic Measurements*
   - *Relation Between Diffusivities from Sorption and Catalytic Measurements*

7. **CONCLUSIONS**

8. **NOTATION**

9. **BIBLIOGRAPHY**

10. **REFERENCES**

1. INTRODUCTION

Diffusion in zeolites has been widely studied over the past fifty years but our understanding of the subject is still far from complete. This is in part because the experimental measurement of intracrystalline diffusion has proved to be a much more difficult and challenging task than might have originally been anticipated. Intracrystalline diffusion, also known as "configurational diffusion"[1] is an activated process, the energy of activation arising largely from steric hindrance. The diffusional activation energy and therefore the diffusivity, for a given sorbate, are therefore directly related to the pore size, as shown schematically in figure 1. Recent experimental evidence shows that, except in situations where steric hindrance is substantial, intracrystalline diffusion is a surprisingly rapid process so that, except when the zeolite crystals are large or the diffusing species are bulky, adsorption/desorption rates are generally controlled by processes other than intracrystalline diffusion. As a result many of the earlier reported diffusivity values derived from uptake rate measurements must be considered suspect.

Intracrystalline diffusion plays an important role in most of the practical applications of zeolites as adsorbents and catalysts. Indeed there are many examples, some of which have been commercialized to large-scale processes, in which the catalyst selectivity depends on differences in diffusion rates between different species[2-4]. The majority of adsorption separation processes depend on differences in equilibrium but some such processes depend on molecular sieving (size exclusion) or differences in sorption rates[5]. These issues are

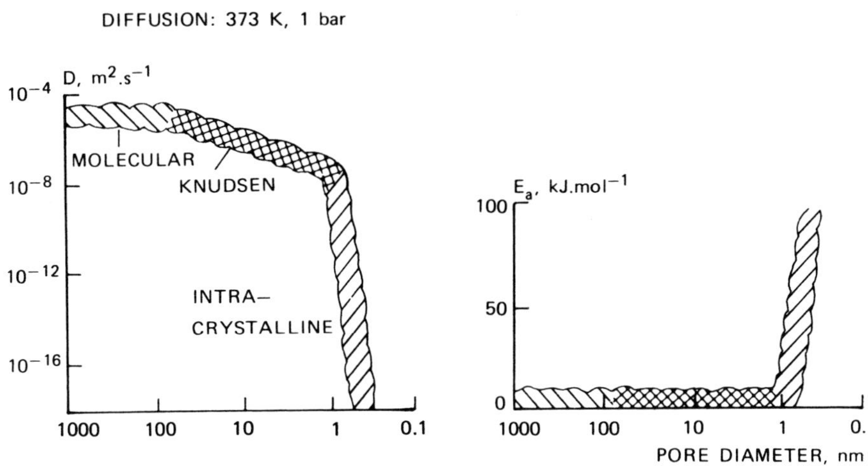

Fig. 1 Effect of pore diameter on diffusivity and diffusional activation energy showing the different regimes.

especially important in zeolite membranes which are likely to see increasing commercialization over the next decade. Diffusion in zeolites has been reviewed in detail in several recent books and articles, a selection of which is given in the "bibliography" presented at the end of this chapter. The aim of the present chapter is to provide a brief introduction to the subject, a review of the various experimental techniques that have been developed for measurement of intracrystalline diffusion and brief comments on some of the issues and approaches currently being followed in the study of this subject.

2. GENERAL PRINCIPLES

2.1 Definitions of Diffusivity[(6)]

Fick's first equation:

$$J = -D(q)\frac{\partial q}{\partial z} \tag{1}$$

provides a convenient starting point for the quantitative analysis of intracrystalline transport in zeolites. This expression can be regarded simply as the definition of the diffusivity (D), which is, at least in principle, a function of the adsorbed phase concentration. The diffusivity defined in accordance with Eq. 1 may be called, more precisely, the *transport diffusivity*, since this quantity measures the ratio of the sorbate flux to the concentration gradient. One may also define a *tracer diffusivity* ($\mathcal{D}$) in accordance with the equation:

$$J_* = -\mathcal{D}\frac{\partial q_*}{\partial z}\bigg|_q \tag{2}$$

This quantity measures the flux of marked molecules under the influence of a concentration gradient of marked molecules at constant total species concentration. In general $D(q)$ and $\mathcal{D}(q)$ will not be the same although one may argue by intuitive physical reasoning that the two quantities should converge in the low concentration limit.

One may also define a self-diffusivity based on the random walk representation of a diffusive process:

$$\mathcal{D} = \frac{1}{2n}\cdot\frac{\overline{r^2}}{t} = \frac{1}{2n}\cdot\frac{\lambda^2}{\tau} \tag{3}$$

where n is the dimensionality of the pore structure, λ is the mean jump distance, τ is the average time between molecular jumps and $\overline{r^2}$ is the mean square displacement over a time period t. It may be shown that the definitions represented by Eqs. 2 and 3 are formally equivalent; we have therefore used the same symbol ($\mathcal{D}$) for both "tracer" and "self" diffusivities.

2.2 Driving Force for Diffusion

Since the true driving force for the transport process is obviously the gradient of chemical potential (μ), rather than the concentration gradient, it is arguably more logical to write, in place of Eq. 1:

$$J = -B(q)\,q \cdot \frac{\partial \mu}{\partial z} \tag{4}$$

The relationship between the mobility (B) and the Fickian diffusivity (D) can be easily derived by considering equilibrium between the adsorbed phase (concentration q) and an ideal vapor phase (pressure p, temperature T)[5]:

$$D(q) = B(q)\,\mathbf{R}T\frac{d\,\ln p}{d\,\ln q} = D_o(q)\frac{d\,\ln p}{d\,\ln q} \tag{5}$$

where $D_o = B\mathbf{R}T$ is the "corrected" or "intrinsic" transport diffusivity. Eq. 5 is commonly referred to as Darken's equation and was developed originally to describe the interdiffusion of two metals to form an alloy[7]. It may also be regarded as the limiting form of the Maxwell-Stefan diffusion equation for single component diffusion.

It is evident from Eq. 5 that, in general, $D \neq D_o$. The thermodynamic correction factor $d\ln p/d\ln q$ depends on loading and can be quite large. For example, for a Langmuir isotherm:

$$\frac{d\,\ln p}{d\,\ln q} = \frac{1}{1 - q/q_s} \quad ; \quad D = \frac{D_o}{1 - q/q_s} \tag{6}$$

so, according to this model, D will approach infinity as $q \to q_s$. In practice a variation in transport diffusivity by more than an order of magnitude between low and high loadings is not uncommon. It is also evident that at low loading within the Henry's Law region, ($d\ln p/d\ln q$) $\to$ 1.0 and $D \to D_o$ so, in the low concentration limit we have $D \cong D_o \cong \mathcal{D}$.

The general relationship between D_o and $\mathcal{D}$ is however more complicated. It may be shown from the principles of irreversible thermodynamics that the relationship must have the form[8, 9]:

$$\frac{D_o}{\mathcal{D}} = 1 + \beta(q)/\alpha(q) \tag{7}$$

where α and β are functions of concentration related to combinations of the straight and cross coefficients in the Onsager formulation. Irreversible thermodynamics provides no evidence as to the magnitude of the factor β/α and experimental information is limited. The study of Barrer and Fender[10] suggests that $\mathcal{D} \cong D_o$ is a valid approximation up to at least 50% of saturation loading. However, accurate measurements of both D_o and $\mathcal{D}$ become difficult at high loadings so while one might anticipate that interference effects will increase as saturation is approached quantitative evidence is limited.

2.3 Non-Isotropic Systems

One of the well-known results of diffusion theory is that for a system in which the diffusional properties are non-isotropic it is always possible to define a new coordinate system in which Fick's equations are applicable[11]. This means that, except in the extreme case of one-dimensional diffusion, the form of the transient sorption/desorption curves will be essentially the same as for an isotropic system although the apparent diffusivity will then be a complex function of the diffusivities in the three principal directions. Detailed studies of diffusion in non-isotropic zeolite crystals such as silicalite can, however, be achieved using oriented crystals with selected faces coated with an impermeable resin, or embedded in an oriented membrane. Such studies show that for n-hexane in silicalite, the ratio of diffusivities in the transverse and longitudinal directions is about three[12, 13] and this is confirmed by PFGNMR measurements.

2.4 Stefan Maxwell Equation: Diffusion in a Binary Adsorbed Phase

The Stefan-Maxwell approach has been extended by Krishna to diffusion in an adsorbed phase by considering the idealized molecular hopping model and regarding the unoccupied sites (vacancies) as the (n + 1) component in the system[14, 15]. The validity of the momentum transfer argument from which the Stefan-Maxwell equation is derived seems questionable in such a situation but the approach may nevertheless be justified on the basis that the results are consistent with experimental observations. With concentrations expressed in terms of fractional occupation of the available sites ($\theta_i = q_i / q_s$), the basic expression becomes:

$$-\frac{1}{\mathbf{R}T}\cdot\frac{\partial\mu_i}{\partial z}=\sum_{j=1}^{n}\frac{\theta_j\left(u_i-u_j\right)}{D_{ij}^1}+\frac{\theta_{n+1}}{D^1_{iv}}\cdot u_i \tag{8}$$

The coefficient D^1_{ij} measures the rate at which components i and j can exchange directly while D^1_{iv} measures the rate of migration to vacant sites, the velocity of which (u_{n+1}) is zero.

For single component diffusion ($\theta_B = 0, N_B = 0$) this reduces to:

$$J=-\frac{D^1_{AV}}{\left(1-\theta_A\right)}\cdot\frac{q_A}{RT}\cdot\frac{d\mu_A}{dz} \tag{9}$$

which is equivalent to Eq. 4 with

$$D_o = B\mathbf{R}T = D^1_{AV}/(1-\theta) \tag{10}$$

as for a Langmuirian system.

For tracer or self-diffusion we have $J_A = -J_B$ and $D^1_{AB} = D^1_A$ which gives:

$$\frac{1}{\mathcal{D}}=\frac{\theta}{D^1_A}+\frac{1-\theta}{D^1_{AV}} \tag{11}$$

showing that with increasing loading, the balance between the two mechanisms should shift in favor of direct exchange.

Habgood and co-workers[16, 17] were the first to provide a rational analysis of sorption rates in a binary adsorbed phase. Using the principle of the chemical potential driving force (Eq. 4) with due allowance for the effect of each component on the chemical potential of the other, they obtained, for a Langmuir system:

$$\frac{\partial \theta_i}{\partial t} = \nabla . \left[D_{eff} \nabla \theta_i \right] \tag{12}$$

$$D_{effA} = \frac{D_{OA}}{(1-\theta_A-\theta_B)} \left[(1-\theta_B) + \theta_A \frac{\nabla \theta_B}{\nabla \theta_A} \right] \tag{13}$$

with a similar expression for D_{effB}. Precisely the same form may be derived from Eq. 8 if the effect of direct exchange of the components is neglected and the second term representing activated jumps to an unoccupied site is assumed dominant. Two variants of this model are plausible; if the jump frequency is independent of surface coverage D^1_{iv} is constant and we obtain Eq. 13 whereas if we assume that the diffusing molecule can successfully execute a jump only when the receiving site is unoccupied we obtain:

$$D_{effA} = D_{OA} \left[(1-\theta_B) + \theta_A \frac{\nabla \theta_B}{\nabla \theta_A} \right] \tag{14}$$

This is in effect the model of Quereshi and Wei[18].

2.5 Single File Diffusion

Self-diffusion in a three-dimensional pore network can be described by the Einstein expression (Eq. 3) from which it is evident that the mean square molecular displacement increases linearly with time. This is true also for a one-dimensional pore system provided that the pore is large enough that diffusing molecules can pass each other. When this condition is not fulfilled we have an entirely different pattern of behavior known as single file diffusion or 'string of pearls' diffusion. The mean square displacement increases with the square root of time[19, 20].

$$z^2(s) = \lambda^2 \left(\frac{1-\theta}{\theta} \right) \sqrt{\frac{2t}{\pi\tau}} \tag{15}$$

which is equivalent to a time varying self-diffusivity given by:

$$\mathcal{D} = \lambda^2 \left(\frac{1-\theta}{\theta} \right) \frac{1}{\sqrt{2\pi\theta t}} \tag{16}$$

This topic has attracted much attention during the last few years. Progress has been reviewed by Kärger[21].

3. THEORETICAL PREDICTION OF INTRACRYSTALLINE DIFFUSIVITIES

It is in principle possible to predict intracrystalline diffusivities *a priori* from transition state theory. However, this requires both a correct physical picture of the transition state and an accurate calculation of the force field so that the entropy and energy of activation can be estimated. Nevertheless useful inferences can sometimes be drawn from transition state theory, especially for diffusion of simple molecules. For example the contribution of the rotational partition function is relatively large so a conclusion concerning the degree to which rotational freedom is restricted in the transition state can sometimes be drawn simply by considering the order of magnitude of the pre-exponential factor in the Arrhenius expression for the temperature dependence of the diffusivity.

A more important and more widely used approach is the *Monte Carlo* calculation which is based on the statistical representation of a diffusion process as a random walk with a factor such as an Arrhenius factor assigned to represent the probability of a successful molecular jump. An early application of this approach was the study of diffusion of n-butane in NaCaA zeolites carried out by Ruthven(22). More recent applications can be found in the works of Dahlke(23), Auerbach(24) and Bell et al.(25) who have applied this approach to study the diffusion of a range of different sorbates in X and Y zeolites and silicalite.

The third approach is a *molecular dynamics* (MD) calculation. Such calculations depend on solving the equations of motion for a molecule (or a set of molecules) confined within the force field of the adsorbent and allowed to move freely subject to the constraints imposed by collisions with the pore walls and with other molecules. The method is, in principle, very powerful but its applicability is limited by two main factors. Firstly the method requires an accurate calculation of the potential field and hence the force on each particle. Secondly the computations are very bulky so, even with modern computers, it is possible to track the movement of molecules only over a very short time interval (several hundred picoseconds) and over molecular displacements of the order of 100 molecular diameters. These constraints restrict the applicability of the MD approach to relatively small and rapidly diffusing molecules.

4. MEASUREMENT OF INTRACRYSTALLINE DIFFUSION

Many commercial adsorbent particles consist of small microporous zeolite crystals formed into a meso/macro porous pellet (fig. 2) generally with the aid of a clay binder. In general such materials may offer three distinct resistances to mass transfer: (i) the micropore resistance of the zeolite crystals (proportional to r_c^2/D_c), (ii) the resistance related to transport through the outer layer of the zeolite crystal (known as the surface barrier resistance, proportional to r_c/k_s) and (iii) the meso- or macropore diffusional resistance of the pellet (proportional to R_p^2/D_p). In addition, a resistance due to the fluid film surrounding the pellet (mass transfer coefficient k_f) may also be present. In this article we emphasize the measurement of intracrystalline diffusion, but it should be noted that the intracrystalline component is by no means always the dominant contribution to the overall mass transfer resistance. In fact commercial adsorbent and catalyst pellets are commonly made from very small zeolite crystals in order to minimize intracrystalline resistance. To make reliable intracrystalline diffusion measurements it is therefore desirable to work with large unaggregated crystals. Since the time constant (r_c^2/D_c) varies with the square of the crystal

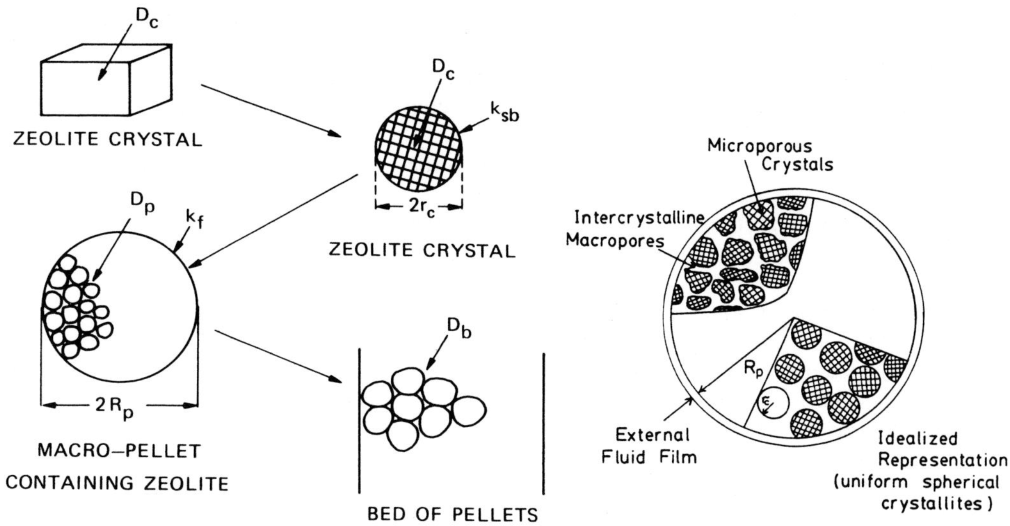

Fig. 2 Schematic diagram showing a composite adsorbent pellet and the principal resistances to mass transfer.

radius, replicate measurements with different crystal size fractions provide a convenient way to confirm intracrystalline diffusion control.

Table 1 lists some of the major experimental methods with their approximate chronology. These methods can conveniently be divided into two broad categories; microscopic and macroscopic depending on whether the technique measures a bulk flux or attempts to track and average the movement of individual molecules. The macroscopic category can be further sub-divided into transient or quasi steady state methods (Table 2).

4.1 Microscopic Techniques: QENS and PFGNMR

The main microscopic methods are quasi-elastic neutron scattering (QENS) and nuclear magnetic resonance (NMR) based techniques. A QENS experiment yields information on molecular mobility at the local (< 10Å) scale. The estimation of long range micropore diffusivities from such measurements therefore depends on estimating the jump length as well as on the implicit assumption that long range diffusion is controlled by the same process that controls the local mobility. This assumption is not always valid.

The earliest NMR measurements were based on estimates of the time between successive molecular jumps as derived from relaxation time measurements[32, 33]. As with a QENS measurement the estimation of a long-range diffusivity from such information requires

TABLE 1

Measurements of Diffusion in Zeolite Crystals: Historical Development

Method	Author
Direct monitoring of transient conc. Profile (H_2O – Heulandite)	Tiselius (1943)[26, 27]
Transient Sorption Rate	
- Volumetric/gravimetric	Barrer (1940)[28, 29]
- Piezometric	Bulow (1970)[30]; Do[31]
NMR Relaxation	Resing (1967)[32, 33]
PFG NMR	Pfeifer, Karger, Lechert (early 1970s)[34-36]
Chromatography (gas)	Haynes, Ma, Ruthven (mid 1970s)[37-39]
Membrane permeation	Hayhurst, Wernick (1983)[40, 41]
Effectiveness Factor	Haag, Post (early 1980s)[42, 43]
ZLC	Eic and Ruthven (mid 1980s)[44, 45]
Frequency Response	Yasuda, Rees (late 1980s)[46, 47]
QENS	Cohen de Lara, Jobic (late 1980s)[48, 49]
Chromatography (liquid)	Awum (1988)[50]
FTIR	Niessen and Karge (1991)[51, 52]
Differential Adsorption Bed	Do (1991)[53]
IR and IR/Freq. Response	Grenier, Meunier, Bourdin (1994)[54, 55]
Tracer ZLC	Hufton (1994)[56]
TAP Reactor	Baerns and Keipert[57]; Nijhuis (1997)[58]

TABLE 2

Classification of Methods for Measuring Intracrystalline Diffusion in Zeolites

Microscopic Methods	QENS		
	NMR - Relaxation		
	- PFG		
Macroscopic Methods	Transient	Sorption Rate	
			Flow – ZLC/TZLC
			Batch – DAB
			- Gravimetric
			- Piezometric
			- FTIR
			- Temp Response
		Chromatographic	
			Gas Phase
			Liquid Phase
			Wall Coated Column
		Freq. Response	
			Pressure
			Pressure/Temperature
	Quasi Steady State	Membrane	
			Wicke Kallenbach
			Single Crystal
			Zeolite Membrane
		Catalyst Effectiveness Factor	

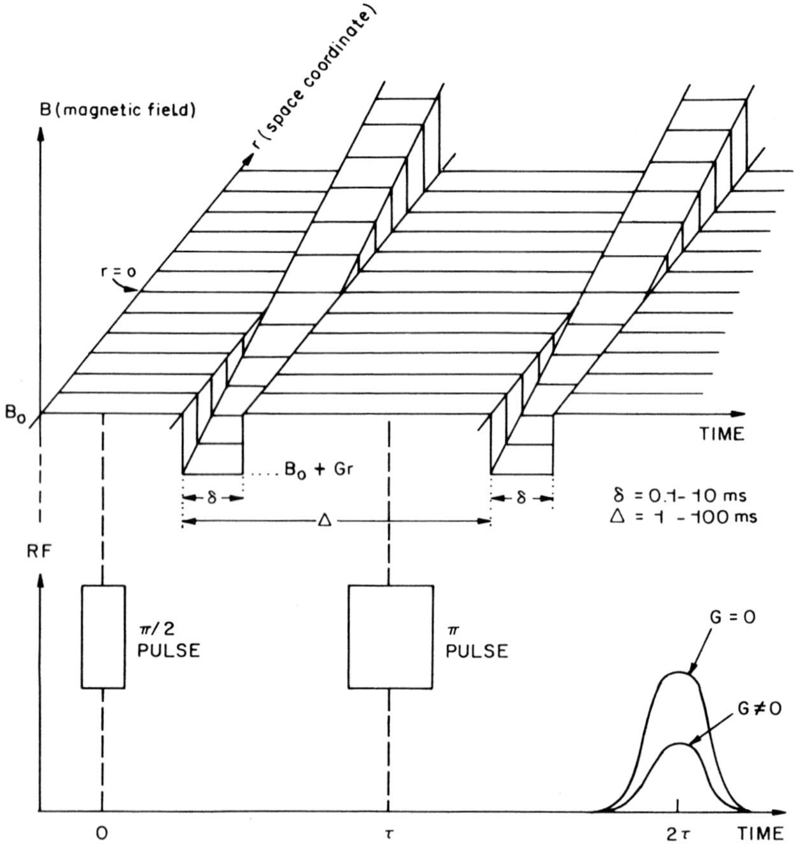

Fig. 3 Sequence of RF and gradient pulses in the PFGNMR technique for measuring intracrystalline self-diffusion.

an assumption concerning the average jump distance. The Pulse Field Gradient NMR (PFGNMR) technique provides a much more reliable approach since although the method is "microscopic" in that the movement of the individual molecules is followed, the method can be applied over macroscopic distances approaching perhaps half the crystal diameter. The method depends on labelling the molecules according to the phase of their nuclear spins and determining the mean square molecular displacement during a known time interval from the attenuation of the spin-echo response signal. The self diffusivity is then estimated directly from the Einstein equation (Eq. 3). It is a "steady state" technique since the measurement is carried out in a system with a uniform concentration of sorbate and yields a direct measurement of the self-diffusivity.

This technique has its origins in the method developed by Stejskal and Tanner[59] for measurement of diffusion in liquids. The development and application to the study of intracrystalline diffusion in zeolites is due largely to the efforts of Pfeifer, Kärger and their collaborators. Full details can be found in the text by Kärger and Ruthven[6] and in the review articles cited in the bibliography. We give here only a very brief introduction aimed at acquainting the newcomer with the essential features of the method and its advantages and limitations.

PFGNMR Technique

In a typical PFGNMR experiment a zeolite sample pre-loaded with (NMR-active) probe molecules is excited with a standard radio frequency (RF) echo pulse sequence in combination with two magnetic field gradient pulses which mark the spatial location of the spins in the sample (Fig. 3). In the absence of the gradient field (i.e., at a constant magnetic field B_o) the spin echo pulse sequence, consisting of a series of two RF pulses of appropriate frequency at $t = 0$ ($\pi/2$ pulse) and $t = \tau$ (π pulse), will lead to an echo signal at $t = 2\tau$, provided that the spin-spin relaxation (characteristic time T_2) is slow compared to τ. The PFG method, however, requires application of both a strong time- (and space-) independent component, B_o, and a pulsed field, Gr, which varies linearly with a space coordinate r following the relation $B = B_o + Gr$. The field gradient (G) is only applied during two short time intervals of identical duration (δ) separated by a time Δ (Fig. 3). After excitation of the nuclear spins with the $\pi/2$ RF pulse, the first field gradient pulse marks the various nuclear spins which precess at different angular velocity determined by the position of the molecule in the zeolite sample. If there were no diffusion, application of the π RF pulse and a second field gradient pulse would completely reverse the accumulated phase incoherence and would lead to a spin echo signal of the same intensity as would be observed in the case where the gradient pulses are absent. If molecular migration takes place in the direction of the gradient (r), however, phase coherence is partially lost and will result in an attenuation of the echo signal. Measurement of the extent of the attenuation as a function of intensity and/or duration of the gradient pulse then provides direct information on the mean square displacement of the diffusing molecules in the zeolite.

The self-diffusion coefficient $\mathcal{D}$, which is related to the molecular mean square displacement $\langle r^2(t) \rangle$ through the Einstein equation (Eq. 3), is related to the ratio of the intensity of the spin echo signal in the presence, A(G), and in the absence, A(0), of a gradient field:

$$\frac{A(G)}{A(O)} = \exp\left(-\gamma^2 G^2 \delta^2 \Delta \mathcal{D}\right) \tag{18}$$

where γ denotes the gyromagnetic ratio and G is the field gradient. Usually, the gradient amplitude G is varied in a series of experiments, while δ and Δ are kept constant. Hence, the self-diffusivity can be obtained by plotting the left-hand side of Eq. 18 against G^2. In a typical PFG experiment, τ and Δ are in the range 1-100 ms in order to satisfy the condition that $\tau, \Delta \ll T_2$. The typical length of the gradient field pulse is of the order of 0.1-10 ms while the magnitude of the gradient field G is normally of the order of 1-10 T/m.

Whether or not the PFGNMR method can be used successfully depends on a number of conditions which are described below.

(i) The most decisive quantity limiting the range of applicability of the PFG technique for slow diffusion processes is the transverse nuclear magnetic (or spin-spin) relaxation time T_2. For molecules adsorbed on zeolites, T_2 is typically of the order of 5-50 ms, which sets an upper limit on the experimental variables Δ and δ in order for the condition $\Delta << T_2$ to be satisfied. If proton spins, having a relatively large gyromagnetic ratio and a high natural abundance, are chosen to probe diffusion, it can be estimated that the lower limit of intracrystalline self-diffusivity that can be measured in a PFG experiment is of the order of 10^{-12} - 10^{-13} $m^2.s^{-1}$.

(ii) The molecular mean square displacement $<r^2(t=\Delta)>$ detected in a NMRPFG experiment must be smaller than the square of the average zeolite crystal radius (r_c^2) in order to ensure that the measured self-diffusivities D are indicative of pure intracrystalline transport. Hence, the upper limit of the intracrystalline self-diffusivity that can be measured in a PFGNMR experiment is given by:

$$(\mathcal{D})_{\text{upper limit}} \approx r_c^2/6\Delta \qquad (19)$$

Assuming the lower limit for Δ to be in the order of 1 ms (which can be easily accomplished at low gradient fields) and an average zeolite crystal radius of 1 μm, the upper limit of $\mathcal{D}$ emerging from eqn. (19) is about 10^{-10} m^2/s. The upper limit can be extended by using larger zeolite crystals.

(iii) If intracrystalline diffusion is too slow to be measured directly using the PFG method, a modification of this technique, known as the NMR Fast Tracer Desorption (FTD) method, has been developed which makes it possible to measure slower diffusion processes. Combination of the PFGNMR and FTDNMR methods has been successfully used to provide direct evidence of a surface barrier. Details of this method are reported elsewhere[60-64].

4.2 Macroscopic Techniques

Uptake Rate Measurements

Direct measurement of the adsorption or desorption rate for a small sample of adsorbent subjected to a well defined change in ambient concentration (or partial pressure) of sorbate is perhaps the most obvious way to attempt to measure intracrystalline diffusion. Gravimetric, volumetric and piezometric methods have all been used to follow the sorption process. Volumetric techniques are generally slow and somewhat cumbersome but gravimetric and piezometric measurements are probably comparable in terms of experimental convenience and limitations arising from the intrusion of extracrystalline rate process. Unfortunately such limitations are often quite severe and, as a result, the reliable application of these methods is restricted to relatively slowly diffusing systems in which the intracrystalline time constant (r_c^2/D_c) is greater than perhaps 100 secs.

The essential components gravimetric and piezometric systems are shown schematically in figure 4. For an isothermal spherical particle subjected to a step change in

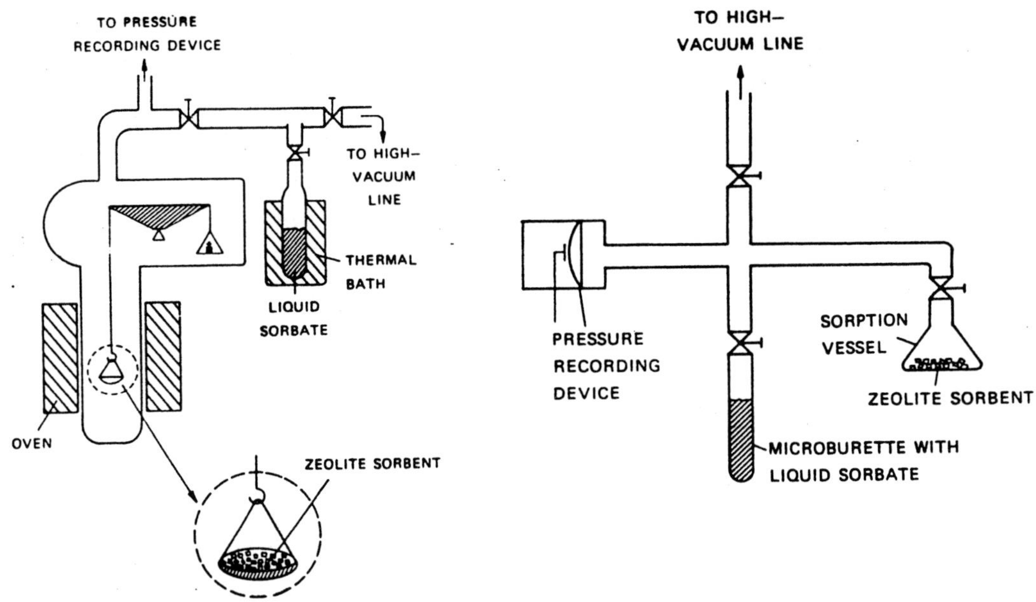

Fig. 4 Experimental set-up for (a) gravimetric and (b) piezeometric measurement of sorption rate.

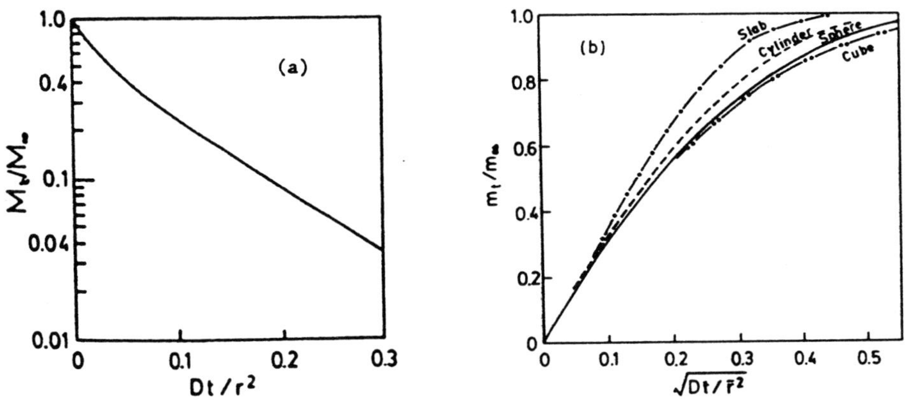

Fig. 5 Theoretical sorption curves for an isothermal diffusion controlled system showing (a) the long time asymptote when plotted on log-linear coordinates (Eq. 22) and (b) the conformity of the initial response to Eq. 21. Note that the initial slope is insensitive to particle shape provided that the equivalent radius is used to define the time constant.

sorbate concentration at the external surface at time zero, the approach to equilibrium, under conditions of diffusion control is given by[11]:

$$\frac{M_t}{M_\infty} = 1 - \frac{6}{\pi^2}\sum_{1}^{\infty}\frac{1}{n^2}\exp\left(-n^2\pi^2 Dt / R^2\right) \tag{20}$$

In this expression for uptake in a crystal $D = D_c$, $R = r_c$ whereas for a pellet $D = D_e$ and $R = R_p$. The short and long time asymptotes are given by:

$$\text{Short Time:} \quad \frac{M_t}{M_\infty} = \frac{6}{R}\sqrt{\frac{Dt}{\pi}} \tag{21}$$

$$\text{Long Time:} \quad \frac{M_t}{M_\infty} = 1 - \frac{6}{\pi^2}\exp\left(-\pi^2 Dt / R^2\right) \tag{22}$$

The form of the response curves is shown in figure 5. The corresponding expressions for other particle shapes are easily derived but there is little numerical difference from the response for a spherical particle of the same external area to volume ratio (i.e. based on an equivalent radius).

In contrast, for the same situation, a particle with surface resistance control follows a simple exponential approach to equilibrium:

$$\frac{M_t}{M_\infty} = 1 - \exp\left(-3k_s / R\right) \tag{23}$$

When mass transfer is rapid and the approach to equilibrium is controlled entirely by heat transfer, the uptake curve obeys[65, 66]:

$$\frac{M_t}{M_\infty} = 1 - \left(\frac{\beta}{1+\beta}\right)\exp\left[\frac{-3ht}{\rho C_p R}\cdot\frac{1}{(1+\beta)}\right] \tag{24}$$

where $\beta = (\Delta H/\rho C_p)(\partial q^*/\partial T)_p$. A plot of log $\left(1 - M_t / M_\infty\right)$vs t thus provides clear evidence concerning the nature of the rate limiting resistance and a convenient way to extract the time constant (R^2/D, $R/3k_s$ or $\rho C_p R/3h$).

The apparent simplicity of this approach is, however, deceptive. For measurement of intracrystalline diffusion the method works well when diffusion is relatively slow (large crystals and/or low diffusivity) but when sorption rates are rapid the uptake rate may be controlled by extracrystalline diffusion (through the interstices of the adsorbent bed) and/or by heat transfer. The intrusion of such effects is not always obvious from the shape of the uptake curve but it may generally be detected by changing the sample quantity and/or the sample configuration. It is in principle possible to allow for such effects in the mathematical model used to interpret the uptake curves (figure 6) and, indeed, the modeling of non-isothermal systems has been studied in considerable detail[65-68]. However, any such intrusion

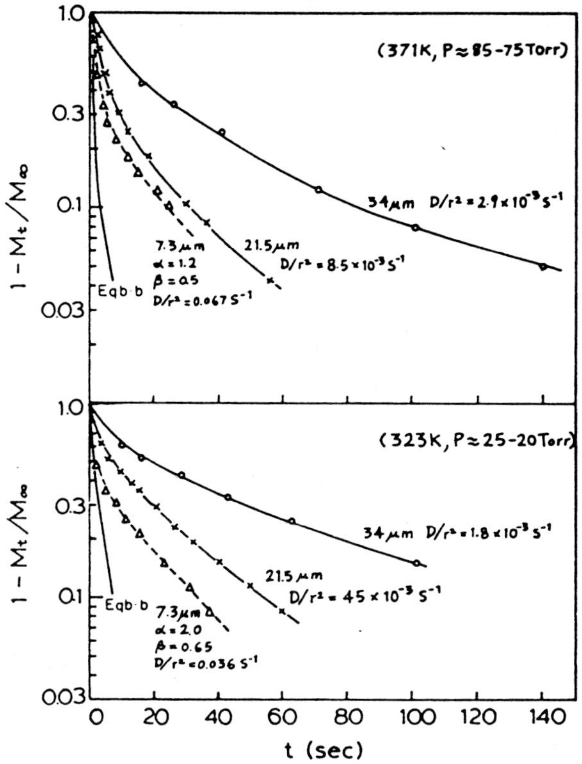

Fig. 6 Experimental uptake curves for CO_2 in 4A zeolite crystals showing near isothermal behavior in large (34 and 21.5 μm) crystals (D_c=9x10^{-9} cm^{2}s^{-1} at 371 K and 5.2x10^{-9} cm^{2}s^{-1} at 323 K). The solid lines are the theoretical curves for isothermal diffusion from Eq. 20 with the appropriate value of D_c/r^2_c. The uptake curves for the small (7.3 μm) crystals show considerable deviation from the isothermal curves but conform well to the theoretical nonisothermal curves with the values of D_c estimated from the data for the large crystals, with heat transfer coefficients from uptake rate measurements for a similar system under conditions of complete heat transfer control. The limiting isothermal curve is also shown by a continuous line with no points. From Ruthven et al.[65].

will obviously diminish the accuracy and confidence with which the intracrystalline diffusivities can be determined.

This technique may also be used to measure effective macropore diffusivities in bi-porous adsorbent pellets. For such a system with a linear equilibrium isotherm and assuming rapid intracrystalline diffusion, the governing diffusion equation is of the same form as for micropore control. The solution is identical to Eq 1 except that R now refers to the particle radius and the diffusivity D is replaced by the effective diffusivity $D_e = \frac{D_p \varepsilon_p}{\varepsilon_p + (1-\varepsilon_p)K}$. Since the equilibrium constant (K) is generally large and varies with temperature according to the van't Hoff equation ($K = K_0 e^{-\Delta H/RT}$) it is clear that a macropore controlled system will generally yield an effective diffusivity that is much smaller than the pore diffusivity and shows a strong (Arrhenius) form of temperature dependence but with an apparent activation energy that approximates the heat of adsorption. This may lead the unwary investigator to conclude that the rate is controlled by an activated (micropore) diffusion process whereas in fact the effect results from the temperature dependence of the capacity of the microparticles. However, a change in the gross particle size can often be used to provide an unequivocal diagnosis of the nature of the controlling resistance.

With vapor phase gravimetric systems it is generally possible to achieve a good approximation to a step change in sorbate pressure although, for strongly adsorbed species this requires a large system volume and a very small adsorbent sample. For piezometric measurements and for liquid phase systems, however, the external concentration (or pressure) is the measured quantity and to achieve accuracy it is therefore necessary to minimize the external system volume. In solving the diffusion equation, the time dependence of the boundary condition at the particle surface must then be considered and this leads to a slightly more complex solution for the uptake curve. The piezometric method has been widely applied(30) but a detailed analysis of this technique reveals that the restrictions on the conditions required to obtain reliable diffusivity data by this method are in fact more severe than had been generally appreciated. (108)

In both gravimetric and piezometric experiments it is generally desirable to make the measurement over a small differential concentration change in order to ensure that the assumption of system linearity is fulfilled. Under these conditions the transient sorption curve (expressed as fractional approach to equilibrium) should be independent of either the step size or direction (adsorption or desorption). Varying the step size and direction thus provides a simple and sensitive experimental test for system linearity.

Uptake rate measurements with large oriented crystals have been used by Caro[12, 13] to demonstrate the non-isotropy of silicalite. Diffusion coefficients for the longitudinal and transverse directions differed by a factor of about three.

An ingenious alternative to the gravimetric method has been developed by Karge and co-workers[51, 52] who tracked the progress of sorption by monitoring the intensity of an IR band characteristic of the adsorbed species. This approach offers the important advantage that, by a judicious choice of the IR wavelength, it may be applied to follow sorption of one (or more) components in a multi-component system.

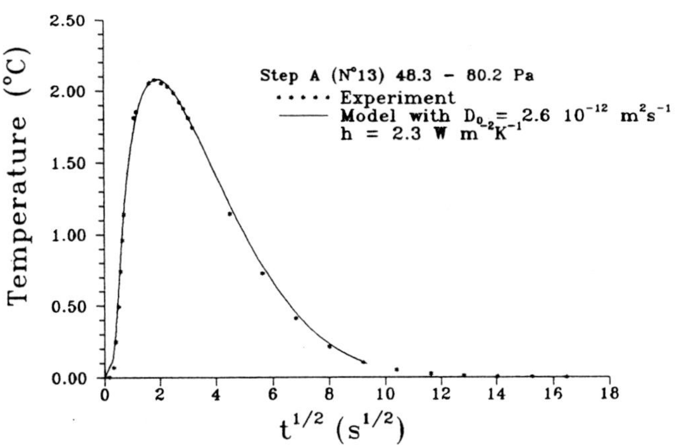

Fig. 7 Transient temperature response for CH_3OH-NaX, pressure step 48-80 Pa showing conformity between experimentally observed temperature and the theoretical curve, calculated from the diffusion model with D_c=2.6x10^{-12} m^2s^{-1}; h=2.3 $Wm^{-2}K^{-1}$. From Grenier et al.[54].

Regardless of the way in which progress of the sorption is followed, all uptake rate measurements are subject to the intrusion of heat transfer resistance and such effects are more severe for strongly adsorbed and rapidly diffusing species. To circumvent this problem Meunier et al[54] introduced a novel approach in which progress of the uptake is followed indirectly by monitoring the temperature of the adsorbent sample, using a sensitive IR detector. A representative response curve is shown in figure 7. A non-isothermal model continuing time constants for both heat transfer and diffusion is needed to interpret the experimental data. However, it emerges that the initial rising portion of the response curve is determined primarily by diffusion while the tail is determined mainly by heat transfer. Both parameters can therefore be determined with confidence, even from a single experimental response curve.

Constant Molar Flow Method

A variant of the usual piezometric system has recently been introduced by Do and co-workers[31]. Instead of subjecting the adsorbent sample to a step change in sorbate pressure the sorbate is introduced into the adsorption vessel at a constant controlled molar flow rate. The pressure in the vessel is monitored continuously on a "Barocel" gauge. Provided that the sorbate flow rate is comparable with the sorption rate the sorption rate may be found directly from a comparison of the measured pressure response with that for a system containing only an inert (non-adsorbing) solid of the same volume. The method is relatively simple and straightforward but is only suitable for relatively slow systems since the transient response is quite sensitive to heat effects.

Frequency Response

In the frequency response method, first applied to the study of zeolitic diffusion by Yasuda[46] and further developed by Rees and co-workers,[47, 69] the volume of a system containing a widely dispersed sample of adsorbent, under a known pressure of sorbate, is subjected to a periodic (usually sinusoidal) perturbation. If there is no mass transfer or if mass transfer is infinitely rapid so that gas-solid mass transfer equilibrium is always maintained, the pressure in the system should follow the volume perturbation with no phase difference. The effect of a finite resistance to mass transfer is to cause a phase shift so that the pressure response lags behind the volume perturbation. Measuring the 'in phase' and 'out of phase' responses over a range of frequencies yields the characteristic frequency response spectrum which may be matched to the spectrum derived from the theoretical model in order to determine the time constant of the mass transfer process. As with other techniques the response may be influenced by heat transfer resistance so, to obtain reliable results, it is essential to carry out sufficient experimental checks to eliminate such effects or to allow for them in the theoretical model. Being a quasi steady state measurement the frequency response technique offers certain advantages over direct measurement of the step response. The form of the frequency response spectrum depends on the nature of the dominant mass transfer resistance and can therefore be helpful in distinguishing between diffusion controlled and surface resistance controlled processes. However, the argument that the cyclic nature of the perturbation eliminates the intrusion of heat effects must be treated with caution. For both p-xylene and 2-butyne in silicalite Shen and Rees(70, 71) observed a bimodal response spectrum and they interpreted the two peaks as indicative of two different transport processes corresponding to diffusion through the straight and sinusoidal channels. Further evidence in support of this hypothesis comes from NMR studies which show that such molecules cannot easily reorient themselves at the channel intersections and from frequency response measurements with silicalite-2, which contains only straight channels of similar dimensions and yields only a single response peak. However, Sun and Bourdin(72) have shown that an alternative explanation based on thermal effects may also be plausible.

In a recent development of the frequency response technique, Bourdin(55) has combined this approach with the IR temperature measurement system. In this experiment the volume of the system is perturbed sinusoidally and both the pressure and temperature responses are measured. It was found that the phase differences between the pressure and temperature were more reliable and reproducible than the phase differences between the pressure and the volume. The explanation seems to be that since the quantity of adsorbent is quite small, a small amount of superficial adsorption on the walls of the apparatus can distort the phase relationship between the pressure and the volume but will not affect the temperature-pressure phase relationship. In the light of this observation there may be a need to re-examine some of the earlier FR results which were derived from the phase difference between the driver piston and the pressure response. The main advantage claimed for this approach is that it provides a clearer discrimination between different mass transfer resistances (internal diffusion/surface barrier).

Chromatographic Measurements

For fast diffusing systems the limitations imposed by extracrystalline resistances to mass and heat transfer make it impossible to derive reliable intracrystalline diffusivity values from direct sorption rate measurements, regardless of the technique used to follow the uptake. Since both heat and mass transfer are enhanced in a flow system the possibility of deriving reliable diffusion values from measurements of the dynamic response of a packed adsorption column has attracted considerable attention. The early models for mass transfer resistance in a chromatographic column were based on the equilibrium stage concept. The Kubin-Kucera model(73, 74) by showing the relationship between the height equivalent to a theoretical plate

(HETP) and the diffusional time constant provided the essential theoretical basis for the chromatographic approach to the measurement of intra-particle diffusivities. The generalization to bi-porous particles was provided by Sarma and Haynes[38].

The chromatographic response is conveniently analyzed in terms of the first and second moments of the pulse response:

$$\mu \equiv \frac{\int_0^\infty c.t\,dt}{\int_0^\infty c\,dt} = \frac{L}{V}\left[1+\left(\frac{1-\varepsilon}{\varepsilon}\right)K\right] \tag{25}$$

where, for a composite particle:

$$K = \varepsilon_P + \left(1-\varepsilon_P\right)wK_C \tag{26}$$

$$\sigma^2 = \frac{\int_0^\infty \left(t-\mu\right)^2 dt}{\int_0^\infty c\,dt} \tag{27}$$

$$\text{HETP} \equiv \frac{\sigma^2}{\mu^2}L = 2\frac{D_L}{v} + \frac{2\varepsilon v}{(1-\varepsilon)}\left[\frac{R}{3k_f} + \frac{R^2}{15\varepsilon_P D_P} + \frac{r_c^{\,2}}{15KD_c}\right]\left[1+\frac{\varepsilon}{(1-\varepsilon)K}\right]^{-2} \tag{28}$$

It is evident that the HETP measures only the overall resistance to mass transfer and cannot provide evidence concerning the nature of this resistance. The chromatographic response is indeed remarkably insensitive to differences in the nature of the mass transfer resistance so, regardless of the way in which the data are analyzed, it is not possible to obtain any such information except by varying critical parameters such as the particle size.

The major difficulty in the analysis of chromatographic data is to separate the axial dispersion and mass transfer contributions since, except for gaseous systems at very low flow rates, the axial dispersion coefficient (D_L) is velocity dependent. For liquid systems D_L varies essentially linearly with velocity so a plot of HETP vs superficial velocity (εv) should be linear with the mass transfer resistance directly related to the slope (see figure 8). For gaseous systems at high Reynolds number this same plot can be used but in the low Reynolds number region a plot of H/v vs $1/v^2$ may be more convenient since in this region D_L is essentially constant and the intercept thus yields the mass transfer resistance[75, 76].

In the application of the chromatographic method to the measurement of intracrystalline diffusivity it is preferable to pack the column directly with unaggregated crystals rather than with composite (pelleted) material since this eliminates the possible intrusion of macropore resistance. The small crystal size of commercial zeolite samples presents a significant practical problem. Early attempts to utilize a column packed directly with such crystals were not very successful. Erroneously small apparent diffusivities were obtained, probably reflecting the anomalously high axial dispersion which is observed in beds of very small particles due to the tendency of such particles to form agglomerates. Columns packed with large synthetic crystals work well[77] and this approach has the advantage that the

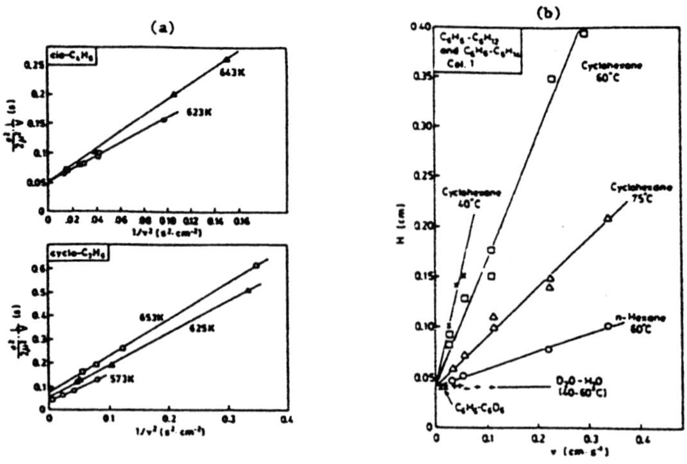

Fig. 8 Chromatographic HETP data. (a) *Vapor Phase.* Plots of H/2v vs $1/v^2$ for cis-butene and cyclopropane in 5A zeolite. From Haq and Ruthven[75]. (b) *Liquid Phase.* Plots of H vs v from liquid chromatographic measurements for benzene/n-hexane in NaX (40 μm crystals). Note that the axial dispersion term ($2D_L/v \cong 0.04$ cm) is essentially the same for C_6H_6-C_6D_6 and for H_2O/D_2O. From Awum et al.[50].

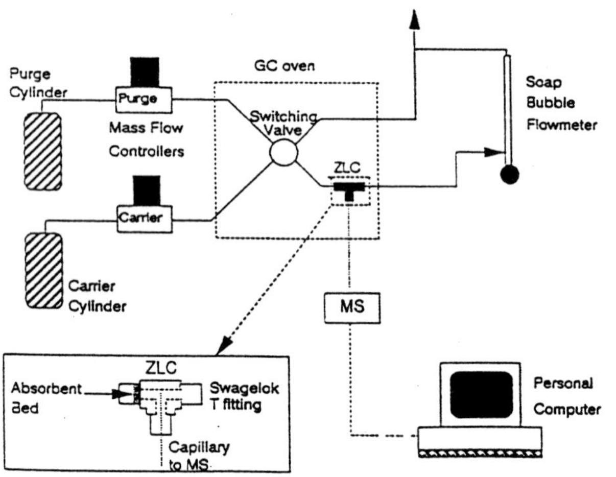

Fig. 9. Schematic diagram showing the experimental set-up for vapor phase ZLC or TZLC measurements.

use of large crystals increases the relative importance of the diffusional resistance (the second term on the right hand side of Eq 28), thus minimizing the impact of any errors in the estimation of the axial dispersion. However, this approach obviously precludes the use of small commercial crystallites. Small commercial crystallites have however been successfully applied in the form of a wall coated column[78] and this may be the best approach when larger crystals are not available.

The Zero Length Column (ZLC) Method

The zero length column (ZLC) method is a chromatographic technique that eliminates the uncertainty due to axial dispersion. It depends on following the desorption of sorbate from a previously equilibrated (small) sample of adsorbent into an inert carrier stream. The ZLC method was introduced, for gas phase adsorption systems, in the late '80s[44, 45] and since then it has found widespread application as a simple and relatively inexpensive way of measuring limiting diffusivities for hydrocarbons and other simple molecules in zeolites and other microporous adsorbents. The method has been extended to the measurement of counter–diffusion in liquid phase adsorption systems by Ruthven and Stapleton[79]. Recently a new variant of the technique, tracer ZLC, has been reported[56], which allows the measurement of self–diffusivities over the entire range of sorbate loadings. A typical experimental set–up for gaseous systems is shown in figure 9.

The actual ZLC column consists of a thin layer of adsorbent material placed between two porous sinter discs. The individual particles (or crystals) are dispersed approximately as a monolayer across the area of the sinter. This minimizes the external resistances to heat and mass transfer, so that the adsorption cell can be considered as a perfectly mixed isothermal, continuous flow cell. Brandani and Ruthven[80] have derived the solution for the general mathematical model for such a system, including the capacity of the fluid phase as well as that of the adsorbent:

$$\frac{c(t)}{c_o} = \sum \frac{2L}{\beta_n^2 + (1 - L + \gamma\beta_n^2)^2 + L - 1 + \gamma\beta_n^2} \exp\left(-\beta_n^2 \frac{D}{R^2} t\right) \tag{29}$$

where the β_n are the positive roots of:

$$\beta_n \cot\beta_n + L - 1 - \gamma\beta_n^2 = 0 \tag{30}$$

with

$$L = \frac{1}{3}\frac{F}{KV_s}\frac{R^2}{D}\ ;\quad \gamma = \frac{1}{3}\frac{V_f}{KV_s} \tag{31}$$

It can be clearly seen that the parameter γ is of the order of 1/K. Therefore the liming case $\gamma = 0$ provides a very useful approximation for gaseous systems where the hold up in the fluid phase can generally be neglected in comparison with the adsorbed phase accumulation. In this situation, Eq. 29 reduces to

$$\frac{c}{c_o} = 2L\sum\frac{\exp\left(-\beta_n^2\frac{D}{R^2}t\right)}{\beta_n^2 + L(L-1)} \tag{32}$$

which is the form obtained from the original ZLC model[44, 45]. It has been shown that eq. 32 can be used with confidence for $\gamma < 0.1$. This model for gas phase adsorption systems has been studied in detail by Brandani and Ruthven[81].

Most of the experimental applications of the ZLC technique have been with gaseous systems and for these systems, the technique may now be regarded as a standard method. Based on our experience it is possible to suggest some guidelines as to how the experiments should be carried out. The key parameter is L, which from its definition, Eq. 31, can be considered as the ratio of the diffusional and washout time constants; R^2/D and KV_s/F. This parameter is also equal to the dimensionless adsorbed phase concentration gradient at the surface of the solid at time zero. From either of these definitions it is evident that L gives an indication of how far removed the system is from equilibrium control. This parameter is proportional to the flow rate, so it can be easily varied and in order to extract a reliable time constant, it is necessary to run the experiment with at least two different flow rates.

A major advantage of the ZLC method is that, for any particular system, the validity of the basic assumptions, under the experimental conditions can be verified directly by a series of simple experimental tests. To establish the validity of the zero length limit, measurements are repeated with columns containing different amounts of adsorbent (see figure 10b). To exclude external mass transfer resistances, experiments may be run with two different carrier gases, typically argon and helium (see figure 10a). Heat effects can be minimized by using high flow rates and system linearity can be checked varying the inlet composition.

If $L>10$ the time constant can be easily extracted from the long time asymptote where Eq. 32 reduces to:

$$\frac{c}{c_o} = 2L\frac{\exp\left(-\beta_1^2\frac{D}{R^2}t\right)}{\beta_1^2 + L(L-1)} \tag{33}$$

A semilog plot of the dimensionless concentration yields L and D/R^2 from the intercept and slope. For $L>10$ the slope should remain constant, while the intercept decreases with increasing L. Possible problems can be encountered if the accuracy of the detector is limited or if the experimental conditions are beyond the range of the Henry law limit. For hydrocarbons a flame ionization detector is typically used. Its high sensitivity allows the response to be followed over a wide concentration range thus facilitating reliable measurement of the long time asymptote. If this asymptote can be measured, the slope should always increase with the flow rate. Since the ZLC response is determined primarily by the limiting behaviour at low concentration, the diffusivity derived in this way corresponds to the limiting value (D_o). The effect of isotherm non-linearity has been studied in detail by Brandani[83]. The method is not well suited to determination of the concentration dependence of diffusivity from integral measurements.

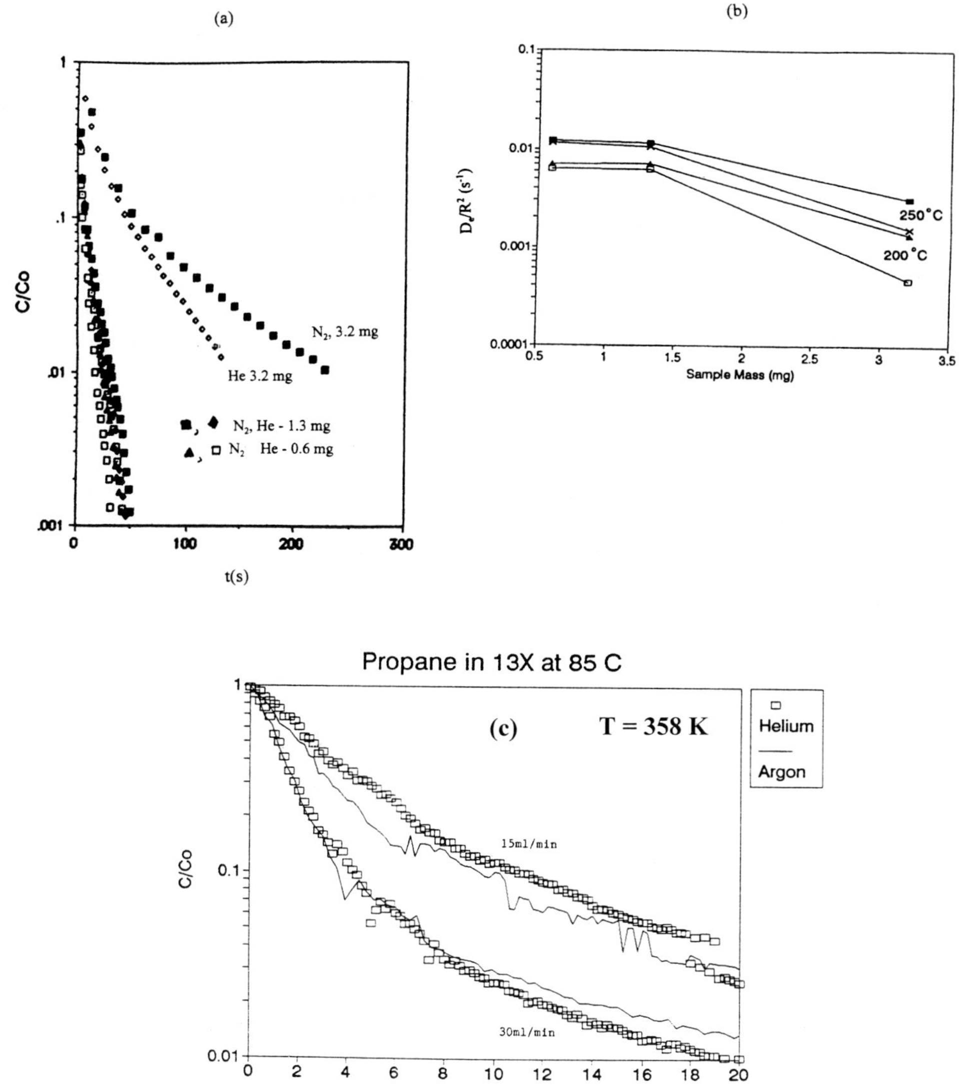

Fig. 10 Effect of sample quantity and nature of purge gas (He or N_2) on (a) the ZLC response curve; (b) the apparent diffusional time constant for benzene-NaX (50 μm crystals at 250°C)[82]; and (c) comparison of ZLC desorption curves for propane-NaX with He and Ar purge[84].

In tracer ZLC (TZLC) the experiment is similar to the standard method but the monitored species is the deuterated form of the sorbate. This introduces an additional constraint on the sorbate and the requirement for a mass sensitive detector such as an on line mass spectrometer. The advantages are the elimination of all possible heat effects, strict linearity of the equilibrium between the fluid phase and the adsorbed phase, and the possibility of measuring directly the tracer diffusivities (which should be the same as the microscopically measured self–diffusivity) over a wide range of concentration. To reduce the costs the carrier may be prepared with a mixture of pure and deuterated hydrocarbons. It has been shown that small imbalances in the concentration of the carrier and the purge streams do not affect the desorption dynamics in any significant way(84). Additional variants of the technique such as the use of partial loading experiments to distinguish between a surface resistance and true intracrystalline diffusion have been discussed elsewhere(81) as well as in the general review article by Brandani and Ruthven(85).

Differential Adsorption Bed (DAB)(53)

Another flow technique that is especially useful for studying multicomponent adsorption kinetics is the differential adsorption bed. In essence a small adsorbent sample is exposed for a known time to a high flow of the sorbate gas mixture in an inert carrier. The bed is then isolated and desorbed by either vacuum or purge. The contents are trapped and analyzed to obtain, after corrections for dead volume, both the composition and total number of moles adsorbed. Repetition of the experiment for different exposure times allows the complete transient sorption curve to be constructed. The method is labor intensive but has the advantages that both kinetic and equilibrium data are obtained and the interpretation of the results is unambiguous. The method has been applied mainly to activated carbon adsorbents but it should be equally applicable to zeolite based materials.

Membrane Measurements

Measurement of the flux through a zeolite crystal membrane under well defined conditions of sorbate pressure on the high and low pressure faces provides an intrinsically attractive approach to the measurement of intracrystalline diffusion. However, in practice this approach is not at all straightforward. The earliest attempts were by Wernick and Osterhuber(41) and Paravar and Hayhurst(40). Both these groups made the membrane from a single large zeolite crystal held in a small hole and sealed with epoxy. Wernick and Osterhuber studied diffusion of n-butane in NaX while Paravar and Hayhurst studied light paraffins in silicalite. Both used the quasi-steady state mode of operation in which a relatively high sorbate pressure is maintained on one side of the crystal with a vacuum on the other side. The flux is then determined from the rate of pressure increase on the low pressure side. The basic system and the form of the pressure response are shown in figure 11. The initial transient leads to a time delay while in the long time region a linear asymptote of pressure vs time is approached. Values for the diffusivity can be derived from both the slope and intercept of such plots: and the consistency of these two values provides a check on the validity of the measurements. The main problem with this approach is associated with the difficulty of properly mounting the small crystal and ensuring that the active faces are not blocked by stray epoxy. More recently measurements have been made with coherent layers of silicalite crystals, grown *in situ* on a stainless steel support(107).

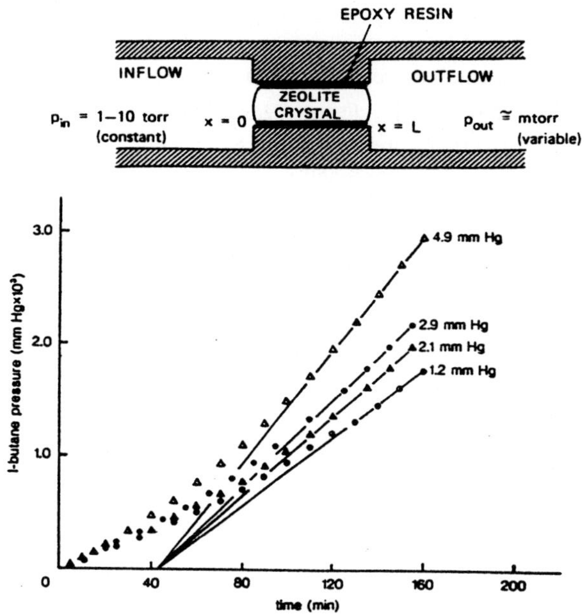

Fig. 11 Permeability measurements for iso-butane through a single (100 μm) silicalite crystal at 334 K. Note the asymptotes all have the same intercept regardless of sorbate partial pressure. This implies that the diffusivity is constant over the relevant range of concentration. From Hayhurst and Paravar[40].

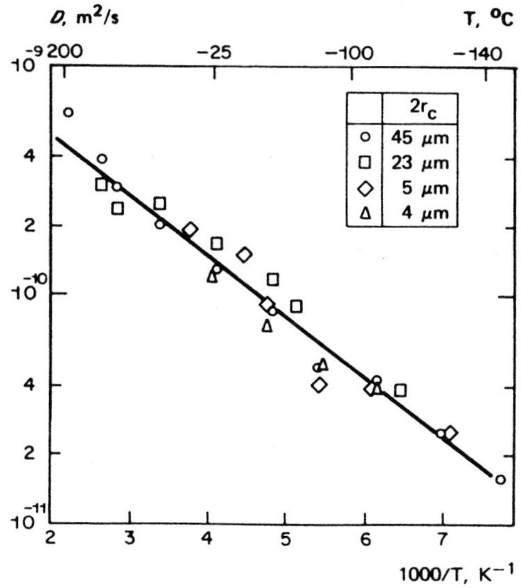

Fig. 12 Effect of crystal size and temperature on PFGNMR self-diffusivity for ethane in NaCaA (5A) zeolite. From Kärger and Caro[86].

Effectiveness Factor

The overall rate of a catalytic reaction, under diffusion limited conditions, depends on the Thiele Modulus $\left(\phi = R\sqrt{k/KD}\right)$. From measurements of reaction rate in a series of particles of different size it is therefore possible to extract both the intrinsic rate constant (k) and the micropore diffusivity. This approach was first suggested by Haag et al[42]. A more complete experimental study (for 2,2 dimethyl butane in HZSM-5) has been reported by Post et al[43] - see section 6.

5. REVIEW OF EXPERIMENTAL DIFFUSIVITY DATA FOR SELECTED SYSTEMS

The experimental methods described above have been applied to study diffusion in a wide range of zeolite systems. For obvious reasons the main focus has been on the practically important zeolites A, X or Y and silicalite/HZSM-5. Diffusion in $AlPO_4$-5 has also attracted attention recently since this material is readily available as large crystals and therefore provides a convenient example of a one-dimensional pore system. Several systems have been studied by both macroscopic and microscopic methods but the "macro" and "micro" diffusivities are not always consistent. This is troubling since it is obvious that one can have confidence in the measured diffusivity values only when the "micro" and "macro" values are in reasonable agreement. This issue is addressed in greater detail below.

5.1 Self-Diffusion

QENS, PFG NMR and tracer exchange measurements yield self-diffusivities directly thus allowing the trends of mobility with such variables as temperature and concentration to be determined without the need for isotherm related correction factors. The temperature dependence in general follow the Arrhenius expression, $\mathcal{D} = \mathcal{D}_\infty e^{-E/\mathbf{R}T}$, with activation energies ranging from a few hundred calories to several k cals per mole, and is therefore conveniently displayed as a plot of log $\mathcal{D}$ vs $1/T$ (see for example figure 12 which shows also the invariance of the diffusivity values with crystal size).

The concentration dependence of the self-diffusivity shows a wide range of different patterns, as illustrated in figures 13 and 14[86-90]. The effect of sorbate-sorbent interaction decreases with loading as the more energetically favorable sites become filled, leading to an increase in mobility with loading. However, sorbate-sorbate interaction increases with loading leading to the opposite trend. On this basis either increasing or decreasing trends with loading can be rationalized. The behavior of methanol in NaX zeolite (figure 14) is of particular interest since it shows a well defined maximum at an intermediate loading. Both the form of the loading dependence and the magnitude of the diffusivity have been confirmed by both PFG NMR and TZLC measurements which, for this system, show excellent quantitative agreement. Further examples of systems showing good agreement between PFG NMR self diffusivities and macroscopically measured values of D_0 derived from ZLC or sorption rate data are shown in figures 15 and 16 while figure 17 shows some examples where there are major discrepancies.

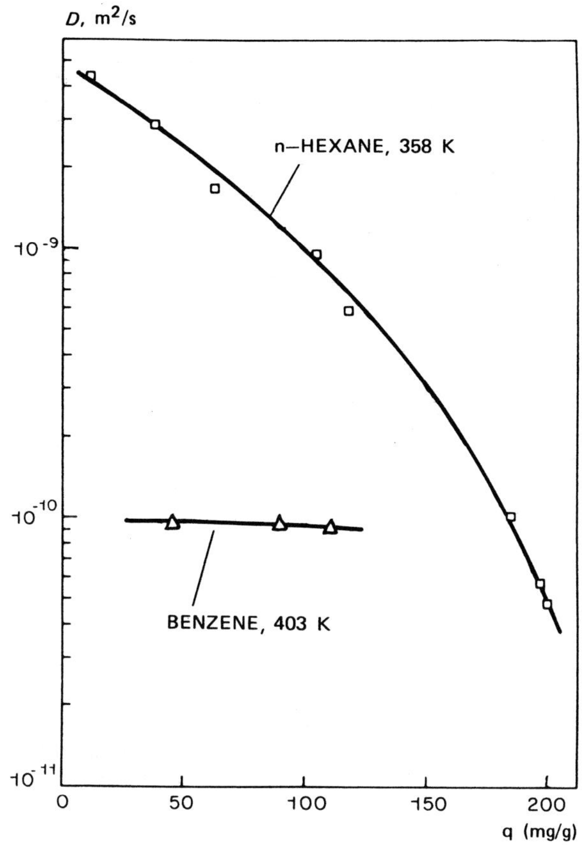

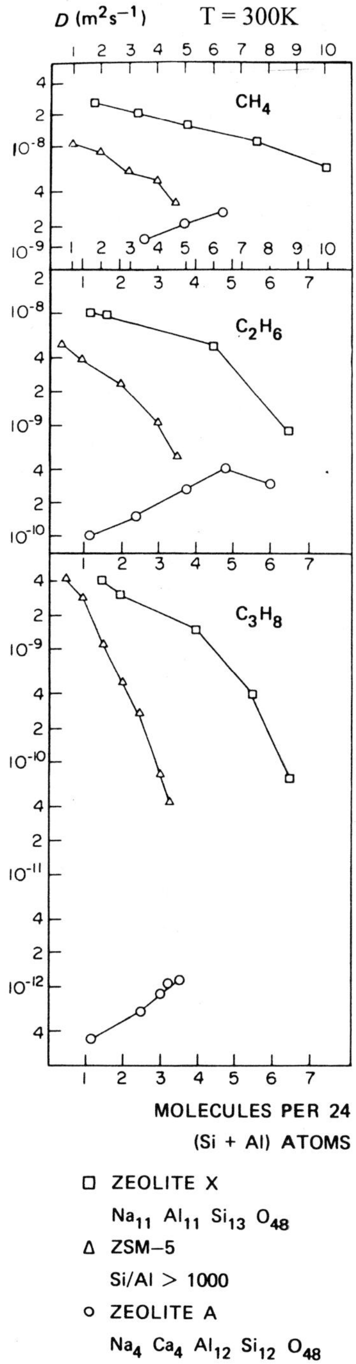

Fig. 13 Concentration dependence of PFG NMR self-diffusivities for (a) n-hexane and benzene in NaX and (b) for light paraffins in NaX (□), silicalite (Δ) and 5A (O). From Kärger and Ruthven[87], Kärger et al.[88] and Pfeifer et al.[89].

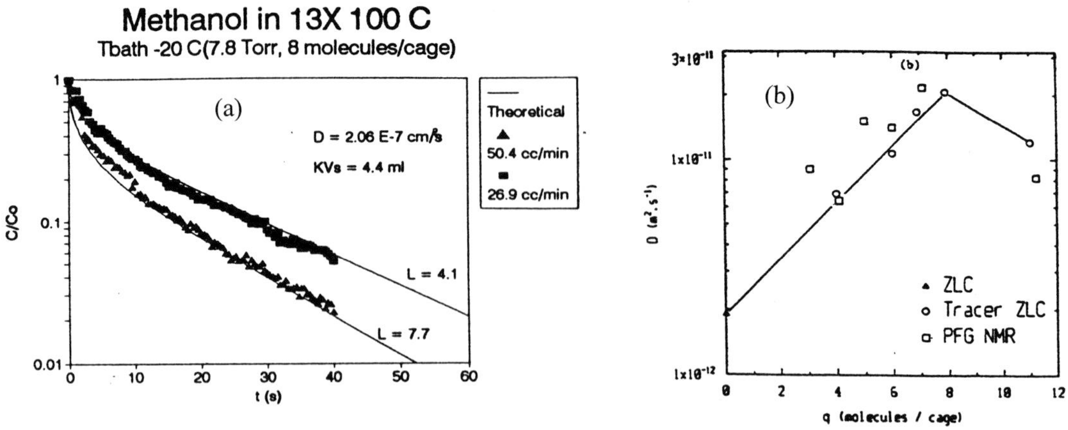

Fig. 14 Diffusion of methanol in large NaX crystals at 100°C. (a) Representative TZLC response curves at two different purge rates and (b) variation of self-diffusivity with loading showing comparison of ZLC, TZLC and PFGNMR data. From Brandani et al.[90].

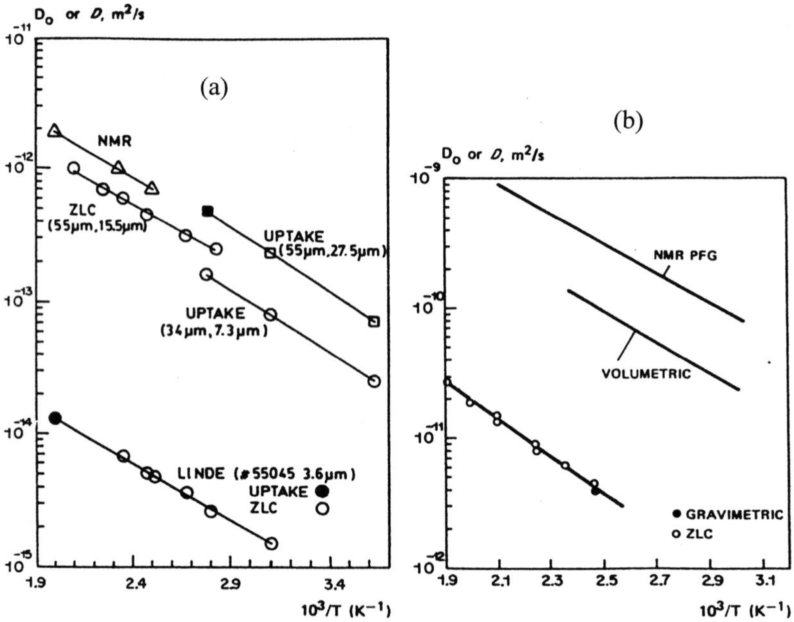

Fig. 15 Temperature dependence of D_o or $\mathcal{D}$. (a) n-butane in 5A zeolite showing correspondence between D_o (from gravimetric measurements) and $\mathcal{D}$ (PFGNMR) for larger crystals. (b) benzene-NaX showing correspondence between gravimetric and ZLC data (for 100 μm and 250 μm crystals). The range of the PFGNMR self-diffusivity values is also shown for comparison. Data from Eic and Ruthven[44, 45], Kärger and Ruthven[87], Germanus et al.[91], and Bülow et al.[92].

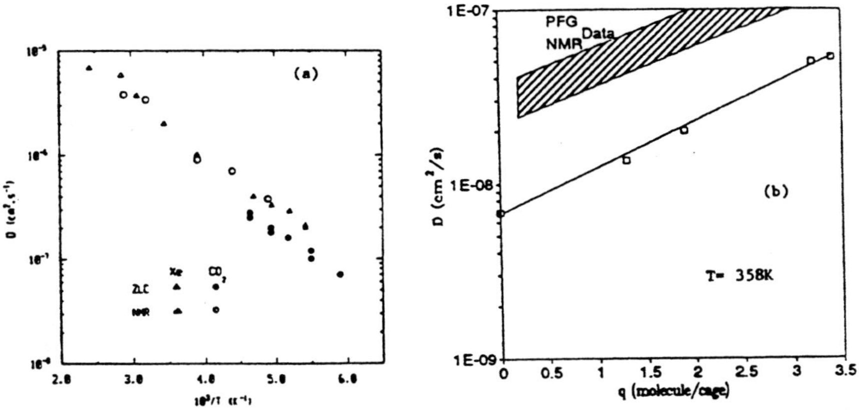

Fig. 16 Comparison of PFGNMR (⊘) and ZLC values for D_o for (a) CO_2 and Xe and (b) propane in large crystals of 5A zeolite. From Ruthven [93] and Hufton et al.[56].

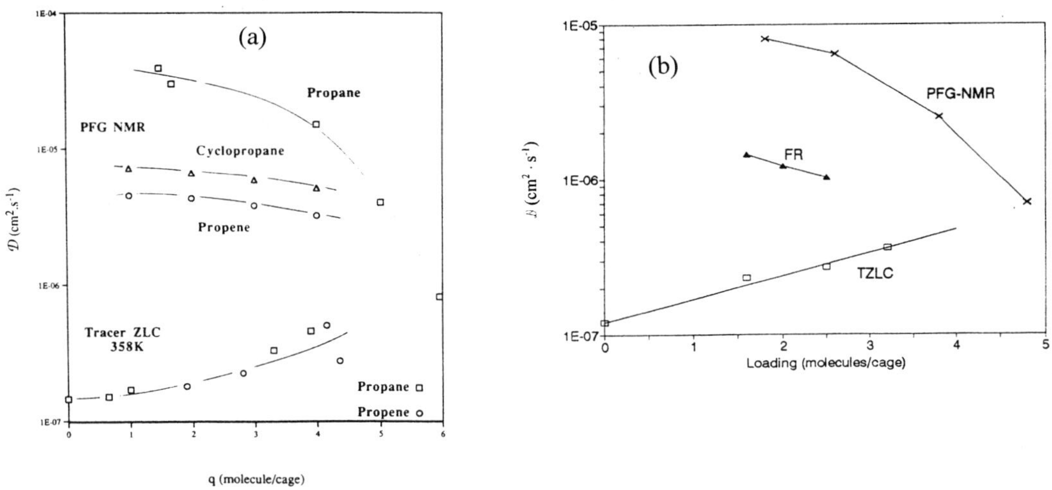

Fig. 17 (a) Comparison of PFGNMR diffusivity data (325 K) and Tracer ZLC data (358K) for propane and propene in NaX zeolite crystals. From Brandani et al.[84]. (b) Comparison of PFGNMR frequency response and tracer ZLC diffusivity data for benzene-NaX at 468 K. From Brandani et al.[82].

5.2 Transport Diffusivities

Consistency of transport diffusivities measured by several different macroscopic experimental techniques (for benzene is silicalite) is illustrated in fig. 18[94]. There appears to be a measurable difference in diffusivity between "silicalite" and the higher Al content HZSM-5 (see below) but for a given material the results obtained by different techniques are quite consistent.

Variation of Crystal Size

In the determination of transport diffusivities by transient techniques the quantity actually measured is the time constant r^2/D. Replicate measurements with different crystal sizes thus provide a straightforward way to investigate the possibility of intrusion of extra crystalline resistances and to confirm the validity of the measured diffusivity values. Figure 19[43] shows a good example of the ideal behavior for diffusion of 2,2 dimethyl butane in HZSM5. r^2/D increases directly with r^2 confirming the constancy of the diffusivity. Further confirmation of the validity of these values comes from fig. 19b which shows consistency of the diffusivity values derived from sorption rate and chromatographic measurements. Similar results for light gases in zeolite 4A are shown in figure 20 in which the values of D_c (rather than r^2/D_c) are plotted against r_c.

In contrast, figure 21 shows that for methane in natural chabazite the apparent intracrystalline diffusivity increases dramatically with crystal size approaching, for very large crystals, the self diffusivity values determined by PFG NMR. The obvious interpretation is that, for this system, the macroscopic transport diffusion measurements in all but the very largest crystals are significantly impacted by external resistances to mass or heat transfer.

Concentration Dependence

Diffusion of the C_8 aromatics in large crystals of NaX is an interesting example. This system has been studied in considerable detail by uptake rate measurements, tracer exchange and ZLC. Some of the data are shown in figures 22a-c. The uptake curves show the expected form with the initial slope varying with r_c in accordance with Eq. 21 so that consistent diffusivity values are obtained for similar pressure steps with different crystal sizes (figure 22a). Differential measurements of the variation of time constant with loading (figure 22b) conform to the simple Darken equation (Eq. 5) and it is evident that the values of D_o/r_c^2 vary inversely with r_c^2 showing the consistency of the rate data. This is shown in figure 22c in which the corrected diffusivity values (D_o) and tracer exchange self diffusivities ($\mathcal{D}$) are seen to be essentially the same and consistent with the limiting values determined by the ZLC method. Remarkably the PFG NMR self diffusivity data show the same general trend ($\mathcal{D}$) (independent of loading over most of the range) but there is a numerical discrepancy of more than an order of magnitude between the two sets of data which appear to converge only in the saturation limit.

A similar discrepancy between macroscopic and microscopic diffusivity measurements has been found for benzene in large crystals of NaX. For this system it has been claimed that piezometric data yield diffusivity values close to the PFG NMR data[92]. However the experimental piezometric system used only one pressure sensor which was connected to the high pressure gas chamber rather than directly to the zeolite chamber. A

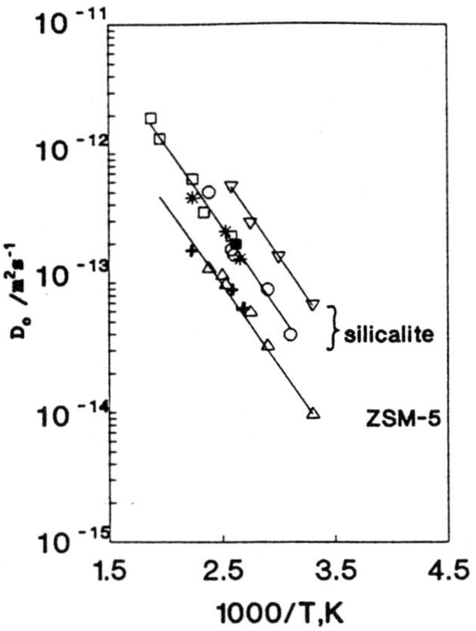

Fig. 18 Comparison of corrected diffusivity of benzene in silicalite-1 and HZSM-5 at low sorbate concentrations. (□ van Den-Begin et al., square wave; (O) Eic and Ruthven, ZLC; (Δ,∇) Zikanova et al., piezometric; (*) silicalite-1, (+) HZSM-6, SSFR; (■) n.m.r. tracer exchange. From Shen and Rees[94].

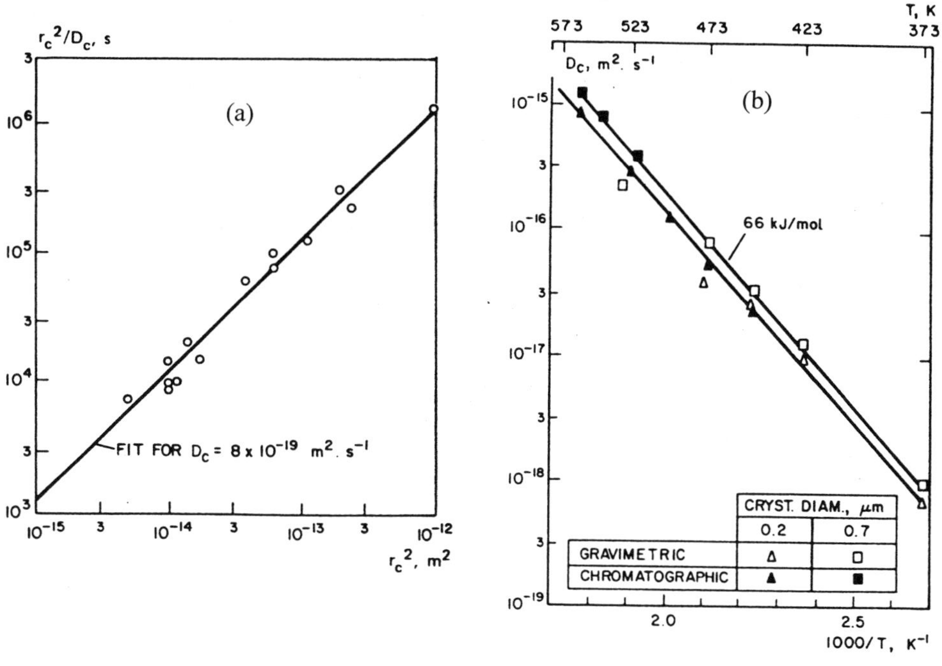

Fig. 19 Diffusion of 2,2 dimethylbutane in silicalite/HZSM5. (a) Variation of (gravimetric) diffusional time constant (r_c^2/D_c) with square of crystal radius showing conformity with the diffusion model. (b) Arrhenius plot showing comparison between gravimetric and chromatographic diffusivities. From Post et al.[43].

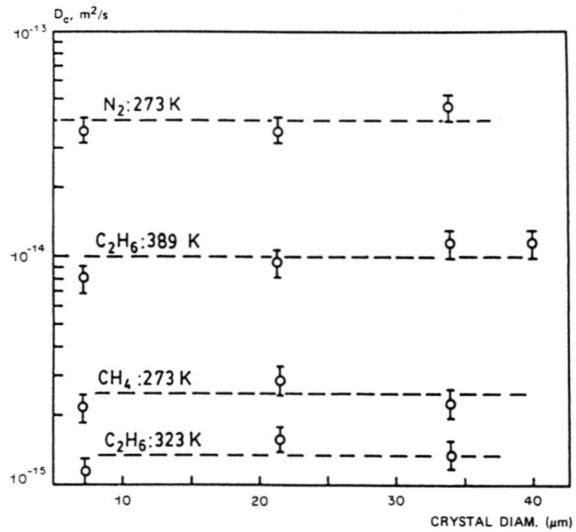

Fig. 20 Gravimetric diffusivity values (D_o) for several gases in 4A zeolite crystals showing consistency of measurements with different crystal sizes. From Yucel and Ruthven[95].

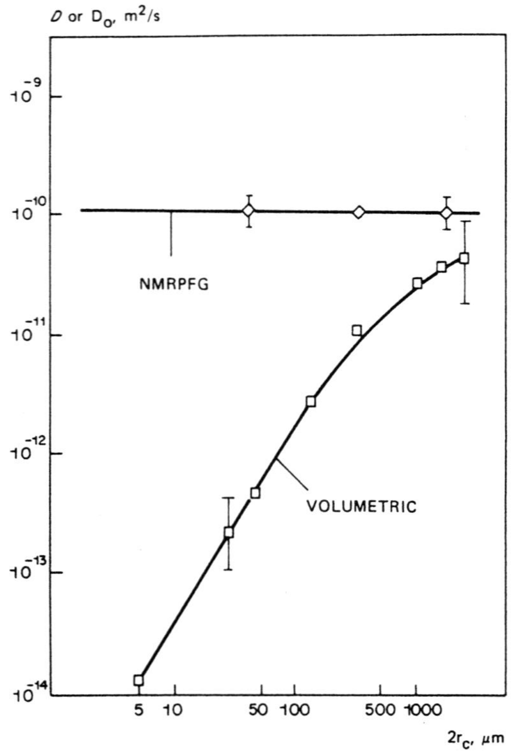

Fig. 21 Comparison of apparent diffusivity values, at 273K, derived from piezometric and volumetric sorption rate measurements for methane in natural chabazite with PFGNMR self-diffusivities. Note that the values converge for large crystals. From Kärger and Caro[86] with permission.

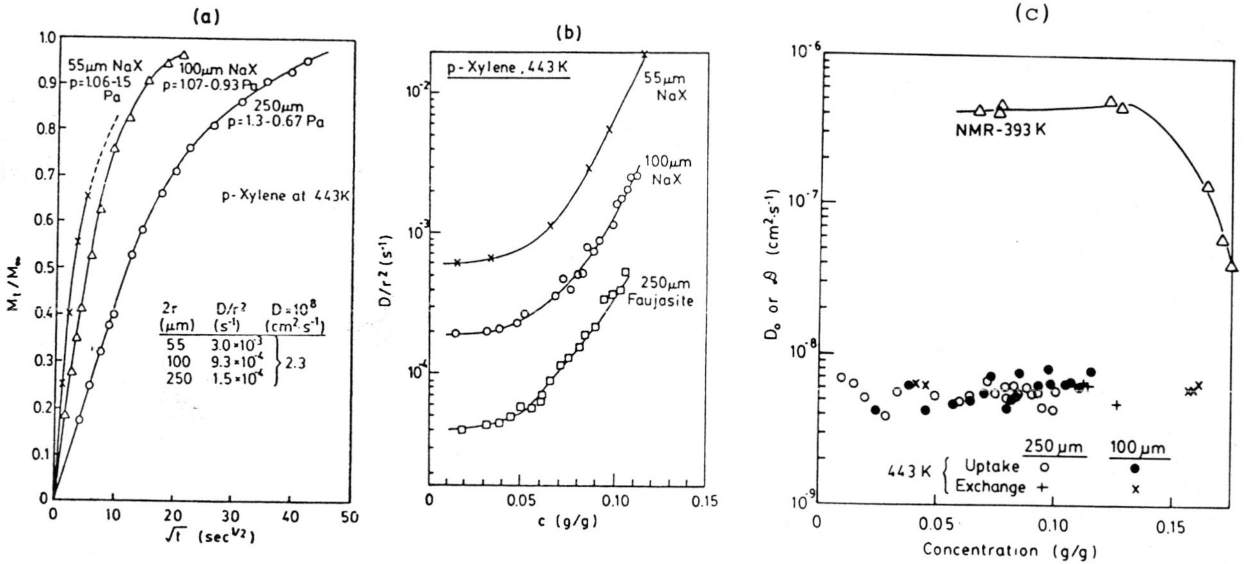

Fig. 22 Experimental sorption curves, diffusional time constants and corrected diffusivities for p-xylene-NaX (55 μm, 100 μm and 250 μm crystals). (a) shows consistency of uptake curves for similar pressure steps with Eq. 21 and constancy of derived diffusionalities for different crystal sizes. (b) shows concentration dependence of the time constants for the three different crystal sizes. (c) shows consistency of derived values of D_o with $\mathcal{D}$ from tracer exchange measurements and deviation from PFGNMR values. From Goddard and Ruthven[97] and Karger and Ruthven[98].

detailed analysis of this experimental system shows that it would probably not have been possible to measure reliable diffusivity values of the magnitude claimed with this experimental setup[108]. More recent macroscopic measurements by the ZLC and TZLC methods[82] have essentially confirmed the earlier large crystal sorption rate data (see figure 17) so the origin of the discrepancy with the PFG NMR remains unresolved[98].

Since the transport diffusivity is in essence the product of the intrinsic mobility and the thermodynamic correction factor (Eq. 5) the variation with loading can be complex. A commonly observed pattern is illustrated in figure 22b. The strong increase of D/r^2 at higher loading levels is due to the thermodynamic correction factor so that values of D_0 calculated according to Eq. 5, are essentially independent of loading (fig. 22c).

For benzene-silicalite at lower temperatures the equilibrium isotherm has an inflexion leading to a maximum and minimum in the diffusivity concentration plot. However this somewhat complex pattern appears to be quantitatively accounted for by the Darken expression so that the corrected difficulty is essentially constant—see fig. 23. These data are quite convincing since the measurements were carried out with several different crystal size fractions with consistent results.

Effect of Chain Length

The effect of chain length on the intracrystalline diffusion of linear paraffins has been studied by several different methods including both macroscopic and microscopic techniques. Representative results are shown in figure 24. In all cases the diffusivities decrease

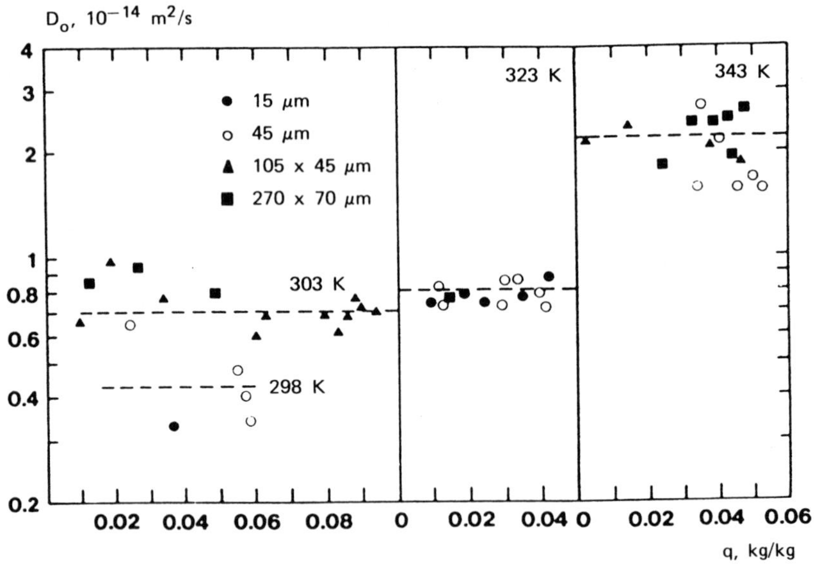

Fig. 23 Benzene in silicalite. Effect of temperature and loading on corrected diffusivity showing consistency of data obtained with different crystal sizes. From Shah and Hayhurst[96] with permission.

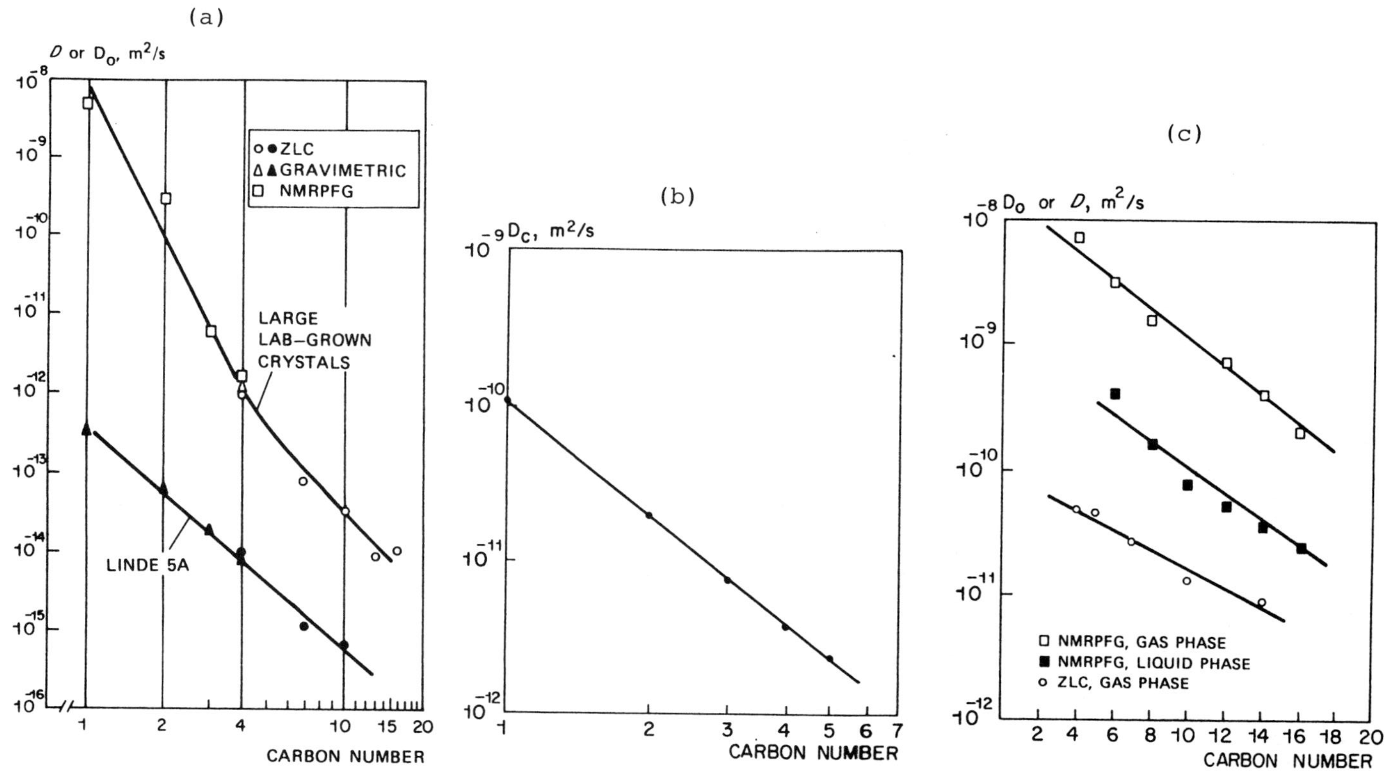

Fig. 24 Variation of diffusivity with carbon number for n-paraffins in (a) 5A zeolite, 473K; (b) silicalite, 334K; and (c) NaX zeolite, 423K. Data from refs. 60, 63, 86-88, 99-105.

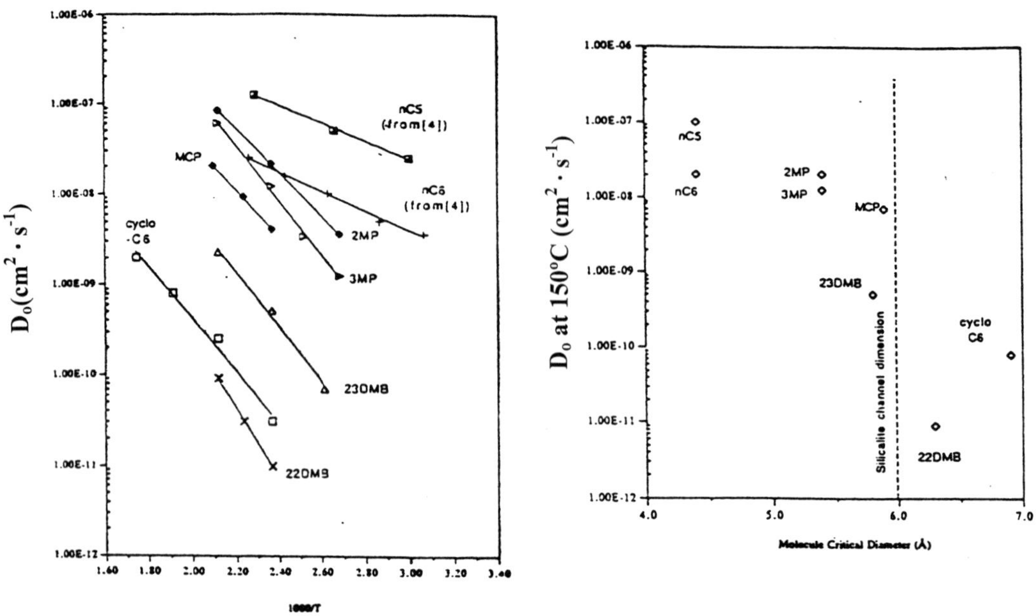

Fig. 25 Diffusion linear and branched paraffins in silicalite (a) Arrhenius plot and (b) variation of D_o at 150°C with critical molecular diameter (gravimetric and ZLC data. From Cavalcante and Ruthven[109].

monotonically with carbon number and there is no evidence to support the existence of a "window effect" as reported by Gorring for diffusion in zeolite T[112] (an intergrowth of erionite-offretite). Indeed a detailed study of diffusion in this system[113] failed to confirm the original results so, in spite of the widespread acceptance of this phenomenon and several attempts to explain it from basic theory we are forced to conclude that the reported behavior was probably simply an artifact of the original measurements and does not represent the true behavior of these systems.

Effect of Molecule Shape

The effect of molecule shape on diffusivity is shown clearly by a recent comparative study of diffusion of the C_6 isomers in large crystals of silicalite[109]. These experiments were carried out with several different crystal sizes so the diffusivity values are probably reliable. The results, shown in figure 25 show that the diffusivities vary between the isomers by about four orders of magnitude. Although there is a small difference in activation energy between the linear and branched isomers the difference in activation energy between singly and doubly branched species and cyclohexane is modest showing that the dramatic reduction in diffusivity must be largely an entropic effect i.e. the requirement for correct orientation of the molecule evidently becomes critical for the more highly branched species. There is a general correlation of diffusivity with critical molecular diameter (the diameter of the cylinder that just circumscribes the molecule in its most favorable equilibrium conformation) but this is clearly not the full story. For example cyclohexane has a larger critical diameter than 2,2 dimethyl butane but the diffusivity is an order of magnitude larger. Clearly the flexibility or compressibility of the molecule must also be important.

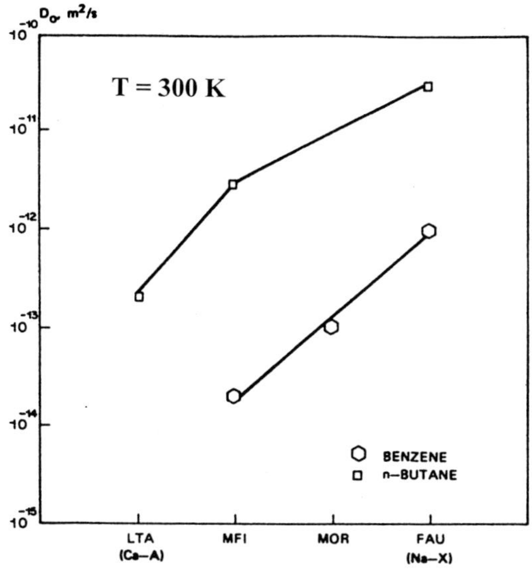

Fig. 26 Comparison of diffusivities (D_o) for n-butane and benzene in several different zeolites showing the effect of pore size. Data are from refs. 96, 100, 110, 111.

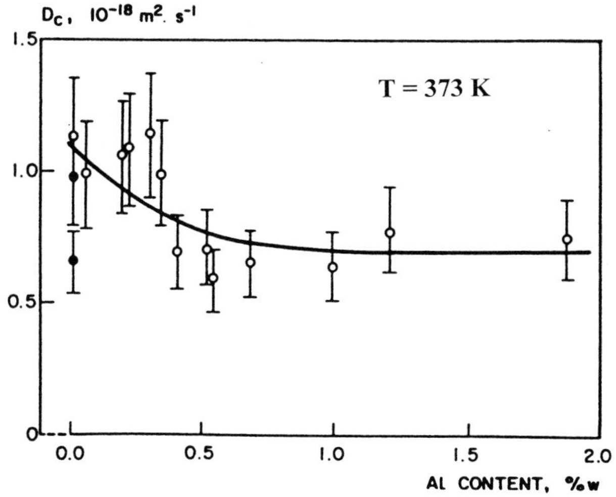

Fig. 27 Effect of Si/Al ratio on the diffusivity for 22 dimethyl butane in HZSM5/silicalite. From Post et al.[(43)].

The effect of pore size is illustrated in fig. 26 for diffusion of benzene and n-butane in several different zeolites.

Effect of Si/Al Ratio

Diffusion in ZSM-5/Silicalite is significantly affected by the Si/Al ratio since the Al centers represent potential cation sites that can act as chemical blocks. This effect is illustrated in Fig. 27[(43)] for 2,2 dimethyl butane (and in fig. 18 for benzene).

The diffusion of 2,2 dimethyl butane has been studied in detail in large crystals of silicalite[(109)] and in much smaller crystals of HZSM5[(43)]. In both these studies the measurements were carried out with a range of differently sized crystals, with consistent results, so there is no reason to doubt the validity of the data. Although the reported activation energies are similar, the difference in diffusivities is surprisingly large (a factor of 20-100) and greatly exceeds the difference normally attributed to the blocking effect of the Al centers (see fig. 27). It seems clear that, for a critically sized sorbate such as 2,2 dimethyl butane, small differences in sample origin and pre-treatment can have a dramatic effect on the diffusivity.

5.3 Diffusion in a Binary Adsorbed Phase

Vapor Phase Systems

Measurements of sorption rates in binary systems under co- and counter-diffusion conditions have been reported by Qreshi and Wei[(18)] and Niessen and Karge[(51, 52)]. The experimental data of Niessen[(115)] for counter-diffusion of aromatics in large crystals of silicalite have been analyzed by van den Broeke[(116)] and shown to be consistent with the chemical potential driving force model (Eq. 13)—see figure 28.

Liquid Phase Systems

Adsorption from the liquid phase involves a counter-diffusion exchange process since the adsorbate displaces solvent in the micropores. Although many industrial scale zeolite based processes are carried out under liquid phase conditions, studies of intracrystalline diffusion in liquid systems are relatively uncommon. The earlier liquid phase sorption rate measurements carried out with small commercial crystals or zeolite pellets were almost certainly controlled by extracrystalline resistances so unreasonably small diffusivity values were obtained. More recently diffusion of benzene in large crystals of NaX has been studied by liquid chromatography[(50)] and by a liquid phase ZLC method[(79)]. The results, shown in figure 29, show remarkable consistency with the extrapolated values of D_o derived from vapor phase measurements at higher temperatures[(80)] but are substantially smaller than the extrapolated PFG NMR self-diffusivities[(91)].

Diffusion of some of the branched alkanes shown in figure 25 has also been studied under liquid phase conditions in large crystals of silicalite by both liquid chromatography and liquid ZLC. For these sorbates the results show a remarkable difference in behavior between gas and liquid systems. Whereas the gas phase data show relatively low diffusivities with large differences between isomers the liquid phase measurements (counter-diffusion vs n-hexane) show higher diffusivity values with less than one order of magnitude difference

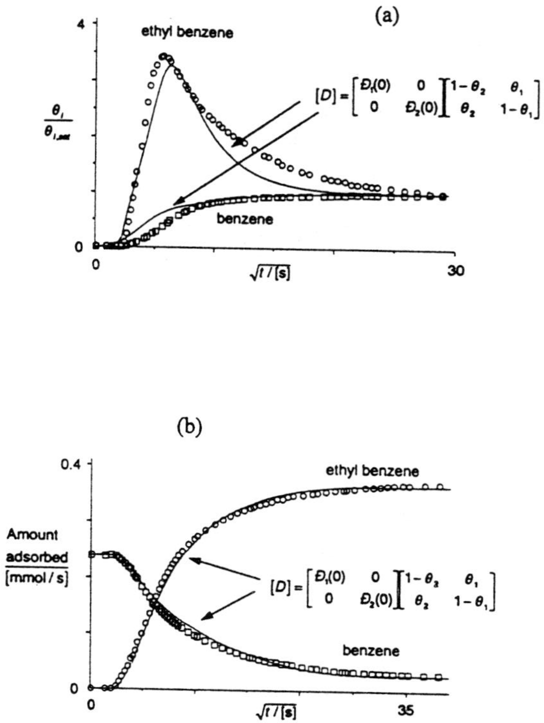

Fig. 28 (a) Co-diffusion and (b) Counter-diffusion transient sorption curves for benzene and ethyl-benzene in silicalite. Pressure 0-115Pa at 375K. For benzene D_o=2.6x10^{-10} cm^2·s^{-1}. These values were derived from the single component measurements. Original experimental data from thesis of Niessen(115); analysis from thesis of van den Broeke(116).

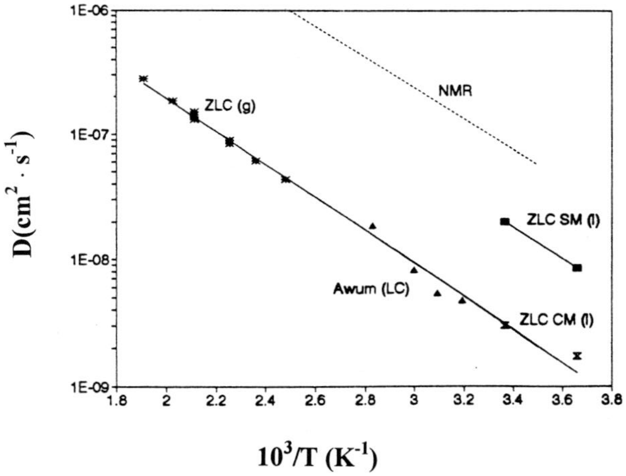

Fig. 29 Liquid phase ZLC measurements for benzene-NaX showing (a) conformity of ZLC response curves with the theoretical model for two different crystal sizes and (b) consistency of the derived liquid phase counter-diffusivities from ZLC measurements with liquid chromatographic data and with the extrapolation of higher temperature vapor phase data. From Brandani and Ruthven(80).

between isomers[114]. The most reasonable explanation seems to be that under fully saturated conditions, the silicalite framework swells slightly so that these species are no longer so severely hindered by interaction with the pore wall, leading to higher diffusivities and smaller differences between isomers.

6. DIFFUSION LIMITATIONS IN ZEOLITE CATALYSIS

In catalytic processes based on zeolites, limitations imposed by the finite rate of intracrystalline diffusion can have a dramatic effect on catalyst performance. If diffusional limitations are involved, variation of the zeolite crystal size will lead to changes in either observed reaction rates or product selectivities (or both). The pertinent literature provides many examples of hydrocarbon conversion reactions where intracrystalline molecular transport has a significant effect on *reaction rates*. Cracking rates of gem-branched dimethyl alkanes on zeolite ZSM-5 have been found to decrease with increasing crystal size, and similar effects have been reported for cracking of 3-methylpentene and n-hexene and n-hexane in differently sized crystal fractions of zeolite ZSM-5. The effect of zeolite crystal size on *selectivity* is probably best exemplified by the generation of para-xylene from either disproportionation of toluene or alkylation of toluene with methanol over zeolite ZSM-5. The desired para-xylene product diffuses in the channel system much more rapidly than the undesired ortho- and meta-xylenes and, consequently, the selectivity to para-xylene increases with zeolite crystal size. Olson and Haag[106] have demonstrated that, for a series of ZSM-5 based catalysts (in toluene disproportionation), a unique relation exists between p-xylene selectivity and diffusion time for ortho-xylene (which is proportional to r_c^2/D_c) determined independently by sorption measurement at lower temperatures.

The extent to which diffusional transport limits the rate of conversion of a single reactant can be quantified using the well-known Thiele relationship[117], giving the effectiveness factor η of a catalyst as a function of the dimensionless Thiele modulus Φ. The effectiveness factor is defined as the ratio between observed and intrinsic rate constants, $\eta = k_{app}/k$. If the reaction is first order in gas phase concentration, the observed rate constant (based on zeolite crystal volume, including the micropores) generally follows from the familiar expression for an integral tubular reactor:

$$k = -\ln(1-\xi)\frac{\varsigma_z * WHSV * \mathbf{R}T}{3.6 * p_o * M} \tag{34}$$

The Thiele modulus for a spherical isotropic zeolite crystallite of radius r_c is given by $\Phi = r_c(k/D_cK_c)^{1/2}$. The reaction rate constant is based on gas phase concentrations, whereas the zeolite diffusivity D_c is defined on the basis of intracrystalline sorbate concentrations. In order to maintain consistency, the adsorption equilibrium constant K_c should therefore be included in the expression for the Thiele modulus (see section 4.2). The relation between η and Φ depends on reaction order, type of zeolite and shape of the zeolite crystals. For a reaction following first-order kinetics in isotropic spherical zeolite crystals, the effectiveness factor function is[118]:

$$\eta = \frac{3}{\Phi}\left\{\frac{1}{\tanh\Phi} - \frac{1}{\Phi}\right\} \tag{35}$$

The above equations apply to diffusion limitations of reactant molecules and can therefore be used to quantify reactant selectivity in shape-selective zeolite catalysis. In the case of product selectivity, a somewhat different approach has been developed.

Determination of diffusivities from catalytic measurements

As has been outlined in the previous section, the effectiveness factors η measured in catalytic experiments using a series of zeolite samples with different crystal size are uniquely related to the proper steady-state diffusivity applicable during catalysis. Determination of the effectiveness factor η yields the quantity Φ. If the intrinsic reaction rate constant k, zeolite crystal size and the equilibrium constant K_c are known, the diffusivity then follows from the value of Φ. Such an approach has been followed by Haag et al., who studied cracking rates of C_6, C_8 and C_9 paraffin and olefin isomers over zeolite ZSM-5 as a function of crystal size.

To evaluate the actual diffusivities, the relation between η and Φ was recalculated to give a relation between η and diffusivity with different values of r_c^2 k as parametric curves (Fig. 30). For each species, sets of diffusivities of a very satisfactory consistency were obtained and it was found that the diffusivities of hydrocarbons in zeolite ZSM-5 decrease by orders of magnitude as the degree of structural branching increases. Since the equilibrium constant K_c was omitted in the Thiele modulus used by Haag et al.[42], the diffusivities reported in Fig. 30 are based on gas phase concentration and should be regarded as K_cD_c, i.e., the product of the equilibrium constant and the conventional intracrystalline transport diffusivity based on intracrystalline sorbate concentration.

Relation between diffusivities from sorption and catalytic measurements

Most attempts to determine transport and self-diffusivities in zeolite crystals are based on a number of physical techniques which have been discussed extensively in Part 4. In the previous section, an alternative method has been discussed allowing elucidation of diffusivities from catalytic studies. The relation between the "physical" and "catalytic" diffusivities, however, is subject to debate since it is not clear to what extent intracrystalline molecular motion prevailing during uptake, chromatographic and PFGNMR studies is representative of counterdiffusion under actual catalytic conditions. Moreover, sorption measurements under catalytically relevant conditions are doomed to fail because the catalytic activity of the zeolite will convert the sorbate.

The relation between diffusion and catalytic reaction of 2,2-dimethylbutane (2,2-DMB) in ZSM-5 zeolite has been studied to bridge the gap between "physical" and "catalytic" diffusivities[43]. Since the intracrystalline mobility of 2,2-DMB determined at lower temperatures (373 K) using conventional uptake studies did not vary significantly with the Si/Al ratio of the zeolite (Fig. 27), ultrapure all-silica MFI samples showing no catalytic activity were used to determine diffusivities at higher temperatures, close to normal catalytic conditions, using the chromatographic technique (Fig. 19) and the results obtained were regarded as being representative of Al-containing MFI zeolite. Simultaneously, the equilibrium constant K_c was determined for a wide range of temperatures.

In the same study, eqn. 34 was used to determine rates of 2,2-DMB and n-hexane cracking over a series of H-ZSM-5 catalysts of different Al content and different crystal size.

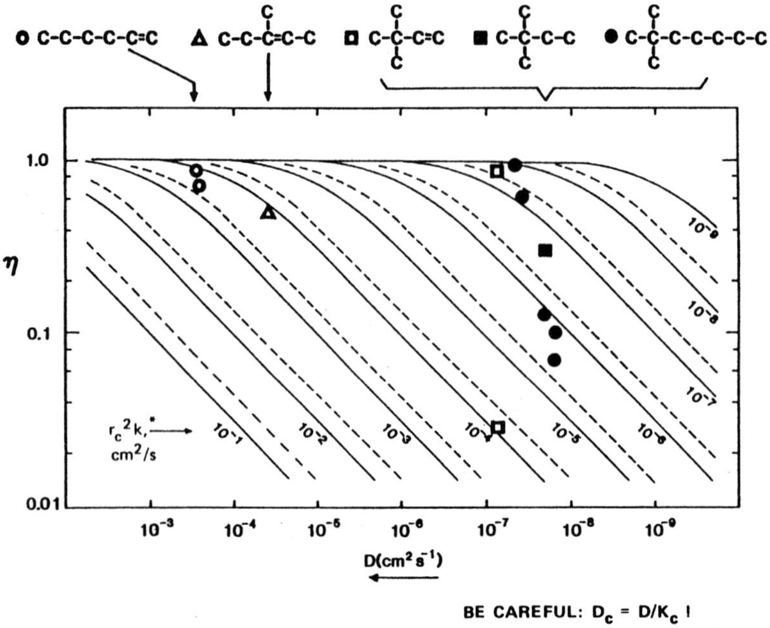

Fig. 30 Relation between effectiveness factor and diffusivity. Showing graphical extinction of diffusivities for C_6 and C_9 hydrocarbons in HZSM5 at 811 K. From Haag et al.[42] with permission.

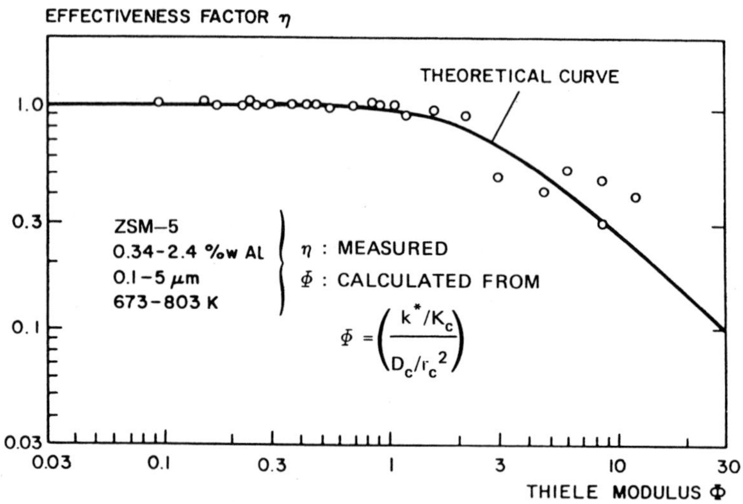

Fig. 31 Experimental verification of the Thiele model for zeolite catalysis. Conversion of 22 Dimethyl butane over HZSM5. From Post et al.[43].

While n-hexane cracking rates were proportional to Al content and independent of crystal size, 2,2-DMB cracking rates in the coarse-crystalline samples were lower than anticipated on the basis of their chemical composition due to diffusion limitations. The Thiele modulus was estimated by extrapolation of the diffusion time constants (D_c/r_c^2) measured at low temperature to actual reaction temperature, using the activation energy of the diffusivity of 2,2-DMB in all-silica MFI materials established with the chromatographic sorption studies. The adsorption constant featuring the Thiele modulus was also obtained by extrapolation of the chromatographic data available for all-silica MFI. Figure 31 shows measured effectiveness factors η for 2,2-DMB conversion over ZSM-5 catalysts as a function of estimated Thiele moduli based on physical sorption studies. The coincidence of the experimentally determined data with the theoretical relation, eqn. (35), indicates that, for the system 2,2-DMB/ZSM-5, diffusivities obtained independently from physical sorption and catalytic studies show excellent agreement. It is not clear yet to what extent the results obtained with this system can be extrapolated to other sorbate/zeolite combinations.

7. CONCLUSIONS

The past three decades have seen an increased emphasis on the study of intra-crystalline diffusivities of various sorbates in zeolite crystals. This interest has been fuelled primarily by the recognition that a thorough knowledge of intracrystalline diffusion phenomena leads to a better understanding of established applications of zeolites in (shape-selective) catalysis and in selective adsorption/separation processes, and may provide the information required for (i) further optimization of existing commercial applications of zeolites and (ii) development of new applications in the above areas. Much progress has been achieved in our understanding of single-component intracrystalline diffusion by the application of a variety of physical techniques and by careful study of the effect of zeolite crystal size. Many of the earlier discrepancies between the results obtained by different techniques have now been resolved and can be traced back to the intrusion of extra-crystalline transport resistances, surface barriers at the outer rim of the zeolite crystal or heat effects. Nevertheless, for a number of sorbate/zeolite systems, substantial differences persist when transport diffusivities obtained from uptake rate measurements and self-diffusivities from the NMR pulse field gradient method are compared.

While real progress has been achieved in single-component diffusion, most technologically important catalytic and adsorption processes involve multicomponent systems. This area, which as yet has received much less attention, should be an important field for future research since it is counterdiffusion rather than single-component diffusion that actually determines zeolite performance in many commercial applications. Nevertheless, evidence has been provided that diffusivities from single-component sorption experiments can be correlated with the catalytic performance of zeolites and can be used to predict catalytic conversions and selectivities.

9. BIBLIOGRAPHY

The following articles provide in depth reviews of one or more aspects of diffusion in zeolites.

Diffusion in Zeolites and other Microporous Solids, by J. Kärger and D.M. Ruthven, John Wiley, New York (1992).

NMR Studies of Diffusion in Porous Solids, by J. Kärger in *Physical Adsorption: Theory and Experiment* p. 297-329, J. Fraissard and C.W. Connor eds. NATO ASI Series C 491. Kluwer, Dordrecht (1997).

Fundamentals of Diffusion in Porous and Microporous Solids, by D.M. Ruthven in *Physical Adsorption: Theory and Experiment*, p. 241-259, J. Fraissard and C.W. Conner eds., NATO ASI Series C 491 Kluwer, Dordrecht (1996).

Measurement of Diffusion in Microporous Solids by Macroscopic Methods, by D.M. Ruthven and S. Brandani, in *Physical Adsorption: Theory and Experiment*, p. 281-290, J. Fraissard and C.W. Conner eds., NATO ASI Series C 491 Kluwer, Dordrecht (1997).

Exciting New Advances in Diffusion of Sorbates in Zeolites, by L.V.C. Rees in *Zeolites and Related Microporous Materials: State of the Art 1994*, p 1133-1150 J. Weitkamp, H.G. Karge, H. Pfeifer and W. Holderich eds., Elsevier, Amsterdam (1994).

On the Comparison between Macroscopic and NMR Measurements of Intracrystalline Diffusion in Zeolites, by J. Kärger in *Molecular Sieves: Science and Technology*, Vol. 7, H.G. Karge and J. Weitkamp eds. Springer, Berlin (1999).

Principles and Applications of Self-Diffusion Measurements by NMR, J. Kärger, H. Pfeifer and W. Heink, *Advan. Magnetic Resonance*, 12, 1-89 (1988).

Molecular Mobility Measurements of Hydrocarbons in Zeolites by NMR Techniques, by J. Caro, H. Jobic, M. Bulow, J. Kärger and B. Zibrowius, *Adv. Catalysis*, 39, 351-414 (1993).

Adsorption Analysis: Equilibria and Kinetics, by D.D. Do, Imperial College Press, London (1998).

The Maxwell-Stefan Approach to Mass Transfer, by R. Krishna and J.A. Wesselingh, *Chem. Eng. Sci.* **52**, 861-911 (1997).

NOTATION

A	intensity of NMR spin-echo signal (Eq. 18)
B	molecular mobility in adsorbed phase
c	sorbate concentration in fluid phase
C_p	heat capacity of adsorbent sample
D	transport diffusivity
D_e	effective diffusivity in macropore = $\varepsilon_p D_p[1+(1-\varepsilon_p)K]$
D_o	corrected transport diffusivity (Eq. 5)
D_p	pore diffusivity (for macroporous pellet)
$\mathcal{D}$	self-diffusivity or tracer diffusivity
F	purge flow rate in ZLC experiment
G	gradient magnetic field
h	heat transfer coefficient (from particle or adsorbent sample)
H	height equivalent to a theoretical plate (HETP)
J	diffusive flux (relative to fixed coordinates)
k	intrinsic (first order) reaction rate constant
k_{app}	apparent first order reaction rate constant
k_s	surface mass transfer coefficient
K	dimensionless Henry's Law constant based on particle volume
K_c	dimensionless Henry's Law constant based on crystal volume
L	dimensionless parameter (Eq. 31)
M_t/M_∞	fractional approach to equilibrium in an uptake experiment
q	adsorbed phase concentration
q_s	saturation limit
r	radial coordinate
r_c	crystal radius
R	particle radius
R	gas constant
t	time
u	diffusive velocity (Eq. 8)
w	volume fraction of zeolite in composite adsorbent particle
z	distance coordinate

Greek Symbols

α, β	functions in Eq 7; dimensionless parameter in Eq. 24
γ	dimensionless parameter (Eq. 31)
δ, Δ	duration of gradient pulse and time interval between gradient pulses in PFG NMR method (see fig. 3)
ε	voidage of column or adsorbent bed
ε_p	porosity of composite adsorbent particle
θ	fractional saturation of adsorbent (q/q_s)
ϕ	Thiele modulus (defined by $\phi = R\sqrt{k/K_c D_c}$ or $R\sqrt{k/D_c}$)
μ	chemical potential; mean retention time
η	effectiveness factor (k_{app}/k)
ξ	extent of reaction
ρ	density of zeolite crystal or adsorbent particle

REFERENCES

1. P.B. Weisz, *Chemtech* **3**, 498-505 (1973)

2. P.B. Weisz and C.D. Prater, *Adv. Catal.* **6**, 143-196 (1954).

3. P.B. Weisz and V.J. Frillette, *J. Phys. Chem.* **64**, 382 (1960).

4. N.Y. Chen and W.E. Garwood, *Catal. Revs. Sci. Eng.* **28**, 185-264 (1986).

5. D.M. Ruthven, *Principles of Adsorption and Adsorption Processes*, John Wiley, New York (1984).

6. J. Kärger and D.M. Ruthven, *Diffusion in Zeolites and Other microporous Solids*, John Wiley, New York (1992).

7. L.S. Darken, *Trans. A.I.M.E.* **175**, 184 (1948).

8. R. Ash and R.M. Barrer, *Surf. Sci.* **8**, 461 (1967).

9. J. Kärger, *Surf. Sci.* **36**, 797 (1973).

10. R.M. Barrer and B.E.F. Fender, *J. Phys. Chem. Solids* **21**, 12 (1961).

11. J. Crank, *Mathematics of Diffusion*, Clarenden Press, Oxford (1956).

12. J. Caro, G. Finger, E. Jahn, J. Kornatowski, F. Marlow, M. Noack, and B. Zibrowius, *Proc. 9th Int. Zeolite Conf.* Montreal (1992), p. 683-691, R. von Ballmoos, J.B. Higgins, M.M.J. Treacy eds., Butterworth-Heinemann, Stoneham, MA (1993).

13. J. Caro, M. Noack, F. Marlow, D. Peterson, M. Griepenstrog, and J.J. Kornatowski, *J.Phys. Chem.* **97**, 13, 685 (1993)

14. R. Krishna, *Chem. Eng. Sci.* **45**, 1779 (1990).

15. R. Krishna, *Chem. Eng. Sci.* **48**, 845 (1993).

16. H.W. Habgood, *Can. J. Chem.* **36**, 1384 (1958).

17. G.F. Round, H.W. Habgood and R. Newton, *Sep. Sci.* **1**, 219 (1966).

18. W.Q. Qureshi and J. Wei, *J. Catal.* **126**, 126 and 147 (1990).

19. P.A. Fedders, *Phys. Revs. B* **40**, 17 (1978).

20. J. Kärger, M. Petzold, H. Pfeifer, S. Ernst and J. Weitkamp, *J. Catalysis* **136**, 283 (1992).

21. J. Kärger, "Single File Diffusion in Zeolites" in *Sorption and Diffusion in Molecular Sieves, Science and Technology* Vol. 7, H.G. Karge and J. Weitkamp eds., Spriger, Berlin (1999).

22. D.M. Ruthven, *Can. J. Chem.* **52**, 3523 (1974).

23. K. Dahlke, G. Emig and E. Aust, *Fourth Int. Conf. On Fundamentals of Adsorption*, Kyoto (1992), Proceedings p. 129-136, ed. M. Suzuki, Kodansha-Elsevier, Tokyo (1993).

24. S.M. Auerbach, N.J. Hanson, A.K. Cheetham and H.I. Metiu, *J. Phys. Chem.* **99**, 10600 (1995).

25. M.C. Coppens, A.T. Bell and A.K. Chakraborty, *Chem. Eng. Sci.* **53**, 2053-2061 (1998).

26. A. Tiselius, *Z. Phys. Chem.* **A169**, 425 (1934).

27. A. Tiselius, *Z. Phys. Chem.* **A174**, 401 (1935).

28. R.M. Barrer, *Trans Faraday Soc.* **37**, 590 (1941).

29. R.M. Barrer, *Zeolites and Clay Minerals*, Academic Press, London, (1978).

30. M. Bülow, J. Kärger, N. van Phat and W. Schirmer, *Z. Phys. Chem. (Leipzig)* **257**, 1205-1216 (1976). See also M. Bulow, et al., *Proc. 6th Internat. Zeolite Conf.* Reno (1983) pp 242-250.

31. D.D. Do, *Chem. Eng. Sci.* **50**, 549-553 (1995), and *Ibid* **53**, 3459-3467.

32. H.A. Resing and J.K. Thompson, *J. Chem. Phys.* **46**, 2876 (1967).

33. H.A. Resing and J.S. Murday, *Proc. 3rd Internat. Zeolite Conf.*, Zurich (1973). Published as *Molecular Sieves*, Adv. Chem. **121**, 414-440, Am. Chem. Soc. Washington (1974).

34. J. Kärger, *Z. Phys. Chem. (Leipzig)* **248**, 27 (1971).

35. H. Pfeifer, "NMR of molecules Adsorbed on Solids" in *NMR-Basic Principles & Process* **7**, 53-153, Springer, Berlin (1972).

36. H. Lechert, K-P. Wittern and W. Schweitzer, "NMR Studies of mobility of Aromatics in Faujasite", *Proc. Zeolite Symp,* Szeged, Hungary Sept. 1978, *Acta Physica et Chemica* **24**, 201-206 (1978).

37. Y.A. Ma and C. Mancel, *Adv. Chem.* **121**, 392 (1974).

38. H.W. Haynes and P.N. Sarma, *Adv. Chem.* **133**, 205 (1974).

39. D.B. Shah and D.M. Ruthven, *AIChE Jl.* **23**, 804 (1977).

40. A. Paravar and D.T. Hayhurst, *Proc. 6th Internat. Zeolite Conf.* Reno, (1983) p. 217-224 D. Olson and A. Bisio eds., Butterworth, Guildford (1984).

41. D.L. Wernick and E.J. Osterhuber, *Ibid* pp 122-130 (1984).

42. W.O. Haag, R.M. Lago and P.B. Weisz, *Faraday Disc.* **72**, 317 (1982).

43. M.F.M. Post, J. van Amstel and H.W. Kouwenhoeven, in *Proc. 6th Internat. Zeolite Conf.*, Reno, 1983 p. 517-527 D. Olson and A. Bisio eds., Butterworth, Guildford, UK (1984).

44. M. Eic and D.M. Ruthven, *Zeolites* **8**, 40-45 (1988).

45. D.M. Ruthven and M. Eic, *Am. Chem. Soc. Symp. Ser.* **368**, 362 (1988).

46. Y. Yasuda, *J. Phys Chem.* **80**, 1867 (1976).

47. N.G. van den Begin and L.V.C. Rees, *Proc. 8th Internat. Zeolite Conf.*, Amsterdam, p. 915-924, P.A. Jacobs and R.A. van santen eds., Elsevier, Amsterdam (1989).

48. H. Jobic, M. Bee and G.J. Kearley *Zeolites* **9**, 312 (1989).

49. E. Cohen de Lara, R. Kahn and F. Mezei, *J. chem. Soc. Faraday Trans* **79**, 1911 (1983).

50. F. Awum, S. Narayan and D.M. Ruthven, *Ind. Eng. Chem. Res.* **27**, 1510 (1988).

51. W. Niessen and H. Karge, *Studies in Surf. Sci. Catalysis* **60**, 213-221 (1991).

52. W. Niessen and H. Karge, *Microporous Materials* **1**, 1-8 (1993).

53. D.D. Do, X. Xu and P.L.J. Mayfield, *Gas Sep. Purif.* **5**, 35-48 (1991).

54. Ph. Grenier, F. Meunier, P.G. Gray, J. Kärger, Z. Xu and D.M. Ruthven, *Zeolites* **14** 242-249 (1994).

55. V. Bourdin, Ph.D. Thesis, Univ. of Paris XI (1996).

56. J.R. Hufton, S. Brandani and D.M. Ruthven, *Proc. 10th Internat. Zeolite Conf.* Garmisch (1994) pp. 1323-1330, J. Weitkamp, H.G. Karge, H. Pfeifer and W. Holderich eds. Elsevier, Amsterdam.

57. M. Baerns and O. Keipert, *Chem. Eng. Sci.* **53**, 3623-3634 (1998).

58. T.A. Nijhuis, L.J.P. van den Broeke et al., *Chem. Eng. Sci.* **52**, 3401-3404 (1997).

59. E.O. Stejskal and J.E. Tanner, *J. Chem. Phys.* **42**, 288 (1965) and **49**, 1768 (1960).

60. R. Richter, R. Seidel, J. Karger, W. Heink, H. Pfeifer, H. Furtig, W. Höse and W. Roscher, *Z. Phys. Chem (Leipzig)* **267**, 1145-1151 (1986).

61. J. Kärger and H. Pfeifer, *Zeolites* **7**, 90-107 (1987).

62. J. Kärger, *AIChE Jl.* **28**, 417 (1982).

63. J. Kärger, H. Pfeifer, M. Rauscher, M. Bulow, M. Samulic, and S.P. Shadnov, *Z. Phys. Chem. (Leipzig)* **262**, 567 (1981).

64. J. Kärger, H. Pfeifer and W. Heink, *Proc. 6th Internal. Zeolite Conf.*, Reno (1983), D. Olson and A. Bisio eds., Butterworths, Guildford (1984).

65. D.M. Ruthven, L.K. Lee and H. Yucel, *AIChE Jl.* **26**, 16 (1980).

66. L-K. Lee and D.M. Ruthven, *J. Chem. Soc. Faraday Trans I* **75**, 2406 (1979).

67. D.M. Ruthven and L.K. Lee, *AichE Jl.* **27**, 654 (1981).

68. R. Haul and H. Stremming, *J. Colloid Interface Sci.* **97**, 348 (1984).

69. L.V.C. Rees and D.M. Shen, *J. Chem. Soc. Faraday Trans I* **86**, 3687 (1990).

70. D.M. Shen and L.V.C. Rees, *Zeolites* **11** 684-689 (1991).

71. D.M. Shen and L.V.C. Rees, *J. Chem. Soc. Faraday Trans I*, **89**, 1063-1065 (1993).

72. L.M. Sun and V. Bourdin, *Chem. Eng. Sci.* **48**, 3783-3793 (1993).

73. M. Kubin, *Coll. Czech. Chem. Commun.* **30**, 1104 and 2900 (1965).

74. E.J. Kucera, *J. Chromatog.* **19**, 237 (1965).

75. N. Haq and D.M. Ruthven, *J. Colloid Interface Sci.* **112**, 164 (1986).

76. C.B. Ching and D.M. Ruthven, *Zeolites* **8**, 68-73 (1988).

77. J.R. Hufton, D.M. Ruthven and R.P. Danner, *Microporous Materials* **5**, 39-52 (1995).

78. M. Delmas, C. Cornu and D.M. Ruthven, *Zeolites* **15**, 45-50 (1995).

79. D.M. Ruthven and P. Stapleton, *Chem. Eng. Sci.* **48**, 89-98 (1993).

80. S. Brandani and D.M. Ruthven, *Chem. Eng. Sci.* **50**, 2055-2059 (1995).

81. S. Brandani and D.M. Ruthven, *Adsorption,* **2**, 133-143 (1996).

82. S. Brandani, Z. Xu and D.M. Ruthven, *Microporous Materials*, **53**, 2791-2798 (1998).

83. S. Brandani, *Chem. Eng. Sci.,* **53**, 2791-2798 (1998).

84. S. Brandani, J.R. Hufton and D.M. Ruthven, *Zeolites,* **15** 624-631 (1995).

85. D.M. Ruthven and S. Brandani, "Measurement of Diffusion in Porous or Microporous Solids by the ZLC Method" in *Recent Advances in Gas Separation by Microporous Materials*, N. N. Kanellopoulos and N. Kekzis eds., Elsevier, Amsterdam (1999).

86. J. Kärger and J. Caro, *J. Chem. Soc. Faraday Trans I*, **73** 1363-1376 (1988).

87. J. Kärger and D.M. Ruthven, *J. Chem. Soc. Faraday Trans I,* **77**, 1485-1496 (1981).

88. J. Kärger, H. Pfeifer, M. Rauscher and A. Walter, *J. Chem. Soc. Faraday Trans I,* **76**, 717-737 (1980).

89. H. Pfiefer, J. Kärger, A. Germanus, W. Schirmer, M. Bulow and J. Caro, *Adsorption Sci. Technol.*, **2**, 229-239 (1985).

90. S. Brandani, D.M. Ruthven and J. Kärger, *Zeolites*, **15**, 494-496 (1995).

91. A. Germanus, J. Kärger, H. Pfeifer, N.N. Samulic and S.P. Shdanov, *Zeolites* **5**, 91-95 (1985).

92. M. Bulow, W. Mietk, P. Struve and P. Lorenz, *J. Chem. Soc. Faraday Trans I*, **79**, 2457-2466 (1983).

93. D.M. Ruthven, *Zeolites* **13**, 594 (1993).

94. D.M. Shen and L.V.C. Rees, *Zeolites* **11**, 666-671 (1991).

95. H. Yucel and D.M. Ruthven, *J. Chem. Soc. Faraday Trans I*, **76**, 60-70 (1980).

96. D.B. Shah, D.T. Hayhurst, G. Evanina and C.T. Guo, *AIChE Jl.* **34**, 1713-1717 (1988).

97. M. Goddard and D.M. Ruthven, *Proc. 7th Int. Zeolite Conf.*, Tokyo (1986) p 467, Y. Murkami et al. eds. Kodansha, Tokyo (1986).

98. J. Kärger and D.M. Ruthven, *Zeolites* **9**, 287 (1989).

99. J. Kärger, M. Bulow, G.R. Millward and J.M. Thomas, *Zeolites* **6**, 146-150 (1986).

100. M. Eic, M. Goodard and D.M. Ruthven, *Zeolites* **8**, 327-331 (1988).

101. D.T. Hayhurst and A.R. Paravar, *Zeolites* **8**, 27-29 (1988).

102. H. Yucel and D.M. Ruthven *J. Chem. Soc. Faraday Trans I* **76**, 71-83 (1980).

103. J. Caro, M. Bülow and J. Kärger, *Chem. Eng. Sci.* **40**, 2169-2170 (1989).

104. D.M. Ruthven, R.I. Derrah and K.F. Loughlin, *Can. J. Chem.* **51**, 3514-3519 (1973).

105. J. Kärger, H. Pfeifer, P. Walther, A. Dyer and C.D. Williams, *Zeolites* **8**, 251-254 (1988).

106. D.H. Olson and W.O. Haag, *Am. Chem. Soc. Symp. Ser.* **248**, 275-307 (1984).

107. F. Kapteijn, W.J.W. Bakker, G. Zheng, J. Pappe and J.A. Moulijn, *Chem. Eng. J.* **57**, 145-153 (1995).

108. S. Brandani, *Adsorption* **4,** 17-24 (1998)

109. C.L. Cavalcante and D.M. Ruthven, *Ind. Eng. Chem. Res.* **34**, 185-191 (1995).

110. M. Eic and D.M. Ruthven, *Zeolites* **8** 472-479 (1988).

111. W. Drachsel and K.A. Becker, *Z. Phys. Chem (Weisbaden)* **122** 910101 (1980).

112. R.L. Gorring, *J. Catalysis* **31**, 13-26 (1973).

113. C.L. Cavalcante, M. Eic, D.M. Ruthven and M. Occelli, *Zeolites* **15**, 293-307 (1995).

114. L. Boulicaut, S. Brandani and D.M. Ruthven, *Microporous Materials* **25**, 81-93 (1998).

115. W. Niessen, Ph.D. Thesis, Fritz Huber Inst. Berlin, Germany (1991).

116. L.J.P. van den Broeke, Ph.D. Thesis, Univ. of Amsterdam, The Netherlands (1994).

117. E.W. Thiele, *Ind. Eng. Chem.* **31**, 916 (1939).

118. C.N. Satterfield, *Mass Transfer in Heterogeneous Catalysis*, MIT Press, Cambridge, MA (1970).

Studies in Surface Science and Catalysis 137
H. van Bekkum, E.M. Flanigen, P.A. Jacobs and J.C. Jansen (Editors)
© 2001 Elsevier Science B.V. All rights reserved.

Chapter 13

Adsorption of gases in zeolite molecular sieves

Lovat V.C. Rees[a] and Dongmin Shen[b]

Introduction to Zeolite Science and Practice - 2nd Completely Revised and Expanded Edition, Elsevier Science Publishers B.V., P.O. Box 103, 1000 AC Amsterdam, The Netherlands

[a]Department of Chemistry, University of Edinburgh, West Mains Road, Edinburgh EH9 3JJ, Scotland, United Kingdom

[b]The BOC Gases Technical Center, 100 Mountain Avenue, Murray Hill, NJ 07974, USA

1. INTRODUCTION

The systematic study of the equilibrium adsorption of gases in zeolite molecular sieves leads to a considerable insight of the state of the guest molecules in the channels and cavities of the host adsorbent. This insight can be used to correlate the behaviour with the structures of the molecular sieves, with the cations, if present, with the framework stability and modification, with intracrystalline electrostatic fields and field gradients, and with the size, electric moments and polarizability of the guest molecule. Zeolite molecular sieves are important adsorbents for fundamental studies of adsorption processes as they are crystalline with well-defined, regular pores and cavities of molecular dimensions defect free over large distances which results in very large areas of reproducible surfaces for adsorption. In many cases the surface of the pores and cavities are precisely defined with cations, if present, in known locations. In many cases the location of the tetrahedral Si, Al, etc. atoms and the framework oxygens have been established from x-ray diffraction and other methods and any changes in the location of these species on adsorption of adsorbates can sometimes be followed. These materials are thus ideally suited for a priori calculations of the adsorption processes especially now that powerful computers and software are readily available.

Early studies involved true zeolites, i.e. aluminosilicates, but in more recent years frameworks with a large variety of framework species have been synthesised and very recently mesoporous aluminosilicates with pores of 3-10nm free diameter have become available. These newer materials have not been studied as extensively as the pure zeolites, however, some of the differences in their adsorptive properties will be included.

In the pure zeolites the Si/Al ratio can be varied from 1.0 to infinity. As each aluminium introduces one negative charge to the framework and requires a unit positive cationic charge to neutralise this charge adsorbents can be prepared which have highly charged hydrophillic surfaces but these surfaces can be altered until, at the limit, pure Si materials can be obtained which have no negative charges on the framework and thus no cations and are, thus, hydrophobic. The adsorptive properties of these materials show transition from surfaces having a high affinity for water and other polar molecules to microporous silicas such as silicalite which will adsorb organic molecules in the presence of water and other polar species.

The molecular sieving properties of these materials is dependent of the free diameter of the windows (rings of framework oxygens) controlling access to the internal surface. Molecular sieves with 6-ring windows of free diameter of 0.28nm which only allow ingress of H_2O and NH_3 up to frameworks with 18-ring windows of ~1.3nm free diameter which allow very large molecules to be adsorbed are now readily available.

In this chapter some attempt will be made to demonstrate as many of these features as possible. However, as sorption processes are used extensively by industry to purify gas streams and to separate mixtures the sorption of mixtures will also be covered. The competition of a second sorbate for the sites on which a first sorbate is sorbed leads to subtle effects when a sorbent is used to separate gas mixtures and these effects have to be studied and understood in the design of separation plants.

Finally, mesoporous materials have recently been developed and the determination of the diameters of these mesopores is an important parameter in the study of such materials. Mesopore diameters can be determined by analysis of the sorption isotherms obtained with these materials. The methods used will be described.

2. THERMODYNAMICS OF GAS ADSORPTION

The thermodynamic approach to the study of phase equilibria is general and can be applied to adsorption processes. Only one general assumption has to be made: the "adsorbed layer" can be treated as a distinguishable phase in the thermodynamic sense.

The thermodynamic principles of adsorption equilibria can be applied to zeolitic systems, if the thermodynamic and geometric properties of the adsorbent can be regarded as independent of temperature and pressure of the surrounding gas and of the concentration of adsorbed molecules. In this case the adsorbent may be considered as being inert, thermodynamically. Under these conditions the adsorbed molecules themselves may be regarded as a distinct phase, and the effect of the zeolite is limited to the creation of a force field, the detailed nature of which need not be specified. However, if the adsorbent is thermodynamically non-inert which is sometimes evident in cases of adsorption on zeolites, solution thermodynamics have to applied.

The condition for equilibrium between an adsorption (a) phase and the co-existing, ambient gas (g) phase is defined as follows:

$$\mu_a = \mu_g \tag{1}$$

since the chemical potentials of the adsorbing species in all phases must be equal. Assuming an ideal gas phase, we may write:

$$\mu_a = \mu_g = \mu_g^o + RT\ln\left(\frac{p}{p^o}\right), \tag{2}$$

where μ_g^o is the standard chemical potential of the gas phase at the reference pressure p^o of 1 atmosphere. Differentiating Equation (2) at constant adsorption phase concentration (n_a) and applying the *Gibbs-Helmholtz* relation

$$\left[\frac{\partial(\Delta\mu / T)}{\partial T}\right] = -\frac{\Delta H}{T^2}, \tag{3}$$

we obtain:

$$\left[\frac{\partial \ln p}{\partial T}\right]_{n_a} = -\frac{\Delta H}{RT^2}, \tag{4}$$

where $\Delta H(= H_a - H_g)$ is the enthalpy change of the adsorbing species due to adsorption. The quantity $-\Delta H(= Q_{st})$ is commonly referred to as the isosteric heat of adsorption. If we assume that the difference of the heat capacity of the adsorbate in adsorption and gas phases is negligible, ΔH is independent of temperature and Equation (4) can directly be integrated to give:

$$\ln p = const. - \frac{-\Delta H}{RT} \tag{5}$$

A plot of $\ln p$ versus $1/T$ at constant concentration will yield a straight line of slope ($-\Delta H/R$). This plot is called the "isostere of adsorption". This situation provides a simple, practical and useful method for determination of the heat and entropy of adsorption. These quantities are usually obtained from adsorption isotherms measured at different temperatures. Since $\mu = H - TS$, it follows from Equation (1) that at equilibrium

$$\frac{H_a - H_g}{T} = S_a - S_g \tag{6}$$

The partial molar entropy of an ideal gas phase is given by

$$S_g = S_g^o - R\ln\left(\frac{p}{p^o}\right), \tag{7}$$

where S_g^o is the molar entropy of the gas in the standard state at the reference pressure $p^o = 1$ atmosphere. A combination of Equations. (6) and (7) leads to

$$S_a - S_g^o = \frac{H_a - H_g}{T} - R\ln\left(\frac{p}{p^o}\right) \tag{8}$$

or

$$\Delta S^o = \frac{\Delta H}{T} - R\ln\left(\frac{p}{p^o}\right) \tag{9}$$

The standard differential entropy of adsorption, ΔS^o, at n_a = const., can be calculated from the adsorption enthalpy for the same value of adsorbate concentration, n_a, and referring to a standard state. The standard *Gibbs* free adsorption energy, ΔG^o, can be calculated as concentration dependence using the following fundamental relationship for the isochoric-isothermic thermodynamic potential:

$$\Delta G^o = \Delta H - T\Delta S^o \tag{10}$$

The standard state of adsorption for a single pure component is defined at a reference pressure of $p^o = 1$ atmosphere and, in addition, at a system temperature T.

2.1 Adsorption Equilibrium

Henry's Law: - Adsorption equilibrium is commonly described in the form of an isotherm, which is a plot of the equilibrium adsorbate concentration, *n*, against gas-phase concentration, *c*, or equilibrium pressure, *p*, at a constant temperature. In general, for physical adsorption on a homogeneous surface at sufficiently low concentration, the isotherm should approach a linear form, which is commonly referred as the Henry's isotherm equation:

$$n = Kc.$$

Since, for an ideal vapour phase, $p = cRT$, the Henry's equation in terms of equilibrium pressure is given by:

$$n = \frac{K}{RT} p = K' p .$$

The limiting slope of the isotherm in the low concentration region is commonly known as the Henry's law constant and the low concentration region as the Henry's law region. The Henry's law constant is a thermodynamic constant, which is temperature dependent and follows the van't Hoff equation:

$$K' = K'_o e^{-\frac{\Delta H_o}{RT}}, \qquad K = K_o e^{-\frac{\Delta E_o}{RT}}$$

where $-\Delta H_o$ and $-\Delta E_o$ represent the differences in enthalpy and internal energy between adsorbed at low coverages and gaseous states, respectively.

In accordance with these relationships plots of ln K' or ln K versus $1/T$ will give $-\Delta H_o$ or $-\Delta E_o$ respectively and these quantities are often found to be essentially linear over a wide temperature range. $(\Delta H_o + RT)$ may be equated to ΔE_o if it is assumed that the volume change per mole adsorbed is -V, the molar volume of the gaseous molecules, so that in $\Delta H_o = \Delta E_o + p\Delta V$, $p\Delta V = -pV = -RT$. It is ΔE_o that is used for interpretation in terms of molecular interactions and statistical thermodynamics.

It is convenient to define a standard state of the adsorbed phase and the corresponding equilibrium constant :

$$\mu_a = \mu_a^* + RT \ln\left(\frac{a_a}{a_a^o}\right) \tag{11}$$

where a_a is the activity of the adsorbed phase and μ_a^* is the standard chemical potential at activity a_a^o. It follows from Equation (2) that

$$\mu_a^* - \mu_g^o = RT \ln\left(\frac{p}{p^o}\frac{a_a^o}{a_a}\right) \tag{12}$$

The equilibrium constant is given by

$$K^* = \frac{p^o}{p}\frac{a_a}{a_a^o} \tag{13}$$

If the standard state is taken as an infinitely dilute ideal adsorbed phase Equation (13) reduces to Henry's law in the low concentration limit and at higher concentrations to a definition of the activity coefficient of the adsorbed phase (γ):

$$\lim_{p\to 0}\frac{n}{p}=\frac{K^{*}a_{a}^{o}}{p^{o}}\equiv K'\,;\quad \text{At higher concentrations, } \gamma=\frac{K'p}{n} \tag{14}$$

2.2 Spreading Pressure

The concept of a spreading pressure π as the appropriate intensive variable for surface adsorption was first introduced by Gibbs. However, the Gibbs formulation implies a two-dimensional adsorbed phase which is unnecessarily restrictive. The same thermodynamic relations may be developed in terms of surface energy and this approach has the advantage that it is equally applicable to two-dimensional adsorption on a surface or three-dimensional adsorption in a microporous solid.

The Gibbs adsorption isotherm provides a general relation between the spreading pressure (or energy) and the adsorbed phase concentration [1]. However, neither the spreading pressure or the energy are amenable to direct experimental measurement. We are more interested in the relation between adsorbed phase concentration and the equilibrium gas pressure.

If the equation of state for the adsorbed phase corresponds to the ideal gas law

$$\pi A = n_a RT \tag{15}$$

where A is the surface area. From the Gibbs isotherm then

$$\left(\frac{\partial\pi}{\partial p}\right)_T=\frac{\pi}{p};\qquad \pi=K'p\;;\quad n=\frac{n_a}{A}=\frac{K'p}{RT} \tag{16}$$

The linear relationship between pressure and adsorbed phase concentration (Henry's law) is equivalent to an ideal gas equation of state for the adsorbed layer in terms of the spreading pressure.

2.3 Langmuir Isotherm

At somewhat higher concentrations the simplest isotherm equation is the Langmuir equation

$$K_1=\frac{\theta}{p(1-\theta)}=\frac{n}{p(n_{sat}-n)} \tag{17}$$

where θ is the fraction of the adsorption sites occupied by adsorbate molecules and n_{sat} is amount of adsorbate required to cover all the adsorption sites.

When the saturation region is approached adsorbate/adsorbate molecule interactions become significant. As a first-order deviation from the Langmuir model one may consider ideal adsorption on a set of localised sites with weak interactions between adsorbed molecules on neighbouring sites. Such a model was suggested by Lacher [2] and also by Fowler and Guggenheim [3]. If the interaction is sufficiently weak that the random distribution of the adsorbed molecules is not affected the resulting expression for the isotherm is

$$K_2 = \frac{\theta}{p(1-\theta)} \exp\left(\frac{2\omega\theta}{RT}\right) \tag{18}$$

where 2ω is the pair interaction energy.

2.4 Mobile Adsorption

Non-polar adsorbates, at least, have been shown to have considerable intracage mobility. For example, for systems such as methane in zeolite 4A, it has been found using neutron scattering that the methane molecules have essentially complete translational freedom within the zeolite cavity with transitions to the neighbouring cages occurring only infrequently. Isotherm equations for such non-localized adsorption have been developed based on the Volmer equation , p(V-b)=RT, and the van der Waals equation, $(p+a/V^2)(V-b)=RT$, where a and b are the coefficients in the van der Waals equation of state.

These are

$$K_3 = \frac{\theta}{p(1-\theta)} \exp\frac{\theta}{1-\theta} \quad \text{and} \tag{19}$$

$$K_4 = \frac{\theta}{p(1-\theta)} \exp\left(\frac{\theta}{1-\theta} - \alpha\theta\right) \quad \text{respectively} \tag{20}$$

where α=2a/bRT.

Barrer and Rees [4] tested an isotherm equation based on the following more accurate equation of state developed by Hirschfelder et al [5]

$$\left(p + \frac{a}{V^2} + \frac{a'}{V^3}\right)\left(V - b + \frac{b'}{V}\right) = RT \tag{21}$$

where a, a′, b and b′ are coefficients. Hirschfelder et al showed that this equation can represent p-V-T relations of gases up to their critical densities (i.e. θ ≈ 0.3) and pressures of several thousand atmospheres. The isotherm equation required two modifications of the Hirschfelder et al equation of state to allow for thermal entropy differences between the fluid in the liquid state as compared with the fluid in the confines of the intracrystalline pores and the smaller number of adsorbate-adsorbate contacts in the adsorbed fluid.

2.5 The Virial Adsorption Isotherm

Equation (21) is inadequate for adsorbed fluids beyond their critical densities. For more dense fluids a virial equation of state can be used. If the gas in the gas phase is considered to be ideal then

$$K' = \frac{n}{p} \exp\left\{2A_1 n + (3/2)A_2 n^2 + (4/3)A_3 n^3 + \ldots\ldots\ldots\ldots\right\} \tag{22}$$

or

$$K = \frac{n}{n_g} \exp\left\{2A_1 n + (3/2)A_2 n^2 + (4/3)A_3 n^3 + \ldots\ldots\ldots\ldots\right\} \tag{23}$$

where p and n_g are the gas phase pressure and concentration, respectively, in equilibrium with n the adsorbed phase concentration in the same units as n_g.

Since from Equation (14)

$$K' = n\,\gamma / p \tag{24}$$

where γ is the activity coefficient of the intracrystalline fluid we can obtain from Equation (22) that

$$\gamma = \exp\left\{2A_1 n + (3/2)A_2 n^2 + (4/3)A_3 n^3 + \ldots\ldots\ldots\ldots\right\} \tag{25}$$

2.6 Statistical Thermodynamic Isotherms

The regular pore structure of zeolites makes them convenient for application of the statistical method. The application of such statistical thermodynamic treatments is fully covered by Brauer et al [6], Barrer [7] and Ruthven[8] and these references should be consulted to see the strengths and weaknesses of the method.

2.7 Other isotherm models

Many isotherm models have been developed in attempts to represent adsorption isotherms of various systems. Some of these models are not thermodynamically consistent and one should, therefore, be careful when dealing with predictions made from these models and this is especially the case when the models are used to predict mixture adsorption isotherms from single component data. Four examples follow.

Freundlich in 1909 introduced an equation with limited application at low/high equilibrium pressures.

$$n = K'p^t \tag{26}$$

This equation was combined with the Langmuir equation to give the Langmuir-Freundlich equation

$$n = \frac{n_{sat}(K'p)^t}{1 + (K'p)^t} \tag{27}$$

The Sips equation [9] was developed initially to describe chemisorption but was later used to describe physisorption

$$n = \frac{n_{sat}K'p^t}{1 + K'p^t} \tag{28}$$

Finally, the Toth isotherm equation was introduced in 1971 [10]

$$n = \frac{n_{sat}p}{(b + p^t)^{1/t}} \tag{29}$$

where t is a constant which is dependent on the spread of adsorption site energies in the adsorbent.

For heterogeneous surfaces multi-site adsorption isotherm models sometime give better correlation with the experimental data. The heats of adsorption on each individual site can be derived from the multi-site models. The multi-site Langmuir isotherm model may be written as:

$$\frac{n}{n_{sat}} = \frac{b_I p}{1+b_I p} + \frac{b_{II} p}{1+b_{II} p} + \ldots + \frac{b_N p}{1+b_N p}$$

where there are N sites of different energies of adsorption.

3. ENERGETICS OF ADSORPTION

The heats or energies of adsorption arise from various interactions. The total interaction energy, ϕ, can include some or all of the following terms

$$\phi = \phi_D + \phi_R + \phi_P + \phi_{F\mu} + \phi_{\bar{F}Q} + \phi_{SP} \tag{30}$$

where ϕ_D is the dispersion energy, ϕ_R is the close range repulsion, ϕ_P is the polarization energy, $\phi_{F\mu}$ is the field-dipole interaction, $\phi_{\bar{F}Q}$ is the field gradient-quadrupole interaction and ϕ_{SP} is the adsorbate-adsorbate (self-potential) interaction.

The first three terms are always present and are referred to as the non-specific components of ϕ. The next two terms, $\phi_{F\mu}$ and $\phi_{\bar{F}Q}$, depend upon the presence of permanent dipole and quadrupole moments respectively in the adsorbate molecules. The final term, ϕ_{SP}, allows for adsorbate-adsorbate interactions and, thus, tends to zero at small coverages.

3.1 Dispersion-Repulsion Energy

The attractive potential arising from dispersion forces between two isolated molecules may be written as

$$\phi_D = -\frac{A_1}{r_{12}^6} - \frac{A_2}{r_{12}^8} - \frac{A_3}{r_{12}^{10}} \tag{31}$$

where r_{12} is the distance between the centres of the interacting molecules and the A terms are constants. The first term, arising from the coupling between instantaneous induced dipoles, dominates. The second and third terms represent the interactions from induced dipole-induced quadrupole and induced quadrupole-induced quadrupole interactions respectively.

The short range repulsive energy arising from the interaction of the electron clouds of the molecules at short distances is often represented semi-empirically by a twelfth-order inverse power expression:

$$\phi_R = \frac{B}{r_{12}^{12}} \tag{32}$$

where B is a constant.

If the two higher order terms in ϕ_D are ignored ϕ_D and ϕ_R can be combined to give the familiar Lennard-Jones 6-12 potential function:

$$\phi = 4\varepsilon\left[\left(\frac{\sigma}{r}\right)^{12} - \left(\frac{\sigma}{r}\right)^{6}\right] \tag{33}$$

The force constant ε and σ have values which are specific for the particular molecular species and values for many common molecules are readily available.

The interaction between two different molecules are generally represented by the arithmetic mean of σ and the geometric mean of ε:

$$\sigma_{12} = \frac{1}{2}(\sigma_1 + \sigma_2) \text{ and } \varepsilon_{12} = \sqrt{\varepsilon_1 \varepsilon_2} \tag{34}$$

For larger molecules where the Lennard-Jones force constants are not available it is customary to replace Equation (33) by an equation of a similar form:

$$\phi = -\frac{A}{r^6} + \frac{B}{r^{12}} \tag{35}$$

where A and B are semi-empirical constants.

There are three commonly used expressions for the calculation of the attractive constant A.

London: $$A = \frac{3}{2}\alpha_1\alpha_2 I, \quad \frac{1}{I} = \frac{1}{I_1} + \frac{1}{I_2} \tag{36}$$

Slater-Kirkwood: $$A = \frac{3eh}{4\pi\sqrt{m}} \frac{\alpha_1\alpha_2}{\left(\sqrt{\alpha_1 / N_1} + \sqrt{\alpha_2 / N_2}\right)} \tag{37}$$

Kirkwood-Muller: $$A = 6mc^2 \frac{\alpha_1\alpha_2}{(\alpha_1 / \chi_1) + (\alpha_2 / \chi_2)} \tag{38}$$

In the above expressions: c = velocity of light; e = charge on the electron; m = mass of the electron; h = Planck's constant; I_1 and I_2 = ionization potentials; α_1 , α_2 = polarizability; χ_1 , χ_2 = magnetic susceptibilities; N_1 , N_2 = number of electrons in outer shells of molecules 1 and 2.

At the equilibrium separation distance r_o, $d\phi/dr$ is zero and B can be obtained in terms of A.

$$B = \frac{Ar_o^6}{2} \tag{39}$$

The value of the repulsion constant B can be estimated by setting r_0 equal to the mean of the van der Waals radii of the two molecules.

Expressions for A_2 and A_3 in Equation (31) corresponding to the Kirkwood-Muller relation for A_1 have been derived [11]

3.2 Electrostatic Energies

In zeolites significant electric fields can exist in the pores and cages if the framework contains appreciable amounts of aluminium. Replacement of a silicon atom by an aluminium atom introduces a unit negative charge which has to be neutralised by a unit positive charge from a cation. The cation/negative framework species can generate a large electric field. Additional contributions to the adsorption energy result from this field. These are given by:

$$\phi_P = -\frac{1}{2}\alpha E^2 \tag{40}$$

$$\phi_{F\mu} = -\mu E \tag{41}$$

$$\phi_{\bar{F}Q} = \frac{1}{2} Q \frac{\partial E}{\partial r} \tag{42}$$

where E is the electric field; α the polarizability, μ the dipole and Q the quadrupole moment of the adsorbate molecule.

4. ADSORPTION OF MIXTURES

Measurement of mixture adsorption equilibria is quite a challenge. There are several methods available to measure mixture adsorption equilibria, but these measurements are time consuming because of large number of variables involved. The prediction of multicomponent adsorption equilibria from single component equilibrium data has attracted a great deal of attention. Such predictions can then be checked against a few experimental mixture data to verify the predictions. For some simple systems, the Langmuir mixing rule or the Ideal Adsorbed Solution Theory (IAST) may apply to give reasonable prediction of multicomponent equilibria, but there is still no universal method which is valid over a wide range of systems.

4.1 Extended Langmuir Model

The Langmuir model for single component adsorption can be easily extended to multicomponent
mixture systems. It is required that the equilibrium data points of the single gas adsorption isotherm are fitted by the *Langmuir* model for single gas adsorption systems. The calculated constants of the single gas adsorption isotherm are used for a prediction of any mixture isotherm at the same temperature. The extended expression for the isotherm is:

$$\frac{n^{(i)}}{n^{(i)}_{sat}} = \frac{b^{(i)} p^{(i)}}{1 + b^{(1)} p^{(1)} + b^{(2)} p^{(2)} + \ldots\ldots} \tag{43}$$

Thermodynamic consistency requires $n^{(1)}_{sat} = n^{(2)}_{sat} = \ldots\ldots = n^{(N)}_{sat}$, but such a requirement is unrealistic for adsorption of molecules of different sizes. This requirement is commonly ignored for an analytic description of adsorption equilibrium within certain concentration range.

4.2 Extended Langmuir-Freundlich Model

Many attempts have been made to improve the correlation of equilibrium data by using more complex expressions, one of the more widely used being the Langmuir-Freundlich model:

$$\frac{n^{(i)}}{n_{sat}^{(i)}} = \frac{b^{(i)} p^{(i)^{t^{(i)}}}}{1 + b^{(1)} p^{(1)^{t^{(1)}}} + b^{(2)} p^{(2)^{t^{(2)}}} + \ldots\ldots} \tag{44}$$

Although it is thermodynamically inconsistent, this expression has been proven to provide a reasonably good empirical correlation of mixture adsorption equilibrium data.

4.3 The Ideal Adsorbed Solution Theory (IAST)

The Ideal Adsorbed Solution Theory (IAST), developed by Myers and Prausnitz [12], describes a calculation of adsorption equilibria for components of a mixture, using only data for their single component adsorption equilibria at the same temperature and on the same adsorbent. The technique is based on the concept of an ideal adsorbed solution and, using classical solution (surface) thermodynamics, an expression analogous to *Raoult*'s law is obtained. The essential idea of the calculation lies in the recognition that for an ideal solution the partial pressure of an adsorbed component is given by the product of its mole fraction in the adsorption phase and the pressure that would be exerted by a single adsorbed component at the same temperature and spreading pressure as those of the mixture. This means, in general, that at $\frac{A\pi}{RT} = const.$ an adsorption equilibrium can be described by means of solution thermodynamics. For an ideal binary system,

$$p^{(1)} = p^{o(1)}(\pi) x_a^{(1)} \tag{45}$$

and

$$p^{(2)} = p^{o(2)}(\pi) x_a^{(2)} \tag{46}$$

If the adsorption isotherms, n = f(p), of the single gases are known, the spreading pressure contributions of the both components to that of the adsorption phase, at the total pressure p are given by

$$\frac{\pi^{(i)} A}{RT} = \int \frac{n^{(i)}}{p^{(i)}} dp \tag{47}$$

and for mixing at constant temperature and spreading pressure:

$$\pi^{(1)} = \pi^{(2)} = \pi^m \tag{48}$$

We also have

$$x_g^{(1)} + x_g^{(2)} = 1 \tag{49}$$

and

$$x_a^{(1)} + x_a^{(2)} = 1 \tag{50}$$

By solving these equations relating to the nine unknowns, (p, $x_g^{(1)}, x_g^{(2)}$, $x_a^{(1)}, x_a^{(2)}$, $p^{o(1)}, p^{o(2)}$, $\pi^{(1)}, \pi^{(2)}$), all other variables can be calculated. The total adsorption phase concentration follows from the principle that for an ideal solution there is no change in the area/molecule on mixing:

$$\frac{1}{n^m} = \frac{x^{(1)}}{n^{o(1)}} + \frac{x^{(2)}}{n^{o(2)}}. \tag{51}$$

A prediction of mixture adsorption isotherms in this way requires experimental single component adsorption isotherm data for the less strongly adsorbed component up to pressures exceeding the total pressure for the binary mixture. For both components, highly accurate adsorption equilibria data in the low-pressure region of single component isotherms are required for the calculation of mixture adsorption equilibria.

4.4 Adsorbate Solution Theory

Surface free energy, $A\pi/m$, which for a microporous materials can be determined from adsorption isotherms as a complex quantity only, without splitting it up into numerical values of specific surface area, A, spreading pressure, π, and number of moles of the sorbent, m, and which should be considered as change of the chemical potential, $\Delta\mu$, of the adsorbent as a result of the adsorption process [13], can be calculated directly from sorption thermodynamic functions. For this purpose, these functions that characterize the adsorption process, were expressed by their polynomial fits with regard to adsorbate concentration, n. The following equations are used:

$$A\pi = \mathrm{F}(n) = \int n \,\mathrm{d}(\Delta G) = \int n \,\mathrm{d}(\Delta H) - T\int n \,\mathrm{d}(\Delta S) \tag{52}$$

$$\Delta H = \mathrm{A} + \mathrm{B}n + \mathrm{C}n^2 + \mathrm{D}n^3 + \ldots\ldots \tag{53}$$

$$\Delta S = \mathrm{a} + \mathrm{b}n + \mathrm{c}n^2 + \mathrm{d}n^3 + \ldots\ldots \tag{54}$$

$$A\pi = \left(\frac{\mathrm{B}}{2}n^2 + \frac{2\mathrm{C}}{3}n^3 + \frac{3\mathrm{D}}{4}n^4 + \ldots\right) - T\left(\frac{\mathrm{b}}{2}n^2 + \frac{2\mathrm{c}}{3}n^3 + \frac{3\mathrm{d}}{4}n^4 + \ldots\right). \tag{55}$$

By combining Equations (52 - 55) with the Gibbs function, Equation (10), this approach being called "Adsorbate Solution Theory" (AST) to differentiate it from the "Ideal Adsorbed Solution Theory" (IAST) [12], one may predict total mixture adsorption thermodynamic functions from single component sorption thermodynamic functions at constancy of the change of chemical potential of the adsorbent, temperature and adsorption phase composition:

$$\exp\left(\frac{\Delta G^m(n^m)}{\mathrm{R}T}\right) = \sum_i^N x^{(i)}\gamma^{(i)} \exp\left(\frac{\Delta G^{o(i)}(n^{o(i)})}{\mathrm{R}T}\right) \tag{56}$$

where the functions $\Delta G^{o(i)}(n^{o(i)})$ denote the concentration-dependent changes of single component Gibbs free adsorption energy at the same values of "surface free energy", as of the binary mixture, *cf.*, Equation (57)

$$(A\pi)^m = (A\pi)^{o(1)} = (A\pi)^{o(2)} \tag{57}$$

From plots of surface free energy, $A\pi/m$, versus adsorbate concentration, n, at a given temperature, total mixture adsorption isotherms at constant sorption phase composition, $x_a^{(i)}$, can be calculated using the following formalism:

$$p^m = \sum_i^N x_a^{(i)} \gamma_a^{(i)} \exp\left(\frac{\Delta G^{o(i)}}{RT}\right) \tag{58}$$

$$\left(\frac{1}{n^m}\right) = \left(\frac{1}{n^m}\right)^e + \sum_{i=1}^N \left(\frac{x_a^{(i)}}{n^{o(i)}}\right) \tag{59}$$

Activity coefficients, $\gamma_a^{(i)}$, of sorbed component i in the sorption phase can also be calculated by means of the following fundamental thermodynamic expressions:

$$\Delta G^e = (\pi\ A)^m \left(\frac{1}{n^m}\right)^e \tag{60}$$

$$\frac{\Delta G^e}{RT} = \sum_{i=1}^N x_a^{(i)} \ln \gamma_a^{(i)} \tag{61}$$

5. EXPERIMENTAL METHODS

Adsorption isotherms can be measured by volumetric or gravimetric techniques. In the volumetric method a dose of adsorbate is measured in a calibrated doser volume using a pressure transducer (e.g. Baratron with a range of zero to 1 atmosphere pressure) to determine the pressure of the dose. From the known volume and pressure the quantity of gas in the doser section can be calculated. The tap to the adsorption bulb containing a known quantity of adsorbent at a controlled temperature, is opened and the system allowed to come to an equilibrium pressure which once again is measured with the Baratron pressure transducer. The amount of gas remaining in the gas phase at equilibrium is calculated from the known volumes of the system at room temperature and the known volume of the adsorption bulb which is at the temperature of the isotherm and thus usually at a temperature different from room. From the difference between the amount of gas in the original dose volume and the gas remaining in the gas phase at equilibrium the amount of gas adsorbed is obtained. The next point on the isotherm is then obtained by introducing another calibrated dose of gas at a pressure somewhat greater than the previous equilibrium pressure from the doser volume and the process repeated as for the first point. Further doses are added until the isotherm is fully scanned.

Desorption points can be determine by removing known amounts of gas from the doser volume and then allowing the system to come to equilibrium again.

Frequently the temperature of the adsorbent is very different from room temperature. In the volumetric method there is usually a tube of small diameter (e.g. 2mm) leading to the adsorbent. When the equilibrium pressure to be measured is quite low it is necessary to apply a correction to the pressure measured at room temperature with the Baratron for the true equilibrium pressure which exists over the adsorbent. The equation for this thermal transpiration correction is described in detail by Liang[14].

In the gravimetric method for the determination of the adsorption isotherm the amount of adsorbate adsorbed is determined from the increase in weight of the adsorbent after exposure to a dose of gas and the system allowed to come to adsorption equilibrium. The weight is measured accurately by means of a sensitive micro-electronic balance as shown schematically in Figure 1. The isotherm is scanned by adding further doses at ever increasing pressures. The desorption isotherm is similarly scanned by removing small incremental amounts of adsorbate until a final low equilibrium pressure is attained.

Usually about 200mg of adsorbent is used although balances can cope with much heavier sample weights. The equilibrium pressure is measured with one or more absolute Baratron pressure transducers which cover the pressure range from zero to 1 atmosphere.

Three corrections may have to be made to the measurements depending on the equilibrium pressure and the temperature difference between sample and the main balance compartment to allow for a) buoyancy effects arising from differences in the volume displaced by the adsorbent and the counterweight; b) condensation on the balance mechanism which can occur if the adsorbate being investigated is below its critical temperature; c) thermal gas motion effects which include thermomolecular flow, slip flow and convection.

The latest gravimetric systems are now fully automated (e.g./ Hiden IGA Systems, Warrington, U.K.). These systems consist of a high-stability microbalance mounted in a vacuum/pressure vessel with gas pressure and sample temperature controllers. These components are manipulated by comprehensive integrated software. These systems allow the user to measure the amount of gas adsorbed in a sample as a function of sample temperature and gas pressure for any predetermined sequence of pressure or temperature steps. The kinetics of the adsorption steps are fully recorded so the rate of adsorption/desorption is readily available. These systems can include an on-line mass-spectrometer for the study of mixtures.

The gravimetric method has several advantages over the volumetric method. For example, non-ideal gas corrections are completely irrelevant so it is easy to use any gas or vapour. Only the sample temperature is important. Temperatures elsewhere in the sensitive volume are usually of no importance. The balance head can therefore be thermostatted to a predetermined temperature which is independent of the sample temperature. Each weight recorded is a direct measurement and does not suffer from the cumulative errors which arise in the multiple dosing of the volumetric method. Mass scales are directly calibrated against standard masses. This calibration is more accurate than calibrations in terms of standard volumes, especially for non-ideal gases. The sample can be connected to wide bore tubing ensuring that outgassing of the sample can be achieved quickly at low pressures. The mass of the sample during the initial outgassing and the mass on outgassing after the isotherm determinations can be monitored.

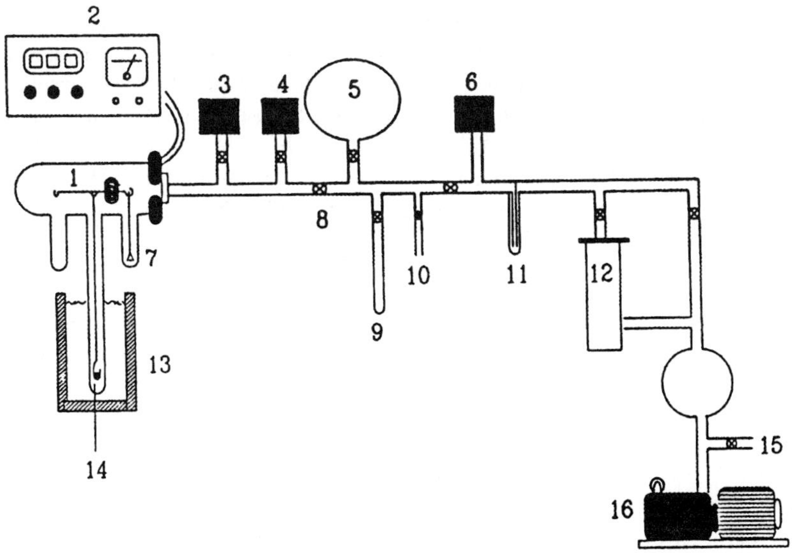

Figure 1. Microsorption Balance

1. Electrobalance
2. Cahn 2000 Electrobalance
3. Baratron pressure transducer
4. Baratron pressure transducer
5. Gas reservoir
6. Pirani gauge
7. Counterweight
8. High vacuum taps
9. Liquid sorbate
10. Gas inlet
11. Cold finger
12. Oil diffusion pump
13. Water bath or tube furnace
14. Zeolite sample
15. Air inlet
16. Rotary pump

The above techniques are used to determine adsorption isotherms. If highly accurate adsorption thermodynamic data are required, i.e. enthalpy, entropy and Gibbs free energy as functions of adsorption phase concentrations up to saturation values, the "isosteric" technique has many advantages over isotherm determinations. The isosteric technique involves a minimum gas phase volume over a large mass of adsorbent. From the isosteres measured isotherms can be readily determined over a wide concentration range. The adsorption saturation value is obtained without presupposing any adsorption isotherm model. The energetic heterogeneity of the adsorbent sites is quantified by characteristic concentration dependencies of the thermodynamic functions. Finally, the isosteric method is extremely useful for determining the thermodynamic functions of mixtures of adsorbates and their partial values i.e. the thermodynamic functions of each component of the mixture in the presence of the other components. The experimental isosteric system is shown in Figure 2.

The basic principle of the method is to maintain a (nearly) constant value of adsorption phase concentration while the equilibrium pressure is measured as a function of temperature. Repeating the procedure for various adsorbate concentrations, n, or, n^m (in case of mixtures), the adsorption thermodynamic functions are obtained as concentration dependencies after subsequent dosing. In cases of mixture isosteric measurements, the adsorption phase composition, $x_a^{(i)}$, is maintained constant in addition to the total adsorbed amount. This requires that for mixtures, doses of identical composition always have to be added. The main features of the isosteric technique are as follows: (*i*) minimum dead volume: large amount of adsorbent; (*ii*) minimum gas-phase-volume to adsorption-phase-volume ratio; (*iii*) low equilibrium pressure; (*iv*) small temperature increments, *c.* 2 - 5 K; (*v*) strongly controlled equilibration criteria.

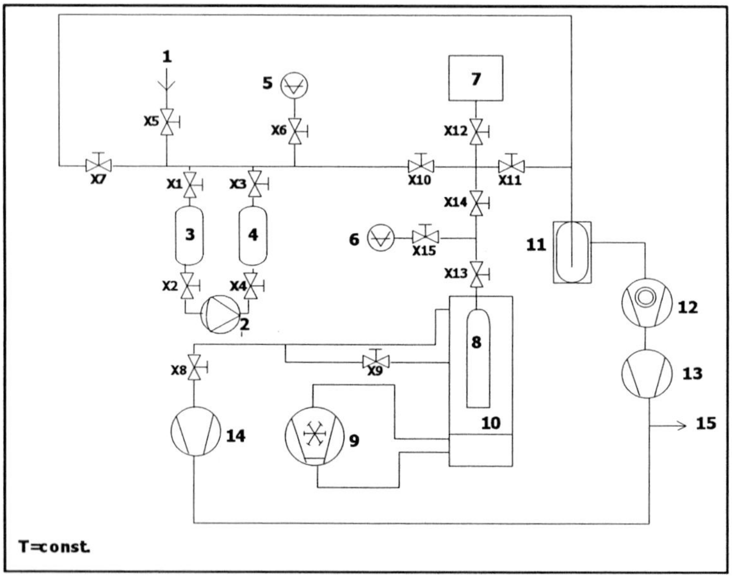

Figure 2. Scheme of the isosteric apparatus. (1. Gas supply; 2. Circulating pump; 3&4. Calibrated doser volumes; 5&6. Pressure sensors; 7. Mass spectrometer; 8. Sample holder; 9&10. Cryostat; 11-15, Vacuum system.)

The isosteric method was first developed by Bering et al [15] and by Schirmer et al [16,17] for single component adsorption and has been extended to investigate binary [18] and ternary mixtures[19] by Bulow et al and has also been successfully used for single components and binary mixtures by Rees et al [20-24]. Thermodynamic analyses of isosteric data and of isosteric heats of multi-component adsorption have been reported by Sircar[25] and Myers et al[26], respectively. Advantages of the latest isosteric technique over other experimental techniques have been described by Shen and Bulow [27] and some of their results will be included in this chapter.

In accordance with the fundamentals of physical adsorption (see Section 2), adsorption isosteres are straight lines at constant adsorbate concentration, n , in Clausius-Clapeyron plots of $\ln p$ vs. $1/T$. From the slopes of these isosteres the isosteric heat and the differential molar heat of adsorption, ΔH, can be calculated;

$$Q_{st} = -RZ(\partial \ln p / \partial T^{-1})_n = H_g - H_a = -\Delta H \qquad (62)$$

where p and T denote, respectively, the adsorbate equilibrium pressure and absolute temperature; R is the universal gas constant, and Z is the compressibility coefficient, $Z = pV/RT$ (with $Z = 1$ for an ideal gas phase and $Z \neq 1$ for a real gas phase). During measurements of the adsorption isosteres, it is essential to check that no adsorption phase transitions occur in the adsorbed phase in the region of physical parameters measured. If a phase change takes place, the isostere does not remain linear, as found for the adsorption of benzene by CaX-type zeolite [28].

If the coexisting gas phase exhibits real gas behaviour, this could be accounted for by the compressibility coefficient, $Z \neq 1$. The heat of adsorption may then become a complex function of pressure and temperature, dependent on the shape of the function $Z = f(p,T)$ even if isosteres remain linear (see [29]).

If isosteres, or data sets (T, p, n) for a given system, are available over the entire adsorbate concentration region, the adsorption equilibrium is comprehensively and fully described. The adsorption thermodynamic functions for the system under consideration can then be obtained from Equations (5), (9) and (10). These equations can be applied to the evaluation of both single component and binary/ternary mixture adsorption isosteres. In the latter case, the mixture isosteres were measured for constant adsorption phase composition by changing, in successive steps, the amount of the chosen gas mixture adsorbed. By analysing the gas phase composition with an on-line mass-spectrometer the mixture isosteres could be deconvoluted into their partial isosteres and the partial thermodynamic functions obtained for each component of the mixture.

The accuracy of the isosteric apparatus has been demonstrated by measuring the sublimation of carbon dioxide (CO_2) in the absence of adsorbent (see Figure 3). The changes of enthalpy, - 25.26 kJ/mol, and entropy, -129.57 J/mol K, typical of CO_2 sublimation as determined by the isosteric method agree well with literature data [30] of - 25.23 kJ/mol and - 129.63 J/mol K, respectively. The isosteric method gives the sublimation energy to an accuracy of ± 70 J/mol.

6. RESULTS AND DISCUSSION

6.1 Equilibrium Studies

The four isotherm equations given in Equations (17-20) have been tested for many different zeolite/adsorbate systems [31-42]. For the system Ar adsorbed in chabazite[32] the Volmer isotherm equation gave the most constant equilibrium constant. However, in most cases this constant and that given by the van der Waals equation represented the data very poorly. This unsatisfactory nature of these equations does not indicate that the adsorbed fluid is not mobile but is a reflection that these models describe the corresponding bulk fluids poorly. If any of these four models do describe the actual isotherms the agreement is best considered as due to compensating effects amongst the various contributing factors.

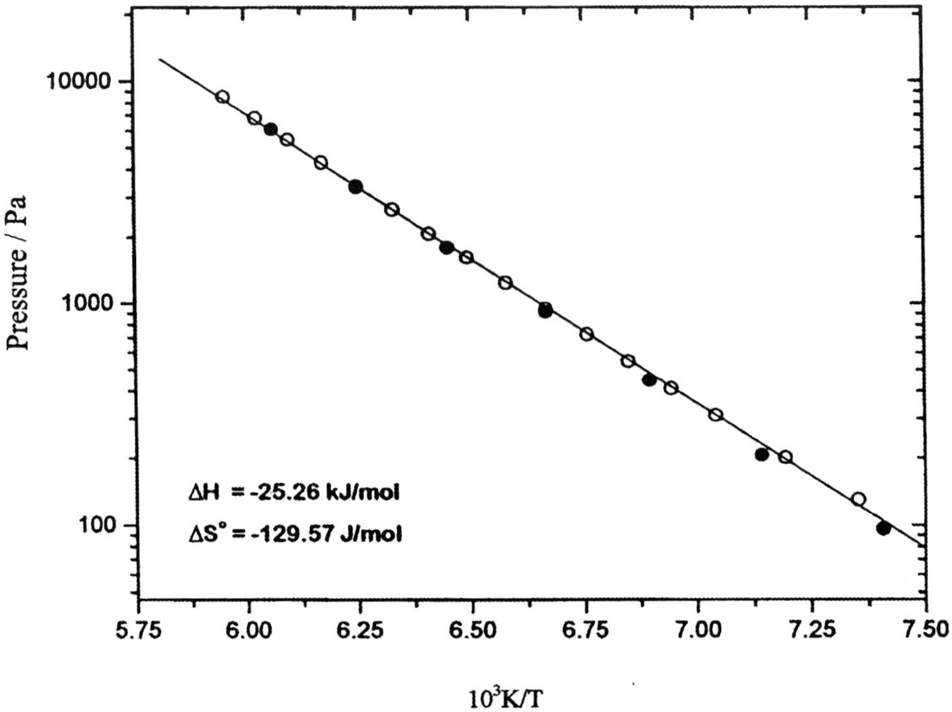

Figure 3. Sublimation curve for carbon dioxide measured isosterically in absence of sorbent (empty and full symbols denote, respectively, experiments performed from low to high temperature and in reverse direction)

Rees and Williams [43] did find that the adsorption of Kr in zeolite NaX followed the Lacher/Fowler and Guggenheim model if only the first 75% of the adsorption was considered. In this region the heat of adsorption was constant initially followed by a small increase due to adsorbate/adsorbate interactions. Over this range of coverage some 10-11 Kr molecules are being adsorbed on the walls of each supercage of this zeolite. The additional 4-5 Kr molecules which are adsorbed at saturation are only adsorbed after the cage walls have been covered and with a much reduced potential. When Equation (18) was used taking only the first 75% of the adsorption into consideration adsorbate/adsorbate interactions resulted in a 2ω value of 2150 J/mol. If the interaction energy of a pair of Kr molecules is calculated from the Lennard-Jones 6:12 potential (see Equation (35)) then a value of $2\omega/Z$ of 833 J/mol is obtained which results in a value of 2.6 for Z.

It is interesting to note that the differential thermal entropy, $\overline{S}_{th}$, is reasonably constant for the molecules adsorbed over the first 75% of the coverage.

Langmuir's and Lacher's isotherm equations each depend upon the concept of a fixed number of sites with one adsorbate molecule per site. This concept is reasonable for systems such as inert gases and methane but as the guest molecule increases in size, for example, as one progresses up the hydrocarbon series then the number of molecules needed to fill, say, a supercage must decrease and the simple concept is inapplicable.

6.2 Isotherms for intracrystalline fluids

Barrer and Rees[4] tested the isotherm equation, based on the more accurate equation of state developed by Hirschfelder et al[5](see Equation (21)). They applied the isotherm equation to the adsorption of Ar in Na- and Ba-X zeolites and of N_2 in K-X. The observed and calculated isotherms for Ar are shown in Figure 4. The agreement is excellent using the bulk values for a, a′, b and b′ and a reduced coordination number of 3.72 and an adjustment to the thermal entropy to allow for the greater restrictions of the adsorbed fluid as compared to the corresponding bulk fluid of the same density.

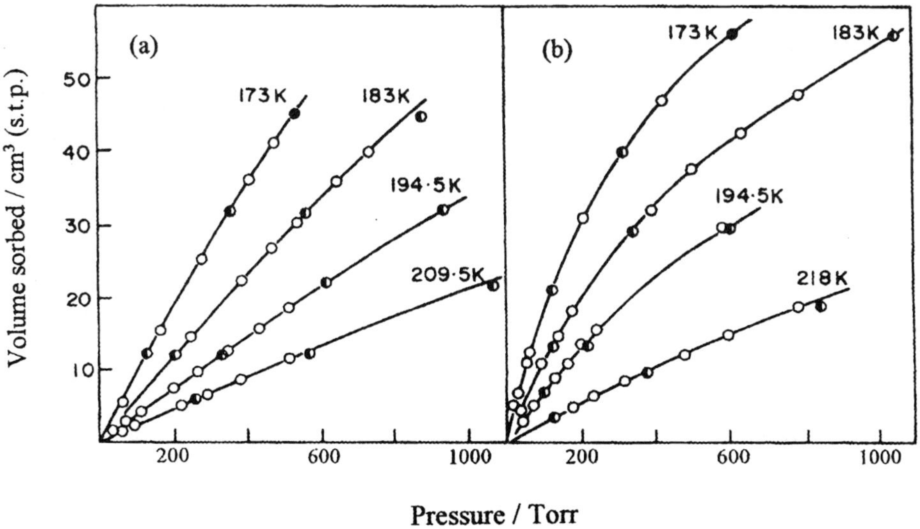

Figure 4. Observed isotherms and isotherms calculated using an equation of state for the sorbed fluid[4]: (a) Ar in Na-X and (b) Ar in Ba-X. ● = point from which argon-argon coordination number was obtained; ◐ = calculated isotherm point; O = experimental isotherm point.

6.3 Virial adsorption isotherm

The virial isotherm equation given by Equation (23) was tested by Barrer and Davies[31] for the systems shown in Figure 5. Not more than three A coefficients were required to give excellent fits of the experimental data even for the strongly curved isotherms. The values of these A coefficients and of K' were obtained from plots of ln (n/p) against n. The temperature coefficients of these A coefficients were determined from the linear functions

Table 1
Virial Constants for various gases in unit volume of H-Chabazite

Sorbate	$10^5 A_1^o$	$10^2 a_1$	$10^8 A_2^o$	$10^2 a_2$	$10^{12} A_3^o$	$10^3 a_3$
Oxygen	10	-0.60	-10.6	-0.61	16.1	-5.3
Nitrogen	24.3	-0.21	7.0	-4.11	-	-
Argon	22.6	-0.51	-17.2	-1.05	44.4	-5.9
Krypton	10.3	0.00	-26.6	-0.49	44.4	-4.4
Xenon	72.0	-0.29	-25.7	-0.35	48.4	-3.9

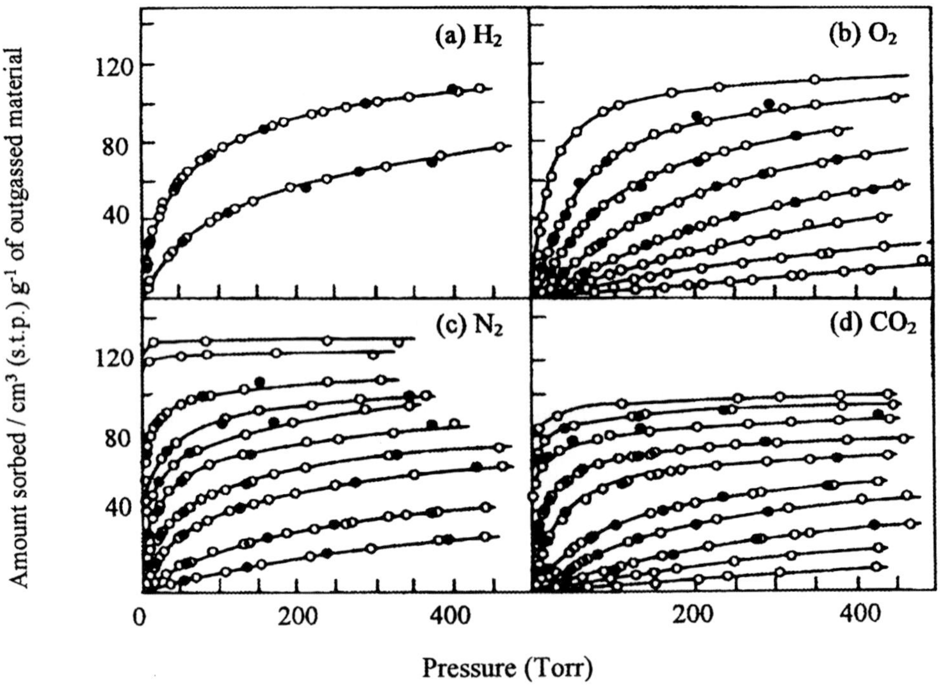

Figure 5. Isotherms for H_2, O_2, N_2 and CO_2 in H-chabazite.[31] O = experimental points; ● = points calculated using Eqn 23, with appropriate values of the A and of K_p.

$$A_i = A_i^o(1 + a_i T) \tag{63}$$

The values of A and the temperature coefficients are given in Table 1. The values of A in this table were determined taking the units of n to be in moles per unit volume of crystal.

Activity coefficients for these systems were obtained from Equation (25) using the values for A listed in Table 1.

6.4 Selected adsorption studies: Hydrocarbons in Zeolites

The adsorption of hydrocarbons in silicalite-1 are examples where only dispersion/repulsion forces are involved (i.e. only the first two terms in Equation (30); the final term involving adsorbate/adsorbate interactions may, also, make a contribution to the total potential at high coverages). The adsorbent has a framework which is constructed only with SiO_4 tetrahedra and, thus, contains no cations which would introduce polarization energies. The adsorption isotherms for ethane, propane, n-butane and n-hexane at various temperatures are shown in Figure 6 [44]. The isotherms for the first three sorbates are all type-1, near rectangular isotherms and give quite good fits with the Langmuir equation (see Equation (17)). Comparison of these isotherms clearly show the stronger interaction between the adsorbate molecule and the adsorbent channel surfaces with increasing chain length.

It is interesting to note, however, that the nitrogen isotherm at 77K in this same adsorbent follows a type-1 isotherm at loadings <24 molecules/unit cell (m/uc) but shows a step for loadings greater than 24 m/uc up to the saturation value of ~34 m/uc. For loadings of <24 m/uc each nitrogen molecule is in close contact with the adsorbent channel surfaces but at higher loadings the adsorbate molecules are located in channel intersections and, therefore, slightly removed from the framework oxygens.

For a regular three-dimensional channel system such as silicalite-1 it is informative to express the adsorption capacity in terms of molecules per unit cell (m/uc) or alternatively the effective total length of adsorbate molecules per unit cell (nm/uc). The latter is the product of chain length of the adsorbate molecule and m/uc. The total channel length of silicalite-1 is given in Table 2 where each unit cell is constructed from four segments of linear (4 x 0.46 nm) and four segments of sinusoidal (4 x 0.66 nm) channels and four channel intersections (4 x 0.54 nm). Also included in Table 2 is the theoretical unit cell adsorption channel length based on two adsorbate molecules being adsorbed per intersection.

Table 2
Silicalite-1 channel lengths

channel type	channel length, nm/uc
sinusoidal channels and intersections	4.83
linear channels and intersections	3.98
sinusoidal and linear channels and intersections (1 molecule/intersection)	6.65
sinusoidal and linear channels and intersections (2 molecules/intersection)	8.81

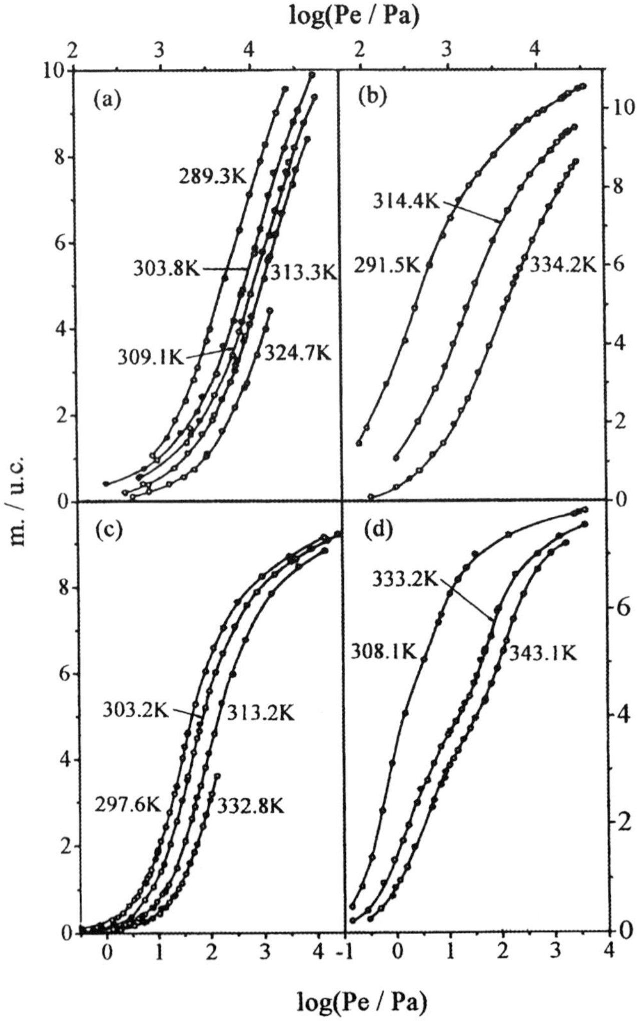

Figure 6. Isotherms of (a) ethane, (b) propane, (c) n-butane, and (d) n-hexane in silicalite-1.

From these channel lengths the theoretical adsorption capacities, n_{theor}, given in Table 3 can be calculated and these capacities may be compared with the experimental capacities, n_{exp}, in column 5 of Table 3. The effective channel length occupied by these molecules (including p-xylene; see later) may be compared with the calculated lengths in Table 2. It can be seen that ethane, propane, n-butane and p-xylene tend to have one adsorbate molecule per intersection whereas n-hexane is tending towards two molecules per intersection (at the equilibrium pressures, P_e, listed in Table 3).

Table 3
Adsorption capacities for various hydrocarbons in silicalite-1

adsorbate	adsorbate length, nm	n_{theor}, m/uc		n_{exp} m/uc	max P_e/P_o	T K	effective channel length, nm/uc
		2 m/int	1 m/int				
ethane	0.526	16.7	12.6	10.58	0.012	289.3	5.57
propane	0.652	13.5	10.2	10.66	0.042	291.5	6.95
n-butane	0.778	11.3	8.5	9.27	0.098	303.2	7.21
n-hexane	1.030	8.6	6.5	7.85	0.138	308.1	8.09
p-xylene	0.98	9.0	6.8	7.63	0.048	313.2	7.48
nitrogen	0.364	24.2	18.3	33.9	0.181	77	12.34

The adsorption isotherms of p-xylene in silicalite-1 are shown in Figure 7 [45]. The adsorption capacity is seen to be 7.63 m/uc. The isotherms of p-xylene display a marked step at 4m/uc and a hysteresis loop at loadings above 4 m/uc. The marked step in adsorption isotherms is frequently found in the adsorption of aromatic molecules in silicalite-1/ZSM-5 and although not so distinct is seen also in the n-hexane isotherms in Figure 6. The hysteresis loop with p-xylene has been reported by many groups in silicalite-1 but tends to disappear when increasing amounts of Al are introduced into the framework to produce ZSM-5 samples. This hysteresis loop is of great interest. Hysteresis is normally associated with capillary condensation which occurs when the adsorbent contains meso- and macro-pores but is difficult to explain for adsorption in micro-pores where multilayer adsorption cannot occur. Several explanations have been proposed for this unusual hysteresis which are based on two different concepts. One is associated with the known phase changes which occur in silicalite-1 when adsorbates are adsorbed/desorbed. For example, the structure changes from $P2_1/n$ to Pnma to $P2_12_12_1$ symmetries under different adsorption conditions/loadings[46,47] but some of these symmetry changes occur at

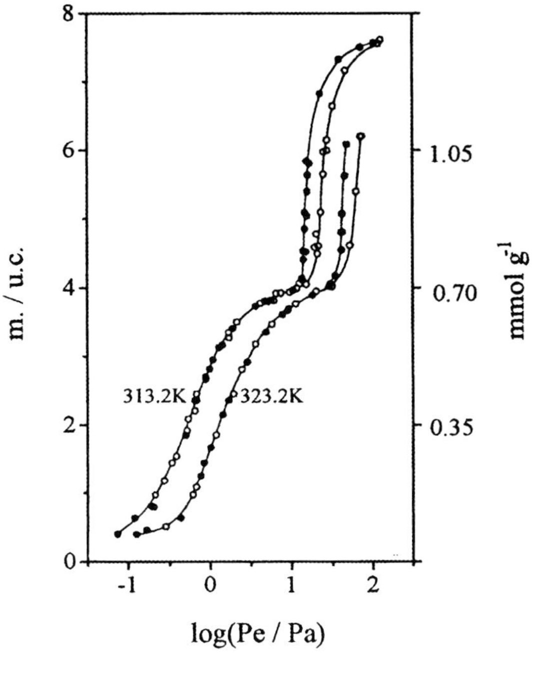

Figure 7. Isotherms of p-xylene in silicalite-1

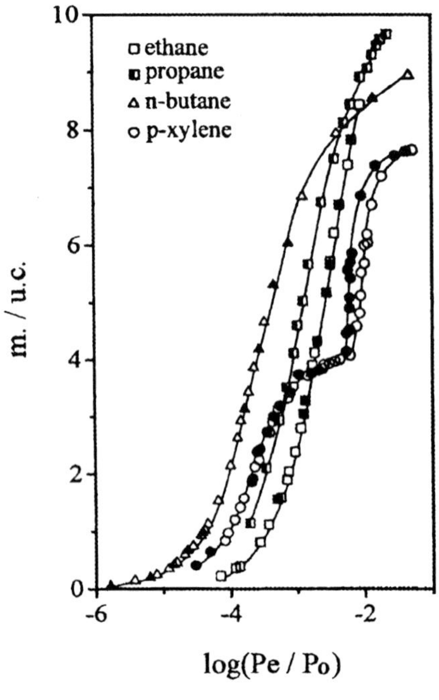

Figure 8. Comparison of plots for hydrocarbons on silicalite-1 at 313.5K.

loadings below 4m/uc whereas the hysteresis loop is always found at loadings above 4m/uc. These symmetry changes are observed with adsorbates where no hysteresis loop is observed. The transformation from orthorhombic to monoclinic symmetry involves only minor displacements of atomic positions within the framework structure which, under appropriate conditions, are highly reversible. This displacement transformation causes only very small changes in the unit cell shape and does not essentially change the size and shape of the channel openings and bend angle of the sinusoidal channel of the silicalite-1/ZSM-5 structure. The fact that these changes are very small is generally overlooked by many workers who continue to explain these changes in the adsorptive properties simply on these very small phase transitions.

The second explanation is due to phase changes/packing differences of the adsorbate on adsorption and desorption and this explanation seems to be more convincing.

In Figure 8 a comparison is made of the affinity, defined as -RT ln (P_e/P_o), where P_o is the saturation vapour pressure of the adsorbed fluid, of silicalite-1 for the adsorption of ethane, propane, n-butane and p-xylene at 313.5K. The affinity can be seen to increase with an increase in chain length of the normal alkane but p-xylene adsorption shows an intermediate affinity between that of n-butane and propane at adsorbate uptake levels of $\leq$ 4 m/uc. However, the adsorbate-adsorbent interaction is drastically reduced at higher adsorbate coverages and the affinity is now less than that of ethane.

Figures 9 and 10 show the adsorption of ethane, ethene, propane and propene adsorbed in NaY (56 Na ions per uc) and the pure silica form of Theta-1 zeolites respectively over a wide range of equilibrium pressures (up to 20 atmospheres) at temperatures of 273, 298 and 323K [48]. The adsorption of these adsorbates in Theta-1 should be similar in behaviour to that found above with silicalite-1 as the channels are of comparable dimensions but NaY has Na cations sited in at least two different sites in the supercages of this zeolite and the effect of large field gradients of the adsorption process can be observed.

The saturation adsorption capacities at P_o are greater in NaY than in Theta-1 because of the much greater free volume of NaY. The initial slopes of the ethane and propane isotherms are steeper in Theta-1 than in NaY due to the enhanced interaction energy in Theta-1 brought about by the closer fit of the molecules in the channels of Theta-1 than in the supercages of NaY. In contrast the initial slopes of the ethene isotherms are much greater in NaY than in Theta-1 because of the additional electrostatic interactions of the π electrons of the double bond with the electric field gradients at the cationic adsorption sites in NaY.

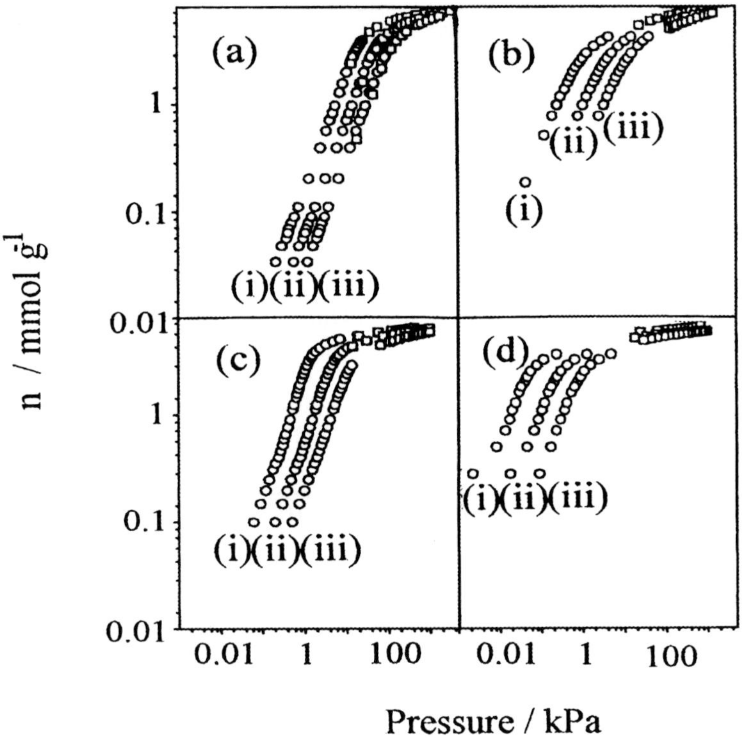

Figure 9. Low (O) and high (□) pressure adsorption isotherms for (a) ethane, (b) ethene, (c) propane, and (d) propene at (i) 273K, (ii) 298K, (iii) 323K on zeolite NaY.

The subtle differences in the adsorption behaviour of these adsorbates in these two zeolites arising from differences in the total free volumes, pore geometries and electric field gradients at the adsorption sites can be clearly seen in Table 4. The saturation capacities, n_{sat}, are some 3.4-3.5 times greater for the C_2 hydrocarbons and some 3.2 times larger for the C_3 hydrocarbons in NaY compared with Theta-1. Since the ratio of the crystallographic free volumes is ~2.5 these experimental ratios demonstrate the more efficient packing of the hydrocarbon molecules in the larger cavities of NaY compared with the one dimensional, narrower channels of Theta-1. The Henry's Law constants K_H in Table 5 show the stronger interaction of ethane and propane molecules with the adsorption sites in Theta-1 due to the enhanced dispersion force interactions of these non-polar molecules with the framework oxygens of Theta-1 compared with those in the NaY cavities. Polarisation of these saturated hydrocarbon molecules by the cations in NaY introduces a cation/induced dipole contribution to the adsorption energy, but this contribution is smaller than the enhanced dispersion force contribution from the close fit of the adsorbate molecules in the smaller channels of Theta-1.

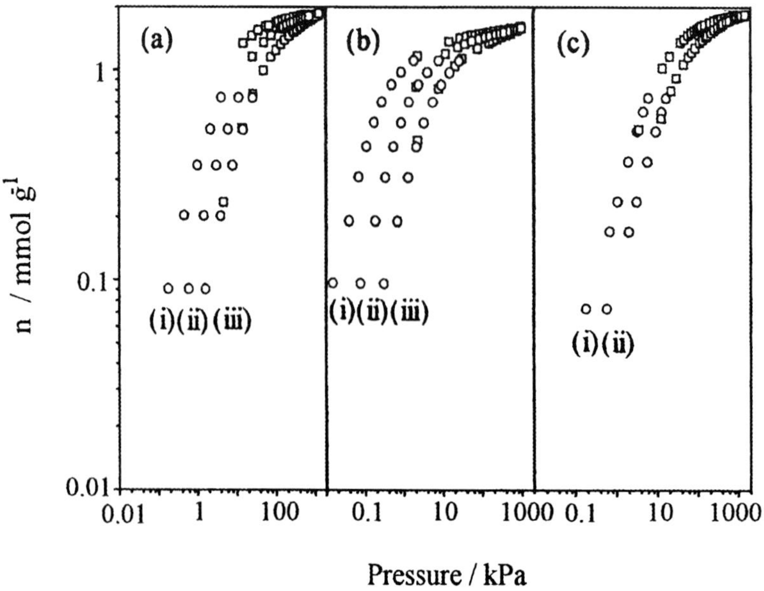

Figure 10. Low (O) and high (□) pressure adsorption isotherms for (a) ethane, (b) propane, (c) ethene at (i) 273K, (ii) 298K, (iii) 323K on zeolite Theta-1.

Table 4
Henry's Law Constants, K_H at 298K, Initial Heats of Adsorption $-\Delta H_K$, Initial Isosteric Heats Q_{st}, and Limiting Adsorption Capacity n_{sat} for the Hydrocarbons in NaY and Theta-1

Zeolite	Sorbate	K_H /mmolg^{-1}kPa^{-1}	$-\Delta H_K$ /kJmol^{-1}	Q_{st} /kJmol^{-1}	$-\Delta H_{liq}$ /kJmol^{-1}	$\Delta H_K/\Delta H_{liq}$	n_{sat} /mmolg^{-1}
NaY	Ethane	0.055	24.97	23.6	14.72	1.69	4.5
	Ethene	0.651	40.94	36.5	13.554	3.02	4.9
	Propane	0.569	35.82	29.6	18.77	1.90	3.8
	Propene	7.948	-	~45	18.40		3.6
Theta-1	Ethane	0.091	29.50	31.9	14.72	2.00	1.33
	Ethene	0.060	27.64	33.0	13.54	2.04	1.41
	Propane	0.700	41.73	42.0	18.77	2.22	1.19
Silicalite-1	Ethane		30.54				
	Propane		37.79				

In comparison the Henry's Law constants for the adsorption of ethene in NaY in Table 5 is an order of magnitude larger than that of ethene adsorbed in Theta-1 and that of propene adsorbed in NaY is over twelve times larger than the corresponding value of ethene adsorbed in NaY. These very large values arise from the large electrostatic interaction of the π electrons of the double bond with the large electric field gradients which are present in NaY.

Table 5
Henry's Law constants K_H in $mmolg^{-1}kPa^{-1}$ for hydrocarbons in NaY and Theta-1

Zeolite	Adsorbate	273K	298K	323K
NaY	Ethane	0.123	0.055	0.027
	Ethene	2.717	0.650	0.225
	Propane	1.581	0.569	0.210
	Propene	32.5	7.948	2.628
Theta-1	Ethane	0.270	0.091	0.036
	Ethene	0.167	0.060	0.025
	Propane	3.274	0.700	0.190

6.5 Energetics of Adsorption of Hydrocarbons

The study of the energetics of adsorption of selected probe molecules is a powerful way of investigating the adsorption properties of molecular sieve adsorbents. For example, in Figure 11 it is shown that the isosteric heat of adsorption of p-xylene in silicalite-1 varies in quite a complex manner with coverage [45]. There is an initial decrease in Q_{st} which is probably associated with the specific interaction of the π electrons of the aromatic ring of p-xylene with the electric field gradients present in the adsorbent due to the very small amounts of Al in the framework and from lattice defects which are introduced when the templating tetrapropylammonium cations used in the synthesis of the adsorbent are removed by calcination of the starting material. A minimum then follows at coverages from 1.5-3.5 m/uc where Q_{st} was approximately constant at ~78kJ/mol which then increases to 81-82 kJ/mol at loadings in excess of 4m/uc.

The differential entropy of adsorption, ΔS^o, for the adsorption of p-xylene in silicalite-1 also falls into two distinct regions. For < 4m/uc, ΔS^o is 157-178 J/mol.K while for >4m/uc, ΔS^o is 201-209 J/mol.K. These large decreases in entropy on adsorption indicate the loss of mobility of the adsorbate molecules on adsorption and the sudden further decrease at > 4m/uc indicates the even greater loss of freedom in the sites which show a larger isosteric heat of adsorption and , thus, a stronger interaction with the framework oxygens. The standard free adsorption energy, ΔG^o, as defined by Equation 10 changes, therefore, from approximately -28kJ/mol at coverages up to 4m/uc to -17kJ/mol at higher coverages. One possible explanation for these thermodynamic parameters is that the first four molecules of p-xylene are adsorbed preferentially in either the straight or sinusoidal channel segments while the last four molecules are adsorbed in the sinusoidal or straight channel segments respectively. It is difficult to decide which channel segments would be preferred by the first four molecules but the straight channel segments is the more likely guess. Similar results have been reported in the literature but the interpretations of the location of the p-xylene molecules vary and depend on phase changes which have been found with silicalite-1/ZSM-5 as a function of temperature, aluminium content of the framework and loading of adsorbate.

Frequency-response diffusion studies at somewhat higher temperatures than the isotherm temperatures in Figure 7, i.e. temperatures where the hysteresis loop tends to disappear, find approximate equal populations of both straight and sinusoidal channel segments even at quite low loadings. The small increase in Q_{st} at coverages greater than 4 m/uc arise from sorbate-sorbate interactions with no preference by the p-xylene molecules for either channel segment.

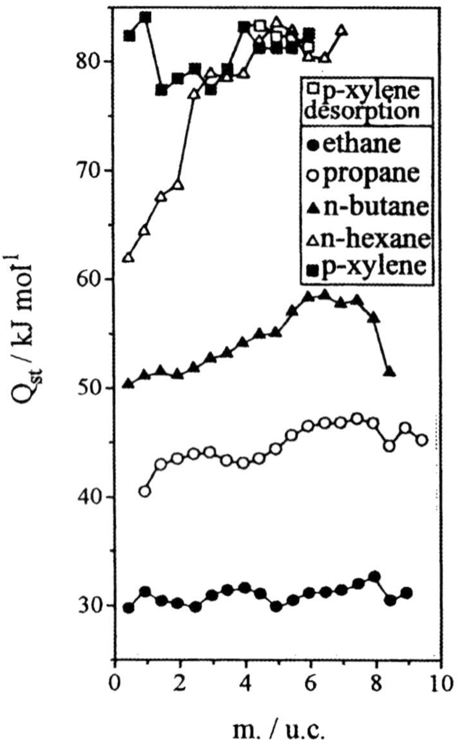

Figure 11. Isosteric heat of adsorption (Q_{st}) on silicalite-1.

This latter interpretation tends to be supported by modelling calculations.

The isosteric heats of adsorption of ethane, propane, n-butane and n-hexane in silicalite-1 are also included in Figure 11 as a function of loading. The heat of adsorption of ethane is near constant over the whole range of coverage. However, as the length of the n-alkane increases adsorbate-adsorbate interactions become more significant until with n-hexane the heat increases from ~60kJ/mol at low coverage to ~80 kJ/mol at loadings of ~8m/uc. Although silicalite-1 has been described as an energetically homogeneous adsorbent one would expect the interaction energy to be different for adsorbate molecules which are completely located in the channel segments and molecules which are longer than the channel segments and have, therefore, to occupy some of the channel intersection volume where the interaction energy will be reduced. To explain the variation of Q_{st} with coverage for n-alkane adsorption a model has been proposed based on the following assumptions: (a) n-alkane ZSM-5/silicalite-1 adsorption is dominated by dispersion forces, ϕ_D; (b) adsorbate molecules would prefer to be located in a channel segment, where the adsorbate-wall interaction is greater than at an intersection; (c) as the n-alkane chain length is increased a greater percentage of the adsorbate length is located in the intersection volume; (d) only end methyl-methyl adsorbate-adsorbate interactions are possible in the channels but side-on adsorbate-adsorbate interactions are possible in the intersections; (e) as the n-alkane chain length is increased beyond the distance between adjacent intersections along a given channel , the model fails, because one molecule can now block two intersections (i.e. chain length >1 nm); (f) when a n-alkane has a molecular length in excess of the linear or sinusoidal channel segment, the molecule will be preferentially located in the sinusoidal channel segment for ≤ 4m/uc, since this allows the maximum adsorbate-wall interaction.

With this model it is now possible to place the first 8m/uc of ethane completely in the four straight and four sinusoidal channel segments. The end methyl groups will be quite far apart up to this loading and adsorbate-adsorbate interactions will be minimal. The first four propane molecules will be located in the sinusoidal channel segments and once again adsorbate-adsorbate interactions will be negligible. However, the next four molecules will be located in the straight channel segments and as the molecule is longer than this segment then the end methyl groups can interact. On increasing the chain length further adsorbate-adsorbate interactions will increase and Q_{st} will show larger and larger adsorbate-adsorbate contributions.

6.6 Energetics of Adsorption of Polar Molecules

When adsorbate molecules contain permanent dipoles or quadrupoles then the energetics of adsorption are enhanced from $\phi_{F\mu}$, field-dipole and $\phi_{\bar{F}Q}$, field gradient-quadrupole interactions, respectively, if the adsorbent contains cations which introduce strong electric fields or electric field gradients at the adsorption surface. These additional components to the adsorption energy are demonstrated in Table 6

Table 6 clearly demonstrates the very large contribution from $(\phi_{F\mu} + \phi_{\bar{F}Q})$ terms when the adsorbate contains a quadrupole moment (N_2 and CO_2) or dipole moment (NH_3 and H_2O) over the contributions from $(\phi_D + \phi_R + \phi_P)$ terms for a zeolite such as NaX which has a large aluminium content and therefore high density of cations.

Table 6
Energetics of adsorption of polar molecules (kJ/mol) [49-52]

Zeolite	Adsorbate	Contribution to Q_{st} ($n \to 0$) from		
		Q_{st} ($n \to 0$)	$(\phi_D + \phi_R + \phi_P)$	$(\phi_{F\mu} + \phi_{\bar{F}Q})$
Na-X	Nitrogen	27.2	13.0	14.2
	Carbon Dioxide	51.0	17.6	33.4
	Ammonia	75.2	15.7	59.6*
	Water	~142	11.1	131*

* There may be a contribution from chemisorption in these initial values

6.7 Adsorption Isosteres and Thermodynamics

The isosteric adsorption technique has been used to study the adsorption of carbon dioxide in three faujasite samples and a micro-/meso-porous carbonaceous adsorbent (M-30) by Bulow and Shen [27]. M-30 (*Osaka Gas Company Ltd.*, Japan) is made from mesocarbon microbeads with a specific surface area of *ca.* 2,450 m^2/g; overall pore volume: 1.8 ml/g (volume fractions of micropores and mesopores represent *ca.* 0.75 and 0.25, respectively); main micropore volume contribution from pores with diameters between 8 and 15 Å; mesopore diameters are smaller than 30 Å; packing density: 0.30 g/ml; primary particle size: *ca.* (10 - 80) μm with a frequency maximum of the distribution curve at *ca.* 30 μm, secondary particle size: *ca.* 1 mm.

The faujasite adsorbents (FAU-I, -II and -III) are in the form of beads ((1.6 - 2.4) mm) that differ mainly in (*i*) content of binder, (*ii*) XRD crystallinity (no account for binder content), and (*iii*) Si / Al ratio (no account for binder content), *cf.*, Table 7. All zeolites contain only alkaline metal cations as monovalent lattice-charge compensating cations in their α-cages. Although not specified the BOC samples most probably contain Li^+ cations. Neither were the zeolite beads checked for extra-framework aluminum-based species although presence of the latter cannot be entirely excluded due to highly complex manufacturing processes [53].

Table 7

Characterization of zeolite adsorbents

Sample	Zeolite type	Origin of material	Binder content, weight %	XRD crystallinity, % *vs.* Internal BOCstandard	Si / Al mole ratio
FAU-I	Faujasite	BOC proprietary material	13	87 ± 5	1.28
FAU-II	Faujasite	Commercial NaX	Proprietary knowledge of manufacturer	Data not available	Proprietary knowledge of manufacturer
FAU-III	Faujasite	BOC proprietary material	17	78 ± 5	1.31

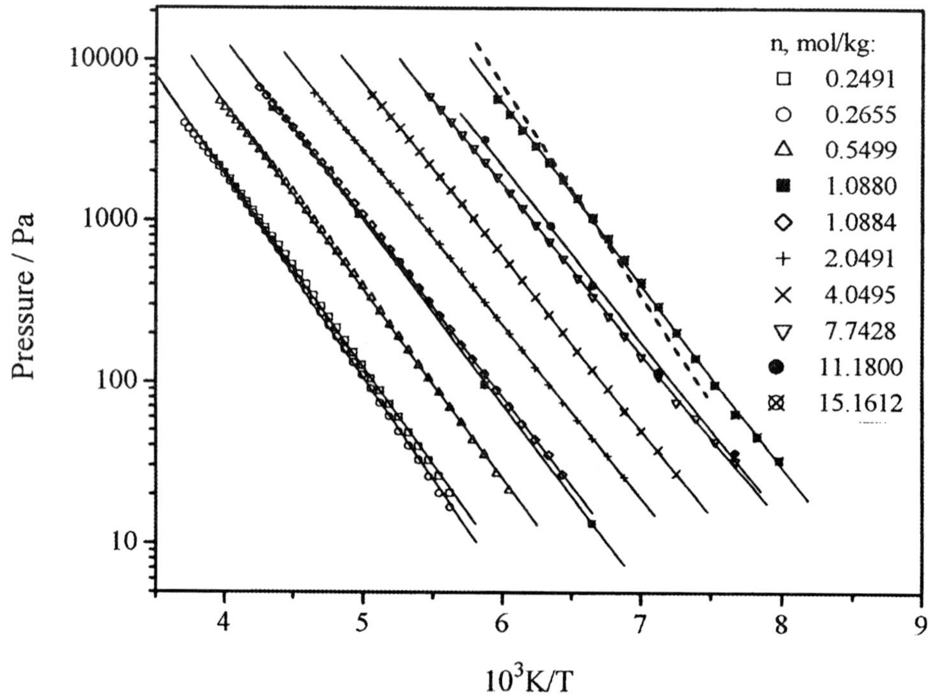

Figure 12. Sorption isosteres of carbon dioxide on sorbent M-30 (solid symbols are data points from another M-30 sample; dashed line is the sublimation curve for carbon dioxide).

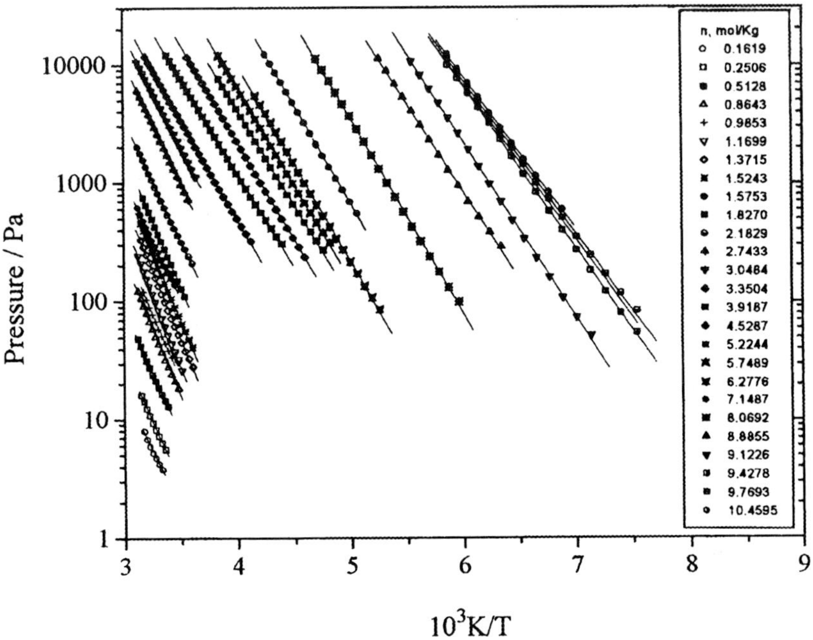

Figure 13. Sorption isosteres of carbon dioxide on FAU-I

Prior to the adsorption experiments, the materials were activated by increasing temperatures from ambient to 673 K *in vacuo* over 24 hours and keeping them at 673 K at *ca.* 10^{-3} Pa for another 120 hours (FAU-II) or 48 hours (FAU-I, FAU-III, M-30).

Sorption isosteres of CO_2 as measured on M-30, FAU-I and FAU-II sorbents in cryogenic temperature regions are presented in Figures 12-14, respectively. Sorption measurements were performed over the entire concentration ranges for zeolitic intracrystalline void volume up to filling the secondary pore volumes of two types of beads, as seen from isosteres. In those figures, each line of symbols represents one isostere measured at the respective sorption phase concentration. As the sorbate concentration approaches saturation, the isosteres approach the CO_2 sublimation curve measured separately, *cf.* Figure 3.

For the carbonaceous material, M-30, the last isostere is intersected with the sublimation curve, the former having a lower slope indicating that isosteric sorption heats and, thus, differential sorption enthalpies, were smaller than the sublimation enthalpy. For FAU-I, the last two isosteres coincide with the sublimation curve showing that saturation capacity was reached. In case of FAU-II, the last isostere proceeds still below the sublimation curve, indicating that total saturation capacity (of beads) was not reached. The coincidence of isosteres with the sublimation curve beyond saturation capacity shows that the isosteric measurements were correct and thermodynamically consistent. For CO_2 sorption in the region of saturation capacity of both related zeolites and the micro-(meso)-porous carbonaceous material, experimental data measured directly is not available from literature.

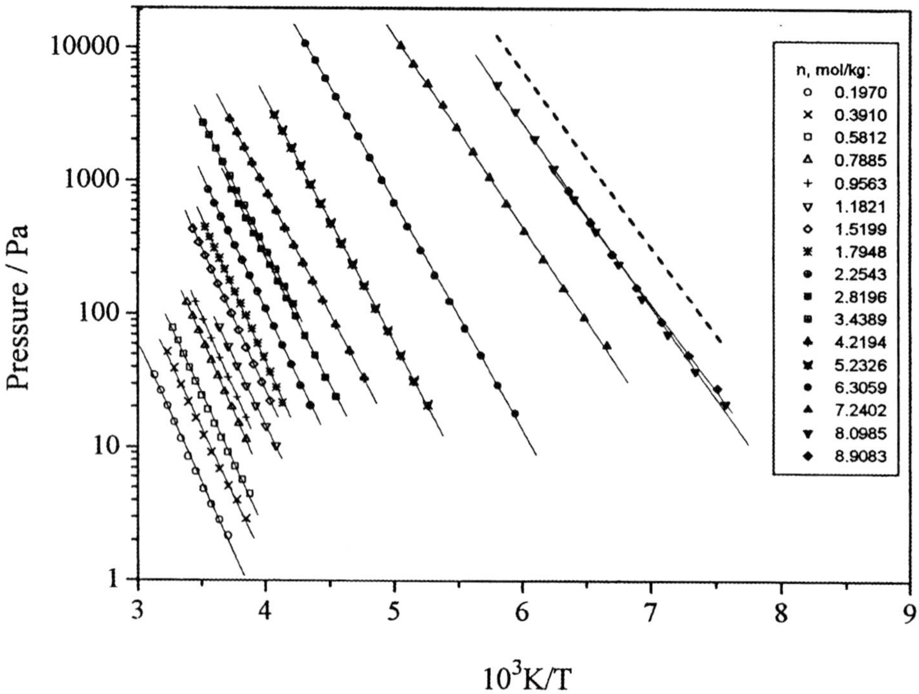

Figure 14. Sorption isosteres of carbon dioxide on FAU-II (dashed line is the sublimation curve)

The differential sorption enthalpy as a function of the CO_2 sorption phase concentration, *cf.*, Figure 15, shows different profiles for the three sorbents. For M-30, the sorption enthalpy decreases initially with increasing concentration, then maintains at *ca.* 21 kJ/mol over a concentration range of 1 ~ 12 mol/kg, and finally increases to approach the enthalpy of CO_2 sublimation, *ca.* 25.23 kJ/mol. The small change in sorption enthalpy, *ca.* 3 ~ 4 kJ/mol, over the entire concentration range, indicates that the sorbent exhibits almost homogeneous sorption sites over a wide pore filling range. A striking and most unusual result is that the sorption enthalpy of CO_2 for the M-30 material is less than the sublimation enthalpy; the reason for this behaviour is still under investigation. The extremely high capacity of M-30 for CO_2 sorption, that reaches *ca.* 15 mol/kg, was confirmed by independent high-pressure sorption measurements.

For FAU-type sorbents, the concentration-dependent differential sorption enthalpy of CO_2 is strongly influenced by their cation composition. Sorption sites in zeolite crystals are apparently energetically heterogeneous with regard to CO_2.

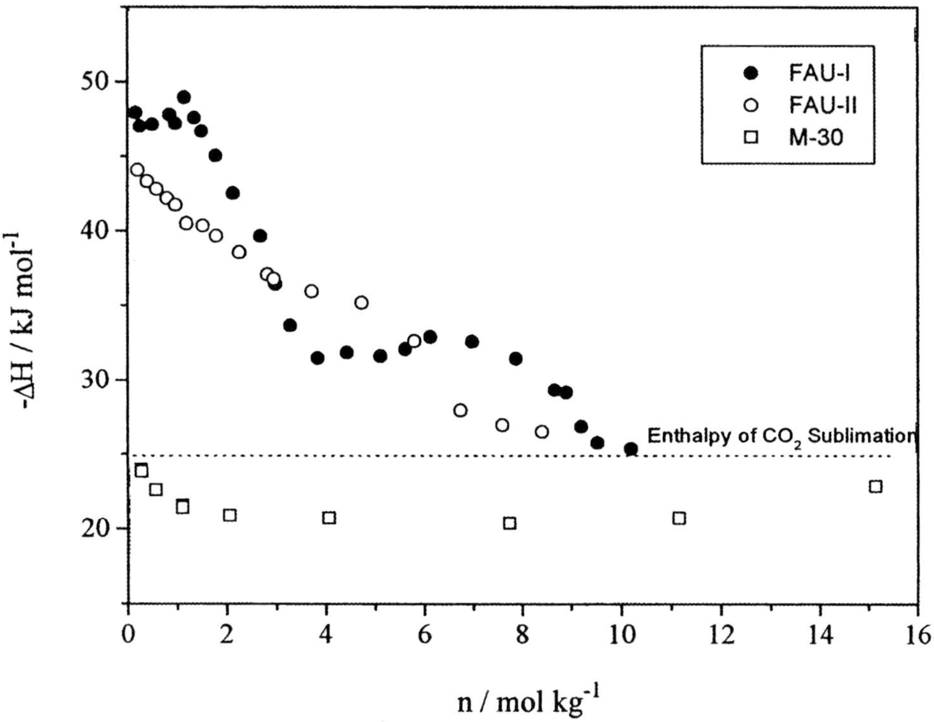

Figure 15. Differential sorption enthalpy for carbon dioxide on sorbents M-30, FAU-I and FAU-II.

Similar results have been reported in the literature dating back to the beginning of the sixties. A few of the important findings should be referred to as follows. Superior CO_2 sorption properties of LiX zeolite (*ca.* 100 % of cations exchanged) compared with those of NaX zeolite and a steady decrease in sorption enthalpy values for CO_2 with increasing concentration were reported [54]. In the case of alkaline metal cations in the zeolites, the initial values of the sorption enthalpy depend on the cation type with sequences both for the enthalpy changes and the extent of energetic heterogeneity as follows: $Li^+ > Na^+ > K^+ > (Rb^+, Na^+) > (Cs^+, Na^+)$, *i.e.*, the smaller the cation the greater is the sorption enthalpy and the energetic heterogeneity (presence of Na^+ in brackets indicates mixed cation compositions).

Sorption enthalpy changes for the two zeolites, FAU-I and FAU-II, amount to > 20 kJ/mol, for concentrations from nearly zero to saturation. For FAU-I, three regions of the dependence of sorption enthalpy on concentration can be seen from Figure 15. The strongest enthalpy change at low concentration characterizes strong specific interaction between CO_2 molecules and cations. At medium concentration, the sorption enthalpy reflects both interaction of CO_2 molecules with zeolitic crystal lattice sites and mutual CO_2-CO_2 interaction. As the concentration approaches pore saturation, the enthalpy change due to sorption reduces further to *ca.* 25.23 kJ/mol. For FAU-II, it decreases gradually from *ca.* 45 kJ/mol to *ca.* 37 kJ/mol with concentration increasing to *ca.* 3 mol/kg, and then it decreases slightly in the concentration range from 3 to 5 mol/kg. Between 5 to 9 mol/kg, the sorption enthalpy reduces finally to a value close to that of the CO_2 sublimation enthalpy. This result is in excellent agreement with data measured for CO_2 on a commercial powder of NaX zeolite, by a calorimeter-volumetric method [55].

By the isosteric method, it is possible not only to obtain CO_2 sorption thermodynamic functions at complete micropore filling but also in the meso- and macropore region up to bulk phase transition of CO_2. The saturation capacities extrapolated for both LiX (122 molecules per unit cell) and NaX zeolites (118 molecules per unit cell) [54] are close to those found experimentally in this work for zeolites FAU-I and FAU-II, where the comparatively high value measured in the case of FAU-I (*ca.* 150 molecules per unit cell) seems to be caused partly by sorption processes and capillary condensation in the secondary pore system of beads.

Calculated values of interaction energies of CO_2 with cations Li^+ and Na^+ that amount to *ca.* 51 and 45 kJ/mol [54], respectively, are close to experimental values for initial sorption enthalpies found on the former materials in a respective sequence. This difference (interaction with Na^+ *vs.* Li^+) is supported by experimental results for differential sorption enthalpies at concentrations that approach zero, as reported in this paper.

The standard sorption entropy, ΔS°, for the three CO_2 sorption systems are calculated as a function of CO_2 concentration in the sorption phase, *cf.* Figure 16. In each case, the variation of ΔS° with concentration is related to the corresponding variation of ΔH. A remarkable entropy loss for sorbed CO_2, compared to the standard gas phase, occurs over the entire concentration range.

ΔS° varies between *ca.* - 65 and - 135 J/mol K. From an entropic point of view, CO_2 molecules are strongly confined in micropores of FAU-I and FAU-II. In M-30 the molecules are more loosely confined, although at the initial concentrations the entropy losses for CO_2 sorbed by the three sorbents, are close to each other. On each of the three sorbents, CO_2 sorption entropy approaches the sublimation entropy as concentration approaches and exceeds its limiting value.

For the well-defined heterogeneous FAU-I sorbent, a wave-like sorption entropy dependence on concentration is found. This pattern corresponds to that of the differential sorption enthalpy as described above, *i.e.*, it is characteristic of a model for occupying several groups of energetically equivalent sorption sites in the sequence of their interaction energies. The wave-like profile in the entropy change is in excellent agreement with computer simulation results for a heterogeneous surface [56].

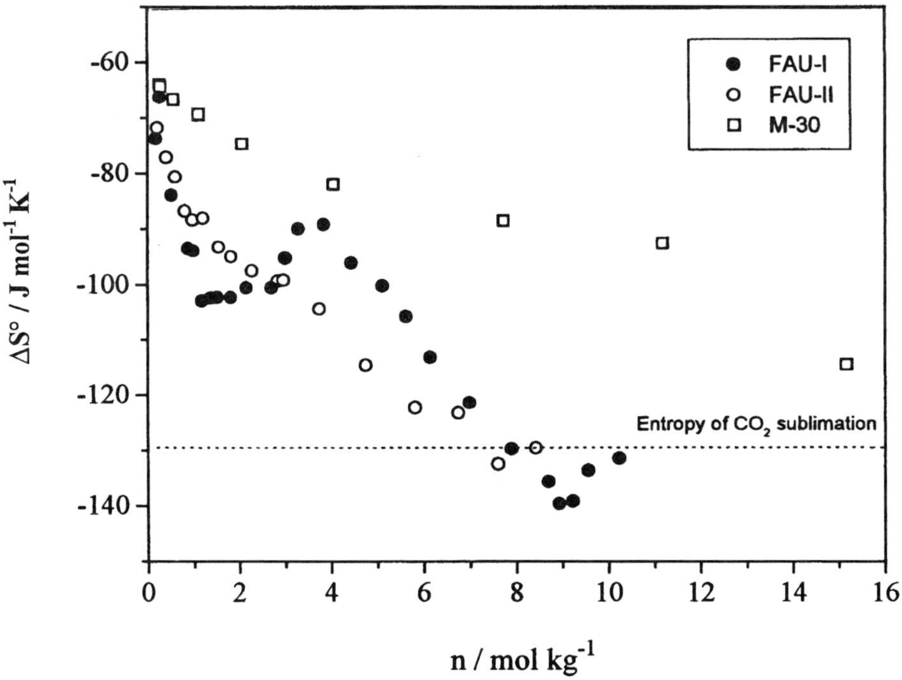

Figure 16. Standard sorption entropy for carbon dioxide on sorbents M-30, FAU-I and FAU-II.

The free energy change, $\Delta G°$, characterizes the natural tendency of a system to spontaneous change. Dependencies of $\Delta G°$ on CO_2 concentration in M-30, FAU-I and FAU-II sorbents at 194.65 K are shown in Figure 17.

In the three systems $\Delta G°$ for CO_2 sorption changes from negative values to zero as the CO_2 sorption phase concentration increases and exceeds saturation capacities demonstrating again the thermodynamic consistency of the experimental data. The larger negative values of $\Delta G°$ in cases of CO_2 sorption on FAU-I and FAU-II, indicate a stronger exothermic sorption process that occurs on zeolites as compared with that on M-30, whose $\Delta G°$ amounts to only about one third of that for FAU-I and FAU-II at initial concentrations. At low temperature, the quantity $T\Delta S°$, is often small, and so $\Delta G° \approx \Delta H$. The shape of the $\Delta G°$ *vs.* n curve becomes, therefore, a mirror image to the dependence of - ΔH *vs.* n.

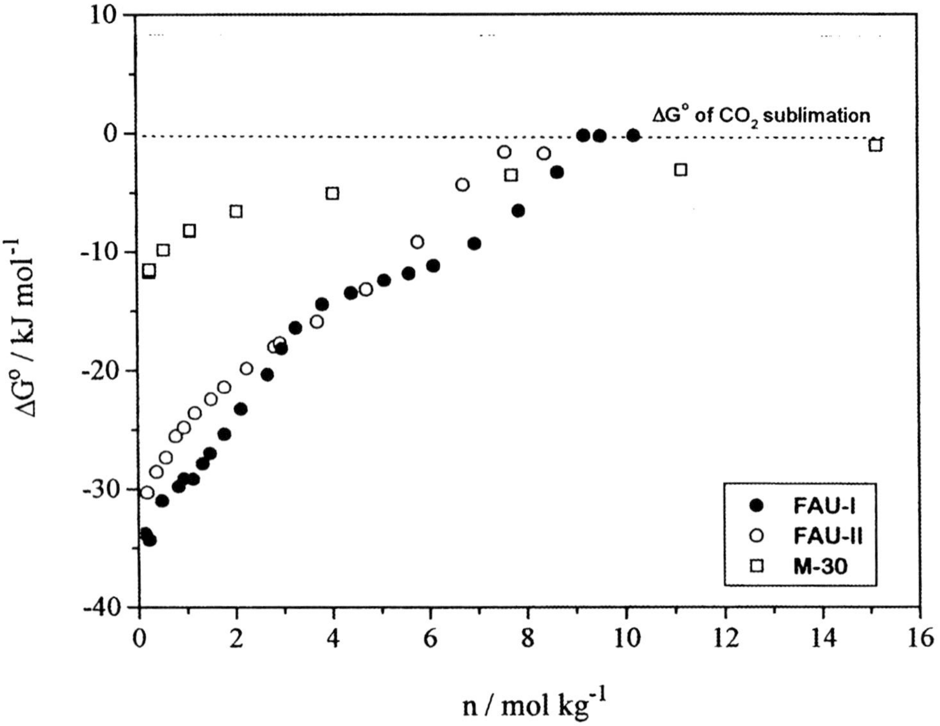

Figure 17. Standard *Gibbs* free energy of carbon dioxide on sorbents M-30, FAU-I and FAU-II, referred to 194.65 K

As known, experimental isosteric data, and, in particular, data for standard free energies as concentration dependencies, allow for both interpolation and extrapolation of sorption isotherms for any physically meaningful regions of temperature and pressure. For the various sorbents, an excellent agreement in isotherms obtained by different experimental methods with those extrapolated from isosteric data was achieved. This is proven by sorption isotherms given in Figure 18 where isotherms derived from isosteric data and measured directly by the microbalance technique are compared.

A related advantage of the isosteric method is that determination of isotherm data at high pressures can be carried out without presupposing an isotherm model equation.

6.8 Adsorption of Mixtures

Starting with data from single-component sorption isosteres of nitrogen and oxygen and total sorption isosteres of their binary mixtures at various sorption phase compositions measured on a FAU-III zeolite sorbent, and utilising thermodynamic relations outlined in the section (4.3), mixture sorption quantities for binary mixtures may be calculated and compared with experiment. The results obtained are given in Figures 19 and 20.

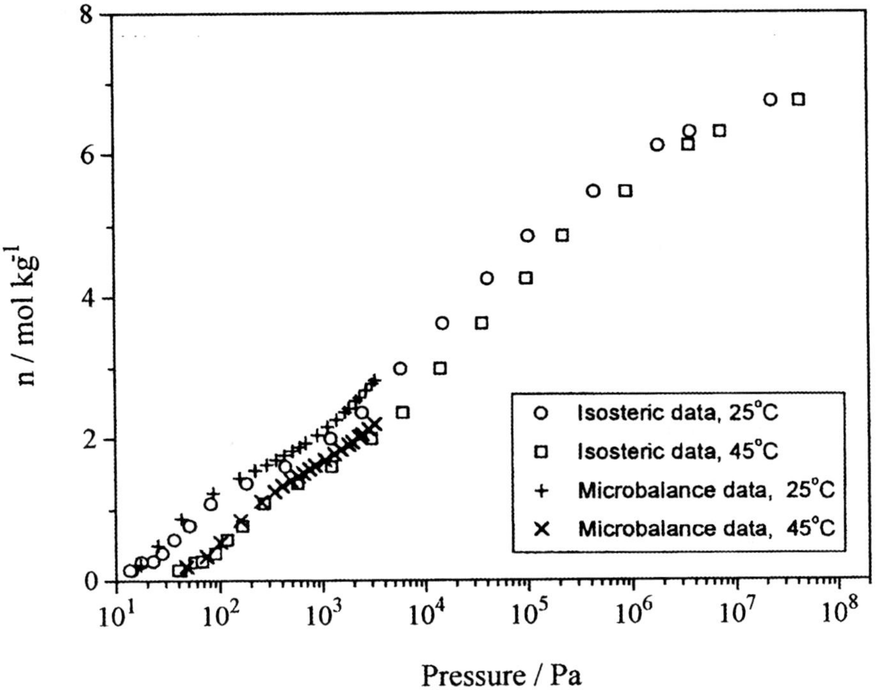

Figure 18. Comparison of carbon dioxide isotherms on FAU-I, obtained from isosteric data and measured by microbalance technique.

Figure 19 shows the free energy sorption data as a function of concentration for the pure gases nitrogen and oxygen and their binary mixtures with reference to a standard temperature of 298 K. A comparison is made for the experimental data and those predicted by means of the "Ideal Adsorbate Solution Theory" (IAST) from the single-component sorption data. For the mixture cases, the mole fractions for nitrogen relate to the sorbed phase composition. Sorption phase concentrations plotted along the *abscissa* comprise sorbed amounts for both single components and total concentration at a given composition.

Figure 20 shows the corresponding total sorption isotherms as functions, $n^m = f(p^m)$, for binary nitrogen-oxygen mixtures at constant sorption phase mole fraction, $x_a^{(i)} = 0.5$. The wave-like character of the isotherms is physically irrelevant but indicates a high sensitivity of the formalism utilized with respect to the accuracy of underlying experimental data (single-component sorption thermodynamic functions). A more detailed discussion of the nitrogen-oxygen mixture / FAU-III system has been published [57].

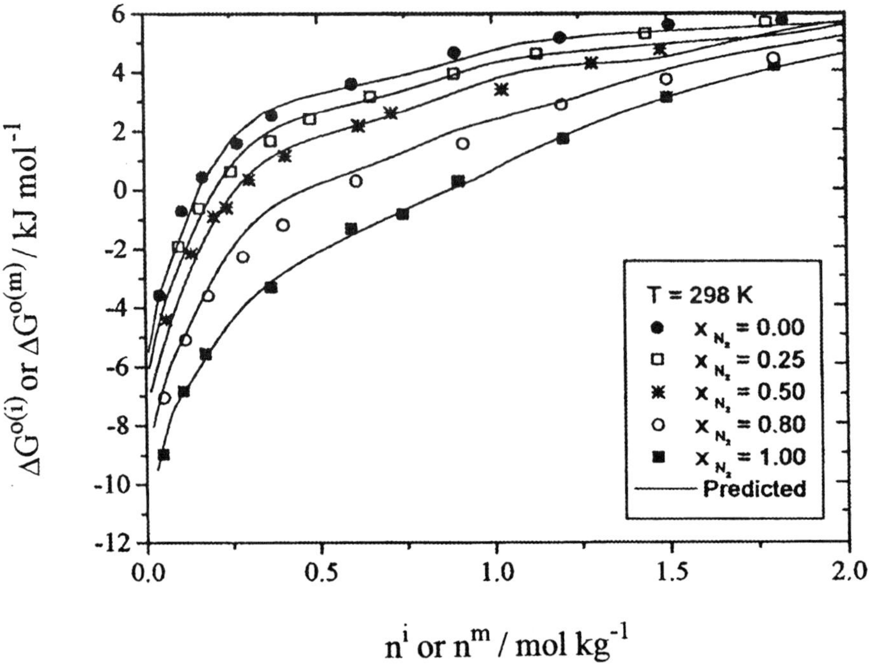

Figure 19. Free energy of sorption for nitrogen and oxygen and their mixtures on sorbent FAU-III at 298 K (symbols: experimental data; curves: predicted by IAST [12]).

The adsorption isotherms for three ethane-propane mixtures adsorbed in silicalite-1 and Na-Y zeolites are shown in Figure 21 along with the pure component isotherms for comparison [22].

The Henry's law constants obtained from the initial slopes of these isotherms at 273, 298 and 323K are listed in Table 8 which also contains the K_H values for the pure components. The initial isosteric heats of adsorption and $\Delta\overline{H}_K$ obtained from the temperature variation of K_H (see section 2.1) are listed for the three mixtures in silicalite-1 and NaY in Table 9.

The ethane-propane mixture isotherms in Figure 21 are well behaved and the Henry's law constants for all mixture isotherms in Table 8 are in good agreement with those predicted from Equation (64) demonstrating that the Henry's law constants for all the mixture isotherms are dependent on the composition of the gas phase

$$K_{H_m} = \sum K_{H_i} Y_i \qquad (64)$$

where K_{H_m} and K_{H_i} are the Henry's law constants for the mixture and pure components, i, respectively and Y_i is the gas-phase mole fraction of component i.

Table 8
Henry's law constants (K_H/ mmol $g^{-1}.kPa^{-1}$) from ethane and propane isotherms (pure and mixture) in silicalite-1 and zeolite NaY at 273, 298 and 323K

T/K	pure propane	ethane-propane 25:75	50:50	75:25	pure ethane
silicalite-1					
273.15	5.11	1.22	0.82	-	0.43
298.15	1.25	0.41	0.27	-	0.14
323.10	0.39	0.16	0.11	-	0.05
zeolite NaY					
273.15	1.95	0.38	0.22	0.15	0.12
298.15	0.57	0.13	0.09	0.06	0.05
323.10	0.20	0.06	0.04	0.03	0.03

Table 9
Mixture differential molar enthalpies calculated from the variation of Henry's law constants with temperature ($-\Delta\overline{H}_K$) for ethane-propane mixtures on silicalite-1 and zeolite NaY at 298.15K compared with the initial isosteric heats calculated from the slopes of the isosteres (Q_{st})

	ethane-propane	Q_{st}/kJ mol^{-1}	$-\Delta\overline{H}_K$ / kJ mol^{-1}
silicalite-1			
	25:75	27	27.41
	50:50	28	27.18
	75:25	-	26.38
zeolite NaY			
	25:75	24	23.93
	50:50	23	22.87
	75:25	22	22.11

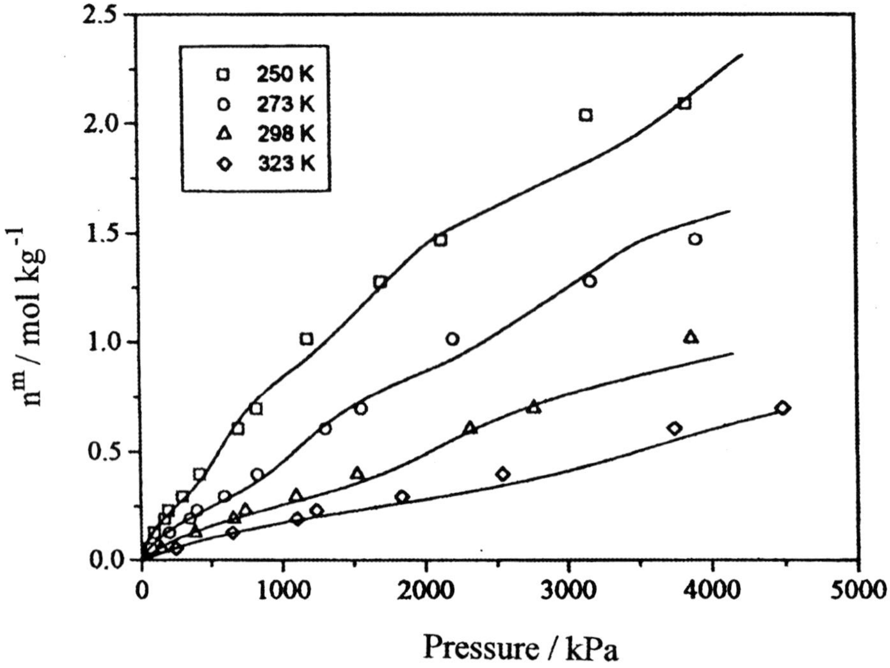

Figure 20. Total sorption isotherms for binary nitrogen-oxygen mixtures at constant sorption phase mole fraction, $x_i^{(s)} = 0.5$ (symbols: experiment; curves: predicted by IAST).

The isosteric heats of adsorption of the mixtures in silicalite-1 when compared with the corresponding heats of the pure components in Figure 22(a) show non-ideal behaviour. The initial heats of the adsorption, Q_{st}, of ~27kJ/mol are in excellent agreement with the differential molar enthalpies obtained from the variation of the Henry's law constants with temperature listed in Table 9. At low coverages the isosteric heats for the 50 and 75 mol % propane mixtures are much smaller than the corresponding heats for the pure ethane. Secondly, there is little difference in the isosteric heats of these two mixtures at low coverages. It is difficult to explain the reasons for this non-ideal behaviour of the isosteric heats when the mixture isotherms are so well behaved.

The corresponding isosteric heats of the three mixtures in NaY in Figure 22(b) show small increases as the proportion of propane increases but for all mixtures the heats are closer to those of pure ethane than pure propane. The isosteric method measures the equilibrium pressure of the adsorbate in the gas phase as a function of temperature at constant composition of the adsorbed phase and these equilibrium pressures are dependent mainly on the less strongly adsorbed species. Thus, the isosteric heat adsorption of mixtures tend to be close to that of the less strongly adsorbed species as has been found for N_2/CO_2 mixtures in silicalite-1 [20].

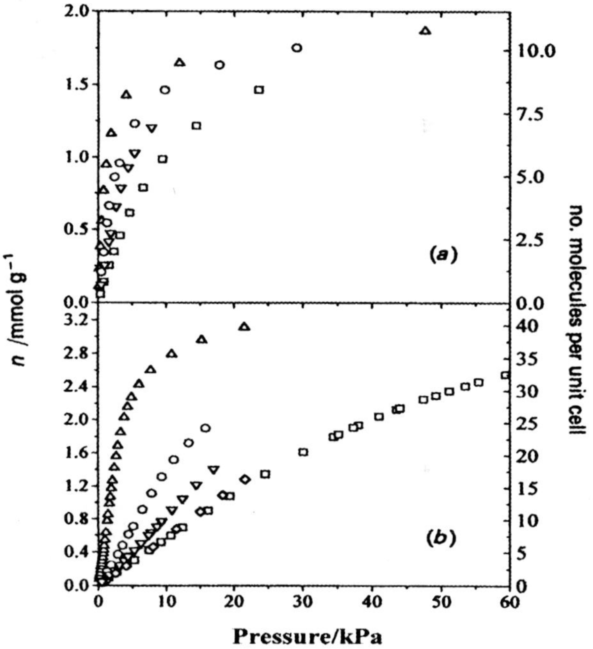

Figure 21. Pure-component and mixture Isotherms for ethane and propane on (a) silicalite-1 and (b) NaY at 298K: Δ, propane; □, ethane; O, 75; ∇, 50; ◊, 25 mol% propane.

Since the isosteric apparatus has an on-line mass-spectrometer which can measure the gas phase composition at every equilibrium point on the isostere and the adsorbed phase composition is known and is constant over the isostere it is possible to deconvolute all mixture isosteres into the partial isosteres of the two components of the mixture. From these partial isosteres the Henry's law constants and the partial thermodynamic parameters of the two components can be calculated and compared with the corresponding quantities obtained from the single component isosteres.

In Table 10 the Henry's law constants at 273, 298 and 323K obtained from the partial isosteres may be compared with the corresponding quantities obtained from the single component isosteres.

The initial isosteric heats of adsorption obtained from the slopes of the partial isosteres as well as $\Delta\overline{H}_K$ obtained from the temperature variation of K_H are listed in Table 11 where it can be seen that there is good agreement.

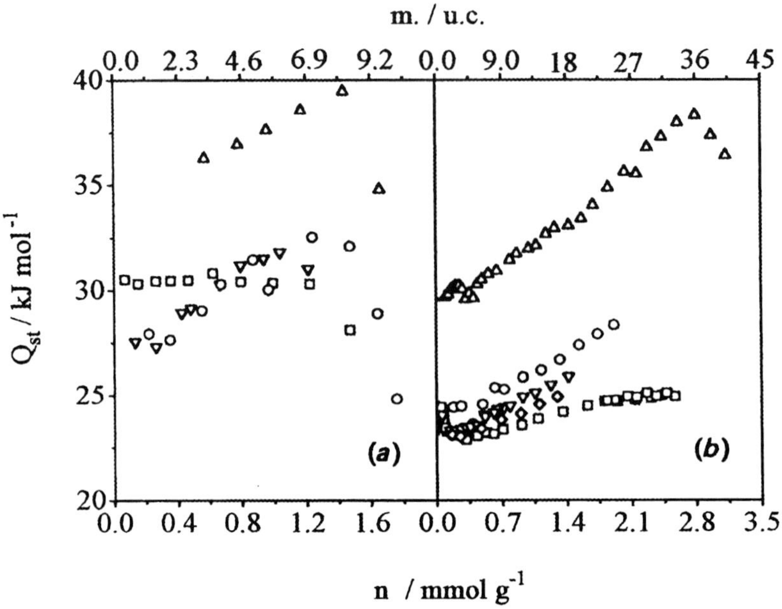

Figure 22. Pure-component and mixture isosteric heats of adsorption as a function of coverage for ethane and propane on (a) silicalite-1 and (b) NaY at 298K. Symbols as in Figure 21.

In Figure 23(a) and (b) the respective partial ethane and propane isotherms for silicalite-1 may be compared with the pure component isotherms at 298 and 323K. Although the initial slopes (i.e. Henry's law constants, see Table 10) are the same for both pure and partial isotherms, the partial isotherms rapidly deviate from the pure isotherms to much lower amounts adsorbed as the equilibrium pressure increases. In the limited channel volume of silicalite-1 the space occupied by the second component quickly affects the adsorption of the first component.

In Figure 23 (c) and (d) the corresponding comparisons can be made for the adsorption in NaY. In this zeolite, which has a much larger adsorption volume, the effect of the second component on the adsorption of the first component is much smaller than that found with silicalite-1. There is a small decrease in the amount adsorbed when the second component is present which probably arises from the competition between the two types of adsorbed molecules for the most energetic Na^+ sites (i.e. SIII) in the supercages of NaY.

Table 10
Henry's law constants (K_H/ mmol g^{-1}. kPa^{-1}) from partial ethane and propane isotherms deconvoluted from the mixture isotherms in silicalite-1 and zeolite NaY at 273, 298 and 323K
(a) Silicalite-1

		partial propane			partial ethane	
T/K	pure propane	25:75[a]	50:50[a]	Pure ethane	25:75[a]	50:50[a]
273.15	5.11	5.01	4.35	0.43	0.37	0.44
298.15	1.25	1.40	1.25	0.14	0.14	0.15
323.10	0.39	0.44	0.44	0.05	0.06	0.06

(b) Zeolite Na Y

		Partial propane				partial ethane		
T/K	pure propane	25:75[a]	50:50[a]	75:25[a]	pure ethane	25:75[a]	50:50[a]	75:25[a]
273.15	1.95	1.16	1.33	1.53	0.12	0.10	0.13	0.11
298.15	0.57	0.46	0.44	0.56	0.05	0.04	0.05	0.05
323.10	0.20	0.17	0.15	0.20	0.03	0.02	0.02	0.02

[a] Ethane : propane ratios in the mixtures.

Table 11
Partial differential enthalpies, calculated from the variation of Henry's law constants obtained from the partial isotherms, with temperature ($\Delta\overline{H}_K$) for ethane-propane mixtures on silicalite-1 and zeolite NaY at 298.15 compared with the initial isosteric heats calculated from the slopes of the partial isosteres(Q_{st})

	Q_{st}/kJ mol^{-1}		$-\Delta\bar{H}_K / kJ.mol^{-1}$	
ethane-propane mixture	ethane	propane	ethane	propane
silicalite-1				
25:75	25	31	25	33
50:50	25	31	26	31
zeolite Na Y				
25:75	20	-	19	26
50:50	21	31	23	29
75:25	22.5	31	22	27

The initial isosteric heats for the pure and partial systems are listed in Table 11. However, a more comprehensive comparison can be made from Figures 24 and 25 where the isosteric heats for the pure and partial systems are shown as a function of coverage for silicalite-1 and NaY, respectively.

The isosteric heats for ethane in silicalite-1 obtained from the partial isosteres at low coverages are smaller than for the corresponding heats for pure ethane, whereas for propane the two sets of heats are similar within the accuracy of the measurements. These results suggest that propane molecules tend to occupy the preferred sites which are most probably the sinusoidal channel segments, forcing the ethane molecules to occupy the shorter straight-channel segments and possibly the even weaker channel intersections at higher propane loadings. The isosteric heat for ethane increases rapidly from 25 to 28 kJ/mol and falls dramatically to 18 kJ/mol at ~2 molecules per unit cell as shown in Figure 24(a) for the 75 mol% propane mixture. When there is less competition by propane as in the 50 mol% mixture, these partial heats do not show a decrease with increasing coverage until a total of ~4 molecules per unit cell are sorbed.

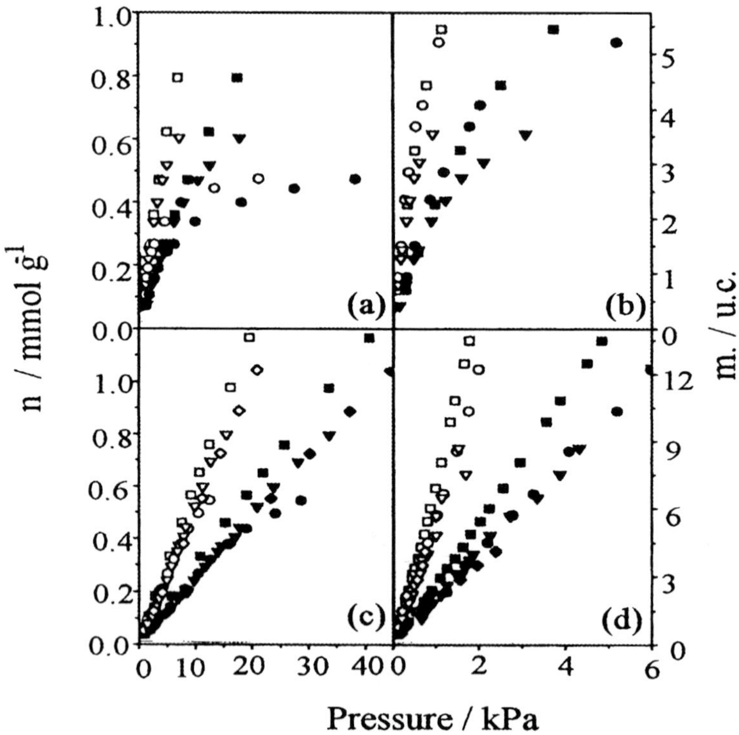

Figure 23. Pure-component and partial isotherms as a function of coverage for (a) ethane and (b) propane on silicalite-1; (c) ethane and (d) propane in NaY at 298K(open symbols) and 323K(filled in symbols). □, pure propane and pure ethane; ○, 75; ∇, 50; ◊, 25 mol% propane.

The partial isosteric heats for ethane in NaY as shown in Figure 25(a) are similar to the corresponding isosteric heats obtained for the pure component sorption isosteres. The initial heats are somewhat smaller and do not show the site heterogeneity found with pure ethane. This result seems reasonable as propane would be expected to compete successfully for those high energy sites. The partial heats for ethane tend to rise rapidly above those for the pure ethane as coverage increases, which probably indicates enhanced adsorbate-adsorbate interactions, but in this case this interaction will be due to ethane-propane interactions. The partial propane isosteric heats are shown in Figure 25(b) to be greater than the corresponding

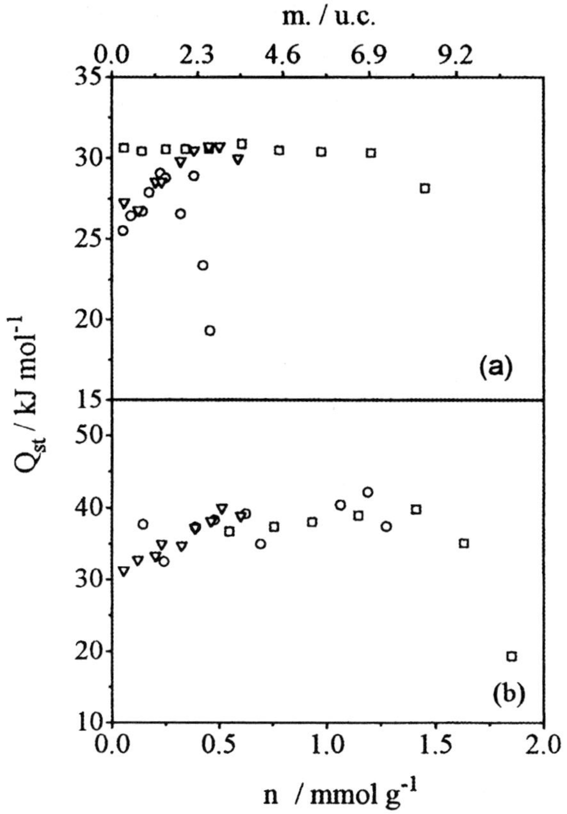

Figure 24. Pure-component and partial isosteric heats of adsorption as a function of coverage for (a) ethane and (b) propane on silicalite-1 at 298K. Symbols as in Figure 21.

pure component heats at all coverages. This may be due to enhanced adsorbate-adsorbate interactions which in this case will be due to propane-ethane interactions and may indicate that the smaller ethane molecules can pack closer to the adjacent propane molecules than the larger propane molecules can in the single component sorption case. These explanations for the differences in the isosteric heats tend to be confirmed when effects of the variations in the compositions of the mixtures on these partial isosteric heats are studied. For example, the partial isosteric heats for ethane increases as the amount of propane in the mixture increases due to enhanced ethane-propane interactions. Similarly, the partial isosteric heats for propane increase as the mixture is enriched in ethane due to a similar increase in propane-ethane interactions.

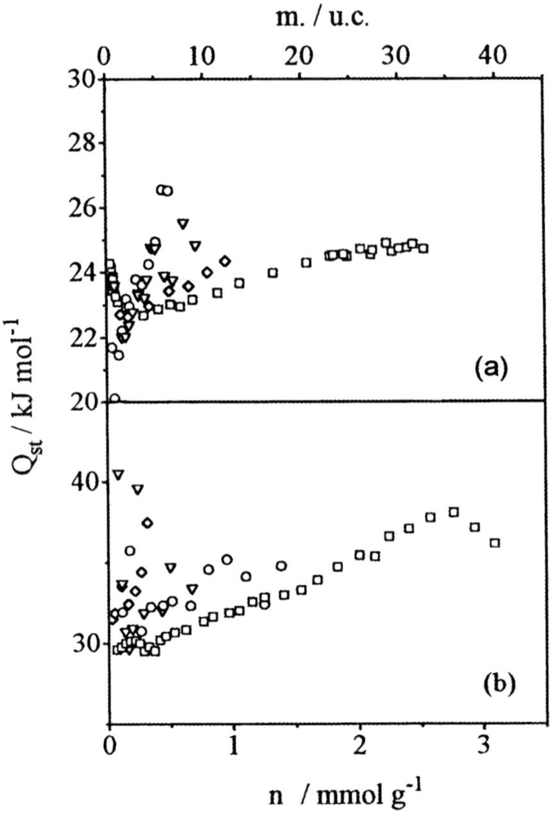

Figure 25. Pure-component and partial isosteric heats of adsorption as a function of coverage for (a) ethane and (b) propane on zeolite NaY at 298K. Symbols as in Figure 21.

6.9 MESOPORE SIZE ANALYSIS

With the recent development of well-defined mesoporous solids, of which MCM-41 is the most widely studied example, there is a need to measure the diameters of the mesopores in these materials. When nitrogen is sorbed at liquid nitrogen temperatures in these materials BET Type IV isotherms result and the analysis of these isotherms leads to both total surface areas and pore sizes.

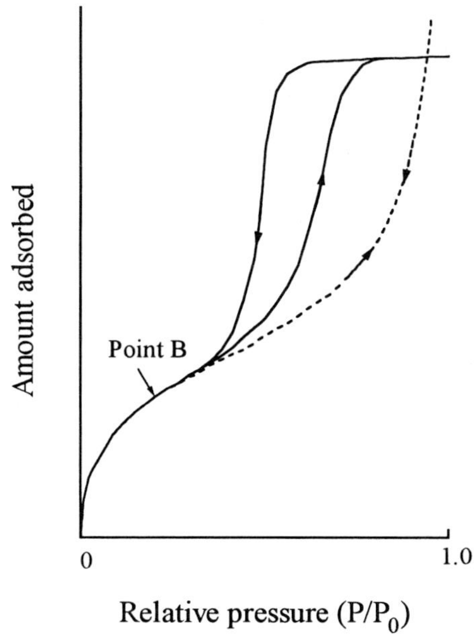

Figure 26. Type IV isotherm (solid lines) showing hysteresis loop. Type II isotherm (dash line) for comparison.

In Figure 26 a typical type IV isotherm obtained with a mesoporous material is shown. Hysteresis loops, i.e. the sorption and desorption branches do not coincide in the higher p/p_o relative equilibrium pressure region, are often found with such isotherms. The amount sorbed is always greater, at any given relative pressure, along the desorption branch than along the sorption branch. At lower relative pressures the isotherm follows the same path as that of the type II isotherm for a non-porous or microporous material. The point of inflection is referred to as the "point B" and the amount sorbed at this point is taken to be that corresponding to the completion of the monolayer and is a simple measure, therefore, of the total surface area of the material.

It is usually assumed that in the hysteresis loop region capillary condensation is taking place. There are several thermodynamic explanations of the physical significance of this region and these are fully discussed by Gregg and Sing[58]. In order to analyse the pore size distribution from the desorption branch of the hysteresis loop of the isotherm Equation(65), the Kelvin equation, is usually employed.

$$\ln \frac{p}{p_o} = -\frac{2\gamma V_L}{RT} \frac{1}{r_m} \tag{65}$$

where p/p_o is the relative pressure of vapour in equilibrium with a meniscus having a radius of curvature, r_m, and γ and V_L are the surface tension and molar volume, respectively, of the sorbate (usually liquid nitrogen). R and T are the gas constant and temperature respectively.

From the Kelvin equation it follows that the vapour pressure over a concave meniscus is less than the saturation vapour pressure p_o at temperature, T. Capillary condensation of a vapour to a liquid will occur at a pressure, p, within all pores or radius, r_m, or less. However, it must be realised that capillary condensation is taking place on pore walls that are already covered with a sorbed film of thickness, t, which is actually dependent on the relative pressure. Consequently, the Kelvin equation leads in the first instance to values of the "core size" rather than the pore size. The values of t for nitrogen at 77.4K have been calculated for p/p_o from 0.40 to 0.95, the normal range of the hysteresis loops for mesoporous materials [59, 60]. The values of t and the corrected pore radius, r_p ,where $r_p = r_m + t$ are given in Table 12.

Table 12
Values of r_p and t as a function of p/p_o for nitrogen at 77.4K

p/p_o	r_p nm	T Nm
0.40	1.560	0.535
0.45	1.740	0.560
0.50	1.945	0.585
0.55	2.185	0.615
0.60	2.490	0.650
0.65	2.870	0.685
0.70	3.370	0.735
0.75	4.050	0.785
0.80	5.070	0.860
0.85	6.75	0.965
0.90	10.19	1.275
0.95	19.9	1.6

The discovery of MCM-41 type materials, with their hexagonal array of uniform, open-ended tubular pores, has led to their use as model mesoporous sorbents. In a recent study the sorption of nitrogen in pure siliceous MCM-41, with various pore widths, has been reported [61]. The sorption isotherms were all type IV in character with sharply rising components in the p/p_o range of 0.14-0.24, 0.19-0.27, 0.29-0.36 and 0.33-0.42, for samples synthesised with dodecyl, tetradecyl, hexadecyl and octadecyltrimethylammonium bromide respectively as the surfactant, indicating fairly uniform, tubular pores of increasing pore diameters in the range 2.4 -3.6 nm. All of the isotherms were reversible i.e. no hysteresis was observed. All isotherms were coincident, within experimental error, over the initial p/p_o region; i.e. less than 0.14, indicating that there was no significant difference in the surface areas of the four samples.

The above observations resulted from the application of the α_s method of isotherm analysis [62]. This method is based on the assumption that for a given gas, e.g. nitrogen, the sorption isotherm on, for example, all silica type surfaces will be similar in shape even though their surface areas may differ greatly. To a first approximation, the different isotherms should be superposable by a simple adjustment of the units of sorption. It should be possible to bring the isotherms into coincidence by normalizing the amount sorbed, e.g. n/A, the amount sorbed per unit surface area. Sing et al [59,63] showed that the sorption of nitrogen on non-porous silicas having surface areas ranging from ~1 to $200m^2g^{-1}$ all fell on to a standard curve (see Table 2.14, ref 62). However, it was found that the standard isotherm for a particular type of material must be determined from solids which are free from micropores. Deviations, then, from this standard isotherm in the higher pressure regions offer a means of detecting and analysing the occurrence of capillary condensation in the mesopores of siliceous type MCM-41 and related materials. There are two convenient methods of detecting deviations from the standard isotherm. There are the "t-plot" [64] and "α_s plot" methods. The latter is a later development of the t-plot method. These plots can be used to detect capillary condensation in mesopores but can, also, determine the volume of the micropores which may be present.

A t-curve, which is the plot of the standard isotherm against t, the statistical thickness of the film, rather than n/n_m, the number of statistical monolayers, is first established (t is obtained from $t=(n/n_m)\sigma$, where σ is the thickness of single monolayer). For nitrogen at 77K σ is taken to be 0.354nm. The isotherm under test is then redrawn as a t-plot, i.e. the amount sorbed is plotted against t rather than p/p_o. The change of the independent variable from p/p_o to t is obtained by reference to the standard t-curve.

If the isotherm under test is identical in shape with the standard, the t-plot will be a straight line which passes through the origin, with slope b_t. Since

$$\frac{n}{n_m} = \frac{t}{\sigma} \quad \text{and thus} \quad n = \frac{t}{\sigma} n_m \tag{66}$$

$$n = b_t t \quad \text{where} \quad b_t = \frac{n_m}{\sigma} \tag{67}$$

As, $A = n_m a_m L$, where A is the surface area, n_m is the amount of sorbate to form a monolayer, a_m is the average area covered by a single sorbate molecule and L is the Avogadro constant, the specific surface area A is related to the slope b_t by

$$A = a_m \sigma L b_t$$

For nitrogen at 77K, with $a_m = 16.2 \times 10^{-20}$ m^2/ molecule, $\sigma = 0.354$ nm and t expressed in nm, we obtain

$$A = 3.45 \times 10^4 b_t \tag{68}$$

If n is expressed in mol g^{-1} then the resultant surface area, in m^2 g^{-1}, is obtained from the t-plot.

If micropores are present in the sorbent the uptake is enhanced in the low pressure region. This results in a t (or α_s , see later) plot which now extrapolates to a positive intercept on the amount sorbed axis instead of passing through the origin and this positive value is equivalent to the micropore volume. The slope of the linear branch at higher t values is now proportional to the external surface area of the sorbent.

If the sorbent contains mesopores, capillary condensation will occur in each pore when the relative pressure reaches a value related to the radius of the pore. The t-plot will then show an upward deviation commencing at the relative pressure corresponding to that of the finest pore present in the sample.

For the purposes of testing the conformity of an isotherm to that of the standard isotherm the numerical thickness of the sorbed layers is irrelevant. It is sufficient merely to replace n_m as the normalizing factor by n_s, the amount sorbed at some fixed relative pressure , $(p/p_o)_s$, which is usually taken as $(p/p_o)_s= 0.4$. The normalized sorption, $n/n_{0.4}(= \alpha_s)$, of the reference sorbent is then plotted against p/p_o to obtain the standard α_s-curve corresponding to the previous t-curve. This α_s-curve can then be used to construct an α_s-plot from the isotherm of the material under test, just as the t-curve was used to construct the corresponding t-plot. If a straight line through the origin results then the test isotherm is identical in shape to that of the standard isotherm and the slope b_α of the linear branch of the α_s-plot will be equal to $n_{0.4}$, just as the slope of the t-plot was equal to n_m/σ. If samples differ only in surface area and not in porosity, their α_s-plots will be straight lines passing through the origin with slope b_α .

The surface area of the sample under test is then obtained as follows:

$$\frac{b_\alpha(sample)}{b_\alpha(reference)} = \frac{A(sample)}{A(reference)} \tag{69}$$

and

$$A(sample) = \frac{b_\alpha(sample)}{b_\alpha(reference)} \times A(reference) \tag{70}$$

The specific surface areas of samples can be determined without reference to their monolayer capacities or to the value of a_m of the sorbate. The method can, therefore, be used for sorbates other than nitrogen.

Standard α_s-curves for both nitrogen and argon on non-porous silica are given in Tables 2.14 and 2.15 of Gregg and Sing [58].

The α_s method was used to analyse the sorption isotherms of nitrogen at 77K in the siliceous MCM-41 samples referred to above. From these analyses the total surface area, A_T, was calculated from the slopes of the α_s-plots in the monolayer region while the corresponding values of the external areas, A_{ext} and the pore volumes, V_p, were calculated from the slope and intercept respectively of the multilayer regions at high p/p_o. The results obtained from these analyses for these four MCM-41 samples are listed in Table 13. All of these samples exhibit high pore volumes and total surface areas but low external surface areas which never exceed 8% of the total area.

The values of the hydraulic pore radius, $r_p(H)$ in Table 13 can be calculated on the assumption of a non-intersecting open ended pore structure from the relation; $r_p(H)=2V_p/(A_T-A_{ext})$. The resulting mean pore widths
(i.e. $2r_p(H)$) were found to be in the range 2.4 to 3.6nm. It is interesting to note that the corresponding values for $r_p(K)$, the corrected Kelvin radius are some 10-30% lower than the $r_p(H)$ values. This difference probably arises from the breakdown of the Kelvin equation at p/p_o values < 0.4. From $r_p(H)$ and the unit cell dimension a_o the wall thickness between the pores can be calculated from; wall thickness = $\sqrt{3}/(2a_o-2r_p(H))$. All samples tended to give similar wall thicknesses of ~0.6nm.

Table 13
Analyses of the nitrogen isotherms of siliceous MCM-41 samples by the α_s method

Sample	Rising p/p_o range	$r_p(H)$ nm	V_p cm^3/g	A_T m^2/g	A_{ext} m^2/g	$r_p(K)$ nm
C12	0.14-0.24	1.18	0.59	1018	18	0.8-1.1
C14	0.19-0.27	1.39	0.70	1038	32	1.0-1.2
C16	0.29-0.36	1.65	0.81	1038	57	1.2-1.4
C18	0.33-0.38	1.81	0.88	1054	83	1.3-1.6

8. ACKNOWLEDGEMENT

We should like to thank Lijuan Song for all her help in the preparation of this chapter

9. REFERENCES

[1] D.M. Ruthven, Principles of Adsorption and Adsorption Processes, John Wiley & Sons, New York, 1984, p 67
[2] J.R. Lacher, Proc. Camb. Phil. Soc. 33 (1937) 518; Proc. Roy. Soc. A161 (1937) 525
[3] R.H. Fowler and E.A. Guggenheim, Statistical Thermodynamics, Cambridge University Press, Cambridge, 1939, p 441
[4] R.M. Barrer and L.V.C. Rees, Trans. Faraday Soc. 55 (1959) 992
[5] J.A. Hirschfelder, R.J. Buehler, H.A. McGee Jr. And J.R. Sutton, Ind. Eng. Chem. 50 (1958) 375
[6] P. Brauer, A.A. Lopatkin and G.Ph. Stepanez, Molecular Sieve Zeolites, Advances in Chemistry Vol. 102, Amer. Chem. Soc., 1971, p 97
[7] R.M. Barrer, Zeolites and Clay Minerals as Sorbents and Molecular Sieves, Academic Press, London 1978, pp 141-148
[8] D.M. Ruthven, Principles of Adsorption and Adsorption Processes, John Wiley & Sons, New York, 1984, pp 78-82
[9] R.J. Sips, J. Chem. Phys. 16 (1948) 490
[10] J. Toth, Acad. Sci. Hungary (Budapest) 69 (1971) 311
[11] R.M. Barrer, Zeolites and Clay Minerals as Sorbents and Molecular Sieves, Academic Press, London 1978, p 191
[12] A.L. Myers and J.M. Prausnitz, AIChE J. 11 (1965) 121
[13] B.P. Bering, A.L. Myers and V.V. Serpinsky, Dokl. Akad. Nauk USSR 193 (1970) 119
[14] C. Liang, J. Phys. Chem. 57 (1953) 910
[15] B.P. Bering, E.G. Zhukovskaja, Ch. M. Rachmukov and V.V. Serpinsky, Z. Chemie (Leipzig) 9 (1969) 13

[16] G. Meinert, A. Grossmann, M. Bulow and W. Schirmer, Adsorbents, Their Synthesis, Properties and Utilization, (Eds. M.M. Dubinin and T.G. Plachenov), Nauka, Leningrad 1971, pp 164-169
[17] W. Lutz, A. Grossmann and W. Schirmer, Chem. Technik (Leipzig) 28 (1974) 739
[18] M. Bulow and P. Lorenz, Fundamentals of Adsorption (Ed A. Liapsis), Engineering Foundation, New York 1987 pp 119-128
[19] M. Bulow, Stud. Surf. Sci. Catalysis Vol 83, Elsevier, Amsterdam 1994, p 209
[20] P. Graham, A.D. Hughes and L.V.C. Rees, Gas Sep. Purific. 3 (1989) 56
[21] L.V.C. Rees, P. Bruckner and J.A. Hampson, Gas Sep. Purific. 5 (1991) 67
[22] J.A. Hampson and L.V.C. Rees, J. Chem. Soc. Faraday Trans. 89 (1993) 3169
[23] P. Bruckner and L.V.C. Rees, Proc. IIIrd Int. Symp. On Separation Technology (Ed. E.F. Vansant), Elsevier, Amsterdam 1994 pp 39-48
[24] Y. Yang and L.V.C. Rees, Microporous Mat. 12 (1997) 117
[25] S. Sircar, Ind. Eng. Chem. Research 31 (1992) 1813
[26] F. Karavias and A.L. Myers, Langmuir 7 (1991) 3118
[27] D. Shen and M. Bulow, Microporous Mesoporous Mater. 22 (1998) 237
[28] H. Blank, M. Bulow and W. Schirmer, Z. physik. Chem. (Leipzig) 260 (1979) 395
[29] A.A. Fomkin, V.V. Serpinsky and B.P. Bering, Izv. Akad. Nauk SSSR, Ser. Chim. 1975 pp 1931-1935
[30] P.W. Atkins, Physical Chemistry (1st Ed), Oxford University Press, Oxford 1978 p 112
[31] R.M. Barrer and J.A. Davies, Proc. Roy. Soc. A320 (1970) 289; A322 (1971) 1
[32] L.A. Garden, G.L. Kington and W. Laing, Proc. Roy. Soc. A234 (1956) 35
[33] W. Schirmer, G. Fiedrich, A. Grossmann and H. Stach, Molecular Sieves, Soc. Chem. Ind., London, 1968 pp 276-288
[34] R.M Barrer and J.W. Sutherland, Proc. Roy. Soc. A237 (1956) 439
[35] R.M. Barrer, F.W. Bultitude and J.W. Sutherland, Trans. Faraday Soc. 53 (1957) 1111
[36] R.M. Barrer and J.A. Lee, Surface Sci. 12 (1968) 354
[37] R.M. Barrer and P.J. Reucroft, Proc. Roy. Soc. A258 (1960) 431
[38] R.M. Barrer and R.M. Gibbons, Trans. Faraday Soc. 59 (1963) 2875
[39] R.M. Barrer and S. Wasilewski, Trans. Faraday Soc. 57 (1961) 1153
[40] R.M. Barrer and B. Coughlan, Molecular Sieves, Soc. Chem. Ind., London, 1968 pp 141-148 and pp 233-241
[41] T. Takaishi, A. Yusa and F. Amakasu, Trans. Faraday Soc. 67 (1971) 3565
[42] P.E. Eberly Jr., J. Phys. Chem. 67 (1963) 2404
[43] L.V.C. Rees and C.J. Williams, Trans. Faraday Soc. 60 (1964) 1973
[44] R.E. Richards and L.V.C. Rees, Langmuir 3 (1987) 335

[45] R.E. Richards and L.V.C. Rees, Zeolites 8 (1988) 35

[46] T. Takaishi, K. Tsutsumi, K. Chubachi and A. Matsumoto, J. Chem. Soc. Faraday Trans. 94 (1998) 601

[47] H. van Koningsveld, H. van Bekkum and J.C. Jansen, Acta Crystallogr., Sect. B 43 (1987) 127

[48] J.A. Hampson and L.V.C. Rees, Stud. Surf. Sci. Catalysis Vol 83, Elsevier, Amsterdam 1994, p 197

[49] R.M. Barrer and G.C. Bratt, J. Phys. Chem. Solids 12 (1959) 146

[50] R.M. Barrer and W.I. Stuart, Proc. Roy. Soc. A249 (1959) 464

[51] R.M. Barrer and B. Coughlan, Molecular Sieves, Soc. Chem. Ind., London, 1968 pp 241-249

[52] R.M. Barrer and R.M. Gibbons, Trans. Faraday Soc. 59 (1963) 2569

[53] F.R. Fitch, M. Bulow and A.F. Ojo, US Patent 5,464,467

[54] R.M. Barrer and R.M. Gibbons, Trans. Faraday Soc. 61 (1965) 948

[55] J. Dunne, R. Mariwala, M. Rao, S. Sircar, R.J. Gorte and A.L. Myers, Fundamentals of Adsorption (Ed. M.D. LeVan), Kluwer Academic Publ., Boston, 1996 pp 277-284

[56] V.A. Bakaev and W.A. Steele, Fundamentals of Adsorption (Ed. M.D. LeVan), Kluwer Academic Publ., Boston, 1996 pp 83-90

[57] M. Bulow and D. Shen, Fundamentals of Adsorption (Ed. F. Meunier), Elsevier, Paris, 1998 pp 87-92

[58] S.J. Gregg and K.S.W. Sing, Adsorption, Surface Area and Porosity 2nd Ed., Academic Press, London, 1982 Chapter 3

[59] M.R. Bhambhani, P.A. Cutting, K.S.W. Sing and D.H. Turk, J. Colloid Interface Sci. 38 (1972) 109

[60] S.J. Gregg and K.S.W. Sing, Adsorption, Surface Area and Porosity 2nd Ed., Academic Press, London, 1982 p 135

[61] M.M.L.Ribeiro Carrott, P.J.M. Carrott, A.J.E. Candeias, K.K. Unger and K.S.W. Sing, Fundamentals of Adsorption (Ed. F. Meunier),Elsevier, Paris, 1998 pp 69-74

[62] S.J. Gregg and K.S.W. Sing, Adsorption, Surface Area and Porosity 2nd Ed., Academic Press, London, 1982 pp 98-100

[63] J.D. Carruthers, P.A. Cutting, R.E. Day, M.R. Harris, S.A. Mitchell and K.S.W. Sing, Chem. and Ind. (1968) 1772

[64] B.C. Lippens and J.H. de Boer, J. Catalysis 4 (1965) 319

Studies in Surface Science and Catalysis 137
H. van Bekkum, E.M. Flanigen, P.A. Jacobs and J.C. Jansen (Editors)
© 2001 Elsevier Science B.V. All rights reserved.

Chapter 14

Introduction to Acid Catalysis with Zeolites in Hydrocarbon Reactions

J.A. Martens and P.A. Jacobs

Centrum voor Oppervlaktechemie en Katalyse, K.U. Leuven,
Kard. Mercierlaan 92, B-3001 Heverlee, Belgium*

1. INTRODUCTION

Catalytic conversion of hydrocarbons over crystalline microporous acidic aluminosilicate catalysts is applied in large petroleum refining processes including catalytic cracking, hydrocracking, isomerization, alkylation and oligomerization. The purpose of this chapter is to provide the reader with a conceptual basis for rationalizing these acid zeolite based catalytic hydrocarbon conversion processes.

It is generally recognized that the zeolite acid strength is more in line with conventional strong acids rather than with superacids [1]. In order to explain the reactivity and the selectivity of the zeolite catalyst, it is insufficient to represent the zeolite catalyst by a proton, by analogy to homogeneous strong BrØnsted acids, without considering the chemical and physical interactions of the hydrocarbon molecules with the zeolite framework. The kinetics of the catalytic conversions are dominated by the strong physical adsorption of the molecules on the catalyst being unique to zeolites [2-4].

The active sites of aluminosilicate zeolites are complex and comprise a hydroxyl group, bridging a Si- and an Al-atom and having strong BrØnsted acid properties, and the oxo bridges, having Lewis base properties [5,6] (Fig.1).

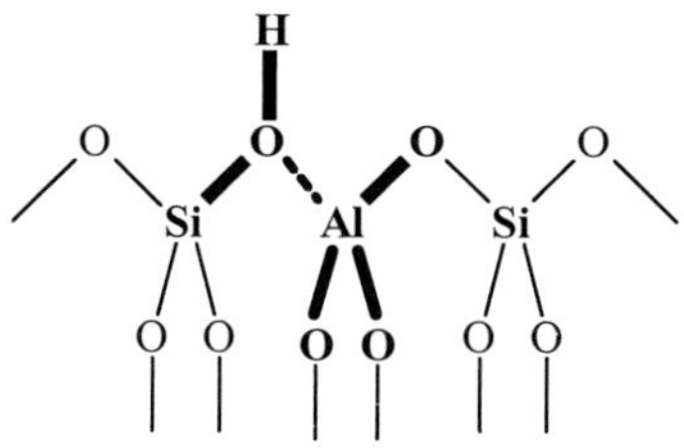

Fig.1. Active sites in hydrogen aluminosilicate zeolites. The bridging hydroxyl group and the oxo ligands of the Al-atom act as bifunctional BrØnsted acid - Lewis base sites.

* This work was carried out in the frame of the GOA research project, sponsored by the Flemish government.

Whereas there is consensus on the nature of the acid sites, there is much less consensus on the reaction mechanisms, even for simple hydrocarbon molecules. It is generally accepted that the reaction mechanisms of hydrocarbon conversions on acid zeolites involve the formation of carbocations. Whether these carbocations occur as reaction intermediates or as transition states is presently a matter of debate, and is likely dependent on the hydrocarbon. Although the experimental observation of carbocations in zeolites remains exceptional [7], the current most successful approach to rationalize the reaction product distributions of model molecules relevant to the industrial applications assumes the occurrence of carbocations as stable reaction intermediates and transformations of these cations known from investigations in liquid superacid solutions [8-11]. The essential features of carbocations and their reactions assumed to occur on zeolites as acid and bifunctional catalysts will be treated in the first two sections of this treatise. In zeolites, the reaction pathways are most of the time altered by molecular shape-selectivity resulting from the confinement of the reacting molecules to the zeolite microvoids offering only a limited space. The topic of shape selective catalysis will be treated in the last section.

2. ALKYLCARBENIUM AND ALKYLCARBONIUM IONS

Among the carbocations of concern for the conversion pathways of hydrocarbons, a distinction has to be made between alkylcarbenium ions and alkylcarbonium ions [12] (Fig.2). Alkylcarbenium ions contain a tri-coordinated positively charged C-atom, the three substituents being alkyl groups or H-atoms. Alkylcarbonium ions contain a penta-coordinated positively charged C-atom, having the same type of substituents. In the alkylcarbonium ions that will be encountered, at least one of the five substituents is a H-atom. The formal representation of alkylcarbonium ions used in Fig.2 does not violate the basic rules of organic chemistry if it is kept in mind that the five bonds of the charged C-atom contain only eight electrons. They can be regarded as σ complexes of a proton and a C-C or C-H σ bond, and denoted as C-alkylcarbonium or H-alkylcarbonium ions, respectively. A few examples of relevant alkylcarbenium and alkylcarbonium ions are given in Table 1. The stability of alkylcarbenium ions depends on the inductive effect of the substituents of the positively charged C-atom. The stability of alkylcarbenium ions decreases in the order:

tertiary (three alkyl substituents) > secondary (two alkyl substituents) >
primary (one alkyl substituent) > methyl cation (no alkyl substituents) (1)

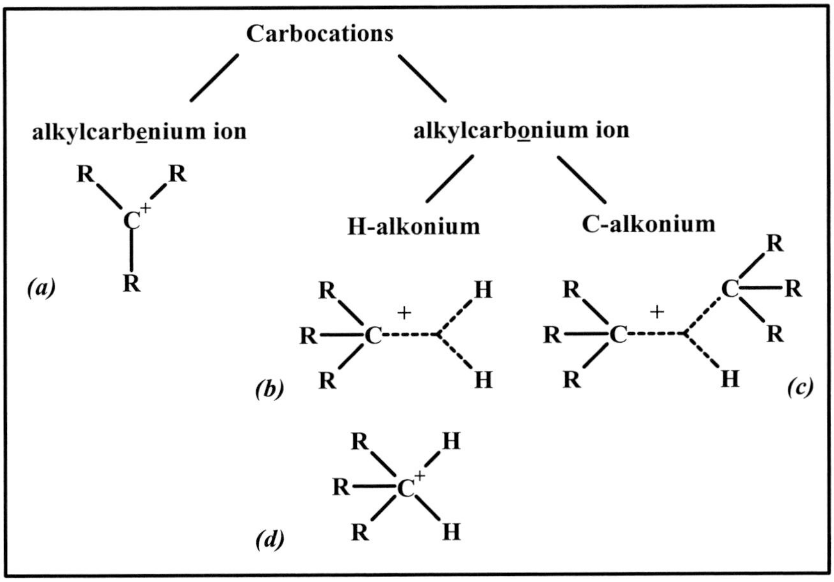

Fig.2. Representation of alkylcarbenium (a) and alkylcarbonium ions (b-d) (R is H or alkylgroup), (b) H-alkonium ion, (c) C-alkonium ion, (d) formal representation of H-alkonium ion.

A methyl substituent has a smaller inductive effect compared to ethyl or larger substituents, the inductive effect being primarily dependent on the number of C-C bonds in the β position to the positively charged C-atom.

The stabilizing effect of alkyl substituents in alkylcarbonium ions is less pronounced than in alkylcarbenium ions and the energy differences between primary, secondary, tertiary, and quaternary alkylcarbonium ions are smaller because of the larger number of substituents and larger charge delocalisation. For cations with 3 to 7 C-atoms, the stability of *sec* H-cycloproponium ions with methylsubstituents on the propane ring is compared to that of *prim*, *sec* and *tert* aliphatic alkylcarbenium ions in Fig.3. The stability of H-cycloproponium ion is situated between that of prop-1-yl and prop-2-yl cations. The stability of the methylsubstituted H-cycloproponium ions increases faster with carbon number than that of the alkylcarbenium ions. The reversal of relative enthalpies between *sec* aliphatic alkylcarbenium ions and *sec* H-cycloproponium ions occurs between carbon number 5 and 6. The curves in Fig.3 suggest that for eight or more C-atoms, the substituted cycloproponium ions may have the lowest enthalpies.

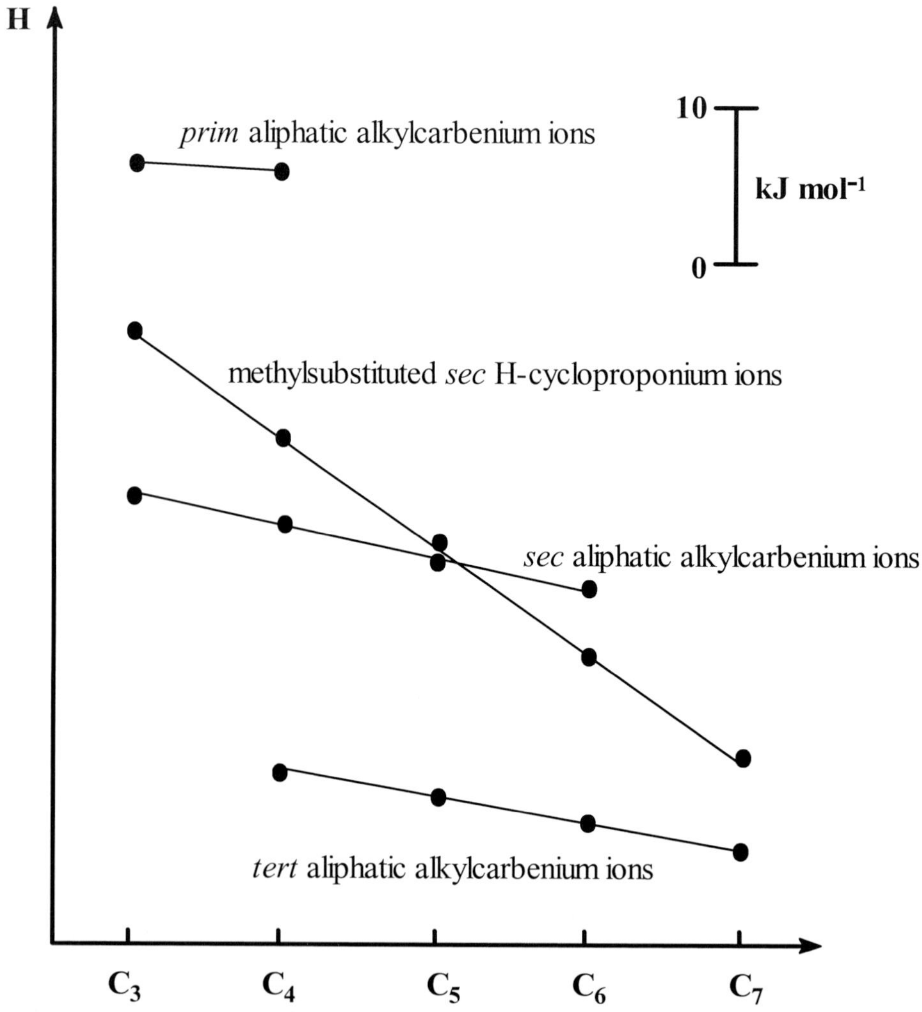

Fig.3. Relative enthalpies of aliphatic alkylcarbenium ions and alicyclic alkylcarbonium ions in the gas phase (C_3 and C_4) and in liquid superacids (C_5-C_7); data from [13].

Table 1
Examples of alkylcarbenium and alkylcarbonium ions

Formula or structure	Name	Formula or structure	Name
Aliphatic alkylcarbenium ions		Aliphatic alkylcarbonium ions	
$CH_3CH_2^+$	ethyl cation	$[CH_5]^+$	methonium ion
$CH_3CH_2CH_2^+$	prop-1-yl cation	$[CH_3CH_4]^+$	H-ethonium ion
$CH_3C^+HCH_3$	prop-2-yl cation	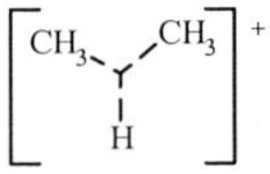	C-ethonium ion
$(CH_3)_3C^+$	2-methylprop-2-yl cation (*tert* butyl cation)	$[CH_3CH_2CH_3CH_3]^+$	2-H-butonium ion
Alicyclic alkylcarbenium ions		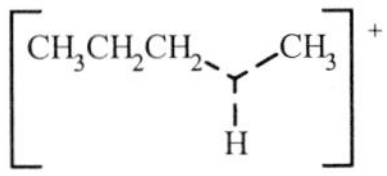	1-C-butonium ion
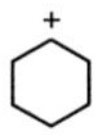	cyclohexyl cation	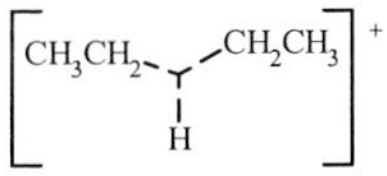	2-C-butonium ion
Aromatic alkylcarbenium ions		Alicyclic alkylcarbonium ions	
	benzenium ion	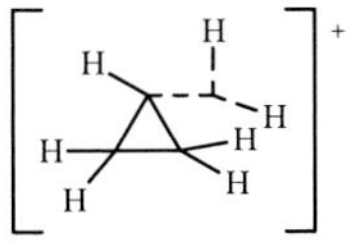	H-cyclopropionium ion (corner-protonated cyclopropane (CPCP))
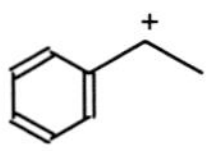	methylphenylcarbenium ion	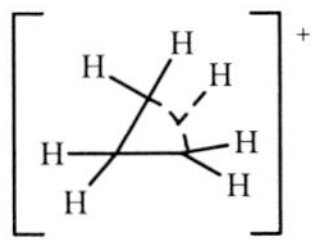	C-cyclopropionium ion (edge-protonated cyclopropane (EPCP))

3. CARBOCATIONS IN ZEOLITES: INTERMEDIATES VERSUS TRANSITION STATES

Ab initio quantum chemical calculations of the entire reaction coordinate of a zeolite catalyzed hydrocarbon conversion process involving large numbers of atoms are computationally very demanding. The published ab initio calculations deal with assumed energy minima and maxima along the reaction coordinate, relatively small hydrocarbons with one to seven C-atoms and a cluster representing the zeolite [14-19]. Calculations revealed that the minimum energy configuration of an aliphatic alkylcarbenium ion and a negatively charged zeolite lattice corresponds to one of the following two situations:

1. formation of a surface alkoxide (alkylsilylether): covalent bonding between positively charged C-atom and Lewis basic O-atom of the zeolite lattice, also referred to as σ-complex:

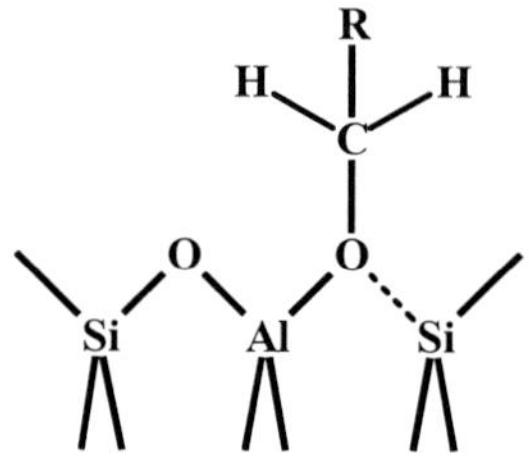

2. donation to the Lewis base site of a proton bound to a C-atom adjacent to the positively charged C-atom. The resulting alkene forms a π-complex with the H-atom of the bridging hydroxyl:

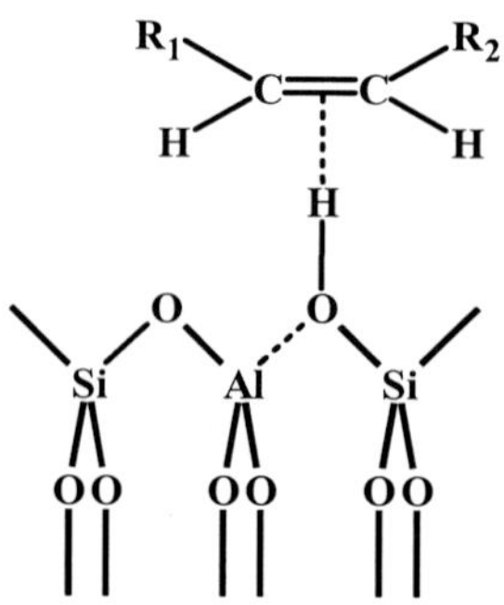

The formation of stable surface alkoxides is encountered in the calculations with primary alkylcarbenium ions [17-19]. The formation of alkoxy groups has been observed by NMR on zeolites contacted with alcohols at low temperatures [20]. According to the calculations, methyl cation, ethyl cation, prop-1-yl, but-1-yl cation, etc., do not occur as stable species and as reaction intermediates in a zeolite, but can occur as transition states, for example in the transformation of a π-complex into a more stable surface alkoxide (Fig.4). This transformation of the adsorbed alkene (π-complex) into the surface alkoxide proceeds via a transition state showing a characteristic 6-ring composed of H-C-C-atoms of the alkyl-fragment and O-Al-O-atoms of the active center. The geometry as well as the electronic structure of the alkyl fragment in the activated complex resemble the free alkylcarbenium ion.

Whereas the surface alkoxide is a more stable species compared to the π-complex with primary alkylcarbenium ions, with small secondary aliphatic alkylcarbenium ions the π-complex is more stable.

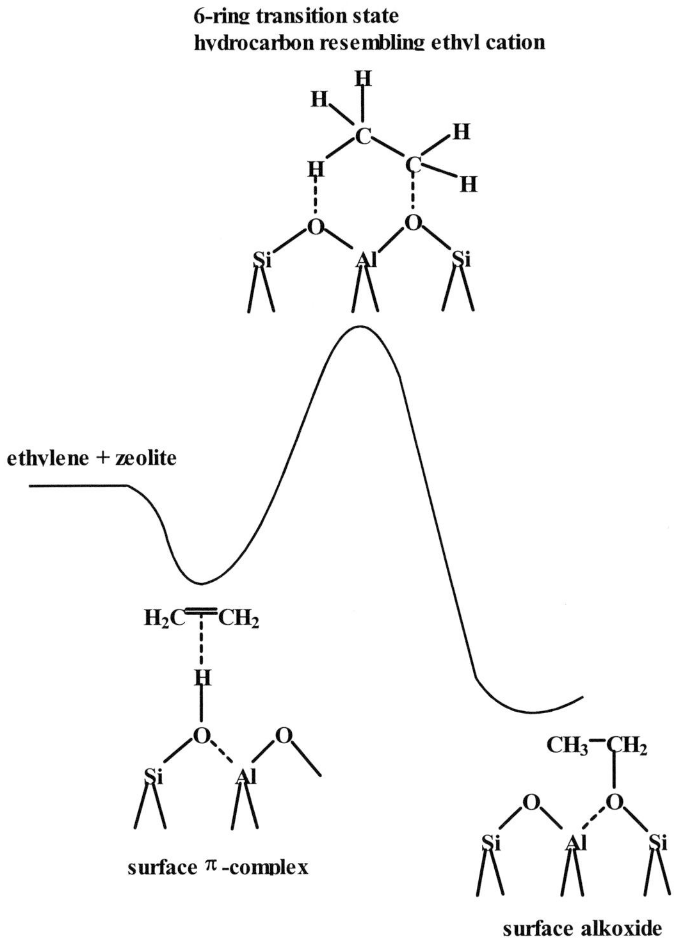

Fig.4. Reaction coordinate of ethoxylation of the bridging hydroxyl, after quantum chemical calculations [1].

A transition in behavior from small hydrocarbon molecules reacting as alkoxides to large hydrocarbons reacting as alkylcarbenium ions is expected to exist.

The formation of a zeolite-alkylcarbenium ion pair:

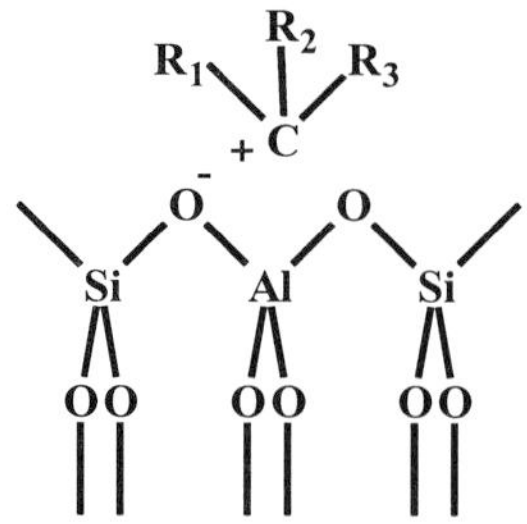

is likely in instances where steric separation of the carbon atom bearing the positive charge and the basic lattice oxygen atoms prevents alkoxide formation [2,21]. But even for the smallest tertiary alkylcarbenium ion, viz. the t-butyl cation, there is experimental evidence for its occurrence as a stable species. Carbon monoxide has an inhibiting effect on the hydrogen-deuterium exchange of perdeuterated isobutane and H-ZSM-5 zeolite, which can only be explained if the tertiary butyl cation is a reaction intermediate undergoing carbonylation [22].

Reaction mechanisms of interest to industrial acid zeolite based hydrocarbon conversion processes are successfully modeled using tertiary alkylcarbenium ions and large secondary alkylcarbenium ions charge balancing the negatively charge on the zeolite framework as reaction intermediates, i.e. local minima along the reaction coordinate [23]. The alkylcarbenium ions assumed to occur in zeolites are the same well-established entities as the organic cations found in superacids, for which it is known that the solvent and the anion play a non-specific role.

3.1. Generation of alkylcarbenium ions on acid zeolites

The formation of alkylcarbenium ions from the feedstock molecules in an acid zeolite can occur according to different mechanisms, depending on the nature of the feedstock and the acidic properties of the catalyst [24]. According to Mechanism 1, a proton from the catalyst is added to an unsaturated molecule. The protonation of an alkene by the acid zeolite (H-Z) leads to the formation of an alkylcarbenium ion:

(Mechanism 1) (2)

Mechanism 2 involves the addition of a proton to a saturated molecule. The protonation of an alkane leads to the formation of an alkylcarbonium ion. The alkylcarbonium ion can be transformed into a smaller alkylcarbenium ion by abstraction of an electroneutral molecule (an alkane or molecular hydrogen), a reaction termed protolytic cracking [25]:

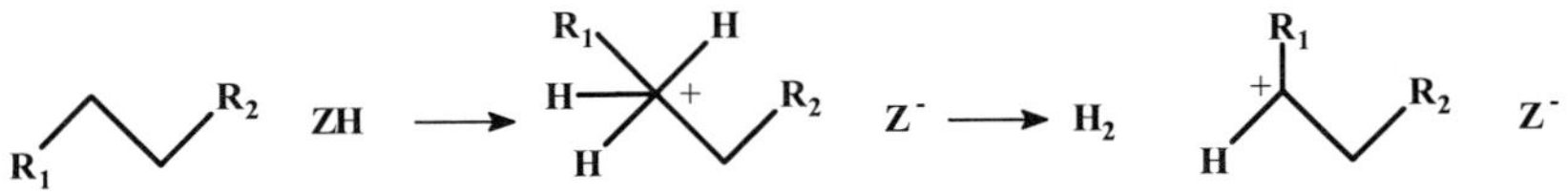

(3)

(Mechanism 2)

Mechanisms 1 and 2 occur on zeolites possessing BrØnsted acidity. Mechanism 1 is much faster than Mechanism 2 and operates at lower temperatures.

Mechanism 3 consists of the abstraction of a hydride anion from an electroneutral molecule. This mechanism involves acid centers of the Lewis type (Z) that bind the hydride:

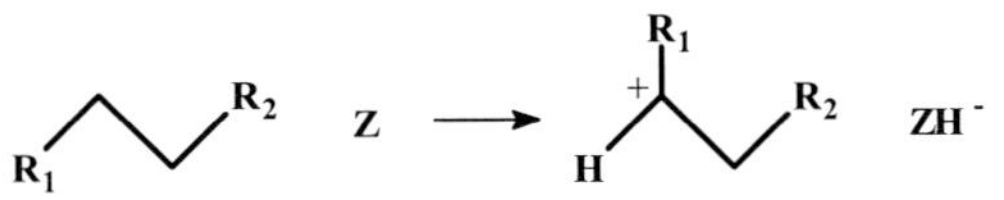

(Mechanism 3) (4)

A variant of Mechanism 3 can occur on a carbenium ion R^+, adsorbed in the zeolite pores, a phenomenon termed bimolecular hydride transfer. In this mechanism, alkylcarbenium ions are chain carriers:

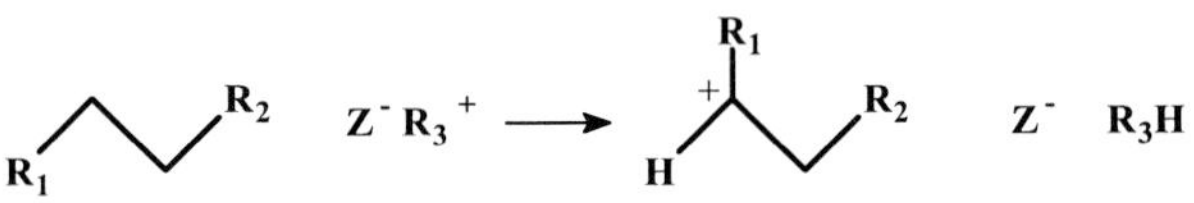

(Mechanism 3*) (5)

Depending on the reaction conditions, the active sites can be covered by molecules blocked in the pores, and behaving like tertiary alkylcarbenium ions. The alkylation of these tertiary alkylcarbenium ions with alkenes gives rise to larger alkylcarbenium ions (Mechanism 4) [26,27] :

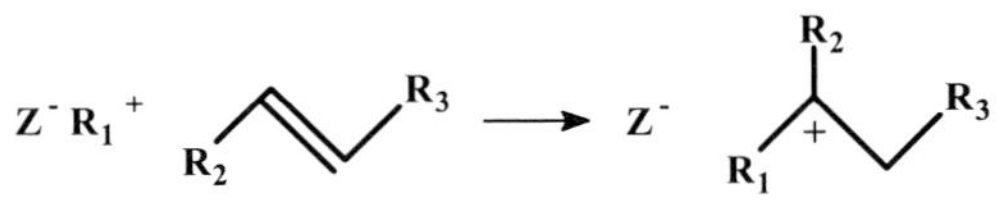

(Mechanism 4) (6)

3.2. Generation of alkylcarbenium ions on bifunctional zeolites

A bifunctional zeolite catalyst is an acid zeolite on which a metal phase is deposited. The function of the metal is to catalyze hydrogenation-dehydrogenation reactions. It is necessary to add hydrogen gas to the hydrocarbon feedstock in order to obtain a high selectivity and good time-on-stream stability of the metal function. The advantage of bifunctional catalysis over monofunctional acid catalysis is that alkylcarbenium ion formation can proceed via Mechanism 1 or 4 instead of Mechanism 2 or 3. The presence of the metal allows the acid zeolite to operate at a reaction temperature that is typically 100 K lower.

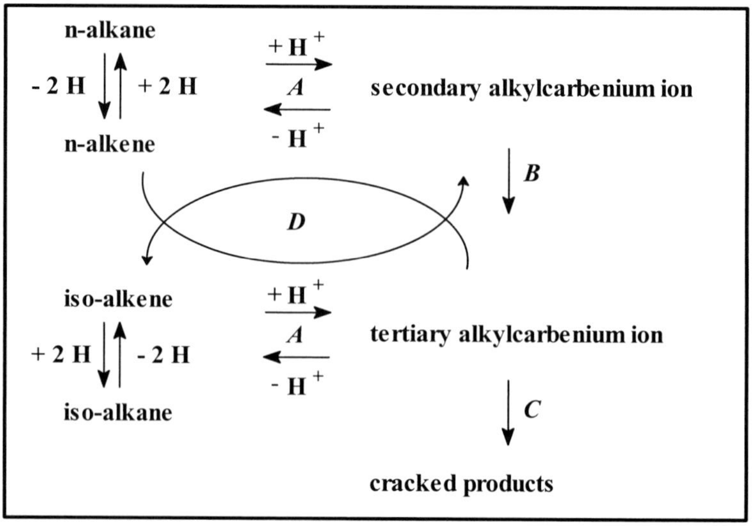

A

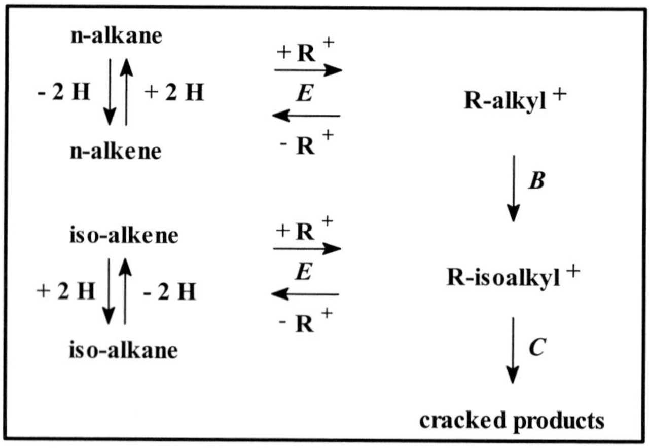

B

Fig.5. Monomolecular (A) and bimolecular (B) bifunctional conversion scheme of n-alkanes [27, 32]

A, protonation-deprotonation reaction on zeolite acid sites;
B, skeletal rearrangements of alkylcarbenium ions;
C, cracking of alkylcarbenium ions;
D, competitive adsorption-desorption of alkenes on acid sites;
E, alkylation-dealkylation on adsorbed *tert* alkylcarbenium ions.

According to the bifunctional reaction scheme for the skeletal isomerization and hydrocracking of a n-alkane [28,29], the metal dehydrogenates the alkane feed molecules into alkenes. The conventional bifunctional reaction scheme involves protonation of these alkenes over the BrØnsted acid sites into secondary alkylcarbenium ions. These cations undergo isomerization and scission reactions, after which they desorb from the acid sites under the form of alkenes. Finally, saturated reaction products are recovered after hydrogenation on the metal phase (Fig.5A). This mechanism is at stake for aluminosilicate zeolites containing platinum or palladium metal, elevated hydrogen partial pressure and hydrocarbons with at least eight C-atoms [30] .With shorter alkenes, there exist alternative bimolecular reaction mechanisms, termed dimerization-isomerization and dimerization-cracking and involving alkylcarbenium ion generation via Mechanism 4 [27] (Fig.5B).

Hydrocarbon conversions on the metal always compete with the bifunctional mechanism. Their contribution becomes important under severe reaction conditions and when the zeolite has weak acidity. The presence of methane and ethane in the reaction products is evidence for the occurrence of metal catalyzed cracking, termed hydrogenolysis [30].

The rearrangements and scission reactions of the alkylcarbenium ion intermediates are the rate limiting steps in the bifunctional reaction scheme. On bifunctional catalysts, the conversion of the feed can be stopped after the primary events by fast desorption of the iso-alkenes (step *D* in Fig.5A). Thus bifunctional catalysis has allowed to gain insight in the complex reaction networks of acid catalyzed conversions of large hydrocarbon molecules. Bifunctional metal-on-acid zeolite catalysis has been very helpful in elucidating the mechanisms of the conversion of aliphatic [10] and alicyclic alkylcarbenium ions [31].

The bifunctional reaction mechanism implies that the mass transport of alkenes, formed on the metal, towards the acid sites and vice-versa has to be efficient. The required intimacy of the two functions in order to obtain the maximum catalytic activity has been quantified and is known as the Weisz Intimacy Criterion [28]:

$$R^2 < 1.2\,10^5 \frac{P_0 D_0}{T\, dN/dt} \tag{7}$$

It determines a maximum average distance, 2R (in m), between the two functions. This distance depends on P_o, the partial pressure of the alkenes (in MPa), T, the reaction temperature (in K) and D_o, the diffusivity of the alkenes in the zeolite pores (in $m^2\ s^{-1}$), and dN/dt, the reaction rate (in mol $s^{-1}\ m^{-3}$). According to Weisz, the intimacy criterion is relatively insensitive to catalyst particle shape and exact kinetics [28]. A comparison between zeolite Y and amorphous silica-alumina catalyst for the conversion of n-alkanes shows that the realization of sufficient intimacy in zeolites is much more critical, due to their high catalytic activity and the low diffusion coefficients of alkanes in the zeolite micropores (Table 2).

Table 2
Maximum average distance, 2R, between metal particles and acid sites in bifunctional silica-alumina [24] and zeolite Y catalysts of the same reactivity

	Silica-alumina	Zeolite Y
D_0 (m^2 s^{-1})	2 10^{-7}	10^{-9}
T (K)	741	473
P_0 (MPa)	10^{-4}	10^{-4}
dN/dt (mol s^{-1} m^{-3})	5	5
2R (m)	50 10^{-6}	4.5 10^{-6}

4. REACTION MECHANISMS OF ACID CATALYZED CONVERSIONS OF LARGE HYDROCARBON MOLECULES IN ZEOLITES

For simplicity, in the following treatment of the reaction mechanisms involving alkylcarbenium ions as reaction intermediates, the anionic zeolite framework will be omitted. In this overview, only those mechanisms will be considered that give rise to chemically distinguishable reaction products. More extensive reviews on the subject are available [8-11].

4.1. Rearrangement and scission reactions of aliphatic alkylcarbenium ions on bifunctional zeolites

There are two types of isomerization reactions of aliphatic alkylcarbenium ions, depending on whether the rearrangement consists of a positional shift of a side chain with respect to the main carbon chain without altering the degree of branching (type A), or of the formation or vanishing of a side chain resulting in an increase or decrease, respectively of the degree of branching (type B) [33]. The degree of branching of an aliphatic carbon skeleton is conveniently derived from the number of $-CH_3$ groups in the molecule:

$$\text{degree of branching} = (\text{number of -CH}_3 \text{ groups}) - 2 \tag{8}$$

Type A isomerization of methylbranched alkylcarbenium ions occurs stepwise through 1,2-methyl shifts. Only subtle rate differences are found between alternative positional shifts. The mechanism of type A isomerization of 2-methylpent-2-yl cation into 3-methylpent-3-yl cation is illustrated in Fig.6. This conversion implies a 1,2-hydride shift, three-ring closure and re-opening, and another 1,2-hydride shift to obtain a tertiary alkylcarbenium ion. The three-ring closure leads to the formation of a H-cycloproponium ion (corner-protonated cyclopropane, CPCP) intermediate, in this case 2-ethyl-3-methyl-1-H-cycloproponium ion. Ethyl and longer alkyl chains can be shifted by one C-atom position along the main chain according to this mechanism.

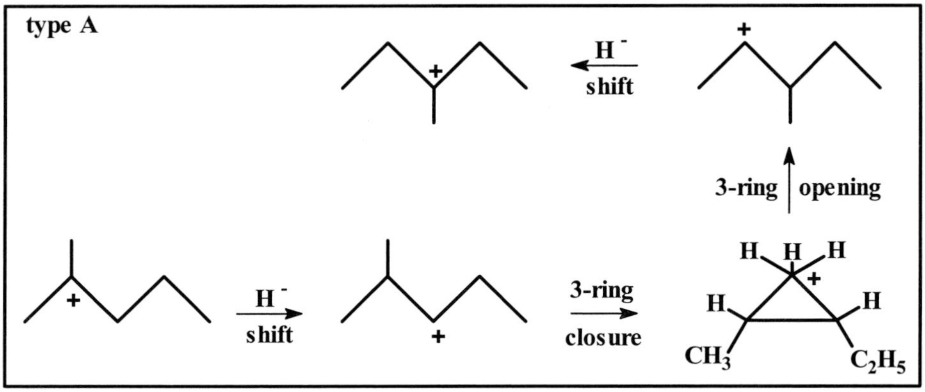

Fig.6. Mechanism of type A isomerization of 2-methylpent-2-yl cation into 3-methylpent-3-yl cation (after ref.10).

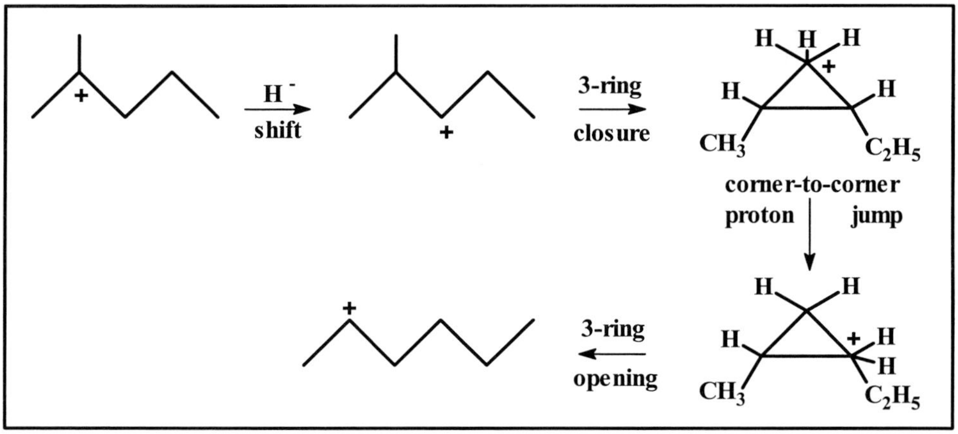

Fig.7. Mechanism of unbranching of 2-methylpent-2-yl cation via CPCP intermediates (after ref.10).

Type B isomerization is responsible for the increase or decrease of the degree of branching with one degree at the time. Methylbranching proceeds over CPCP intermediates. The mechanism of vanishing of the methylbranching in 2-methylpent-2-yl cation is explained in Fig.7. With respect to the 1,2-methyl shift mechanism, detailed in Fig.6, branching involves an additional step, viz. a corner-to-corner jump of a proton on the cyclopropane ring [13]. This proton jump involves the transformation of a H-cycloproponium ion over a C-cycloproponium ion into a different H-cycloproponium ion. This proton jump is the slow step of a type B rearrangement, explaining why type B rearrangements are slower than type A isomerizations. Jumps of the proton to corner C-H bonds with a lower number of alkyl substituents on the C-atom result in an increase of the branching degree upon ring opening, while jumps towards more substituted corners lead to a decrease of the branching degree.

Type B isomerization via H-cycloalkonium ions with rings that are larger than three-rings accounts for the direct formation of ethyl- and larger n-alkyl side chains [10]. The ethylbranching and butylbranching mechanisms of tridec-4-yl cation are shown in Figs.8 and 9, respectively. The rate of a branching rearrangement decreases the larger the size of the ring involved [10].

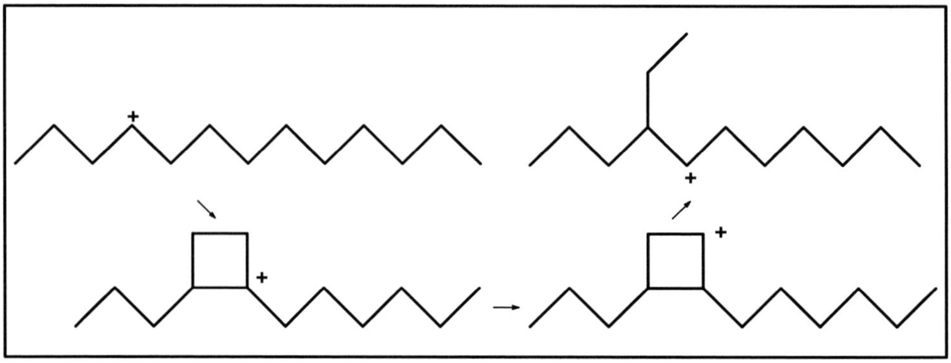

Fig.8. Ethylbranching mechanism of tridec-4-yl cation via substituted corner-protonated cyclobutanes (CPCB). (The formal representation of C-H bonds in the CPCB's is omitted).

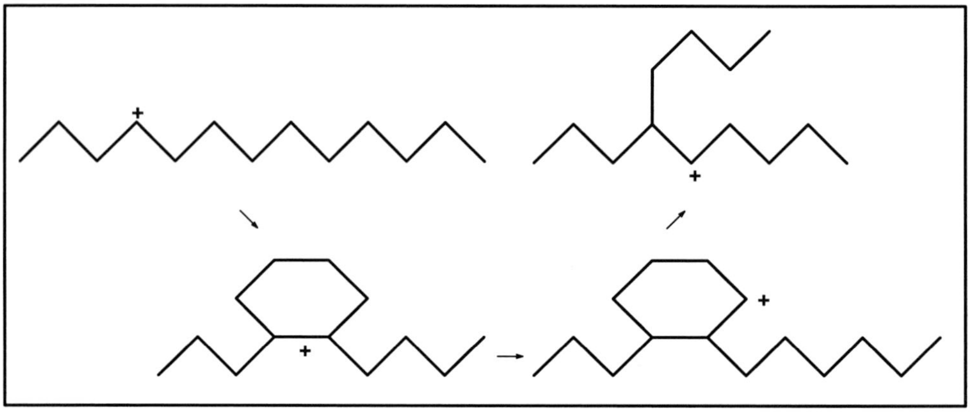

Fig.9. Butylbranching mechanism of tridec-4-yl cation via substituted corner-protonated cyclohexanes (CPCH). (The formal representation of C-H bonds in the CPCH's is omitted).

Cracking of alkylcarbenium ions occurs through β-scission. For long-chain alkylcarbenium ions, containing eight or more C-atoms, there exist four modes of β-scission, denoted as A (*tert* → *tert*), B_1 (*sec* → *tert*), B_2 (*tert* → *sec*), C (*sec* → *sec*), respectively [32,34]. The distinction between the mechanisms is based on the position of the side chains relative to the charged C-atom (Fig.10). These specific configurations are α,γ,γ-tribranched, γ,γ-dibranched, α,γ-dibranched and γ-branched for A, B_1, B_2 and C, respectively.

The rate of the β-scission reactions decreases in the order:

$$A >> B_1,B_2 > C \tag{9}$$

type	ions involved	example
A	*tert* → *tert*	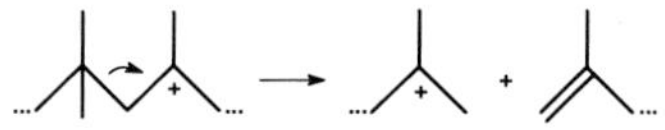
B_1	*sec* → *tert*	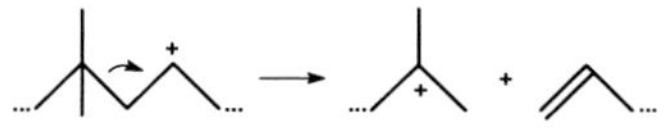
B_2	*tert* → *sec*	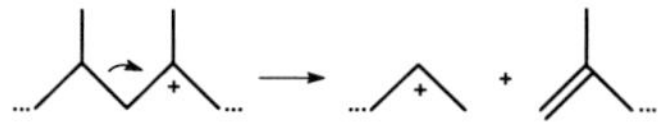
C	*sec* → *sec*	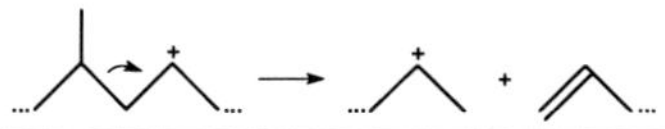

Fig.10. Possible β-scission mechanisms on secondary and tertiary alkylcarbenium ions [34].

The susceptibility of an alkylcarbenium ion to β-scission increases with increasing degree of branching, while that for the generation of additional branchings remains constant. This explains why skeletal branching and cracking are consecutive reactions and why the yield of branched isomers from n-alkanes exhibits a maximum at a certain degree of conversion (Fig.11).

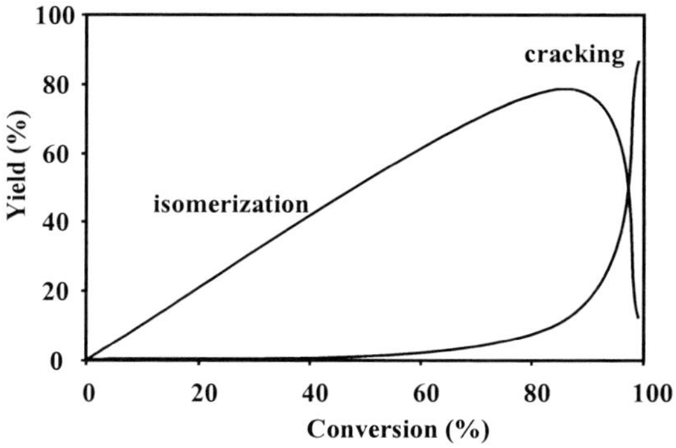

Fig.11. Evolution of the yield of skeletal isomers and cracked products with conversion of a long n-alkane on a bifunctional zeolite catalyst.

4.2. Rearrangement and scission reactions of alicyclic alkylcarbenium ions on bifunctional zeolites

The mechanisms of the different types of rearrangements of alicyclic alkylcarbenium ions are very similar to those of aliphatic alkylcarbenium ions [10], though there are more pathways available. The degree of branching of an alicyclic carbon skeleton can be obtained by counting each tertiary C-atom for one branching, and each quaternary C-atom for two branchings.

External alkyl shifts correspond to a positional shift of an alkyl substituent of the ring. They proceed according to the type A isomerization mechanism explained for the aliphatic alkylcarbenium ions and do not alter the degree of branching. The reaction steps involved in isomerization of 1-ethyl-4-methylcyclohex-1-yl cation into 1-ethyl-3-methylcyclohex-1-yl cation are shown in Fig.12.

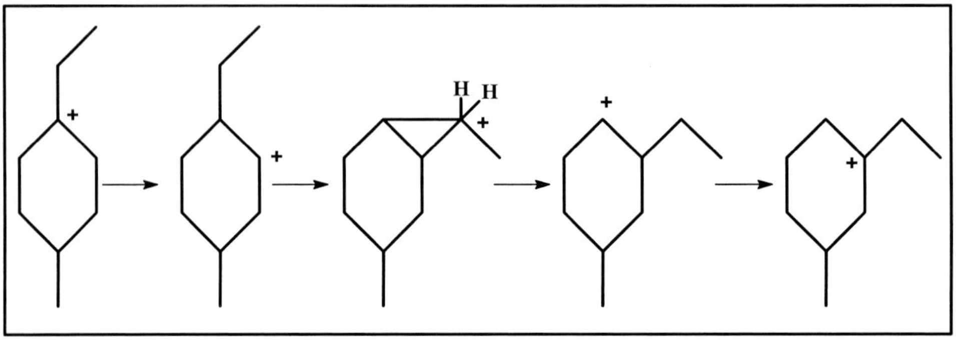

Fig.12. Mechanism of 1,2-ethylshift in 1-ethyl-4-methylcyclohex-1-yl cation.

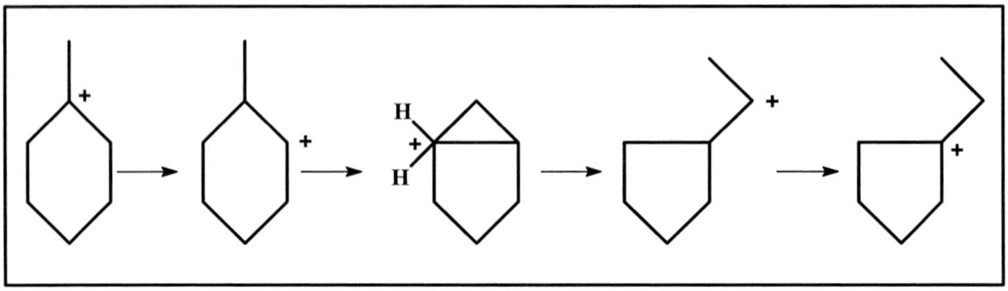

Fig.13. Mechanism of side-chain elongation through ring contraction in methylcyclohex-1-yl cation.

Internal alkyl shifts result in the elongation or shortening of existing side chains with one C-atom and concomitant contraction or expansion of the ring. Internal alkyl shifts proceed through similar reaction steps as the external alkyl shifts and do not alter the degree of branching. The mechanism of the elongation of a methyl side chain in methylcyclohex-1-yl cation is shown in Fig.13. In this type of rearrangement, the three C-atoms of the cycloproponium intermediates involved belong to the cycloalkane ring, whereas in external alkyl shifts, only two ring C-atoms are involved (Fig.12).

Generation and vanishing of side chains of the ring of substituted cycloalkyl cations coincides with ring contraction and ring expansion, respectively. The formation of a methyl side chain proceeds via CPCP intermediates. An essential difference between this type of reaction and the internal and external alkyl shifts is the necessity of a corner-to corner proton jump on the CPCP structure. The analogy with the type A and B isomerization mechanisms of the aliphatic alkylcarbenium ions is obvious. The conversion pathway of 1-methylcyclohex-1-yl cation into 1,2-dimethylcyclopent-1-yl cation is shown in Fig.14.

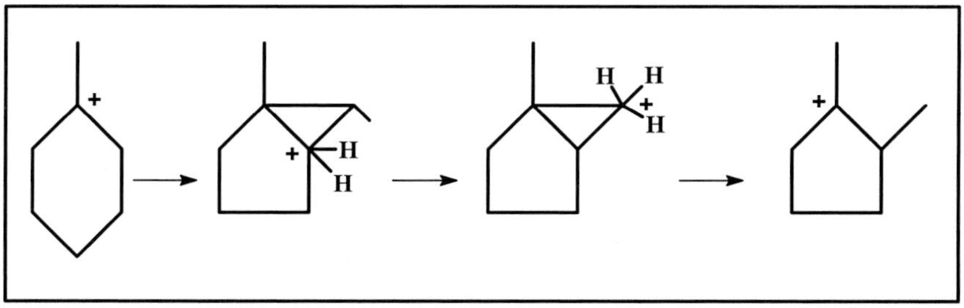

Fig.14. Mechanism of methyl side-chain generation through ring contraction in 1-methylcyclohex-1-yl cation.

In alkylcycloalkyl cations with a long alkyl substituent on the ring, ring branching and internal and external alkyl shifts on the ring are much faster than chain branching [31]. Mechanistically, the ring undergoes methylbranching, and the methyl side chain is transferred from the ring to the long side chain via external alkyl shifts (Fig.15).

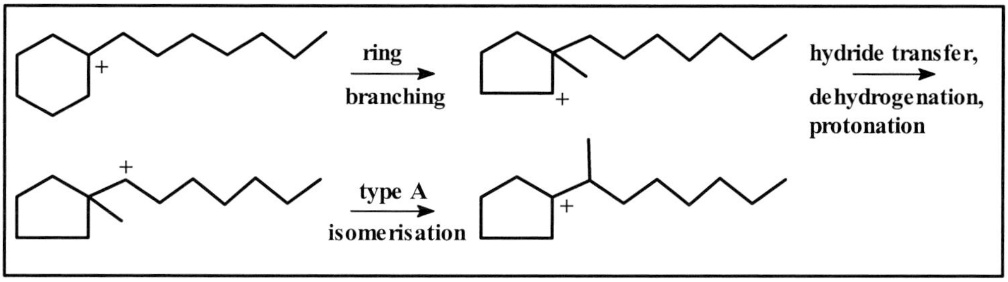

Fig.15. Mechanism of ring-to-chain transfer of a methylbranching in cycloalkyl cations with long n-alkyl substituent on the ring [31].

Internal ring alkylation reactions occur in unsaturated cycloalkyl cations such as cyclooct-4-ene-1-yl and cyclodec-4-ene-1-yl cation. As an example, the transformation of cyclooct-4-ene-1-yl cation into bi [330]cyclooct-1-yl cation is given in Fig.16.

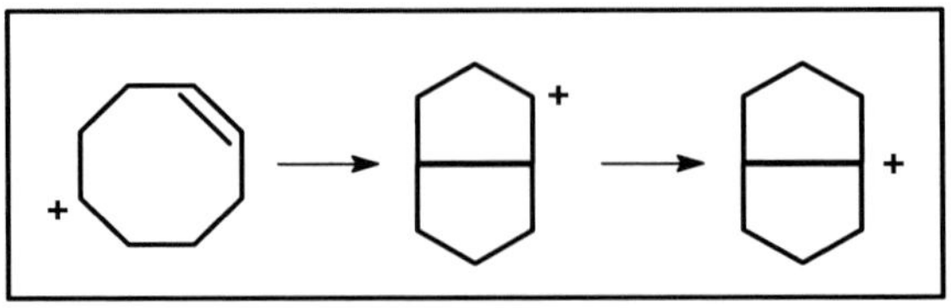

Fig.16. Mechanism of internal ring alkylation in cyclooct-4-ene-1-yl cation.

Starting from alicyclic hydrocarbons with large rings such as a ten-ring, numerous parallel reaction routes are available (Fig.17). Their interpretation based on alkylcarbenium ion chemistry is straightforward.

The β-scission mechanisms of aliphatic alkylcarbenium ions can also be operative on alicyclic alkylcarbenium ions. Examples are shown in Fig.18.

Owing to an unfavorable orientation of the orbitals in five- and six-rings, endocyclic β-scissions are very unfavorable [33].

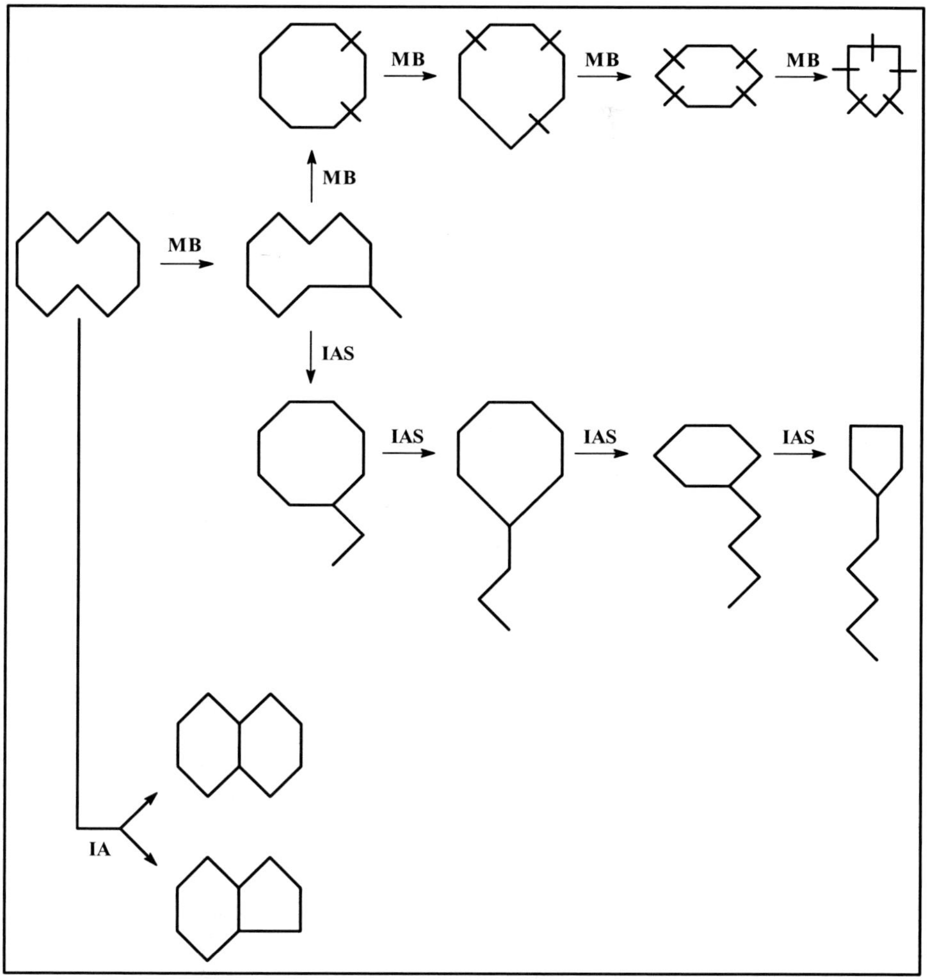

Fig.17. Conversion pathways of cyclodecane on bifunctional zeolites.
MB: methylbranching
IAS: internal alkyl shift
IA: internal ring alkylation

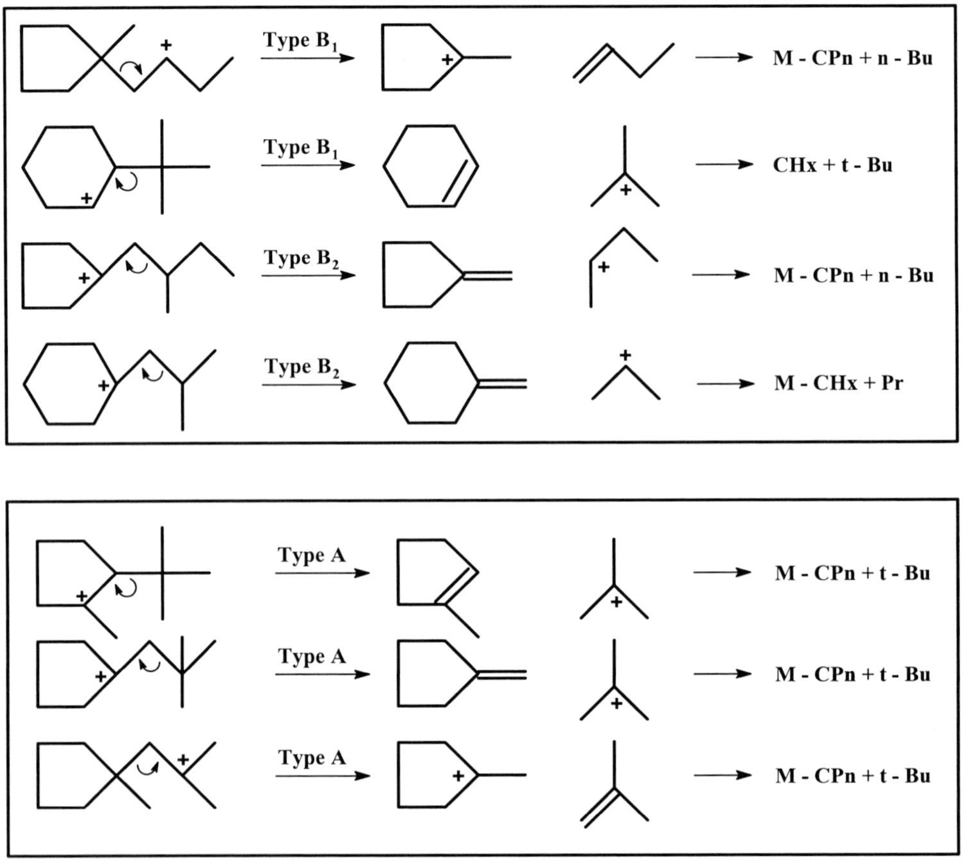

Fig.18. Examples of β-scission of alicyclic alkylcarbenium ions [35].

4.3. Monofunctional conversions of aliphatic and alicyclic hydrocarbons

In the absence of a terminating function, which in bifunctional catalysis is a noble metal dispersed in the zeolite pores, the propagation in monofunctional catalytic cracking of aliphatic and alicyclic hydrocarbons occurs via bimolecular hydrogen transfer reactions (Mechanism 3*, section 3.1). Most authors agree on a chain propagation mechanism in cracking of saturated hydrocarbons on an acid zeolite catalyst (Fig.19) [20,36-38].

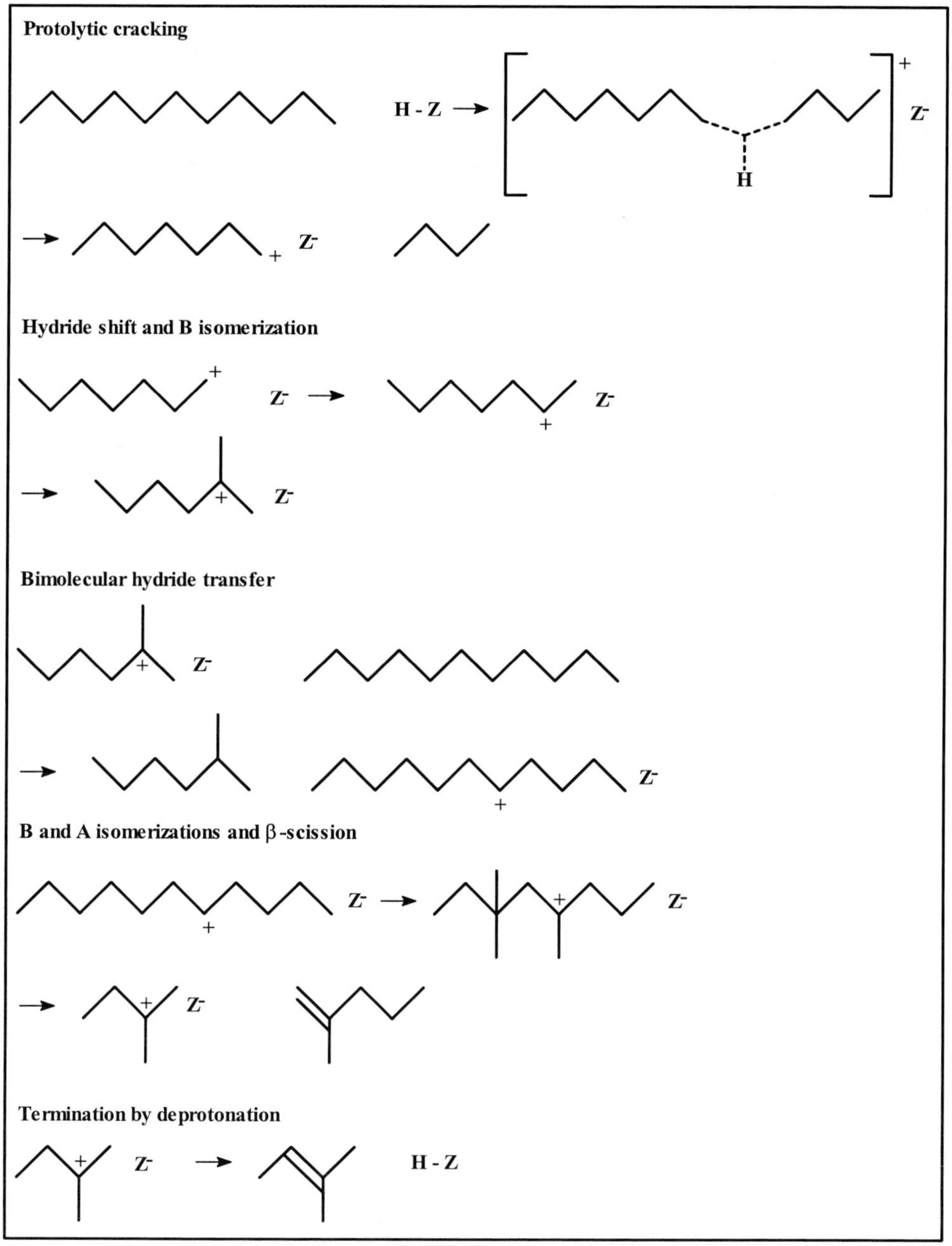

Fig.19. Chain mechanism at stake in the catalytic cracking of aliphatic hydrocarbons.

The initial step in the cracking of a n-alkane consists in the generation of alkylcarbenium ion intermediates through protolytic cracking of alkylcarbonium ions upon release of molecular hydrogen or a short alkane (Mechanism 2, section 3.1). Protolytic cracking is favored at high reaction temperatures, and at low hydrocarbon partial pressures and conversions [25]. The alkylcarbenium ions undergo the same transformations as those depicted for bifunctional catalysis comprising intramolecular hydride shifts, skeletal isomerization and β-scission, competing, however, with bimolecular hydride transfer reactions. Intermolecular hydride ion transfer is generally the slow step and requires the strongest acid sites [39], which consequently are prone to fast deactivation. It has been estimated that hydride transfer from a tertiary C-atom in an alkane is sixteen times faster than from a secondary C-atom [40]. Thus the formation of saturated reaction products is the result of such hydride transfer reactions. A major difference between hydrocracking (bifunctional metal-acid) and catalytic cracking (monofunctional acid) is the formation of C_1 and C_2 products in the latter. In hydrocracking, the formation of primary alkylcarbenium ions must not be invoked to explain experimental product distributions. Given the average higher reaction temperatures in catalytic cracking, the occurrence of primary alkylcarbenium ions has often been invoked, especially in the older literature [39], although more recent literature explains the formation of short fragments based on protolytic cracking [41,42].

Alkylation of alkylcarbenium ions with alkenes, resulting in the formation of longer alkylcarbenium ions and recracking is an important reaction pathway, especially with short alkanes. The major reactions which intervene in the conversion of butane are listed in Fig.20.

It is evident that with C_2-C_5 substrates which initially do not allow the β-scission mechanisms to occur, alkylation or oligomerization will occur prior to the β-scission.

From all this it follows that hydride transfer is a key reaction in acid catalyzed conversions of saturated hydrocarbons, as it determines (1) the reactivity of an alkane substrate (2), the yield of alkanes in the cracked products (3) and the high ratio of iso- to n-alkanes in the C_4 and C_5 cracked product fraction [39].

Ring opening in alicyclic hydrocarbons occurs mainly through endocyclic protolytic cracking [43]. Exocyclic protolytic cracking corresponds to a scission in a side chain [44,45]. Examples are shown in Fig.21.

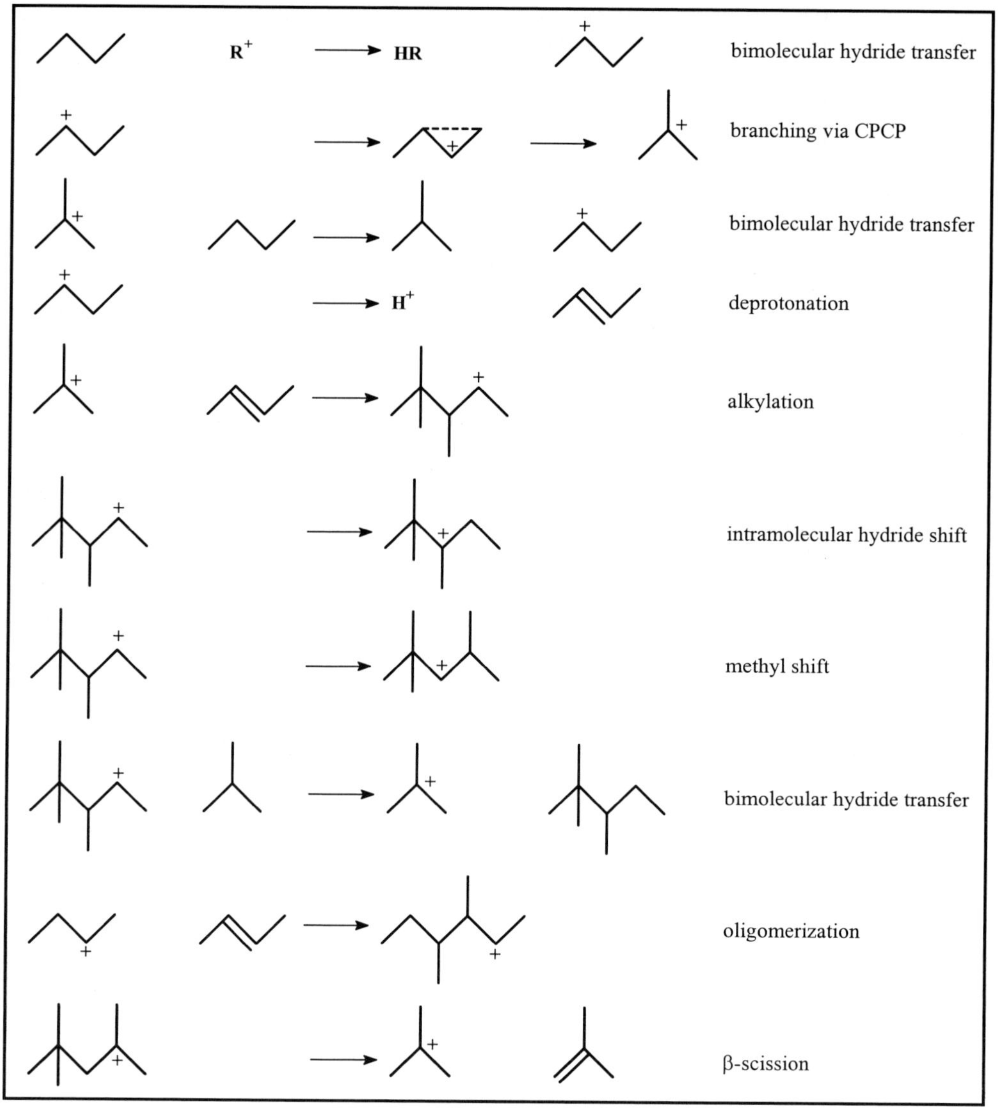

Fig.20. Alkylcarbenium ion reactions occurring in the conversion of butane on an acid zeolite catalyst.

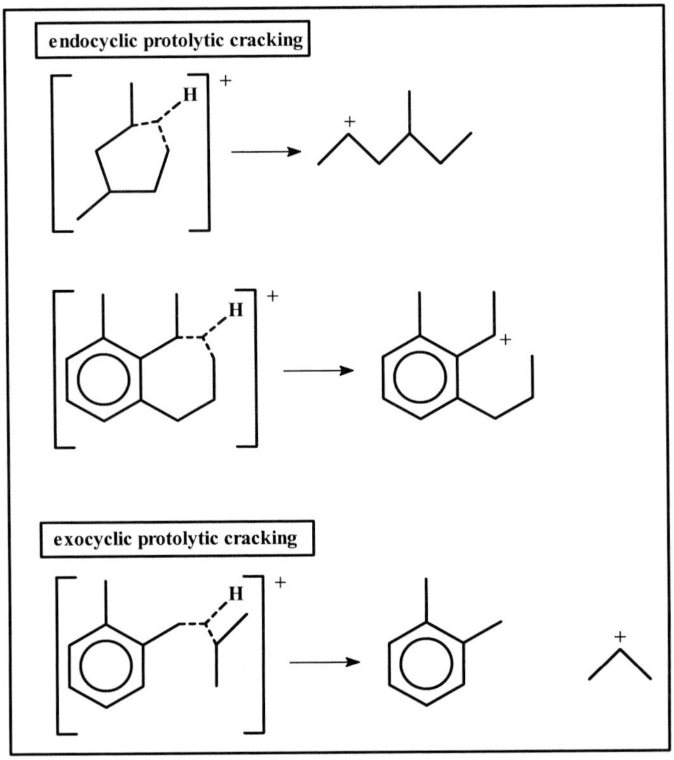

Fig.21. Examples of endocyclic and exocyclic protolytic cracking reactions of alicyclic hydrocarbons.

4.4. Reactions of saturated hydrocarbons with unsaturated hydrocarbons

A key conversion in Fluid Catalytic Cracking is the aromatization of alicyclic hydrocarbons by hydrogen transfer to alkenes:

$$\text{cycloalkane} + 3 \text{ alkene} \dashrightarrow \text{aromatic} + 3 \text{ alkane} \qquad (10)$$

This aromatization process involves a series of hydride transfer and deprotonation processes an example of which is elaborated in Fig.22.

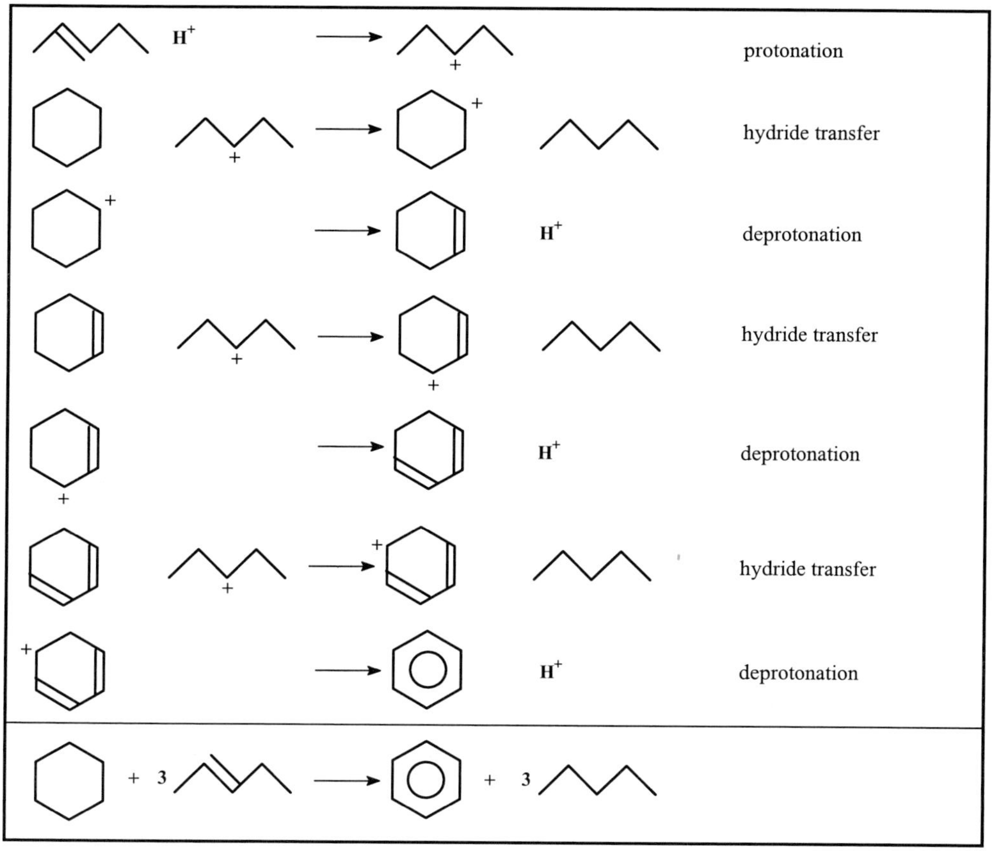

Fig.22. Aromatization mechanism of alicyclic hydrocarbons through hydrogen transfer to unsaturated hydrocarbons.

Isoalkanes can be alkylated with alkenes at low reaction temperatures. The reaction network of isoalkane/alkene alkylation is very complex and can be subdivided into reaction schemes of true alkylation, self-alkylation, destructive alkylation, multiple alkylation and recracking [46] (Fig.23A-C).

When butenes are alkylated with isobutane, it is possible to distinguish true from self-alkylation based on the nature of the trimethylpentane reaction products, provided secondary type A isomerization is slow. Butane formation reveals the presence of self-alkylation. The distinction of true and self-alkylation can be made based on the carbon number of the alkylate obtained using an isoalkane and an alkene of a different carbon number [47]. On H-zeolites, true alkylation predominates. The formation of alkylate requires a donation of a hydride from the isoalkane. Proton loss gives rise to the formation of an alkene. In case of deprotonation, the overall reaction corresponds to an oligomerization, while in case of hydride transfer, the overall reaction network is one of alkylation.

Acid zeolites show substantial deactivation in these processes [48,49]. Coke molecules are formed through a combination of hydride transfers, intermolecular and intramolecular alkylation reactions and oligomerizations leading to heavy, unsaturated cyclic and acyclic compounds.

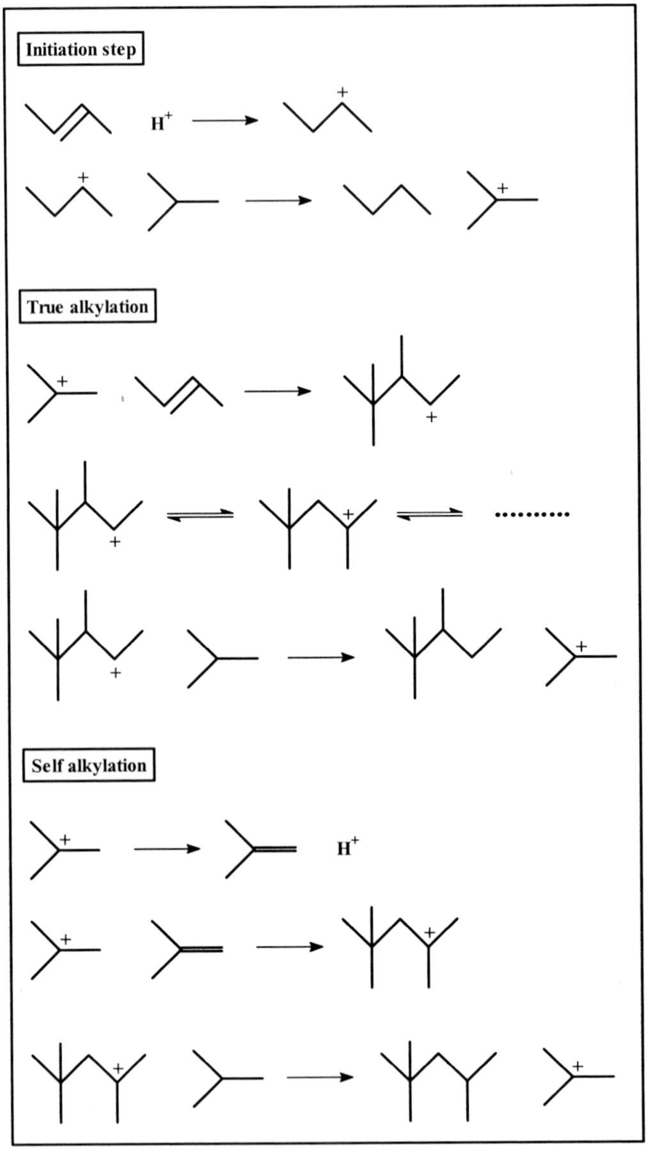

Fig.23A. Mechanism of true alkylation and self-alkylation.

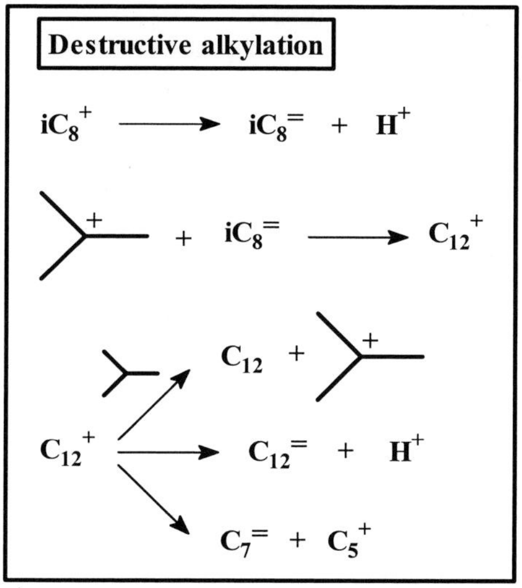

Fig.23B. Destructive alkylation.

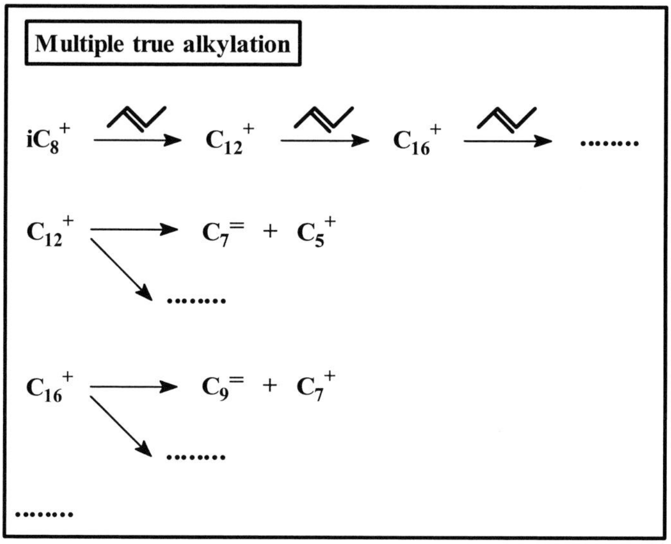

Fig.23C. Multiple true alkylation and cracking.

4.5. Monofunctional conversions of alkenes

The generally accepted mechanism for the alkylation of aromatics with alkenes on H-zeolites is depicted in Fig.24. Dealkylation of an alkylaromatic proceeds according to the reverse sequence of reaction steps [50].

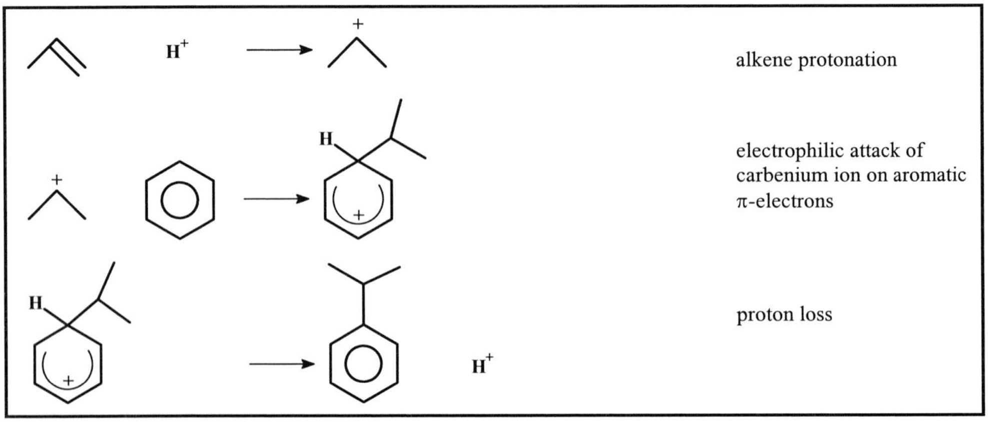

Fig.24. Mechanism of the acid catalyzed isopropylation of benzene.

Protonation of the alkene on the BrØnsted acid site gives the corresponding secondary alkylcarbenium ion, which exerts an electrophilic attack on the π-electrons of the (alkyl)aromatic ring, resulting in a benzenium type cation which then re-aromatizes upon deprotonation.

True oligomerization of light alkenes can be rationalized based on six types of reaction steps (Fig.25). The intermolecular proton transfer (step 6) is in competition with hydride transfer, causing the formation of saturated products and hydrogen deficient coke precursor molecules.

The following reactivity order holds for alkene oligomerization [52]:

$$\text{isobutene} \gg \text{butenes} > \text{propene} > \text{ethene} \qquad (11)$$

This order can be readily explained by the relative stabilities of the alkylcarbenium ions involved.

Under more severe reaction conditions, oligomers undergo cracking and the alkene fragments co-oligomerize. Alicyclic and aromatic hydrocarbons and coke are formed through a combination of reaction steps already explained.

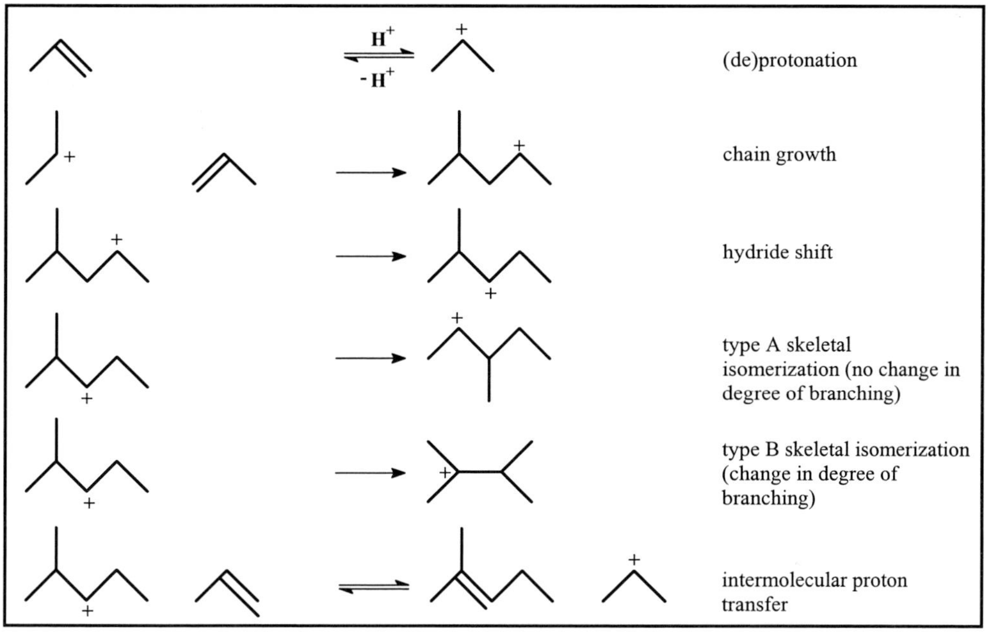

Fig.25.Mechanism of true oligomerization of propene [51].

4.6. Isomerization and disproportionation of alkylaromatics

The isomerization of alkylbenzenes can be rationalized using an intramolecular mechanism: during protonation, a benzenium ion is formed, followed by a single (or successive) 1,2-alkyl shifts in the benzenium ion [53], the alkyl shift being the rate limiting step (Fig.26).

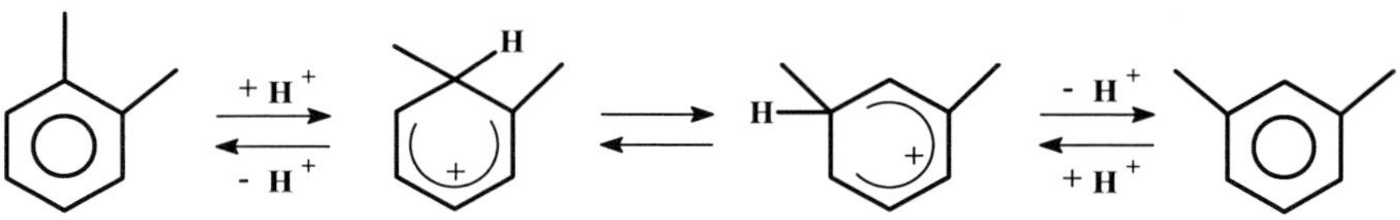

Fig.26. Mechanism of isomerization of o-xylene into m-xylene.

There is ample experimental evidence that the disproportionation of alkylbenzenes also is a BrØnsted acid site catalyzed reaction [54]. Mechanistically, a distinction has to be made between methyl- and ethylsubstituted alkylaromatics, and aromatics with larger alkyl groups. For the latter, a sequence of dealkylation-realkylation is possible, at least when the formation of primary alkylcarbenium ions can be avoided. For short alkyl substituents there is much evidence for a mechanism via diarylmethyl cation intermediates (Fig.27) [54].

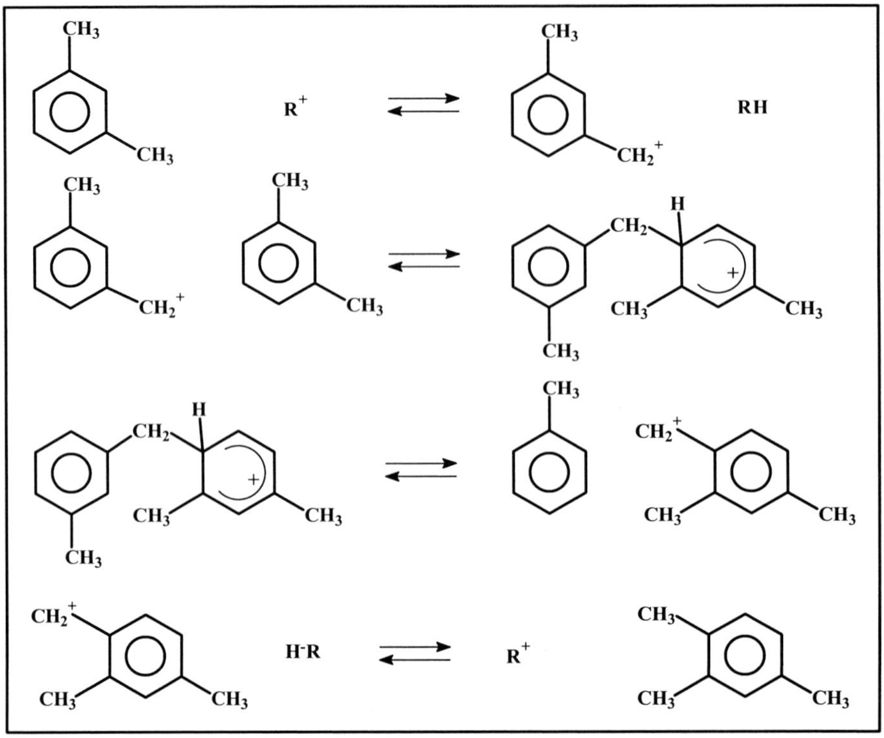

Fig.27. Mechanism of disproportionation of m-xylene.

The presence of molecular hydrogen decreases the selectivity for toluene disproportionation on acid mordenite [55], which is explained by the following equilibrium:

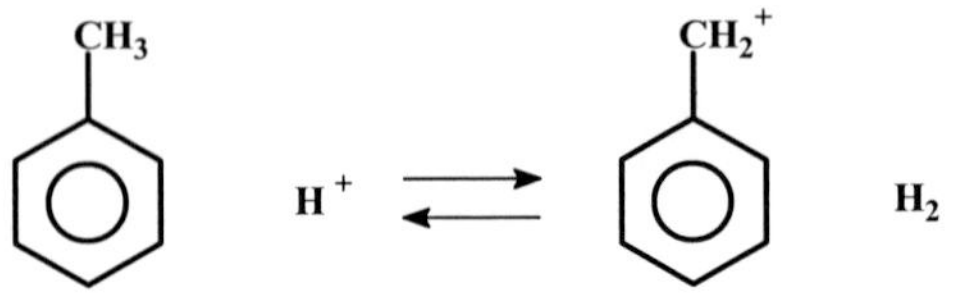

(12)

The addition of alkanes that are capable to form *tert* alkylcarbenium ions slows down the o-xylene disproportionation without affecting the isomerization. This is indicative of the different nature of intermediates for isomerization and disproportionation [55]. Between benzylic cations and isoalkanes, the following hydride transfer reaction is possible [55]:

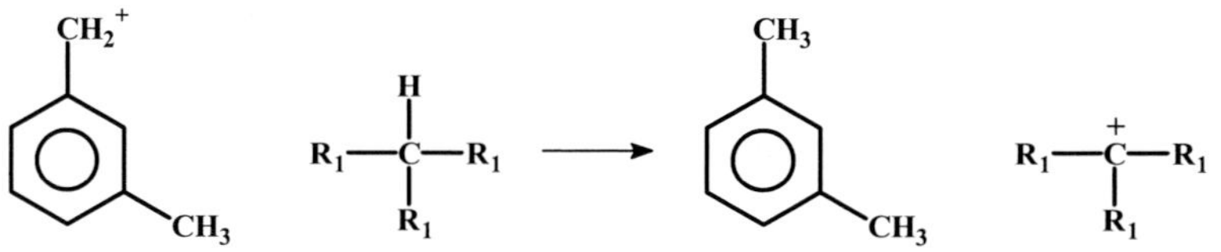

(13)

5. SHAPE-SELECTIVE CATALYSIS IN ACID ZEOLITES

5.1. Molecular shape-selective catalysis, pore mouth catalysis and key-lock catalysis

Molecular shape-selectivity, or simply shape-selectivity, is the reaction specificity arising from the presence of a sterically confined environment in which the molecules are converted. Shape selectivity consists in often a subtle matching of size and shape of reactants, transition states and products with the size and shape of micropores of the zeolite. To obtain a net shape selective effect, it is required that the shape-selective environment is repeated many times in the solid. This is the reason why shape-selectivity is unique to zeolites. The first paper on shape-selective catalysis appeared in 1960 [56]. Since then the research activity in this area has grown significantly and the subject is very well documented. In the course of time, there have been significant changes in the rationalization of shape-selective effects. In the older, the emphasis was on diffusion and transition state effects, whereas in the more recent work, the importance of physisorption is much more emphasized.

This overview will be limited to an explanation of the different concepts and mechanisms that have been proposed. For further details and examples, the reader is referred to the review articles [57-64].

5.1. Molecular shape-selective catalysis, pore mouth catalysis and key-lock catalysis

To discuss shape-selectivity, pore mouth and key-lock catalysis, it is convenient to group zeolite pore architectures into three categories: (1) cage and window structures, (2) intersecting tubes and channels with connecting apertures and (3) tubular pores (Fig.28, a). In the first two categories of structure types, the exterior crystal surface exposes interrupted cages and incomplete channel intersections with apertures that are substantially larger than the windows to the underlying cavities. Bulky molecules with kinetic diameters larger than the windows can penetrate into these cavities, whereas they cannot migrate deeper into the bulk of the zeolite crystal. Another possibility is a partial penetration of a molecule into the micropores, for example in a pore mouth (Fig.28,b), or with two or more groups simultaneously into several pore openings with the bulky moiety of the molecule remaining outside, a situation designated with the term key-lock adsorption [65](Fig.28, c). The possibility of pore mouth and key-lock catalysis is available in any zeolite structure type (Fig.28). Shape-selective catalysis occurs in the bulk of the zeolite crystals.

Considering simply a cubic zeolite crystal with a size of 2R (μm) and with interconnected microcavities of 1 nm, the body of such crystal contains of the order of $8\ 10^9\ R^3$ microcavities. The fraction of the cavities located at the crystal boundary is $24\ 10^6\ R^2$. The percentage of cavities available for pore mouth catalysis corresponds to $0.3\ R^{-1}$. For crystals of 1 μm, the number of cavities at the crystal boundary is 0.6%, for crystals of 100 nm, it is already 6%.

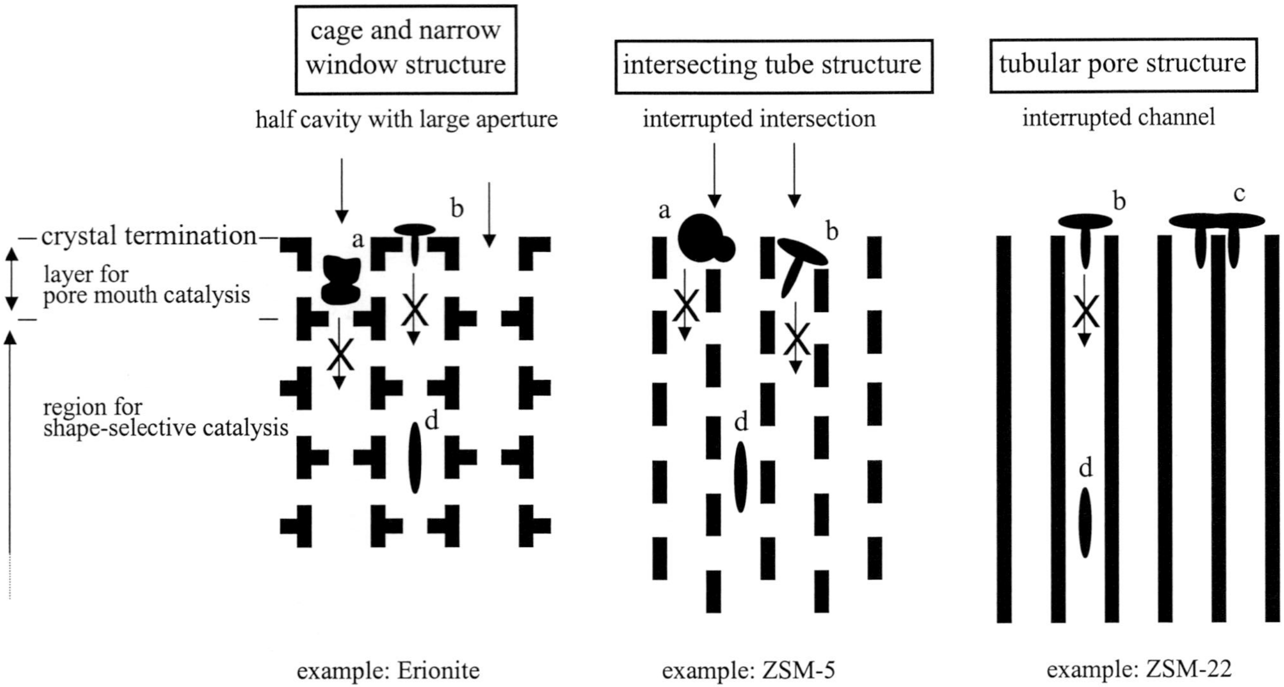

Fig.28. Shape-selective environments in the different zeolite structure types: (a) large molecules have access to the interrupted cavities and channel intersections of the layer for pore mouth catalysis; (b) molecules are plugged into the pore apertures; (c) molecules are converted in multiple pore mouths according to key-lock catalysis; (d) molecules are converted in the intracrystalline shape-selective environment (after ref.66).

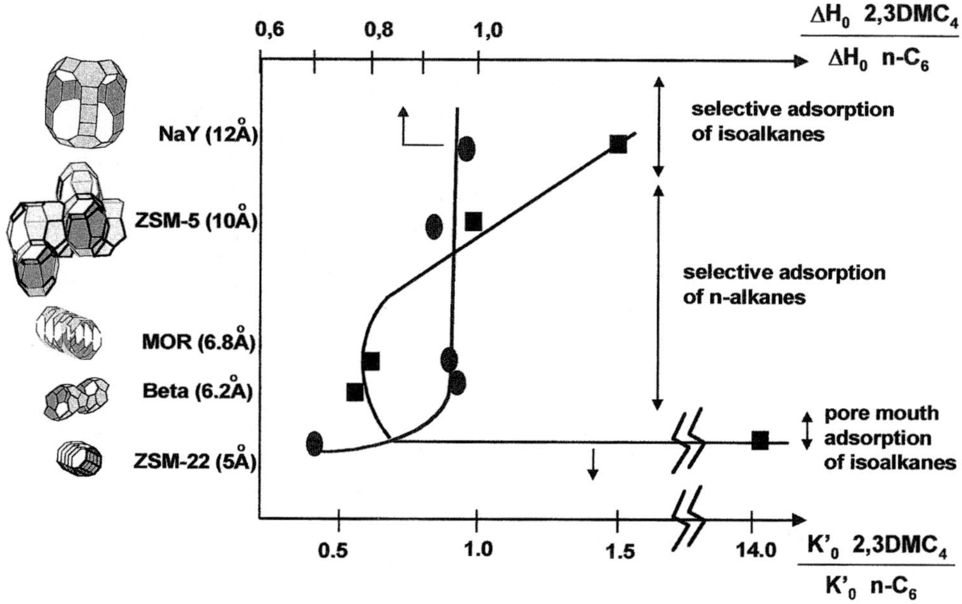

Fig.29. Ratio of adsorption enthalpy (ΔH°) and of pre-exponential factors of van't Hoff equation (K'∘) of 2,3-dimethylbutane and hexane on large- and medium pore zeolites in the Henry region and temperature of 200-375°C [66].

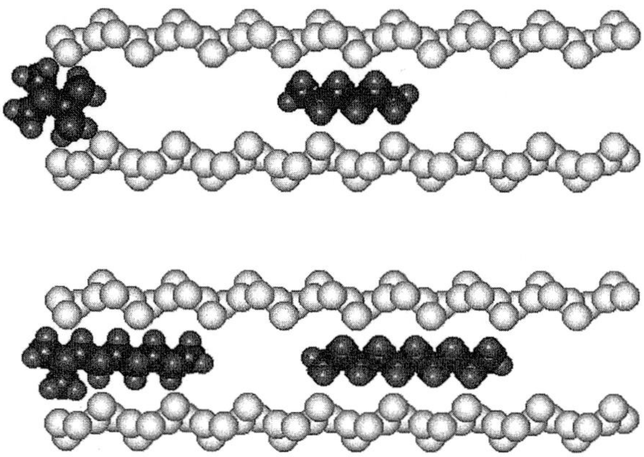

Fig.30. Representation of pore mouth adsorption of 2,3-dimethylbutane and 2-methylnonane and intracrystalline adsorption of hexane and decane on the ZSM-22 zeolite structure type [66].

An example of the pore mouth versus bulk adsorption of alkanes depending on branching and zeolite structure is elaborated in Fig.29 based on adsorption data for 2,3-dimethylbutane and hexane in the Henry region at 200 to 375°C [67]. In zeolites with pore sizes of 0.6 nm and wider, the adsorption enthalpies of the iso- and the n-alkane are very similar. The selectivity is dependent on the entropy factor (pre-exponential factor of the van 't Hoff equation), being more favorable for the isoalkane in Y zeolites, and for the n-alkane in large pore zeolites with smaller cross-section (mordenite and beta) and in intersecting medium pore zeolites (ZSM-5). On the tubular medium pore zeolite ZSM-22, the isoalkane has a lower adsorption enthalpy and a considerably higher pre-exponential factor compared to the n-alkane. The isoalkane adsorbs selectively in the pore mouths where it looses less entropy upon adsorption compared to the n-alkane that is trapped inside a micropore (Fig.30).

Framework terminating silanols and bridging hydroxyl groups on the external zeolite surface are much weaker Brönsted acids compared to the acid sites inside the micropores [68]. Molecules adsorbing in pore mouth and key-lock modes (Fig.28, a and b) can reach the strong Brönsted acid sites present in the first atom layers of the subsurface and be converted at high turnover frequencies and with peculiar selectivity sterically imposed by the pore mouth structure [65].

5.2. Mechanisms of molecular shape-selective catalysis

Most of the shape-selective phenomena encountered belong to one of the following three categories:

- **Reactant selectivity**: it takes place when zeolite catalyst acts as a molecular sieve and excludes certain molecular sizes and structures from the intracrystalline voids, while other less bulky molecules are able to enter; a whole variety of zeolite types with different sizes (and shapes) of the pore orifices are available so that this critical exclusion limit can be varied over a wide range of molecular sizes;

- **Product selectivity**: it is the result of discrete diffusivities of the different reaction products in the pores of the zeolite crystals; typical examples are the isomerization reactions of alkylaromatics. This selectivity is not only dependent on the pore size but also on the crystal size of the catalyst particles. The existence of some kind of "molecular traffic control" has been invoked to describe qualitatively the transport of molecules with different shape and/or size in the intracrystalline volume of zeolites with two discrete sets of pores, as is the case in pentasil-type zeolites [69] and mordenite [70].

- **Transition state shape selectivity**: it occurs when the spatial configuration around a transition state or a reaction intermediate located in the intracrystalline volume is such that only certain configurations are possible. Transalkylation reactions of alkylaromatics are textbook examples of this kind of selectivity. In other cases, it is not always easy to distinguish between product and transition state shape selectivity. If the catalyst can be synthesized with crystal sizes which are significantly different, the distinction is easily made as only the former selectivity is dependent on crystal size. Transition shape-selectivity is traditionally linked to the suppression of undesired side reactions such as coke formation. Santilli et al. proposed to use the term "inverse shape selectivity" in those instances where the transition state shape-selectivity arises from a positive discrimination of specific transition states [71]. Skeletal branching of hexane in large pore zeolites is a typical example (Fig.31). The highest selectivity for multibranched isohexanes was observed in zeolites with pores with very specific diameters and with optimum interaction with the desirable isomers. A special manifestation of molecular shape-selectivity in zeolites is the cage and window effect, observed in hydrocracking of long n-alkanes. A cage or window effect occurs when the reactivities and/or diffusivities do not change monotonically within a homologous series of compounds [57]. A well known example of a cage effect reported by Gorring [72] is the hydrocracking of hexadecane on a zeolite with erionite topology. The distribution of the cracked products exhibits a minimum at C_7-C_9, and maxima at C_4 and C_{12}. The regular hydrocracking pattern in the absence of molecular shape selectivity shows a maximum at C_7-C_9 (central cracking of the molecule) and gradually less cracking at C-C bonds more to the end of the carbon chain [73]. Erionite contains cylindrical cavities with 8-ring windows (8-membered rings of oxygen atoms delineate the cage apertures). C_7-C_9 alkanes fit neatly in an erionite cage. The original explanation of the cage and window effect offered by Gorring was that C_7-C_9 hydrocarbons are preferentially trapped in the erionite cavities and undergo extensive cracking. The erionite cage effect in hydrocracking is then a special case of product shape-selectivity. Later, binodal or even trinodal product distributions were observed with many zeolites, provided a n-alkane with sufficient chain length was hydrocracked [74,75]. It was interpreted in terms of transition state selectivity and suppression of cracking at specific C-C bonds. Provided the adsorbed feed molecule stretches across two neighboring cavities, separated by a window, the latter becomes a steric barrier for alkyl shifts towards carbons positioned in the window (Fig.32). Since the C-C bonds that crack preferentially are those between a tertiary or quaternary carbon and a secondary carbon, the cracking of C-C bonds positioned in zeolite windows is suppressed.

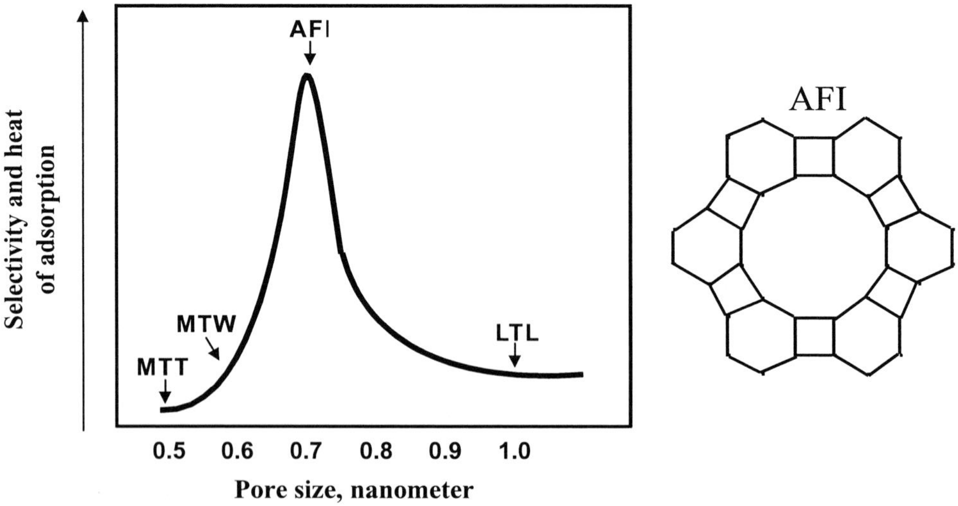

Fig.31. Inverse shape selectivity in the skeletal multibranching of hexane in large pore zeolites. AFI zeolites represent the optimum pore diameter, reflected in the highest heat of adsorption of multibranched isohexanes, after ref.71.

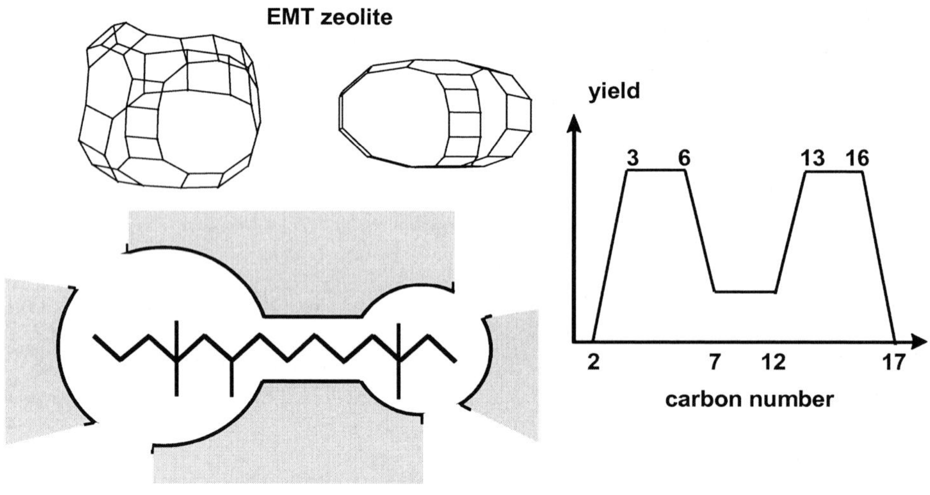

Fig.32. Isononadecane molecule trapped in two cages of EMT type zeolite. The formation of cracked products with seven to twelve carbon atoms requiring scission of C-C bonds located in the window between these cages is suppressed.

REFERENCES

1. J. Sommer, M. Hachoumy, F. Garin and D. Barthomeuf, J. Am. Chem. Soc., 116 (1994) 5491.
2. J. A. Lercher and K. Seshan, Current Opinion in Solid State & Material Science, 2 (1997) 57.
3. J. Wei, Chem. Eng. Sci., 51 (1996) 2995.
4. J.F. Denayer, G.V. Baron, G. Vanbutsele, P.A. Jacobs and J.A. Martens, Chem. Eng.Sci., 54 (1999) 3553.
5. V.B. Kazansky, Stud. Surf. Sci. Catal., 85 (1994) 251.
6. F. Haase and J. Sauer, J. Am. Chem. Soc., 117 (1995) 3780.
7. T. Xu, D.H. Barich, P.W. Goguen, W. Song, Z. Wang, J.B. Nicholas and J.F. Haw, J. Am. Chem. Soc., 120 (1998) 4025.
8. H. Pines, The Chemistry of Catalytic Hydrocarbon Conversions, Academic Press, New York, 1981.
9. A. Corma, Chem. Rev., 95 (1995) 559.
10. J.A. Martens and P.A. Jacobs, in Theoretical Aspects of Heterogeneous Catalysis, J.B. Moffat (ed.), Van Nostrand Reinhold, New York, 1990, p.52.
11. D.S. Santilli and B.C. Gates, Handbook of Heterogeneous Catalysis, G. Ertl, H. Knözinger and J. Weitkamp (eds.), VCH, 1997, p.1123.
12. G.A. Olah and P. von R. Schleyer, Carbonium Ions, Willey, New York, 1968, 342.
13. P. Vogel, "Carbocation Chemistry", Stud. Org. Chem., Elsevier, Amsterdam, 21 (1985).
14. G.J. Kramer, R.A. van Santen, C.A. Emeis and A.K. Novak, Nature, 363 (1993) 529.
15. G.J. Kramer and R.A. van Santen, J. Am. Chem. Soc., 117 (1995) 1766.
16. A.M. Rigby, G.J. Kramer and R.A. van Santen, J. Catal., 170 (1997) 1.
17. I.N. Senchenya, N.D. Chuvilkin and V.B. Kazansky, Kinet. Katal., 26 (1985) 1073.
18. I.N. Senchenya and V.B. Kazansky, Kinet. Katal., 28 (1987) 566.
19. V.B. Kazansky and I.N. Senchenya, J. Catal., 119 (1989) 108.
20. W.E. Farneth and R.J. Gorte, Chem. Rev., 95 (1995) 615.
21. J.A. Martens and P.A. Jacobs, in 'Handbook of Heterogeneous Catalysis', G. Ertl, H. Knözinger and J. Weitkamp (eds.), VCH, 1997, p.1137.
22. J. Sommer, D. Habermacher, R. Jost, A. Sassi, A.G. Stepanov, M.V. Luzgin, D. Freude, H. Ernst and J. Martens, J. Catal., 181 (1999) 265.
23. G.D. Svoboda, E. Vynckier, B. Debrabandere and G.F. Froment, Ind. Eng. Chem. Res., 34 (1995) 3793.
24. B.W. Wojciechowski and A. Corma, Catalytic Cracking, Catalysts, Chemistry and Kinetics, Marcel Dekker, New York, Basel, 1986, p.9.
25. W.O. Haag and R.H. Dessau, in Proceed. 8th International Congress on Catalysis, Dechema, Frankfurt am Main, 1984, Vol II, p. 305.
26. P. Andy, N.S. Gnep, M. Guisnet, E. Benazzi and C. Travers, J. Catal., 173 (1998) 322.
27. E. Blomsma, J.A. Martens and P.A. Jacobs, J. Catal., 155 (1995) 141; ibidem 159 (1996) 323.
28. P.B. Weisz, Adv.Catal., 13 (1962) 137.
29. M.L. Coonradt and W.E. Garwood, Ind. Eng. Chem. Prod. res. Dev., 3 (1964) 38.

30. J. Weitkamp, W. Gerhardt and P.A. Jacobs, Proceed. Int. Symp. Catal. Zeolites, Siofok (Hungary), May 1985, p.261.
31. W. Souverijns, R. Parton, J.A. Martens, G.F. Froment and P.A. Jacobs, Catal. Lett., 37 (1996) 207.
32. J.A. Martens, P.A. Jacobs and J. Weitkamp, Appl. Catal., 20 (1986) 239.
33. D.M. Brouwer and H. Hogeveen, Rcl. trav. Chim. Pays-Bas, 89 (1970) 211.
34. J. Weitkamp, P.A. Jacobs and J.A. Martens, Appl. Catal., 8 (1983) 123.
35. J. Weitkamp, S. Ernst and C.Y. Chen, Stud. Surf. Sci. Catal., 49B (1989) 1115.
36. J. Abbot and J.D. Head, J. Catal., 125 (1990) 187.
37. S.J. Collins and P.J. O'Malley, J.Catal., 153 (1995) 94.
38. A. Corma, P.J. Miguel and A.V. Orchilles, J. Catal., 145 (1994) 171.
39. J.A. Rabo, in "Zeolites: Science and Technology", F.R. Ribeiro, A.E. Rodrigues, L.D. Rollmann and C. Naccache (eds.), Martinus Nijhof, The Hague, Boston, Lancaster, 1984, p.291.
40. N.Y. Chen and W.O. Haag, in "Hydrogen Effects in Catalysis", Z. Paal and G. Menon (eds.), Marcel dekker, New York and Basel, 1988, p.695.
41. S.J.Collins and P.J. O'Malley, Chem. Phys. Lett., 246 (1995) 555.
42. S.J.Collins and P.J. O'Malley, J. Catal., 160 (1995) 94.
43. A. Corma, F. Mocholi and V. Orchilles, Appl. Catal., 67 (1991) 307.
44. A. Corma, P.J. Miguel, A.V. Orchilles and G.S. Koermer , J. Catal., 135 (1992) 45.
45. A. Corma, P.J. Miguel, A.V. Orchilles and G.S. Koermer , J. Catal. ,145 (1994) 181.
46. L. Schmerling , Friedel-Crafts and Related Reactions, G.A. Olah (ed.), Interscience Publishers, 1964.
47. F. Cardona, N.S. Gnep, M. Guisnet, G. Szabo and P. Nascimento, Appl. Catal., 128 (1995) 243.
48. K.P. de Jong ,C.M.A.M. Mesters, D.G.R. Peferoen, P.T.M. van Brugge and C. de Groot, Chem. Eng. Sci., 51 (1996) 2053.
49. M.F. Simpson, J.Wei and S. Sundaresan, Ind. Eng. Chem. res., 35 (1996) 3861.
50. C.J.A. Mota and R. Rawet, Ind. Eng. Chem. Res., 34 (1995) 4326.
51. V.N. Ipatieff and H. Pines, Ind. Eng. Chem., 28 (1936) 684.
52. G.E. Langlois, Ind. Eng. Chem., 45 (1953) 1470.
53. A. Corma, A. Cortes, I. Nebot and F. Tomas, J. Catal., 57 (1979) 444.
54. M. Guisnet, Stud. Surf. Sci. Catal., 20 (1985) 273.
55. M. Guisnet and N.S. Gnep, in "Zeolites: Science and Technology", F.R. Ribeiro, A.E. Rodrigues, L.D. Rollmann and C. Naccache (eds.), Martinus Nijhof, The Hague, Boston, Lancaster, 1984, 571.
56. P.B. Weisz and V.J. Frilette, J. Phys. Chem., 64 (1960) 382.
57. S.M. Csicsery, ACS Monogaph, 171 (1976) 680.
58. S.M. Csicsery, Zeolites, 4 (1984) 202.
59. P.B. Weisz, Pure Appl. Chem., 52 (1980) 2091.
60. E.G. Derouane, in "Intercalation Chemistry", M.S. Wittinghton and A.J. Jacobson (eds.), Academic Press, 1982, p.101.
61. E.G. Derouane, Stud. Surf. Sci. Catal., 4 (1980) 5; ibidem 19 (1984) 1.
62. J. Dwyer, Chem. Ind., 7 (1984) 229.
63. J. Weitkamp and S. Ernst, Catal. Today, 19 (1994) 107.

64. J.A. Martens, W. Souverijns and P.A. Jacobs, in "Comprehensive Supramolecular Chemistry", vol. 7, "Solid-State Supramolecular Chemistry: Two- and Three-dimensional Inorganic Networks", G. Albert and T. Bein (eds.), Pergamon, 1996, Vol. 9, p.621.
65. J. Martens, W. Souverijns, W. Verrelst, R. Parton, G. Froment and P. Jacobs, Angew. Chem. Int. Ed. Eng. 34 (1995) 2528-2530.
66. A.M. Goossens, G. Vanbutsele and J.A. Martens, in "Fundamentals of Adsorption 6", F. Meunier (ed.), Elsevier, Paris, 1998, p.31.
67. J.F. Denayer, G.V. Baron, W. Souverijns, P.A. Jacobs and J.A. Martens, J. Phys. Chem. B, 102 (1998) 4588.
68. M. Trombetta, G. Busca, M. Lenarda, L. Storaro and M. Pavan, Appl. Catal.A, 182 (1999) 225.
69. E.G. Derouane and Z. Gabelica, J. Catal., 65 (1980) 486.
70. G.S. Lee, J.J. Maj, S.C. Rocke and J.M. Garcés, Catal. Lett., 2 (1989) 243.
71. D.S. Santilli, T.V. Harris and S.I. Zones, Microporous Mater., 1 (1993) 329.
72. R.L. Gorring, J. Catal., 31 (1973) 13.
73. J. Weitkamp, in "Hydrocracking and Hydrotreating", J.W. Ward and S.A. Quader (eds.), ACS Symp. Ser., 20 (1975) 1.
74. J.A. Martens, M.Tielen and P.A. Jacobs, Stud. Surf. Sci. Catal., 46 (1989) 49.
75. J.A. Martens, G.M. Vanbutsele and P.A. Jacobs, in "Proc. 9th IZC", von Ballmoos, J.B. Higgins and M.M.J. Treacy (eds.), Butterworth-Heinemann, Stoneham, 2 (1993) 335.

Studies in Surface Science and Catalysis 137
H. van Bekkum, E.M. Flanigen, P.A. Jacobs and J.C. Jansen (Editors)
© 2001 Elsevier Science B.V. All rights reserved.

Chapter 15

PREPARATION OF ZEOLITE CATALYSTS

Herman W.Kouwenhoven(1), Bas de Kroes (2)

(1) retired, c/o Technisch Chemisches Laboratorium, ETH-Zentrum, 8092 Zuerich, Switzerland.

(2)AKZO NOBEL Chemicals B.V., Research Centre Amsterdam, P.O.Box 37650, 1030 BE Amsterdam, The Netherlands.

1.INTRODUCTION

Zeolite catalysts are successfully applied in hydrocarbon processing and in chemical conversion processes. Properties which make zeolites versatile catalyst components are:

Size and shape of the crystallites are variables in catalyst design and are determined by the synthesis conditions.

The internal surface area is high and thermally very stable compared to those of the amorphous silicas and aluminas used traditionally in catalyst preparation.

The pores are well defined and of molecular dimensions, the pore size is determined by the crystallographic structure.

The pore openings ruling the accessibility of the internal surface may be subtly changed by ion-exchange, postimpregnation or chemical reaction with specific compounds, (pore size engineering).

Charge-compensating cations can be removed by ion-exchange.

T-atoms are in most structures highly accessible, their nature determines the polarity of the surface and its catalytic activity.

The chemical composition of the lattice is dependent on the synthesis conditions. In many cases the Si/Al ratio may be varied within wide limits and for some structures also the nature of the T-atoms is a variable.

The chemical composition of the lattice may be changed by the application of the various techniques for dealumination- and T-site substitution reactions (secondary synthesis), as a side effect a profound change in the texture of the zeolite may occur by the formation of mesopores.

The catalytic activity of the surface of a zeolite can be modified by a treatment with a reactive chemical and this modification may be limited to the outer-surface of the crystals by selection of a reagent, which due to its size, cannot reach the internal surface.

Catalyst components such as a hydrogenation/dehydrogenation function can be introduced via ion-exchange with a precursor compound, and active materials, such as metallo-phthalocyanines may be synthesized in situ in the zeolite pores.

A large number of variables has accordingly to be evaluated during an investigation of the catalytic activity of a zeolite for a given reaction. For large scale petrochemical processes the economic incentive of an attractive product composition justifies the expensive and lengthy research effort for reaching optimum catalyst performance. For the smaller scale fine chemicals manufacturing processes such an effort is more difficult to justify and the number of applications of zeolites in these processes is accordingly presently relatively small. Existing and potential applications of zeolite catalysts in fine chemicals manufacture are discussed in Chapter 18 by Hölderich and van Bekkum and in (ref. 1 and 2).
We will discuss the preparation of zeolite catalysts in general and highlight the preparation of catalysts for: fluid catalytic cracking; shape selective hydrogenation, and for shape selective production of dimethylamine. Special attention will be given to the following aspects : zeolite activation; T-site substitution reactions; metal emplacement, the preparation of zeolite supported noble metal catalysts and catalyst shaping. Detailed information on the various aspects is presented in the references.

2 CATALYST PREPARATION PROCEDURE

2.1 Zeolite Activation

2.1.1 Removal of the template

The zeolite sample used as a catalyst component may contain template molecules which have to be removed before further processing. Normally these molecules are strongly sorbed in the zeolite lattice and are often located in positions having a limited accessibility. Compared to "free" template molecules their counterparts occluded in the zeolite lattice are stabilized and require a much higher temperature for complete oxidation during calcination in air. Moreover the latter process is strongly exothermic and burning away e.g. 1%w CH_2 in a zeolite results in an adiabatic temperature rise of about 600 K. It is therefore advisable to study the thermal decomposition both in an inert and in an oxidizing atmosphere by thermal analysis in order to establish the reaction pattern. Generally it is advantageous to first remove the bulk of the organic material by thermal decomposition in an inert medium and to eliminate remaining carbonaceous residues by subsequent oxidation. Although most zeolites have a high thermal stability such a staged calcination procedure may also prevent unwanted T-atom migration out of the zeolite lattice due to stabilization reactions during the oxidative template removal. A detailed discussion of template removal procedures is presented in (ref.3).

2.1.2 Ion-exchange for preparation of solid acids

Removal of the charge-compensating cations (usually Na) by ion-exchange will normally follow the template removal step. Ion-exchange of zeolites is reviewed in Chapter 11 by Townsend and Coker. For the preparation of H-zeolites the most effective and gentle method starts with an ammonium-exchange. The ammonium-exchanged zeolite is subsequently calcined to convert it into the H-form. On a small scale a staged procedure in inert gas followed by an inert gas/air mixture after reaching a temperature of about 720 K is preferable; on a large scale calcination in air is applied. A study of the reaction by thermal analysis is very useful in order to be aware of the exothermic effects which may accompany the removal of ammonia (ref.4). Deep removal of Na by conventional ion-exchange at about 300 K, from zeolites having a structure which consists of various types of cages differing in accessibility, such as zeolite Y, is extremely time consuming. For instance Maesen (ref. 5) reported that removal of 87% of the Na from Na-Y by ion-exchange at 293 K, using 20 ml of 1M NH_4NO_3/g Na-Y, requires about 50 successive one day exchanges. Under hydrothermal conditions at 433 K, the rate of ion-exchange of Na-Y with an ammonium nitrate solution is appreciably enhanced and using 20 ml of 1M NH_4NO_3/ g Na-Y 87% removal of the Na contained in Na-Y may be reached in one step in 30 minutes. Deep removal of Na from Na-Y (> 95%) is usually carried out by repeated ion-exchange at about 373 K, combined with calcination. As a result of the latter Na-atoms located in double six rings and sodalite cages are redistributed over the zeolite surface. At the same time solid state reactions occurring in the zeolite remove Al-atoms from the lattice. This phenomenon, which is referred to as stabilization, is discussed later in some detail in relation to the manufacture of catalytic cracking catalysts. In a broader context stabilization is reviewed in (ref.6) and in Chapter 6 by Szostak.

Zeolites having a Si/Al ratio of 5 or higher are stable enough to allow direct exchange of Na^+ by H^+ ions by acid treatment without loss of crystallinity. Lattice dealumination is an inescapable side effect of acid treatment and depends on the zeolite structure: it is easy to dealuminate Beta and difficult to dealuminate ZSM-5. Acid leaching of mordenite is for instance discussed in (ref.7).

Table 1
Catalytic activity of Pt-H-zeolite Y for n-pentane hydroisomerization.
Effect of sodium content.

Crystallinity X-Ray %	Na_2O (% w)	Temp.for 30% conversion, K
90	2.02	578
80	0.27	573
80	0.02	523

Conditions: WHSV: 1 h^{-1}; 30 bar; H_2/C_5 molar ratio:2.5.
0.5%w Pt on zeolite Y.

Acid (H) zeolites are widely applied in catalytic conversion processes, see for instance in this book Chapter 14 by Jacobs and Martens, and Chapter 17 by Stork and Maxwell. The catalytic activity of H-zeolites is strongly dependent on the degree of Na removal, Table 1 (ref.8). For maximum Brønsted acid activity deep removal of sodium and/or other bases is essential. The deactivating effect of Na ions on acid activity is similar to that of small amounts of adsorbed NH_3. Deactivation of Brønsted acidity by sodium and ammonium ions is for instance discussed in (ref.9) and (ref.10).In the absence of Na or other bases, the catalytic activity of H-zeolites depends both on the number and on the quality of the acid sites. The number of acid sites is equal to the number of Al atoms in T sites. The effect of lattice Si/Al ratio on the catalyst performance in the isopropylation of biphenyl of some mordenite samples is shown in Fig. 1 (ref.11).

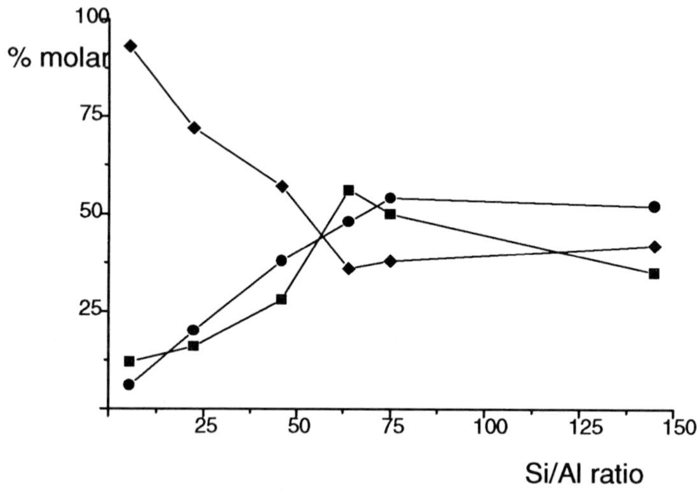

Fig.1. Alkylation of biphenyl (BP) with propene over modified mordenite catalysts (ref. 11). Effect of Si/Al ratio on product composition.at runhour 4.
♦ : selectivity IBP; ■ :BP conversion; ● :selectivity 4,4'-DIBP.
Conditions: Solvent: mesitylene; Temp.: 453 K; WHSV(BP): 1.09 h^{-1};
BP/C3 = 1(mol); BP concentration: 3.5 mol/ l solvent

For each particular zeolite structure the acidity per site reaches its maximum value at a specific Si/Al ratio. This ratio is equal to or higher than that required for a surface configuration in which each Al atom is isolated and has no next nearest T sites occupied by Al-atoms. For zeolites having a Si/Al ratio lower than about 9, the strength of the acid sites will not be a constant per site. The width of the curve representing the acid-site strength distribution is related to the Si/Al ratio; the nature and the ordering of the trivalent atoms in the structure determine the acid strength per site. This is illustrated by the NH_3 desorption temperature and examples of two reactions given in Table 2 for samples which have the MFI structure and contain the T atoms: B, Fe, Ga, or Al. The relations between acid strength and Si/Al ratio for various zeolite structures are discussed in (ref.12).

Table 2
Nature of T atom, acidity and activity of MFI silicates.

	NH_3 Desorption (ref.13)		C_4 Cracking (ref.14)		Rearrangement of 2 phenylpropanal to phenylpropanone at 573 K (ref.15)			
T Atom	Si/T	T Max (K)	Si/T	k^* 773 K	Si/T	WHSV (h^{-1})	Conv.	Sel.
B	170	423			94	0.8	63	97
Fe	196	603			36	2.0	98	95
Ga	396	623	138	7.1				
Al	140	653	140	45.5				

*)k=ml/g/min

Hydroxyl groups on the outer (and inner) surface may be deactivated by reaction with alkylsilyl compounds :

$$(C_nH_{2n+1})_3SiH + HO\text{-Surface} \rightarrow (C_nH_{2n+1})_3Si\text{-O-Surface} + H_2$$

The reaction product is oxidized by calcination and the active hydroxyl groups of the product are covered with a layer of SiO_2.

2.1.3. Introduction of other elements, shape selective hydrogenation

Most elements may be introduced by direct cation-exchange of their aqueous salt solutions with a zeolite. Well known examples are rare earth elements in zeolite X and Y; Ca or K in zeolite A ; Cu in ZSM-5 etc. Upon further activation by reduction or metal compound formation the zeolite may be (partially) converted into the H-form. In case acidic activity is unwanted the acid-sites have to be neutralised before the material can be applied as a catalyst.
The preparation of Ni on zeolite A catalysts for the shape selective hydrogenation of acetylene in its mixtures with butadiene and ethene is discussed by Corbin et al. in (ref.16). Catalysts were prepared by ion-exchange of 3A(KA), 4A(NaA) and 5A(CaA) with a $Ni(NO_3)_2$ solution and some products were subsequently exchanged with alkali carbonate solutions in order to tune their pore size for enhanced selectivity in hydrogenation.
ESCA results showed that most of the Ni was deposited on the outside of the zeolite crystals and not in the pores. Removal of acetylene from steam cracker light ends represents a significant improvement in the production of polymer grade ethene, since it allows the recovery of butadiene from an acetylene free gas mixture by cryogenic processing. For high acetylene selectivity the hydrogenation activity must

be located in the zeolite pores, however, and catalytic sites at the outer surface have to be deactivated. Most efficient poisoning of the Ni activity outside the pores in Ni on zeolite A catalysts was obtained by the addition of thiophene, which is too large to enter the pores and thus selectively deactivates the remaining Ni activity on the outside of the zeolite crystals, cf Table 3.

Table 3
Shape selective hydrogenation of acetylene mixed with butadiene and ethylene.

	Conversion			Adsorption	
Catalyst	C_2H_2	C_4H_8	MeOH	EtOH	n-PrOH
Ni-Na-A (III)	100	100	19.5	16.3	9.6
Ni-K-A (VII)*	100	17	17.5	3.0	0.4
Ni-K-A TEOS*	100	11	17.1	5.0	0.6

* Thiophene added during reaction.
Conditions: Temp.: 475 K, Pressure: atmospheric, Gas composition:i in % vol. H2=28, C2H2 = 0.35; C2H4 = 32, C4H8 = 0.65, He = 39, Contact time 0.5-1.0 sec, 1 ml of catalyst.

Selectivity was further enhanced by treating the materials with liquid tetraethyl orthosilicate (TEOS), followed by a calcination in air at 700 K. Since TEOS is too large to enter the pores of zeolite A, it was assumed that a layer of SiO_2 was formed on the outer surface of the TEOS modified material.

2.1.4 Preparation of noble metal catalysts

Noble metal containing catalysts are usually prepared by ion-exchange of metal-ammine complexes with the zeolite. Because of the cost of noble metal, the metal loading should be low and the exposed metal surface should be as high as possible. This means that the metal must be well dispersed and have an even distribution over the available zeolite surface. The type of metal complex, the conditions of ion-exchange, the calcination procedure and the reduction conditions each have a strong effect on the dispersion of the metal in the finished catalyst. The subject is discussed in detail in (refs. 17, 18 and 19), which also review preceeding literature, in the following paragraph we will present some additional points.
Two types of zeolite supported noble metal catalysts may be distinguished. The first comprises bifunctional catalysts, consisting of an acidic function and a metallic hydrogenation/ dehydrogenation function (ref.17). This class of catalysts is applied in processes, characterized by an acid catalyzed reaction, which is in many cases triggered by reactive, unsaturated hydrocarbons, which are present in low concentrations in the reactant mixture. These reactive species may be generated by the metallic hydrogenation-/dehydrogenation function. At the same time the metal function stabilizes conversion by lowering the concentration of unsaturated coke precursors which are adsorbed on the acidic catalyst surface, by hydrogenation of the same. In Fig.2 it is for example shown that both the introduction of platinum in

the zeolite or the application of a higher hydrogen partial pressure results in a lower deactivation rate. The conversion, however, is inversely related to the hydrogen partial pressure and the activity level of H-mordenite at which the rate of deactivation is close to zero is very low and not practicable. In most hydrocarbon reactions use of a hydrogen atmosphere and introduction of a hydrogenation function will appreciably lower the rate of deactivation and thus result in a more stable conversion level.

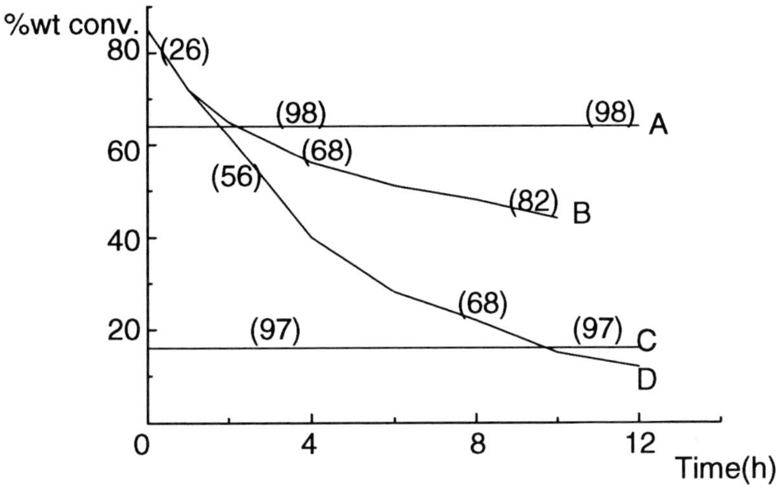

Fig. 2 n-Pentane isomerisation activity of H-mordenite (H-M).
Catalyst deactivation and stabilisation of conversion by Pt
(x) Numbers : selectivity to C_5
Conditions: Temp. 423 K; WHSV: 1 h^{-1}; Pressure nC_5 : 8.6 kg/cm^2;
Pressure H_2: A = 20 kg/cm^2, (Pt on H-M) , B= 17.2 kg/cm^2(H-M),
C= 43 kg/cm^2 (H-M) D = 8.6 kg/cm^2 (H-M);

The second type is represented by monofunctional catalysts, consisting of a metal supported on a neutral (ref.18) catalytically inactive zeolite. The supporting zeolite may be a hydrophobic zeolite (ref.19) or a mixed oxide material which was neutralized before use in hydrogenation.

Apart from its functions as metal support and acidic catalyst, the zeolite component may also induce shape selectivity. Zeolite-supported noble metal catalysts are best prepared by ion exchange using an aqueous solution of a cationic metal ammine complex, such as $Pt(NH_3)_4X_2$, provided that the pores are large enough for the $Pt(NH_3)_4^{2+}$ cations to enter the zeolite structure. In general metal ammine complexes have a strong interaction with the zeolite. As a result the metal will be deposited near the outer surface of the crystals, forming a thin outer layer of high metal concentration. A homogeneous metal distribution may, however, be obtained by the use of an ammonium zeolite and the addition of a competing cation, such as NH_4^+ to the ion-exchange solution and a long equilibration time. The ion- exchange procedure is in some detail discussed in (refs. 19 and 20), which mention also the effect of a competing cation. Using Al free silicalite as support, hydrophobic Pt catalysts were prepared for D_2/H_2O exchange. The ion-exchange capacity of silicalite is strongly dependent on the pH of the exchanging solution. At a pH of 9.5 a Pt loading of about 8%w could be prepared, whereas at pH=5 the uptake capacity was about 0.8%w. This shows that Si-OH groups in silicalite participate in the ion-exchange at a pH above about 7. The preparation is summarized in Table 4.

Table 4
Preparation of Pt on silicalite by ion-exchange (ref.19)

	Silicalite	Commercial,type 115 from UCC
	Metal emplacement	Cation-exchange with $Pt(NH_3)_4Cl_2$
	pH ion-exchange	5 → Pt about 0.8%w
		9.5 → Pt about 8%w(addition NH_4OH)
	Drying	400 K in air
	"Calcination"	573 K in vacuo
	Reduction	573 K in H_2
Remarks		
	Use of commercial zeolite alleviates reproducibility problems. At high pH high silica zeolites have an enhanced ion-exchange capacity which is in excess of the number of non Si framework atoms. Pt dispersion depends on ion-exchange conditions: High pH → good dispersion (high H/Pt) Low pH → poor dispersion (low H/Pt)	

In commercial catalyst manufacture the application of competitive ion-exchange is technically feasible. Costwise its application is problematic, however, since the metal uptake by the zeolite is not complete and the noble metal accordingly has to be recovered from the spent ion-exchange solution. Calcination of the noble metal ion-exchanged zeolite presents some problems in large scale operations. It has been firmly established that calcination in an oxidizing atmosphere is a prerequisite for obtaining a well dispersed noble metal on Al-containing zeolite supports. Thermal analysis of the calcination yields basic data for the optimal calcination procedure. Results for $Pt(NH_3)_4$-NH_4-zeolite Y in air are shown in Fig.3.
Comparison with TPD data of NH_4-zeolite Y in air in Fig.4 (ref.21) shows that unless proper measures are taken, the calcination may easily result in extensive thermal damage due to its highly exothermic character. The calculated adiabatic temperature rise due to the burning of one %wt of NH_3 is about 250 K. Judicious choice of a temperature programme for the calcination is in some cases sufficient to prevent temperature excursions. In other cases it may be necessary to develop an alternative preparation procedure to circumvent problems during calcination. Reduction of the calcined material is generally simple and the reaction occurs at relatively low temperature. Care should, however, be taken that the partial pressure of water is low since well-dispersed noble metals may sinter in mixtures of H_2O and H_2 at temperatures above 720 K.

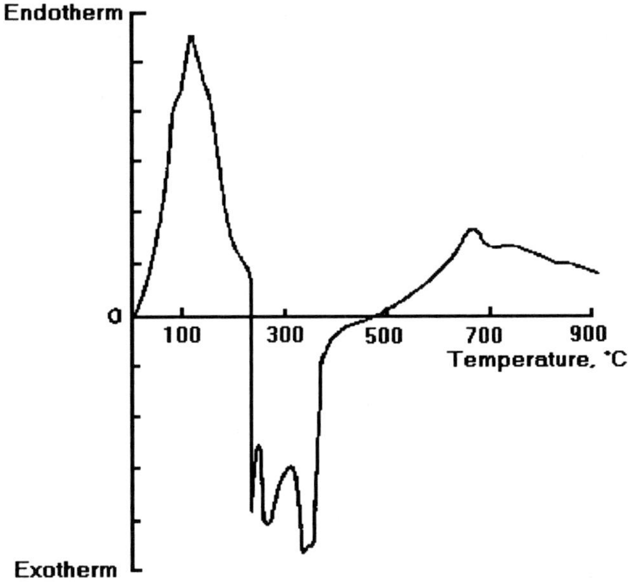

Fig.3 Thermal analysis of $Pt(NH_3)_4$ exchanged NH_4-zeolite Y in air. Strong exothermic effect.

Finishing of a silicalite-based catalyst prepared by competitive ion-exchange in the presence of NH_3 is possible without an intermediate calcination (ref.19). The presence of chloride ions during reduction of a $Pt(NH_3)_4^{2+}$ loaded silicalite has a detrimental effect, however, on the Pt dispersion of the finished catalyst.
A report of the preparation of Pt on zeolite Y hydrogenation catalysts is presented in (ref.22). It appears that the preparation method was not optimal. The products described in this article have a relatively poor metal dispersion and some of the metal is deposited on the outside of the crystals and not in the zeolite pores. In case the $Pt(NH_3)_4^{2+}$ cation (0.54 nm) is too large to enter the pores of the zeolite it is sometimes possible to synthesize the zeolite in a solution which contains some $Pt(NH_3)_4^{2+}$ compound such as $Pt(NH_3)_4Cl_2$. This method was first described by Weisz et al. (ref.23) for the preparation of shape selective Pt hydrogenation catalysts supported on zeolite-A. Most of the metal was located in the zeolite cavities and the materials were ion-exchanged with a $CaCl_2$ solution until >98% of the Na^+ ions were exchanged with Ca^{2+}. During the deep ion-exchange with the $CaCl_2$ solution the $Pt(NH_3)_4^{2+}$ cations located on the outer surface of the zeolite crystals were removed. Catalysts were dried and calcined at 723 K , prereduced with H_2 and used in the selective hydrogenation of the n-olefins in mixtures of isobutene with n-butene or propene. A high selectivity for hydrogenation of the n-olefin was obtained, provided that no Pt was present outside the zeolite pores.

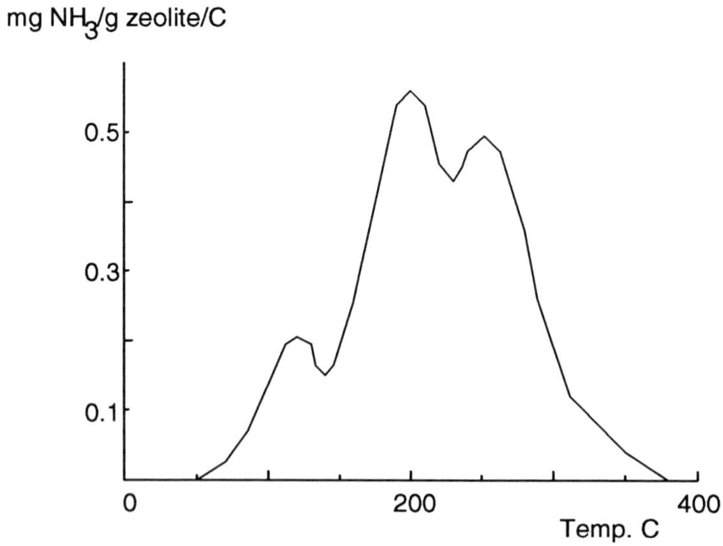

Fig.4 NH_3 TPD of NH_4-zeolite Y.(ref. 21)

2.1.5 Ion-exchange with solid salts

After early exploration of high temperature ion-exchange of zeolites with solid salts around 1970 the subject was not pursued and it lasted until about 1985 that research was resumed simultaneously by several groups. In a review article by Karge and Beyer (ref.24) it is shown that solid state ion-exchange is an attractive method for the introduction of elements in zeolites and that in one step high degrees of exchange may be reached which are difficult to obtain by conventional methods. In most studies the zeolite is in the Na, H, or NH_4 form and mostly oxides or chlorides of the in-going elements are applied. Usually finely dispersed powders of the zeolite and the salt (or oxide) of the in-going cation are mixed and heated, the required end-temperature depends on the nature of the cations and anions involved in the exchange. In most cases the exchange is fast during the initial period and sometimes exchange occurs already at ambient conditions.

2.1.6 Catalyst preparation by impregnation

An alternative method for the introduction of other elements is impregnation with a solution using the so-called "incipient wetness" technique. The volume of the solution containing the required weight of salt should be equal to the pore volume of the zeolite to be impregnated. The available pore volume of the zeolite must be measured with the same solvent as used in preparing the solution. It is preferable to apply nitrates or salts of organic acids such as acetates, since these are easily decomposed by heating without formation of residual materials. Bifunctional catalysts are applied in many petrochemical and oil conversion processes and their preparation is well known. Pore volume impregnation is for instance applied in the manufacture of hydrocracking catalysts, which contain mostly stabilized Y as acidic component and a Ni/Mo/S or Ni/W/S hydrogenation function.

2.1.7. Pore size engineering, preparation of shape selective methylamine catalysts.

The conventional method for the manufacture of methylamines from methanol and ammonia consists of an acid catalyzed reaction using silica-alumina or other acidic solids as catalysts. The product composition which is obtained in this process depends on the reaction temperature and is in thermodynamic equilibrium. At 623 K the product consists of 15% monomethylamine (MMA) , 23% dimethylamine (DMA) and 62% trimethylamine (TMA). Commercial demand for these products is different: the most valued compound is DMA, followed by MMA and the least valued is TMA. It was found that for instance mordenite (ref. 25) and zeolite rho (ref. 26) could be converted into DMA-selective catalysts by pore size engineering and poisoning of outer surface activity. Next to the methylamines, some dimethyl ether (DME) was formed as a byproduct. The reaction of ammonia and methanol over the Na^+- and the H^+-forms of mordenite and erionite was investigated by Kogelbauer et al. (ref. 27), who reported that the molar ratio of the adsorbed reactants on alkali exchanged zeolites was methanol/ ammonia = 4mol/mol. On the acidic materials ammonia adsorption was most abundant and methanol adsorption was minor. Both forms of the zeolites were active catalysts, the proton form, however, was one order of magnitude more active. Over the H-zeolite formation of a C-N bond required breaking of a N-H bond of an adsorbed ammonia by methanolysis. Over the Na-zeolite a C-N bond was formed by ammonolysis of a C-O bond of adsorbed methanol. It was assumed that the rate of formation of DME was lower over H-zeolites because of the high surface concentration of ammonium and the lower activation energy of the methanolysis reaction. The effect of a high temperature (973 K) treatment with gaseous $SiCl_4$ on the selectivity for DMA formation of various forms of mordenite was investigated by Segawa and Tachibana (ref.28). Catalysts were characterized by ^{29}Si- and ^{27}Al-MAS-NMR, chemisorption, XPS and IR spectroscopy.

Table 5
Si/Al ratio and adsorption properties of H-mordenite.
Effect of $SiCl_4$ treatment. (ref.28)

	Si/Al ratio			Adsorption * (mmol/g)				
Catalyst	bulk	framework	surface	MMA	DMA	TMA	MeOH	NH_3
H-M**	9.7	9.3	10.5	1.5	1.4	1.2	1.5	1.8
H-SC-M**	10.5	11.4	13.8	1.0	0.9	0.05	1.4	1.3
SC-H-M**	21.1	58	∞	0.4	0.4	0.1	0.7	0.7

* Measured at 373 K, MeOH measured at 300 K. **: H-M: H-mordenite, Si/Al = 10.
H-SC-M : Na-M treated with $SiCl_4$ at 973 K, ion-exchanged with NH_4NO_3,
SC-H-M : H-M treated with $SiCl_4$ at 973 K.

The $SiCl_4$ treatment resulted in a 50% increase in the surface Si/Al ratio of Na-mordenite and had a minor effect on the bulk Si/Al ratio of Na-mordenite. On the other hand for H-mordenite the $SiCl_4$ treatment led to an appreciable aluminum loss from lattice positions, with formation of extra framework alumina in the zeolite pores and complete removal of aluminum from the zeolite surface, cf Table 5. In general reaction with $SiCl_4$ led to the deposition of a layer of SiO_2 on the outer surface of the finished catalysts and to a change in mordenite pore diameter, which made the material unaccessible for TMA, see also Table 6, here H-SC-M is the best catalyst. Similarly Bergna et al. (ref.29) studied zeolite rho, ZK 12, chabazite and pollucite, which were treated with various chemicals to deposit a layer of alumina or silica, thus changing both the pore diameter and the outer surface chemistry. Results were in ageement with those obtained by Segawa and Tachibana (ref.28). A recent review of methylamines synthesis is presented in (ref 75).

Table 6
Selectivity of the reaction of ammonia and methanol over H-mordenite
Effect of $SiCl_4$ treatment. (ref.28)

Catalyst	Reaction rate MeOH $mol.g^{-1}h^{-1}\ 10^{-2}$	Product composition,% wt DME	MMA	DMA	TMA
H-M	4.5	21	21	20	33
H-SC-M	3.9	1	33	65	1
SC-H-M	8.6	48	9	13	30

Conditions: Temp: 653 K, Pressure: atmospheric, Gas composition:
NH_3 / MeOH / N_2 = 1 / 1 / 31, WHSV: not given. For catalysts, see Table 5.

Another example of pore size engineering is for instance the well-known ZSM-5 based catalyst used in alkylation of aromatics and related reactions. ZSM-5 is impregnated with a Mg salt to change and adapt its effective pore diameter and increase its selectivity for p-xylene formation in toluene disproportionation (ref.30).
In general, preparation of shape selective catalysts requires a close control of the location of the catalytic function, and catalytic sites outside the zeolite pores must be deactivated to prevent losses in shape selective reactions. It is moreover clear that the pore opening limits the size of molecules both when entering and when leaving the zeolite structure. In the latter case, too bulky molecules remain enclosed in the zeolite structure and participate in further reactions.

2.1.8 Preparation of basic catalysts

Applications of zeolites in base catalyzed reactions are relatively rare. Various methods for the preparation of basic zeolites are discussed below:
In the as synthesized form high silica zeolites such as ZSM-5 or beta, having a Si/Al ratio above about 35, usually contain intracrystalline Si-OH groups, which may be associated with alkali ions. After calcination for template removal these materials behave similar as alkali silicates in contact with water and their basicity is evident by

a pH of about 10. Dessau (ref.31) described the synthesis of high silica ZSM-5 samples having a Si/Al ratio of 300 and 13,000 and a Na/Al ratio of 7.6 and 300 respectively. The calcined samples were applied as catalysts in the base catalyzed formation of methylcyclopentenone by ring closure of acetonylacetone. In comparison the acidic form of the sample having a Si/Al ratio of 300 converted acetonylacetone with high selectivity into 2,5-dimethylfuran. In both reactions the conversion at 623 K was about 95% at a selectivity above 90%. No evidence was presented that migration of alkali to the outer surface during calcination could be excluded, and catalytic sites on the outer surface were not deactivated before the catalysis experiments. Hathaway and Davis (ref.32) prepared basic faujasites by impregnation of Y and X with cesium acetate. The calcined materials were applied as basic catalysts, for instance in the side chain alkylation of toluene. This preparation method was further investigated by Rodriguez et al. (ref.33), who impregnated various amounts of cesium acetate on Cs-exchanged X. After impregnation materials were dried and calcined at 823 K. A series of catalysts was thus obtained containing nominally 0 to 26 additional Cs atoms per unit cell. Materials were characterized by XRD, N_2 adsorption isotherms and CO_2 Temperature Programmed Desorption (TPD). The Knoevenagel condensation of benzaldehyde with ethyl cyanoacetate in the solvent dimethyl sulfoxide was applied as a test reaction. According to the XRD data no loss in crystallinity occurred during preparation, and the sorption data showed that both the surface area and the micropore volume decreased steadily with increasing cesium loading. For each sample the TPD of CO_2 had a maximum at about 573 K and was minimal at 773 K. Above about 823 K further significant amounts of CO_2 were developed from the samples containing > 8 cesium oxide species per unit cell, which was ascribed to carbonate decomposition. It was concluded that Cs-X can theoretically accomodate up to 1 cesium oxide species per supercage, which is reached by impregnation of 16 molecules of cesium acetate per unit cell. The correlation between CO_2 desorption data and the rate of the Knoevenagel reaction indicated that 6 - 8 cesium oxide species were incorporated into each unit cell and that remaining Cs_2CO_3 which was deposited outside the zeolite pore system, participated in the catalytic reaction. The authors suggested that some of this Cs_2CO_3 dissolved and acted as a homogeneous catalyst.

Ionic alkali-metal clusters such as $(Na_4)^{3+}$ are formed in zeolite Y by impregnation with an ammoniacal NaN_3 solution and are applied in reduction and in base-catalyzed reactions (refs.34,35). The nature of these clusters has been investigated by spectroscopic techniques and it was confirmed that $(Me_4)^{3+}$ units are formed (ref.36). Formation of $(Na_4)^{3+}$ clusters in zeolite X and Y by reaction at ambient conditions with a solution of a Li-alkyl or of a Li-aryl compound was reported in (ref.37). The preparation of alkali-clusters in zeolites is reviewed by Srdanov et al. in (ref.38), and a general review of basic zeolites is presented by Barthomeuf in (ref.39).

2.2 T-Site Substitution Reactions

It was mentioned earlier that the framework composition of a zeolite can be

modified by steaming at high temperature (>730 K) or by acid extraction. In both cases Al atoms are removed while the zeolite retains its crystallinity. The vacancies created in the lattice by steaming are generally repaired by Si migration. As a result of the substitution of Al by Si in zeolite Y its unit-cell size will decrease, because the atomic radius of Si is smaller than that of Al. The relation between unit-cell size and Si/Al ratio for zeolite Y is the subject of many reports, see for instance (ref.40). High temperature steaming is widely applied on an industrial scale, the process requires strict process control for reproducible results. Acid leaching which is also industrially applied for changing the Si/Al ratio, however, will leave lattice positions vacant which were formerly filled by Al atoms. Accordingly the stability of the lattice is in many cases affected, unless the structural damage is repaired.
Several other methods have been developed to change the T-site population in an existing zeolite such as reaction with aqueous $(NH_4)_2SiF_6$ (ref.41), reaction with an anhydrous halide in boiling CCl_4 (ref.42) and gasphase reaction with anhydrous halides such as $SiCl_4$ at high temperatures (ref.43). The reactions with aqueous $(NH_4)_2SiF_6$ and with $SiCl_4$ in CCl_4 show that removal of Al from well crystallised zeolite Y by chemical reactions at temperatures <373 K is relatively easy. Treating zeolite Y with aqueous solutions including the introduction of other elements is discussed in (ref.44), and it is shown that also the reverse reaction occurs i.e. the introduction of Al atoms in a dealuminated zeolite or a high silica material. An early report of this reaction (ref.45) presented evidence that extrusion of a B-silicate having the ZSM-5 structure using alumina as a binder, resulted in a final material containing lattice aluminum and having a much higher catalytic activity than the original B-silicate. (Re-)insertion of aluminum atoms into the faujasite framework has been reported in (ref.46). The framework composition was followed by solid state MAS NMR measurements, which showed clearly that Al reinsertion is possible using hot, basic aluminate solutions. In a recent paper Lutz et al. (ref.72) present evidence that the NMR measurements rather show that an extra framework Al-rich aluminosilicate is formed as a result of an alkaline treatment of DAY dealuminated by steaming. Beyer et al. (ref.47) studied the solid state reaction of ammonium zeolites and $(NH_4)_2SiF_6$ at elevated temperatures. During reaction ammonia was evolved, Si was incorporated into T sites formerly occupied by Al atoms and $(NH_4)_3AlF_6$ and NH_4HF_2 were formed. After removal of NH_4HF_2 by sublimation and of $(NH_4)_3AlF_6$ by washing with water, fully crystalline zeolites with an increased Si/Al ratio were obtained.
A review of the dealumination of zeolite Y with gaseous $SiCl_4$ at elevated temperatures is presented by Martens et al. (ref. 48). In the studies discussed in this review zeolite Y was dealuminated by reaction with $SiCl_4$ at a temperature between 423 and 623 K, followed by N_2 stripping at a temperature between 623 and 923 K. The mechanism of the dealumination reaction involves two consecutive reaction steps, in the first step $SiCl_4$ reacts with the zeolite at a temperature between 423 and 623 K with rupture of one Si-Cl bond, formation of the chloride of the charge compensating cation, and formation of a Si-O bond. In the second step, which occurs at a temperature above 623 K, silicon-aluminum exchange occurs with substitution of Si into the lattice and formation of the $AlCl_4^-$ complex of the charge compensating cation. The number of Al atoms, which may be substituted by Si depends on the number of charge compensating cations, which is available for

formation of an $AlCl_4^-$ complex, and was about equal to the number of charge compensating cations in the supercages of the zeolite structure. The temperature stability of the $AlCl_4^-$ complexes depends on the nature of the cation: in zeolite Y Li^+ complexes decompose at about 735 K and Na^+ complexes at about 785 K. Cs^+ complexes decompose at a higher temperature than the Na^+ complexes. H^+ complexes decompose at a lower temperature than the Li^+ complexes and were not detected in this work.
The pore structure of the dealuminated zeolite depends strongly on the dealumination technique. In general it appears that "chemical" dealumination by treatment with e.g. gaseous $SiCl_4$ or an aqueous solution of NH_4SiF_6 does not result in appreciable formation of secondary pores and the pore size distribution of thus treated zeolites is similar to that of the starting materials. On the other hand dealumination by high temperature steaming followed by acid extraction leads to appreciable mesopore formation. Some examples of this phenomenon are e.g. presented in (ref.49) for mordenite and in (ref.50) for zeolite Y. The extent of the formation of mesopores by high temperature steaming followed by acid extraction depends on the nature of the starting material and on experimental details of the applied procedures and is accordingly difficult to reproduce. For commercial application in conversion processes the secondary pore structure generated by dealumination is an important parameter, since it may strongly improve the accessibility of the catalytically active inner surface of the zeolite crystals and thus has a strong influence on the resulting catalyst performance.
T-site substitution reactions change the nature of the zeolite surface chemistry appreciably, their application has been compared to a secondary synthesis step (ref.46).

2.2.1 T-site substitution and catalyst deactivation

In many applications of zeolite catalysts the conversion level is not a constant at constant reaction conditions. Rather, a steadily decreasing conversion is observed. In general various deactivating surface reactions may occur and one of the most frequently observed mechanisms is that of a surface gradually being covered by a layer of polymeric species. An example of a deactivating catalyst was shown in Fig. 2. In order to increase the active life of the catalyst, measures may be taken depending on the mechanism of the deactivating reaction. When the deactivation is due to poremouth plugging as a result of coke formation on the outer surface of the zeolite particles, a possible remedy is the selective removal of the active sites on the outer surface. Many examples of such a measure have been reported in the literature, an example is the deactivation of the outer surface of mordenite by reaction with $SiCl_4$ and formation of a SiO_2 layer, discussed before in 2.1.7 for H-mordenite. The outer surface deactivation was in this case due to the removal of protonic sites by the substitution of Al- by Si-atoms.
Deactivation by formation of carbonaceous residues on the inner surface of the zeolite is mostly due to condensation reactions of reactive intermediates. The rate of this reaction may be lowered by increasing the Si/Al ratio of the zeolite and thus increasing the separation of the protonic sites. The optimal performance as a function of catalyst composition has accordingly to be determined. Preferentially this

is studied on samples obtained in direct synthesis, possible amongst others for ZSM-5 and zeolite Beta. In other cases secondary synthesis may be applied to change the Si/Al ratio of the zeolite.

2.3 Effect of Crystallite Size

For many zeolites the size and shape of the crystallites are dependent on the synthesis conditions. Since reactions in zeolites are frequently diffusion limited, crystallite site has a strong effect on the activity and selectivity of the catalyst. This aspect is dealt with in the chapter by Post and Ruthvenin this book and in (ref.51). In relation to the preparation of zeolitic catalysts one must be aware of the fact that crystallite size is important for catalyst performance and that its effects should be considered in catalyst development. The influence of crystallite size on reaction rate and selectivity is for instance discussed by van der Pol et al. (ref.52) for the hydroxylation of phenol using titanium-silicalite-I catalysts.

2.4 Catalyst Shaping

2.4.1. Granules, particle size <0.1 mm

Granulated catalysts consisting of particles smaller than 0.1 mm are usually manufactured by spray-drying of a slurry which contains all the active components of the final catalyst and a binder. This unit operation cannot be scaled down to the usual laboratory type of batch size, because the dimensions of the equipment determine, amongst other properties, to a large extent the maximum size of the particles which can be made. In general the zeolite component of a spray-dried catalyst is in an activated form and the spray-dried product is ready for delivery, and is usually applied as a catalyst after an in situ thermal activation. Apart from their catalytic activity, the physical and mechanical properties of spray-dried powders are important parameters which are mainly determined by the binder system and the manufacturing process.

2.4.2. Catalyst pellets

Various methods are applied for the preparation of catalyst pellets, see Table 7, and zeolitic catalysts do not require special methods. By far the most frequently applied method in commercial catalyst manufacturing is extrusion, using an hydraulic binder such as clay, alumina, silica, or a combination of these. Although the technique is straightforward, manufacturing of dense extrudates requires special equipment. Simple equipment is available for the laboratory extrusion of experimental catalysts. It is, however, difficult to prepare representative batches of catalyst on a laboratory scale as far as catalyst texture and strength are concerned. These aspects have a large effect on catalyst performance and may even influence the design of the reactor and the configuration of the catalyst bed(s). In many cases the extrusion mixture will contain all the components of the finished product. The extrusion mixture is plasticised by intensive mixing in a high shear mixer or mixer-muller and the resulting dough is shaped using an auger extruder. The finished catalyst is obtained after drying and calcination of the "green" extrudates. It is advantageous that the calcined product is the end product. Further operations in which shaped particles are contacted with aqueous solutions are usually complicating and costly.

Table 7
Methods for the manufacture of shaped zeolites (> 0.1 mm).

Procedure	Binder	Remarks
Agglomeration	Clay	Use in separation processes.
Extrusion	Alumina	Use in catalysis and most other processes.
Compaction		May damage zeolite.
Hydrothermal conversion of shaped precursor.		Preparation of FCC catalysts, method for the preparation of "self-bonded" shaped zeolites.

2.4.3 Structured Zeolite Catalysts.

The use of catalysts and adsorbents in the treatment of large gas volumes for removal of minor components and air pollution abatement is rapidly expanding. Generally these applications require a reactor configuration which is tailored for an efficient contact between gas and solid and a low pressure drop over the bed of active material. Structured zeolites have been considered in the literature for applications as membranes, adsorbents, catalyst components and in electronic devices (ref.73). The wide range of well defined pore sizes and the adaptability of the surface chemistry offered by zeolites make them attractive base materials for new developments in these areas. The main methods for the preparation of structured zeolites are :

Deposition of zeolites by dipcoating.

Dipcoating is an efficient and well proven technique for the deposition of thin layers of an oxidic material on existing supports. The dipcoating technique is very versatile, since it may be applied to most support materials, however complicated their shape. Both the support and the material to be deposited should avail of reactive surface groups. It appears that condensation reactions between SiOH groups on the outer surface of the zeolite crystals and hydroxyl groups on oxidic and metallic supports usually result in the formation of an adhering zeolite layer upon curing of a zeolite coated support. The interaction is not strong however and the thickness of the layer is moreover very limited. Accordingly a precursor of an hydraulic binder is usually added to the zeolite slurry, which upon curing assists in the formation of a strongly bonded zeolite layer on the support. The thickness of this layer depends on the the zeolite content of the slurry and the number of times that the dipcoating is repeated. The zeolites may be tailored for a specific application before the coating operation. Structured zeolites prepared by the dipcoating technique are applied in adsorption and in catalysis.

Growth of zeolite layers by in situ synthesis.

Total coverage of an existing support with a closed layer of zeolite crystals will hardly ever be obtained by dipcoating and is not relevant for applications other than in membranes. Research on the growth of closed layers of zeolite crystals on existing supports by in situ synthesis is presently emphasized because of the promising possibilities for applications in membrane technology. Porous metallic or oxidic supports consisting of aggregates of powders and/or thin wires, such as Knudsen diffusors or filter plates are generally applied and covered by deposition of zeolite crystals from the synthesis mixture. Growth of closed layers of large or small mutually oriented zeolite crystals on supports might allow additional applications in the manufacture of electronic devices. Both in the open literature and in patents many publications concerning the preparation procedure have recently appeared. Main problems are the formation of a dense layer of crystals, their mutual orientation and the prevention of the occurrance of non covered support areas.

2.5 Analysis

Table 8
Specified properties of shaped catalysts.

Property	Powder	Extrudates (granules)
Specific surface area, m^2/g	+	+
Pore size distribution	+	+
Pore volume ml/g	+	+
Compacted bulk density	+	+
Particle density	+	+
Skeletal density	+	+
Particle size distribution	+	+
Average length mm, (n particles)		+
Average diameter mm, (n particles)		+
Maximum/minimum diameter, mm		+
Maximum/minimum length, mm		+
Side crushing strength,N/mm (n particles)		+
Attrition %w	+	+
Chemical composition	+	+
Loss on ignition at ...K	+	+
Zeolite content %w	+	+
Zeolite unit cell size	+	+
Zeolite crystallinity	+	+
Metal dispersion/distribution	+	+
Manufacturing method	+	+
Performance test		

Finished catalysts must meet a number of chemical and physical specifications. In an ideal case the specifications agreed upon between manufacturer and user describe the material sufficiently accurate to obviate extensive catalyst testing. Since a large number of variables influences the final performance of zeolitic catalysts it is of utmost importance that both the starting materials and the products of the various catalyst preparation steps are fully characterized and documented. Moreover, whenever commercial samples are available their use in catalyst development is strongly advised since these materials are generally prepared on a large scale and their use may alleviate the appreciable reproducibility problems, which are usually faced during research on the application of zeolitic catalysts using experimental zeolite samples. Analytical techniques for zeolite characterization are described in Chapter 8 by van Hooff, Roelofsen and Lercher.
A number of properties of catalysts are measured using standardized methods, many of these were developed in joint projects of catalyst manufacturers and users under the auspices of the ASTM (ref.53). Other analytical methods are agreed between manufacturer and user. Properties which are frequently agreed upon to serve as product specifications are collected in Table 8.

3 FLUIDIZED CATALYTIC CRACKING (FCC) CATALYSTS

3.1 FCC Process.General Data

The FCC process is the most important refinery process for the production of gasoline components. Typical FCC feedstocks such as for instance gas oils, vacuum tower bottoms, atmospheric tower bottoms, coker gas oils and lube extracts (ref. 54) are converted into lighter products by cracking reactions. A typical product composition is given in Table 9. Valuable products are LPG, gasoline and light cycle oil; fuel gas and residue are byproducts and coke on catalyst represents a loss which is removed by coke burning in the regenerator.
FCC is a continuous regenerative process, a process scheme for a riser cracker is e.g. presented in Chapter 17 by Stork and Maxwell. The preheated feed is injected at the bottom of the riser reactor and mixed thoroughly with the freshly regenerated, hot catalyst. The catalyst to oil weight-ratio in the mixture is 5 to 6 and its temperature is about 773 K. The residence time in the riser reactor is about 3 seconds and after this very short reaction time the catalyst is completely deactivated by coke deposition. Deactivated catalyst and hydrocarbons are separated in the steam stripper vessel. Coke on catalyst is subsequently removed by burning in air at 1030 K in the regenerator reactor.The residence time in the regenerator is about 10 min. The regenerated catalyst is mixed with some fresh catalyst and transported to the reactor.
An average FCC unit has a catalyst inventory of about 300 - 1000 tons, with 400 tons being typical. Throughput ranges from 15.000 - 130.000 barrels/day, i.e. 2250 - 19500 tons/day, with a typical FCC unit handling 7500 tons/day.Because of the extreme reaction conditions the FCC catalyst must have a high attrition resistance and a very high physical strength, moreover its thermal stability must be very high in order to withstand the regeneration process.

Table 9
Typical Product Composition FCC Process
Feedstock:Vacuum gasoil + Residue

Component	% w
Fuel gas (H_2;C_1;C_2)	3
LPG (C_3;$C_3^{=}$;C_4;$C_4^{=}$)	17
Naphtha	52
Light cycle oil	16
Bottoms	5
Carbon on catalyst	6

Some typical data on the process conditions are collected in Table 10.

Table 10
Typical Data of an Average FCC Unit

Catalyst inventory	400 tons
Fresh catalyst rate	3 - 4 tons/day
Feed rate	7500 tons/day
Cat/oil ratio w/w	5-6
Linear velocity in riser reactor	10 m/sec
Temperature in riser	800 K
Temperature of feed	530 K
Temperature in regenerator	1000 K

A review of the distribution of the 445 actual and planned world wide FCC units is given in Table 11. About 40% of the active FCC units are located in North America, which is the largest gasoline market (demand of around 8.000.000 barrels/day).

Table 11
Number of FCC Units per Area

Unit Status	North America	South America	Asia Pacific	Eurafrica	Total
Active	147	33	83	105	368
Construction	0	0	3	0	3
Inactive	12	8	24	24	68
Planned	0	0	6	0	6
Grand Total	159	41	116	129	445

World wide FCC catalyst consumption is about 1230 tons/day or 450.000 tons/year, about half of which is consumed in North America.

3.2 FCC Catalyst Production Capacity, Main Catalyst Producers.

Worldwide the installed production capacity for FCC catalyst is about 550.000 tons/year, see Table 12.

Table 12
Worldwide Production Capacity

Worldwide Production Capacity	ktons/year
USA	350
Europe	100
Japan,Brazil,China	100

The appreciable overcapacity results in a very competitive situation amongst the producers. The accompanying total capacity for Y-zeolites is about 150.000 tons/year, since the average percentage of zeolite in the FCC catalyst is in the 25 - 30 % range. The main FCC catalyst producers and the location of their plants are listed in Table 13.

Table 13
Main FCC Catalyst Producers

Company	Plants Location
Grace	USA/Europe
Engelhard	USA/Europe
Shokubai-Kasei	Japan
Akzo Nobel	USA/Europe/South America

3.3 Future Developments in FCC.

The very competitive market situation and economical and environmental requirements in FCC have led to many catalyst improvements in recent years. Even further improvements in product performance have to be substantiated in order to meet future demands anticipated for the FCC process. Expected economic demands are:

- Higher gasoline yield.
- Higher gasoline octane number.
- Higher yield by lower gas- and coke formation.
- Expanded boiling range of feedstock (resid cracking).
- Improved metal (Ni,V) tolerance of FCC catalyst.
- Lower catalyst losses by dust emissions
- Lower SO_2 emissions.

These goals are most probably reached at lowest cost by tailoring the FCC catalyst rather than by adaptation of the FCC process proper. In most of the areas for process improvement mentioned above appreciable progress has been made in recent years due to improvements in catalyst performance.

3.4 Composition of the FCC Catalyst

3.4.1 Matrix

The two main components of a FCC catalyst are the active zeolite and the matrix material, the latter acts as a continuous binder phase between the zeolite crystallites. As such the matrix determines the physical properties of the catalyst. In recent years the function of the matrix has been expanded and includes now catalytic activity to crack large molecules into fragments small enough to enter the zeolite pore system. In addition the properties of the matrix include activity to trap contaminants that are present in the feedstocks. As heavier feeds are processed and catalysts are operated at higher metal contaminant levels it is increasingly important for the matrix to passivate Ni, thus preventing excessive coke and hydrogen production, and to trap V and Na to prevent zeolite destruction (V) and/or poisoning (Na) (refs. 55; 56; 57) Components in the matrix precursor mix are alumina,silica-alumina gel and pretreated clays.

3.4.2 Zeolite Component.

The main zeolite component in modern FCC catalysts is based on zeolite Y. In this book the structure of zeolite Y is e.g. presented in Chapter 3 by Baerlocher and McCusker. The zeolite has a pore opening of about 0.8 nm, it is activated by ion exchange with NH_4 ions or trivalent rare earth ions. Acid sites are generated by calcination of the ion exchanged zeolite and these acid sites are active in the catalytic cracking process.

3.4.2.1 Acid Site Configuration

Important factors related to the acid site configuration are:

- The number of acid sites
- The strength of the acid sites
- The distribution of the acid sites in the lattice.

The acid site configuration determines to a large extent the ratio of the various chemical reactions occurring during the cracking process. In this respect an important factor is the degree of isolation of the acid sites, illustrated in Fig. 5. This illustration is simplified and shows only four of the actual nine next nearest T-atom (NNN)sites.

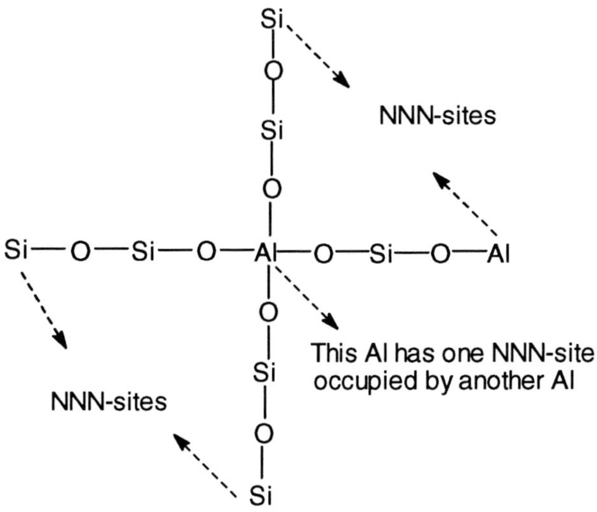

Fig. 5. Acid-site configuration.

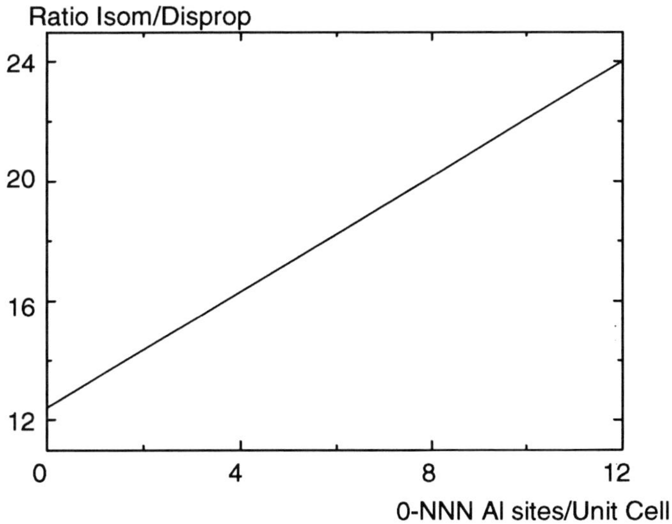

Fig.6. o-Xylene conversion over zeolite H-Y. Effect of acid-site configuration on isomerization/disproportionation ratio.(ref. 62)

Isomerization: o-xylene→ m- xylene + p-xylene

disproportionation o-xylene→ toluene + trimethylbenzenes

Acid strength increases as the number of Al atoms in NNN sites decreases from 9

to 0. A maximum in acidity is reached at a Si/Al ratio (SAR) of 9 -12, when all Al atoms in a unit cell have no Al atoms in NNN sites (ref. 56). Further removal of Al results in a decrease in overall acidity because of the Al loss (refs.59; 60; 61).
The effect of the degree of isolation of the acid sites (number of Al sites without an Al atom on a NNN position, i.e. 0 NNN sites) on a chemical reaction is illustrated for reactions of o-xylene in Fig. 6 (ref. 62).
The ratio between the rate of isomerisation and that of disproportionation depends on the number of 0-NNN sites per unit cell in the zeolite. The effects of the acid site configuration on catalyst performance are an active field of research. For FCC catalysts the situation is complicated due to the fact that the configuration changes under the operating conditions in a FCC unit. The change in active site configuration may be studied in the laboratory by the application of simulated deactivation procedures. It appears that both the rate of deactivation and its mechanism are dependent on the quality of the zeolite used in the manufacture of the catalyst and on the hydrothermal stability of the zeolite component. Changes in acid site configuration occur as a result of exposure to high temperature in the presence of steam, conditions which are common in a FCC unit.

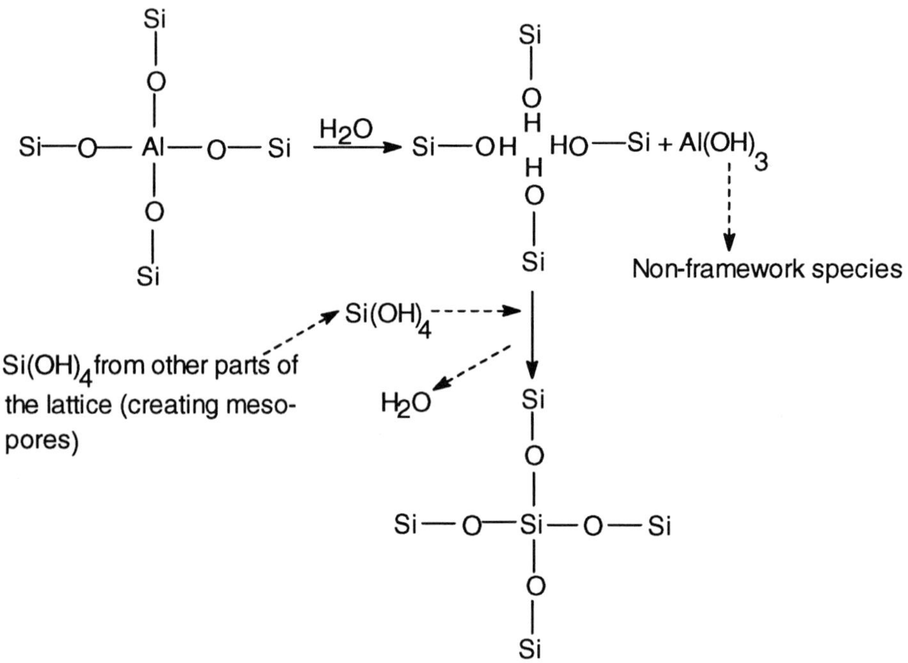

Fig. 7. Solid state reactions in zeolite Y during steaming; dealumination and repair

Al is removed from T sites and forms extra (non)framework alumina as a separate phase (ref.63). The vacated T sites are partly filled by silica from other parts of the lattice (Fig.7) some structural damage will, however, always remain.

Recent investigations have shown that extra framework alumina is also catalytically active. During the alumina migration a transition state occurs where alumina is partly bound to the zeolite structure. It appears that this transition state alumina has a strong effect on the catalytic activity (refs. 64; 65; 66; 67).

3.4.3 Manufacture of Y Zeolite Component

It has been stated earlier that both the quality and the activation procedure of the zeolite used in the manufacture of the FCC catalyst have a strong influence on its performance. The activated zeolite is one of the raw materials in the FCC catalyst manufacture.

3.4.3.1 Synthesis of Na-Zeolite Y

A simplified block diagram of the synthesis process is presented in Fig. 8. A silica source, an alumina source and caustic soda are mixed in water and sometimes a seeding mixture is added in order to accelerate the crystallization process.

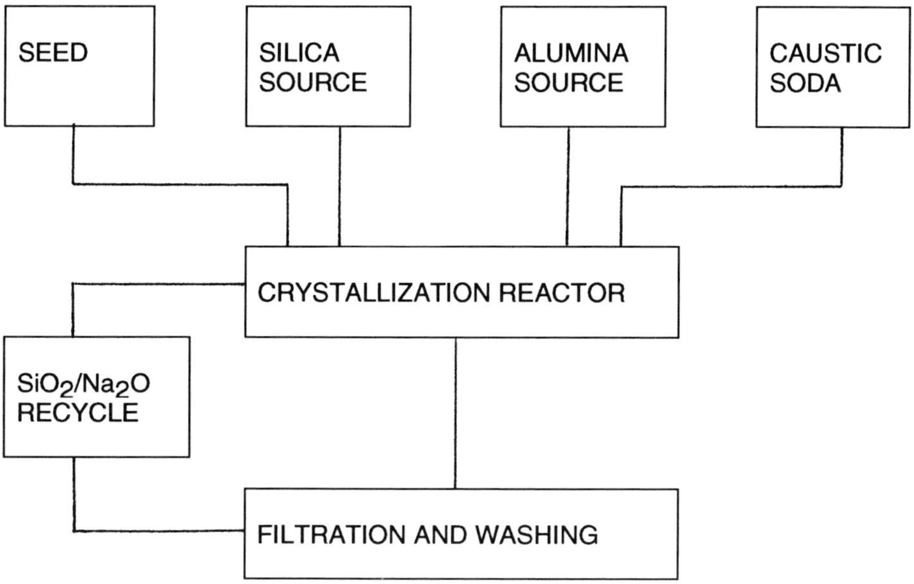

Fig.8. Industrial synthesis of zeolite Na-Y.

Upon mixing the ingredients an amorphous silica-alumina gel is formed. The crystallization process is carried out at 373 K and lasts 10-50 h, depending on the desired Si/Al ratio in the final product. The mechanism of the crystallization of zeolite Y from the amorphous gel is not yet well understood. Moreover it appears

that zeolite Y is one out of several crystalline structures which can be formed under the reaction conditions (temp. about 375 K).
Fig. 9 shows that for example the composition of the reaction mixture has a strong

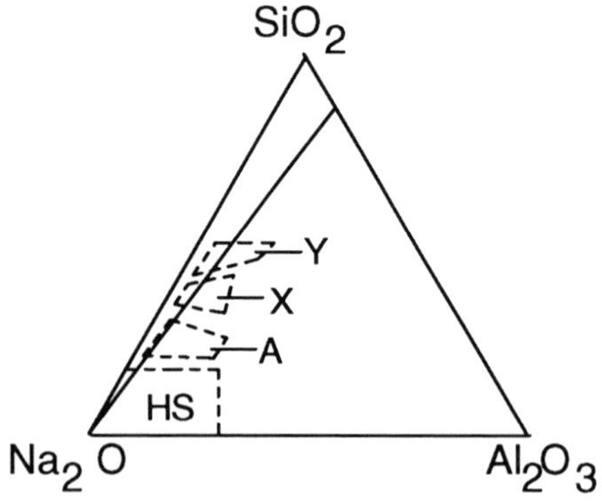

Fig. 9. Synthesis of zeolite Na-Y. Product composition depends on molar ratio of the components.(HS is hydroxysodalite)

influence on the end product. This illustrates that the homogeneity of the gel is an important factor. An inhomogeneous gel may easily result in locally different crystalline structures. These impurities accelerate the recrystallization of an eventually formed zeolite Y into other phases, see Fig.10.

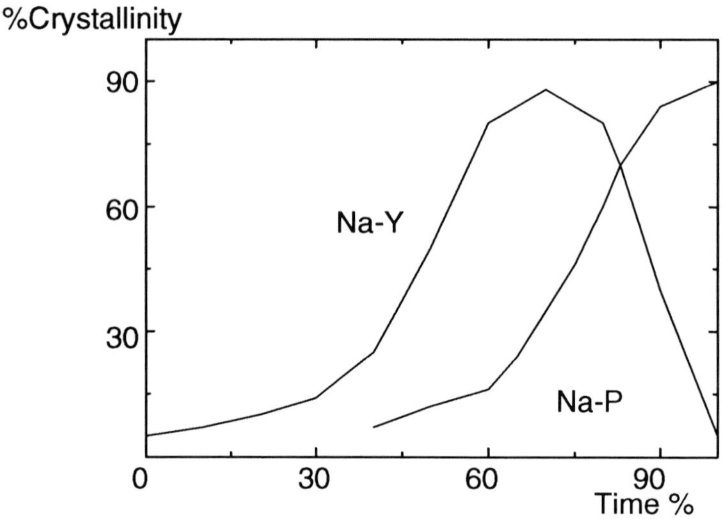

Fig. 10.Recrystallization of Na-Y into Na-P.

Scale up of laboratory recipes is accordingly an important step in the commercialization of zeolites. Additional problems arising during scale up of zeolite synthesis are discussed below.

The crystallization process consists of two stages as is illustrated in Fig.11. It is generally accepted that during the induction time small nuclei are formed, which upon reaching a critical concentration show an accelerated growth to the final crystals (ref.68).

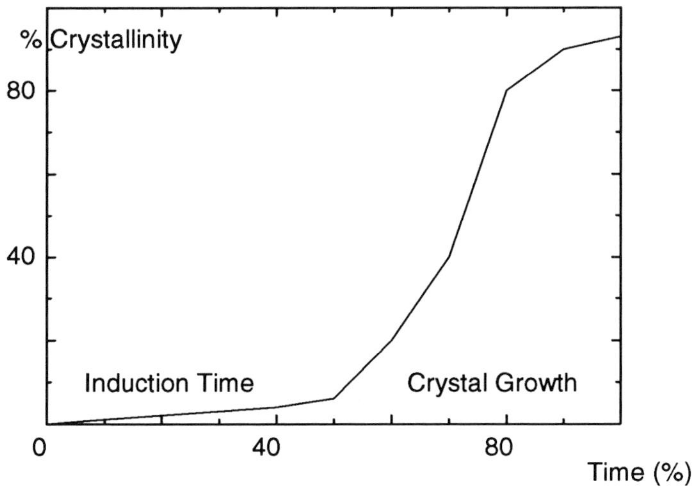

Fig. 11. Synthesis of zeolite Na-Y. Product purity depends on conditions during induction time.

Practical experience shows that for zeolite Y the induction period represents a very delicate process stage which can easily be disturbed by agitation or in general by shear forces. The nucleation process accordingly is strongly influenced by the homogeneity of the starting Si/Al gel, the way of agitating and the temperature. Seeding of the synthesis mixture alleviates some of the above mentioned problems. It results in an end product of higher purity which, however, contains very small crystals and is difficult to handle. A general problem of zeolite Y slurries is presented by their pseudoplastic behaviour. A settled zeolite Y slurry will form a massive and very hard sediment that is extremely difficult to handle. On the other hand a "solid" filtercake is transformed into a milky liquid under the influence of shear. This property of zeolite Y slurries has important consequences for the lay-out of a production unit.

3.4.3.2 Ion-Exchange and Zeolite Activation

During this process step NaY is converted into the catalytically active material. A simplified block diagram of the activation procedure is given in Fig. 12. About 70% of the Na ions are replaced by NH_4^+ or rare earth (RE^{3+}) ions during the first ion exchange and an intermediate heating step is applied to convert the zeolite into the H-form. The temperature of the intermediate calcination is usually between 823 and

1050 K, the presence of steam during this process step is essential. During heating Na is redistributed over the zeolite surface and is thus made available for further ion exchange to about 70% replacement by NH_4. After the second ion exchange the Na_2O content is less than 1 %w. During the calcination some Al is removed from the zeolite lattice, the product is usually described as stabilized Y. This product has a higher Si/Al ratio in its framework than the NaY raw material from which it was made and contains some non framework alumina (NFA).

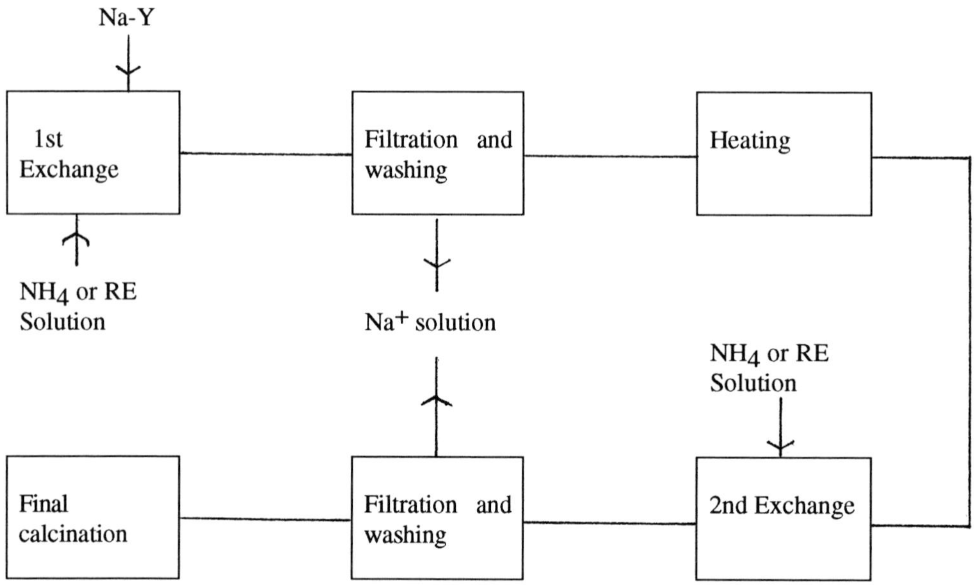

Fig.12. Ion-exchange of zeolite Na-Y for sodium removal. In each ion-exchange step about 70% of the sodium is removed.(RE:Rare Earth)

The nature of NFA and its role in the catalytic activity of the various forms of stabilized Y sieve materials is a widely investigated research subject. It is now believed that activity and selectivity of Y zeolites in FCC catalysts are determined by an interaction between framework and non-framework aluminum species. Strict control of the conditions of the stabilization process is necessary to maintain an homogeneous Al distribution (ref. 67). Important parameters of the production process are :

The Si/Al ratio of the raw Na-Y,
The Na_2O content before the heating step,
Temperature,residence time and H_2O partial pressure during the heating step,
The Na_2O content after the second ion exchange step.

With the exception of the heating step, zeolite activation does not present difficult technological problems. Control of process conditions during calcination in large scale equipment is the most difficult aspect in scale up to commercial operation.
The latest trend in FCC catalyst manufacture is to subject the zeolite component to further activation procedures.Such a process step may consist of a repeated intermediate heating step followed by an ion exchange. More Na is removed and the product has an higher Si/Al ratio in its framework. Alternate T-site substitution reactions may be applied on the Na-Y material using the procedure described in (ref.14). Recent work has, however, shown that reaching high degrees of T-site substitution is difficult and may only be achieved at the expense of zeolite crystallinity (ref. 69).

3.4.4 Effect of ZSM-5 Addition

Product quality of FCC units is changed by addition of other zeolites to the FCC catalyst (ref.74). Addition of ZSM-5 is most frequently applied since it results in a better gasoline quality albeit at the cost of a lower gasoline yield. In the narrow pores of ZSM-5 monobranched and n-paraffins are preferentially adsorbed. These molecules, which form an appreciable part of the FCC gasoline fraction, have a relatively low octane number and are removed by shape selective cracking in the

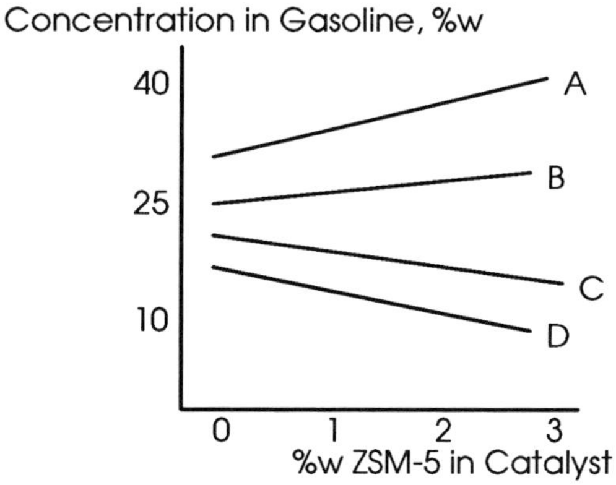

Fig. 13 . Effect of the addition of ZSM-5 on gasoline composition.
A : aromatics, B : branched olefins, C : branched paraffins, D : n-olefins

narrow pores of ZSM-5, lowering the gasoline yield but improving its octane number (cf Fig. 13). The cracked products are olefinic and may be applied in the manufacture of high octane components or as chemical feedstock. The pore structure of FCC catalyst is schematically presented in Fig. 14. Upon contacting the catalyst resid molecules are cracked and demetallized on the outer surface and in the widest pores of the catalyst particles. The lightest fraction of the gasoil, which is formed in the macro- and mesopores is further cracked in the zeolite pores and the quantity and quality of the gasoline in the final product depends amongst others on

the nature of the zeolites present in the catalyst.
It is clear that FCC catalyst composition may be varied widely since the number of parameters is large. The total number of different commercial catalysts which is

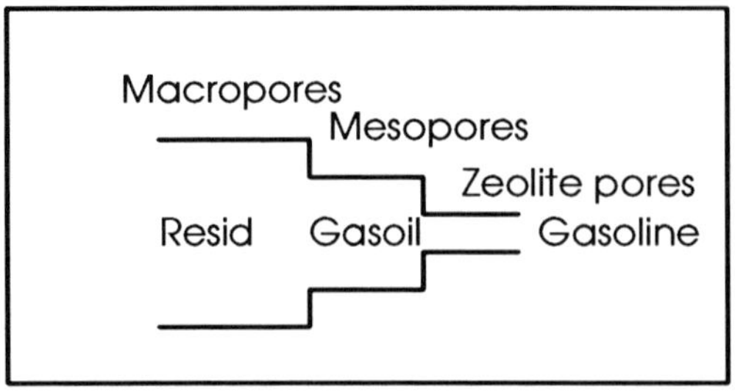

Fig. 14. Pore struccture of FCC catalyst

available on the market is comparable to that of the total number of FCC units, and the design of hardly any of the latter is identical to that of another unit. This calls for customized catalysts and rationalizes the large number of different materials.

3.4.5 Manufacture of the Composite FCC Catalyst

The last step in the manufacture of FCC catalyst is its shaping into a fluidizable powder. To this end an aqueous slurry is prepared containing the matrix precursors and the activated zeolite component. The slurry is spray-dried in a large continuous unit and the product is shipped to the consumers. Some typical properties are collected in Table 14.

Table 14
FCC catalyst,typical properties.

Pore Volume,H_2O,ml/g		0.20-0.35
Surface area,m^2/g		250-350
App. bulk density,g/ml		0.7-0.8
Zeolite content,%w		20-30
Particle size distribution	μm	%w
	< 149	97
	< 105	80
	< 80	65
	< 40	20
	< 20	2

4 CONCLUSION

In the preparation and commercial manufacture of zeolite catalysts results of fundamental studies in zeolite chemistry are very often quickly applied. Especially in large applications such as FCC and hydrocracking even a small improvement in catalyst selectivity often has a considerable economic value. Because of the strong research effort in the area of zeolites, new catalysts having an improved performance are produced at a high rate. This lively interaction between fundamental and applied science makes research into the use of zeolite catalysts very rewarding.
For additional reading excellent reviews are found in (refs. 51 and 70). A review of relevant patents is given in (ref.71).

REFERENCES

1 W. Hölderich, M. Hesse, F.Näumann, Angew.Chemie 100, (1988), 232
2 H.van Bekkum, H.W.Kouwenhoven, Recl.Trav.Chim.Pays Bas 108, (1989), 283
3 M.Soulard, S.Bilger, H.Kessler, J.L.Guth, Zeolites, 7, (1987), 463.
4 C.Dimitrov,Z.Popova,S.Mladenov,K,H.Steinberg, H.Siegel in D. Kallo, Kh.H. Minachev (Eds.), "Catalysis on Zeolites", Akad.Kiado Budapest 1988, p. 135
5 Th.L.M.Maesen, Thesis, Delft University of Technology , 1990, J.Chem.Soc Faraday Trans.,87, (1991), 787
6 H.Bosaceck and V.Patzelova in D. Kallo, Kh.H. Minachev (Eds.), "Catalysis on Zeolites", Akad.Kiado Budapest 1988, p. 169.
7 B.L.Meyers,T.H.Fleisch,G.J.Ray,J.T.Miller, J.B.Hall, J.Catal. 110, (1988), 82.
8 H.W. Kouwenhoven in W.M. Meier, J.B.Uytterhoeven (Eds), "Molecular Sieves", Adv.in Chem.Series 121, Washington, 1973, p.529.
9 W.K.Hall, J.Engelhardt, G.A.Sill in P.A.Jacobs, R. A. van Santen (Eds), "Zeolites,Facts, Figures,Future", Elsevier, Amsterdam 1989, Stud.Surf.Sci. Catal.49, p. 1253.
10 P.O.Fritz, J.H.Lunsford, J.Catal. 118, (1989), 85.
11 D.Vergani, R.Prins, H.W.Kouwenhoven, Appl. Catal. 163, (1997), 71
12 J.D. Barthomeuf, Materials,Chemistry and Physics,17,(1987), 49
13 C.T.W.Chu, C.D.Chang, J.Phys.Chem. 89, (1985), 1569.
14 D.K.Simmons, R.Szostak, P.K.Agrawal, T.L.Thomas, J.Catal.106,(1987),287.
15 W.F.Hölderich, Pure & Appl.Chem. 58, (1986), 1383.
16 D.R.Corbin, L.Abrams, C.Bonifaz, J.Catal. 115, (1989), 420.
17 M.Guerin,C.Kappenstein,F.Alvarez,G.Gianetto, M.Guisnet, Appl.Catal.45, (1988), 325.
18 S.T. Homeyer, W.M.H. Sachtler in P.A. Jacobs, R.A.van Santen (Eds), "Zeolites, Facts, Figures, Future", Elsevier, Amsterdam, 1989, Stud. Surf. Sci.Catal.49, p. 975.
19 H.A. Rangwala, J.A. Szymura, S.E. Wanke, F.D. Otto, Can.Journ Chem. Eng., 66,(1988),843.

20 H.W.Kouwenhoven, H.J.A.van Helden Ger.Offen 1.816.822, (1968) Shell Research

21 R.J.Mikovski, J.F.Marshall,J.Catal. 44,(1976),170.

22 H.Bergem, E.A. Blekkan, A.Holmen in "Hydrotreatment and Hydrocracking of Oil Fractions", G.F.Froment, B.Delmon, P.Grange, (Eds.), Elsevier, Amsterdam, 1997, p 391

23 P.B. Weisz, V.J. Frilette, R.W. Maatman, E.B. Mower, J.Catal. 1, 1962, 307.

24 H.G.Karge, H.K.Beyer in "Zeolite Chemistry and Catalysis", P.A.Jacobs, N.I.Jäger, L.Kubelková, B.Wichterlová, (Eds.), Elsevier, Amsterdam, 1991, p 43

25 Y.Ashina, T.Fujita, M. Fukatsu, K.Niwa, Y.Yagi, in "Proceedings of the 7th Intern. Zeolite Conf.", Tokyo, 1986, Y.Murakami, A. Iijima, J.W.Ward, (Eds.), Kodansha-Elsevier, Tokyo, 1988, p1960.

26 H.E.Bergna, D.R.Corbin,G.C.Sonnichsen, US Pat. 4,752,596, 1988, Dupont

27 A.Kogelbauer, Ch. Gründling, J.A.Lercher, in "Zeolites and Related Microporous Materials ; State of the Art 1994", J.Weitkamp, H.G.Karge, H.Pfeifer, W.Hölderich (Eds.),Stud.Surf.Sci. Catal., 84, Elsevier, 1994, p.1475

28 K.Segawa, H.Tachibana, J. Catal.131, (1991), 482

29 H.E.Bergna, M.Keane, D.H.Ralston, G.C.Sonnichsen, L Abrams, R.D.Shannon, J.Catal. 115, (1989), 148

30 N.Y.Chen, W.W.Keading, F.G.Dwyer, J.Am.Chem.Soc.101, (1979), 6783.

31 R.M.Dessau, Zeolites, 10,(1990),205.

32 P.E.Hathaway, M.E.Davis, J.Catal., 116, (1989), 263, 279

33 I.Rodriguez, H. Cambon, D. Brunel, M. Laspéras, P. Geneste, in "Heterogeneous Catalysis and Fine Chemicals III" ,M. Guisnet et al. (Eds.), Elsevier, Amsterdam, 1993, p623.

34 L.R.M. Martens,P.J.Grobet, P.A.Jacobs, Nature,315,(1985),568.

35 R.E.H. Breuer, E.de Boer, G.Geismar, Zeolites 9,(1989),336.

36 K.B.Yoon, J.K.Kochi, J.Chem.Soc.Chem.Commun. 1988, 510.

37 M.W. Anderson, J. Klinowski, J.Chem.Soc.Far.Trans.I 82, (1986), 1449.

38 V.I.Srdanov, W.P.Blake, D.Markgraber, H.Metin, G.D.Stucky, in " Advanced Zeolite Science and Applications", J.C.Jansen, M.Stöcker,H.G.Karge, J.Weitkamp, (Eds.), Stud.Surf. Science and Catal.,85, Elsevier, 1994, p. 130.

39 D.Barthomeuf, Cat.Reviews, Science and Engineering, 38, (1996), 521

40 J.Dwyer, F.R.Fitch, F.Machader, G.Qin, S.M.Smyth, J.C.Vickerman, J.Chem.Soc. Chem. Commun., 422 (1981)

41 G.W. Skeels, D.W. Breck in D. Olson, A. Bisio (Eds.), Proc.Sixth Int.Zeol. Conf. 1983, Butterworths,U.K.,1983, p.87.

42 L.V.C. Rees, E.F.T. Lee, PCT, WO 88/01254 ,(1988).

43 H.K. Beyer, I.M. Belenykaja, F. Hange, M. Tielen, P.J. Grobet, P.A. Jacobs, J.Chem.Soc.Far.Trans.I, 81,(1985),2889.

44 P. Fejes, I. Kirisi, I. Hannus, Gy. Schöbel in D. Kallo, Kh.H. Minachev (Eds.), "Catalysis on Zeolites"Akad.Kiado Budapest 1988, p. 135 ref 4, p. 205.

45 C.T.W. Chu, J. Catal. 93, (1985), 451.

46 H. Hamdan,B. Sulikowski, J. Klinowski, J.Phys.Chem 93, (1989), 350.

47 H.K. Beyer, G. Borbély-Pálné, J. Wu,in "Zeolites and Related Microporous Materials, State of the Art 1994", J. Weitkamp, H.G. Karge, H. Pfeifer,H. Hölderich (Eds.), Stud. Surf.Sci. Catal. 84, Elsevier, Amsterdam, 1994, p.933
48 J.A. Martens, P.J. Grobet, P.A. Jacobs in "Preparation of Catalysts V", G. Poncelet, P.A. Jacobs, P. Grange, B. Delmon (Eds.), Elsevier, Amsterdam, 1991, p355.
49 M.M. Olken, J.M. Garces in "Proc. Ninth Int. Zeol. Conf.", R. von Balmoos, J.B. Higgins, H. M. J. Treacy, (Eds.), Butterworth-Heinemann, Stoneham, 1993, p. 559.
50a U.Lohse, M.Mildebrath, Z.anorg.allg.Chem. 476,(1981), 126.
50b R.Fang, H.W.Kouwenhoven, R.Prins in "Zeolites and Related Microporous Materials: State of the Art 1994", J.Weitkamp, H.G.Karge, H.Pfeiffer, W. Hölderich (Eds.), Stud. Surf. Sci. Catal. 84, Elsevier, Amsterdam,1994, p. 1441.
51 W.O. Haag, N.Y. Chen, in "Catalyst Design, Progress and Perspectives" L.L. Hegedüs (ed.), J.Wiley & Sons, 1987, p. 163.
52 A. J. H. P. van der Pol, A. J. Verduyn, J. H. C. van Hooff in "Proc. Ninth Int. Zeol. Conf.", R. von Ballmoos, J. B. Higgins, H. M. J. Treacy, (Eds.), Butterworth-Heinemann, Stoneham, 1993, p. 607.
53 ASTM Standards on Catalysts, ASTM, Philadelphia, PA, e.g. 1988.
54 A.Avidan in J.S.Magee, M.M.Mitchell, Jr (Eds.), "Zeolites, Fluid Catalytic Cracking: Science and Technology", Elsevier, Amsterdam, 1993, Stud. Surf. Sci. Catal. 76, p.1.
55 M.L.Ocelli, Catal. Rev. Science Eng. 33, 241, 1991
56 J.L.Sanchez, J.P.Hager, in EDP Congress 1993, Proc. from the Symp., TMS Annual Meeting, Minerals, Metals and Materials Society, PA, 1993, p. 665.
57 C.A.Trujillo, U.N.Uribe, P.P.Knops-Gerrits, L.A.Oviedo, P.A.Jacobs, J.Catal. 168, (1997) 1.
58 A.Humphries, D.Harris, P.O'Connor in J.S.Magee, M.M.Mitchell, Jr (eds), "Zeolites, Fluid Catalytic Cracking: Science and Technology", Elsevier, Amsterdam, 1993, Stud. Surf. Sci. Catal. 76, p.41.
59 Q.L.Wang, G.Giannetto, M.Guisnet, J. Catal. 130,(1991), 471.
60 P.V.Shertukde, W.K.Hall, J.M.Dereppe, G.Marcelin, J.Catal.139,(1993), 468.
61 A.Corma, F.V.Melo, D.Rawlence, Zeolites 12 (1992) 261.
62 E.van Broekhoven, H. Wijngaards, Proc. Ketjen Catalysts Symposium 1988 F8, p. 1.
63 J. Scherzer in " Catalytic Materials", ACS, Washington, 1984,p. 157.
64 L. Kubelkova, S. Beran, A. Malecka, V.M.Mastikhin, Zeolites 9 (1989), 12.
65 G.Garralon, A.Corma, V.Fornes, Zeolites 9 (1989), 84.
66 E.H.van Broekhoven, S. Daamen, R.G. Smeink, H. Wijngaards,J. Nieman in P.A. Jacobs, R.A.van Santen (Eds.), "Zeolites, Facts, Figures, Future", Elsevier, Amsterdam, 1989, Stud.Surf.Sci.Catal.49, p.1291.
67 R.M. Barrer, Zeolites 1, (1981), 130.
68 A.P.Matharu, L.F.Gladden, S.W.Carr in H.K.Beyer, H.G.Karge, I.Kiricsi, J.B.Nagy, (Eds.), "Catalysis by Microporous Materials", Elsevier, Amsterdam, 1995, Stud.Surf.Sci.Catal. 94, p.147.

69 A.Humphries, S.Yanik, L.A.Gerritsen, P.O'Connor, P.Desai, Hydrocarbon Processing. 1991, 69
70 R.Szostak."Molecular Sieves", Van Nostrand Reinhold, New York, 1989
71 P.Michiels, "Molecular Sieve Catalysts", EPO Applied Technology Series, vol. 9, Pergamon Press, Oxford, 1987.
72 W. Lutz, W. Gessner, D. Müller, Zeolites 19,(1997), 209.
73 H.van Bekkum, E.R.Geus, H.W.Kouwenhoven in J.C.Jansen, M.Stöcker, H.G.Karge, J.Weitkamp, (Eds.),"Advanced Zeolite Science and Applications", Amsterdam 1994, Elsevier, Stud. Surf. Sci.Catal. 85, p.509.
74 S.T.Sie in J.C.Jansen, M.Stöcker, H.G.Karge, J.Weitkamp, (Eds.), "Advanced Zeolite Science and Applications", Amsterdam 1994, Elsevier, Stud. Surf. Sci.Catal. 85, p. 587.
75 D.R.Corbin, S.Schwarz, G.C.Sonnichsen, Catal. Today 37, (1997), 71.

Studies in Surface Science and Catalysis 137
H. van Bekkum, E.M. Flanigen, P.A. Jacobs and J.C. Jansen (Editors)
© 2001 Elsevier Science B.V. All rights reserved.

Chapter 16

COKE FORMATION ON ZEOLITES *

H.G. KARGE

Fritz-Haber-Institut der Max-Planck-Gesellschaft, Faradayweg 4-6, 1000 Berlin 33 West (FRG)

ABSTRACT

A number of techniques and methods suitable for investigation of coke formation on acidic zeolite catalysts is presented and their application illustrated by pertinent results. Such tools for studying coked zeolite catalysts are: various spectroscopic techniques (IR, NMR, ESR, UV-VIS); gravimetry (TGA, also combined with GC and MS) for the determination of the amount of coke deposits and their H/C ratio; extraction of the coked zeolite and chemical analysis (GC, MS) of the extracts; adsorption measurements. Particular attention is paid to the problem of the nature of coke components and, in this context, some emphasis is laid on the distinction between different types of coke. Important phenomena related to coke lay-down such as the effect of acidity, shape selectivity, mechanism and kinetics and location of coke deposition are also discussed.

INTRODUCTION

Coke formation is the most frequent cause of catalyst deactivation in acid-catalyzed hydrocarbon reactions. Therefore, a great number of studies had been carried out already on coke formation on the classic cracking catalysts, i.e. amorphous silica/alumina which was employed before the advent of zeolite catalysts [1]. Also, coke formation was investigated on so-called bifunctional catalysts which exhibit, besides a hydrogenation function, i.e. usually a metal component, an acidic function as well. Examples are platinum, palladium, or nickel supported by acidic alumina or silica/alumina [2]. Similarly, coke formation on desulfurization and denitrogenation catalysts was studied [3].

It is interesting to note, however, that in the first decade of zeolite catalysis little attention was paid to the phenomenon of coke formation, with the exception of a few studies such as the pioneering work by Eberly et al. [4-5]. It was only in the late seventies that some papers appeared, dealing with coke formation on zeolites [6-9]. Since then we have to notice an ever increasing interest in this topic, on the aspects of both industrial application and fundamental research on zeolite catalysis.

The economical point of view of coke formation in industrial hydrocarbon processing cannot be overestimated: Even though the activity of a coked zeolite catalyst may be recovered by coke combustion, regeneration is frequently incomplete and, in any case, requires additional investment.

According to a current view of coke formation this is usually visualized as the build-up of very bulky, polyaromatic systems with even graphite-like structure [7, 10]. How-

*An updated version on this subject will appear in Vol.5 of Mol.Sieves Science and Technology, Eds.H.G.Karge and J.Weitkamp, to be published by Springer-Verlag, Berlin.

ever, this view might be too narrow in many cases of deactivation of zeolite catalysts caused by deposition of carbon-containing residues during hydrocarbon reactions. Frequently, such reactions are carried out at relatively low temperatures where the exclusive or preferential formation of polyaromatic residues is unlikely. Such reactions are, for instance, the alkylation of benzene by olefins [11], or the alkylation of isobutane by butenes [12-13], over acidic zeolite catalysts. Nevertheless, catalyst deactivation, sometimes even a very rapid one, is observed in such reactions as well. It must be ascribed, then, to carbonaceous deposits which are olefinic or to some extent even paraffinic in nature.

Therefore, a more general definition of coke, underlying tentatively the present study, is the following one: *"Coke consists of carbonaceous deposits which are deficient in hydrogen compared with the coke-forming reactant molecule(s)".*

However, not only the nature of those deposits called "coke" is a matter of debate and the subject of many investigations. There are numerous problems related to coke and coke formation on acidic zeolites: very important questions concern the effects of reaction temperature and pressure, time on stream, reactant composition, catalyst properties (such as structure, acidity and binder), location and kinetics of coke deposition. Even though a large body of data and results has been accumulated on coke formation during the last decade, many questions are still open and controversially discussed. Therefore, the aim of the present review is not, primarily, to report on general results of research in the field of coke formation. Rather, particular attention is paid to the methods and techniques which were developed and employed to tackle the above-mentioned aspects and problems related to coke on zeolite catalysts.

The great variety of techniques comprises, inter alia, gravimetric or thermogravimetric analysis(TGA); mass spectrometry (MS); coke combustion, combined with gas chromatography (GC), TGA or MS; spectroscopic methods like infrared (IR), electron spin resonance (ESR), nuclear magnetic resonance (NMR, e.g. ^{1}H NMR, ^{13}C NMR, ^{129}Xe NMR) with and without high resolution techniques (magic angle spinning or MAS, cross polarisation or CP, high field decoupling), ultra violet and visible spectroscopy (UV-VIS), photoelectron spectroscopy (PES). Besides spectroscopic methods, which usually do not require destruction of the coke/zeolite system in order to liberate coke-constituting species, also treatment of the coked catalysts with mineral acids, followed by extraction and chemical analysis, was employed. Adsorption measurements with various adsorbates on the fresh and coked samples were carried out especially to elucidate the locus of coke deposition, i.e. internal pore structure vs. external surface of the zeolite crystallites.

This enumeration of tools for coke investigation is certainly not exhaustive, and new methods are likely to appear in the near future. However, the above-mentioned methods and techniques seem to be the most important and frequently employed ones. Thus, they will be briefly presented in the following paragraphs and their application illustrated by suitable examples and significant results. In conclusion, an attempt will be made to summarize some aspects of coke formation which seem to be, at the present state of the art, of general validity.

METHODS OF COKE INVESTIGATION AND RESULTS

1. Determination of the H/C ratio

The H/C ratio of carbonaceous deposits is an important parameter in characterising the coke, even though the composition of coke may be complex and comprise constituents with different hydrogen content. The determination of the H/C ratio (via oxidation and analysis of the H_2O and CO_2 formed) is, in the case of zeolites, by no means trivial. The main source of error results from H_2O adsorption from the ambient. Surprisingly, this is scarcely discussed in the literature, and frequently H/C ratios of coke deposits on zeolite catalysts are reported without any indication of the method of determination.

To solve the principal problem one additional measurement is necessary. Thus, Entermann et al. [14] monitored the oxygen consumption in a recirculation apparatus during coke combustion.

Weitkamp et al. [15-16], developed another procedure which minimizes the experimental error due to adsorbed or structural water of the coked zeolite catalyst and, thus, provides reliable results. The principle of this method can be visualized from the scheme of Figure 1 (for experimental details see Refs. 15 and 16). The procedure starts

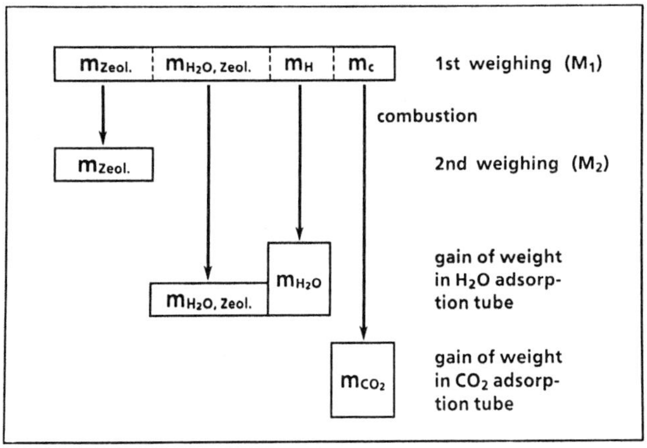

Fig. 1. Scheme for the determination of the H/C ratio in coked zeolite catalysts [15, 16]

with weighing of the coked zeolite containing the mass of the zeolite matrix, m_{zeol}, the mass of adsorbed and structural water, $m_{H_2O,zeol}$, and the masses of hydrogen, m_H, and carbon, m_C, of the coke. The data obtained by this first weighing are $M_1 = m_{zeol} + m_{H_2O,zeol} + m_C + m_H$. After combustion, the increase in mass of the adsorption tubes for H_2O (filled with magnesium perchlorate) and for CO_2 (filled with sodium asbestos) is determined, i.e. ($m_{H_2O,zeol} + m_{H_2O}$) and m_{CO_2}. Finally, the weight of the dry zeolitic material, generally converted to the amorphous state, is determined via "back weighing". This step is, in fact, the crucial one. However, if readsorption of water is carefully excluded this step provides $M_2 = m_{zeol}$. The evaluation of the H/C ratio follows from equ. (1) through (7), where equ. (7) contains only the measured data, viz. M_1, M_2, m_{H_2O} and m_{CO_2} (atomic weights rounded).

$$\Delta M = M_1 - M_2 = m_{H_2O,\ Zeol.} + m_c + m_H \tag{1}$$

$$m_{CO_2} = \frac{44}{12} m_c \tag{2}$$

$$m_C = \frac{12}{44} m_{CO_2} \tag{3}$$

$$m_{H_2O} = m_{H_2O,\ Zeol.} + 9\, m_H \tag{4}$$

$$m_{H_2O} - \Delta M = 8 m_H - m_c = 8\, m_H - \frac{12}{44} m_{CO_2} \tag{5}$$

$$m_H = \frac{1}{8}\left\{ m_{H_2O} - \Delta M + \frac{12}{44} m_{CO_2} \right\} \tag{6}$$

$$H/C = \frac{m_H}{1/12\, m_c} = \frac{3}{2}\, \frac{m_{H_2O} - \Delta M + \frac{12}{44} m_{CO_2}}{12/44 m_{CO_2}} \tag{7}$$

Weitkamp et al. checked the procedure with several test hydrocarbons and determined the H/C ratio of coke formed upon reaction of various hydrocarbons (n-heptane, 2,2,4-trimethylpentane, 1-hexene, methylcyclohexene, toluene) over LaNaY zeolite.

More recently, Karge et al. [17-18] combined TGA (using a temperature-programmed microbalance, Perkin Elmer, TGS 2) and GC. The scheme of the experimental set-up is depicted in Figure 2. Prior to combustion of the coke in a stream of oxygen (if necessary

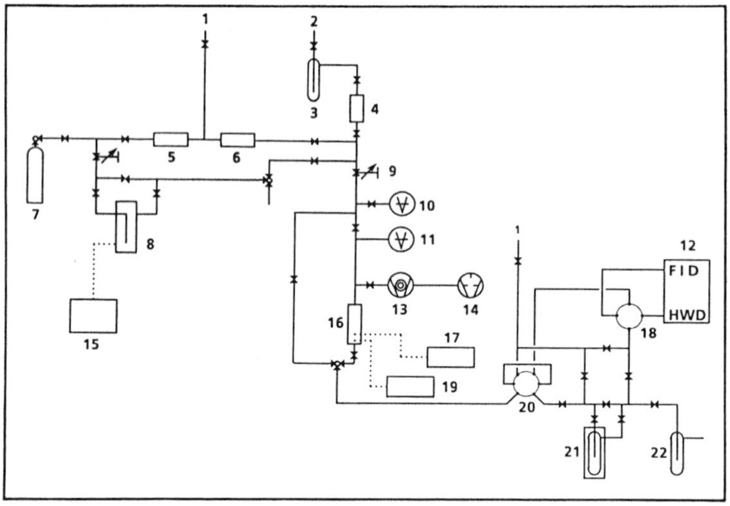

Fig. 2. Scheme of the experimental set-up for the determination of the H/C ratio in coked zeolite catalysts

1: He; 2: synthetic air, 3: molecular sieve trap; 4: gas drying, 5, 6: gas cleaning; 7: feed; 8: saturator; 9: fine adjustment valve, 10: baratron; 11: ionization gauge; 12: GC; 13: turbo molecular pump; 14: roughing pump; 15: thermostat; 16: microbalance; 17: temperature controller; 18: four-way valve, 19: microbalance controller; 20: gas sampling valve; 21: cooling trap; 22: bubbler.

in contact with CuO as a catalyst), adsorbed water and desorbable hydrocarbons are removed from the coked zeolite in a stream of dried N_2 at e.g. 400 K. The desorbed compounds are trapped at 100 K. When the desorption from the catalyst is completed, the trap is rapidly heated to 500 K and the trapped compounds purged into the GC for analysis. Similarly, the products of the subsequent coke combustion (H_2O,CO_2) are trapped and analyzed. Tests with measured amounts of adsorbed water and with suitable hydrocarbons (e.g. coronene, H/C = 0.5) confirmed the reliability of the TGA/GC technique. The weight loss measured during desorption or combustion is compared with the weight of the desorbed materials and combustion products evaluated from the GC analysis. Not only the H/C ratio of coked samples obtained from catalytic runs in separated reactors could be determined with the help of the apparatus and procedure described, but the coke formation could be carried out in the microbalance as well, when a feed stream was passed over the catalyst pretreated inside the balance. In this case, the amount of deposited carbonaceous materials was monitored via the weight gain. Subsequently, the H/C ratio was determined as outlined. In this way, the H/C ratio of coke deposits formed for instance through reaction of ethylene or methanol over H-ZSM-5 catalysts was measured as a function of reaction temperature and time on stream. Examples of results are presented in Figures 3 and 4.

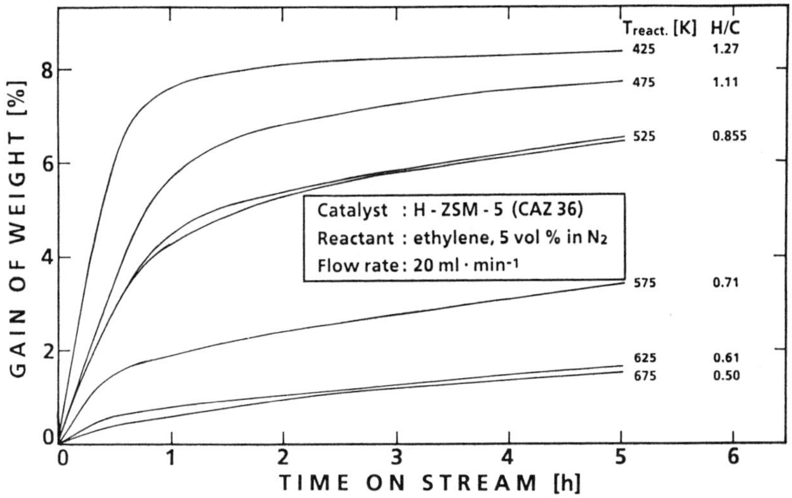

Fig. 3. TGA measurements of coke formation upon reaction of ethylene as a function of temperature.

Figure 3 demonstrates the weight gain of a H-ZSM-5 catalyst (lot label: CAZ36) due to coke deposition during the reaction of ethylene at reaction temperatures (T[react]) increasing from 425 to 675 K. Also, the H/C ratios after 6 h on stream are indicated. Two periods of coke formation are observed: a rapid initial deposition of carbonaceous materials (during the first 30 to 60 minutes) and a subsequent period of much slower coking. The amount of coke deposited after a given time (e.g. 5 h) decreases with increasing reaction temperature. However, the lower the reaction temperature the shorter the time on stream after which the weight gain levels off. The H/C ratio significantly decreases with increasing T[react].

The situation is even more complicated because the H/C ratio decreases not only with increasing reaction temperature as indicated in Figure 3 but also with time on stream.

This is shown for one particular reaction temperature in Figure 4. The "aging" of coke which is demonstrated by Figure 4 depends, in turn, on the reaction temperature and may affect the type of up-take curves, such as depicted in Figure 3, in a rather complex manner.

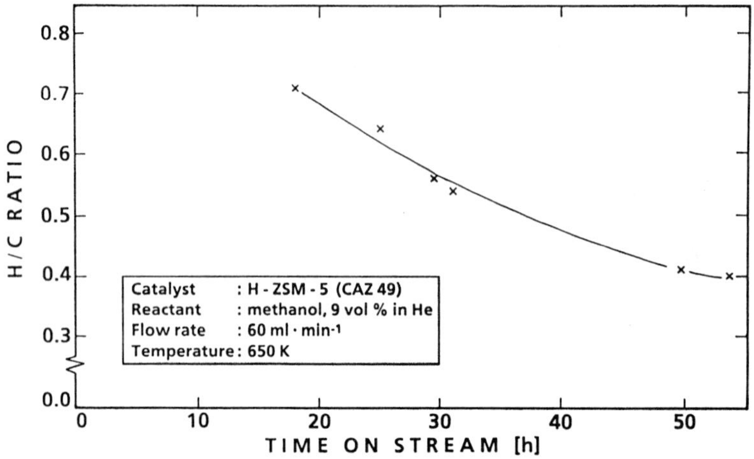

Fig. 4. H/C ratio of coke as a function of time on stream.

Table 1 summarizes results of coke formation obtained via TGA/GC for two different ZSM-5 catalysts (lot labels CAZ 36 and CAZ 49) on stream of (i) ethylene/nitrogen and (ii) methanol/nitrogen at various reaction temperatures. The main differences between the two catalysts were as follows: CAZ 36 possessed a higher number of acidic OH groups per gramme, lower acidity strength and larger crystallite sizes (Ø ≈ 4 µ) compared to CAZ 49 (Ø ≈ 0,4 µ) (see also Ref. 19).

Table 1 Coke formation on two H-ZSM-5 catalysts

Temperature	CAZ 36 ethylene		CAZ 36 methanol		CAZ 49 ethylene		CAZ 49 methanol	
[K]	wt%	H/C	wt%	H/C	wt%	H/C	wt%	H/C
425	8.3	1.27			7.2	1.81		
475	7.8	1.11			5.8	1.50		
525	6.6	0.86			4.8	1.38		
575	3.6	0.71	2.9	1.26	2.5	1.35	1.0	1.52
625	1.6	0.61	2.4	1.12	1.1	1.12	0.9	1.42
675	1.5	0.50	2.0	0.83	0.5	1.02	0.8	1.32

Feed: 5 vol% ethylene in N_2 or 9 vol% methanol in He; flow rate: 20 ml · min^{-1} in a thermobalance, TGS 2; time on stream: 6 h

Results similar to those reported here were arrived at by other authors: Magnoux et al. [20], Guisnet et al. [21] and Schulz et al. [22] who also observed a decrease of the H/C ratio of coke as a function of reaction temperature, *viz.* with coke formed upon propene transformation over ultrastable H-Y catalysts (USH-Y) [20] and upon methanol conversion over H-ZSM-5, RE-Y, or H-MOR [22]. The GC measurements of the H/C ratio by Magnoux et al. [20] refer to the so-called soluble coke. Also, aging of the carbonaceous deposits, i.e. declining H/C ratio with time on stream was found by Schulz et al. [22]. Finally, it was reported in several cases that the rate of coke deposition was particularly high during the initial period of the coke forming reaction. Thus, Neuber et al. [23] found rapid deactivation in a first stage of the reaction due to coke formed by rapid cracking of 1-methylnaphthalene over H-Y and ZSM-20 at 573 K. Similar results were reported by Solinas et al. [24] on coke deposition during methylnaphthalene isomerization over H-Y, whereas Schulz et al. [22] measured a high initial deactivation upon methanol conversion over RE-Y only for a high reaction temperature (753 K).

2. Infrared Spectroscopic Investigations of Coke Formation

IR spectroscopy was one of the first techniques employed to investigate the nature of carbonaceous deposits laid down on zeolite catalysts during hydrocarbon reactions. Eberly Jr. [5] observed upon reaction of n-hexene over H-Y the appearance of a prominent IR band in the CH deformation region around 1600 cm^{-1}. He ascribed this band to highly unsaturated carbonaceous residues, most likely polyolefins, as was suggested by comparison with spectra typical of such species which had been reported earlier [25-26].

Reaction of olefins over acidic mordenite catalysts (H-MOR, Be-MOR) was studied via IR by Karge et al. [27]. At low temperatures (about 350 K to 375 K) instantaneous polymerization upon contact of the olefin with the zeolite surface was observed. IR bands of saturated hydrocarbons (ν_{as} [CH_3] = 2960 cm^{-1}; ν_{as} [CH_2] = 2930 cm^{-1}; ν_s [CH_3] = 2860 cm^{-1}) predominated in the spectra (compare also Refs. 6 and 28) which were obtained in an IR cell for static measurements [29]. On successive heating to higher temperatures in high vacuum the coke band (around 1585 cm^{-1}) appeared. However, no CH bands at wavenumbers higher than 3000 cm^{-1}, which would be indicative of aromatics, were observed. Simultaneous MS measurements of the species desorbed into the high vacuum indicated fragments $[CH]_n$ with n = 2, 3 . . . 7 [6]. Conversion measurements conducted separately in a microflow reactor but under similar conditions revealed parallel formation of coke (polyolefins?), gaseous olefinic polymers and paraffins, indicating significant hydrogen transfer. Both IR and conversion experiments with catalysts pretreated at various temperatures in high vacuum suggested that complete removal of the acidic OH groups, i.e. dehydroxylation as evidenced by IR, resulted in a loss of any olefin conversion including coke formation [6]. The dehydroxylated catalysts exhibited high densities of so-called Lewis sites [30] but they regained the capability for coke formation only when rehydroxylation occurred either due to traces of water present in the feed or formed by oxidation of the hydrocarbons; oxidation may be caused by oxygen from the zeolite lattice and/or oxygen strongly adsorbed on the catalysts. Thus, the presence of Lewis sites, as indicated by IR after pyridine adsorption, provided no activity for generation of coke; rather, the presence of Brønsted sites was the indispensable requirement for this process to occur (see also Refs. [12, 31]. This is at variance with results reported by Weeks et al. [28] who claimed that polymerization of 1-butene took place on dehydroxylation sites of deammoniated H-Y.

IR spectroscopy has not only the advantage of enabling us to investigate coke formation without destruction of the zeolite matrix containing the residues (*vide infra,* section 7). It can also be successfully employed for studying coke formation *in situ*, i.e. observing the deposition of carbonaceous materials on the working catalyst. This requires the use of a suitable IR flow reactor cell which may be connected to a device (MS, GC) for analysis of the product stream. Several types of such cells are described in the literature, see for instance Refs. [32-35].

Eisenbach and Gallei [7] studied the formation of coke upon conversion of n-hexane and n-hexene over acidic faujasite-type catalysts. They found that the presence of olefins remarkably favoured coke formation, as indicated by the growth of the coke band. These authors, however, ascribed the coke band around 1600 cm^{-1} to graphite-like deposits.

Systematic studies of *in-situ* IR and conversion measurements in order to follow the formation of coke on working zeolite catalysts were conducted by Karge et al. [36-37] and Lange et al. [38]. IR cells, complementary equipment and procedure used by these authors are described in detail in Ref. 33. Some of the results obtained are reviewed in the following paragraphs.

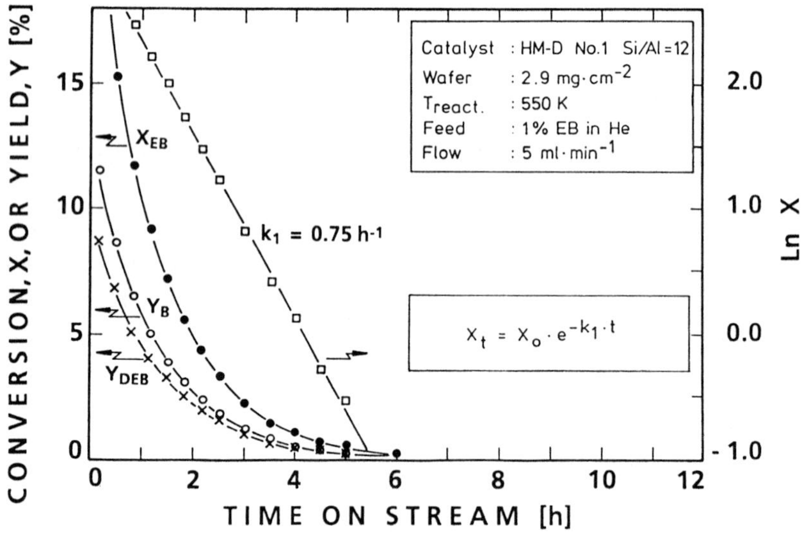

Fig. 5 Decrease in activity of an acidic mordenite upon coke deposition during conversion of ethylbenzene as a function of time on stream [36].

Figure 5 shows the decrease in activity of a slightly dealuminated hydrogen mordenite catalyst during conversion of ethylbenzene (EB) under conditions where, besides the main reaction, *viz.* disproportionation to benzene (B) and diethylbenzenes (DEB's), dealkylation occurs to a small extent. The olefin formed gives rise to coke formation, which results in deactivation of the catalyst. Deactivation, measured via the decrease in ethylbenzene conversion, X, with time on stream, t, follows a first order relationship, *viz.*

$X_t = X_o \cdot e^{-k_1 \cdot t}$, where X_o is the initial conversion and k_1 the rate constant; k_1 characterizes the different deactivation behaviour of various mordenites with systematically varied Si/Al ratios [36].

Spectra obtained from the working catalyst during ethylbenzene conversion are displayed in Figure 6. The most interesting feature of the set of spectra obtained in the region of stretching modes of acidic OH groups (active Brønsted sites) around 3600 cm^{-1} is the constancy of their integrated absorbance (dashed areas); the deviations observed are within the limit of experimental error. This means that the active sites were not poisoned or consumed even though the activity decreases (similar to the behaviour shown in Figure 5) and, consequently, the coke band around 1600 cm^{-1} (see middle part of Figure 6) ceased to grow.

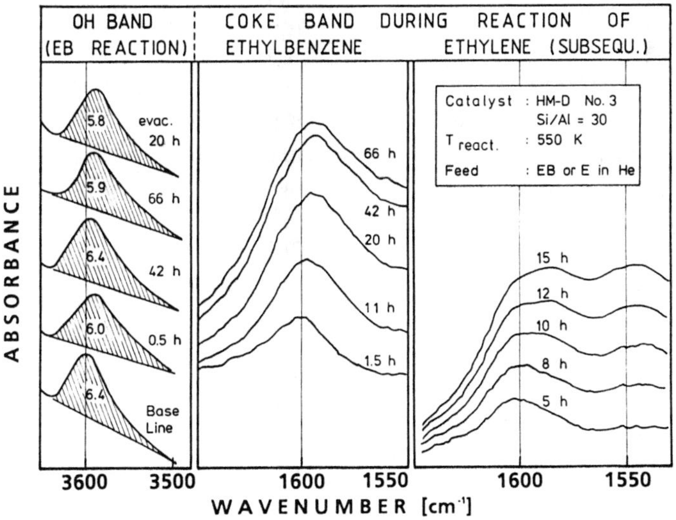

Fig. 6. In-situ IR spectra of a working hydrogen mordenite catalyst during conversion of ethylbenzene [36]

However, not only did the intensity of the band of acidic OH groups remain unaffected during the ethylbenzene reaction but these sites were also still active. This was confirmed when, after 66 h under a stream of ethylbenzene in helium, the feed stream was changed to ethylene in helium. Immediately, a conversion of ethylene was measured. Prior to scanning the spectra under ethylene/helium stream a new base line was established or, in other words, the last coke band spectrum of the ethylbenzene reaction (obtained after 66 h) was subtracted from all the following ones measured after the change of the feed. The result is demonstrated by the right hand part of Figure 6: the coke band started again to grow.

In conclusion, deactivation of the mordenite catalyst during the reaction of ethylbenzene was caused by deposition of coke blocking the active sites for the relatively large ethylbenzene molecules whereas the smaller ethylene molecules still had access to the active centres. Similar observations were made with H-ZSM-5 catalysts.

However, Bibby et al. [39] in their study of the formation of coke on and its removal from H-ZSM-5 MTG catalysts claimed that the acidic OH groups (band at 3610 cm^{-1})

were almost completely eliminated with increasing build-up of carbonaceous deposits up to about 5 wt%. Subsequently, a reinforced intensity of the 3610 cm^{-1} band was observed at ca. 7.5 wt% followed by another fall at 10 wt%. Simultaneously, the band of the non-acidic OH groups (3740 cm^{-1}) continuously declined. These features are difficult to understand and may originate, to a certain extent, from spectroscopic difficulties with the heavily coked samples.

In contrast, the OH groups of H-Y were significantly more affected by coke deposition upon olefin reaction [37]. Eisenbach and Gallei [6] reported analogous behaviour of Ca-Y during n-hexane conversion over Pt/Ca-Y and even claimed irreversible consumption of OH groups due to coke formation.

The intensity of the coke band around 1600 cm^{-1} proved to be a suitable measure for the amount of coke deposited [36, 40-42], even though the nature of the coke might be rather complex. In this context it is worth noting that the position of the so-called coke band around 1600 cm^{-1} in all *in-situ* IR experiments of coke formation shifted to lower wavenumbers when the time on stream was extended. This may indicate that the structure of the main components of the coke changed with time on stream, likely in the direction of more unsaturated compounds, e.g. aromatic species [5]. This corresponds to the aging of the coke (compare section 1).

A plot of the integrated absorbance of this band vs. the amount of coke deposited, measured for instance by a balance, provides usually straight lines [40-41]. Sometimes, however, a distinct change in the slope of such plots is observed indicating a marked alteration in the nature of the coke due to aging [*vide supra*, Figure 4 and Ref. 43].

Analogous *in-situ* IR studies as described above with ethylbenzene were carried out by using an ethylene/helium stream from the beginning of the experiment [37-38]. Results are shown in Figures 7 and 8.

Upon contact with the olefin at low temperatures (300 - 450 K) the intensity of the band of acidic OH groups at 3605 cm^{-1} decreased. Simultaneously a broad band around 3500 cm^{-1} appeared indicating a shift of a part of the 3605 cm^{-1} band to lower wavenumbers due to hydrogen bonding with adsorbate species (Figure 7). The overall absorbance was almost constant.

In the region of the CH stretching modes again the bands of paraffinic species developed (ν_{as} [CH_3]=2958 cm^{-1}; ν_{as} [CH_2]=2926 cm^{-1}; ν_s [CH_3]=2872 cm^{-1}; ν_s[CH_2]= 2856 cm^{-1}). The region between 1500 and 1300 cm^{-1} (Figure 8) exhibited the corresponding CH deformation bands, *viz.* δ_{as} [CH_3]=1485 cm^{-1}; δ_{as}[CH_2]=1468 cm^{-1}; δ_s[CH_3]= 1382 cm^{-1} and δ_s[CH_2]= 1359 cm^{-1}.

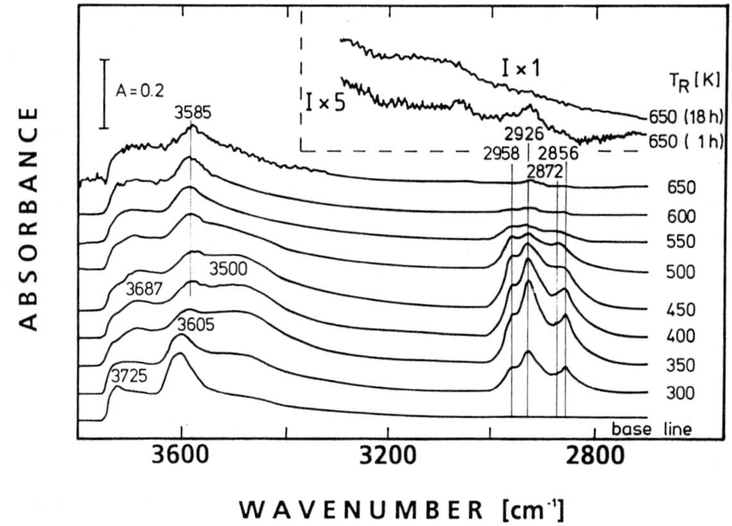

Fig. 7. IR spectra of carbonaceous deposits formed by reaction of ethylene over hydrogen mordenite [38].

When the reaction temperature was increased above 500 K, the situation significantly changed: the band of acidic OH's was essentially regenerated; the CH bands were generally weakened whereby the ratio CH_3/CH_2 seemed to increase, indicating substantial cracking and branching of the carbonaceous residues on the surface. Finally, at relatively high reaction temperatures and/or extended time on stream, a weak band around 3080 cm^{-1} (Figure 7, insert) and bands around 1600 cm^{-1} and 1540 cm^{-1} were observed

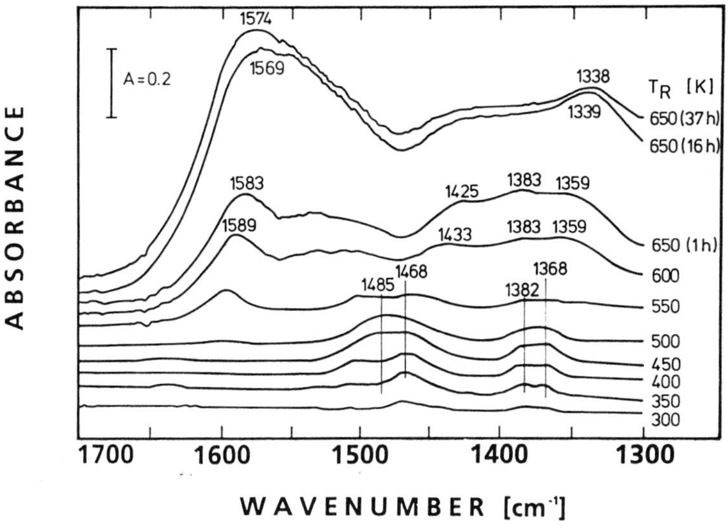

Fig. 8. IR difference spectra of carbonaceous deposits formed by reaction of ethylene over hydrogen mordenite [38].

(Figure 8). The first band was most likely indicative of CH stretching modes of aromatics while the 1600 cm^{-1} band ("coke" band) could be ascribed to polyalkenes and/or aromatic species. The band at 1540 cm^{-1} was also found by Fetting et al. [41] under similar conditions and had been assigned to alkylnaphthalenes. However, it may be also typical of polyphenylene structures [44]. After prolonged time on stream at high temperatures (650 K), both bands around 1600 and 1540 cm^{-1} collapsed into one broad band with its centre at about 1570 cm^{-1}. Similar experiments and, in general, coincident results were reported in Refs. [40, [45] and [46]. Interestingly, Haber et al. [46] reported formation of aromatic deposits upon reaction of olefins such as but-1-ene and dec-1-ene over H,Na-Y at relatively low temperatures, *viz.* 398 and 363 K, respectively.

The above-described *in-situ* IR results suggest discrimination between at least two types of coke, viz. a so-called coke I (low temperature coke, "soft coke", "white coke") constituted mainly of paraffinic and, to a lesser extent, olefinic or polyolefinic species (weak band around 1600 cm^{-1}) and a so-called coke II (high-temperature coke, "hard coke", "black coke") which consists of polyalkenes, but predominantly of alkylaromatics and polyaromatics (intense band below 1600 cm^{-1}, band above 3000 cm^{-1}).

3. Nuclear Magnetic Spin Resonance Spectroscopy

^{1}H-NMR spectroscopy has been successfully employed for chemical analysis of hydrocarbon extracts from coked zeolites (*vide infra*, section 7). Also, proton magnetic resonance spectroscopy provided valuable elucidation of problems related to the location of the carbonaceous deposits on coked zeolites (*vide infra*, section 13). The new technique of high resolution NMR, in particular MAS NMR of ^{13}C [47], turned out to be a powerful tool for the investigation of coke and coke precursors. This method was elegantly applied, for instance by Derouane et al. [48], van den Berg et al. [49], Neuber et al. [23] and Maixner et al. [50] when investigating coke formation upon reaction of alkenes over hydrogen mordenite or H-ZSM-5 and conversion of polynuclear aromatics. In the study of Neuber et al. [23] samples of H-Beta, H-ZSM-5 and H-Y were deactivated during conversion of 1-methylnaphthalene at 453 K. Their CP/MAS ^{13}C NMR spectra were obtained, in fact, after contact with the ambient atmosphere and showed prominent signals at 19 ppm and 127 ppm vs. TMS. These signals are typical of methyl groups attached to the aromatic ring and to carbon from the aromatic nuclei, respectively [51]. Increase of the temperature of reaction over H-Beta to 673 K resulted in a slight decrease of the CH_3 intensity relative to that of the 127 ppm signal due to a small extent of dealkylation. Thus, the coke structure formed upon 1-methylnaphthalene conversion over H-Beta, H-ZSM-5 and H-Y was interpreted by the authors to be partially demethylated aromatics with a limited number of aromatic rings.

^{13}C-MAS NMR spectra can be much improved if feed hydrocarbons for the coking reaction are used which are enriched in ^{13}C. In Figure 9 sets of spectra are displayed which were obtained from H-MOR catalysts activated at 700 K and subsequently coked in an atmosphere of 1 kPa of 1,2 - ^{13}C - C_2H_4 [38]. The series of experiments started with a sample contacted with ethylene at 300 K. Prior to each further heating step the

1,2 - ^{13}C-C_2H_4 atmosphere was renewed, since previous measurements [52] had proven that this procedure closely simulates on-stream conditions. However, immediately before carrying out the MAS NMR measurements the samples had to be contacted with the ambient atmosphere, *viz.* when they were placed into the rotor tube. For comparison, CP measurements were therefore conducted without MAS under exclusion of ambient atmosphere. Even though the quality of the spectra was lower than in the MAS case no substantial difference with respect to the main signals was encountered.

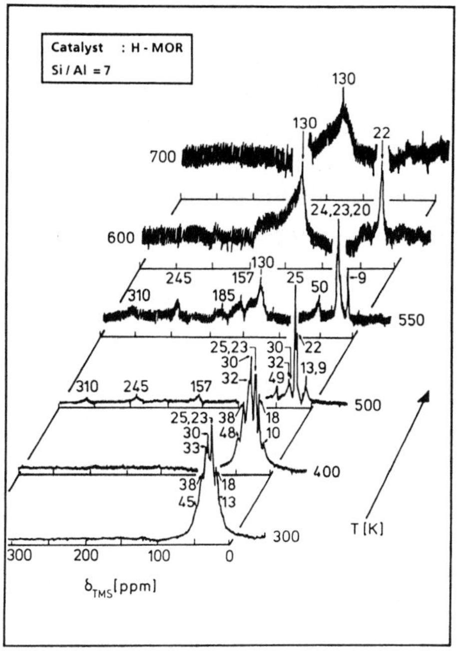

Fig. 9. ^{13}C-MAS NMR spectra of carbonaceous deposits formed by reaction of ethylene over hydrogen mordenite [38].

At low temperatures, i.e. up to 500 K, signals were observed which gave evidence for the formation of branched and possibly also linear alkanes. The carbons of paraffinic chains, CH_3 - CH_2 - $(CH_2)_n$ - , may contribute to the NMR lines observed at 13, 25, and 30 to 33 ppm (Figure 9). A large fraction of branched chains was indeed indicated by the signal around 40 ppm and an intensity ratio for the lines at 13 and 25 ppm lower than 1 : 1 [47]. Interestingly, a line was observed at 48 to 50 ppm which was possibly due to formation of alkoxide, i.e. species such as CH - CH_2 - O - zeolite. The oligomeric deposit remained nearly unchanged upon heating to 450 K. Upon heating up to 550 K, the carbonaceous species isomerized and subsequently cracked whereby smaller molecules were formed. This is derived from the change in the spectra between 0 and 40 ppm, where the lines around 18 and between 30 and 40 ppm significantly decreased, leaving lines at 20, 23, 24 and 50 ppm and, concomitantly, causing the appearance of lines at higher shifts. At high coking temperatures (700 K) the signal of highly unsaturated species (130 ppm) absolutely predominated.

Around 500 K, various carbocations developed which gave rise to well-defined NMR lines at 130, 157, 185, 245 and 310 ppm. The line at 130 ppm is ascribed to highly unsaturated species, most probably aromatics; that at 157 ppm together with the 245 ppm line is likely to indicate acyclic and cyclic carbocations. The line at 185 ppm, which appears after heating the deposit to 550 K, together with the line of "high-temperature" coke species at 130 ppm, could be indicative of protonated benzene, alkylbenzenes or polynuclear aromatics [47, 53-54]. The protonated carbon of the benzene ring would then be responsible for the signal around 50 ppm.

The existence of polyenic carbocations is not clearly supported by the present results, since this would require not only the appearance of the 130 ppm line but also the presence of a line in the range between 80 and 100 ppm [55]. The occurrence of uncharged polyenes, however, is not excluded (compare section 5).

Thus, the results of these ^{13}C MAS NMR investigations are in good agreement with the discrimination of two types of coke as derived from IR results (compare previous section as well as sections 4 and 5).

More recently, Anderson et al. [56] developed a very promising technique permitting ^{13}C MAS NMR studies of coke formation under conditions which are even closer to those of real catalysis. They kept a series of small tubes, containing the catalyst H-ZSM-5, under the feed stream (methanol) at reaction temperature (650 K) and followed the conversion of methanol to gasoline by on-line GC. At certain intervals the tubes with used catalyst samples were sealed off. The sealed tubes could be investigated without intermittent contact of the coked catalyst samples with the ambient atmosphere. Thus, the authors were able to follow the build-up of coke precursors and coke by ^{13}C MAS NMR as a function of time on stream and other reaction parameters. Up to about 30 hydrocarbon species held at the surface of the working catalyst were identified. Surprisingly, also an intense signal was observed which was ascribable to CO.

CP ^{13}C MAS NMR spectroscopy was also applied for the investigation of coke formation upon conversion of n-hexane (not enriched in ^{13}C) over H-ZSM-5 catalysts at 813 K. The results showed formation of aromatic ($\delta = 128$ ppm) and, to a lesser extent, most likely olefinic carbonaceous residues [57]. Finally, the same technique was used by Andersen et al. in their work on retained and desorbed products from reaction of 1-hexene over H-ZSM-5 [58].

4. Electron Spin Resonance Spectroscopy

It is a general experience that upon coke formation on zeolites free radicals appear. These can be investigated by ESR. The experiments can be carried out in X or Q band (8.2 to 12.4 or 33 to 50 GHz, respectively; Ref. [59]), using static EPR cells of fused silica [52] or even tiny flow reactors for *in-situ* measurements [60-61]. Leith [62] and Kucherov and Slinkin [63-66] were the first to publish interesting ESR spectra obtained after contact of activated acidic zeolites with olefins under static conditions.

The particular features of the ESR signals provide information about the type of coke formed. Thus, Leith [62] and Kucherov and Slinkin [63-64] observed a well-resolved 15-line ESR signal with a hyperfine splitting constant of a = 0.8 mT and a g-value of 2.0028 when they adsorbed propylene on hydrogen mordenite. The multiplicity of lines proved that the olefin must have oligomerised and, indeed, the observed spectrum could be ascribed to oligomeric olefinic or allylic species [62-64]. In similar experiments by Lange et al. [52] this signal was well reproduced. At higher reaction temperatures (350-450 K) under static conditions but with renewed ethylene atmosphere prior to each heating step, an ESR spectrum developed with seven hyperfine lines (a = 1.6 mT) and a g-value of 2.0025. This spectrum was almost identical to that obtained from γ-irradiated glassy allylchloride [67] and similar to the spectrum shown in Figure 11 for under-stream conditions at 430-470 K [61]. At even higher temperatures (above about 500 K), however, the hyperfine splitting of the spectrum vanished and the signals collapsed into a single line with a g-value close to that of free electrons, indicating species with a highly delocalized π-electron system.

When the numbers of spins per gram of coked catalyst were plotted as a function of reaction temperature with all the other conditions unchanged, curves resulted as shown in Figure 10. This figure demonstrates the clearly separated two regions of two types of coke, *viz.* the oligomeric olefinic/allylic deposits (below 500 K, multiple line spectra, a = 1.6 mT) vs. highly unsaturated hydrocarbons (above ca. 550 K, single line spectrum, g = 2.0023). The spin concentration of the latter coke species steeply increased with temperature.

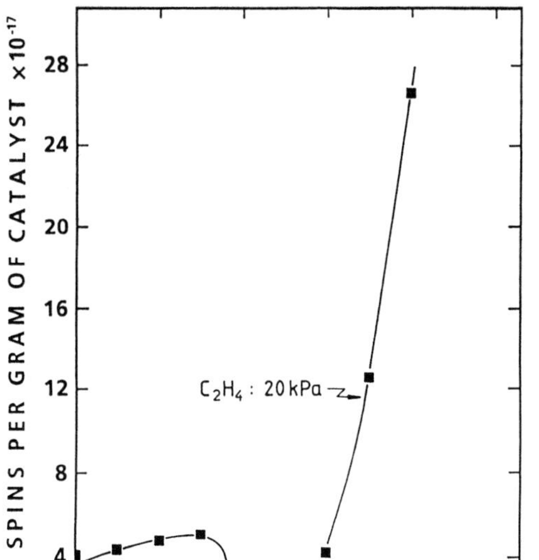

Fig. 10. Intensity of the EPR spectra (spin density) after reaction of ethylene over hydrogen mordenite as a function of temperature; ethylene atmosphere renewed after each heating step [52].

Similar results were obtained by in-situ measurements in an ESR flow reactor [61], as is illustrated by Figure 11, where a stream of 8 vol% ethylene in helium was passed at successively higher reaction temperatures over a catalyst of hydrogen mordenite which was activated at 700 K in a flow (15 ml · min^{-1}) of dry air (1 h) and nitrogen (1 h) prior to reaction.

The number of radicals typical of the low-temperature coke (multiple-line ESR spectra) decreased with decreasing Al content of the zeolite, i.e. decreasing number of acidic sites (Brønsted sites) which are required for formation of oligomeric species. This is demonstrated by Figure 12. In the series used, the density of Lewis sites, which might be responsible for the stabilization of these radicals, decreased in the same sequence as the density of the Brønsted sites. However, the effect of the zeolite acidity (type, number and strength of the acidic sites) of dealuminated mordenites used in the experiments of Ref. [52] on the formation of "high-temperature coke" (single line ESR spectra) was much more complex and requires further investigation.

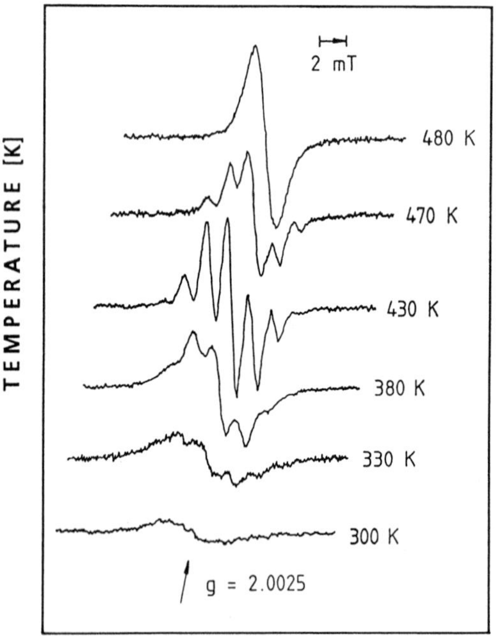

Fig. 11. In-situ EPR spectra of radicals formed during coke formation upon ethylene reaction over hydrogen mordenite [61].

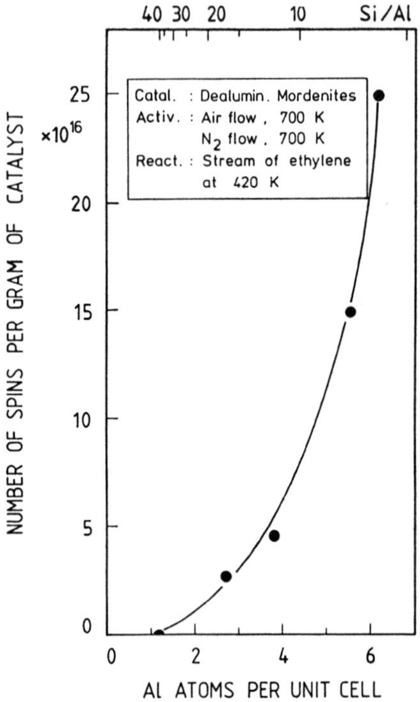

Fig. 12. Density of the radicals (spins) formed during ethylene reaction over hydrogen mordenites as a function of their Al content (acidity) [61].

Even though only a small fraction of the carbon atoms of the coke deposits corresponded to radicals (about 0.1 to 1%), the ESR results can be regarded as representative for the formation of carbonaceous residues on the zeolite catalysts. This is suggested by the results presented in Figure 13. A close (linear) correlation was found between the concentration of radicals of high-temperature coke (determined via ESR) and the total amount of coke formed on hydrogen mordenite at 573 K (determined via TGA). Interestingly, the maximum deposition of coke was 12 wt-%, irrelevant of what sample of the series of dealuminated mordenites (compare Figure12) was employed for the ethylene reaction. This seems to suggest that under the conditions of this reaction the maximum deposit of coke is related to the primary pore structure (*vide infra*, section 13). This differs from the results obtained when the coke-forming reaction was the dealkylation of the more bulky ethylbenzene (compare Ref. 36). In this case deactivation was caused by site blockage; the ultimate loading with coke depended on the dealkylation activity and decreased with decreasing Al content (number of active Brønsted sites).

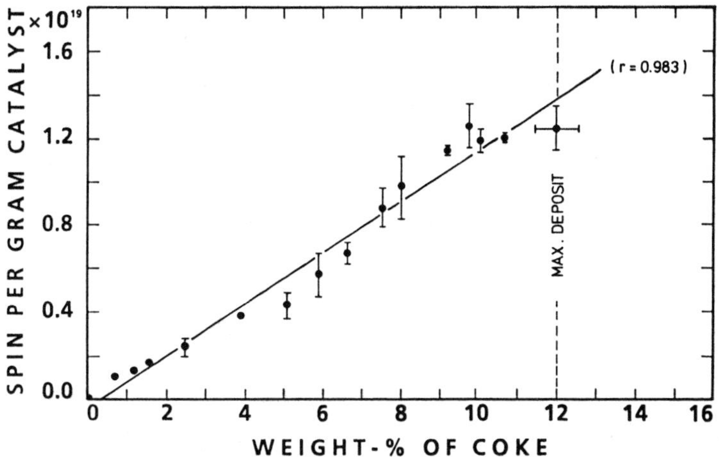

Fig. 13. Correlation between the number of radicals and the amount of high-temperature coke deposited during the reaction of ethylene over hydrogen mordenite at 573 K [61].

5. Ultraviolet-visible Spectroscopy

Even though zeolite samples usually exhibit low transparency and, moreover, the transmittance is strongly reduced upon coke deposition, coking of zeolites can also be studied by UV-VIS spectroscopy in transmittance mode. This would even enable us to carry out spectroscopic investigations with the very same samples in other spectroscopic ranges, e.g. in IR, ESR etc. The techniques are described in detail elsewhere [68-69]. However, measurements in reflection mode are also possible [70]. This is particularly noteworthy, since it may happen that the spectroscopic transmittance of the samples is indeed too low.

UV-VIS spectroscopic studies related to coke formation were carried out by only a few authors [19, 70-72]. It turned out that UV-VIS spectroscopy is rather sensitive and capable of providing valuable information about intermediates, coke precursors and final coke structures.

As an example, a set of difference spectra (i.e. with the background spectrum of the zeolite matrix having been subtracted) is depicted in Figure 14. These spectra were obtained under static conditions with a H-ZSM-5 sample (total Si/Al ratio 33.5). The transparent wafer (12.6 mg · cm^{-2}) was pretreated in high vacuum (10^{-5} Pa, 10 h) and subsequently loaded at room temperature with methanol (0.5 mmol per g zeolite).

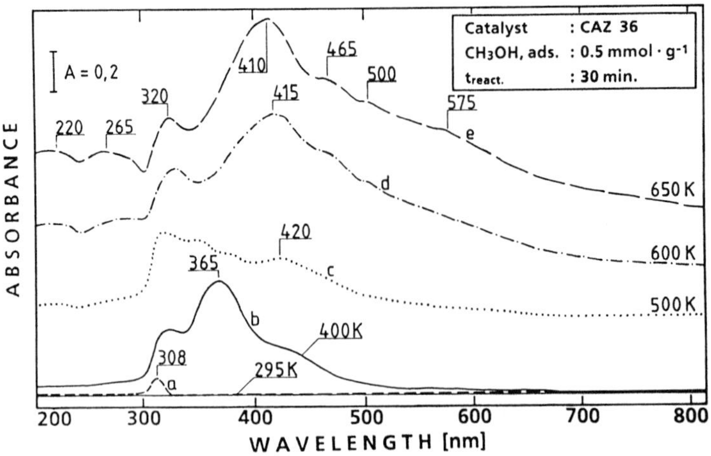

Fig. 14. UV-visible difference spectra of a methanol-loaded H-ZSM-5 catalyst (CAZ 36) after heating at successively higher temperatures [19].

Immediately after contact with methanol a small band at 308 nm appeared (difference spectrum a); absorbance in this range (around 310 nm) is usually observed after interaction of olefins with acidic adsorbents and originates from π-allylic carbocations [73-75]. Subsequent stepwise heating to 400, 500, 600 and 650 K of the adsorbate/catalyst system in a closed UV-VIS cell led to the spectra b, c, d and e, respectively, which exhibit a number of bands typical of intermediates and carbonaceous residues.

At low temperature, e.g. 400 K, the most prominent band is that around 365 nm. UV bands in the range 360-385 nm are ascribable to polyenylic cations such as $[CH_2 \cdots C(Me) \cdots CH \cdots CH \cdots CH_2]^+$, see Refs. [71-73, 76]. The broad band around 320 nm may still contain some contributions of π-allylic cations but mainly arises from cyclohexenyl cations (315 nm), since such cyclization products of olefin interaction on acidic catalysts are indeed indicated by UV bands in the range of 275-330 nm [70, 75]. The shoulder at 420 nm is most probably due to more bulky species such as diphenylcarbenium ions [77] or polyalkylaromatics and condensed aromatic compounds [70, 73] which, to a small extent, may even form at low temperatures.

Signals of such species, however, become predominant at higher reaction temperatures at the expense of the polyenic species which are weakened and, finally, vanish upon the transformation of "low-temperature coke" into "high-temperature coke". Pertinent assignments of the bands which appear at higher reaction temperatures are as follows: a pair of bands (320 and, more intense, 410-420 nm) is indicative of bulky aromatic species (diphenyl or polyphenyl carbenium ions, polyalkylaromatics, condensed aromatic ring systems, see Refs. 70, 77). Evidence of formation of such species at high temperatures (600-700 K) is also provided by bands at 500 and 575 nm [70, 77]. Bands observed at 220, 265 and 465 nm may be ascribed to dienes, cyclohexadiene and/or benzene and cations of substituted benzenes, respectively.

Similar experiments with olefins (ethylene, propylene, butene, butadiene, hexatriene etc.) resulted in spectral features which were, at least to a large extent, essentially the same as described above [78]. However, the fact that in UV-VIS studies polyenylic cations were observed is at some variance with the ^{13}C MAS NMR results (see section 2.3), but this may indicate that the UV-VIS technique is very sensitive and only a small fraction of the polyalkenes need to be present as carbocations in order to be identified in the 360-385 nm region.

In agreement with the preceding considerations of the IR, NMR and ESR results, the UV-VIS investigations illustrate again that two types of coke should be distinguished (coke I and coke II, see section 4); at low temperatures, olefinic/polyolefinic constituents predominate whereas at higher temperatures the main components of the carbonaceous deposits are bulky aromatic and polyaromatic species.

However, it could be shown that the preferential occurrence of one type of coke or the other depends not only on the reaction temperature but on properties of the catalysts as well. Thus, for instance, small crystallites of H-ZSM-5 (average diameter 0.4 μm) with fewer but more strongly acidic sites than those used in the experiments of Figure 14 favoured the formation of coke I even at relatively high reaction temperature [19]. This is illustrated by Figure 15 comparing the spectra of two H-ZSM-5 catalysts which differ in their properties as outlined above.

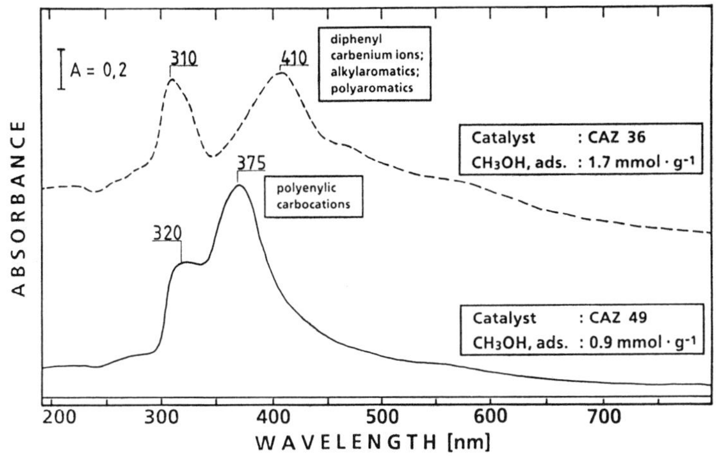

Fig. 15. UV-visible difference spectra of two different H-ZSM-5 catalysts (CAZ 36, CAZ 49) loaded with methanol and heated at 650 K.

Differences in behaviour with respect to formation of carbonaceous deposits, as demonstrated by the two examples of Figure 15, may have significant consequences in the catalytic performance with respect to the desired reaction (e.g. methanol-to-gasoline conversion) as well as for the catalyst regeneration [19, 78].

6. Photoelectron Spectroscopy

Only recently, photoelectron spectroscopy, i.e. X-ray photoelectron spectroscopy (XPS) and Auger electron spectroscopy (AES), was applied in studying coke formation [79, 80]. Determination of the C/Si ratio as a function of coke loading was used to distinguish between deposition in the interior pore system and on the external surface of the zeolite crystallites (Ref. [79], see section 13). Fleisch et al. [80] claimed that their investigations by core electron energy loss spectroscopy (CEELS) enabled them to distinguish clearly between coke compounds with sp^3 and sp^2 hybridisation. They reported good agreement of the results obtained by this technique and CP MAS NMR measurements.

The interpretation of XPS spectra of coked zeolite catalysts, however, is difficult. Assignments of signals might be erroneous because of severe conductivity problems; the disturbance caused by charging effects depends on the degree and topology of the surface coverage with coke. Nevertheless, photoelectron spectroscopy is a promising and attractive method for coking studies, all the more as modern spectrometer devices permit essentially *in-situ* measurements, i.e. investigation of the coked surface without any intermittent contact with ambient atmosphere [81].

Auger Electron Spectroscopy (AES) enables us to avoid disturbances of the above-mentioned type because of the high electron energy used. Figure 16 shows an example of a carbon Auger electron spectrum [82]. It was obtained from an H-ZSM-5 (CAZ 49)

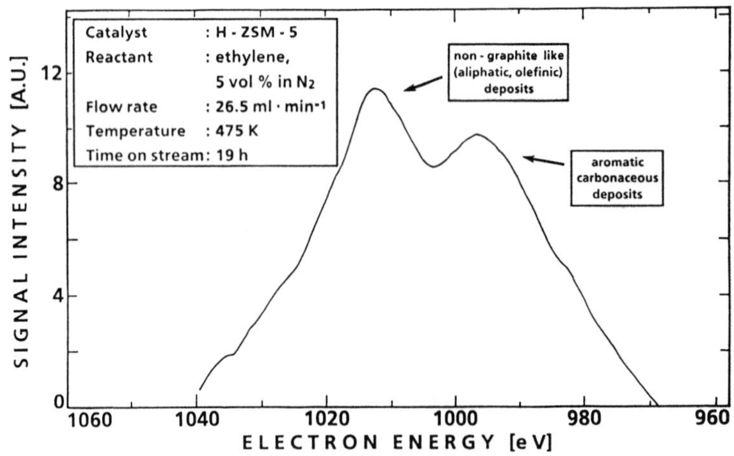

Fig. 16. Carbon Auger spectrum of an H-ZSM-5 catalyst (CAZ 49) after coking in an ethylene stream at 475 K.

sample which was covered with coke after 19 h exposure to a stream of 5 vol% ethylene in N_2 (total pressure 100 kPa) at 475 K. The spectrum exhibits two broad peaks, *viz.* at 995 eV and around 1010 eV. The former peak is ascribable to highly aromatic carbonaceous deposits ("carbon black" type, see Ref. [83]). These species are probably very bulky and, therefore, their presence restricted to the external surface of the crystallites. The high energy peak around 1010 eV is indicative of deposits richer in hydrogen and aliphatic or olefinic in character. However, a distinction between sp^3 and sp^2 hybridisation is not possible. Although quantitative estimations about the relative abundance of the two types of coke constituents cannot be derived from these results, the appearance of both sufficiently resolved signals in the spectrum of Figure 16 is in qualitative agreement with the spectroscopic results described in sections 2 through 5.

A sample, which was coked in a feed stream of 12 vol% methanol in He at 650 K for 19 h, provided AES results similar to those shown in Figure 16 [82].

Spectra obtained by XPS from the same H-ZSM-5 samples (coked with ethylene or methanol) still exhibited the Si and Al signals at 103.2 eV and 75.0 eV, respectively. However, the peak areas of the coked samples were reduced by about 85 % compared to the corresponding peak areas of the fresh catalysts. Since the depth of observation of the XPS method is restricted to a few monolayers, a coverage of the external surface of the zeolite grains with a thick homogeneous layer of coke would result in a complete suppression of the Si and Al signals. Therefore, one has to conclude that the external zeolite surface was not completely covered by a homogeneous coke layer; rather, a topologically heterogeneous coverage is suggested.

This has been confirmed by more recent work using a significantly improved technique which allows even for "in-situ" experiments [81]. Results obtained so far refer to coking of an H-ZSM-5 catalyst (activated in high-vacuum at 675 K prior to reaction) under the stream of diluted ethylene (5 vol% in He) at 650 K. In view of well-established XPS data for carbon-containing species [84] the results may be interpreted as follows:

(i) coke deposition in the form of a thin (1 to 2 monolayers), complete and conductive film of carbonaceous materials starts to occur already in the very first stage of the experiment (0.4 to 0.8 wt% coke, obtained by coking with ethylene);

(ii) upon longer exposure to the feed stream thicker carbon-containing islands grow on top of the first layer, exhibiting electrical charging effects and rendering the coke deposition topologically heterogeneous;

(iii) in agreement with earlier results (IR, NMR, ESR, UV-VIS, AES), two types of carbonaceous deposits can be distinguished, *viz.* paraffinic or olefinic, hydrogen-rich ones (corresponding to coke of type I) and residues similar to "carbon black" but still non-graphite species (corresponding to coke of type II).

Photoelectron spectroscopy will certainly allow for further valuable and very detailed *in-situ* investigations of the nature of coke and the mechanism of its formation.

7. Extraction of coked zeolite catalysts

Venuto et al. [85] were the first to dissolve coked zeolite catalysts and analyse the extracts of the solution by GC. Since they used RE-X catalysts, a relatively mild acid treatment was sufficient to destroy the zeolite matrix. Coke forming reactions were, *inter alia,* alkylation of benzene by olefin (at 485 K) and olefin polymerisation. After destruction of the coked zeolites (H/C ratios of coke about 1.3) the acidic suspensions were extracted with $CHCl_3$. The black tar extracted from the RE-X catalysts (deactivated during the alkylation of benzene by olefin) was a complex mixture of high-molecular-weight aromatic compounds (highly alkylated benzenes, polyalkylnaphthalenes and higher condensed polycycles). Interestingly, essentially the same species were found trapped in the pore system of the zeolite catalyst when ethylene alone was passed over RE-X at 485 K. This suggests that, at least to a great extent, the reactions of ethylene and intermediates formed therefrom are responsible for deactivation of the RE-X catalyst employed for the alkylation of benzene. Becker et al. [11] arrived at the same conclusion when they studied the deactivation of hydrogen mordenite during alkylation of benzene with ethylene or propylene. However, Walsh and Rollmann [86] reported on an interesting effect of the feed composition on coke formation. They used radiotracer methods and found that the coke was mainly produced from the aromatics when a mixture of paraffins and aromatics was passed over Y- and mordenite-type catalysts at 635 K.

According to Venuto et al. [31, 85], formation of polynuclear aromatics inside the zeolite pore structure is due to a complex series of side reactions of ethylene and/or polyalkylbenzenes, including polymerisation, hydrogen transfer and dehydrocyclisations.

The method of acid treatment and extraction of coked zeolite catalysts was further developed and extensively used by Guisnet and co-workers in a series of systematic studies on coke formation [20-21, 87-92].

The techniques used by these authors have been described in Refs. [20, 87]. In a first step the coke components, trapped in the channels and cavities of a zeolite pore system, are liberated by dissolution of the aluminosilicate matrix in hydrofluoric acid (40%) at room temperature. Subsequently the soluble components are extracted by CH_2Cl_2 as a solvent; in some cases "insoluble coke" remains in the form of black particles (*vide infra*). The extracts are analysed by GC, HPLC, ^{1}H-NMR and MS.

A serious problem related to the extraction method, however, is whether or not changes in the original chemical nature of the coke occur upon the rigorous treatment with mineral acids. Guisnet and co-workers [20, 87] have carried out a series of tests which seem to confirm that the procedure employed does not generate any artifacts. They loaded inert carriers (SiO_2) with, for instance, phenanthrene or 1-tetradecene and submitted these systems to the same treatment as was applied to coked catalysts. No chemical modification of the hydrocarbon test materials was observed. However, it is still not certain that this also holds for highly reactive species (e.g., polyenic compounds etc.) when in contact with the active surface of the catalyst. Concerted experiments, using spectroscopic tools and the extraction method, are being carried out to clarify this point [93].

Guisnet's group studied coke formation upon propene conversion over ultrastabilised H-Y, (US H-Y), H-MOR and H-ZSM-5 [20], and on n-heptane cracking over H-Y, hydrogen mordenite (H-MOR), H-ZSM-5, hydrogen offretite (H-OFF) and hydrogen erionite (H-ERI) [21, 88-92].

As an illustration, we describe some results obtained by Magnoux and Guisnet when investigating the nature of coke deposits and the mode of deactivation on n-heptane cracking over H-ERI [90]. The coke loading was varied through the variation of the time on stream.

The authors found that at low coke loadings the coke was "non-polyaromatic"; the aromaticity increased with severity of coke deposition. Evidence for this general trend was provided, *inter alia*, by ^{1}H-NMR (Figure 17). Particularly with high coke loadings

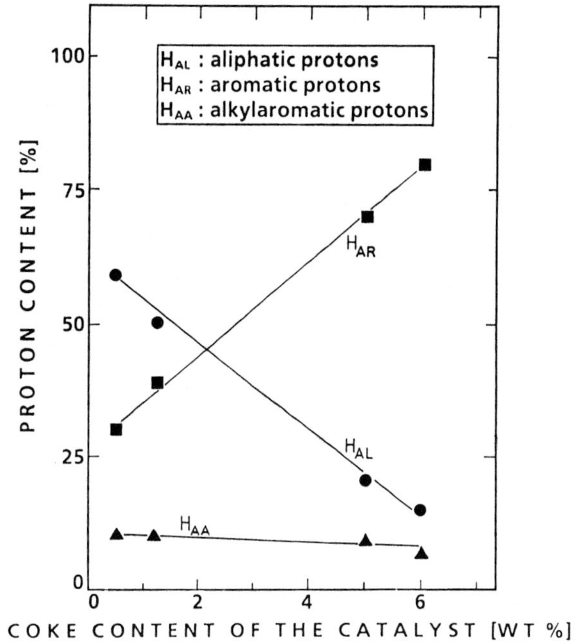

Fig. 17. ^{1}H NMR analysis of the soluble fraction of coke (formed upon n-heptane reaction over H-ERI); details see text [90]

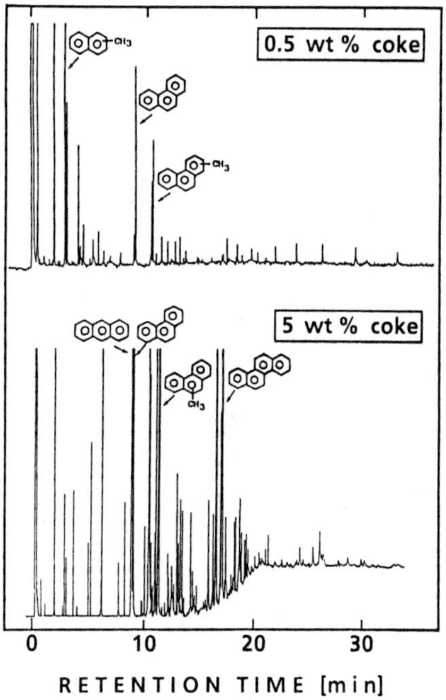

Fig. 18. Gas chromatographic analysis of extracts of coke formed upon n-heptane reaction over H-ERI [90]; details see text.

(5 to 6 wt%), polyaromatic compounds in the form of small black particles (H/C < 0.5) were detected. The "soluble coke", however, was a rather complex mixture. Depending on the coke loading, i.e. time on stream (aging), a number of main families of compounds constituting the carbonaceous deposits were deduced from the GC and MS results. The upper part of Figure 18 shows the main compounds forming the carbonaceous residues after a short time on stream or low coke loading (0.5 wt%). Some of the main families of coke constituents determined at higher loadings are indicated in the gas chromatogram in the lower part of Figure 18. At even higher coke content (6 wt%), accumulated during n-heptane cracking over H-ERI, the composition of the soluble coke was less complex; it was mainly comprised of anthracene, phenanthrene and chrysene. Similar results were obtained for coke formed over H-ZSM-5 and H-MOR.

The method of extraction was also applied by Andersen et al. [58] in their study on retained and desorbed products which occur upon reaction of 1-hexene over H-ZSM-5 catalysts. Their attempt to derive a detailed scheme of the hydrocarbon chemistry, involved in coke formation during this reaction, is based to a large extent on the results obtained by the extraction technique. Evidence was provided of a key role of cyclopentadienes as important intermediates in the route to coke formation.

8. Adsorption measurements

Comparison between the adsorption capacity of fresh and coked zeolite catalyst samples is frequently used to characterize the deposition of carbonaceous materials in the void volume of the zeolite structure. More specifically, adsorption measurements are used to clarify whether the coke is predominantly laid down in the zeolite pores or on the outer surface.

There is a relatively large number of studies of this type [see, e.g., Refs. 20, 39, 89-90, 92-97]. Adsorbate species, which are very often used in this context, are nitrogen, water, ammonia, trimethylamine and small paraffin molecules. However, also more bulky molecules such as pyridine, methyldiisopropylamine, substituted benzenes, methylene blue, which had no access to the particular internal volume, were used in order to probe the external surface of the zeolite crystallites [96-97].

As an example, measurements by van Hooff et al. [94] were used to compare different zeolite structures (H-Y, H-MOR, H-ZSM-5) with respect to pore filling. Cracking of n-hexane at 573 K was employed for coke formation. The void volumes of the fresh and coked catalysts were probed by n-butane adsorption. Results are presented in Table 2. From these data the authors concluded that in the case of H-Y and H-MOR coke deposition proceeded preferentially inside the pore structure whereas this occurred only to a minor extent in the channels of H-ZSM-5.

Similarly, Karge et al. [98] determined the BET analogous surface area of a series of dealuminated mordenite samples prior and subsequent to coking by ethylene (m[cat] : 0.015 g activated at 675 K in flowing N_2 (100 ml · min^{-1}); feed: 5 vol% ethylene in nitrogen; v : 16 ml·min^{-1}; T[react] : 650 K; time on stream (TOS) : 100 h). The measurements were carried out with an "Omnisorb 360" apparatus [99] or with a "Quantasorb" (Quanta Chrome Comp., N.Y.). The uncoked samples were degassed at 675 K, the coked samples at 625 K prior to the measurements of the surface area. The results showed that 75-90% of the BET analogous internal surface was lost upon a coke loading of about 12 wt%.

Table 2. Integral conversion, coke deposition and decrease of pore-volume after 300 min reaction at 573 K for the 3 different zeolite catalysts [94]

Catalyst	Integral Conversion gC / gCat.	Coke Deposition wt%	Coke Selectivity %	Pore Volume (PV) ml/g		
				initial	final	ΔPV
H-ZSM-5	1.82	1.93	1.1	0.145	0.134	0.011
H-Mord.	0.34	4.75	14	0.099	0.022	0.077
H-Y	0.22	2.59	12	0.141	0.084	0.057

Bibby et al. [96] used basic probe molecules (ammonia, pyridine) for monitoring the change of the adsorption capacity of H-ZSM-5 catalysts caused by coke deposition. From their results they simultaneously derived conclusions concerning the available acidic centres before and after coke loading (see also section 11) and the locus of coke deposition (section 13).

Hydrocarbon molecules with different effective molecular diameters (n-hexane, 3-methylpentane) were employed by Guisnet and co-workers [92] in order to determine not only the change in adsorption capacity as a function of coke deposition but also to elegantly discriminate between the various localizations of the deposited coke components in, for instance, the different cavities of hydrogen offretite (see section 13).

9. Temperature-programmed oxidation

Temperature-programmed oxidation should be a suitable tool to discriminate between different types of coke deposited on a catalyst. It has been, *inter alia*, successfully applied to coke investigation on bifunctional catalysts. Although bifunctional zeolite catalysts are usually employed under hydrogenating conditions which remove, to a large extent, polymerisable, unsaturated coke-forming species, coke formation is frequently encountered with this type of catalyst as well. One example was already mentioned (see section 2) as studied *via* in-situ IR by Eisenbach and Gallei [7].

Barbier [2], in fact, investigated coke formation on reforming catalysts which used Al_2O_3 as a support (Pt/Al_2O_3). He studied the coke deposits formed upon conversion of cyclopentane/cyclopentene mixtures and was able to discriminate, *via* temperature-programmed combustion of the deposits, between coke laid down on the metal and coke located on the support. Analogous investigations should be possible with coked bifunctional catalysts containing acidic zeolites as a support.

PARTICULAR PHENOMENA RELATED TO COKE FORMATION

10. Selectivity of coke formation

The idea that coke formation in zeolites is a shape-selective reaction was first advanced by Rollmann [8] and Rollmann and Walsh [9]. In their studies they systematically investigated the correlation between structure properties, sorptive behaviour, catalytic selectivities and coking (aging) tendencies of a series of acidic zeolites (erionite, ferrierite, TMA-offretite, mordenite, zeolite L, zeolite Y and H-ZSM-4). It was shown that those catalysts (ferrierite, erionite) which exhibited a high (reactant) selectivity in the conversion of a five-component feed (2,3-DMB, 3-MP, NC6, benzene, toluene) were also characterised by low coke yields and low aging rates. By contrast, the large pore zeolites (H-MOR, H-L, H-Y, H-ZSM-4) did not selectively convert the feed mixture and, correspondingly, suffered from severe coking and aging (see Table 3). TMA-offretite represented a unique case in the group of zeolites investigated. It contains both 12- and 8-ring channel systems, both accessible by n-hexane. However, only the 12-ring channels are capable of accommodating the larger cyclohexane or benzene molecules. This intermediate position of TMA-offretite is reflected in its selectivity and coking behaviour.

Table 3. Coke Selectivities[a] [8]

Catalyst	Selectivity	
	Coke[b]	Normalized[c]
Erionite	0.14	0.16
Ferrierite	0.03	0.04
TMA offretite	0.7	0.8
Mordenite	0.3	0.4
Zeolite L	0.4	0.5
ZSM-4	0.4	0.5
Zeolite Y	2.2	2.9

[a] 5-Component feed; 3.3 days on stream; all catalysts treated under a standard cycle with WHSV = 3 h^{-1}; T = 589-811 K; p = 1.4 · 10^6 Pa; H_2/HC = 3; [b] grams/100 g conversion, observed. [c] grams/100 g conversion, normalized to 50% conversion using first-order approximation.

Already Rollmann [8] suggested in his work that the formation of more bulky structures such as cycloparaffins ("coke precursors") is sterically hindered by the restrictive pore systems of reactant or product shape-selective catalysts, e.g. ferrierite and erionite. In a subsequent paper Rollmann and Walsh [9] confirmed their view of shape selectivity in coke formation (Figure 19) and also included the investigation of the new prominent ZSM-5 catalyst with its remarkably low propensity to form coke. Furthermore, the concept of geometrical restrictions for the formation of spaceous transition states of coke building species ("transition state selectivity") was discussed [10].

However, a modified hypothesis for the low-coking tendency of H-ZSM-5 in comparison with H-Y and H-MOR was advanced by Schulz et al. [100]. These authors assumed that only the first stage of formation of carbonaceous deposits is governed by shape-selective properties of the surrounding of the acidic sites inside the pore structure: on H-ZSM-5 this, indeed, results in low initial deposition of carbonaceous residues. The further growth of coke, however, does not proceed, according to Schulz et al. [100], via a carbenium ion mechanism but via direct methylation of the carbonaceous deposits (by the feed, e.g. olefin or methanol) followed by elimination of methane. For stoichiometric reasons at least one carbon atom is added to the coke when one CH_4 molecule is evolved. Thus, evolution of CH_4 is seen as a suitable indicator for growth of coke (compare also Ref. [101]).

In view of this concept one would expect more or less pronounced induction periods for coke formation on shape--selective catalysts such as H-ZSM-5. But this is not the case, at least not always, and the mechanism and kinetics of coke formation seem to depend on various factors (see, e.g. Figure 8 in Ref. 19). Therefore, more work is needed, employing

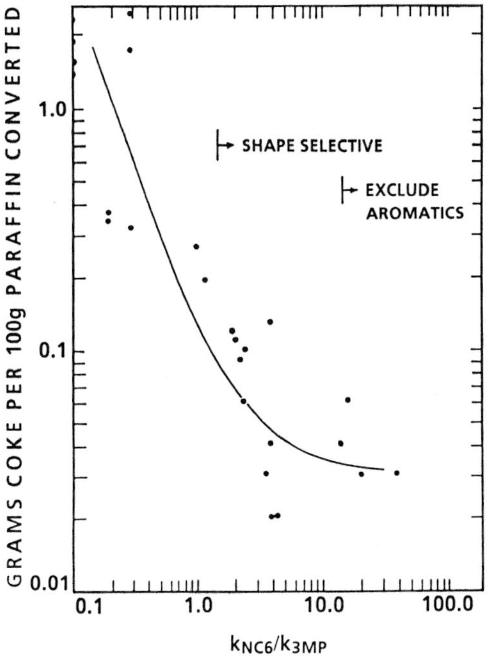

Fig. 19. Coke yield vs. shape selectivity of paraffin conversion over acid zeolite catalysts [8]; the shape selectivity is measured through the ratio of the first order rate constants, k_{NC6}/k_{3MP}, for disappearance of the respective paraffin isomers (n-hexane, 3-methylpentane) at 700 K.

various techniques, to check the interesting hypothesis by Schulz et al. In any case it is obvious that the effect of shape selectivity with respect to coke formation has to be considered primarily in the case of the so-called type II coke (see section 5) and does not play a major role, if any, with type I coke (low-temperature or white coke).

11. Effect of acidity

The role of acidity (nature of acidic sites, i.e. Brønsted vs. Lewis centres, number and strength of sites) was already mentioned several times in preceding sections. However, the relationship between coke formation and acidity is far from evident, and it seems too early to make an attempt of proposing general correlations between both phenomena.

As mentioned in section 4 (Figure 12), the yield of low-temperature coke continuously decreased with the Al content of dealuminated mordenites, whereas the effect on high-temperature coke, although existent and measurable, depended in a rather complex manner on the Si/Al ratio [61]. The coking behaviour of a series of mordenites with systematically varied (framework) Si/Al ratio during hydroisomerization of n-hexane was studied by Haas et al. [102]. These authors found that synthetic mordenites steamed at 725 K or strongly dealuminated by leaching with hydrochloric acid (Si/Al up to 39) exhibited improved activity, stability and regeneration properties. Also, Kubelková et al. [45] did observe an effect of acidity (strong Brønsted sites and electron accepting centres, i.e. cationic Al species) on coke formation when ethylene was reacted over ZSM-5 or modified Y-type catalysts. Temperature-programmed desorption of ammonia from fresh and coked catalysts seemed to indicate that the most acidic sites were preferentially poisoned by coke.

In contrast, Rollmann [8] did not observe any influence of the modulus SiO_2/Al_2O_3 on the coking behaviour of two pairs of differently dealuminated hydrogen mordenite and H-Y catalysts. Similarly, Schulz et al. [100] claimed that the acidic sites are required only for the initial slow stage of formation of coke precursors in the methanol-to-gasoline reaction at 650 and 750 K and that the acid sites are not involved in further coke build-up.

On the other hand, Bibby et al. [103] reported that in fact the overall deposition of coke during methanol conversion over H-ZSM-5 was higher the lower the Si/Al ratio, whereas the initial coke formation was not related to the aluminium content. Furthermore, it was deduced from ammonia and pyridine adsorption measurements on H-ZSM-5 samples that during the initial period of coke deposition (up to about 4 to 5 wt% of coke during methanol reaction at 650 K) a rapid loss of acid Brønsted sites occurred. This was followed by a much slower decrease in the number of acidic OH groups. However, after a total coke deposition of about 15 wt% the number of acidic sites, indicated by the bases, dropped to zero and the activity of the catalysts had completely vanished. This is at some variance with the observations by Karge et al. [36] who found via *in-situ* IR and conversion measurements that the number of acidic OH groups in H-MOR and H-ZSM-5 remained essentially unaffected in spite of severe coking and loss of activity. This was explained by deactivation through blocking rather than consumption or poisoning of sites.

Although there seems to be noticeable disagreement about the role of the number (and sometimes also the strength) of Brønsted sites with respect to coke formation over acidic zeolites, most authors agree that the presence of acidic OH groups is the indispensible requirement for coking to occur (at least in the initial stage of the process). Only a few studies seem to suggest a predominant effect of Lewis sites on the propensity of the zeolite to form coke [28]. In general, however, these investigations have not excluded that Brønsted centres were present on the catalyst surface; they may form, for instance, through a (limited) rehydroxylation in which Lewis sites are involved [104].

12. Mechanism and Kinetics

The current view of the pathways of coke formation is presented in the scheme of Figure 20. Many intermediates and precursors of coke, suggested by this scheme, were identified (see sections 2-7) as well as constituents of the "white" coke or type-I coke, generally formed at lower temperatures (sections 2-6). It has been shown that similar species occur during deposition of carbonaceous materials, regardless of whether the feed contains paraffins, olefins or methanol. However, the actual network of the reactions depends most probably on the specific situation (catalyst, feed, temperature etc.) and has not yet been experimentally established and unambiguously evaluated in a particular case.

The kinetics of coke deposition is frequently described by the, in fact, empirical Voorhies equation [105]

$$A = k_c \cdot t^{n_c}\ ,\quad n_c \geq o \tag{8}$$

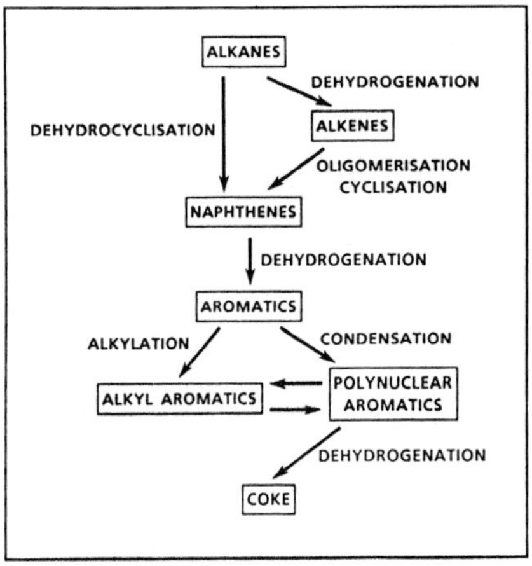

Fig. 20. Current view about coke formation on zeolites.

where A = amount of coke deposited [wt%], k_c = constant [wt% · min$^{-n_c}$], t = time on stream [min], n_c = parameter. Voorhies obtained a value of n_c close to 0.5 and, therefore, suggested that coke deposition is related to diffusion, the coke itself being the diffusion barrier. However, subsequent studies of a number of workers (compare Ref. [1]) have shown that values of n_c occurred which were significantly higher than 0.5. Moreover, it turned out that both k_c and n_c were affected by the nature of the feed as well as by reaction conditions, e.g. space velocity. More elaborate relations for deactivation and coke deposition are, for instance, provided in Refs. [106-108].

On the basis of the *in-situ* IR and conversion measurements upon ethylbenzene reaction over dealuminated mordenites (presented in section 2) it was found that both aging (activity decrease) and coke deposition could be approximated by a relationship of first order in time [36, 61] (Figure 21).

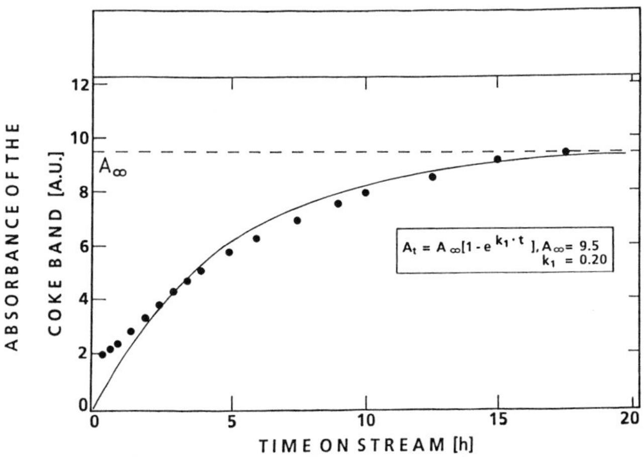

Fig. 21. Coke formation (measured via the absorbance of the coke band) upon ethylbenzene reaction over hydrogen mordenite (Si/Al=12) as a function of time on stream; first order relationship.

The results allowed elimination of the variable t and presentation of the loss of activity as a function of coke deposited (Figure 22). The slopes of the straight lines in Figure 22 provide a measure for the sensitivity of the catalyst to coke formation: the steeper the straight line the lower the amount of coke required to effect a given decrease in conversion.

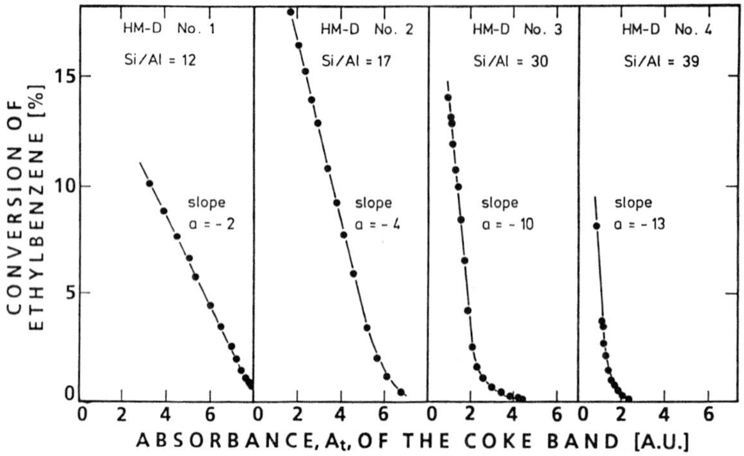

Fig. 22. Sensitivity of hydrogen mordenites (with various Si/Al ratios) to coke formation upon conversion of ethylbenzene (see text) [36].

Another approach, relating the coke yield not simply to the time on stream but taking into account the different times of exposure of the catalyst, in order to obtain a distinct state of deactivation, was advanced by Schulz et al. [22,100]. These authors defined a normalized time on stream (NTOS) which is the real time on stream divided by the time elapsed until unconverted feed (methanol) was observed due to deactivtion of the catalyst. This facilitated the comparison of various catalysts with respect to their catalytic performance.

13. Localization of Coke Deposition

An interesting and important problem related to coking zeolite catalysts is the localization of the coke deposits. Usually the question arises whether the coke is preferentially formed inside the zeolite pore system or on the external crystallite surface. In some cases, even more detailed information is required when various types of cavities of a zeolite structure are possible loci of coke deposition.

In Figures 23 and 24 the coke deposition upon the reaction of a relatively small coke-forming molecule (ethylene) is compared for three catalysts, *viz.* H-Y, H-MOR and H-ZSM-5. The measurements both via the absorbance of the IR band around 1600 cm^{-1} (coke band, see section 2) and via the weight gain (TGA) seem to suggest that the final

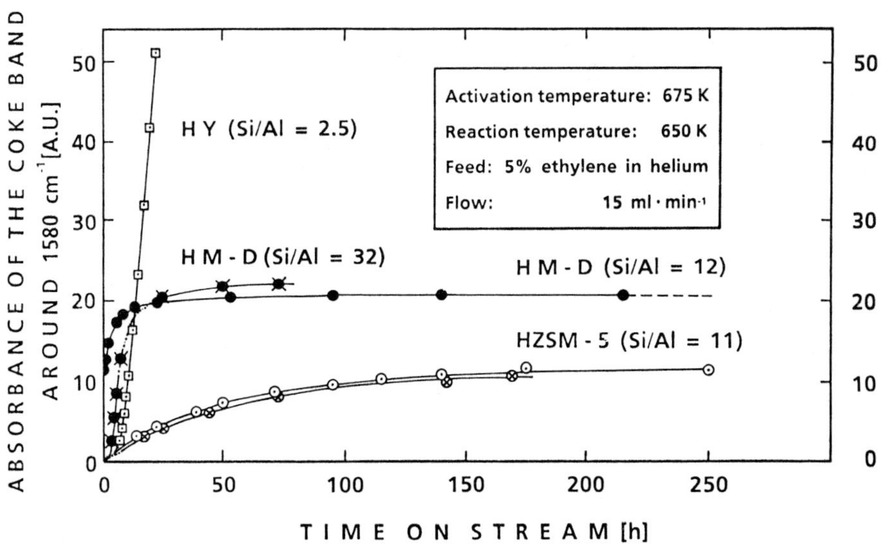

Fig. 23. Coke deposition (measured via IR) during ethylene reaction over three types of zeolites (H-Y, H-MOR, H-ZSM-5) with different void volumes [37].

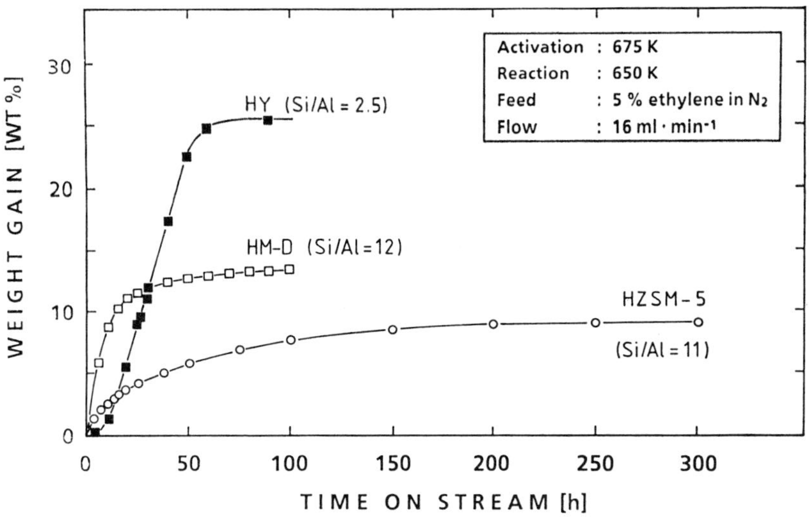

Fig. 24. Coke deposition (measured via TGA) during ethylene reaction over three types of zeolites (H-Y, H-MOR, H-ZSM-5) with different void volumes [37].

coke levels in Figures 23 and 24 (26 wt%, 12 wt% and 8 wt%) reflect the sequence of the void volumes of H-Y, H-MOR and H-ZSM-5: 0.38 (supercages only), 0.20 or 0.13 (total or main channel volume), and 0.16 $cm^3 \cdot g^{-1}$, respectively. The final coke level of 8 wt% in the case of H-ZSM-5 (see Figure 24) is in remarkably good agreement with the value found by Sexton et al. [79] for internal coke formation, i.e. channel filling (*vide infra*). The IR results (Figure 23) show indeed the same sequence (and a good reproducibility) of the results as the TGA measurements; in fact, for H-Y the final coke level was not observed due to the very low transmittance of a heavily coked H-Y sample. If one assumes an average density of 1.5 $g \cdot cm^{-3}$ of the coke, the above figures for mordenite correspond to a 50% filling of the total void volume [109] (0.20 $cm^3 \cdot g^{-1}$) or a 75% filling of the main channel system (0.13 $cm^3 \cdot g^{-1}$). Thus, it is likely that, under the reaction conditions applied, a significant fraction of the amount of the final coke loading is accommodated inside the pore system.

On the basis of his data, Rollmann [8] arrived at the same conclusion. The situation is similar to that observed with H-Y and H-ZSM-5. Moreover, the decrease of the internal adsorption capacity measured for water and nitrogen (see section 8) after coke formation also suggests that at least a large fraction of the coke was deposited in the internal pore system of H-MOR and H-ZSM-5.

Similarly, Bibby et al. [39] derived from their ammonia and pyridine adsorption studies on fresh and coked H-ZSM-5 samples that at least the initial period of coke formation occurred in the interior of the zeolite crystallites where the acidic active sites are located.

In their XPS work on coking of H-ZSM-5 under a stream of methanol in nitrogen at 643 K, Sexton et al. [79] also derived from the C/Si ratios that up to 8 wt% coke the filling of the pores predominated. The C/Si ratio increased linearly with the amount of coke deposited. Formation of coke on the external surface occurred at higher loadings and was indicated by an exponential increase of the C/Si ratio (Figure 25). The results were in reasonable agreement with the theory advanced by these authors.

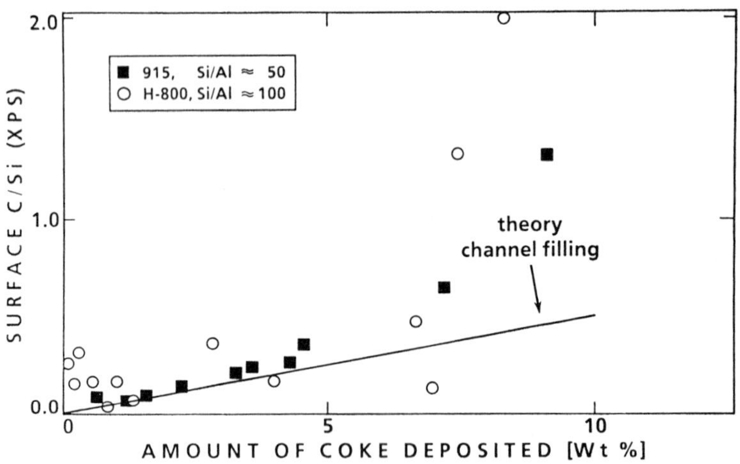

Fig. 25. C/Si ratio of coke deposited from a stream of methanol in nitrogen at 645 K on H-ZSM-5 catalysts [79].

Already Rollmann [8] pointed out that even in the structure of narrow- or medium-sized channel systems such as TMA-erionite (or ZSM-5) bulky coke species may form if the intersections of the channels provide sufficient space to accommodate these residues or, if necessary, the transition states through which they form.

However, in many cases H-ZSM-5 catalyst samples bear more than 20 wt% coke after complete deactivation in, for instance, a stream of methanol. In such events it is obvious that not all of the coke could be deposited in the internal pore system. Rather, a great part of the coke must be located on the external surface and fill the intercrystalline volume.

As an example, reference is made to the work of Bibby et al. [103]. For a first period of coking these authors observed indeed that the sorption capacity or the void volume of their H-ZSM-5 catalysts decreased linearly with the amount of coke deposited during methanol reaction at 635 K. They concluded that internal coke formation occurred intimately associated with the active sites. Internal deposition was confirmed by the X-ray diffraction (XRD) patterns which showed similar changes of the lattice parameters as were observed upon occlusion of large organic molecules or bulky ions. However, it was also observed that the maximum amount of coke deposited (up to 26.7%) exceeded the measured decrease of the void volume, indicating formation of external coke as well. In a subsequent study, Bibby et al. [96] confirmed their results via adsorption experiments with trimethylamine (TMA) and ethyldiisopropylamine (EDA). TMA had access to the pore volume of H-ZSM-5 and probed the total number of acidic sites whereas EDA provided information only about the external surface. Again, the authors arrrived at the conclusion that the reduction of the sorption capacity was caused by simple pore filling.

Andersen et al. [58] showed that, after deacidification of the external surface of H ZSM-5 by interaction with $SiCl_4$, the fraction of higher aromatics decreased when 1-hexene was reacted over this catalyst. Hence, in the case of the non-modified H-ZSM-5 those bulky species (naphthalenes, phenanthrenes, anthracenes) seem to form preferentially outside the pore system. According to Andersen et al., coke constituents initially formed in the interior of a (non-modified) H-ZSM-5 crystal might, at higher temperatures (T[react] ≥ 600 K) migrate to the external zeolite surface where, at acidic sites, conversion to more bulky, involatile residues is possible.

Applying transmission electron microscopy (TEM) in combination with electron energy loss spectroscopy (EELS), Gallezot et al. [110] were able to locate external coke deposits formed upon n-hexane cracking over ultra-stabilized H-Y, H-OFF and H-ZSM-5. Their techniques provided also information about the structure of the coke constituents, *viz.* coronene- or pentacene-like species.

An elegant technique to reliably discriminate between coke deposited in the intracrystalline and the intercrystalline volume was developed by Bülow et al. [111]. They carried out ^1H-NMR measurements on H-ZSM-5 samples coked by n-hexane cracking at 775 K or mesitylene reaction at 800 K. By means of NMR pulsed field gradient techniques the intracrystalline self-diffusion coefficient in solid samples, D_i, was measured for methane and propane. NMR tracer desorption experiments provided data about the fraction γ of molecules which are able to leave the crystallite during the observation time; γ is related to the effective diffusion coefficient D_d. While D_i is affected only by carbonaceous deposits inside the pore system and not by external coke layers, D_d is sensitive to both types of coke localization.

From the evaluation of D_i and D_d Bülow et al. [111-112] were able to conclude that (i) upon n-hexane cracking two stages of coke deposition occur: during the initial period coke is laid down in the intracrystalline channel network, during the final periods coke is mainly formed near the external surface; (ii) coke formed from mesitylene is, even at the beginning of the reaction, predominantly deposited close to the external crystallite surface.

Finally, another method should be mentioned which appears to be a suitable tool for determination of the coke distribution, *viz.* the ^{129}Xe NMR technique [113]. This technique rendered possible, for instance, measurements of the free volume of a H-Y zeolite accessible for Xe atoms before and past coking upon reaction of n-hexane or propylene. Also, it allowed for the discrimination between internally and externally deposited residues.

Frequently, the sizes of channels and cavities of a particular zeolite architecture are not uniform. As an example offretite was already mentioned (section 10). This structure contains two types of pores, *viz.* linear cylindrical channels (12-membered rings, diameter 0.63 nm) and gmelinite cages (8-membered rings, diameter 0.45-0.50 nm) which interconnect the channels. Guisnet and co-workers [92] studied the localization of coke deposits formed upon n-heptane cracking on hydrogen offretite. Combining their determination of the coke composition (conducted analogously to the example of section 7) and adsorption measurements using four differently sized adsorbates (i.e. nitrogen, n-hexane, NC6, and 3-methylpentane, 3MP) these authors were able to localize the various coke species related to their period of formation. (i) In the initial stage, i.e. at low coking, small compounds are formed which are trapped in the gmelinite cages. This reduces the total adsorption capacity, e.g. for N_2, but does not hinder NC6 or 3MP from entering the pores. (ii) At higher coke loadings (0.8 through 4.5 wt%), the coke comprises polyaromatic molecules overflowing into the large channels. They limit or block access to the large channels, in particular for 3MP. (iii) For even greater coke content (> 4.5 wt%) very bulky polyaromatic compounds form in the pores close to the external surface preventing also N6C from penetrating the large channels and the gmelinite cages.

Systematic and detailed adsorption experiments such as described by Guisnet and co-workers [90, 92], employing adsorbates with suitably selected sizes, shapes and adsorptive properties, promise to provide better information as to where the deposition of carbonaceous residues occurs when coke is formed on a working zeolite catalyst. Most likely, such measurements can be advantageously combined with spectroscopic investigations. *In-situ* studies, i.e. adsorption intermittent with periods of reaction but without contact to the ambient, would be desirable.

SUMMARY AND CONCLUDING REMARKS

It has been shown that there exists a variety of techniques which are suitable for investigating coke formation during hydrocarbon reactions over acidic zeolite catalysts and related phenomena. In many cases they were indeed successfully applied. It appears from the large body of results accumulated in this field so far that some general conclusions may be derived. For instance, the following features seem to be well established.

(i) At least two classes of coke may be distinguished by several criteria, *viz.* coke I (white or soft coke, preferentially formed at low reaction temperatures) and coke II (black or hard coke, mainly occurring at higher temperatures, e.g. at T[react.] > 500 K), even though the borderline between the two types of coke may not always be very sharp.

(ii) The so-called coke II, to which much more attention usually has been paid, is constituted mainly of polyaromatics, frequently of condensed ring systems.

(iii) Not only higher reaction temperatures but also extended time on stream ("aging of coke") favour the formation of more bulky, polyaromatic coke components with a low H/C ratio.

(iv) Coke formation is a shape-selective process and is significantly influenced by the architecture of the particular zeolite catalyst.

However, there are other aspects of coke formation, whose investigation seems to lead to conflicting or, at least, ambiguous results. Examples are the role of acidity, i.e. nature, number and strength of acidic sites on coke formation, and the locus of coke deposition, in particular with narrow- or medium-pore-size zeolites. However, the effect of acidity, the location of coke deposition, kinetics and mechanism of coke formation etc. most probably depend on a number of specific conditions of the process under consideration. Therefore, even more systematic and detailed studies may be required to obtain a consistent and satisfying description of these phenomena.

NOTE (added after preparation of the manuscript)

Very recently, a review article by Guisnet et al. [114] appeared in which the authors, after a brief overview of the pertinent literature, present an excellent and concise report of their studies on coking and deactivation of zeolites and the effect of zeolitic pore structures. It is based, inter alia, on Refs. [20-21] and Refs. [87-92]. This article may be very helpful for a reader entering the field of coke formation on zeolite catalysts.

ACKNOWLEDGEMENTS

The author thanks those colleagues and co-workers who contributed to this review. Particularly, he thanks Professor J. Weitkamp (University of Stuttgart) for helpful discussion and criticism. Financial support by the Bundesminister für Forschung und Technologie (BMFT), Projects 03C 111 1 and 03C 231 1, is gratefully acknowledged.

REFERENCES

[1] E.E. Wolf and F. Alfani, Catal. Rev.-Sci. Eng. 24 (1982) 329-371.

[2] J. Barbier, in: "Catalyst Deactivation 1987", Proc. 4th Int. Symp., Antwerp, Belgium, Sept. 29-Okt. 1, 1987 (B. Delmon and G.F. Froment, Eds.), Elsevier, Amsterdam 1987; Studies Surf. Sci. Catalysis 34 (1987) pp.1-19.

[3] A. Arteaga, J.L.G. Fierro, F. Delannay and B. Delmon, Appl. Catal. 26 (1986) 227-249.

[4] P.E. Eberly, Jr., C.N. Kimberlin, Jr., W.H. Miller and H.V. Drushell, Ind. Eng. Chem., Process Des. Dev. 5 (1966) 193-198.

[5] P.E. Eberly, Jr., J. Phys. Chem. 71 (1967) 1717-1722.

[6] H.G. Karge and J. Ladebeck, Proc. Symp. Zeolites, Szeged, Hungary, Sept. 11-14, 1978 (P. Fejes, Ed.); Acta Universitatis Szegediensis, Acta Physica et Chemica, Nova Series 24 (1978) pp. 161-167.

[7] D. Eisenbach and E. Gallei, J. Catal. 56 (1974) 377-389.

[8] L.D. Rollmann, J. Catal. 47 (1977) 113-121.

[9] L.D. Rollmann and D.E. Walsh, J. Catal. 56 (1979) 139-140.

[10] S. Csicsery, in: "Shape Selective Catalysis, Route to Chemical Fuels", Proc. of a Symp., ACS Meeting, Seattle, USA, March 20-25, 1983, pp. 116-141.

[11] K.A. Becker, H.G. Karge and D. Streubel, J.Catal. 28 (1973) 403-413.

[12] J. Weitkamp, Proc. 5th Int. Conference on Zeolites, Naples, Italy, June 2-6, 1980 (L.C.V. Rees, Ed.) Heyden, London, 1980, pp. 858-865.

[13] J. Weitkamp, in: "Catalysis by Zeolites", Proc. Int. Symp., Ecully (Lyon), France, Sept. 9-11, 1980 (B. Imelik et al., Eds.) Elsevier, Amsterdam, 1980; Studies Surf. Sci. Catalysis 5 (1980) 65-75.

[14] W. Entermann and H.C.E. van Leuven, Anal. Chem. 44 (1972) 589-590.

[15] J. Weitkamp, in: "Innovation in Zeolite Materials Science", Proc. Int. Symp., Nieuwport (Belgium), Sept. 13-17, 1989 (P.J. Grobet et al., Eds.) Elsevier, Amsterdam, 1988; Studies Surf. Sci. Catalysis 37 (1988) pp. 515-534.

[16] C. Chen, Diploma Thesis, Engler-Bunte-Institut der Universität Karlsruhe, 1983.

[17] H.G. Karge, H. Darmstadt, R. Amberg, Z. Zheng and M. Laniecki, Applied Catalysis, submitted for publication.

[18] H. Darmstadt, Diploma Thesis, Fritz-Haber-Institut der Max-Planck-Gesellschaft / Technische Hochschule Darmstadt, 1988.

[19] H.G. Karge, M. Łaniecki, M. Ziołek, G. Onyestyak, A. Kiss, P. Kleinschmit and M. Siray, in: "Zeolites: Facts, Figures, Future", Proc. 8th Int. Conf. Zeolites, Amsterdam, The Netherlands, July 9-14, 1989 (P.A. Jacobs and R. van Santen, Eds.) Elsevier, Amsterdam, 1989; Studies Surf. Sci. Catalysis 49 (1989) pp. 1327-1337.

[20] P. Magnoux, P. Roger, C. Canaff, V. Fouche, N.S. Gnep and M. Guisnet, in: "Catalyst Deactivation 1987", Proc. 4th Int. Symp., Antwerp, Belgium, Sept. 29 Okt. 1, 1987 (B. Delmon and G.F. Froment, Eds.), Elsevier, Amsterdam 1987; Studies Surf. Sci. Catalysis 34 (1987) pp. 317-330.

[21] M. Guisnet, P. Magnoux and C. Canaff, in: "New Developments in Zeolite Science and Technology", Proc. 7th Int. Zeolite Conf., Tokyo, Japan, August 17-22, 1986 (Y. Murakami, A. Iijima, and J.W. Ward, Eds.) Kodansha, Tokyo and Elsevier, Amsterdam, 1986, pp. 701-707.

[22] H. Schulz, W. Böhringer, W. Baumgartner and Zhao Siwei, in: "Catalyst Deactivation 1987", Proc. 4th Int. Symp., Antwerp, Belgium, Sept. 29-Okt. 1, 1987 (B. Delmon and G.F. Froment, Eds.), Elsevier, Amsterdam 1987; Studies Surf. Sci. Catalysis 34 (1987) pp. 479-492.
[23] M. Neuber, S. Ernst, H. Geerts, P.J. Grobet, P.A. Jacobs, G.T. Kokotailo and J. Weitkamp, in: "Catalyst Deactivation 1987", Proc. 4th Int. Symp., Antwerp, Belgium, Sept. 29-Okt. 1, 1987 (B. Delmon and G.F. Froment, Eds.), Elsevier, Amsterdam 1987; Studies Surf. Sci. Catalysis 34 (1987) pp. 567-577.
[24] V. Solinas, R. Monaci, E. Rombi and L. Forni, in: "Catalyst Deactivation 1987", Proc. 4th Int. Symp., Antwerp, Belgium, Sept. 29-Okt. 1, 1987 (B. Delmon and G.F. Froment, Eds.), Elsevier, Amsterdam 1987; Studies Surf. Sci. Catalysis 34 (1987) pp. 493-500.
[25] R.N. Jones and C. Sandorfy, in: "Chemical Applications of Spectroscopy", Interscience Publishers, Inc., New York, N.Y. 1956.
[26] E.R. Blout, M. Fields and R. Karplus, J. Am. Chem. Soc. 70 (1948) 194-198.
[27] H.G. Karge, Proc. 4th Int. Zeolite Conf., (Molecular Sieves II), Chicago, Ill., USA, April 18-22, 1977 (J. Katzer, Ed.); Am. Chem. Soc., Washington D.C.; ACS Symp. Series 40 (1977) pp. 584-595.
[28] T.J. Weeks, Jr. and A.P. Bolton, Proc. 3rd Int. Conf. Molecular Sieves, Zürich, Switzerland, Sept. 3-7, 1973; Recent Progress Reports (J.B. Uytterhoeven, Ed.) Leuven University Press, Leuven, 1973. pp. 426-431.
[29] H.G. Karge, Z. physik. Chem. [N.F.] 122 (1980) 103-116.
[30] H.G. Karge, Z. physik. Chem. [N.F.] 76 (1971) 133-153.
[31] P.B. Venuto and P.S. Landis, Advances in Catalysis 18 (1968) 259-371.
[32] H.G. Karge and K. Klose, Z. physik. Chem. [N.F.] 83 (1973) 92-99.
[33] H.G. Karge, W. Abke, E.P. Boldingh and M. Łaniecki, Proc. 9th Iberoamerican. Symp. Catalysis, Lisbon, Portugal, July 16-21, 1984 (M.F. Portela, Ed.) Jorge Fernandes, Lisbon, 1984, pp. 582-593.
[34] E. Gallei and E. Schadow, Rev. Sci. Instrum. 45 (1974) 1504-1506.
[35] W.C. Hecker, A. Bell, R.F. Hicks, C.S. Kellner and B.J. Savatsky, J. Catal. 71 (1981) 216-218.
[36] H.G. Karge and E.P. Boldingh, Catalysis Today 3 (1988) 53-63.
[37] H.G. Karge and E.P. Boldingh, Catalysis Today 3 (1988) 379-386.
[38] J.-P. Lange, A. Gutsze, J. Allgeier and H.G. Karge, Appl. Catalysis 45 (1988) 345-356.
[39] D.M. Bibby, G.D. McLellan and R.F. Howe, in: "Catalyst Deactivation 1987", Proc. 4th Int. Symp., Antwerp, Belgium, Sept. 29-Okt. 1, 1987 (B. Delmon and G.F. Froment, Eds.), Elsevier, Amsterdam 1987; Studies Surf. Sci. Catalysis 34 (1987) pp. 651-658.
[40] Kredel, Dissertation, Technische Hochschule Darmstadt, 1988.
[41] F. Fetting, E. Gallei and P. Kredel, Ger. Chem. Eng. 7 (1984) 32-38.
[42] F. Fetting, E. Gallei and P. Kredel, Chem.-Ing.-Tech. 54 (1982) 606-607.
[43] H.G. Karge and M. Łaniecki, to be published.
[44] DMS Arbeitsatlas der Infrarot-Spektroskopie, Verlag Chemie, Weinheim, 1972.
[45] L. Kubelková, J. Nováková, M. Tupá and Z. Tvarůzková, Proc. Int. Symp. Zeolite Catalysis, Siófok, Hungary, May 13-16, 1985 (P. Fejes and D. Kalló, Eds.) (Acta Physica et Chemica Szegediensis) Petöfi Nyomada, Kecskemét, 1985, pp. 649-657.

[46] J. Haber, J. Komorek and T. Romotowski, Proc. Int. Symp. Zeolite Catalysis, Siófok, Hungary, May 13-16, 1985 (P. Fejes and D. Kalló, Eds.) (Acta Physica et Chemica Szegediensis) Petőfi Nyomada, Kecskemét, 1985, pp. 671-679.
[47] E. Breitmeier and W. Völter, "^{13}C-NMR Spectroscopy, Methods and Application in Organic Chemistry", Wiley Interscience, New York, 1972, pp. 221-223.
[48] E.G. Derouane, J.P. Gilson and J.B. Nagy, ZEOLITES 2 (1982) 42.
[49] J.P. van den Berg, J.P. Wolthuizen, A.D.H. Claque, G.R. Hays, R. Huis and J.H.C. van Hooff, J. Catal. 80 (1983) 130-144.
[50] S. Maixner, C.Y. Chen, P.J. Grobet, P.A. Jacobs and J. Weitkamp, in: "New Developments in Zeolite Science and Technology", Proc. 7th Int. Zeolite Conf., Tokyo, Japan, August 17-22, 1986 (Y. Murakami, A. Iijima, and J.W. Ward, Eds.) Kodansha, Tokyo and Elsevier, Amsterdam, 1986, p. 693-700.
[51] J.B. Stothers: "Carbon-13 NMR Spectroscopy" in "Organic Chemistry", - A Series of Monographs, Vol. 24, Academic Press, New York, London 1972, pp. 90-100.
[52] J.-P. Lange, A. Gutsze and H.G. Karge, J. Catal. 114 (1988) 136-143.
[53] G.A. Olah, J.S. Staral, G. Asencio, G. Liang, D.A. Forsyth and G.D. Mateescu, J. Am. Chem. Soc. 100 (1973) 6299-6308.
[54] G.L. Nelson and E.A. Williams, Progr. Phys. Org. Chem. 12 (1976) 229-342.
[55] R.N. Young, Progr. Nucl. Magn. Reson. Spectrosc. 12 (1979) 261-286.
[56] M.W. Anderson and J. Klinowski, in: "Recent Advances in Zeolite Science"; Proc. of the Annual Meeting of the British Zeolite Association (BZA), Cambridge, UK, 1989, (J. Klinowski and P.J. Barrie, Eds.) Studies Surf. Sci. and Catalysis, Elsevier, Amsterdam, in press.
[57] D. Neuhaus, Diploma Thesis, Karl-Marx-Universität Leipzig, Sektion Physik, 1989.
[58] J.R. Andersen, Y.-F. Chang and R.J. Western, J. Catal. 118 (1989) 466-482.
[59] J.E. Wertz and J.R. Bolton. "Electron Spin Resonance in Elementary Theory and Practical Applications", McGraw Hill, New York, N.Y., 1972.
[60] H.G. Karge, E.P. Boldingh, J.-P. Lange and A. Gutsze, Proc. Int. Symp. Zeolite Catalysis, Siófok, Hungary, May 13-16, 1985 (P. Fejes and D. Kalló, Eds.) (Acta Physica et Chemica Szegediensis) Petőfi Nyomada, Kecskemét, 1985, pp. 639-647.
[61] H.G. Karge, J.-P. Lange, A. Gutsze and M. Laniecki, J. Catal. 114 (1988) 144-152.
[62] I.R. Leith, J. Chem. Soc. Commun. (1972) 1282-1283.
[63] A.V. Kucherov and A.A. Slinkin, Kinet. Katal. 23 (1982) 1172; Engl. Translat.: 997-1003.
[64] A.V. Kucherov and A.A. Slinkin, Kinet. Katal. 24 (1983) 947;. Engl. Translat.: 804-810.
[65] A.V. Kucherov and A.A. Slinkin, in: "Structure and Reactivity of Modified Zeolites"; Proc. Int. Conf., Prague, ĆSSR, (P.A. Jacobs et al., Eds.) Elsevier, Amsterdam 1984; Studies Surf. Sci. Catalysis 18 (1984). pp. 77-84.
[66] A.V. Kucherov, A.A. Slinkin, D.A. Kondratyev, T.N. Bodarenko, A.M. Rubinstein and Kh.M. Minachev, J. Mol. Catal. 37 (1986) 107-115.
[67] P.B. Ascough and H.E. Evans, Trans. Faraday Soc. 60 (1964) 801-808.
[68] H.G. Karge, S. Trevizan de Suarez and I.G. Dalla Lana, J. Phys. Chem. 88 (1984) 1782-1784.
[69] H.G. Karge, M. Ziolek and M. Łaniecki, J. Catal. 109 (1988) 252-262.

[70] J.C. Vedrine, P. Dejaifve and E.D. Garbowski, in: "Catalysis by Zeolites", Proc. Int. Symp., Ecully (Lyon), France, Sept. 9-11, 1980 (B. Imelik et al., Eds.) Elsevier, Amsterdam 1980; Studies in Surface Sci. Catalysis 5 (1980) pp. 29-37.
[71] H. Förster, S. Franke and J. Seebode, J. Chem. Soc. Faraday Trans. I 79 (1983) 373-382.
[72] M. Łaniecki and H.G. Karge, Proc. 6th Int. Symp. Heterogeneous Catalysis, Sofia, Bulgaria, July 13-17, 1987 (D. Shopov et al., Eds.) Publ. House Bulgarian Academy of Science, Sofia 1987, pp. 129-134.
[73] E.D. Garbowski and H. Praliaud, J. Chim. Phys. Phys. Chim.-Biol. 76 (1979) 687-692.
[74] E.D. Garbowski, J.-P. Candy and M. Primet, J. Chem. Soc. Faraday Trans. I 79 (1983) 835-844.
[75] G.N. Asmolov, O.V. Krylov, O.V. Bragin and D.B. Furman, Kinet. Katal. 18 (1977) 730; Engl. Translat. 18 (1977) 609-613.
[76] T.S. Sørensen, J. Am. Chem. Soc. 87 (1965) 5075-5084.
[77] G.A. Olah, C.U. Pittmann, Jr., R. Waack and M. Doran, J. Am. Chem. Soc. 88 (1966) 1488-1495.
[78] H.G. Karge, M. Ziołek and H. Darmstadt, in preparation.
[79] B.A. Sexton, A.E. Hughes and D.M. Bibby, J. Catal. 109 (1988) 126-131.
[80] Th.H. Fleisch, G.W. Zajac, B.L. Meyers, G.J. Ray and J.T. Miller, in "Catalysis: Theory to Practice", Proc. 9th Int. Congr. Catalysis, Calgary, Canada, June 26-July 1, 1988 (M.J. Phillips and M. Ternan, Eds.) The Chemical Institute of Canada, Ottawa, Ontario, 1988, pp. 483-490.
[81] M. Muhler, R. Schlögl and G. Ertl, Surface Sci. 189 (1987) 69-79.
[82] R. Schlögl, H.G. Karge, H. Darmstadt and M. Wesemann, in preparation.
[83] E. Tegler, G. Wiech and H. Faessler, J. Atomic Mol. Phys. 14 (1981) 1273-1282.
[84] R. Schlögl, Surface Sci. 190 (1987) 861-872.
[85] P.B. Venuto and L.A. Hamilton, I & EC Product Research and Development 6 (1967) 190-192.
[86] D.E. Walsh and D. Rollmann, J. Catal. 49 (1977) 369-375.
[87] M. Guisnet, P. Magnoux and C. Canaff, in: "Chemical Reactions in Organic and Inorganic Constrained Systems" (C.R. Setton, Ed.) Nato Series C 165, Reidel, Dordrecht, 1986, pp. 131-140.
[88] P. Magnoux, P. Cartraud, S. Mignard and M. Guisnet, J. Catal. 106 (1987) 235-241.
[89] P. Magnoux, P. Cartraud, S. Mignard and M. Guisnet, J. Catal. 106 (1987) 242-250.
[90] P. Magnoux and M. Guisnet, ZEOLITES 9 (1989) 329-335.
[91] P. Magnoux, M. Guisnet, S. Mignard and P. Cartraud, J. Catal. 117 (1989)495-502.
[92] S. Mignard, P. Cartraud, P. Magnoux and M. Guisnet, J. Catal. 117 (1989) 503-511.
[93] H.G. Karge and H. Darmstadt, in preparation.
[94] J.G. Post and J.H.C. van Hoof, in: "New Developments in Zeolite Science and Technology"; 7th Int. Zeolite Conf., Tokyo, Japan, August 17-22, 1986 (Japan Assoc. of Zeolite, Ed.) Tokyo, 1986, Preprints of Poster Papers, Paper 1D-20, pp. 249-250.
[95] D.M. Bibby, in: "New Developments in Zeolite Science and Technology"; 7th Int. Zeolite Conf., Tokyo, Japan, August 17-22, 1986 (Japan Assoc. of Zeolite, Ed.) Tokyo, 1986, Preprints of Poster Papers, Paper 1D-11, pp. 231-232.

[96] D.M. Bibby and C.G. Pope, J. Catal. 116 (1988) 407-414.
[97] G.P. Handreck and T.D. Smith, J. Chem. Soc. Faraday Trans. I 84 (1989) 645-654.
[98] H.G. Karge, R. Fiedorow, J. Schütze and R. Schlögl, unpublished results.
[99] W.J.M. Pieters and A.F. Venero, in: "Catalysis on the Energy Scene", Québec, Canada, Sept. 30-Oct. 3, 1984 (S. Kaliaguine and A. Mahay, Eds.) Elsevier, Amsterdam 1984; Studies Surface Sci. Catalysis 19 (1984) pp. 155-163.
[100] H. Schulz, W. Böhringer, W. Baumgärtner and Zhao Siwei, in: "New Developments in Zeolite Science and Technology"; Proc. 7th Int. Zeolite Conf., Tokyo, Japan, August 17-22, 1986 (Y. Murakami, A. Iijima and J.W. Ward, Eds.) Kodansha, Tokyo and Elsevier, Amsterdam, 1986, pp. 915-922.
[101] A. Corma and B.W. Wojciechowski, Catal. Rev.-Sci. Eng. 27 (1985) 29-150.
[102] J. Haas, F. Fetting and L. Gubicza, Proc. Int. Symp. Zeolite Catalysis, Siófok, Hungary, May 13-16, 1985 (P. Fejes and D. Kalló, Eds.) (Acta Physica et Chemica Szegediensis) Petöfi Nyomada, Kecskemét, 1985, pp. 659-669.
[103] D.M. Bibby, N.B. Milestone, J.E. Patterson and L.P. Aldridge, J. Catal. 97 (1986) 493-502.
[104] H.G. Karge, Z. phys. Chem. [N.F.] 99 (1975) 241-254.
[105] A. Voorhies, Ind. Eng. Chem. 37 (1945) 318-322.
[106] G.F. Froment and K.B. Bischoff, Chem. Eng. Sci. 16 (1961) 189-201.
[107] G.F. Froment and K.B. Bischoff, Chem. Eng. Sci. 17 (1962) 105-114.
[108] M.B. Ajinka, W.H. Ray and G.F. Froment, Ind. Eng. Chem., Process Des. Dev. 13 (1974) 107-112.
[109] D.W. Breck and R.W. Grose, Advances in Chemistry Ser. 121 (1973) 319-329.
[110] P. Gallezot, C. Leclerq, M. Guisnet and P. Magnoux, J. Catal. 114 (1988) 100-111.
[111] M. Bülow, J. Caro, J. Völter and J. Kärger, in: "Catalyst Deactivation 1987", Proc. 4th Int. Symp., Antwerp, Belgium, Sept. 29-Okt. 1, 1987 (B. Delmont and G.F. Froment, Eds.) Elsevier, Amsterdam 1987; Studies Surf. Sci. Catalysis 34 (1987) pp. 343-354.
[112] J. Kärger, H. Pfeifer and W. Heink, Proc. 6th Int. Zeolite Conf., Reno, USA, July 10-15, 1983 (D. Olson and A. Bisio, Eds.) Butterworths, Guildford, Surrey, UK, 1984, pp. 184-200.
[113] T. Ito, J.L. Bonardet, J. Fraissard, J.B. Nagy, C. André, Z. Gabelica and E.G. Derouane, J. Appl. Catal. 43 (1988) L5-L11.
[114] M. Guisnet and P. Magnoux, Appl. Catalysis 54 (1989) 1-27.

Studies in Surface Science and Catalysis 137
H. van Bekkum, E.M. Flanigen, P.A. Jacobs and J.C. Jansen (Editors)
© 2001 Elsevier Science B.V. All rights reserved.

Chapter 17

Hydrocarbon processing with zeolites

I.E. Maxwell[a] and W.H.J. Stork[b]

[a]Avantium Technologies B.V., Zekeringstraat 29, 1014 BV Amsterdam

[b]Shell Research and Technology Centre, Amsterdam, Badhuisweg 3, 1031 CM Amsterdam, The Netherlands

This chapter on hydrocarbon processing with zeolites covers both existing and new catalytic applications of zeolites in oil refining and gas conversion. By way of introduction some structural aspects related to these industries are discussed to provide some background in which to relate the current and future developments. Further, the unique opportunities and limitations of zeolites are discussed in order to put the applications in hydrocarbon processing into perspective.

Some more evolutionary type developments are covered where the introduction and continued improvement of zeolite catalysts have had a significant impact on existing process technologies. Examples of such technologies in this category include catalytic cracking, hydrocracking, paraffin isomerization and olefin oligomerization.

Newer process technologies in which the application of zeolitic catalysts has led to new process concepts are also included. The unique shape selective properties of zeolites are shown to play a dominant role in this developing field of applications. Examples include catalytic iso-dewaxing, LPG to aromatics (CYCLAR), selective oxidation, methanol-to-olefins (MTO), deep catalytic cracking (DCC) and selective catalytic NOx removal (SCR).

Finally, some general trends are discussed in terms of zeolite catalysis and how these might be expected to have a further impact on hydrocarbon process technology in the future. Significant opportunities are believed to exist for further developments related to both existing and emerging processes. The discovery and application of new synthetic molecular sieves, leading to new or rejuvenated processes, also holds promise for the future. In addition, advances in understanding are being achieved by the recent emphasis on theoretical studies, while the emerging combinatorial synthesis and high-speed screening technologies represent a great increase in experimental power over the time-consuming and empirical approaches employed hitherto.

1. INTRODUCTION

1.1. Background

Zeolites are currently being used industrially on a very large scale. Table 1 gives an estimate of the present synthetic zeolite production. The greater part of this production is intended for ion-exchange applications in detergents, using mainly zeolite A, but large amounts are also being produced for adsorption (drying; physical separations of hydrocarbons) and catalytic applications. In terms of volume most of the zeolitic catalysts are used in oil refining, where especially zeolite Y is employed on a large scale in catalytic cracking; applications in other refining processes, gas conversion, chemical processes and environmental catalysis, although growing, represent much smaller volumes. The turnover of zeolite use for catalysis has doubled since 1986: altogether an impressive growth.

Table 1
Applications of synthetic zeolites.

Application	tonnes/annum
Ion exchange	>1.000.000
Adsorption	~75.000
Catalysis	~120.000 (of which >90% Y-zeolite for catalytic cracking)

In this survey the applications of zeolites in oil refining, petrochemicals processes and gas-conversion-processes will be reviewed. Before embarking on a detailed discussion, however, the characteristics that make zeolites so valuable in their catalytic applications are briefly summarized.

The basics of oil refining are also discussed, i.e. the conversion and upgrading of the various oil fractions into transportation fuels boiling in the gasoline and middle-distillate range. The important applications of zeolites in oil refining are discussed, the various processes being reviewed in detail. This is followed by a discussion of the new emerging gas conversion processes which make use of zeolite catalysts. After a brief introduction to the basics of gas conversion, the processes for the conversion of methanol, synthesis gas, light paraffins and olefins are dealt with separately.

This chapter does not pretend to provide a comprehensive review of all the studies that have been carried out on each topic, but is rather intended to present a selection of the more relevant topics to provide the reader with a general impression of the "state of the art".

1.2. Zeolite properties

On the basis of their present use in applied catalysis, various important aspects of zeolites can be identified. It is, furthermore, noteworthy that many crucial catalytic properties can often be tailored to suit a particular application. These properties can be summarized as follows:

(i) Zeolites, especially the silica-rich materials, often exhibit a high thermal and hydrothermal stability . This allows their use as catalysts at high temperatures, and their frequent oxidative regeneration even in the presence of some steam.

(ii) Zeolites are crystalline microporous solids, allowing a relatively high concentration of reactant molecules inside the zeolite cages; thus, in the presence of a zeolite the reaction will proceed as if it were carried out at higher partial pressure compared witt most other catalytic solid surfaces. This so-called cage effect has been nicely illustrated by Rabo [1], as shown in Table 2 for n-hexane cracking over KY.

Table 2
Zeolites: Non-acid catalysis n-hexane over KY, 500 °C (from ref. 1).

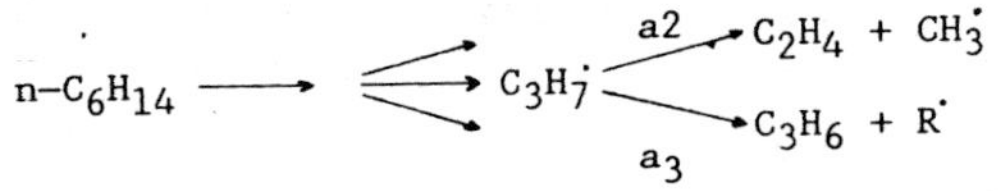

	a_2	a_3 (%)
Thermal	44	11
KY	18	39

KY results in: - 5x higher cracking
- different selectivity (less C_1, C_2)

KY favours bimolecular reactions

(iii) Zeolites can be synthesized with a wealth of different structures. Thus, until recently, one could classify zeolites as those with pore openings consisting of 8-, 10-, or 12-membered rings. Recently, molecular sieves with larger openings, e.g. with 14 or 18-membered rings (such as in UTD-1 , SSZ-44 , AlPO-8 and VPI-5) and even with wider pores such as in cloverite (20 rings) and in MCM-41, MCM-48 and related ordered mesoporous materials have been synthesized, as discussed in detail elsewhere in this volume. Moreover, the pore system can be one, two or three dimensional, it can contain pores of different sizes, and consist of tubes and/or cages, etc. In recent years the importance of the dimensionality of the pore system, and of the presence or absence of cavities as opposed to straight channels has become clearer. The fact that the pore openings are of similar dimensions to hydrocarbon molecules, i.e. from about 0.3 to 1.2 nm, is the characteristic feature that forms the basis of the catalytic shape-selective behaviour of zeolites. However, more recently, also the mesoporous system in zeolites, produced by structural defects, has received more attention. By steaming or acid extraction part of the zeolite can be dissolved or restructured leaving a mesoporous system in addition to the microporous one, which may be of importance in the transport of reactants or products and thus the conversion of relatievely large molecules which may be present in a feedstock.

(iv) Zeolites and related materials can nowadays be prepared with a wide variety of chemical compositions. Originally, zeolites were typically aluminosilicates with a Si/Al atomic ratio between 1 and 5; however, with the advent of templated synthesis and controlled modification procedures (e.g. ultrastabilization) higher ratios (up to infinity) have become attainable, such as in ZSM-5, and in many other zeolites.

(v) Isomorphous replacement of especially Al (e.g. by Ga, Fe, B, Ti) in zeolites is well established, although the number of commercial applications of such isomorphously substituted zeolites still remains rather limited. In the 1980's, researchers at UOP discovered a new family of molecular sieves based on aluminophosphates, where two Si atoms in a zeolite had been replaced by one Al and one P atom [2]. Currently, also SAPOs, MeAPOs, MeAPSOs, etc. are available, in which further elements have been isomorphously introduced in an AlPO structure. In general, these materials, after e.g. Si introduction, have only a moderate activity which, for many applications, has so far been a severe handicap.

(vi) Zeolites generally have ion-exchange properties, which is the reason for their use in detergents. For this application gismondine (zeolite P) has been successfully developed giving faster rates of ion exchange than the traditional zeolite A [2a]. For catalytic use ion exchange can be used to introduce metals in a highly dispersed manner but also. via ammonium-ion exchange followed by calcination, to introduce acidity. The acidity of zeolites is, in general, very high, being about 1000 times higher than that of amorphous silica-alumina, and is directly related to the framework aluminium content, as can be seen from Figure 1 [3, 4, 5]. The concentration of the acid sites can be adjusted through the framework Al content, while the intrinsic acidity of the sites can be modified, for example, by isomorphous replacement of Al by Ga, Fe, and B, with the acidity decreasing in this order, as is illustrated in Figure 2 [6, 7]. It is difficult, on the other hand, to increase the acidity of zeolites beyond that of the normal bridging OH group. "Superacidic" sites in which an extra framework Al is associated with the bridging OH have been reported [7a] and [7b] specific functions may be anchored on to (all-silica) zeolite structures.

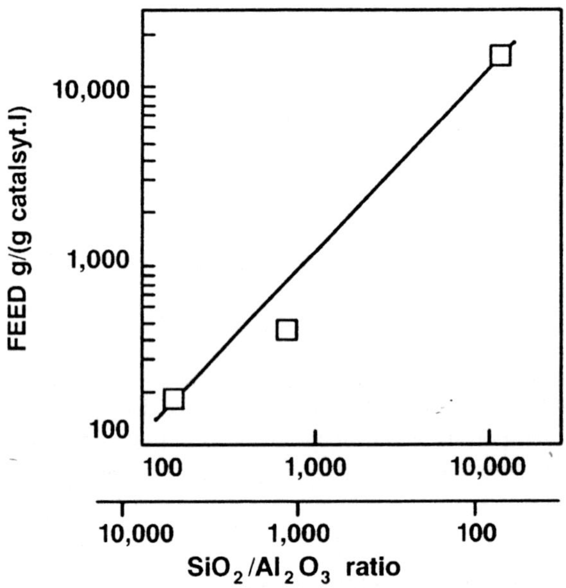

Catalytic conversion (activity per A1 site) is the same for any silica/alumina ratio when the feed rate is adjusted proportionately

Figure 1. Zeolites: activity in acid catalysis.

(vii) The near-molecular pore dimensions of zeolites result in a wide range of diffusivities (spanning some 10 orders of magnitude), which impart the shape-selective properties to these materials. At least three types of shape-selective catalytic behaviour have been identified to date:
- reactant selectivity
- product selectivity
- restricted transition state selectivity.

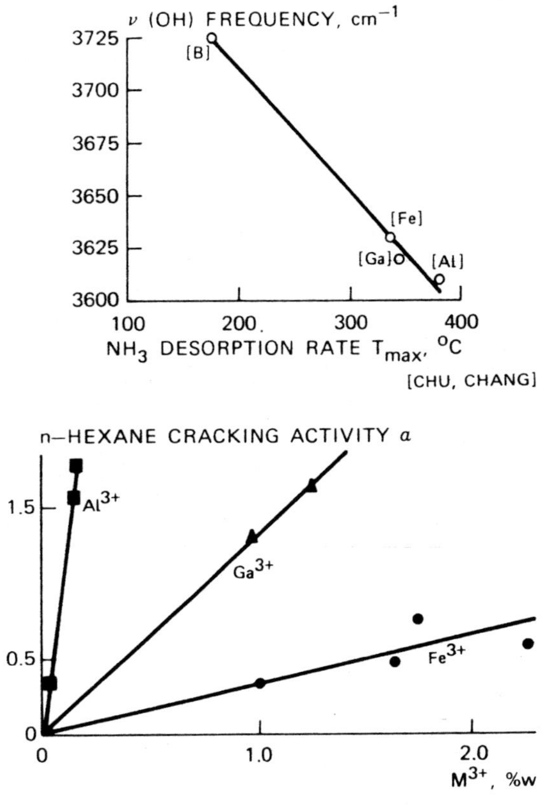

Figure 2. Zeolites: acidity and isomorphous substitution (from refs. 6, 7).

These different forms of shape selectivity are illustrated in Figure 3 [8, 9]. In many cases combinations of these may be operating for a particular reaction.

(viii) Despite this rich variety of properties, although at present over 100 different zeolite structure types are known, only about 10-15 of those have found commercial application. This is primarily due [9a, 9b] to the fact that, for the commercialization of a new zeolite, an extensive and costly effort is required to scale up for the manufacture of commercial quantities of the material in a way that is well-controlled, environmentally acceptable, economically justified (given the application and the required volumes), and, most importantly, reproduces the performance of the original material. For this reason, a significant performance incentive is required to use a new rather than an already commercially available (or a modification of an existing) zeolite. This obviously presents a severe "barrier to entry" to bring completely new zeolites into commercial use; nevertheless there such examples, for instance in aromatics alkylation, where the conventional Friedel-Crafts methods were first substituted by zeolite-based processes and, subsequently, by still newer processes using a second generation of zeolite catalysts.

The applications of zeolites are discussed with emphasis on the special features of these materials as compared to conventional catalyst systems.

2. OIL-CONVERSION PROCESSES

2.1. Basics of oil refining

The refining of crude oil is a major industry in which worldwide some 40 million barrels (one barrel contains 159 litres) of crude oil are being processed daily in numerous refineries. The basic principles involved in oil refining remain unchanged but the trend is towards increasing complexity.

Crude oil is first split into various fractions in a primary atmospheric distillation step. In this way the raw materials for the main oil products are recovered, which include naphtha (transportation fuel and petrochemical feedstock with boiling range: C_5 to about 180°C) to produce gasoline; and middle distillates to produce kerosine (aviation fuel with boiling range: about 130 to 300°C) and diesel/gas oil (transportation and domestic fuel; 150-370°C). In addition, base oils for lubricants can be recovered (see Figure 4). These fractions are the primary oil products (i.e. transportation fuels and naphtha for chemical feedstock), but many of these primary products need to be improved by further purification and upgrading before they can meet specification and be used as final commercial products.

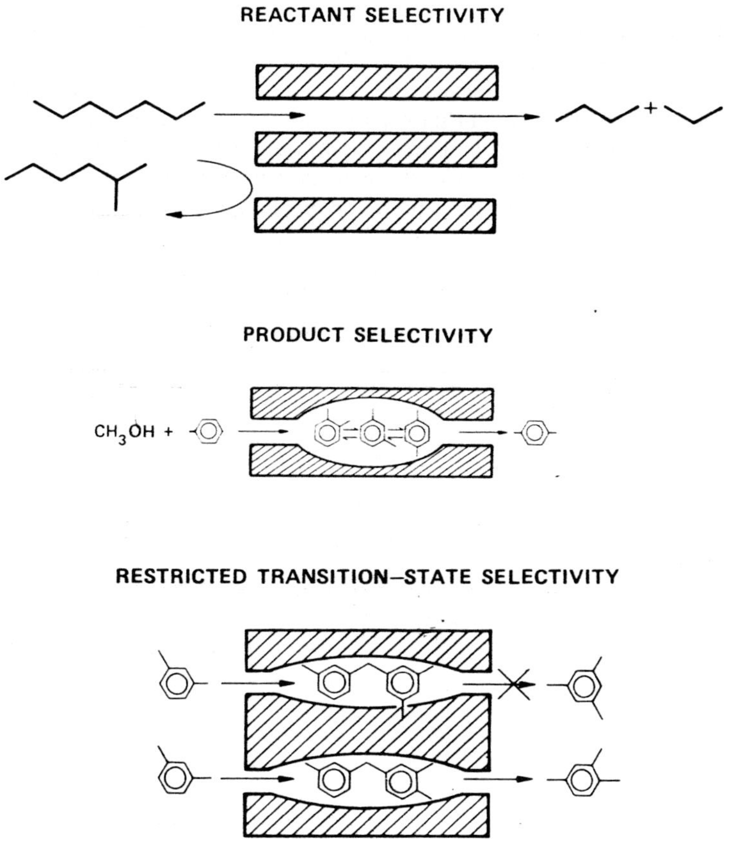

Figure 3. Different types of shape selectivity (from ref. 8).

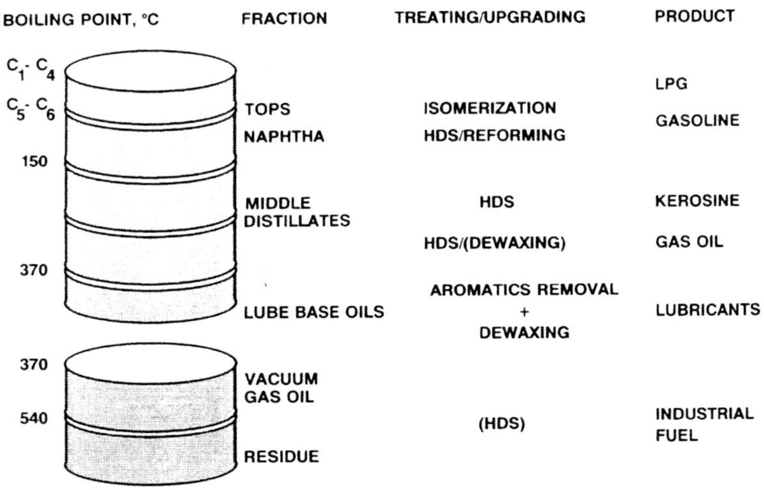

Figure 4. A simple oil refinery (basics).

In fact, the further upgrading of these primary oil products is becoming increasingly important as environmental legislation becomes more stringent. Sulfur, for example, is removed from oil fractions by means of catalytic hydrodesulfurization (HDS) using Co(Ni)/Mo/alumina catalysts. Combustion properties are important for the transportation fuels which require upgrading to meet product quality requirements. For example, the C_5-C_6 tops fraction may need to be isomerized to branched paraffins to increase the octane number, while the octane of the naphtha fraction needs to be increased by aromatization in the catalytic reforming process. For gas oils the wax content may impede the cold flow properties and thus require a dewaxing step which, in a modern refinery, may well be catalytic, utilizing shape-selective zeolites.

This simple (so-called "hydroskimming") refinery produces the required boiling point fractions, but utilizes in this way only the lighter part of the crude oil. The heavier fractions of the barrel (boiling point above about 370°C) can only be used as low-value industrial fuel. The more modern so-called "complex refinery" has additional processing capabilities (see Figure 5) which enable increased amounts of transportation fuels to be recovered by (partial) conversion of this heavy residue fraction of the barrel of oil. This is often achieved by first separating the residue in a vacuum distillation tower into fractions boiling below and above 540 °C, and then converting the lighter fraction (vacuum gas oil) into gasoline and middle-distillate products in a catalytic cracking unit or in a hydrocracking unit, respectively.

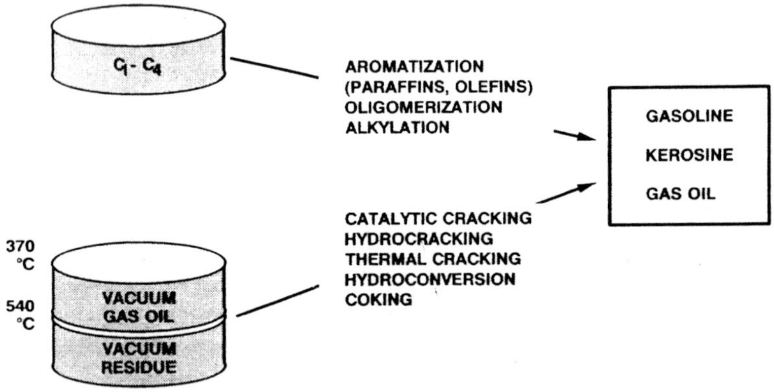

Figure 5. Conversion options in a complex refinery.

The remaining heavy vacuum residue can be converted into more valuable lighter fractions (as shown in Figure 5) by means of thermal processes (i.e. visbreaking or coking) or catalytically in the presence (i.e. hydroconversion) or absence of hydrogen (residue cat. cracking). These two options are fundamentally different in the way they achieve the necessary increase in hydrogen to carbon ratio of the products, i.e. by adding hydrogen or by removing carbon (i.e. as coke).

The overall refinery yield of transportation fuels can be further increased by converting the light gases associated with the crude oil, or those produced in the refinery, into higher molecular-weight hydrocarbons. These "synthesis" processes include: alkylation of propene/butene/i-butane mixtures to yield high-octane gasoline-range products, and oligomerization of olefins. These various processing options can be translated into many different refinery schemes and it usually depends on many factors, such as local crude supply, desired product slate, product quality requirements and desired flexibility, etc., as to which scheme will be optimal in a particular situation.

Having outlined the various processing schemes in refineries to achieve the desired products, and also the correct boiling ranges, it is worthwhile to consider the molecular structure of the crude oil fractions and the desired products. Crude oil consists of a very large number of components, and hence one always limits oneself to overall compositions, albeit that a separation in groups of components is standard. Within a given crude, with increasing boiling point, the ring content (aromatics + naphthenes) of the molecules increases, as does the heteroatom (S, N, O, and trace amounts of metals such as Ni and V) content, while the hydrogen content decreases. The molecules consist of (a number of) rings, with paraffinic side chains; usually the heteroatoms form part of the ring (certainly these are the most difficult to remove). These generalizations hold for all crude oils but clear differences exist between various crudes, some being much more paraffinic (few rings), others being more naphthenic or aromatic, while large differences in boiling point are found.

As regards the molecular structure of the desired products: for naphtha to be used in gasoline a high octane number is required, which can be achieved by isoparaffins, (iso) olefins and aromatics; for kerosine and gasoil on the other hand a high hydrogen content is generally attractive. In some cases n-paraffins can be troublesome because they may crystallize as wax, especially at low temperatures. Thus, the ideal lubricant molecule, to achieve its low dependance of viscosity on temperature (high "viscosity index") is a long paraffin with only few and short side chains. Increasingly stringent specifications to the transportation fuel are at present the most important factor dictating changes in the refinery, both for gasoline and for middle distillates production. The specifications for the near future as currently envisaged are given in ref. 8a; if anything, the tightening of the specification is accelerating. Throughout this chapter it will be clear that these changes in specifications affect most refinery processes.

A discussion of a modern oil-refining complex is not complete without considering the petrochemical industry, since this uses significant amounts of oil products as feedstock. Basically, the chemical industry requires as building blocks components such as synthesis gas, olefins and aromatics (benzene, toluene and xylenes, BTX). These are usually manufactured from naphtha, which is converted to olefins and aromatics in a thermal cracking unit; currently, there is a trend to use also heavier oil fractions as feedstocks for this unit. Synthesis gas can be manufactured from naphtha in a steam reformer, but also from coal or heavy oil residues by gasification. Aromatics (BTX) can also be recovered from the high-octane product from a catalytic reformer. Given the increasing amounts of oil products required by the petrochemical industry as feedstock - for example, ethylene capacity will reach 100 million tonnes/annum in 2001 - these outlets can significantly influence the optimal refinery scheme.

Now that the various refining processes and options have been outlined, it is of interest to see where zeolite catalysts are being applied. This has recently been surveyed by Degnan [8b]. Figure 6 shows the various crude-oil boiling fractions, which are now considered as feedstocks, and indicates for which processes zeolite catalysts are either already being used commercially or are likely to be applied in the near future. It is encouraging from the viewpoint of the zeolite chemist to observe that the list of applications where zeolitic catalysts are competitive, and often much more attractive compared with other catalyst systems, appears to be continuously growing. In the following sections of this chapter the various applications of zeolites in oil-refining processes will be discussed in more detail.

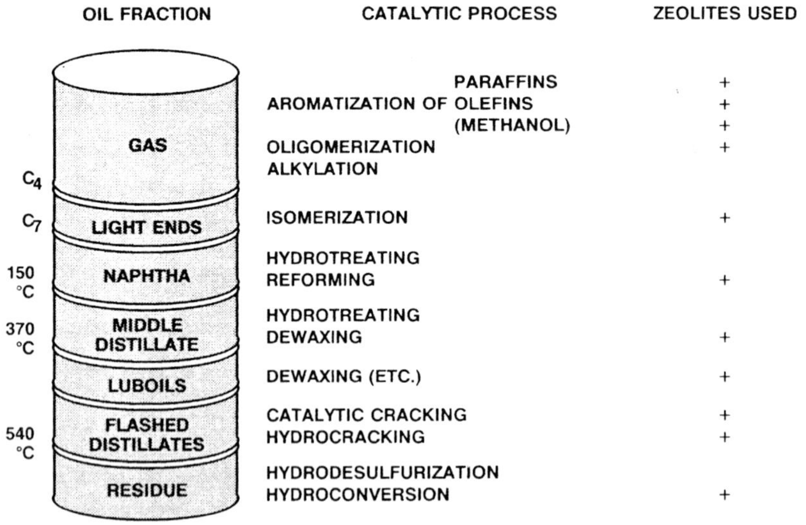

Figure 6. Zeolite catalysis in oil refining.

2.2. Paraffin hydroisomerization

Environmental considerations have brought about a rapid phase-out of the lead addition to gasoline, which has been further stimulated by the fact that automobile exhaust catalysts are rapidly poisoned by lead additives. Since lead was added to gasoline to increase the octane number, other means to achieve this must be employed.

In Table 3 the various hydrocarbon components of a typical US gasoline pool are listed together with their octane numbers [10]. Gasoline octane can, for example, be boosted by increasing the catalytic reformer severity (i.e. removing low-octane paraffins, but often also increasing the aromatics content), by increasing paraffin alkylation (yielding highly branched, high octane paraffins, see below), by adding methyl *tert*-butyl ether, MTBE (see below) or ethanol, and by replacing straight-run tops (C_5-C_6 paraffins) with isomerized products. Such paraffin isomerization can bring about a substantial increase in the octane number, as can be seen from Table 4. For example, *n*-pentane and especially *n*-hexane have much lower octane numbers than their corresponding branched isomers (in particular the dibranched hexanes). It is not surprising, therefore, that the capacity of paraffin isomerization has increased over the years, a recent figure being 1.35 million barrels per day [10a].

Table 3
Typical composition of U.S. gasoline pool (from ref. 10).

	%vol	Octane range
Butanes	6	91 - 93
Lt. straight run	4.5	55 - 75
Cat. Crackate	36	84 - 89
Hydrocrackate	3	85 - 87
Coker	1	60 - 70
Alkylate	11	90 - 94
Reformate	34	86 - 96
MTBE	1	106 -110
Isomerate	3.5	80 - 90

Table 4
Paraffin isomerization. Research octane number of C_5 and C_6 paraffins.

	RON-O
Pentanes	
n-Pentane	62
Methylbutane	93
2,2-dimethylpropane	83
Hexanes	
n-Hexane	29
2-Methylpentane	78
3-Methylpentane	76
2,2-Dimethylbutane	92
2,3-Dimethylbutane	104

This paraffin skeletal isomerization conversion is an equilibrium-limited reaction, as shown in Figure 7, with a low reaction temperature favouring isomerization and thus high-octane products. A very low reaction temperature of some 150°C can be employed by using chlorided alumina type catalysts containing a noble metal such as platinum [10b]. However, these catalyst systems suffer from the disadvantage of being very sensitive to trace impurities in the feed such as water. By contrast, zeolite-based catalysts which operate at higher temperatures (i.e. 250°C) are more robust and can withstand low levels of impurities such as sulfur and water in the feedstock. Interestingly, given the general trend toward increasing zeolite penetration, a new commercial amorphous catalyst was recently announced (by UOP, based on earlier work by Cosmo Oil and Mitsubishi Heavy Industries), that is clearly more active (by about 80°C) than the zeolitic ones, and which is supposedly not very sensitive to poisons [10c]. Recently Sie has reviewed paraffin isomerization catalysts and processes extensively [10d].

Zeolitic catalysts for paraffin isomerization were first developed for the Shell Hysomer process using platinum on H-mordenite. At the normal reaction temperatures the equilibrium is such that some normal paraffins are not converted to branched products, as can be seen from Figure 7, and it can therefore be attractive to combine the Hysomer isomerization process with the ISOSIV iso/normal separation process, developed by Union Carbide (now UOP).

In the ISOSIV process the normal paraffins are selectively absorbed by zeolite 5A (Ca-A), while the branched isomers cannot enter the pores of the molecular sieve. The effluent of the Hysomer process can thus be separated in the ISOSIV step and the normals recycled back to the catalytic section and so complete conversion to iso-paraffins is achieved. (It would be even more attractive to recycle also mono-branched hexanes, but although the use of ferrierite for separation of mono and dibranched paraffins has been described [10e], to our knowledge this has so far not been applied commercially for a variety of reasons). The combination of the ISOSIV and the Hysomer processes is the basis of the joint Shell/UCC(UOP) process, TIP (Total Isomerization Process), as shown in Figure 8, which results in an increase of some 9-10 octane (RON-0) points, compared with the single-step Hysomer process [10]. The Hysomer and TIP processes are being extensively applied; although the first unit was started up in 1969, the real increase in capacity was in the eighties as a result of the growth in octane demand. More recent process improvements of the TIP process have resulted in a 10-15 % cost reduction [10[. It has been estimated that the current installed capacity of the Shell Hysomer process is about 25 million tons/year [10f].

Hydroisomerization is a typical acid-catalysed reaction, the mechanism is believed to be as shown in Figure 9. In this mechanism carbenium ions are formed through protonation of alkenes; such ions can then rearrange and desorb as isomers, or be cracked through beta-scission. The intermediate in the rearrangement is assumed to be a protonated cyclopropane (PCP) species, as shown in Figure 9. This mechanism also explains the bifunctional requirements (the hydrogenation and acidic functions) of the catalyst. Nevertheless, it should be stressed that bare H-mordenite also has a significant isomerization activity [11], as shown in Figure 10. However, to obtain optimal selectivity and stability the bifunctional catalyst is much to be preferred [11a].

Weitkamp has published (ref. 12) results on cracking and isomerization over a bifunctional zeolite catalyst for alkanes of various chain lengths, as shown in Figure 11. Higher molecular- weight alkanes are much more reactive than lower ones and clearly the isomerization proceeds at lower temperatures than the hydrocracking. While C_6 paraffins can be isomerized rather selectively, this selectivity decreases with increasing chain length. This can be readily understood on the basis of the mechanisms outlined above: for C_7, C_8 and higher paraffins the hydrocracking can proceed via beta-scission, a step that for C_5/C_6 cannot be incorporated easily into the outlined mechanism [12a, 13a]. The PCP intermediate mechanism does not allow for an easy isomerization of n-C_4, and indeed, this reaction requires much higher temperatures. In ref. 10a it is postulated that the reaction proceeds through a primary carbenium ion, explaining the low reactivity, while Sachtler et al. [12aa] have recently again emphasized the possibility of a reaction path via dimerization. It is noteworthy that Blomsma [12b] has recently shown that, also in C_7 isomerization over Pd or Pt BETA catalysts, the dimerization isomerization path has a significant contribution.

These facts explain why paraffin isomerization so far has concentrated on C_5-C_6. In the case of feedstocks with wider boiling ranges, isomerization of the lighter fractions tends to be accompanied by excessive cracking of the heavier components. Furthermore, isomerization of the heavier alkanes is also less selective due to competitive hydrocracking reactions.

Selective isomerization of n-C_7 into single branched isomers has been reported over ferrierite [12c] but, as the formation of dibranched isomers was limited, the gain in octane in this operation was small. In this context it should be mentioned that the isomerizarion of n-C_7 over MoC catalysts as described by Ledoux [12d] supposedly proceeds according to a different mechanism and does not exhibit much cracking.

Clearly, considerable effort has been devoted to the development of high-performance hydroisomerization catalysts by industrial laboratories. The mordenite catalyst has to be converted from the as-synthesized Na form into the active highly acidic form, which involves not only ion exchange of Na for H, but also an adjustment of the silica/alumina ratio. Figure

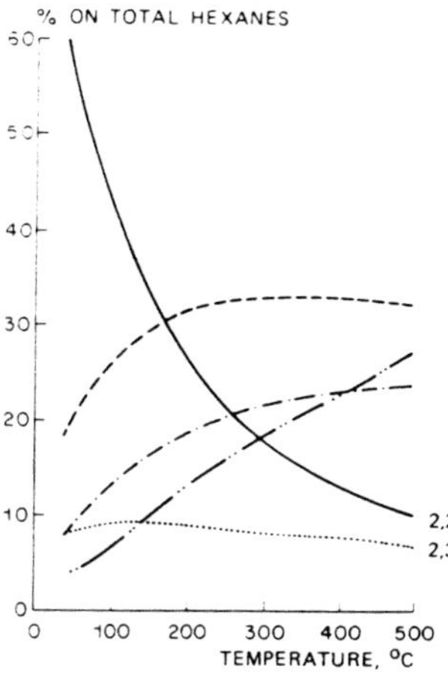

Figure 7. Thermodynamic equilibria for hexane hydroisomerization.

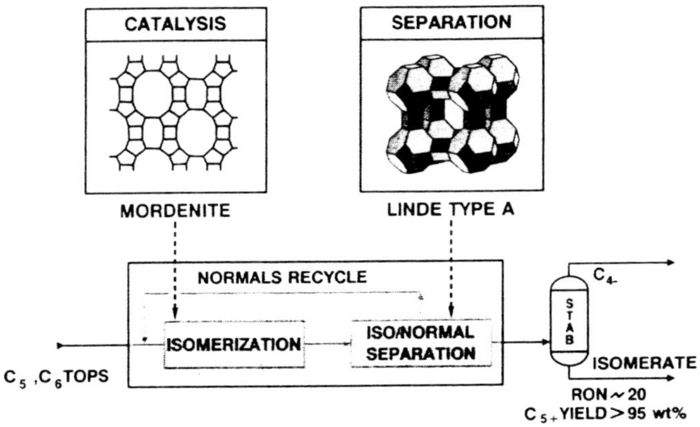

Figure 8. Paraffin Total Isomerization Process (TIP).

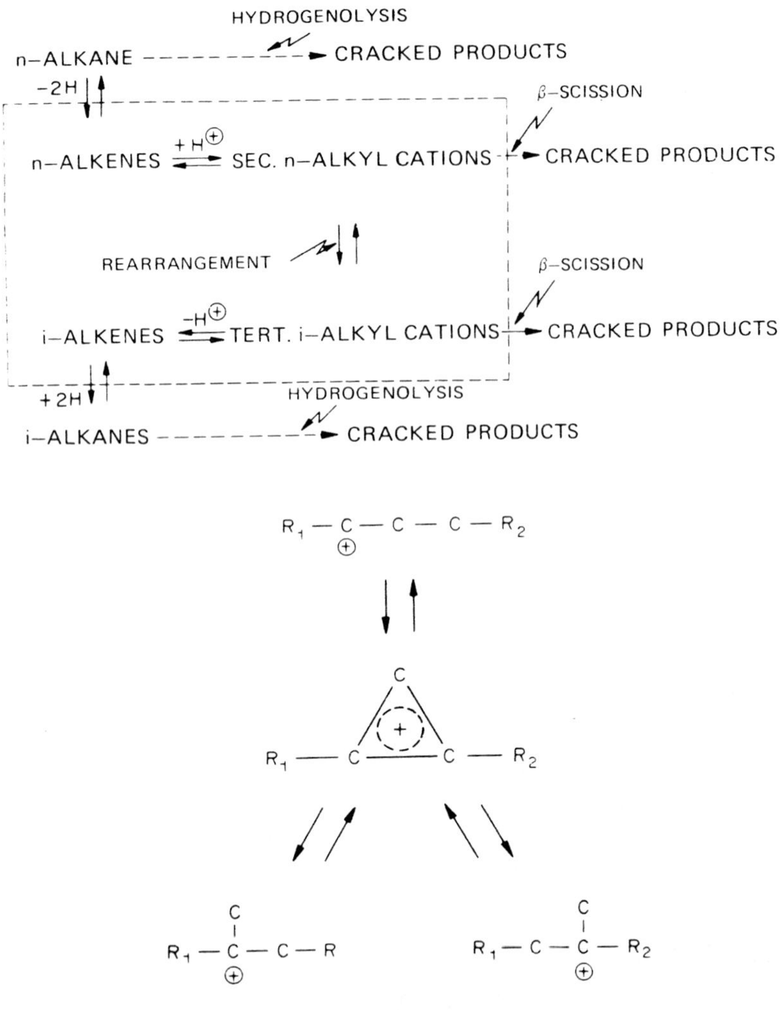

Figure 9. a) Mechanism for n-alkane hydroisomerization.
b) Protonated cyclopropane intermediate.

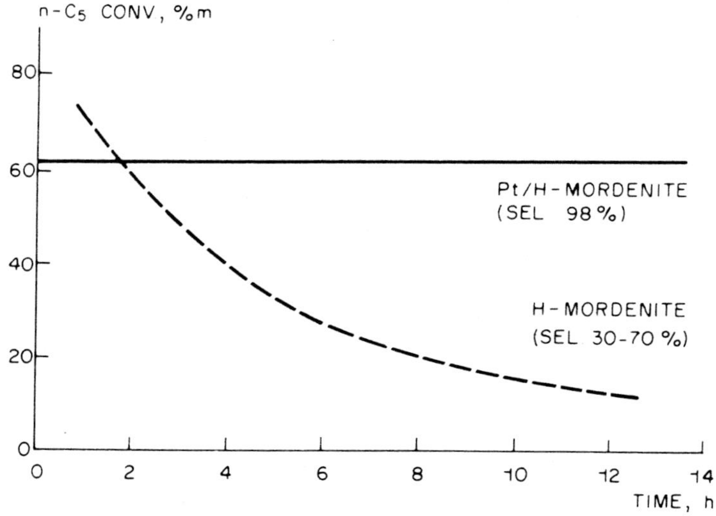

Figure 10. Influence of metal function in pentane hydroisomerization (from ref. 11).

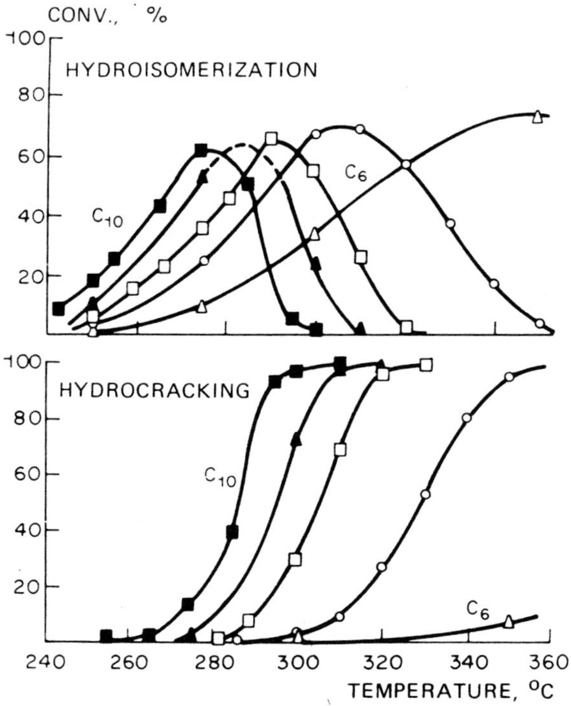

Figure 11. Comparison of hydroisomerization and hydrocracking for C_6-C_{10} paraffins.

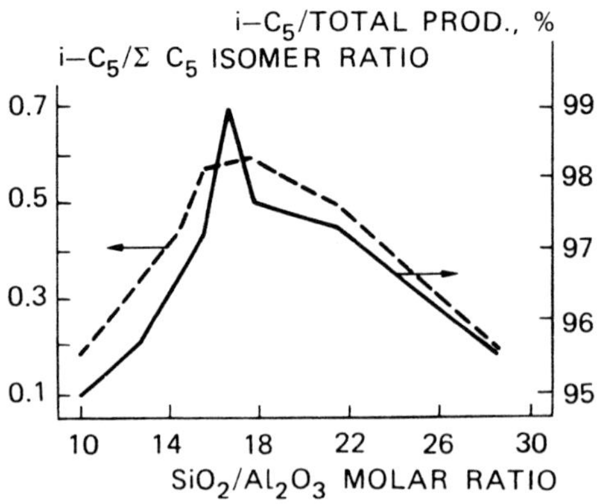

Figure 12. Effect of mordenite SiO_2/Al_2O_3 ratio for pentane hydroisomerization (from ref. 13).

12 gives an example of the influence of the Si/Al ratio on activity and selectivity [13]; solid-state NMR has proved to be a powerful technique to monitor the zeolite changes during such treatments [14]. The optimum Si/Al atomic ratio is about 10, which is very close to the value that gives the highest acidity according to Barthomeuf [14a] on the basis of the density of acidic sites with Si as next nearest neighbours. Another catalyst manufacturing step which has been shown [15] to influence performance is the calcination, as shown in Figure 13.

Research on improved isomerization catalysts continues, as is demonstrated by the growing patent literature, and indeed Figure 14 shows that significant potential benefits may be obtained from catalysts with improved activity, as has been calculated, for example, using a hydroisomerization process model developed by Shell [10]. Zeolites beta and omega [15a] have been regularly studied for this application, omega apparently giving a higher yield of (the high- octane) dibranched isomers. Datema et al. [15b] have attempted a direct synthesis of mordenite with Si/Al atomic ratio of 10, so far without success, however.

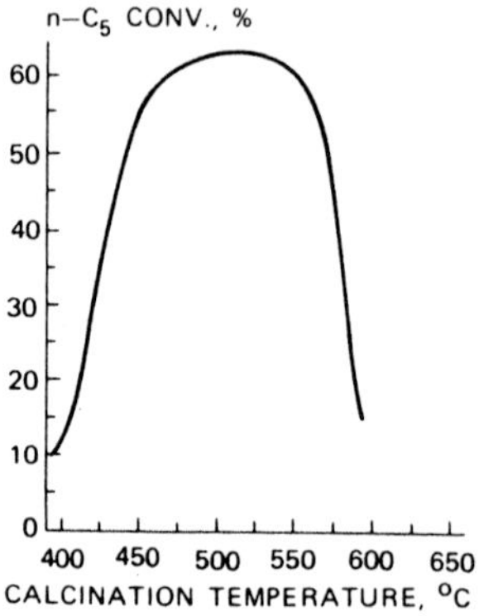

Figure 13. Effect of mordenite calcination temperature on pentane hydroisomerization activity (from ref. 15).

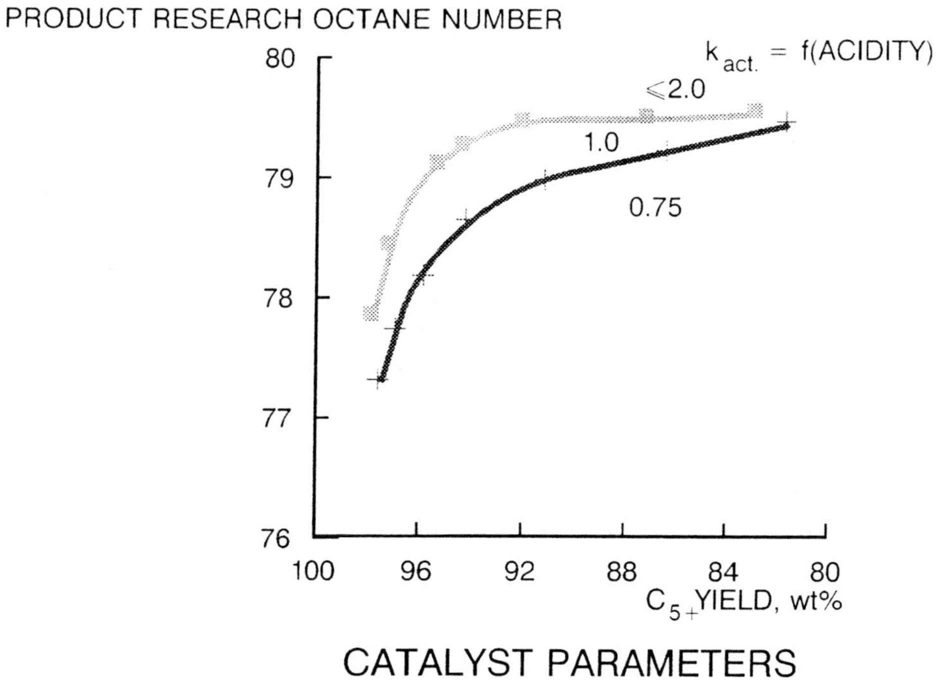

Figure 14. Hydroisomerization catalyst modelling (from ref. 10).

2.3 Skeletal isomerization of olefins

In recent years the skeletal isomerization of olefins has gained much attention, since in this way additional amounts of iso-butene, which is the feedstock for MTBE, can be produced. MTBE is a well known gasoline octane booster, now becoming disfavoured for environmental reasons, but a few years ago a definite shortage of this chemical was predicted. A number of processes have therefore been developed for the conversion of n-butene into iso-butene. In august 1991 Texas Olefins company started the first commercial scale development unit (228 mT/d) for isomerization of butene, using the so-called SKIP process, which uses an inexpensive catalyst with a lifetime of a few hours, operated in a swing-reactor system [15c]. IFP has also devoted a considerable effort to developing a catalyst for the skeletal isomerization of butene or pentene, and developed a proprietary alumina-based catalyst, again with a short lifetime, apparently of the order of 100 hours. This is to be operated in, for example, a continuous reformer-type unit [15d].

The stability of the zeolite catalysts introduced more recently for this application is dramatically better, as can be estimated from Figure 14a [15e]. Industrial research work at Shell [15e] and Texaco [15f] has shown ferrierite to be a very attractive catalyst for this application. Comparison with other medium-pore zeolites such as TON and MFI suggests ferrierite to be superior in selectivity, giving less C_5+ by-product, as shown in figure 14b [15e, 15f]. Mobil and BP have jointly developed the ISOFIN process, and improved on the earlier BP catalyst [ref. 15g]. Lyondell has put the earlier Shell work into commercial use, also employing further improved ferrierite catalysts [15h].

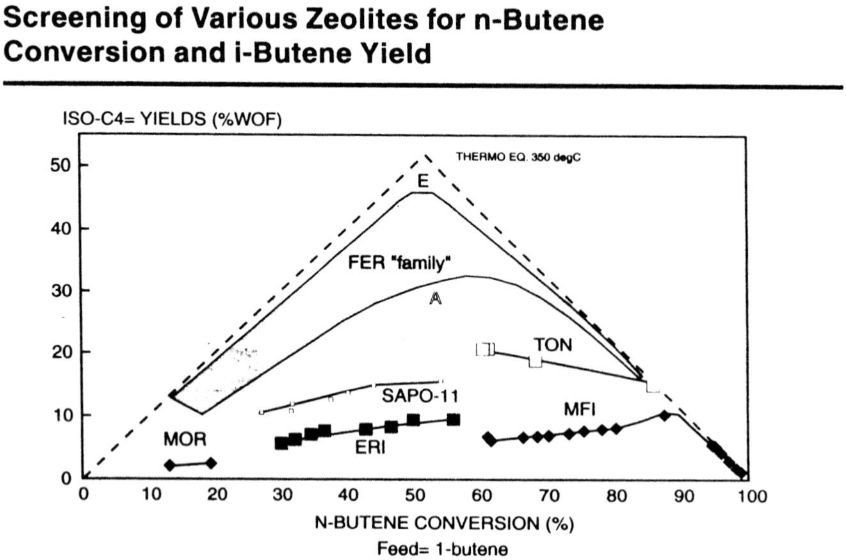

Figure 14a. Screening of various zeolites for n-butene conversion and i-butene yield.

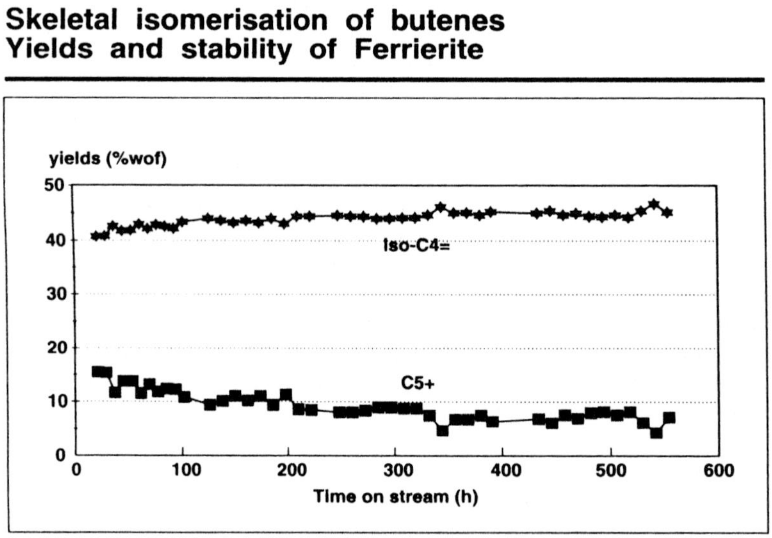

Figure 14b. Selectivity and stability of ferrierite in olefin isomerization (ref. 15e).

The catalytic chemistry is of interest since, for thermodynamic reasons, low temperatures are desired, while the main non-selective reaction is the dimerization of butene (or pentene) . Mooiweer et al. [15e] have suggested that the reaction proceeds via dimerization inside the zeolite cages, followed by isomerization and selective cracking to isobutene, the octene intermediates being unable to escape the zeolite. Guisnet [15i]) has pointed out that the operating catalyst has a high level of carbonaceous species, and that possibly the butene is attached to these, and isomerized in that way, in fact at the crystal surface of the zeolite. Also, Ponec et al. [15j] have opposed the bimolecular mechanism, arguing that a higher production of propene and pentene would be expected with such a mechanism. For recent work on the kinetics of the isomerization of linear butenes over H-FER and on the role of the acid site location, see ref. [16a,b].

2.4 Olefin/paraffin alkylation

In many refineries olefins and paraffins are converted into highly branched octanes over liquid acids such as HF or sulfuric acid. In principle, the alkylation reaction of butene with iso-butane is a very favourable way to increase the gasoline pool with high-octane components (currently, around 10 vol.% of gasoline consists of alkylate); however, in practice the reaction is not easy to achieve. For a selective alkylation to be favoured over undesired olefin oligomerization, high excesses of iso-butane have to be used in the feed and recycled, while catalyst life in general is short. In plants with sulfuric acid this leads to large amounts of spent acids, while HF gives rise to environmental concern. This area has been reviewed from a catalytic viewpoint by Corma [15k] and Weitkamp et al [15kk].

Obviously a heterogeneous catalyst, and preferably an environmentally friendly one, would be desirable. In recent years, therefore, a number of heterogeneous-catalyst based processes have been taken into development, using catalysts such as supported BF_3, SbF_5, CF_3SO_3H [15l], etc. As these often still involve environmentally unfriendly chemicals, research into zeolite based alkylation catalysts has also continued [15k, 15m] . Early work in this direction had been carried out, for example, at the Sun laboratory. It early became clear that the main problem in this area is catalyst stability; after a relatively short time the catalyst loses its alkylation activity, retaining only olefin oligomerization activity.

Mesters et al. [15n] have shown that, in faujasites, a high density of acid sites is preferred for catalyst life, but that zeolite beta gives an even better performance. The reaction is preferably carried out in slurry operation, while the regeneration of these catalysts is achieved through hot hydrogen treatment, particularly effective by the promotion of the catalyst by small amounts of noble metals. Nevertheless, it appears that, at present, processes using heterogeneous catalysts based on supported super-acids are leading the process developments, which has at least reached the pilot plant stage [15kk].

2.5. Gasoline-upgrading processes

Two different types of gasoline-upgrading process can be distinguished which utilize zeolite-based catalysts. The first type involves a shape-selective cracking mechanism and the second type makes use of shape-selective aromatization reactions.

2.5.1 Shape-selective cracking

This process makes use of the shape-selective properties of medium-pore zeolites such as ZSM-5 to preferentially crack paraffinic components in gasoline to light gaseous hydrocarbons. An octane increase is achieved, since the most reactive n-paraffin molecules have the lowest research octane number and vice versa [16], as is demonstrated for ZSM-5 in Figure 15. The first commercial process of this type was the so-called selectoforming process from Mobil [17], which was introduced in the mid-1960s and later improved as the M-forming process [16], whereby the more shape-selective ZSM-5 zeolite was introduced as catalyst. Gulf researchers [18] were also active in this field and reported selective upgrading of reformates and straight-run gasolines using the medium-pore zeolite, ferrierite.

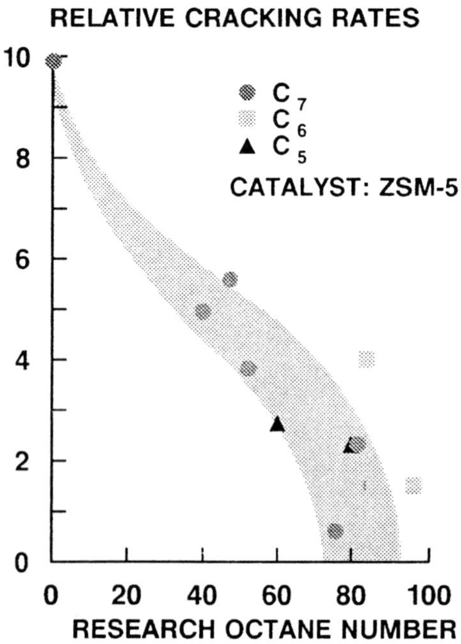

Figure 15. Cracking rates vs octane number: ZSM-5 catalyst (from ref. 16).

The hydroprocessing conditions typically applied for gasoline upgrading via shape-selective cracking are mild pressures (30 - 50 bar) and temperatures in the range 300 - 350°C. The commercial success of these processes has, however, been rather limited due to the relatively high loss in gasoline yield compared to the octane gain. In addition, the saturated by-product light gases have, in general, a relatively low value.

More recently, this basic concept has, however, been applied in catalytic cracking technology, where a shape-selective zeolite cracking component (e.g. ZSM-5) is introduced as a co-catalyst to increase gasoline octane. This will be discussed in more detail in the section dealing with catalytic cracking.

2.5.2 Desulfurization of naphtha ex fluid catalytic cracking

Naphtha obtained from an FCC unit usually contains a high concentration of sulfur compounds (unless the feedstock for the FCC unit has been first desulfurized), and also a high concentration of olefins, the latter compounds being favourable for the octane number (see above). In conventional desulfurization over (Ni/Mo or Co/Mo) HDS catalysts, olefins are significantly hydrogenated and, under deep desulfurization conditions, a clear loss in octane number results. In some cases this may be detrimental in the overall refinery scheme and, for such cases, processes were developed whereby, after desulfurization, the low octane components are selectively cracked to gas, using ZSM-5 type catalysts, at the expense of a lower yield of naphtha. Examples of such processes are Octgain by Mobil [18a] and ISAL by Intevep/UOP [18b].

2.5.3. Aromatization reactions.

An interesting, more recent development has been the application of zeolites to gasoline upgrading by means of selective aromatization reactions. Aromatization is an important catalytic reaction in a modern refinery/petrochemical complex to produce both high-octane gasoline and benzene/toluene/xylenes (BTX). The technologies currently applied are Fluid Catalytic Cracking (FCC) and Catalytic Reforming. In ("conventional") catalytic reforming, using bimetallic Pt/X/Cl/alumina catalysts, the advent of continuous reformers has allowed refineries to operate at low hydrogen partial pressures, giving clearly higher C_5+ yields than in the past with the semi-regenerative units.

A new approach which has been developed by Chevron workers [19] addresses the fact that conventional reforming catalysts have relatively poor aromatization selectivities for C_6 and C_7 aliphatic components, as shown in Figure 16. By contrast, a novel catalyst based on

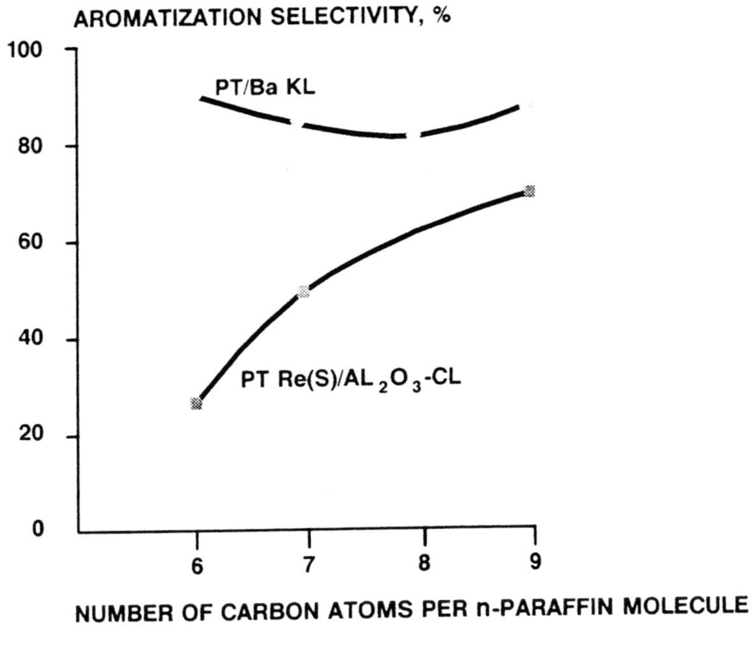

Figure 16. Catalytic reforming with zeolite L (from ref. 19).

zeolite L has been shown to exhibit high selectivities towards aromatics formation, including low-molecular weight paraffins such as hexanes and heptanes (see Figure 16). The catalyst is composed of highly dispersed platinum clusters in barium and/or potassium-exchanged potassium zeolite L. A significant feature of the catalyst is the lack of acidity, with the aromatization properties being attributed to a combination of the platinum clusters and the shape-selective properties of the zeolite. Lane et al. [19a] conclude, from a comparison between Pt/K L and Pt/K Y zeolites, that the high selectivity in zeolite L is indeed consistent with a confinement model [19b], in which *n*-hexane is adsorbed as a 6-ring pseudocycle.

Derouane, however, reported that similar aromatization properties were shown by Pt supported on stabilized magnesia [19c], which would suggest that shape selectivity is not an important factor. Recently, Chevron researchers published results on a number of alternative systems such as Pt/Cs/B/(Al)beta [19d].

On the basis of the Pt / L catalyst system, Chevron has developed a new process called AROMAX [20], which is primarily intended to be applied for aromatization of the lighter feedstocks and is thus complementary to conventional catalytic reforming. A major drawback of the catalyst system, however, appears to be the high sensitivity to sulfur contaminants. This generally requires an extra feedstock pretreatment step, which may have limited the application of this process route to date.

Undoubtedly, new process routes to achieve gasoline octane upgrading with high selectivity will continue to be of interest in the future as the phase-out of lead in gasoline continues on a global scale.

2.6. Middle distillate upgrading processes: hydrogenation

The past few years have witnessed a drive towards high cetane (low aromatics) diesel fuels, primarily stimulated by government regulations in Sweden and California. Obviously, a strong economic incentive exists to manufacture such products in existing, that is low-pressure, distillate-hydrotreating reactors. Because of thermodynamic constraints, this implies that the operating temperatures have to be low as well, and therefore catalysts have to be used that are highly active, even in the presence of significant amounts of S and N. Noble metals, supported on acidic zeolites, were found to be excellent catalysts for this application, the zeolite enhancing the tolerance to sulfur of the noble metals (Pt, Pd). (It is interesting to compare this with the catalysts in section 2.5.3, where basic zeolites imparted a very high sulfur-sensitivity to the noble metals. Clearly, the role of the zeolite in these systems is different from what was seen so far: rather than directly exerting (shape-selective) acid catalytic activity, here the zeolite serves to modify the properties of the supported noble metals.) The preferred line-up is such that the middle distillate is first passed over a conventional Ni/Mo/alumina hydrotreating catalyst, and then, mostly after gas/liquid separation, with "pure" hydrogen over the dedicated noble-metal catalyst, as shown in Figure 16a [20a]. The process line-up can vary according to circumstances; for further references see ref. 20b. At present, high cetane number middle distillates are manufactured commercially in, for example, Sweden and California, using such catalysts [20c, 20d].

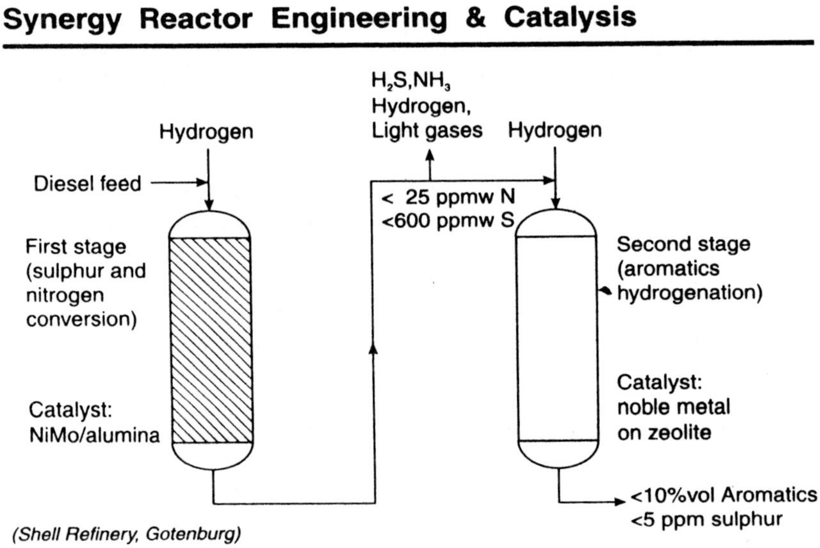

Figure 16a. Line-up of Gothenburg unit.

Middle-distillate upgrading through deep desulfurization and hydrogenation are often discussed together, since both involve severe processing. While the conventional approach to deep desulfurization is through dedicated, highly active Co/Mo/alumina catalysts, here also "second-stage" zeolite-containing catalysts have been proposed. At present these do not yet appear to have found commercial application but, if the trend towards lower sulfur specifications in diesel fuel continues, this could well change.

2.7 Catalytic dewaxing

The presence of wax, which essentially consists of normal paraffins, is often a problem in terms of product quality for gas oils and lubricating base oils, because of wax crystallization at low temperatures. The cold-flow properties of these products are sensitive to even quite low concentrations of waxy components and therefore processing is required to selectively remove the n-paraffin molecules.

If the wax is selectively removed from a gas oil, for example, a heavier fraction may be included in this product, which can often be attractive economically. This is illustrated by the fact that some 20 MDDW (Mobil distillate dewaxing) units are reported to be operational [20f], while also the AKZO/FINA process has been licensed to a number of companies [20g]; in addition, CFI (Cold-Flow Improved) catalysts from suppliers such as Suedchemie are being used commercially. The chemistry of lubricating base oils is quite complex: a crucial parameter here is the viscosity index, VI (the temperature dependence of the viscosity, a high VI indicating a low temperature dependence); *n*-paraffins have very high values, whereas branched molecules and ring systems have lower values. Since *n*-paraffins crystallize as wax, an ideal lube oil consists of saturated molecules with some ring structures and slightly branched paraffin chains which do not easily crystallize. Thus, from straight-run oil fractions, first the aromatic components are removed by furfural extraction or by hydroprocessing (which also removes the heteroatoms S and N, which are potential poisons for the dewaxing catalyst, and is increasingly applied), and finally the wax (highly linear paraffin molecules) . Dewaxing is conventionally carried out by solvent extraction, but nowadays catalytic dewaxing is becoming increasingly popular. It has been estimated that gasoil dewaxing and cold-flow improvement is applied worldwide in some 40 units, while the dewaxing capacity for lubricating base oils amounts to about 5 million tonnes p.a. (i.e. some 15 % of the total base capacity).

The catalytic dewaxing process is carried out in trickle-flow operation over a (bifunctional) shape-selective zeolite catalyst, under a hydrogen flow. The hydrogenation function (often noble metals, though also Ni and Ni/W have been described) improves the catalyst selectivity and stability. Depending on the type of feedstock (i.e. whether or not hydrotreated) the catalyst life can vary from weeks to years; regeneration is often achieved through hot hydrogen stripping. The technology used to be dominated by Mobil, but recently also Chevron, AKZO-FINA , Criterion/Zeolyst/Lyondell and others offer catalytic dewaxing technology and catalysts.

In catalytic dewaxing one uses the shape-selective properties of a zeolite to selectively convert the normal paraffins, either by (hydro)cracking to lighter products or, preferably, by isomerization to branched paraffins. This shape selectivity is illustrated in Figure 17, taken from ref. 21, for ZSM-5: clearly the cracking rate decreases with increasing degree of branching. The selective removal of the normal paraffins in gas oil- dewaxing is nicely illustrated in Figure 18, taken from ref. 22, where the sharp peaks of the normal paraffins in the gas chromatograph of the feed are clearly absent in the product after catalytic dewaxing. Figure 19 shows that with increasing depth of dewaxing and thus decreasing pour point of a lube oil, the alkanes are selectively converted (ref. 23). The medium-pore-size (10-ring) zeolites appear to be particularly suitable to obtain a high selectivity, as is demonstrated in the comparison between ZSM-5 and mordenite (ref. 23), (the first generation BP catalyst), as shown in Table 5.

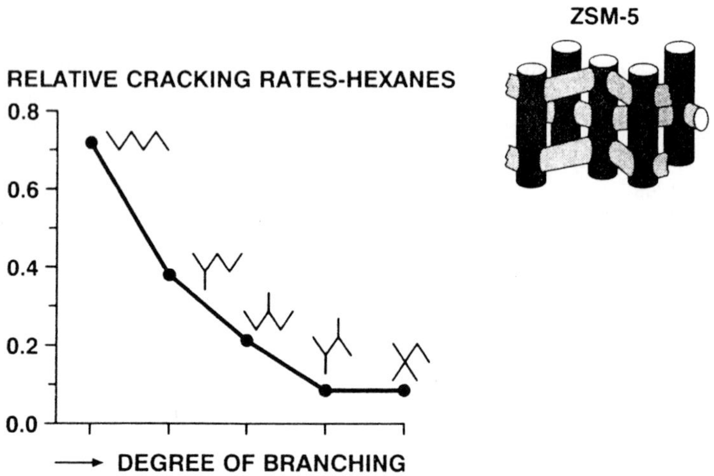

Figure 17. Shape-selective catalytic dewaxing (from ref. 21).

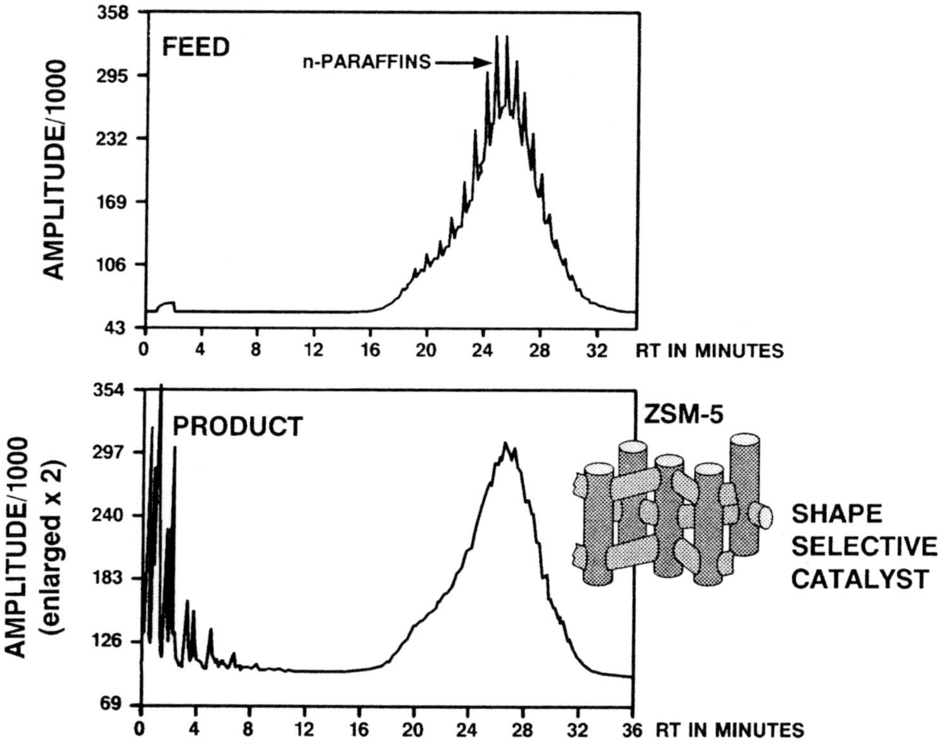

Figure 18. Shape-selective catalytic dewaxing (from ref. 22)

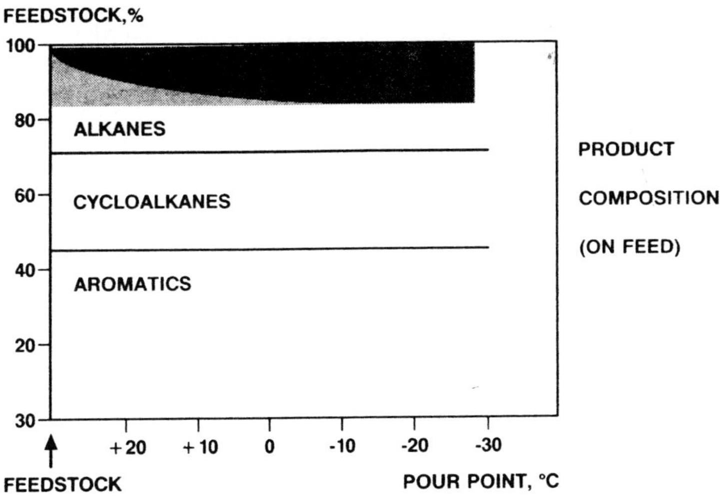

Figure 19. Molecular composition of lube oil as a function of pour point (from ref. 23).

Table 5
Dewaxing with ZSM-5 vs. mordenite (from ref. 23).

Catalyst		Feed	ZSM-5		Pt/mordenite	
Conversion,	%w	0	20	23	27	36
Product pour point,	°C	+35	-12	-26	-12	-26
Viscosity index		-	94.3	89.4	89.6	77.6

Table 6
Dewaxing over ZSM-5 and ZSM-23 (from ref. 23).

		ZSM-5	ZSM-23
Pore size,	nm	0.54 x 0.56	0.45 x 0.56
Conversion,	%wt	15	11
Product pour point,	°C	-12	-12
Viscosity index		101.0	108.7

However, also within the medium-pore size zeolites, optimization is possible, as is illustrated in Table 6, where ZSM-5 and ZSM-23 are compared [23]. ZSM-23 which has a somewhat smaller pore size, gives a higher lube-oil yield, of a higher quality (viscosity index, VI) than ZSM-5 at equal product pour point. In addition to zeolite structure, crystallite size and aluminium content are also important variables to further optimize catalyst performance. As shown in Figure 20 [24] there is a significant advantage in activity and stability for the smaller zeolite particles, while Figure 21 [25] suggests that the cracking activity can also be optimized by controlling the aluminium content and steaming conditions [25a]. According to Ratnasamy et al. [23a] isomorphous substitution in ZSM-5 of Al by B, Fe (or Ga) results in catalysts with an increased selectivity, albeit at the expense of activity, in line with the phenomenon described in Section1.

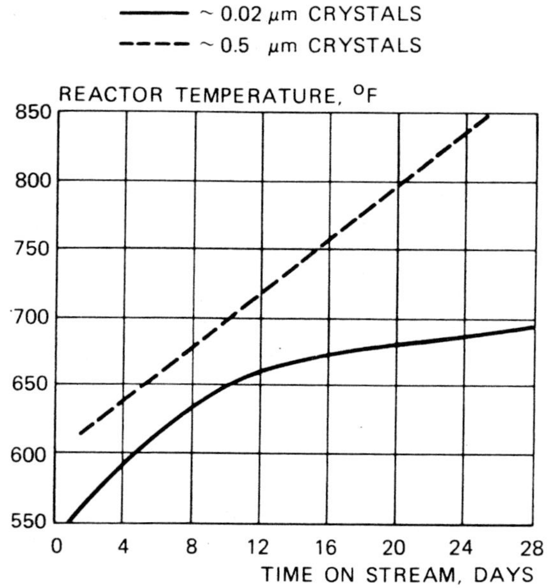

Figure 20. Influence of ZSM-5 crystallite size (from ref. 24)

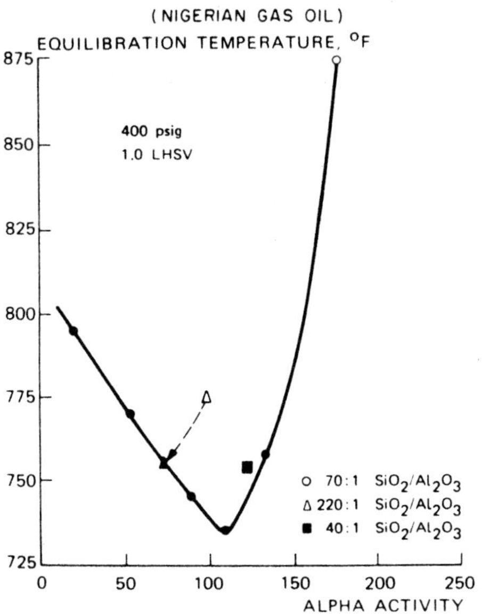

Figure 21. Effect of alpha activity (adjusted through steaming) on catalytic dewaxing activity (from ref. 21)

Table 7
Isomerization activity of ZSM-5 and beta-based catalysts (from ref. 26)

	Isomerization of raw light gas oil		
	Example No.		
	8 (Pt/Beta)	9 (Ni/ZSM-5)	10 (Zn/Pd/ZSM-5)
Reaction pressure kPa	6996	5272	6996
Temperature, °C	402	368	385
LHSV	1	2	2
Products percent:			
C_{1-4}	2.3	8.6	15.9
C_5–165 °C	16.1	11.4	19.8
165 °C +	81.6	79.1	64.3
Total liquid product pour point, °C	-53	-34	-54

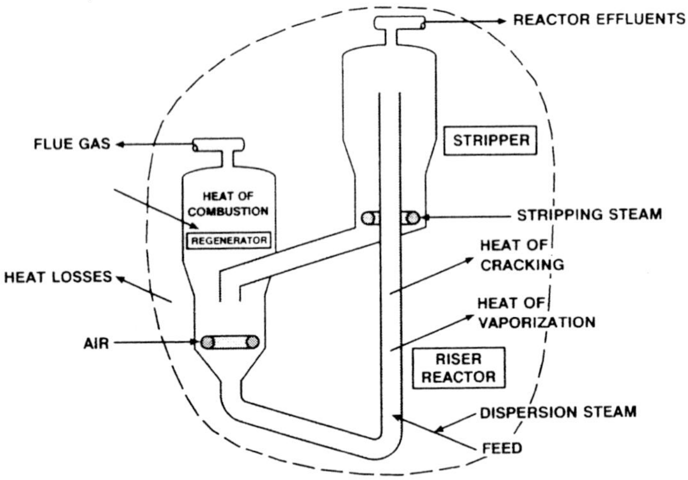

Figure 22. Simplified overall FCC configuration and heat balance.

2.8. Catalytic Cracking

Catalytic cracking is a moving-fluidized-bed process for the conversion of vacuum distillates and residues into olefinic gases, high-octane gasolines and diesel oil fractions over an acidic catalyst (see Figure 22). Fluid catalytic cracking has been a principal factor in refining since the late 1940s; a prime reason for the durability and perhaps the most significant characteristic of the cracking process is the flexibility in terms of the variety of feedstocks from a wide range of crudes which can be processed [27-29]. This is accomplished by cracking the vaporized feed over a solid acid catalyst (i.e. zeolite Y) at a temperature between 470 and 540°C and a pressure between 0.5 and 1.5 kg/cm2 gauge in a riser reactor, with contact times of 1-3 seconds. The catalyst promotes a reaction mechanism which can to a large degree be explained by carbonium-ion chemistry [28].

After catalyst/vapour separation at the end of the riser, the catalyst is transported to the regeneration section, where about 1 %w carbon on catalyst and a small amount of hydrogen present as hydrocarbon on the catalyst is burnt off. The reactions in the regenerator are exothermic, and produce temperatures of about 700°C and a 20 % steam environment. The hot catalyst circulates between the reactor and the regenerator, which acts as a heat carrier.

Conventional FCC using vacuum distillate feeds operates in heat-balanced operation, where the heat released in the regenerator is sufficient to vaporize and catalytically crack the feed.

However, the present trend towards heavier-residue-containing feedstocks results in high catalyst coke makes, which can cause excessive regenerator temperatures in heat-balanced operation, leading, for example, to hydrothermal catalyst deactivation and low catalyst circulation rates (i.e. cat/oil ratios). To solve these problems, new developments such as regenerator catalyst cooling have been introduced to remove the excess heat and give increased operational flexibility [27]. Hence, modern residue FCC units with this configuration employ heat-balanced operation.

2.8.1 Zeolite Y catalysts.

The introduction of zeolite Y base catalysts, composed typically of zeolite (10-50%), matrix and binder (see Chapter 15) in the mid-1960s had a dramatic impact on the performance of FCC units. This is illustrated in Figure 23, where the pronounced increase in gasoline yield is evident.

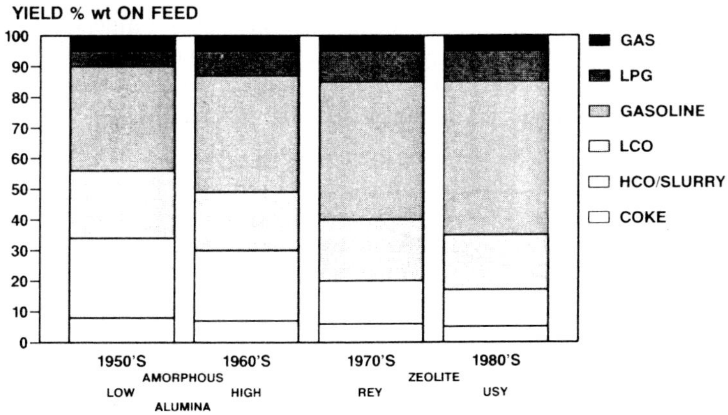

Figure 23. FCC catalyst development history (from ref. 1).

Further, the high activity and relatively low coke make of zeolite catalysts enabled the reactor technology to advance from dense fluidized beds to short-contact-time risers with an associated improvement in performance (see Figure 24).

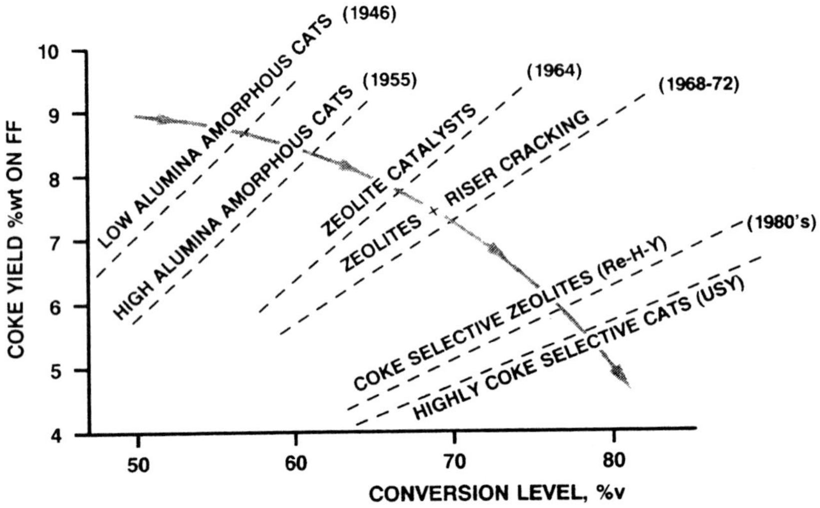

Figure 24. Trends in catalytic cracking catalyst performance (from ref. 1).

However, the shift towards heavier residue-containing feedstocks and the need for increased gasoline octane (due to the removal of lead additives for environmental reasons) continue to pose major challenges to FCC catalyst manufacturers.

The first generation of zeolite FCC catalysts [28] involved the use of acidic rare-earth-ion exchanged zeolite Y (i.e. REY zeolites). These catalysts were not only much more active for cracking reactions than the previous amorphous systems but they also influenced the composition of the gasoline. In fact, as shown in Figure 25, gasoline octane (i.e. both MON and RON) was significantly reduced. This can be attributed to an increased rate of hydrogen transfer reactions between olefins and naphthenes giving paraffins and aromatics.

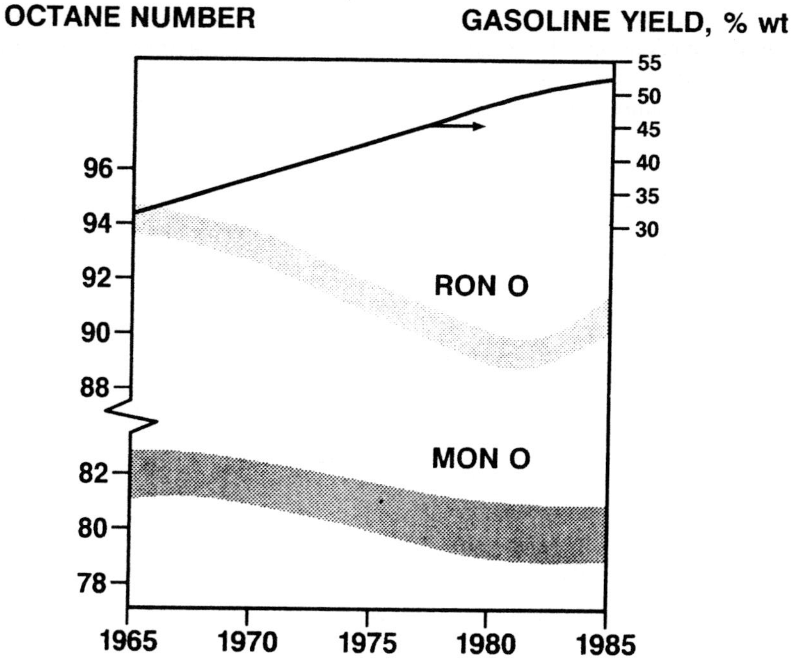

Figure 25. FCC gasoline octane and yield development histroy (from ref. 1).

It is this increased paraffin content of the gasoline, which is not sufficiently compensated for by the aromatics formation, that results in decreased gasoline octane. This problem has led to the introduction of so-called ultrastable zeolite Y (i.e. USY zeolites) materials [30,31], which exhibit lower unit-cell parameters and thus reduced acid-site density. Since hydrogen transfer is bimolecular this reaction path is suppressed in the zeolite Y based catalysts with lower acid-site density. This effect is illustrated in Table 8, which shows the FCC gasoline composition as a function of zeolite type [32].

Table 8
Catalytic cracking: comparison of FCC gasoline composition as a function of catalyst type (from ref. 32).
Conditions: reaction temperature = 520°C, conversion = 67%, contact time = 1.2 s.

Catalyst		Rare-earth Y	USY	Partial rare-earth USY
Composition C_5-C_{10}				
Fraction of gasoline, %w				
Saturates	normal	2.1	2.0	2.3
	branched	21.5	14.7	16.4
Olefins	normal	12.1	11.6	13.1
	branched	19.4	23.1	23.6
Cyclic	saturates	11.5	11.8	9.7
	olefins	6.6	12.0	7.8
Aromatics		26.8	24.8	27.1
Total olefins		38.1	46.7	44.5
Calculated				
RON		89	94	94
MON		78	79	82

The enhanced olefin make due to hydrogen-transfer suppression is also shown to result in an increase in gasoline octane, particularly RON. This effect is also reflected in a marked decrease in paraffin/olefin ratio, for example, in the C_3-C_5 carbon range as a function of decreasing unit cell size.

An additional benefit of the USY zeolites has been reduced coke make, which has enabled the processing of heavier feedstocks with retention of heat-balanced operation. These benefits led to a quite dramatic shift away from REY- to USY-based catalysts as shown, for example, in Figure 26 for North America.

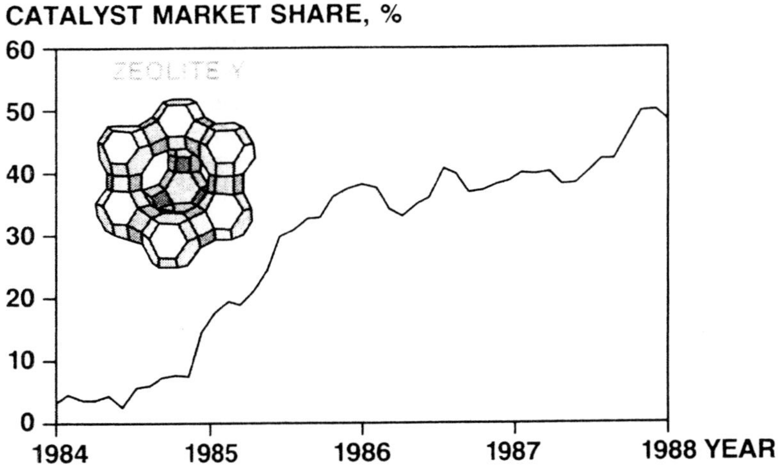

Figure 26. Application of ultrastable zeolite Y in catalytic cracking in North America (from ref. 32).

However, more recently some of the shortcomings of USY-type catalysts have been recognised and these include both low activity and insufficient gasoline MON. This has led to the hybrid-type catalysts whereby a balance is achieved between the REY and USY zeolites in order to optimize the advantages of both types of catalyst. As shown in Table 8, these partial rare-earth USY-type catalysts exhibit a better balance in terms of hydrogen-transfer reactions whereby a significant improvement is realized in terms of gasoline MON quality [32]. In addition, these new-generation catalysts make use of more modern techniques to manufacture USY, which resulting in a more nearly defect-free zeolite Y structure and an associated improvement in performance, particularly in terms of hydrothermal stability. It is also noteworthy that, under the hydrothermal conditions in the regenerators of FCC plants, the unit cell sizes (UCS) of the zeolite components of the catalysts tend to reduce further and reach so-called equilibrium values.

This further shrinkage of the UCS under operational conditions needs to be taken into account during the catalyst design and development phase. For this reason, fresh catalysts are exposed to a hydrothermal treatment prior to testing in the laboratory, in order to simulate the effects of equilibration in the commercial unit.

Despite these quite extensive catalyst developments FCC gasoline quality remains a problem in the modern refinery complex, which is further exacerbated by the poorer quality of feedstocks which are being processed. This gasoline octane problem is reflected in the so-called "octane dip", which is shown in Figure 27, where the MON and RON show a marked minimum in the mid-boiling range (i.e. C_8 molecular weight).

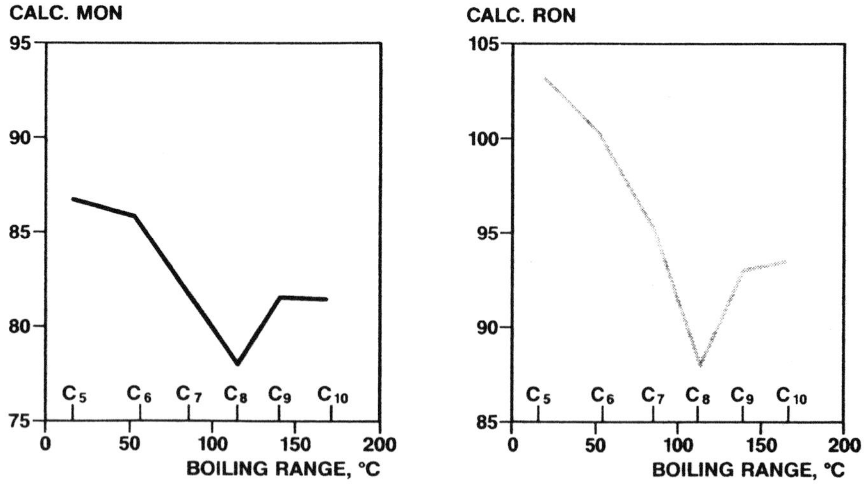

Figure 27. FCC gasoline octane as a function of boiling range.

This can be largely attributed to the presence of low octane saturated cyclic components in this carbon number range [33], as shown in Figure 28.

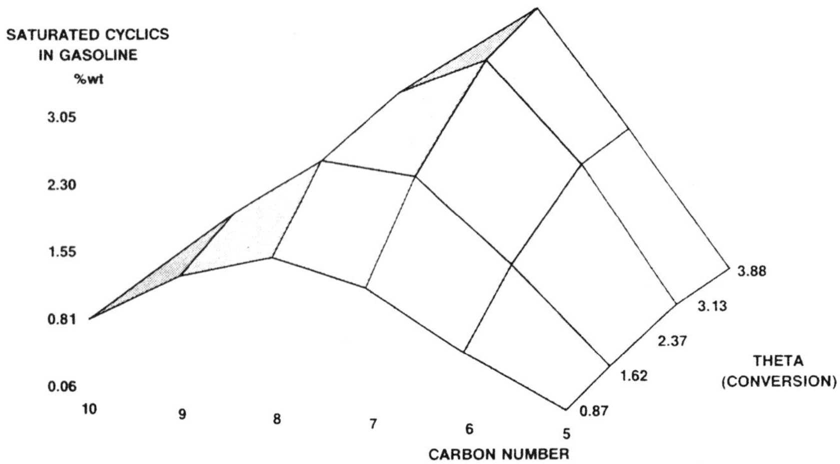

Figure 28. FCC gasoline composition as a function of conversion (from ref. 33).

Thus, there is scope for further improvement of FCC catalysts, which will require a good understanding of zeolite structure/catalytic performance relationships.

2.8.2 ZSM-5 additives

Several petroleum companies, viz. Exxon, Mobil and Gulf [34], have studied shape-selective zeolite catalytic processes for improving the octane quality of reformate and gasoline streams. The basic concept is similar to catalytic dewaxing, which involves the application of reactant shape-selectivity by which n-paraffin molecules are preferentially removed by selectively cracking to light (gaseous) hydrocarbons. The octane increase is achieved because the most reactive n-paraffin molecules have the lowest research octane number and vice versa for a medium-pore-size pentasil zeolite, such as ZSM-5.

In FCC the ZSM-5 has been used primarily in the form of a separate particle additive; however, it has also been successfully employed as a composite,i.e. a catalyst containing ZSM-5 and the faujasite cracking component in the same particle [35]. As discussed above, the purpose of ZSM-5 is to selectively centre-crack n-paraffins to produce primarily propylene and butylene as by-products, both of which are suitable for use as alkylation feedstock. Hence, if alkylation capacity is available, the alkylate yield more than compensates for the cracked gasoline loss, resulting in a net increase in gasoline plus alkylate.

Up to the present, the ZSM-5 additive has been mainly used in conjunction with REY catalysts, where the gasoline fraction is rich in *n*-paraffins. More recent studies [36], where ZSM-5 has been applied with USY-type FCC catalysts, have shown that low-octane components of the gasoline are, in this case also, preferentially cracked out to C_3 and C_4 olefins but with an associated loss in gasoline yield (see Figure 29).

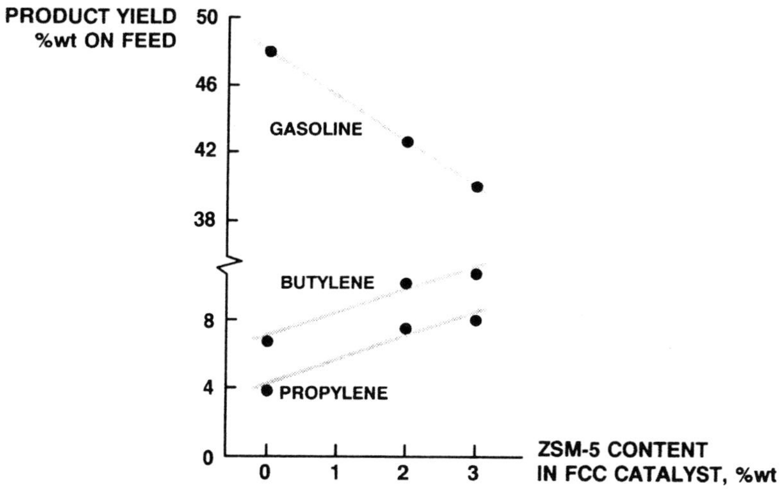

Figure 29. Product yields: ZSM-5 FCC co-catalyst (from ref. 36).

A corresponding increase in (calculated) RON and MON is observed, as shown in Figure 30.

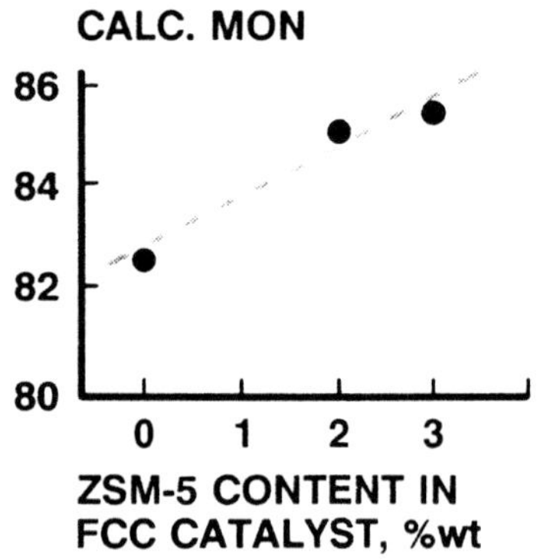

Figure 30. Calculated motor octane number: ZSM-5 FCC co-catalyst (from ref. 36).

An interesting finding of this study [36] was that, under FCC conditions, no preferential cracking of straight-chain relative to branched components occurred, as had been expected on the basis of catalytic dewaxing technology. The octane improvement was thus shown to be primarily due to an increase in concentration of aromatics in FCC gasoline (see Table 9). The observed cracking of branched paraffins likely reflects the reduced shape-electivity of the ZSM-5 zeolite at the relatively high operating temperatures applied in FCC (i.e. 480-520°C) compared with catalytic dewaxing (300-350°C),

This concept of using zeolite co-catalysts to modify the performance of a generic FCC catalyst system (for example zeolite Y) can, in principle, significantly increase the product flexibilty of the catalytic cracking unit. In fact, an extension of this technology whereby the shape selective zeolite (ZSM-5) is used as the primary rather than the co-catalyst has led to a process termed Deep Catalytic Cracking (DCC) where lower olefins are the major product and gasoline is a by-product (see Tables 9a and 9b).

Table 9
Catalytic Cracking Gasoline composition with ZSM-5 FCC co-catalyst (from ref. 36).

Added ZSM-5,		%wt	0	2	3
Conversion,		%wt	68.5	71.1	70.3
Gasoline yield,		%wt	48.2	42.5	39.8
Gasoline,		%wt			
Saturates	normal		2.0	1.7	1.8
	branched		17.9	14.8	14.0
Olefins	normal		12.7	10.1	8.3
	branched		26.3	26.5	28.0
Cyclics	saturated		5.5	8.1	7.5
	olefinic		4.9	2.7	2.4
Aromatics			30.6	36.1	37.8
Calculated					
RON			95.9	98.7	100.1
MON			82.7	85.2	86.2

Table 9a
Deep catalytic cracking process conditions.

Item	DCC	FCC
Typical residence time, sec	10 - 60	1 - 30
Cat / oil	9 - 15	5 - 10
Steam, wt% of feed	10 - 30	1 - 10
Reactor temperature, °F	950 - 1070	950 - 1020
Pressure, psig	10 - 20	15 - 30
Catalyst	ZSM-5	Zeolite Y

Table 9b
Deep catalytic cracking process: typical yields.

Olefin	wt%
Ethylene	3.5
Propylene	22.9
Butylene	17.3
Isobutylene	6.6

(Sinopec, 1997)

2.8.3. Catalytic cracking to lower olefins

The use of ZSM-5 as an additive [36a, 36b] in FCC units has increased steadily in recent years, particularly in the US, to meet the growth in demand for isobutene which has been driven by the increased application of methyl *tert*-butyl ether (MTBE) as a high octane gasoline additive.

The Deep Catalytic Cracking (DCC) technology, which involves the use of ZSM-5 as the primary catalyst in a FCC type of moving bed reactor system to maximize lower olefins production with gasoline being a by-product, has been developed primarily by Sinopec researchers in China [36c]. In addition to deploying ZSM-5 as a single catalyst system, different process conditions compared with conventional FCC are employed, such as shorter contact times and higher temperatures to further maximize lower olefin yields. Some typical lower olefins product yields under DCC operating conditions are given in Table 9b. Normal FCC feedstocks such as vacuum distillates can be used in the DCC process but paraffinic feeds tend to give higher lower olefin yields.

Sinopec have commercialized this technology in China [36d] and have recently licensed it for the construction of a DCC plant in Thailand. Application of DCC technology will likely be attractive in situations where there is a relatively high demand for propylene compared to ethylene and where lower olefin pipeline networks have not yet been developed.

2.8.4 Residue catalysts

With the continued growth of residue processing in FCC units, increasing attention is being given to improving the performance of catalysts for this type of operation. Some catalyst aspects of importance in this respect are:

- coke make
- bottoms conversion
- metals resistance
- gas make

Both the zeolite and the matrix play an important role in the optimal performance of a residue catalyst [37]. Some properties of the zeolite structure which are considered to be important are:

- mesopore structure for bottoms conversion
- unit cell parameter for coke make

- defect structure for hydrothermal stability
- resistance of zeolite structure to vanadium attack

The resistance to performance decline due to nickel and vanadium contamination is one of the most challenging problems in the development of modern residue FCC catalysts. Approaches to this problem include the use of nickel passivation agents (e.g. antimony or bismuth) to suppress the undesirable dehydrogenation reactions which are catalysed by nickel [38]. For vanadium metal, traps such as basic alkaline earth oxides, or metals such as strontium, are often applied, which act as scavengers and thus prevent the vanadium from destroying the zeolite structure. However, although many of these techniques are effective they are generally limited to relatively low levels of contamination. This has led to the development of catalyst rejuvenation techniques which are claimed not only to remove metals but also to substantially restore catalyst activity [39,40].

Many challenging problems remain to further improve the performance of FCC catalysts, as the requirements in terms of both feedstock (i.e. poorer quality) and products (i.e. better quality) become ever more demanding. It is vital to the future of the FCC process that these challenges are met, and improved catalysts and, in particular, modified zeolites will likely play an important role in these developments. In recent years companies such as Akzo Nobel, Engelhard, Grace Davison, Intercat and Cosmo Oil have all developed new catalysts for bottoms upgrading. The approaches tend to vary from physical characteristics such as pore size to changing the composition by, for example, adding Lewis-acid components such as alumina.

2.9. Hydrocracking

Hydrocracking is an oil-conversion process of growing importance due to a number of structural trends in the refining and petrochemicals industries. These include a shift towards an increased demand for middle-distillate product (i.e. kerosine, diesel and gas oil) relative to gasoline, which is particularly evident in the Pacific Basin and the Indian sub-continent (see Figure 31).

Another trend which is evident on a global scale is the increasing environmental legislation. Hydrocracking ,which inherently produces oil products which have low concentrations of heteroatoms (i.e. sulfur and nitrogen) and aromatic components is, from an environmental viewpoint, an attractive processing route. Recently a useful introduction to the hydrocracking process has been published by Scherzer and Gruia [40a].

In contrast to the situation in catalytic cracking, catalysts for hydrocracking need to have a significant life; depending on the unit conditions, roughly between 0,5 year (mild hydrocracking, at low hydrogen partial pressure) and 5 years (two-stage operation at high hydrogen pressures). To this end, the feedstock is generally first denitrogenated, mostly over NiMoP/alumina catalysts, since the organic nitrogen compounds are severe catalyst poisons [49]. The denitrogenated feedstock is then passed over the hydrocracking catalyst, which can be done according to a number of different process line-ups, as outlined below. In all cases the hydrocracking catalyst operates according to the principles of bifunctional catalysis (see also under Section 2.2 Hydroisomerization), the (de)hydrogenation function being provided by noble metals or base metals, and the acid function by zeolites, or amorphous silica-aluminas, etc. The feedstock molecules are dehydrogenated to active intermediates that adsorb on the acid sites of the catalyst. These intermediates are subsequently cracked, and then, after hydrogenation, desorb again from the catalyst. "Ideal" hydrocracking refers to the situation in which, after one cracking event, the product molecules are hydrogenated and desorb again; this results in a high yield of primary products, avoids overcracking, and hence minimizes gas make. "Ideal" hydrocracking has been studied in detail by Weitkamp et al.[40b]. Clearly the hydrogenation function has to be sufficiently strong to hydrogenate the intermediate to avoid further cracking; " in balance" therefore with the cracking function.

The demands on the hydrogenation function hence depend on the process conditions; when cracking is carried out in the presence of ammonia (see below, "series-flow"), the acid function of the catalyst is poisoned, and hence requires higher operating temperatures, also enhancing the activity of the hydrogenation function. Under these conditions the activity of the acid function will be limiting. In the "two-stage" process line-up where cracking is carried out in the absence of ammonia, temperatures are much lower, and it is the activity of the hydrogenation function that is generally limiting. Examples of this have been given by Ward [41]. In all cases, the intrinsic reactivities of the hydrocarbon molecules in the oil are such that first the side-chains of the aromatic molecules are cracked from the aromatic nuclei, after which the naphthenes are cracked, and only lastly the paraffins. This implies that, with increasing conversion, the unconverted product on average is increasingly paraffinic, as a consequence of which bottom products from partial conversion hydrocrackers can be very useful as feedstocks for ethylene crackers, for catalytic crackers and for lubricant base oil. Although in the latter case a dewaxing step may be necessary, it was outlined in Section 2.8 that "nearly" normal paraffins are the ideal lubricating base oil components. At present it is estimated that 15 % of all paraffinic lubricating base oils are manufactured through hydrocracking [41a].

The intrinsic reactivity sequence holds in the absence of any steric constraints, e.g. for wide pore catalysts such as amorphous silica alumina`s; when using zeolitic catalysts this sequence may be modified in that heaviest molecules and/or the multi-ring molecules encounter too strong diffusion limitations, resulting in lower effective reactivity; this is shown in Figure 31a [41b]. Obviously these effects are the strongest for heavy feedstocks, at low operating temperatures. In figure 31b [41] these effects are illustrated by Field-Ion Mass Spectroscopic analyses of the unconverted products obtained, in two-stage operation, over fully amorphous, fully zeolitic, and composite catalysts (composite catalysts contain both a zeolitic and an amorphous cracking function). The composite catalysts do not only have merits in balancing the conversion of the paraffinic and naphthenic molecules; also with feedstocks that have components over a wide boiling range these catalysts can balance the conversion of the lighter and the heavier components [41d].

The multicomponent nature of the feedstock, combined with the preferential conversion of some species relative to others can lead to special effects in hydrocracking recycle operation to (almost) complete extinction. When a catalyst is particularly suitable for conversion of e.g. the lower boiling molecules in the feedstock, the heavier molecules will build up in recycle, and, in order to achieve the desired high overall conversion, the lighter molecules will be overcracked [43]. For a successful high-conversion recycle operation, the conversion of the various feedstock components needs to be reasonably well balanced through proper catalyst design, or by deploying a combination of catalysts.

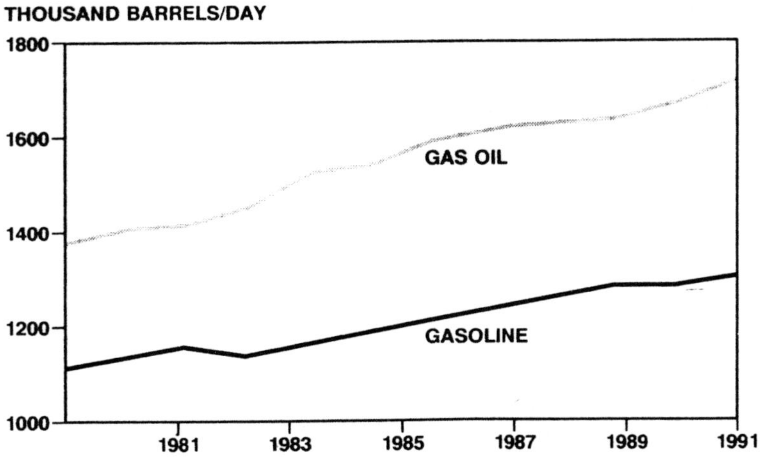

Figure 31. Demand for gas oil and gasoline in the Pacific basin.

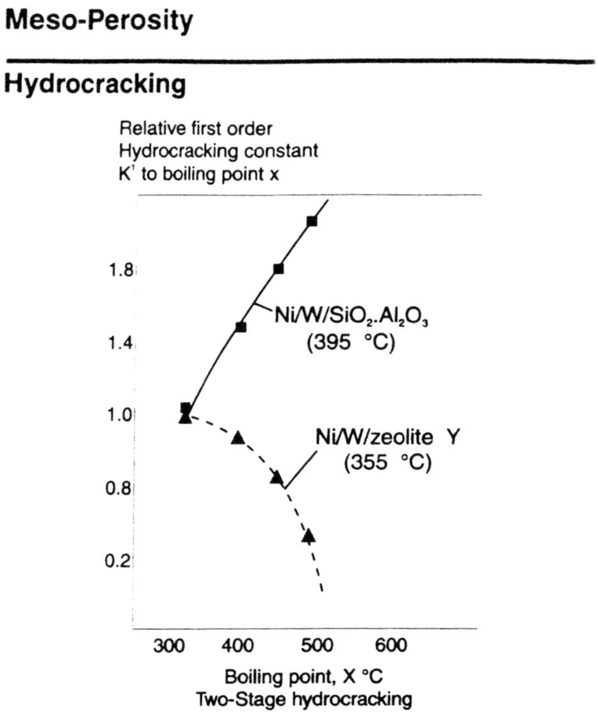

Figure 31a. Relative hydrocracking rates for $SiO_2.Al_2O_3$ and zeolite Y based catalysts.

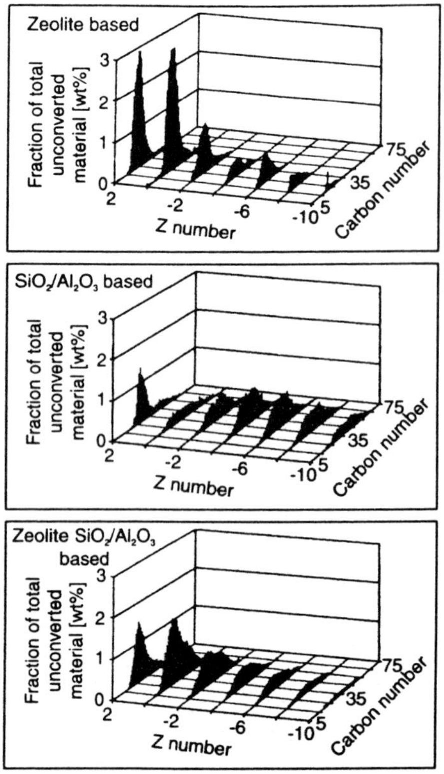

Figure 31b. Hydrocracking catalysts: FIMS spectrometry of unconverted product.

At least three types of hydrocracking process configurations can be distinguished. The simplest, so-called single-stage, configuration is that in which both the hydrotreating and the hydrocracking reactions are carried out in a single reactor (see Figure 32). Alternatively, the hydrotreating and hydrocracking steps may be performed in separate reactors. The most conventional, so-called two-stage, configuration consists of two reactors with interstage product removal (see Figure 33), whereby gaseous by-products such as H_2S and NH_3 are not carried through to the second-stage reactor. A more modern and cost-effective process is the series-flow configuration, which has no such interstage product separation (see Figure 33) and thus requires more robust second-stage catalyst systems such as those based on zeolites.

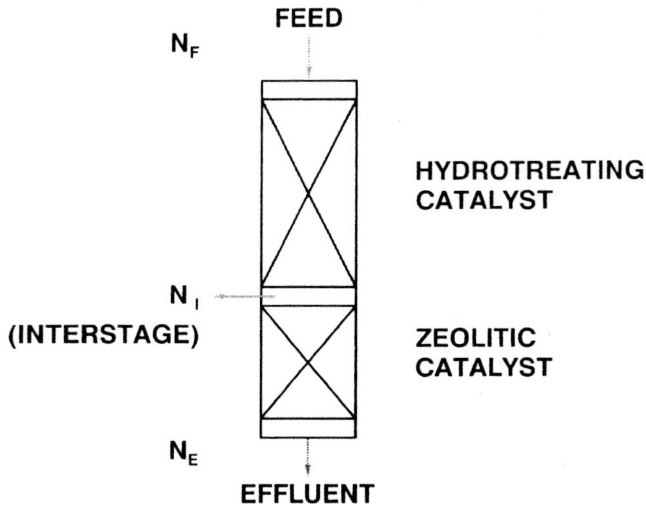

Figure 32. Stacked-bed reactor configuration.

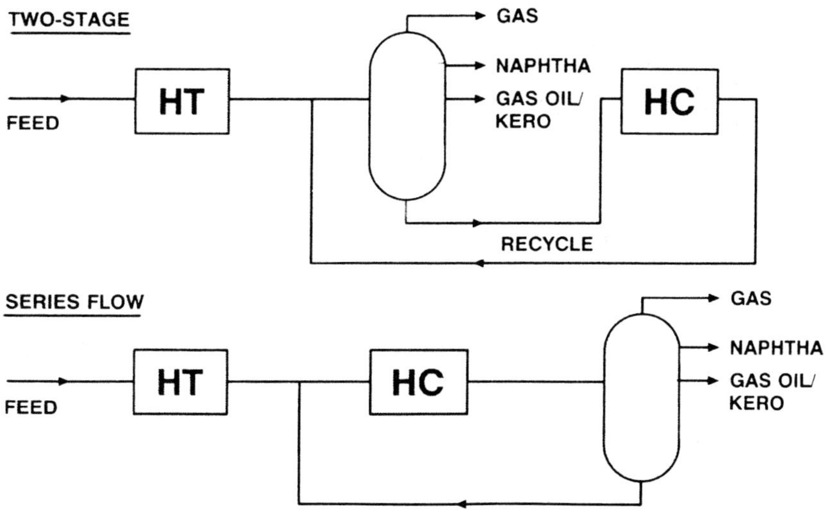

Figure 33. Hydrocracking process configurations.

A typical feature of the hydrocracking process is the product yield flexibility, as is demonstrated in Figure 34, whereby the primary product can be, for example, naphtha, kerosine or gas oil. The process conditions and the catalyst type applied can be used to shift the product yield as required [41].

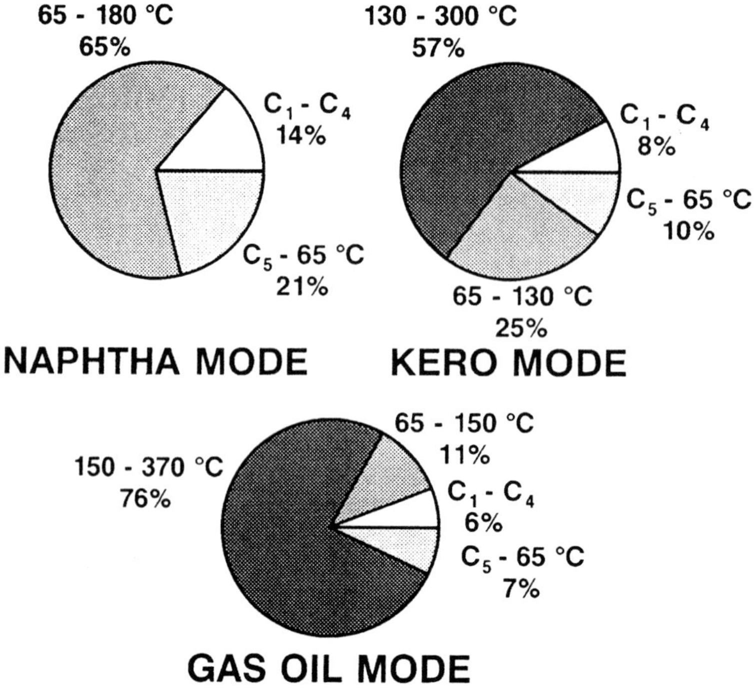

Figure 34. Hydrocracking process product flexibility (from ref. 42).

2.9.1 Zeolite-based catalysts

Although the first commercial hydrocrackers made extensive use of amorphous inorganic oxides such as aluminosilicates as catalysts, it was soon realised that zeolites with their high acidity and well-defined pore structure can offer substantial advantages [42]. One of the major advantages of zeolite catalysts is their markedly reduced coking tendency, which results in significantly improved catalyst stability. This is particularly important when heavy feedstocks are processed, where high rates of catalyst deactivation can become a major process constraint. The wide application of zeolite catalysts, for example, within the Shell Group hydrocrackers [42] has been an important factor in achieving a significant increase in feedstock heaviness and associated economic benefits (see Figure 35).

Another important characteristic of zeolite catalysts is their uniform and relatively high concentration of Bronsted acid sites, which results in high levels of hydrocracking activity. This is reflected in lower initial reactor temperatures and, together with reduced rates of coke deactivation, results in enhanced catalyst life. Typical performance data for amorphous aluminosilicate and zeolite catalysts are shown in Figure 36 where the benefits of the zeolite catalyst in terms of both initial activity and stability are apparent.

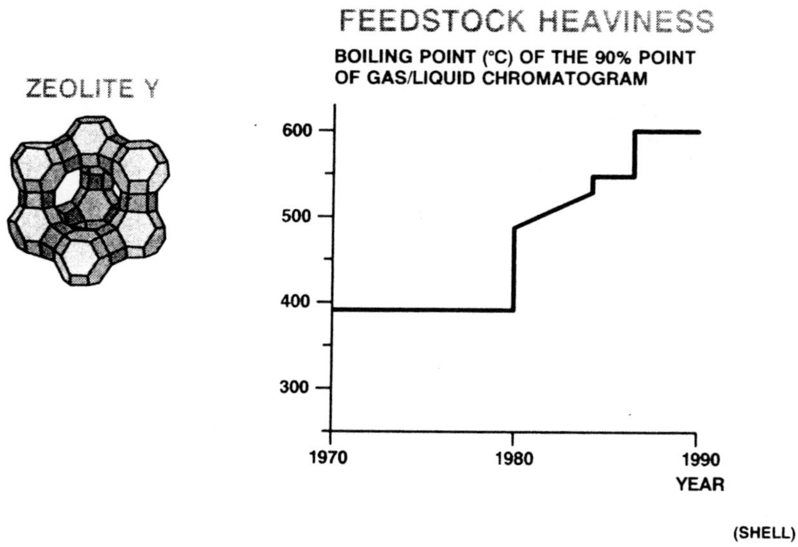

Figure 35. Hydrocracking application of zeolite Y (from ref. 42).

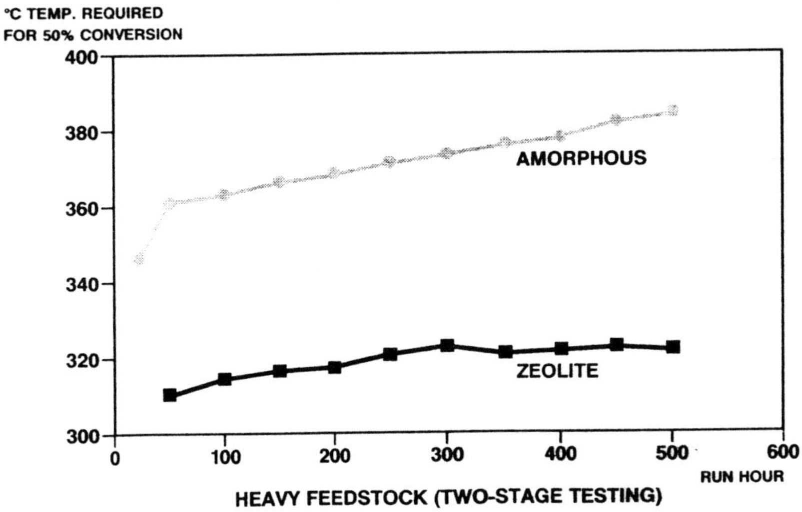

Figure 36. Hydrocracking catalysts (from ref. 44).

As regards selectivity, the zeolite-based catalysts also exhibit different behaviour compared with amorphous systems [43]. This can often be traced back to an imbalance between hydrogenation and cracking activity, caused by the strength of the cracking function of zeolites; in addition, the diffusion of the products out of a zeolite is slower than out of an amorphous material, thus enhancing secondary cracking. First-generation zeolitic hydrocracking catalysts are generally more naphtha- and thus less middle-distillate-selective than their amorphous counterparts. In North America this has been an advantage since in this region hydrocrackers are normally operated in a naphtha-selective mode. However, as previously mentioned, in the developing countries where there is a high demand for middle-distillate product, naphtha-selective zeolite catalysts are less applicable.

Another disadvantage of first-generation zeolite-Y-based catalysts was their relatively poor stability with respect to selectivity maintenance (increased gas make) with time on stream; this is particularly apparent when operating with heavy feedstocks and recycle to extinction [43,44]. This led to a significant research effort by a number of companies [41, 45-47] into the development of new zeolite-based catalysts which are more middle-distillate-selective. Studies at Shell [44], for example, have shown that the product selectivity of zeolite-Y-based hydrocracking catalysts can be profoundly influenced by varying the unit cell dimension (see Figure 37). This parameter can be regarded as a measure of the catalyst acid-site density. First-generation zeolite-Y-based catalysts typically exhibit high acid-site densities. As shown in Figure 37 it has been found that, by decreasing this unit cell parameter and thereby lowering the acid site density, the selectivity towards middle-distillate products can be markedly improved without a major activity penalty. Furthermore, this shift towards heavier products also results in reduced gas make and thus increased liquid yields.

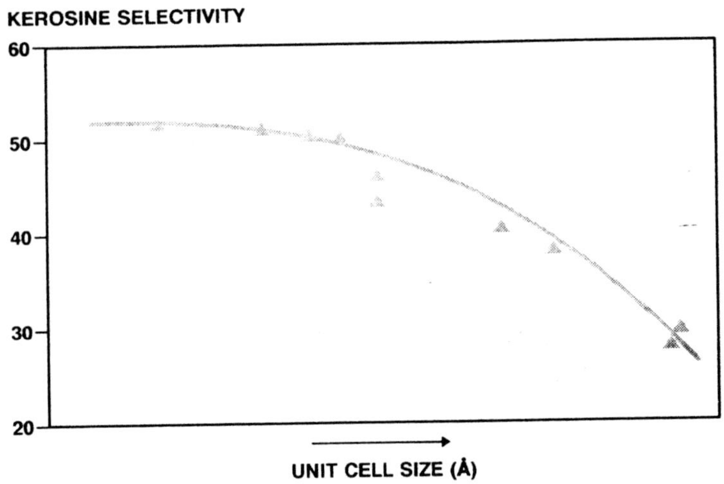

Figure 37. Zeolitic hydrocracking catalysts (from ref. 44).

It should be noted, however, that in addition to reduced acid-site density, lower-unit-cell zeolite Y materials are known to possess other characteristics such as a mesopore stucture and extra-lattice alumina [47], which most likely also contribute to the improved catalyst performance. In fact, low-unit-cell zeolite Y-based catalysts have been extensively studied for catalytic cracking and are currently widely applied to achieve reduced coke make and improved gasoline octane [31]. Although it is tempting to draw analogies it should be emphasized, however, that due to the markedly different catalytic chemistry involved, direct comparisons between catalyst properties and performance for these two processes should be treated with caution.

In addition, the stability with respect to selectivity during catalyst ageing is also markedly improved for low unit-cell-zeolite Y catalysts. This has been demonstrated in the laboratory by means of an accelerated ageing test (see Figure 38), where the low- and high-unit-cell zeolite Y-based catalysts are compared under the same process conditions. These improved performance features of low-unit-cell zeolite-Y-based hydrocracking catalysts have now also been demonstrated in commercial practice under recycle conditions [44]

2.9.2 Stacked-bed configuration.

In recent years the performance of single-stage and first stage hydrocracking units has been substantially improved by the introduction of so-called stacked beds [48]. As shown in Figure 32, this involves combining both hydrotreating and zeolitic catalysts within the same reactor configuration. The major advantage of such stacked-bed systems is that the conversion can be significantly increased without a penalty in deactivation rate.

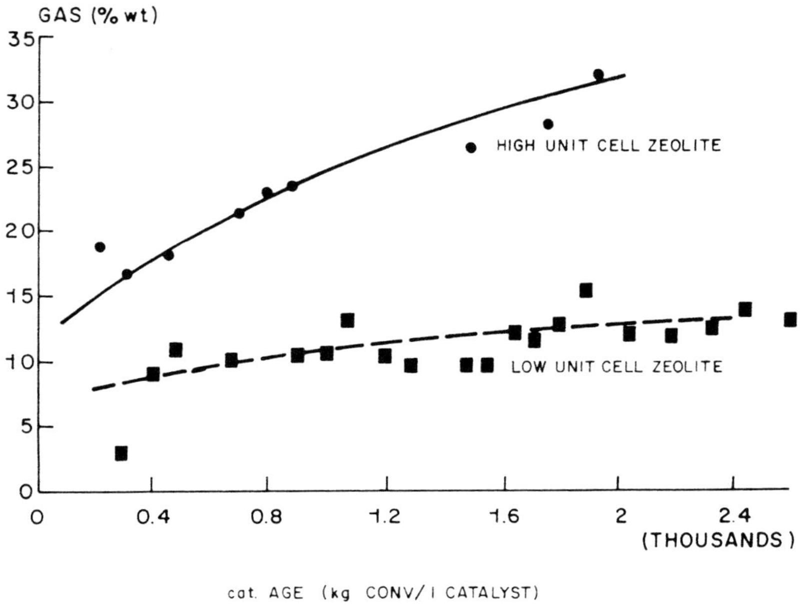

Figure 38. Zeolitic hydrocracking catalysts: accelerated ageing test (from ref. 44).

It is the high cracking activity of zeolites, in the presence of relatively high concentrations of organic nitrogen compounds, which makes these types of catalysts particularly useful for this application. The very substantial activity gains that can be typically achieved by the use of such stacked beds are shown in Table 10. Furthermore, the activity advantages of the stacked-bed systems increase with temperature, the apparent activation energies of these dual-catalyst systems being higher than that of the single hydrotreating catalyst [48].

However, one of the potential limitations of these stacked beds in the past has been the reduced selectivity to middle-distillate products, caused by the introduction of the zeolite component. The recent development of more highly middle-distillate-selective zeolite catalysts has now substantially reduced this disadvantage with retention of the activity gain [48], as is demonstrated in Figure 39 and Table 10.

The above discussion pertains to the relatively high pressures (e.g. 100-150 bar pressure) which prevail in conventional hydrocracking. However, the stacked-bed systems using zeolite catalysts can equally well be applied [49] under mild hydrocracking conditions, which are typified by lower operating pressures (for example 50-80 bar pressure).

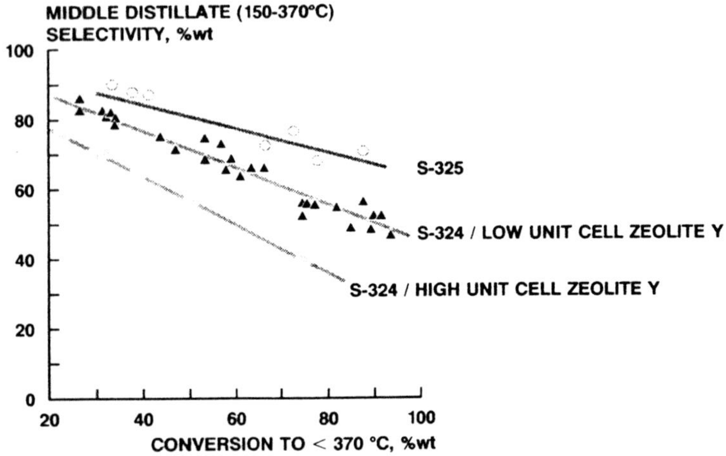

Figure 39. Middle distillate selectivity: hydrocracking catalyst stacked beds (from ref. 48).

Table 10
Hydrocracking Process: Comparison product of selectivities for single/stacked beds (from ref. 48).

Catalyst	S324		S324/LOW UCZ		S324/HIGH UCZ
Conversion to <370 °C, %wt	50	70	50	70	70
Gas	4	5	3	4	5
Naphtha/gasoline	23	28	25	45	56
Kerosine	21	24	33	25	20
Gas oil	52	43	39	26	19
Activity T req, °C	T	T+14	T−35	T−30	T−32

high unit cell zeolite: HIGH UCZ
low unit cell zeolite : LOW UCZ

2.9.3. Series-flow/two-stage configurations

New-generation zeolite catalysts have been successfully applied in conventional two-stage and series-flow modes of operation [44]. In particular, as previously mentioned, the high activity, stability and nitrogen resistance of zeolite catalysts have made the lower-cost, series-flow, mode of operation (see Figure 33) for middle distillate products also [49a] a technically feasible option which is likely to grow in importance.

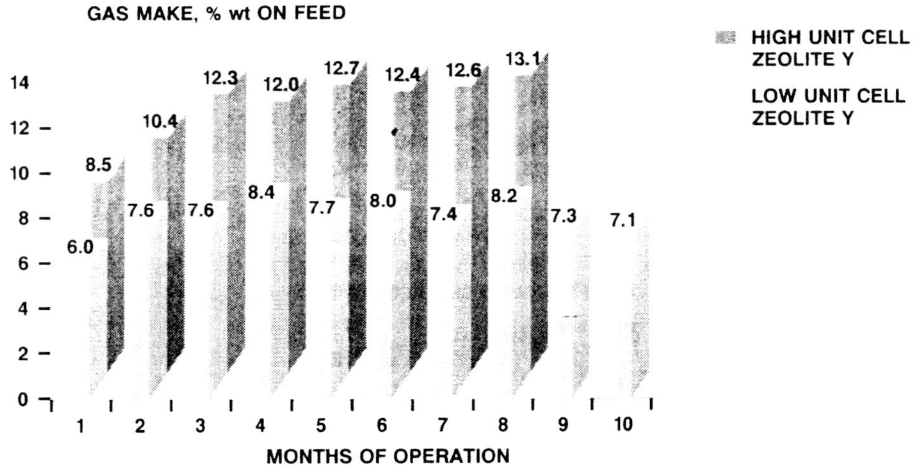

Figure 40. Zeolitic hydrocracking catalysts: commercial performance (series flow) (from ref. 44).

An example of the markedly improved performance experienced in commercial series-flow operation of new-generation zeolitic hydrocracking catalysts is shown in Figure 40. Clearly, the new low-unit-cell zeolite-Y-based catalyst results in a marked reduction in gas make, compared to the previously applied high-unit-cell zeolite-based analogue. Laboratory data [44] indicate that this improved performance is particularly enhanced when processing heavier feedstocks. Since this heavy feedstock trend is expected to continue, it is likely that these types of zeolite catalysts will find increasing application in hydrocrackers in the future. Furthermore, in view of the increasing importance of hydroprocessing routes and the significant advances that have already been made in applying zeolite-based hydrocracking catalysts, it is most likely that this field of research will continue to receive considerable attention in the future.

2.9.4 Middle distillate selective catalysts containing zeolites

The above discussion has emphasized the variations in zeolites which may be considered in designing hydrocracking catalysts. However, for processing very heavy feedstocks, zeolitic catalysts will always tend to be less selective than amorphous catalysts, because the reactivity distribution is shifted more towards lighter molecules than with amorphous catalysts as a result of the size-exclusion effect, as shown in Figure 40a [49b]. Therefore zeolitic catalysts are more active for cracking the products than the feedstock, which will intrinsically tend to result in overcracking heavy feedstocks. This indicates there is an incentive to use zeolitic materials with wider pores, but so far attempts with e.g. VPI-5 [49c], or MCM-41 [49d] have not given convincing positive results. To nevertheless use some of the advantages of zeolites in hydrocracking, composite catalysts containing both zeolites and amorphous silica alumina have been developed. With these catalysts selectivities approaching those of fully amorphous systems can be achieved, whilst maintaining the advantages of zeolites in terms of stability. In particular when such catalysts are used in combination with the stacked beds, as outlined in Section 2.9.3, very attractive overall performances, with very stable yield patterns can be realized, as is reported [49e] for the performance of a number of commercial hydrocrackers.

2.9.5. The Paragon process

Another interesting zeolite-catalyst-related development in the field of hydrocracking is the so-called Paragon process developed by Chevron [50]. This process concept couples a conventional high-pressure hydrocracker with a low-pressure step, which makes use of a shape-selective zeolite catalyst such as ZSM-5 (see Figure 41). In this last reactor, normal and slightly branched paraffinic hydrocarbons are selectively cracked to olefins in the C_3-C_8 range (see Figure 42), which can thus significantly increase the gasoline yield and octane quality compared with a conventional hydrocracker. However, although the concept is elegant the process does suffer from a number of disadvantages, such as the fact that the organic nitrogen level of the feed to the shape-selective zeolite catalyst may not exceed 1 ppm to avoid rapid catalyst deactivation. Furthermore, the combination of high- and low-pressure reactors within a process is expensive in terms of compression-decompression costs. These disadvantages would appear, at least to date, to have retarded commercial development.

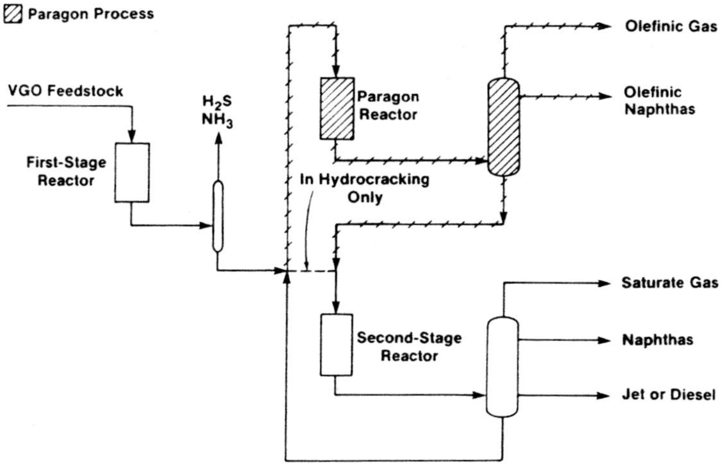

Figure 41. The Paragon process (from ref. 50).

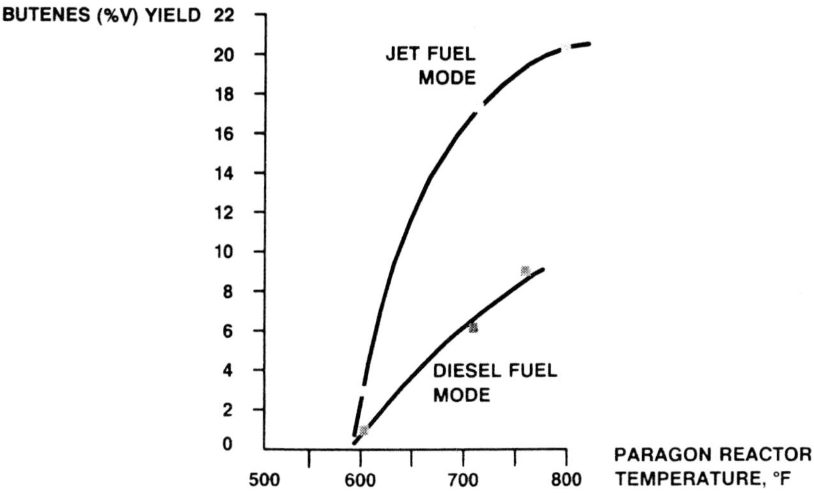

Figure 42. Paragon process olefin yields (from ref. 50).

2.10. Residue hydroprocessing

As described in the Introduction, atmospheric and vacuum-distilled residues, which are the oil fractions in which the heaviest molecules of a crude oil have been concentrated, are in themselves not very attractive products. The various contaminants in a crude oil, such as sulfur and nitrogen compounds, metals such as nickel and vanadium, and large aromatic molecules (asphaltenes) which are easily transformed into coke, are concentrated in these fractions. The highest concentrations of these contaminants are found, of course, in the vacuum residue. In addition, the properties of the various residues depend strongly on the crude origin, some examples of which are given in Table 11.

Table 11
Typical properties of heavy oil fractions.

Feedstock		Arabian Heavy Vacuum Distillate	Arabian Heavy Atmospheric Residue	Arabian Heavy Vacuum Residue	Maya Vacuum Residue
Yield on crude,	%wt	19	54	35	40
Sulfur,	%wt	2.9	4.5	5.4	5.3
Vanadium,	ppmw	0.5	97	150	635
Nickel,	ppmw	0.7	31	48	127
C_5 asphaltenes,	%wt*	–	12	20	34
H/C ratio		1.70	1.55	1.44	1.37

*The material which flocculates when mixing the residue with n-pentane is called pentane (C_5) asphaltenes.

The material which flocculates when mixing the residue with n-pentane is called pentane (C_5) asphaltenes.

The traditional outlet of residual oil has been industrial and marine heavy fuels but, with the increasingly stringent environmental standards, these residues can no longer be blended as such into these fuels and need to be desulfurized first. Thus, in the seventies a large number of residue hydrodesulfurization units were built (particularly in Japan), reducing the sulfur content in the residue by some 80-90 %.

However, these low-sulfur residual fuels remain less valuable than gasoline or middle-distillate oil products and thus an economic driving force exists not only to desulfurize in these units but also to achieve (by a proper choice of catalysts and conditions) an ever-increasing conversion of residue into distillates. This is illustrated, for example, in Figure 43, where the increasing conversion in Shell residue units is plotted versus time [51]. It has been reported that, in Shell`s Pernis (vacuum residue) HYCON unit, a conversion of 580°C+ components of 70% is achieved [51a]. Both Amoco and Veba have reported [51b, 51c] on ways to achieve 90% conversion in their process units: ebulliating bed and “Combi Cracking”, respectively, when operated for maximum conversion.

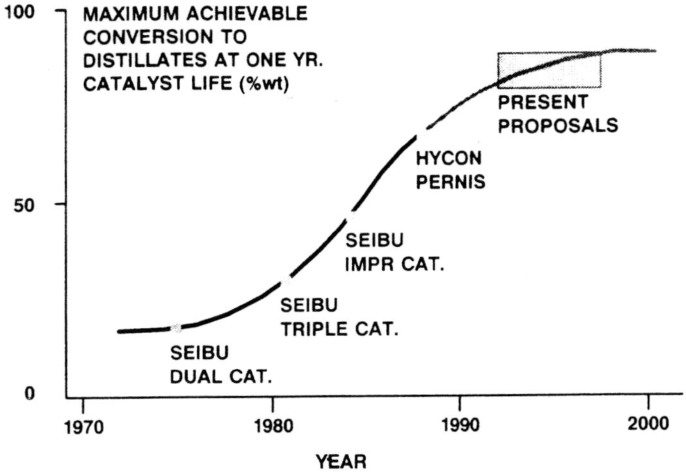

Figure 43. Residue hydroprocessing conversion development in Shell (from ref. 51).

Given the trend, in distillate hydrocracking, to process increasingly heavy feedstocks, it is not surprising that hydrocracking type catalysts, including zeolitic catalysts, have also been studied for residue hydroprocessing. Obviously, the residual feedstocks, with their high contaminant levels and large refractory molecules, are more difficult to process than the distillate feedstocks discussed earlier. On the other hand, even modest levels of conversion are economically attractive, so that complete conversion is by no means necessary.

In Japan, where there is a large interest in residue hydroprocessing, a national project was started by the formation of RAROP, the Research Association for Residual Oil Processing, in which oil and chemical companies cooperated with major government funding; this group appears to have taken the lead in the development of zeolitic catalysts for this purpose and has presented several publications on this subject [52-56].

It would appear that the starting point of these investigations has been a search for a catalyst that is active for cracking in the presence of H_2S, as could be judged from toluene disproportionation and cumene cracking experiments. Several metal-loaded ultrastable Y-sieves were studied, from which an Fe-loaded USY, prepared by treatment of USY with ferric nitrate at pH 1.6-2.1, proved to be particularly active [52, 53]. A detailed study was carried out to characterize this catalyst, from which it was concluded that ferric complexes are bonded to framework vacancies created by acid-induced dealumination (see Figure 44). These iron complexes then apparently grow into small iron oxide particles of about 1 nm size, due to the pH increase accompinying the zeolite dealumination step [52]. The acidic properties seem to result from the adsorption of H_2S on the supported Fe species.

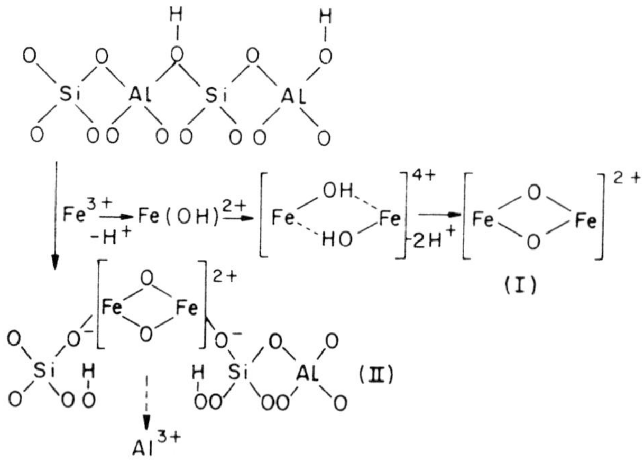

Figure 44. Ferric complex induced zeolite framework dealumination (from refs. 52, 53).

The Fe-containing zeolite was incorporated into a catalyst by extrusion with alumina and impregnation with cobalt and molybdenum. This catalyst was then tested with a Kuwait atmospheric residue in trickle-flow operation at 13.5 MPa, space velocity 0.3 h^{-1}, and 400-410°C. Surprisingly, the catalyst exhibits a cracking activity which increases at 410°C during the first 1500 hours. This is attributed to a graphitization of the carbonaceous species deposited on the catalyst. Ultimately, the catalyst gives about 65 %w conversion to material boiling above 343°C, with a high selectivity to naphtha, under the quoted conditions [54]. No data have been published that indicate whether the lighter residue fractions are converted preferentially, as might be expected if the interior of the zeolite crystals still plays a dominant role in the conversion reactions. If the outer surface of the zeolites were completely coked, rendering the inside inaccessible, no such selectivity would be expected. Possibly the iron particles have another role, apart from inducing acidity, and are involved in reducing the coke-forming tendency. Itoh [57] has published a study on clay catalysts that have been promoted with iron, in which the iron appears to halve the coke formation in what is essentially a thermal cracking regime.

The zeolite-based catalysts ("R-HYC") have been used commercially for some years, in combination with residue HDS catalysts, in the hydroprocessing of atmospheric residues, operating under the conditions as indicated above and, in line with the experience in distillate hydrocracking, at relatively high temperatures. Other catalyst companies have also put such catalysts into commercial use in fixed-bed units. Under ebulliating-bed unit conditions, some laboratory results have been published that were not entirely favourable, in terms of flocculation behaviour of the product, which is always an important aspect in residue [57a]. Thus, it would appear that there is indeed scope to apply zeolite-based catalysts to upgrade even the heaviest fractions of the crude oil barrel, but it is clear that there is still space for more development and performance improvement.

3. GAS-CONVERSION PROCESSES

3.1. Basics of gas conversion

The evolution of proven world oil and gas reserves up to 1990 (see Figure 45) showed a dramatic increase in the latter and a levelling off the former. This trend is expected to continue, though in a less steep fashion (see, for example, the Internet pages of BP-Amoco), which will most likely result in a gradual shift towards the use of gas as a feedstock for the manufacture of transportation fuels and even petrochemicals. Recognition of this situation has led to increased global interest in emerging technologies which can efficiently convert natural gas into liquids and higher-added-value products. The incentive to produce liquids from gas also stems from the desire to employ such process technology directly, at remote natural gas field sites, to save on transportation costs and to avoid gas flaring at oil recovery sites. Such new technology could thus potentially compete very favourably with pipelines and LNG tankers which are the present prime means of gas transportation to consumers. Some examples of such process routes are shown schematically in Figure 46, where both methane and LPG can be employed as feedstocks.

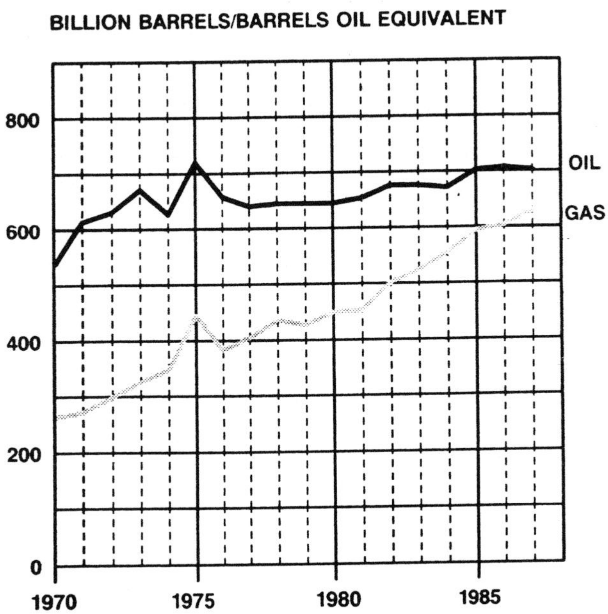

Figure 45. Evolution of proven world oil and gas reserves.

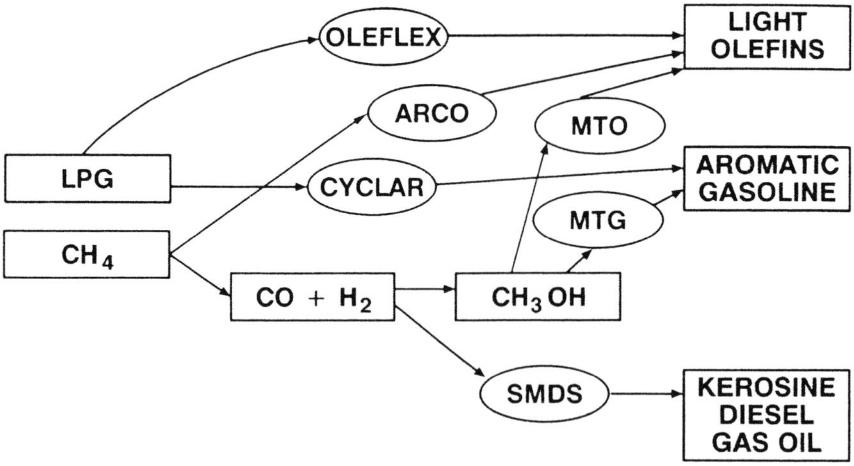

Figure 46. Catalytic gas upgrading processes.

It is evident from this diagram that some routes involve direct conversion from natural gas to liquids, such as ARCO's direct oxidative coupling [58], while others make use of synthesis gas (for example, the Shell Middle Distillate Synthesis [59]) and methanol (for example the Mobil Methanol to Gasoline, MTD process [60]) as intermediates. Some of these process routes make use of zeolite-based catalysts, which will be discussed individually.

3.2. Conversion of Methanol to Gasoline

In the Mobil methanol-to-gasoline (MTG) process, as shown in Figure 47, methanol is initially converted into olefins over ZSM-5 followed by oligomerization and cyclization reactions, and hydrogen transfer, which lead to the selective production of gasoline-range aromatics and light paraffins.

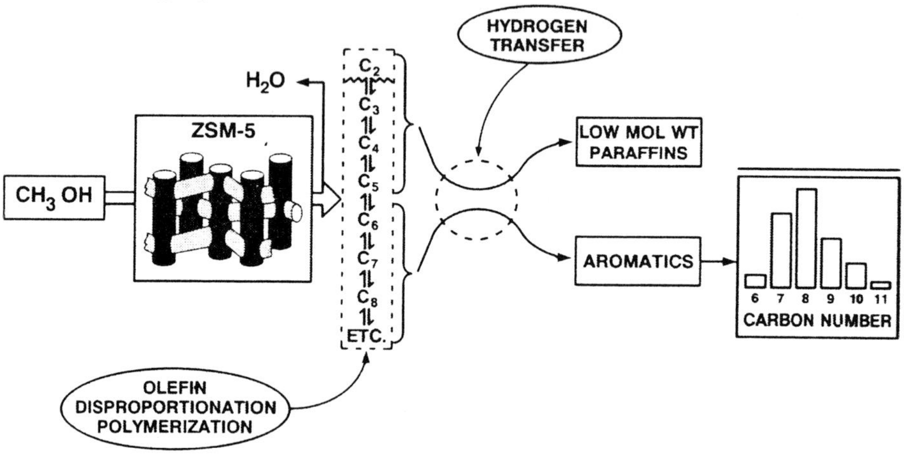

Figure 47. Methanol to gasoline reaction mechanism (from ref. 61).

Historically, the MTG process was first developed in the seventies and early eighties at the peak of the energy crisis and high oil prices. The technology was initially developed for fixed-bed reactors but this was later followed by the demonstration of fluid-bed operation on a pilot-plant scale, first for MTG in 1984 and then for methanol to olefins, (MTO) in 1985.

The effect of zeolite structure on yield is shown in Table 12, where the reaction products of methanol conversion over various zeolites at 570 C are shown [62]. The small-pore zeolites cannot easily accommodate aromatics and therefore yield mainly lower olefins, while the medium-pore zeolites yield primarily aromatics and lower paraffins (a combination required by the overall H/C mass balance) and some non-aromatic C_5+ product, depending on the conversion level. Wide-pore zeolites, however, give high concentrations of heavy aromatics, as shown also in Figure 48 from ref. 62.

Over the medium-pore zeolites the largest aromatic molecule produced is essentially durene (1,2,4,5-tetramethylbenzene) and although the wide-pore sieves permit further alkylation, they also exhibit a high selectivity to coke formation, which results in a short catalyst life (see Table 13). Table 12 and Figure 49 illustrate that profound differences in product selectivities are found for zeolites with different pore sizes. Other catalyst parameters that have been studied include the effect of Al content, impregnation with phosphorus, and addition of Zn, Mn, etc. [62]. Such catalyst variations can be used to direct the operation either towards the MTO or the MTG mode. Alternatively, this can be achieved through reduction of the methanol partial pressure by means of steam, for example, or by increasing the temperature and space velocity, which both favour olefins rather than aromatics.

Table 12
Methanol conversion to hydrocarbons over various zeolites (370°C, 1 atm., 1 LHSV) (from ref. 69).

	Hydrocarbon distribution (%wt) in				
	Erionite	ZSM-5	ZSM-11	ZSM-4	Mordenite
C_1	5.5	1.0	0.1	8.5	4.5
$C_2^=$	0.4	0.6	0.1	1.8	0.3
C_2	36.3	0.5	0.4	11.2	11.0
$C_3^=$	1.8	16.2	6.0	19.1	5.9
C_3	39.1	1.0	2.4	8.7	15.7
$C_4^=$	5.7	24.2	25.0	8.8	13.8
C_4	9.0	1.3	5.0	3.2	9.8
C_5+ aliphatic	2.2	14.0	32.7	4.8	18.6
A_6	-	1.7	0.8	0.1	0.4
A_7	-	10.5	5.3	0.5	0.9
A_8	-	18.0	12.4	1.3	1.0
A_9	-	7.5	8.4	2.2	1.0
A_{10}	-	3.3	1.5	3.2	2.0
A_{11}+	-	0.2	-	26.6	15.1

Table 13
Dimethyl ether conversion: effect of zeolite type (from ref. 67).

Catalyst	Residual Activity, %	Coke, %wt	Coke/hydro-carbons x 10^3
HZSM-5	90	1.5	0.3
HY	8	9.3	40
H-Erionite/Chabazite	0.15	8.6	90
H-Mordenite	0.15	6.8	200

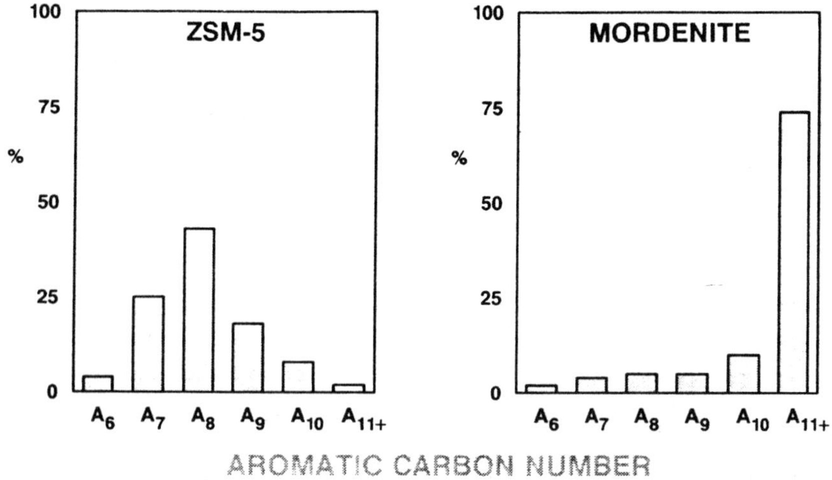

Figure 48. Methanol to gasoline: effect of zeolite pore size (from ref. 62).

In the scale-up to commercial operation two factors are of particular importance, namely catalyst deactivation and heat production. Despite the relatively high stability of the ZSM-5 catalyst, deactivation in the MTG operation is pronounced. The catalyst is progressively coked and the total catalyst inventory is completely deactivated after about one month under these conditions. The heat production in the MTG reaction is high, leading to an adiabatic temperature rise of some 650°C [61].

Two process setups have been developed to cope with this heat release. In the older fixed-bed system, the methanol is first dehydrated to dimethyl ether, releasing 20 % of the heat, and is then in a second stage reacted further to gasoline. In the second stage, a recycle of light gases is applied, not only to dissipate the heat, but also to increase the gasoline selectivity. The commercial MTG plant in New Zealand, with a capacity of 14500 bbl/d of gasoline, started up in 1985/86, and has the setup described above. The second stage was carried out in five reactors in parallel (four for the reaction and one for regeneration). The New Zealand complex further contains a product treatment step, to reduce the durene content of the gasoline to 2 % by weight.

Fluid-bed operation was an obvious alternative to the multi-reactor fixed bed operation. This was developed in a 100 b/d demonstration unit in Germany, Wesseling, which ran in the MTG mode of operation during 1984. The temperature in such a fluidized bed is, of course, uniform and part of the catalyst is continuously withdrawn for regeneration. By varying the amount of this stream, the average catalyst activity can be controlled. The fluid-bed performance has considerable advantages over the fixed bed in terms of thermal efficiency, gasoline yield and octane quality [62, 63].

Finally, in the context of the methanol-to-gasoline route, the combination of MTO plus MOGD (Mobil Olefins to Gasoline and Diesel) should be mentioned. This, according to Mobil [63], has the advantage of offering a high degree of product flexibility.

3.3. Methanol conversion to lower olefins

The zeolite-based technology for the production of lower olefins employing methanol as the feedstock is often termed Methanol to Olefins (MTO). The Mobil MTO process [59] can be considered as an intermediate step of their methanol-to-gasoline (MTG) process [60]. The product distribution of methanol conversion over ZSM-5 as a function of reciprocal space velocity (contact time) is shown in Figure 49, from which it can be seen that lower olefins are important intermediate products. By a judicious choice of process conditions (e.g. temperature, space velocity, pressure and methanol partial pressure) the product distribution can be maximized towards the production of light olefins [61, 64].

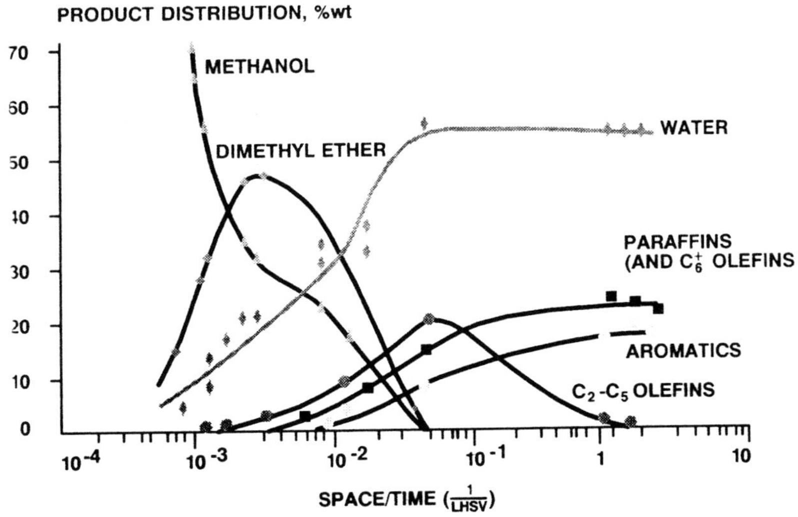

Figure 49. Methanol to hydrocarbons reaction path: ZSM-5 Catalyst (371°C) (from ref. 61).

A particularly interesting feature of this zeolitic catalytic chemistry is the C-C bond coupling step, which is effected starting from a C_1 hydrocarbon (methanol). Although there have been extensive mechanistic studies [65-68] the detailed chemistry of this quite remarkable catalysis is not as yet fully understood. The shape-selective properties of the ZSM-5 zeolite structure play an important role in the catalytic performance. For example, the relatively high selectivity to lower olefins (see Table 14), which is achieved by suppression of cyclization and hydrogen-transfer reactions, can be attributed not only to an appropriate choice of process conditions but also to transition-state shape selectivity. The relatively low coke make (see Table 14) is also a reflection of the shape selective behaviour of this catalyst system.

Table 14
Methanol conversion to olefins (from ref. 67).

Number of zeolites oxygen atoms/ring	"Small pore" 8	"Medium pore" 10
Typical catalysts	erionite zeolite T chabazite ZK-5 ZSM-34 SAPO-34	ZSM-5 modified ZSM-5
Typical yields, %wt		
$C_2^=$	25-55	5-25
$C_3^=$	20-50	20-40
Total olefins	50-95	50-80
C_5+	1-15	25-60
Coke	0.5-5	less than 0.5

A variety of "small"- and "medium"-pore zeolites other than ZSM-5 have also been studied [63] for methanol conversion, as shown in Table 14. Interestingly, the "small"-pore zeolites such as ZSM-34 and SAPO-34 give higher yields of lower olefins and, in particular, ethylene. This can likely be attributed to a greater degree of shape- selectivity and thus further suppression of consecutive reactions. However, this is not entirely consistent with the higher coke makes which apparently occur when using these smaller pore materials.

UOP have developed an MTO catalyst [ref. 68, 69] based on the molecular sieve SAPO-34 (chabazite structure) which is highly selective for converting methanol to ethylene and propylene. Selectivities to these lower olefins of up to 85 % are claimed at 100 % methanol conversion (see Table 15). It has also been shown that adding water to the methanol feed has a positive effect on lower olefin selectivity and reduces coke make on the catalyst [69a]. For a study of the effect of crystal size of SAPO-34 on the selectivity and deactivation in the MTO reaction see ref. 69. UOP together with Norsk Hydro are offering nowadays SAPO-34 based MTO technology, for the conversion of natural gas - via methanol - to ethylene and propylene [68].

Table 15
Methanol conversion to olefins with "medium pore" silicoaluminophosphate zeolite, SAPO 34 (Union Carbide, from ref. 68).

Methanol conversion	100	100
Product yields, %wt		
Ethylene	61.1	43.0
Propylene	27.4	41.8
Butenes	5.4	10.8
Pentenes	-	-
Total olefins	93.9	95.6

There have also been a number of studies [69b, 69c, 69d, 69e], on various zeolites and modifications to enhance the selectivity towards lower olefins production from methanol. The techniques employed include cation exchange, pore modification by silica deposition, dealumination and phosphorus impregnation. All these studies point to the importance of both acidity and pore structure in determining optimal catalytic performance.

This technology has been in various stages of development for many years; in 1996 UOP and Norsk Hydro announced [69f] that they were jointly licensing MTO technology with the intention to build a first plant based on this process. The technology is apparently based on a second-generation catalyst system utilizing the molecular sieve ALPO-34, which was originally developed by UOP. This represents the first application of an aluminium phosphate type molecular sieve catalyst system.

It is conceivable that a number of factors such as cheap sources of natural gas, improved synthesis gas technologies combined with the continuous incremental improvements in methanol technologies that MTO could indeed become an economically viable route to produce lower olefins from natural gas, particularly in locations where this is plentiful and cheap.

3.4. Conversion of synthesis gas to gasoline

In New Zealand the MTG plant was combined with a methane steam reformer to produce synthesis gas and a methanol synthesis plant to overall produce gasoline from natural gas [70]. With such a combination of processes one attempts to improve the economics through a close integration of the various steps. In the Haldor Topsoe integrated gasoline synthesis process (TIGAS) the methanol synthesis and the MTG reaction are integrated, without methanol being isolated as an intermediate product. A multifunctional catalyst was developed which permitted production of oxygenates (methanol and DME) from synthesis gas at moderate pressures, which thus improves the process integration. The process has been developed in an industrial pilot plant. Compared with the Mobil MTG version, the hydrogenation of the olefinic gases over the oxygenate catalyst is probably disadvantageous [71].

A more rigorous approach towards integration is to directly convert synthesis gas into aromatic gasoline in a single step. This necessitates combining the CO hydrogenation function and the aromatization function into a single catalyst, possibly also including a CO-shift function. From the patent literature it would appear that this approach has indeed been extensively studied. Mobil, for example, has worked on ZnO/Cr203-ZSM-5 [72] catalyst systems, and has also studied combinations of Fischer-Tropsch functions and ZSM-5. Shell also has quite an extensive patent coverage on direct gasoline synthesis, and both BP and UOP have been active in this area. As mentioned above, apart from the routes involving methanol-type intermediates, Fischer-Tropsch-based routes will convert synthesis gas into distillate-range products. The Shell SMDS (Shell Middle Distillate Synthesis) process is an example of such a route. However, these process routes do not typically make use of zeolitic catalysts.

3.5. Light Petroleum Gases (LPG) conversion to aromatics

The direct conversion of LPG into aromatic components is an interesting route to produce both high-octane gasoline components and benzene/toluene/xylenes (BTX) as petrochemicals. BP and UOP have jointly developed [73-75] the CYCLAR process to convert C_3-C_5 paraffins and olefins into aromatics-rich products. This process makes use of a gallium-loaded ZSM-5 type catalyst, as shown in Figure 50.

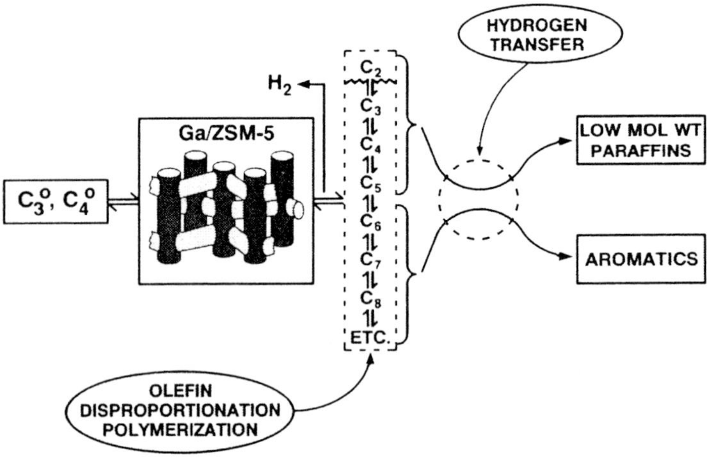

Figure 50. LPG to aromatics: CYCLAR process (UOP/BP).

The mechanism involves the initial dehydrogenation of paraffins to form lower olefins and hydrogen, followed by olefin oligomerization and cyclization reactions to produce aromatics, in which hydrogen transfer plays an important role. In many respects this mechanistic scheme resembles that used to describe the MTO and MTG processes (see Figure 47).

There have been several studies [76] aimed at finding a substitute for gallium as the dehydrogenation/cyclization catalyst component. Although zinc silicates, for example, have been shown to be active for aromatization the zinc component is unfortunately too volatile under the relatively high process temperatures employed. Platinum has also been shown to further enhance the aromatization activity [76] of, for example, the zinc silicate catalyst system.

A typical product yield from the CYCLAR process is shown in Table 16, from which it can be seen that the aromatic product is very rich in benzene and toluene. This particular distribution of aromatics is significantly different from that obtained in the continuous platforming process. Further, the product distributions are similar for propane and butane feedstocks. The resultant octanes (RON and MON) of the gasoline fractions are high, reflecting the selectivity towards aromatic products (see Table 16). The process conditions employed include relatively high temperatures (475-575°C) and low pressures (2-10 bara), which thermodynamically favour the formation of aromatics. The rate of catalyst deactivation is such that moving-bed reactor technology has to be employed with continuous catalyst regeneration, similar to that applied in the low-pressure Continuous Catalytic Reforming (CCR) process. A demonstration unit for the CYCLAR process was operated in the past at the BP Grangemouth refinery.

Table 16
LPG to aromatics (CYCLAR): Typical products yields from ref. 75).

Overall composition, %wt		Aromatics composition, %wt	
Total aromatics	64	Benzene	31
Fuel gas	30	Toluene	41
Hydrogen	6	Xylenes/ethylbenzene	17
		C_9+	11
	100		100

Reactor temperature	: 475-575 °C
Pressure	: 2 - 10 bara
Catalyst contact time	: 5 - 20 s
LHSV	: 2-4
Catalyst	: Ga/ZSM-5
Feedstock	: propane

In the short term, current feedstock / product price structures may somewhat limit the commercial potential of LPG-to-aromatics conversion technology. However, in niche situations where the LPG price is low, hydrogen is in short supply and aromatics are required for petrochemicals, this zeolite-catalyst based process technology can prove an attractive option, fo example, in the Middle East where LPG is available at low cost.

3.6. Olefin oligomerization

Olefin oligomerization has long been known and used to upgrade catalytic cracker off-gas streams in refineries, an example being the CATCON polygasoline process [77] developed by UOP. Improvements are obtained by the application of zeolite catalysts. In particular, both Mobil [78-80] and Shell [81] have developed zeolite-catalyst-based processes: the MOGD and SPGK processes, respectively. Perhaps the main advantage of the zeolite catalysts is the yield flexibility, whereby via an appropriate adjustment of process conditions a switch in primary product between gasoline and middle distillates is possible.

The MOGD process makes use of a ZSM-5 zeolite as the shape-selective catalyst system which, when operated at relatively low temperatures, produces mainly middle-distillate product via olefin-oligomerization reactions. By increasing the operating temperature and adjusting the fractionation conditions, gasoline can be produced as the primary product (see Table 17). This process has been applied on a semi-commercial scale at the Mobil refinery complex in Paulsboro [80].

Table 17
Olefin Oligomerization: Distillate and Gasoline modes for the Shell Polygasoline Kero Process (SPGK) (from ref. 81).

	Max distillate mode	Gasoline mode
C_1–C_2	–	–
C_3	1	4
C_4	2	5
C_5–165 °C gasoline	18	–
165 °C+ distillate	79	–
C_5–204 °C gasoline	–	84
204 °C+ distillate	–	7

$C_3^=/C_4^=$ feed

The Shell SPGK process [81] is based on a modified zeolite catalyst system, which enables both gasoline and middle-distillate products can be produced as primary products by appropriate adjustment of the operating conditions (see Figure 51).

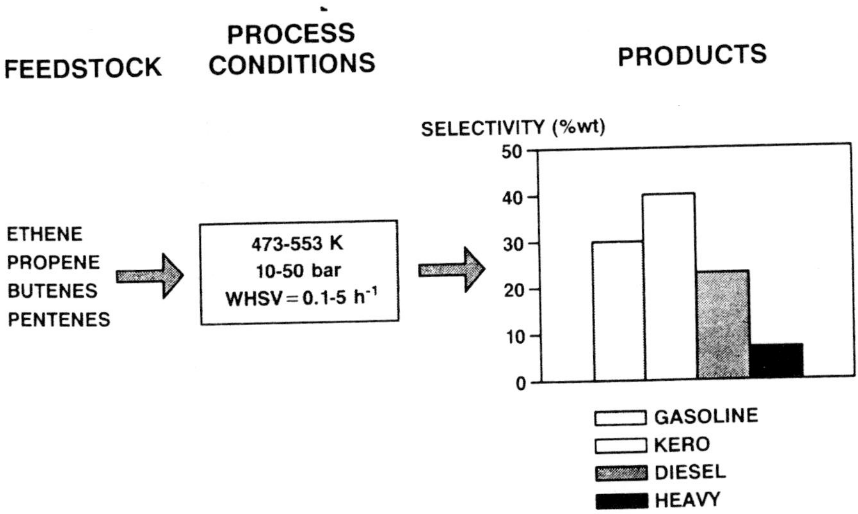

Figure 51. Olefin oligomerization: Shell SPGK process (from ref. 82).

The attractiveness of olefin oligomerization to produce transportation fuels depends very much on the individual situation in each refinery. In modern large, fully integrated refinery complexes, this type of technology could potentially be exploited to produce, for example, high-quality middle-distillate products. However, this particular utilization of lower olefins also has to compete with alternative applications, e.g. as feedstocks to petrochemical processes, which in some cases may be more attractive. An advantage of the zeolite-based olefin-oligomerization routes is that normally only minimal olefin feed purification is required.

Future developments in this field could possibly be towards applicat:ions related to the manufacture of petrochemicals rather than that of transportation fuels, given the relatively high prices of lower olefins. In this regard there may be many new opportunities to develop more sophisticated non-acid zeolite catalysts [82], rather than the present predominantly acid-catalysed systems.

4. CONCLUSIONS AND FUTURE DEVELOPMENTS

A perception of the different phases of maturity of zeolite based catalytic technologies is given in the form of a S-curve (see Figure 52). This approach attempts to classify the different technologies in terms of emerging, developing and maturing technologies. It is interesting to observe that there is quite a balanced distribution of zeolite catalytic technologies across all sectors of the S-curve. This is indicative of a dynamic area of technology with new applications continually entering the front end of the S-curve while other processes are maturing and moving into second-generation catalyst systems.

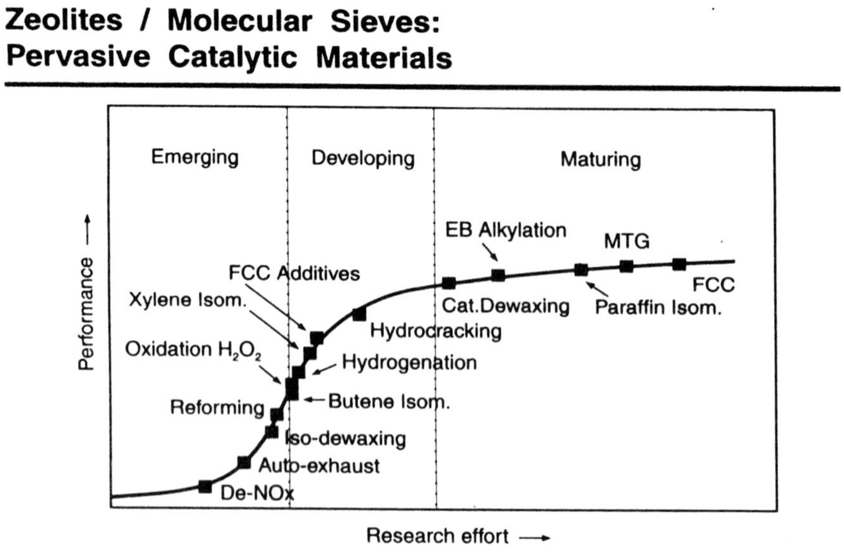

Figure 52. Maturity of zeolite-based catalytic technologies.

There would appear to be a number of forces at play which are driving this relatively high level of innovation, such as incremental improvements to catalysts for existing processes, new emerging catalytic reaction types of industrial interest, and applications in new areas.

Many incremental improvements to existing processes have involved the development of new-generation zeolite-based catalyst systems. Examples include: new- generation zeolites for applications in hydrocracking; iso-dewaxing catalysts for wax removal from lubeoils; extended and improved zeolite catalysts for aromatics alkylation and isomerization; and continuous improvement in FCC catalysts, particularly for residue processing.

Newer application areas of zeolite catalysis include: high temperature DeNOx for power generation emissions control (HT-SCR), transportation emissions control (LB-SCR), petrochemicals (DCC for lower olefins; AROMAX and CYCLAR for paraffins to aromatics) and fine-chemicals catalysis such as oxidation and acylation. Most of these newer application areas have been driven by novel catalytic reaction types such as hydrocarbon based DeNOx and oxygen transfer using oxidants such as H_2O_2 and N_2O. The catalyst systems involved in these new catalytic chemistries are primarily based on transition-metal ions exchanged into particular zeolites such as ZSM-5 and beta which may be indicative of a rather novel type of active site.

Further along the S-curve, where the technologies are maturing, innovation and process improvement is mainly driven by the development of next generation catalyst systems, or a merging of catalyst and related technologies such as novel reactors or combinations of both. Important innovations arose from the use of ultra-stabilized Y (hydrogenation and hydrocracking), zeolite beta, and MCM-22 (aromatics processing). A novel shape-selective zeolite developed by Chevon researchers enabled a paradigm shift from dewaxing to iso-dewaxing in lube-oil processing technology.

To date, the application of non-zeolitic microporous (for example AlPOs) and mesoporous (for example MCM-41) structures still appears to be quite limited, despite perhaps high initial expectations. However, particularly for the mesoporous materials it may be too early to draw conclusions on catalytic applicability and the gestation period may turn out to be rather long.

Nevertheless, the field of zeolite catalysis continues to be dynamic and there is every indication that catalytic innovation utilizing these remarkable microporous materials will continue at the same relatively high rate or even accelerate into the foreseeable future.

5. ACKNOWLEDGEMENTS

The authors would like to express their sincere gratitude to their many colleagues for the stimulating discussions on the subject of zeolite catalysis. Particular thanks are due to the following people in relation to the present review article: Drs. J. P. van den Berg, J. Biswas, A. A. Esener, J. W. Gosselink, A. Hoek, T. Huizinga, A. Klazinga, M. Manton, J. K. Minderhoud, G-J. den Otter, M. F.M. Post, H. Schaper, B.D. Murray, L. Huve, T.L.M. Maesen, C.M.A.M. Mesters, and A. Nowak. The authors are most grateful to Dr. R.S. Downing who performed the difficult task of putting the various sections together and finalizing the chapter.

REFERENCES

1 J. Rabo, in: Zeolite Science and Technology, F.A. Ribeiro, A.E. Robriques, L.D. Rollman and C. Naccache (Eds.), Martinus Nijhoff, The Hague, 1984,p, 291.

2. E.M. Flanigen, R.L. Patton and S.T. Wilson, Structural, Synthetic and Physicochemical Concepts in Aluminophosphate Based Molecular Sieves, **in:** P.J. Grobert, W.J. Mortier, E.F. Vansant and G. Schulz-Ekloff, (Eds.) Innovation in Zeolite Materials Science, Proc. Nieuwpoort Symposium, (1987), Elsevier, 1988, pp. 13-29.

2a C.J. Adams, A.Araya, S.W. Carr, A.P. Chapple, K. Franklin, P. Graham, A.R. Minihan, T.J. Osinga and J.A. Stuart, in Progress in Zeolite and Microporous Materials, Studies in Surface Science and Catalysis, Elsevier Science B.V.,Vol. 105, (1997), p. 1667

3 J.N. Miale, N.Y. Chen and P.B. Weisz, J. Catal.. 6, (1966) 278-287.

4 W.O. Haag, R.M. Lago and P.B. Weisz, Nature, 309, (1984) 589-591.

5 P.B. Weisz, Chem Tech, 1987, p. 368.

6 C.T.W. Chu and C.D. Chang, J.Phys.Chem., 89, (1985) 1569-1571.

7 M.F.M. Post, T. Huizinga, C.A. Emeis, J.M. Nanne and W.H.J. Stork, An Infrared and Catalytic Study of Isomorphous Substitution in Pentasil Zeolites, in: H.G. Karge and J. Weitkamp (Eds.), Zeolites as Catalysts, Sorbents and Detergent Builders: Applications and Innovations, Proc. Wurzburg (1988) Symposium, Elsevier, 1989, pp. 365-377.

7a. R.M. Lago et al., Stud. Surf. Sci. and Catal. , 28 (1986), p. 677 (IZC Tokyo 1986)

7b. M.E. Davis, e.g. presentation at the Van Bekkum Symposium, Delft, 1997.

8 S.M. Csicsery, Zeolites, 4 (1984), 202.

8a EU Environmental specifications for market fuels for year 2000, proposed by the Council of Environmental Ministers 19.6.97

8b Thomas F. Degnan, Jr., Topics in Catal. 13 (2000) 349-356.

9 I.E. Maxwell, J. Inclusion Phenom., 4 (1986) 1-29.

9a T. F. Degnan, paper presented at the World Technology congress, San Diego, California, august 1996
9b J.E. Naber, K.P. de Jong, W.H.J. Stork, H.P.C.E. Kuipers and M.F.M. Post , (1994) in Zeolites and related microporous materials: State of the art 1994, J.E. Weitkamp, H.G. Karge, H. Pfeifer and W. Hoelderich, (eds.), 1994, Elsevier Science BV., p. 2197-2219
10 I.E. Maxwell and G.J. den Otter, Eur. Pat. Appl. E.P. 463850.
10a C. Gosling, R. rosin, P. Bullen, T. Shimizu, and T. Imai, paper presented at the World conference on refining, Vehicle Technology and Fuel Quality, Brussles, Belgium, may 21-23, 1997
10b P.J. Kuchar, J.C. Bricker, M.E. Reno and R.S. Haizmann, Fuel Processing Technol. 35, (1993) 183-200
10c Oil Gas Journal, april 7, 1997, p. 41
10d S.T. Sie, in Handbook of Heterogeneous Catalysis, eds. G. Ertl, H. Knoezinger and J. Weitkamp, VCH/Wiley, 1997, p. 1998
10e e.g. Patent US 505974
11 H.W. Kouwenhoven, Molecular Sieves, ACS Advances in Chemistry Series 121, 1973, pp, 529-539. 11a M.A. Lanewala, P.E. Pickert and A.P. Bolton, J. Catal. 9 (1967) 95-97
12 J. Weitkamp, Hydrocracking and Hydrotreating, in: J.W. Ward (Ed.), ACS Symposium Series 20, American Chemical Society, 1975, pp. 1-27.
12a S.T. Sie, Ind. Eng. Chem. Res. 31, 1993, 1881-1889; ibid. 32, 1993, 403-408
12aa H. Liu, G.D. Lei and W.M.H. Sachtler, Applied Catalysis, A, (1996), 146 (1), 165-180; see also W.M. H. Sachtler, to be published in Topics in Catalysis
12b E. Blomsma, J.A. Martens and P.A. Jacobs, J. Catal. 155, 1995, 141-147; ibid. 159, 1996, 323-331. E. Blomsma, J.A. Martens and P.A. Jacobs, in Proceedings of the 11th Int. Zeolite Conference, Seoul, South Korea, 1996, Eds. H. Chon, S.-K. Ihm and Y.S. Uh
12c e.g. EP-A-653399
12d M.J. Ledoux, J.L. Guille, P.H. Cuong, and S. Marin, Eur. Pat. Appl. EP 474570 A1 92031113
13 P.B. Koradia, J.R. Kiovsky and M.Y. Asim, J. Catal., 66 (1.980) 290-293.
14 G.R. Hays, W.A. van Erp, N.C.M. Alma, P.A. Couperus, R. Huis and A.E. Wilson, Zeolites, 4 (1984) 377-383.
14a D. Barthomeuf , in Catalysis 1987, p. 177, Elsevier, 198815J.A. Gray and J.T. Cobb, J. Catal., 36 (1975) 125-141.
15a M. Boulet, E. Bourgeat-Lami, F. Fajula, T.des Courieres and E. Garrone, in Proceedings of the 9th Int. Zeolite Conf. , Montreal 1992, Eds. R. von Ballmoos et al., Buttersworth- Heinemann, 1993 p. 389-396; F. Fajula, M. Boulet, B. Coq, V. Rajaofanova, F. Figueras and T. des Courieres, in L.Guczi et al.,Eds, New Frontiers in Catalysis, Proceedings of the 10 th Int. Congress on Catalysis, july 19-24, 1992, Budapest, Elsevier Science Publishers, 1993, p. 1009-1016 D. McQueen, B.H. Chiche, F. Fajula, A. Auroux, C. Guimon, F. Fitoussi and P. Schulz, J. Catal. 161 (1996) 587-596
15b H. Jongkind, K.P. Datema, S. Nabuurs, A. Seive and W.H.J. Stork , Microporous materials, 10 (1997) p.149
15c T.P. John and S. Thomas, paper AM-93-47 presented at the 1993 NPRA annual mtg, march 21-23,1993, San Antonio

15d R. M. Gialella, L. Mank, J.-P.Burzynski, D. Duee and C. Travers, paper AM-93-15, presented at the 1993 NPRA annual meeting , march 21-23, 1993, San Antonio

15e H. H. Mooiweer, K.P. de Jong, B. Kraushaar-Czarnetzki, W.H.J. Stork and B.C.H. Krutzen, in Zeolites and Related Microporous Materials: State of the Art 1994, J. Weitkamp, H.G. Karge , H. Pfeifer and W. Hoelderich (Eds.) Studies in Surface science and Catalysis, vol 84, p. 2327-2334, Elsevier Science B.V. 1994 K.P. de Jong, Catalysis Today 29 (1996) 171-178; J.G. Buglass, K.P. de Jong and H.H. Mooiweer, paper at the Int. Symposium on Coking and Mitigation, August 20-24, 1995, Chicago, USA

15f R.A. Sawicki, R.J. Pellet, E.J. Kuhlmann, H.-M. Huang, C.-L. O`Young, R.V. Kessler and D.G. Casey, paper AM-95-45 presented at the 1995 NPRA annual mtg, march 19-21, 1995, San Francisco W.-Q. Xu, Y.-G. Yin, S.L. Suib, J.C. Edwards and C.-L. O`Young, J. Catal. 163, 232-244 (1996), and references therein.

15g S. Kumar Kunchal, F.G. Dwyer and C.P. Ambler paper AM-93-45 at the 1993 NPRA annual meeting, march 21-23, 1993, San Antonio

15h J. Barin Wise and D. Powers, in ACS Symposium Series, vol. 552, Environmental Catalysis, (1994) J.N. Armor, editor, p. 273, Everett, J. Barin Wise and D. Powers, Presentation at Solid Acid Conference, november 16, 1993, session 4, Houston 15i M. Guisnet, P. Andy, N.S. Gnep, C. Travers and E. Benazzi, J. Chem. Soc. , Chem. Commun. 1995, p. 1685; P. Andy, N.S. Gnep, M. Guisnet, E. Benazzi and C. Travers, J. Catal. 173, (1998), p 322

15j J. Houzvicka, O Diefenbach and V. Ponec, J. Catalysis, 164 (1996) 288-300

15k A. Corma and A. Martinez, Catal. Rev.-Sci. Eng. 35 (4) (1993) 483

15kk J. Weitkamp and Y. Traa (1997) in G. Ertl, H. Knoezinger and J. Weitkamp (eds.), Handbook of heterogeneous catalysis, chapter 3.14, Wiley-VCH, p. 2039

15l A.K. Rhodes, Oil &Gas Journal, Aug. 22 (1994) p.52

15m A. Corma, V. Gomez and A. Martinez, Applied Catal. A119 (1994) 83

15n K.P. de Jong, C.M.A.M. Mesters, D.G.R. Peferoen, P.T.M. van Brugge and C. de Groot, Chemical Engineering Science , vol. 51, no 10, pp. 2053-2060, 1996

16 P.B.Weisz, Pure Appl.Chem., 52 (1980) 2091-2103

16a L. Domokos, L. Lefferts, K. Seshan and J.A. Lercher, J. Catal. 197 (2001) 68

16b L. Domokos, L. Lefferts, K. Seshan and J.A. Lercher, J. Mol. Catal. 162 (2000) 147

17 S.D. Burd, Jr. and J. Mazuik, Hydrocarbon Process., 51 (1972) 97-102.

18 J.P. Grannetti and A.J. Perrotta, Ind. Eng. Chein. Process Des. Dev., 14 (1975) 86-92.

18a WO 93/04146 world patent application; T.L. Hilbert, P.P. Durant and M.S. Sarli, Hydrocarbon Asia, november/december 1996, p. 56

18b J.A. Salazar, C. Badra, J.A. Perez, E. Palmisano, W. Garcia and B. Solari, Symposium on the removal of aromatics, sulfur and olefins from gasoline and diesel, presented at the 212th Natl. ACS meeting, Division of petroleum chemistry, Orlando, Fl. August 25-29, 1996

19 T.R. Hughes, W.C. Buss, P.W. Tanun and R.L. Jacobson, in: Y. Murakami, A. Lijima and J.W. Ward (Eds.), 7th Int. Zeolites Conf. Proc., Kodansha/Elsevier, Tokyo, 1986, pp. 725-732.

19a G.S. Lane, F.S. Modica and J.T. Miller, J. Catal. 129 (1991) 142

19b E. Derouane, D.J. VanderKerken , Applied Catalysis, 45 (1988) L15

19c R.J. Davis, and E. Derouane, Nature 349 (1991) 313

19d S. Zones paper S7b2, Symposium Advanced Catalytic Materials, MRS Fall Meeting, Boston, USA, december 2-5, 1996

20 D.V. Law, P.W. Tamm and C.M. Detz, Energy Progress, Vol.7, N4, December 1987, pp. 215-222. 20a J. P. van den Berg, J.P. Lucien, G. Germaine, and G.L.B. Thielemans, Fuel Processing Technol. 35 (1993) 119-137

20b W.H.J. Stork, in Hydrotreatment and hydrocracking of oil fractions, eds. G.F. Froment, B. Delmon and P. Grange, Elsevier Science BV, 1997, p. 41-67

20c B.D. Murray, B.H.C. Winquist, L.J. Sanborn, S.N. Milam, R.M. Foley, W.F. Johns and D.P. Schofield, paper 243d at the AIChE mtg, San Francisco, USA, november 13-18, 1994

20d P.N. Hannerup and B. Cooper, Oil Gas- European Magazine 1 (1995) p.25

20e I. Mochida, K. Sakanishi, X. Ma, S. Nagao, and T. Isoda, Catalysis Today, 29 (1996) 185-189

20f A.Appetiti, F. Simola, and B. Binarco Paper AM 90-43 at the 1990 NPRA Annual Mtg, San Antonio, Texas, USA. F.A. Smith and R.W. Bortz, Oil Gas J. 1990, aug. 13, p. 51

20g H.W. Homan Free, T. Schockaert and J.W.M. Sonnemans, Fuel Processing Technol. 35 (1993) 111-118

21 K.W. Smith, W.C. Starr and N.Y. Chen, Oil Gas J. 78 (1980) May 26, 75-84.

22 N.Y. Chen, R.L. Gorring, H.R. Ireland and T.R. Stein, Oil Gas J. 75 (1977) June 6, 165-170.

23 J.G. Bendoraitis, A.W. Chester, F.G. Dwyer and W.E. Garwood, in Y. Murakami, A. Lijima and J.W. Ward (Eds.), 7th Int. Zeolites Conf. Proc., Kodansha/Elsevier, Tokyo, 1986, pp. 669-675.

23a S.Sisavanker, K.J. Waghmare, K.M. Reddy and P. Ratnasamy, in Proceedings of the 9th International Congress on Catalysis, Calgary 1988, p. 120

24 US patent 3,968,024 (1976).

25 CB patent 2,027,742 (1980).

25a S.Sisavanker, K.J. Waghmare, K.M. Reddy, A.N. Kothasthane and P. Ratnasamy, J. Chem. tech. Biotechnol. 48 (1990) 261 H.K. Beyer, P.Feher and K. Jakob, Hungarian journal of Ind. Chem. 14 (1986) 345

25b C.L. Baker and M.P. McGuiness, paper at the 1995 NPRA annual meeting, march 19-21, 1995, San Francisco, USA

26 US patent 4,419,220 (1977).

26a S.J. Miller ACS meeting Div. Petr. chem. august 1993, Chicago p. 788 S.J. Miller, in J. Weitkamp, H.G. Karge, H. Pfeifer and W. Hoelderich (eds.), Zeolites and Related Microporous Materials: State of the Art 1994, Elsevier Science, p. 2319 WO 92/01657, (1990) to Chevron Research and Technology Company US patent 5 135 638 to Chevron research and Technology Company

27 J.E. Naber, P.H. Barnes and M, Akbar, The Shell Residue Fluid Catalytic Cracking Process, Japan Petroleum Refining Conference, Tokyo, October 19-21, 1988.

28 B.W. Wojciechowski and A. Corma, Catalytic Cracking, Marcel Dekker, New York, 1986.

29 L. Upson, I. Dalin and W.R. Wichers, Heat balance - The Key to Cat. Cracking, 3rd Katalistiks FCC Symposium, Amsterdam, May 26-27, 1982,

30 B. de Kroes, C.J. Groenenboom and P. O'Connor, New Zeolites in FCC, Ketjen Catalysts Symposium 1986, Scheveningen, May 25-28, 1986.

31 L.A. Pine, P.K. Maher and W.A. Wachter, **J.** Catal., 85, (1984) 466.

32 J. Biswas and I.E. Maxwell, The Relationship Between Generic Catalyst Types and FCC Performance, to be published, Applied Catalysis.
33 J. Biswas and I.E. Maxwell, Octane Enhancement Induced By Zeolites in Fluid Catalytic Cracking, Paper 8th International Zeolite Conference, Amsterdam, July 10-14, 1989
34 J.M. Maselli and A.W. Peter, Catalytic Reviews Science Engineering, 26, (1984) 525-554.
35 H.F. Henz, V.M. de Marco Meniconi and J.M. Fuoco, Petrobas Experience with Octane Enhancement in Resid Cat. Cracking, Ketjen Catalyst Symposium '86, Scheveningen 1986.
36 J. Biswas and I.E. Maxwell, Octane Enhancement in Fluid Catalytic Cracking. Part 1, to be published, Applied Catalysis
36a Adewuyi, Y.G., Klocke, D.J. and Buchanan, J.S., Appl. Catal., A (1995), 131 (1), 121-133
36b Buchanan, J.S. and Adewuyi, Y.G., Appl. Catal., A (1996), 134 (2), 247-262
36c Li, Z.T., Shi, W.Y., Pan, R.N. and Jiang, F.K., Prepr. Am. Chem, Soc., Div. Pet. Chem., (1993), 38,(3), 581-583
36d Fu, A., Hunt, D., Bonilla, J.A. and Batachari, A., Oil Gas J., (1998), 96 (2), 49-53
37 J. Biswas and I.E. Maxwell, Recent Process and Catalyst-Related Developments in Fluid Catalytic Cracking, submitted to Applied Catalysis.
38 A.R. English and D.C. Kowalzyk, Oil Gas J. 82 (1984) July 16, 127.
39 F.J. Elvin, NPRA Meeting March 1986, Paper AM-86-41.
40 F.J. Elvin, NPRA Meeting March 1987, Paper AM-87-44.
40a J. Scherzer and A.J. Gruia, Hydrocracking Science and Technology, Marcel Dekker, Inc, New York,1996
41 J.W. Ward, Design and Preparation of Hydrocracking Catalysts, G. Poncelet and P.A. Jacobs (Eds.), Preparation of Catalysts III, Elsevier Science Pub., Amsterdam, 1983.
J.W. Ward, Fuel Processing Technol. 35 (1993) 55-85
41c A. van Dijk, A.F. de Vries, J.A.R. van Veen, W.H.J. Stork and P.M.M. Blauwhoff, Catal. Today 1991, 11, p. 129
41d I.E. Maxwell, J.K. Minderhoud, W.H.J. Stork and J.A.R. van Veen, in Handbook of Heterogeneous Catalysis,, eds. G. Ertl, H. Knoezinger, J. Weitkamp, VCH/Wiley, 1997, p.2017
42 I.E. Maxwell, Catalysis Today, 1 (1987) 385-413.
43 T. Yan, Ind, Eng. Chem. Process Des. Dev., 22 (1983) 154-160.
44 I.E. Maxwell, T. Huizinga, A.A. Esener, A. Hoek, F. van der Meerakker, W.H.J. Stork and 0. Sy, Oil Gas J., 1991, april 22, p. 77-82
45 M.E. Reno, B.L. Schaefer, R.T. Penning and B.M. Wood, 1987 NPRA Annual Meeting, Paper AM-87-60.
46 P.J. Nat, 1988 NPRA Annual Meeting, Paper AM-88-75.
Relations between Structure and
47 J. Scherzer, in: Catalytic Materials;
Reactivity, ACS Symposium Series 248, T.E. Whyte et al. (Eds.), 1984, p. 157.
48 A.A. Esener and I.E. Maxwell, "Studies in Surface Science and Catalysis, Hydrotreating Catalysts", (Eds.) M.L. Occelli and R.G. Anthony, vol. 50, pp. 263-271.
49 J.W. Gosselink, A. van de Paverd and W.H.J. Stork, Mild Hydrocracking, Optimization of Multiple Catalyst Systems for Increased VGO Conversion, Catalysts in Petroleum Refining Conference, Kuwait, 1989,

49a P.J. Nat, J.W.F.M. Schoonhoven and F.L. Plantinga, in: Catalysts in Petroleum Refining, D.L. Trimm et al., eds., Elsevier Amsterdam 1989, p.399

49c B. Kraushaar-Czarnetzki, R.J. Dogterom, W.H.J. Stork, C.A. Emeis, and J.P. van Braam Houckgeest, J. Catal. 141 (1993) p. 140

49d A. Corma, A. Martinez, V. Martinez-Soria and J.B. Monton, J. Catal. 153 (1995), p. 25

49e T. Huizinga, J.M.H. Teunissen, J.K. Minderhoud, J.A.R. van Veen and H.J. Springett, paper at the AIChE meeting, San Francisco, november 15, 1994

50 D.J. O'Rear, Ind. Eng. Chem. Res. 26, (1987), 2337-2344.

51 J.M. Oelderik, S.T. Sie and D. Bode, Applied Catalysis, 47 (1989) 1-24.

51a. B. Scheffer, M.A. van Koten, F.C. de Boks, W. Rozema and K.W. Roebschlaeger, paper at the ACS Symposium on Advances in Catalysis and Processes for Heavy Oil Conversion, San Francisco, april 13-17, 1997, and accepted for publication in Catalysis Today

51b. e.g. J.F. Mosby, R.D. Buttke, J.A. Cox and C. Nikolaides, CHEM. Eng. Science, 41, 1986, 989-995, and other AMOCO publications

51c. e.g. K. Niemann and F. Wenzel, Fuels Processing Technology, 35, (1993), p. 1

52 S. Hikada, A. Ino, K. Nita, Y. Maeda, K. Morinaga, and N. Yamazoe, Modification of Y Zeolite by Ferric Nitrate Solution Proc. 7th Int. Zeolite Conf., Tokyo, 1986, pp. 329-336.

53 S. Hikada, A. Ino, S. Nakai, H. Shimakawa, T. Mibuchi and K. Nita, Acidic Properties of FeHY-zeolites in the Presence of H2S, paper H-9 of the Ketjen Catalysts Symposium, Amsterdam, 1986

54 K. Nita, S. Nakai, S. Hidaka, T. Mibuchi, H. Shimakawa, Ken-Ichi and K. Inamura, Deactivation and Reactivation Behaviour of Zeolite Hydrocracking Catalyst for Residual Oil, p. 501 in: B. Delmon and G.F. Froment, (Eds.) Catalyst Deactivation, Elsevier Science Publishers, 1987.

55 H. Sue, The Idemitsu (Kosan Co. Ltd.) Hydrocracking Process produces Distillate from Residual Oils, 3rd UNITAR Heavy Crude Tarsands int. Conf., Long Beach, California, Augtist 1985.

56 H. Sue and M. Fujita, Idemitsu (Kosan Co.) R-HYC (residuum hydrocracking) Catalvsts for Resid Conversion, 51st API Refining Dep. Mid-vear Meeting San Diego, California, December 1986.

57 T. ltoh, New Hydrocracking Catalyst for Heavy Gil Upgrading, Symposium on Advances in Resid Upgrading, Division of Petroleum Chemistry, ACS Denver Meeting, April 5-10, 1987.

57a E.P. Dai and C.N. Campbell, in Catalytic Hydroprocessing of petroleum and distillates, M.C. Oballa and S.S. Shih, eds., Marcel Dekker, 1994, p. 127-141

58 N,W. Green and R.V. Ramanathan, Conversion of Natural Gas to Transport Fuels, AICHE Spring Meeting, March 6-10, New Orleans, 1988, Paper 70B.

59 M.J. van de Burgt, C.J. van Leeuwen, J.J. Dell'Amico and S.T. Sie, The Shell Middle Distillate Synthesis Process, paper Tll, at the Methane Conversion Symposium, University of Auckland, New Zealand, 27 April - 1 May, 1987.

60 H.H. Gierlich, W. Dolkmeyer, A. Avidan and N. Thiagarajan, Chem.-Ing.-Tech. 58, Heft 3 (1986), MS 1462-1486.

61 C.D. Chang, Catal. Rev.-Sci. Eng., 25 (1) (1983) 1-118.

62 C.D. Chang and A.J. Silvestri, CHEMTECH, October 1987, 624-631.

63 A.A. Avidan, Gasoline and Distillate from Methanol, Methane Conversion Symposium, University of Auckland, New Zealand, 27 April - 1 May, 1987.

64 C.D. Chang, Catal. Rev. - Sci. Eng., 26 (334) (1984) 323-345.
65 J.P. van den Berg, J.P. Wolthuizen and J.H.C. van Hooff, Proc. of the Fifth Conference on Zeolites, L.V.C. Rees (Eds.) Heyden, London, 1980, p. 649.
66 T. Mole, J. Catal., 84 (1983) 423.
67 G.A. Olah, H. Doggwieler, J.D. Felberg, S. Frohlich, M.J. Grdina, R. Karples, T. Keumi, S. Inaba, W.M. Ip, K. Lammertsma, G. Salein end D.C. Tabor, J. Am. Chem. Soc., 106 (1984) 2143.
68 S.T. Wilson and P. Barger, Microporous Mesoporous Mater. 29 (1999) 117.
69 D. Chen, K. Moljord, T. Fuglerud and A. Holmen, Microporous Mesoporous Mater. 29 (1999) 191.
69a Marchi, A.J. and Froment, G.F., Appl. Catal., (1991), 71 (1), 139-152
69b Dehertog, W.J.H. and Froment, G.F., Appl. Catal., (1991), 71 (1), 153-165
69c Prakash, A.M., Unnikrishnan, S. and Rao, K.V., Appl. Catal., (1994), A, 110 (1), 1-10
69d Nawaz, S., Kolboe, S., Stocker, M., Stud. Surf. Sci. Catal., (1994), 81, (2nd Natural Gas Conversion Conference) , 393-398
69e Hutchings, G.J., Johnston, P., Lee, D.F., Warwick, A., Williams, C.D. and Wilkinson, M., J. Catal., (1994), 147 (1), 177-185
69f European Chem. News, (16 Dec. 1996), 66, (1745), 25
70 Oil Gas J. 83 (34) August 1985, p. 38-39.
71 J. Topp-Jorgensen and J.R. Rostrup-Nielsen, Oil Gas J. 84 (1986), May 19, 1986.
72 C.D. Chang, H.W. Lang and R.L. Smith, J. Catal., 56 (1979) 169-173.
73 Pet, Times, 91, Aug. 1987, 2213-2214,
74 R.E.Bolan, PEP Review N85-3-3, SRI International 1987.
75 US Patent 4,528,412 (1985).
76 T. Inui, Y. Makino, F. Okazumi and A. Miyamoto, in: P.J. Grobet, W.J. Mortier, E.F. Vansant and G. Schulz-Ekloff (Eds.), Innovation in Zeolite Materials Science, Proc. Int. Sympos., Nieuwpoort (Belgium), Elsevier, 1988, pp. 487-494.
77 V.N. lpatieff and G. Egloff, Oil Gas J., 33 (52) 1934, 31.
78 W.E. Garwood, Prepr, Am. Chem. Soc. Div. Pet. Chem., 27 (2) (1982) 573.
79 US Patent 3,960,778 (1974).
80 S.A. Tabak and J.H. Beech, Mobil Olefin to Gasoline and Distillate (MOGD) Process, paper presented at the AICHE Spring National Meeting, Houston, April 2-6, 1989.
81 J.P. van den Berg, K.H.W. Roebschlaeger end I.E. Maxwell, The Shell Polygasoline and Kero (SPGK) Process, paper presented at the llth North American Meeting of the Catalysis Society, Dearborn, USA, May 7-11, 1989.
82 I.E. Maxwell, Advances in Catalysis, 31, (1982), 1-76.
83 M.E. Davis, C. Saldarriaga, C. Montes, J, Garces, C. Crowder, Nature, 331, (1988) 698.

Studies in Surface Science and Catalysis 137
H. van Bekkum, E.M. Flanigen, P.A. Jacobs and J.C. Jansen (Editors)
© 2001 Elsevier Science B.V. All rights reserved.

Chapter 18

Zeolites and related materials in organic syntheses. Brönsted and Lewis Catalysis

W.F. Hölderich[a] and H. van Bekkum[b]

[a] University of Technology RWTH Aachen, Department of Chemical Technology and Heterogeneous Catalysis, Worringerweg 1, 52074 Aachen, Germany

[b] University of Technology Delft, Laboratory of Organic Chemistry and Catalysis, Julianalaan 136, 2628 BL Delft, The Netherlands

1. INTRODUCTION

Present-day main applications of synthetic zeolites include the use as cation exchangers in detergent formulations, as selective adsorbents in drying and separations, and as catalysts in oil refining and petrochemistry.
In the latter areas by far the greatest use is in catalytic cracking, other important areas being hydrocracking, hydroisomerization and aromatics processing (cf. Chapter 17). In a not too far future the methanol-to-lower-olefins (MTO) process may constitute another important application.

Compared to the successful use of zeolites in hydrocarbon processing, their use in the synthesis of organic intermediates and fine chemicals is in a relatively early state of development. Two reasons for this are

- many of the target organic systems to be synthesized are too bulky to be built in or to desorb from the zeolite pore systems
- the average synthetic organic chemist is not sufficiently acquainted with zeolites, their handling and tuning and their potential, other than their use as drying agent.

One of the first to draw attention to the broad potential of the combination of acidity and shape selectivity offered by zeolites in organic reactions was Venuto (1). The tendency to utilize this potential for specific, highly selective syntheses in the field of intermediates and fine chemicals has continued steadily as reviewed by the present authors (2) in 1991 and by Venuto (3) in 1994. The most recent reviews are given in ref. (4).

Zeolites may be modified in many ways; they can be tuned over a wide range of acidity and basicity, and of hydrophylicity and hydrophobicity, many cations can be introduced by ion exchange and isomorphous substitution is possible also allowing build-in of isolated redox centers in the lattice. Moreover metal crystallites and metal complexes can be entrapped within the microporous environment (5). Altogether rich opportunities are provided of designing and employing heterogeneous catalysts that are tailored to suit the reactions desired. In this respect the experience gained in petrochemistry offers valuable guidance.

This chapter contains:

- a short discussion on general aspects of zeolites which are of relevance to their use in organic syntheses;
- a paragraph on non-catalytic uses of zeolites in organic preparative procedures;

- a main section on zeolite catalysis in organic syntheses with focus on the following areas
  - Isomerization reactions
  - Electrophilic substitution of arenes
  - Cyclization reactions including heterocyclic ring formation
  - Nucleophilic substitution and addition
  - MPV reduction and Oppenauer oxidation
  - Zeolite-catalyzed two- and multi-step syntheses

For zeolite-catalyzed oxidation reactions the reader is referred to Chapter ....

2. GENERAL ASPECTS

2.1. Effects of zeolite structure

As mentioned in earlier chapters a distinction is made between small-, medium- and large-pore zeolites. The entrance to molecules of these three categories is limited by rings of 8, 10 or 12/14 oxygens, respectively, as shown in Figure 1. For more detailed geometric information on a number of important zeolites, see Chapter 3, the IZA Atlas (5a) and the Appendix. The crystallographic accessibilities given in the Appendix may be increased by about 10% to obtain the limiting sizes of entering reactants and desorbing product molecules.

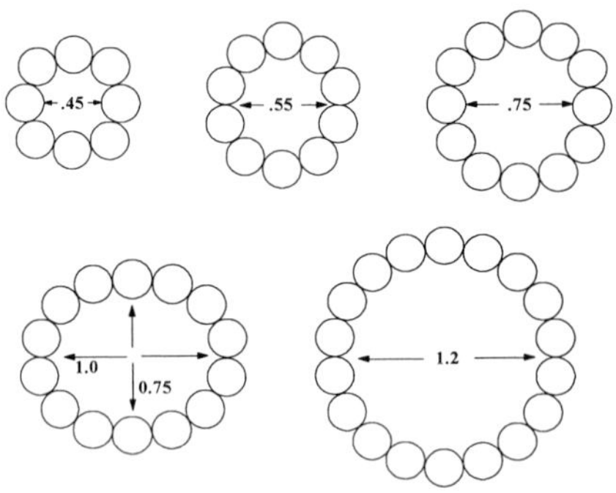

Figure 1. Rings of oxygen atoms as present in zeolites A, ZSM-5, Faujasite (X/Y), UTD-1 and the super large pore aluminophosphate VPI-5; diameters in nm.

Table 1 gives some further examples of the three zeolite classes, also indicating whether the pore system is one-dimensional (parallel channels, no interconnections, code I), or three-dimensional (interconnected channels or cages, code III). Furthermore the Si/Al ratio or minimum Si/Al ratio is given.

In zeolites which are closely related with respect to composition (Si/M range, M = Al, Ga, Fe, B, Ti) and building units, quite different channel arrays can exist. Thus, in ZSM-5 and ZSM-11, zig-zag and/or orthogonal connections are present, respectively, and also intergrowths of these endpoints exist, whereas TON and ZSM-12 can be visualized as a parallel bundle of macaroni.

Table 1
Zeolites and zeotypes classified according to accessibility of and connections in pore system.

Small pore	Si/Al	Dim.	Medium pore	Si/Al	Dim.
A	1	III	ZSM-5	>10	III
Chabazite	2	III	Silicalite-1	∞	III
SAPO-34	< 0.2	III	B-MFI[a]	∞	III
MAP	1	III	TS-1[b]	∞	III
Erionite	3	III	ZSM-11	> 10	III
			Ferrierite	> 5	III
			MCM-22		III
			TON	> 10	I
			AlPO-11[c]	0	I
			SAPO-11	< 0.2	I
		Large pore	Si/Al	Dim.	
		Faujasite/X/Y	> 1	III	
		SAPO-37	< 0.2	III	
		Beta	> 5	III	
		Ti-Beta[b]	∞	III	
		ITQ-7		III	
		Mordenite	> 5	II	
		Offretite	> 3.5	I	
		L	> 3	I	
		Omega	> 2.6	I	
		ZSM-12	> 11	I	
		AlPO-5[c]	0	I	
		SAPO-5	< 0.2	I	
		CIT-5		I	
		UTD-1		I	

[a] Si/B > 20. [b] Ti-MFI, Si/Ti > 30. [c] Al/P = 1.

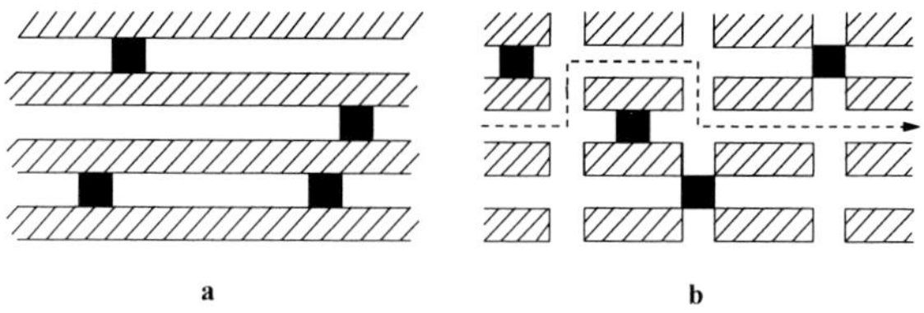

Figure 2. Coke-formation in parallel channel systems (a) and in three-dimensional systems (b).

Such structural differences influence the effective lifetime of zeolitic catalysts. Catalyst deactivation takes place largely as a result of coke deposition (with a spectrum ranging from polycyclic aromatics, especially at higher temperatures, to long-chain, wax-like paraffinic deposits, particularly at lower temperatures). In general, zeolites with a three-dimensional structure will deactivate - at a given amount of coke - less rapidly than zeolites with a one-dimensional framework. In three-dimensional zeolites a channel which is blocked at one point can still be utilized for a catalytic reaction via the interconnections, as schematically shown in Figure 2.

Furthermore it seems that the larger the connecting cages or crossings are, the larger the ease of coke-formation and deactivation is. Thus X- and Y-zeolites with supercages with a diameter of 1.3 nm generally deactivate faster than medium pore pentasil zeolites with crossings of about 0.9 nm diameter (6). Another important variable in coke formation presumably is the Si/Al ratio which determines the number and nature of the acidic sites. See Chapter 16 for a full account of coke formation.

Zeolites can display different types of shape selectivity (7-13) in organic reactions (*cf.* Scheme 12). *Reactant* selectivity allows only part of the reacting molecules to enter the catalyst pores. In *product* selectivity, only compounds with the correct dimensions can diffuse out of the pores. Only those reactions can take place of which the transition state fits within the internal zeolite pores; this may lead to transition state selectivity.

It may be noted that transition state selectivity belongs to a family of reaction selectivities having in common that the surrounding zeolite affects the relative importance of parallel and/or consecutive reactions. Of particular importance in organic synthesis often is the so-called "restricted growth" type selectivity in which the formation of higher addition, substitution or oligomerization products is suppressed due to the dimensions of these products.

Shape selectivity can only operate when the reaction occurs within the zeolite pore framework. There are, however, cases in which it is uncertain whether reaction takes place on both the inner and the outer surfaces or, on the other hand, only on the outer surface of the zeolite. Furthermore, examples are known in which the molecular dimensions would certainly permit reaction within the zeolite framework, but where the course of the reaction is apparently not determined by shape selectivity but instead by other parameters such as thermodynamic considerations.

In order to determine the contribution of the outer surface of the zeolite crystals bulky molecules may serve as probes (see Chapter 21).

2.2. Modification of zeolites

The properties of zeolites and thus their catalytic behaviour can be varied within certain limits by modifying the zeolites either during or after the actual synthesis (cf. Chapter 6).

Variations in synthesis formulation and in synthesis conditions allow

- to synthesize a particular type of zeolite;
- to establish a certain Si/Al ratio, often taking into account some limits (cf. Table 1) pertaining to the zeolite in question;
- to introduce other T-atoms (*e.g.* B, Ga, Fe, Ge, Ti), again often a maximum build-in percentage of such T-atoms applies (cf. Table 1);
- to synthesize relatively large or small crystals of the zeolite under consideration. The crystal size has a direct bearing on the inner/outer surface area ratio and on the average length of the intracrystalline diffusion paths of reacting molecules. Generally, activity decreases (14) with increasing crystal size, whereas selectivity can increase or decrease depending on the reaction studied;
- to modify the morphology of the zeolite crystals. Systems having parallel channels, such as zeolite L or mordenite, show the fastest crystal growth in the direction of the channels. By manipulation of the growth rates, morphologies may be obtained (15,16) which represent more efficient catalysts.

2.2.1. Introduction of Brönsted and Lewis acidity

The number of counter ions in zeolites, which is directly related to the content of trivalent T-atoms, can be adjusted by building-in different ratios of tetravalent (Si) and trivalent T-atoms. In general, zeolite synthesis yields the neutral sodium form. By means of ion-exchange with NH_4-salts and subsequent calcination one obtains the proton *i.e.* the Brönsted acidic form of the zeolites.

Factors affecting the acid strength per site include the Si/T ratio and the type of T-atom. The acidity of the well-studied ZSM-5 zeolites is known to fall according to the T-atom sequence Al > Ga > Fe >> B. In this way the acidity spectrum ranges from the mildly acidic borosilicates to the strongly acidic sites in alumino- and gallosilicates, with high Si/T ratio. As a consequence, any organic reaction subject to proton catalysis can, in principle, also be catalyzed by a zeolite. Some advantages over conventional Brönsted catalysts will be listed in a subsequent paragraph. In some cases the conversion into the proton form may also be carried out by direct ion-exchange with an aqueous mineral acid (HCl, HNO_3). The strong zeolite ZSM-5 is an example. For many other zeolites a treatment with mineral acid will also modify the properties of the zeolite, *e.g.* by extracting T-atoms out of the framework.

Ion-exchange from aqueous solution also allows the introduction of many types of cations, *e.g.* transition metals, lanthanides, into zeolites. In several cases existing knowledge of exchange equilibria allows one to achieve a desired cation loading in a zeolite. Again, any type of homogeneously cation-catalyzed reaction can, in principle, be catalyzed by a zeolite carrying this cation, provided the reactants have access to the zeolite pores.

The zeolite lattice, acting as a macro-ligand, often offers the cation several oxygen surroundings, which can impose restrictions to the cation's coordination chemistry and its catalytic activity. For instance in zeolites X and Y (see Figure 3) the cation positions range from locked-in and inaccessible for reactants (positions S_I and $S_{I'}$) to the active and mobile S_{III} position in the supercages. Careful studies (17) provide a detailed picture of the ion-exchange of La^{III} and Ce^{III} at various temperatures in the zeolites Na-Y and Na-X. It appears that at ca. 80°C La^{III} and Ce^{III} enter the sodalite units and hexagonal prisms. The latter cation positions have a high-oxygen coordination. In order to be able to penetrate the 6-ring windows of the sodalite unit and the hexagonal prism, the La and Ce water coordination has to be strongly adapted. Obviously, it is important to know the distribution of the active cations and their mobility under reaction conditions.

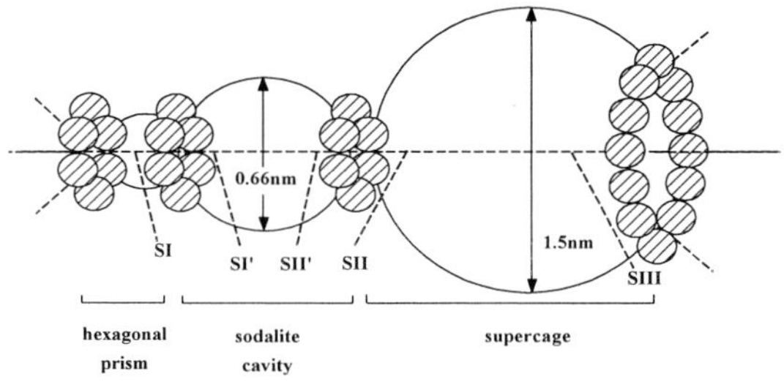

Figure 3. Cation positions in the faujasite lattice (adapted from ref. 18).

2.2.2. Basic zeolite catalysts

The main emphasis of zeolite catalysis has been upon reactions catalyzed by strong or weak acids. If, however, complete ion-exchange with alkali metal ions such as K, Rb or Cs is carried out, it is possible not only to neutralize the Brönsted acid centers, but also to prepare weakly basic zeolites. This is particularly true if these neutral zeolites are then impregnated with alkali hydroxides.

For some examples the reader is referred to a paper of Corma *et al.* (19) reporting on condensations of benzaldehyde with *e.g.* diethyl malonate over alkali metal exchanged X- and Y-zeolites. The catalytic acitivity was found to increase with decreasing Si/Al ratio and with increasing size of the alkali cation. The reactions are assumed to involve ionization of the activated C-H bond followed by attack of the negatively charged reactant on the carbonyl group of benzaldehyde.

In another study on base-catalysis by alkali-modified X- and Y-zeolites (20) the conversion of isopropanol (towards propene on acid sites, towards acetone on basic sites) is taken as a probe reaction.

In another approach Martens *et al.* (21) developed sodium cluster-containing X-, Y- and L-zeolites by thermal decomposition of adsorbed sodium azide. These systems were able to catalyze 2-butene isomerization and aldol condensations.

For reviews on basicity and basic catalytic properties of zeolites we refer to ref. (21a and b). Ordered mesoporous materials of the MCM-41 type may be modified towards basic catalysts in several ways (*cf.* Chapter 22). In the sodium form such systems already exhibit basicity. The basic character can be increased by coating the inner walls with Cs_2O or $CsLaO_2$ (21c). Hybrid basic catalysts can be obtained by tethering organic amines to the MCM-41 wall (21d and e).

2.2.3. Introduction of metal(0) and of metal complexes

Metal ion or metal-complex exchange, followed by a proper reduction procedure, leads to metallic particles, *e.g.* Pt, Pd, inside the zeolite (5).

Such metal entities may activate C-H bonds and may serve as catalyst in (de)hydrogenation, aromatization and oxidation reactions. They also exert a stabilizing effect in several reactions and catalyze oxidative reactivation of used zeolites.

Generally the noble metal is introduced into the zeolite by ion exchange with an aqueous solution of a cationic metal-ammine complex ($Pt(NH_3)_4^{2+}$ or $Pd(NH_3)_4^{2+}$) while performing the ion-exchange in competition with NH_4-ions (*cf.* Chapter 15). The subsequent calcination (NH_3-removal) and reduction (H_2) has been studied in great detail by Sachtler *et al.* (22) for the systems Pt-in-Y and Pd-in-Y. The location (supercage or sodalite cage, *cf.* Figure 3) of the naked noble metal ion prior to reduction has a large effect on the ease of reduction and the degree of association after reduction. For an analysis of the Pt(II) reduction process and the effect of hydrogen partial pressure on the outer/inner surface division see ref (22a).

Small metal complexes can be exchanged like the above-mentioned noble metal-ammine complexes or adsorbed into zeolites; larger complexes, sized up to the diameter of cages or channel crossings, can be constructed inside the zeolite. Various complexes ranging from metal carbonyl complexes to phthalocyanines in - mainly Y - zeolites have been prepared and characterized.

Illustrative work (23) on metal carbonyl complexes involves adsorption of $Ru(CO)_5$ (diam. .63 nm) into zeolite NaY, transformation in the zeolite to $Ru_3(CO)_{12}$ (diam. .92 nm) and subsequently towards a Ru hydrogenation catalyst.

The entrapment of metal phthalocyanines into zeolites (23a) also implies the heterogenization of homogeneous catalysts. Here the zeolite imposes factors leading to very interesting regioselectivity and reactant selectivity in the field of oxidation. Several reports (24) mention the catalytic use of zeolite-encapsulated chiral complexes, constructed by adding chiral ligands to metal ions inside the zeolite. For a review on metal complexes in zeolites, see ref. (18) and (5).

2.2.4. High temperature treatments

The calcination of zeolites serves in the first place to create the H-form by decomposition of the ammonium-form or by removing organic template molecules incorporated into the zeolite framework during the synthesis. At temperatures exceeding 400°C, dehydroxylation may occur leading to transformation of Brönsted acid sites into Lewis acid centers.

Since this means that the catalytic properties are affected, it is necessary to carry out the tempering under carefully controlled conditions.

Calcination also serves regeneration of deactivated (coked) zeolitic catalysts.
The tempering of zeolites in the presence of water is a well-known method for preparing thermally stable zeolites with a long active lifetime. This so-called steaming brings about dealumination (alumina migrates out of the zeolite framework into the cages) with partial curing of the lattice by insertion of Si. The products obtained in this way are known as ultrastable zeolites *e.g.* US-Y.

In another dealumination procedure, in which the zeolite is treated with $SiCl_4$ in the gas phase Al is replaced by Si without developing defect sites in the framework. Liquid phase dealumination techniques include treatment with $(NH_4)_2SiF_6$ (24) and the use of a solution of $SiCl_4$ in carbon tetrachloride (25).

2.2.5. Tuning accessibility of zeolites

As has already been mentioned in an earlier chapter, fine tuning of the effective pore width of zeolites, subsequent to their synthesis, can be achieved by ion-exchange with cations of different sizes.

A further possibility is by means of CVD (chemical vapour deposition) methods, in which Si- or Ge-alkoxides are deposited at the mouth of the zeolite pore and then subjected to heat treatment. In this way the pore opening is artificially narrowed (26).

By reacting zeolites with diborane or silane Vansant *et al.* (26a) were able to alter the adsorption properties of zeolites such as H-mordenite profoundly.

2.2.6. Inertization of external surface of zeolites

The active centers located at the outer surface of zeolites do not display any shape selective behaviour. This negative effect is particularly marked in the case of small crystallites.

Accordingly the aim is to reduce the Al content on the external surface as far as possible. This can be achieved either by lowering drastically the Al concentration in the reaction solution towards the end of the zeolite synthesis resulting in the formation of an SiO_2 coating (27-29) or alternatively, by neutralizing or poisoning the acid centers of the outer surface subsequent to the synthesis. Bulky nitrogen bases such as 4-methylquinoline (30) or silanes bearing bulky substituents such as triphenylchlorosilane (31,32) are suitable for the two latter purposes. Covering by a thin silica layer may be achieved by a treatment with tetraethyl orthosilicate (32a).

2.3. Shaping zeolitic catalysts. Binders

With few exceptions zeolites in the pure state, in particular the Si-rich organophilic types, are difficult to formulate into extrudates, tablets or microspheres for use in fluidized beds. It is therefore necessary to add a binder material for this purpose.

Suitable binders are various aluminum oxides, preferably boehmite, amorphous aluminosilicates, finely divided SiO_2, mixtures of SiO_2 and Al_2O_3, TiO_2, ZrO_2, and clays. After molding, the extrudates or tablets are for example dried at 110°C/16 h and calcined at 500°C/16 h. It is also possible to obtain advantageous catalysts by molding the isolated zeolite immediately after drying and subjecting it to calcination only after molding.

When extruding binder-free often peptization aids are used, being for example ethyl cellulose, stearic acid, potato starch, formic acid, oxalic acid, acetic acid, nitric acid, ammonia, amines, silica esters and graphite or mixtures thereof (32a). Part of these aids have the ability to form chemically new interfaces between particles, others work physically as glue. Upon calcination most of these materials will be removed. In this way any side-reactions caused by the binder material are avoided.

When zeolite-catalyzed reactions are conducted in the liquid phase use can of course also be made of the pure zeolite crystals denoted as zeolite powder.

2.4. Use of zeolites in the liquid phase

As mentioned in a preceding paragraph some high silica zeolites (Si/Al > 10) are stable in aqueous acid medium at moderate temperatures. Generally low silica zeolites (A, X, Y) are not stable under conditions of low pH, and care should be taken when such zeolites are used in a low pH slurry technique as catalyst or adsorbens. The same applies when acid-unstable zeolites are used in reactions producing inorganic acid, e.g. aromatic halogenation, or in formulations containing strongly Al-coordinating ligands.

It is recommended in case some doubt arises as to the stability of the zeolite under the conditions applied, to check the crystallinity of the spent zeolite by X-ray analysis and/or adsorption capacity.

Finally it should be recalled that in liquid phase zeolite-catalyzed reactions the solvent should be carefully chosen. Because of its own adsorption the solvent has an effect on the reactant (and product) concentrations inside the zeolite. Rate and selectivity can be profoundly influenced in this way (33).

2.5. Some general remarks

The broad range of techniques which can be employed during the zeolite synthesis as well as the modifications which can be made at a subsequent stage afford an extremely large number of possibilities for optimizing the catalyst for a particular reaction. This is known as catalyst design or catalyst tailoring. Each individual measure which is adopted can influence activity, selectivity, stability and life time of the zeolite catalysts.

Zeolite catalysis is obviously best served by a balanced strength of adsorption of reactants together with just weak adsorption of products. When relatively strongly adsorbing product molecules are involved in liquid-phase batch experiments, progressive deactivation will occur, whereas in gas-phase flow work, pore filling with reactants and products will reach a steady state. It should be mentioned in this connection that, in the micro-domains of the zeolite crystals, the rate and selectivity of a given reaction are governed by the local "concentrations" of reactants and catalytic sites whereas the products, as observed outside the zeolite - in the bulk gas or liquid – also reflect the diffusion rates of reactants and products, the probability of consecutive reactions and also contributions of the outer zeolite surface – if not made inert - and of the binder material when present.

2.6. Advantages of zeolites

The easily reproducible production of well defined zeolite surfaces and of catalyst prepared from them are important advantages.

Major applications of zeolites are in reactions catalyzed by proton acids and Lewis acids, where the change from a homogeneous to a heterogeneous procedure brings advantages with respect to easy separation and disposal of the catalyst, avoidance of corrosion etc. In this regard, their shape selectivity often has an advantageous effect on the composition of the product.

As carriers for active components, zeolites make it possible to increase the activity and stability of the catalyst towards steam and high temperature, *e.g.* by doping with rare earths. Doping with suitable metals enables hydrogenation and oxidation reactions to be carried out. Bifunctional zeolitic systems can be easily designed and prepared. In all these cases the shape selectivity of the support is an additional factor controlling the reaction.

The thermal stability of the zeolites permits them to be used at relatively high temperatures. They are therefore advantageous for reactions in which the thermodynamic equilibrium requires high temperatures. At the same time carrying out the reaction at a high temperature level makes it possible to recover process heat effectively.

Zeolite catalysts can be regenerated readily with air merely by burning off the coke, which is frequently responsible for their deactivation; after this treatment they generally regain their initial activity.

3. NON-CATALYTIC USE OF ZEOLITES IN ORGANIC SYNTHESIS

Application of zeolites in organic synthesis includes non-catalytic and catalytic uses. The first category comprises
- drying and purification of reactants and media;
- separation of products;
- application as re-usable reagent;
- use as slow release carrier and as selective reaction host;
- use as reactant concentrator;
- use as scavenger, *e.g.* in shifting equilibria.

Some examples may illustrate these methods.

3.1. Drying and purification of reactants and media

Important variables in the use of zeolites as selective adsorbents (34) are the pore size (accessibility), the Si/Al ratio and the counterions. Zeolites with low Si/Al (type A, type X, Si/Al = 1, type MAP) are highly hydrophilic due to the high cation content. Zeolites with high Si/Al ratio (*e.g.* silicalite-1, ZSM-5 with Si/Al $\rightarrow \infty$) are organophilic and will selectively adsorb organic compounds such as ethanol, 1-butanol and phenol from an aqueous solution.

Drying of liquids and gases constitutes one of the early and well-established uses of zeolites. Especially the hydrophilic zeolite A (NaA, KA) is used for this purpose. Alternatively, the more acid-stable small-pore zeolite chabazite can be used. At low water pressures or concentrations zeolites are superior to conventional sorbents like silica. The high affinity for water is coupled with a fairly high saturation capacity (25 wt %).

When drying organic liquids size and nature of the organic compound have of course a bearing on the drying efficiency; the more hydrophobic the compound, the deeper the drying. Table 2 lists some examples (35). For drying of small molecules, *e.g.* acetonitrile, zeolite KA is the desiccant of choice. KA is also applied industrially for drying of bioethanol.

Table 2
Drying of solvents (ref. 35) with zeolite NaA (5% wt/v) at 30°C.

Solvent	Residual water, ppm
Toluene	.01
Benzene	.03
Dichloromethane	.07
Diethyl ether	2
1,4-Dioxane	13
Tetrahydrofuran	28

In drying the enzyme invertase zeolite NaA proved to be somewhat more effective than the classical drying agent P_2O_5.

Recently zeolite membranes became commercial (35a) for solvent dehydration. The tubular membranes consist of a zeolite layer (presumably zeolite A) on porous ceramics or stainless steel. The membranes operate at 95-120°C and show > 98% water selectivity in the permeate while working with 5% water-containing organic solvents.

Zeolites can be useful in keeping reaction formulations dry during the reaction. Thus zeolite NaA proved to be (36) an essential component for selectivity and activity in the stereoselective Sharpless epoxidation of allylic alcohols (Figure 4). Primarily the role of the zeolite appears to be the protection of the Ti catalyst from traces of water present or formed in the reaction medium.

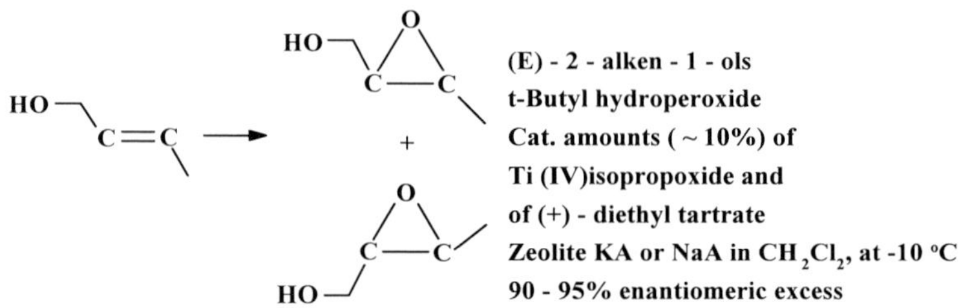

Figure 4. Zeolite A as drying agent in enantioselective epoxidation (ref. 36).

Another chiral Ti-catalyst serves the addition of 1,1-disubstituted olefins to methyl glyoxylate (37). Again the presence of zeolite NaA is reported to be required for obtaining high stereoselectivity. Its precise role is not clear, as yet.

Traces of thiophene can be selectively adsorbed from benzene onto zeolite ZSM-5 (37a). Finally we mention in this paragraph the use of zeolite NaX for the selective adsorption (38) of hydroperoxide impurities from ether solvents or alkene reactants. The zeolite is more effective than the conventional alumina adsorbents. When using a Co^{II}-exchanged zeolite X selective decomposition of peroxide impurities can be achieved.

3.2. Separation of products

Zeolites are used widely as selective adsorbents in a range of large scale (39) as well as small scale separation processes.

Separations may be a result of differences in size and shape of the molecules to be separated or may be due to different affinity of the zeolite towards the compounds to be separated. Table 3 lists some known separations between or within classes of organic compounds and indicates the separation principle and the overall class of zeolites applied.

Table 3
Adsorptive separations by zeolites.

Class(es) of compounds	Separation due to	
	Size	Affinity
Linear branched aliphatics	small pore zeolite	
Functionalized aliphatics		large pore
Disubstituted aromatics (o,m,p)	medium pore	large pore
Trisubstituted aromatics	medium pore (1,2,4 *vs* 1,2,3/1,3,5)	large pore
Tetrasubstituted aromatics	large pore	
Disubstituted naphthalenes	large pore	
Monosaccharides		large pore
Oligosaccharides	large pore	

Well-established separations *by shape* - a form of exclusion chromatography - include for instance the separation of linear and branched alkanes over CaA, and the separation of 1,2,4-trisubstituted benzenes from their larger 1,2,3- and 1,3,5-substituted isomers over medium pore zeolites (*e.g.* ZSM-5, AlPO-11).

Separations by size on large pore (X, Y) zeolites are amongst others the separation of 2,6- and (the larger) 2,7-disubstituted naphthalenes and separation of some 1,2,4,5-tetrasubstituted benzenes from their larger 1,2,4,6-isomers. A fine example is the complete separation of 2,5-dinitro-1,4-di-t-butylbenzene from its more bulky 2,6-dinitro isomer over zeolite CaX (from cyclohexane solution at 80°C) (39a).

The discovery of the ordered mesoporous materials substantially enlarged the potential of zeolitic materials in separation of organic compounds.

UOP developed under the name Sorbex a series of affinity-based advanced separation processes (40) applying simulated-moving-bed technology. An example is the Parex process, world-wide applied on a large scale to separate p-xylene from mixed C_8-aromatics (o-, m- and p-xylene and ethylbenzene). The p-xylene-selective adsorbent is an ion-exchanged X-zeolite.

Other Sorbex separation processes include the Olex process (linear olefins from linear paraffins), the MX process (m-xylene from C_8-aromatics) and the Cresex process (m- and/or p-cresol from xylenols).

In addition to application in separating regio-isomers zeolites may become useful in the separation of stereoisomers. In the carbohydrate field diastereomeric monosaccharides have been separated over X- and Y-zeolites with cation variation over the alkali and alkaline earth metals (41). For instance glucose/fructose separation is achieved over CaY or CaX offering an alternative for the commonly used sulfonated polystyrene resin (in the Ca-form).

Mannose/glucose separation can be performed over BaX or BaY. Differences in the strength of the cation-carbohydrate complexes are assumed to play an important role in these separation processes.

Disaccharides are to be considered - in view of their dimensions – as boundary cases for adsorption into the FAU pore system. Consequently the temperature may be an important variable (42).

So far no examples are known of separation of enantiomers over zeolites. This would require a chiral zeolite or a zeolite loaded in a stable and homogeneous way with a chiral adsorbate. Here excellent opportunities are foreseen for ordered mesporous materials with a chiral entity, *e.g.* a cyclodextrin, attached to the walls of the pores.

An example of selective adsorption of organics from an aqueous medium by a zeolite is the room temperature adsorption of caffeine from aqueous solutions using dealuminated, steam-treated zeolite Y, having a Si/Al ratio of ≥ 13. The adsorbed caffeine may be recovered by ethanol extraction. It is claimed that the coffee flavour components are not affected by this extraction (43).

3.3. Application as re-usable reagent

Zeolites may serve as a carrier for oxidizing cations (Figure 5). Thus CeY can be applied (44) to oxidation reactions, *e.g.* pinacol cleavage, and the zeolitic reagent can be subsequently regenerated by high-temperature air oxidation. In this way, some conventional oxidations by metal compounds might be turned into clean procedures, provided that metal-ion extraction does not occur during the process.

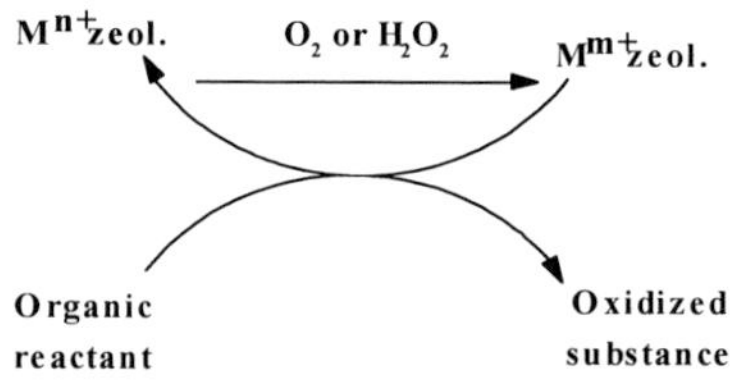

Figure 5. Regenerable zeolite oxidant.

The high cation-exchange capacity of zeolites has been applied in the use of Ag-exchanged zeolites A, X, and Y to promote (45) the coupling of protected monosaccharides. In view of the size of the reactants, the reaction has to take place at the outer zeolite surface (Figure 6), which means that particle size and surface roughness are important factors. The precise way of action and fate of the Ag^{I} ions have to be settled yet.

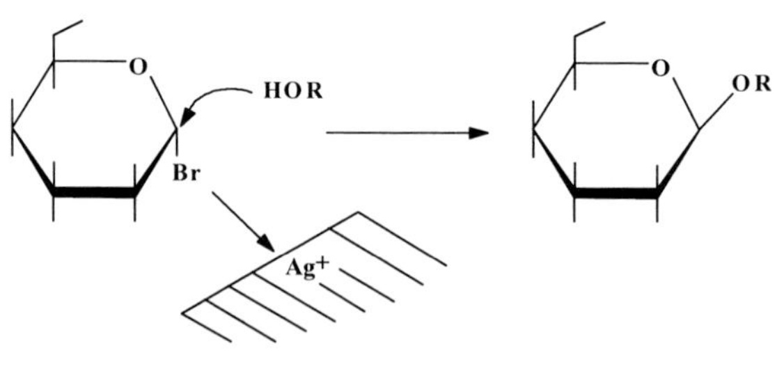

Figure 6. Carbohydrate coupling by Ag zeolite.

3.4. Use as slow release carrier and as selective reaction host

Chemicals can be adsorbed in zeolites either from the gas phase or by using solutions in appropriate solvents, up to complete pore volume saturation if required. The charged zeolite can be used in slow-release techniques - *e.g.* in reactions which are difficult to control as to regioselectivity and consecutive reactions. An example is the use of bromine-loaded zeolite CaA for para-selective bromination of aniline and the toluidines (46).

Interesting effects of microporous solids (present in near-stoichiometric amount) on yield and regioselectivity of aromatic halogenation and nitration by activated reagents were reported by Smith *et al.* (47). Thus para-selective chlorination of monosubstituted benzenes is obtained using t-butyl hypochlorite and zeolite NaHX in acetonitrile. The enhanced p-selectivity is ascribed to steric factors (shape selectivity).

Another interesting example of a zeolite acting as a host lattice for reacting molecules is the photochlorination of n-alkanes, adsorbed into a ZSM-5 zeolite (48). The selectivity for primary C-H conversion of dodecane, S_t, relative to internal chlorination varies from 1.2 to 7.8 compared to S_t = 0.4 for photochlorination of liquid dodecane. The selectivity for terminal chlorination reaches a maximum at loadings of 1-2% which corresponds to 10-20% filling of the zeolite pore system. At these low loadings the alkane will prefer (48a) the straight channels of the zeolite, and the chlorine atoms might arrive through the sinusoidal channels (Figure 7).

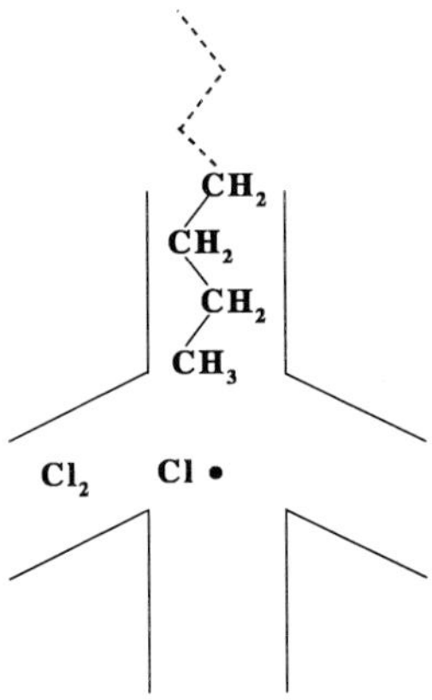

Figure 7. Possible mechanism of terminal photochlorination.

3.5. Zeolites as reactant concentrator

The principle of physical catalysis was elegantly demonstrated by Dessau (49) who studied the cyclodimerization of butadiene to yield vinylcyclohexene over several large-pore Na zeolites and other porous materials. It appeared that Na zeolites, having pores > 0.55 nm and a carbon molecular sieve with pores of 1.3 nm, showed catalytic activity. Smaller pore zeolites, such as ZSM-5 and an amorphous silica with pores >> 1.5 nm, showed hardly any catalytic activity for this reaction. The porous material (zeolite, carbon molecular sieve) concentrates the reactant for the bimolecular reaction and the effect is comparable to the results obtained under high pressure conditions.

3.6. Use as scavenger, e.g. in shifting equilibria

The ability of (hydrophilic) zeolites to selectively adsorb water from a reaction mixture was already mentioned in Section 3.1. Zeolites of the A and X type (*e.g.* NaA, NaX) are also known to be efficient hydrochloric and hydrobromic acid scavengers which can be of use in organic syntheses.

The yield of common organic reactions is often limited by a relatively low equilibrium conversion. Such equilibria, *e.g.* acetalization, esterification, enamine formation, transacetalization and transesterification, can be brought to essential completion by selective adsorption of the small side-product (water, methanol) into a suitable zeolite.

For Brönsted-acid-catalyzed water-forming reactions - acetalization, esterification - an obvious combination would seem, at first sight, to be a conventional homogeneous proton catalyst together with zeolite KA. Problems may arise, however, from cation exchange leading to catalyst deactivation. Preferred combinations (*cf.* Figure 8) consist of KA together with a solid acid catalyst: a sulfonic acid resin, silica-alumina, or a large-pore zeolite such as HY or rare earth (RE)-Y (50). In the last method (II), the reactants should fit in the Y zeolite, which moreover should be chosen with a relatively high Si/Al ratio, *i.e.* relatively hydrophobic in order to prevent water accumulation in the catalyst. Also a high Si/Al mordenite has been studied (51) as esterification catalyst.

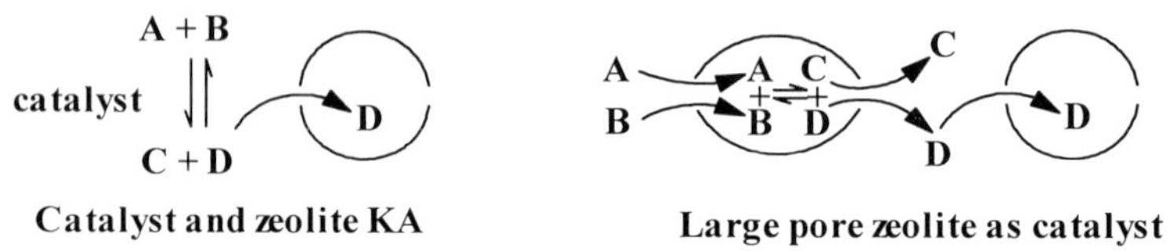

Figure 8. Shifting equilibria by selective side-product removal.

In another water-producing reaction, the enamine synthesis, an amine and a carbonyl compound are involved and the preferred catalyst is expected to be of the Lewis acid type. In this case, a good adsorbent/catalyst combination is zeolite KA and alumina (52).

Finally we mention the use of zeolite KA as water scavenger in the recently discovered direct amidation of oleic acid (53). The enzyme *Candida antarctica* lipase B is the catalyst, the NH_3 source is ammonium carbamate and the reaction is conducted under very mild conditions (35°C).

4. ZEOLITE-CATALYZED ISOMERIZATIONS

4.1. General remarks

Zeolites have a high potential for catalysis of isomerization reactions. In some industrial isomerization processes non-regenerable catalysts (H_2SO_4, $AlCl_3$) are used and zeolite catalysis may lead to clean technology. An interesting option for several products is the combination isomerization/separation: a zeolite-catalyzed isomerization is followed by a chromatographic separation over zeolites and recycle of the unwanted isomers, thus avoiding organic waste products. Established integrated processes of this type include the TIP Process for total isomerization of alkanes (see Chapter 17) and several processes providing di- or trisubstituted benzenes, of which some examples will be given.

4.2. Double bond isomerization of olefins

Double bond isomerizations can be performed using acidic catalysts via a carbenium ion mechanism or using basic catalysts via a carbanion mechanism at the allylic proton (54,55). The advantage of the basic over the acidic catalysts for this reaction is that cracking and skeletal isomerization are suppressed.

One of the most extensively investigated isomerization reactions is the conversion of butenes;

in the steam cracking of naphtha, after butadiene and isobutene have been removed, a C_4-fraction containing a varying proportion of 2-butene to 1-butene is obtained and this ratio can be changed if desired by isomerization. For this isomerization, as well as for the isomerization of longer-chain olefins, acid or basic zeolites can be applied and this is subject of numerous publications (56).

For example, the reaction can be controlled to give the desired isomer by adjusting the acidity of the catalyst and the temperature. At 175-200°C, using an acid B-MFI zeolite, 88% of the 1-butene is converted into 2-butene in a trans/cis ratio of 5:3. Na-B-MFI is suitable for isomerization in the reverse direction; at 450°C trans-2-butene is converted in a yield of 60% into 1-butene and cis-2-butene in a ratio of 7:5 (57-60).

4.3. Double bond isomerization of olefins with functional groups

In the isomerization of 2-alkylacroleins into 2-methyl-2-alkenals according to Reaction 1 a B-MFI zeolite migrates the terminal double bond without reduction in selectivity due to skeleton isomerization or reactions at the aldehyde group. Thus Ce-boron zeolite is an effective heterogeneous catalyst which, at 300°C, gives an 82% selectivity of tiglaldehyde (Reaction 1, R = CH_3) at a conversion of 20% (61); conventional catalysts such as S-poisoned Pd catalyst in the presence of H exhibit the disadvantage of considerable formation of hydrogenation products.

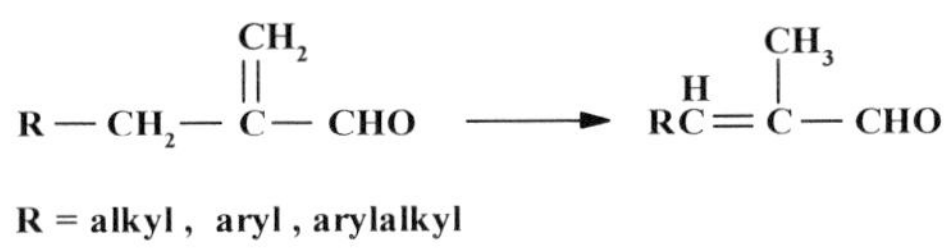

(1)

The isomerization of pentenoic acid esters is possible over noble metal or transition metal doped Y (62,63) and pentasil zeolites (64,65) without attack or cleavage of the ester group. For example (64) when using B-MFI impregnated with 0.8 wt % Pd methyl 3-pentenoate at 180°C and WHSV = 1,4 h^{-1} is converted in a product mixture of 87% 3-, 8% 4-, 4% trans-2-pentenoate.

If the olefin is substituted by an acetoxy group, the pentasil zeolites are not only capable of isomerizing the double bond, but they also produce a rearrangement through a simultaneous shift of the functional group (66). Thus 1,4-diacetoxybutene - see Reaction 2 - is converted into 3,4-diacetoxybutene on boron and aluminum pentasils at 300°C, at a conversion of over 50%. The selectivity is 50-60%. This transformation is reversible; if 3,4-diacetoxybutene is passed over boron zeolite at 200°C, the 1,4-product is obtained with a selectivity of 80%. Hitherto, allyl rearrangements of this type have only been carried out in the liquid phase using SeO_2 or Pt or Pd halides in the presence of O_2. The use of pentasil zeolites in the gas phase offers an alternative to this.

(2)

The aim of selective double bond isomerization is only attained to a limited extent, since, mixtures of isomers are formed. The target product can, however, be obtained by removing, by separation and recycling, the undesired isomers.

4.4. Skeletal isomerization of saturated hydrocarbons

In the chemistry of intermediates, sometimes a selective rearrangement of the carbon skeleton is required. Skeletal isomerization reactions of alkanes take place under surprisingly mild conditions over strongly acidic zeolites. Thus, n-butane is converted into isobutane at 200°C on H-Y- or La-Y-zeolite (67).

Bifunctional Pt-H-mordenite at 250°C is applied for alkane isomerization in the TIP process (Chapter 17); for a discussion of the reaction mechanism including the role of Pt in reducing coke formation and cracking, see ref. (68).

Alkanes containing only one side methyl group are formed as the primary products when long-chain alkanes are isomerized using bifunctional catalysts. Thus, n-undecane is isomerized on Pt/Ce-Y-zeolite in a stream of hydrogen at 275°C at conversions of up to 40% into a mixture of isoundecanes, without cracking processes manifesting themselves (69). Reactions of this type can also be used to characterize shape-selective zeolites (see Chapters 8 and 14).

An interesting skeletal rearrangement is the conversion of tetrahydrodicyclopentadiene into adamantane, reported by Honna *et al.* (70,2d) to take place over rare earth exchanged Y-zeolites in a H_2/HCl atmosphere at 250°C. Initially, an *endo-exo* isomerization takes place (Reaction 3). A stable catalyst is obtained by adding a hydrogenation component (Pt-Re-Co). Selectivities up to 50% have been observed.

On the basis of a mechanistic study, including H-D exchange experiments, Lau and Maier conclude (71) that strongly acidic zeolite sites first form pentacoordinate carbonium ions, which lose H_2 upon transformation into carbenium ions. A sequence of 1,2-shifts then leads to adamantane.

Intermolecular hydride and proton transfer are assumed to occur.

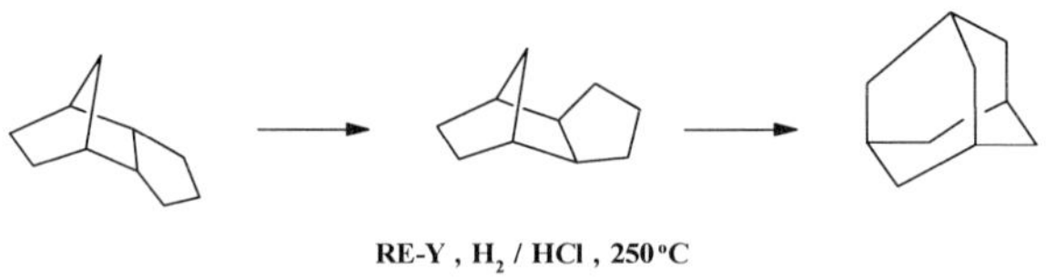

(3)

Several workers have reported on the use of zeolites in the isomerization of terpene hydrocarbons. Here, proton-acid and non-proton catalysis generally lead to quite different product mixtures. For the isomerization of β-pinene to myrcene (Reaction 4), which is of industrial interest, a maximum selectivity of 83% was reported (72) using a zeolite coded TSZ-642.

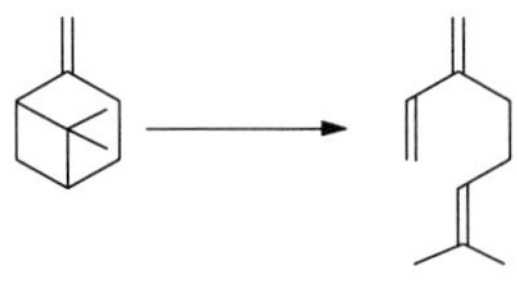

(4)

4.5. Isomerization of functionalized saturated systems

4.5.1. Aldehyde-ketone rearrangement

(5)

Preparation of ketones from aldehydes (Reaction 5) would be an attractive route, as the latter are easily accessible via oxo synthesis, for example. This type of isomerization is known, *e.g.* with tin-, molybdenum- and copper-containing mixed oxide catalysts or using cerium oxide on alumina. Disadvantages of these processes are the low selectivities obtained with satisfactory conversions and the fact that steam is needed in order to get high selectivity and long life time. Because of this, unsymmetrical ketones are generally prepared in industry by the condensation of various organic acids followed by decarboxylation. The unavoidable formation of symmetrical ketones and carbon dioxide are disadvantages in this process.

It has been found by Hölderich *et al.* at BASF that ketones can be obtained advantageously by isomerization of aldehydes using B- or Fe-MFI zeolites (73). A plus point of this method is that the addition of steam is not necessary. The results are summarized in Table 4. As to the priority of the migrating group it is found that the methyl group migrates exclusively if phenyl, benzyl and methyl groups are present. This could be caused by the high shape selectivity of the pentasil zeolites.

Table 4
Aldehyde-ketone rearrangement (73) over MFI-type zeolites.

Feed	Zeolite	Condit.[d]	Product	Conv.	Select.
2-Methyl-propanal	boron[a]	400°C $2\ h^{-1}$	Butan-2-one	42%	90%
Pival-aldehyde	boron[a]	360°C $2\ h^{-1}$	3-Methyl-butan-2-one	92%	85%[e]
2-Phenyl-propanal	boron[a]	400°C $0.8\ h^{-1}$	1-Phenyl-propan-2-one	63%	97%
2-Phenyl-propanal	iron[b]	400°C $2\ h^{-1}$	1-Phenyl-propan-2-one	98%	95%
2-Phenyl-propanal	iron[c]	400°C $2\ h^{-1}$	1-Phenyl-propan-2-one	100%	87%
3-Phenyl-2-methyl-propanal	boron[a]	400°C $2\ h^{-1}$	4-Phenyl-butan-2-one	59%	85%

[a] SiO_2/B_2O_3 = 47, without binder. [b] SiO_2/Fe_2O_3 = 18, without binder. [c] Extruded with boehmite as a binder in the ratio 60:40 wt %. [d] Under isothermal conditions in a tube reactor; WHSV h^{-1} = g feed/g catalyst and hour. [e] Plus 12% isoprene.

The iron-zeolite is more active than the boron-zeolite, as is shown by the conversion of 2-phenylpropanal. This tendency is found *e.g.* in the conversion of methanol to olefins too (74).

The iron-zeolite formed into extrudates with boehmite as a binder is less selective than the pure zeolite catalysts. Using the alumino-zeolite ZSM-5 only low selectivity is obtained at high conversion rate even when steam is used (75), showing that the zeolite's Brönsted acidity is not a major factor in this reaction.

It was found (75a) that ω-phenyl alkanals rearrange over $CsLaO_2$ coated MCM-41 and HMS materials towards phenyl alkyl ketones. Conditions are mild: 100°C in the liquid phase (solvent free). As with the above-discussed high temperature isomerization of branched aldehydes the mechanism of the rearrangement is as yet unclear.

4.5.2. Pinacol rearrangement

The pinacol rearrangement is an acid-catalyzed dehydration of vicinal diols leading to carbonyl compounds (Reaction 6, showing the conversion of pinacol into pinacolone). In a few cases unsaturated alcohol and conjugated dienes are obtained too. This applies regardless of the use of a homogeneous or a heterogeneous catalyst. Both reactions of the 1,2-diols take place on zeolites (76).

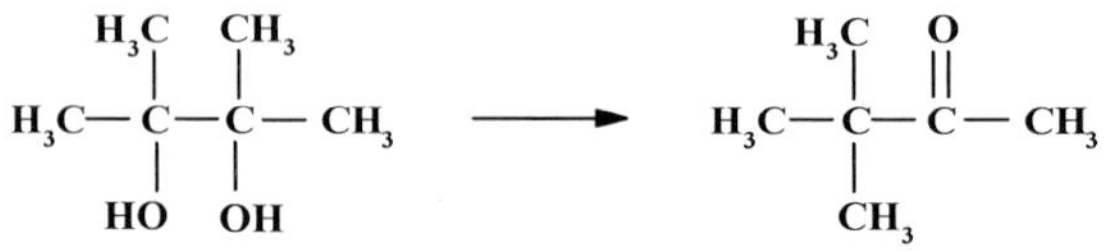

(6)

When using zeolites of the faujasite and pentasil type yields between 41 and 83% are achieved at 105°C, the best performance being obtained with H-Y zeolite.
Within the faujasite group it was found (77) that differences in Brönsted acidity (promoting the pinacol arrangement) and in basicity (promoting β-H elimination) result in differences in selectivity for ketone and diene formation, respectively. If the temperatures are raised to 400°C or higher, dienes are preferably formed.

Also some SAPOs (SAPO-5 and SAPO-37) turned out (77a) to be good catalysts in the conversion of pinacol towards pinacolone. The same holds for MAPO-5 molecular sieves (M = Fe, Ni or Cu) applied at 110°C in a batch reactor (77b). Low polarity solvents, especially 2-octanone, assumed to be able to displace the product ketone from the inner surface, gave the best results here.

1,2-Diols containing a primary OH group are rearranged, after dehydration, to give aldehydes. For example, phenylglycols, which can contain a wide variety of substituents in the aromatic nucleus, are converted (Reaction 7) into phenylacetaldehydes on aluminum and boron MFI zeolites at 250-300°C. The selectivity at complete conversion is between 88 and 96% (78). In this reaction, the zeolites are superior to the amorphous metal-doped or undoped aluminum silicates which give yields of 50-86%. Like phenylglycol itself, ethers and esters of the latter can be isomerized in high yields (91-96%) (78). These reactions are only of academic interest and will probably not be used in industry, since the 1,2-diols are usually prepared from the corresponding epoxides, which undergo rearrangement by zeolites to give such aldehydes too (*cf.* Section 4.5.3).

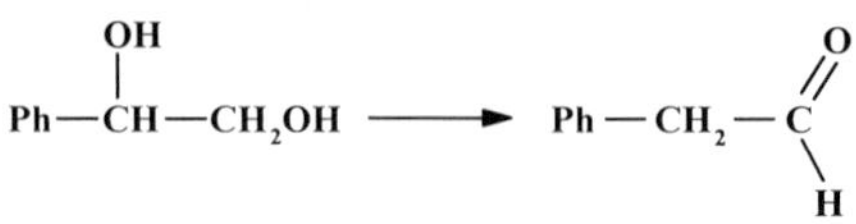

(7)

4.5.3. Rearrangement of epoxides

Activation of epoxides for ring opening reactions can be achieved by Brönsted acid catalysts (addition of a proton to the epoxide oxygen) as well as by Lewis acid catalysts (coordination of the epoxide oxygen to a multivalent cation (78a). As to the regioselectivity of the ring opening electronic as well as steric factors can play a role. These general considerations apply to the opening of the epoxide ring to give aldehydes using heterogeneous catalysts, such as zeolites, in the liquid or gas phase. Indeed, zeolites in their proton form (H-ZSM-5) as well as the essentially neutral Ti-silicalite (TS-1) and Ti-Beta are found to catalyze this type of reactions. The reaction of styrene oxide or alkyl- or alkoxy-substituted styrene oxides using titanium zeolites in acetone or methanol at 30-100°C results in phenylacetaldehydes, according to Reaction 8 with selectivities higher than 90% at conversions between 90 and 100% (79). In a subsequent patent (80) the Enichem workers report on the direct conversion of vinylbenzene compounds (4-methoxystyrene, 4-methylstyrene and α-methylstyrene) to phenylacetaldehydes over TS-1 using aqueous H_2O_2 in *e.g.* methanol as solvent. At approximately 100% conversion selectivities are over 90%, the epoxide is assumed to be the intermediate (*cf.* Chapter 19).

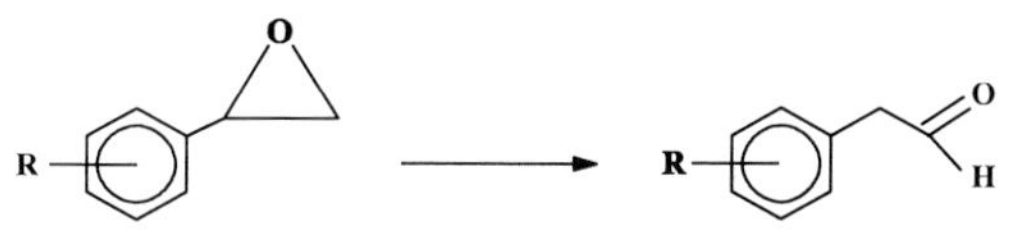

R = alkyl, aryl, arylalkyl, halogen, haloalkyl, alkoxy, alkylthio

(8)

It is also possible to prepare phenylacetaldehydes containing a very wide variety of substituents on zeolites in the gas phase, without using solvents. Particularly attractive features of the acidic boron-, iron- and aluminum-pentasils are their high service lives and yields (> 90%) (78,81).

The high regioselectivity observed in the isomerization of styrene oxides towards phenylacetaldehydes may be related to the stabilization of the developing benzylic cation by the adjacent phenyl ring. In a similar way the selective conversion of phenylglycol into phenylacetaldehyde can be understood.

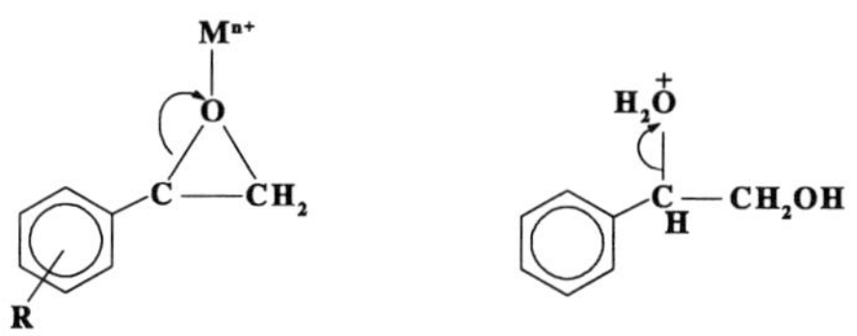

Analogous to the rearrangement of styrene oxide compounds more complicated substances such as phenyl-pyruvic acid methyl ester derivatives can be synthesized from readily available glycidic acid esters as shown in Reaction 9. These esters can be used as intermediates for herbicides as for example the triazinones and for the synthesis of L-amino acids.

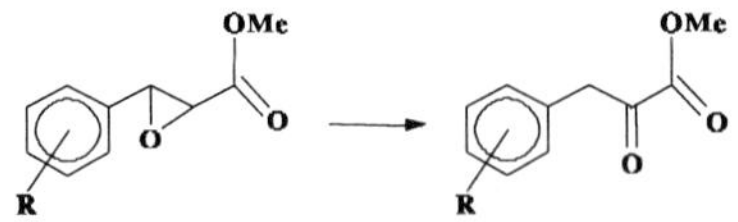

(9)

In the gas-phase reactions zeolites are superior compared to 'non-zeolitic' molecular sieves, metal phosphates and metal oxides. Over the mildly acidic B-MFI (Si/B = 24) the synthesis of phenyl-pyruvic acid methyl ester gives yields up to 94% at 200°C and WHSV = 3 h^{-1}. In the case of the cesium doped material (0.6 wt %) 95% of product yield could be achieved (81a). With these concepts α-ketocarboxylic acid esters can be prepared from readily available feedstocks while conventional methods require the use of costly and environmentally problematical Grignard reactions.

In zeolite-catalyzed isomerization of aliphatic epoxides the observed regioselectivities are relatively low. Also rearrangements have been reported.

Thus, 2-methyl-2,3-epoxybutane is converted at 150°C on MFI type zeolites to 52% methyl isopropyl ketone, 40% pivalaldehyde and 7% isoprene (82). The pivalaldehyde results from a pinacol-type methyl shift in the intermediate ring-opened carbenium ion.

1,5-Dioxospiro-(2,6)-octane is readily converted, over a zeolite catalyst to 4-formyltetrahydropyrane, a valuable intermediate for the production of fragrances, pharmaceuticals and agrochemicals (Reaction 10). This compound is hardly obtainable by known synthesis routes (Rosenmund reduction). Using aluminum or boron containing MFI type zeolites at 300°C in the gas phase the 4-formyltetrahydropyrans are readily produced in high yields of over 97% (82a,b). This reaction can also be carried out successfully in the presence of solvents such as THF resulting in similar yields under the same conditions (82c).

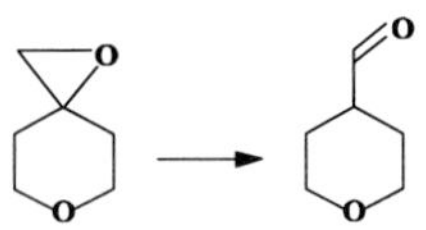

(10)

Terpene epoxides are very reactive compounds. Some products are formed conventionally by isomerization using homogeneous catalysts such as BF_3•Et_2O. The rearrangement of isophorone oxide (Reaction 11) yields the keto aldehyde (**2**), which is an intermediate for fragrance chemicals. If the reaction is performed in the presence of zeolites, the keto aldehyde (**2**), along with the α-diketone (**3**), is mainly formed (82d,e).

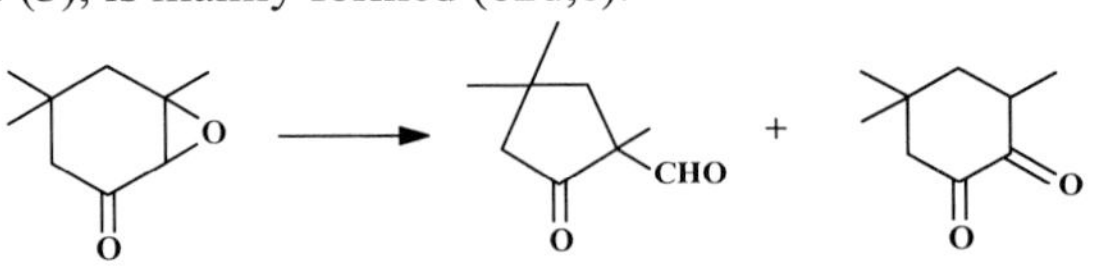

(1) (2) (3)

Isomerization of isophorone oxide

(11)

Hölderich *et al.* obtained high yields, up to 81%, of keto aldehyde (**2**) over H-FER after 6 h in the liquid phase at 110°C using toluene as the solvent (82d). H-US-Y catalysts show high performance in the formation of (**2**). Especially a H-US-Y (Si/Al = 48) catalyst, modified by treatment with diluted HCl (pH = 2) is remarkably active, also at high weight hourly space velocities. The performance of this catalytic system is comparable to the system described by Sheldon *et al.* (82e). They found that a reaction in benzene at 80°C over 2 h over dealuminated H-mordenite gave 100% conversion and 85% selectivity to (**2**).

Many materials, including zeolites have been tested as catalysts for the isomerization of a-pinene oxide towards the fragrance intermediate campholenic aldehyde (Reaction 12). This is the key molecule for the synthesis of various sandalwood-like fragrances.

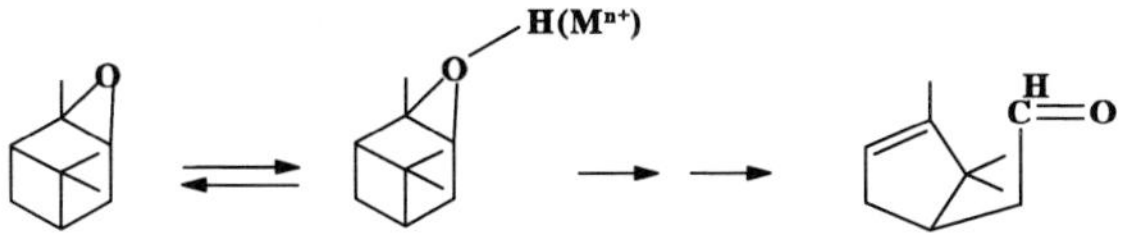

(12)

It appears that certain Lewis acid cataysts are less active but more selective than strong Brönsted acids in this isomerization. The present industrial process applies $ZnBr_2$ in benzene as the catalyst. Selectivity is about 85% which is not easy to beat.

Good results (up to 76% selectivity at 98% conversion) have been obtained by Hölderich *et al.* using HCl-treated H-USY (83). Recently the van Bekkum group showed (84) that Lewis acidic zeolites such as Ti-Beta give excellent results in the rearrangement of α-pinene oxide in both liquid and gas phases. Experiments conducted in the liquid phase with 1,2-dichloroethane as solvent resulted in 81% selectivity for campholenic aldehyde at 29% conversion after 24 h. Gas-phase tests resulted in further improvements in the catalytic performance of this system. With methylcylcohexanone as co-adsorbate 93% selectivity at an initial conversion of 100% was achieved at 90°C. Although the conversion decreases linearly after 6 h on stream, the catalyst can be completely regenerated up to 100 times by a burn-off in air at 480°C.

For mechanistic considerations of the rearrangement, see ref. (84a).

4.5.4. Rearrangement of cyclic acetals

Another example of C-O bond cleavage over zeolites is the conversion of cyclic acetals (85) leading to aldehydes rather than to olefinic compounds. Acetals of neopentyl glycol (2,2-dimethyl-1,3-propanediol) are isomerized, particularly on pentasil zeoliltes, at 250-400°C to neoalkanals with > 90% selectivity at 50-80% conversion (Reaction 13).

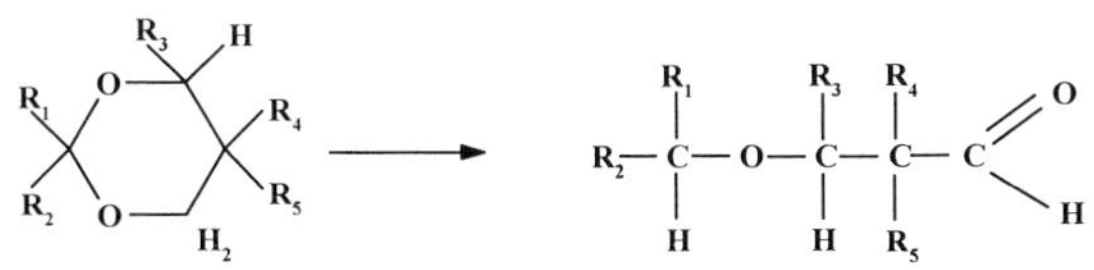

R_1, R_2, R_4, R_5 = H, alkyl, alkenyl, aryl, arylalkyl, alkylaryl, arylalkenyl, alkenylaryl, heterocyclic residue ; R_3 = H, alkyl

(13)

4.5.5. Ring enlargement reactions

Several zeolite-catalyzed ring enlargement reactions exist which are basically of the Wagner-Meerwein type, *i.e.* protonation of the reactant is followed by a dehydration/migration step and subsequently the intermediate enlarged C-containing ring system is stabilized by a final reaction step.

In the following sections the formation of O- and N-containing heterocyclic rings by such a ring enlargement will be exemplified.

4.5.5.1. Conversion of tetrahydrofuran derivatives into dihydropyran systems

Zeolites (*e.g.* X) are reported (86) to be better catalysts than alumina for the rearrangement of tetrahydrofurfuryl alcohol at approximately 350°C to give 2,3-dihydro-2H-pyran. 1,2,5-Pentanetriol can also be used as a starting material for the latter compound.
It is known that alkyl-substituted 1,2,5-pentanetriols can be converted in the presence of p-toluenesulfonic acid to 2-hydroxymethyltetrahydrofurans (95% yield), which compounds can, in turn, be transformed in a second step by a gas phase reaction on Al_2O_3 to 2,3-dihydropyrans (Reaction 14). The disadvantages are two reaction steps and the poor yield of the rearrangement (50%).

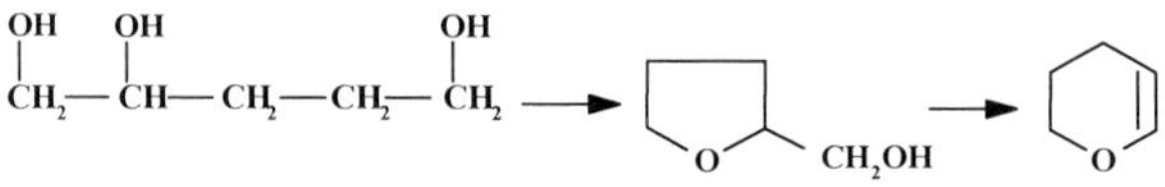

(14)

The dehydration of 1,2,5-pentanetriol on zeolite catalysts proceeds to a large extent directly to 2,3-dihydropyran. When the reaction is carried out on a boron zeolite (H-form) at 350°C, 73% conversion and 70% selectivity are achieved. A by-product is the reaction intermediate 2-hydroxymethyltetrahydrofuran. If the catalyst is doped, *e.g.* with W (3.1% by wt) both activity and selectivity are increased, and 2,3-dihydropyran is obtained with a selectivity of 85% at 100% conversion (87). So by employing a bifunctional zeolite catalyst it is possible to bring about dehydration and subsequent rearrangement in a single step.

An example of ring enlargement of lactones is the zeolite-catalyzed conversion of dihydro-5-(hydroxymethyl)-2-furanones into 3,4-dihydro-2-pyrones (88), which are intermediates, *inter alia*, for insecticides.

4.5.5.2. The Beckmann rearrangement

The most important industrial example of the Beckmann rearrangement is the reaction of cyclohexanone oxime to ε-caprolactam (Reaction 15), which is the starting material for Nylon-6. The classical synthetic route involves the oximation of cyclohexanone with hydroxylamine-sulfate and the subsequent rearrangement of the oxime in concentrated sulfuric acid. Selectivity is high but approximately 2 t $(NH_4)_2SO_4$ per t caprolactam are inevitably obtained as co-product. Further problems encountered include handling a large amount of fuming sulfuric acid and corrosion of the apparatus. In order to eliminate these problems, attempts have been made for many years to switch from a homogeneous to a heterogeneous process and many types of catalyst systems have been tested.

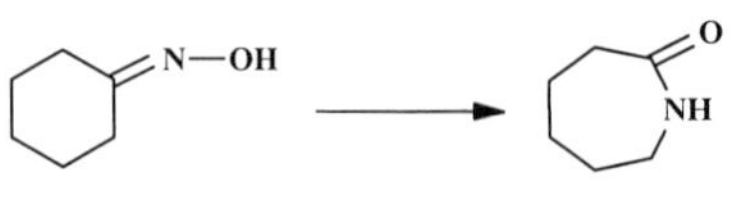

(15)

As long ago as the 1960's Venuto and Landis used zeolites for this purpose (1). X- and Y-zeolites as well as mordenite in the H-form or doped with rare earth or transition metals are employed; *e.g.* cyclohexanone oxime (30 wt % dissolved in benzene) is converted over HY at 380°C and WHSV = 1.2 h^{-1} to ε-caprolactam with 76% selectivity and 85% conversion during the first two hours. The principal by-product is 5-cyanopent-1-ene. As the reaction is continued, the overall conversion decreased to about 30% after 20 h with a drop to 50% selectivity for caprolactam.

Selectivity and catalyst service time were recently improved by Tatsumi *et al.* (89) who introduced alcoholic solvents, particularly 1-hexanol. Thus applying a molar ratio of cyclohexanone oxime/1-hexanol/N_2 of 1/9/5.9 at 623 K in the gas phase over H-USY (Si/Al 62) a selectivity to caprolactam of 94.6% at 99.9% conversion was obtained after 6 h on stream. Weakly acidic ordered mesoporous silica aluminas (MCM-41, FSM-16) also gave selectivities up to 95% provided 1-hexanol was applied as the solvent (89a).

As to mechanistic considerations, first a Brönsted acid catalyzed reaction was assumed: protonation of the OH group followed by concerted dehydration/migration and subsequently water addition. As shown by the selective phenyl migration in (E)-acetophenone oxime (90) the group trans with respect to the OH group migrates. This is also observed in the homogeneous Beckmann rearrangement.

Due to the results of Sato *et al.* (91-93) the focus shifted to medium pore zeolite catalysts with reduced acidity particularly on the outer surface and to other weakly acidic microporous materials.

Sato *et al.* showed for instance that a high Si/Al ratio and inertization of the outer surface - where the reaction was shown to take place - are beneficial in the case of H-ZSM-5 catalyst. Table 5 shows the effect of external surface silanation.

Table 5
Comparison of silanated with nonsilanated H-ZSM-5 in the reaction of cyclohexanone oxime[a] to ε-caprolactam (ref. 93).

Catalyst	Time on stream (h)	Conversion (%)	Selectivity (%)
silanated[b]	3.3	100	95.0
	31.0	98.2	95.0
nonsilanated	3.3	100	79.7
	27.0	95.8	89.4

[a] Reaction conditions: 8 wt % solution of oxime in benzene, 350°C, WHSV = 11.7 h^{-1}, 1 atm, CO_2 as carrier gas molar ratio oxime/CO_2/benzene = 1/5.6/18.3. [b] H-ZSM-5 with Si/Al = 1600, treated with chlorotrimethylsilane at 350°C for 4 h.

Persuing the idea of reducing the acidity of the zeolite in order to get high selectivity and long life time of the catalyst, also TS-1 (93a), B-MFI (93b) and SAPO-11 (94) were used for the Beckmann rearrangement. The activity of B-MFI has been studied in much detail (93b). By applying ethanol and water as diluents catalyst life time and selectivity could be improved. Working at 300°C in the gas phase selectivities up to 95% were obtained. Comparison of B-MFI crystals of various sizes indicated the reaction to take place at the outer surface, presumably by silanol nests. Complete regeneration could be achieved by calcination at 500°C. Using SAPO-11, a 5 wt % solution of cyclohexanone oxime in acetonitrile is converted at 350°C, atmospheric pressure and WHSV = 10.8 h^{-1} to ε-caprolactam with 95% selectivity and 98% conversion. A pilot plant (UOP) has been running (94a) using SAPO-11 as the catalyst.

For a recent review on heterogeneous catalyzed Beckmann rearrangements see ref. (94b). Finally it may be mentioned that a successful zeolite-catalyzed rearrangement of cyclohexanone oxime would be the final bridge in a new process sequence leading to ε-caprolactam (see Figure 9). Two other reaction steps will be discussed later.

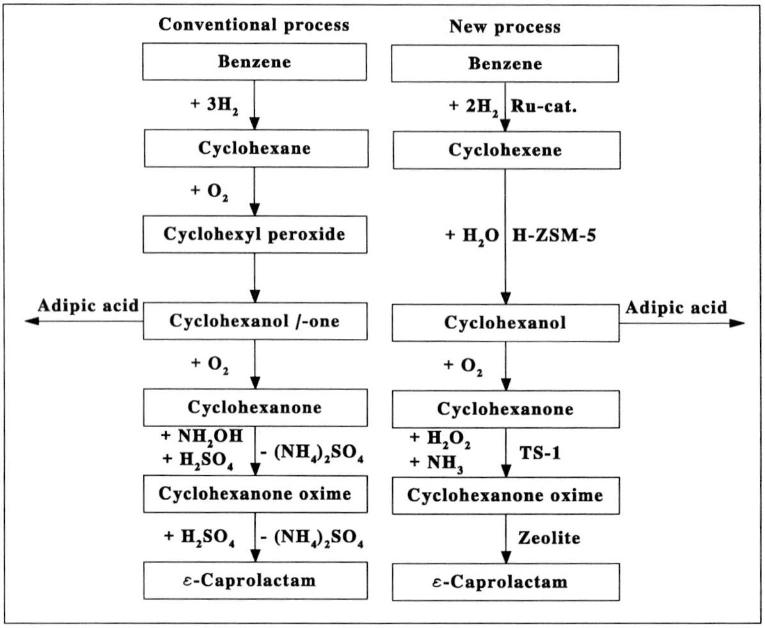

Figure 9. Synthesis routes of ε-caprolactam and adipic acid (93b).

4.5.6. Benzamine rearrangement

The synthesis of substituted pyridines by rearrangement of aminobenzenes in the presence of ammonia constitutes an interesting reaction. Thus aniline was converted (95) at 510°C over H-ZSM-5 with 52% selectivity into α-picoline (2-methylpyridine). Later (95a) 85% selectivity was achieved by optimalization of the NH_3 pressure. In the absence of NH_3 diphenylamine is the main product.

In a similar way 1,3-diaminobenzenes rearrange (96) to a mixture of 2- and 4-aminopyridines (Reaction 16) and α-naphthylamine to 1-methylisoquinoline.

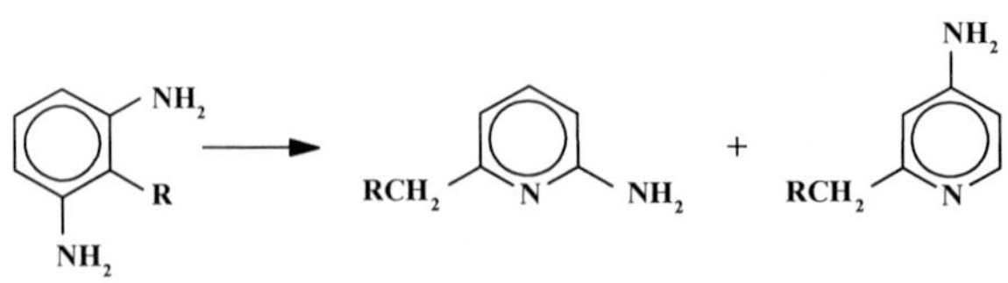

(16)

A mixture of 1,3-diaminobenzene and NH_3 (molar ratio 1:60) is converted at 350°C and 190 bar on H-ZSM-5 to 2-amino-6-methylpyridine with 83% selectivity at 43% conversion. A comparison with silica-alumina or alumina under the same reaction conditions shows the superior properties of the zeolite over other acid catalysts not possessing zeolite structure. Zeolites show the best performance at around 600 K and 10 MPa. This is a useful new route for aminopyridines, compounds which were hitherto only available by the complicated Tschitschibabin-reaction of sodium amide with pyridine.

The mechanism of these exciting reactions has been elucidated by Prins (95b), and is assumed to involve addition of NH_3 to the arylamine, ring opening to an enaminoimine intermediate through a reverse aldol reaction followed by ring closure and NH_3 elimination.

This benzamine rearrangement opens new synthetic routes but selectivities need improvement.

4.6. Isomerization of arenes

4.6.1. Skeletal isomerization of alkylsubstituted arenes

The conversion of substituted arenes into their isomers is of industrial interest and frequently serves as a model reaction for the shape selectivity of zeolites (32). The isomerization of xylenes is one of the processes carried out on a large industrial scale using zeolites as catalysts. It forms the subject of numerous publications. Here the shape selectivity of the zeolites is utilized to increase the proportion of p-xylene in C_8-arene fractions.

The mechanism of xylene interconversion is believed to involve proton addition followed by a 1,2-methyl shift.

Recently cyclohexyltoluene isomerization has been applied in a new synthetic approach of 3-methylbiphenyl (96a).

Polynuclear aromatic compounds can also be isomerized using zeolites. 1-Methylnaphthalene is converted (40%) into 2-methylnaphthalene with a selectivity of 95% using H-Na-Y-zeolite at 270°C in a stream of hydrogen, but there is a loss of activity of ca. 50% in the course of 10 h (97). For a study of this isomerization over H-ZSM-5 and the large pore zeolites H-ZSM-12 and HY, see ref. (98).

4.6.2. Isomerization of arenes containing functional groups

The principle of the xylene isomerization is applicable to a number of toluene derivatives carrying functional groups such as hydroxyl, amino, nitrilo and halogen. As shown, *e.g.* by Weigert (99,100), zeolites of the ZSM-5 type are particularly suited for such equilibrations. The highly acidic H-ZSM-5 is able to protonate the aromatic nucleus, to form the intermediate carbenium ion, also when strongly deactivating groups as cyano are present. Thus a contact time of 3 s at 500°C suffices (99) to establish the three component equilibrium of 46% ortho-, 34% meta- and 20% para-tolunitrile over H-ZSM-5.

A zirconium-containing MFI zeolite is able to isomerize o- and/or p-toluidine into mixtures of the o-, m- and p-isomers and is superior to the H-ZSM-5 catalyst in terms of both product yield and catalyst lifetime (101). At 430°C o-toluidine is converted on the zirconium zeolite to an o-, m-,p-mixture in a weight ratio of 37:45:15. Aniline can be applied as a diluent (100). Conditions are less severe than applied in the benzamine-to-picoline isomerization. Also 2-ethylaniline can be isomerized.

Toluidine is produced by the hydrogenation of nitrotoluene mixtures of approximate composition 63% o-, 4% m- and 33% p-isomers. However, the m-isomer is of great importance as an intermediate for dyestuffs and agrochemicals. In a subsequent isomerization process (Reaction 17) on an MFI-type zeolite a toluidine mixture rich in the m-isomer is obtained. The components can be separated and the o-/p-isomers are recycled.

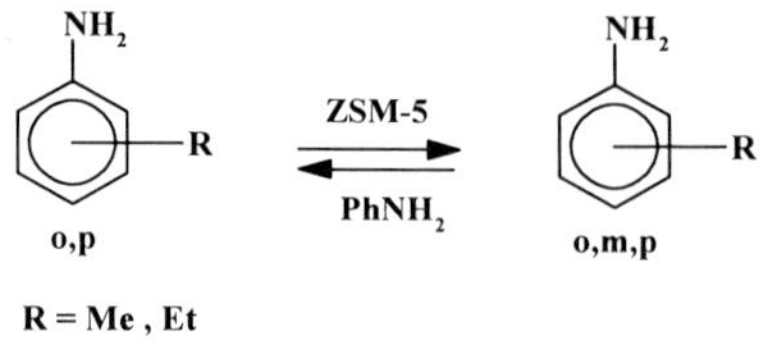

(17)

It may be noted that here the equilibrium mixture is obtained and not as with the xylenes a mixture in which the p-isomer dominates. Apparently diffusion rate differences do not prevail in the toluidine isomerization; factors are the higher reaction temperature and the fact that amines diffuse slower - presumably more jumpy due to protonation/deprotonation - through the H-ZSM-5 channels than hydrocarbons of the same size. For a comparison of results on toluidine isomerization as well as a proposed reaction mechanism, see ref. (102).

Several, mainly patent publications, deal with isomerization (and separation) of chlorinated benzenes such as chloro- and dichlorotoluenes, over zeolites. Often integrated processes can be designed combining zeolite-catalyzed isomerization with separation over zeolites or AlPO molecular sieves and including recycle of unwanted isomers.

The isomerization of the dichlorotoluenes may serve as an example. Upon direct chlorination 2,4- and 2,5-dichlorotoluene are the main products. The 2,6-, 2,3- and the 3,4-isomer are present in low amounts and the 3,5-isomer is absent in such mixtures. Isomerization covering all dichlorotoluenes (Figure 10) is achieved over the large pore zeolites Beta, omega and mordenite in the H-form at temperatures 300-350°C in hydrogen (103,104). By doping the zeolite with Re, Ag or Ni a stable catalyst is obtained (105). Thus AgH-mordenite at 350°C remained completely stable over 150 h as measured by the amount of 2,6-dichlorotoluene formed (8.6%).

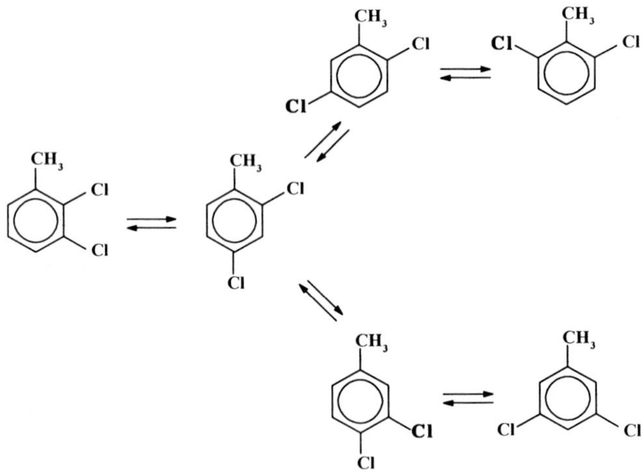

Figure 10. Isomerization of dichlorotoluenes on a large pore zeolite.

Specific isomers can be separated from dichlorotoluene mixtures by adsorptive separation using a simulated moving bed, wherein a faujasite type zeolite is used as the adsorbent (106). The specific isomer is for example 2,6-dichlorotoluene which is the starting material for the herbicide 2,6-dichlorobenzonitrile (Casoron).

When using an MFI-type zeolite, Re-doped (105) or Zr-doped (106), the attainable equilibrium is limited to 2,4-, 2,5- and 3,4-dichlorotoluene (Figure 11). The slightly larger 2,3-., 2,6- and 3,5-isomers are assumed to take part in the equilibrium at the crossings in the zeolite but are too bulky to diffuse through the channels and leave the zeolite crystal. A fine case of product selectivity.

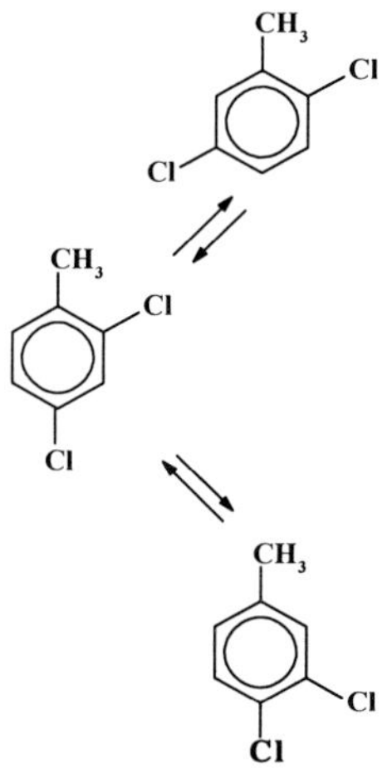

Figure 11. Isomerization of dichlorotoluenes on a medium pore zeolite.

In accordance with the above equilibrium limitation the medium pore molecular sieve $AlPO_4$ –11 is able to separate 2,6-dichlorotoluene from its 2,4- and 2,5-isomers; for the latter two compounds the sieve is accessible.

Finally it may be noted in this paragraph on chlorobenzenes that the classical isomerization catalyst, $AlCl_3$, (i) will give all six isomers, (ii) is corrosive, (iii) cannot be regenerated.

4.6.3. Isomerization of substituted heteroaromatics

Isomerization reactions of heteroaromatic compounds, *e.g.* alkylthiophenes and halogenothiophenes are also catalyzed by zeolites. Thus 2-methylthiophene in admixture with steam is converted at 330°C on H-ZSM-5 to a product mixture containing 53% 2- and 42% 3-methylthiophene (107). 2-Chlorothiophene can be converted into 3-chlorothiophene (84%) with minor dechlorination (5%) taking place. The 2-chloro-isomer is formed preferentially in the chlorination of thiophene. Taking into account the easy separation of 2- and 3-chlorothiophene zeolite-catalyzed isomerization provides a good route to the 3-isomer. For a review on isomerization of halothiophenes, see ref. (108).

4.6.4. The Fries rearrangement

The Fries rearrangement is the isomerization of phenyl esters towards o- and p-acylphenols, and is usually performed using a relatively large amount of $AlCl_3$ as a catalyst. As the aryl ketone forms a stable 1:1 molar adduct with the $AlCl_3$ more than stoichiometric amounts are required. The obvious interest in the development of a heterogeneously catalyzed process has resulted in studies on the use of solid acid, especially zeolites, as catalysts.

Most of these studies concern the synthesis of hydroxyacetophenones either by Fries rearrangement of phenyl acetate or by acetylation of phenol with acetic acid or acetic anhydride. Especially the catalysis groups at Poitiers and at Zürich have been very active here. Most studied catalyst is zeolite H-ZSM-5. The reaction has been conducted in the gas as well as in the liquid phase. For a recent review see (109).

Upon reacting phenol or a substituted phenol and a carboxylic acid or anhydride in the proper temperature range over an acid catalyst, ester formation followed by Fries rearrangement is to be expected (Reaction 18). A set of reactions exist - in principle as equilibria - but catalyst and temperature determine which part of the network is in operation.

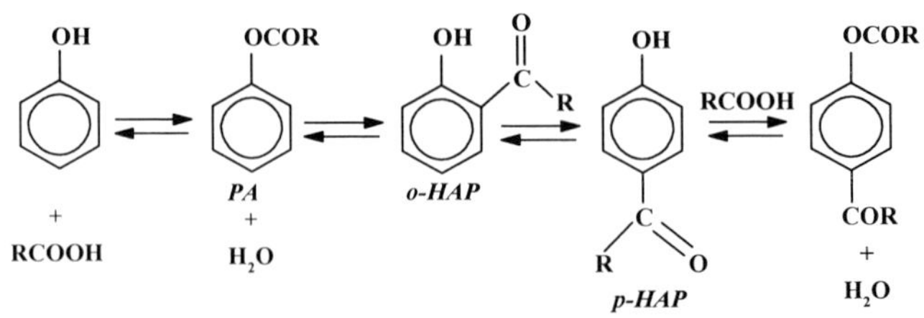

(18)

Thus relatively high proportions of o-hydroxyacetophenone (o-HAP) are observed in the direct acetylation of phenol over ZSM-5 type catalysts. On a H-ZSM-5 (Si/Al = 480) at 300°C with a residence time of 8 s, phenol reacts with acetic acid to yield 2-hydroxyacetophenone with 47% selectivity at 67% phenol conversion (110). The para-isomer is only formed to an extent of 2% and 1% of 4-acetylphenyl acetate is present. The remainder of the reaction product consists largely of phenyl acetate (32%). In fact, on H-ZSM-5 (Si/Al = 24) at approximately 250°C the ortho-isomer is formed with almost 99% selectivity at 15% phenol conversion.

The ortho-selectivity might stem from an intramolecular rearrangement towards the ortho-isomer with stabilization by chelation (Al), followed by a slow intermolecular ortho-para rearrangement. Also direct acylation of phenol is possible (109). The p-hydroxyacetophenone (p-HAP) is thought to be mainly formed in H-ZSM-5 by transacylation of phenol by phenyl acetate (PA). Other zeolites as ZSM-12 (110a) and H-Beta (110b) have been tested as well. Particularly H-Beta gave good results, with p/o HAP ratios up to 4.7.

Regioselectivity simplifies in the Fries rearrangement when using resorcinol as the reactant. Here excellent results have been obtained (110c) in the benzoylation towards 2,4-dihydroxybenzophenone (Reaction 19), a precursor of a well-known sunscreen agent. H-Beta is the catalyst of choice, benzoic acid is the acylating agent. Yields of almost 90% have been achieved working in the liquid phase.

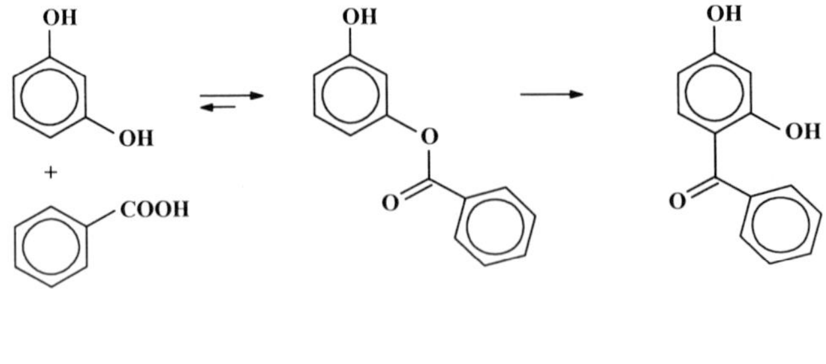

(19)

The present industrial process involves the reaction of benzotrichloride ($PhCCl_3$, obtained by chlorination of toluene) with resorcinol in the presence of $FeCl_3$ and produces polluting and corrosive waste streams.

The limited space offered by zeolite Beta becomes apparent when 2-methyl- and 2,6-dimethylbenzoic acid are applied as reactants (110c) together with resorcinol. With 2-methylbenzoic acid the conversion to the benzophenone is accelerated due to electronic effects. With 2,6-dimethylbenzoic acid the reaction slows down strongly because the intermediate ester is too bulky to be formed inside Beta. Here the mesoporous MCM-41 material in the H-form is a good catalyst (110d). The same holds for the reaction of 1-naphthol and 2,6-dimethylbenzoic acid.

5. ELECTROPHILIC SUBSTITUTION OF ARENES

In electrophilic aromatic substitution often metal halides ($AlCl_3$, $FeCl_3$, $ZnCl_2$, $SnCl_4$, BF_3) or mineral acids (H_2SO_4, H_3PO_4, HF) are applied as the catalyst. Regeneration of these conventional catalysts generally not occurs.

Zeolites and related solid acids are taking over the catalytic jobs. Zeolites are environmentally benign, they reduce waste disposal profoundly. Moreover they are noncorrosive.

5.1. Aromatic alkylation

In the field of alkylation of benzene with lower olefins (ethene, propene) zeolite-based processes are dominating now. For ethylbenzene this is already the case for almost 20 years; the question in case of a new plant is not whether a zeolite-based process will be selected but which one. For cumene the transition towards zeolite catalysis takes place since a few years.

Some excellent reviews exist on benzene alkylation (111-113). Moreover, some articles provide an accurate comparison of zeolite's performance in benzene alkylation with lower olefins, including variations in Si/Al ratio, zeolite crystal size etc.

Thus Bellussi *et al.* (114) compared zeolites Beta and USY in benzene alkylation with ethene and propene whereas Corma *et al.* (115) did so for zeolites MCM-22, Beta and ZSM-5.
Zeolite Beta seems to be the most active catalyst from these studies whereas MCM-22 shares a good activity with an excellent stability. The latter quality may be a result of the alkylation taking place (115) on the outer surface of the plate-shaped MCM-22 crystals where half-cages (cups) are present, with suitable dimensions for benzene alkylation.

5.1.1. Ethylbenzene and cumene

The conventional routes to ethylbenzene are the liquid phase $AlCl_3$-based process (< 130°C) with improved version operating at higher temperature (160-180°C) and the gas phase process operating at 300°C and 60 bar over BF_3-on-alumina as the catalyst.

A true breakthrough was the Mobil-Badger process (116-119) applying H-ZSM-5 as the catalyst. Ethene is converted to ethylbenzene, with 99% selectivity, at about 400°C, under approximately 20 bar, at a WHSV of 3 h^{-1} (based on ethene) and a benzene/ethene molar ratio of 6 to 7. Benzene conversion is about 20%.

The pentasil zeolite is shape selective; the channels allow the transport of benzene, ethylbenzene and diethylbenzenes but prevent the emerging of polyalkylated products. Diethylbenzene, which is undesirable, can be converted to ethylbenzene with benzene over the same catalyst.

With respect to process technology, energy savings and environmental protection, the advantages of this heterogeneous catalyst over the conventional homogeneous $AlCl_3$-catalyst are:
- alkylation at 400°C permits as much as 95% of the heat to be recovered

- the catalyst is non-corrosive and need not be separated off or processed
- there are no disposal problems as in the case of $AlCl_3$ hydrolysis
- and deactivated catalysts can be easily regenerated.

Many plants all over the world practise or have practised the Mobil-Badger process. In recent years other zeolite catalysts (Y, MCM-22) came to the fore. Especially MCM-22 is a favorite catalyst nowadays.

The classical catalyst for the cumene synthesis is kieselguhr supported phosphoric acid (SPA). This system is first subjected to calcination with formation of polyphosphoric acid and silicophosphoric acid.

Several zeolite-based cumene processes are today in operation (112). The CD Tech and Lummus-Unocal processes make use of a modified Y-zeolite. The Enichem process applies zeolite H-Beta and the Mobil-Raytheon process uses zeolite MCM-22.

At Dow-Chemical a de-aluminated mordenite has been developed (120) by calcining H-mordenite and treating the zeolite with mineral acids (HNO_3, HCl). The severity of the thermal and acid treatments, the number and sequence of the treatments and the type of binder material define the properties of these catalysts that are coded 3-DDM: 3-dimensional dealuminated mordenite. Mesopores are introduced enhancing the access (*cf.* schematic picture below) to the 1-D channels of 3-DDM greatly. See also ref. (120a) for a description of 3-DDM preparation.

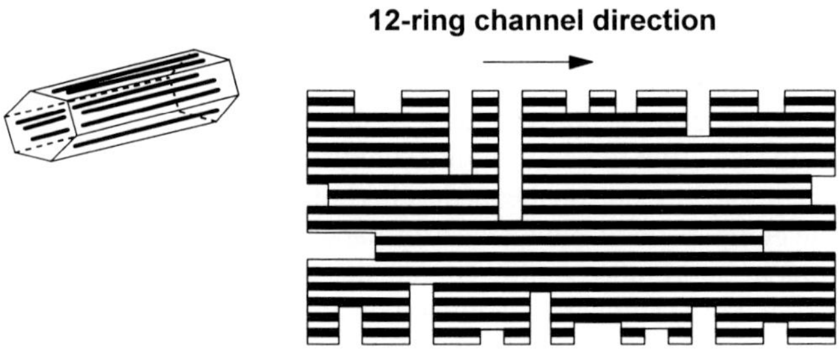

3-DDM, cross-section

Finally a cumene process called Q-Max has been developed by UOP. Here presumably a large pore silicoaluminophosphate molecular sieve with a small amount of Mg in the framework, MgSAPO-31 is the catalyst.

5.1.2. Linear alkylbenzenes

Linear alkylbenzenes are important precursors of biodegradable anionic surfactants (LAS, linear alkylbenzenesulfonates). The conventional catalyst is HF. The first solid acid catalyst, a fluorided silica-alumina, was introduced some 5 years ago (DETAL process). The DETAL process is safer than the HF process and also more cost-effective because no special metallurgy is required and no calcium fluoride waste stream exists (112).

Zeolites such as Beta and ZSM-12 may come to the fore here (121) because they display higher selectivity to the desired 2-phenyl isomers.

5.1.3. Mechanism

The mechanism of alkylation of benzene on H-zeolites is assumed to involve formation of alkylated surface sites which attack the aromatic nucleus as depicted in Reaction 20 for ethylation.

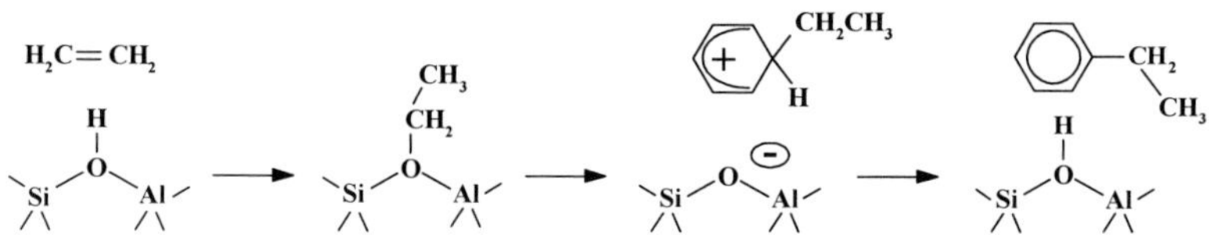

(20)

When subjecting a substituted aromatic system ArX to zeolite-catalyzed alkylation the initial regioselectivity will be determined by the relative stabilities of the 3 transition states $ArX(R)^+$ leading to the o-, m- and p-isomers. It will be clear that the zeolite micro domain may add electronic (adsorption) and steric factors to such factors inherently pertaining to the $ArXR^+$ system. Taking into account the small differences in the inherent stability of o- and p-ArXR for X = alkyl, OH, halogen and other o,p-directing X, such effects imposed by the zeolite may have a substantial influence on the o/p ratio.

When the reaction conditions are such (*e.g.* high temperature) that isomerization occurs, differences in diffusion rates may play a role and *e.g.* crystal size may become a factor. Summarized:

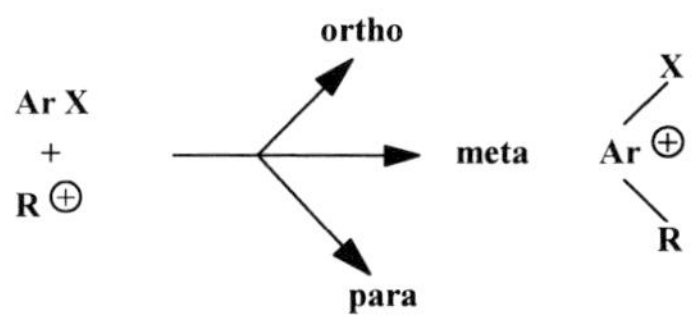

o Inherent electronic + steric factors

Transition state selectivity

o Factors imposed by zeolite micro-domain

Product size selectivity

o Under isomerization conditions
ArXR diffusion rates crystal size effects

As to consecutive reactions in aromatic alkylation such compounds may - depending upon their size and the pore size of the zeolite used - show up in the product mixture or just exist in the larger cages and crossings and take part in equilibria. For a convincing direct NMR observation of the latter phenomenon in aromatic methylation, see ref. (122).

As to regioselective alkylation the focus has been on selective p-alkylation. In a comparative study (123) of the zeolites H-Y, H-mordenite and H-ZSM-5 in toluene alkylation - under assumed equilibrium conditions – just H-ZSM-5 showed pronounced p-selectivity, especially when large crystals were used with homogeneous Al-distribution.

This important feature of the pentasil zeolite may be further improved by modification, *e.g.* by adding Mg- and/or P-compounds. Mobil has commercialized a process to produce p-ethyltoluene – precursor of p-methylstyrene – using these types of catalysts.

5.1.4. Alkylation of dinuclear aromatics

Many papers and patents deal with the alkylation - with methanol or lower olefins - of the dinuclear aromatics naphthalene and biphenyl. Gas phase as well as liquid phase techniques have been applied.

Thus the industrially important methylation of naphthalene has been carried out (124-127) over H-Y, H-mordenite and H-ZSM-5. Mixtures of 1- and 2-methylnaphthalene are obtained with H-ZSM-5 exhibiting the highest selectivity towards the 2-isomer (and in the consecutive reaction towards 2,6-dimethylnaphthalene).

These results raised discussions on *outer surface catalysis*. Thus Fraenkel *et al.* (125) assumed that methylated naphthalenes do not fit into the ZSM-5 pore system and that accordingly the shape selective methylation of (Me) naphthalene over H-ZSM-5 occurs on external Brönsted acid sites. These external surface sites are located in "half" channel intersection cavities which can accommodate well the mono- and dimethylnaphthalene products. Derouane (128) supported this idea while introducing the general term "molecular nesting". In the recent work of Corma *et al.* (128a) on delaminated MCM-22 systems this has become reality. Here large outer surfaces containing half-cages exist as catalytic sites.

Weitkamp *et al.* (126) measured the adsorption of the methylnaphthalenes onto ZSM-5; at 100°C only the 2-isomer is adsorbed and in a quantity (6.5 wt %) well exceeding the amount calculated for outer surface adsorption. It is assumed, however, that at the reaction temperature, 400°C, also 1-methylnaphthalene can enter the pores of ZSM-5.

at 100 °C

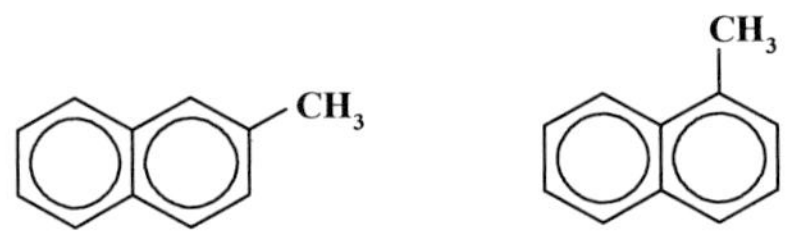

adsorbed **not adsorbed onto ZSM-5**

For the production of 2,6-diisopropylnaphthalene mordenite-based catalysts, 3-DDM type, have been developed by Dow, Catalytica and Monsanto (112). The conventional $AlCl_3$-catalyst leads to a mixture of di-, tri- and tetraisopropylnaphthalenes (128b). As shown by the Dow group and *e.g.* by Kikuchi *et al.* 3-DDM catalysts with high Si/Al ratio (> 50) lead to selective formation of 2,6- and 2,7-diisopropylnaphthalenes with 2,6:2,7 ratio of about 3.

2,6-Diisopropylnaphthalene can be converted to naphthalene-2,6-dicarboxylic acid and to 2,6-dihydroxynaphthalene, two monomers for specialty polyesters.

Another example in this area is the selective p,p'-isopropylation of biphenyl (Reaction 21) over modified mordenites in the liquid phase. Reported selectivities towards 4,4'-diisopropylbiphenyl are 84% (at 69% conversion) over fluorided mordenite at 150°C and 20 bar (129) and 74% (at 98% conversion) over highly dealuminated mordenite at 250°C (130). For a recent review see ref. (120).

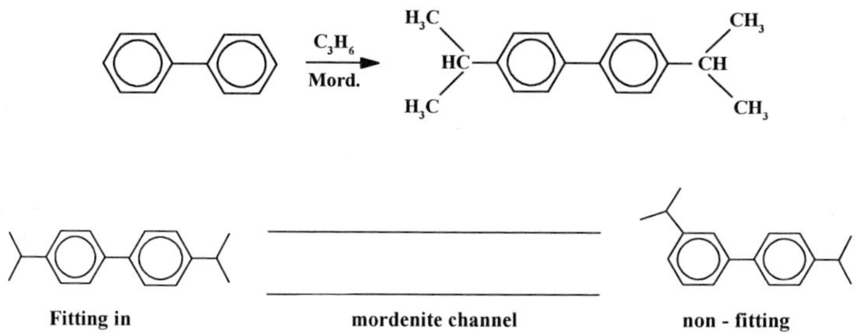

(21)

5.2. Alkylation of phenols and aromatic amines

Arenes containing functional groups such as a hydroxyl or amino group can also be alkylated on zeolitic catalysts. These conversions, however, are more complex than the alkylation of alkylbenzenes, because attack at the nucleus as well as at the functional group can take place, *i.e.* in the case of phenol not only carbon (C-) but also oxygen (O-) alkylation is possible. For recent reviews on this topic see refs. (111,131). Often such reactions are carried out in the vapour phase at atmospheric pressure and at temperatures between 200 and 350°C using pentasil and faujasite-type zeolites. A broad spectrum of products, consisting essentially of cresols, xylenols, anisoles, methylanisoles and diphenyl ether, is obtained when phenol is methylated using zeolites. Both Brönsted and Lewis acid catalysis can play a role in phenol alkylation; Brönsted sites will activate the alkylating agent by protonation, Lewis acid sites, *e.g.* Al^{III}, will activate both phenol, by coordination/deprotonation, and the alkylating reactant, by coordination. Obviously when phenol and alkylating molecule coordinate to one and the same Lewis site only O-alkylation or ortho-alkylation can (initially) occur. This is indeed observed in the methylation of phenol over alumina at 260°C (*cf.* ref. (119)).

Generally kinetic control is observed in the zeolite-catalyzed alkylation of phenol, yielding o- and p- (and 2,4- and 2,6-) isomers whereas meta-isomers, which can be accommodated by the zeolites tested, and which are the thermodynamically most stable isomers, virtually do not show up. Apparently more severe conditions are required for the o,p-to-m-isomerization.

Thus at 400°C over ZSM-5 the reaction of phenol with ethylene yielded ethylphenols with a p:m:o ratio of 17:52:31 (132). Note that an ethyl group will migrate more easily along the benzene nucleus than a methyl group.

The difficulty to obtain m-cresol can be approached (133) by a reaction sequence starting with the isopropylation of toluene over H-ZSM-12 yielding – under optimized conditions – a mixture of isopropyltoluenes (cymenes) containing 64% of the m-isomer (equilibrium composition). This mixture is shape-selectively cracked (de-alkylated) over a modified ZSM-5 catalyst which cracks the p- and the o-isomer and leaves the m-isomer essentially unaffected. m-Cymene is subsequently converted to m-cresol by oxidation/rearrangement.

When alkylating phenol with small alkylating agents generally the o-isomers predominate (119)). The o/p-picture changes upon applying more bulky alkylating molecules. A more bulky transition state is required and this may result in zeolites with increased shape selectivity for p-products. Examples include phenol t-butylation over NaH-Y (134) and alkylation of phenol with 1-octanol (135) over RE-Y, H-mordenite and H-ZSM-12.

In the latter case an additional selectivity plays a role: the initially formed primary carbenium ion 1-C_8^+ (*cq.* the 1-octyl alkylated lattice oxygen) will rapidly isomerize to the secondary 2-C_8^+ which can undergo further H-shifts towards 3-C_8^+ and 4-C_8^+ before alkylating phenol. Over RE-Y, as well as on Amberlyst-15, all three octyl structures are formed whereas with the less spacious mordenite and ZSM-12 just the 2- and 3-octyl systems are observed (see Figure 12).

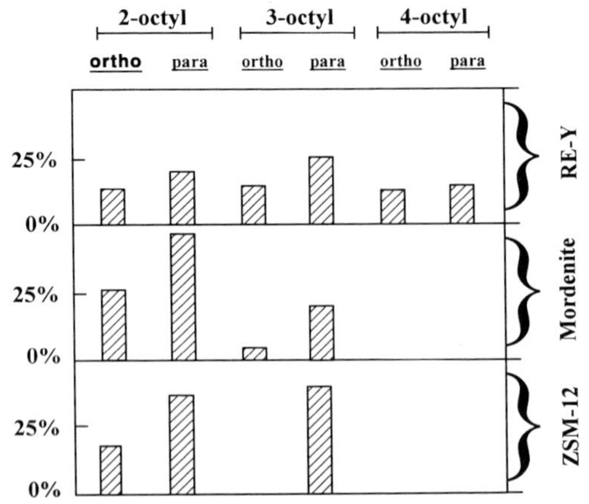

Figure 12. Selectivity and p/(p+o)-ratio in the alkylation of phenol with 1-octanol (135).

An early case of understanding zeolite catalysis is the ethylation of phenol with ethylene as reported by Venuto and Landis (1). Benzene is ethylated faster than the activated phenol over HY zeolite. This is explained by the non-balance of adsorption of phenol (strongly adsorbed) and ethylene onto HY.

For a rationalization of solvent effects in the cyclohexylation of phenol over USY we refer to the work of Espeel *et al.* (135a). An optimum in solvent polarity exists where both reactants are well adsorbed in the zeolite.

The alkylation of aromatic amines over zeolites includes the reaction of aniline with methanol and with olefins. In principle, reaction can take place at the N-containing group forming N-alkylated compounds or at the nucleus forming C-alkylated compounds. The methylation of aniline, for example, yields toluidines, N-methylaniline and N,N-dimethylaniline. All are useful intermediates for dyestuffs, agrochemicals and drugs as well as for the organic synthesis.

As to the aniline/methanol reaction it should be mentioned that alumina and MgO are highly selective catalysts in providing N,N-dimethylaniline and N-methylaniline, respectively. Zeolites could play an important role in C-alkylation.

Upon methylating aniline over zeolites H-Y and H-ZSM-5 generally mixtures of N- and C-methylated products are observed (see *e.g.* ref. 136). Ione *et al.* demonstrated (137) that the product composition obtained over ZSM-5 type catalysts is strongly affected by the reaction temperature. With increasing temperature from 275-450°C the content of N-alkylated products decreases from 100 to 0.3% and the C-methylated products increase.

Within the C-alkylate the regioselectivity depends on temperature, WHSV and on zeolite chemical composition. Using H-ZSM-5 the p-isomer of the toluidines is favoured in the temperature range between 275 and 350°C and its content is about 80%. Under these conditions no m-toluidine is found. Below 275°C the o-isomer and above 425°C the m-isomer is favoured. At 285°C the o/p-ratio increases with increasing WHSV indicating the p-isomer to be a secondary product. At 450°C the thermodynamic equilibrium of the isomers is formed on H-ZSM-5. By modification of the zeolite diffusional effects become operational and the p-isomer predominates again.

Burgoyne *et al.* have studied large pore H-zeolites, amorphous silica-alumina and alumina as catalysts in the alkylation of anilines with propene and isobutene (138). Reactions were performed either cofeeding the reactants over the solid catalyst in a fixed-bed reactor or by reacting the reagents and catalyst in a stirred autoclave. The temperature is again an important variable: with increasing temperature N-alkylated, o-alkylated and p-alkylated aniline can become the major product when alkylating aniline with isobutene.

The catalyst activity order observed is H-Y > silica-alumina > alumina. Also dealuminated H-mordenite (139) and SAPO-37 (140) proved active catalysts.

Under proper reaction conditions high yields of o-alkylated products may be obtained. The proposed mechanism (138) involves concerted cycloaddition of alkene to protonated arylamine. In the prime commercial process (138a) for arylamine ethylation homogeneous aluminum anilide catalysis is applied. This process is highly selective for ortho-ethylation but the rate falls off sharply with higher alkenes.

The t-butylation of toluene-2,6-diamine over H-Y shows shape-selectivity as it leads (138) essentially to the mono-t-butyl compound whereas on silica-alumina di-t-butylation is observed (Reaction 22).

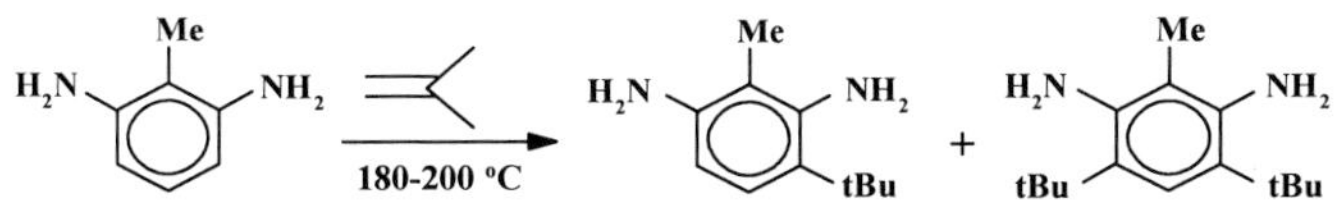

Silica-alumina	conv. 57%	75 : 15	
Zeolite H-Y	conv. 70%	82 : 2	

(22)

5.3. Alkylation of heteroarenes

Only a few references are dealing with zeolite-catalyzed alkylation of heteroaromatics perhaps because these systems are apparently intrinsically more complex.

Thiophene can be alkylated with methanol over H-ZSM-5 yielding, at 450°C, a product containing 10% of 2-methylthiophene, 10% of 3-methylthiophene and 9% of dimethylthiophenes (141,142).

The alkylation of pyridine over faujasite type catalysts (143) with methanol takes place primarily at the aromatic nucleus. In subsequent reactions, *cf.* Reaction 23, the picolines formed can either undergo further ring alkylation to lutidines or side chain alkylation to yield ethyl- or vinylpyridines. When HY, LiY, SrY or BaY are employed, ring methylation occurs almost exclusively to give picolines and lutidines (Table 6), with the alkaline earth doped Y-zeolites leading to higher yields than HY and LiY.

When the two latter zeolites are applied the β-position (3-position) is preferentially attacked whereas the use of Sr- and Ba-Y leads to preferential α- and γ-methylation. Isomers of lutidine (dimethylpyridine) are formed as by-products.

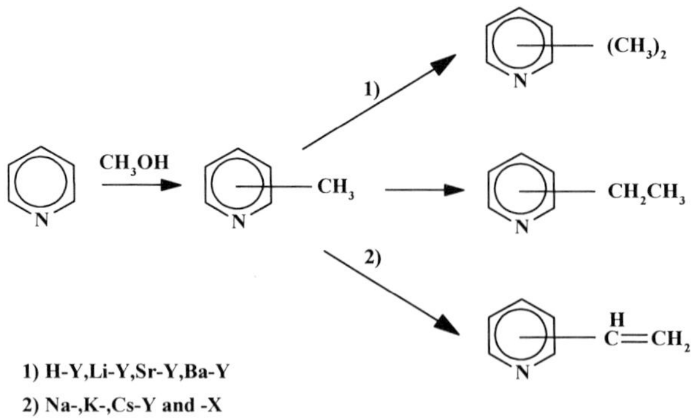

(23)

Table 6
Product spectrum of the reaction of pyridine and methanol.

Catalyst	H-Y	Li-Y	Sr-Y	Ba-Y	Na-Y[a]	Cs-Y[a]
conversion (%)	32	31	62	63	65	82
yield (%):						
picolines	18	13	31	35	13	15
lutidines	6	6	22	22	3	2
ethylpyridines	1	2	/	/	13	27
vinylpyridines	/	/	/	/	5	7

Reaction conditions: 400°C, LHSV = 1.3 h^{-1}, molar ratio methanol/pyridine = 8.
[a] 450°C.

On the other hand, when pyridine is treated with methanol in the presence of X- and Y-zeolites ion-exchanged with alkali metals (not Li), side chain alkylation prevails as a consecutive reaction (Table 6). The principal products are ethylpyridine and vinylpyridine as well as isomers of picoline and lutidine.

5.4. Acylation of arenes

Aromatic acylation processes are widely applied in the fine chemical industry as reaction steps in the synthesis of pharmaceuticals and fragrances. Below some examples of acylated precursor molecules are given together with the final functional molecule. Traseolide is a musk-type fragrance, manufactured by Quest Int., ibuprofen and naproxen (S-enantiomer) are well- known pharmaceuticals.

Acylated aromatic	Precursor of
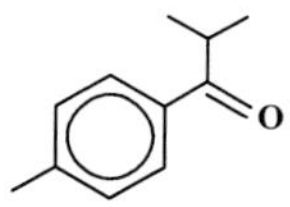	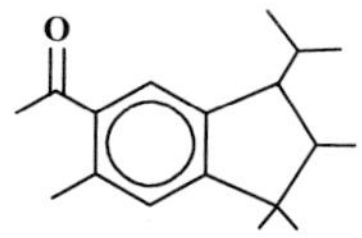 traseolide
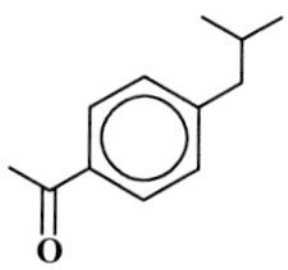	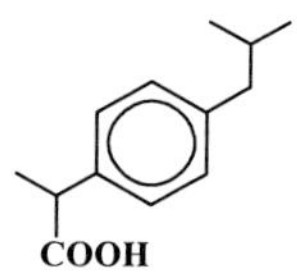 ibuprofen
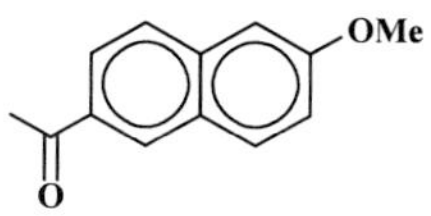	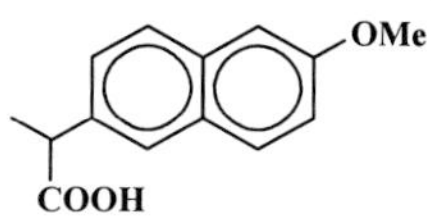 naproxen

Present industrial acylation of aromatics generally involves acid chlorides as the acylating reactants together with more than stoichiometric amounts of metal chloride ($AlCl_3$, $TiCl_4$, $FeCl_3$) "catalysts". This because of strong coordination of the product ketone to the Lewis-catalyst. In this way a substantial waste problem arises. Other acylating agents include acid anhydrides which exhibit similar disadvantages due to an even higher catalyst demand. Moreover, the combination metal chloride catalyst and hydrochloric acid formed in the reaction gives rise during work-up to highly corrosive media.

For the above reasons a direct and clean truly catalytic route involving the free acids or the anhydrides as acylating agents would be most attractive. Zeolite catalysts show great promise in this respect.

As a first successful approach (Reaction 24) the direct acylation of toluene and p-xylene by straight chain aliphatic acids (C_2-C_{22}) using zeolite NaCe-Y (70% exchange) as the catalyst was reported (144) by Chiche *et al.* A stirred batch autoclave was used, and an excess of aromatic hydrocarbon was applied. The amount of zeolite catalyst was relatively high.

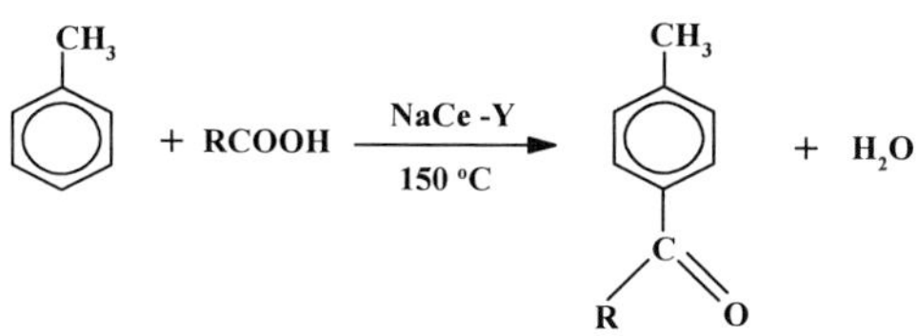

(24)

The acylation of toluene proceeds with increasing yield as the chain length of the acylating agent increases, and reaches a maximum of 96% with dodecanoic acid. Perhaps here the best balance of adsorption exists. It is to be expected - in view of the hydrophilic nature of the zeolite used - that the water formed will remain mainly in the zeolite pores and will deactivate the acidic sites. This will limit the turnover number.

Also acid clays, particularly Al^{III}-exchanged montmorillonite, were found (145) to catalyze the direct acylation of toluene.

As to the regioselectivity of the toluene acylation, in all NaCeY-catalyzed reactions more than 93% of the p-isomer is obtained. This is only partly due to the shape selectivity of the zeolite since acylation on the Al-montmorillonite or with the homogeneous catalyst $AlCl_3$ also leads to a preponderance of the p-isomer (see Table 7). However, the zeolite is superior in this respect.

Table 7
Comparison of regioselectivity among clay, zeolite and $AlCl_3$ in the acylation of toluene at 130°C (ref. 145).

Acylating agent	Isomeric composition (% over								
	NaCeY			Al mont. mor.			$AlCl_3$		
	o	m	p	o	m	p	o	m	p
octanoic acid	3	3	94	5	9	86	4	16	80
dodecanoic acid	3	3	94	5	11	84	4	15	81
palmitic acid	0.5	1	98.5	5	10	85	4	12	84

The relatively low reactivity of short chain carboxylic acids in the liquid phase over NaCe-Y is in line with a report (146) on the gas phase acylation of toluene and ethylbenzene by acetic acid over H-ZSM-5. At 250°C low conversions - with high regioselectivity - are observed.

Modest yields (12%) were also obtained by Toray workers (147) in the direct acylation of benzene with propionic acid over several dealuminated zeolites at 250-270°C in the gas phase. Improved results were reported by Singh and Pandey who applied H-ZSM-5 (Si/Al 41) in the acylation of benzene with acetic acid, also at 250°C in the gas phase. Applying a benzene:acetic acid molar ratio of 2:1 a selectivity of 91% at 43% conversion (HOAc) was obtained.

Quite recently Davis *et al.* (148,149) studied the acetylation of isobutylbenzene with acetic anhydride over H-Beta samples at 100°C in the liquid phase (1,2-dichloroethane as solvent). Low catalyst activity together with high (> 99%) p-selectivity was observed. Note that the acetic acid formed may act as inhibitor. A patent to Uetikon (149a) operates the reaction over H-Beta at 140°C and claims a yield of 80% with 96% p-selectivity.

As to the adsorption preference of reactants and products the following sequence is expected (and partly proven) to hold onto zeolites with Si/Al say up to 20.

$$RCOOH > ArCOR > RCOCl > ArH$$

Compared with the order in conventional $MHal_3$ catalysis:

$$ArCOR >> RCOCl >> ArH$$

the differences are less pronounced in zeolite catalysis. Moreover by selecting zeolites with high Si/Al ratio the adsorption of ArH will improve which might more than compensate the lower amount of acid sites.

The mechanism might involve – on the analogy of aromatic alkylation – an acylated surface site, see below. Initially free carboxylic acids are assumed to adsorb in a symmetrical way onto an acidic zeolite site.

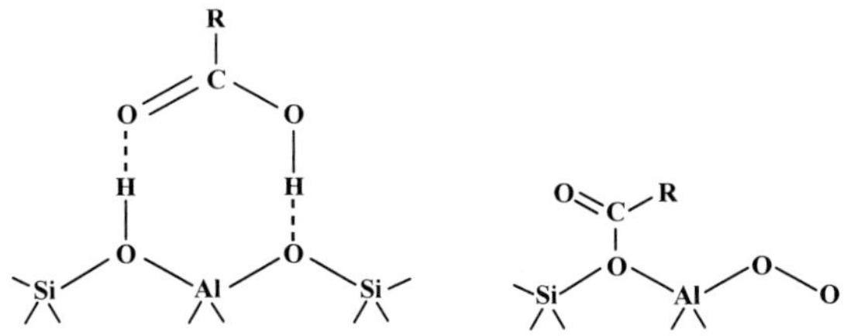

Finally we mention in this section the zeolite catalyzed synthesis of anthraquinone, which compound is an important intermediate. Zeolite NaCe-Y is used (150) at 550°C to convert benzene and phthalic anhydride (molar ratio 25:1) towards anthraquinone (Reaction 25). Using CO_2 as the carrier gas a conversion of 65% together with a selectivity of 92% is mentioned.

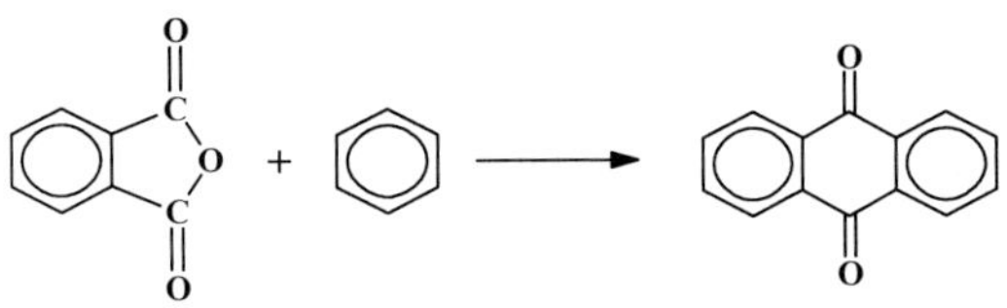

(25)

For recent reviews on aromatic acylation see ref. (151) and (151a).

5.5 Acylation of aromatic ethers: anisole and 2-methoxynaphthalene

In zeolite-catalyzed acylation aromatic ethers have two advantages compared with benzene and alkylbenzenes:

(i) the ethers are activated for electrophilic substitution

(ii) the ethers are in better competition with the acylating agents for adsorption into the zeolite.

Following studies of Corma *et al.* (152) and Prins *et al.* (153) on acylation of anisole, Rhodia (at that time Rhône-Poulenc) workers started a systematic investigation (154) into the industrially relevant acetylation of anisole and veratrole (Reactions 26).

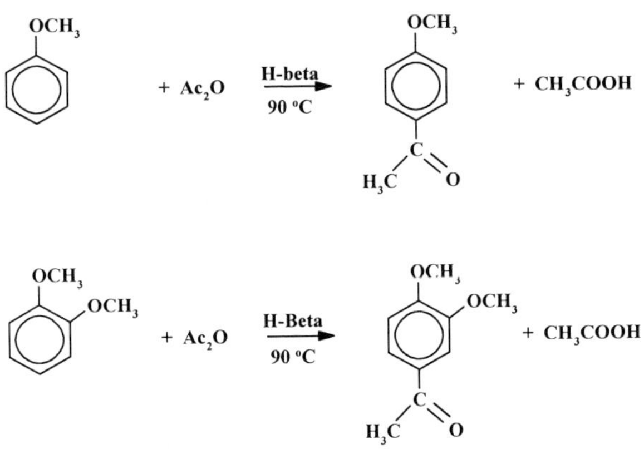

(26)

Many variables were studied. See Table 8 (154) for a comparison of catalysts. Zeolites HY and H-Beta show by far the best performance in the case of anisole, whereas HY is the catalyst of choice in the acetylation of veratrole.

The authors rationalise these outcomes by comparing the sizes of the postulated acylation intermediates (see below) with the intersection volumes of the zeolites H-ZSM-5, H-Beta and HY which amount to 150, 310 and 1700 (supercage) $Å^3$, respectively.

Table 8.
Acylation of anisole and veratrole by acetic anhydride[a].

Reactant	Catalyst	Yield (%)[b]
Anisole	H-ZSM-5	12
Anisole	H-Mordenite	29
Anisole	H-Beta	70
Anisole	HY	69
Anisole	exchanged clay	14
Anisole	Al-clay	16
Veratrole	H-ZSM-5	12
Veratrole	H-Mordenite	25
Veratrole	H-Beta	53
Veratrole	HY	95

[a] Reaction conditions: molar ratio reactant: acetic anhydride 5, 10 wt % catalyst, 90°C, 6 h.
[b] Selectivity to 4-acetyl compound always > 98%.

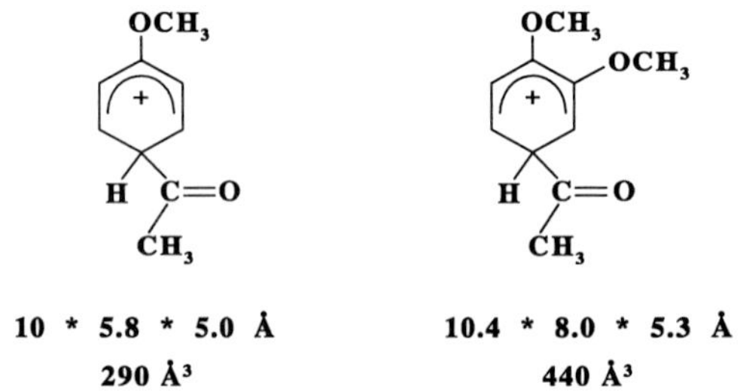

Dimensions of acylation intermediates

A solvent-free process was developed with reactants continuously passing a fixed bed of catalyst (H-Beta). In the mean time both processes are in industrial operation which means a real breakthrough in aromatic acylation. The new process for p-acetylanisole is a clean (green) process and also brings along a substantial process simplification (151a) with respect to the conventional $AlCl_3$-catalyzed process. That multi-stage process used 1,2-dichloroethane as the solvent and involved a hydrolysis step, several washings and separations, and water and solvent recycle. Two recent studies (154a,b) deal with product inhibition exerted by the p-acetylanisole in the acetylation reaction of anisole.

Compared to the above success story the zeolite-catalyzed selective 6-acylation of 2-methoxynaphthalene – required for the synthesis of the drug Naproxene – is in a much earlier state of development.

Here the kinetically favoured substitution is at the 1-position and accordingly reaction of 2-methoxynaphthalene and acetic anhydride over a mesoporous catalyst of the MCM-41 type leads (155) to > 98% 1-acetyl substitution (liquid phase 100°C).

Then two approaches can be studied:

(i) formation of the 1-isomer followed by isomerization to the thermodynamically more stable 6-isomer

(ii) application of a zeolite with a pore system too narrow for the 1-isomer to form.

Recently reported evidence (149,156) suggests that inside zeolite Beta only 6-substitution occurs. This requires full passivation of Beta's outer surface.

Outer surface activity can be measured by reacting bulky probe molecules (157) that will not enter the Beta pores at not too high temperatures. Below two examples are shown.

11.0 * 10.8 * 5.8 Å

A promising method to passivate outer zeolite surfaces is treatment with tetraethyl orthosilicate (TEOS) which leads upon hydrolysis and condensation to a thin and porous silica layer (158,159).

When applying the method to Beta (relatively large crystals) a strongly enhanced selectivity to 6-acetyl-2-methoxynaphthalene (up to 92%) was observed (149).

Recently zeolite ITQ-7 was shown to be a promising catalyst for selective 6-acylation of 2-methoxynaphthalene (159a).

5.6. Acylation of heteroarenes

Hölderich *et al.* reported (160) that it is possible to perform the acetylation of heteroaromatics such as thiophene, furan and pyrrole in the gas phase on zeolite catalysts with high selectivity according to Reaction 27.

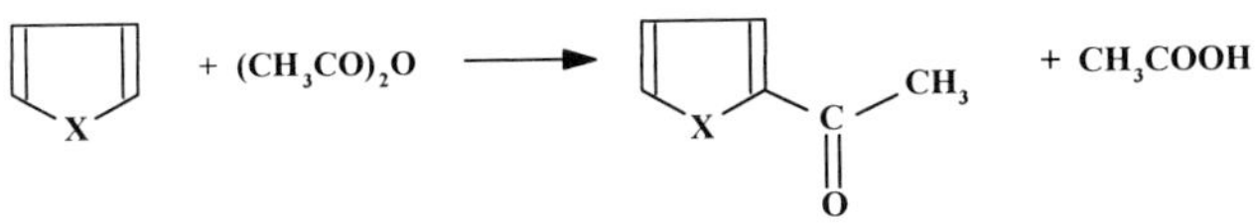

X = O, S, NH

(27)

The reaction of thiophene with acetic anhydride at 250°C on a boron ZSM-5 zeolite leads to 2-acetylthiophene with 99% selectivity at 24% conversion. On a Ce-doped boron-ZSM-5 at 200°C 2-acetylfuran is obtained with 99% selectivity at 23% conversion. On the other hand, in the case of pyrrole, which tends to polymerise, the acidity of the catalyst and also the temperature must be reduced in order to achieve a high selectivity. A boron zeolite doped with 0.2% by wt. Cs yields 2-acetylpyrrole with 98% selectivity and 41% conversion at 150°C. As in the homogeneously catalysed reactions, the acylation on zeolites proceeds largely at the 2-position of the heteroaromatics so sterical differences, if any, between substitution at the 2- and 3-position are overruled by the electronic effect.

More recently thiophene was also acylated in the liquid phase. Thus thiophene was acylated with butyryl chloride over several zeolites at 100°C in chlorobenzene as the solvent (161). As expected there was a correlation between acylation rate and number of Lewis sites in the zeolite catalysts and not with the number of Brönsted sites.

A logical picture seems:

acylating agent	activation by
RCOCl	Lewis sites
RCOOH	Brönsted sites
$(RCO)_2O$	Lewis and Brönsted sites

For a study on the acetylation of the very reactive molecule benzofuran over USY the reader is referred to ref. (162). Reaction (with acetic anhydride) occurs under mild conditions (60°C) but the catalyst deactivates rather rapidly.

Direct C-acylations of imidazoles and pyrazoles by means of Friedel-Crafts reactions were unknown, and it was predicted that this reaction is not possible at all; one was therefore forced to adopt more complicated processes. Surprisingly, with the aid of zeolite catalysts at high temperature it turned out to be possible to carry out direct acylation (163). Thus upon passing a mixture of 2-methylimidazole and acetic acid or acetic anhydride over a boron-MFI zeolite at 400°C, a conversion of 63% and a selectivity of 85% to 2-methyl-4-acetylimidazole is observed. In conclusion the use of zeolites in aromatic acylation has provided new and interesting procedures with already industrial application. Noteworthy is the environmental friendliness of the new methods compared with the conventional ones.

5.7. Aromatic nitration

Present-day technical nitration of benzene, toluene and chlorobenzene involves reaction of the aromatic in the liquid phase with mixtures of HNO_3-H_2SO_4-H_2O. The nitration, which is strongly exothermic, is conducted batchwise or continuously. For reviews of these well-established processes see *e.g.* ref. (164-166).

Benzene nitration is executed at 50-90°C and the yield on benzene is 95-98% (w/w). In current industrial practice continuous adiabatic nitration is preferred to batch processing. Environmental regulations have led to improved methods for product recovery and regeneration of spent sulfuric acid. The latter is commonly performed off-site by specialized firms.

Nitration of toluene for production of mononitrotoluenes is conducted at somewhat milder conditions (25-40°C) because of the activating methyl group. When dinitrotoluenes are required the reaction temperature is 65-70°C.

A heterogeneous catalyst, replacing the sulfuric acid in the aromatic nitration would certainly be welcomed. Several groups have tackled this challenge. Thus Prins *et al.* studied (167) vapour phase nitration (170°C, 65 wt % aq. HNO_3) of benzene over various zeolites. Best results as to selectivity and stability were obtained over clinoptilolite. It is assumed that the nitration takes place at the outer surface of the zeolite crystals.

Recently (168) Choudary *et al.* applied azeotropic removal of (product) water in the liquid phase nitration of toluene and other monosubstituted benzenes over various zeolites. Here zeolite H-Beta gave the best performance. It is to be expected, however, that Al-leaching occurs under the conditions applied.

Kogelbauer and Kouwenhoven have recently reviewed (169) the state of the art and potential of heterogeneous catalysis in aromatic nitration. These authors are not optimistic about future replacement of the present bulk nitration processes.

In small scale operations the method developed by Smith *et al.* (170) may be of interest. Here acetyl nitrate acts as nitrating agent and is formed *in situ* from 90% nitric acid and acetic anhydride. Zeolite Beta is assumed to catalyze both reactions. High yields and p-selectivities are obtained for monosubstituted benzenes (see Table 9). The catalyst is not completely stable. Furthermore, safety considerations will obstruct large scale use of the method.

Table 9

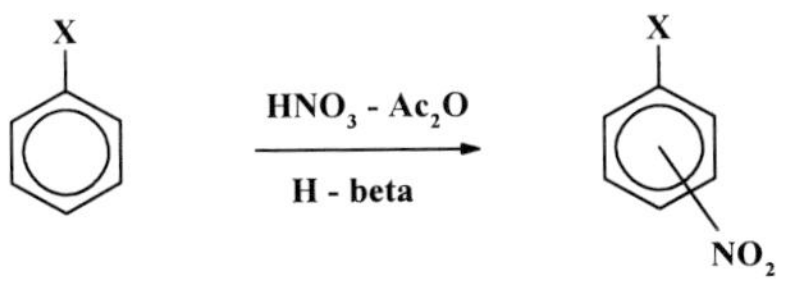

The effect of X on the nitration of PhX at room temperature

X	t/min	Yield (%)	Product proportions (%)		
			ortho	meta	para
F	30	>99	6	0	94
Cl	30	>99	7	0	93
Br	5	>99	13	0	87
H	30	>99	-	-	-
Me	30	>99	18	3	79
Et	10	>99	15	3	82
iPr	30	>99	9	3	88

5.8. Aromatic halogenation

Chloro- and bromo-substituted arenes are prepared industrially in the presence of Lewis acids, especially $FeHal_3$. The use of zeolite catalysts in the gas or liquid phase could reduce corrosion and waste disposal problems. Systems carrying an activating substituent such as OH, OR, NH_2, do not require a catalyst. So reports on zeolite-catalyzed halogenation of such compounds, *e.g.* anisole, should be considered with caution.

The abilities of zeolites in - high temperature - isomerization and in separation of halobenzenes have already been mentioned in Section 4.6.2.

A review on heterogeneously catalyzed halogenation has been compiled by Ratnasamy *et al.* (171). This review covers much papers on zeolite catalysis in aromatic halogenation. Another quite recent chapter of Ratnasamy and Singh (172) is also devoted to heterogeneous catalysis in halogenation.

5.8.1. Chlorination of arenes

Upon chlorinating benzene in the gas phase at 175°C over various zeolites (molar ratio $Cl_2:C_6H_6 = 5$) the composition of the reaction product was found (173) to depend strongly on the nature of the zeolite catalyst (see Figure 13).

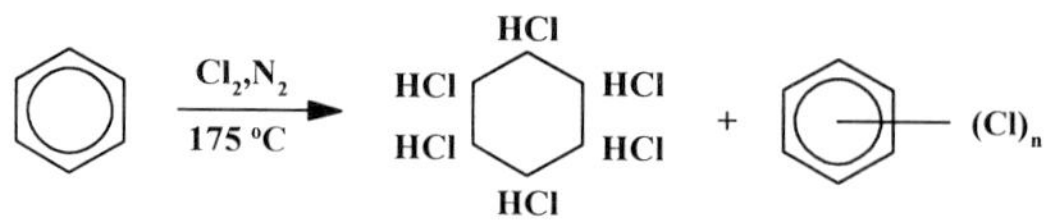

	Benzene conv. %	Addition:Substitution
Silicalite-1	97	96:4
NaH-ZSM-5	98	90:10
H-Mordenite	91	84:16
NaH-ZSM-11	86	72:28
KL	86	11:89
RE-Y	100	3:97
LA-LPV (13% alum.)	94	7:93

Figure 13. Gas phase chlorination of benzene.

On the high silica zeolites addition and not substitution (Figure 13) prevails; apparently a radical reaction is induced which - in view of the dimensions of the hexachlorocyclohexanes formed (0.8 nm) - mainly takes place at the outer surface and/or in the pore mouth openings. Free radicals were detected (ESR) upon adsorption of o-dichlorobenzene on zeolite H-Beta (174).

Over Y and L zeolites the gas phase benzene chlorination mainly leads to substitution with an ionic mechanism involved.

Zeolites of the X-, Y- and L-type have been employed, particularly by Japanese workers, in the liquid phase chlorination of toluene (175,176), of halobenzenes (177-179), and of halotoluenes (180). Mild temperatures are applied and often p/o-ratios are somewhat higher than obtained with conventional $FeCl_3$-catalysis. At 100°C in the liquid phase over Na-ZSM-5 chlorination proceeds (181) to the dichloro stage whereas over NaY also tri- and tetrachlorobenzenes are formed.

Further improvement of p-selectivity, *e.g.* in p-dichlorobenzene synthesis, has been explored by impregnation with salts (181) and/or by adding quaternary ammonium compounds (182) or lower aliphatic acids (183) to the liquid phase. The latter type of addition (*e.g.* chloroacetic acid) also improves (184) the 4,4'-selectivity in the chlorination of biphenyl over KL. The background of this effect is as yet unclear. A complicating factor is the fact that the acid additive can also exert a catalytic action in the halogenation.

An important series of patents (171) and papers (*e.g.* 185-187) on zeolite-catalyzed aromatic chlorination reflects the studies of Singh *et al.*, working at the NCL at Pune, India.

These authors investigated the liquid phase chlorination of several aromatics over various zeolite catalysts. Here zeolite KL turned out to be the catalyst of choice, showing amongst others the highest regioselectivities. Thus in toluene chlorination at 80°C zeolite KL gives (185) a p/o ratio of 4.05 whereas the other K-exchanged zeolites give a p/o ratio < 1.

In the o-xylene chlorination (186) (80°C, 1,2-dichloroethane as the solvent) KL gave a much higher 4-Cl/3-Cl ratio than the other K-zeolites, see Figure 14.

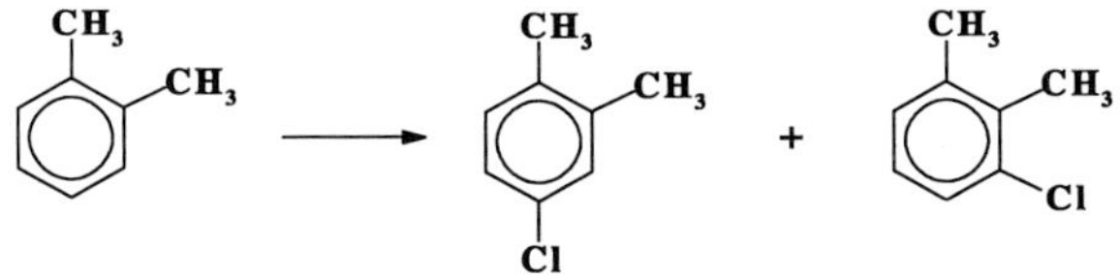

Catalyst :		
	KL	**11.7 : 1**
	KX	**1.76 : 1**
	K-Beta	**1.63 : 1**
	K-Mordenite	**1.55 : 1**
	K-Y	**1.45 : 1**
	K-ZSM-5	**1.44 : 1**
	$FeCl_3$ (homog.)	**1.00 : 1**

Figure 14. Liquid phase chlorination of o-xylene.

The differences in regioselectivity are as yet not fully understood. The cross-section of the 4-isomer is smaller than that of the 3-isomer, which might be a clue in the high regioselectivity of the 1-dimensional KL catalyst compared to the 3-dimensional systems (X, Y, Beta). However, the 1-D K-mordenite is then expected to display also higher regioselectivity, which was not observed.

As mentioned in Section 3.4 very high p-selectivities have been obtained by Smith *et al.* (47) in the chlorination of chloro- and bromobenzene. Applying t-butyl hypochlorite as reactant and NaX as the host material a para/ortho ratio of 32 was found.

Anisole chlorination proceeds uncatalyzed towards 2,4-dichloroanisole in CCl_4 at 25°C. Upon adding NaCeY-62 as a catalyst subsequent chlorination towards 2,4,6-trichloro- and 2,3,4,6-tetrachloroanisole takes place (188).

Similarly a third chloro-substituent can be introduced in phenyl acetate (Reaction 28).

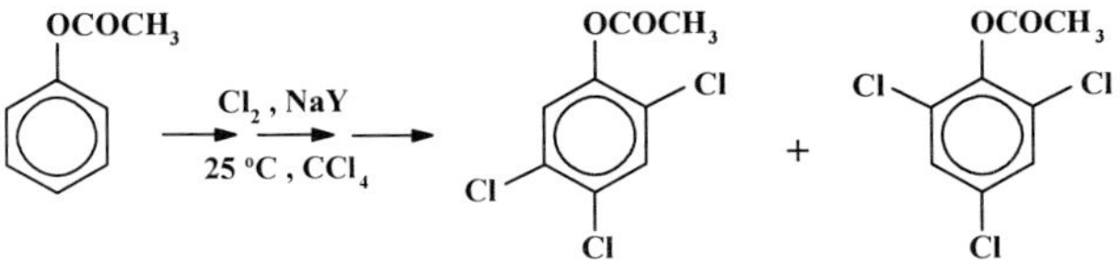

(28)

The 2,4,5-trichlorophenyl acetate is found to prevail whereas with $FeCl_3$-catalysis only the 2,4,6-isomer is observed. Apparently shape selectivity is involved. It may be noted that the o,p-directing power of the $OCOCH_3$-group is less than that of OH or OCH_3. Furthermore competitive adsorption experiments showed 2,4,5-trichlorophenyl acetate to be stronger adsorbed onto a Y-zeolite than the 2,4,6-isomer; the better accommodation of the 2,4,5-isomer might be perceptible in the transition state of chlorination.

Finally, the ad- and desorption of 2,4,6-trichloro-anisole and –phenyl acetate in/from zeolite Y are noteworthy in view of the diameter of these compounds which is almost 0.9 nm. This underlines the margin in accessibility mentioned in Section 2.1.

As to scavenging of the HCl generated in the chlorination and probably damaging the zeolite, see the next section.

5.8.2. Bromination of arenes

A paper of one of the authors *et al.* (189) gives details on the liquid phase bromination of halobenzenes over various Y-zeolites. Ref. (190) contains a study on the zeolite Y catalyzed bromination of toluene.

Using the bromination of bromobenzene in CCl_4 at 25°C towards o- and p-dibromobenzene as a test reaction (189) the following order of activity is found:

KY < NaY << CaX < RE-Y, CeY < HY, NaMgY, NaCaY

Activity as well as p/o selectivity are found to increase with increasing degree of cation exchange of NaY. Regioselectivity obtained in bromination of bromo-, chloro- and fluorobenzene is substantially higher than observed with $FeBr_3$-catalysis.

From adsorption isotherms (CCl_4, 25°C) the following order of adsorption strength onto NaCaY-82 can be deduced (189):

HBr > o-diBr > p-diBr, mono-Br benzene > Br_2

Hydrogen bromide, generated during the reaction is held responsible for the decay of the catalyst which manifests itself in reduced p/o-selectivity in the course of an experiment and upon re-use of a zeolite Y catalyst.

Attempted remedies by scavenging HBr include the addition of $NaHCO_3$ and KA (189), of KA and of propylene oxide (191). With the latter method a high (98:2) initial p/o-ratio in the bromination of toluene over NaY has been reported. The catalyst is deactivated completely however, after some 10% conversion.

The order of reactivity found in zeolite-Y-catalyzed liquid phase bromination is

toluene > benzene > fluoro- > chloro- > bromobenzene

which suggests an electrophilic attacking species. A proposed mechanism (Figure 15) involves activation of bromine - which is already in interaction with the aromatic nucleus - by zeolitic Brönsted sites. Also bromine activation by cations, *e.g.* Na^I, may be envisaged, as evidenced by Raman spectroscopy (192).

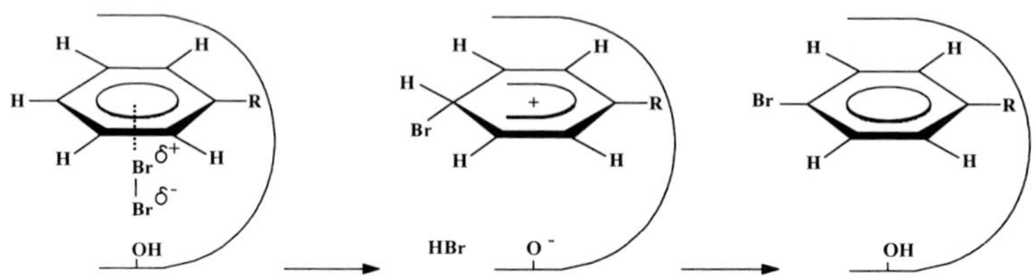

Figure 15. Mechanism of zeolite-catalyzed bromination.

Reported p-dibromobenzene syntheses over zeolites at higher temperature include the bromination (193) of bromobenzene or benzene over Li-faujasite at 200°C (p-selectivity 92%) and the oxidative bromination (194) using Br_2 and/or HBr/O_2 at 170°C over CuY.

5.8.3. Aromatic iodination

Direct iodination requires - due to the reductive power of HI – oxidative conditons. Zeolite-catalyzed oxyiodination of benzene (195), naphthalene (196) and several substituted arenes (197) has been described. Iodine, air and the aromatic substrate are fed in the gas phase at atmospheric pressure over the catalyst. Non-acidic faujasite, *e.g.* KX, sometimes impregnated with K-salts, and divalent metal exchanged pentasil zeolites, *e.g.* Ca-ZSM-5, are recommended as catalysts. Good conversions and high selectivities are reported.

5.9. Hydroxyalkylation of arenes

As is the case with aromatic acylation, aromatic hydroxyalkylation is developing to a successful field for zeolite catalysis.

Hydroxyalkylation of aromatics is known to be a difficult reaction because the consecutive reaction towards diarylmethane systems usually is fast. The limited space in zeolites might prevent the second step. However, early work of Venuto and Landis (198) using the large pore HY zeolite as the catalyst at 182°C showed invariably bisarylmethane formation (*cf.* m-xylene/formaldehyde, shown below). An exception was the reaction of phenol and hexafluoroacetone where the electron-attracting fluoro atoms will oppose carbenium ion formation (Reaction 29).

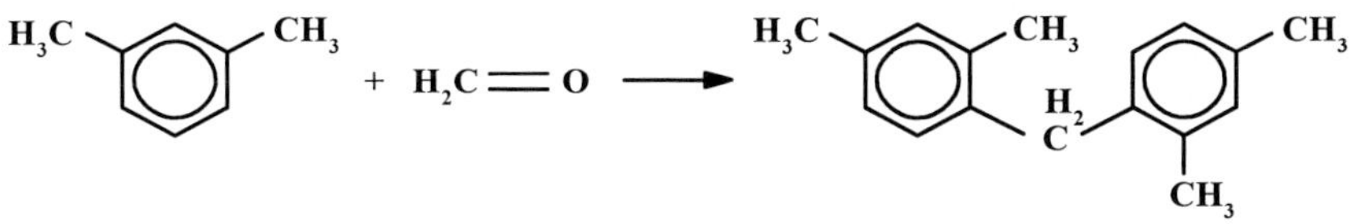

Select. > 99%
Conv. 70%

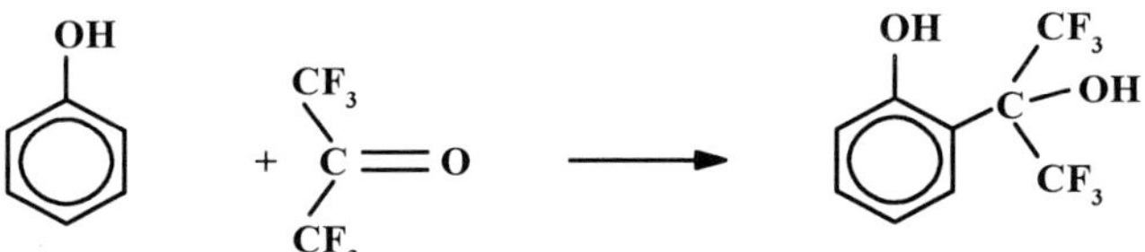

Select. > 99%
Conv. 50%

(29)

Also reaction of aniline with formaldehyde (199) was reported to yield bis(4-aminophenyl)-methane with a selectivity of 85% at 40% conversion over Y-type zeolites. Similarly the reaction of toluene and trioxane over several molecular sieves, *e.g.* SAPO-11, of the AlPO family (200) leads to bis(4-methylphenyl)methane. Studies of the Corma group (201) applying zeolites USY, Beta and Omega, and of the van Bekkum group (202) using SAPO molecular sieves also mainly gave rise to bisarylmethane formation.

On the other hand two patents (203) claim successful hydroxymethylation and hydroxyethylation of phenol by reacting phenol at 150°C with the cyclic aldehyde trimers trioxane and paraldehyde, respectively, over a H-Y type catalyst in the liquid phase. Remarkably high selectivities (up to 88%) for the p-hydroxy(m)ethylphenol are mentioned.

In a patent to Asahi Chemical (204) a basic zeolite, CsNaX, is applied under mild conditions (60°C) in the reaction of phenol and paraformaldehyde. Here o- and p-hydroxymethylphenol (in a ~ 1:1 ratio) are obtained with a selectivity of 95% at 66% conversion.

A research cooperation between the French groups of Rhodia and of CNRS at Montpellier led to the successful application (also at low temperature) of dealuminated mordenite in aromatic hydroxymethylation (205-207). This method is applied in the first industrial example of zeolite-catalyzed hydroxymethylation: the reaction of guaiacol and formaldehyde (Reaction 30) which is an important step in a new and clean vanillin process being operated by Rhodia.

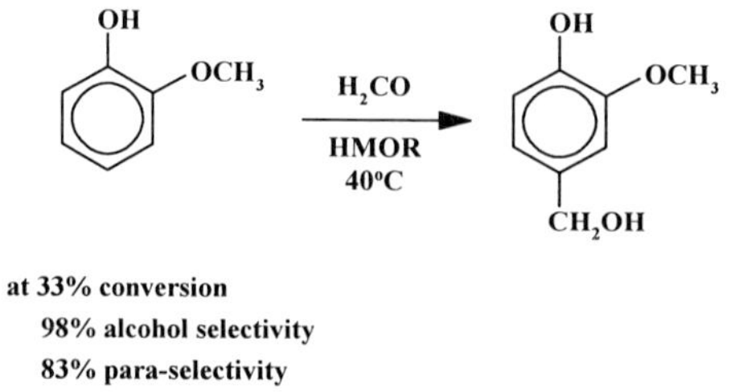

(30)

In another interesting approach (208) the aldehyde is reacted in the form of an acetal. Thus acetaldehyde dimethyl acetal and phenol (molar 1:10) react over H-ZSM-5 at 250°C in the gas phase to give 2-(1-methoxyethyl) phenol in 59% selectivity (Reaction 31).

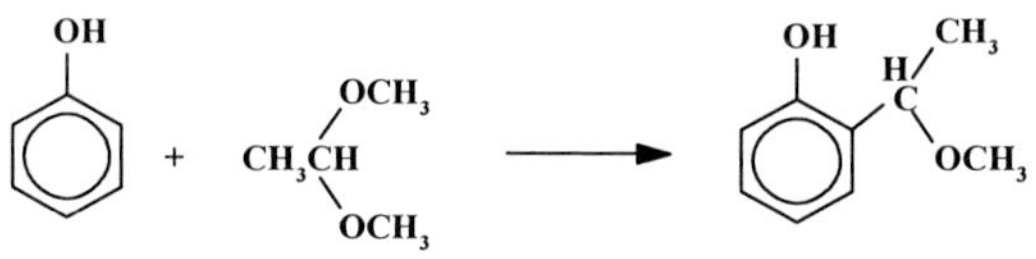

(31)

A challenge remains to react the non-activated benzene with formaldehyde, as a salt-free route towards benzyl alcohol. Nowadays benzyl alcohol is manufactured by chlorination of toluene followed by hydrolysis of benzyl chloride.

In a recent paper (209) a selectivity to benzyl alcohol of 22% at 35% conversion is reported by reacting benzene and formaldehyde (in the presence of some water) over H-ZSM-5 (Si/Al = 40) at 240°C. Interestingly, the main product is benzyl methyl ether (selectivity 58%) which compound might be formed *via* a Cannizzaro reaction affording methanol, then formation of the dimethyl acetal and finally aromatic alkylation as shown above in Reaction 31.

For a complete review on heterogeneously catalyzed aromatic hydroxyalkylation the reader is referred to ref. (210).

Recently successful hydroxypropylation was reported (210a) by reacting 2-methoxynaphthalene and propylene oxide over Ti- and Zr-Beta. Oligomerization of propylene oxide is a competing reaction. The desired product 6-(1-hydroxy-2-propyl)-2-methoxynaphthalene (precursor of the drug naproxen) was obtained with selectivities up to 20%.

6. ZEOLITE-CATALYZED CYCLIZATIONS

Some ring-forming reactions - *e.g.* anthraquinone formation - and some ring transformations - *e.g.* the benzamine rearrangement - were mentioned already.

For the well-known methanol-to-gasoline (MTG) process, which is in fact a distribution of H and C (ratio 2:1) over aliphatic and aromatic hydrocarbons, the reader is referred to Chapter 15 and to ref. (211).

The analogous ethanol-to-gasoline conversion over H-ZSM-5 (212) proceeds also in the presence of excess of water which enables integrated conversion (213) of carbohydrates to hydrocarbons by consecutive fermentation and dehydration/aromatization.

In addition to the reaction of lower alcohols (and olefins) also light alkanes, *e.g.* propane, can be converted over ZSM-5 catalysts into aromatics; preferably a dehydrogenating component (Ga (214,215) or Zn (216)) is incorporated or added.

Cyclization/aromatization of hexane and higher n-alkanes is selectively attained (217) using Pt-loaded neutral L-zeolites. For instance n-hexane is converted over 0.6% Pt/K-L at 460°C and WHSV 2 h^{-1} in 80% selectivity to benzene. n-Octane yields over Pt/BaK-L at 460°C 30% aromatics of which 88% consists of ethylbenzene and o-xylene (and 12% m- and p-xylene). Mechanistic insight in this reaction is provided by Derouane *et al.* (218).

Some additional zeolite-catalyzed cyclization reactions will be dealt with in this section.

6.1. Carbocyclic ring formation

6.1.1. Diels-Alder cycloadditions

In paragraph 3.5 physical catalysis was shown to operate in a [4 + 2] cyclodimerization reaction over microporous solids. By adding a suitable cation a chemo-catalytic component can become active too in several ways.

Thus, highly selective cyclodimerization of butadiene to vinylcyclohexene over $Cu^{I}Y$ (Reaction 32) has been reported by Maxwell *et al.* (4a,219). By contrast, in homogeneous catalysis, a mixture of cyclic dimers and trimers is normally obtained. The selectivity of the zeolite is attributed to reaction selectivity, since the Cu^{I} intermediate required to form vinylcyclohexene is less space-demanding than the intermediates necessary for the formation of the other oligomers. The stability of the $Cu^{I}Y$ catalyst is found to be strongly dependent on its way of preparation.

A one-step process is nowadays available to convert vinylcyclohexene into styrene, so together cyclization/dehydrogenation opens the way to convert cheap butadiene into styrene.

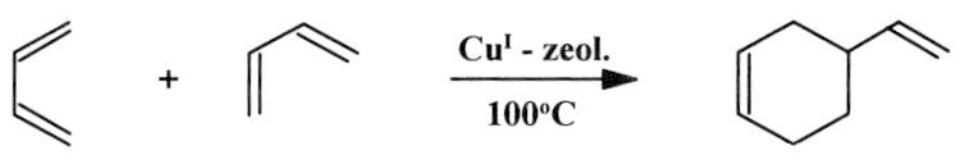

(32)

Another example (Reaction 33) is the $Cu^{I}Y$ catalysis of the Diels-Alder reaction of furan and α,β-unsaturated carbonyl compounds such as methyl vinyl ketone (220) at the required low temperature (0°C). Essentially, the same *exo/endo* product ratio (2.5) is observed as found using Fe^{III}-doped clay (bentonite) as the catalyst (221). Here, the dienophile might well be activated by cation coordination to the carbonyl oxygen.

Accordingly Cu(II)Y appeared to be more active than Cu(I)Y in the related cyclo-addition of cyclopentadiene and methyl vinyl ketone (222). The activity of HY and of HUSY can be traced back to proton activation or coordination of the dienophile to defect Al.

In more recent work of Lechert *et al.* (223) also mesoporous materials of the MCM-41 type appeared to be good catalysts in the Diels-Alder reactions of isoprene with methyl vinyl ketone and with 1,4-naphthoquinone. MCM-41 displays a distinctly higher activity than HY though its acidity is lower. Again Al-coordination might play a role.

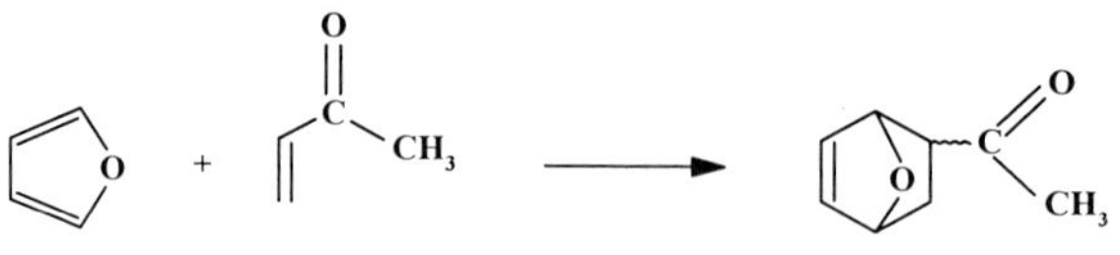

(33)

Another mechanism of activation by zeolites has been postulated in a study (224) on cyclodimerization of 1,3-cyclohexadiene and 2,4-dimethyl-1,3-pentadiene on NaX. Electron transfer to the zeolite is assumed to take place leading to a reactive cation radical.

These authors also observed zeolite (NaX) catalysis in the [2 + 2] cyclodimerization of the electron-rich styrene derivative 4-(1-propenyl)anisole (anethole) yielding a cyclobutane system. Here the question arises of the contributions of inner and outer zeolite surface.

Another interesting zeolite-catalyzed [2 + 2] cyclo-addition is Reaction 34, the conversion of cyclopropene towards tricyclo[3.1.0.0]hexane (225). This cyclodimerization proceeds with high selectivity over NaA and KA (yields of 95 and 97%, respectively), whereas on zeolites with pore size larger than .4 nm, such as CaA, NaX and NaY polymerization takes place. Similarly 1- and 3-methylcyclopropene can be selectively cyclodimerized to the corresponding tricyclohexanes.

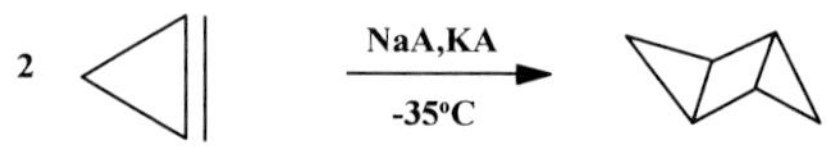

(34)

This new method of cyclodimerization of cyclopropenes is superior to known procedures of preparing tricyclohexanes. As to the mechanism the authors propose a stepwise ionic mechanism in which spatial restrictions imposed by the small pore zeolite prevent approach of a third cyclopropene molecule.

6.1.2. Cyclopropanation

Copper-exchanged X and Y zeolites are active catalysts for the decomposition of diazo compounds (226,227) leading to carbenoid intermediates. Cyclopropanation of various olefins has been carried out (228) by the application of NaCuX as a catalyst for the decomposition of ethyl diazoacetate, see Reaction 35.

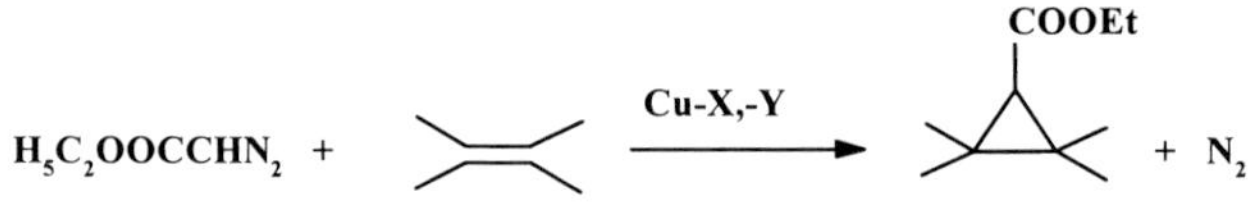

(35)

Typically the ethyl diazoacetate is gradually added to a suspension of NaCuX in the olefin at 80°C. The reaction is followed by monitoring the nitrogen evolution. Side products are the dimeric compounds diethyl fumarate and maleate and polymeric material assumed to be formed at the outer surface of the zeolite. Compared to conventional copper catalysts, the zeolite catalysts give rise to relatively low amounts of polymeric side-products. Important variables are the degree of copper exchange and the water content of the NaCuX. Some dienes were found to give mono-cyclopropane systems. For instance, 1,1-dichloro-4-methyl-1,3-pentadiene can be efficiently monocyclopropanated.

Some chiral complexes were constructed inside the zeolite, *e.g.* the neutral Cu(L-alanine)$_2$ and the cationic Cu(R-1,2-diaminopropane)$_2$. When used as a catalyst, these systems induce just a small asymmetric cyclopropanation of 1,1-dichloro-4-methyl-1,3-pentadiene.

Similarly recently (229) chiral Cu and Rh complexes were heterogenized on a modified USY zeolite. As test reactions the cyclopropanations of styrene and 1,2-dihydropyrane with ethyl and t-butyl diazoacetate were studied. The Cu catalysts were more active than the Rh-systems. Reactivity and selectivity (cyclopropanation *vs* dimerization) did not deviate significantly from those obtained by the homogeneous complexes. The advantage of the zeolite-based catalysis is that they can be easily recovered and re-used without loss of activity.

6.1.3. Anthracene formation

When passing (229a) benzyl alcohol over alumina, fluorided alumina, zeolite H-Y or heteropolyacids at temperatures of 300-500°C anthracene is formed (Reaction 36). Side products are toluene and benzaldehyde. The authors assume o-benzylbenzyl alcohol to be an intermediate. Zeolite H-Y gives the best performance. In a preparative example a solution of benzyl alcohol in benzene was passed over a bed of H-Y at 400°C and WHSV .95 h^{-1} to give 64.5% (on benzyl alcohol) of anthracene.

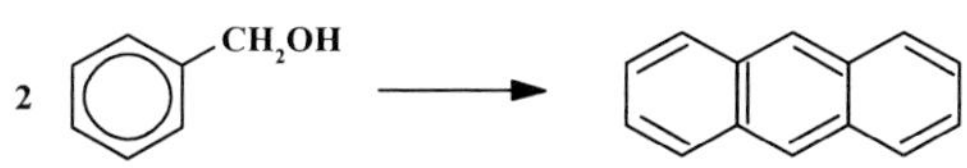

(36)

6.2. Ring closure towards heterocyclic compounds

Heterocyclic compounds play an important role as intermediates and end-products in fine chemistry. For instance many bio-active molecules - pharmaca and agrochemicals - contain a heterocyclic ring. Ono *et al.* have demonstrated (230) the great potential of zeolite catalysis in heterocyclic-compound syntheses. Applications of zeolites in heterocyclic-ring (trans)formations include:
- cyclization reactions;
- heteroatom substitution (O $\rightarrow$ N);
- ring-size isomerization (5 $\rightarrow$ 6);
- aromatization reactions.

In this section some examples will be given on zeolite-catalyzed formation of oxygen- and nitrogen-containing heterocycles.

6.2.1. Oxygen-containing ring systems

6.2.1.1. Formation of cyclic ethers

Several examples exist of zeolite-catalyzed dehydration of diols to form five-membered ring heterocycles. For instance, trans-1,4-cyclohexanediol is transformed at 215°C into 7-oxabicyclo[2.2.1]heptane (Reaction 37) over a specially treated NaA zeolite (231). Selectivity is 71% at 98% conversion; apparently outer surface catalysis is involved.

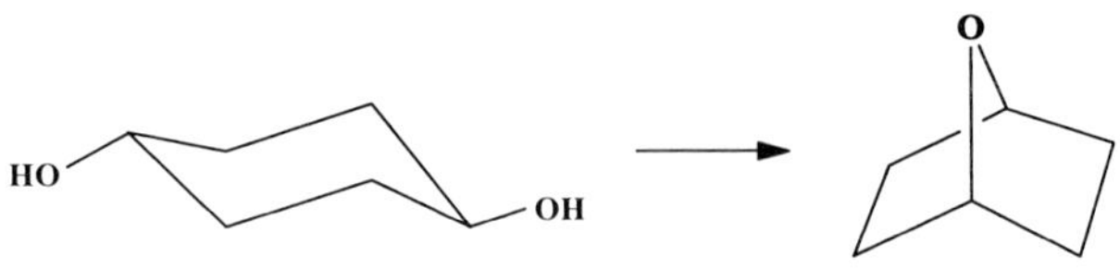

(37)

One of the present authors studied the cyclodehydration of 1,4-butanediol and 1,4-but-2-enediol towards tetrahydrofuran and dihydrofuran, respectively. Highly selective conversion of 1,4-butanediol is obtained (232) *e.g.* over a boron- or iron-zeolite of the MFI type at 200 or 300°C. When passing 1,4-but-2-enediol over various boronsilicates (233) crotonaldehyde (2-butenal) showed up as a major side product beside dihydrofuran.

Recently the conversion of tetritols (erythritol and threitol) to the 1,4-anhydro-systems was reported (234) to take place in the presence of KA zeolite. Here too, catalysis should be due to outer zeolite surface and/or binder material.

We mention here also the dehydration of 1,4-diketones over H-zeolites (235), leading to 2,5-disubstituted furans according to Reaction 38. For example, 2,5-dimethylfuran (a compound with a blending octane number of > 200) is obtained with 98% yield from acetonylacetone (R = CH_3) over H-ZSM-5 (Si/Al 140) at 300°C. The reaction may involve intramolecular acetalization of the mono-enolized diketone followed by dehydration.

On basic Na-ZSM-5 under identical conditions acetonylacetone enters into an intramolecular Claisen-condensation leading to 3-methyl-2-cyclopenten-1-one in 89% yield. In this way the acetonylacetone conversion is of diagnostic value for the nature of the active sites present in medium pore zeolites.

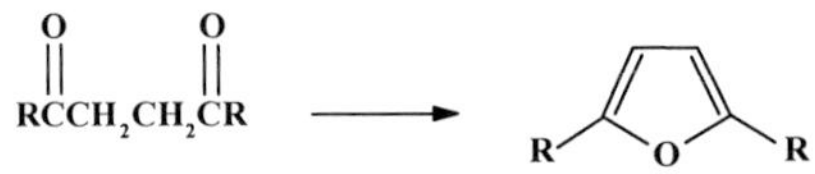

(38)

Interesting cyclic ether formations by 2-step Ti-catalyzed epoxidation/ring closure reactions have been reported by Corma *et al.* (236) and by Tatsumi *et al.* (237).

Thus linalool was converted (236) by treatment with t-butyl hydroperoxide over Ti-Beta and Ti-MCM-41 to a mixture of five- and sixmembered cyclic ethers.

In the more limited space of TS-1 (Ti-MFI) essentially complete regioselectivity to the fivemembered ring ethers was achieved in the 2-step reaction of 4-pentene-1-ol and 4-hexen-1-ol with hydrogen peroxide (Reaction 39). These reactions (237) represent fine examples of a zeolite acting as bifunctional and regioselective catalyst.

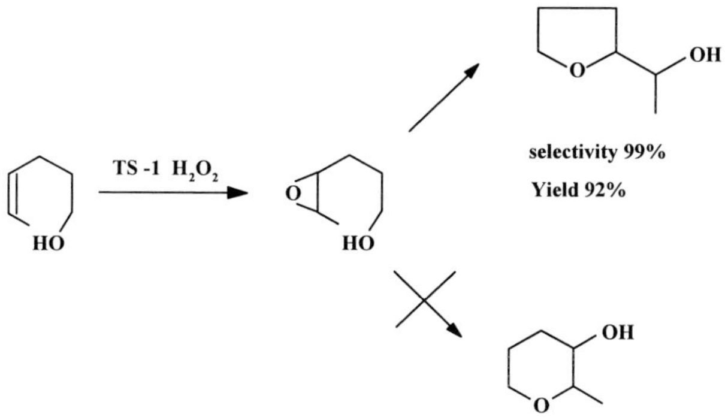

(39)

6.2.1. 2. Synthesis of coumarin derivatives

By reaction of phenols with bifunctional molecules, which are able to alkylate and to esterify, several coumarin derivatives have been prepared using H-Beta as the catalyst (238,239).
Thus resorcinol and propynoic acid react at 150°C in p-chlorotoluene as the solvent towards the perfumery ingredient umbelliferone (7-hydroxycoumarin) in good yield (Reaction 40). H-Beta catalyzes the esterification as well as the ring closure alkylation step. With acrylic acid dihydrocoumarins are obtained.

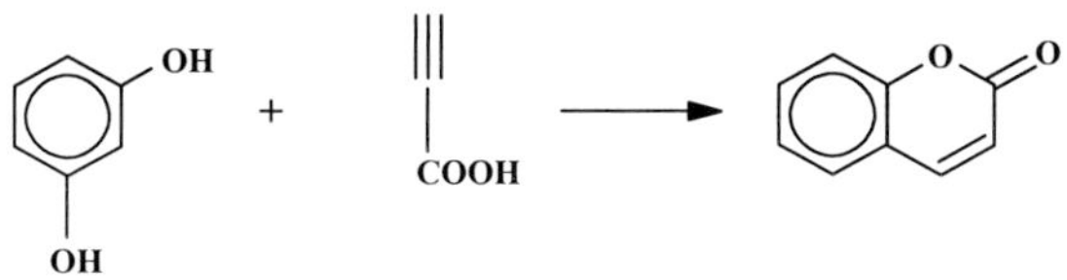

(40)

Another route to coumarins is the Pechmann condensation of phenols with β-ketoesters. The reaction proceeds *via* transesterification and intramolecular hydroxyalkylation, followed by dehydration, and is conventionally conducted using sulfuric acid as catalyst. Resorcinol and ethyl acetoacetate (Reaction 41) reacted solvent-free over H-Beta at 150°C to give 7-hydroxy-4-methylcoumarin (78% isolated yield). This is another perfumery ingredient which is also an intermediate for the insecticide Hymecromone.

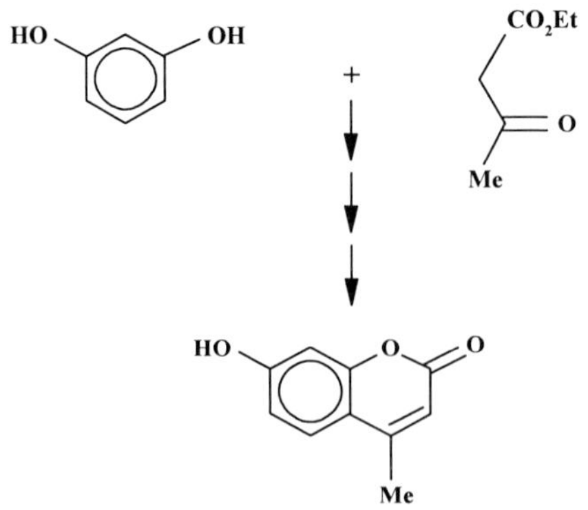

(41)

6.2.2. Nitrogen-containing ring systems

For a review on the use of zeolites as catalysts in the synthesis of N-containing compounds the reader is referred to ref. (240) and for a recent review on heterogeneous catalysis in heterocyclic synthesis to ref. (240a).

6.2.2.1. O/N replacement in cyclic compounds

Zeolites offer an interesting alternative to conventional acid non-shape-selective catalysts for the replacement of O by N in heterocyclic compounds.

Tetrahydrofuran (THF) and NH_3 (molar ratio 1:7) react at 350°C over H-L to give pyrrolidine - Reaction 42 - with a selectivity of 91% at 53% conversion and over H-Y with a selectivity of 82% at 61% conversion. Alumina and silica-alumina are also active in this reaction but the selectivity is poor. The alkali forms of zeolites L and Y are not active at all, indicating that Brönsted acid sites exert the catalytic action (241). A logical sequence would seem: oxygen protonation/-nucleophilic substitution/hydroxyl protonation/intramolecular nucleophilic substitution. In view of the presence of (excess of) ammonia and pyrrolidine just a small part of the protons is available for catalysis.

(42)

Primary amines also enter into zeolite-catalyzed O/N substitution of THF. Thus THF reacts with propylamine ($R = C_3H_7$) over Al-doped H-Y at 360°C to give 1-propylpyrrolidine with 75% selectivity at 61% conversion (230).

The corresponding six-membered ring, tetrahydropyran, undergoes O/N replacement towards piperidine over dealuminated HY- and HL-zeolites which give a better performance than the parent H-zeolites (242). The alkali forms are inactive whether dealuminated or not, so Brönsted acidity seems to be required.

Five- and six-membered lactones are converted to the corresponding lactams (Reaction 43) over various Y-zeolites.

(43)

For instance, γ-butyrolactone (n = 3) and NH_3 (molar ratio 1:5) react over CuY at 260°C to give 2-pyrrolidinone with 80% selectivity at 31% conversion (243).

By contrast caprolactone (n = 5) and NH_3 are found to give 5-hexenenitrile as the main product upon reaction over pentasil type or HY zeolites (244) in contrast to the expected caprolactam. This may be related to the more difficult ring closure of the supposed acyclic intermediate compared to 5- and 6-membered ring formation. For a full discussion of the different behaviour of the lactones, see ref. (240).

6.2.2.2. Cyclocondensation with ammonia

Aldol condensations of aldehydes and ketones on zeolites have been extensively described. Such condensation reactions, which are preferably carried out in the gas phase, lead in the presence of NH_3 over acid catalysts to pyridine and alkylated pyridines (240). Various aldehyde/aldehyde and aldehyde/ketone combinations were reacted (245) together with NH_3 over H-ZSM-5 (Si/Al 200-400).

In some cases high selectivities are obtained. A fine example is the synthesis of β-alkylpyridines (246) by reaction of acrolein, an alkanal and ammonia over MFI zeolites (Reaction 44).

(44)

When passing a mixture of ammonia acrolein and butanal (molar ratio 3:1:1) over a HF-treated B-MFI-zeolite at 400°C and WHSV 3 h^{-1} one obtains β-ethylpyridine with 72% selectivity. For R = C_4H_9 and C_6H_{13} the selectivities are 78% and 90%, respectively. In all cases conversion is complete and catalyst lifetime > 48 h. The increasing selectivity with increasing chain length indicates shape selectivity; the zeolite might induce the long chain alkanals to adopt - on average - favourable positions with respect to the other reactants.

In industry pyridine and β-picoline (3-methylpyridine) are generally coproduced by condensing formaldehyde, acetaldehyde and ammonia (Reaction 45). Also small amounts of α- and γ-picoline and of higher alkylated pyridine are formed.

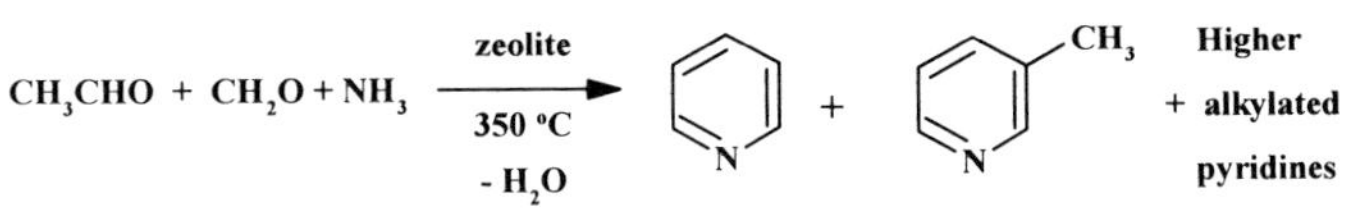

(45)

The conventional amorphous silica-alumina catalysts have been substituted here by zeolites, especially of the H-ZSM-5 type (247,248). Higher yields and higher pyridine/β-picoline ratios are obtained with zeolite catalysis.

The micropores will reduce the formation of higher alkylated pyridines. The zeolites can be further improved by incorporating metal oxides (*e.g.* Pb, Tl, Co) or noble metals (see ref. 240a for a review). Frequent reactivation of the catalyst (fixed bed) or adding regularly fresh catalyst in a fluid-bed method remains necessary.

As an example, a Pb-MFI catalyst, operated at 450°C in a fixed bed reactor at 450°C and fed with $CH_2O/CH_3CHO/NH_3$ in a 1.0:2.0:4.0 molar ratio at WHSV = 1000 gave 79% total pyridines with pyridine/β-picoline ratio of 7.5. Also zeolites MCM-22 (249) and Beta (250) perform well in combined pyridine/β-picoline synthesis.

Pyridines and picolines are manufactured on a fairly large scale (together over 50.000 t/a) whereas 3,5- and 2,6-dimethylpyridines (lutidines) are made on a 10-100 t/a scale.

The combination acetone/formaldehyde/NH_3 leads to 2,6-lutidine. Zeolites ZSM-5 (251) and Beta (250) have been studied as catalysts here. Instead of formaldehyde also methanol can be used. Over the zeolite dehydrogenation (*cf.* 252) is assumed to occur. When feeding a mixture of acetone, ^{13}C-labelled methanol, ammonia and water (molar ratio 2:1:4:13.7) to H-ZSM-5 (Si/Al 96) at 450°C at WHSV (acetone) 0.3 h^{-1}, 2,6-lutidine is formed (13% selectivity) which is exclusively labelled at the 4-position (253) (Reaction 46).

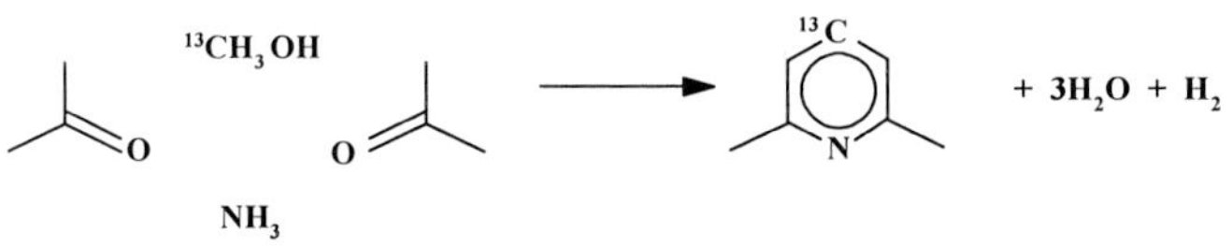

(46)

3,5-Lutidine can be prepared from propionaldehyde, formaldehyde and ammonia over a zeolitic catalyst, H-ZSM-5 (254) or H-Beta (250), but in practice this chemical is isolated from the higher methylpyridines in the pyridine/β-picoline manufacture.

By reacting an arylamine with lower aldehydes (255) quinoline derivatives can be obtained (Reaction 47). Various Brönsted acids, including large pore zeolites can be applied as catalysts.

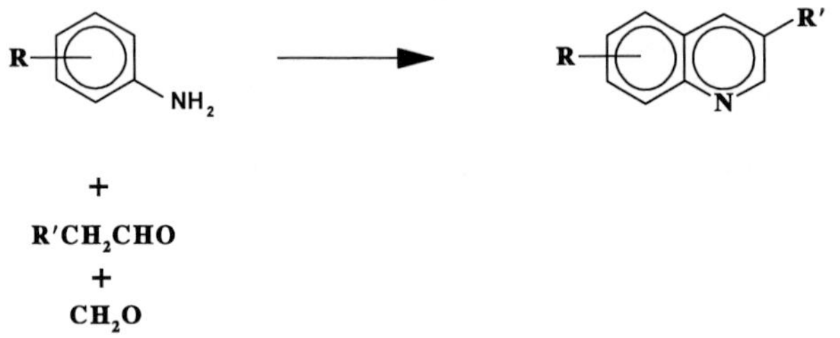

(47)

6.2.2.3. Oxidative pyridine synthesis

The reaction of ethanol with ammonia on zeolite catalysts leads to ethylamine (240). If, however, the reaction is carried out in the presence of oxygen, then pyridine is formed (256,257). B-MFI or Fe-containing MFI type catalysts are particularly suitable for this purpose. Thus, a mixture of ethanol, NH_3, H_2O and O_2 (molar ratio 3:1:6:9) reacts on B-MFI at 330°C and WHSV - 0.17 h^{-1} to yield pyridine with 48% selectivity at 24% conversion. At 360°C the conversion is

81% but there is increased ethylene formation at the expense of pyridine. Further by-products include diethyl ether, acetaldehyde, ethylamine, picolines, acetonitrile and CO_2. When applying H-mordenite, HY or silica-alumina as the catalyst under similar conditions pyridine yields are very low and ethylene is the main product. The one-dimensional zeolite H-Nu-10 (TON) turned out to be another pyridine-forming catalyst (258). A mechanism starting with partial oxidation of ethanol to acetaldehyde followed by aldolization, reaction with ammonia, cyclization and aromatization can be envisaged. An intriguing question is why pyridine is the main product and not methylpyridines (picolines) (258). It has been suggested in this connection that zeolite radical sites induce C_1-species formation.

Improved yields of pyridine and picolines are obtained (258) when instead of ammonia methylamine is applied as the N-reactant in the reaction of ethanol over H-TON in the presence of oxygen.

6.2.2.4. Thiazole synthesis

Researchers at Merck developed a new zeolite process for 4-methylthiazole which is a precursor of an important fungicide (thiabendazole). The 2-step process (259,260) starts from cheap chemicals (Reaction 48) and is an impressive improvement compared to the existing salty multistep synthesis of 4-methylthiazole (starting with chlorination of acetone). The catalyst is a composite of Cs_2SO_4, silica and H-ZSM-5, so base-catalysis is involved. A selectivity of 62% at 86% conversion and a catalyst lifetime of 2 weeks are reported.

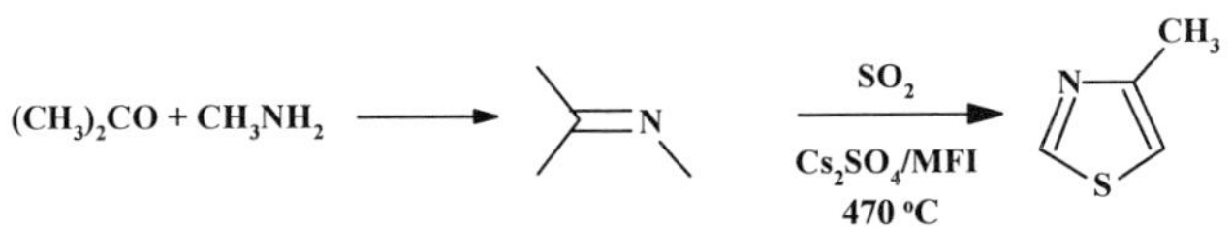

(48)

6.2.3. Formation of bicyclic compounds

Zeolites give higher selectivities than existing industrial catalysts in the synthesis of diazabicyclo[2.2.2]octane (DABCO) from precursors such as N-hydroxyethyl- and N-aminoethylpiperazine. At 400°C 10% conversion and 87% selectivity are obtained (261) over H-ZSM-5 (Si/Al 35-55). Conventional catalysts, *e.g.* alumina, are more active but less selective leading to difficultly separable byproducts. When using a high silica zeolite, the only by-product is piperazine.

It is also possible (262) to use ethanolamine as the starting material for DABCO. At 400°C and WHSV 10 h^{-1} a 64% yield of DABCO is claimed using a ZSM-5 catalyst (Reaction 49).

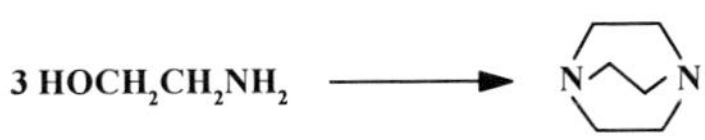

(49)

The well-known Fischer indole synthesis provides a versatile method for synthesizing substituted indoles from the arylhydrazones of aldehydes or ketones. In general it is not necessary to isolate the arylhydrazone; the reaction may be conducted as a one-pot procedure starting from the carbonyl compound and arylhydrazine. The enehydrazine tautomer of the arylhydrazone undergoes (263) a rapid [3,3]-sigmatropic rearrangement, followed by elimination of ammonia to yield the indole product. With unsymmetrical ketones two isomeric indoles (as shown in Reaction 50) result and zeolites might show shape selectivity in this reaction.

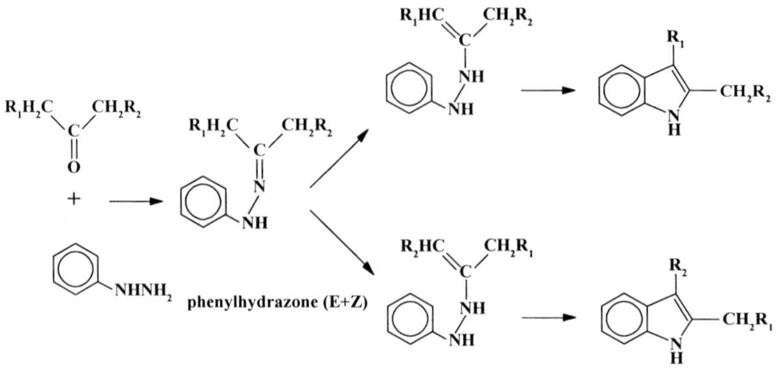

(50)

The traditionally employed, homogeneous Brönsted and Lewis acid catalysts require neutralization during product work-up and thus generate salt and/or metal-containing wastes. An early, non-shape-selective, application of zeolites in the Fischer indole synthesis was reported by Venuto and Landis (264). As catalysts, they employed calcium- and rare-earth-exchanged zeolite X under continuous-flow conditions at 150°C. Good yields of indoles were obtained with both acetone and cyclohexanone phenylhydrazones; *e.g.* 72.5% in the latter case.

The phenylhydrazine/cyclohexanone reaction was recently further investigated (265) over various zeolites in the H-form and also including methyl-substituted reactants (thus increasing product size). The HY sample appeared by far the most active catalyst.

Prochazka *et al.* screened (in liquid-phase batch reactions, with refluxing cyclohexane solvent fourteen different zeolites with five different unsymmetrical ketones, both in one-pot reactions and with isolated hydrazones as substrates (266). With large-pore zeolites, yields of 80-90% were obtained from the phenylhydrazones of 3-hexanone, 3-undecanone, 5-methyl-3-heptanone and 1-phenyl-2-butanone.

The last substrate appeared to afford a striking example of shape selectivity. According to Prochazka *et al.*, the use of mordenite as catalyst led to the selective (93%) formation of the least bulky isomer, 2-benzyl-3-methylindole, from 1-phenyl-2-butanone phenylhydrazine, whereas only 2-ethyl-3-phenylindole was formed when the reaction was performed in acetic acid. However, in attempts by Rigutto *et al.* to reproduce these findings, mixtures of both isomers were invariably obtained (267) and it was found that molecular sieves possessing a one-dimensional channel system, such as mordenite, were ineffective in altering the indole isomer ratio. In the cyclization of 1-phenyl-2-butanone, Rigutto *et al.* found that amongst three-dimensional zeolites H-Beta showed the highest preferential formation (75%) of 2-benzyl-3-methylindole (267).

Of the various possibilities for realising preferential formation of a single isomer, *product shape selectivity* is unlikely in the present case, since the indole products do not interconvert, which would rapidly result in plugging of the pores by the bulkier, non-desorbing isomer. On the basis of adsorption data of the isomeric indoles from 1-phenyl-2-butanone and phenylhydrazine, Rigutto *et al.* suggested that *restricted transition-state selectivity* is the most likely mechanism for determining regioselectivity.

In attempts to disentangle the various factors contributing to regioselectivity, the studies of Rigutto *et al.* were extended by Kunkeler *et al.* (268). In addition to 1-phenyl-2-butanone these workers also employed as substrates 1-phenyl-2-propanone (R_1 = Ph, R_2 = H in Reaction 50) and 3-heptanone (R_1 = Et, R_2 = n-Bu).

Of several active zeolite catalysts tested, H-Beta was found to be the most selective heterogeneous catalyst for the synthesis of the 'linear' isomer (75% selectivity) from 3-heptanone and 1-phenyl-2-butanone phenylhydrazones while, in contrast, H-Y gave about equal amounts of both isomers. Over both zeolites 1-phenyl-2-propanone gave fully selective the more bulky isomer: 2-methyl-3-phenylindole showing that non-sterical factors dominate here.

Although adsorption experiments with H-Beta gave results in accordance with the preference for the linear product from 3-heptanone phenylhydrazone, a bulky/linear adsorption ratio of 28/72 being found (268), and similar results were obtained for the indole isomers from 1-phenyl-2-butanone, the total adsorption was small. Furthermore, when a sample of Beta, the external surface of which had been passivated by silylation (158), was employed, no activity was found. This suggests that it is the external surface, rather than the internal pore walls, which is responsible for the observed activity, so that the observed variations in isomer distribution should not be ascribed to pore-induced shape selectivity.

For a recent review on heterogeneous catalysis in the Fischer indole synthesis see ref. (269). The parent compound, indole, can be prepared by gas phase reaction of aniline and ethylene glycol over H-ZSM-5 (Si/Al 180) at 300°C in a H_2 atmosphere (270). The reaction is assumed to involve subsequently nucleophilic substitution, electrophilic substitution and dehydrogenation. The yield is moderate (54%) but again the procedure seems environment-friendly compared to the existing method.

Zeolites and clays have been postulated as prebiotic in the formation of biomolecules. In this connection, the observed formation (271) of adenine and guanine from C_1 precursors in zeolite NaX is interesting.

Reaction 51 formulates the formation of adenine (1%) from ammonium cyanide. Note the C:N ratio of 1 in the product molecule.

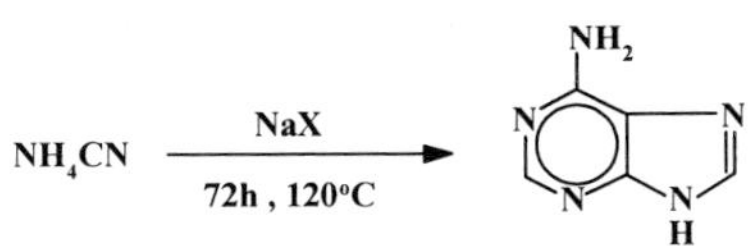

(51)

7. NUCLEOPHILIC SUBSTITUTION AND ADDITION

In nucleophilic substitution - a type of reaction frequently encountered in aliphatic chemistry - zeolites can play their role by activating the "leaving group" through protonation or metal ion coordination and by displaying shape selectivity *e.g.* limiting consecutive reactions.
In this paragraph some examples will be given of nucleophilic substitution together with some addition reactions in which a zeolite-activated olefin reacts with an O- or N-nucleophile.

7.1. Reactions of alcohols with ammonia

Amination reactions are of considerable interest in the chemical industry. Especially methylamines and ethylamines play important roles as intermediates. For a recent review on amination the reader is referred to (272).

Two important ways to aliphatic amines are amination of alkenes and of alcohols. The latter amination can be executed over metal catalysts (the alcohol is dehydrogenated and a reductive amination follows) or over solid acid catalysts.

Thus methylamines are produced by the reaction of methanol with ammonia on acidic heterogeneous catalysts, *e.g.* silica-aluminas. Generally, however, a product mixture is desired which differs from the equilibrium composition, in that it should contain as high as possible a fraction of mono- or dimethylamine (MMA or DMA) (273). Various zeolite catalysts have been studied (274-278) in order to minimize the formation of trimethylamine (TMA). For a review on methylamines synthesis see ref. (279). The Nitto-process for the production of di- and monomethylamine from methanol and ammonia by gas phase reaction applies a modified ion-exchanged mordenite preventing the formation of trimethylamine. The selectivity for dimethylamine is about 65% and that for trimethylamine less than 5% at 320°C. Therefore, in contrast to the conventional production, there is no excess of trimethylamine which has to be recycled. According to SRI International's evaluation (279a), the Nitto-process can increase the capacity by about 30-50% and reduce the energy consumption by 40-50% in existing installations (less distillation, no recycling) and can require about 30-40% less capital investment in new plants. The product shape selectivity of the modified mordenite is effected by adjusting its acidity and pore opening (Figure 16) and by poisoning the outer surface. This Nitto-process is commercially applied by Nitto and ICI and constitutes a typical industrially successful example of utilizing the shape selectivity of zeolites. Du Pont as well as Air Products are in a process to develop alternative approaches to the Nitto-technology which might involve zeolite RHO and chabazite.

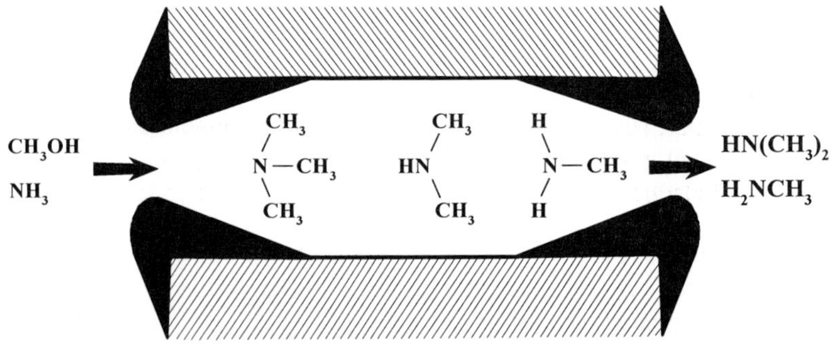

Figure 16. Product selectivity in methylamines synthesis.

The mechanism of methanol amination is intriguing because protonation of ammonia on a Brönsted site would not seem in favour of the reaction. Lercher *et al.* (280) advanced a mechanism (Figure 17) for the first steps in the H-zeolite catalyzed reaction. The subsequent desorption of methylamine from the active site is a slow step and requires the assistance of ammonia.

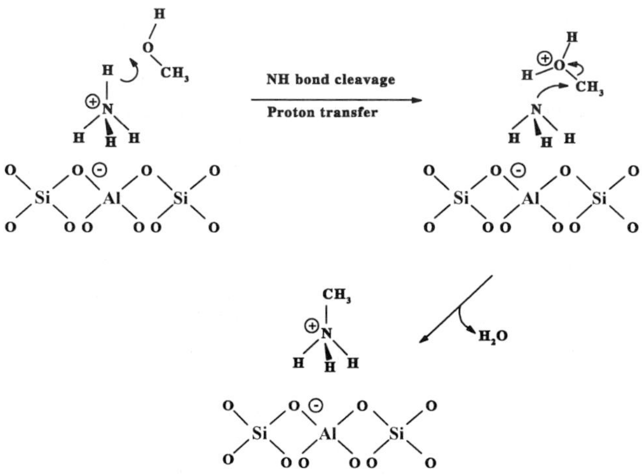

Figure 17. Mechanism of methylamine formation (280).

In the amination of ethanol (281) and higher alcohols over H-zeolites elimination towards alkenes will always be a competing reaction.

Shape selective C_2 to C_4 alcohol amination by dehydrogenation/reductive amination is described in a patent (282). Reactions are performed in the presence of hydrogen using cobalt- or nickel-exchanged mordenite or zeolite Y at 300°C. Selectivity towards monoalkylamines is substantially higher on the zeolite catalysts than on a cobalt on silica catalyst.

The amination of alcohols on zeolite catalysts is not restricted to monofunctional molecules. It has been shown, for instance, that nucleophilic substitution of the OH-group in ethanolamine with NH_3 yields ethylenediamine (283). A dealuminated (Si/Al = 7.2) rare earth or H-exchanged mordenite is used as the catalyst. With ethanolamine and ammonia in a molar ratio of 1:4 at 300°C ethylenediamine is obtained with 84% selectivity at 15% conversion. As by-product aminoethylethanolamine is formed exclusively. This behaviour contrasts with that of many conventional catalysts such as Co, Ni, Cu/Cr, Pt on supports where selectivity to ethylenediamine is sacrificed in favour of conversion.

7.2. Addition of ammonia and amines to olefins

Another approach towards amines is the direct addition of ammonia or alkylamines to olefins. Various zeolites have been studied as catalysts: HY and rare earth Y (284-287), H-offretite (284,286,290), H-clinoptilolite, H-chabazite-erionite and H-erionite (284,286,288), H-mordenite (284-287,289) as well as pentasil zeolites (284,290-292). The amination is favoured by low temperature, high pressure and high ammonia-to-olefin ratio. However, a minimum reaction temperature is required to activate the olefins and this treshold temperature depends on the olefin structure.

Ethylene is aminated, at a minimum temperature of 320°C using HY, H-erionite and H-mordenite. A 4:1 molar NH_3 olefin ratio is applied (284,286). The conversion increases with temperature, *i.e.* from approximately 2% conversion at 320°C to about 12% conversion at 380°C over HY. In the presence of H-erionite 2.5% conversion is achieved at 320°C and 13.5% at 380°C. Mono- and diethylamines (wt ratio > 9:1) are the main products and the selectivity for these compounds is > 98%. At temperatures above 380°C the formation of nitriles and higher olefins is observed.

In the case of the amination of propene the minimum temperature is 300°C to achieve 2% olefin conversion over HY. The molar ratio of mono- to diisopropylamine is 93 to 7.

Isobutene reacts with NH_3 over HY to t-butylamine (TBA, Reaction 52) already at 220°C (285). At 300°C the equilibrium conversion (9%) is reached. Over this temperature range > 90% selectivities are obtained using a 2:1 ammonia/isobutene molar feed ratio.

However, the Y-catalyst suffers from rapid deactivation. BASF has developed an MFI-type zeolite which shows not only more than 99% selectivity, but also affords commercially acceptable catalyst life. The absence of inorganic coproducts as well as extremely toxic starting materials and intermediates in this process provides evidently advantages over the traditional HCN-based Ritter route to TBA starting from isobutene and hydrogen cyanide where the resulting formamide is hydrolysed. Thus, the zeolite-based process is worth to be called environmentally safe and friendly. According to the Ritter-reaction, 4.5 t starting material are needed to produce 1 t TBA and 3 t waste are produced per t TBA (E-factor is 3). In the case of the BASF-process, which is in operation in 2 plants at Antwerpen, 1 t starting material yields almost 1 t TBA.

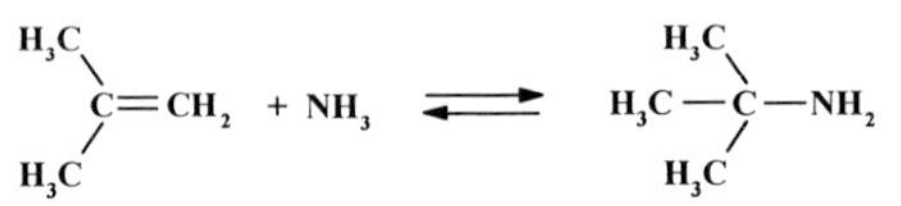

(52)

Concerning the reaction mechanism it is found (284-286) that strongly acidic sites are necessary and the amination occurs via protonated intermediates – possibly with alkoxy structure - as evidenced by the relative ease: isobutene > propene > ethene.

A recent kinetic study (293) on the amination of isobutene (and 1-butene) over a series of H-zeolites confirmed that MFI is the best catalyst for the TBA synthesis. Moreover, it was found that ammonia and isobutene show the same strenght of adsorption onto H-MFI (Si/Al = 14) and onto H-Beta (Si/Al = 15).

Upon zeolite-catalyzed addition of a diamino-compound such as ethylenediamine to isobutene only mono-alkylation is observed (294), even when the olefin is present in large excess. Thus at 300°C and 300 bar N-t-butyl ethylenediamine is obtained in 20.5% yield in the presence of a B-MFI zeolite.

The amination of dienes over zeolites is possible too (295), and suitable dienes can give rise to N-containing ring systems. For example 2,5-dimethyl-1,5-hexadiene - containing two active double bonds - reacts with NH_3 (Reaction 53) over Ce-doped borosilicate at 300°C to give 2,2,5,5-tetramethylpyrrolidine with 25% selectivity at 24% conversion. The noncyclic mono-addition product is obtained with 31% selectivity.

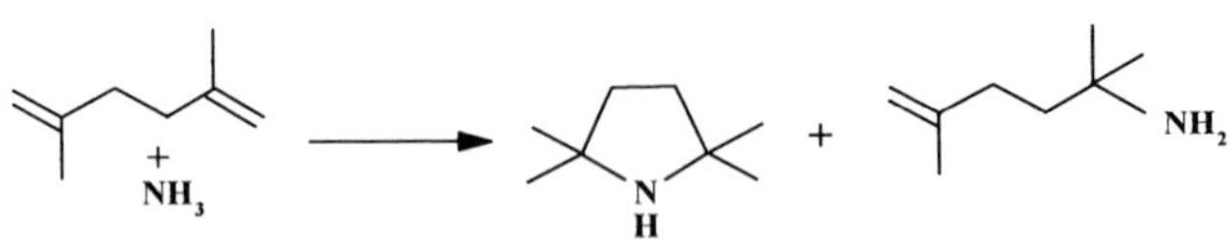

(53)

7.3. Synthesis of anilines

The synthesis of anilines is of industrial interest. Catalytic hydrogenation of aromatic nitro-compounds can often be applied but other existing methods are not without disadvantages and a new clean zeolite-based technology would be welcomed. Reported routes involve amination/dehydrogenation of alicyclic systems and nucleophilic substitution of aromatic compounds with NH_3.

The preparation by reaction of alicyclic alcohols like cyclohexanol, or ketones like cyclohexanone with ammonia is carried out in the presence of a ZSM-5 type catalyst which contains a metal having dehydrogenation activity (296). Using NiHZSM-5 at 480°C and 200 psig only 16.2% aniline selectivity is observed. Dimeric compounds such as diphenylamine and carbazole are preferentially formed.

Since amination and simultaneous dehydrogenation proceed with only moderate yield, it is preferable to choose phenol as starting material for the synthesis of aniline. Here ammonia is assumed to add to the tautomeric ketonic form of phenol. Over H-ZSM-5 in the gas phase at 510°C and 28 atm a phenol conversion of 94% and an aniline selectivity of 91% (95,297) are achieved. Main side product is 2-methylpyridine, formed in a consecutive reaction (cf. Section 4.5.6). Over Na-ZSM-5 at lower WHSV the selectivity to aniline raises to 96.5%. The reaction is also catalyzed by Y-, X- and mordenite zeolites, but these catalysts display a markedly shorter lifetime than H-ZSM-5, and in the case of X- and Y-zeolites the conversion is lower (297,298). By-products such as diphenylamine and carbazole, which are obtained in the presence of non-zeolitic silica-alumina according to the Halcon-Scientific Design process are suppressed or eliminated by virtue of the shape selectivity of the ZSM-5 catalyst.

Several groups have reported (299-302) on the reaction of chlorobenzene and ammonia in the presence of zeolite catalysts (Reaction 54). First particularly Cu-exchanged faujasites have been tested, which seemed a logical choice. However, conversions are modest and benzene shows up as a major side product.

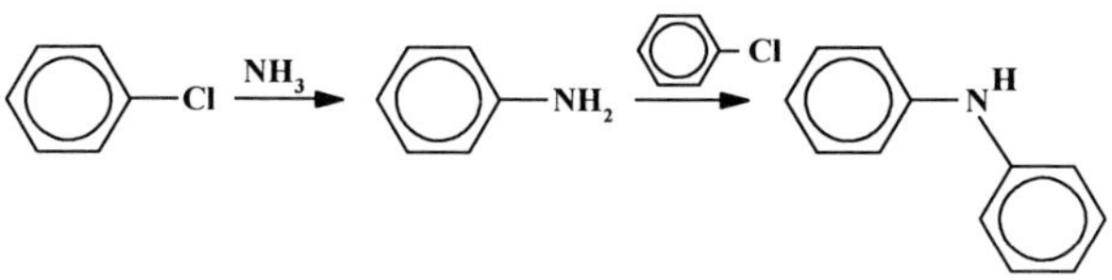

(54)

It has been found that significantly higher selectivities can be obtained by using copper zeolites with smaller pore sizes; copper-silica-alumina was inactive under these conditions, presumably because of a rapid deactivation. On the zeolites copper-L (channel diameter 7.1 Å), copper-mordenite (6.5 x 7 Å) and copper-ZSM-5 (5.5 Å), nucleophilic substitution prevailed.

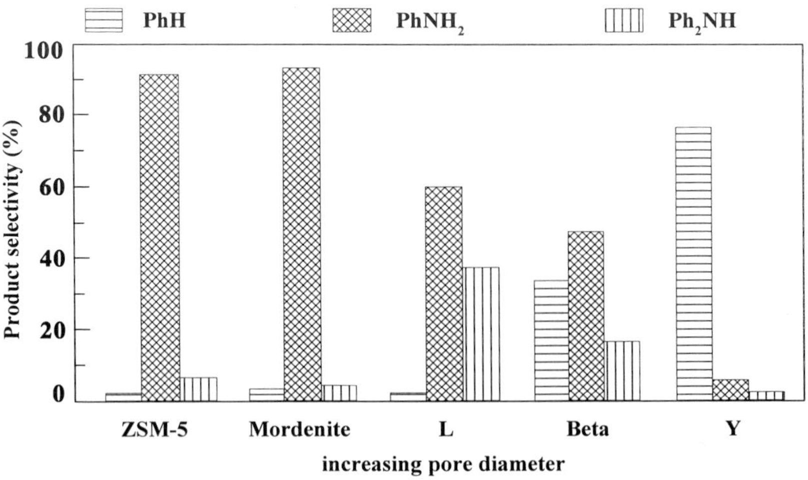

Figure 18. Selectivities in the chlorobenzene conversion with ammonia, catalyzed by different copper-exchanged zeolites (all containing 3 ± 0.3 wt % copper) at 400°C (adapted from ref. 302).

With copper-mordenite and copper-ZSM-5, aniline selectivities of 90-95% were obtained; with copper-L, the secondary substitution product diphenylamine was also formed at significant levels. The benzene formation with these three catalysts was very low. In contrast, reduction towards benzene was pronounced on copper-Beta and especially on copper-Y (302) (see Figure 18). The activity sequence (Y ≈ L >> mordenite > Beta > ZSM-5) indicates that a high activity is related to a low Si:Al ratio.

As with zeolite Y, the catalytic activity of mordenite exchanged with different metal ions was found to decrease in the order: Cu > Ni >> Ag > Zn. Likewise, the substitution selectivity was found to decrease in the order: Cu > Ni > Ag (303).

Copper ZSM-5 was also found to perform well in the chlorobenzene to phenol conversion with water. Best results were obtained (304) with a Cu-ZSM-5 sample prepared by solid-state ion exchange: phenol was produced at 50% initial conversion with 97% selectivity.

For a recent review on heterogeneously catalyzed nucleophilic aromatic substitution reactions, see ref. (305).

7.4. Nucleophilic addition to epoxides

The addition of polar compounds such as H_2O and NH_3 to epoxides on zeolite catalysts yields the correponding diol or alcoholamine, respectively. In the ring opening of asymmetrical epoxides with amines, higher catalytic activity is achieved with weakly acidic or weakly basic (NaY, NaX) than with strongly acidic (HY) or strongly basic (CsY) zeolites (306). An interesting feature is the regioselectivity of this addition which turns out to be dependent on the type of zeolite and counterion.

For instance, during the reaction of 1,2-epoxyoctane with aniline in benzene at 80°C according to Reaction 55, R = C_6H_{13}, addition takes place at both the 1- and the 2-position, *i.e.* on the less or more hindered site of the epoxide ring. The ratio of the two isomers 1/2 is dependent upon the zeolite catalyst employed. On both NaX and NaY the two isomers are obtained in 90% yield, but on NaX the ratio 1/2 is 36 and on NaY 73.

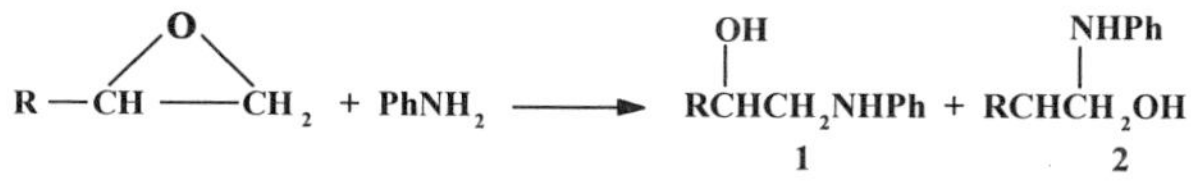

(55)

In the reaction of styrene oxide (R = C_6H_5) Brönsted acid type Y zeolites (HY, CaY) give a 1/2 ratio of < .01 whereas the use of NaY and KY leads to 1/2 ratios of 0.16 and 8.3, respectively. The high preference for 2 formation when H^+-catalysis is applied indicates some developing positive charge on the α-carbon which is stabilized by the phenyl substituent. In cation catalysis steric factors apparently dominate.

In the reaction of ammonia and ethylene oxide zeolites are able to restrict the products to the linear mono- and diethanolamine.

Addition of an azido group to 2,3-epoxyalcohols leads to 2-azido-1,2-diols (307). When a CaY-zeolite loaded with sodium azide is used as reagent, a much higher regioselectivity is achieved than with conventional systems such as NaN_3/NH_4Cl in aqueous methanol, or when Al_2O_3 or SiO_2 loaded with NaN_3 are employed. Ring opening to yield the isomer in which the azido group is adjacent to the long carbon chain is favoured. In the case of 2,3-epoxy-1-octanol, the ratio is 94:6 at 85% yield on NaN_3/CaY, but only 66:34 at 65% yield on NaN_3/Al_2O_3 and 76:24 at 88% yield with NaN_3/NH_4Cl. The regioselectivity is explained by assuming bidentate coordination of the substrate to Ca^{II}.

These examples illustrate the potential of zeolite catalysts to direct the regioselectivity in organic syntheses.

7.5. Addition of oxygen-compounds to alkenes

Numerous reports exist (1-3) on the addition of hydroxyl-compounds (water, alcohols, carboxylic acids) to olefins. Generally the first step is proton addition followed by addition of the nucleophilic OH-compounds and transfer of a proton. Sometimes zeolites do offer distinct advantages over conventional Brönsted catalysts in these reactions. We will confine ourselves to some comments and examples.

Water addition. Alcohols are prepared by the acid-catalyzed hydration of olefins. When zeolites are used it is also necessary to accept low conversions (< 20%), in order to achieve high selectivities.

For the hydration of C_2 to C_4 olefins we refer to some comparative studies (308-310) in which several types of zeolites including the effect of the Si/Al ratio have been investigated. Ferrierite and MFI-type catalysts give the best performance.

New zeolite-based hydration technology has been developed by Asahi Chemical workers (311). Cyclohexene - available now through partial hydrogenation of benzene (also Asahi technology) - is hydrated in a 3-phase system, initially consisting of cyclohexene/water/zeolite ZSM-5 in a 2.4/2.7/1 wt ratio. The zeolite catalyst resides in the water phase. This hydration is a step in a new low-salt synthesis route to caprolactam (*cf.* Section 4.5.5.2).

Zeolites have been studied as catalysts in several terpene hydrations of industrial relevance (312-314). A recent successful example is the hydration of dihydromyrcene towards dihydromyrcenol (2,6-dimethyl-7-octen-2-ol) (Reaction 56).

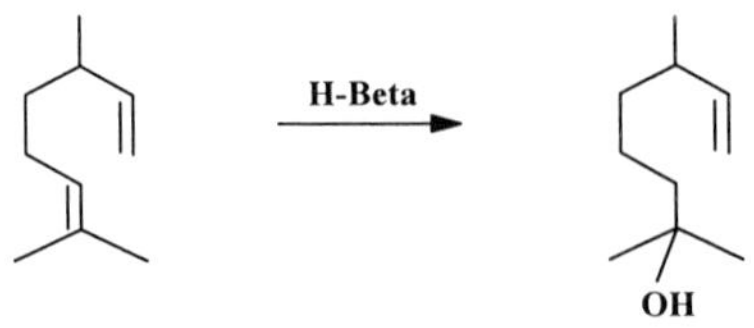

(56)

The conventional industrial method involves the use of concentrated sulfuric acid. Yield is high but dilute sulfuric acid is produced as a waste stream. Here the Corma group found (315) that H-Beta's with low defects (and thus high hydrophobicity) and with Si/Al *e.g.* 27, gave good conversions together with excellent selectivities towards dihydromyrcenol.

Medium and large pore zeolites might be tuned to serve further challenging hydrations in the terpene field, such as the isomerization/hydration of α-pinene towards (iso)borneol (316,317).

Addition of *alcohols*. Ethers can be obtained by the acid-catalyzed addition of alcohols to olefins. This reaction is of commercial interest especially for the production of the octane boosters, MTBE and ETBE, from isobutene and methanol (Reaction 57) or ethanol, respectively. When zeolite H-ZSM-5 (316) or Nu-10 (317) are used as catalysts, then conversions of approximately 35% and a MTBE selectivity of 95% are achieved. At approximately 100°C and 35 bar a weakly acidic B-MFI zeolite affords MTBE in 86% yield (318). Industrially, MTBE continues to be produced with the aid of acidic ion exchangers, since zeolite catalysts are at present less effective. For environmental reasons MTBE is becoming seriously under pressure.

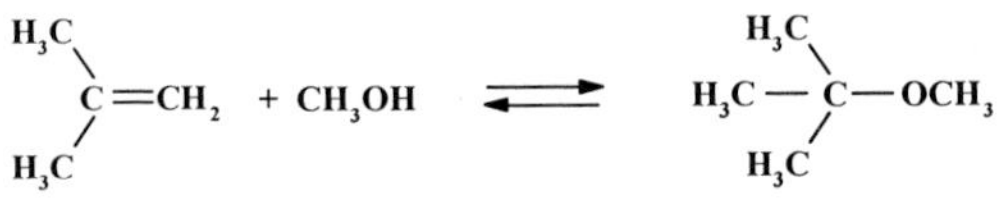

(57)

Several groups have studied (319) the industrially relevant hydromethoxylations of limonene and α-pinene over zeolite catalysts in order to substitute the conventional polluting catalysts. Best results have been obtained by the Aachen group who applied a H-Beta catalyst (fixed bed) in gas phase reaction at mild temperatures.

Thus limonene conversion with methanol (Reaction 58) occurred with 95% selectivity to the monomethyl ether at 60% conversion at 40°C. The catalyst deactivates slightly during 8 h on stream. The high selectivity of the side chain methoxylation might be due to less hindered alkoxy formation on the zeolite active site.

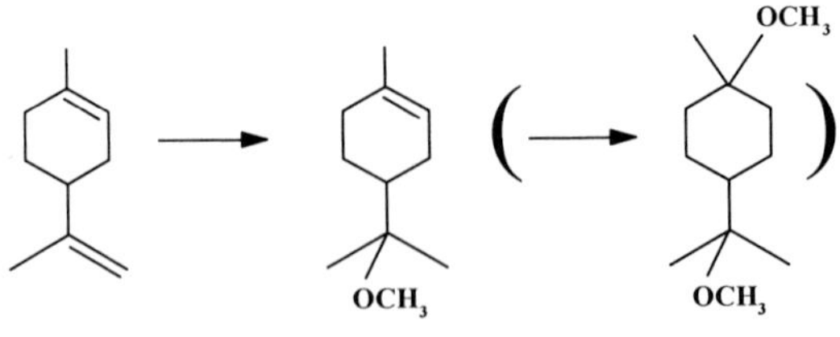

(58)

The mesoporous H-MCM-41 material turned out to be an excellent catalyst in the addition (320) of various alcohols, including the bulky cholesterol to the double bond of dihydropyran.

Addition of *carboxylic acids*. The zeolite-catalyzed addition of acids (in most cases acetic acid) to olefins - Reaction 59 - is described in several patents (321-323).

As esterification method this addition has the advantage that no water is produced. Consequently hydrophilic catalysts can be applied without the problem of product (water) inhibition. Conversions of only 30% were achieved with H-ZSM-5 and H-ZSM-12 at 150-200°C and 25 bar. In addition, mixtures of isomers are formed if asymmetrical olefins are used. 1-Olefins afford the 2-carboxylates with a selectivity of 80% (323). The use of layered silicates, in which replacement has been carried out with trivalent ions, offers sometimes better results than zeolites. Yields of up to 90%, based on olefins, have been reported (324,325).

$$R{-}CH{=}CH_2 + R_1COOH \longrightarrow \underset{\displaystyle OCOR_1}{\underset{|}{RCHCH_3}} \qquad \mathbf{(59)}$$

Zeolites, especially of the MFI type, as well as phosphates, catalyze the intramolecular addition of carboxylate groups in unsaturated acids (326). Thus a mixture of 2-, 3- and 4-pentenoic acid esters reacts in the presence of water to yield 5-methylbutyrolactone. Using HY at 180°C 93% selectivity to the lactone at 49% conversion of the esters is obtained.

7.6. Esterification and acetalization

The nucleophilic addition of alcohols to carbonyl compounds as aldehydes/ketones and carboxylic acids is the major reaction step in acetalization and esterification, respectively. As water is formed in these reactions, a zeolite catalyst should be relatively hydrophobic. Accordingly, the use of acid-leached mordenite (Si/Al 40-150) as the catalyst in the esterification of propionic acid and 1-butanol towards butyl propionate has been reported (327). In recent studies (328,329) zeolite Beta samples with Si/Al 5-100 were applied in liquid and gas phase esterification.

In liquid phase (140°C) esterification of long chain alcohols and carboxylic acids a Beta with high Si/Al is preferred presumably because of more easy water removal from the active sites (328). With shorter chain reactants (*i.e.* more hydrophilic) this effect is much less pronounced. This is also the case for short chain reactants entering esterification in the gas phase (fixed bed H-Beta). Liquid and gas phase esterification were found to differ (328,329) in that in liquid phase esterification at 140°C a substantial autocatalytic component is operating. Some authors have overlooked this phenomenon.

Another interesting catalyst in esterification is the hybrid system obtained by modifying MCM-41 material by attaching sulfonic acid groups to the pore walls (330-332; Chapter 20). In our hands this catalyst was completely stable in esterification at 140°C (328).

This was not the case when heteropoly acids adsorbed into MCM-41 were used in esterification (333). In liquid phase as well as in gas phase esterification at 110°C sintering of the heteropoly acids was observed leading to outer surface located particles. Presumably re-dispersion is possible.

The equilibrium mixtures obtained in the esterification reactions can be shifted towards the desired product by distilling off the water azeotropically, by selective water adsorption (*cf.* Section 3.6) or by passing the reaction mixture along a water selective membrane (328) which may contain microporous silica or zeolite A as separating layer. An example is the esterification

of fatty acids with ethanol (334) with high yields in the presence of KA or NaA. Zeolites A can also serve as HCl or HBr scavenger when reacting alcohols with acid chlorides. An example is the regioselective benzoylation of 1,6-anhydroglucose in the presence of NaA (335).
Transesterification using zeolites, *i.e.* the formation of a new ester starting from another one, is described analogously to the synthesis of esters (336,337). For example, yields of approximately 50% are obtained at 150°C in the preparation of methyl acetate from acetic acid and methyl formate on zeolite H-ZSM-5.

Alcoholysis of esters, the other type of transesterification, is best performed with a basic catalyst (alkoxide). Zeolites can be of help by adsorbing the displaced alcohol. Thus dimethyl terephthalate and t-butyl alcohol are quantitatively converted into di-t-butyl terephthalate in the presence of NaA or CaA (338).

Finally we mention in this section on esterification the interesting ring closure (333) of ω-hydroxyalkanoic acids towards macrocyclic lactones (Reaction 60). Dealuminated HY (Si/Al = 7) is applied as the catalyst. For instance 15-hydroxypentadecanoic acid gives the lactone in a yield of 55% which is much higher than would be obtained with homogeneous H^+-catalysis with the reactant concentration applied. The authors assume the reaction to take place at the outer (and mesoporous) surface of the catalyst.

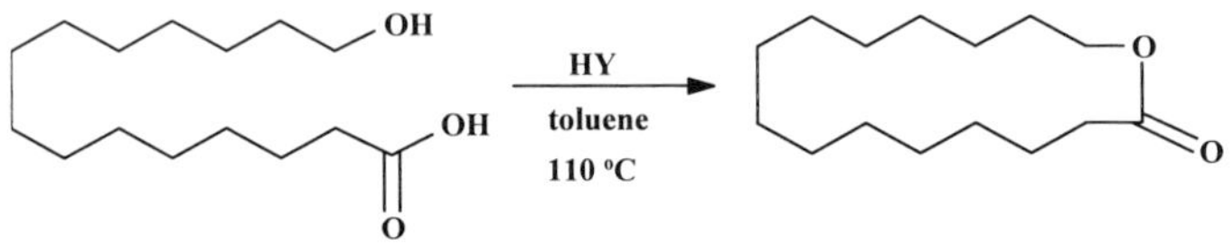

(60)

Acetals, prepared from carbonyl compounds and alcohols, are important intermediates in fine chemistry. Zeolites and related materials are excellent catalysts for acetal formation. Activation of the carbonyl group for attack by the hydroxyl group of the alcohol can be achieved by interaction of the carbonyl oxygen with a protonic zeolite site, with a counter cation or with a defect Al-site.

Consequently, zeolites can catalyze acetalization in several ways and indeed zeolites have been shown (50,339,340) to be good acetalization catalysts. In the case of bulky reactants/products also the mesoporous materials of the MCM-41 type (341) and ITQ-2 dealuminated zeolites (342) appear to be excellent catalysts.

In recent years the groups at Valencia and at Delft have devoted much attention to catalytic alkylation of carbohydrates leading to a new class of biofriendly surfactants. Here a hemi-acetal, as is present in the cyclic structures of, *e.g.* glucose and fructose, is converted into a full acetal. The usually applied homogeneous - organic sulfonic acid catalyzed - alkylation is accompanied by colour formation, and a heterogeneous catalyst would be welcomed.

In the alkylation of gluose with 1-butanol zeolites Y and Beta appeared (343) to be good catalysts. Due to the large difference in hydrophilicity/hydrophobicity of glucose and the alcohol careful tuning of the zeolite's Si/Al ratio is recommended. This has been nicely illustrated (344) in the case of H-Beta as the catalyst. Interestingly the alkyl glucofuranosides (5-ring systems) appear first as products though the two furanose forms are < 1% in the glucose mutarotation equilibrium. Several explanations can be advanced.

In fructose alkylation MCM-41 is the catalyst of choice (345). With lower alcohols quantitative conversions are obtained, whereas with long chain alcohols a two-step process - *via* butyl fructoside - is preferred. With strong acid catalysts, like sulfonic acids, homogeneous or as a resin, or with H-zeolites, yields of alkyl fructosides are low and dehydration to 2-hydroxymethylfurfural prevails. H-MCM-41's mild acidity suffices to activate fructose, especially its two furanose, five-membered forms (*cf.* Reaction 61) which total about 30% in the equilibrium. Consequently in the initial reaction mixture the alkylated furanose forms (2 and 3) dominate. The effect of the Si/Al ratio of the MCM-41material was small within the range Si/Al = 30-100.

The difference in reactivity between glucose and fructose is ascribed to the more stable intermediate oxycarbenium ions directly attainable in the case of fructose.

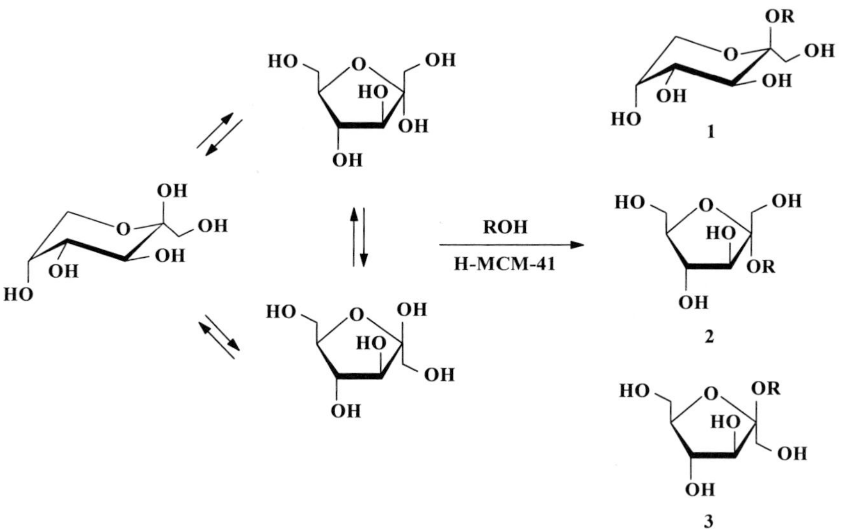

(61)

Also fructose-containing disaccharides such as leucrose, isomaltulose and lactulose can be alkylated (346) over H-MCM-41. Importantly, the glycosidic bond connecting the two sugar units is not broken.

For a complete overview of carbohydrate alkylation, see a recent chapter of Corma and Iborra (347).

7.7. Addition of S-H and P-H compounds to olefins

The acid-catalyzed reaction of olefins with H_2S furnishes thiols and thioethers (348-350). Non-doped X- and Y-zeolites catalyze the conversion of branched olefins into tertiary thiols in high yields (351,352) but only give moderate yields in the case of linear olefins. Recourse is made in this case to RE-Y- and Cd-X-zeolites, which give conversions of > 99% and selectivities > 90% (348,349). The reactions are carried out at 100-250°C and under elevated pressure (up to 250 bar). Compared with conventional acidic catalysts, zeolites give better yields and service lifes.

As discovered by one of the authors and his co-workers (353) zeolites catalyze the addition of phosphine and mono- and dialkylphosphines to olefins (Reaction 62). Olefin conversion is much higher than in the case of amination, while maintaining high selectivity.

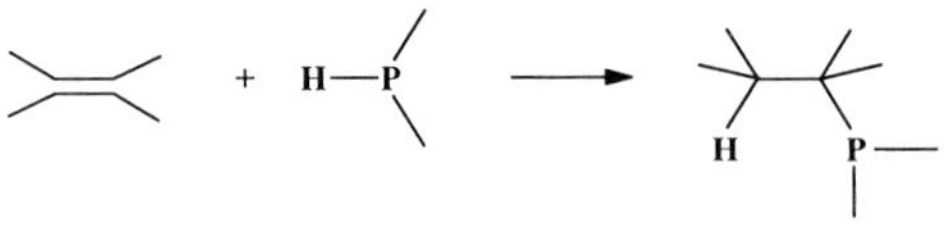

(62)

At 100°C and 8 bar in a glass ampoule isobutene reacts with PH_3 (mol ratio 1:2) on a boron MFI- type zeolite to give t-butylphosphine with 98% selectivity at 42% conversion. When the temperature is raised to 200°C the conversion increases but the selectivity falls to 85%. Under the same conditions at 100°C the reaction of isobutene with methylphosphine yields 92% selectivity of methyl-t-butylphosphine at 41% conversion.

Also cyclic olefins (cyclopentene, cyclohexene) and dienes (*e.g.* 1,4-cyclohexadiene) have been successfully converted (353) with PH_3 into phosphines.

8. MPV REDUCTION AND OPPENAUER OXIDATION

The Meerwein-Ponndorf-Verley (MPV) reduction of aldehydes and ketones and its reverse, the Oppenauer oxidation of alcohols (together denoted as MPVO reactions) are highly selective hydrogen-transfer reactions. Carbon-carbon double bonds, carbon-halogen bonds and nitro groups are not reduced. The hydrogen donors are easily oxidisable secondary alcohols (2-propanol, 2-butanol) while the oxidants are simple ketones as acetone or cyclohexanone.

The MPVO reactions are usually homogeneously catalyzed by metal alkoxides such as $Al(OPr^i)_3$. The activity of these catalysts is related to their Lewis acidic character in combination with ligand exchangeability. The MPVO reactions proceed *via* a cyclic six-membered transition state in which both the reductant and the oxidant are coordinated to the metal centre of the metal alkoxide catalyst (Figure 19). The alcohol reactant is coordinated as alkoxide. Activation of the carbonyl by coordination to Al(III) initiates the hydride transfer reaction from the alcoholate to the carbonyl. The formed alkoxide may leave the catalyst *via* an alcoholysis reaction with the bulk alcohol (354).

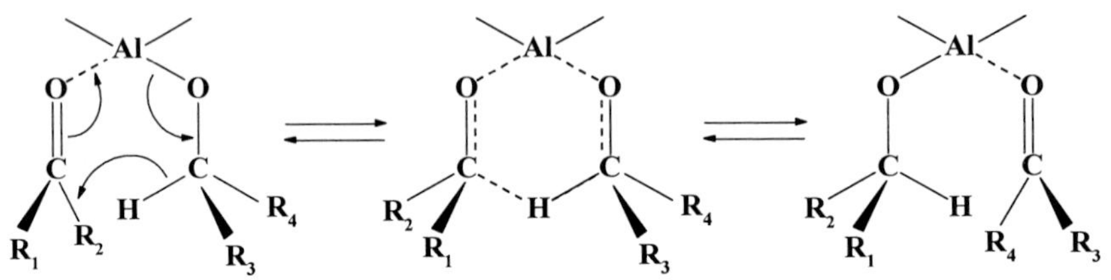

Figure 19. Reaction mechanism for the MPVO reaction.

Only a few examples exist of zeolite-catalyzed MPVO reactions. Shabtai *et al.* (355) studied the potential of alkali or alkaline earth exchanged X-type zeolites in the gas phase (100-180°C) MPV reduction of various saturated and unsaturated aldehydes and ketones using isopropanol and reducing agent. In the reduction of linear aldehydes over NaX a gradual decrease in the reduction rate was observed, upon increasing chain length, which was attributed to increasing diffusional limitations in the micropores. Selectivities to the corresponding 1-alcohols were generally high (> 95%). Application of Lewis-acidic CaX gave acetalisation of the aldehydes as an important side-reaction. The X-type zeolite/isopropanol system was not capable of reducing α,β-unsaturated aldehydes. Shape-selectivity was found in the selective conversion of citronellal under MPV conditions. In NaX there was enough space for the substrate to perform an intramolecular ring closure to isopulegol whereas over CsX reduction to the linear citronellol was observed (Reaction 63).

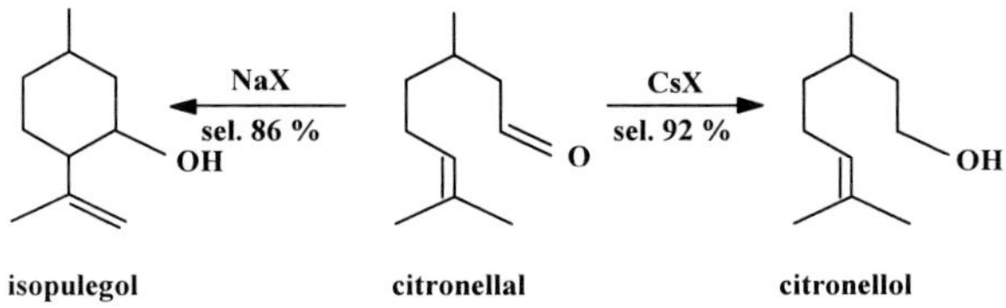

(63)

In the reduction of methylcyclohexanone isomers (100°C) it was observed that the 4-isomer reacted relatively fast and gave a thermodynamically determined product distribution (*cis:trans* = 24:76).

Recently, Creyghton *et al.* have reported (356) the application of zeolite Beta in the stereoselective (> 95%) reduction of 4-*tert*-butylcyclohexanone to *cis*-4-*tert*-butylcyclohexanol (Reaction 64). This zeolite-based catalyst proved to be fully regenerable without loss in activity or stereoselectivity. This is of industrial relevance, as the *cis*-isomer is a fragrance intermediate. Other active solid catalysts, including zeolite HY, amorphous silica-alumina and MCM-41 material, invariably gave the thermodynamically more stable *trans*-isomer in excess (~90%).

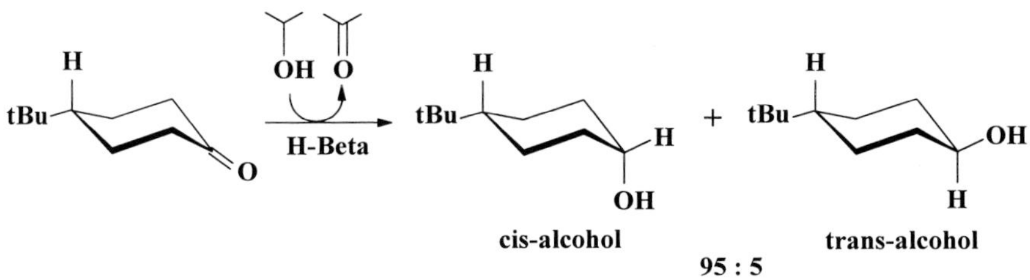

(64)

The activity of the zeolite Beta catalyst was found to increase upon increasing activation temperature together with mild steaming (357). Infrared results indicated a relation between the catalytic activity and the amount of partially hyrolyzed framework aluminium as reflected by the 3780 cm^{-1} band. For the assignment of this high frequency band see ref. (358) and (359).

The transition states leading to the *cis*- and *trans*-alcohols (see below) differ substantially in the way in which they can be accommodated in the pores of zeolite Beta. That for the *cis*-isomer is linearly aligned with the pore axis and can easily be accommodated within the straight channels of zeolite Beta. The transition state for the formation of the *trans*-isomer is more or less perpendicular to the channel wall and cannot be well accommodated within the micropores.

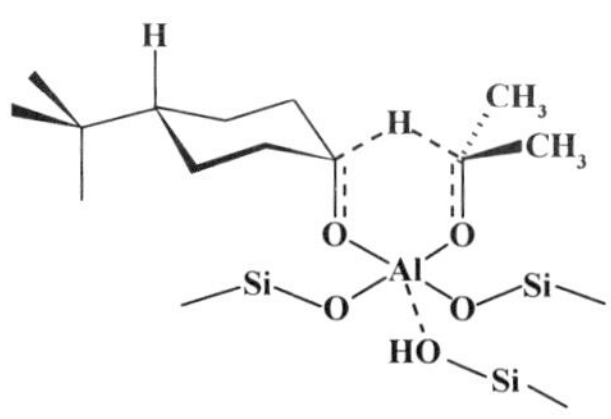

TS to cis

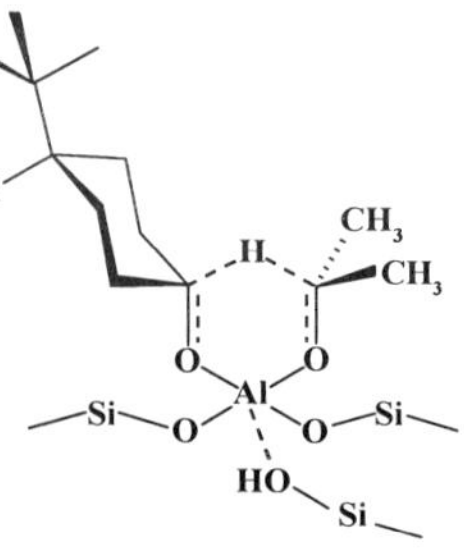

TS to trans

The dynamic character of Al in zeolite Beta and its ability to cycle between the tetrahedral full-lattice configuration and lattice-grafted forms was first demonstrated by Fajula (360) using ^{27}Al-NMR, and recently confirmed (361) by ^{27}Al-MAS-NMR and Al K-edge XANES studies.

Zeolite BEA was also found (356) to be an active and stereoselective catalyst in the Oppenauer oxidation of 4-*tert*-butylcyclohexanol to 4-*tert*-butylcyclohexanone, the reverse reaction. In a competitive experiment, using an almost equimolar mixture of *cis*- and *trans*-4-*tert*-butylcyclohexanol, and acetone as the oxidant, the *cis*-isomer was almost exclusively converted to the corresponding ketone, the *trans*-isomer remaining essentially unchanged.

Since the MPVO reactions can generally be catalyzed by Lewis acids, Ti-Beta and various other titanium-containing materials were tested in the reduction of 4-*tert*-butylcyclohexanone. Ti-Beta (Si/Ti = 69) is indeed (362) able to catalyze the MPV reduction, although with a lower TOF than observed for Al-Beta. Other titanium-containing materials, *i.e.* Ti-MCM-41 (Al-free), rutile (TiO_2) and TiO_2-deposited-on-silica, were found to be completely inactive in the MPV reduction. The high selectivity (98%) of Ti-Beta towards the interesting thermodynamically less favorable *cis*-alcohol and the non-reactivity of dissolved titanium isopropoxide, indicate that the reaction takes place over framework titanium and not over leached titanium species and further confirms that the titanium site in Ti-Beta has considerable Lewis acidic properties.

In the Oppenauer oxidation (butanone as the oxidant, 85°C) of a mixture of *cis*- and *trans*-4-t-butylcyclohexanol Ti-Beta selectively converted the *cis*-isomer. Compared to Al-Beta less side-products were formed (362).

Ti-Beta was also studied (363) in gas phase MPV reduction at 100°C using 4-methylcyclohexanone as the substrate and various secondary alcohols as reductant. Dehydration to 4-methylcyclohexene is a side reaction here. Conversion and stereoselectivity depended strongly on the structure of the reducing alcohol. The molar ratio reducing alcohol: 4-methylcyclohexanone inside the zeolite appears to be an important variable. For a full account see ref. (363).

9. ZEOLITE-CATALYZED TWO- AND MULTI-STEP SYNTHESES

In the field of organic synthesis increasing attention is being paid to multifunctional catalysis, in which chemical conversions consisting of a number of individual reactions are brought about in the minimum number of steps. Zeolite catalysts enable several catalytic steps to be combined in mutually optimized fashion. Examples which have already been mentioned include

- the epoxidation/ring closure of linalool and other unsaturated alcohols towards cyclic ethers;
- the reaction of caprolactone with ammonia involving addition, ring cleavage and dehydration to yield 5-hexenenitrile;
- the amination/rearrangement of phenol with ammonia leading to 2-methylpyridine;

- the condensation/imination/cyclization/dehydrogenation of acrolein, aldehyde and ammonia to yield 3-alkylpyridines;
- the dehydrogenation/condensation/imination/cyclization/dehydrogenation of ethanol and ammonia or methylamine to yield pyridines;
- the condensation/rearrangement of phenylhydrazine and ketones towards 2,3-alkylated indoles;
- the esterification/alkylation of phenols and α,β-unsaturated carboxylic acids towards coumarins and dihydrocoumarins;
- the esterification/hydroxyalkylation/dehydration of phenols and β-ketoesters towards coumarins;
- the isomerization/hydration of α-pinene towards isoborneol.

Some further examples of such two-step or multistep zeolite-catalyzed conversions will be considered in the following.

9.1. Dehydration and hydrogenation

Zeolites, particularly alumino- and boro-MFI-type zeolites are suitable catalysts (364) for the dehydration of α-hydroxyketones to α,β-unsaturated ketones in high yield. If a hydrogenating component such as Pd or Cu is added, then saturated unsymmetrical ketones can be obtained according to Reaction 65.

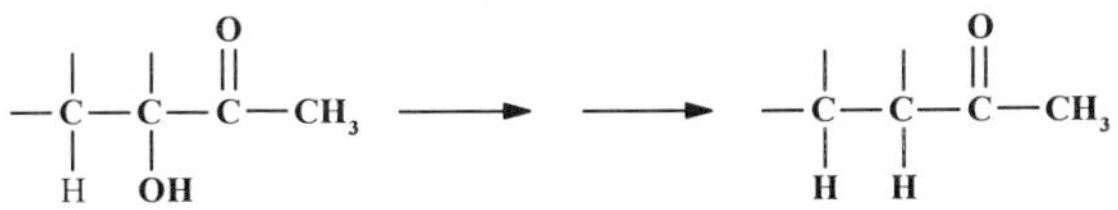

(65)

Thus, using a Ce/Pd-doped B-MFI zeolite (Si/B 94, 2.3 wt % Ce, 0.5 wt % Pd) 3-hydroxy-2-methylbutan-2-one is converted quantitatively at 375°C under H_2 to methyl isopropyl ketone. Under similar conditions non-zeolitic catalysts such as Pd (0.5 wt %) on alumina give relatively poor results.

9.2. Aldol condensation, dehydration and hydrogenation

Another route to α,β-unsaturated ketones is by means of the aldol condensation, which proceeds smoothly on acidic zeolite catalysts (365,366). The aldol condensation of acetone on acidic zeolites yields mesityl oxide, isobutene, phorones, mesitylene and alkylphenols. In the presence of H-ZSM-5 (Si/Al = 34) acetone reacts at 200-250°C in a sealed ampoule under autogeneous pressure to give mesityl oxide with 90% selectivity at 25% conversion (367).

Basic zeolites, such as NaX, CsNaX and NaKL, induce acetone oligomerization/dehydration through a carbanion mechanism which also leads to the dimeric mesityl oxide, the trimer phorone which compound can cyclize to isophorone and to higher oligomers. Martens *et al.* have shown (368) that zeolites give rise to smaller amounts of heavy products than alumina does. The high potential of Na- and Cs-exchanged MCM-41 materials in aldol condensations was reported by Kloetstra *et al.* (369). Thus benzaldehyde entered aldol condensation over Na-MCM-41 in a solvent-free procedure at 150°C with several ketones giving good yields. For instance the benzaldehyde/α-tetralone combination gave the aldol condensation product in 98% selectivity at 60°C conversion.

Under hydrogenating conditions the aldol condensation yields saturated ketones (370-373). Thus acetone reacts in a H_2-stream at 180°C on Pd-ZSM-5 (Si/Al - 24, 0.5 wt % Pd) according to Reaction 66 to give methyl isobutyl ketone with 98% selectivity at 29% conversion (373).

(66)

Under identical reaction conditions but using Pd-Y-zeolite one achieves the same conversion but a selectivity of only 30%. In this case hydrogenation of the acetone to isopropanol occurs to a much greater extent.

9.3. Acetalisation and aldol condensation towards Jasminaldehyde

A new process to α-pentylcinnamaldehyde (Jasminaldehyde), a well-known fragrance, involves acetalisation of heptanal towards the dimethyl acetal followed, in the same pot, by slow hydrolysis and aldol condensation with benzaldehyde (374), see Reaction 67, MCM-41 material in the H-form catalyzes all three consecutive steps and is the catalyst of choice.

First heptanal is refluxed in methanol, at around 80% formation of dimethylacetal, methanol is distilled off in vacuum, benzaldehyde is added and the mixture is heated at the desired temperature. In this way the heptanal concentration is controlled and undesired self condensation is suppressed. At 125°C and conversion 99% a selectivity to Jasminaldehyde of 68% was achieved.

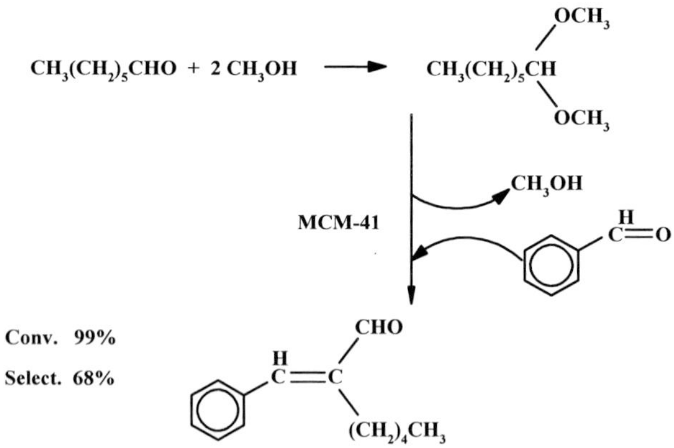

(67)

9.4. Hydroformylation and ketone formation

Rh^{III}-exchanged zeolites NaX and NaY are found (375) to catalyze the reaction of propene, CO and H_2 towards C_7 ketones according to Reaction 68.

(68)

Typical reaction conditions are 150°C, 1 atm, 1.25 s residence time over a fixed catalyst bed containing RhNaY (3.5 wt % Rh). Butyraldehyde and isobutyraldehyde are assumed to be intermediates. A mechanism for the second step is not given. The product ketones, 4-heptanone and 2-methyl-3-hexanone are formed in a ratio 2:1.

For a comparison of homogeneous and zeolite-supported Rh-catalysts in hydroformylation of various C_6-olefins the reader is referred to ref. (376).

9.5. Amidation and dehydration

Aliphatic dinitriles, such as adipodinitrile, n = 4, are prepared from dicarboxylic acids and ammonia (Reaction 69) on a large scale using fluidized and fixed bed processes in the gas phase with heterogeneous catalysts. Amides would seem logical intermediates in this reaction. Selectivity-reducing cyclization to cyclopentanone and cyanocyclopentanonimine occurs in this reaction using conventional catalysts.

$$\mathbf{HOOC(CH_2)_nCOOH} \longrightarrow \longrightarrow \mathbf{NC(CH_2)_nCN} \qquad \mathbf{(69)}$$

These side reactions can be substantially reduced if a B-MFI-zeolite is used (377) which is charged with 5.6 wt % Na and 7.6 wt % P, as the comparison with conventional catalysts based on SiO_2 shows (Table 10). The transition state selectivity of the zeolite catalyst, which does inhibit cyclization, is clearly responsible for this.

Table 10
Adipic acid amidation/dehydration.

Products (mol%)	Catalyst	
	Na/P-B-zeolite[b] (%)	Na/P-SiO_2 (%)
Adiponitrile	94.0	83.4
Cyanovaleric acid	0.7	3.0
Cyanovaleramide	2.8	0.6
useful products	97.5	87.0
Cyanocyclopentanonimine	0.5	2.4
Cyclopentanone	0.9	4.3

[a] Fluidized bed reactor, 400°C, 200 g adipic acid/500 g catalyst. [b] Si/B = 94.

9.6. Hydrolysis and hydrogenation

An interesting example of a single-stage zeolite-catalyzed method for a two-step reaction is the direct conversion of polysaccharides of the glucan type, especially starch, towards D-glucitol (sorbitol). In Reaction 70 this is formulated for the amylose component of starch.

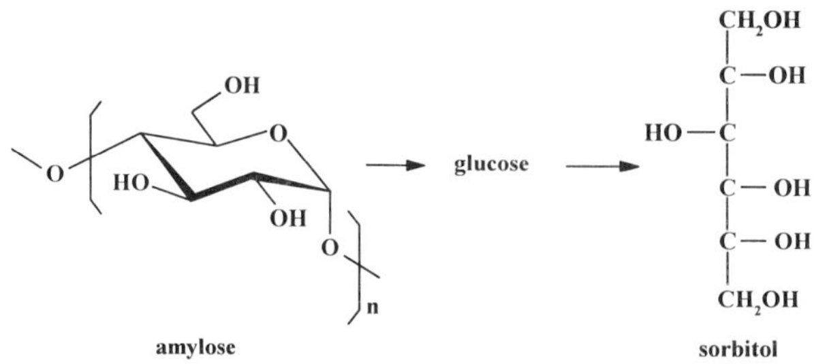

(70)

The usual industrial process requires purification of the intermediate glucose because the enzymatic hydrolysis does not reach completion.

A zeolite-based and recently commercialized process (378) combines hydrolysis and hydrogenation by using Ru-loaded H-USY (3 wt % Ru) as a dual-function catalyst. The outer zeolite surface would seem to provide the Brönsted acidity required for the hydrolysis of the polymeric substrate. Surface roughness and crystal size are expected to be important factors. Pressure surprisingly accelerate hydrolysis as was recently found (379) in the hydrolysis of inulin over H-Beta. The Ru hydrogenation component of the catalyst can exert its action at the inner as well as at the outer surface of the zeolite as the Y pore system is accessible to glucose.

Typical reaction conditions are: 180°C, batch autoclave, 5.5 MPa H_2, starch concentration 30 wt %, Ru/starch wt/wt 0.002. With this formulation a reaction time of 1 h suffices to obtain essentially quantitative conversion. The selectivity to sorbitol is > 95%. Just minor amounts of mannitol and pentitols are formed. The catalyst can be re-used many times.

Similar excellent results are also obtained (378) by combining a 5% Ru-on-carbon catalyst with an acidic zeolite catalyst (H-USY, H-mordenite or H-ZSM-5).

Altogether a fine example of zeolite catalysis demonstrating that zeolites also show promise in the conversion of polymeric materials.

FINAL REMARKS

The rapid development of zeolite catalysis in the synthesis of various types of organic compounds is reflected in the considerable body of recent literature including many patents and patent applications.

As mentioned in the introduction of this chapter the size of the molecules to be converted or synthesized seems limited by the pore dimensions of zeolites. However, a growing number of examples related to outer surface catalysis, or to selective scavenging of side products or impurities or to slow release of reagents illustrates that zeolites can be of much value also in the synthesis or conversion of large molecules or materials. Moreover, the recently discovered super-large pore molecular sieves will enlarge the spectrum of molecules having access to zeolites considerably. For the ordered mesoporous materials essentially no limits exist with respect to reactant and product size. A further recently discovered class is that of the dealuminated zeolites (128a) where the zeolitic layer surface exerts the catalytic action.

A limiting factor sometimes mentioned is the relatively rapid decay of zeolite catalysts due to coke formation. Here many examples exist - as shown in the foregoing - of stabilizing zeolites by adding components or by adapting the reaction mixture, e.g. by adding water or carbon dioxide. Also process conditions - fluidized bed with continuous regeneration - can be applied to overcome this trouble. Moreover and very importantly, zeolites can generally be completely reactivated by calcination.

Advantages - as outlined amongst others in Section 2.6 - are manifold and justify the expectation that a further exponential growth of the use of zeolites and related materials will take place. The present authors are convinced that zeolite catalysis and technology will be future cornerstones of a clean environmentally friendly organic chemicals industry.

1 P.B. Venuto, ChemTech., 215 (1971); P.B. Venuto and P.S. Landis, Adv. Catal., 18 (1968) 259.

2 W.F. Hölderich and H. van Bekkum, in Introduction to Zeolite Science and Practice, Eds. H. van Bekkum, E.M. Flanigen and J.C. Jansen, Elsevier (1991) (Stud. Surf. Sci. Catal. 58), p. 631.

3 P.B. Venuto, Microporous Mater. 2 (1994) 297.

4a A. Corma, Chem. Rev. 97 (1997) 2373.

4b Handbook of Heterogeneous Catalysis, Eds. G. Ertl, H. Knözinger and J. Weitkamp, Wiley-VCH (1997), Vol. 5, various chapters.

4c Catalysis and Zeolites. Fundamentals and Applications, Eds. J. Weitkamp and L. Puppe, Springer-Verlag (1999), several chapters, especially Chapter 6 by P. Espeel, R. Parton, H. Toufar, J. Martens, W.F. Hölderich and P.A. Jacobs, p. 377.

4d A. Corma and H. Garcia, Catal. Today 38 (1997) 257.

4e M.G. Clerici, Topics in Catal. 13 (2000) 373.

5 For a review, see M. Ichikawa, Platinum Metals Rev. 44 (2000) 3.

5a W.M. Meier, D.H. Olsen and Ch. Baerlocher, Atlas of Zeolite Structure Types, 4th revised Ed. (1996), Elsevier.

6a I.E. Maxwell, Adv. Catal., 31 (1982) 1.

6b J.M. Thomas and C.R. Theocharis, in Modern Synthetic Methods (Ed. R. Scheffold), 5 (1989), p. 249.

6c G. Perot and M. Guisnet, J. Mol. Catal., 61 (1990) 173.

7 J. Weitkamp, S. Ernst, H. Dauns and E. Gallei, Chem.-Ing.-Tech., 58 (1986) 623.

8 S.M. Csiscery, ACS Monograph, 171 (1976) 680; Zeolites, 4 (1984) 202.

9 W.O. Haag, R.M. Lago and P.B. Weisz, Faraday Disc., 72 (1982) 317.

10 N.Y. Chen, ACS Symposium Series, 368, 29 (1988) 468.

11 S.M. Csiscery, J. Catal., 108 (1987) 433.

12 P.A. Jacobs and J.A. Martens, Stud. Surf. Sci. Catal., 28 (1986) 23.

13 E.G. Derouane, J. Catal., 100 (1986) 541.

14 C. Herrmann, J. Haas and F. Fetting, Appl. Catal., 35 (1987) 299.

15 L.D. Rollmann and E.W. Valyocsik, US Patent 4.205.052 (1980), Mobil Oil Corp.

16 J.P. Verduyn, Eur. Pat. Appl. 219.354 (1986), Exxon Co.

17 E.F.T. Lee and L.V.C. Rees, Zeolites, 7 (1987) 446, 473.

18 J.H. Lunsford, Rev. Inorg. Chem., 9 (1987) 1.

19 A. Corma, V. Fornes, R.M. Martin-Aranda, H. Garcia and J. Primo, Appl. Catal., 59 (1990) 237.

20 P.E. Hathaway and M.E. Davis, J. Catal., 116 (1989) 263.

21 L.R.M. Martens, P.J. Grobet, W.J.M. Vermeiren and P.A. Jacobs, Stud. Surf. Sci. Catal., 28 (1986) 935; 31 (1987) 531.

21a H. Hattori, Chem. Rev. 95 (1995) 537.

21b D. Barthomeuf, Catal. Rev.-Eng. Sci. 38 (1996) 521.

21c K.R. Kloetstra, M. van Laren and H. van Bekkum, J. Chem. Soc., Faraday Trans. 93 (1997) 1211.
21d *e.g.* A. Derrien, G. Renard and D. Brunel, Stud. Surf. Sci. Catal. 117 (1998) 445.
21e D.J. Macquarrie, J. Chem. Soc. Chem. Commun. 1996, 1961.
22 S.T. Homeyer and W.M.H. Sachtler, J. Catal., 118 (1989) 266; Stud. Surf. Sci. Catal., 49 (1989) 975, and references cited herein.
22a E.J. Creyghton, A.C.T. van Duin, J.C. Jansen, P.J. Kooyman, H.W. Zandbergen and H. van Bekkum, J. Chem. Soc. Faraday Trans. 92 (1996) 4609.
23 W.R. Hastings, C.J. Cameron, M.J. Thomas and M.C. Baird, Inorg. Chem., 27 (1988) 3024.
23a N. Herron, G.D. Stucky and C.A. Tolman, J. Chem. Soc. Chem. Commun. 1986, 1521.
24 H.K. Beyer, I.M. Belenykaja, F. Hange, M. Tielen, P.J. Grobet and P.A. Jacobs, J. Chem. Soc. Far. Trans. I, 81 (1985) 2889; G.W. Skeels and D.W. Breck, in Proc. 6th Int. Zeolite Conference, Reno, Butterworth (1983) p. 87.
25 L.V.C. Rees and E.F.T. Lee, PCT, WO 88/01254 (1988).
26 T. Hibino, M. Niwa, Y. Murakami, M. Sano, S. Komai and T. Hanaichi, J. Phys. Chem., 93 (1988) 7847.
26a A. Thys, G. Peeters, E.F. Vansant and I. Verhaert, J. Chem. Soc. Faraday Trans. I, 79 (1983) 2821, 2835.
27 P.G. Rodewald, US Patent 4.477.583 (1984), Mobil Oil Corp.
28 G.H. Kuehl, US Patent 4.482.531 (1984), Mobil Oil Corp.
29 E.W. Valyocsik, US Patent 4.490.342 (1984), Mobil Oil Corp.
30 J.R. Anderson, K. Foger, T. Male, R.A. Rajadhyaksha and J.V. Sanders, J. Catal., 58 (1979) 114.
31 W.O. Haag and D.H. Olson, US Patent 4.117.026 (1978), Mobil Oil Corp.
32 J.A. Martens, M. Tielen, P.A. Jacobs and J. Weitkamp, Zeolites, 4 (1984) 98.
32a W.F. Hölderich, L. Riekert, M. Kotter and U. Hammon, Ger. Offen DE 3.231.498 (1984), BASF AG.
33 For an example see: Th.M. Wortel, W. van den Heuvel, N.A. de Munck and H. van Bekkum, Acta Phys. Chem. (Szeged), 24 (1978) 341.
34 For a review see: D.M. Ruthven, Chem. Eng. Progr., (2) (1988) 42.
35 D.R. Burfield, G.H. Gan and R.H. Smithers, J. Appl. Chem. Biotechnol., 28 1978) 23. See for drying of other solvents by zeolites: D.R. Burfield et al., J. Org. Chem., 42 (1977) 3060; 43 (1978) 3966; 46 (1981) 629.
35a Leaflet of Mitsui Engineering & Shipbuilding Cy. Tokyo (1999).
36 Y. Gao, R.M. Hansen, J.M. Klunder, H. Masamune and K.B. Sharpless, J. Am. Chem. Soc., 109 (1987) 5765.
37 K. Mikami, M. Terada and T. Nakai, J. Am. Chem. Soc., 111 (1989) 1940.
38 Th.M. Wortel and H. van Bekkum, J. Org. Chem., 45 (1980) 4763.
39 For a review on bulk separations see: R.V. Jasra and S.G.T. Bhat, Separ. Sc. and Technol., 23 (1988) 945.
39a A.J. Hoefnagel, J.H.A.J. Nunnink, A. van Veen, P.E. Verkade and B.M. Wepster, Recl. Trav. Chim. Pays-Bas 88 (1969) 386.
40 For a recent review on UOP's separation processes, see J.D. Sherman, Proc. Natl. Acad. Sci. USA, 96 (1999) 3471.
41 J.D. Sherman and C.C. Chao, in Proc. 7th Int. Zeolite Conference, Kodansha (1986), p. 1025.

42 M.C. Heslinga, H. Pluim and H. van Bekkum, in "Zeolites for the nineties", 8th Int. Zeolite Conference (1989), p. 347.
43 T.P.J. Izod, Eur. Pat. Appl. 13.451 (1980), Union Carbide Corp.
44 M. Floor, A.P.G. Kieboom and H. van Bekkum, Recl. Trav. Chim. Pays-Bas, 108 (1989) 128.
45 C.A.A. van Boeckel, T. Beetz, A.C. Kock-van Dalen and H. van Bekkum, Recl. Trav. Chim. Pays-Bas, 106 (1987) 596.
46 M. Onaka and Y. Izumi, Chem. Lett., (1984) 2007.
47 K. Smith, Bull. Soc. Chim. France, (1989) 272; K. Smith, M. Butters and B. Nay, Synthesis (1985) 1157.
48 N.J. Turro, J.R. Fehlner, D.. Hessler, K.M. Welsh, W. Ruderman, D. Fernberg and A.M. Braun, J. Org. Chem., 53 (1988) 3731.
48a B. Smit and T.L.M. Maesen, Nature 374 (1995) 42.
49 R.M. Dessau, J. Chem. Soc. Chem. Commun., (1986) 1167.
50 Th.M. Wortel, W.H. Esser, G. van Minnen-Pathuis, R. Taal, D.P. Roelofsen and H. van Bekkum, Recl. Trav. Chim. Pays-Bas, 96 (1977) 44.
51 R. Huang, Q. Chen and Z. Yu, C.N. 85.102.395, Chem. Abstr., 108 (1988) 133817.
52 D.P. Roelofsen and H. van Bekkum, Recl. Trav. Chim. Pays-Bas, 91 (1972)605.
53 M.J.J. Litjens, A.J.J. Straathof, J.A. Jongejan and J.J. Heijnen, Tetrahedron 55 (1999) 12411.
54 H. Pines, "The Chemistry of Catalytic Hydrocarbon Conversions", Academic Press (1981) 123.
55 K. Tanabe, M. Misono, Y. Ono and H. Hattori, Stud. Surf. Sci. Catal., 51 1989) 215.
56 *E.g.* L.R. Martens, P.J. Grobet, W.J. Vermeiren and P.A. Jacobs, Stud. Surf. Sci. Catal., 28 (1986) 935.
57 W. Hölderich, A. Lindner, W.D. Mross, M. Schwarzmann, K. Volkammer and U. Wagner, Eur. Pat. 129.899 (1985), BASF AG.
58 A. Lindner, U. Wagner, K. Volkamer, W. Hölderich, W. Hochstein and W. Immel, Eur. Pat. 129.000 (1985), BASF AG.
59 W. Hölderich, A. Lindner, W.D. Mross and M. Strohmeyer, DBP 3.427.979 1986), BASF AG.
60 D.L. Sikkenga, US Pat. 4.503.282 (1985), Standard Oil Co.
61 R. Fischer, W. Hölderich, W.D. Mross and H.M. Weitz, Eur. Pat. 0.167.021(1986), BASF AG.
62 H.W. Schneider and R. Kummer, DEP 3.317.163 (1984), BASF AG.
63 W. Hölderich, R. Fischer, K.D. Malsch and H. Lendle, DEP 3.521.380(1986), BASF AG.
64 W. Hölderich, W. Richter, H. Lendle and K.D. Malsch, DEP 3.521.381(1986), BASF AG.
65 W. Hölderich, H. Aichinger, F. N„umann and R. Fischer, DEP 3.638.011(1988), BASF AG.
66 W. Hölderich, R. Fischer, W.D. Mross and H.M. Weitz, Eur. Pat. 0.132.737 (1985), BASF AG.
67 M. Stoecker, P. Hemmersbach, H. Raeder and J.K. Grepstad, Appl. Catal., 25 (1986) 223.
68 H.W. Kouwenhoven, Adv. Chem. Ser., 121 (1973) 529.
69 J. Weitkamp, W. Gerhard and P.A. Jacobs, Acta Phys. Chem. (Szeged),(1985) 261.
70 K. Honna, M. Sugimoto, N. Shimizu and K. Kurisaki, Chem. Lett., (1986)315.

71 G.C. Lau and W.F. Maier, Langmuir, 3 (1987) 164.
72 M. Nomura and Y. Fujihara, Nippon Nogei Kagaku Kaishi, 57 (1983) 1227; CA, 101 (1984) 7440f.
73 W. Hölderich, F. Merger, W.D. Mrosz and R. Fischer, Eur. Pat. 0.162.387(1985), BASF AG.
74 K.G. Ione, L.A. Vostrikova and V.M. Mastikin, J. Mol. Catal., 31 (1985)355.
75 H.C. Linstid and G.S. Koermer, US Pat. 4.537.995 (1985), Celanese Corp.
75a K.R. Kloetstra, J. van den Broek and H. van Bekkum, Catal. Lett. 47 (1997) 235.
76 For a review see: A. Molnar, in Fine Chemicals through Heterogeneous Catalysis, R.A. Sheldon and H. van Bekkum, Eds., Wiley-VCH (2001), p. 232.
77 A. Molnar, S. Bucsi and H. Bartok, Stud. Surf. Sci. Catal., 41 (1982)203.
77a F.A. Jabur, V.J. Penchev and C.P. Bezoukhanova, J. Chem. Soc. Chem. Commun. (1994) 1591.
77b B.-Y. Hsu and S. Cheng, Microporous and Mesoporous Mater. 21 (1998) 505.
78 W. Hölderich, N. Götz, L. Hupfer, R. Kropp, H. Theobald and B. Wolf, DOS 3.546.372 (1987), BASF AG.
78a For a recent review on rearrangement of epoxides, see W.F. Hölderich and U. Barsnick, in Fine Chemicals through Heterogeneous Catalysis, R.A. Sheldon and H. van Bekkum, Eds., Wiley-VCH (2001), p. 217.
79 C. Neri and F. Buonomo, Eur. Pat. 100.117 (1983), Enichem Anic SpA.
80 C. Neri and F. Buonomo, Eur. Pat. 102.097 (1984), Enichem Anic SpA.
81 M. Orisaku and K. Sano, Jap. Pat. 61.112.040 (1986), Mitsubishi Gas Chem.
81a W.F. Hölderich and N. Götz, 9th IZC, Montreal (1992), R. von Ballmoos *et al.*, Eds., Butterworth-Heinemann, Vol. II, p. 309.
82 W. Hölderich, N. Götz, L. Hupfer and H. Lermer, DEP 3.632.529 (1988), BASF AG.
82a N. Götz , W.F. Hölderich and L. Hupfer, US Pat. 4.968.831 (1988), BASF.
82b W. Spiegler, W.F. Hölderich, N. Goetz , L. Huper and J. Wild, US Pat. 4.824.973 (1986), BASF.
82c W.F. Hölderich, in Acid-Base Catalysis, K. Tanabe, H. Hattori, T. Yamagchi and T. Tanaka, Eds., Kodansha-VCH (1989) 1.
82d C. Meyer, W. Laufer and W.F. Hölerich, Catal. Lett. 53 (1998) 131.
82e R.A. Sheldon, J.A. Elings, S.K. Lee, H.E.B. Lempers and R.S. Downing, J. Mol. Catal. A 134 (1998) 129.
83 A. Liebens, C. Mahaim and W.F. Hölderich, Stud. Surf. Sci. Catal. 108 (1997) 587.
84 P.J. Kunkeler, J.C. van der Waal, J. Bremmer, B.J. Zuurdeeg, R.S. Downing and H. van Bekkum, Catal. Lett. 53 (1998) 135.
84a G. Carr, G. Dosanjh, A.P. Millar and D. Whittaker, J. Chem. Soc. Perkin Trans 2 (1994) 1419.
85 W. Hölderich, F. Merger and R. Fischer, Eur. Pat. 199.210 (1986), BASF AG.
86 Z. Ore, Z. Chen and O. Jiang, Kexue Tongbao, 32 (1987) 462.
87 W. Hölderich, R. Fischer and W. Mesch, DOS 3.636.430 (1987), BASF AG.
88 H.G. Schmidt, Eur. Pat. 75.680 (1986), Dynamit Nobel AG.
89 L.X. Dai, K. Koyama, M. Miyamoto and T. Tatsumi, Appl. Catal. A: General 189 (1999) 237.
89a L-X. Dai, K. Koyama and T. Tatsumi, Catal. Lett. 53 (1998) 211.
90 P.S. Landis and P.B. Venuto, J. Catal., 5 (1966) 245.
91 H. Sato, N. Ishii, K. Hirose and S. Nakamura, Stud. Surf. Sci. Catal., 28(1986) 755.

92 H. Sato, K. Hirose, N. Ihii and Y. Umada, Eur. Pat. 234.088 (1987), Sumitomo Chem. Co.

93 H. Sato, K. Hirose, M. Kitamura, H. Tojima and N. Ishii, Eur. Pat. 236.092 (1987), Sumitomo Chem. Co.

93a A. Thangaraj, S. Sivashankar and P. Ratnasamy, J. Catal. 137 (1992) 252.

93b W.F. Hölderich, J. Röseler, G. Heitmann and A.T. Liebens, Catal. Today 37 (1997) 353 and 38 (1997) 223.

94 K.D. Olson, Eur. Pat. 251.168 (1988), UCC.

94a J. A. Cusumano, ChemTech (1992) 482.

94b T. Tatsumi, in Fine Chemicals through Heterogeneous Catalysis, R.A. Sheldon and H. van Bekkum, Eds., Wiley-VCH (2001), p. 185.

95 C.D. Chang and P.D. Perkins, Zeolites, 3 (1983) 298.

95a A. Tschumper and R. Prins, Appl. Catal. A: General 172 (1998) 285.

95b R. Prins, Catal. Today 37 (1997) 103; and R. Prins, in Fine Chemicals through Heterogeneous Catalysis, R.A. Sheldon and H. van Bekkum, Eds., Wiley-VCH (2001), p. 205.

96 H. LeBlanc, L. Puppe and K. Wedemeyer, DE 3.332.687 (1985), Bayer AG.

96a A.J. Hoefnagel and H. van Bekkum, to be published.

97 V. Solinas, R. Monaci, B. Marongiu and L. Forni, Appl. Catal., 9 (1984)109.

98 M. Neuber, H.G. Karge and J. Weitkamp, Catal. Today, 3 (1988) 11.

99 F.J. Weigert, J. Org. Chem., 51 (1986) 2653.

100 F.J. Weigert, J. Org. Chem., 52 (1987) 3296.

101 K. Eichler, E. Leupold, H.J. Arpe and H. Baltes, DE 3.420.707 (1985), Hoechst AG.

102 R.H. Hardy and H. Burton, J. Catal., 111 (1988) 146.

103 K. Iwayama, Y. Magatani and K. Tada, Jap. Pat. 6.393.739 (1988), Toray Ind.

104 K. Tada, Y. Imada and K. Iwayama, Jap. Pat. 6.393.738 (1988), Toray Ind.

105 K. Iwayama, Y. Magatani and K. Tada, Eur. Pat. Appl. 0.278.729 (1988); Jap. Pat. 01.254.633 (1989), Toray Ind.

106 K. Eichler, H.J. Arpe, H. Baltes and E. Leupold, Eur. Pat. Appl. 164.045 (1985), Hoechst AG.

107 H. Litterer and E.J. Leupold, DBP 3.537.288 (1987), Hoechst AG.

108 U. Dettmeyer, K. Eichler, K. K hlein, E.J. Leupold and H. Litterer, Angew. Chem., 99 (1987) 470; Angew. Chem. Int. Ed., 26 (1987) 468.

109 M. Guisnet and G. Perot, in Fine Chemicals through Heterogeneous Catalysis, R.A. Sheldon and H. van Bekkum, Eds., Wiley-VCH (2001), p. 211.

110 B.B.G. Gupta, US Pat. 4.668.826 (1987), Celanese Corp.

111 J.S. Beck and W.O. Haag, in Handbook of Heterogeneous Catalysis, G. Ertl, H. Knözinger and J. Weitkamp, Eds., Wiley-VCH (1997), Vol. 5, p. 2123.

112 J.A. Horsley, ChemTech, October 1997, p. 45.

113 Hydrocarbon Processing, March 1999, p. 104.

114 G. Bellussi, G. Pazzuconi, C. Pereo, G. Girotti and G. Terzoni, J. Catal. 157 (1995) 227.

115 A. Corma, V. Martinez-Soria and E. Schnoeveld, J. Catal. 192 (2000) 163.

116 F.G. Dwyer, in W.R. Moser (Ed.), Chemical Industries, Vol. 5, Marcel Dekker (1981) p. 29.

117 J. Weitkamp, Acta Phys. Chem., 31 (1985) 271.

118 W.F. Hölderich and E. Gallei, Ger. Chem. Eng., 8 (1985) 337; Chem.-Ing.-Tech., 56 (1984) 908.

119 R.F. Parton, J.M. Jacobs, D.R. Huybrechts and P.A. Jacobs, Stud. Surf. Sci. Catal., 46 (1989) 163.
120 G.R. Meima, G.S. Lee and J.M. Garces, in Fine Chemicals through Heterogeneous Catalysis, R.A. Sheldon and H. van Bekkum, Eds., Wiley-VCH (2001), p. 151.
120a D. Vergani, R. Prins and H.W. Kouwenhoven, Appl. Catal. A: General 163 (1997) 71.
121 S. Sivasankar, A. Thangaraj, R.A. Abdulla and P. Ratnasamy, Stud. Surf. Sci. Catal. 75 (1993) 397.
122 M.W. Anderson and J. Klinowski, J. Am. Chem. Soc., 112 (1990) 10.
123 D. Fraenkel and M. Levy, J. Catal., 118 (1989) 10.
124 K. Eichler and E. Leupold, DBP 3.334.084 (1985), Hoechst AG.
124a R.J. Sampson, Ch.B. Hanson and J.P. Candlin, Eur. Pat. 202.752 (1986), ICI.
124b O. Takahashi, Jap. Pat. 62.29.536 (1987), Idemitsu Kosan Co.
125 D. Fraenkel, M. Cherniavsky, B. Ittah and J. Levy, J. Catal., 101 (1986) 273.
126 M. Neuber and J. Weitkamp, in "Zeolites for the nineties", 8th Int. Zeolite Conference (1989), p. 425.
127 J. Weitkamp, M. Neuber, W. Höltmann and G. Collin, DEOS 37.23.104 (1989), Rütgerswerke AG.
128 E.G. Derouane, J. Catal., 100 (1986) 545.
128a A. Corma, V. Fornes, J. Martinez-Triguero and S.B. Pergher, J. Catal. 186 (1999) 57.
128b J.M. Thomas and K.I. Zamaraev, Angew. Chem. Int. Ed. 106 (1994) 316.
129 K. Tanaguchi, M. Tanaka, K. Takahata, N. Sakamoto, T. Takai, Y. Kurano and M. Ishibashi, Eur. Pat. Appl. 288.582 (1988), Mitsui Petrochem. Ind.
130 G.S. Lee, J.J. Maj, S.C. Rocke and J.M. Garces, Catalysis Letters, 2 (1989) 243.
131 P. Espeel, R. Parton, H. Toufar, J. Martens, W. Hölderich and P.A. Jacobs, in Catalysis and Zeolites. Fundamentals and Applications, J. Weitkamp and L. Puppe, Eds., Springer Verlag (1999) p. 377.
132 T. Yamagashi, T. Idai and E. Takahashi, US Pat. 4.927.979 (1990) Maruzen Petrochemical.
133 W.W. Keading, M.M. Wu, L.B. Young and G.T. Burress, US Pat. 4.197.413 (1980) Mobil Oil Corp.
134 A. Corma, H. Garcia and J. Primo, J. Chem. Research, (1988) 40.
135 L.B. Young, Eur. Pat. 29.333 (1981), Mobil Oil Corp.
135a P.H.J. Espeel, diss. Leuven (1994); P.H.J. Espeel, K.A. Vercruysse, M. Debaerdemaker and P.A. Jacobs, Stud. Surf. Sci. Catal. 84 (1994) 1457.
136 P.Y. Chen, M.C. Chen, H.Y. Chu, N.S. Chang and T.K. Chuang, Stud. Surf. Sci. Catal., 28 (1986) 739.
137 K.G. Ione and O.V. Kikhtyanin, Stud. Surf. Sci. Catal., 49 (1989) 1073.
138 W.F. Burgoyne, D.D. Dixon and J.P. Casey, ChemTech, (1989) 690, and references cited therein. W.F. Burgoyne and D.D. Dixon, Appl. Catal., 62 (1990) 161 and 63 (1990) 117.
138a W. Keim, in Ullmann's Encylopedia of Industrial Chemistry, 5th Ed., VCH, Weinheim (1985), Vol. A1, p. 197.
139 R. Agrawal, S. Auvil and M. Deeba, Eur. Pat. 240.018 (1987), Air Prod. Chem. Inc.
140 R. Pierantozzi, Eur. Pat. 245.797 (1987), Air Prod. Chem. Inc.
141 K. Eichler and E. Leupold, Eur. Pat. 175.969 (1986), Hoechst AG.
142 V. Solinas, R. Monaci, G. Longu and L. Forni, Acta Phys. Chem., 31 (1985) 291.
143 H. Kashiwagi, Y. Fujiki and S. Enomoto, Chem. Pharm. Bull., 30 (1982) 404, 2575.

144 B. Chiche, A. Finiels, C. Gauthier, P. Geneste, J. Graille and D. Pioch, J. Org. Chem., 51 (1986) 2128.
145 B. Chiche, A. Finiels, C. Gauthier, P. Geneste, J. Graille and D. Pioch, J. Mol. Catal., 42 (1987) 229.
146 B.B.G. Gupta, Eur. Pat. 239.383 (1987), Celanese Corp.
147 M. Morimoto *et al.*, Jap. Pat. 97.151.154, 97.151.155 and 97.176.081, Toray Ind.
148 A.P. Singh and A.K. Pandey, J. Mol. Catal. A: Chemical 123 (1997) 141.
149 P. Andy, J. Garcia-Martinez, G. Lee, H. Gonzalez, C.W. Jones and M.E. Davis, J. Catal. 192 (2000) 215.
149a A. Vogt and A. Pfenninger, EP 0701987 A1 (1996) Uetikon AG.
150 Jap. Pat. 81.142.233 (1981), Mitsui Toatsu Chemicals Inc.
151 H.W. Kouwenhoven and H. van Bekkum, in Handbook of Heterogeneous Catalysis, G. Ertl, H. Knözinger and J. Weitkamp, Eds., Wiley-Vch (1997), Vol. 5, p. 2358.
151a P. Metivier, In Fine Chemicals through Heterogeneous Catalysis, R.A. Sheldon and H. vanBekkum, Eds., Wiley-VCH (2001), p. 161.
152 A. Corma, M.J. Climent, H. Garcia and J. Primo, Appl. Catal., 49 (1989) 109.
153 G. Harvey, A. Vogt, H.W. Kouwenhoven and R. Prins, Proc. 9th Int. Zeolite Conf., Vol. II, R. von Ballmoos, J.B. Higgins and M.M.J. Treacy, Eds., Butterworth, Boston (1992), p. 363.
154 M. Spagnol, L. Gilbert and D. Alby, in The Roots of Organic Development, J-R. Desmurs and S. Ratton, Eds., Industrial Chemistry Libray, Elsevier (1996)Vol. 8, p. 29; M. Spagnol, L. Gilbert, E. Benazzi and C. Marcilly, WO 96/35655 (1996) Rhodia.
154a E.G. Derouane, C.J. Dillon, D. Bethell and S.B. Derouane-Abd Hamid, J. Catal. 187 (1999) 209.
154b D. Rohan, C. Canaff, E. Fromentin and M. Guisnet, J. Catal. 177 (1998) 296.
155 E.A. Gunnewegh, S.S. Gopie and H. van Bekkum, J. Mol. Catal. A 106 (1996) 151.
156 H.K. Heinichen and W.F. Hölderich, J. Catal. 185 (1999) 408.
157 E.J. Creyghton, J.A. Elings, R.A. Sheldon and H. van Bekkum, Microporous Mater. 5 (1996) 299.
158 P.J. Kunkeler, D. Moeskops and H. van Bekkum, Microporous Mater. 11 (1997) 313.
159 R. Weber, K.P. Moller and C.T. O'Connor, Microporous Mesoporous Mater. 35-36 (2000) 533.
159a P. Botella, A. Corma and G. Sastre, J. Catal. 197 (2001) 81.
160 W.F. Hölderich, H. Lermer and M. Schwarzmann, DOS 3.618.964 (1987), BASF AG.
161 Y. Isaev and J.J. Fripiat, J. Catal. 182 (1999) 257.
162 F. Richard, H. Carreyre and G. Perot, J. Catal. 159 (1996) 427.
163 H. Lermer, W.F. Hölderich, T. Dockner and H. Koehler, DEOS 3.724.035 (1987), BASF AG.
164 K. Schofield, Aromatic Nitration, Cambridge University Press, London (1980).
165 L.V. Malysheva, E.A. Paukshtis and K.G. Ione, Catal. Rev.-Sci. Eng. 37 (1995) 179.
166 R.L. Adkins, in Kirk-Othmer's Encyclopedia of Chemical Technology, 4th Ed., Vol. 17 (1996) p. 133.
167 L.E. Bertea, H.W. Kouwenhoven and R. Prins, Stud. Surf. Sci. Catal. 84 (1994) 1973.
168 B.M. Choudary, M. Sateesh, M. Lakshmi Kantam, K. Koteswara Rao, K.V. Ram Prasad, K.V. Raghavan and J.A.R.P. Sarma, Chem. Commun. (2000) 25.
169 A. Kogelbauer and H.W. Kouwenhoven, in Fine Chemicals through Heterogeneous

Catalysis, R.A. Sheldon and H. van Bekkum, Eds., Wiley-VCH (2001), p. 123.
170 K. Smith, A. Musson and G.A. DeBoos, J. Org. Chem. 63 (1998) 8448.
171 P. Ratnasamy, A.P. Singh and S. Sharma, in Handbook of Heterogeneous Catalysis, G. Ertl, H. Knözinger and J. Weitkamp, Eds., Wiley-VCH (1997), Vol. 5, p. 2348.
172 P. Ratnasamy and A.P. Singh, in Fine Chemicals through Heterogeneous Catalysis, R.A. Sheldon and H. van Bekkum, Eds., Wiley-VCH (2001), p. 133.
173 T. Huizinga, J.J.F. Scholten, Th.M. Wortel and H. van Bekkum, Tetrahedron Lett., (1980) 3809.
174 J. Pardillos, D. Brunel, B. Coq, P. Massiani, L.C. de Ménorval and F. Figueras, J. Am. Chem. Soc., 112 (1990) 1313.
175 Y. Higuchi and T. Suzuki, Eur. Pat. 112.722 (1984), Ihara Chem. Ind.
176 K. Sekizawa, T. Hironaka and Y. Tsutsumi, Eur. Pat. 171.256 (1986), Toyo Soda Mfg.
177 T. Miyake, K. Sekizawa, T. Hironaka and Y. Tsutsumi, US Pat. 4.861.929 (1989), Tosoh Corp.
178 K. Sekizawa, T. Miyake, T. Hironaka and Y. Tsutsumi, US Pat. 4.849.560 (1989), Toyo Soda Mfg.
179 T. Suzuki and Y. Higuchi, US Pat. 4.822.933 (1989), Ihara Chem. Ind.
180 Y. Higuchi and T. Suzuki, Eur. Pat. 118.851 (1984), Ihara Chem. Ind.
181 T. Miyake, K. Sekizawa, T. Hironaka, M. Nakano, S. Fujii and Y. Tsutsumi, Stud. Surf. Sci. Catal., 28 (1986) 747.
182 T. Hironaka, K. Sekizawa and Y. Tsutsumi, DEOS 32.20.391 (1987), Toyo Soda Mfg.
183 T. Suzuki and C. Komatsu, Eur. Pat. 154.236 (1985), Ihara Chem. Ind.
184 A. Botta, H.J. Buysch and L. Puppe, DEOS 38.09.258 (1989), Bayer AG.
185 A.P. Singh, S.B. Kumar, A. Paul and A. Raj, J. Catal. 147 (1994) 360.
186 S.B. Kumar and A.P. Singh, J. Catal. 150 (1994) 430.
187 A.P. Singh, S. Sharma and S.M. Kale, J. Mol. Catal. A: Chemical 127 (1997) 101.
188 J.C. Oudejans, A.C. Kock-van Dalen and H. van Bekkum, unpublished results.
189 Th.M. Wortel, D. Oudijn, C.J. Vleugel, D.P. Roelofsen and H. van Bekkum, J. Catal., 60 (1979) 110.
190 F. de la Vega and Y. Sasson, Zeolites, 9 (1989) 418.
191 F. de la Vega and Y. Sasson, J. Chem. Soc., Chem. Commun., (1989) 653.
192 R.P. Cooney and P. Tsai, J. Raman Spectrosc., 8 (1979) 195, 236.
193 S. Kai, Jap. Pat. 60.224.644 (1985), Asahi Chem. Ind.
194 S. Kai, Jap. Pat. 60.224.645 (1985), Asahi Chem. Ind.
195 G. Paparatto and M. Saetti, Eur. Pat. 181.790 and 183.579 (1986), Montedipe SpA.
196 G.C. Tustin and M. Rule, US Pat. 4.778.939 (1988), Eastman Kodak Co.
197 M. Rule, G.C. Tustin, D.W. Lane and T.H. Larkins, US Pat. 4.778.940 (1988), Eastman Kodak Co.
198 P.B. Venuto and P.S. Landis, J. Catal. 6 (1966) 237.
199 Y. Kiso, T. Takai and T. Hayashi, Eur. Pat. 329.367 (1989), Mitsui Petrochem. Ind.
200 K.D. Olson, Eur. Pat. 251.169 (1987), Union Carbide Corp.
201 M.J. Climent, A. Corma, H. Garcia, S. Iborra and J. Primo, Appl. Catal. A: General 130 (1995) 5; J. Catal. 130 (1991) 138; Appl. Catal. 51 (1989) 113.
202 M.H.W. Burgers and H. van Bekkum, Stud. Surf. Sci. Catal. 78 (1993) 567.
203 T. Kiyora, Jap. Pat. 63.307.834 and Jap. Pat. 63.307.835 (1988), Mitsui Toatsu Chem. Inc.
204 H. Ishida and Y. Takamatsu, Jap. Pat. 04275246 (1992) Asahi Chemical Cy.
205 C. Moreau, A. Finiels, S. Razigade, L. Gilbert and F. Fajula, WO 9637452 (1996)

Rhodia.
206 C. Moreau, R. Jacquot and M. Spagnol, in Catalysis in Organic Reactions, Marcel Dekker, Inc., New York (1998) p. 51.
207 J. Lecomte, A. Finiels, P. Geneste and C. Moreau, J. Mol. Catal. A: Chemical 140 (1999) 147.
208 B.B.G. Gupta, US Pat. 4.694.111 (1987), Celanese Corp.
209 X. Wu and R.G. Anthony, J. Catal. 184 (1999) 294.
210 P. Metivier, in Fine Chemicals through Heterogeneous Catalysis, R.A. Sheldon and H. van Bekkum, Eds., Wiley-VCH (2001), p. 173.
210a A. Brait and M.E. Davis, Appl. Catal. A: General 204 (2000) 117.
211 C.D. Chang and A.J. Silvestri, ChemTech, (1987) 624 and references cited therein.
212 See *e.g.* a. J.C. Oudejans, P.F. van den Oosterkamp and H. van Bekkum, Appl. Catal., 3 (1982) 109; b. G.A. Aldridge, X.E. Veryklos and R. Mutharasan, Ind. Eng. Chem. Process Des. Dev., 23 (1984) 733.
213 P.A. de Boks, T.F. Huber, W.J. van Nes, P.F. van den Oosterkamp, H. van Bekkum, G.C. van Eybergen, J.H. van den Hende, N.W.F. Kossen, C.I. Mooring, J.C. Oudejans, A.C. Snaterse and J.A. Wesselingh, Biotechnology Lett., 4 (1982) 447.
214a H. Kitagawa, Y. Sendoda and Y. Ono, J. Catal., 101 (1986) 12.
214b N.S. Gnep, J.Y. Doyemet and M. Guisnet, Stud. Surf. Sci. Catal., 46 (1989) 153.
214c L. Petit, J.P. Boursonville and F. Raatz, Stud. Surf. Sci. Catal., 49 (1989) 1163.
214d T. Inui, Y. Ishihara, K. Kamachi and H. Matsuda, ibid., 1183.
214e P. Meriaudeau, G. Sapaly and C. Nacchae, *ibid.*, 1423.
214f C.R. Bayense and J.H.C. van Hooff, Recent Res. Reports, 8th Int. Zeol. Conf. Amsterdam, 1989, p. 181.
215 For a recent review on Ga-containing zeolites see: R. Fricke, H. Kosslick, G. Lischke and M. Richter, Chem. Rev. 100 (2000) 2303.
216 T. Mole, J.R. Anderson and G. Greer, Appl. Catal., 17 (1985) 141.
217 a. J.R. Bernard, in L.V.C. Rees (ed.), Proc. 5th Int. Zeol. Conf., Naples, Heyden, London, 1980, p. 686; b. T.R. Hughes, W.C. Buss, P.W. Tamm and R.L. Jacobson, in Proc. 7th Int. Zeol. Conf., Tokyo, Kodansha-Elsevier, 1986, p. 725.
218 E.G. Derouane and D.J. Vanderveken, Appl. Catal. 45 (1988) L15.
219 I.E. Maxwell, R.S. Downing, J.J. de Boer and S.A. van Langen, J. Catal., 61 (1980) 485, 493.
220 J. Ipaktschi, Z. Naturforsch., 416 (1986) 496.
221 P.Laszlo and J. Lucchetti, Tetrahedron Lett., (1984) 4387.
222 D. Hochgraeber-Paetow and H. Lechert, Proceedings 9th IZC, R. von Ballmoos *et al.*, Eds. (1993), p. 483.
223 D. Genske, K. Bornholdt and H. Lechert, Stud. Surf. Sci. Catal. 117 (1998) 421.
224 S. Ghosh and N.L. Bauld, J. Catal., 95 (1985) 300.
225 A.J. Schipperijn and J. Lukas, Recl. Trav. Chim. Pays-Bas, 92 (1973) 572.
226 J.C. Oudejans, J. Kaminska and H. van Bekkum, Recl. Trav. Chim. Pays-Bas, 102 (1983) 537.
227 M. Onaka, H. Kita and Y. Izumi, Chem. Lett., (1985) 1895.
228 J.C. Oudejans, J. Kaminska, A.C. Kock-van Dalen and H. van Bekkum, Recl. Trav. Chim. Pays-Bas, 105 (1986) 421.
229 M.J. Alcon, A. Corma, M. Iglesias and F. Sanchez, J. Mol. Catal. A: Chemical 144 (1999) 337.
229a K. Ganesan and C.N. Pillai, J. Catal., 119 (1990) 8.

230 For an early review, see Y. Ono, Heterocycles, 16 (1981) 1755.
231 D.J. Sikkema, P. Hoogland, J. Bik and P. Lam, Polymer, 27 (1986) 1441.
232 W.F. Hölderich and M. Schwarzmann, Eur. Pat. 309.911 (1989), BASF AG.
233 W.F. Hölderich and M. Schwarzmann, Eur. Pat. 309.906 (1989), BASF AG.
234 M. Kurszewska, E. Skorupa, R. Kasprzykowska, P. Sowinski and A. Wisniewski, Carbohydr. Res. 326 (2000) 241.
235 R.M. Dessau, Zeolites, 10 (1990) 205; US Pat. 4.658.045 (1987), Mobil Oil Corp.
236 A. Corma, M. Iglesias and F. Sanchez, Chem. Commun. (1995) 1635.
237 A. Bhaumik and T. Tatsumi, Chem. Commun. (1998) 463.
238 A.J. Hoefnagel, E.A. Gunnewegh, R.S. Downing and H. van Bekkum, Chem. Commun. (1995) 225.
239 E.A. Gunnewegh, A.J. Hoefnagel, R.S. Downing and H. van Bekkum, Recl. Trav. Chim. Pays-Bas 115 (1996) 226.
240 W.F. Hölderich, Stud. Surf. Sci. Catal., 46 (1989) 193.
240a C.H. McAteer and E.F.V. Scriven, in Fine Chemicals through Heterogeneous Catalysis, R.A. Sheldon and H. van Bekkum, Eds., Wiley-VCH (2001), p. 275.
241 Y. Ono, K. Hatada, K. Fujita, A. Halgeri and T. Klii, J. Catal., 41 (1976) 322.
242 Y. Ono, A. Halgeri, M. Kanedo, K. Hatada, ACS Symp. Ser., 40 (1977) 596.
243 K. Hatada and Y. Ono, Bull. Chem. Soc. Japan, 50 (1977) 2517.
244 W.F. Hölderich and M. Schwarzmann, Eur. Pat. 267.438 (1988), BASF AG.
245 D. Feitler, W. Schimming and H. Wetstein, US Pat. 4.675.410 (1987), Nepera Inc.
246 W.F. Hölderich, N. Götz and G. Fouquet, Eur. Pat. 263.464 (1988), BASF AG.
247 C.D. Chang and W.H. Lang, US Pat. 4.220.783 (1980) Mobil Oil Corp.
248 H. Sato, S. Shimizu, N. Abe and K. Hirose, Stud. Surf. Sci. Catal. 84 (1994) 1951.
249 P.J. Angevine, C.T-W. Chu and T.C. Potter, US Pat. 5.395.940 (1995), Mobil Oil Corp.
250 C.H. McAteer, D.C. Brown and R.D. Davis, US Pat. 5.780.635 (1998) Reilly Industries.
251 A.V. Rama Rao, S.J. Kulkarni, R. Ramachandra Rao and M. Subrahmanyam, Appl. Catal. 111 (1994) L101.
252 Y. Matsumura, K. Hashimoto and S. Yoshida, J. Catal., 100 (1986) 392.
253 F.J. van der Gaag, R.J.O. Adriaansens, H. van Bekkum and P.C. van Geem, Stud. Surf. Sci. Catal., 52 (1989) 283.
254 S.J. Kulkarni, R. Ramachandra Rao, Y.V. Subba Rao, M. Subrahmanyam and A.V. Rama Rao, Appl. Catal. 136 (1996) L1.
255 C.H. McAteer, R.D. Davis and J.R.Calvin, US Pat. 5.700.942 (1992) Reilly Industries.
256 F.J. van der Gaag, F. Louter, J.C. Oudejans and H. van Bekkum, Appl. Catal., 26 (1986) 191.
257 F.J. van der Gaag, F. Louter and H. van Bekkum, Proc. 7th Int. Zeol. Conf., Tokyo, Kodansha, Elsevier (1986), p. 763.
258 R.A. le Fêbre, A.J. Hoefnagel and H. van Bekkum, Recl. Trav. Chim. Pays-Bas 115 (1996) 511.
259 B.S. Besty, F.P. Gortsema, G.T. Wildman and J.J. Sharkey, US Pat. 5.231.187 (1993) Merck & Co.
260 F.P. Gortsema, B. Besty, J.J. Friedman, D. Matsumoto, J.J. Sharkey, G. Wildman, T.J. Blacklock and S.H. Pan, in Catalysis of Organic Reactions, J.R. Kosak and T.A. Johnson, Eds., Chem. Industries Series, No. 53, Marcel Dekker (1994), p. 445.

261 R.A. Budnik and M.R. Sandner, Eur. Pat. 158.139 (1985), Union Carbide Corp.
262 H. Sato and M. Tsuzuki, Int. Appl. WO 8.703.592 (1987), Idemitsu Kosan Co.
263 J. March, Advanced Organic Chemistry, 4th Ed., Wiley-Interscience (1992), p. 1141.
264 P.B. Venuto and P.S. Landis, Adv. Catal. 18 (1968) 345.
265 D. Bhattacharya, D.W. Gammon and E. van Steen, Catal. Lett. 61 (1999) 93.
266 M.P. Prochazka, L. Eklund and R. Carlson, Acta Chem. Scand. 44 (1990) 610; M.P. Prochazka, Thesis Umea University, Sweden, 1990.
267 M.S. Rigutto, H.J.A. de Vries, S.R. Magill, A.J. Hoefnagel and H. van Bekkum, Stud. Surf. Sci. Catal. 78 (1993) 66.
268 P.J. Kunkeler, M.S. Rigutto, R.S. Downing, H.J.A. de Vries and H. van Bekkum, Stud. Surf. Sci. Catal. 105 (1997) 1269.
269 R.S. Downing and P.J. Kunkeler, in Fine Chemicals through Heterogeneous Catalysis, R.A. Sheldon and H. van Bekkum, Eds., Wiley-VCH (2001), p. 178.
270 H. Sato and M. Tsuzuki, Jap. Pat. 63.196.561 (1988), Idemitsu Kosan Co.
271 F. Seel, M. von Blon and A. Dessaur, Z. Naturforsch., 37b (1982) 820.
272 T. Mallat and A. Baiker, in Handbook of Heterogeneous Catalysis, Eds. G. Ertl, H. Knözinger and J. Weitkamp, Wiley-VCH, 1997, Vol. 5, p. 2334.
273 F.J. Weigert, J. Catal., 103 (1987) 20.
274 Y. Ashina, T. Fujita, M. Fukatsu, K. Niwa and J. Yagi, Proc. 7th Int. Zeol. Conf., Tokyo, 1986, Kodansha, p. 779.
275 R.D. Shannon, M. Keane, L. Abrams, R.H. Staley, T.E. Gier and G.C. Sonnichsen, J. Catal., 113 (1988) 367; 114 (1988) 8; 115 (1989) 79.
276 H.E. Bergna, M. Keane, D.H. Ralston, G.C. Sonnichsen, L. Abrams and R.D. Shannon, J. Catal., 115 (1989) 148.
277 H.E. Bergna, D.R. Corbin and G.C. Sonnichsen, US Pat. 4.683.334 (1987) and 4.752.596 (1988), DuPont.
278 K. Segawa, A. Sugiyma, M. Sakaguchi and K. Sakurai, in K. Tanabe et al., Eds., Acid-Base Catalysis, Kodansha, 1989.
279 D.R. Corbin, S. Schwarz and G.C. Sonnichsen, Catal. Today 37 (1997) 71.
279a W.S. Fong, SRI-International, PEP Review, no. 89-3-4 (1991).
280 C. Gründling, V.A. Veefkind, G. Eder-Mirth and J.A. Lercher, Stud. Surf. Sci. Catal. 105 (1997) 591.
281 V.A. Veefkind, PhD Thesis, University of Twente (1998).
282 M. Deeba, Eur. Pat. Appl. 311.900 (1989), Air Prod. Chem. Inc.
283 M. Deeba, M.E. Ford and T.A. Johnson, Eur. Pat. 252.424 (1988), Air Prod. Chem. Inc.
284 M. Deeba, M.E. Ford and T.A. Johnson, Stud. Surf. Sci. Catal., 38 (1987) 221.
285 M. Deeba and M.E. Ford, J. Org. Chem., 53 (1988) 4594.
286 M. Deeba, M.E. Ford and T.A. Johnson, J. Chem. Soc. Chem Commun., 8 (1987) 562.
287 J. Peterson and H.S. Fales, US Pat. 4307250 (1981); Eur. Pat. 39918 (1984), Air Prod. Chem. Inc.
288 M.E. Deeba, W.J. Ambs, Eur. Pat. 77016 (1985), Air Prod. Chem. Inc.
289 J. Peterson and H.S. Fales, US Pat. 4375002 (1983), Air Prod. Chem. Inc.
290 W.J. Ambs, M.E. Deeba and J.F. White, Eur. Pat. 10192 (1986), Air Prod. Chem. Inc.
291 V. Taglieber, W. Hölderich, R. Kummer, W.D. Mross and E. Saladin, Eur. Pat. 133938 (1985) and Eur. Pat. 132736 (1985), BASF AG
292 W. Hölderich, V. Taglieber, H. Pohl, R. Kummer and K.G. Baur, DBP 36342747 (1987), BASF AG.

293 M. Lequitte, F. Figueras, C. Moreau and S. Hub, J. Catal. 163 (1996) 255.
294 M. Hesse, W. Hölderich and M. Schwarzmann, Eur. Pat. 296.495 (1988), BASF AG.
295 W. Hölderich, M. Hesse and M. Schwarzmann, Eur. Pat. 297.446 (1989), BASF AG.
296 C.D. Chang and W.H. Lang, US Pat. 4.434.299 (1984), Mobil Oil Corp.
297 C.D. Chang and W.H. Lang, Eur. Pat. 062.542 (1982); C.D. Chang and P.D. Perkins, Eur. Pat. 082.613 (1986), Mobil Oil Corp.
298 K.K. Pitale and R.A. Rajadhyaksha, Curr. Sci., 54 (1985) 447.
299 M.G. Warawdekar and R.A. Rajadhyaksha, Zeolites, 7 (1987) 574.
300 D.G. Jones and P.S. Landis, US Pat. 3.231.600 (1966).
301 K. Hatada, Y. Ono and K. Tominaga, in Adv. Chem. Ser. ACS, 121 (1973) 501.
302 M.H.W. Burgers and H. van Bekkum, J. Catal. 148 (1994) 68.
303 M.H.W. Burgers, PhD Thesis, Delft University of Technology (1995).
304 H. Nakajima, M. Koya, H. Ishida, H. Minoura and Y. Takamatsu, Microporous Mater. 2 (1994) 237.
305 M.H.W. Burgers and H. van Bekkum, Handbook of Heterogeneous Catalysis, Eds. G. Ertl, H. Knözinger and J. Weitkamp, Wiley-VCH, 1997, Vol. 5, p. 2364.
306 M. Onaka, M. Kawai and Y. Izumi, Chem. Lett., (1985) 779.
307 M. Onaka, K. Sugita and Y. Izumi, Chem. Lett., (1986) 1327; and in K. Tanabe et al., Eds., Acid-Base Catalysis, Kodansha, 1989, p. 33.
308 E. Kikuchi, T. Matsuda, K. Shimomura, K. Kawahara and Y. Morita, Stud. Surf. Sci. Catal., 28 (1986) 771.
309 M. Iwamoto, M. Tajima and S. Kagawa, J. Catal., 101 (1986) 195.
310 K. Eguchi, T. Tokiai and H. Arai, Appl. Catal., 34 (1987) 275.
311 M. Misono and T. Inui, Catal. Today 51 (1999) 369.
312 J.C. van der Waal, H. van Bekkum and J.M.Vital, J. Mol. Catal. A: Chemical 105 (1996) 185.
313 H. Valente and J.M. Vital, Stud. Surf. Sci. Catal. 108 (1997) 555.
314 M. Nomura, T. Tada, I. Inoue and Y. Fujihara, Chem. Exp. 8 (1993) 689.
315 P. Botella, A. Corma, J.M. Lopez Nieto, S. Valencia, M.E. Lucas and M. Sergio, J. Catal., in the press.
316 M. Nomura and Y. Fujihara, Kinki Daigaku Kokagubu Kenkyu Hokoku, 19 (1985) 1.
316 P. Chu and G.H. Kuehl, US Pat. 4.605.787 (1986), Mobil Oil Corp.
317 Q. Chen, Zh. Yu and Zh. Liu, CN 85.102.398 (1986), Guangzhu Chem. Int.
317 J.A. Daniels and A. Stewart, Eur. Pat. 055.045 (1986), ICI.
318 M.R. Klotz, US Pat. 4.584.415 (1986), Amoco.
319 K Hensen, C. Mahaim and W.F. Hölderich, Stud. Surf. Sci. Catal. 105 (1997) 1133 and references cited therein.
320 K.R. Kloetstra and H. van Bekkum, J. Chem. Res. (1995) 26.
321 L.B. Young, US Pat. 4.365.084 (1982), Mobil Oil Corp.
322 Jap. Pat. 56.092.833 (1979), Mitsubishi Gas Chem. Ind.
323 S. Haruhito, DBP 3.149.979 (1982), Idemitsu Kosan Co.
324 R. Gregory, D.J. Westlake and J.J. Harry, Eur. Pat. 045.618 (1986), BP Co.
325 J.A. Ballantine, M. Davies, E.M. O'Neil, I. Patel, J.H. Purnell, M. Ravanakorn and K.J. Williams, J. Mol. Catal., 26 (19840 57.
326 F. Näumann, W. Hölderich and M. Schwarzmann, DPB 2.802.980 (1988) BASF AG.
327 R. Huang, Q. Chen and Z. Yu, CN 85.102.395 (1986).
328 M.J. Verhoef, PhD Thesis, Delft University of Technology (2000).
329 R. Koster, PhD Thesis, University of Amsterdam (2001).

330 W.M. van Rhijn, D.E. De Vos, B.F. Sels, W.D. Bossaert and P.A. Jacobs, Chem. Commun. (1998) 317.

331 W.M. van Rhijn, D.E. De Vos, W.D. Bossaert, J. Bullen, B.H. Wouters, P.J. Grobet and P.A. Jacobs, Stud. Surf. Sci. Catal. 117 (1998) 183.

332 J. Diaz, C. Marquez-Alvarez, F. Mohino, J. Perez-Pariente and E. Sastre, J. Catal. 193 (2000) 295.

333 M.J. Verhoef, P.J. Kooyman, J.A. Peters and H. van Bekkum, Microporous Mesoporous Mater. 27 (1999) 365.

334 J. Wang, Y. Li and Y. Zhou, Riyong Huaxue Gongye, 2 (1988) 52.

335 T.B. Grindley and R. Thangarasa, Carbohydr. Res. 172 (1988) 311.

336 L.A. Hamilton, US Pat. 3.328.439 (1967), Mobil Oil Corp.

337 H. Eckhardt, K. Halbritter, W. Hölderich and W.D. Mross, DBP 3.506.632 (1985), BASF AG.

338 D.P. Roelofsen, J.W.M. de Graaf, J.A. Hagendoorn, H.M. Verschoor and H. van Bekkum, Recl. Trav. Chim. Pays-Bas, 89 (1970) 193.

339 T. Ookoshi and M. Onaka, Tetrahedron Lett. 39 (1998) 293.

340 A. Corma, M.J. Climent, H. Garcia and J. Primo, Appl. Catal., 59 (1990) 333.

341 M.J. Climent, A. Corma, S. Iborra, M.C. Navarro and J. Primo, J. Catal. 161 (1996) 783.

342 I. Rodriguez, M.J. Climent, S. Iborra, V. Fornes and A. Corma, J. Catal. 192 (2000) 441.

343 A. Corma, S. Iborra, S. Miquel and J. Primo, J. Catal. 161 (1996) 713.

344 M.A. Camblor, A. Corma, S. Iborra, S. Miquel, J. Primo and S. Valencia, J. Catal. 172 (1997) 76.

345 A.M. van der Heijden, F. van Rantwijk and H. van Bekkum, J. Carbohydr. Chem. 18 (1999) 131.

346 A.T.J.W. de Goede, I.G. van der Leij, A.M. van der Heijden, F. van Rantwijk and H. van Bekkum, PCT Int. Appl. WO 96.136640 (1996).

347 A. Corma and S. Iborra, in Fine Chemicals through Heterogeneous Catalysis, R.A. Sheldon and H. van Bekkum, Eds., Wiley-VCH (2001), p. 257.

348 H.E. Fried, Eur. Pat. 122.654 (1984), Shell Int. Res.

349 D. Kallo, G. Onyestyak and J. Papp, in Proc. 6th Int. Zeol. Conf., Reno, Butterworth, 1984, p. 444.

350 G. Onyestyak, D. Kallo, J. Papp and J.E. Detrekoy, Hung. Pat. 29.972 (1984).

351 G.T. Kerr and P.S. Landis, US Pat. 3.312.751 (1967), Mobil Oil Corp.

352 B. Buchholz, US Pat. 4.102.931 (1978); E.J. Dziera and B. Buchholz, Eur. Pat. 082.289 (1983), Pennwalt Corp.

353 W.F. Hölderich, M. Hesse and E. Sattler, in Proc. 9th Int. Congr. Catal. Calgary, 1988, Vol. 1, p. 316.

354 For a review see: C.F. de Graauw, J.A. Peters, H. van Bekkum and J. Huskens, Synthesis 10 (1994) 1007.

355 J. Shabtai, R. Lazar and E. Biron, J. Mol. Catal. 27 (1984) 35.

356 E.J. Creyghton, S.D. Ganeshie, R.S. Downing and H. van Bekkum, Chem. Commun. (1995) 1859; J. Mol. Catal. A: Chemical 115 (1997) 457.

357 P.J. Kunkeler, B.J. Zuurdeeg, J.C. van der Waal, J.A. van Bokhoven, D.C. Koningsberger and H. van Bekkum, J. Catal. 180 (1998) 234.

358 I. Kiricsi, C. Flego, G. Pazzuconi, W.P. Parker, Jr., R. Millini, C. Perego and G. Bellussi, J. Phys. Chem. 98 (1994) 4627.

359 A. Vimont, F. Thibault-Starzyk and J.C. Lavalley, J. Phys. Chem. 104 (2000) 286.
360 F. Fajula, in " Zeolites, a refined Tool for Designing Catalytic Sites", L. Bonneviot and S. Kaliaguine, Eds., Elsevier, 1995, p. 133.
361 J.A. van Bokhoven, PhD Thesis, Utrecht University (2000); J.A. van Bokhoven, D.C. Koningsberger, H. van Bekkum and A.P.M. Kentgens, J. Am. Chem. Soc. 122 (2000) 12842.
362 J.C. van der Waal, E.J. Creyghton, P.J. Kunkeler, K. Tan and H. van Bekkum, Topics In Catal. 4 (1997) 261.
363 J.C. van der Waal, P.J. Kunkeler, K. Tan and H. van Bekkum, J. Catal. 173 (1998) 74.
364 W.F. Hölderich, K. Schneider and L. Hupfer, DPB 3.632.530 (1988), BASF AG.
365 G.P. Hagen, US Pat. 4.433.174 (1984), Standard Oil Co.
366 Y.I. Isakov and Kh.M. Minachev, Russ. Chem. Rev., 51 (1982) 1188.
367 C.J. Plank, E.J. Rosinski and G.T. Kerr, US Pat. 4.011.278 (1977), Mobil Oil Corp.
368 L.R.M. Martens, W.J. Vermeiren, D.R. Huybrechts, P.J. Grobet and P.A. Jacobs, in Proc. 9th Int. Congr. Catal., Calgary, 1988, Vol. 1, p. 420.
369 K.R. Kloetstra and H. van Bekkum, Chem. Commun. (1995) 1005.
370 Y. Isakov, N.Y. Usachev, T. Isakova and K. Minachev, Izv. Akad. Nauk SSR Ser. Khim., (1985) 1965 and (1986) 299.
371 Y. Isakov, K. Minachev and T. Isakova, Dokl. Akad. Nauk SSR, (1985) 284.
372 G.I. Golodets, N.V. Pavlenko, L.F. Korzhova, K.M. Vaisberg and Y.I. Churkin, Kinet. Katal., 25 (1984) 1015.
373 T.J. Huang and W.O. Haag, US Pat. 4.339.606 (1982) and Eur. Pat. 112.821(1986), Mobil Oil Corp.
374 M.J. Climent, A. Corma, R. Guil-Lopez, S. Iborra and J. Primo, J. Catal. 175 (1998) 70.
375 E. Rode, M.E. Davis and B.E. Hanson, J. Chem. Soc. Chem. Commun., 1985, 716.
376 M.E. Davis, P.M. Butler, J.A. Rossin and B.E. Hanson, J. Mol. Catal., 31 (1985) 385.
377 W. Hölderich, H. Lendle, P. Magnussen, H. Leitner, J.H. Mangegold and W. Leitenberger, Eur. Pat. 196.554 (1986), BASF AG.
378 P.A. Jacobs and H. Hinnekens, Eur. Pat. Appl. 329.923 (1989), Synfina-Olefina S.A.
379 A.M. Heinen, PhD Thesis, Delft University of Technology (2000).

Studies in Surface Science and Catalysis 137
H. van Bekkum, E.M. Flanigen, P.A. Jacobs and J.C. Jansen (Editors)
© 2001 Elsevier Science B.V. All rights reserved.

Chapter 19

Metal Ions Associated to Molecular Sieve Frameworks as Catalytic Sites for Selective Oxidation Reactions

G.Bellussi [a] and M.S. Rigutto [b]
[a] *ENIRICERCHE S.p.A., Via, F. Maritano 26, 20097 San Donato, Milano, Italy*
[b] *Shell Global Solutions International b.v., P.O. Box 38000, 1030 BN Amsterdam, The Netherlands*

1. Introduction

Selective oxidation reactions are applied throughout the chemical industry to produce chemicals ranging from bulk scale oxygenates to pharmaceuticals.[1] For obvious economical and environmental reasons, most bulk scale oxidation processes are catalytic and most employ air or oxygen as the oxidant. Catalytic methods are for the last two decades also gaining ground in fine chemicals production, albeit at a fairly slow pace, because it is often more difficult to justify the significant research efforts needed to find and develop them.

Researchers have started to explore zeolites as selective oxidation catalysts in the late 1960's, not long after their value as acid catalysts in refining and petrochemicals manufacture had started to become well-established.[2] Much of the early research was focused on the use of the low-silica zeolites available at that time (*e.g.* A, X, Y, chabazite, mordenite), impregnated or ion-exchanged with transition metals, as catalysts in (mostly gas-phase) oxidation reactions. With the advent of high silica zeolites in the course of the 1970's came the possibility to synthesise materials containing transition metal ions in their covalent frameworks - which in the early 1980's led to what is still the most succesful example of oxidation catalysis by molecular sieves: titanium silicalite as a catalyst for oxygenations with dilute hydrogen peroxide.

In the following section, we shall very briefly discuss some interesting examples of selective oxidation catalysis with ion-exchanged systems, before going to what is the main theme of this chapter: oxidation catalysis by framework species in molecular sieves. Zeolite-entrapped metal complexes with organic ligands, the so-called "ship-in-a-bottle" catalysts, are the topic of a separate chapter elsewhere in this volume.

1.1. Oxidation catalysis by transition metal ion-exchanged and impregnated zeolites

The composition of zeolites is easily varied by ion exchange or impregnation treatments, and it is thus often relatively straightforward to prepare materials in which transition metal species are well-dispersed, well-mixed and highly accessible to reactants. A vast amount of rather diverse research into oxidation catalysis by such materials has been carried out mostly in the 1970's and early 1980's.[2] Some of the -in our view- most interesting examples are listed in Table 1.

Table 1. Overview of oxidation reactions and peroxide decomposition reactions catalysed by ion-exchanged and impregnated zeolites. Adapted from de Vos et al.[3]

Substrate	Oxidant (+ co-substrate)	Catalyst	Products	Known analogue	Ref.
Ethane	O_2 (+ NH_3)	Co-Beta	Acetonitrile	-	4
Ethene	O_2	CuPd-Y	Acetaldehyde	homog.	5
Propene	O_2	CuPd-Y	Acetone	homog.	5
	O_2 (+ HOAc)	CuPd-Y	Allyl Acetate	heterog.	6
	O_2	Cu-Y	Acrolein	(heterog.)	7
Alkenes	O_2	(Cu^{2+},VO^{2+})-Y	2,3-Epoxyalcohols	homog.	8
Butadiene	O_2	($V_xP_yO_z$) on various zeolites	Maleic Anhydride, Furan	heterog.	9
2,6-Dialkylphenol	*t*-BuOOH	Co-X	Quinones, (Diphenoquinones)	homog. (diff. selectivity)	10
Cumyl hydroperoxide	-	H-ZSM-5	Phenol + acetone	homog.	11
t-Butyl hydroperoxide	-	Co-Y, Ce-Y, La-Y	*t*-butanol	homog.	12
Benzene Fluorobenzene Toluene	N_2O	H-ZSM-5	Phenol Fluorophenol Cresols	-	13

One might say that zeolites, and particularly the ion-exchanged low-silica materials, have a number of properties which do not always seem to make them particularly attractive as (carriers for) oxidation catalysts. They tend to sorb oxygenated products much more strongly than the substrate (usually a hydrocarbon) - a tendency one expects to be less pronounced for other (supported) metal oxides; they can restrict the diffusion of products which are often more sensitive to oxidation than the substrate and thus promote unwanted consecutive reactions; furthermore, di- and trivalent transition metal ions may hydrolyse, which can give rise to (in most cases undesirable) acidity. Indeed, in much of the early work, ion-exchanged low-silica zeolites turned out to be good catalysts for deep oxidation.

Of the more successful examples of zeolite-catalysed oxidation reactions mentioned in Table 1, most can be viewed as heterogeneous analogues of known homogeneous systems, in which the zeolite takes the role of (co-)solvent and weakly coordinating anion at the same time. It appears that zeolites are best at catalysing reactions whose solution phase counterparts are best carried out in polar, acidic media. An instructive and well-documented example is the Wacker oxidation of ethylene to acetaldehyde catalysed by zeolite Y exchanged with copper and palladium ions,[5,14] in which case it has been shown that the mechanism is similar to that of the homogeneous reaction. An advantage is that the zeolite catalyst allows for gas-phase operation and thus avoids the need for expensive corrosion-resistant reactor equipment as well as an expensive catalyst separation step.

In some particular cases, as in the oxidation of substituted phenols to *p*-benzoquinones, shape selectivity may be exploited to suppress the formation of (bulky) coupling products,[10] *i.c.* diphenoquinones. Shape selectivity can in principle also help prevent unwanted reactions of catalytically active metal species, like oligomerisation and disproportionation. Ship-in-a-bottle systems may in some cases display unique transition metal chemistry thanks to that effect.

As with many of the framework-substituted materials (as will be discussed later on in the chapter), one can expect that application of ion-exchanged zeolites in liquid-phase will lead to leaching of transtition metal species. This aspect has not been given much attention in the earlier literature.

1.2. Catalysis by framework species

Molecular sieves containing isolated transition metal ions, demonstrably or supposedly incorporated into or bound to their frameworks, have received considerable attention during the past fifteen years. Following the discovery by Taramasso et al.[15,16] and other ENI workers[17,18] of titanium silicalite-1 (TS-1) and its unique properties as a catalyst in oxygenations using hydrogen peroxide (TS-1 will be discussed in some detail in section 2 of this chapter), many such systems have been synthesised and studied, and some have shown promise as catalysts for selective oxidation reactions. As we will be discussing the catalytic chemistry of such materials, we will also either implicitly refer to or explicitly discuss the coordination chemistry of "isomorphously substituted framework species"[19] which appears to be somewhat of a contradiction. Als we will see in the course of the chapter, and indeed in quite some detail for titanium species in titanium silicalites in section 2, it is not.[20] The catalytic systems that will be reviewed include titanium, vanadium and chromium silicalites, as well as cobalt-, vanadium-, and chromium-substituted AlPO's. The metals in this list are known to catalyse a host of different oxidation reactions in which very different mechanisms operate, some with electron transfer steps and autoxidation chains, others with heterolytic or homolytic oxo or peroxo oxygen transfer steps.[21] Different as these mechanisms are, they all of course involve coordinative interactions of some kind.

When trying to understand what can be specific for catalysis by molecular sieves containing transition metal framework sites, one needs to address the following topics:

i. The local structure of the site. The case where the metal occupies one of the regular tetrahedral framework sites (and is tetrahedrally coordinated) is called *isomorphous framework substitution*.

ii. The interaction of the site with reactants, products and solvents: its coordination chemistry. In one extreme view, it is dominated by the requirements of an only slightly flexible molecular sieve framework. In another extreme view, a flexible framework adapts to the coordination chemistry of the ion.[22]

iii. Specific properties of the surrounding framework, such as its hydrophilicity/hydrophobicity characteristics, or its shape selective properties.

The first two items describe properties of the site which are expected to depend on the metal in question, so we will discuss them in the appropriate sections; one can mention one property which different framework-connected metal ions are supposed to have in common, which is their monomeric nature.

Concerning the last item, it is possible to make a few general remarks before treating specific catalysts. Shape selectivity effects are well known and often well understood in acid catalysis,[23] but clear examples in oxidation reactions are rare. Since oxidations are mostly irreversible, and usually not accompanied by reactions that can effect the interconversion of isomeric products, only *reactant selectivity* or *restricted transition state selectivity* can be of practical importance. Slow diffusion of, for instance, some undesired isomer, will not lead to an improved selectivity, but only to low reaction rates.

Specific sorbent -or solvent- properties of zeolites can be of great significance. The pronounced hydrophobicity of titanium silicalites is undoubtedly one of their key properties, and it allows them to be used as catalysts for selective oxidations with aqueous hydrogen peroxide. Selective sorption by molecular sieves can also give rise to unexpected solvent effects. Although selective sorption by molecular sieves is a well-established phenomenon,[24] we feel that its importance in liquid-phase catalytic reactions is often still underestimated.

Finally, it should be mentioned that, at least for metal ions that substitute isomorphously, the surrounding lattice can be regarded as a pre-organised polydentate ligand, from which complete solvolysis is difficult.

2. Titanium-containing silicalites

Among the metal-containing molecular sieves, titanium-containing silica-based materials are the most intensively studied ones. The reason for this is that one member of this family, titanium silicalite-1 (usually denoted as TS-1), catalyses a broad range of selective oxidations with hydrogen peroxide (see Fig 1).[17,18] Its catalytic properties are of great interest, both commercially and scientifically. Mechanistic aspects of the oxidation reactions catalysed by TS-1 will be discussed in section 2.2. Before that is done, a brief review of the literature on the synthesis of titanium-containing silicalites is given (section 2.1). Finally, an effort is made to draw some conclusions concerning the structure of the titanium sites in these materials, and their ability to react with various molecules including hydrogen peroxide (section 2.3).

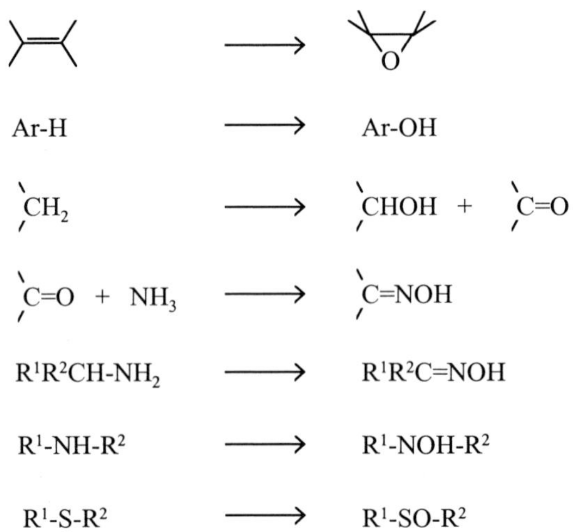

Figure 1. Conversions catalysed by TS-1.

Most of this section is devoted to TS-1, which has the MFI structure, and its closest relatives, which are also compositional variants of known high-silica zeolites and contain only a small amount of titanium. As we will see in section 2.3., there is now overwhelming evidence for tetrahedral framework siting of titanium in TS-1, substituting for silicon, but the situation is not always as clear for other materials. A few titanium-rich molecular sieves are known; they have unique structures composed of infinite chains of titanium octahedra connected to silicon tetrahedra and actually constitute a separate family of materials.[25] We only mention them once here, since they seem to be of little relevance to oxidation catalysis. A likely reason for their reported poor catalytic performance in oxidations[26] is the simple fact that framework titanium ions are completely shielded by silicon tetrahedra and hence inaccessible from the pores.

2.1. Synthesis

Tetravalent titanium usually assumes an octahedral coordination, and has a strong tendency to polymerize in aqueous systems, usually resulting in precipitation and eventual formation of one of the very stable forms of titanium dioxide. One could therefore expect the synthesis of materials containing isolated tetrahedral titanium sites to be rather difficult, and this expectation proves to be true: some specific features of the synthesis of TS-1 are now understood to a certain extent, and it is known to involve a number of conditions which are highly critical. The synthesis of zeolitic materials with other structures (MEL, BEA, ZSM-48), containing titanium in framework positions, has been described, but with the exception of [Ti]-BEA,[27] these materials have not yet been deeply investigated.

Synthesis of TS-1

In the patents of Taramasso et al.[15] two methods are described for synthesizing TS-1. Both are specific variations on general recipes for the preparation of silicalite-1, and both use the same template, the tetrapropylammonium cation. The methods differ mainly in the way in which the precursor mixture for crystallization is prepared: following the first method, it is obtained by controlled hydrolysis of tetraethoxytitanium(IV) and tetraethoxysilane; we shall label this method the "mixed alkoxide" method. Following the second method, the precursor is obtained by mixing colloidal silica and a solution containing tetrapropylammonium hydroxide, hydrolysed titanium alkoxide, and hydrogen peroxide. (The peroxide forms complexes with titanate species, thus suppressing their tendency to polymerize.) We shall label this second method the "dissolved titanium" method. The composition of the mixtures and the conditions that can be used to crystallize TS-1 from them are roughly the same.[16] An alternative 'dry' procedure to synthesize TS-1 by impregnating silica with a titanium precursor dissolved in aqueous TPAOH was developed first by Padovan et al.[28] Later modifications of this method employed TiO_2/SiO_2 co-gels.[29]

Many workers have experienced that the preparation of TS-1 *via* the mixed alkoxide method is highly critical. Van der Pol et al.[30] showed, that deviations from the optimum mixing temperature, the hydrolysis temperature and the speed of hydrolysis easily lead to unwanted precipitate formation in the early stage of gel preparation. It might be of importance to note that the tetrapropylammonium ion seems to act like a phase transfer catalyst during the hydrolysis of the alkoxide mixture, and allows it to proceed smoothly. Less reactive titanium sources, such as the tetrabutoxide,[31] have been used in an attempt to make the preparation procedure less critical, but the effects of this approach have been strongly questioned.[32] Kraushaar[33] proposed a simple model, according to which, prior to hydrolysis, all titanium should be present as single titanium alkoxide units in mixed oligomers with the silicon alkoxide. Acceptor properties of the titanium alkoxides towards dioxane[34] suggest that in $Si(OEt)_4$, titanium ethoxide could form solvated species like *e.g.* $Ti(OEt)_4[(\mu^2\text{-}OEt)Si(OEt)_3]_2$ or $Ti(OEt)_4(\mu^2\text{-}OEt)_2Si(OEt)_2$, under conditions where other titanium alkoxides would be either unsolvated or would remain as oligomers containing

Ti(μ-OR)Ti units. Unassociated monomers or oligomers would then give rise to aggregated oxidic phases upon hydrolysis. As isolated titanium species are expected to be metastable relative to such phases, the formation of the latter can be considered irreversible and should be prevented. Gao et al.[35] showed that an alternative approach is to hydrolyse $Si(OEt)_4$ with aqueous $TiCl_3$, $Ti^{3+}_{(aq)}$ being less prone to oligomerisation.

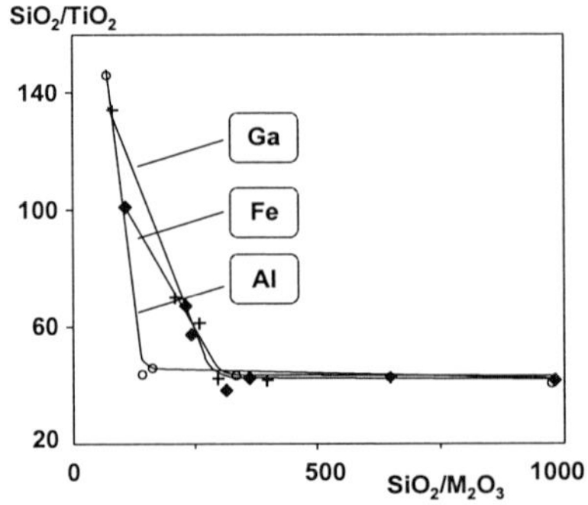

Figure 2. Competition between trivalent ions and titanium in the crystallisation of [M^{III},Ti]-MFI from synthesis mixtures with constant Si/Ti ratios and varying Si/M^{III} ratios.[37]

Another possible pitfall in TS-1 synthesis is the presence of alkali metal ions in the synthesis mixture. Several authors[17,36,37] have pointed out that a small amount of sodium or potassium, which can for instance originate from a commercial tetrapropylammonium hydroxide solution, suffices to prevent the insertion of titanium into the molecular sieve framework. Instead, titanium species aggregate to form an amorphous phase, and ultimately anatase.[37] The role of alkali metal ions in the synthesis of high-silica zeolites is not very clear;[38] they seem to catalyse the equilibration of silicate species, presumably by promoting hydrolysis and recondensation reactions of Si-O-Si bonds, stabilizing terminal $Si-O^-$ groups by complexation.[39] A similar reasoning might apply to Si-O-Ti bonds. However, Khouw et al.[40] reported that sodium present during TS-1 synthesis is not necessarily detrimental, but leads to a silicalite containing both titanium and sodium, from which sodium can be removed by an acid wash after calcination, and whose performance is then comparable to a TS-1 obtained by conventional sodium-free synthesis. A very different effect is caused by the addition of salts of trivalent metals like Al^{3+}, Ga^{3+} or Fe^{3+} to the synthesis mixture, which also suppresses Ti incorporation to a certain extent,[37] but which does not cause segregation of titanium. As a consequence, it is possible to synthesize doubly substituted analogues of TS-1 (see Fig. 2), like e.g. [Ti,Al]-MFI,[37,41] which can act as bifunctional catalysts. Further study of such competitive crystallizations might give clues about the gel chemistry involved in titanium incorporation.

The use of fluoride ions in TS-1 synthesis was first reported by Guth and coworkers[42] and Qiu et al.[43] In an effort to synthesize TS-1 in the presence of fluoride, Kooyman et al.[44] observed anatase impurities. In addition, lower activities and selectivities were found in phenol hydroxylation and alkane oxygenation, compared to a conventional TS-1 sample. Characterisation by EXAFS has confirmed that the use of fluoride in synthesis leads to the formation of octahedral, extraframework titanium.[45]

Crystallization of TS-1 can be performed in a fairly wide temperature range, as can be seen in Fig. 3(a). The results shown were obtained with the mixed alkoxide method. The effectiveness of titanium incorporation increases with temperature, up to the point where anatase formation starts to occur, at 200 °C. Crystallisation at low temperature (100 °C) of sols prepared by the 'mixed alkoxide' method favours the formation of colloidal TS-1 crystals with an average size in the 80-110 nm range.[46] (Interestingly, such small particles show a peculiar tendency to stack at neutral *p*H, *e.g.* during washing, which can even give rise to cm-long fibres under certain conditions.[47]) At a fixed temperature, the titanium content of the crystalline product is always lower than that of the precursor mixture, as is shown in Fig. 3(b). This means that the solution is enriched in titanium during synthesis, and that unnecessary long crystallization times ultimately lead to the deposition of extraframework titanium phases.[28]

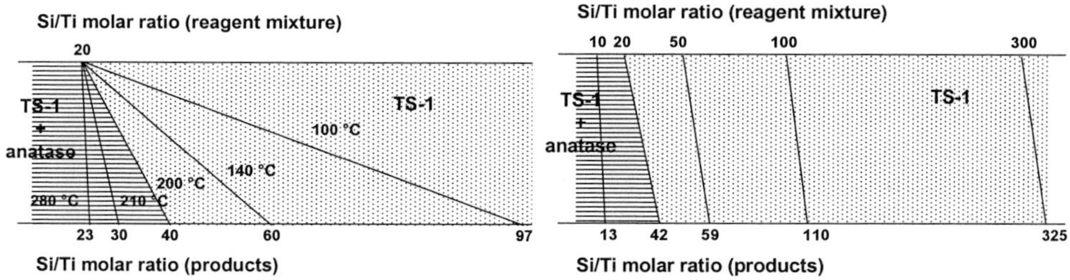

Figure 3. Influence of (a, left) the crystallisation temperature and (b, right) the gel composition on Ti contents in TS-1.[16]

Perego et al.[48] pointed out, that titanium could be incorporated in the MFI framework up to a Ti/(Ti+Si) molar ratio of 0.025. This apparent compositional ceiling is probably not a thermodynamic one and, as TS-1 is already a metastable phase, it might be possible to break through it. For instance, TiO_2-SiO_2 glasses can contain titanium substituting for silicon in tetrahedral sites up to a Ti/(Si+Ti) ratio of 0.07 - 0.09.[49] In fact, Thangaraj et al.[31a] claimed the preparation of TS-1 with much higher titanium contents using an alternative synthesis method (*vide supra*), and found the upper limit of the Ti/(Ti+Si) ratio to be 0.10. However, Millini et al.[32] observed that the XRD data reported in that paper did not agree with the results of Perego et al.,[48] and concluded from a detailed re-investigation by Rietveld analysis that in the materials of Thangaraj, a large fraction of the titanium was present as extraframework material. Further discrepancies in ^{29}Si MAS NMR and IR spectra[31,48] confirm this conclusion. The XRD data of van der Pol et al.,[30] who has claimed an upper limit for Ti/(Ti+Si) of 0.04, also disagree with the data of Perego et al,[48] although to a lesser extent.

Several studies have been aimed at replacing the rather expensive alkali-free TPAOH by other templating agents. Hexapropyl-1,6-hexanediammonium and the much less expensive 1,6-hexanediamine were found to be effective although the latter promotes the formation of large crystals, elongated along the c- or b-axis, depending on titanium content.[50] TS-1 and a boron-containing analogue have been synthesised in the presence of (alkali-free) methylamine and TPA bromide.[51]

Synthesis of other medium-pore materials: TS-2, [Ti]-ZSM-48

Because the MFI and MEL structures are related,[52] and because the silicalites having these structures (*i.e.* silicalite-1 and silicalite-2) can be prepared in similar ways using somewhat different templates, it is perhaps not surprising that a titanium-containing equivalent of silicalite-2, denoted as TS-2, can be synthesized.[53,54] In addition, Bellussi et al.[53] (and later Tuel et al.[55]) reported the synthesis of MFI-MEL structural intergrowths containing titanium, which they labeled TS-3. In fact, materials classified as having the MEL structure are in reality often disordered intergrowths of MEL and MFI.[52] Reddy et al.[54] reported the possibility to synthesize TS-2 with rather high titanium contents using the method of Thangaraj et al.[31] (*vide supra*). The synthesis of TS-2 is not as well studied as that of TS-1, but one can expect it to respond in similar ways to the parameters discussed in the above.

The synthesis of [Ti]-ZSM-48 has been reported by Serrano et al.[56] and later by Tuel et al.[57] A variant of the dissolved titanium method, with added hydrogen peroxide, was used to prepare materials with Ti/(Ti+Si) ratios up to 0.037 (see section 2.3 for a discussion of the purity of these materials). [Ti]-ZSM-48 has non-intersecting pores with 10-membered ring apertures: probably for this reason, it proved quite inactive as a catalyst in low-temperature liquid-phase reactions.

Synthesis of titanium-containing (ultra-)large-pore materials: [Ti]-beta, [Ti]-ZSM-12 and Ti-MCM-41

TS-1 is a medium pore material, which makes it suitable as a catalyst for the conversion of small molecules, typically up to C_8. A need for clean oxidation methods however also exists in the realm of fine chemical synthesis, where the use of stoechiometric reagents like percarboxylic acids or chromium compounds is still widespread. A catalyst with a scope as wide as that of TS-1 but also capable of converting larger molecules is clearly a desirable aim for research in selective oxidation, and the work done so far on large-pore analogues of TS-1 seems to be a firm step in the good direction.

The synthesis of zeolite beta containing titanium in its framework was first reported by Camblor et al.[58] Here, again the dissolved titanium method was used, now without added peroxide, with alkali-free tetraethylammonium hydroxide (TEAOH) as the template. At first, the addition of aluminium to the synthesis mixture was found to be necessary for crystallisation of beta, yielding products with Al/(Al+Si) ratios of 0.010 or higher. This apparently does not suppress titanium incorporation as strongly as it does in the case of TS-1 synthesis.[37] As will be discussed in more detail in later sections, the presence of aluminium (and silanol groups associated with lattice defects) in the material is undesirable in most cases. Rigutto et al.[59] prepared aluminium-free zeolite beta containing framework titanium by reacting borosilicate beta with $TiCl_4$. Formation of clustered extraframework titanium species could be minimised though not entirely suppressed by employing a multistep procedure depicted in Fig. 4.

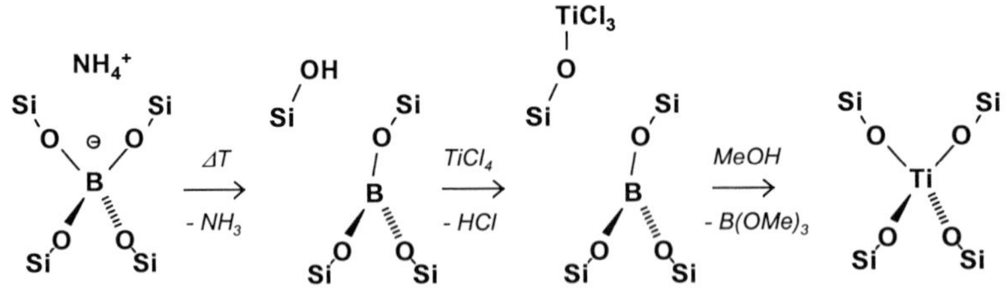

Figure 4. Route for replacing boron in borosilicate frameworks with titanium.[59]

Direct crystallisation of aluminium-free [Ti]-beta turned out to be a more difficult feat, if only because at the time, the synthesis of all-silica beta had not been achieved. Soon after, it was found that sodium-free gels containing dibenzyldimethylammonium hydroxide (DBDMAOH) yield the latter[60] but not the former, mainly because of the longer synthesis times required in the presence of titanium combined with the limited stability of DBDMAOH under hydrothermal conditions.[61] The first to succeed were Saxton et al., who used a novel diquaternary template (see Fig. 5).[62] Camblor et al., after having made the observation that [Ti,Al]-beta continued to crystallise after preferential consumption and incorporation of all the aluminium from the synthesis gel, managed to prepare aluminium-free materials by seeding a mixed alkoxide-derived gel containing TEAOH with dealuminated crystals of zeolite beta.[63] Van der Waal et al.[64] used the saturated, more stable analogue of their DBDMAOH template, bis(cyclohexylmethyl)diammonium hydroxide (see Fig. 5) to synthesise aluminium-free materials with faily large (~1×1×0.2 μm) crystals.

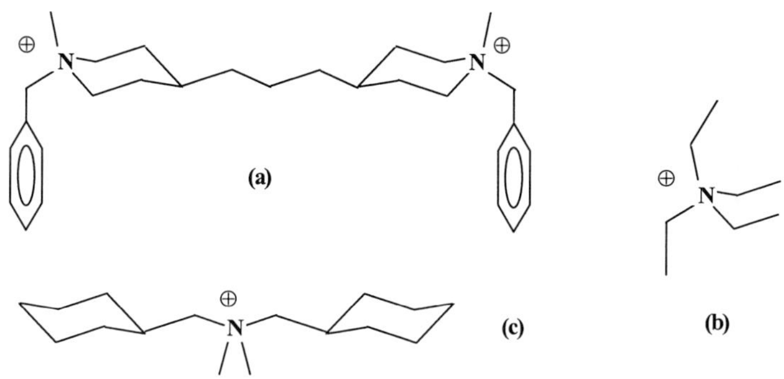

Figure 5. Templates suitable for the synthesis of aluminium-free [Ti]-beta: (a) 4,4'-trimethylene-bis(N-benzyl-N-methyl piperidinium); (b) tetraethylammonium; (c) di(cylclohexylmethyl) dimethylammonium.

As mentioned, hydrophobicity is considered an important characteristic of TS-1. The [Ti]-beta materials mentioned above were all found to contain significant concentrations of silanol groups associated with lattice defects of unknown structure, rendering them still fairly hydrophilic. By employing tetraethylammonium fluoride, Blasco et al.[65] managed to prepare truly hydrofobic, defect-free [Ti]-beta materials.

[Ti]-ZSM-12 could be prepared by a mixed alkoxide synthesis employing $Ti(OBu)_4$ and alkali-free aqueous bis(diethylammonium) hydroxide.[66] Titanium-containing SSZ-33, a material with intersecting 10- and 12-membered ring pores, was synthesised from a borosilicate precursor using $TiCl_4$.[67] [Ti]-UTD-1, a material with 14-membered ring pores, was synthesised by Balkus et al.[68] with the same cobalticenium template used for the all-silica material.

After the discovery of MCM-41, several authors described the synthesis and properties of titanium-containing mesoporous silica materials belonging to the M41S family. The synthesis of titanium-containing MCM-41 was first reported by Corma et al.[69] Similar materials were later described by others.[70] MCM-41 is not a crystalline material in the same sense as zeolites are: its pore walls are disordered and its internal pore surface is ridden with silanol groups.[71] The use of the term 'isomorphous substitution' to describe these materials is hence not very meaningful. One would expect a titanium-containing MCM-41 material to behave much like a titanium-on-amorphous-silica catalyst rather than as a TS-1 analogue, in terms of both surface chemistry and catalysis. This is indeed what is observed.[72,73]

Synthesis: conclusive remarks

Although the synthesis of TS-1 has been studied rather intensively in the past decade, many aspects of the specific titanium chemistry involved are still unclear. The stabilizing interactions that allow the formation of neutral, tetrahedral titanium species -even in the presence of water- are worth identifying in their own right, but they might give some important clues about the catalytic properties of the tetrahedral complex as well. The deleterious effect on titanium incorporation of mobilising agents such as fluoride ions, trace levels of alkali metal cations, or of mobilising conditions like or high temperatures, as well as the competition by trivalent metals, need a better understanding. This would at least be of great help in the development of synthesis methods for novel titanium-substituted materials.

2.2. Reactions catalysed by titanium-containing silicalites

The scope of TS-1 catalysed reactions and the advantages of TS-1 over other catalysts have been reviewed extensively,[17,18,48,74] so we will limit our treatment to the data and trends that are of interest for mechanistic considerations. In all reactions mentioned below, hydrogen peroxide was used as the oxidant unless specified otherwise.

Hydroxylation of aromatics

The hydroxylation of phenol to dihydroxybenzenes (1,2-DHB or catechol and 1,4-DHB or hydroquinone) is currently the most important application of TS-1, and a many studies have been devoted to this reaction.[17,18,36,44,74,75,76,77,78] It is often used as a test reaction, because of its sensitivity to the quality of the catalyst: the presence of extraframework titanium significantly lowers the reaction selectivity by catalysing the decomposition of hydrogen peroxide[17] and the formation of quinones and coupling products.[79] Acetone is the solvent of choice, but methanol or water give similar activities. The solvent does influence selectivity: acetone promotes formation of catechol, whereas methanol gives higher selectivities to hydroquinone. The tentative explanation is that intraporous reactions are more *para*-selective and that acetone sustains *ortho*-selective catalytic reactions on the external surface of TS-1 crystals by efficient dissolution of tars that otherwise poison the surface sites.[76,80] The use of acetonitrile gives somewhat lower rates, and concomitantly, hydrogen peroxide decomposition becomes a relatively important side reaction.[81] The reaction is probably diffusion-limited as observed rates strongly depend on crystal size.[81] The selectivity for *para*-hydroxylation is higher with TS-1 than with Brønsted acids or Fenton-type catalysts[17] which is commonly interpreted as a shape selectivity effect. It also indicates that, for the TS-1 catalysed reaction, coordination of phenol to the site is not part of the reaction sequence, in contrast to some other cases.[82,83] This in turn might be one of the reasons (in addition to the hydrophobicity of TS-1) why further oxidation of dihydroxybenzenes to quinones is relatively slow on TS-1.

TS-1 also catalyses the hydroxylation of benzene itself and other aromatics that are more electron-rich than benzene.[48,74,84,85] Chlorobenzene, nitrobenzene, benzoate esters and other substrates bearing electron-withdrawing substituents do not react. Shape selectivity manifests itself in the fact that substrates carrying an otherwise activating bulky alkyl group, e.g. isopropylbenzene or *t*-butylbenzene, are also unreactive. Only a few other cases of aromatics hydroxylation involving group 4 - 6 metals and hydroperoxides have been reported: relevant examples are the electrophilic, heterolytic hydroxylation of naphthols by $Ti(O^iPr)_4$/*t*-butyl hydroperoxide or by $[MoO(O_2)_2]$·py·HMPT,[86] and the hydroxylation of benzene and benzene derivatives by the vanadium peroxo picolinate complex $VO(O_2)(Pic)\cdot 2H_2O$.[87] According to Mimoun et al. and Bonchio et al.[87] oxidations by the vanadium reagent involve homolysis of the O-O bond and a radical chain pathway. Many other examples of the reaction involving homolytic[82] or heterolytic[83] oxygen transfer are known. For the case of TS-1, both pathways can be envisaged. Reactivity trends and isomer ratios could lead to the conclusion that the reaction involves an heterolytic, electrophilic attack on the aromatic ring,[82,83,87] but similar results could also be obtained through a pathway involving an electrophilic radical attack. On the other hand, benzylic oxidation, usually a homolytic reaction involving radicals, prevails over hydroxylation of the nucleus when ethylbenzene is the substrate.[74,84,88] This seems to imply that the same species can perform (or initiate) homolytic as well as heterolytic oxygen transfer. C-H bond activation at saturated carbon atoms is discussed further on in this section.

Recently, the conversion of acetophenone to a mixture of 2- and 4-hydroxyacetophenone and phenyl acetate catalysed by TS-1 was reported.[89] Formation of the ester was explained by a catalytic Bayer-Villiger oxidation, so this reaction is somewhat different from the direct hydroxylations described in the above.

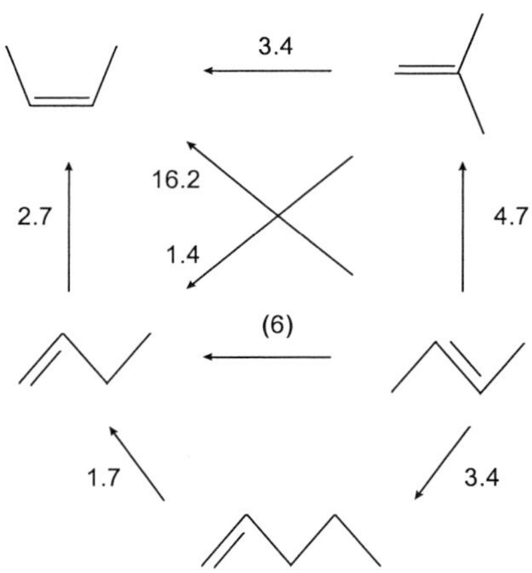

Figure 6. Relative reactivities of lower alkenes in epoxidation catalysed by TS-1. The arrows point to the more reactive alkene.[90]

Epoxidation

TS-1 is a particularly active and selective catalyst for the epoxidation of alkenes,[74] and allows this reaction to be carried out at temperatures as low as 0 °C.[90,91] Even electron-poor alkenes, such as allyl chloride, allylic ethers and alcohols, are epoxidized with high yields and selectivities.[92,93] Epoxidation occurs with retention of configuration,[94] evidencing a heterolytic mechanism. Relative reactivities among alkenes depend mainly on the nucleophilicity of the double bond, but steric factors sometimes dominate,[90] as in the case of cyclohexene, which reacts two orders of magnitude slower than 1-hexene.[74] Relative reactivities of some lower alkenes are shown in Fig. 6. In contrast, homogeneously catalysed epoxidation rates are mainly governed by electronic factors, while steric factors only have a minor influence.[95] Restricted transition state selectivity as well as differences in sorption strength and diffusion rates are thought to cause these differences.

The similar reactivities of allyl alcohol and allyl ethers and esters observed with TS-1,[74] indicate that the mechanism does not involve coordination of the substrate in the form of a surface alkoxide.[96] This was also concluded by Adam et al.[97] who determined diastereoselectivities for a number of different allylic alcohols using different catalytic systems (TS-1, [Ti]-beta, $Ti(O^iPr)_4$/TBHP, $VO(acac)_2$/TBHP, CH_3ReO_3/ROOH) as well as stoechiometric reagents (*m*-chloroperbenzoic acid and dimethyldioxirane) to obtain clues about the mechanism.

Allylic oxidation is usually not observed as a side reaction,[74] although work by Tatsumi et al.[98] on the oxidation of allylic alcohols shows that it can sometimes compete with epoxidation, especially for the bulkier substrates. The authors attributed this effect to a lack of reactivity of the double bond due to steric hindrance. Coordination of the alkene to titanium as part of the reaction sequence was invoked by Huybrechts et al.[99] to explain the inhibition of *n*-hexane oxidation by 1-octene. This view was countered by Clerici et al.[90] who did not observe such inhibition with what was claimed to be a purer catalyst.

Styrene and related compounds are epoxidised rather than undergoing hydroxylation of the aromatic ring. Styrene itself is rapidly rearranged to 2-phenylacetaldehyde with high selectivity.[72,100] Further oxidation involving C-C bond cleavage has also been reported.[101] Adam et al. observed that in the case of α-methylstyrene, cleavage is preceded by the formation of 2-phenyl-1-hydroperoxy-2-propanol which further reacts to acetophenone.[102] Curiously enough the formation of the hydroperoxyalcohol is promoted by non-protic solvents like acetonitrile.

With TS-1, epoxidation rates depend on the solvent used[90,91] in a way that is very different from the solvent effects observed for homogeneously catalysed epoxidations.[103] The use of methanol or ethanol gives higher rates than the use of aprotic solvents: e.g., in methanol, which is the solvent of choice, the rate is an order of magnitude higher than in acetonitrile. The addition of fluorides or alkali metal acetates inhibits the reaction.[90] Clerici et al.[90] and Bellussi et al.[104] proposed a mechanism in which provides an explanation for both the solvent effects and the inhibitory effects. This mechanism contains a proposal for the structure of the actual oxygenating titanium complex and will therefore be discussed in section 2.3.

In 1993 Camblor et al.[58] reported the use of [Ti,Al]-beta as an epoxidation catalyst, less active than TS-1 for substrates like 1-hexene,[105] but capable of converting larger molecules like cyclododecene.[106] Under the conditions applied in the early work on [Ti,Al]-beta, Brønsted acidity due to the aluminium in these materials unfortunately caused efficient solvolysis of the initially formed epoxides. Also, it was observed that aluminium had a negative effect on activity,[107] which was later recognised to be due to the hydrophilicity of the acid sites. The bifunctional nature of [Ti,Al]-beta was exploited in the conversion of linalool, which was found to give cyclic ethers by acid-catalysed rearrangement of initially formed epoxide, with selectvities close to that of the enzyme epoxidase.[108] The most straightforward remedy for the selectivity problem was to neutralise the acidity by ion-exchange:[109] lithium-exchanged [Ti,Al]-beta was found to much more selective towards the epoxide (76 % at 7 % conversion) with lower but still appreciable activity. Larger cations caused more activity loss, presumably by increasing the hydrophilicity of the catalyst, and were not as effective in inhibiting solvolysis.

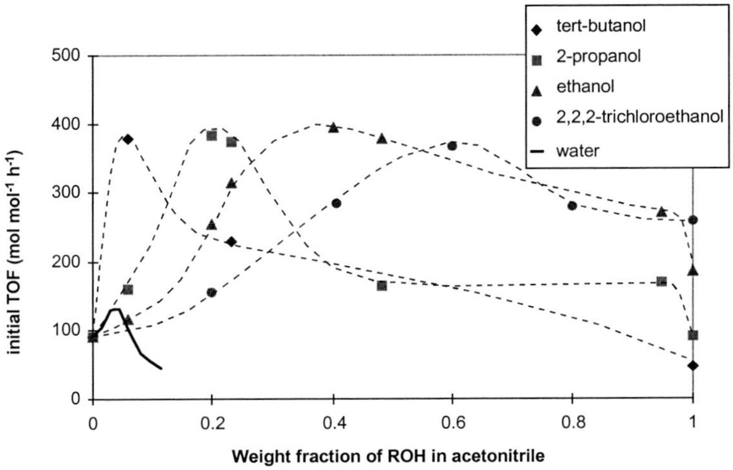

Figure 7. Solvent effects in [Ti]-beta-catalysed epoxidation of 1-octene: effect on the rate of the concentration of various cosolvent alcohols and water in mixtures with acetonitrile.[112]

Later Corma et al.[110] observed that acetonitrile was very effective at surpressing epoxide solvolysis/hydrolysis reactions in [Ti,Al]-beta without causing any loss of activity, presumably by succesfully competing with epoxide molecules for coordination to the acidic proton. With the synthesis of aluminium-free materials,[62,63,64] and later of truly hydrophobic [Ti]-beta materials that could really be regarded as TS-1 analogues,[65] it became clear that differences in intrinsic per-site acitivity between [Ti]-beta and TS-1 were not as large as they had been percieved to be earlier[63] if only conditions were properly optimised for the different sorption characteristics of the large-pore materials. Aluminium-free materials were found to be capable of epoxidising acid-sensitive substrates like terpenes, although in the conversion of camphene, the highly strained epoxide was still found to isomerise efficiently to camphyl aldehyde. [61,111] No results of the use of the defect-free materials in such reactions have yet been reported, but one can expect them to perform even better.

Van der Waal. et al.[112] studied solvent effects in [Ti]-beta-catalysed epoxidations in some detail. They found that in alcohol-acetonitrile mixed solvent systems, epoxidation activity always goes through a maximum (see Fig. 7), the maximum activity being independent of the alcohol chosen but occuring at an alcohol concentration which depends on the polarity of the alcohol. Most remarkably, similarly water (in acetonitrile) was also found to accelerate the reaction rate.[61] This was interpreted as evidence that alcohols or water participate in the reaction mechanism, as in the case of TS-1.

Ammoximation of cyclohexanone and related reactions

Another reaction catalysed by TS-1 is the ammoximation of ketones. The application to cyclohexanone is commercially important: its reaction with ammonia and hydrogen peroxide in *t*-butanol gives cyclohexanone oxime with high selectivity (>95 % at 80 - 90 %

conversion).[113] Acetophenone derivatives also give oximes but this can be accompanied by cleavage.[114] Amorphous titanium-on-silica with a high titanium loading (9.8 wt %)[115,116] and phosphotungstic acid[117] also catalyse ammoximation of ketones, but they do so with somewhat lower selectivities and less efficient use of hydrogen peroxide.

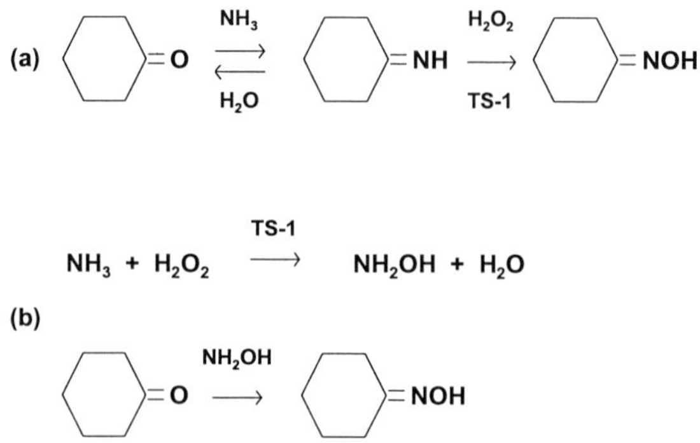

Figure 8. Reaction pathways for the ammoximation of cyclohexanone.

Two different reaction sequences have been proposed for the TS-1-catalysed reaction (see Fig. 8). The first scheme involves (uncatalysed) intermediate formation of the imine, followed by an oxygenation step: it was proposed by Thangaraj et al.[118] and Reddy et al.[119] to explain the by-products formed in the ammoximation reaction. Interaction of cyclohexanone with TS-1 and formation of imine upon addition of ammonia, as shown by infrared spectroscopy, was interpreted by Tvaruzkova et al.[120] as evidence for this reaction sequence. In addition, imine oxygenation by hydrogen peroxide is a known reaction and is catalysed by molybdenum and tungsten complexes.[21] However, Zecchina et al.[121] made the following observations which disfavour this first sequence: (i) bulky substrates that are not sorbed by the catalyst, like cyclododecanone and 4-*t*-butylcyclohexanone, are nevertheless converted, reacting at relative rates equal to those observed for oximation with hydroxylamine; (ii) TS-1 catalyses the formation of hydroxylamine from ammonia with good selectivity; (iii) UV-Vis spectra of TS-1 contacted with H_2O_2/NH_3 indicate that first a $[(SiO)_3L_2Ti(\eta^2\text{-}O_2)]^-NH_4^+$ complex is formed[122] which can react to form hydroxylamine. The second reaction sequence, which involves catalytic formation of hydroxylamine, is therefore more likely.

TS-1 also catalyses the oxidation of dialkylamines to the corresponding hydroxylamines,[123] of primary amines[124] and benzyl- and allyl amines[125] to oximes, of aniline to nitrosobenzene and azoxybenzene[126] and of sulfides to sulfoxides.[127] The homogeneously catalysed analogues of these reactions are well studied, and are known to involve a nucleophilic attack of the substrate on a peroxo oxygen, followed by O-O bond heterolysis.[128]

Oxidation of alcohols

TS-1 catalyses the oxidation of both primary and secondary alcohols at moderate temperatures.[74,129,130] Glycols may be oxidised to α-hydroxyketones.[131] With primary alcohols, high selectivities to aldehyde can be obtained at < 30 % conversion of the alcohol. Most secondary alcohols react much faster than the primary alcohols. Shape-selective effects are also observed in this reaction: a remarkable one is the difference in reactivity between 2-

pentanol and 3-pentanol, the latter reacting approximately ten times slower. Some acid catalysis occurs and leads to the formation of side products (acetals, esters). Also, unselective reactions occur on the external surface that can be suppressed by addition of di-*t*-butylphenol.[130] Alcohol oxidation has not received as much attention as some of the other reactions catalysed by TS-1, and not much is known about its mechanism. A radical process can be envisaged as well as a two-electron process; the relatively high selectivities to aldehydes indicate the latter. It is interesting to note that alcohol oxidation by neutral or anionic tungsten or molybdenum peroxo complexes is probably a two-electron process[132,133] not unlike Oppenauer oxidation[134] but with a peroxo ligand as the hydride acceptor in stead of a coordinated ketone. The molybdenum picolinate n-oxido complex $MoO(O_2)_2(pico)^-$ is especially selective to aldehydes.[132] It is unclear to which extent heterolytic alcohol oxidation by peroxo complexes can still be considered an "electrophilic" reaction;[135] it is commonly said that less electrophilic reagents are needed to effect alcohol oxidation compared to alkene epoxidation. For alcohol oxidation catalysed by [Ti]-beta, a mechanism has been proposed involving a complex similar to the epoxidising species proposed by Clerici et al. (see section 2.3).[136]

Oxygenation of alkanes

Examples of alkane oxygenation where good selectivities to alcohols and ketones are obtained at appreciable conversions of the substrate are rare. Huybrechts et al.[137] and Clerici[138] discovered independently that TS-1 catalyses the oxidation of alkanes to secondary alcohols and ketones. Under the somewhat different conditions used, somewhat different results were obtained; a comparison is made in Table 1. The most important difference is the fact that Huybrechts et al.[137] observed a statistical distribution of oxygenate isomers, whereas Clerici[138] found a preference for the 2-position, which he interpreted as due to steric and inductive effects. It is not unlikely that under the severe conditions used by Huybrechts et al., the participation of radicals and molecular oxygen in the process becomes more important.

Clerici observed that in methanol at 55 °C, addition of 2,6-di-*t*-butyl-4-methylphenol, carbon tetrachloride, chloroform or dichloromethane did not influence reaction rates or product distributions. Free-radical *chain* mechanisms can therefore be considered less likely under these mild conditions. Evidence for the electrophilic nature of the process was obtained from the oxidation of 1-chlorohexane and methyl heptanoate: in both cases oxygenation occurred preferentially on the remote secondary C-H bonds.[137]

Table 2. *n*-Hexane oxidation catalysed by TS-1.

Reference	137	138
Solvent	Acetone (two-phase)	Methanol (one-phase)
Temperature	100 °C	55 °C
H_2O_2 conversion	> 90 %	98 %
Sel. Based on converted H_2O_2	70 %	91 %
(2-ol+2-one)/(3-ol+3one)	1.1	2.3

Recently the possibility was reported to prepare methyl *t*-butyl ether (MTBE) in a single step from isobutane and methanol by using a bifunctional [Ti,Al]-ZSM-5 material which catalysed both oxidation and etherification.[139]

The fact that TS-1 is able to oxidize alkanes in the presence of an oxidizable alcohol such as methanol is remarkable. It could be due to an intrinsic property of the active complex, or perhaps to selective sorption properties of the silicalite.[24,74] Solvent effects similar to those observed for TS-1 catalysed epoxidation are also found in this reaction, although there are a few differences. Methanol gives the best selectivities, but in *t*-butanol the reaction is faster and selectivities are only slightly lower. On the other hand, epoxidation in *t*-butanol is much slower than in methanol.[94] This difference has been explained in terms of steric hindrance of the site active in olefin epoxidation, formed upon the interaction of a titanium hydroperoxo complex with a solvent molecule (see section 2.3).[94] Another common feature of epoxidation and alkane oxygenation is the fact that both reactions are inhibited by alkali metal salts, especially the acetates, and by fluorides.[138]

It is as to yet unclear what the mechanism of the oxygen insertion reaction is, and whether it consists of consecutive steps involving radicals or of a single "oxenoid" two-electron step. A vanadium peroxo complex is known that can effect alkane oxygenation, but it is different from TS-1 in the sense that it also oxygenates primary C-H bonds and operates *via* a radical chain mechanism.[132] A few strongly electrophilic species, such as ozone or the $H_3O_2^+$ ion in superacidic media,[140] can effect heterolytic oxygen insertion into C-H bonds. Such reactions are supposed to involve formation of a three-center two-electron bond by attack on a σ-bond ("oxenoid" insertion). Reactions involving a hydrogen abstraction step followed by a fast recombination of radicals in a so called *solvent cage* are also known.[141] Such solvent cage reactions are also not affected by radical chain inhibitors or chloroalkanes. Clerici et al.[138] preferred a solvent cage mechanism on the basis of the large isotope effect observed for TS-1 ($k_H/k_D = 4.1$). The observations that hydroxylation of tertiary C-H bonds is accompanied by loss of configuration, and that in the oxidation of alkylcyclopropanes ("radical clocks") no ring opening occurs, when taken together strongly support this interpretation.[142]

2.3. Structure and reactivity of the titanium site

Influences on the surrounding lattice

The original identification of titanium in TS-1 as a true tetrahedral framework atom had been based on indirect evidence, obtained from infrared spectroscopy, X-ray powder diffraction, EDX microprobe analysis and ^{29}Si MAS NMR measurements.[48] It was shown that the unit cell volume of calcined TS-1 increases linearly with titanium content. Provided that neither the site itself nor the lattice is strongly distorted, the unit cell volume V can be expected to depend on the degree of substitution x = Ti/(Ti+Si) following the equation:

where V(x) is the observed unit cell volume, V(0) the unit cell volume of the pure all-

$$V(x) = V(0) \cdot (1 + [(d_{Ti\text{-}O} / d_{Si\text{-}O})^3 - 1] \cdot x) \qquad \text{(Eq. 1)}$$

silica material and d is an average bond length. This is known as *Vegard's law*,[143] and it is usually obeyed when substitution is both isomorphous and isovalent. Fitting of X-ray data of TS-1 with eqn. 1 (assuming a value of 1.59 Å for $d_{Si\text{-}O}$[144]) gives a titanium-oxygen bond length of 1.78 Å, which closely agrees with bond lengths found in other compounds containing tetrahedral titanium in oxidic environments.[145] Linear dependence of unit cell expansion on titanium content was also observed for other titanium-containing materials. In an effort to treat the phenomenon in a quantitative fashion, we have summarized some of the data from literature in Table 3.

Table 3. Summary of XRD data for titanium-substituted materials.

Material	R[a)]	hypothetical $d_{Ti\text{-}O}$ (Å)
TS-1[48]	0.395	1.78
TS-1[31a]	0.113	1.65
TS-2[31b]	0.308	1.74
[Ti]-ZSM-48[56]	0.230	1.70
idem	0.337[b)]	1.75[b)]
[Ti]-beta[65]	0.147	1.66

[a)]R according to $V(x)=V(0)(1+Rx)$, $R=(d_{Ti\text{-}O}/d_{Si\text{-}O})^3-1$; $d_{Si\text{-}O}=1.59$ Å (for symbols, see eqn. 1). [b)]Values obtained when the point corresponding to the highest Ti content ($x=0.037$) is ignored (ref. 56).

The shortest known average Ti-O bond length for titanium in a tetrahedral environment is 1.764 Å.[145] Apart from the first entry, the bond lengths calculated from XRD data therefore seem to be out of range. If eqn. 1 is valid for all the above materials, one possible way to explain the data is to assume that these materials contain some extraframework titanium.[32] Molecular dynamics (MD) simulations agree closely with TS-1 XRD data and indicate the validity of Vegard's law for that case.[146] On the other hand, XRD data for the [Ti]-beta materials described by Blasco et al.[65] do not follow eqn. 1; nevertheless the materials have been very well characterised and there seems to be no indication for the presence of extra-framework titanium impurities. When looked at in more detail, lattice expansion is anisotropic in a complex way that depends on titanium siting.[147] It seems that the phenomenon is even more complex for [Ti]-beta, where expansion is not only anisotropic but where the anisotropy also changes with titanium content. Njo et al. found, by means of Monte Carlo/ Molecular Mechanics calculations, that titanium in TS-1 has a strong preference for two specific crystallographic sites (T2 and T12); similar complications can be expected in [Ti]-beta.[151] In any case, it seems advisable to excercise some caution in interpreting unit cell expansion data, and to acquire independent evidence to ascertain the purity of new molecular sieves materials containing titanium.

An observation which is difficult to understand is the decrease in the temperature of transition to orthorhombic symmetry that is caused by incorporation of titanium.[148] Such a decrease in transition temperature has also been observed for substitution of silicon by boron and aluminium,[149] whereas substitution by germanium effects a strong *increase*.[150] Little can be said about the cause of these phenomena without detailed calculations.[151] Distortions of the lattice effected by the presence of framework titanium sites are also apparent from ^{29}Si NMR spectra of a series of TS-1 or [Ti]-beta samples: the well-resolved features of the spectrum of the all-silica material already become blurred at low titanium contents.[48,65]

Direct spectroscopic methods for probing the site geometry

The most easily accessible method for direct probing of titanium(IV) environments in solids is diffuse reflectance UV-Vis spectroscopy. The spectra one obtains are charge transfer spectra, which are somewhat difficult to interpret. Pure, as-synthesized as well as calcined, dehydrated samples of TS-1 show a single absorption maximum around 48,000 cm^{-1}. By comparison of spectra with those obtained from model compounds, Boccuti et al.[152] were able to assign the absorption to tetrahedral titanium(IV) species. In addition, the use of UV-Vis spectroscopy was shown to allow the detection of octahedral (supposedly extra-framework) titanium impurities with good sensitivity.[153]

EXAFS and XANES measurements have confirmed the tetrahedral geometry of titanium sites in TS-1.[45,145,154,155] The strongest evidence comes from the average Ti-O bond lengths, which were determined to be 1.83±0.02 Å,[45] 1.80±0.01 Å[145] and 1.81±0.01 Å,[154] respectively (Note that these values are consistent with the above interpretation of the XRD data). In addition, Pei et al.[145] were able to conclude from their data that a second shell of oxygen and/or silicon atoms was present within a radial distance range of 3.1-3.8 Å, which is consistent with the environment of a framework site. Previous EXAFS and XANES data indicated the presence of multiple five- and six-coordinated species in TS-1.[156] Pei et al.[145] and Bordiga et al.[154] showed that this could have been due to the presence of extraframework material, while hydration of the titanium site might also have played a role.[45,154] (We will discuss site hydration separately).

Alternative interpretations[157] of EXAFS/XANES data have been brought up but have since been convicingly refuted.[145,158]

The 960 cm^{-1} band and its interpretation

As-synthesized and calcined, dehydrated samples of TS-1 show a characteristic band at approximately 960 cm^{-1} in their infrared[48] and Raman spectra.[159] The infrared feature is shared by TS-2, [Ti]-ZSM-48 and [Ti]-beta (For these materials, no Raman data in the 800-1100 cm^{-1} range have been published). The interpretation of this band has been a matter of debate: originally, it had been attributed to a stretching vibration of the silicon-oxygen bond in Si-OTi bridges,[48,152] but later titanyl groups (ν Ti=O)[17,99] and silanol groups (ν Si-OH),[160] which are expected to resonate in the same frequency region, have also been invoked. In addition to what has been mentioned earlier in this section, there is substantial evidence for the original assignment which can be summarized as follows:

- The position of the 960 cm^{-1} band is not changed by treatments with D_2O.[104] In contrast, a clear secondary isotope effect was found for Si-O vibrations in silanols on silica.[161]
- A titanium-oxygen double bond would give rise to a charge transfer band in the visible region (25,000-30,000 cm^{-1}): such a band is not observed. Instead, a band is found near 48,000 cm^{-1} which can be adequately explained as originating from tetrahedral titanium.[152]
- EXAFS results show that the range of titanium-oxygen bond lenghts for titanium sites in TS-1 is very narrow,[145] which renders the presence of a titanium-oxygen double bond short enough to resonate at 960 cm^{-1} unlikely. Three groups independently found values close to 1.81 Å for the Ti-O bond length, which is close to what one would expect for a symmetrical tetrahedral oxygen coordination.
- No ready exchange with $^{18}O_2$ occurs, not even at elevated temperatures.[152]

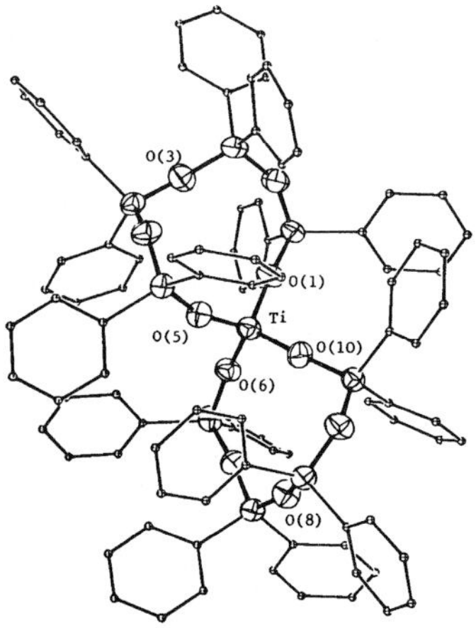

Figure 9. The tetrahedral titanium site in the spiro compound $Ti[O(OSiPh_2)_4]_2$[164]

Detailed treatment of the Si-OTi vibrational mode and of some of the arguments listed here were given by Boccuti et al.[152], Astorino et al.,[162] and as part of wider computational studies by Smirnov et al.[146] and de Man et al.[163] The latter, an extensive *ab initio* study of molecules representing possible titanium geometries in titanium silicalites, favored isomorphous substitution of titanium for silicon over other possibilities.

In summary, we think that for (dry, calcined) TS-1, the model according to which titanium occupies tetrahedral framework positions, substituting for silicon, is the only model that can consistently explain all the present data. We could expect the site to look somewhat like the central tetrahedral titanium atom in the siloxide whose structure is rendered in Fig. 9.[164] The question which then remains to be answered is: what can be the reactivity of such a tetrahedral framework atom, and what are the interactions that give it its catalytic properties ?

On the reactivity of the titanium site

Boccuti et al.[152] studied the interaction of protic molecules with TS-1 using infrared and DR UV-Vis spectroscopy. It was shown that adsorption of water, ammonia and methanol broadened and shifted the 960 cm^{-1} band somewhat and lowered its intensity. Adsorbed water also causes a shift of the UV-Vis adsorption maximum from 48,000 to 42,000 cm^{-1} which indicates an increase in the coordination number of titanium. Adsorbed ammonia causes an additional resonance to appear at 38,000 cm^{-1}, which is ascribed to a $NH_3 \rightarrow Ti(IV)$ ligand-to-metal charge transfer.[154] The changes in infrared and UV-Vis spectra are reversed on outgassing. When TS-1 is contacted with labeled water and subsequently degassed, a shift of the 960 cm^{-1} band to 952 cm^{-1} ($H_2{}^{17}O$) or 937 cm^{-1} ($H_2{}^{18}O$) is observed.[104] In addition, $H_2{}^{17}O$ treatment of TS-1 causes the appearance of a narrow signal at 360 ppm in its ^{17}O MAS NMR spectrum. In view of the observed shift of the 960 cm^{-1} band, it is likely that this signal arises from a Si-^{17}O-Ti bridging oxygen. Taken together, these results show that expansion of the titanium coordination shell is accompanied to some degree by hydrolysis and recondensation reactions of Si-O-Ti bonds (see Fig. 10).[165]

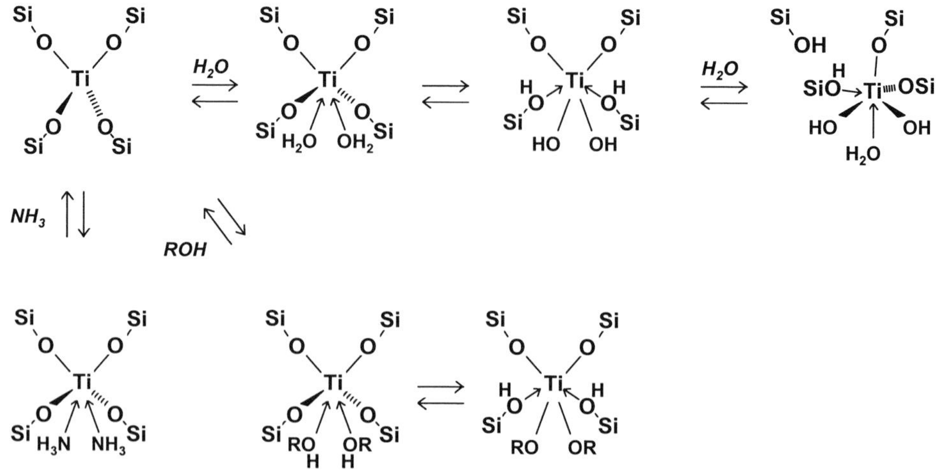

Figure 10. Some of the possible reactions of the titanium site in TS-1 with protic adsorbates.

The use of XANES and EXAFS spectroscopy[154] allowed detailed analysis of the interaction between the titanium site and adsorbed ammonia: the coordination number of titanium was shown to increase from four to six, while the average titanium-ligand bond length increased from 1.81 Å to 1.84 Å, accompanied by an increase in bond length variation. Outgassing at 400 K largely reversed these changes. Similar results were observed by Lopez et al.[45] for water adsorption on TS-1 using XANES, EXAFS, ^{17}O MAS NMR and $^{47,49}Ti$ NMR. $^{47,49}Ti$ NMR signals were only observed for hydrated samples, which might have been due to the averaging effect of water ligand exchange. A summary of the above is the reaction scheme displayed in Figure 10.

A remarkable aspect of the site's coordination chemistry, and an important one in terms of catalysis, is its resistance to more extensive hydrolysis. In contrast, isolated titanium species on amorphous silica are hydrolysed by ambient moisture.[159] Tetrakis-siloxides of titanium are particularly resistant to hydrolysis,[166] which appears to be more than a mere kinetic effect. For example, the compound depicted in Fig. 9 is synthesized from $Ti(OR)_4$ and $Ph_2Si(OH)_2$ with simultaneous formation of water; its stability is presumably due to O $\rightarrow$ Ti π-bonding.[164] Compared to such simple siloxide complexes, the structure surrounding the titanium site in TS-1 can moreover be viewed as a pre-organised ligand, which gives additional energetic and entropic stability towards complete hydrolysis. The pronounced hydrophobic character of the TS-1 pores should provide additional kinetic stability. It is reasonable to assume that the primary driving force for hydrolysis of the titanium site is the expansion of its coordination shell (rather than the formation of Ti-OH bonds *per se*), and that further hydrolysis cannot take place unless Ti-O-Ti bridges are formed.

Of most direct interest with respect to catalysis is the nature of the interaction of the site with hydrogen peroxide. A limitation in studying surface (hydro-) peroxo complexes of TS-1 is their low stability at ordinary temperatures, which makes them less suitable for X-ray spectroscopy or solid state NMR studies. Using DR UV-Vis spectroscopy, Geobaldo et al.[167] observed a new charge transfer band at 26,000 cm^{-1} when TS-1 was contacted with aqueous hydrogen peroxide. A band at 26,000 cm^{-1} with the same shape was also observed for the

$[TiF_5(O_2)]^{3-}$ ion, and it was assigned to a $O_2^{2-} \rightarrow Ti^{4+}$ ligand-to-metal transfer. Zecchina et al.[122] observed a similar band but at much higher wavenumbers (38,000 cm^{-1}) for a sample which had been evacuated after H_2O_2 exposure. Huybrechts et al.[36] observed a broadening and shift of the infrared 960 cm^{-1} band (*vide supra*) and the simultaneous appearance of a weak band at 880 cm^{-1} upon contacting TS-1 with hydrogen peroxide. These results allow one to conclude that hydrogen peroxide is directly coordinating to the titanium site forming a peroxo or hydroperoxo surface complex which is probably five-coordinated or octahedral, depending on the solvent loading. The binding mode of the (hydro-) peroxo anion is less obvious. In the $[TiF_5(O_2)]^{3-}$ ion, as in most stable titanium peroxo complexes, the peroxo species is bonded side-on, and the spectral similarities suggest side-on bonding for the case of TS-1 as well. However, the TS-1 complex is a reactive species, and it might have a geometry which is different from known titanium peroxo compounds, which are rather inert.[168] The bonding mode will be discussed in more detail later in this section.

When hydrogen peroxide is present, TS-1 acts as a Brønsted acid, catalysing the hydrolysis or alcoholysis of epoxides.[104] This finding suggests that the complex contains a hydroperoxo rather than a peroxo ligand,[169] or alternatively that the presence of a peroxo ligand enhances the acidity of another protic ligand.[170] Moreover, the rate of these acid-catalysed reactions is solvent dependent (in much the same way as the epoxidation reaction, see section 2.2), which suggests involvement of a solvent molecule in the complex. In a recent study, Clerici et al.[171] were able to obtain stable alkali metal containing peroxo complexes of TS-1 by treatment with alkaline hydrogen peroxide solutions (these can be considered the ion-exchanged forms of the above Brønsted-acidic complex). The complexes show an additional charge transfer band at 33,000 cm^{-1} in the UV-Vis spectrum, and a weak infrared band at 866 cm^{-1}. They do not act as oxidizing agents unless they are reacidified. The authors concluded that the complex probably contained a peroxo ion bonded side-on to a framework titanium species (see Fig. 11).

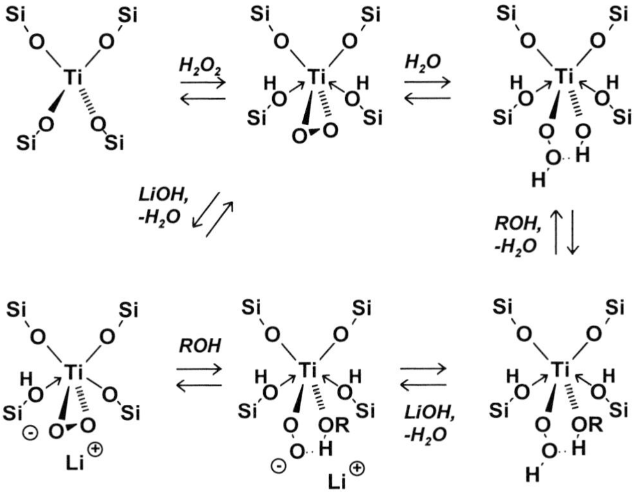

Figure 11. Some of the possible reactions of the titanium site in TS-1 with hydrogen peroxide.

About the acitive complex, and the mechanism of electrophilic oxygenations catalysed by TS-1

Discussions on mechanism in heterogeneous catalysis often rely on analogies with homogeneous systems, which are usually easier to characterize, and allow one to do kinetic studies in a straightforward way. In the case of TS-1, it is difficult to find suitable homogeneous analogues, since soluble titanium complexes do not catalyse oxidations with hydrogen peroxide. The related titanium-on-silica system, which catalyses the epoxidation of alkenes with alkyl hydroperoxides,[172] does not offer many clues, since it has not been studied in great detail.

To simplify the discussion, let us focus on the TS-1 catalysed epoxidation reaction, which is in several aspects similar to its homogeneous transition metal-catalysed counterpart (see section 2.2). Looking for clues, one is confronted with the many unsolved problems that even surround the homogeneous case.[103,173,174] Nevertheless, consensus exists on the following points that can be relevant to our case:

i. Complexes that catalyse the heterolytic epoxidation of simple alkenes with hydrogen peroxide or alkyl hydroperoxides are almost always Lewis acids.

ii. Complexes containing a transition metal with an empty d-shell (e.g. Mo(VI), Ti(IV), Re(VII) in CH_3ReO_3) are much more effective than other Lewis acids.

iii. In all known stable peroxo complexes of the group 4 - 7 transition metals, the peroxo moiety is bonded side-on (η^2); in alkylperoxo complexes η^2-bonding also occurs but is asymmetric, the bond between the alkyl-bearing oxygen and the metal being considerably longer than the other metal-oxygen bond. Efficient epoxidation is thought to require smooth oxygen transfer without disruption of metal-oxygen bonding, as depicted in Fig. 12(a) (so-called O^1,O^2-migration of the metal).[175]

iv. Three-membered rings (Fig. 12(a)) are not the only structures from which oxygen can be transferred smoothly: organic peracids epoxidize *via* a very different mechanism involving a five-membered ring (see Fig. 12(b)). Note that also in this case, no bonds are disrupted during the transfer reaction.

v. Homogeneous metal-catalysed epoxidation and epoxidation by peracids experience similar solvent effects: polar solvents enhance the reaction rate, except for strongly coordinating or protic solvents which slow down or completely inhibit the reaction. With the exception of TS-1, this also applies to homogeneous and heterogeneous catalysis by titanium.

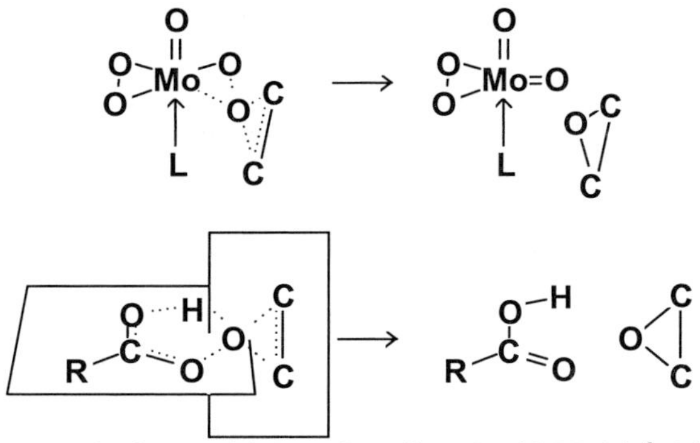

Figure 12. Transition states in the epoxidation of an alkene by (a) $MoO(\eta^2-O_2)_2 \cdot L$ and (b) a percarboxylic acid.

In addition, a proposal for the mechanism of epoxidation catalysed by TS-1 has to account for the following observations:

vi. None of the mechanisms postulated for homogeneously catalysed epoxidations *per se* can adequately explain the solvent effects observed for TS-1, and in particular the beneficial influence of methanol and ethanol (see section 2.2).

vii. Stable η^2-O_2 titanium peroxo complexes are known, but unlike their $MoO(\eta^2-O_2)_2L$ relatives, they are more stable and less reactive than hydrogen peroxide itself, and none of them acts as an oxygen transfer agent. In one known case, such an unreactive $Ti(\eta^2-O_2)$ unit can even be converted into a Ti(OH)(OOR) unit which *does* effect epoxidation of alkenes.[176]

viii. The active species in TS-1 possesses Brønsted acidity and ion exchange properties (see also section 2.3).

Oxygen transfer from a $Ti(\eta^2-O_2)$ species was proposed by Huybrechts et al.[62] and Notari et al.[18] and this does as such seem to be a most plausible mechanism. It does not *per se* explain the solvent effects (*vi*), but those could arise from selective sorption by the silicalite matrix, as suggested by Romano et al.,[74] or one could explain them by assuming that a certain degree of solvolysis of TiOSi bridges is necessary to make the peroxo moiety easily approachable. Perhaps more important is the lack of reactivity of known $Ti(\eta^2-O_2)$ complexes (*vii*). Ab initio calculations[177] show that for *hydro*peroxides, the peroxo moiety in $Ti(OH)_3(OOH)$ has a tendency to bind as η^1-OOH, and predict that this complex is capable of oxidising alkenes *via* an η^2-OOH transition state. This is an indication that the "η^2 mechanism" is plausible but that it might not involve a stable η^2-OOH intermediate. It would be interesting if such calculations could be done on systems that are closer to the actual structure of the active complex (*i.e.* including siloxide ligands and solvent molecules).

Clerici et al.[90] rather interpreted the solvent effects (*iv*) as evidence for the participation of solvent molecules in the reaction. They formulated a mechanism for TS-1 catalysed epoxidation which involves a hydroperoxo rather than a peroxo species, and coordination of an alcohol molecule to the site (see Fig 13).

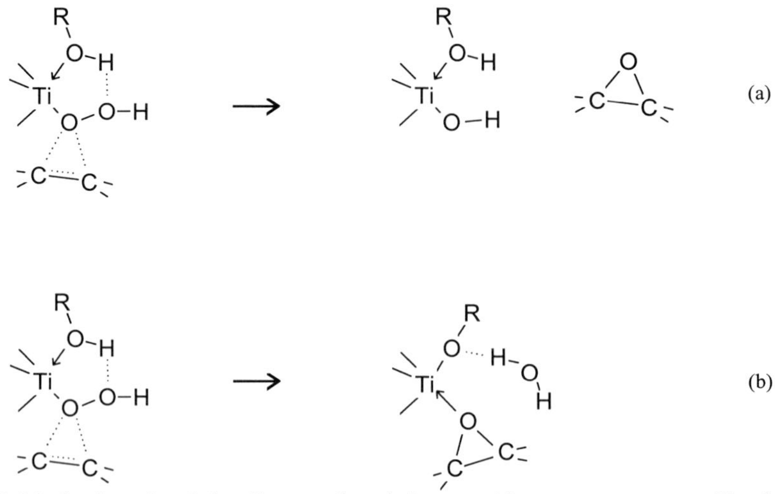

Figure 13. Mechanisms involving five-membered ring transition states as proposed by (a) Clerici et al.[90] and (b) Neurock et al.[178]

Note that pathway (a) in Fig. 13 is in fact not very different from the "η^2 mechanism" in that it also involves a metal 1,2-shift; pathway (b) involves a transition state similar to that of epoxidation by a percarboxylic acid, and has been identified as a plausible mechanism by means of DFT calculations, even though a barrier height was not computed.[178] In either case, coordination of the alcoholic solvent is expected to become increasingly difficult with increasing bulkiness, which would explain the rate decrease in the series: methanol > ethanol > *t*-butanol. In aprotic solvents, water can take over the role of the alcohol, but is apparently less effective. The five-membered ring thus formed should then allow for smooth oxygen transfer, analogous to the mechanism of oxygen transfer from peracids (*vide supra*). The scheme also accounts for the acidity of the TS-1/H_2O_2 complex. According to the "percarboxylic acid" description, the catalytic action of the framework titanium(IV) ion is essentially that of a neutral, oxophilic Lewis acid, and no specific chemical property of titanium, such as its d^0 configuration (see *ii* and *iii*), is invoked.

Synthesis and isolation of titanium(IV) (hydro-)peroxo complexes capable of effecting epoxidation and more detailed calculational studies would surely help to further elucidate the mechanism of TS-1 catalysed epoxidation. For the oxygen transfer steps in other reactions catalysed by TS-1 the situation is less clear: some plausible hypotheses have been sketched (we briefly discussed some possibilities in section 2.2), but as to yet they lack convincing experimental support.

3. Vanadium-containing silicalites

The encouraging results obtained in selective oxidations by the use of TS-1 gave rise to a growing interest in the isomorphous substitution of transition metals other than Ti in silicalites. The well known catalytic properties of supported vanadium oxides for selective oxidations and ammoxidations of organic compounds,[179] as well as some similarities between vanadium and titanium in their catalytic peroxo chemistry have prompted a number of studies on the possibility to insert vanadium in the framework of crystalline microporous silicas and aluminosilicates. In this section we will briefly review the synthesis and characterization of V-containing silicalites having the MFI or the MEL structure. Some other V-containing materials (ZSM-48 and NCL-1) are also briefly discussed. Finally, we will try to describe some aspects of the catalytic chemistry of V-containing silicalites.

3.1 Synthesis

For a better understanding of the problems related to the synthesis of V-containing silicalites, it is useful to briefly recall some of the relevant aqueous chemistry.[180] The most common oxidation states of vanadium are III, IV and V. In basic solutions (*p*H>9), vanadium(III) and vanadium(IV) species are both easily oxidized by air to give vanadate(V) anions. In fluoride-containing solutions, vanadium forms octahedral air-stable anions in oxidation states IV and V, predominantly of the type $VOF_n(H_2O)_m^{p-}$. All of these air-stable anions are highly soluble and do not seem to interact readily with silicate species.[181]

The lower oxidation states seem to provide better starting points for synthesis as their oxides are less acidic and a better interaction with silica is expected. Little is known about their aqueous chemistry in basic solutions: vanadium(IV) can form vanadate ions (most of them not well-characterized) whose composition and degree of polymerization strongly depend on *p*H and concentration. The main species in dilute, basic (*p*H>11) solutions is reportedly monomeric $VO(OH)_3^-$.[180b] The aqueous chemistry of vanadium(III) in basic solutions is very poorly documented.

The known coordination chemistry of vanadium does not justify the expectation that vanadium ions will be able to occupy tetrahedral lattice positions in a silica. A truly stable and more or less symmetrically bound tetrahedral coordination is only found for vanadium(V) in vanadates of metals that form basic or amphoteric oxides. On silica, tetrahedral monooxo or dioxo surface complexes are formed,[182] but the interaction is weak and the species are easily hydrolysed.[183] In this respect, vanadium(V) chemistry is similar to that of phosphorus(V).

The usual coordination geometry of vanadium(IV) is square pyramidal or distorted octahedral, and compounds containing tetrahedral vanadium(IV) are rare. In most compounds of vanadium(IV), the vanadyl(IV) unit (formally VO^{2+}) is present: one of the V-O bonds is very short and has essentially the character of a double bond. Tetrahedral coordination is also rarely found for vanadium(III), one possibly relevant example being the VCl_4^- ion that occurs discretely in some salts.[180]

Synthesis of vanadium-containing MFI and MEL type zeolites.

A summary of the results of hydrothermal syntheses of V-containing MFI- and MEL-type silicalites that have been reported in literature is given in Table 4. Because of the widely different gel compositions and crystallization conditions used by different authors, it is difficult to distinguish trends in these results. We will nevertheless try to make some general remarks.

Table 4. Overview of recipies for the preparation of V-containing silicalites and zeolites

N°	Reagent Mixture				Crystal. Solid obtained			Ref.
	Source of V	Templ. agent$^{(1)}$	Si/V$^{(2)}$	M^+/Si$^{(2,3)}$	Si/V$^{(2)}$	Si/Al$^{(2)}$	Crystal. phase	
1	VCl_3	HMDA	53	0.5	63	120	MFI	184
2	$VOSO_4$	TPA-Br	(4)	(4)	174	123	MFI	185
3	VCl_3	TPA-Br	(4)	(4)	178	132	MFI	185
4	NH_4VO_3	TPA-Br	42	(4)	-	106	MFI	185
5	VCl_3	HMDA	(4)	(4)	21.7	198	MFI$^{(7)}$	186
6	VCl_3	HMDA	(4)	(4)	21.6	455	ZSM-48$^{(7)}$	186
7	VCl_3	TBA-OH	(4)	(4)	30.8	669	MEL$^{(7)}$	186
8	$VO(C_2O_4)$	TPA-Br	5	~0.5	42	-	MFI	187
9	$VOSO_4$	TPA-OH	50	(5)	98	770	MFI	188
10	M^+VO_3	TPA-OH	(4)	(4)	~85	-	MFI	189
11	VCl_3	TPA-OH	31	-	50	-	MFI	190
12	VCl_3	TPA-OH	31	0.3	28	-	MFI$^{(7)}$	190
13	VCl_3	TPA-OH	15	-	58	-	MFI	191
14	VCl_3	TPA-OH	15	-	118	-	MFI	191
15	$VOSO_4$	TBA-OH	20	-	41	-	MEL	192
16	$VOSO_4$	TBA-OH	40	-	78	-	MEL	192
17	NH_4VO_3	TBA-OH	40	-	182	-	MEL	192
18	NH_4VO_3	TPA-Br	40	1.0$^{(6)}$	152	-	MFI	193

(1) HMDA=hexamethylenediamine, TPA=tetrapropylammonium, TBA=tetrabutylammonium; (2) Molar ratios; (3) M^+=alkali cation; (4) alkali metal are present in the starting mixture but it is not possible to determine the amount; (5) 0.11 equivalents of NH_3 were used; (6) M^+=NH_4+ and the synthesis is performed in the presence of F^- at pH=7.9; (7) a transformation to crystobalite occurs upon calcination in air at 550°C.

The importance of choosing the proper vanadium source was pointed out by Miyamoto et al.[185] who showed that the use of V(III) or V(IV) salts as the vanadium source, but not that of NH_4VO_3, resulted in appreciable levels of vanadium in the product. Later, this result was largely confirmed by others. The use of vanadium(IV) salts in combination with fluoride at neutral *p*H was also found to lead to very low vanadium contents.[194]

Some authors[186,190] reported that vanadium-containing zeolites and silicalites that had been prepared using VCl_3, underwent a structural collapse leading to the formation of crystobalite upon calcination in air at 550°C; the same was not observed when the samples were calcined at 550°C in a H_2 atmosphere. Such behaviour was in fact only observed when sodium ions were present in the starting reaction mixture.[190] The structural collapse has been tentatively attributed to the oxidation of V(III) or V(IV) species very well dispersed in the zeolite matrix. Since air had not been excluded in any of the reported preparations using V(III), the vanadium source must have been, at least partially, oxidized to V(IV) and V(V) already during the preparation of the precursor gel or the subsequent hydrothermal treatment.[185,190]

The use of vanadyl(IV) sulfate as the vanadium source in the absence of sodium was found to lead to thermally stable materials with more reproducible vanadium contents in the final product[188] compared to the use of V(III) sources,[190] provided that the oxidation of vanadium(IV) during gel preparation or crystallization was avoided. This approach was shown to be applicable to the synthesis of MEL-type materials as well.[192] In contrast to the compositional ceiling of Si/V≈90 found for MFI-type materials,[188] Hari Prasad Rao et al.[192] reported that the V content of the MEL-type materials increased linearly with the V content of the precursor gel when the latter was in the range 20<Si/V<160; the vanadium uptake efficiency was reported to be around 50%. On the basis of semiquantitative ESR spectra, the same authors concluded that the V(IV) content of MEL type silicalite decreased when the Si/TBAOH molar ratio in the precursor gel was chosen higher than 5.0, and that during crystallization, the V(IV) content of the solids increased with an increasing degree of crystallinity.

Synthesis of other vanadium-containing zeolites.

The synthesis of a V-containing ZSM-48 was reported by Habersberger et al.[186] from a gel containing VCl_3 as source of vanadium, hexamethylenediamine as the structure-directing agent and sodium hydroxide (see Table 4). The crystalline material thus obtained was transformed into cristobalite upon calcination in air. Tuel et al.[195] described the preparation of a V-containing ZSM-48, starting from $VOSO_4$, 1,8-diaminooctane, fumed silica and water. The zeolite was stable towards calcination in air, confirming that as for the V-containing MFI silicalite, the presence of sodium ions is responsible for the instability to thermal treatment in an oxidative atmosphere.[190] Again, the Si/V molar ratio in the crystalline solid was about twice that in the precursor gel.

The first example of a large pore V-containing zeolite was reported by Reddy et al.,[196] who described the synthesis of a V-containing NCL-1 silicalite from a reaction mixture containing fumed silica, $VOSO_4$, NaOH and hexamethylenebis(triethylammonium) bromide. The authors claim NCL-1 to be a new molecular sieve and have classified it as a large pore material on the basis of its sorption properties.[197] Synthesis of vanadium-containing zeolite beta has also been reported.[198]

3.2. Characterization of vanadium species

Most characterization studies on vanadium-containing molecular sieves have focussed on V(IV) or V(V) species. No effort has been described to characterize vanadium(III) species in the materials prepared from vanadium(III) salts. Centi et al.[191] reported that MFI-type materials prepared from vanadium(III) mainly contain vanadium(IV) and vanadium(V). They found at least three different species, one of which was an undefined vanadium-rich oxidic phase. Washing an air-calcined sample with ammonium acetate at room temperature was reported to remove most (>80%) of the vanadium. These and other complications arising from the use of vanadium(III) salts (see also section 3.1) do not seem to make them very suitable precursors.

Table 5. ESR parameters of V-containing silicalites and zeolites (from X-band spectra at R.T.)

Material[1]	V-source in initial gel	(2)	$A_{//}$(G)	$g_{//}$	A_{-} (G)	g_{-}	(3)	Ref.
MFI (2)	Tab.3	H_2	196	1.957	91	2.015	-	185
MFI (3)	Tab.3	H_2	191	1.927	89	1.994	-	185
MFI (8)	Tab.3	A.S.	186	1.949	72	1.990	sq.pyr.	186
MFI (9)	Tab.3	A.S.	199	1.933	74	1.991	sq.pyr.	188
MFI (10)	Tab.3	A.S.	192.5	1.939	87.8	1.961	sq.pyr.	189
MFI (11)	Tab.3	A.S.	191	1.95	55	2.00	sq.pyr.	190
MFI (12)	Tab.3	A.S.	191	1.93	89	1.97	sq.pyr.	190
MFI (13)	Tab.3	air[4]	203		80		d.octa.	191
MFI (14)	Tab.3	H_2[4]	154	1.911	68	1.963	d.tetra.	191
MEL (15)	Tab.3	A.S.	185	1.932	72	1.981	-	192
MFI (18)	Tab.3	H_2	189	1.925	84	1.997	-	193
ZSM-48	$VOSO_4$	A.S.	188	1.936	70	1.99	-	195
NCL-1	$VOSO_4$	A.S.	197	1.929	72	1.973	-	196

(1) The numbers between brackets refer to the entry numbers in Table 4; (2) Synthesis post-treatment: A.S. = as synthesized, air = calcined in air, H_2 = calcined in air then reduced in H_2; (3) Proposed coordination of the V(IV) species; (4) spectrum recorded at 77 K.

Most of the V-containing zeolites described in the literature are reported to contain predominantly vanadium(IV) after crystallization, even if the materials had not been prepared from a vanadium(IV) salt. ESR spectra of these V(IV)-containing samples show anisotropic hyperfine splitting caused by the ^{51}V nucleus with or without the presence of an underlying broad singlet. The singlet is thought to arise from vanadium-rich phases or oligomeric species, while the hyperfine pattern originates from isolated ions. A proper analysis of the spectra should include pattern simulation and estimation of the contribution of the singlet. A quantitative approach is preferred, because line broadening can make vanadium(IV) oligomers difficult to detect.

Linewidths in the hyperfine pattern are usually not reported, although they can be used to calculate a lower limit for the average spin-spin distance in a sample and thus indirectly the homogeneity of the vanadium distribution.[199] Using electron probe microanalysis, Rigutto and van Bekkum[188] showed that in large crystals of vanadium-containing MFI, the metal is distributed more or less homogeneously. In fact, concentration depressions even coincided with grain boundaries in intergrown crystals.

In Table 5, the ESR parameters are listed for V(IV) species in several V-containing zeolites. The values reported for g and hyperfine coupling tensors generally agree with the presence of square pyramidal or distorted octahedral $(VO)O_4$ vanadyl species.[200] In one case a low value of $A_{//}$ (154 G) was arguably ascribed to tetrahedral V(IV).[191] Obviously, a vanadyl species cannot be located at a regular tetrahedral framework site, and other possible situations of vanadium, e.g. as a *framework satellite*, have to be considered. We will come back to this later.

Several papers report the possibility to oxidize and reduce the vanadium species reversibly.[185,187,188,192,193,194,195] Upon calcination in air, V(IV) is oxidized to V(V); the former species is restored (judged from ESR spectra) by thermal treatment in hydrogen or a hydrocarbon reductant. Although nonquantitative, these results were also taken to evidence the lack of mobility of vanadium species undergoing red/ox cycles at high temperatures. Immobility, on a micrometer scale, was also evidenced by microprobe analysis.[188]

Table 6. ^{51}V NMR chemical shifts for V-containing silicalites and zeolites

Material	(1)	Pretreat. (2)	$\delta_{max}^{(3)}$ (ppm)	$\delta_{iso}^{(4)}$ (ppm)	(5)	Ref.
V-MFI	8	A.S.	-	-680	tetrah.	187a
V-MFI	9	C.	-500	-500	tetrah.	188
V-MFI	10	C.	-	-557	tetrah.	189
V-MFI	11	C.	-385	-	octah.	190
V-MFI	12	A.S.	-570	-570	tetrah.	190
V-MFI	13	C.	-480	-	tetrah.	191
V-MEL	16	C.	-	-573	tetrah.	202
V-ZSM-48	-	C.	-530	$-575^{(6)}$	tetrah.	195

(1) Reference to entry number of materials listed in Table 4; (2) A.S.= as synthesized, C. = calcined in air; (3) location of the maximum absorption intensity in the static spectrum; (4) location of the center band in MAS spectra; not corrected for second order quadrupolar shifts (5) attribution of the vanadium coordination made in the refs.; (6) value obtained from simulation of the powder pattern. All shifts are relative to $VOCl_3$.

The local symmetry of the V(V) species present in calcined samples has been investigated by ^{51}V NMR spectroscopy; Table 6 lists some results reported in the literature. The many ^{51}V NMR spectra of reference vanadium compounds that have been published in the last few years[182,201] have greatly facilitated the interpretation of the usually poorly resolved spectra obtained from solid catalysts. From the data shown in Table 6, it appears that in most calcined vanadium-containing silicalites, V(V) is present in tetrahedral coordination. In some cases, the MAS spectra display poorly resolved spinning sideband patterns due to chemical shift dispersion; in all of these materials, multiple tetrahedral vanadium(V) species are present. Very narrow MAS signals are observed for some other materials,[188,195,202] probably due to reaction of the site with water; a fast hydrolysis/condensation equilibrium is expected to give rise to anisotropy averaging. Recent data show that in completely dehydrated V-MFI samples the main species is a tetrahedral monomer with low symmetry (C_{2v} or lower: σ_{11}=-370 ppm,σ_{22}=-560 ppm,σ_{33}=-950 ppm).[203] ^{51}V spectra also showed that the species reacts more or less readily with ambient water and alcohols, depending on the hydrophobicity of the molecular sieve matrix. Similar observations have been made for vanadium-containing ZSM-12.[204]

Several authors observed a band at about 960 cm^{-1} in the IR[189,190,192,195,196] or Raman[203] spectra of calcined V-containing silicalites that was not present in the spectra of the corresponding silicalites prepared in the abscncc of vanadium. As was discussed in section 2.2 of this chapter, in TS-1 an IR or Raman band at about 960 cm^{-1} may best be attributed to an asymmetric stretching mode of framework SiO_4 tetrahedra bound to another ion (*in casu* Ti). By analogy, the IR band at 960 cm^{-1} has been taken as evidence for the presence of V-O-Si bridges in vanadium-containing silicalites.[192,195] A direct correlation between vanadium content and unit cell parameters has been reported for V-containing MEL[192] and V-containing ZSM-48,[192] while for MFI-type materials such a correlation was absent.[188,189] For

V-containing MEL type silicalite in particular, the relative intensity of 960 cm^{-1} IR band and the unit cell volume were reported to increase linearly with the vanadium content.[192b]

Adsorption of ammonia, pyridine and acetonitrile revealed the presence of Lewis and weak Brønsted sites on V-MFI[191] and V-MEL[202] that were not observed in the corresponding silicalites. In agreement with the presence of weak Brønsted acidity, an ion exchange capacity was observed for V-containing MFI-type silicalites.[188]

On the basis of the results mentioned above, hypotheses on the environment of V(IV) and V(V) were given by some authors. Figure 14 depicts the local vanadium environments proposed by Rigutto et al.[188] and Centi et al.[191] For comparison, the structure of $VO(OSi)_3$ species on amorphous, dehydrated silica[205] and the vanado(IV)silicate anion chain in the mineral haradaite,[206] $Sr_2[(VO)_2Si_4O_{12}]$ are also shown.

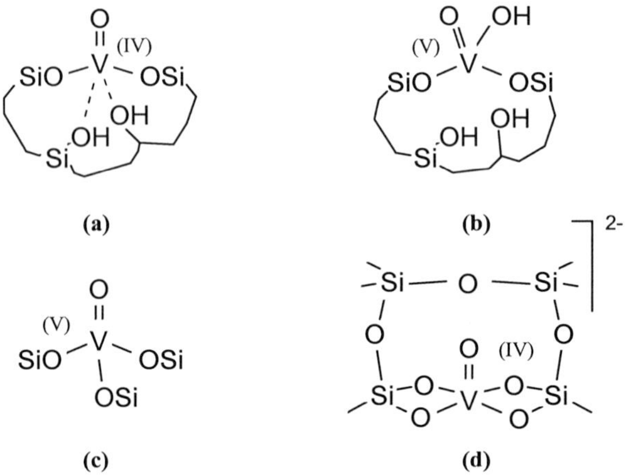

Figure 14. Structures for the vanadium site in V-containing silicalites (a and b), in V-on-amorphous-SiO_2 catalysts (c) and in the mineral haradaite (d).

Structure of the vanadium site: Conclusive remarks

The results reported in literature suggest that it is possible to prepare vanadium-containing zeolites and silicalites in which isolated vanadium species are probably connected to the framework at defect sites forming so-called *framework satellites*. There is no tendency for vanadium to adopt a symmetrically binding tetrahedral coordination, and no evidence is found in the literature for *isomorphous* substitution. One can only speculate about the nature of the defect sites to which vanadium binds. The structures shown in Fig. 14 depict a lattice vacancy as the binding site. Although silicalites are often found to contain lattice defects, the presence of vacancies is a matter of controversy;[207] one nevertheless could imagine their formation to be induced by the presence of vanadium. Dessau et al.[207a] proposed a different structure for the defects in silicalites, consisting of a hydrolysed Si-O-Si bond in a four ring; such a site could also bind a vanadyl ion. The structure of the vanadium site in the mineral haradaite (see Fig. 14) provides us with an indication of the coordinative requirements of vanadium(IV) in a similar environment. The satellite species in vanadium-containing silicalites is accessible to molecules that can enter the micropores, and can apparently change its coordination and oxidation state reversibly. The oxidized vanadium(V) species is both a weak Lewis acid and a weak Brønsted acid.

The formation of vanadium-rich impurities during crystallization as well as the, apparently related, phenomenon of thermal instability can best be prevented by using a vanadium(IV) salt as the precursor while avoiding the presence of alkali metal ions[188,192] and oxygen during synthesis.[188]

3.3 Catalytic reactions

Vanadium-containing silicalites have been tested as oxidation catalysts in the presence of different kinds of oxidants: O_2 and N_2O have been used in gas phase reactions,[187,208] and H_2O_2 in liquid phase reactions.[202,209,210]

Gas phase reactions

In the oxidative dehydrogenation of lower alkanes, a V-containing-silicalite showed a lower activity but a higher selectivity towards alkenes[208a] and aromatics[187] compared to H-ZSM-5. In the oxidative dehydrogenation of propane, Zatorski[208a] et al. observed a very high selectivity to propene with either O_2 or N_2O, using a catalyst with a very low vanadium content (see Table 7). The observed behaviour for V-silicalite and other modified silicalites was at least partially attributed to the presence of so called "defect sites" but no hypothesis was given on the reaction mechanism. On the basis of an extensive characterization of the same catalyst, Bellussi et al.[208b] later suggested the following reasons for its remarkable selectivity: (*i*) the presence of isolated tetrahedral vanadium(V) as the sole site of oxygen activation; (*ii*) a limited tendency for oxygen insertion or alkene activation at allylic positions due to the presence of stable V-O-Si bonds containing non-electrophilic bridging oxygen ions, and a stable coordination of vanadium; (*iii*) the absence of strong Brønsted acid sites.

In methanol oxidation, vanadium-containing MFI-type silicalites prepared from Ludox and $Si(OEt)_4$ and a 1 % vanadium-on-silica catalyst were found to show similar selectivities to formaldehyde (> 90%).[203] Higher activities were found for the silicalites, neatly proportional to the Henry constants for methanol adsorption: intrinsic rate constants were found to differ by a factor of two at most.

Table 7. Oxidative dehydrogenation of propane of metal-containing silicalites[208]

Catalyst	Si/M	Oxidant	T (°C)	Conv. (%)	Selectivity (%)				
					C_1+C_2	C_3H_6	C_4+C_5	arom.	COx
H-ZSM-5	254	O_2	450	18.5	5	45	5	19	26
Fe-MFI	250	O_2	450	20.1	3	40	2	25	30
B-MFI	36	O_2	450	5.6	2	74	-	-	24
V-MFI	270	O_2	450	12.5	2	73	-	-	25
H-ZSM-5	254	N_2O	350	18.9	4	39	5	32	20
B-MFI	36	N_2O	350	2.1	3	84	-	-	13
V-MFI	270	N_2O	350	12.5	1	95	-	-	4

(1) Performed in a plug flow reactor, space velocity 564 h^{-1}, feed: 2.3% propane, 1.65% O_2 (or 9.6% N_2O), balance He, atmospheric pressure, activities measured after 1h time on stream.

Liquid phase reactions

Most of the research on oxidations with H_2O_2 in the presence of V-containing silicalites has been performed by the research group of the National Chemical Laboratory in Puna (India).[202,209] The catalysts used were V-containing MEL-type silicalites (see Table 4, entries 15 and 16). The results reported in alkane oxidation and phenol hydroxylation are comparable to those of a reference TS-2, but selectivities, activities and hydrogen peroxide yields are lower than what has been reported for TS-1 (Tables 7 and 8). Results similar to those obtained with V-MEL have been reported for V-containing ZSM-48.[195]

Table 8. Oxyfunctionalisation of hexane catalysed by Ti- and V-containing silicalites

Catalyst (1)	Conv. (%)	H_2O_2 conv. (%)	H_2O_2 sel. (%)(2)	Product selectivity (%)							Ref.
				1-ol	2-ol	3-ol	1-al	2-one	3-one	other	
V-MEL (3)	14.6	100	57.1	3.7	9.2	8.2	7.2	26.3	25.0	21.4	209b
TS-2 (3)	15.9	100	58.6	-	19.1	17.1	-	23.7	23.0	16.6	209b
TS-1 (4)	27.6	98	86.0	-	32.1	25.9	-	39.8	2.0	-	138

(1) V-MEL, Si/V=79; TS-2, Si/Ti=77; TS-1, Si/Ti=46; (2) H_2O_2 used for monofunctional product formation; Reaction conditions: (3) cat=0.1 g, n-hexane=5 g, n-hexane/H_2O_2 (mol)=3, solvent (acetonitrile)=20 g, T=373 K, reaction duration=8 h; (4) cat=0.856 g, n-hexane=2.6 g, n-hexane/H_2O_2 (mol)=2.19, solvent (methanol)=50 g; T=328 K, reaction duration=1h.

The main reported differences in the behaviour of Ti-containing silicalites and V-containing silicalites are the higher activity of the latter in the oxyfunctionalization of primary carbon atoms in alkanes and also of toluene,[209,210] indicating a more pronounced homolytic tendency of the active complex. Following a proposal by Mimoun et al.[87] for the mechanism of hydroxylation by oxoperoxovanadium(V)picolinate complexes, Ramaswamy et al.[209b] assumed the involvement of vanadium(IV) superoxo complexes in these reactions for the case of V-MEL as well.

Table 9. Phenol hydroxylation catalysed by Ti- and V-containing silicalites

Catalyst (1)	Conv. mol (%)	H_2O_2 conv.(%)	H_2O_2 (2) sel.(%)	Product Distribution (%)			Ref.
				pBQ(3)	HQ(4)	Catechol	
V-MEL (5)	24.3	100	55.7	3.9	52	44.1	209b
TS-2 (6)	20.8	100	70.0	0.9	49.8	49.3	209b
TS-1 (7)	27.0	100	82.0	1.0	50.0	49.0	74

(1) V-MEL, Si/V=79; TS-2, Si/Ti=29; TS-1, Si/Ti=34; (2) H_2O_2 used for monofunctional product formation; (3) *p*-benzoquinone; (4) hydroquinone; (5) cat.=0.1g, phenol=1g, phenol/H_2O_2 (mol)=3, T=333K, reaction duration=8h, solvent=water; (6) cat=0.1g, phenol=1g, phenol H_2O_2 (mol)=3, T=342K, reaction duration=29h, solvent=acetone; (7) cat=0.72g, phenol 20.7g, phenol/H_2O_2(mol)=3.8, T=373K; reaction duration=1h, solvent=water/acetone.

Rigutto et al. observed similar selectivity patterns in V-MFI-catalysed reactions but generally with much lower peroxide efficiencies (20 % or lower) than found for V-MEL.[211] Moreover, for V-MFI as well as V-MEL,[212] partial (~5 %) leaching of vanadium was observed, perhaps not sufficient to explain the observed activity but a worrying indication of lack of stability. It is difficult to see why V-MFI and V-MEL would behave differently given the fact that the vanadium sites in both materials have similar spectroscopic signatures (see section 3.2). De Vos et al.[3] remarked that in liquid phase oxidations with hydrogen peroxide, vanadium-containing materials give rise to selectivities that are very much like those of the homogeneously catalysed reactions. It seems justified to question whether vanadium silicalites indeed act as truly heterogeneous catalysts in liquid-phase reactions.

4. Other transition metal-containing silicalites

Ferrisilicate analogs of many different high-silica zeolites have been described and framework siting of iron(III) ions in these materials is well-established.[213] Framework iron(III) species in silicalites do however not seem to have interesting redox properties, with the interesting possible exception of iron-silicalite-catalysed oxidation of benzene to phenol with N_2O in the gas phase[214] and the non-catalytic oxidation of methane to methanol with reactive oxygen species of unknown nature generated by decomposition of N_2O on iron-silicalite.[215] The role of transition metal species in the catalysis of oxygen transfer from N_2O is a matter of debate: an earlier interpretation invokes acid sites and intermediacy of N_2OH^+ ions as possible donors of incipient "OH^+".[13]

Some interesting results in the catalytic oxidative cleavage of unsaturated compounds to aldehydes with hydrogen peroxide have recently been reported for chromium-containing MFI-type silicalites prepared from chromium(III) salts.[216,217] *E.g.*, oxidation of methyl methacrylate with non-alkaline hydrogen peroxide was reported to give methyl pyruvate in 80% selectivity at 95% conversion.[216] In homogeneous systems, chromium is not a practical catalyst because it efficiently decomposes hydrogen peroxide.[218] Catalysis in reactions with *t*-butyl hydroperoxide (TBHP) has also been reported,[219,220] but in view of the fact that TS-1 does not catalyse oxidations with this reagent, it is quite possible that the chromium-containing silicalite catalysts in question may have been subject to leaching.

The chromium species in these silicalites have not been characterized in any great detail; Sheldon et al.[217] claimed isomorphous framework substitution by chromium, but evidence was not provided. Tetrahedral coordination of chromium(III) is very rare and the reported green colour of as-synthesized materials[217] suggests octahedral coordination. For MFI-type silicalites prepared from $Si(OEt)_4$, $Cr(NO_3)_3$ and alkali-free Pr_4NOH, low chromium incorporation efficiencies were observed;[221] no effect on unit cell parameters was found, which led the authors to conclude that the chromium species is not a framework species. Similar conclusions were drawn later by van der Puil et al.[222] Spoto et al.[223] prepared a chromium-containing silicalite by grafting of chromic acid onto internal silanol groups at 973 K followed by reduction, and concluded from infrared studies that a framework-connected chromium(II) species is formed which readily interacts with NO and CO.

5. Transition metal-containing aluminophosphates

Szostak[224] already remarked that in general, $AlPO_4$'s seem more flexible in their ability to incorporate low levels of foreign ions into their structures than silica-based materials. This generalization does not seem to apply to transition metals, for which there are only three well-established cases of isomorphous substitution for each class of materials: iron, cobalt and zinc for aluminophosphates, and titanium, iron and zinc for silica-based molecular sieves. There is however reason to believe that aluminophosphate frameworks are more tolerant towards non-tetrahedral ions. Some structures are known to contain non-tetrahedral aluminium bearing hydroxide ions (*e.g.* $AlPO_4$-17[225]) or water molecules (*e.g.* VPI-5[226]). Several studies indicate that reversible hydration of framework aluminium, while only rarely observed for zeolites, is a common feature of $AlPO_4$'s.[20,225] In this section, one case of isomorphous substitution (CoAPO) and two supposed cases of non-isomorphous substitution (VAPO and CrAPO) are treated.

5.1. CoAPO-5, CoAPO-11 and other cobalt-containing $AlPO_4$'s

Low levels of cobalt(II) can be incorporated into the framework of $AlPO_4$-5[227] and $AlPO_4$-11,[228,229] substituting for aluminium on tetrahedral framework sites, as evidenced by (quantitative) diffuse reflectance UV-Vis and near-infrared spectra (DREAS), ESR and ^{31}P NMR measurements. There is quite some disagreement among the authors cited here about the maximum attainable content of framework cobalt, but cobalt incorporation up to a level of Co/(Co+Al+P)=0.005 (molar ratio) is apparently feasible. According to Schoonheydt et al.,[227a] octahedral cobalt(II) (which they assumed to be non-framework material) is very difficult to detect by DR uv-vis spectroscopy when much more strongly absorbing tetrahedral species are also present. Still lower maximum contents of framework cobalt have been reported for CoAPO-16 and CoAPO-34.[228,230] In view of the very high framework cobalt content of CoAPO-50 (Co/(Co+Al+P)=0.19) these apparent compositional ceilings are somewhat difficult to explain.[231]

Upon calcination in air or oxygen, both tetrahedral cobalt(III) and strongly distorted tetrahedral cobalt(II) species are formed in CoAPO's.[229,230] The oxidized species are reduced back to tetrahedral cobalt(II) by treatment with hydrogen at 773 K[229] or even with methanol (not a particularly good one-electron donor) at room temperature.[228,230] Infrared studies indicate that tetrahedral framework cobalt(II) ions in CoAPO's (in the protonated form) are capable of coordinating acetonitrile.[232]

CoAPO-11 and CoAPO-5 have both been used as catalysts in the autoxidation of p-cresol to p-hydroxybenzaldehyde in methanolic sodium hydroxide.[233] A selectivity of 90 % at 90 % conversion was claimed. In a further examination of this reaction, Peeters et al.[234] found leaching of cobalt from CoAPO's and questioned whether CoAPO acts as a truly heterogeneous catalyst. Several workers[235,236] reported CoAPO-5 to be an active and moderately selective catalyst for the autoxidation of cyclohexane and *n*-alkanes in acetic acid. Selectivities to adipic acid up to 45 % were reported at conversions in the range 30-40%. The cobalt(III) ion is a very strong oxidant; its regeneration from cobalt(II) in the above reactions is believed to be effected by homolytic reduction of the hydroperoxide intermediates formed in the autoxidation cycle,[236] as in the homogenous case. Here also, the question arises whether the observed catalysis has been truly heterogeneous. Analysis of spent catalyst employed in acetic acid medium indicated leaching and reprecipitation of cobalt.[237] It appears however that in acid-free media, cobalt species in CoAPO are stable and CoAPO-5 and -11 display stable activity in cyclohexane autoxidation.[237]

5.2. Vanadium-containing $AlPO_4$-s: VAPO-5 and VAPO-11

The synthesis of pure VAPO-5 using V_2O_5 as the vanadium source has been reported by Montes et al.[238] and Jhung et al.[239] In both cases, vanadium was largely reduced during synthesis, and the crystalline product was shown by ESR to contain mainly well-dispersed monomeric vanadyl(IV) species, which were oxidized to vanadium(V) upon calcination at 773 K. Using quantitative ESR measurements, Jhung et al[239] showed that up to 80% of the vanadium(V) could be reduced to monomeric vanadyl(IV) species with ESR parameters nearly identical with those of the species present in the as-synthesized materials, by treatments with reductants like xylenes or trimethylbenzenes at 373 K. The presence of well-dispersed, immobile vanadium species was taken to evidence framework siting of the materials. Both Jhung et al.[239] and Montes et al.[238] made the assumption that vanadium(IV) substituted for phosphorus(V), mainly on the basis of elemental analyses and a generalization of the predictions by Flanigen[240] on the incorporation of tetravalent and pentavalent elements. Rigutto et al.[241] found compositions consistent with substitution for aluminium rather than phosphorus for VAPO-5 synthesized in the presence of fluoride using vanadyl(IV) sulfate as the vanadium source. In addition, ^{51}V NMR showed that at low vanadium contents, the dominant species in calcined materials is a square pyramidal or distorted octahedral vanadyl(V) species, consistent with a phosphate environment but difficult to explain when a second coordination shell of aluminium ions is assumed. In any case, the materials are fairly complex and materials from different preparations can show significantly different spectroscopic signatures.

VAPO-5 was found to catalyse oxidations with *tert*-butyl hydroperoxide (TBHP), such as the epoxidation of allylic alcohols (>95% selectivity at 50% conversion of the substrate) and benzylic oxidations.[241] Shape selective poisoning indicated that the reactions occured inside the pores of the molecular sieve. Somewhat lower activities and selectivities were obtained with VAPO-11.[242] Based on ^{51}V NMR studies on the interaction of vanadium species in VAPO-5 with TBHP and some other sorbates[241] a model was proposed for the catalytic site including partial solvolysis to form peroxo complexes. Also for VAPO materials, leaching has been observed, and clear indications have been found that homogeneously catalysed reactions do occur at least in some cases.[243] Again, caution is justified.

Concepcion et al.[244] have investigated the use of VAPO-5 as a catalyst in the gas-phase oxidative dehydrogenation of ethane and propane. It was observed that in materials with additional framework-incorporated magnesium, deep oxidation is suppressed.

5.3. Chromium-containing $AlPO_4$'s

Although several papers report the synthesis of chromium-containing $AlPO_4$'s,[245] not much effort has been made to characterize these materials in detail. Helliwell et al.[246] concluded from a single crystal study of a chromium-containing aluminophosphate with the AlPO-14 structure that chromium was occupying 6% of the octahedral framework sites (AlPO-14 contains tetrahedral, five-coordinated and octahedral gallium sites[247]). Non-tetrahedral framework-connected ions (cf. vanadyl species in VAPO's) might also occur in other chromium-containing AlPO's. Diffuse reflectance UV-Vis spectra seem to indicate that octahedral chromium(III) is present in as-synthesized CrAPO-5 which is oxidized to tetrahedral chromium(VI) upon calcination.[245b]

Chen et al.[220,248,249,250,251,252] reported chromium-containing AlPO-5 to be an active and selective catalyst in the decomposition of hydroperoxides towards ketones (*e.g.* 86% selectivity to cyclohexanone from cyclohexyl hydroperoxide at 87% conversion), in benzylic oxidations and in the (aut-)oxidation of secondary alcohols and of alkanes, with TBHP and

TBHP/oxygen. Lempers et al.[253] found that chromium-containing AlPO's catalyse the oxidation of alkenes to α,β-unsaturated ketones, but attributed a significant share if not all of the catalytic activity in these and other reactions involving hydroperoxides to dissolved chromium species that had been leached from the catalyst.[254]

6. Conclusive remarks

The majority of studies on the chemistry of framework transition metal sites in molecular sieves has been devoted to titanium silicalite-1. These studies show that a tetrahedral framework ion can have considerable freedom in reversibly coordinating small molecules, and that the framework, even though it acts as a pre-organised ligand, is sufficiently flexible to allow this. The apparent resistance of the titanium site to extensive hydrolysis, essential to catalyst stability, at the same time imposes restraints on the accessibility of the site; as a consequence, it appears that the site remains sufficiently electron-deficient to be an active catalyst for oxygen transfer from hydrogen peroxide. Together with those properties, site separation and selective sorption by the hydrophobic molecular sieve matrix can explain most of the important features of the catalytic chemistry of titanium silicalites. So far, this knowledge has been applied with success in the research of several groups that has led eventually to hydrophobic [Ti]-beta, a true large pore analogue of TS-1 capable of converting large molecules.

Furthermore, we have seen several examples of molecular sieve materials containing non-tetrahedral transition metal species chemically bound to their frameworks. Such materials are less-well investigated than isomorphously substituted materials. Most transition metals do not fulfill the requirements for isomorphous substitution, but might nevertheless yield interesting catalysts when anchored to the framework of a molecular sieve. Caution should however be excercised in applying such materials as catalysts in liquid phase reactions, particularly in those cases involving solubilising agents like hydroperoxides or acetic acid, as many such materials seem to be prone to leaching.

References

1. R.A. Sheldon, J.K. Kochi, Metal-Catalysed Oxidations of Organic Compounds, Academic Press, New York, 1981; J.E. Lyons, Appl. Ind. Catal., **3**, (1984), 131.
2. For a review of the early work on metal-ion exanged zeolites see: D.B. Tagiev, K.M. Minachev, Russ. Chem. Rev., **50**, 1981, 1009.
3. For a thorough review see: D.E. de Vos, Ph.L. Buskens, D.L. Vanoppen, P.P. Knops-Gerrits, P.A. Jacobs, in "Comprehensive Coordination Chemistry", Vol. 7, "Solid-State Supramolecular Chemistry: Two- and Three-Dimensional Networks", G. Alberti, Th. Bein Eds., Pergamon, (1996), p. 647.
4. Y. Li, J.N. Armor, Chem. Commun., **1997**, 2013.
5. T. Kubota, F. Kumada, H. Tominaga, T. Kunugi, Int. Chem. Eng., **13**, (1973), 539; H. Arai, T. Yamashiro, T. Kubo, H. Tominaga, Bull. Jpn. Pet. Inst., **18**, (1976), 39; P.H. Espeel, G. de Peuter, M.C. Tielen, P.A. Jacobs, J. Phys. Chem., **98**, (1994), 11588.
6. Kh. Minachev, V.V. Kharlamov, Acta Phys. Chem., **31**, (1985), 497.
7. I. Mochida, S. Hayata, A. Kato, T. Seiyama, J. Catal., **19**, (1970), 405.
8. O. Dittmann, K.L. Schiff, A. v. Fuener, P. Strehl, Ger. Offen. 2,431,034, 1976; J.E. Lyons, U.S. Pat. 4,182,722, 1980.
9. G.C. Centi, Z. Tvaruzkova, F. Trifirò, P. Jiru, L. Kubelkova, Appl. Catal., **13**, (1984), 69.
10. J.C. Oudejans, H. van Bekkum, J. Mol. Catal., **12**, (1981), 149.
11. See e.g.. C.D. Chang, B.P. Pelrine, Eur. Pat 125,065, 1984 (to Mobil Oil).
12. M. Floor, A.P.G. Kieboom, H. van Bekkum, Recl. Trav. Chim. Pays-Bas, **107**, (1988), 362.
13. M.H. Gubelmann, J.M. Popa, P.J. Tirel, Eur. Pat. 406,050, 1991 (to Rhône Poulenc); P.J. Tirel, M.H. Gubelmann, J.M. Popa, in "Recent Progress Reports at the 9th International Conference on Zeolites" (Montreal), 1992, p. 61; E. Suzuki, K. Nakashiro, Y. Ono, Chem. Lett., **1988**, 953.
14. P.H. Espeel, M.C. Tielen, P.A. Jacobs, J. Chem. Soc., Chem. Commun., **1991**, 669; P.H. Espeel, G. De Peuter M.C. Tielen, P.A. Jacobs, J. Phys. Chem., **98**, (1994), 11588.
15. M. Taramasso, G. Perego, B. Notari, U.S. Patent No. 4,410,501 (1983), to Snamprogetti S.p.A; M. Taramasso, G. Manara, V. Fattore, B. Notari, U.S. Patent No. 4,666,692 (1987), to Snamprogetti S.p.A.
16. G. Bellussi, V. Fattore, Stud. Surf. Sci. Catal., **69**, (1991), 79.
17. B. Notari, Stud. Surf. Sci. Catal., **37**, (1987), 413 and refs. cited therein.
18. B. Notari, Stud. Surf. Sci. Catal., **60**, (1991), 343 and refs. cited therein.
19. R.M. Barrer, in D.H. Olson and A. Bisio (Eds.), Proc. 6th Int. Zeol. Conf., Butterworths, 1983, 870; J.C. Vedrine, Stud. Surf. Sci. Catal., **69**, (1991), 25.
20. Aluminium species in the framework of zeolites and $AlPO_4$'s can also expand their coordination shell: R. Meinhold, N. Tapp, J. Chem. Soc., Chem. Commun., **1990**, 219; E. Bourgeat-Lami, P. Massiani, F. Di Renzo, P. Espiau, F. Fajula, T. Des Courières, Appl. Catal., **71**, (1991), 139; A.S. Medin, V.Yu Borovkov, V.B. Kazansky, A.G. Pelmetschikov, G.M. Zhidorimov, Zeolites, **10**, (1990), 668.
21. R.A. Sheldon, J.K. Kochi, Metal-Catalyzed Oxidations of Organic Compounds, Academic Press, New York, 1981.
22. Some highly relevant model studies have been performed on transition metal siloxides. See: F.J. Feher, J.F. Walzer, Inorg. Chem., **30**, (1991), 1689, and refs. cited therein.
23. S.M. Csicsery, Zeolites, **4**, (1984), 202.
24. R.M. Dessau in W.H. Frank (Ed.), "Adsorption and Ion Exchange with Synthetic Zeolites", ACS Symp. Ser., **135**, (1980), 123; for examples of solvent effects in zeolite-catalyzed reactions see *e.g.* Th.M. Wortel, W. v.d. Heuvel, N.A. de Munck, H. v. Bekkum, "Proceedings of the Symposium on Zeolites", Szeged, Hungary, 1978, p.341; P.H.J. Espeel, K.A. Vercruysse, M. Debaerdemaeker, P.A. Jacobs, Stud. Surf. Sci. Catal., **84 B**, (1994), 1457.

25. D.M. Chapman, A.L. Roe, Zeolites, **10**, (1990), 730; S.M. Kuznicki, K.A. Trush, WO Patent No. 91/18833, (1991), to Engelhardt Corp.
26. P.J. Kooyman, unpublished results.
27. T. Blasco, M.A. Camblor, A. Corma, J. Pérez-Pariente, J. Am. Chem. Soc., **115**, (1993), 11806.
28. M. Padovan, F. Genoni, G. Leofanti, G. Petrini, G. Trezza, A. Zecchina, Stud. Surf. Sci. Catal., **63**, (1991), 431.
29. M. Uguina, D.P. Serrano, G. Ovegero, R. van Grieken, M. Camacho, Appl. Catal. A, **124** (1995) 391; D.P. Serrano, M. Uguina, G. Ovegero, R. van Grieken, M. Camacho, Microporous Mater., **4** (1995) 273; *idem, ibid.,* **7** (1996) 309.
30. A.J.H.P. van der Pol, J.H.C. van Hooff, Appl. Catal. A, **92**, (1992), 93.
31. (a) A. Thangaraj, R. Kumar, S.P. Mirajkar, P. Ratnasamy, J. Catal., **130**, (1991), 1; (b) A. Thangaraj, M.J. Eapen, S. Sivasanker, P. Ratnasamy, Zeolites, **12**, 943.
32. R. Millini, E. Previde Massara, G. Perego, G. Bellussi, J. Catal., **137**, (1990), 503.
33. B. Kraushaar, Ph.D. Thesis, Eindhoven Univ. of Technology, 1989.
34. C.G. Barraclough, R.L. Martin, G. Winter, J. Chem. Soc., **1964**, 758.
35. H. Gao, J. Suo, S. Li, J. Chem. Soc., Chem. Commun., 1995, 835.
36. D.R.C. Huybrechts, I. Vaesen, H.X. Li, P.A. Jacobs, Catal. Lett., **8**, (1991), 237.
37. G. Bellussi, A. Carati, M.G. Clerici, A. Esposito, Stud. Surf. Sci. Catal., **63**, (1991), 421.
38. M. Goepper, H.-X. Li, M.E. Davis, J. Chem. Soc., Chem. Commun., **1992**, 1665.
39. R.K. Iler, "The Chemistry of Silica", John Wiley, New York, 1979, p. 150.
40. C.B. Khouw, M.E. Davis, J. Catal., 151 (1995) 77.
41. A. Thangaraj, R. Kumar, S. Sivasanker, Zeolites, **12**, (1992), 135.
42. J.L. Guth, H. Kessler, J.M. Higel, J.M. Lamblin, J. Patarin, A. Seive, J.M. Chezeau, R. Wey, in M.L. Occelli, M.E. Robson (Eds.), "Zeolite Synthesis", ACS Symp. Ser., **398**, (1989), 176; J.M. Popa, J.L. Guth, H. Kessler, Eur. Patent Appl. No. 292,363, (1988).
43. Qiu Shilun, Pang Wenqin, Yao Shangqing, Stud. Surf. Sci. Catal., **49A**, (1989), 133.
44. P.J. Kooyman, J.C. Jansen, H. van Bekkum, in R. von Ballmoos, J.B. Higgins, M.M.J. Treacy (Eds.), Proc. 9th Int. Zeolite Conf., Butterworth-Heinemann, Boston, 1992, p. 505.
45. A. Lopez, M.H. Tuilier, J.L. Guth, L. Delmotte, J.M. Popa, J. Solid State Chem., **102**, (1993), 480.
46. G. Zhang, J. Sterte, B. Schoeman, J. Chem. Soc., Chem. Commun., 1995, 2259.
47. K.T. Jung, J.H. Lee, J.H. Hyun, D.S. Kim, J.G. Kim, Y.G. Shul, Stud. Surf. Sci. Catal. **105**, (1997), 2241.
48. G. Perego, G. Bellussi, C. Corno, M. Taramasso, F. Buonuomo, A. Esposito, in Y. Murakami, A. Iijima, J.W. Ward (Eds.), "Proceedings of the 7th International Conference on Zeolites", Kodansha Elsevier, Amsterdam, 1987, p 129.
49. R.B. Greegor, F.W. Lytle, D.R. Sandstrom, J. Wong, P. Schultz, J. Non-Crystalline Solids, **55**, (1983), 27; F. Lytle, R.B. Greegor, in F.L. Galeener, D.L. Griscom, M.J. Weber, "Defects in Glasses", Mater. Res. Soc. Symp. Proc., **61**, (1985), 259.
50. A. Tuel, Y. Ben Taârit, Zeolites, **14**, (1994), 594; A. Tuel, Zeolites, **16**, (1996), 108; A. Tuel, Stud. Surf. Sci. Catal., **105**, (1997), 261.
51. M. Shibata, J. Gerard, Z. Gabelica, Stud. Surf. Sci. Catal., **105**, (1997), 261.
52. G.T. Kokotailo, P. Chu, S.L. Lawton, W.M. Meier, Nature, **275**, (1978), 119; G. Perego, M. Cesari, G. Allegra, J. Appl. Cryst., 17 (1984), 403; G. Perego, G. Bellussi, A. Carati, R. Millini, V. Fattore, in "Zeolite Synthesis", M.L. Occelli, H.E. Robson (eds.), ACS Symp. Ser., **398**, (1989), 360.
53. G. Bellussi, A. Carati, M.G. Clerici, A. Esposito, R. Millini, F. Buonuomo, Bel. Patent No. 1,001,038, (1989), to Eniricerche S.p.A. and Enichem Synthesis S.p.A.
54. J.S. Reddy, R. Kumar, Zeolites, **12**, (1992), 95.
55. A. Tuel, Y. Ben Taarit, Microporous Mater., **1**, (1993), 179; A. Tuel, Y. Ben Taarit, C.

Naccache, submitted for publication in Zeolites.
56. D.P. Serrano, H.-X. Li, M.E. Davis, J. Chem. Soc., Chem. Commun., **1992**, 745.
57. A. Tuel, Y. Ben Taârit, Zeolites, **15**, (1995), 164.
58. M.A. Camblor, A. Corma, J. Pérez-Pariente, Zeolites, **13**, (1993), 82.
59. M.S. Rigutto, R. de Ruiter, J.P.M. Niederer, H. van Bekkum, Stud. Surf. Sci. Catal., **84**, (1994), 2245.
60. J.C. van der Waal, M.S. Rigutto, H. van Bekkum, J. Chem. Soc., Chem. Commun., **1994**, 1241.
61. J.C. van der Waal, Ph.D. thesis, Delft University of Technology, Delft University Press, 1998.
62. R.J. Saxton, US Pat. 5,453,511, (1995), to ARCO.
63. M.A. Camblor, M. Constantini, A. Corma, L. Gilbert, P. Esteve, A. Martinez, S. Valencia, J. Chem. Soc., Chem. Commun., **1996**, 1339.
64. J.C. van der Waal, P. Lin, M.S. Rigutto, H. van Bekkum, Stud. Surf. Sci. Catal., **105**, (1997), 1093.
65. T. Blasco, M.A. Camblor, A. Corma, P. Esteve, J.M. Guil, A. Martínez, J.A. Perdigón-Melón, S. Valencia, J. Phys. Chem. B, **102**, (1998), 75.
66. A. Tuel, Zeolites, **15**, (1995), 236.
67. C.B. Dartt, M.E. Davis, Appl. Catal. A, **143**, (1996), 53.
68. K.J. Balkus, Jr., A. Khanmamedova, A.G. Gabrielov, S.I. Zones, Stud. Surf. Sci. Catal., **101**, (1996), 1341.
69. A. Corma, M.T. Navarro, J. Pérez-Pariente, J. Chem. Soc., Chem. Commun., **1994**, 147.
70. See *e.g.*: P.T. Tanev, M. Chimbwe, T.J. Pinnavaia, Nature, **368**, (1994), 321.
71. A. Monnier, F. Schüth, Q. Huo, D. Kumar, D. Margolese, R. Maxwell, G. Stucky, M. Krishnamurty, P. Petroff, A. Firouzi, M. Janicke, B. Chmelka, Science, **261**, (1993), 1299; J.S. Beck, J.C. Vartuli, W.J. Roth, M.E. Leonowicz, C.T. Kresge, K.D. Schmitt, C.T-W. Chu, D.H. Olson, E.W. Sheppard, S.B. McCullen, J.B. Higgins, J.L. Schlenker, J. Am. Chem. Soc., **114**, (1992), 10832.
72. A. Corma, M.T. Navarro, J. Pérez-Pariente, F. Sánchez, Stud. Surf. Sci. Catal., **84 A**, (1994), 69; presented at the 10th International Zeolite Conference in Garmisch-Partenkirchen, Germany, 1994.
73. R. Hutter, T. Mallat, D. Dutoit, A. Baiker, Topics Catal., **3**, (1996), 421.
74. U. Romano, A. Esposito, F. Maspero, C. Neri, M.G. Clerici, Stud. Surf. Sci. Catal., **55**, (1990), 33; U. Romano, A. Esposito, F. Maspero, C. Neri, M.G. Clerici, Chim. Ind., **72**, (1990), 610.
75. B. Kraushaar-Czarnetzki, J.H.C. van Hooff, Catal. Lett., **2**, (1989), 43; P.J. Kooyman, P. van der Waal, P.A.J. Verdaasdonk, J.C. Jansen, H. van Bekkum, Catal. Lett., **13**, (1992), 229.
76. A. Tuel, S. Moussa-Khouzami, Y. Ben Taarit, C. Naccache, J. Mol. Catal., **68**, (1991), 45; J.A. Martens, Ph.L. Buskens, P.A. Jacobs, A.J.H.P. van der Pol, J.H.C. van Hooff, C. Ferrini, H.W. Kouwenhoven, P.J. Kooyman, H. van Bekkum, Appl. Catal. A, **99**, (1993), 71.
77. A.J.H.P. van der Pol, A.J. Verduyn, J.H.C. van Hooff, Appl. Catal. A, **92**, (1992), 113.
78. G. Bellussi, C. Perego, in "Handbook of Heterogeneous catalysis", G. Ertl, H. Knozinger, J. Weitkamp Eds., Wiley-VCH, 1997, Vol. 5 p.2329-2334.
79. M. Constantini, E. Garcin, M. Gubelmann, J.-M. Popa, Eur. Patent Appl. 385,882, (1990), to Rhône Poulenc Chimie.
80. A. Tuel, Y. Ben Taârit, Appl. Catal. A, **102**, (1993), 69.
81. A.J.H.P. van der Pol, A.J. Verduyn, J.H.C. van Hooff, Appl. Catal. A, **96**, (1993), L13.
82. E.A. Kharakanov, S.Y. Narin, A.G. Dedov, Appl. Organomet. Chem., **5**, (1991), 445.
83. R.O.C. Norman, R. Taylor, "Electrophilic Substitution in Benzenoid Compounds", Elsevier, Amsterdam, 1965, p. 110.

84. A. Thangaraj, R. Kumar, P. Ratnasamy, Appl. Catal., **57**, (1990), L1.
85. A. Bhaumik, P. Kumar, J. Chem. Soc., Chem. Commun., **1995**, 349.
86. K. Krohn, K. Brüggmann, D. Döring, P.G. Jones, Chem. Ber., **125**, (1992), 2439.
87. H. Mimoun, L. Saussine, E. Daire, M. Postel, J. Fischer, R. Weiss, J. Am. Chem. Soc., **105**, (1983), 3101; M. Bonchio, V. Conte, F. Coppa, F. Di Furia, G. Modena, Stud. Surf. Sci. Catal., **66**, (1991), 497.
88. N.K. Mal, A.V. Ramaswamy, Appl. Catal. A, **143**, (1996), 75.
89. A. Bhaumik, P. Kumar, R. Kumar, Catal. Lett., **40**, (1996), 47.
90. M.G. Clerici, P. Ingallina, J. Catal., **140**, (1993), 71.
91. M.G. Clerici, G. Bellussi, U. Romano, J. Catal., **129**, (1991), 159.
92. H. Gao, G. Lu, J. Suo, S. Li, Appl. Catal. A, **138**, (1996), 27.
93. G.J. Hutchings, D.F. Lee, A.R. Moinihan, Catal. Lett., **39**, (1996), 83.
94. M.G. Clerici, G. Bellussi, Eur. Patent Appl. 315,248, (1992).
95. J. Sobczak, J.J. Zio_kowski, J. Mol. Catal., **13**, (1981), 11.
96. T. Itoh, K. Jitsukawa, K. Kaneda, S. Teranishi, J. Am. Chem. Soc., **101**, (1979), 159.
97. W. Adam, A. Corma, A. Martínez. C.M. Mitchell, T. Indrasena Reddy, M. Renz, A.K. Smerz, J. Mol. Catal. A, **117**, (1996), 357.
98. T. Tatsumi, M. Yako, M. Nakamura, Y. Yuhara, H. Tominaga, J. Mol. Catal., **78**, (1993), L41.
99. D.R.C. Huybrechts, Ph.L. Buskens, P.A. Jacobs, J. Mol. Catal., **71**, (1992), 129.
100. S.B. Kumar, S.P. Mirajkar, G.C.G. Pais, P. Kumar, R. Kumar, J. Catal., **156**, (1995), 163.
101. J.S. Reddy, U.R. Khire, P. Ratnasamy, R.B. Mitra, J. Chem. Soc., Chem. Commun, **1992**, 1234.
102. W. Adam, A. Corma, A. Martinez, M. Renz, Chem. Ber., **129**, (1996), 1453.
103. K.A. Jørgensen, Chem. Rev., **89**, (1989), 431.
104. G. Bellussi, A. Carati, M.G. Clerici, G. Maddinelli, R. Millini, J. Catal., **133**, (1992), 220.
105. M.A. Camblor, A. Corma, A. Martinez, J. Pérez-Pariente, S. Valencia in V. Cortés Corberán, S. Vic Bellón (Eds.), Preprints of the 2nd World Congress on New Developments in Selective Oxidation, Benalmádena, Spain, 1993, paper G.4.
106. A.Corma, M.A. Camblor, P. Esteve, A. Martínez, J. Pérez-Pariente, J. Catal., **145**, (1994), 151.
107. M.A. Camblor, A. Corma, A. Martínez, J.Pérez-Pariente, S. Valencia, Stud. Surf. Sci. Catal., **82**, (1994), 531.
108. A. Corma, M. Iglesias, F. Sanchez, J. Chem. Soc., Chem. Commun., **1995**, 1635.
109. T. Sato, J. Dakka, R.A. Sheldon, Stud. Surf. Sci. Catal., **84**, (1994), 1853.
110. A. Corma, P. Esteve, A. Martínez, J. Catal., **161**, (1996), 11.
111. J.C. van der Waal, M.S. Rigutto, H. van Bekkum, Applied Catal. A, **167**, (1998), 331.
112. J.C. van der Waal, H. van Bekkum, J. Mol. Catal. A, **124**, (1997), 137; J.C. van der Waal, Ph.D. Thesis, Delft University of Technology, Delft University Press, 1998.
113. P. Roffia, G. Leofanti, A. Cesana, M. Mantegazza, M. Padova, G. Petrini, S. Tonti, P. Gervasutti, R. Varagnolo, Chim. Ind., **72**, (1990), 598.
114. J. Le Bars, J. Dakka, R.A. Sheldon, Appl. Catal., **136**, (1996), 69; K.G. Davenport, R.A. Sheldon, J. Le Bars, W.H. Mueller, U.S. Pat. Appl. 5,466,869, 1995, to Hoechst Celanese Corp.
115. P. Roffia, G. Leofanti, A. Cesana, M. Mantegazza, M. Padova, G. Petrini, S. Tonti, P. Gervasutti, Stud. Surf. Sci. Catal., **55**, (1990), 43.
116. T. Tatsumi, N. Jappar, J. Catal., **161**, (1996), 570.
117. S. Tsuda, Chem. Econ. Eng. Rev., **1970**, 39.
118. A. Thangaraj, S. Sivasanker, P. Ratnasamy, J. Catal., **131**, (1991), 394.
119. J. Sudhakar Reddy, S. Sivasanker, P. Ratnasamy, J. Mol. Catal., **69**, (1992), 383.

120. Z. Tvaruzkova, M. Petras, K. Habersberger, P. Jiru, Catal. Lett., **13**, (1992), 117.
121. A. Zecchina, G. Spoto, S. Bordiga, F. Geobaldo, G. Petrini, G. Leofanti, M. Padovan, M. Mantegazza, P. Roffia, Stud. Surf. Sci. Catal., **75**, (1992), 719.
122. A. Zecchina, S. Bordiga, C. Lamberti, H. Ricchiardi, C. Lamberti, G. Ricchiardi, D. Scarano, D. Scarano, G. Petrini, G. Leofanti, M. Mantegazza, Catal. Today, **32**, (1996), 97.
123. S. Tonti, P. Roffia, A. Cesana, M.A. Mantegazza, M. Padovan, Eur. Patent Appl. 314,147, (1989), to Montedipe S.p.A.
124. J.S. Reddy, P.A. Jacobs, Catal. Lett., **37**, (1996), 213.
125. R.J.T. Ravindranathan, A. Sudalaj, Tetrahedron Lett., **36**, (1996), 1903.
126. S. Goutier, A. Tuel, Appl. Catal. A, **118**, (1996), 173; T. Selvam, A.V. Ramaswamy, J. Chem. Soc., Chem. Commun., **1996**, 1215.
127. R.S. Reddy, J.S. Reddy, R. Kumar, P. Kumar, J. Chem. Soc., Chem. Commun., **1992**, 84; A.V. Ramaswamy, S. Sivasanker, Catal. Lett., **22**, (1993), 239; V. Hulea, P. Moreau, F. Di Renzo, J. Mol. Catal. A, **111**, (1996), 325; A. Corma, M. Iglesias, F. Sanchez, Catal. Lett., **39**, (1996), 153; M. Crocker, C.A. Emeis, R.H.M. Herold, J. Mol. Catal. A, **110**, (1996), L7.
128. See e.g. O. Bortolini, F. Di Furia, P. Scrimin, G. Modena, J. Mol. Catal., **7**, (1980), 59 and refs. cited therein.
129. F. Maspero, U. Romano, J. Catal., **146**, (1994), 476.
130. E.V. Spinacé, H.O. Pastore, U. Schuchardt, J. Catal., **157**, (1995), 631.
131. R.A. Sheldon, J. Dakka, Erdöl, Erdgas, Kohle, **109**, (1993), 520.
132. V. Conte, F. Di Furia, in G. Strukul (Ed.), "Catalytic Oxidations with Hydrogen Peroxide as Oxidant", Kluwer, Dordrecht, 1993, p. 223.
133. S.E. Jacobson, D.A. Muccigrosso, F. Mares, J. Org. Chem., **44**, (1979), 921.
134. C.F. de Graauw, J.A. Peters, H. van Bekkum, J. Huskens, *Synthesis*, **10**, (1994), 1007.
135. For an explanation of the electrophilic nature of metal peroxo complexes see R.D. Bach, G.J. Wolber, B.A. Coddens, J. Am. Chem. Soc., **106**, (1984), 6098, and refs. cited therein.
136. A. Corma, P. Esteve, A. Martinez, Appl. Catal. A, **143**, (1996), 87.
137. D.R.C. Huybrechts, L. de Bruyker, P.A. Jacobs, Nature, **345**, (1990), 240.
138. M.G. Clerici, Appl. Catal., **68**, (1991), 249.
139. R. van Grieken, G. Ovejero, D.P. Serrano, M.A. Uguina, J.A. Melero, J.Chem. Soc., Chem. Commun., (1996), 1145.
140. G. Olah, Acc. Chem. Res., **20**, (1987), 422.
141. See e.g. M. Faraj, C.L. Hill, J. Chem. Soc., Chem. Commun., **1987**, 1487.
142. C.B. Khouw, C.B. Dartt, J.A. Labinger, M.A. Davis, J. Catal., **149**, (1994), 195.
143. To be more precise, it is a (very close) linear approximation of Vegard's law. It should be noted that, to the knowledge of the authors, it has not been verified extensively for cases where one of the pure end members does not exist. See for example: L.V. Azároff, "Introduction to Solids", McGraw-Hill, New York, 1960, p. 290.
144. H. van Koningsveld, J.C. Jansen, H. van Bekkum, Zeolites, **10**, (1990), 235.
145. S. Pei, G.W. Zajac, J.A. Kaduk, J. Faber, B.I. Boyanov, D. Duck, D. Fazzini, T.I. Morrisson, D.S. Yang, Catal. Lett., **21**, (1993), 333 and refs. cited therein.
146. K.S. Smirnov, B. van de Graaf, Microporous Mater., **7**, (1996), 133.
147. Y. Oumi, K. Matsuba, M. Kubo, T. Inui, A. Miyamoto, Microporous Mater., **4**, (1995), 53.
148. A. Tuel, Y. Ben Taarit, J. Chem. Soc., Chem. Commun., **1992**, 1578.
149. D.G. Hay, H. Jaeger, J. Chem. Soc., Chem. Commun, **1984**, 1433.
150. Z. Gabelica, J.L. Guth, Stud. Surf. Sci. Catal., **49A**, (1989), 421.
151. S.L. Njo, H. van Koningsveld, B. van de Graaf, J. Phys. Chem. B, **101**, (1997), 10065.
152. M.R. Boccuti, K.M. Rao, A. Zecchina, G. Leofanti, G. Petrini, Stud. Surf. Sci. Catal., **48**, (1989), 133.
153. A. Zecchina, G. Spoto, S. Bordiga, A. Ferrero, G. Petrini, G. Leofanti, M. Padovan, Stud. Surf. Sci. Catal., **69**, (1991), 251.

154. S. Bordiga, F. Boscherini, F. Buffa, S. Coluccia, F. Genoni, C. Lamberti, G. Lefanti, L. Marchese, G. Petrini, G. Vlaic, A. Zecchina, J. Phys. Chem., accepted for publication.&&
155. S. Bordiga, S. Coluccia, C. Lamberti, L. Marchese, A. Zecchina, F. Buffa, F. Genoni, G. Leofanti, G. Petrini, G. Vlaic, J. Phys. Chem., **98**, (1994), 4125; R.J. Davis, Z. Liu, J.E. Tabora, W.S. Wieland, Catal. Lett., **34**, (1995), 101; C. Cartier, C. Lortie, D. Trong On, H. Dexpert, L. Bonneviot, Physica B, **208/209**, (1995), 653; L. Le Noc, D. Trong On, S. Solomykina, B. Echchahed, F. Beland, C. Cartier, D. Moulin, L. Bonneviot, Stud. Surf. Sci. Catal., **101**, (1996), 611.
156. P. Behrens, J. Felsche, S. Vetter, G. Shultz-Ekloff, N.I. Jaeger, W. Niemann, J. Chem. Soc., Chem. Commun., **1991**, 678; E. Schultz, C. Ferrini, R. Prins, Catal. Lett., **14**, (1992), 221.
157. D. Trong On, A. Bittar, A. Sayari, S. Kaliaguine, L. Bonneviot, Catal. Lett., **16**, (1992), 85.
158. A. Jentys, C.R.A. Catlow, Catal. Lett., **22**, (1993), 251.
159. G. Deo, A.M. Turek, I.E. Wachs, D.R.C. Huybrechts, P.A. Jacobs, Zeolites, **13**, (1993), 365.
160. M.A. Camblor, A. Corma, J. Pérez-Pariente, J. Chem. Soc., Chem. Commun., **1993**, 557.
161. F. Boccuzzi, S. Coluccia, G. Ghiotti, C. Morterra, A. Zecchina, J. Phys. Chem., **82**, (1978), 1298.
162. E. Astorino, J.B. Peri, R.J. Willey, G. Busca, J. Catal., **157**, (1995), 482.
163. A.J.M. de Man, J. Sauer, J. Phys. Chem, **100**, (1996), 5025.
164. M.B. Hursthouse, A. Hossain, Polyhedron, **3**, (1984), 95; V.A. Zeitler, C.A. Brown, J. Am. Chem. Soc., **79**, (1959), 4618.
165. L. Bonneviot, D. Trong On, A. Lopez, J. Chem. Soc., Chem. Commun., **1993**, 685.
166. F. Schindler, H. Schmidbaur, Angew. Chem., **79**, (1967), 697.
167. F. Geobaldo, S. Bordiga, A. Zecchina, E. Giamello, G. Leofanti, G. Petrini, Catal. Lett., **16**, (1992), 109.
168. H. Mimoun, M. Postel, F. Casabianca, J. Fischer, A. Mitschler, Inorg. Chem., **21**, (1982), 1303.
169. A.F. Ghiron, R.C. Thompson, Inorg. Chem., **29**, (199), 4457.
170. J.D. Lydon, L.M. Schwane, R.C. Thompson, Inorg. Chem., **26**, (1987), 2606.
171. M.G. Clerici, P. Ingallina, R. Millini, in R. von Ballmoos, J.B. Higgins, M.M.J. Treacy (Eds.), Proc. 9th Int. Zeolite Conf., Butterworth-Heinemann, Boston, 1992, p. 445.
172. R.A. Sheldon, J. Mol. Catal., **7**, (1980), 358.
173. H. Mimoun, Angew. Chem., **94**, (1982), 750.
174. R.D. Bach, G.J. Wolber, B.A. Coddens, J. Am. Chem. Soc., **106**, (1984), 6098, and refs. cited therein.
175. See *e.g.* G. Boche, F. Bosold, J.C.W. Lohrenz, Angew. Chem. Int. Ed. Engl., **33**, (1994), 1161.
176. H.J. Ledon, F. Varescon, Inorg. Chem., **23**, (1984), 2735.
177. E. Karlsen, K. Schöffel, Catal. Today, **32**, (1996), 106.
178. M. Neurock, L.E. Manzer, Chem. Commun. **1996**, 1133.
179. (a) G. Centi, F. Trifirò, J.R. Ebner, V. Franchetti, Chem. Rev., **28**, (1989), 400; (b) G.C. Bond, A.J. Sarkany, G.D. Parfitt, J. Catal., **57**, (1979), 476; (c) G.C. Bond, P. König, J. Catal., **77**, (1982), 309; P.J. Gellings, Catalysis, **7**, (1985), 105.
180. (a) R.J.H. Clark, in J.C. Bailar et al. (eds.), "Comprehensive Inorganic Chemistry", Pergamon, Oxford, 1973, p. 491; (b) L.V. Boas, J.C. Pessoa in G. Wilkinson (ed.), "Comprehensive Coordination Chemistry", Pergamon Press, Oxford, Vol. 3, 453; (c) A. Butler in D. Chasteen (ed.), "Vanadium in Biological Systems", Kluwer Academic, Dordrecht, 1990, p. 25.
181. R.K. Iler, "The Chemistry of Silica", John Wiley, New York, 1979, p. 665.
182. N. Das, H. Eckert, H. Hu, I.E. Wachs, J. Walzer, F. Feher, J. Phys. Chem., **97**, (1993), 8240.
183. G. Deo, I.E. Wachs, J. Phys. Chem., **95**, (1991), 5889.

184. L. Marosi, J. Stabenow, M. Schwarzman, DE Pat. 2,831,631, (1980).
185. A. Miyamoto, D. Medhanavyn, T. Inui, Appl. Catal., **28**, (1986), 89.
186. K. Habersberger, P. Jiru, Z. Tvaruzkova, G. Centi, F. Trifirò, React. Kintet, Lett., **39**, (1989), 95.
187. (a) P. Fejes, J. Halasz, I. Kiricsi, Z. Kele, I. Hannus, C. Fernandez, J.B. Nagy, A. Rockenbauer, Gy. Schöbel, in "New Frontiers in Catalysis", L. Guczi et al. (eds.), Akademiai Kiado, Budapest, 1993, part A, 421; (b) P. Fejes, I. Marsi, I. Kiricsi, J. Halasz, I. Hannus, A. Rockenbauer, Gy. Tasi, L. Korecz, Gy. Schöbel, Stud. Surf. Sci. Catal., **69**, (1991), 173.
188. M.S. Rigutto, H. Van Bekkum, Appl. Catal., **68**, (1991), L1.
189. J. Kornatowski, M. Sychev, V. Goncharuk, W.H. Baur, Stud. Surf. Sci. Catal., **65**, (1991), 581.
190. G. Bellussi, G. Maddinelli, A. Carati, A. Gervasini, R. Millini, in R. Von Ballmoos et al. (eds.), Proc. 9th Int. Zeol. Conf., Butterworth-Heinemann, Boston, 1992, Vol.I, 207.
191. G. Centi, S. Perathoner, F. Trifirò, A. Aboukais, C.F. Aissi, M. Guelton, J. Phys. Chem., **96**, (1992), 2617.
192. (a) P.R. Hari Prasad Rao, R. Kumar, A.V. Ramaswamy, P. Ratnasamy, Zeolites, **13**, (1993), 663, (b) P.R. Hari Prasad Rao, A.V. Ramaswamy, P. Ratnasamy, J. Catal., **137**, (1992), 225.
193. S.B. Hong, C.G. Kim, Y.S. Uh, Y.K. Park, S.I. Woo, Korean J. Chem. Eng., **9**, (1992), 16.
194. M.S. Rigutto, unpublished work.
195. A. Tuel, Y. Ben Taarit, Appl. Catal. A. General, **102**, (1993), 201.
196. K.R. Reddy, A.V. Ramaswamy, P. Ratnasamy, J. Chem. Soc., Chem. Commun., **1992**, 1613.
197. R. Kumar, K.R. Reddy, A. Raj, P. Ratnasamy, Proc. 9th Int. Zeol. Conference, R. Von Ballmoos et al. (eds.), Butterworth-Heinemann Publ., Montreal 1992, Vol. 1, 189.
198. T. Sen, M. Chatterjee, S. Sivasanker, J. Chem. Soc., Chem. Commun., **1995**, 207.
199. H. Takahashi, M. Shiotany, H. Kobayashi, J. Sohma, J. Catal., **14**, (1969), 134.
200. S.S. Eaton, G.R. Eaton, in D. Chasteen (ed.), "Vanadium in Biological Systems", Kluwer Academic, Dordrecht, 1990, p. 199.
201. See *e.g.*: H. Eckert, I.E. Wachs, J. Phys. Chem., **93**, (1989), 6796 and refs. cited therein.
202. P.R. Hari Prasad Rao, A.A. Belhekar, S.H. Egde, A.V. Ramaswamy, P. Ratnasamy, J. Catal., **141**, (1993), 595.
203. M.S. Rigutto, H. van Bekkum, G. Deo, N. Arora, I.E. Wachs, H. Eckert, manuscript in preparation.
204. I.L. Moudrakowski, A. Sayari, C.I. Ratcliffe, J.A. Ripmeester, K.F. Preston, J. Phys. Chem., **98**, (1994), 10895.
205. N. Das, H. Eckert, H. Hu, I.E. Wachs, J.F. Walzer, F.J. Feher, J. Phys. Chem., **97**, (1994), 8240.
206. F. Liebau, "Structural Chemistry of Silicates", Springer Verlag, Berlin, 1985, p. 195.
207. (a) R.M. Dessau, K.D. Schmitt, G.T. Kerr, G.L. Woolery, L.B. Alemany, J. Catal., **104**, 484; (b) R.M. Dessau, K.D. Schmitt, G.T. Kerr, G.L. Woolery, L.B. Alemany, J. Catal., **109**, 472; (c) K. Yamagishi, S. Namba, T. Yashima, **95**, (1991), 872.
208. (a) L.W. Zatorski, G. Centi, J. Lopez Nieto, F. Trifirò, G. Bellussi, V. Fattore, Stud. Surf. Sci. Catal., **49B**, (1989), 1243; (b) G. Bellussi, G. Centi, S. Perathoner, F. Trifirò, in S. Oyama, W. Hightower (eds.), "Catalytic Selective Oxidation", ACS Symp. Ser., **523**, 1993, p. 281.
209. (a) P. Ratnasamy, R. Kumar, Catal. Lett., **22**, (1993), 227; (b) A.V. Ramaswamy, S. Sivasanker, Catal. Lett., **22**, (1993), 239; (c) P.R. Hari Prasad Rao, K.R. Reddy, A.V. Ramaswamy, P. Ratnasamy, Stud. Surf. Sci. Catal., **78**, (1993), 385; (d) P.R. Hari Prasad Rao, A.V. Ramaswamy, Appl. Catal. A. General, **93**, (1993), 123; (e) P.R. Hari Prasad Rao, A.V. Ramaswamy, J. Chem. Soc., Chem. Commun., (1992), 1245.
210. C. Marchal, A. Tuel, Y. Ben Taarit, Stud. Surf. Sci. Catal., **78**, (1993), 447.
211. M.S. Rigutto, Ph.D. Thesis, Delft University of Technology, 1996.

212. P.R. Hari Prasad Rao, A.A. Belhekar, S.G. Hedge, A.V. Ramaswamy, P. Ratnasamy, J. Catal., **141**, (1993), 595.
213. P. Ratnasamy, R. Kumar, Catal. Today, **8**, (1991), 329.
214. G.A. Sheveleva, A.S. Kharitonov, G.I. Panov, V.I. Sobolev, N.L. Razdobarova, E.A. Paushktis, V.N. Romannikov, Neftekhimiya, **33**, (1993), 530.
215. V.I. Sobolev, K.A. Dubkov, O.V. Panna, G.I. Panov, Catal. Today, **24**, (1995), 251.
216. M. Kawai, T. Kyora, Japan. Pats. 0356,439 and 0358,954, (1991), to Mitsui Toatsu Chemicals.
217. R.A. Sheldon, J.D. Chen, J. Dakka, E. Neeleman, Preprints of the IInd World Congress on Selective Oxidation, Belmaneda, Spain, paper G.1.
218. J. Muzart, Chem. Rev., **92**, (1992), 113.
219. N.B. Barhate, M. Sasidharan, A. Sudalai, R.D. Wakharkar, Tetrahedron Lett., **37**, (1996), 2067.
220. J.D. Chen, H.E.B. Lempers, R.A. Sheldon, Stud. Surf. Sci. Catal., **85**, (1995), 75.
221. U. Cornaro, P. Jiru, Z. Tvaruzkova, K. Habersberger, Stud. Surf. Sci. Catal., **69**, (1991), 165.
222. N. van der Puil, Widyawati, J.C. Jansen, H. van Bekkum, Stud. Surf. Sci. Catal., **84 A**, (1994), 210.
223. G. Spoto, S. Bordinga, E. Garrone, G. Ghiotti, A. Zecchina, J. Mol. Catal., **74**, (1992), 175.
224. R. Szostak, "Molecular Sieves; Principles of Synthesis and Identification", Van Nostrand Reinhold, New York, 1989.
225. C.S. Blackwell, R.L. Patton, J. Phys Chem., **88**, (1984), 6135.
226. L.B. McCusker, C. Baerlocher, E. Jahn, M. Bülow, Zeolites, **11**, (1991), 308;
227. (a) R.A. Schoonheydt, R. de Vos, J. Pelgrims, H. Leeman, Stud. Surf. Sci. Catal., **49 A**, (1989), 559; (b) C. Montes, M.E. Davis, B. Murray, M. Narayana, J. Phys. Chem., **94**, (1990), 6425.
228. B. Kraushaar-Czarnetzki, W.G.M. Hoogervorst, R.R. Andrea, C.A. Emeis, W.H.J. Stork, J. Chem. Soc., Faraday Trans., **87**, (1991), 891.
229. M.P.J. Peeters, L.J.M. van de Ven, J.W. de Haan, J.H.C. van Hooff, Colloids Surf., **72**, (1993), 87.
230. L.E. Iton, I. Choi, J.A. Desjardins, V.A. Maroni, Zeolites, **9**, (1989), 535.
231. J.M. Benett, B.K. Marcus, Stud. Surf. Sci. Catal., **37**, (1988), 269.
232. J. Jänchen, M.P.J. Peeters, J.H.M.C. van Wolput, J.P. Wolthuizen, J.H.C. van Hooff, U. Lohse, J. Chem. Soc., Faday Trans., **90**, (1994), 1033.
233. J. Dakka, R.A. Sheldon, Netherl. Pat. 9,200,968, (1992).
234. M.P.J. Peeters, M. Busio, P. Leijten, J.H.C. van Hooff, Appl. Catal. A, **118**, (1994), 51.
235. B. Kraushaar-Czarnetski, W.G.M. Hoogervorst, W.H.J. Stork, Stud. Surf. Sci. Catal., **84**, (1994), 1869; B. Kraushaar-Czarnetski, W.G.M. Hoogervorst, Eur. Pat. 519,569 (1993) to Shell Research B.V.
236. S.S. Lin, H.S. Weng, Appl. Catal. A, **105**, (1993), 289; S.S. Lin, H.S. Weng, J. Chem. Eng. Jpn., **27**, (1994), 211.
237. D.L. Vanoppen, D.E. De Vos, M.J. Genet, P.G. Pouxhet, P.A. Jacobs, Angew. Chem., Int. Ed. Engl., **34**, (1995), 560.
238. C. Montes, M.E. Davis, B. Murray, M. Narayana, J. Phys. Chem., **94**, (1990), 6431.
239. S.H. Jhung, Y.S. Uh, H. Chon, Appl. Catal., **62**, (1990), 61.
240. E.M. Flanigen, R.L. Patton, S.T. Wilson, Stud. Surf. Sci. Catal., **37**, (1987), 13.
241. M.S. Rigutto, H. van Bekkum, J. Mol. Catal, **81**, (1993), 77; M.S. Rigutto, H. van Bekkum, Extended Abstracts of the 9th Int. Zeolite Conf., Montreal, 1992, abstract no. RP54.
242. M.J. Haanepen, J.H.C. van Hooff, DGMK Tagungsbericht, **9204**, (1992), 227.
243. M.J. Haanepen, Ph.D. Thesis, Eindhoven University of Technology, 1996.

244. P. Concepcion, J.M. Lopez-Nieto, J. Pérez-Pariente, Catal. Lett., **19**, (1993), 333; *idem*, *ibid.*, **28**, (1994), 9.
245. (a) X. Yu, P.J. Maddox, J.M. Thomas, Polyhedron, **8**, (1989), 819; (b) B.Z. Wan, K. Huang, T.C. Yang, C.T. Tai, J. Chin. Inst. Chem. Eng., **22**, (1991), 17; (c) S. Hocevar, J. Batista, V. Kaucic, J. Catal., **139**, (1993), 351.
246. M. Helliwell, V. Kaucic, G.M.T. Cheetham, M.M. Harding, B.M. Kariuki, P.J. Rizkallah, Extended Abstracts of the 9th Int. Zeolite Conf., Montreal, 1992, abstract no. RP203.
247. J.B. Parise, Acta. Crys. C, **42**, (1986), 670.
248. J.D. Chen, J. Dakka, E. Neeleman, R.A. Sheldon, J. Chem. Soc., Chem. Commun., **1993**, 1379; J.D. Chen, J. Dakka, R.A. Sheldon, Appl. Catal. A, **108**, (1994), L1; J.D. Chen, M.J. Haanepen, J.H.C. van Hooff, R.A. Sheldon, Stud. Surf. Sci. Catal., **84**, (1994), 973.
249. J.D. Chen, Ph.D. Thesis, Delft University of Technology, 1995.
250. R.A. Sheldon, J.D. Chen, J. Dakka, E. Neeleman, Stud. Surf. Sci. Catal., **83**, (1994), 407.
251. J.D. Chen, H.E.B. Lempers, R.A. Sheldon, J. Chem. Soc., Faraday Trans., **92**, (1996), 1807.
252. J.D. Chen, R.A. Sheldon, J. Catal., **153**, (1995), 1.
253. H.E.B. Lempers, R.A. Sheldon, Appl. Catal. A, **143**, (1996), 137.
254. H.E.B. Lempers, R.A. Sheldon, Stud. Surf. Sci. Catal., **105**, (1997), 1061.

Studies in Surface Science and Catalysis 137
H. van Bekkum, E.M. Flanigen, P.A. Jacobs and J.C. Jansen (Editors)
© 2001 Elsevier Science B.V. All rights reserved.

Chapter 20

Zeolite-based Supramolecular Assemblies

Dirk E. De Vos and Peter A. Jacobs

Center for Surface Science and Catalysis, Katholieke Universiteit Leuven, 3001 Heverlee, Belgium

Zeolite materials are nowadays familiar to every chemistry graduate as ion exchangers, adsorbents and catalysts. The economic success of numerous zeolite processes has moreover resulted in substantial funding for research into new applications of these materials. This combination of financial resources and scientific maturity gave the zeolite community all the assets to successfully enter the realm of supramolecular chemistry. Since the eighties, research teams have therefore conducted all conceivable organic or inorganic reactions and syntheses in zeolite environments. Two properties of zeolites are particularly attractive. First, their pore size is close to the dimensions of many molecules, and thus one often deals with isolated molecules or at most with a few molecules at a time. This site isolation prevents formation of large aggregates. Secondly the pore system allows to spatially organise molecules. Such compartmentalisation of the reaction mixture bears full resemblance to natural macromolecular structures, which explains the continuous interaction between supramolecular zeolite science and biomimetic chemistry.

In this chapter we discuss a disparate array of zeolite-based supramolecular assemblies. In all cases, molecules are encapsulated inside the zeolite or adsorbed at the outer surface, and as a result they start to behave differently than they would in solution or as bulk materials. Often chemical reactions or compounds are obtained which would have been impossible without the specific interaction between the zeolite host and its guest molecules. The chapter contains following topics:

- assembly of host-guest systems:
  - via adsorption of the guest into the host,
  - by synthesis of the zeolite host around the guest,
  - by *in situ* synthesis of the guest inside the host, with formation of new covalent or coordinative bonds,
  - via covalent anchoring of guest species on the surface,
- catalysis with zeolite-associated metal complexes, *e.g.* in oxidation, hydrogenation or hydroformylation,
- interaction of supramolecular zeolite systems with light,
- zeolite-confined clusters of semi-conductors or alkali metals,
- polymerisation in zeolites.

From our overview, we exclude those systems that are merely precursors to catalytic metal particles, such as many carbonyl complexes. Carbonyl chemistry in zeolites has been reviewed in detail [1], and a more general discussion of zeolite-based catalyst preparation is given in Chapter 15 of this book.

1. SUPRAMOLECULAR ASSEMBLY

Depending on the zeolite pore size, the solubility and the charge of the guest, direct exchange or adsorption of the guest compound into the zeolite may be possible. Examples are the exchange of Ru-hexammine and Cu amino acid complexes into faujasites [2,3], or the adsorption of neutral metallocenes into zeolite Y [4]. Alternatives are *(i)* the synthesis of the zeolite around an entrapped compound (**1.1**), or, vice versa, *(ii)* synthesis of the guest inside the existing zeolite matrix (**1.2** and **1.3**; Figure 1).

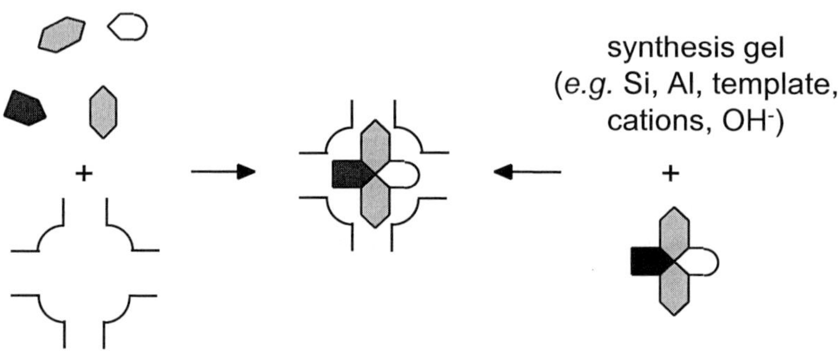

Figure 1. Schematic representation of (*left*) ship-in-a-bottle synthesis of a guest inside a zeolite and (*right*) zeolite crystallisation around an entrapped guest.

1.1. Zeolite synthesis around a guest compound

Guest compounds can be added to a zeolite synthesis gel. Depending on charge and solubility, small amounts of the guest molecules may be entrapped in the final synthesis product [5]. This method is especially popular for synthesis of zeolites X or Y, in which the use of a template is only optional. For instance phthalocyanines (Pc), perfluorophthalocyanines and thioindigo have been incorporated in zeolite X [6-8]; methylene blue was encapsulated during the synthesis of FAU, AFI and CHA structures [9-11]. The entrapment of (substituted) ZnPc in the AlPO-5 structure proves that these guest molecules can reside in self-created structural defects [12].

In some cases, metal complexes with a rigid structure have been used as true templates in zeolite synthesis. The cobalticinium ion ($CoCp_2^+$), in association with OH^- or with F^-, functions as a template for structures with NON topology (nonasil and

ZSM-51) [13,14]. The $CoCp_2^+$ fits tightly inside the ellipsoidal NON cages and, because of the strong hydrogen bonding, can only rotate around its fivefold symmetry axis at temperatures above 485 K [15]. Moreover, $CoCp_2^+$ is a template for the cage-containing structure AlPO-16. AlPO-5 formation has also been observed in the presence of the same template [16].

With the permethylated variant $CoCp^*{}_2^+$ as a template, Balkus and co-workers obtained the new structure UTD-1, which is the first 14-MR high-silica zeolite. The structure has now been solved, and appears to contain two polymorphs with a faulting probability of 20 % [17]. The template can be removed via HCl washing.

Finally, one may work with the thermal lability of some frameworks. For instance, after adsorption of methylene blue in VPI-5, the aluminophosphate is hydrothermally transformed into AlPO-8, which contains entrapped methylene blue dimers [18]. In a related study, it has been observed that the entrapment of phthalocyanines inside VPI-5 stabilises this labile structure towards hydrothermal degradation. This was ascribed to propping of the pores with stacks of encapsulated phthalocyanines [19].

1.2. Synthesis of the guest inside the host: Formation of new molecules

The classical example of a ship-in-a-bottle synthesis is the intrazeolitic formation of the aromatic phthalocyanine ring from four molecules of dicyanobenzene. The dicyanobenzene easily enters the 12-membered ring pores of zeolite Y, and via a complex cyclisation at 473-573 K, phthalocyanines are formed. These are far too large to be removed from the zeolite, and are entrapped in the supercage [20-24].

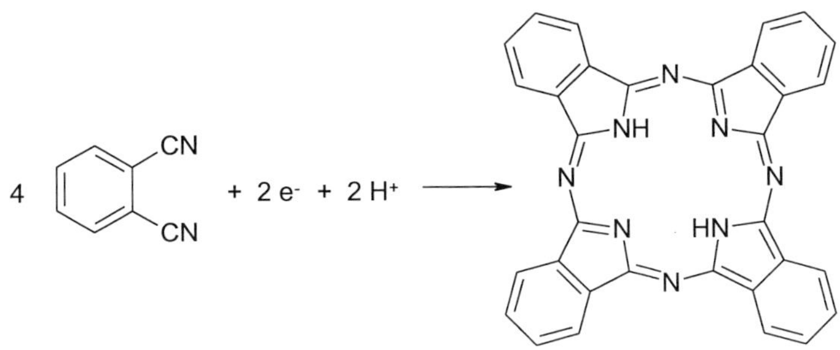

(1)

The overall process is a two-electron reduction, and therefore the procedure works best with slightly hydrated zeolites. The phthalocyanines can be metallated by starting from a metal-exchanged zeolite, or by adding a metal source such as a metallocene into the cyclisation. The latter procedure minimises the amount of unchelated metal ions in the eventual product [23]. The *in situ* synthesis of phthalocyanines results in much higher concentrations of the complex in the framework than the zeolite synthesis methods of the previous paragraph. The *in situ* synthesis has been successfully applied to NaX, NaY, VPI-5 and hexagonal faujasite [19, 21, 23, 25].

There are several new examples of *in situ* synthesis of dye components in zeolites [26-31]. Azo compounds are obtained when N,N-dimethylaniline is added to a zeolite impregnated with the diazonium salt $(RN_2)_2ZnCl_4$. Especially Y zeolites with pre-existing mesoporosity function well as hosts. Even zwitterionic dyes, with R = $PhSO_3H$ have been prepared in US-Y [26]:

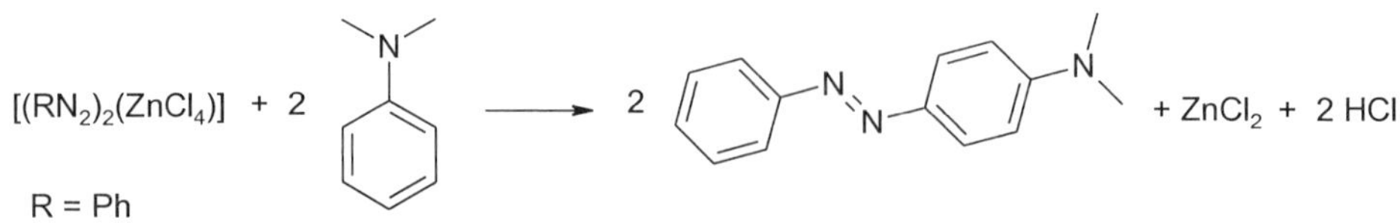

(2)

A recent development is the synthesis of the triphenylmethyl cation in zeolite Y. The acidity of the dehydrated framework catalyses the Friedel Crafts reaction of benzene and CCl_4. Thus a stable carbocation is formed in the faujasite supercage.

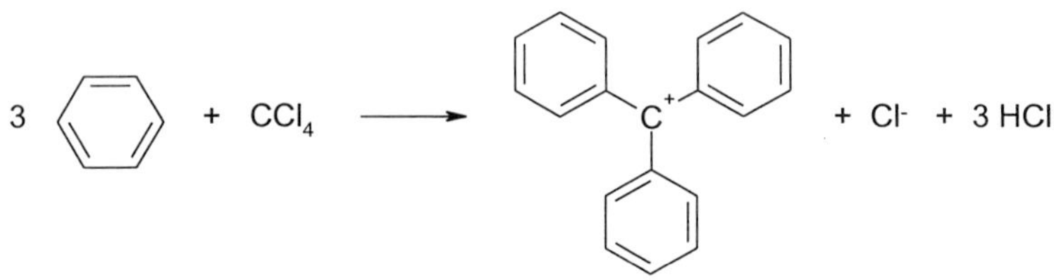

(3)

Exposure to water or to hydride yields Ph_3COH or Ph_3CH respectively [27].

In a similar zeolite-catalysed Friedel-Crafts reaction, triphenylmethyl (trityl) cations are obtained from the reaction of one molecule benzaldehyde with 2 molecules of a *para*-substituted aromatic compound, followed by hydride abstraction by the zeolite [28].

Triphenylpyrilium cations have been synthesised by intrazeolite condensation of acetophenone and chalcone. The reaction stoichiometry requires a 2-electron oxidation by a supplementary chalcone molecule, and a proton transfer from the zeolite [29-31]:

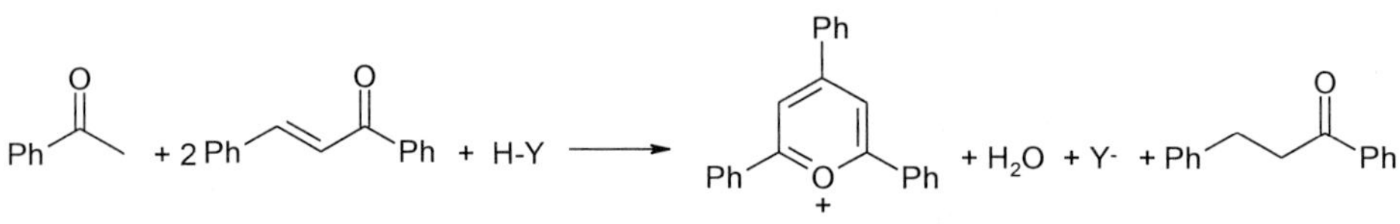

(4)

Finally, the strong base N,N',N"-tricyclohexylguanidine has been claimed to be entrapped in NaY by condensation of cyclohexylamine and N,N'-dicyclohexylcarbodiimide [32]. However, water contact results in hydrolysis of the guanidine and formation of dicyclohexylurea.

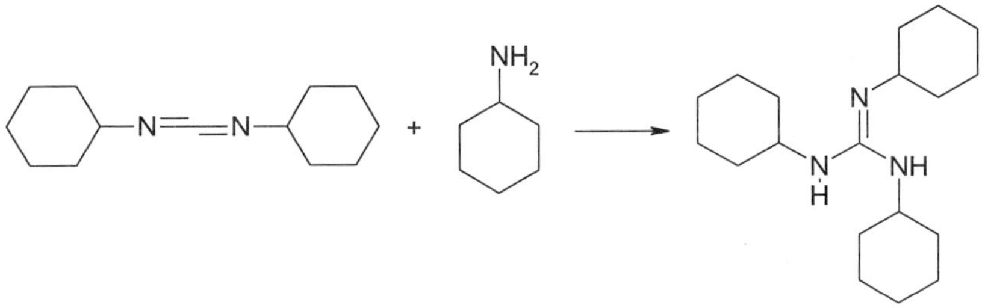

(5)

1.3. The in situ complexation

In the previous paragraph, large encapsulated entities were obtained through formation of covalent bonds. If one desires to have entrapped metal complexes, one may as well create new coordinative bonds. Intrazeolite complexations of cations in framework sites are well known, especially in zeolite Y. Employed monodentate ligands include ammonia [33], alkylamines [34], H_2S [35], CH_3NC [36], pyridine [37] and phosphines [38]. Literature also reports on the use of polydentate ligands such as Schiff bases [39,40], polypyridines [41], and polyamines [42]. Intrazeolite complex chemistry has been reviewed in detail [1, 5, 43, 44].

In most cases, the zeolite is first exchanged with the transition metal ion, and partially or fully dehydrated. This treatment may cause motion of metal ions to less accessible sites. However, the cation mobility is sufficiently high for this process to be reversible. The zeolite is then exposed to the ligand in its pure gaseous or liquid form, or to a solution of the ligand in an inert solvent. Depending on the ligand excess, and on the post-synthesis treatment (*e.g.* evacuation) one may obtain different arrangements of metal, ligand and zeolite.

If enough ligand is available, and if the metal has a strong affinity for the external ligand, highly coordinated complexes are formed. Examples include $Co(NH_3)_5^{2+}$ [33], $Co(ethylenediamine)_3^{2+}$ [45], $Ru(bipyridine)_3^{2+}$ [46] or $Co(terpyridine)_2^{2+}$ in zeolite Y [47]. Obviously, the eventual number of donor atoms surrounding the metal ion depends on the individual preference of the metal ion. Cu^{2+} prefers a square-planar coordination, and the planar complexes $Cu(NH_3)_4^{2+}$ or $Cu(pyridine)_4^{2+}$ therefore correspond to a full complexation [37, 48].

The coordination sphere of the metal ion may also be shared by the zeolite and the external ligand(s). In a reaction of PMe_3 with NiNaY, a pseudo-tetrahedral complex is formed with three lattice oxygen atoms and one phosphine around nickel [49]. A similar $Me(O_L)_3L$ arrangement is obtained after controlled decomposition of $Cu(NH_3)_4^{2+}$ in zeolite Y (Figure 2) [50].

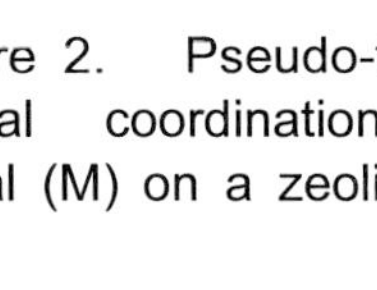

Figure 2. Pseudo-tetrahedral coordination of metal (M) on a zeolite 6-ring.

When the zeolite participates in the coordination, it acts mostly as a tridentate chelating ligand via the six-membered rings. The competition between this surface and the ligand sometimes leads to differences between complexations in solution and in a zeolite. For instance, the solution reaction of Co^{2+} with cyclam, salen, acacen or other tetradentate Schiff bases yields square planar complexes. In a dehydrated zeolite CoNaY, these planar complexes are only formed to a very limited extent; and detailed analysis of the ESR parameters for these planar intrazeolite complexes proves that the axial positions are 'empty' [42, 51]. As the zeolite is not capable of binding as a monodentate ligand, the coordinatively unsaturated square planar complexes are hardly formed. The majority of the Co resides in a shared zeolite-ligand environment (Figure 3).

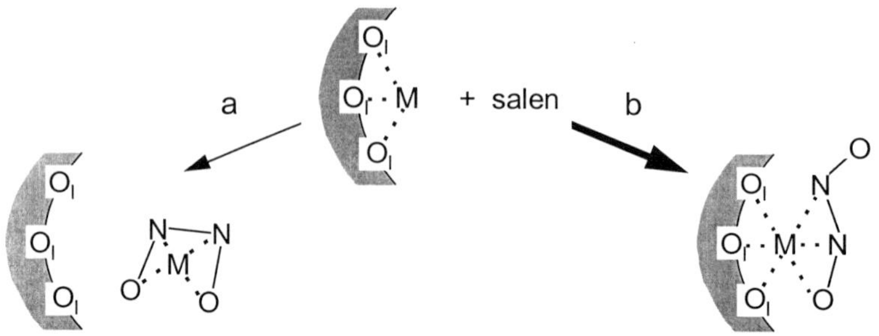

Figure 3. Intrazeolite chelation of a metal ion by salen mainly yields a shared zeolite-ligand coordination (b). Square planar complexes (a) are hardly formed.

Careful ligand design can improve the yield of the desired complexes in the intrazeolite complexation. For instance, if the tetradentate Schiff base salen is replaced with a pentadentate ligand containing an extra coordinating amine group, the zeolite is easily expelled from the coordination sphere, and a large amount of O_2 binding, pentacoordinate complexes is formed [52].

Heteroleptic complexes have been obtained by simple exposure of the zeolite to a ligand mixture, as with $Co(bpy)(terpy)^{2+}$ [53]. In an alternative two-step procedure, the ligands are added consecutively to the metal zeolite:

$$Ru(NH_3)_6^{3+} + 2\ bpy \rightarrow Ru(bpy)_2^{2+} + 4\ NH_3 + N_2H_4 \quad (6)$$

$$Ru(bpy)_2^{2+} + L' \rightarrow Ru(bpy)_2(L')^{2+} \quad (7)$$

This elegant route has led to an array of heteroleptic systems, with L' = bipyrazine, dimethylbipyridine or diazafluorene [54].

1.4. Location of guests in the host

For many of the syntheses described in the previous paragraphs, there is considerable debate whether the guest is truly inside the intact host structure, and whether the dispersion over the porous framework is homogeneous.

In some cases, there is evidence that the synthesis of nanostructures leads to a partial degradation of the framework. The resulting mesopores have been proposed to host part of the guests. For instance, mesopore formation has been observed in the exposure of Cd^{2+} zeolites to H_2S [55], or in the *in situ* synthesis of phthalocyanines in zeolite Y [56]. Also in the synthesis of azo dyes in faujasites, the retention of the dye in the material has been related to the occurrence of mesopores [26].

Even if stability of the guest to extraction is not a conclusive proof for occlusion, it certainly is an important issue if the host-guest systems are to be used in suspension. An illustrative case is that of the neutral Me(salen) complexes in zeolite Y. The original idea of steric entrapment in the faujasite structure has been experimentally contradicted; with a polar solvent such as ethanol, the larger part of Me(salen) complexes can be extracted from the structure [51, 57-59].

Spectroscopic or diffraction techniques often give indirect evidence for the intrazeolite location of host species. Rietveld refinement of XRD data for triphenylmethylium species in zeolite Y has detected 'some' electron density intensity in the supercages, although attempts to interpret the electron density peak directly without assumption of the guest structure have failed [28]. For a VPI-5 with occluded FePc in its channels, the intensity of the ^{27}Al NMR lines is strongly enhanced by cross-polarisation with the immobile protons of the entrapped species [19].

1.5. Physicochemical characterisation

For zeolite chemists, the characterisation of occluded species has often been the testing ground to introduce novel techniques or to improve their applicability to highly dispersed materials. The following examples do not constitute a comprehensive enumeration but intend to illustrate specific aspects of zeolite-guest interactions.

Vibrational spectroscopy in the IR or Raman mode is a sensitive way to detect deformations in guest compounds. Thus the adsorption and oxidation of metallocenes in zeolite Y were analysed in detail [4]. The symmetry lowering induced by the interaction with the zeolite gives IR intensity to normally IR forbidden vibrations. IR and Raman arguments have been used in the discussion of the possible deformation of phthalocyanines in the faujasite framework. Raman can be particularly useful to observe vibrations below 1200 cm^{-1}. In the latter domain, zeolites are strong IR absorbers but only weak Raman scatterers [60]. For instance, Raman allows to distinguish between the different binding modes (end-on superoxo, μ-peroxo, μ-superoxo) of O_2 on intrazeolite Co-amine complexes [61]. With transient Raman spectroscopy, excited states have been identified, *e.g.* the $bpy^{\cdot -}$ anion in the photo-excitation of intrazeolite $Ru(bpy)_3^{2+}$ [62].

Electronic spectra of metal zeolites can be recorded from the UV over the visible into the near infrared domain, and are helpful in elucidating the metal coordination, even if the variability of the extinction coefficients and the considerable line widths represent major drawbacks [43]. Intra- and intermolecular charge transfer phenomena are easily monitored in the diffuse reflectance mode [63]. Other light-related phenomena are discussed in section 3.

Electron paramagnetic resonance (EPR) spectroscopy is generally used to investigate metal complexes or clusters with unpaired electrons. Of all spectroscopic techniques, it yields by far the most detailed fingerprints of metal coordination. EPR even detects small amounts of unusual redox states such as Ni^{3+}, Ni^{1+}, Mn^{4+} or Pd^{1+} [64,65]. A classical example is the binding of O_2 on monomolecular Co^{2+} centers, with primary formation of superoxo $Co^{3+}(L)_n.O_2^-$ complexes. Depending on the relative sizes of the zeolite cage and the complex, the formation of binuclear complexes is retarded, and the reversible character of the O_2 binding is enhanced [33,34]:

$$Co^{2+}(L)_n + O_2 \rightleftharpoons Co^{3+}(L)_n.O_2^- + Co^{2+}(L)_n \quad (8)$$

$$Co^{3+}(L)_n.O_2^- + Co^{2+}(L)_n \xrightarrow{slow} Co^{3+}(L)_n.O_2^{2-}.Co^{3+}(L)_n \quad (9)$$

EPR can also detect changes in spin multiplicity. For $Co(bpy)_3^{2+}$ in zeolite Y, Lunsford and co-workers detected a surprising temperature-dependent transition from the EPR-invisible high-spin form (with three unpaired electrons) to the EPR-detectable low-spin complex with a single unpaired electron. The stabilisation of the low-spin form with its shorter Co-N bonds was ascribed to the space limitations imposed by the zeolite Y supercage [66]. In the specific case of Fe, similar information regarding valence, spin multiplicity and coordination can be obtained from Mössbauer spectroscopy [41, 67, 68].

XPS probes the coordination states and valences in the outer layer of the zeolite structure. Particularly in combination with elemental analysis of the bulk material, XPS is a precise method to detect guest accumulation or depletion at the border of the zeolite crystals [39,40]. EXAFS studies have provided detailed insights into the structures of more or less extended intrazeolite clusters, such as Cd selenides [69].

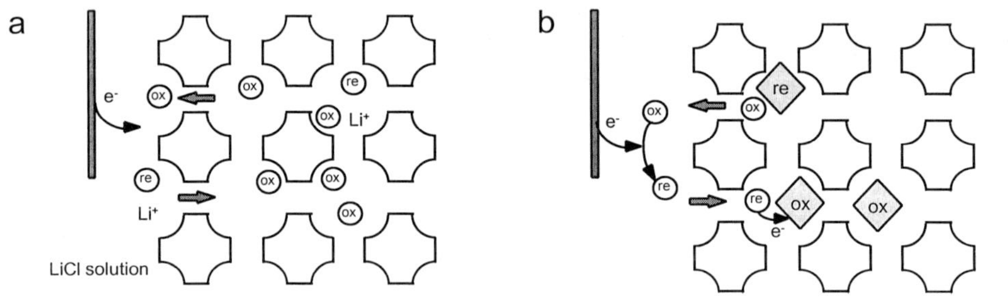

Figure 4. Zeolite electrode processes: (a) extrazeolitic redox of mobile ions at the electrode, (b) intrazeolitic redox reaction of immobilised species, mediated by mobile electron carriers.

Electrochemically addressing zeolite-entrapped species has been the focus of a major debate over the last years [57, 58, 70, 71]. The electrode and sample preparations have large effects on the results. When the guest species are mobile, for instance zeolite-exchanged cations in an electrolyte-containing suspension, the electroactive species migrate to the electrode surface and the actual electron transfer is extrazeolitic (Figure 4a) [72]. The work of Mallouk has demonstrated that intrazeolite electron transport is possible via hopping over mobile vehicles such as metallocenes or viologens (Figure 4b) [73]. Up to now, there is no proof at all for intrinsic, non-mediated electroactivity of immobilised species in the zeolite interior. A very limited response may be generated by metal complexes on the outer zeolite surface [57, 58, 74].

1.6. Covalent linking

An organic moiety can be attached to an (alumino)silicate matrix via alkoxysilyl reagents. This requires that the supports have a sufficient density of -OH groups on the (outer) surface. A suitable support is US-Y, which may be further modified to increase its mesoporosity, *e.g.* via steaming and a citric acid treatment. Corma *et al.* have prepared a series of chiral heterogeneous catalysts by anchoring triethoxysilylated prolineamide derivatives on US-Y [75-77]:

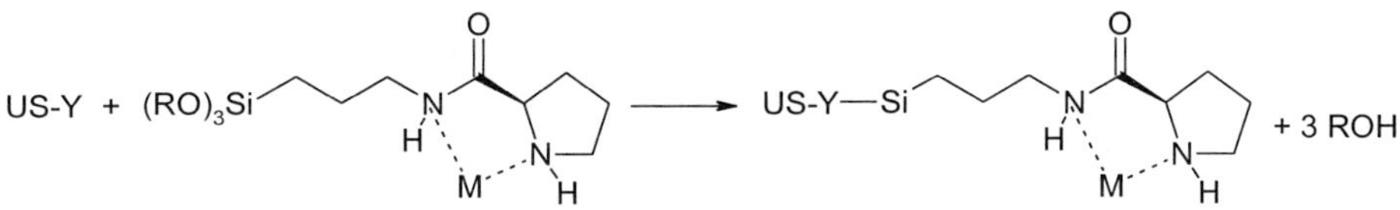

(10)

Such methods are also easily applied to the mesoporous molecular sieves of the MCM-41 family. Via acid treatment, one may increase the amount of reactive surface -OH groups. Often two-step procedures are used. The surface is first functionalised with a simple reactive group, *e.g.* -Cl, $-NH_2$ or an epoxide:

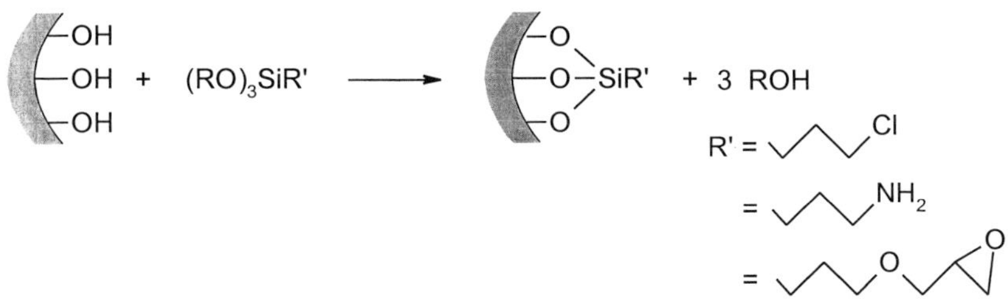

(11)

Next the ligand or the complex is reacted with this surface. The epoxide opening in particular can be achieved in mild conditions. Such methods have been applied in heterogenisation of Schiff base and triazacyclononane type ligands [78-79]. In a variation on this theme, a strong base solid catalyst was obtained by covalent anchoring of the guanidine base 1,5,7-triaza-[4,4,0]-bicyclodecene on an oxirane-functionalised MCM-41 surface [80] (12):

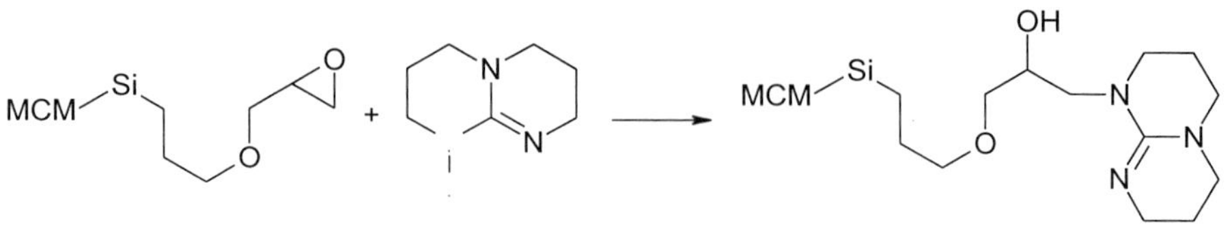 (12)

2. CATALYSIS WITH SHIP-IN-A-BOTTLE SYSTEMS

2.1. Oxidation with single oxygen atom donors

Phthalocyanines. Herron, Stucky and Tolman first used FePc complexes, immobilised in zeolite Y (FePcY) as catalysts for hydrocarbon oxygenation. The original FePcY catalysts were synthesised starting from Fe-exchanged zeolite Y. As the complexation of this Fe by the Pc ligands is never complete, a considerable amount of Fe remains in lattice exchange positions [22]. This free Fe catalyses the decomposition of peroxides, and as a result common oxidants such as tBuOOH cannot be used. With PhIO as an oxidant, only low turnover numbers (< 10) were observed.

An alternative synthesis of the catalyst was elaborated by Romanovskii, and parallelly by Jacobs and Parton [23, 81]. The use of ferrocene minimises the amount of residual free Fe in the zeolite, and the use of a peroxide oxidant now becomes highly advantageous. With FePcY and tBuOOH, n-octane is oxidised to a mixture of alcohols and ketones. In a single run at room temperature, over 6000 catalytic cycles are obtained in alkane oxidation [82, 83].

The activity per complex is further enhanced by introducing electron-withdrawing nitro or fluoro groups in the phthalocyanine ring [84,85]. With Ru perfluorophthalocyanines and tBuOOH, both high activities and peroxide efficiencies have been observed in the cyclohexane oxidation. This facile tuning of the catalytic site clearly illustrates the benefits of a mixed organic-inorganic catalyst design.

While dissolved phthalocyanines are subject to oxidation by other active complexes, the faujasite-entrapped complexes seem essentially stable to oxidative degradation [83]. The mechanism of the FePcY catalyzed alkane oxidations has been investigated via *e.g.* kinetic isotope effects and by competitions between secondary and tertiary C-H bonds. Based on such experiments, the oxygen transfer mechanisms of FePcY and its natural model cytochrome P-450 seem highly analogous [24, 86].

Polarity and sorption are of extreme importance for the FePcY catalyst. For the oxidation of cyclohexane with tBuOOH, the adsorption of the reactants and products on the catalyst has been quantified [83]. Adsorption of polar compounds on the inorganic structure appeared as the major source for catalyst deactivation. A simple extraction with acetone removes these products and restores the catalytic activity.

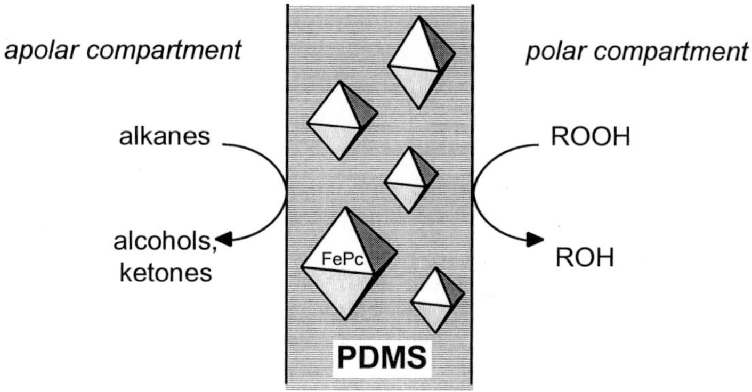

Figure 5. Embedding the FePcY catalyst in a polydimethylsiloxane (PDMS) membrane. The membrane separates two reactant compartments.

In a more elaborate design, the FePcY catalyst was incorporated into a polydimethylsiloxane membrane (Figure 5). This changes the lipophilic-hydrophilic balance around the zeolite crystals, and favors the apolar alkane substrate in its adsorption close to the active site. This ultimately yields a further multiplication of the turnover numbers, especially upon application in membrane type reactors [24].

Schiff base complexes. Moderate results have been obtained in epoxidations catalysed by Mn(salen) complexes in zeolites [87-88]. With PhIO, turnover numbers are low, and with tBuOOH, allylic oxidation dominates. More favourable product selectivities are obtained in alkane oxidations, for instance from cyclohexane to mainly cyclohexanone. In the oxidation of alkanes, the stability of the catalyst can be multiplied by substitution of the phenol groups of the salen ligand with pyridine groups [88]:

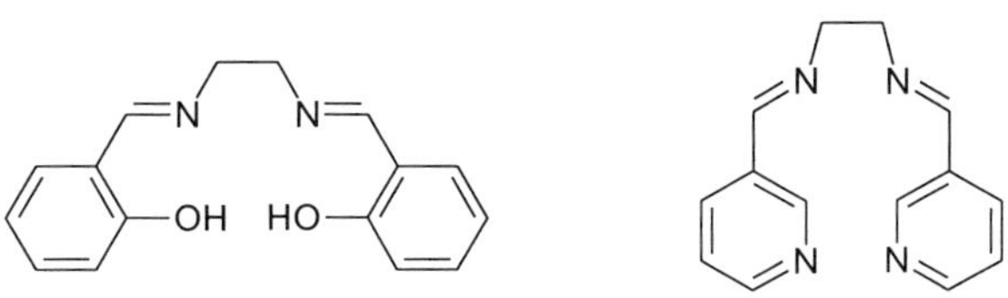

(13)

In view of the lability of zeolite-immobilised Me(salen) complexes to extraction from zeolite Y, it is a necessity to use a highly apolar reaction medium. There is a recent report on the synthesis of a chiral Mn Schiff base complex in hexagonal faujasite [89]. The entrapment seems to preserve both the activity and the chiral induction in the epoxidation of *e.g.* styrenes with commercial bleach.

Bipyridine complexes. In solution, mixtures of Mn salts and bipyridine ligands have only limited catalytic activity. This is due to the formation of poorly soluble polynuclear clusters, the structure of which is dependent on the charge-compensating anion. These clusters catalyse peroxide decomposition rather than selective oxyfunctionalisation [90]. When *bis* complexes of Mn and bipyridine are assembled in zeolite Y cages, the zeolite framework not only functions as the charge-balancing anion, but also sterically impedes the formation of bi- or polynuclear clusters. In these conditions, the mononuclear intrazeolite $Mn(bpy)_2^{2+}$ catalyses the oxygenation of olefins to the corresponding epoxides (Figure 6). Depending on the composition of the framework and the co-cation population, these epoxides may be further hydrolysed and oxidised. Thus the epoxide is the main product with $Mn(bpy)_2^{2+}$ in NaX; but with the same complex in zeolite Y, diols and even diacids are formed [91-93].

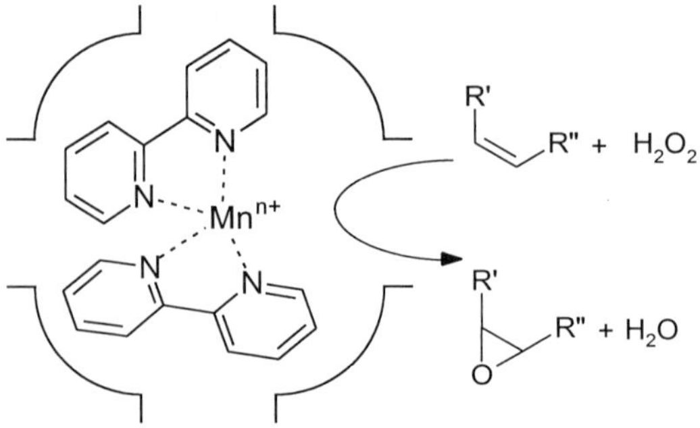

Figure 6. Entrapment of $Mn(bpy)_2^{2+}$ in a zeolite Y supercage prevents formation of oligonuclear complexes. The mononuclear Mn bipyridine complex is a catalyst for selective epoxidation with aqueous H_2O_2.

Polyamine ligands. Mn complexes of 1,4,7-trimethyl-1,4,7-triazacyclononane are highly active catalysts for olefin epoxidation, at the condition that they are used at low temperature and in a ketone solvent [94]. These cationic complexes can be easily heterogenised in the zeolite Y framework, and retain an important activity with several hundreds of catalytic cycles *e.g.* in styrene epoxide formation. Under oxidative conditions, this relatively small complex may form dimers inside the zeolite structure, as is proved by the characteristic 16-line EPR signal of a Mn^{III}-Mn^{IV} dimer (14). The precise nature of the oxygen transferring agent remains however to be elucidated.

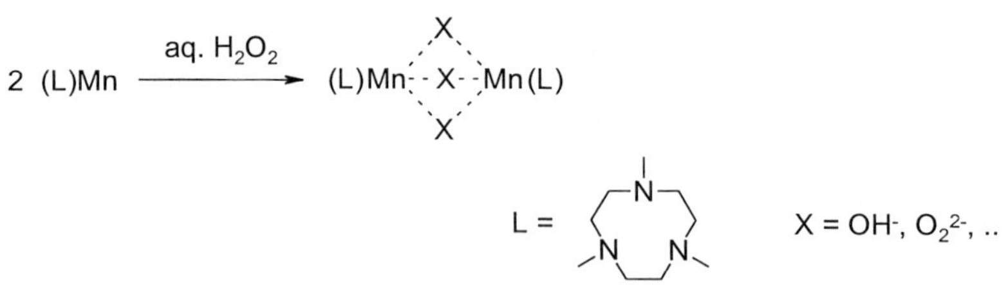

(14)

2.2. Oxidations with O_2

Cobalt phthalocyanines catalyse the oxidation of mercaptans to disulfides with O_2 [95-96]. As the specific activity of these complexes increases with dispersion, they have been heterogenised on different organic or inorganic materials. The success of these materials not only depends on their ability to maintain a high catalyst dispersion, but also on the polarity of the thiol substrate [96]. Thus the rather hydrophobic US-Y is the best support for oxidation of mercaptoethanol in an aqueous phase, while for dodecanethiol in hexane, the best activity was obtained with CoPc on zeolite Y.

Direct activation of O_2 for oxidation of other substrate classes seems far more difficult. Triphenylphosphine has been oxidised to its oxide by zeolite-hosted $(bpy)_2Ru^{IV}{=}O^{2+}$, but the reaction is only stoichiometric [97]. There are reports on the autoxidation of ethylbenzene, catalysed by zeolite-supported Co phthalocyanines. In such reactions the Co functions as a Haber-Weiss catalyst for the decomposition of hydroperoxides. However, even highly aromatic systems such as phthalocyanines lack long term stability in the high temperature, free radical conditions of these reactions [25].

A particular type of O_2 activation was described by Fox et al.. $Ru(bpy)_3^{2+}$, exchanged on the outer zeolite surface, functions as a photosensitiser and converts $^3O_2$ into $^1O_2$. This active form of dioxygen then reacts with specific selectivities with olefins such as 1-methylcyclohexene, yielding hydroperoxides as the primary product [98]:

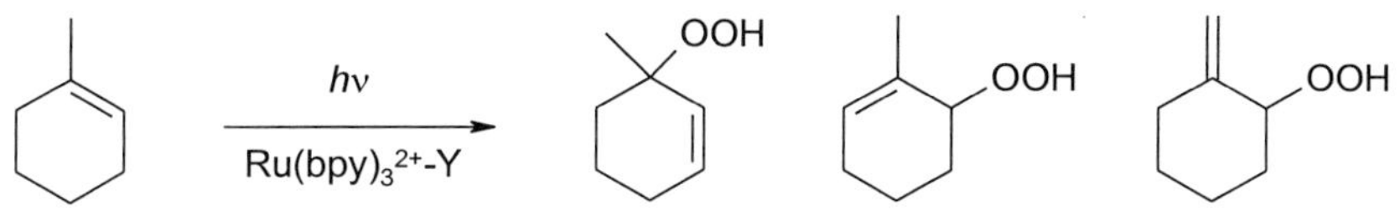

(15)

2.3. Reactions in a reducing atmosphere

Whenever zeolite-supported complexes of Pt, Pd, Rh, Ni or other metals are exposed to H_2 or CO, the complexes might be transformed into particles of zerovalent metal. The advantage of a supramolecular design for hydrogenations therefore is not as much an increased activity as an enhanced selectivity. An added ligand may increase the dispersion of the metal phase and thus limit multiple hydrogenations. For instance, Pd(salen)-NaY displays a higher selectivity for the monohydrogenation of cyclooctadiene than a corresponding PdNaY [59, 99]. Substrate shape selectivity has been observed with Pd(salen)-NaY, or with a zeolite Y in which small Rh entities are introduced by deposition of $Rh(allyl)_3$ [100-101]. The zeolite might also simply function as a support for large metal particles on the outer surface. Chiral modifiers such as cinchona alkaloids can be adsorbed on the extraporous Pt particles to enable enantioselective reduction of for instance pyruvates [102-103].

There are some reports on hydroformylations with zeolite-supported Rh compounds, sometimes in the presence of phosphine ligands. Invariably leaching of the Rh from the support is experienced as a major problem. Nevertheless clear

effects of the zeolite structure have been observed, *e.g.* in the linear : branched ratio [104-107].
Numerous studies have been devoted to the preparation of catalysts with zeolite-supported metal complexes as precursor materials. The scope of these reactions encompasses *i.a.* Fischer-Tropsch and watergas shift reactions, with metals such as Fe, Ru or Os [1, 108, 109].

3. INTERACTION OF SHIP-IN-A-BOTTLE SYSTEMS WITH LIGHT

The photophysical properties of simple guest molecules depend strongly on the characteristics of the host, such as framework polarity, exchanged cations and pore architecture. This is illustrated by following examples:

- The emission spectra of zeolite-adsorbed pyrene reflect the lattice polarity, zeolite X being the most polar environment known [110]. The intensity of the emission from the excited dimer (or excimer) is a function of the zeolite's water content [111].
- The zeolite pore system influences the aggregation of the light-absorbing molecules. Examples are pyrene and thionine, which might exist as monomers or as dimers in the pores of L zeolites and faujasites [112-113].
- Heavy cations (Cs^+, Rb^+, Tl^+) enable observation of phosphorescence from alkenes, aromatics or aromatic ketones that normally mainly display fluorescence [114].

Host effects are also important in the photochemical behaviour of adsorbed molecules:

- The light-induced homolytic cleavage of benzylketones A(C=O)B into A* and B* radicals leads to varying products, depending on the space available for the radical intermediates [115]. Similar observations have been made for photo-Fries reactions [116].
- When faujasite-adsorbed cycloalkanones or alkylphenylketones are subjected to photolysis in Norrish type reactions, the outcome depends largely on the exchanged cations (*e.g.* Li^+), because of the potential interaction between the carbonyl group and light cations in particular [117-118]. When the zeolite matrix is modified by adsorption of a chiral molecule, such as (-)ephedrine, a moderate *ee* is observed in the products of the type II photoreaction of some alkylphenylketones [119] (Figure 7).
- Effects of heavy atoms are observed in the dimerisation of acenaphthylene, or in the rearrangements of β,γ-unsaturated ketones [120-121].

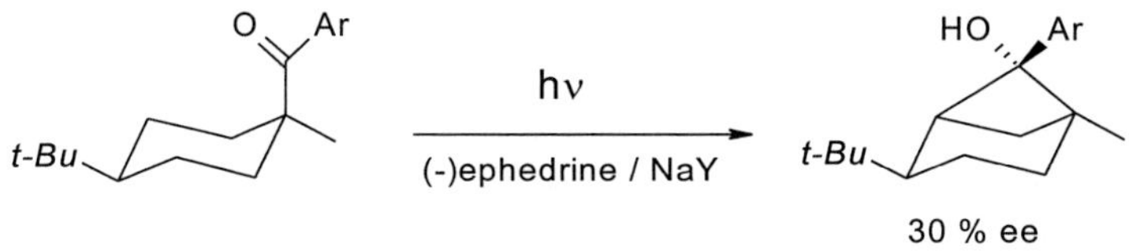

Figure 7. Enantioselective photochemistry in a NaY zeolite, impregnated with a chiral inductor.

While these examples illustrate some of the essential findings in the field, the reader should consult recent specialised reviews for a more complete discussion [122-125]. In the rest of this section, we will discuss four application-directed approaches into the design of light-interacting zeolite host-guest assemblies.

3.1. Photostabilisation of dyes

The stability of organic dyes against photobleaching can be raised considerably by encapsulation in NaY or AlPO-5. As singlet oxygen is often involved in dye degradation, one of the roles of the framework may be to quench the $^1O_2$ which is formed via energy transfer from the excited dye [9].

3.2. Spectral hole burning

The future opto-electronic memory devices will be composed of individually switcheable light-sensitive spots. In order to add a frequency dimension to the spatial dimensions, it is desirable to address dye molecules at different laser frequencies within the broad absorption band. If the absorption properties of the dye are permanently changed by the irradiation, *e.g.* by alteration of its position in the inorganic matrix, the absorption in the narrow frequency range of the laser excitation is much lower: a hole has been burnt into the spectrum (Figure 8) [126].

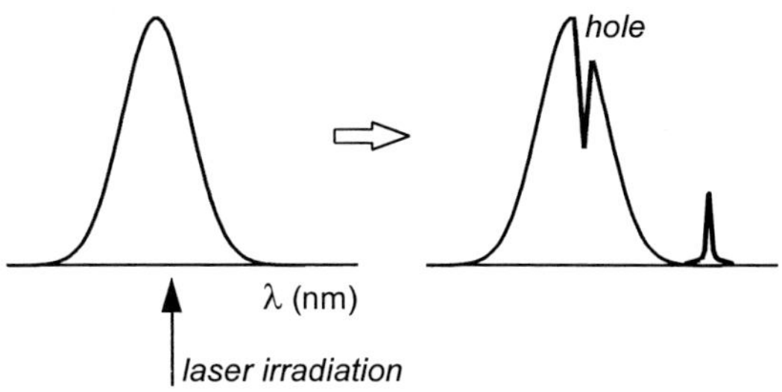

Figure 8. Principle of spectral hole burning.

Such spectral hole burning has been observed with substituted ZnPc's in AlPO-5, and for thionine in faujasite [12, 127]. Some of these holes are persistent up to 80 K. This resistance of the irradiated dye towards decay to its original state is ascribed to a stabilisation of the changed dye by the surrounding frozen solvent shell inside the pores.

3.3. Non-linear optical effects

When a hyperpolarisable molecule is irradiated with strong laser light, the emitted light contains not only the frequency of the incident light but also the double frequency. This phenomenon is called frequency doubling, or second harmonic generation (SHG), and is a non-linear optical effect because the dependence of the polarisation on the electric field contains higher order terms. For this effect to be observable in macroscopic systems, one of the important conditions is that the molecules be organised in a non-centrosymmetric way, *i.e.* without an inversion center. Moreover, the organised domains should be sufficiently large [126].

All conditions for SHG are fulfilled for *para*-nitroaniline (PNA), adsorbed in AlPO-5, VPI-5, AlPO-11 or ZSM-5, and for dimethylaminobenzonitrile in AlPO-5 [128-129]. The PNA in AlPO-5 system has been studied with powders and with large oriented single crystals. Thus the non-linear optical response, the electronic and vibrational spectra have been measured as a function of the angle between the incident light and the AlPO-5 channel direction [130-131]. The PNA molecules form head-to-tail assemblies held together by hydrogen bonding between the NO_2 and the NH_2 groups; the crystal contains large domains with identical dipole orientations.

3.4. Charge transfer processes

Light can be used to induce transfer of electrons either within a molecule, or from a photo-excited donor molecule D to an acceptor A:

$$D + h\nu \rightarrow D^* \tag{16}$$

$$D^* + A \rightarrow D^{\bullet+} + A^{\bullet-} \tag{17}$$

The latter process is a separation of charges, and therefore a conversion of light into chemical energy. Similar phenomena take place in nature in the light-driven electron transfer chains of the photosynthesis. The yield of the process critically depends on avoiding the back electron-transfer reaction, which undoes the charge separation:

$$D^{\bullet+} + A^{\bullet-} \rightarrow D + A \tag{18}$$

In nature this is realised by a sophisticated spatial organisation of the redox partners, *e.g.* in a membrane. As zeolites can discriminate and organise molecules depending on size and charge, they are obvious candidates to use as a matrix for artificial light-driven electron transfer chains [63, 132, 133].

Charge transfer phenomena have been observed for numerous donor-acceptor pairs in zeolites. In case of the interaction between aromatics and viologens, the absorption spectrum (*i.e.* the energy difference between the ground state and the charge transfer state) is practically the same for pairs in solution and in the zeolite [63]. However, the lifetime of the excited state increases tremendously, *e.g.* in the photoexcitation of naphthalene - methylviologen (MV^{2+}), from 30 ps in acetonitrile to 6.8 μs in a zeolite:

$$\text{Naphthalene} + MV^{2+} \underset{}{\overset{h\nu}{\rightleftharpoons}} \text{Naphthalene}^{\bullet+} + MV^{\bullet+} \quad (19)$$

This was ascribed to a reduced mobility of both cationic products in the anionic lattice. Both partners are much more localized than in solution, and this reduces the probability of a back-electron transfer.

The most investigated model systems are those involving $Ru(bpy)_3^{2+}$ or related complexes in faujasites [46, 62]. The photophysics of zeolite-entrapped $Ru(bpy)_3^{2+}$

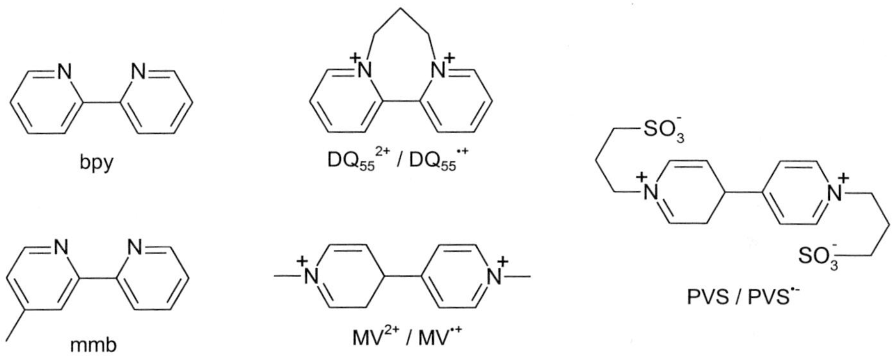

Figure 9. Components of zeolite-based electron-transfer chains

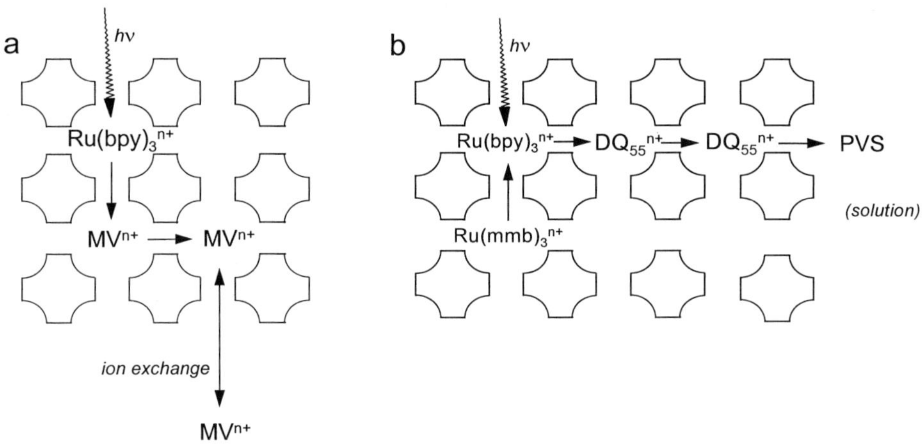

Figure 10. Zeolite-based electron transfer chains. (a) Light-induced electron transport from $Ru(bpy)_3^{2+}$ to methylviologen. The charge separation is maintained by electron hopping to adjacent cages, and by ion exchange. (b) The redox chain $Ru(bpy)_3^{n+}$ / DQ_{55}^{n+} / PVS. $Ru(mmb)_3^{2+}$ reduces the initially formed $Ru(bpy)_3^{3+}$.

have been characterised by its absorption and emission, and via the Raman spectra of the ground and excited states. In the dehydrated zeolite, the lifetime of the triplet charge transfer excited state (^{3}CT) is much lower than in the hydrated zeolite [62]. The proposed explanation was that in the ^{3}CT state, the excited electron is in the bipyridine ligand, and this causes an unfavourable electrostatic interaction with the lattice. In hydrated zeolites, an increased occupation of the supercages by $Ru(bpy)_3^{2+}$ causes a decrease of the emission intensity via concentration quenching [134].

Via a two-step synthesis, faujasite-entrapped heteroleptic complexes are obtained, in which one of the bpy ligands is replaced by a methylated bipyridine, bipyrazine, or diazafluorene. In comparison with the solution complexes, the excited state lifetimes seem increased. This has been ascribed to a destabilisation of the ligand field (LF) state. This LF state corresponds to a Ru-N bond lengthening and thus, in solution, to dissociation of one of the ligands. The zeolite sterically impedes population of this state, as well as the ensuing size increase of the complex and eventual deligation [135-136].

Electron transfer from zeolite-entrapped $Ru(bpy)_3^{2+}$ has been studied in combination with *e.g.* zeolite-exchanged viologens (Figures 9 and 10). The location of the viologen in a neighbouring supercage, and the ensuing electron hopping towards adjacent MV^{2+} molecules gives rise to a durable charge separation. The reduced $MV^{\cdot+}$ is easily observed with Raman spectroscopy [137-138]. Even better results are obtained when more components are added to the transfer chain, or when viologens with different reduction potentials are used [133]. The redox potentials of these acceptors (*e.g.,* benzylviologen) might even vary due to steric effects imposed by the zeolite on the viologen conformation.

In order to 'harvest' the light energy, the reduced $MV^{\cdot+}$ can be ion-exchanged out of the zeolite by fresh MV^{2+} (Figure 10 a) [139]. Alternatively, the electron can be relayed to the zwitterionic propylviologensulfonate (PVS), which once reduced, is in its anionic form charge-excluded from the negative zeolite lattice (Figure 10 b) [140]. Finally, it has been proposed to stabilise the photo-generated $Ru(bpy)_3^{3+}$ by reduction with a $Ru(mmb)_3^{2+}$ in an adjacent supercage (mmb = 5-monomethyl-2,2'-bipyridine). The Ru center is then no longer available for the back electron transfer [141]. It is evident that such complicated schemes are highly dependent on the careful assembly of the components inside the zeolite.

A different approach has been followed by Mallouk et al.. Instead of entrapping the $Ru(bpy)_3^{2+}$ inside the zeolite, the outer surface of zeolites such as Y, L or mordenite was decorated with Ru complexes in which one of the bipyridines has a viologen covalently attached to it (Figure 11). This arrangement slows down the forward electron transfer, but even far more the back reaction. The electrons can be transmitted from the covalently attached acceptor to ion-exchanged methylviologens in the zeolite interior [142].

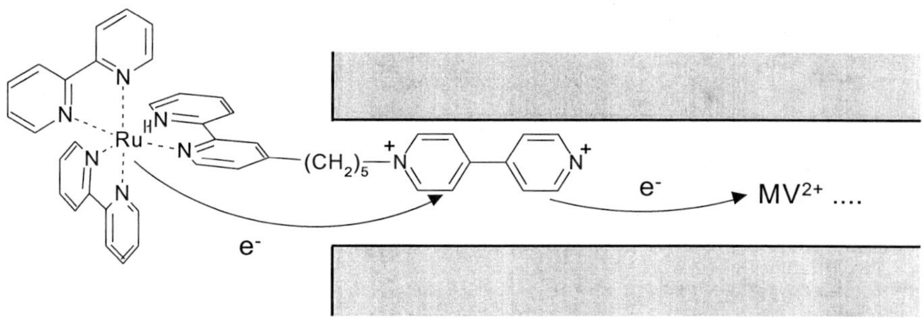

Figure 11. Electron transfer from a size-excluded, MV^{2+}-functionalised Ru-bipyridine complex to methylviologens in the interior of zeolite L.

Moreover, the $Ru(bpy)_3^{n+}$ species at the outer zeolite surface can be reduced by other electron donor species in solution, which again lengthens the duration of the charge separation [143].

There have been attempts to integrate these partial electron transfer chains into the real life water splitting into O_2 and electrons. Downstream the electrons have been transformed into H_2 with Pt particles in zeolite L as catalysts [144]. Upstream the reaction of $Ru(bpy)_3^{3+}$ with water has been studied [145]. Evolution of oxygen has been claimed, with HO*, H_2O_2 and O_2^- as intermediates. A most critical step seems the diffusion of HO* radicals from one Ru center to an adjacent one. This process might well cause oxidative damage to the complexes.

The spatial organisation approach is of course not limited to Ru photosensitisers. For instance, electron transport has been realised in the chain Fe^{2+} - methylacridinium - 7,7,8,8-tetracyanoquinodimethane, with photosensitisation of the acridinium dye [146].

4. CLUSTER FORMATION IN ZEOLITE ENVIRONMENTS

4.1. Alkali metal clusters

Just as electrons can dissolve in liquids, zeolites can accommodate excess electrons. Typically, the extra electron is shared by several alkali ions, and the electron is said to be trapped in an ionic cluster. There are various methodologies to create this electron surplus:

- The pioneers in this field used γ-irradiation and produced low concentrations of paramagnetic Na_4^{3+} centers [147]:

$$4\ Na^+ + e^- \rightarrow Na_4^{3+} \qquad (20)$$

  Other forms of ionising radiation, *e.g.* high-frequency UV, can also be used [148];
- Photo-excited host compounds, for instance anthracene or pyrene, can donate excess electrons inside alkali-exchanged zeolites. Evidently, the results are strongly dependent on the Si : Al ratio of the zeolite, and on the presence of transition ions or adsorbates [149];
- Higher amounts of doping can be achieved by distilling Na into the framework, even if it is difficult to realise a homogeneous reaction of the Na vapor over the zeolite;
- Chemical agents may be used, for instance Li, dissolved in primary amines, or butyllithium in an alkane solvent. An alternative and well-controllable procedure uses Na azide, which is adsorbed from alcoholic solutions and thermally decomposed [150-152].

These methods give access to a wide range of excess electron concentrations, and different techniques must be used to gain insight into the structure of the formed clusters. The detected species strongly reflect the techniques applied. For instance, diffuse reflectance spectroscopy can evidence occurrence of Na_4^{3+}, which is red while larger clusters give blue hues. Electron paramagnetic resonance (EPR) is highly suitable to detect low concentrations of species with a single unpaired electron. In the tetrahedral Na_4^{3+} cluster, one electron experiences the hyperfine interaction of four Na nuclei with I=3/2, and a 13-line spectrum is obtained. However, superimposition of this signal on the signals of Na_5^{4+} and Na_6^{5+} with 16 and 19 lines respectively leads to complicated patterns, which can only be interpreted through careful spectral simulation [153]. A more serious disadvantage is that EPR systematically overlooks diamagnetic species, such as Na_5^{3+}, Cs_4^{2+} or Na^-. More or less direct evidence for these spin-coupled species has been obtained from ^{23}Na MAS NMR, neutron diffraction or from measurements of the overall magnetic susceptibility of the material. Spin pairing, as in Na_5^{3+}, seems to be a general event except in materials with a very low electron excess [154].

The locations of the clusters have been intensively studied, largely via diffraction work in combination with spectroscopic methods. In sodalite, but also in LTA and FAU zeolites, tetrahedral Na_4^{3+} clusters are hosted by the sodalite cages [154]. The sodalite cages of A and X zeolites can also host K_3^{2+} ions [155]. Structures of Rb and Cs clusters are probably very different. For instance the Cs_4^{2+} cluster linearly crosses the sodalite cage of zeolite A [156].

There have been hopes that at high alkali metal loadings, a continuous conducting metallic phase would be formed within the insulating zeolite lattice. At higher Na concentrations, the materials become black, and the EPR spectra coalesce into a singlet at *g*=2. It is clear now that such spectroscopic features can be explained by an increased exchange between clusters in close proximity, for instance in adjacent sodalite cages of SOD or FAU materials [157].

A less futuristic application of these metal clusters is strong base catalysis. The isomerisation of butenes has been studied to prove the carbanionic nature of the intermediates. Thus *cis*-2-butene isomerises mainly to 1-butene and not to *trans*-2-butene (Figure 12). Materials prepared by decomposition of NaN_3 in NaY catalyse the side chain alkylation of *e.g.* toluene, or the aldol condensation of acetone [152, 158]:

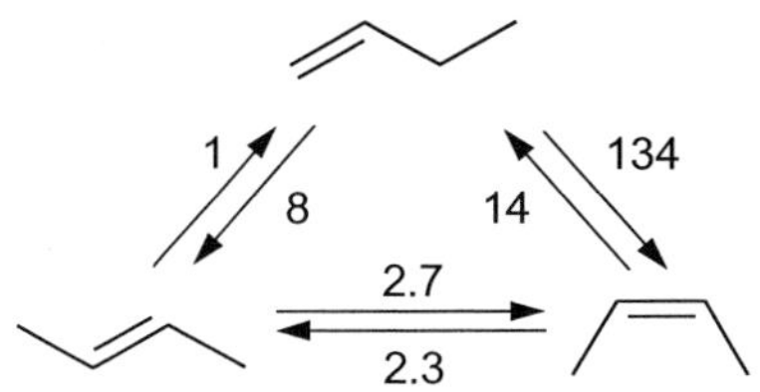

Figure 12. Relative rate constants for butene isomerisation with Na metal on zeolite Y.

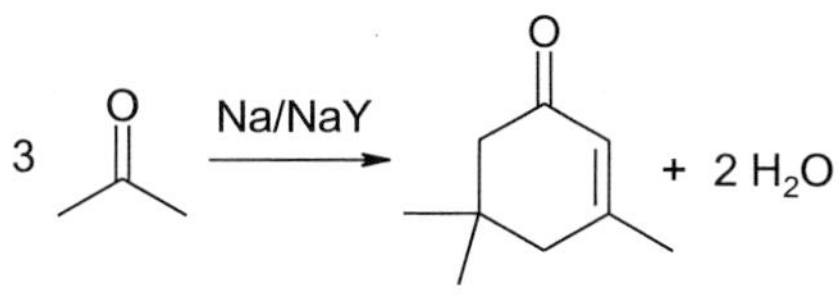

(21)

4.2. Semiconductor clusters

A major focus in the design of advanced materials is the synthesis of extremely small semiconductor particles [159]. Exposure of Cd zeolites to H_2S vapors results in Cd sulfide nanoparticles. Evidence from Rietveld refinements and EXAFS studies led to the picture of the cubane cluster Cd_4S_4, encapsulated in the sodalite cage of faujasites [160]. Of particular interest are the reflectance spectra of these materials. Bulk CdS is a semiconductor, and the band gap corresponds to the onset of the absorption in the electronic spectrum. When only a small fraction of the sodalite cages is occupied with cubane Cd_4S_4 clusters, this absorption is strongly blue-shifted (*i.e.* to the UV), indicating that the band gap is larger for this quantum-confined CdS particle. When the Cd concentration is raised, the band gap gradually decreases. This effect was interpreted as arising from interactions between Cd_4S_4 clusters in neighbouring sodalite cages of zeolite Y. Such interactions are much weaker in zeolite A.

The picture of each sodalite cage being occupied by a Cd_4S_4 cluster may be idealized. The H_2S treatment of zeolite Y can create mesopores, and large CdS clusters have been observed outside the framework [55]. Nevertheless, the spectroscopic properties, such as the blue shift upon decrease of the CdS concentration, seem independent on the precise location of the CdS aggregates.

Other compounds of potential interest for semi-conduction have been dispersed over zeolite matrices. Upon insertion in zeolites, elemental Se builds structures which are influenced by the lattice topology [161]. Chains are formed in MOR zeolites, while Se_8 rings form in zeolite A. Similar work has been performed on Te, while Ge and Si have been deposited as (di)germane and disilane [126]. WO_x clusters are formed by photolysis of adsorbed $W(CO)_6$ in the presence of O_2 [162, 163]; GaP can be assembled from $(CH_3)_3Ga$ and PH_3 [164]. A particular advantage of the use of metalalkyls such as $(CH_3)_3Ga$ or $(CH_3)_2Zn$ is that these compounds react selectively with the Brønsted acid sites inside the zeolites.

5. POLYMERS IN ZEOLITES

Adsorption of monomers such as acetylene in certain zeolites easily leads to polymerisation [165-166]. For instance, the average length of the conjugated polyacetylene fragments in zeolite channels has been determined based on the frequency of the C=C stretching vibration in the Raman spectrum [167].

There has been a major effort in the synthesis of conducting polymers in the channels of zeolites or mesoporous hosts. Such systems are anticipated to lead to miniaturisation of electronic devices. The polymerisation might or might not require addition of oxidants. In the polymerisation of pyrrole or thiophene in zeolite Y or mordenite, exchanged Cu^{2+} or Fe^{3+} are the oxidants [168-170]. Overoxidation of the polypyrrole introduces conductivity, which is associated with quinoidal structures (Figure 13).

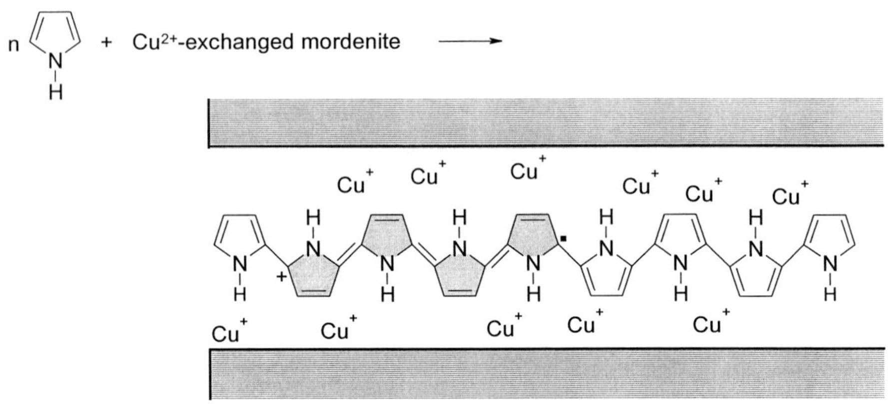

Figure 13. Polymerisation of pyrrole and oxidation to conducting polymeric structures in the pores of mordenite.

Another oxidant, aqueous peroxodisulfate, is used in the oxidative polymerisation of aniline adsorbed in mordenite or zeolite Y [171-172]. The conductivity of the polyaniline is a function of the oxidation and protonation levels, and is therefore influenced by the structure of the zeolite host. In an alternative approach, acrylonitrile is polymerised with radical initiation in MCM-41, and after pyrolysis, encapsulated conducting carbon wires are obtained [173-174].

ACKNOWLEDGMENTS

The continuous interest of the Center for Surface Science and Catalysis of the K.U.Leuven in the field of zeolite Supramolecular Chemistry and Catalysis is funded by the Belgian Federal Government in the frame of a IUAP project. DDV is a post-doc researcher of FWO (Belgium).

REFERENCES

1. T. Bein, in Comprehensive Supramolecular Chemistry, eds. J.L. Atwood, J.E.D. Davies, D.D. Macnicol, F. Vögtle, Volume 7, p. 579.
2. W. De Wilde, G. Peeters and J.H. Lunsford, J. Phys. Chem. 84 (1980) 2306.
3. B.M. Weckhuysen, A.A. Verberckmoes, I.P. Vannijvel, J.A. Pelgrims, P.L. Buskens, P.A. Jacobs and R.A. Schoonheydt, Angew. Chem. Int. Ed. Engl., 34 (1995) 2652.
4. G.A. Ozin and J. Godber, J. Phys. Chem., 93 (1989) 878.
5. K.J.J. Balkus and A.G. Gabrielov, J. Inclusion Phen. Mol. Recognition, 21 (1995) 159.
6. K.J.J. Balkus, C.D. Hargis and S. Kowalak, ACS Symp. Series 499 (1992) 347.
7. A.G. Gabrielov, K.J.J. Balkus, S.L. Bell, F. Bedioui and J. Devynck, Microporous Materials, 2 (1994) 119.
8. R. Hoppe, G. Schulz-Ekloff, D. Wöhrle, C. Kirschhock and H. Fuess, Langmuir, 10 (1994) 1517.
9. R. Hoppe, G. Schulz-Ekloff, D. Wöhrle, C. Kirschhock and H. Fuess, Stud. Surf. Sci. Catal., 84 (1994) 821.
10. S. Wohlrab, R. Hoppe, G. Schulz-Ekloff and D. Wöhrle, Zeolites, 12 (1992) 862.
11. R. Hoppe, G. Schulz-Ekloff, J. Rathousky, J. Starek and A. Zukal, Zeolites, 14 (1994) 126.
12. M. Ehrl, F.W. Deeg, C. Bräuchle, O. Franke, A. Sobbi, G. Schulz-Ekloff and D. Wöhrle, J. Phys. Chem. 98 (1994) 47.
13. K.J.J. Balkus and S. Shepelev, Microporous Mater. 1 (1993) 383.
14. G. van de Goor, B. Lindlar, P. Behrens and J. Felsche, J. Chem. Soc. Chem. Commun. (1995) 2559.
15. P. Behrens, G. van de Goor, C.C. Freyhardt, Angew. Chem. 107 (1995) 2895.
16. K.J.J. Balkus, A.G. Gabrielov, S. Shepelev, Microporous Materials, 3 (1995) 489.

17. C.C. Freyhardt, M. Tsapatsis, R.F. Lobo, K.J.J. Balkus and M.E. Davis, Nature 381 (1996) 295.
18. Y. Jin and H. Chon, Chem. Commun. (1996) 135.
19. R.F. Parton, F. Thibault-Starzyk, R.A. Reynders, P.J. Grobet, P.A. Jacobs, C.P. Bezoukhanova, W. Sun and Y. Wu, J. Mol. Catal. A., 97 (1995) 183.
20. A. Zakharov and B. Romanovskii, Vest. Mosk. Univ. Ser. Khim., 20 (1979) 94.
21. G. Meyer, D. Wöhrle, M. Mohl and G. Schulz-Ekloff, Zeolites, 4 (1984) 30.
22. N. Herron, G. Stucky and C. Tolman, J. Chem. Soc. Chem. Commun. (1986) 1521.
23. R.F. Parton, L. Uytterhoeven and P.A. Jacobs, Stud. Surf. Sci. Catal., 59 (1991) 395.
24. R.F. Parton, I.F.J. Vankelecom, C.P. Bezoukhanova, M. Casselman, J. Uytterhoeven and P.A. Jacobs, Nature 370 (1994) 541.
25. S. Ernst, Y. Traa and U. Deeg, Stud. Surf. Sci. Catal. 84 (1994) 925.
26. C. Schomburg, D. Wöhrle and G. Schulz-Ekloff, Zeolites 17 (1996) 232.
27. T. Tao and G.E. Maciel, J. Am. Chem. Soc. 117 (1995) 12889.
28. M.L. Cano, A. Corma, V. Fornés, H. Garcia, M.A. Miranda, C. Baerlocher and C. Lengauer, J. Am. Chem. Soc. 118 (1996) 11006.
29. A. Corma, V. Fornés, H. Garcia, M.A. Miranda, J. Primo and M. Sabater, J. Am. Chem. Soc. 116 (1994) 2276.
30. A. Corma, V. Fornés, H. García, M.A. Miranda and M.J. Sabater, J. Am. Chem. Soc. 116 (1994) 9767.
31. M. L. Cano, F.L. Cozens, H. García, V. Martí and J. C. Scaiano, J. Phys. Chem. 100 (1996) 18152.
32. R. Sercheli, A.L.B. Ferreira, M.C. Guerreiro, R.M. Vargas, R.A. Sheldon and U. Schuchardt, Tetrahedron Lett. 38 (1997) 1325.
33. E.F. Vansant and J.H. Lunsford, Adv. Chem. Ser. 121 (1973) 441.
34. R. Howe and J.H. Lunsford, J. Am. Chem. Soc. 97 (1975) 5156.
35. N. Herron, Y. Wang, M.M. Eddy, G.D. Stucky, D.E. Cox, K. Moller and T. Bein, J. Am. Chem. Soc. 111 (1989) 530.
36. J.H. Lunsford and E.F. Vansant, J. Chem. Soc. Faraday Trans. II 69 (1973) 1028.
37. P.S.E. Dai and J.H. Lunsford, Inorg. Chem. 19 (1980) 262.
38. R.A. Schoonheydt, D. Van Wouwe and H. Leeman, J. Chem. Soc. Faraday Trans. I 76 (1980) 2519.
39. N. Herron, Inorg. Chem. 25 (1986) 4714.
40. K.J.J. Balkus, A. Welch and B. Gnade, Zeolites 10 (1990) 722.
41. W. Quayle, G. Peeters, G. De Roy, E. Vansant and J.H. Lunsford, Inorg. Chem. 21 (1982) 2226.
42. D.E. De Vos, D.L. Vanoppen, S. Libbrecht, Y. Bruynseraede, P.P. Knops-Gerrits, X.Y. Li and P.A. Jacobs, Chem. Eur. J. 1 (1995) 144.
43. W.J. Mortier and R.A. Schoonheydt, Prog. Solid State Chem. 16 (1985) 1.

44. D.E. De Vos, P.P. Knops-Gerrits, R.F. Parton, B.M. Weckhuysen, P.A. Jacobs and R.A. Schoonheydt, J. Inclusion Phen. Mol. Recognition, 21 (1995) 159.
45. R. Schoonheydt and J. Pelgrims, J. Chem. Soc. Dalton Trans. (1981) 914.
46. W. De Wilde, G. Peeters and J.H. Lunsford, J. Phys. Chem. 84 (1980) 2306.
47. K. Mizuno, S. Imamura and J.H. Lunsford, Inorg. Chem. 23 (1984) 3510.
48. D. Flentge, J.H. Lunsford, P.A. Jacobs and J.B. Uytterhoeven, J. Phys. Chem. 79 (1975) 354.
49. R.A. Schoonheydt, D. Van Wouwe and M. Vanhove, J. Colloid Interface Sci. 83 (1981) 279.
50. W. De Wilde, R.A. Schoonheydt and J. Uytterhoeven, ACS Symp. Ser. 40 (1977) 132.
51. D.E. De Vos, E.J.P. Feijen, R.A. Schoonheydt and P.A. Jacobs, J. Am. Chem. Soc. 116 (1994) 4746.
52. D.E. De Vos, F. Thibault-Starzyk and P.A. Jacobs, Angew. Chem. Int. Ed. Engl. 33 (1994) 431.
53. S. Imamura and J.H. Lunsford, Langmuir 1 (1985) 326.
54. K. Maruszewski, D.P. Strommen, K. Handrich and J.R. Kincaid, Inorg. Chem. 30 (1991) 4579.
55. M. Wark, G. Schulz-Ekloff and N.I. Jaeger, Catal. Today 8 (1991) 467.
56. E. Paez-Mozo, N. Gabriunas, F. Lucaccioni, D.D. Acosta, P. Patrono, A. La Ginestra, P. Ruiz and B. Delmon, J. Phys. Chem. 97 (1993) 12819.
57. C. Senaratne, J. Zhang, M.D. Baker, C.A. Bessel and D.R. Rolison, J. Phys. Chem. 100 (1996) 5849.
58. C.A. Bessel and D.R. Rolison, J. Phys. Chem. B 101 (1997) 1148.
59. D.E. De Vos and P.A. Jacobs, Proc. 9th Int. Zeolite Conf. II (1992) 615.
60. P.P. Knops-Gerrits, D.E. De Vos, E.J.P. Feijen and P.A. Jacobs, Micropor. Mater. 8 (1997) 3.
61. P.K. Dutta and R. Zaykoski, J. Phys. Chem. 93 (1989) 2603.
62. J.A. Incavo and P.K. Dutta, J. Phys. Chem. 94 (1990) 3075.
63. K.B. Yoon, Chem. Rev. 93 (1993) 321.
64. H. Liu, W. Shen, W.H. Quayle and J.H. Lunsford, Inorg. Chem. 23 (1984) 4553.
65. L. Kevan, Acc. Chem. Res. 20 (1987) 1.
66. K. Mizuno and J.H. Lunsford, Inorg. Chem. 22 (1983) 3484.
67. Y. Umemura, Y. Minai and T. Tominaga, J. Chem. Soc. Chem. Commun. (1993) 1822.
68. Y. Umemura, Y. Minai, N. Koga and T. Tominaga, J. Chem. Soc. Chem. Commun. (1994) 893.
69. K. Moller, M.M. Eddy, G.D. Stucky, N. Herron and T. Bein, J. Am. Chem. Soc. 111 (1989) 2564.
70. D.R. Rolison, Stud. Surf. Sci. Catal. 85 (1994) 544.
71. F. Bedioui, J. Devynck and K.J.J. Balkus, J. Phys. Chem. 100 (1996) 8607.

72. C. Senaratne and M.D. Baker, Anal. Chem. 64 (1992) 697.
73. Z. Li and T. Mallouk, J. Phys. Chem. 91 (1987) 643.
74. D.R. Rolison, C.A. Bessel, M.D. Baker, C. Senaratne and J. Zhang, J. Phys. Chem. 100 (1996) 8610.
75. A. Corma, M. Iglesias, C. Del Pino and F. Sanchez, J. Organomet. Chem. 431 (1992) 233.
76. A. Carmona, A. Corma, M. Iglesias, A. San Jose and F. Sanchez, J. Organomet. Chem. 492 (1995) 11.
77. A. Corma, M. Iglesias, M.V. Martin, J. Rubio and F. Sanchez, Tetrahedron: Asymmetry 3 (1992) 845.
78. P. Sutra and D. Brunel, Chem. Commun. (1996) 2485.
79. Y.V.S. Rao, D.E. De Vos and P.A. Jacobs, Chem. Commun. (1997) 355.
80. Y.V.S. Rao, D.E. De Vos and P.A. Jacobs, Angew. Chem., in press (1997).
81. A. Zakharov, Mendeleev Commun. (1991) 80, and references in [5].
82. D.R.C. Huybrechts, R.F. Parton and P.A. Jacobs, Stud. Surf. Sci. Catal. 60 (1991) 225.
83. R.F. Parton, G.J. Peere, P.E. Neys, P.A. Jacobs, R. Claessens and G.V. Baron, J. Mol. Catal. A 113 (1996) 445.
84. K.J.J. Balkus, M. Eissa and R. Lavado, J. Am. Chem. Soc. 117 (1995) 10753.
85. R.F. Parton, C.P. Bezoukhanova, J. Grobet, P.J. Grobet and P.A. Jacobs, Stud. Surf. Sci. Catal. 83 (1994) 371.
86. I.F.J. Vankelecom, R.F. Parton, P.A. Jacobs, J. Catal. (1996). COMPLETE
87. C. Bowers and P.K. Dutta, J. Catal. 122 (1990) 271.
88. D.E. De Vos, P.P. Knops-Gerrits, D.L. Vanoppen and P.A. Jacobs, Supramolecular Chemistry 6 (1995) 49.
89. S.B. Ogunwumi and T. Bein, Chem. Commun. (1997) 901.
90. J.E. Sarneski, D. Michos, H.H. Thorp, M. Didiuk, T. Poon, J. Blewitt, G.W. Brudvig, R.H. Crabtree, Tetrahedron Lett. 32 (1991) 1153.
91. P.P. Knops-Gerrits, D.E. De Vos, F. Thibault-Starzyk and P.A. Jacobs, Nature 369 (1994) 543.
92. P.P. Knops-Gerrits, F. Thibault-Starzyk and P.A. Jacobs Stud. Surf. Sci. Catal. 84 (1994) 1411.
93. P.P. Knops-Gerrits, I.F.J. Vankelecom, E. Béatse and P.A. Jacobs, Catal. Today 32 (1996) 63.
94. D.E. De Vos, J.L. Meinershagen and T. Bein, Angew. Chem. Int. Ed. Engl. 35 (1996) 2211.
95. V. Iliev, A. Andreev, D. Wöhrle and G. Schulz-Ekloff, J. Mol. Catal. 66 (1991) 5.
96. F. Thibault-Starzyk, M. Van Puymbroeck, R.F. Parton and P.A. Jacobs, J. Mol. Catal. A 109 (1996) 75.
97. P.K. Dutta and S.K. Das, J. Am. Chem. Soc. 119 (1997) 4311.
98. T.L. Pettit and M.A. Fox, J. Phys. Chem. 90 (1986) 1353.
99. S. Kowalak, R. Weiss and K.J.J. Balkus, Chem. Commun. (1991) 57.

100. T.N. Huang and J. Schwartz, J. Am. Chem. Soc. 104 (1982) 5244.
101. D.R. Corbin, W.C. Seidel, L. Abrams, N. Herron, G.D. Stucky and C.A. Tolman, Inorg. Chem. 24 (1985) 1800.
102. K. Morgenschweis, E. Pohlkien and W. Reschetilowski, Stud. Surf. Sci. Catal. 108 (1997) 167.
103. W. Reschetilowski, U. Böhmer and J. Wiehl, Stud. Surf. Sci. Catal. 84 (1994) 2021.
104. M.E. Davis, P.M. Butler, J.A. Rossin and B.E. Hanson, J. Mol. Catal. 31 (1985) 385.
105. J.M. Andersen and A.W.S. Currie, Chem. Commun. (1996) 1543.
106. L. Alvila, T.A. Pakkanen, T.T. Pakkanen and O. Krause, J. Mol. Catal. 75 (1992) 333.
107. A. Fukuoka, L. Rao, N. Kosugi, H. Kuroda and M. Ichikawa, Appl. Catal. 50 (1989) 295.
108. S.L. Suib, K.C. McMahon, L.M. Tau and C.O. Bennett, J. Catal. 89 (1984) 20.
109. L.E. Iton, R.B. Beal and D.T. Hodul, J. Mol. Catal. 21 (1983) 151.
110. K.K. Iu and J.K. Thomas, Langmuir 6 (1990) 471.
111. V. Ramamurthy and D.R. Sanderson, J. Phys. Chem. 97 (1993) 13380.
112. B.H. Baretz and N.J. Turro, J. Photochem. 24 (1984) 201.
113. V. Ramamurthy, D.R. Sanderson and D.F. Eaton, J. Am. Chem. Soc. 115 (1993) 10438.
114. V. Ramamurthy, J.V. Casper, D.F. Eaton, E.V. Kuo, D.R. Corbin, J. Am. Chem. Soc. 114 (1992) 3882.
115. N.J. Turro and Z. Zhang, Tetrahedron Lett. 30 (1989) 3761.
116. K. Pitchumani, M. Warrier and V. Ramamurthy, J. Am. Chem. Soc. 118 (1996) 9428.
117. V. Ramamurthy, D.R. Corbin, N.J. Turro and Y. Sato, Tetrahedron Lett. 30 (1989) 5829.
118. V. Ramamurthy, X.G. Lei, N.J. Turro, T.J. Lewis and N.J. Scheffer, Tetrahedron Lett. 32 (1991) 7675.
119. M. Leibovitch, G. Olovsson, G. Sundarababu, V. Ramamurthy, J.R. Scheffer and J. Trotter, J. Am. Chem. Soc. 118 (1996) 1219.
120. V. Ramamurthy, D.R. Corbin, C.V. Kumar and N.J. Turro, Tetrahedron Lett. 31 (1990) 47.
121. B. Borecka, A.D. Gudmundsttir, G. Olovsson, V. Ramamurthy, J.R. Scheffer and J. Trotter, J. Am. Chem. Soc. 116 (1994) 10322.
122. V. Ramamurthy and N.J. Turro, J. Inclusion Phen. Mol. Recognition, 21 (1995) 239.
123. V. Ramamurthy and M.A. Garcia-Garibay, in Comprehensive Supramolecular Chemistry, eds. J.L. Atwood, J.E.D. Davies, D.D. Macnicol, F. Vögtle, Volume 7, p. 693.
124. V. Ramamurthy, D.F. Eaton and J.V. Caspar, Acc. Chem. Res. 25 (1992) 299.

125. J.K. Thomas, Chem. Rev. 93 (1993) 301.
126. P. Behrens and G.D. Stucky, in Comprehensive Supramolecular Chemistry, eds. J.L. Atwood, J.E.D. Davies, D.D. Macnicol, F. Vögtle, Volume 7, p. 721.
127. M. Ehrl, H.W. Kindervater, F.W. Deeg, C. Bräuchle and R. Hoppe, J. Phys. Chem. 98 (1994) 11756.
128. S.D. Cox, T.E. Gier, G.D. Stucky and J. Bierlein, J. Am. Chem. Soc. 110 (1988) 2986.
129. F. Marlow, J. Caro, L. Werner, J. Kornatowski and S. Dähne, J. Phys. Chem. 97 (1993) 11286.
130. L. Werner, J. Caro, G. Finger and J. Kornatowski, Zeolites 12 (1992) 658.
131. F. Marlow, W. Hill, J. Caro and G. Finger, J. Raman Spectr. 24 (1993) 603.
132. P.K. Dutta, J. Inclusion Phen. Mol. Recognition, 21 (1995) 215.
133. P.K. Dutta and M. Ledney, Prog. Inorg. Chem. 44 (1997) 209.
134. W. Turbeville, D.S. Robbins and P.K. Dutta, J. Phys. Chem. 96 (1992) 5024.
135. K. Maruszewski, D.P. Strommen and J.R. Kincaid, J. Am. Chem. Soc. 115 (1993) 9345.
136. K. Maruszewski and J.R. Kincaid, Inorg. Chem. 34 (1995) 2002.
137. P.K. Dutta and J.A. Incavo, J. Phys. Chem. 91 (1987) 4443.
138. P.K. Dutta and W. Turbeville, J. Phys. Chem. 96 (1992) 9410.
139. P.K. Dutta and M. Borja, J. Chem. Soc. Chem. Commun. (1993) 1568.
140. P.K. Dutta and M. Borja, Nature 362 (1993) 43.
141. M. Sykora and J.R. Kincaid, Nature 387 (1997) 162.
142. J.S. Krueger, J.E. Mayer and T.E. Mallouk, J. Am. Chem. Soc. 110 (1988) 8232.
143. E.H. Yonemoto, Y.I. Kim, R.H. Schmehl, J.O. Wallin, B.A. Shoulders, B.R. Richardson, J.F. Haw and T.E. Mallouk, J. Am. Chem. Soc. 116 (1994) 10557.
144. Z. Li, C.M. Wang, L. Persaud and T.A. Mallouk, J. Phys. Chem. 92 (1988) 2592.
145. M. Ledney and P.K. Dutta, J. Am. Chem. Soc. 117 (1995) 7687.
146. S. Fukuzumi, T. Urano, T. Suenobu, Chem. Commun. 213 (1996).
147. P.H. Kasai and R.J. Bishop, in Zeolite Chemistry and Catalysis, ACS Monograph 171, ed. J.A. Rabo (1976) 350.
148. X. Liu, K.K. Iu and J.K. Thomas, Chem. Phys. Lett. 224 (1994) 31. (UV)
149. X. Liu, K.K. Iu and J.K. Thomas, J. Phys. Chem. 98 (1994) 13720.
150. P.A. Anderson, D. Barr and P.P. Edwards, Angew. Chem. 103 (1991) 1511.
151. K.B. Yoon and J.K. Kochi, J. Chem. Soc. Chem. Commun. (1988) 510.
152. L.R.M. Martens, P.J. Grobet and P.A. Jacobs, Nature 315 (1985) 568.
153. P.A. Anderson and P.P. Edwards, J. Chem. Soc. Chem. Commun. (1991) 915.
154. P.P. Edwards, P.A. Anderson and J.M. Thomas, Acc. Chem. Res. 29 (1996) 23.
155. T. Sun and K. Seff, J. Phys. Chem. 98 (1994) 10156.

156. N.H. Heo and K. Seff, J. Am. Chem. Soc. 109 (1987) 7986.
157. N.P. Blake and G.D. Stucky, J. Inclus. Phen. Mol. Recogn. 21 (1995) 299.
158. L.R. Martens, W. Vermeiren, D.R. Huybrechts and P.A. Jacobs, Proc. 9th Int. Congress on Catalysis, Eds. M.T. Philips and M. Ternan, 1 (1988) 420.
159. N. Herron, J. Inclus. Phen. Mol. Recogn. 21 (1995) 283.
160. Y. Wang and N. Herron, J. Phys. Chem. 92 (1988) 4988.
161. J.B. Parise, J.E. MacDougall, N. Herron, R. Farlee, A.W. Sleight, Y. Wang, T. Bein, K. Moller and L.M. Moroney, Inorg. Chem. 27 (1988) 221.
162. G.A. Ozin, R.A. Prokopowicz and S. Ozkar, J. Am. Chem. Soc. 114 (1992) 8953.
163. G.A. Ozin, S. Ozkar and R.A. Prokopowicz, Acc. Chem. Res. 25 (1992) 553.
164. J.E. Mac Dougall, H. Eckert, G.D. Stucky, N. Herron, Y. Wang, K. Moller, T. Bein and D. Cox, J. Am. Chem. Soc. 111 (1989) 8006.
165. C. Pereira, G.T. Kokotailo and R.J. Gorte, J. Phys. Chem. 95 (1991) 705.
166. S.D. Cox and G.D. Stucky, J. Phys. Chem. 95 (1991) 710.
167. P. Tsai, R.P. Cooney, J. Heaviside and P.J. Hendra, Chem. Phys. Lett. 59 (1978) 510.
168. T. Bein and P. Enzel, Angew. Chem. 101 (1989) 1737.
169. J.R. Blanking, G.J. Millar, G.A. Bowmaker and R.P. Cooney, J. Raman Spectr. 24 (1993) 523.
170. P. Enzel and T. Bein, J. Chem. Soc., Chem. Commun. (1989) 1326.
171. P. Enzel and T. Bein, J. Phys. Chem. 93 (1989) 6270.
172. C. Wu and T. Bein, Science, 264 (1994) 1757.
173. P. Enzel, J.J. Zoller and T. Bein, J. Chem. Soc. Chem. Commun. (1992) 633.
174. C. Wu and T. Bein, Science, 266 (1994) 1013.

Studies in Surface Science and Catalysis 137
H. van Bekkum, E.M. Flanigen, P.A. Jacobs and J.C. Jansen (Editors)
© 2001 Elsevier Science B.V. All rights reserved.

Chapter 21

The use of bulky molecules as probes for investigating the contributions of the external and internal pore-wall activities of zeolite catalysts

P.J. Kunkeler[a], R.S. Downing[b] and H. van Bekkum[c]

[a]Akzo PQ Silica, Ankerkade 111, 6222 NL Maastricht, The Netherlands

[b]Avantium Technologies B.V., Zekeringstraat 29, 1014 BV Amsterdam, The Netherlands

[c]Laboratory of Applied Organic Chemistry and Catalysis, Delft University of Technology, Julianalaan 136, 2628 BL Delft, The Netherlands.

The micropores of zeolites have diameters comparable to molecular dimensions, which makes possible the phenomenon of shape selectivity in catalytic reactions. In order to take advantage of this, it is essential that the reaction actually takes place at sites on the internal-pore walls and not on the external surface. For determining the relative reactivities of the external and internal zeolite surfaces, probe molecules may be employed, either as model substrates in catalytic reactions, or for physico-chemical study of their interaction with the surface. The probe will typically be a bulky molecule incapable of penetrating the micropores. In some cases, the probe itself may be able to enter the internal pore system but gives a mixture of reaction products of differing bulk, the product distribution then forming the diagnostic test of the relative extents of internal- and external-surface activity. Probe molecules are found particularly useful for investigating the effectiveness of treatments applied for the purpose of deactivating the external surface in order to suppress non-shape-selective reactions. In this survey, the factors determining the choice of probe molecules for various applications are discussed, with examples of specific applications, and a protocol for the use of probes to check the effectiveness of an external-surface passivation treatment is proposed.

1. INTRODUCTION

The internal micropore walls of zeolites form extensive surface areas available for catalytic activity. However, the contribution of the external surface of the crystallites cannot be ignored in interpreting catalytic phenomena on zeolites. Some zeolites, for example Beta [1], are known to have extensive and reactive external surfaces which can contribute substantially to the overall reaction. For a number of reasons it may be desirable to distinguish the separate contributions of the external and internal micropore wall surface to the overall activity:

1. The possession by zeolites of channels and cages with diameters comparable to those of organic molecules gives rise to the valuable property of shape selectivity (**Figure 1**) [2]. In order to take full advantage of this, it is essential that the reaction takes place within the

internal pore structure of the zeolite and not at its external surface. Especially in zeolite-catalyzed reactions giving mixtures of isomeric products of varying size, there is a temptation to interpret selectivity changes as due to micropore-induced shape selectivity, whereas other factors may also play a role. An example is provided by the finding, by Schultz-Eklof *et al.* [3], of almost exclusively *para*-products from the oxidation of phenol oxidation over non-microporous Ti-MCM-41. Since the mesopores of this catalyst cannot induce shape selectivity, the authors proposed preferential adsorption and polymerization of the *o*-isomers as the explanation of the observed *para*-selectivity. Only if the external surface has been eliminated as the reaction site can a claim of micropore-induced shape selectivity be substantiated.
2. The rigorous interpretation of the kinetics of zeolite-catalyzed reactions may require knowledge of whether the reaction takes place in the micropores or on the external surface.
3. The selection of the appropriate catalyst morphology for a particular reaction is facilitated by knowledge of whether the reaction occurs truly within the internal pore structure or on the crystallite surfaces.
4. Various post-synthesis treatments are applied to zeolites to reduce external-surface acidity or to modify pore-mouth dimensions [4]. Probe experiments are valuable for monitoring the effects of such treatments.

To obtain information about the distribution of catalytically active sites between the external and internal pore-wall surfaces, probe molecules may be employed which are capable of interacting with the active sites and of a size such that they are excluded from pores below a certain threshold diameter. The present chapter provides an overview of the use of this bulky probe approach, the considerations and precautions involved in selecting and using suitable molecules, and some examples of applications to various catalytic reactions.

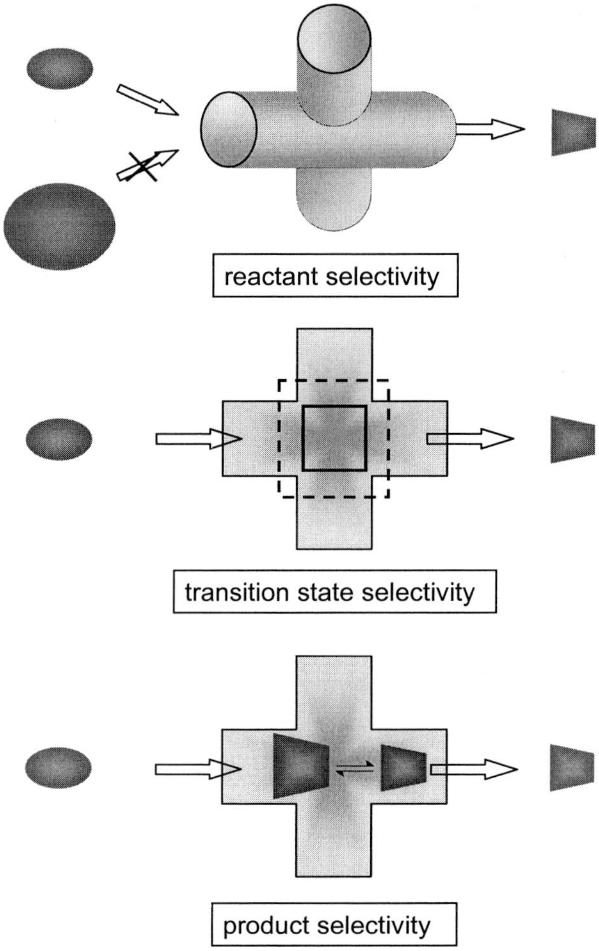

Figure 1. Shape selectivities based on the access dimensions and internal pore structure of zeolites.

There are two ways in which probe molecules can be used. Either their reversible interactions with the outer surface catalytic sites can be studied by physico-chemical techniques, or they can form the substrates in catalytic reactions. The latter method gives more direct information on the relative contributions of the external and internal surfaces to the total catalytic activity and is therefore generally to be preferred. However, a suitable analytical method, based on physico-chemical measurements and preferably calibrated against the results from a catalytic test reaction, can result in considerable savings in time. Frequently, both methods are employed to yield complementary information. For example, in catalytic cracking, kinetic experiments require elaborate gas-phase reactors operating at high temperatures and interpretation of the kinetics is complicated by rapid deactivation as a result of coke formation and the complexity of the reactions. In such cases analytical tests are of value.

The majority of the published work with probe molecules concerns their use to monitor the effects of post-synthesis treatments applied to the zeolite. Types of treatment applied include:

a) pretreatment by deposition of a silica coating, from halo- or alkoxysilanes too large to enter the pores, in order to passivate the outer surface or to reduce the size of pore openings [4]
b) pretreatment by selective removal of active material (for example, aluminium in the case of acid catalysts) from the surface [5,6];
c) the use of bulky poisons, such as amines or phosphines, to deactivate acid sites [7,8] by (reversible) outer surface adsorption; the interaction of these molecules with sites at the pore entrances may cause some restriction of access.

The first two are applied as pretreatments, while the poisons of the third method are usually employed *in-situ* in a reaction mixture. Treatments a and c will, in general, affect the adsorption rate of reactants and the desorption rate of products. A general protocol for employing probe molecules to monitor the effects of applied surface treatments is depicted in Scheme 1. Only by the use of such a rigorous approach can both the decrease in external surface activity and the maintenance of the internal activity be confirmed.

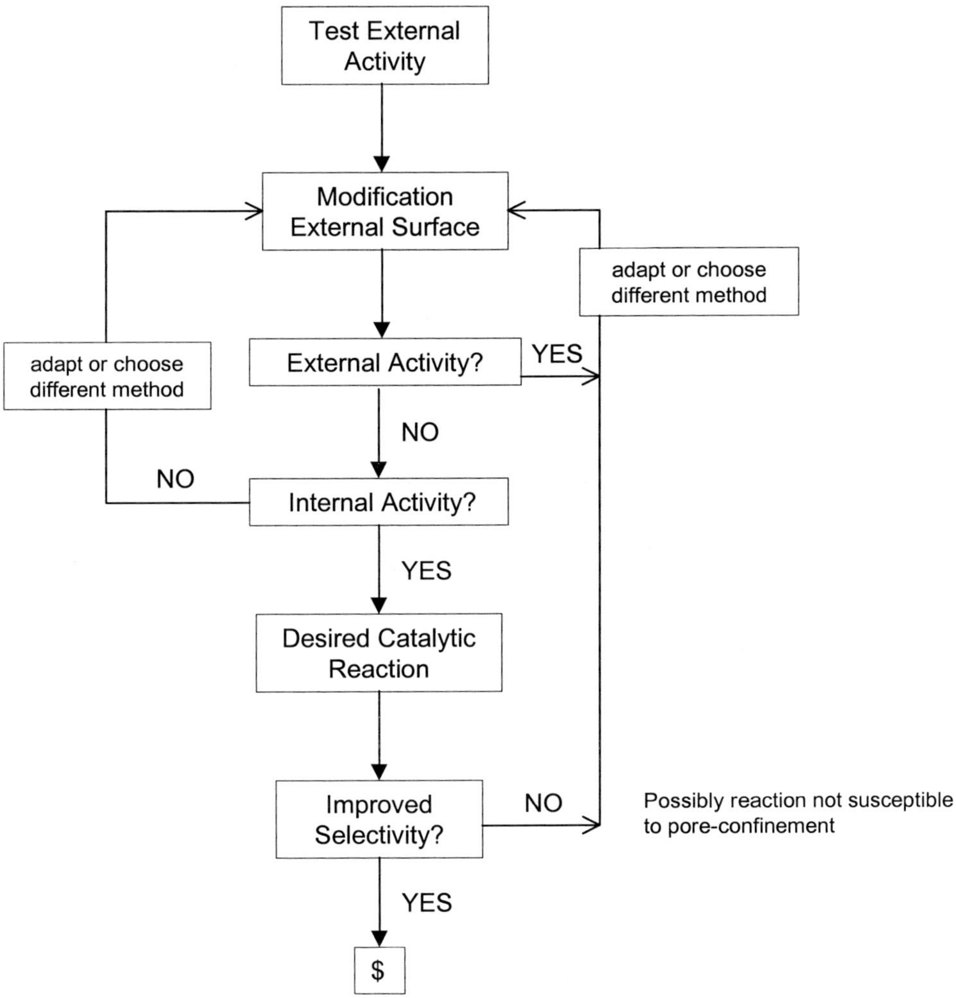

Scheme 1. The path for optimizing the shape selectivity in zeolite catalysed reactions by external surface passivation.

The effectiveness of such treatments requires monitoring by means of appropriate probe molecules. Suitably chosen probes may reveal, for example, whether passivation of the external surface is complete, or whether treatment designed to remove external-surface aluminium may bring about other effects, such as the creation of mesoporosity.

In some cases, the above treatments are themselves applied for diagnostic purposes; for example, to investigate the possibility of reaction on the external surface of a zeolite, one of the treatments described above may be applied and the catalytic performance compared with that of the untreated zeolite. However, unless a check on the effectiveness of the treatment is conducted, misleading results can be obtained and this approach is less rigorous than the use of a suitable probe.

2. SELECTION OF A PROBE MOLECULE

The first consideration when selecting a probe molecule is the type of activity under study; acid, base, hydrogenation and oxidation catalysts have all been investigated using probes but the majority of the published examples refer to Brønsted acid catalysis. The choice is also influenced by the type of study planned. For a catalytic experiment, the bulk of the probe should exclude it from the entire micropore system. On the other hand, for physico-chemical studies, a probe of a similar size to a substrate molecule of interest may be required.

It is important to realise that the crystallographic pore diameter does not represent a suitable threshold molecular diameter, since molecules 20% larger than this diameter are readily admitted, especially at elevated temperatures. For example, naphthalene (9.1 x 6.8 x 3.4 Å) is readily taken up at 100°C by H-ZSM-5 zeolite (pores 5.1 x 5.5 and 5.3 x 5.6 Å) [9].

The probe will generally be chosen as an appropriate analogue of the substrate of interest, although the analogy between model and substrate is never perfect. The introduction of bulky substituent groups will, in general, alter the reactivity of the model with respect to the substrate of interest. This needs to be taken into account when interpreting rate differences between the two compounds in their reactions on the outer surface of the zeolite.

3. USE OF PROBE MOLECULES FOR PHYSICO-CHEMICAL INVESTIGATION

Adsorption experiments with molecules of differing sizes can be used to estimate the relative extents of the external and internal surface areas. However, to determine the distribution of catalytically active sites between the two, probe molecules capable of interaction with the active sites are required. Techniques that have been reported for the study of surface-probe interactions include infra-red (IR), UV-visible (UV-vis) and NMR spectroscopy, thermogravimetry (TGA) and temperature-programmed desorption (TPD).

Infra-red studies of the interaction of probe molecules with acidic oxides and zeolites has been reviewed comprehensively by Lercher *et al.* [10], who provide a number of valuable criteria to be considered in selecting probe molecules. Probes for investigating the distribution of acid sites between the external and internal surfaces include amines, pyridine and substituted pyridines (see Figure 2), and aromatic hydrocarbons of varying bulk. The probe should be stable; tertiary amines as probes for acid sites may be unsuitable since they may undergo disproportionation on acid sites. In probing steric constraints around acid sites with basic probe molecules of various sizes, it is important to realise that the protons of Brønsted-acid sites can be attracted by the base molecule from sites which are sterically well shielded, posing a problem for the determination of site accessibility.

The infra-red spectroscopy of adsorbed quinoline was employed by Corma *et al.* [11] to study Brønsted- and Lewis-acid sites in Y zeolites. Quinoline (9.1 × 6.8 x 3.5 Å) was chosen as a suitable model for the bulky gasoil molecules forming the feed in catalytic cracking since, although not completely excluded from the internal pore structure (cage opening 7.4 Å), its penetration is strongly diffusion-limited. The same authors also found 2,6-di-*tert*-butylpyridine (11.0 x 6.0 x 5.6 Å) to be a useful probe for the external-surface acidity of both medium-pore (e.g. ZSM-5) and for unidirectional large-pore zeolites such as Mordenite, although it was able to penetrate easily into the pore system of 3-dimensional large-pore Beta. They demonstrated a good correlation between the intensity of the $N\text{-}H^+$ stretching band of adsorbed 2,6-di-*tert*-butylpyridine at 3370 cm^{-1} and the cracking reactivity of large reactant molecules unable to penetrate the pore system [12]. The use of this probe illustrates how increasing substitution of a molecule can result in a change in its interaction with the surface.

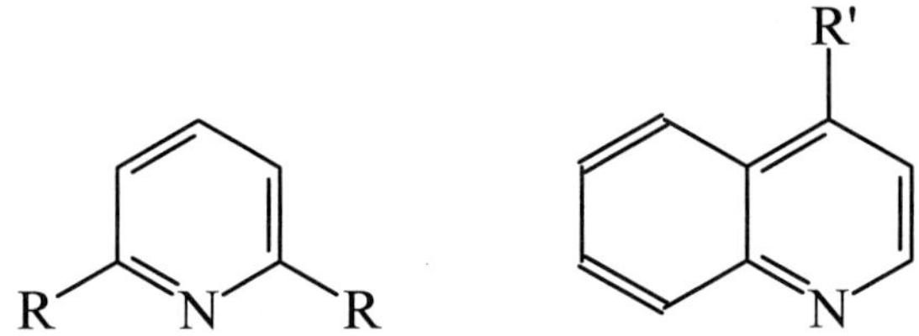

Figure 2. Basic molecules employed as probes for zeolite acid sites (R = H: pyridine; CH_3: 2,6-dimethylpyridine; *tert*-C_4H_9: 2,6-di-*tert*-butylpyridine. R' = H: quinoline; CH_3: 4-methyl-quinoline).

Steric shielding ensures that the principal adsorption mode of 2,6-di-*tert*-butylpyridine (with pK_b 5.02, compared with the pyridine value of 5.23) is via the aromatic ring, rather than the electron lone pair of the nitrogen atom as in pyridine itself [13].

Weber *et al.* employed TPD of ammonia, pyridine and 4-methylquinoline [14] to monitor the effects of surface-passivation treatments applied to H-ZSM-5 zeolite. Ammonia and pyridine were employed as probes for total acidity and 4-methylquinoline for external-surface acidity. These enabled the authors to show that both a silica coating, applied by chemical vapour deposition (CVD), and a silicalite coating were equally effective in rendering the external sites inert, but that the silica coating was preferable in reducing the total acidity to a smaller extent. As an alternative probe for the external acidity of H-ZSM-5, 2,6-dimethylpyridine was employed by Melson *et al.* [15], although its size suggests it to be a boundary case, to be employed only at low temperatures.

Probe molecules for basic sites on zeolites and oxides have been reviewed by Lavalley [16]. Although bulky probes capable of distinguishing internal and external sites are not specifically addressed, the use of pyrrole as probe molecule is noted, suggesting that bulky pyrrole derivatives might be suitable for this purpose.

To compare the surface polarities of silicas exposed to various treatments, Tavener *et al.* [17] employed UV-visible spectroscopy to measure the energy of the $\pi \rightarrow \pi^*$ transition of a bulky adsorbed dye molecule, Reichardt's dye. This could also be applied to zeolites, though it cannot be used with strong acids, which limits its usefulness.

4. PROBE MOLECULES AS SUBSTRATES IN CATALYTIC REACTIONS

Probe molecules are employed in catalytic reactions in various ways. The application of a substrate molecule too large to penetrate the micropore structure makes use of the phenomenon of reactant shape-selectivity; the probe can undergo reaction only on the external surface. In addition, a substrate able to enter the pores can give information if its products include molecules of different sizes, so that their distribution is subject to product or transition-state shape selectivity (see **Figure 1**). Examples of both kinds are given below.

4.1. Bulky substrate molecules as probes

A reaction of great commercial importance in the petroleum refining industry is the catalytic cracking of gasoils over acid zeolites. In order to identify the relative contributions to the activity of the internal and external zeolite surface and the non-zeolitic matrix, kinetic experiments with probe molecules have been conducted. For large-pore zeolites, such as faujasite, mordenite and beta, 1,3,5-tri-*tert*-butyl- and 1,3,5-tri-*iso*propylbenzene (TTBB and TIPB), *cf.* Figure 3, which have diameters (CPK models) much larger (11.0 × 11.0 × 6.3 Å for TTBB and 9.2 × 9.2 × 5.0 Å for TIPB) than the pore-mouth of 12-ring aperture zeolites (approx. 7.4 Å), have been employed [5,18-21]. For example Aguiar *et al.*, [20] compared the cracking rates of *n*-heptane and TIPB over rare-earth Y (REY) zeolites of differing crystallite sizes and hence external surface area. No influence of external area was found for *n*-heptane, which can penetrate the zeolite pores, but a strong inverse correlation between size and cracking rate was found for TIPB, indicating that the reaction is confined to the outer surface. The fact that even the largest of these probes, TTBB, has been shown to be taken up, albeit slowly, by zeolite NaX at 180°C [22], means that the above findings result from the fact that the cracking reaction proceeds much more rapidly than diffusion into the pore system.

For medium-pore zeolites, smaller molecules have been employed. To test the external activity of ZSM-5, the cracking of 1,3-di-*iso*-propylbenzene at 400 °C [23] and the isomerization of 1,2,4-trimethylbenzene to 1,3,5-trimethylbenzene at 450 °C [21] were applied.

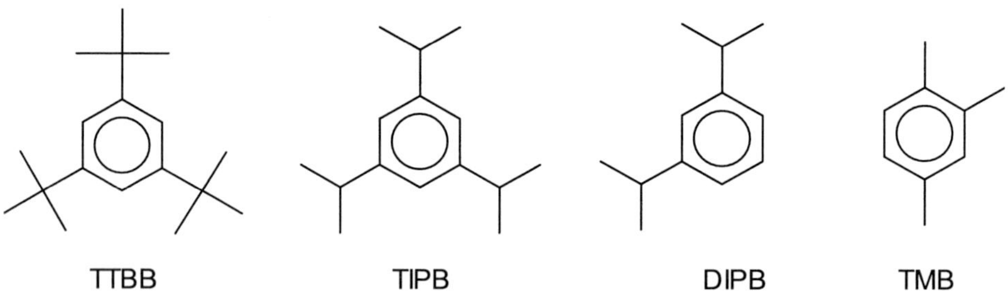

Figure 3. 1,3,5-tri-*tert*-butylbenzene (TTBB), 1,3,5-tri-*iso*-propylbenzene (TIPB), 1,3-di-*iso*-propylbenzene (DIPB) and 1,2,4-trimethylbenzene (TMB).

For the determination of the external activity of ZSM-5, and of composite materials consisting of layers of this zeolite on various supports, the dehydration of 3,3-dimethyl-2-butanol was developed by Van der Puil [24], see Figure 4. This reaction can take place on mild Brønsted or Lewis acid sites, while cracking of the molecule was only observed for the strong Brønsted acidic catalysts. Thus, the product distributions gave additional information on the composite materials prepared in terms of acid strength. Reactions were carried out in a batch-wise gas-phase microreactor at 175 °C under atmospheric pressure.

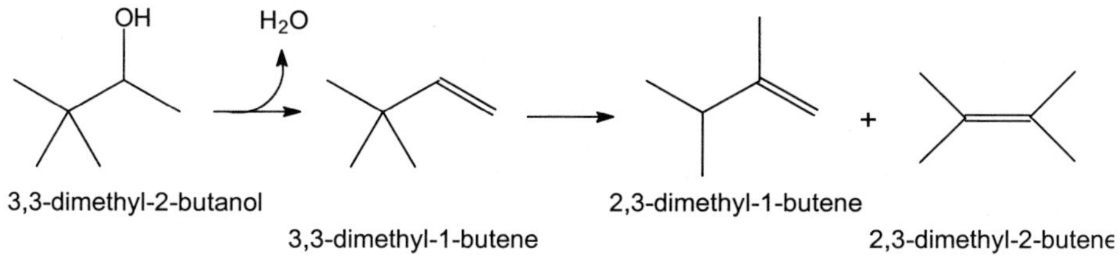

Figure 4. Dehydration of 3,3-dimethyl-2-butanol into 3,3-dimethyl-1-butene and consecutive isomerisation to 2,3-dimethylbutenes.

In addition, the isomerisation of 2,4,4-trimethyl-1-pentene (Figure 5) was used to verify the results obtained with the dehydration reaction of 3,3-dimethyl-2-butanol and the cracking of 1-heptene. Here, also, cracking was only observed in the presence of strong acid sites.

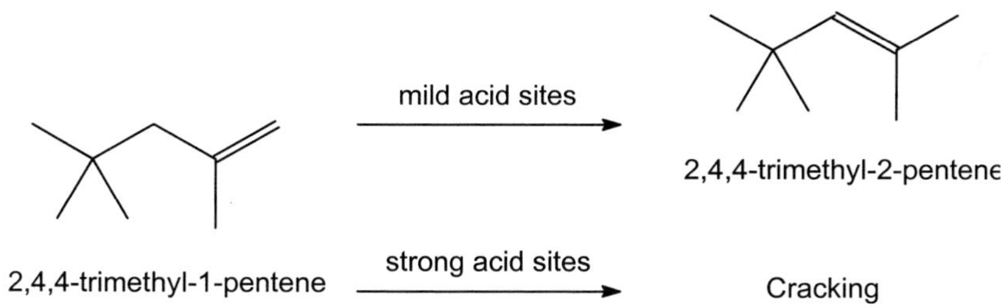

Figure 5. Isomerisation of 2,4,4-trimethyl-1-pentene.

Large-pore zeolites are increasingly employed in the synthesis of fine chemicals. Creyghton *et al.* [7] developed a family of probe molecules suitable for studying the external activity of 12-ring aperture zeolites under the mild conditions (<100°C) used in these liquid phase syntheses. The parent compound, allyl 3,5-di-*tert*-butylphenyl ether (Figure 6), may be employed as probe for both acidic and hydrogenation sites. Over H-Beta and H-Mordenite, it undergoes Brønsted-acid-catalyzed Claisen rearrangement at 80 °C to give 2-allyl-3,5-di-*tert*-butylphenol. This in turn isomerizes to 4,6-di-*tert*-butyl-2-methyldihydrobenzylfuran, which is obtained in a yield of greater than 96%; the reaction is a clean one, without formation of heavy by-products which might lead to catalyst deactivation. The probe was employed to compare the effect of different external-surface passivation treatments of H-Mordenite [7] and H-Beta [25]. For H-Mordenite, it was demonstrated that triphenylphosphine produced a total deactivation, while the weak base triphenylamine gave only a partial activity reduction. Dealumination by nitric acid treatment reduced the external activity to about 7% of the untreated H-Mordenite but a control experiment with allyl phenyl ether, which is able to penetrate the micropores, showed the activity of the treated and untreated catalysts to be similar, indicating that the nitric acid treatment had improved the accessibility of the internal pore system.

In the case of H-Beta, the probe allyl 3,5-di-*tert*-butylphenyl ether was employed to monitor the reduction of the external activity by silica-coating, using TEOS in hexane. It was found that this treatment was only effective for Beta in the form of relatively large, well-formed crystals [25] with particle size 2-3 μm. The external activity of microcrystalline batches of Beta could not be reduced to any significant extent by the silica-coating procedure.

The double bond of this compound also makes it valuable as a probe of hydrogenation functionality. The hydrogenation of the probe molecule (Figure 6) was conducted over a series of platinum catalysts on microporous (Beta) and non-microporous carbon and alumina supports [7]. From the kinetics obtained with reference catalysts of known Pt dispersion it could be shown that the rates per unit mass of exposed Pt were similar and showed little influence of the support. This information could then be used to determine the amounts of Pt exposed on the external surface of different Pt/Beta catalysts and thus compare their suitability for shape-selective hydrogenation.

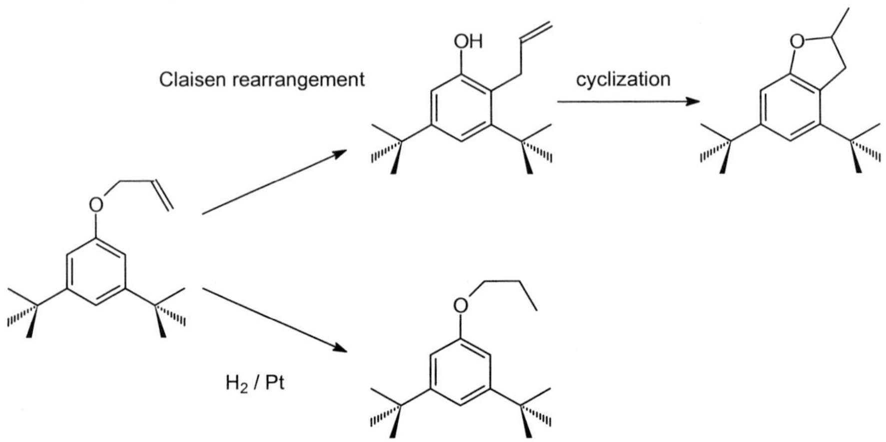

Figure 6. Allyl 3,5-di-*tert*-butylphenyl ether as outer surface H^+ probe in the Claisen rearrangement and as outer surface Pt probe in hydrogenation.

A derivative of the above molecule was developed by Creyghton *et al.* for the purpose of probing the active site of zeolite Beta for the Meerwein-Ponndorf-Verley (MPV) reduction, a Lewis acid catalyzed reaction [26]. Beta was found to show a remarkable stereoselectivity in the reduction of 4-*tert*-butylcyclohexanone with 2-propanol, giving almost exclusively the *cis*-alcohol product. The probe molecule 1-(3,5-di-*tert*-butylphenoxy)propan-2-one (see Figure 7.), which is too bulky to enter the micropores of Beta, underwent no MPV reduction, although it reacted readily, though non-stereoselectively, over several non-microporous catalysts and USY zeolite. This showed that the external surface of Beta is inactive for the reaction, as required for the proposed micropore-induced transition-state shape selectivity.

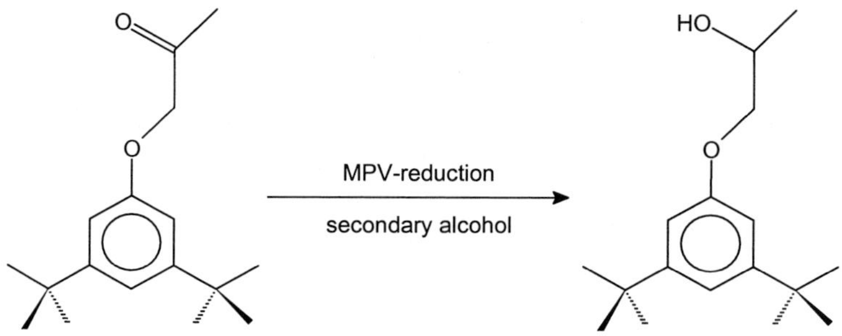

Figure 7. Reduction of 1-(3,5-di-*tert*-butylphenoxy)propan-2-one to 1-(3,5-di-*tert*-butylphenoxy)propan-2-ol using 2-propanol as the reductant.

The above compounds are not commercially available but their syntheses have been published [7,26]. It will be clear that other derivatives of the parent compound could be developed as probes for a variety of organic reactions. Other di-*tert*-butyl aromatic compounds with potential as probe molecules include 2,4-di-*tert*-butylphenol, the Friedel-Crafts alkylation of which with cinnamyl alcohol over various solid-acid catalysts was investigated by Armengol *et al.* [27]. Untreated H-Y zeolite proved to be inactive for the reaction but an H-Y sample, treated to increase the mesopore (2 – 6 nm) content, by steaming and citric acid digestion, was found to be active, as were mesoporous MCM-41 and non-microporous silica-alumina. Since 2,4-di-*tert*-butylphenol is adsorbed fairly readily in the pores of untreated Y zeolite, it is presumably the blocking of the pores by the bulky product which prevents the reaction from proceeding. The compound 2,6-di-*tert*-butylphenol could be of value as an oxidation probe, undergoing oxidation with H_2O_2 to the quinone over Ti-MCM-41 [28] but being too large to enter zeolite pores at low to moderate temperatures.

4.2. Self-probing reactions

The applications described above employ probes which are too large to enter the micropores of zeolites and thus represent examples of reactant shape selectivity. It is also possible to employ substrates which are themselves able to enter the pores but react to give products of different sizes; reaction in the micropores may favour the formation of the smaller or more linear (generally the more desirable) product, while the product ratio from reaction at the external surface will show no influence of shape selectivity. In these cases the probe makes use of the property of transition-state or product shape selectivity. In the former, the the less bulky transition state is favoured, leading to kinetic control and formation of the more linear product. In the latter case, both products are formed but diffusion control ensures that only the more linear is able to desorb; if the reaction is non-reversible, the reaction will be halted by the accumulation of the bulkier product in the pores.

Reactions of this type could be described as self-probing and thus fulfill Lercher's criterion [10] that, where possible, the substrate itself forms the best probe. They include xylene isomerization, olefin oligomerization, and Friedel-Crafts alkylation and acylation reactions of naphthalene and biphenyl derivatives. Shape selectivity is crucial for the production of the linear isomers and many of these are studied intensively owing to their commercial importance as sources for polymer and drug intermediates.

In the *iso*-propylation of naphthalene, for example, the linear 2,6- isomer forms the starting material for several speciality monomers. Conventional catalysts such as $AlCl_3$ give this isomer in the thermodynamic 1:1 ratio with the undesired 2,7-isomer (see Figure 8). Chu and Chen found [29] that 12-ring zeolites give higher 2,6-isomer selectivity than non-microporous catalysts but that H-ZSM-5 gave a low 2,6/2,7-isomer ratio as a result of non-shape-selective reaction at the external surface. Kim *et al.* [30] showed that, in the *iso*-propylation of naphthalene over H-Mordenite catalyst, substantial improvement of the 2,6-/2,7-isomer ratio could be achieved by selective deactivation of the acid sites on the external surface with cerium. To confirm the passivation of the surface acidity, they employed a second probe reaction, the cracking of TIPB. A process has been developed by Catalytica which employs mordenite as catalyst to increase the ratio of 2,6- to 2,7-di-*iso*-propyl-naphthalene to 2.9 [31]. An example of the diagnostic use of this reaction is provided by O'Donovan and O'Connor [32], who used the ratio of 2,6- to 2,7-di-*iso*-propyl isomers to compare the external-surface acidity of Mordenite cracking catalysts prepared by a variety of hydrothermal and acid treatments.

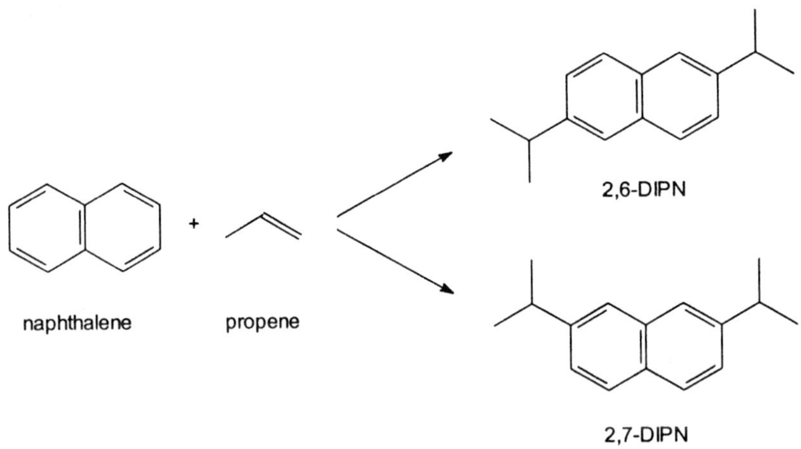

Figure 8. Isopropylation of naphthalene with propene.

The methylation of 2-methylnaphthalene with methanol over H-ZSM-5 catalyst was studied by Inui *et al.* [33], who employed a mechano-chemical method for selectively neutralizing the acid sites on the external surface. The zeolite was ball-milled with barium oxide, which was shown to lead to an improvement in the selectivity to ββ'- (2,6- plus 2,7-dimethyl) isomers. Here again, the authors employed an independent probe reaction to monitor the decrease of external activity; they showed that the activity for TIPB cracking, compared with that for *iso*-propylbenzene, was reduced over the BaO-treated catalyst.

A rather direct method of demonstrating the difference between the internal and external surfaces of H-Mordenite catalyst with respect to shape selectivity was employed by Sugi *et al.* [34]. The reaction investigated was the *iso*-propylation of biphenyl. Following the catalytic experiments, the catalyst was dissolved, using aqueous HF, in order to obtain the product encapsulated in the micropore structure. Analysis of this showed the 4,4'-di-*iso*-propylbiphenyl content of the material recovered from the micropores to be substantially higher than that of the bulk reaction product, which also contained non-linear isomers from the non-shape-selective reaction at the external surface.

The acylation of 2-methoxynaphthalene (see Figure 9) over zeolite H-Beta was investigated by Harvey *et al.* [1]. Adsorption experiments under reaction conditions showed that the linear product, 2-acetyl-6-methoxynaphthalene, can penetrate the 12-ring channels of Beta, while the 1-acetyl-2-methoxy-isomer is excluded. The acylation reaction was conducted over H-Beta catalyst modified by various acid treatments. The 1,2-/2,6-isomer ratio in the products showed some correlation with decreasing external-surface activity but the effect was far from linear. Probably in this case a number of different factors are operating, including the distributions of Lewis and Brønsted sites between the internal and external surfaces.

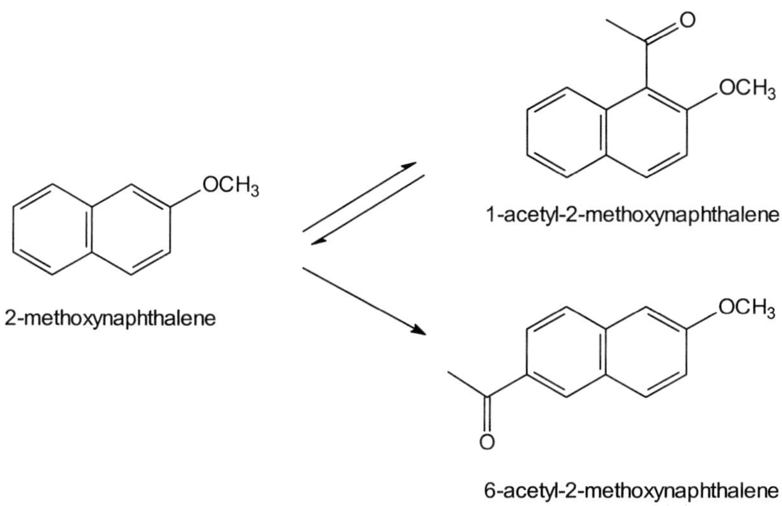

Figure 9. Acylation of 2-methoxynaphthalene using acetic anhydride as the acylation reactant and chlorobenzene as the solvent under reflux conditions over H-Beta [1].

The shape-selectivity of H-ZSM-5 catalyst in the production of *para*-dialkylbenzenes has also received considerable attention owing to its industrial relevance [35]. There has been considerable debate regarding whether the occurrence of ortho- and meta-isomers is the result of external-surface acidity or the relation between intrinsic reactivities (including isomerization) and diffusion coefficients of the three isomers in the micropores. Melson and Schüth [15] showed that, in the disproportionation of ethylbenzene over ZSM-5 samples of differing Si/Al ratio and crystallite size, the *para*-selectivity showed a good correlation with the external acidity determined by titration with 2,6-dimethylpyridine.

The oligomerization of propene over HZSM-5 catalyst has been investigated with a view to minimizing the branching of the product. Wilshier *et al.* [36] showed that reaction on the external surface of the zeolite predominated, yielding highly branched products. The silylation of the surface employing hexamethyldisylazane, and poisoning of the external surface acid sites with 4-methylquinoline, both reduced the total activity and gave improved shape selectivity. Chen and Bridges [37] employed the viscosity index (VI) of the C20+ product fraction as a measure of the linearity of the propene oligomers. Treatment of the catalyst with

2,6-di-*tert*-butylpyridine gave a substantial increase in the product VI and hence in the linearity, while ^{13}C-NMR confirmed the reduction in chain-branching.

Finally, an investigation which well illustrates the use of various techniques in combination was conducted by Chen and Schramm [18] on the external-surface acidity of zeolites H-ZSM-5 and H-BEA. Triphenylphosphine and triarylphosphonium salts were employed to passify the external acidity and quantitative ^{31}P-NMR spectroscopy was used both to obtain an estimate of the relative strength of the acid sites and to determine the percentage of the total acid sites residing on the surface, which was found to be 10.6% in the case of BEA, compared with only 2.7% for H-ZSM-5. The cracking of bulky TTBB was used to demonstrate the loss of external-surface activity as a result of the passivation, while propene oligomerization showed reduced activity but increased linearity (VI) of the lube-oil product as a result of the suppression of the non-shape-selective reaction on the external surface. Consistent with the higher external-acidity percentage and larger pore size of BEA, the oligomerization activity loss with this catalyst was greater and the product linearities lower.

5. CONCLUDING REMARKS

From the foregoing, it will be clear that a rigorous investigation of external-surface activity may require the use of complementary physico-chemical and catalytic experiments. The treatment of a zeolite to moderate its external-surface activity needs to be monitored by probe studies of both the external- and internal-surface activities of the modified zeolite.

REFERENCES

1. G. Harvey, G. Binder and R. Prins, Stud. Surf. Sci. Catal. 94 (1995) 397.
2. P.A. Jacobs and J.A. Martens in: H. van Bekkum, E.M. Flanigen and J.C. Jansen, Eds., Introduction to Zeolite Science and Practice, Stud. Surf. Sci. Catal. 58 (1991) Elsevier Amsterdam etc., Chapter 12.
3. G. Schulz-Ekloff, J. Rathousky and A. Zukal,., Preprints, Div. Petrol. Chem., Amer. Chem. Soc., 1995, 39, 221-224,
4. E.F. Vansant, Pore Size Engineering in Zeolites, Wiley, 1993.
5. R. Szostak, in reference 2, Chapter 5.
6. M.R. Apelian, T.F. Degnan and A.S. Fung, US Patent 5,234,872 (1993) to Mobil Corp.
7. E.J. Creyghton, J.A. Elings, R.S. Downing, R.A. Sheldon and H. van Bekkum, Microporous Mater., 5 (1996) 299.
8. P.J. Kunkeler, J.A. Elings, R.A. Sheldon and H. van Bekkum, P182, Book of Abstracts, 12th IZC, Baltimore, USA, 1998.
9. H. van Koningsveld and J.C. Jansen, Microporous Mater., 6 (1996) 159-167.
10. J.A. Lercher, C.Gründling and G. Eder-Mirth, Catal. Today, 27 (1996) 353-376.
11. A. Corma, V. Fornés and F. Rey, Zeolites, 13 (1993) 56-59.
12. A. Corma, V. Fornés, L. Forni, F. Marqués, J. Martínez-Triguero and D. Moscotti, J. Catal., 179 (1998) 451-458.
13. H. Knözinger, H. Krietenbrink and P. Ratnasamy, J. Catal. 48 (1977) 436.

14. R.W Weber, J.C.Q. Fletcher, K.P. Möller and C.T. O'Connor, Microporous Mater. 7 (1996) 15.
15. S. Melson and F. Schütz, J. Catal. 170 (1997) 46.
16. J.C. Lavalley, Catal. Today, 27 (1996) 377-401.
17. S.J. Tavener, J.H. Clark, G.W. Gray, P.A. Heath and D.J. Macquarrie, J. Chem. Soc., Chem. Commun., (1997) 1147.
18. C.S. Hsia Chen and S.E. Schramm, Microporous Mater. 7 (1996) 125.
19. J.H. Kim, M. Okajima and M. Niwa, Stud. Surf. Sci. Catal. 105 (1997) 1965.
20. E.F.S. Aguiar, M.L.M. Valle, M.P. Silva and D.F. Silva, Zeolites 15 (1995) 620.
21. H.P. Röger, K.P. Möller and C.T. O' Connor, Microporous Mater. 8 (1997) 151.
22. Eric N. Coker, Dik P. Roelofsen, Richard M. Barrer, Jacobus C. Jansen and Herman van Bekkum, Microporous and Mesoporous Mater., 22 (1998) 261-268.
23. P. Tynjälä and T.T. Pakkanen, J. Mol. Catal. A 122 (1997) 159.
24. N. van der Puil, *PhD. Thesis,* (1997) Delft University of Technology, ISBN 90-9010583-2
25. P.J. Kunkeler, D. Moeskops and H. van Bekkum, Microporous Mater.11 (1997) 313-323.
26. E.J. Creyghton, S.D. Ganeshie, R.S. Downing and H. van Bekkum, J. Chem. Soc., Chem. Commun., (1995) 1859, see also *PhD. Thesis* E.J. Creyghton, Delft University of Technology, ISBN 90-407-1382-0.
27. Elvira Armengol, Maria L. Cano, Avelino Corma, Hermenegildo Garcia and Maria T. Navarro, J. Chem. Soc., Chem. Comm., (1995) 519-520.
28. P.T. Tanev, M. Chibwe and T.J. Pinnavaia, Nature, 368 (1994) 321-323.
29. S.-J. Chu and J.-W. Chen, Appl. Catal. A: General, 123 (1995) 51-58.
30. J.H. Kim, Y. Sugi, T. Matsuzaki, T. Hanaoka, Y. Kubota, X. Tu, M. Matsumoto, S. Nakata, A. Kato, G. Seo and C. Pak, Appl. Catal. A 131 (1995) 15.
31. James A. Cusumano, CHEMTECH (August 1992) 482-489.
32. A.W. O'Donovan and C.T. O'Conner, Stud. Surf. Sci. Catal. 105 (1997) 949.
33. T. Inui, Shu-Bin Pu and Jun-ichiro Kugai, Appl. Catal. A: General, 146 (1996) 285-296.
34. Y.Sugi, T. Matsuzaki, T. Hanaoka, Y. Kubota, T. Nakajima, A. Igarishi and K. Kunimori, Collected Abstr., 3rd Internat. Symposium on Acid and Base Catalysis, Rolduc, The Netherlands, April 1997.
35. D. Fraenkel, Ind. Eng. Chem. Res. 29 (1990) 1814-1821.
36. K.G. Wilshier, P. Smart, R. Western, T. Mole and T. Behrsing, Appl. Catal. 31 (1987) 3391.
37. C.S. Hsia Chen and R.F. Bridger, J. Catal. 161 (1996) 687.

Studies in Surface Science and Catalysis 137
H. van Bekkum, E.M. Flanigen, P.A. Jacobs and J.C. Jansen (Editors)
© 2001 Elsevier Science B.V. All rights reserved.

Chapter 22

BEYOND TWELVE MEMBERED RINGS

Eelco T.C. Vogt[*], Charles T. Kresge[‡], and James C. Vartuli[#]

[*] Akzo Nobel Chemicals, Research Center Amsterdam, PO Box 37650, 1030 Amsterdam, The Netherlands
[‡] Present Address: The Dow Chemical Company, Corporate R&D, 1776 Building, Midland, Michigan 48674, United States of America
[#] Mobil Technology Company, Paulsboro Technical Center, Paulsboro, New Jersey 08066, United States of America.

This chapter deals with a subject that did not yet exist in the literature at the time of the first edition of this book, only 9 years ago. In these nine years, the field of ultra-wide pore molecular sieves has attracted considerable attention. A large number of new materials has been and still is invented, encompassing zeolites, aluminophosphates, and ordered mesophases. We will attempt to give an overview of the developments of this field in the last decade. We will limit ourselves to molecular sieves, and will not discuss layered structures in great detail, although some of the mesoporous phases we will discuss are closely related to layered materials. A number of reviews has been published on this subject in recent years, each of them covering different aspects [1,2,3].

We will need to start off by defining a number of terms. Following the IUPAC definitions, we will consider materials with accessible pore openings smaller than 2 nm to be microporous, and materials with accessible pore openings larger than 2 nm mesoporous. Zeolites, microporous aluminosilicates, have generally been described in terms of small-, medium- or large- (sometimes wide-) pore depending on the number of tetrahedral atoms circumscribing the pores. Small-pore zeolites, such as Zeolite A (LTA), Sodalite (SOD), and Zeolite P (GIS), have pores made up of 8 tetrahedral atoms, and about 0.4 nm in diameter. Medium pore zeolites, like ZSM-5 (MFI), have pores made up of 10 tetrahedral atoms, and about 0.55 nm in diameter. Wide-pore, or large pore zeolites, like Zeolite Y (FAU), Mordenite (MOR), and Beta (-BEA) have pores made up of 12 tetrahedral atoms, measuring about 0.75 nm in diameter. Following Casci and others [3,4], we would call materials with pores circumscribed by more than 12 tetrahedral atoms ultra-wide or ultra-large pore. However, following the above IUPAC nomenclature, materials with pores larger than 2 nm should be called mesoporous, and hence do not fall under the definition of zeolites. Assuming perfectly cylindrical pores, zeolites would have a maximum of about 20 T-atom pores (a 2 nm diameter leads to a 6.3 nm circumference, the average T-T distance in zeolites is about 0.31 nm). Materials like MCM-41, although definitely ordered and porous molecular sieves, should not be referred to as zeolites, but rather as mesoporous ordered materials.

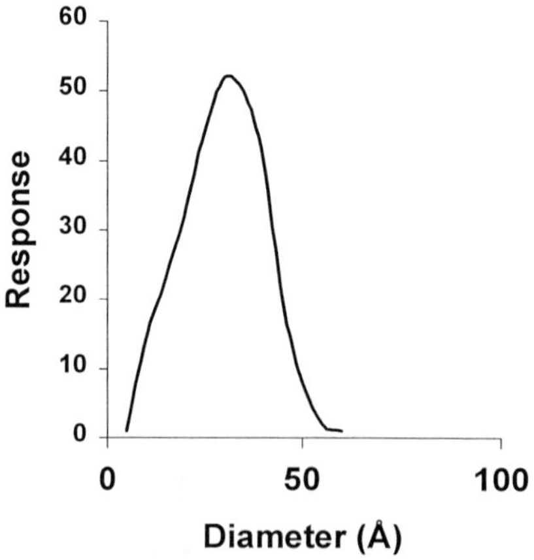

Figure 1: Molecular size distribution in a typical Vacuum Gas Oil as observed with Size Exclusion Chromatography

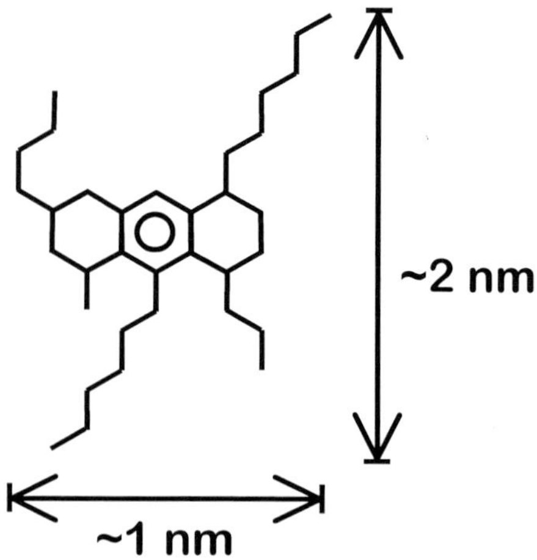

Figure 2: Hypothetical large hydrocarbon molecule, carbon number = 35, boiling point ~ 540°C

The ability to discriminate molecules on size and shape (the definition of molecular sieving) is by no means limited to zeolites alone. Various materials with chemical compositions ranging from carbons to aluminophosphates to clays, silicas and aluminas are known to exhibit molecular sieving. Among these materials, the ordered three dimensional structure found in materials with zeolite-type crystal structures (the 'zeotypes') is special, mainly because it leads to a very well defined, narrow pore size distribution. The well-defined pore system of zeotypes has prompted their widespread use in ion exchange, catalysis, and gas separation. Especially from the field of catalysis, where the main application of zeolites is in refinery applications like Fluid Catalytic Cracking and Hydrocracking, came the desire to make zeotypes with pores larger than 0.75 nm, which would allow larger molecules to access the active pore system of the zeotypes. 12-membered ring zeotypes like Zeolite Y will allow access to molecules not much larger than 0.75 nm in kinetic diameter. Actually, due to thermal motion of the zeolite crystal that can expand the pore mouth, molecules slightly larger than the formal pore diameter have been observed to enter zeolite pore systems.

To illustrate the size of molecules in realistic feedstocks, we can turn to size exclusion chromatography. Although the diffusion processes governing size exclusion chromatography are quite different from those in zeolite pores, an attempt can nevertheless be made to estimate the size distribution in Vacuum Gas Oil, a major feed for FCC- and Hydrocracking-processes [5]. As can be seen from Figure 1, for the specific feed that was investigated, the average molecule size is about 3.5 nm. A hypothetical large hydrocarbon molecule of carbon number 35 and an estimated boiling point of 540°C is depicted in Figure 2. In Figure 3 we have tried to depict the size of an even heavier hypothetical Resid molecule compared to the size of the pores of MCM-41. As a model for MCM-41, we have used the hypothetical structure net 81(7), described by Smith and Dytrich [6]. In the true MCM-41, the pore walls are at least twice as thick as depicted here, and most probably not nearly as ordered. Nevertheless, it can be easily observed that the heavy Resid molecule, with a molecular weight of about 3500, easily fits in the pores.

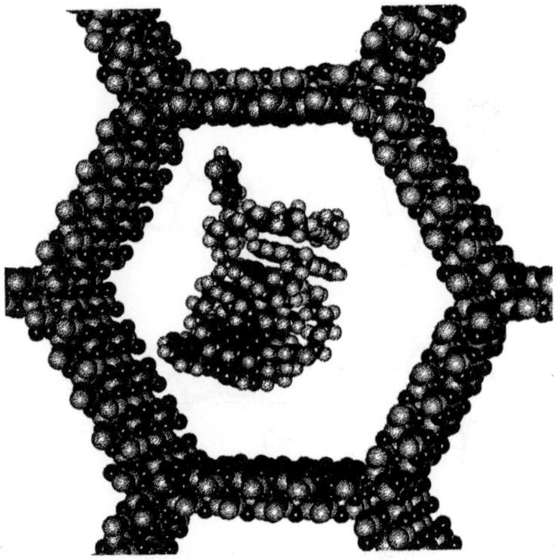

Figure 3: A hypothetical heavy resid molecule in the pore of Smith and Dytrich's hypothetical structure 81(7) (ref. 6)

HISTORY

The history of ultra-wide pore molecular sieves starts with the discovery of VPI-5, an 18-membered ring alumino-phosphate, by Davis and coworkers in 1988 [7,8,9]. It was later demonstrated that d'Yvoire had prepared a material with the same framework topology, called H1, already in the early 60's [10]. However, Davis and coworkers solved the structure [11], and made a pure material in larger quantities [12,13] (H1 was made only in very limited quantities in dilute systems, and apparently was impure). Derouane and von Ballmoos prepared a material called MCM-9 also before Davis' publication [14], but the structure of this material (a mixture of silicon-containing AlPO4-11 and VPI-5) was not recognized until later [15]. Figure 4 depicts the framework structure of VPI-5.

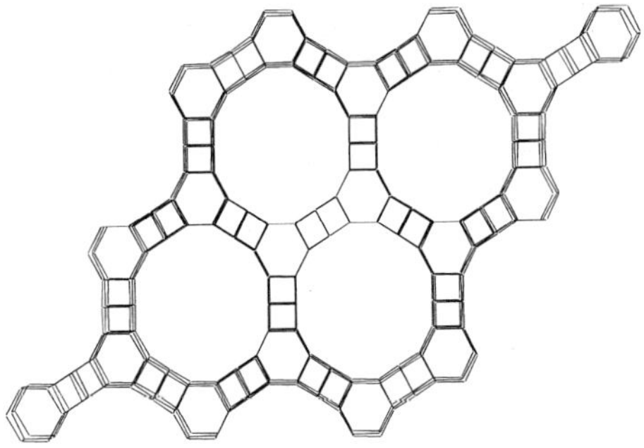

Figure 4: The framework structure of VPI-5, structure code VFI

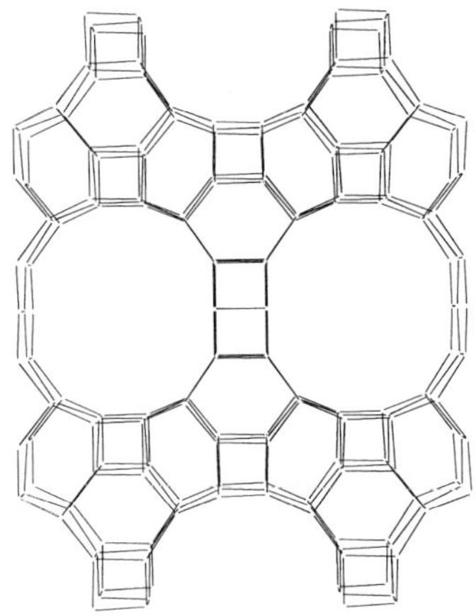

Figure 5: The framework structure of $AlPO_4$-8, structure code AET

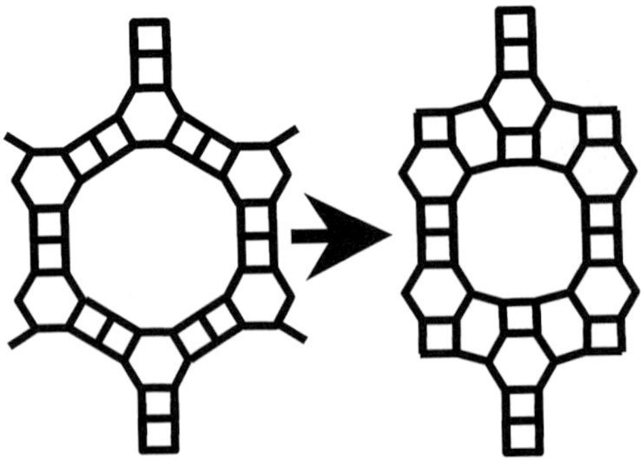

Figure 6: The transformation of VPI-5 to $AlPO_4$-8 according to Vogt and Richardson [19]

VPI-5 is unstable towards moist air and steam in the temperature range around 70-100°C. Various authors report on ways to enhance the thermal stability of VPI-5 [e.g.,16,17]. The material converts to $AlPO_4$-8 (see figure 5), which was already described in the original Union Carbide $AlPO_4$-patents before Davis' papers [18].

It was later shown simultaneously by workers at Akzo Nobel and Mobil [19,20], that $AlPO_4$-8 actually was also an ultra-wide pore material, exhibiting a one-dimensional 14-membered ring pore system (see figure 5 and 6). Although the one-dimensional pore of $AlPO_4$-8 is about 0.7-0.9 nm wide, there is hardly any report of substantial accessible pore volume for $AlPO_4$-8. This is believed to be caused by the fact that the conversion of VPI-5 to $AlPO_4$-8 can actually take place in three different directions. This probably leads to extensive faulting in the $AlPO_4$-8 crystals, and effective blockage of the one-dimensional pore at the pore-mouth.

Szostak et al. reported in 1990 on a naturally occurring ultra-wide pore molecular sieve, the iron-aluminophosphate Cacoxenite, which was previously also described by Moore and Shen [21,22]. The material, which has a one-dimensional irregular 24-membered ring system, is extremely unstable towards moist air at increased temperature, and has therefore not been investigated further.

In 1991, Estermann et al. described a galliumphosphate containing an intricate three-dimensional pore system consisting of clover-leaf shaped 30 membered ring pores, giving access to giant 3 nm wide cages [23,24,25]. The material, prepared from a fluoride-containing medium, retains fluoride ions in double-four ring cages, charge compensated by inorganic cations, and the framework is incompletely coordinated in certain lattice positions (see figure 7). At those spots, most notably the incisions between the

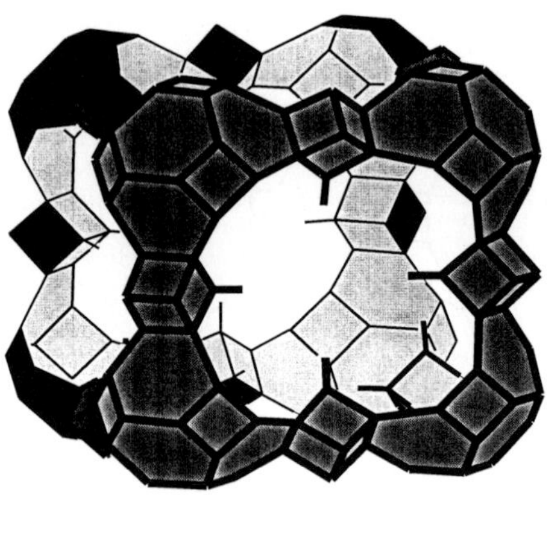

Figure 7: The framework structure of Cloverite. Reprinted from [23] by permission. Copyright 1991, Macmillan Magazines Ltd.

clover "leafs", the framework is terminated by hydroxyls. Like the materials described before, Cloverite is also extremely sensitive to moisture at increased temperature, which is why hardly any application has been described up to the present time.

JDF-20, an aluminophosphate with a one-dimensional 20-membered ring system was first reported in 1992 by Huo et al. [26,27,28]. Like VPI-5 and cloverite, the material is extremely unstable towards moisture. JDF-20 collapses to amorphous material upon calcination.

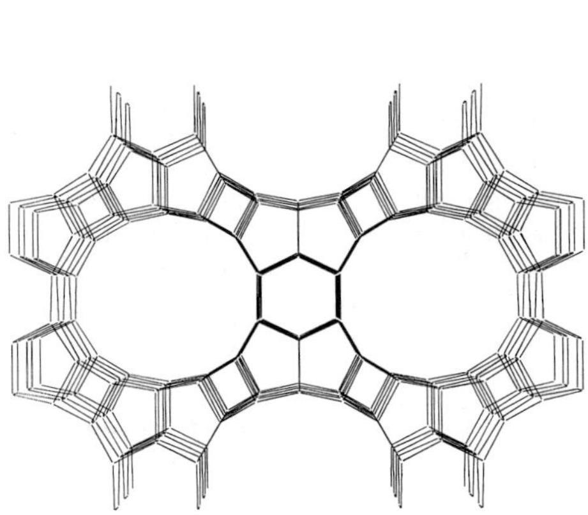

Figure 8: The framework structure of UTD-1

Apart from the aluminophosphate-based structures described above, there is also a number of silicate based ultra-wide pore materials. For instance, In recent years, the family of ultra-wide pore molecular sieves has been expanded with the discovery of UTD-1 by the group of Balkus of the University of Texas at Dallas [29]. This material, again a one-dimensional molecular sieve (see figure 8), is prepared using a fairly unconventional template system, namely di- (pentamethyl-cyclopentadienyl) cobalt salts. The template molecules apparently stack like a roll of oreo-cookies to form a cylindrical mold for the UTD-1 structure, which contains 14-membered ring pores. UTD-1 appears to be the first completely crystalline aluminosilicate with pores larger than 12-membered rings to date.

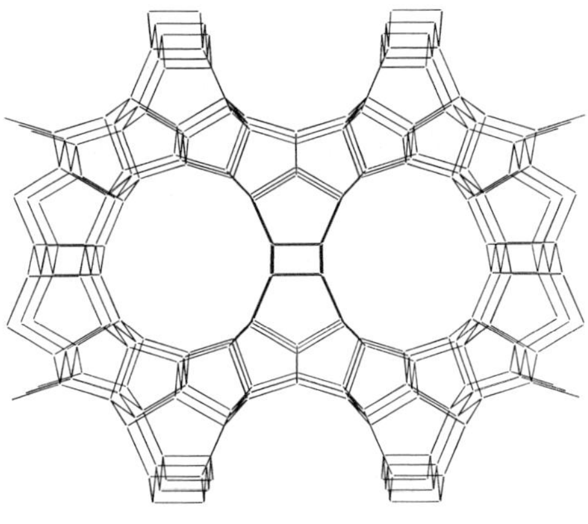

Figure 9: Framework structure of CIT-5

Recently, Wagner and coworkers have reported the discovery of another 14-membered silica based zeolite, CIT-5 (see figure 9). This material, prepared using N(16)-methylsparteinium and lithium, is also composed of one-dimensional channels. This material appears to have good thermal stability, retaining crystallinity to temperature >900°C [30].

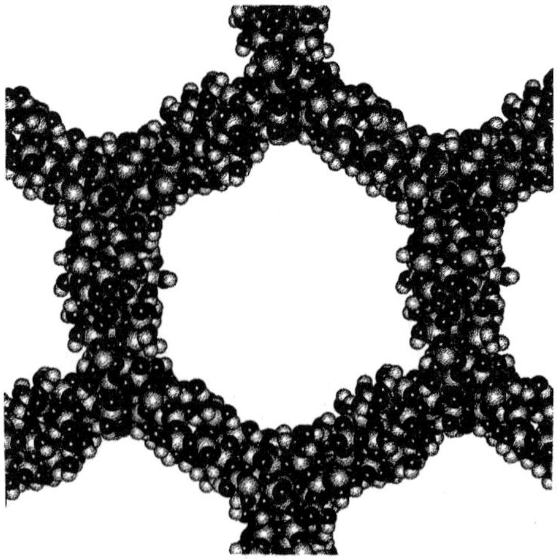

Figure 10: Framework structure model for MCM-41. Model provided by Robert Bell, Royal Institution of Great Britain [69]

A major boost to the field of ultra-wide pore molecular sieve research was given by the discovery by Mobil researchers of a family of materials prepared from surfactant-containing synthesis gels. The members of this family, MCM-41, MCM-48, and MCM-50, were first described in patents published in 1991 [31]. In subsequent papers, the Mobil researchers showed that MCM-41 (see figure 10) was a molecular sieve with a regular one-dimensional pore system with pores of originally around 3 nm in diameter [32,33,34]. Adjusting the length of the apolar tails of the surfactant molecules pores [34], typically C_{12}-C_{16} trimethyl-ammonium salts, was found to control the diameter of the pores. The synthesis of the materials appears to be controlled by the micelle-formation of the surfactant molecules or their silicate salts [35]. Adding expander molecules, like 1,3,5-trimethylbenzene, which prefer to be in the hearts of the micelles, also leads to wider pores. MCM-41, the hexagonally ordered one-dimensional molecular sieve, is formed under conditions where cylindrical micelles are favored in the surfactant-silicate system. MCM-50 is a series of layered materials, forming under conditions where laminar surfactant complexes are more favored, whereas MCM-48 is formed under conditions where spherical micelles are more stable.

Although the Mobil materials have received considerable attention [36,37], at the same time, workers from Toyota and Waseda University described the preparation of similar materials [38,39]. The workers, who studied the recrystallization of a layered silicate called Kanemite in the presence of exactly the same surfactants as used by the Mobil workers, found they could prepare fairly ordered mesoporous molecular sieves, which apparently lost the structural features of Kanemite in the transition. The material, later dubbed FSM-16, did exhibit a single large peak at low angle in the XRD-pattern. Further study proved the materials behaved essentially like amorphous silica-alumina.

After the discovery of the Mobil- and Toyota/Waseda-materials, various groups have created synthetics pathways to mesoporous materials that structurally resemble the M41S family. The Mobil work started from a cationic surfactant S^+ (Cetyltrimethylammonium, CTMA) and an anionic inorganic species I^-(silicate).

In addition to the S^+I^- route, there have been reports on S^-I^+ routes, $S^+X^-I^+$ (X^-=Cl^-, Br^-), and $S^-M^+I^-$ (M^+=Na^+, K^+) routes. In the last two routes, small ions with charge opposite to the surfactant mediate between the surfactant and the inorganic building block. In yet another important route, Pinnavaia and coworkers [40] make use of S^0I^0 combinations, where both the surfactant and the inorganic species are uncharged. Typical reactants are primary amines with long alkyl tails (S^0) and TEOS (I^0). It can be assumed that the interaction between the template and the inorganic species is based on hydrogen bridging with the lone pair of the amine head groups.

The aluminum source used in the synthesis of MCM-41 apparently influences the acidity of the resulting structure. Reddy and Song [41] used various sources of aluminum in the synthesis of MCM-41materials with varying silica/alumina ratios. The authors conclude that aluminum isopropoxide is the preferred source, leading to improved crystallinity and acidity.

PREPARATION

Ultra-wide pore molecular sieves have been prepared with a variety of templates, and some even in the absence of templates. For instance, the original $AlPO_4$-H1 described by d'Yvoire [10] was made in a template free system, and the structure of VPI-5 seems to be stabilized by triple helices of water molecules in the pores [42]. There does not appear to be a clear indicator of what an organic template should look like for it to succeed in the synthesis of ultra-wide pore molecular sieves. Although the micelle-type structures formed by surfactants in the synthesis of materials seem to allow the preparation of a range of pore sizes, pore diameters between 0.75 nanometer and 2 nanometer seem difficult to attain. Incidental successes like VPI-5, UTD-1, and CIT-5 do not presently seem to be predictable based on template geometry or synthesis conditions. Various authors have reflected on the design of zeolite synthesis. See for instance Chapter 5 in this handbook, Chapter 4 in the original edition of this handbook [43], the work of Flanigen [44], Lowe [45], Jacobs [46], Davis and Zones [47,48,49], Harris [50], and Blum and Hyde [51]. In addition, the group of Thomas at the Royal institution in London has made great progress in using computer modeling to design templates for new zeolite structures [52]. Still, however, the complete design of a synthesis procedure for a new molecular sieve has not been demonstrated. This, in our opinion, remains one of the great challenges for zeolite synthesis.

ANALYSIS

The techniques used to study ultra-wide pore molecular sieves are no different from the techniques normally applied to molecular sieves, as we will describe in the next paragraphs.

Electron Microscopy

Ultra-wide pore molecular sieves basically are normal molecular sieves with slightly wider pores, which makes it possible to study the pore systems with standard resolution electron microscopy, as opposed to the high-resolution transmission electron microscopy needed for more conventional molecular sieves.

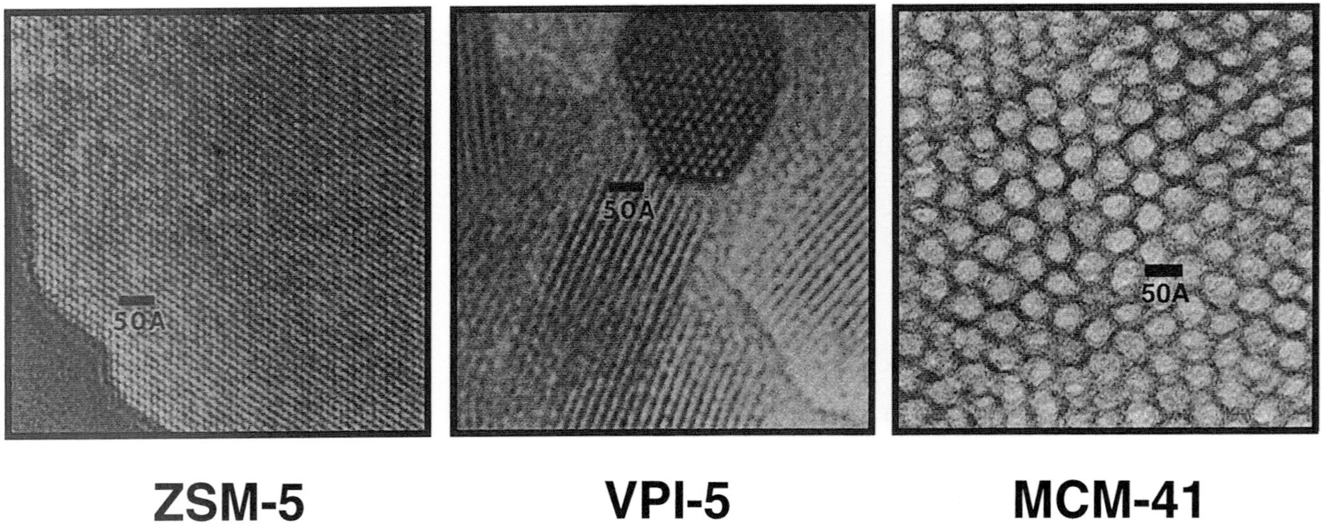

Figure 11: An electron microscopy study of the structures of ZSM-5, VPI-5, and MCM-41 shows the differences in pore sizes

Figure 11 is a good indication of the difference in pore sizes for ZSM-5, VPI-5 and MCM-41, captured with transmission electron microscopy.

Alfredsson and Anderson [53] have used transmission electron microscopy to study the structure of MCM-48. They conclude that the structure can be described using the gyroid minimal surface, one of the set of infinite periodic minimal surfaces. Previous work has already been successful in describing the structures of other zeolites (among others A and Y) with this type of mathematical curved structures.

The structure is described as the result of combining sine waves in three dimensions. The resulting mathematical structure, which is cubic, is classified in spacegroup Ia-3d, the same as found for MCM-48. Alfredsson and Anderson [53] compared XRD-patterns and TEM images of MCM-48 with those predicted by model calculations on the gyroid surface, and found a surprising resemblance.

Adsorption studies
The large pore systems in the ultra-wide pore molecular sieve materials sometimes give rise to unexpected problems in the interpretation of adsorption isotherms. Most materials show a sharp step in the sorption isotherms, which is ascribed to capillary condensation within uniform mesopores. The effect was noted by Davis et al. in VPI-5 with argon sorption [54], and also repeatedly found with MCM-41 and similar materials.

Wu et al. [55] studied the formation of MCM-41 and its pore structure. They concluded that a more defined structure is formed at lower surfactant concentrations. Nitrogen sorption was used to probe the various pores formed in the structure. Incomplete growth or packing problems in MCM-41 gives rise to slit-like mesopores, which were also observed in the Scanning Electron Microscope.

Upon transformation to $AlPO_4$-8, VPI-5 appears to lose most of it's pore volume (up to 90%), which is surprising given the fact that $AlPO_4$-8 is also an ultra-wide pore molecular sieve. However, we have to take into account the fact that both $AlPO_4$-8 and VPI-5 have one-dimensional pores in which blocking at the pore-entrance will effectively block the entire pore. If we combine that with the fact that the transformation reaction can take place in three different directions in the crystal plane, this effect is more understandable. Figure 12 demonstrates the effective blocking of the pore system of $AlPO_4$-8 by superimposing three layers of $AlPO_4$-8

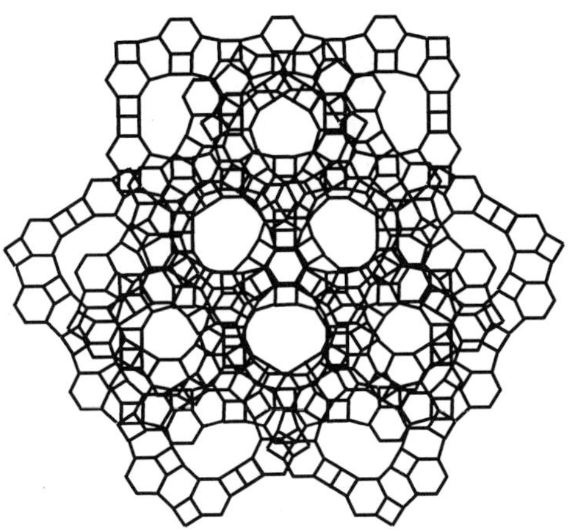

Figure 12: Pore blockage in AlPO4-8 caused by transformation in three different crystallographic directions

under an angle of 120°, which is the angle under which these layers could form in the actual process.
Kloetstra et al. [56] use thermoporometry to measure the pore size distribution of MCM-41. Thermoporometry is a calorimetric procedure, in which the solidification of a liquid adsorbed in the pores of a structure is followed in a calorimeter. The solidification temperature is dependent on the pore dimensions, and the pore volume can be determined from the energy involved. The technique appears very suitable especially to investigate the pores of mesoporous materials.

NMR
In the characterization of ultra-wide pore molecular sieves, NMR has mainly been applied to study the hydrophilicity of VPI-5 [57], the positioning of water molecules in the VPI-5 pores [58,59], and the crystallinity of the materials from the M41S family.

^{29}Si-MAS-NMR on MCM-41 usually indicates the presence of a large number of silanol groups, considered to be evidence for the amorphous nature of the pore walls. ^{27}Al-MAS-NMR in as-synthesized MCM-41 has been observed to yield proof of mainly tetrahedral coordination of the alumina. After high temperature treatment, however, and especially in the presence of steam, framework alumina appears to become dislodged. The nature of the synthetic pathway is related to the amount of surface hydroxyls (Q^3 silica) found. The amount of Q^3 silicons can be as high as 50%.

XRD
Peaks in the low-angle diffraction range dominate the XRD-pattern of many of the ultra-wide pore molecular sieves. In some cases, like many reports of MCM-41 synthesis, only one peak is actually found. In materials with a more crystalline structure, like VPI-5, which is also an two-dimensional hexagonally ordered material, there is more structure in the pattern.

The structure of VPI-5 was first solved by Davis et al. [11], and later refined by various others. McCusker et al. [42] used high-resolution synchrotron powder diffraction to study the structure, and were able to locate an intricate structure of water molecules that appeared to be part of the structure of VPI-5. The water molecules were ordered in a triple helix, and part of the helix structure appeared to be bound to one of the aluminum atoms in the asymmetric unit, which was thus octahedral. The exact location of this octahedral aluminum atom has been the subject of quite some debate. Grobet et al. in 1989 proposed a structure where one of the aluminum atoms in the six-ring of VPI-5 was octahedral [59, 60]. Richardson et al. [61], who studied $AlPO_4$-54, the structural analog of VPI-5, also in 1989 described an increased thermal ellipse, which could be interpreted as an indication of deviation from the normal tetrahedral coordination for the aluminum atom in the center of the fused four-ring. Maistriau et al. [58] performed solid state MAS-NMR on VPI-5 and $AlPO_4$-8, and described the role of the hydrated aluminum atom in the transition of VPI- to $AlPO_4$-8. They posed a model in which the aluminum atom in the center of the fused four ring is octahedral, and also proposed that the structural reorganization of VPI-5 starts at this site. Figure 6 shows the proposed recrystallization process, as it was originally described by Richardson et al. [62].

Computer modeling
Kitao et al. [63,64] used computer modeling to study the hydrophilicity of VPI-5, and found that only the aluminum atom in the center of the fused four-rings can adsorb water with an attractive interaction energy. This study, based on Ab Initio Quantum Mechanical calculations on small clusters with the geometry of the various Al-sites in the VPI-5 structure, concludes that only this specific site can adsorb water, and only if it is already hydrated. The other Al-sites (hydrated and dehydrated) do not appear to be capable of sorbing water.

In recent years, Cheetham and Harding [65] performed a single crystal X-ray diffraction study on VPI-5, and confirmed all of the framework positions and the positions of the water molecules as originally proposed by McCusker et al. [42].

Richardson et al. described a series of hypothetical nets with 18 ring-channels that all have the same projection as VPI-5 [61]. These structures were later used by De Man et al. [66] to study the intrinsic stability of ultra-wide pore materials. They concluded that net 520, the one describing the actual structure found in VPI-5, appeared to be the most stable. Many other authors have described hypothetical frameworks for ultra-wide pore materials. It is, for instance, interesting to note that the structure for VPI-5 had already been predicted by Smith and Dytrych in 1984 [6]. Li et al. [67], describe a structure called MIR-28, a hypothetical 28-ring structure, which they claim should be stable, based on energy minimizations.

Feuston and Higgins used molecular dynamics studies to come up with a number of model structures for MCM-41 [68]. The authors built structures adhering to the general hexagonal structure found for MCM-41, with different wall thickness and pore diameter, and studied their dynamic behavior. Based on this study, the authors conclude that amorphous pore walls with a thickness of about 1.1 nm are in best agreement with the observed XRD pattern for MCM-41 with lattice constant of about 4.5 nm. Recent work on amorphous models for MCM-41 was done by Bell [69] (see figure 10).

Keffer et al. [70] calculated potential maps for Xenon in various zeolites and molecular sieves. The potential maps can be used to study the accessible pore volume and localized adsorption sites, but also to study diffusion mechanisms. The ultra-wide pores studied ($ALPO_4$-8 and VPI-5) of course are both unidimensional, so little difference is observed in their characteristics, except for the more rectangular shape of the accessible pore.

Catalytic tests
A number of catalytic tests have traditionally been used to establish the effective pore width of zeolites.

The Constraint Index, developed by Mobil workers in the late 1980's [71], is based on the comparison of the rates of acid catalyzed cracking of n-hexane and

3-methylpentane. Since the Constraint Index is conceptually based on the reactant shape selectivity for the branched versus the normal hexane, it is not too difficult to understand that the method has difficulty distinguishing pores in the range larger than about 0.6 nanometer. The Constraint Index, which uses the ratio of the cracking rates of n-hexane and 3-methylpentane, is found to be 1-12 for zeotypes with medium pores, like ZSM-5. For larger pore zeolites, like Beta and Y, it is typically smaller than unity. There is no difference between the constraint index for zeolite beta, and for amorphous silica-alumina. It is thus obvious that the constraint index can not be used to determine the exact pore size of ultra-wide pore molecular sieves, although it can of course give an indication when the structure of a molecular sieve is unknown.

An improved method is described by Weitkamp et al. [72], who developed the Spaciousness Index, based on the bifunctional hydrocracking of a C_{10}-naphthene, like butylcyclohexane or pentylcyclopentane. The Spaciousness Index is defined as the ratio of the yields of isobutane and n-butane in the cracking reaction. The method relies on the fact that the rearrangement leading to the formation of isobutane from the C_{10}-naphthene requires a rather bulky transition state, which is hindered in smaller pore zeolites. Hence, where the Constraint Index is reactant shape selective, the Spaciousness Index is transition state selective. The Spaciousness Index gives a better indication of the internal pore size for larger pored materials. However, its applicability ends at about the cage size of zeolite Y, or about 1.2 nm, since at that size the transition state is no longer hindered. Thus, the spaciousness index is also of limited value to establish the pore size of ultra-wide pore molecular sieves.

From the comparison of the standard catalytic techniques to assess the pore size, we must conclude they are basically not suitable to study ultra-wide pore molecular sieves. Therefore, we must resort to test-reactions with larger molecules, basically probing the pore system with increasing reactant molecule sizes.

Martens et al. [73] describe a method based on the hydroconversion of n-decane. The bi-functional cracking and isomerization gives a number of different ratios that can be used to determine the structure and dimension of zeolite pores. This technique seems to be more useful than the others, because there is still quite a lot of differentiation even in twelve membered structures. This technique thus seems a likely candidate to probe larger pore molecular sieves.

A well-known molecule that is larger than the window size of zeolite Y is 1,3,5-tri-isopropylbenzene. For instance, Davis et al. [54] probed the pores of VPI-5 with 1,3,5-tri-isopropylbenzene in sorption experiments. A problem with using tri-isopropyl benzene as a probe in cracking reactions is the fact the molecule fairly easily cracks on the outer surface of the molecular sieve crystal, which makes interpretation of the test results complicated.

Chen and Schramm [74] circumvented this problem by deactivating the outer surface of the crystals with bulky organophosphorous compounds, which allowed the authors to study the intrinsic cracking process of 1,3,5-tri-tert-butylbenzene in ZSM-5 and zeolite beta. Of course, the molecule is too large to enter the ZSM-5 pores, but it can be used to study the cracking on the external surface of the crystals. The technique can conceivably also be used for molecular sieves with larger pores, although probably not on the extremes like MCM-41, where the pores would allow access to most of the conventional bases.

Therefore, it appears that only catalytic tests with real petroleum feedstocks can give an indication of the accessible pore size of the ultra-wide pore molecular sieves. Corma et al. [75] used NiMo loaded MCM-41 to study mild hydrocracking reactions, and compared the results with those for similar catalysts based on Zeolite Y and amorphous silica alumina. The authors conclude the availability of uniform pores in the mesoporous range is the reason for superior performance of the MCM-41 based catalyst, especially in single stage hydrocracking. When comparing the same aluminosilicates without metal loading, Corma et al. [76] found the Y-zeolite to be the most active in heptane cracking, by a factor of about 140 over MCM-41. However, in gasoil cracking, MCM-41 and zeolite Y differed only a factor 10 in activity. The authors claim that the sites in Zeolite Y are much more acidic than those in MCM-41, but the increased accessibility of the sites in MCM-41 compensates in the case of gasoil cracking. The authors note that steaming has a detrimental effect on the activity of MCM-41, the structure apparently collapses, which would make the material incompatible with FCC operation.

Mokaya et al. [77] used thermogravimetric desorption of cyclohexylamine and cumene (isopropylbenzene) cracking to probe the acidity of MCM-41, and found that both the number of acid sites and their strength is lower in MCM-41 than it is in zeolite Y.

Stability - Theory

An intriguing question concerning ultra-wide pore materials is their intrinsic acidity and stability. These issues have been the subject of various theoretical studies. There appear to be two different schools of thought concerning the intrinsic activity of larger pored materials as compared to smaller pored materials. O'Malley et al. [78] claim that the acidity of molecular sieves is related to the T-O-T angle in the bridging OH in the aluminosilicate structure. The angle is obviously related to the hybridization of the oxygen *p*-orbitals, and it is understandable that changes in this hybridization should affect the strength of the OH bond, and hence, its acidity. O'Malley and others have investigated the acidity as a function

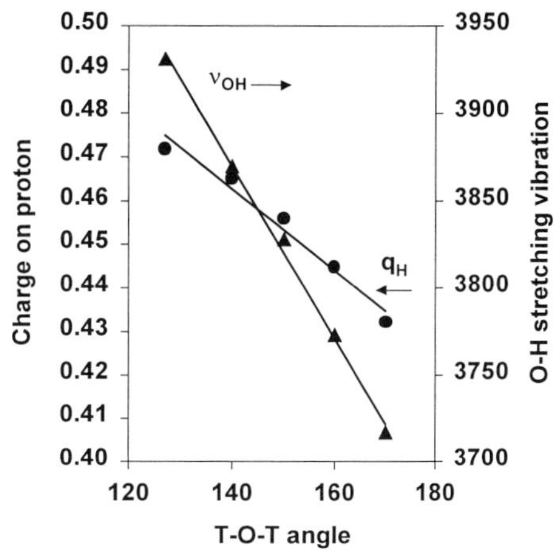

Figure 13: Acidity of zeolite proton as a function of the T-O-T angle

of T-O-T angle, and have found a minimum acid strength at angles of 127° (see Figure 13). Larger pore systems, that more closely resemble amorphous materials, could have more relaxed T-O-T angles, and hence lower acidity. The relation between acid strength and T-O-T angle has been demonstrated by other authors as well [79]. However, we feel there is no reason to assume that the average T-O-T angle for ultra-wide pore materials should be any wider than that of medium pore or large pore zeolites. Therefore, we feel another theory may be required.

Derouane and co-authors have published an extensive series of papers on the so-called curvature effect [80,81,82]. The basis of this theory is the assessment that the Van der Waals interaction between molecule and pore wall will be optimal when the molecule radius and the pore radius coincide. The zeolite pore system thus acts as a solvent, stabilizing the molecule in the pore system to the extent of the sorption energy, which will be highest for combinations of matching pore size and molecule size.

It has been demonstrated that actually the activity of various zeolites in cracking reaction can be found to be dependent of the adsorption energy of the molecule in the zeolite pore system. Furthermore, this sorption energy, and thus the cracking activity can be predicted by a simple formula.

In Figure 14, we have plotted the turnover frequency for hexane cracking at 538°C (1000°F) for various molecular sieves of similar acid site content. It can be seen that the TOF declines in a smooth way with increasing pore size, and that the activity observed closely matches the predicted activity. Thus, it appears that within the framework of this theory the activity of large pore molecular sieves in cracking reactions can be predicted fairly accurately by accounting for the amount of interaction between the curved pore wall and the molecule. For hexane cracking, the highest interaction, and hence the highest activity (usually interpreted as the highest acidity), is thus found for medium pore zeolites like ZSM-5. It should be noted that the TOF for hexane cracking can vary three decades within a fairly narrow pore size range.

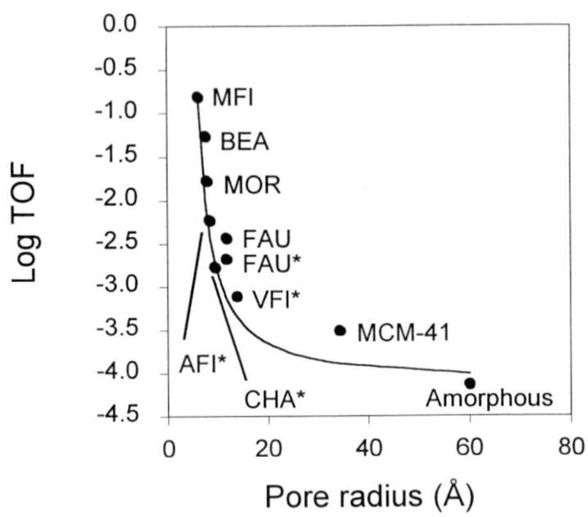

Figure 14: The turnover frequency in hexane cracking at 538°C as a function of pore diameter in various molecular sieves

It should also be noted, that the activity for MCM-41 is predicted to be similar to that for amorphous silica-alumina, which is indeed observed.

Various authors have claimed that acid sites in SAPOs are less active than those in zeolites. This statement is based on NH_3 adsorption measurements, but also on comparison of cracking activities for small molecules. Theoretic considerations based on simple electronegativity arguments by Sauer [83], however, indicate there is no reason for differences in acid strength. More detailed studies by Sauer et al. on this subject are in progress [84,85]. The measured curve and the predicted curve for both zeolites and silicoaluminophosphates are combined in Figure 14. The overlap of both curves is striking. The results indicate that on a macroscopic scale, the activities of zeolites and SAPOs of similar pore diameter are similar. This is most obvious for the isostructural zeolite Y and SAPO-37. This indicates that for cracking of small hydrocarbons, the differences in heat of sorption induced by structural differences are the major contributing factor to the activity.

In recent literature, Corma and coworkers have reported on a theory they called "electronic confinement" [86,87]. The authors performed quantum-mechanical calculations and conclude that the interaction between the outer orbitals of the pore-lining atoms and the outer orbitals of the molecule sorbed in the zeolite pores leads to a compression of the outer orbitals of the molecule. This compression then, leads to higher energy levels for the molecular orbitals, and hence, an activation of the molecule. Although the authors have demonstrated the effect for various molecules in various zeolites, and the effect is certainly real, we believe the magnitude of the effect is highly overestimated by the computational methods applied by Corma et al. [86,87]. We thus believe that electronic confinement does not offer an explanation for the diminishing activity observed in larger pore zeotypes.

Various authors have calculated the lattice enthalpies of molecular sieves with force-field methods. Invariably, a dependence of the lattice enthalpy of the framework density is found [66,88,89,90]. To illustrate this statement, we have plotted in Figure 15 the lattice enthalpies for various silicates with different lattice densities. It can easily be observed that there appears to be a linear dependence of the lattice enthalpy with the framework density, with the less dense, more open structures, having the highest enthalpy, and hence the lowest stability. A similar effect was found for silicates.

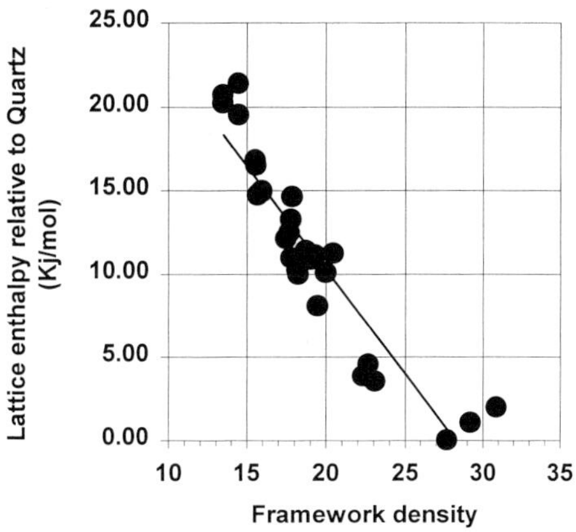

Figure 15: Calculated lattice enthalpies for selected silicates as a function of framework density, after the work of Henson et al. [90].

The decrease in stability that is generally predicted by theoretical work is actually also observed in experiments, as also indicated by Henson et al. [90]. Navrotsky et al.

[91,92,93] measured the enthalpy of formation for various silicates by several calorimetric techniques and found correlations very similar to those described before.

Recently, Garcés of Dow Chemical [94] proposed a theory relating void fraction and structure in porous solids. The author reduces the pore system of zeotype structures to single voids, and the dense framework components as a single dense shell. The author then continues to show that most zeolites can be described either as having spherical voids, cylindrical voids, or as layered solids. Garcés correlated the void fraction of various zeolites to the calculated void space diameter for the three basic structures [94]. He notes that for most well known zeolite structures the void structure predicted by the model is in good agreement with the known structure. MCM-41, however, is found to deviate, and the author claims that his model predicts the pore structure of MCM-41 should be spherical rather than cylindrical. However, a simple calculation shows that if we assume the shell to be two or three monolayers thick (which is much more realistic for MCM-41 with its 1 nm pore walls) the adapted model actually correctly predicts the structure of MCM-41 to be cylindrical (see Figure 16).

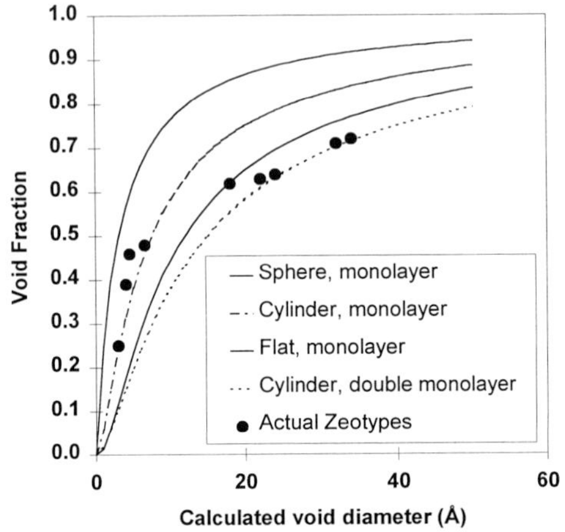

Figure 16: Void fraction calculations following Garcés' model [94] with adaptations as described in the text

APPLICATIONS

Although initially most researchers seemed to be interested in MCM-41 and related structures for conversion of large hydrocarbons in refinery applications, that field of applications seems to have been abandoned in the last years. Although a number of papers [75,76] and patents [95,96] have appeared on the application of MCM-41 in major refinery process, the majority of applications seem to concentrate on applications that are closer to fine chemicals.

Reactions of large molecules with the mesoporous materials, either in acid form (via aluminum incorporation) or via impregnation with catalytically active metals/complexes have received much attention. MCM-41 materials have demonstrated the ability to convert bulky molecules that are too large to react within the classical microporous molecular sieve pore structure. For example, it was reported that aluminosilicate MCM-41 was a suitable catalyst for the alkylation of the bulky 2,4-di-tert-butylphenol with cinnamyl alcohol [97]. It was shown that while this alkylation does not occur in the restricted environment of an HY zeolite (pore size about 0.75 nm), the primary alkylation product 6,8-di-tert-butyl-2-phenyl-2,3-dihydro[4H]benzopyran is formed using the large pore diameter mesoporous molecular sieve.

Other work on metals incorporation included the introduction of heteropolyacids such as $H_3PW_{12}O_{40}$ [98]. Here it was shown that the acid incorporation allowed higher selectivity (to less bulky alkylation products) in the alkylation of 4-tert-butyl-phenol with styrene. It was shown that the less bulky products were more comparable to the internal pore dimensions than the larger products normally formed over completely open amorphous substrates containing the same acid species.

Selective partial oxidation of molecules over MCM-41 materials has also received much attention. For example, titanium-containing mesoporous materials (incorrectly referred to as Ti-zeolites), isomorphous with hexagonal MCM-41, and synthesized with titanium-tetraethoxide as the titanium source, were found to selectively oxidize 1-hexene to the resulting epoxide with H_2O_2 as the oxidant [99]. It seems to make sense to differentiate between Ti-containing mesoporous materials prepared by co-condensation of the Ti-source (as above), as opposed to post-synthesis addition of the Ti-species (grafting). The materials prepared with the different preparation techniques behave different, as is exemplified for instance in their resistance to leaching by H_2O_2 and Tert-Butyl hydroperoxide (TBHP). Large molecule oxidation was also studied. It was found that selective epoxidation of norbornene could be achieved with TBHP as oxidant. The same reaction could not be achieved over Ti-ZSM-5 because of the bulky nature of the reactants. Unfortunately, the work gives no comparative data for an amorphous Ti/silica analog which would contribute to the shape-selectivity claims ascribed to the mesoporous sieve. Further work shows that alpha-terpineol can be epoxidized with these catalysts [100]. In other papers, a Ti-containing mesoporous molecular sieve was capable of the shape-selective conversion of 2,6-di-tert-butyl phenol to the corresponding quinone [40,101]. The sum of these works shows that the oxidations of aromatics and olefins can be effected with these Ti-mesoporous materials, but the more elusive shape-selective oxidations of alkanes and alcohols remain the exclusive domain of the microporous Ti-containing molecular sieves.

Rao et al. [102] heterogenized active manganese triazacyclononane epoxidation catalyst. The authors covalently anchored Mn-1,4,7-triazacyclononane complexes onto the MCM-41 surface, which was first reacted with 3-glycidil-oxypropyl)trimethoxysilane. The triazacyclononane (tacn) is then added, and it reacts with the glycidiloxypropyl-residue. The *tacn* is then further functionalized, and doped with Mn^{2+}. Epoxidation of styrene with this catalyst shows reasonable turnovers, but the authors report that similar activities can be found with amorphous silica-supports. Thus, there seems to be no specific advantage of using MCM-41 as the support.

Parton et al. [103] incorporated iron-phthalocyanine complexes in zeolite Y and VPI-5. They prepared these complexes by in-situ reaction of sorbed ferrocene with 1,2-dicyanobenzene. The catalysts were tested in the oxyfunctionalization of alkanes with

TBHP. Turnover numbers for the complex sorbed in VPI-5 surpass those found for the free complexes. Furthermore, incorporation in the VPI-5 pores protects the iron complexes from oxidative destruction.

Mesoporous silicates have also been used for the controlled polymerization of styrene, methyl methacrylate and vinyl acetate [104]. In the study, 2.5 and 4 nm MCM-41 materials were used as hosts. It was observed that the properties of the polymerized guests varied with the degree of confinement. Thus, an increase in the average molecular mass of polymethylmethacrylate was observed with an increase in confinement on polymerization within 4 and 2.5 nm pores, respectively. The authors suggest that the increase in confinement leads to a smaller quantity of growing polymer chains, which are unable to interact, resulting in the formation of longer chains. A decrease of 20 Kelvin in the glass transition temperature of polyvinyl acetate formed in 2.5 nm pore diameter material was observed. It was suggested that the decrease leads to a suppression of the cooperative motion of the polymer.

Van Looveren et al. [105] describe the copolymerization of ethylene and propylene using $[\{C_2H_4(1\text{-indenyl})_2\}Zr(CH_3)_2]$ supported on methylaluminoxane-MCM-41. The methylaluminoxane cocatalyst is anchored to the internal pore walls and is generated in-situ by hydrolysis of trimethylaluminum, and the Zr-complex is adsorbed into the pores. Using this system, the Zr-complex is found to be active at much lower Al/Zr-ratios than needed for the homogeneous systems.

Another area that the mesoporous structures have received a great deal of attention in is the area of separation and adsorption. The uniform pore structure within the mesopore range and the resultant high pore volume offer materials for separations that vary from the removal of organic and inorganic contaminants in waste streams to chromatographic media. Early work demonstrated the extraordinary high sorption capacity of the M41S materials for hydrocarbon molecules such as benzene [32,33]. It was also shown that modification of the pore walls by post functionalization could substantially alter the sorption capacity and behavior of these materials [33]. Replacing the pore wall silanol groups with trimethylsilyl groups created a more hydrophobic environment within the pore structure. The resulting material demonstrated a substantial reduction in sorption capacity of polar molecules such as water.

A recent article illustrated the use of functionalized MCM-41 for the removal of heavy metals from waste streams. These researchers functionalized the pore walls of a silica MCM-41 with a mercaptopropylsilane [106]. The resultant material demonstrated a high affinity to extract mercury and other heavy metals from both aqueous and nonaqueous waste streams. Furthermore these researchers also demonstrated that this functionalized MCM-41 material could be regenerated and reused. Another recent article indicates that enzymes could be attached to the pore walls of MCM-41 [107]. The authors of this paper were able to adsorb globular enzymes, cytochrome C and trypsin to siliceous MCM-41.

Ultimately, uniform mesostructures synthesized as membranes would have great utility for separation applications ranging from chemicals to pharmaceuticals. Several research groups are actively pursuing the synthesis protocol for this type of material. A recent article detailed the use of magnetic fields to align the pore structure of the resultant mesoporous silica illustrating the techniques used to tackle this difficult synthesis problem [108]. Attachment of chiral molecules to the pore walls could create the interesting possibility of separating pharmaceutically interesting enantiomers.

Turning to more novel applications, it has been speculated that MCM-41 materials could show utility in a variety of electronic, optical and other advanced applications. It has been demonstrated that 2,4,6-triphenylpyrylium (TP^+) ion incorporated in MCM-41 can be used as a highly efficient electron-transfer material [109]. In this paper, it was shown that the TP^+ ion, when so incorporated, could be used as a sensitizer (in the presence of long wave radiation) to convert cis-stilbene to trans-stilbene. In a comparative study, TP^+ in MCM-41 was much more effective in this conversion than $TP^+BF_4^-$ or TP^+ incorporated into the smaller Y zeolite or TP^+ on amorphous silica-alumina. In summary, it was shown that the mesoporous hosts are promising for photoinduced electron transfer reactions.

Finally the mesoporous materials were used as nanosized reactors to form polyaniline within the pore system of copper or iron exchanged MCM-41 by adsorption of aniline vapor followed by oxidative polymerization [110,111]. The resulting encapsulated polyaniline was examined for electrical conductivity. Results showed that the conductivity of the encapsulated polyaniline is greatly reduced compared to polyaniline prepared in the bulk, supporting the supposition of the materials location inside the channel system and not on the external silicate surface. When the polyaniline is extracted, the conductivity increases significantly as compared with bulk samples.

Heterogenized sulfonic acid catalysts have been prepared by the functionalization of MCM-41 with 3-mercaptopropyl-trimethoxysilane, and subsequent oxidation of the anchored SH groups [112]. The catalysts were tested in the synthesis of 2,2-bis(5-methylfuryl)propane, an intermediate for macromolecular chemistry. The anchored sulfonic acid catalysts show good activity, and do not deactivate as quickly as normal zeolites, possibly because they are more hydrophobic.

Using similar concepts, Lasperas et al. [113] prepared solid basic catalysts by anchoring amine functions to the pore walls. These catalysts were tested in the Knoevenagel condensation of benzaldehyde with ethyl cyanoacetate. The differences between primary and tertiary amine groups were investigated, and the immobilized primary amines were found to be more active, probably because they allow the formation of an imine intermediate.

Kloetstra et al. [114,115] prepared basic catalysts by impregnation of MCM-41 with mixtures of Cesium and Lanthanum salts. Small, mildly basic, intraporous $CsLaO_x$ clusters are formed upon thermal decomposition. Their catalytic application was tested in Knoevenagel condensation of enolates to benzaldehyde and the Michael addition of ethyl cyanoacetate to ethyl acrylate.

It was found that the catalysts exhibit good activity, and can be re-used after a solvent flush and high-temperature cycle without apparent loss of activity.

CONCLUSION

Molecular sieves with pores beyond twelve membered rings are in a relatively new field, in which exciting new developments are reported with high frequency. Commercial applications for some of the new materials appear on the horizon. There is still a challenging, relatively virgin field of pore diameters between 14 and 20-membered rings (roughly between 1 and 2 nm), in which new crystalline materials would be very welcome for various applications. The authors are confident that the ever-increasing interest for this field will eventually lead to discoveries in this area.

ACKNOWLEDGMENTS

The authors would like to thank Mobil Oil and Akzo Nobel Chemicals for permission to publish this review paper.

REFERENCES

1 A. Sayari, in Recent advances and horizons in zeolite science and technology, Studies in Surface Science and Catalysis **102**, Chon et al., eds., Elsevier, Amsterdam, 1996, 1-46.
2 X.S. Zhao, G.Q. Lu, G.J. Millar, Ind. Eng. Chem. Res., **35**, 1996, 2075-2090.
3 J.L. Casci, in Advanced Zeolite Science and Applications, Studies in Surface Science and Catalysis, vol. **85**, Jansen, Stöcker, Karge and Weitkamp, eds., Elsevier, Amsterdam, 1994, 329-356.
4 D.W. Breck, in Zeolite Molecular Sieves: Structure, chemistry and use, Wiley, London, 1974.
5 P. O'Connor, F. van Houtert, Proceedings Ketjen Catalyst Symposium 1986, Akzo Chemie Amersfoort, The Netherlands, paper F-8.
6 J.V. Smith, W.J. Dytrich, Nature, **309**, 1984, 607-608.
7 M.E. Davis, C. Saldarriaga, C. Montes, J.M. Garcés, C.E. Crowder, Nature, **331**, 1988, 698-699.
8 M.E. Davis, C. Saldarriaga, C. Montes, J.M. Garcés, C.E. Crowder, Zeolites, **8**, 1988, 362-366.
9 M.E. Davis, C. Montes, J.M. Garcés, ACS Symp. Ser., **398**, 1989, 291-304.
10 F. d'Yvoire, Bull. Soc. Chem., 1961, 1762.
11 C.E. Crowder, J.M. Garcés, M.E. Davis, Adv. X-ray Analysis, **32**, 1988, 503-510.
12 M.E. Davis, C. Montes, P.E. Hathaway, J.P. Arhancet, D.L. Hasha, J.M. Garcés, J. Am. Chem. Soc., **111**, 1989, 3919-3924.
13 M.E. Davis, D. Young, Tokyo Meeting Microporous Crystals, 1990.
14 E.G. Derouane, R. von Ballmoos, US Patent 4,673,559, 1989.

15 E.G. Derouane, L. Maistriau, Z. Gabelica, A. Tuel, J.B. Nagy, R. von Ballmoos, Appl. Catal., **51**, 1989, L13-L20.
16 J.R. Anderson, W.R. Jackson, D. Hay, Z. Yang, A. Campi, Zeolites, **16**, 1996, 15-21.
17 A. Karlsson, D, Akporiaye, M. Stocker, E.M. Myhrvold, Catalysis in Microporous Materials, Studies in Surface Science and Catalysis, vol. **94**, Beyer, Karge, Kirisci, and Nagy, eds., Elsevier, Amsterdam, 1995, 501-508.
18 S.T. Wilson, B.M. Lok, E.M. Flanigen, U.S. Patent 4,430,440, 1982.
19 E.T.C. Vogt, J.W. Richardson, Jr., J. Solid State Chem., **87**, 1990, 469-471.
20 R.M. Dessau, J.L. Schlenker, J.B. Higgins, Zeolites, **10**, 1990, 522-524.
21 R. Szostak, R. Kuvadia, J. Brown, T.L. Thomas, Zeolites: Facts, Figures, Future, Proceedings 8th International Zeolite Conference, Studies in Surface Science and Catalysis, vol. **49**, Jacobs and van Santen, eds., Elsevier, Amsterdam, 1989, 439.
22 P.B. Moore, J. Shen, Nature, **306**, 1983, 356.
23 M. Estermann, L.B. McCusker, C. Baerlocher, A. Merrouche, H. Kessler, Nature, **352**, 1991, 320-323.
24 A. Merrouche, J. Patarin, H. Kessler, M. Soulard, L. Delmotte, J.L. Guth, Zeolites, **12**, 1992, 226.
25 M.E. Davis, Nature, **352**, 1991, 281-282.
26 Q. Huo, R. Xu, S. Li, Z. Ma, J.M. Thomas, R.H. Jones, A.M. Chippendale, J. Chem. Soc. Chem. Commun., 1992, 875.
27 R.H. Jones, J.M. Thomas, J. Chen, R. Xu, Q. Huo, S. Li, Z. Ma, A.M. Chippindale, J. Solid State Chem., **102**, 1993, 5605.
28 Q. Huo, R. Xu, S. Li, Y. Xu, Z. Ma, Y. Yue, L. Li, in Proceedings from the ninth International Zeolite Conference, Montreal, 1992, R. von Ballmoos et al., eds., Butterworth-Heinemann, 1993, 279-286.
29 C.C. Freyhardt, M. Tsapatsis, R.F. Lobo, K.J. Balkus, Jr., M.E. Davis, Nature, **381**, 1996, 295-298.
30 P. Wagner, M. Yoshikawa, M. Lovallo, K. Tsuji, M. Tsapatsis, and M.E. Davis, J. Chem. Soc., Chem. Commun. 1997, 2179.
31 J.S. Beck, C.T. Chu, I.D. Johnson, C.T. Kresge, M.E. Leonowicz, W.J. Roth, J.C. Vartuli, Int. Pat. Appl. WO 91/11390, August 8, 1991.
32 C.T. Kresge, M.E. Leonowicz, W.J. Roth, J.C. Vartuli, and J.S. Beck, Nature, **359,** 1992, 710-712.
33 Beck, J.C. Vartuli, W.J. Roth, M.E. Leonowicz, C.T. Kresge, K.D. Schmitt, C.T.W. Chu, D.H. Olson, E.W. Sheppard, S.B. McCullen, J.B. Higgins, and J.L. Schlenker, J. Am. Chem. Soc., **114**, 1992, 10835.
34 J.S. Beck, J.C. Vartuli, G.J. Kennedy, C.T. Kresge, W.J. Roth, S.E. Schramm, in Zeolites Science 1994, Recent progress and discussions, Studies in Surface Science and Catalysis, vol. **98**, Karge and Weitkamp, eds., Elsevier, Amsterdam, 1995, 15.
35 G.D. Stucky, Q. Huo, A. Firouzi, B.M. Chmelka, S. Schacht, I.G. Voigt-Martin, F. Schüth, in Progress in zeolite and microporous materials, Studies in Surface Science and Catalysis, vol. **105**, Chon, Ihm, and Uh, eds., Elsevier, Amsterdam, 1996, 3.

36 M.E. Davis, Nature, **364**, 1993, 391-393.
37 X. Zhao, G.Q. Lu, G. Millar, X.S. Li, Catal. Lett., **38**, 1996, 33-37.
38 T. Yanagisawa, K. Kuroda, C. Kato, Bull. Chem. Soc. Jpn., **61**, 1988, 3743-3745.
39 T. Yanagisawa, K. Kuroda, C. Kato, Reactivity of Solids, **5**, 1988, 167-175.
40 P.T. Tanev, M. Chibwe, and T.J. Pinnavaia, Nature, **368**, 1994, 321.
41 K.M. Reddy, C. Song, Catal. Lett., **36**, 1996, 103-109.
42 L.B. McCusker, Ch. Baerlocher, E. Jahn, M. Bulow, Zeolites, **11**, 308-313, 1991.
43 J.C. Jansen, S.T. Wilson, in Introduction to Zeolite Science and Practice, Chapter 4, Studies in Surface Science and Catalysis, vol. **58**, van Bekkum, Flanigen and Jansen, Editors, Elsevier, Amsterdam, 1989, 77-152.
44 E.M. Flanigen, Adv. Chem. Ser. **121**, 1973, 119-139.
45 B.M. Lowe, in Innovation in Zeolite Materials Science, Proceedings of an International Symposium, Nieuwpoort, 1987, Studies in Surface Science and Catalysis, vol. **37**, Grobet, Mortier, Vansant, and Schulz-Ekloff, Editors, Elsevier, Amsterdam, 1987, 1-12.
46 E.J. Feijen, J.A. Martens, P.A. Jacobs, In Zeolites and Related Microporous Materials: State of the Art 1994, Proc. 10th International Zeolite Conference, Garmisch-Partenkirchen Germany, Studies in Surface Science and Catalysis, vol. **84**, Weitkamp, Karge, Pfeifer, and Hölderich, eds., Elsevier, Amsterdam, 1994, 3-21.
47 M.E. Davis, Chemtech, September 1994, 22-26.
48 S.I. Zones, D.S. Santilli, in Proceedings from the ninth International Zeolite Conference, Montreal, 1992, von Ballmoos et al., eds., Butterworth-Heinemann, Boston, 1993, 171-179.
49 M.E. Davis, S.I. Zones, Proceedings, Symposium on Synthesis of Zeolites, Layered Compounds and other Microporous Solids, ACS 209th National Meeting, Anaheim, CA, 1995, 233-234
50 T.V. Harris, S.I. Zones, In Zeolites and Related Microporous Materials: State of the Art 1994, Proc. 10th International Zeolite Conference, Garmisch-Partenkirchen Germany, Studies in Surface Science and Catalysis, vol. **84**, Weitkamp, Karge, Pfeifer, and Hölderich, eds., Elsevier, Amsterdam, 1994, 29-36.
51 Z. Blum, S. Hyde, Chem. Mater., **2**, 1990, 312-314.
52 D.W. Lewis, R.A. Catlow, J.M. Thomas, D.J. Willock, G.J Hutchings, Nature, **382**, 1996, 604.
53 V. Alfredsson, M.W. Anderson, Chem. Mater., **8**, 1996, 1141-1146.
54 M.E. Davis, C. Montes, P.E. Hathaway, J.P. Arhancet, D.L. Hasha, J.L. Garcés, J. Am. Chem. Soc., **111**, 1989, 3919-3924.
55 C.N. Wu, T.S. Tsai, C.N. Liao, K.J. Chao, Microporous Mater., **7**, 1996, 173-185.
56 K.R. Kloetstra, H,W. Zandbergen, M.A. van Koten, H. van Bekkum, Catal. Lett., **33**, 1995, 145-156.
57 J.O. Perez, P.J. Chu, A. Clearfield, J. Phys. Chem., **95**, 1991, 9994-9999.

58 L. Maistriau, Z. Gabelica, E.G. Derouane, E.T.C. Vogt, J. van Oene, Zeolites, **11**, 1991, 583-592.

59 J.A. Martens, E. Feijen, J.L. Lievens, P.J. Grobet, P.A. Jacobs, J. Phys. Chem., **95**, 1991, 10025-10031.

60 P.J. Grobet, J.A. Martens, P.A. Jacobs in :Zeolites for the nineties: Recent Research Reports presented during the 8th International Zeolite Conference, Amsterdam, July 10-14, 1989, 195-196.

61 J.W. Richardson, J.V. Smith, J.J. Pluth, J. Phys. Chem., **93**, 1989, 8212-8219.

62 J.W. Richardson Jr., E.T.C. Vogt, Zeolites, **12**, 1992, 13-19.

63 O. Kitao, K.E. Gubbins, Chem. Phys. Lett., **227**, 1994, 545-550.

64 O. Kitao, K.E. Gubbins, J. Phys. Chem., **100**, 1996, 12424-12430.

65 G. Cheetham, M.M. Harding, Zeolites, **16**, 1996, 245-248.

66 A.J.M. de Man, R.A. van Santen, E.T.C. Vogt, J. Phys. Chem., **96**, 1992, 10460-10466.

67 B. Li, P. Sun, Q. Jin, D. Ding, Zeolites, **17**, 1996, 424-427.

68 B.P. Feuston, J.B. Higgins, J. Phys. Chem., **98**, 1994, 4459-4462.

69 R.G. Bell, Proceedings of the 12th International Zeolite Conference, Baltimore, 1998, to be published by the Materials Research Society.

70 D. Keffer, V. Gupta, D. Kim, E. Lenz, H.T. Davis, A.V. McCormick, J. Mol. Graphics, **14**, 1996, 108-116.

71 V.J. Frillette, W.O. Haag, R.M. Lago, J. Catal., **67**, 1981, 218-222.

72 J. Weitkamp, S. Ernst, R. Kumar, Appl. Catal., **27**, 1986, 207-210.

73 J.A. Martens, M. Tielen, P.A. Jacobs, J. Weitkamp, Zeolites, **4**, 1984, 98.

74 C.S. Hsia Chen, S.E. Schramm, Microporous Mater. , **7**, 1996, 125-132.

75 A. Corma, A. Martínez, V. Martínez-Soria, J.B. Monton, J. Catal., **153**, 1995, 25-31.

76 A. Corma, M.S. Grande, V. Gonzales-Alfaro, A.V. Orchilles, J. Catal., **159**, 1996, 375-382.

77 R. Mokaya, W. Jones, Z. Luan, M. Alba, J. Klinowski, Catal. Lett., **37**, 1996, 113-120.

78 P.J. O'Malley , J. Dwyer, J. Phys. Chem., **92**, 1988, 3005-3007.

79 A.G. Pelmenchikov, V.I. Pavlov, G.M. Zhidomirov, S. Beran, J. Phys. Chem., **91**, 1987, 3325.

80 E.G. Derouane, J.M. Andre, A.A. Lucas, J. Catal., **110**, 1988, 58-73.

81 E.G. Derouane, J.B. Nagy, C. Fernandez, Z. Gabelica, E. Laurent, P. Maljean, Appl. Catal., **40**, 1988, L1-L10.

82 E.G. Derouane, in Guidelines for Mastering the properties of molecular sieves, D. Barthomeuf, ed., Plenum Press, New York, 1990, 225-239.

83 J. Sauer, in Recent Advances in Zeolite Science, Studies in Surface Science and Catalysis, Vol. **52**, Klinowski and Barrie, eds., Elsevier, Amsterdam 1989, 73-90.

84 J. Sauer, K.-P. Schroeder and V. Termath, to be published in Coll. Czech. Chem. Commun.

85 M. Braendle and J. Sauer, J. Am. Chem. Soc., **120**, 1998, 1556.

86 C.M. Zicovich-Wilson, A. Corma, P. Viruela, J. Phys. Chem., **98**, 1994, 10863-10870.

87 A. Corma, H. Garcia, G. Sastre, P. Viruela. J. Phys. Chem. B, **101**, 1997, 4575.
88 E. de Vos-Burchart, E.T.C. Vogt, B. van de Graaf, J. Chem. Soc., Faraday Trans., **88**, 1992, 2761-2769.
89 R.A. van Santen, G. Ooms, C.J.J. den Ouden, B.W. van Beest, M.F.M. Post, in Zeolite synthesis, ACS symposium series **398**, M.L Ocelli and H.E. Robson, eds., American Chemical Society, 1989, 617-633.
90 N.J. Henson, A.K. Cheetham, J.D. Gale, Chem. Mater., **6**, 1994, 1647-1650. Figure 15 was adapted from this reference, and is reprinted with permission of the copyright holder. Copyright 1994 American Chemical Society.
91 A. Navrotsky, I. Petrovic, Y. Hu, C. Chen, M.E. Davis, Microporous Mater., **4**, 1995, 95.
92 I. Petrovic, A. Navrotsky, C. Chen. M.E. Davis, In Zeolites and Related Microporous Materials: State of the Art 1994, Proc. 10th International Zeolite Conference, Garmisch-Partenkirchen Germany, Studies in Surface Science and Catalysis, vol. **84**, Weitkamp, Karge, Pfeifer, and Hölderich, eds., Elsevier, Amsterdam, 1994, 677.
93 A. Navrotsky, I. Petrovic, A. Navrotsky, M.E. Davis, S.I. Zones, Chem. Mater., **5**, 1993, 1805.
94 J.M. Garcés, Adv. Mater., **8**, 1996, 434-437.
95 J.J.L. Heinerman, E.T.C. Vogt, Int. Pat. Appl. WO 94/26846, 1993.
96 P.J. Nat., E.T.C Vogt, Int. Pat. Appl. WO 94/26847, 1993.
97 E. Armengol, M.L. Cano, A. Corma, H. García, and M.T. Navarro, J. Chem. Soc., Chem. Commun., 1995, 519-520.
98 I.V. Kozhevnikov, A. Sinnema, R.J.J. Jansen, K. Pamin, and H. van Bekkum, Catal Lett, **30**, 1995, 241-252.
99 A. Corma, M.T. Navarro, and J.P. Pariente, J. Chem. Soc., Chem. Commun, 1994, 147.
100 A. Corma, M.T. Navarro, J. Pérez-Pariente, and F. Sánchez, In Zeolites and Related Microporous Materials: State of the Art 1994, Proc. 10th International Zeolite Conference, Garmisch-Partenkirchen Germany, Studies in Surface Science and Catalysis, vol. **84**, Weitkamp, Karge, Pfeifer, and Hölderich, eds., Elsevier, Amsterdam, 1994, 69.
101 A. Sayari, R.J. Sudhakar, and A. Dicko, Symp. on Synthesis of Zeolites. Layered Compounds and other Microporous Solids, Div. of Petrol. Chem. 209th National Am. Chem. Soc. Meeting, Anaheim, CA, 1995.
102 Y.V.S. Rao, D.E. de Vos, T. Bein, P.A. Jacobs, J. Chem. Soc., Chem. Commun., 1997, 355.
103 R.F. Parton, L. Uytterhoeven, P.A. Jacobs, in Heterogeneous Catalysis and Fine Chemicals III, Studies in Surface Science and Catalysis, vol. **78**, Guisnet et al, eds., Elsevier, Amsterdam, 1991, 395-403.
104 P.L. Llewellyn, U. Ciesla, H. Decher, R. Stadler, F. Schüth, and K. Unger, In Zeolites and Related Microporous Materials: State of the Art 1994, Proc. 10th International Zeolite Conference, Garmisch-Partenkirchen Germany, Studies in Surface Science and Catalysis, vol. **84**, Weitkamp, Karge, Pfeifer, and Hölderich, eds., Elsevier, Amsterdam, 1994, 2013.

105 L. van Looveren, D.F. Geysen, K.A. Vercruysse, B.H. Wouter, P.J. Grobet, P.A. Jacobs, Angewandte Chemie, Int. Ed., **37**, 1998, 517-520.

106 X. Feng, G.E. Fryxell, L.-Q. Wang, A.Y. Kim, J. Liu, and K.M. Kemner, Science, **276**, 1997, 923-26.

107 J.F. Diaz and K.J. Balkus, J. Mol. Catal. B: Enzyme, **2,** 1996, 115-26.

108 S.H. Tolbert, A. Firouzi, G.D. Stucky, and B.F. Chmelka, Science, **278,** 1997, 264-69.

109 A. Corma, V. Fornés, H. García, M.A. Miranda, and M.J. Sabater, J. Am. Chem. Soc., **116**, 1994, 9767-9768.

110 C.G. Wu and T. Bein, Chem. Mater., **6**, 1994, 1109-1112.

111 C.B. Wu and T. Bein, Science, **264**, 1994, 1757-1759.

112 W.M. van Rhijn, D.E. de Vos, B.F. Sels, W.D. Bossaert, P.A. Jacobs, J. Chem. Soc., Chem. Commun., 1998, 317.

113 M.Lasperas, T. Llorett, L. Chaves, I. Rodriguez, A. Cauvel, D. Brunel, Heterogeneous Catalysis and Fine Chemicals IV, Blaser, Baiker, Prins, eds., Elsevier, Amsterdam, 1997, 75-82.

114 K.R. Kloetstra, M. van Laren, H. van Bekkum, J. Chem. Soc., Faraday Trans., **93**, 1997, 1211-1220.

115 K.R. Kloetstra, H. van Bekkum, J. Chem. Soc., Chem. Commun., 1995, 1005.

Appendix

Zeolite structures

This appendix presents a number of structural characteristics of some important zeolites and of $AlPO_4$-5. The zeolite structure types are listed in alphabetic order according to their IUPAC memonic codes.
The figures labeled (a) show a schematic representation of the channel system pictured as it is oriented within a crystal of typical morphology. The framework structures are labeled (b); T-O-T links in these figures are represented as straight lines.
some of the framework structures have been reproduced from W.M.Meier, Ch. Baerlocher and D. Olson, Atlas of Zeolite Structure Types, IZA Structure Commission, 1998. Crystallographic and other data on a large number of zeolite structures not listed here can also be found in the Atlas and on the IZA Website: www.IZA-structure.org./databases/

Contents:
AFI
BEA
FAU
FER
LTA
LTL
MFI
MOR

AFI

(a)

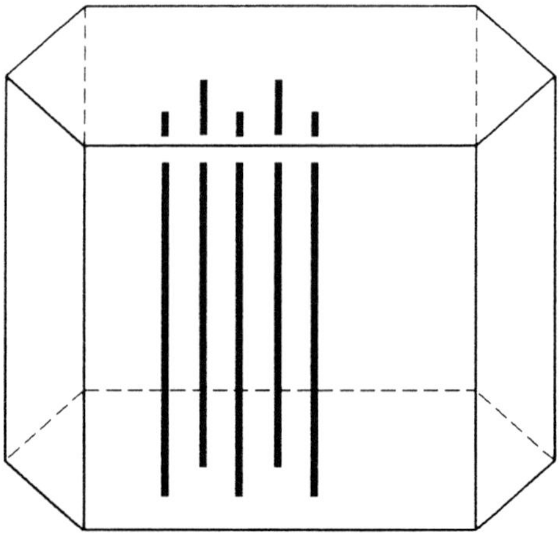

(b)

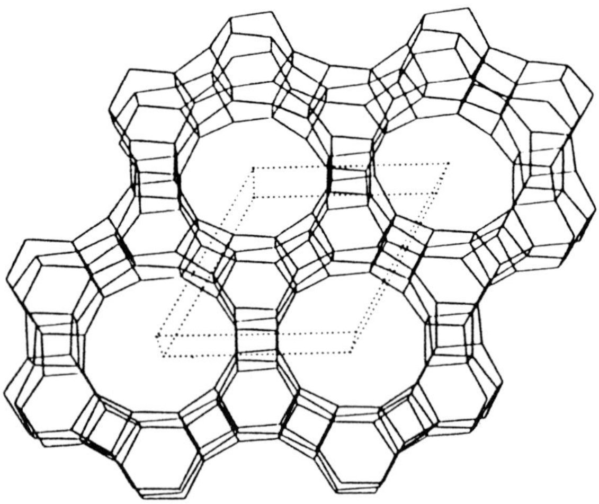

IUPAC name: AFI ($AlPO_4$-5)

Composition: $[Al_{12}P_{12}O_{48}] \cdot (C_3H_7)_4NOH \cdot x\ H_2O$

Important structural isotypes: SAPO-5; SSZ-24 (all-silica)

Channel system: 1-D; straight circular 12-ring 7.3 Å channels

BEA

(a)

(b)

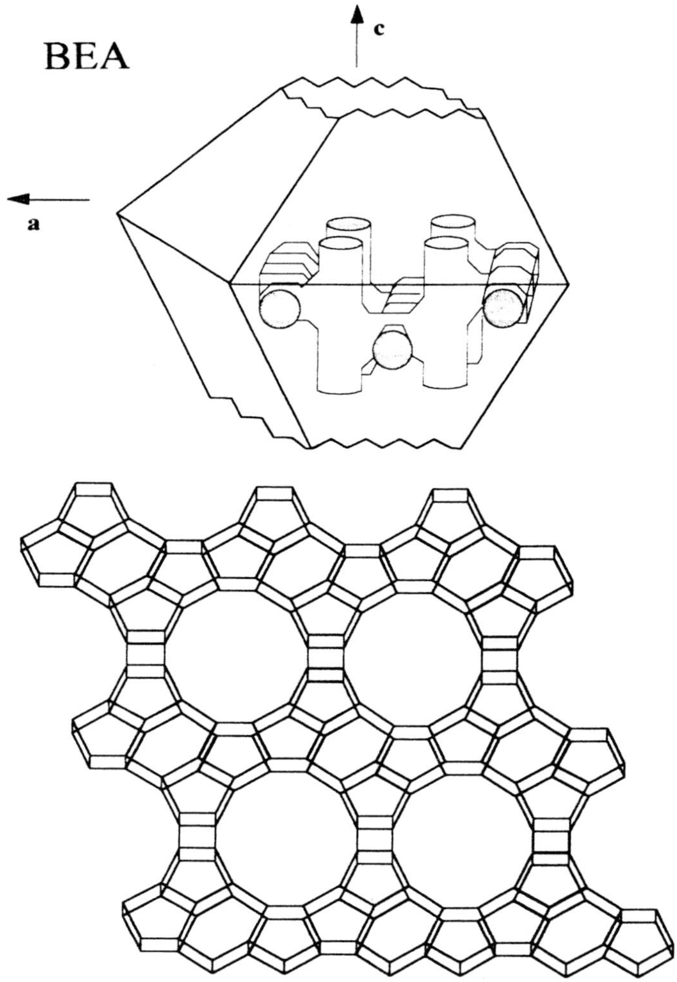

IUPAC: BEA (Beta)

Composition: $Na_n[Al_nSi_{64-n}]O_{128}$ with $n < 7$

Inportant structural isotype: Tschernichite

Channel system: Pore system is 3-D , 12-ring in c-direction 7.6x6.4 Å and two 12-ring channels in a direction perpendicular to c-direction with pores of 7.6x6.4 and 5.5x5.5 Å, resp.

FAU

(a)

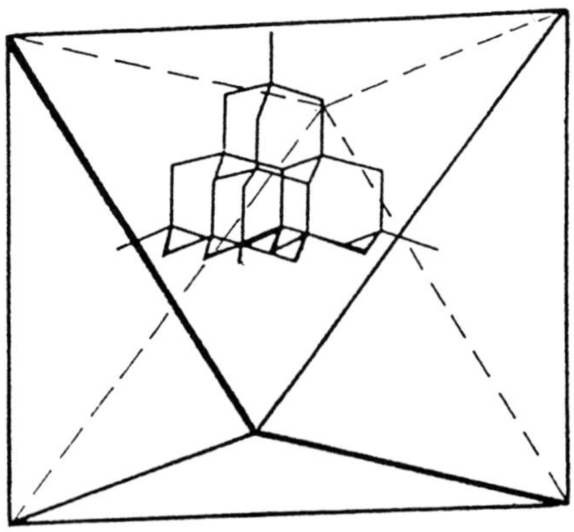

(b)

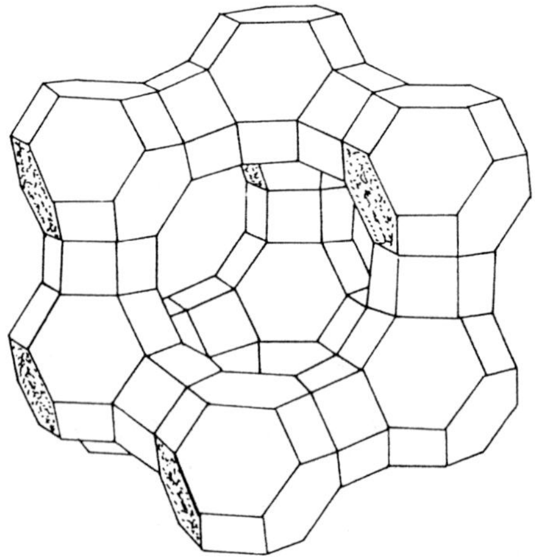

IUPAC name:	FAU (Faujasite)
Composition:	$Na_{58}[Al_{58}Si_{134}O_{348}] \cdot 240\ H_2O$ (NaY)
Important structural isotypes:	X ($1 \le Si/Al \le 1.5$); Y ($Si/Al \ge 2.5$); SAPO-37
Channel system:	3-D; circular 12-ring 7.4 Å windows connecting ~ spherical 11.8 Å cavities (supercages)

FER

(a)

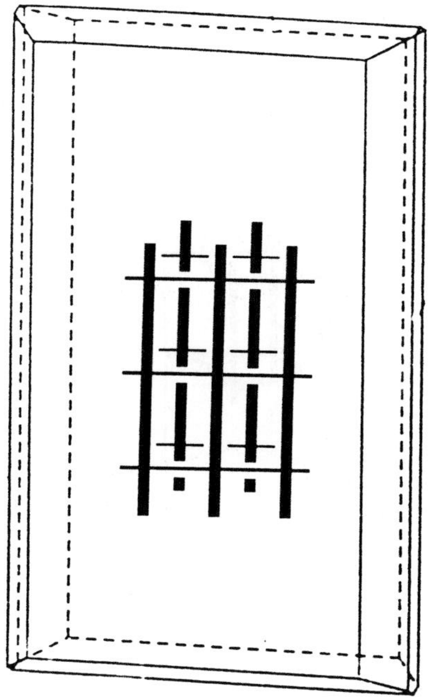

(b)

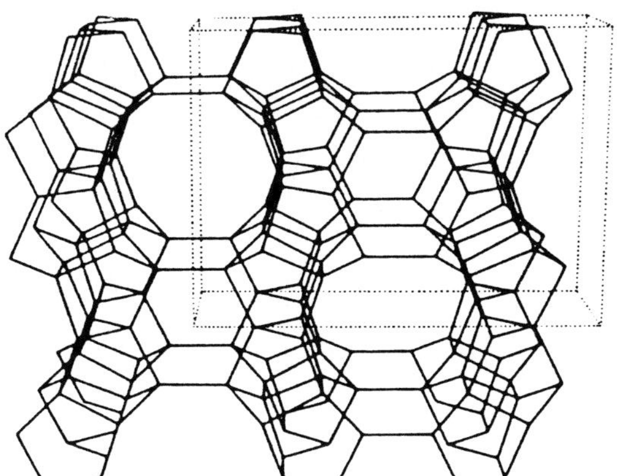

IUPAC name: FER (Ferrierite)

Composition: $Na_2Mg_2[Al_6Si_{30}O_{72}] \cdot 18\ H_2O$

Important structural isotypes: high silica types (e.g. ZSM-35, NU-23; $5 \leq Si/Al \leq 25$)

Channel system: 2-D; straight 10-ring 4.2×5.4 Å channels connected by straight 8-ring 4.8×3.5 Å channels

LTA

(a)

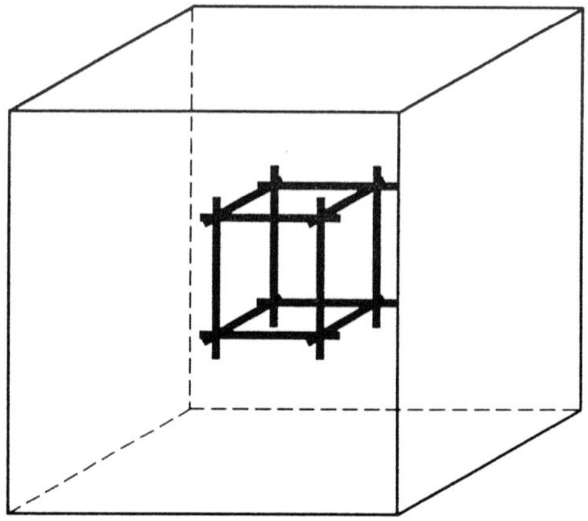

(b)

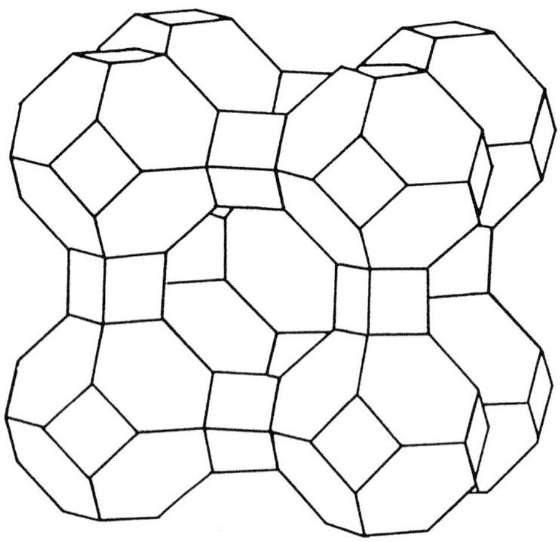

IUPAC name:	LTA (Linde Type A)
Composition:	$Na_{12}[Al_{12}Si_{12}O_{48}] \cdot 27\ H_2O$
Important structural isotypes:	SAPO-42; high silica types ($1 \leq Si/Al \leq 3$)
Channel system:	3-D; circular 8-ring 4.1 Å windows connecting ~ spherical 11.4 Å cavities

LTL

(a)

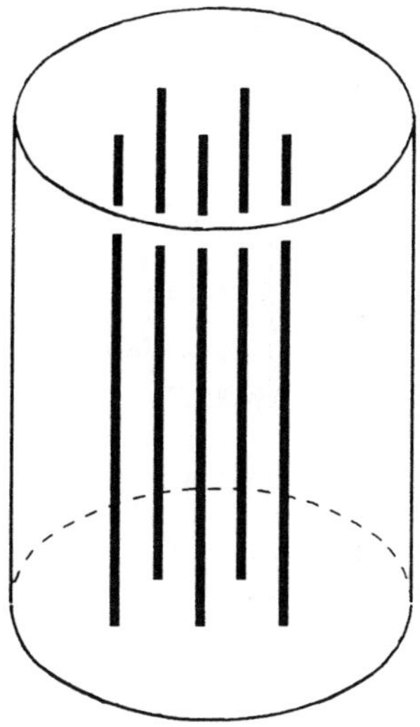

(b)

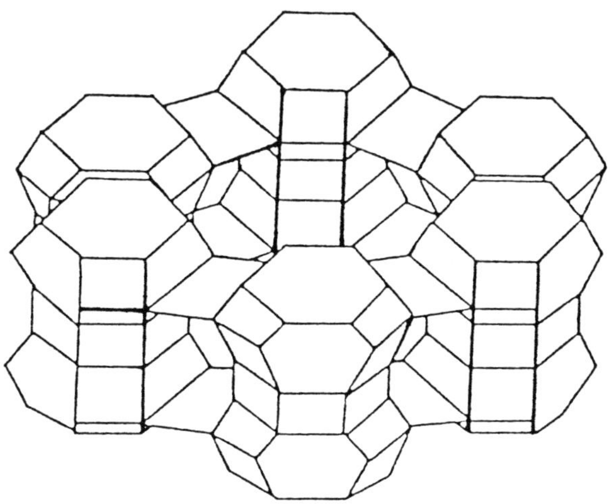

IUPAC name:	LTL (Linde Type L)
Composition:	$K_6Na_3[Al_9Si_{27}O_{72}] \cdot 21\ H_2O$
Channel system:	1-D; straight circular 12-ring 7.1 Å channels

MFI

(a)

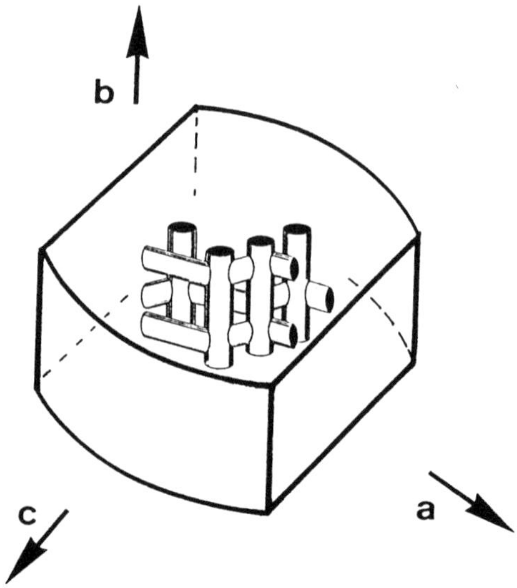

(b)

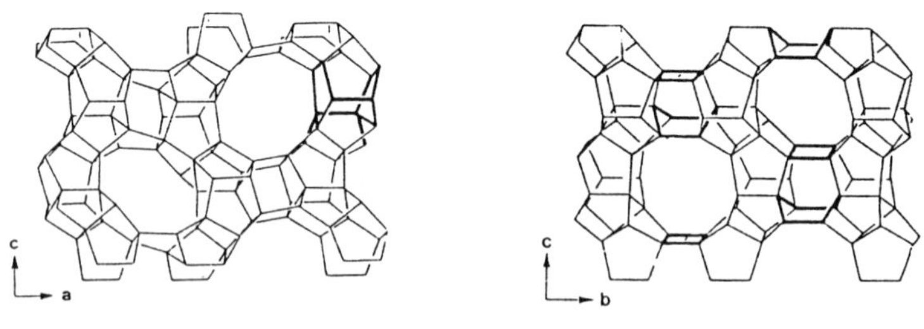

IUPAC name:	MFI (ZSM-5)
Composition:	$Na_n[Si_{96-n}Al_nO_{192}] \cdot 16\ H_2O$ ($n \leq 8$)
Important structural isotypes:	Silicalite-1 (Si/Al = ∞); Boralite; TS-1; (Si,Ge)-MFI
Channel system:	3-D; straight 10-ring 5.2×5.7 Å channels connected by sinusoidal 5.3×5.6 Å channels; intersection cavities ~ 9 Å

MOR

(a)

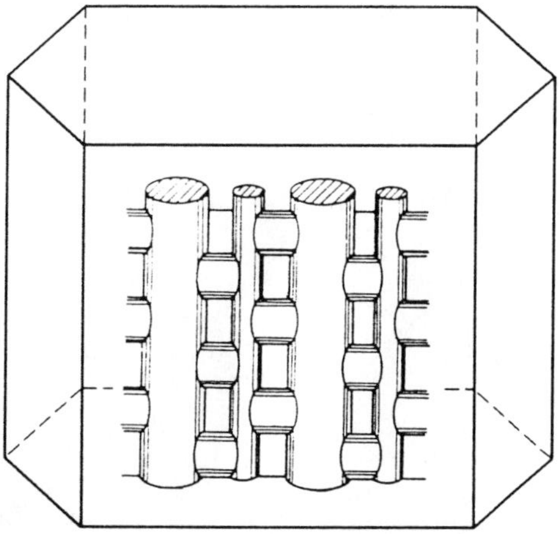

(b)

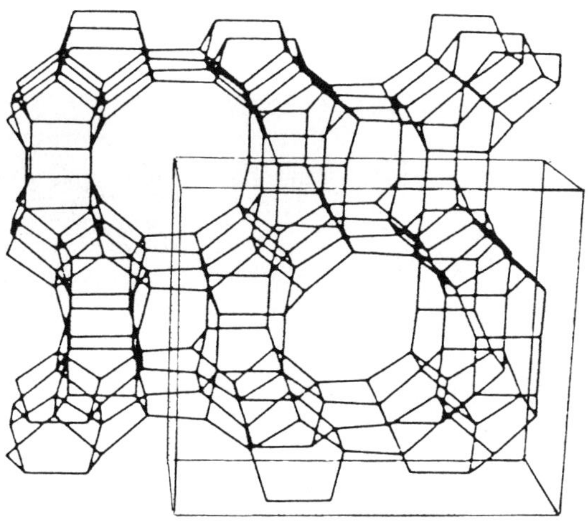

IUPAC name:	MOR (Mordenite)
Composition:	$Na_8[Al_8Si_{40}O_{96}] \cdot 24\ H_2O$
Important structural isotypes:	high silica types ($5 \leq Si/Al \leq 20$)
Channel system:	2-D; straight 12-ring 7.0×6.5 Å channels connected by short alternating 8-ring channels (~3 Å)

Edith Flanigen, lecturing at the IZA Pre-Conference School

Peter Jacobs and Koos Jansen in serious discussion but glasses in the hand

Herman van Bekkum and Paul Venuto discussing zeolite application in organic conversions

Participants of the 1989 Pre-Conference School held at the Dutch National Soccer Centre.
Various authors were present, e.g. sitting no 2-4 from right: Martin Post, Ian Maxwell and Rosemary Szostak

Keyword index

A

A-zeolites 834
 acid treatment 263
 crystals 180, 221
 in detergent formulations 3
 production 3, 4
 structure 43, 55
Accessibility 822, 827
Acetal isomerization 841
Acetalization 834, 887, 894
Acid catalysis 633
Acid site configuration 633
Acid sites, strength 695
Acid-leached FAU, TEM of 350
Acidity 369
 and TOT angle 1015
 isomorphous substitution 751
 of Si, Al cluster 750
 theory 420
Activated carbon 12
Activation of zeolites 674
Acylation of arenes 856
 acylated surface 859
 acylation intermediates 860
 adsorption preferences 858
 of anisole 860
 of heteroarenes 861
 of 2-methoxynaphthalene 861, 999
 of phenol 848
 of toluene 857
 of veratrole 859
 Rhodia processes 859
Adamantane 836
Addition of hydrogen sulfide 889
Addition of phosphines 889
Addition to olefins
 alcohol addition 886
 carboxylic acid addition 887
 water addition 844, 885
Adenine formation 879
Adsorbent consumption 16
Adsorption 579
 adsorbate solution theory 590
 alkanes/alkenes on NaY 603
 alkanes/alkenes on Theta-1 604
 argon in Na-X and Ba-X 597
 capacities for hydrocarbons 601
 comparison of adsorbents 12
 computer simulation 436
 energetics of adsorption 586, 605
 enthalpies 665
 ethane/propane on NaY 619, 620, 624
 ethane/propane mixtures on silicalite-1 617, 618, 622, 623
 experimental methods 591
 extended Langmuir model 588
 equilibrium 582
 free energy of sorption for N_2 and O_2 616
 gases in Chabazite 598
 gravimetric uptake curves 538-543
 Henry's law constants 617
 ideal adsorbed solution theory (IAST) 589
 isosteric heats of adsorption on silicalite-1 606
 isosteric method 594
 isotherm of p-xylene in silicalite-1 601
 isotherms 19, 551, 463
 kinetics 538
 Langmuir isotherms 583
 Mobile adsorption 584
 n-alkanes in silicalite-1,adsorption isotherms 600
 N_2/O_2 mixtures on FAU 618
 nitrogen on MCM-41 626
 octane *vs* iso-octane 14
 of gases 579
 of mixtures 588, 614
 of nitrogen bases 993
 of organics from water 829
 of polar molecules 607
 sorption entropy of CO_2 613
 sorption isosteres of CO_2 608-610
 statistical thermodynamic isotherms 585
 theory 424, 580
 thermodynamics of gas adsorption 580
 virial adsorption isotherm 585, 598
Aerosil 200 183
Affinity separations 831
AFI ($AlPO_4$-5) 51, 120, 125
Air separation 16
Al Keggin ion 322-325
^{27}Al NMR
 general 405
 non-framework Al 406
 SAPO's 414
 zeolite Y 407
Al, solution chemistry of 322
Alcoholysis of esters 888
Aldol condensation 893
Alfa-cavity (α-cage) 39
Alkali metal clusters 975
 butene isomerization 977
 from sodium azide 976
 in LTA and FAU 976
 ^{23}Na MAS NMR 976
 side chain alkylation of toluene 977
Alkali removal 278
Alkane conversion schemes
 on bifunctional zeolites 642
Alkane oxidation 925, 942, 966, 967
Alkene oligomerization 660
Alkylation 765, 849
 of framework 638
 of fructose 889
 reactions 658-660

Alkylcarbenium ions
  generation of 640
  on bifunctional zeolites 641
  reactions of 655
Alkylcarbonium ions, structures 637
All silica molecular sieves 18, 211, 212
Alpha activity 283
AlPO-5 240, 246, 668
AlPO-8 52, 1005, 1006
AlPO-11 235, 379
AlPO-31 235
AlPO-34 240, 247
AlPO-39 235
AlPO-41 235
AlPO-52 238
$AlPO_4$ molecular sieves, families 231
  ABC family 129
  AEI 136
  AEL 119
  AET 122, 125, 126
  AFG 132
  AFO 123
  AFT 133
  AFX 132
  AHT 121, 126
  $AlPO_4$-H1 1009
  APC 117
  APD 120
  ATO 98
  ATS 90
  ATT 103, 116
  ATV 119, 125, 126
  AWW 137
  bonding concepts 234
  fluoride in synthesis 247
  historical 229
  large crystals 240
  mixed solvents in synthesis 242
  NMR of 414
  sheet structures 245
  structures in IZA Atlas 232, 233
  synthesis 235
  templating 236
Alumina
  as adsorbent 12
  as Al source 182, 183
Alumination 284
Aluminium, dynamics of 892
Aluminium, extra-framework 276
Aluminium non-framework 406
Aluminophosphate 20, 21
Amination
  of alcohols 879
  of dienes 882
  of olefins 881
  of phenol 883
Ammonia methylation 879
Ammonia-TPD 379
Ammonium fluoro silicate method 275, 280
Ammoximation 923
Amoco 798
Analcime 180
Anilines, synthesis by nuclophilic substitution 883
Anthracene formation 871
Anthraquinone 859
Applications 748
  of zeolites 4, 30
Argon adsorption 382
Aromatic alkylation
  cumene synthesis 850
  4,4'-diisopropylbiphenyl 852
  2,6-diisopropylnaphthalene 852
  ethylbenzene synthesis 849
  linear alkylbenzenes 850
  mechanism 851
  of dinuclear systems 851
  of heteroarenes 855
  of phenols and anilines 853-855
  zeolite based processes 849
Aromatic halogenation
  bromination 865
  chlorination 863
  gas phase chlorination 864
  iodination 866
  o-xylene 864
Aromatic separation 831
Aromatization 657, 767
  mechanism 657
AROMAX process 768
Atlas of Zeolite Structure Types 14, 15, 20, 31, 38, 42, 69
Autoclaves 187

B

β-cage (sodalite unit) 39
B-MFI as catalyst 837, 843, 941
Barrer, R.M. 5, 11, 15, 31, 263, 317, 474, 529
Base catalysis 825, 1021
BASF 837
Basicity 369, 381
BEA (Beta) 5, 50, 58, 210, 147-150, 848, 849, 885-887 891, 1031
Beckmann rearrangement 842-844
Benzaldehyde condensation 894
Benzamine rearrangement 894
Benzene isopropylation 660
Benzyl alcohol, conversion to anthracene 871
Beryllophosphate 24, 251
Beryllosilicate 24
BET equation 353
Bicyclic compounds 877
Bifunctional catalysis
  in hydrocarbon conversion 643, 759
  in starch conversion 895
Binder materials 689, 827
Bis-quats as templates 218

Biphenyl
  alkylation 676, 852
  chlorination 864
Boron replacement by Ti 918
Boron zeolites 823, 837, 843
Boron-11 NMR 413
BP 808
Breck, D.W. 11, 14, 16, 31
British Zeolite Association (BZA) 31
Bromination 865
  mechanism 866
Bronsted acid catalysis 821
Bronsted acid site 425
Bronsted acidity 824
Building unites 70
Building zeolites 69
Bulky probe molecules 996, 997
Butane cracking 285, 368
t-Butyl hydroperoxide in oxidation 945, 966, 967, 1019
t-Butylamine 882

C

Cacoxenite 1006
Caffeine, adsorption of 832
Calorimetry 380
Campholenic aldehyde 841
CAN 98, 130
Cancrinite cage 39
Caprolactam synthesis 842
Carbenium ions 634-637
Carbocations in zeolite catalysis 638
Carbocations in zeolites
  intermediates *vs* transition states 638
Carbohydrate coupling 832
Carbohydrate separations 831
Carbon dioxide adsorption 386
Carbon molecular sieves 12
Carbon-13 NMR 412
Carbonium ions 634-637
Carbonyl complexes in zeolites 824
Catalysis in hydrocarbon reactions 633
Catalyst granules 688
Catalyst pellets 688
Catalyst shaping 688
Catalysts, FCC 691
Catalytic cracking 691-694
Catalytic dewaxing 769
CATCON process 809
Cation exchange 473
  equilibria 476, 482
Cation positions 56, 473, 499
  in faujasite 825
Cation sites, by ^{23}Na NMR 412
Cation templating theory 201
Ce^{III}-exchange 825
Ce-Y as re-usable reagent 832
Ce-Y in aromatic acylation 857
Cesium-133 NMR 393
CFI 93
CHA 131, 136
Chabazite 15, 16, 21, 46, 508
Chains in zeolite structures 40
Characterization techniques 345
Chevron 768, 770, 796
Chiral catalysts 965
Chiral complexes 826, 871
Chlorination 863
  of n-alkanes 833
  of anisole 858
  of benzene 864
  of phenyl acetate 865
Chromatographic method
  to determine diffusivities 544
Chromium-containing systems
  Cr-MFI 943
  CrAPOs 945
CI (Constraint Index) 359
Cinnamyl alcohol, alkylation with 1018
CIT-5 25, 823, 1006, 1007
Clathrasils 212
Clays 299
  acidity 317
  benzidine reaction 318
  cation exchange capacity 306
  in acylation 860
  IR frequencies 312
  NMR data 313-316
  sorption 308
  swelling 308
Clinoptilolite 17, 263, 508
Cloverite 24, 25, 30, 52, 237, 247, 1006
Cluster approximation 423
Co-APOs 231, 241, 243
Co-catalyst 781
Co-complexes in Y 961
Co-X, Co-Y 912
Co^{II}-X 830
CoAPOs 944
  in p-cresol oxidation 944
  in cyclohexane oxidation 944
  leaching 944
Coke
  adsorption measurements 730
  3-D *vs* 1-D systems 823
  deposition 737
  effect of acidity 733
  ESR spectroscopy 720-723
  external coke 739
  extraction of 727, 729
  formation 707, 731
  formation of polynuclear aromatics 728
  from ethylbenzene 735, 736
  from ethylene 711, 717, 719, 722, 737
  from methanol 711, 724, 725, 738
  H/C ratio determination 709, 712

investigation 709-731
IR spectroscopy 713-718
localization 736
mechanism of formation 735
NMR spectroscopy 718-720
selectivity of formation 731
temperature-programmed oxidation 731
UV/Vis spectroscopy 723-725
^{129}Xe NMR 740
XPS and AES spectroscopy 726
Computational studies 421
Concentration effect 833
Condensation 196
Constraint index 363, 1013
Consumption of zeolites 4
catalysts 16
regional 4
Cracking activity 366, 368
Cracking mechanisms 653
Crankshaft chains 111
Cronstedt 5, 15
Crystal growth 199
Crystal size distribution 203
Crystal size effects 566, 688, 773
Crystals 180
Crystallization steps 202
Crystallization, zeolites 198, 522
Crystallographic pore diameter 992
Cu-beta 884
Cu-L 884
Cu-mordenite 883
Cu-X 870
Cu-Y 869, 870, 883, 912
CuPd-Y 912
Curvature effect 1016
CYCLAR process 808
Cyclizations 869
Cyclocondensation towards pyridines 875
Cyclodimerization 870
Cyclohexane oxiation 944, 966
Cyclohexanone, ammoximation 844
Cyclohexanone oxime 842
Cyclopropanation 870
Cyclopropene dimerization 870

D

DABCO synthesis 877
DAC 94, 108, 110
DAF-1, -3 237
Darken relation 497, 501
DD-3R 212
3-DDM 850
Dealuminated mordenite 850
Dealuminated Y zeolites 265, 696
Dealumination
by chemical treatment 266, 280, 281
by hydrothermal treatment 266, 279, 696
loss of crystallinity 263
mechanism 264
NMR spectra 401
of zeolite Y 265, 266, 270, 696
Deep catalytic cracking 783
Dehydration/hydrogenation 893
Deprotonation energy 425
Desulfurization 766
Detergency
criteria for a zeolite builder 506
Na/Ca exchange in zeolite A 505
Na/Ca/Mg-zeolite A system 504
Na/Mg exchange in zeolite A 505
zeolite A as builder 503
Detergent formulations
developments 507
zeolite A *vs* MAP 516-518
zeolites in 4, 30, 506
Di-acetoxybutene, isomerization 835
Diamines as templates 219
Diborane, in tuning accessibility 827
Dichlorotoluenes, isomerization 846
Diels-Alder cycloadditions 869
Diffusion
benzene in NaX 549, 553, 555, 565
benzene in ZSM-5 557, 560
branched paraffins in silicalite 562
butane in NaCaA 554
carbon dioxide in theta-1 and silicalite-1 441, 443
classification of methods 535
CO_2 in CaA 555
CO_2 in NaA 541
co-diffusion of benzene and ethylbenzene 565
comparison of benzene and butane 563
computer simulation 440
counter-diffusion of benzene and ethylbenzene 565
2,2-dimethylbutane in ZSM-5 557, 563, 568
driving force 529
effect of molecular shape 562
ethane in NaCaA 551
general principles 528
hexane in NaX 553
C_6-C_9 hydrocarbons in ZSM-5 568
in large crystals 554
isobutane in silicalite 551
liquid phase conditions 564
MD studies 440
methane, ethane, propane 553
methane, ethene in NaA 558
methane in chabazite 558
methane in MFI and MEL 445
methane in silicalite 441
methanol in NaX 543, 554
n-paraffins in NaCaA 561
n-paraffins in NaX 561
n-paraffins in silicalite 561

propane in CaA 555
propane in NaX 549
propane, propene in NaX 555
regimes 527
self-diffusion 552
single file 531
Stefan Maxwell Equation 530
Xe in CaA 555
p-xylene in NaX 559
Diffusion limitations in zeolite catalysis 566
Diffusivities
chromatographic method 544
effect of crystal size 556
effect of pore diameter 527
effect of Si/Al ratio 564
effect of sorbate concentration 556
from catalytic experiments 567
gravimetric method 539
macroscopic methods 538
membrane technique 550
microscopic methods 533
NMR Pulse Field Gradient Method 536
n-paraffins, effect of chain length 560
piezometric measurements 539
Qens mehod 534
survey of methods 535
theoretical prediction 532
uptake rate measurements 538
ZLC method 547
Dihydropyran formation 832
1,4-Diketone conversions 872
Dimensionality 823
Dimethylamine 880
1,2-Diol isomerization 838
Disaccharide separation 831
Disproportionation of alkylaromatics 662
Dodecasils 166, 213, 214
Drying of reactants and solvents 829

E

EAB 131
EDAX 363
EDI 127, 128
EDTA-method 274, 275, 279
EDX 363
Effectiveness factor 568
Electron transfer material 1021
Electronic confinement 1017
EMC-2 25, 58
EMT 25, 45, 58, 138, 668
Enamine formation 834
Enantiomer separation 832, 1021
Enantioselective epoxidation 830
Enantioselective ketone cyclization 971
Eniricerche 5
Enzyme, drying of 829
Enzymes, in MCM-41 1020
EPI 94, 108, 110
Epoxidation 921, 933, 968, 1019
Epoxides
addition of aniline 884
addition of azide 885
formation 830
isomerization 839
ERI 131
Erionite 17, 21, 265, 664, 823
ESR-spectroscopy of coke 720
Esterification 834, 887
Ethylation of surface 639
ETS-4 25
ETS-10 25
EU-1 219
EU-2 219
EU-4 219
EUO 95
Evolution 17, 18
Expander molecules in synthesis ordered mesoporous materials 1008
Extra-framework Al species 276, 277
Extra-framework species 56
Extrudates (granules) 690
Exxon 14, 781

F

FAU 58, 138
Si, Al ordering schemes 398
Faujasite zeolites 5
structure 21, 44, 350, 1032
TEM 5, 656, 691, 694, 701, 774
FCC 777
catalyst development 775, 776
cocatalyst 781-783
gasoline 780
process 691-694
Fe-MFI
in butane cracking 285
in oxidation 943
Ferrierite 106, 110, 210, 212, 219, 763, 1033
FEZA (Federation of European Zeolite Associations) 31
Fibrous zeolites 127
Fick's equation 528
FIMS spectrometry 788
Fischer indole synthesis 877
Fluid catalytic cracking 774
Fluoride
in synthesis 210, 216, 247
methods 275
Framework
densities 1017
flexibility 474
Friedel-Crafts alkylation 849-855
Fries-rearrangement 847-849
Fructose, alkylation of 889

FSM-16 26, 1008
Functional group determination 369

G

Ga-MFI 285
Gallium
  in MFI 916
  substitution of Al 286
Galloarsenate 24
Gallophosphate 24, 246, 1006
Gas oil
  cracking 1015
  demand 787
  molecular size distribution 1004
Gas reserves 801
Gas-upgrading processes 802
Gel formation 191
Germanosilicate 24
Gismondine 21, 38, 41, 46, 63, 206, 207, 247, 116
Glucose, alkylation of 888
GME 117, 130
Grace 16
Gulf 781

H

Halogenation, aromatic 863
Heats of adsorption of alkanes 457
Hectorite 305, 312
Heptane, hydroisomerization 277
Heterocyclic ring formation 871
Heterogenization of functions in MCM-41 1021
Heteropoly acids, in MCM-41 1019
Heulandite 3
Hexane adsorption isotherms 19
Hexane cracking 263, 282, 366, 1016
Hexane in silicalite loading 460
HF 275
HI (VPI-5) 1005
History 15
Hoffman degradation of template 197
HY in organic synthesis 841, 843
Hydrobromic acid adsorption 866
Hydrocarbon reactions 633, 747
  acid-catalyzed 644
  mechanism of 644-647
Hydrochloric acid adsorption 865
Hydrocracking 785-796
  accelerated ageing test 793
  commercial performance 795
  kerosine selectivity 792
  process configurations 789
  product flexibility 790
  single/stacked beds 795
Hydroformylation 894
Hydrogen peroxide 914
Hydrogen-1 NMR 409
  chemical shift ranges 409
Hydroisomerization 675, 679, 756, 759
Hydrolysis/hydrogenation 895
Hydroperoxides
  in oxidation 945, 966, 1019
  selective adsorption 830
Hydrophilicity 829
Hydroxyalkylation 867, 920
Hydroxylation 920, 942

I

Indole synthesis 877
Inertization of external surface 827
Infrared spectroscopy 373-382
Intergrowth 58
International conferences, history 31
International Zeolite Association (IZA) 2, 31
International Zeolite Conferences 1, 31
Internet sites IZA 38
Iodination 866
Ion exchange 467, 825
  affinity and selectivity 479
  ammonia removal 509
  concentration scales and ideal behaviour 488
  different exchange sites 473
  diffusion within the micropores 496
  effect of temperature 498
  equilibrium aspects 476, 482
  exchange sites 473, 499
  experiment *vs* prediction 486, 502
  film transfer resistance 494
  framework flexibility 474
  hierarchical model 493
  hysteresis 474
  in high-silica zeolites 512
  isotherms, types of 477-479
  kinetics 492, 500
  mass action quotient 481
  Na/K exchange in ZSM-5 478
  Na/K/Cd-exchange in zeolite X 486
  Na/Li exchange in zeolite P 478
  Na/Pb exchange in MOR 479
  nuclear waste treatment 507
  prediction of 483
  prediction procedures 484
  separation factor 480
  solid state exchange 514
  theoretical models and experimental test 486, 502
  thermodynamic equilibrium constant 481
  transport processes 495
  wastewater treatment 509
  zeolite stability 475
Ion sieving 472
IR spectra, calculated 447
IR spectroscopy 351, 713
Iron, in MFI synthesis 916

Iron phthalocyanine
in VPI-5 1019
in Y 966, 1019
Iron-containing silicalite
in benzene to phenol synthesis 943
in methane to methanol conversion 943
Iron-loaded USY 780
Isobutene from n-butene, zeolite catalysts 764
Isolation of zeolite product 204
Isomerization
general 834
of aldehydes 835, 837, 838
of alkylaromatics 661
of anilines 844
of cyclic acetals 841
of cyclohexanone oxime 842
of dichlorotoluenes 846
of diols 838
of epoxides 839-841
of 1-methylnaphthalene 845
of olefins 763, 834
of phenyl esters 847
of β-pinene 836
of saturated hydrocarbons 836
of substituted heteroaromatics 847
of toluidines 845
of xylenes 695, 845
Isophorone oxide 840
ISOSIV process 16, 758
Isotherms, types 353
ITE/RTH family 144-146
ITQ-7 823

J

Jasminaldehyde 894
JBW 103
JDF-20 246, 1007

K

Kanemite, precursor of mesoporous ordered materials 1008
KFI 136

L

La^{III}-exchange 825
Lactone
conversion 875
formation 888
Lactonization 888
Lattice
constant 364
dynamics 444
enthalpies 1017
stability 1017
LEV 132
Lewis acid catalysis in MPV reduction 892
Lewis acidity 825, 892
LIO 132
Liquid phase reactions 828
Lithium-7 NMR 736
Loewenstein's rule 405
LOS 131
LOV 156-159
Low temperature reaction mixture 190
LPG to aromatics 808
LTA 55, 137, 446, 1034
LTL 104, 263, 767, 1035
Lube oil 771

M

M41S 26
Macrocyclic lactones 888
Manganese complexes
in MCM-41 1019
occluded in Y 967, 968
MAP in detergents 507, 516
Market 4, 16
MAS-NMR
of SAPOs 249
spectroscopy 364
Matrix materials 694
MAZ 105
McBain 15, 26
MCM-1 to -9 242
MCM-22 25, 49, 850
MCM-41 26, 29, 373, 626, 826, 849, 887, 889, 918, 965, 1005, 1008-1010, 1012, 1015-1022
MCM-48 26, 1008, 1011
MCM-50 1008
MeAPO 22
MeAPSO 23
Mechanisms of hydrocarbon isomerization 645-650
MEL (ZSM-11) 48, 57, 141-144
Melanophlogite 213, 214
Membrane method, to determine diffusivities 408
Membranes, zeolite-based 209, 830
MER 118
Mesoporous aluminophosphates 250
Mesoporous ordered materials 26, 1003
Mesoporous pillared layered silicates 330-332
Metal complexes 826
Metal ions, associated to framework 911
Metal particles in zeolites 675, 679, 681
Metal-in-zeolite in bifunctional catalysis 759, 895
Metal(O) introduction 826
Methane H/D exchange 432
Methane in silicalite-1, calculated IR 442
Methanol

adsorption modes 430, 431
to gasoline 802-805
to olefins 805-807
Methylamine and dimethylamine, selective synthesis of 880
Methylamine 683, 880
Methylnaphthalenes 845
Methylviologen attached to $Ru(bpy)_3$ 975
MFI 5, 47, 141-144, 1036
MFS 106
Mica 302, 329
Microwave in synthesis 196
Milton, R.M. 11, 14, 16, 27, 31
MnAPO 231
Mobil-Badger process 849
Mobil Corp 5, 6, 26, 781, 809, 1008
Modelling 423
Modification 824
Molecular dynamics (MD) 434, 532
MON 162
Monosaccharide separation 831
Monte-Carlo method 452-458
Montmorillonite
in acylation 860
pillared 329
Montmorillonites 309, 318, 334, 336
Mordenite (MOR) 17, 48, 107, 110, 222, 265, 379, 399, 1037
Morphology, modification of 824
MPV reduction 890, 996
MTG process 802-805
MTO process 805-807, 821
MTW 91
Mutinaite 25
MWW 25, 49

N

Na-Y ^{17}O NMR 408
Na-Y ^{29}Si chemical shifts 395
NaY, industrial synthesis 697
Naphthalene methylation 852
α-Naphthylamine, rearrangement of 844
NAT 127, 128
Natural zeolites 3, 16, 175
applications 17
prices 17
Natrolite 180
NaX NMR 396
NES 96
Nickel on A catalysts
in hydrogenation 678
preparation 677
Nitration 862
Nitrile synthesis 895
Nitrogen adsorption 511
Nitrogen-containing ring syntheses 874
^{27}Al-MAS NMR 405-407
^{29}Si-MAS NMR 393-405, 515
^{27}Al NMR 405-407
^{1}H NMR 409
^{23}Na-NMR 411
^{31}P NMR 415
^{129}Xe NMR 358
NMR
cation sites 412
chemical shift anisotropy 389
2D techniques 391
dipolar coupling 388
important parameters 391
in situ studies of reactions 414
multiple-pulse sequences 389
non-framework Al 407
of coke 718-720
Si chemical shifts 394-400
suitable nuclei 393
templates 412
NMR, liquid state in zeolite synthesis 403
NMR of coke 718-720
NMR PFG technique in determining diffusivities 535
NMR, solid state
^{11}B 393, 409
^{13}C 412
cations 412
^{133}Ce 412
cross-polarization 390
dealuminated Y 401
experimental techniques 389
^{7}Li 393, 412
line broadening 388
magic angle spinning 389
nuclear spin interactions 388
oxygen-17 393, 407
principles 388
quadrupolar interaction 389
^{29}Si chemical shifts 394-400
siliceous zeolites 399
sodium-23 393, 411
Noble metal catalysts 678, 680
NON 95
Non-catalytic use 829
Non-linear optical effects 972
Nu-10 (TON) 877
Nucleation 198
Nucleophilic addition to epoxides 884
Nucleophilic substitution 879, 883
NW-3 349

O

Octane numbers 757, 766, 777, 780
Offretite
^{29}Si chemical shifts 395
^{29}Si MAS NMR spectrum 399
Offretite (OFF) 57, 103, 130, 395, 399
OH vibration frequencies 370, 371

Oil-conversion 752-756
Olefin
  isomerization 764
  oligomerization 809
Olefin/paraffin alkylation 76
Olefins, from methanol 805
Oligomerization mechanisms 661
Omega 206, 395, 399, 823
Omega, ^{29}Si chemical shifts 395
  ^{29}Si MAS NMR spectrum 399
Oppenauer oxidation 892
Optipur 183
Organic syntheses 821
Organophilicity 829
OSI 92
Outer surface
  selective dealumination 990
  silica coating of 990
  use of bulky poisons 990
Overgrowth, X onto A 208
Oxidation
  general 911
  of alcohols 924
  of alkanes 925, 942, 966, 967
  of dialkylamines 924
  of glycols 924
  overview 912
  Wacker-type 912
  with alkyl hydroperoxides 945, 966, 1019
  with ammonia and hydrogen peroxide 924
  with hydrogen peroxide 914
  with nitrous oxide 912, 941
  with oxygen 912
  with ship-in-a-bottle systems 966
Oxidative dehydrogenation 941
Oxidative pyridine synthesis 876
Oxygen adsorption isotherms 19
Oxygen containing ring synthesis 871
Oxygen-17 NMR 407
  of NaY 408

P

Paraffin hydroisomerization 759
Paragon process 796
Patents on zeolites 4
Pd and Pt introduction 826
Pd/Cu-zeolites
  in olefin oxidation 912
Pentasil unit 39
Peroxide decomposition 912
Phase change, upon ion-exchange 474
Phenol
  alkylation 853
  from benzene 943
  hydroxylation 920, 942
Phenyl acetaldehyde 839, 922
PHI 118
Photochemistry in zeolites 970-975, 833
Photochlorination 833
Photoelectron spectroscopy of coke 726
Photostabilization of dyes 971
Phthalocyanines in zeolites 958
Phyllosilicates 299
Physical catalysis 833
Picoline formation 844
PILCS characterization 326
Pillared clays 300, 317, 326, 332-337
  in catalysis 335
  in separation 336, 337
Pillaring with
  metal cluster cations 321
  metal oxides 321
  organics 317
  organometallic 320
  polyoxycation 320
  Zr complex 333
Pinacol cleavage 832
Pinacol rearrangement 838
α-Pinene oxide isomerization 841
β-Pinene isomerization 836
Platinum-BaKL 767
Platinum-HMOR 679
Platinum-HY 675, 681
Platinum-in-zeolite
  in aromatization 767
  in isomerization 759, 836
Poisoning external surface 827, 990
Polymerization
  ethene/propene 1020
  methyl methacrylate 1020
  over Zr-complex in MCM-41 1020
  styrene 1020
  vinyl acetate 1020
Polymers in zeolites 978
Polypyrrole in mordenite 978
Pore diameters
  and activity in cracking 1016
  and size of molecules 356, 992
Pore mouth adsorption 665
Pore mouth catalysis 663
Pore size engineering 683
Pore size flexibility 992
Pore sizes of zeolites 13
Pore volume
  of clays 309
  of pillared clays 327
  of zeolites 354
Post-synthesis modification 261
Powder diffraction pattern 58
Preparation of zeolite catalysts 673
  by impregnation 682
  degree of Na-removal 676, 700
  dipcoating 689
  ion exchange 675, 682, 699
  removal of template 674
  shaping 688

$SiCl_4$ treatment 683
Probe molecules 861
allyl 3,5-di-t-butylphenyl ether 996
criteria 374-376
1-(3,5-di-t-butylphenoxy)propan-2-one 997
2,6-di-t-butylpyridine 993
in IR spectroscopy 374
4-methylquinoline 993
quinoline 992
triarylphosphonium salts 1000
1,3,5-triisopropylbenzene 994, 1014
triphenylphosphine 1000
1,3,5-tri-t-butylbenzene 994, 1014
Processing, hydrocarbon 633
Product adsorption 858
Product selectivity 662, 989
Protocol for optimizing shape selectivity 991
Protonation of isobutene 433
Pseudoboehmite 183
Pt-containing zeolites 675, 679, 767, 836
Publications, number of 4
Publications regarding zeolites 4
Pyridine
adsorption on MOR 380
alkylation 855
as probe molecule 374
formation 875-877
Pyrrolidine formation 874
P-zeolite 16, 47, 63, 698

Q

Quantum chemistry
calculation of surface ethylation 639
of zeolite acidity 424
Quinoline as probe molecules 992

R

Radical reaction 864
Radionuclide separation 507
Raman spectroscopy 351
RCI 360
Re-usable reagent 832
RE-Y 263, 834, 836, 853
Reactant selectivity 666, 913, 989
Reactants 181
Reaction parameters, in synthesis 205
Rearrangement reactions of hydrocarbons 648
Rectorite, pillared 329
Redox properties 912
Refined Constraint Index 360
Refinery 752, 753
Regeneration 829
Regioselectivity 863, 873, 884
Residue catalysts 784
Residue hydroprocessing 797
Restricted growth selectivity 824
RHO 136, 137
Rietveld refinement 64, 349
Ring closure
coumarin synthesis 873
epoxidation/ring closure 872
oxygen containing rings 871
towards heterocyclic compounds 871
RON 757, 766, 777, 780
RSN 160
RT1, RT2, RT3 170
RTH 144-146
Ru catalyst complexes 826
Ru-H-USY 896

S

Saccharides separation 831, 832
SAPO 22
acidity 1016
MASNMR of 249
SAPO-5 22, 240, 242, 249, 823, 842
SAPO-11 229, 335, 379, 823, 843, 844
SAPO-17 415
SAPO-34 229, 240, 247, 248, 286
SAPO-37 237, 238, 248, 249, 823, 838
SAPO-40 238
SAPO-44 415
SAPO-LTA 238
Saponite 305, 312
SAT 133
Scavenger, zeolites as 833
Scission reactions in hydrocarbons 648
Secondary building units 69
Secondary synthesis methods 261
Selective scission 668
Selectivity types 666, 913, 989
Self-diffusion 552
of light paraffins 553
SEM 346
Separation by affinity 831
Separation factor in cation exchange 479, 480
Separations by size 831
Shape selective catalytic dewaxing 770
Shape selective cracking 765
Shape-selectivity 663, 664, 753, 989
Shell 758, 798, 809
Shifting equilibria 833, 834
Ship-in-a-bottle systems 958
alkali metal clusters 975
bipyridine complexes 967, 969
characterisation 963
charge transfer processes 972-975
covalent linking of ligands 965
entrapped tricyclohexylguanidine 960
examples of in situ complexation 961
FePcY catalyst 966
in PDMS membrane 967
in situ synthesis of dyes 960

interaction with light 970
Mn-complexes 967, 968
non-linear optical effects 972
Pd(salen)-NaY 969
phthalocyanines from 1,2-dicyanobenzene 959
polyamine ligands 968
polymers in zeolites 978
Schiff base complexes 967
semiconductor clusters 977
spectral hole burning 971
synthetic approaches 958
triphenylmethyl cation 960
zeolite as a ligand 962
Si, Al ordening in X and Y 398
Si MAS NMR 364, 365, 515
SI (Spaciousness Index) 361
^{29}Si-MAS NMR 393-405
Si/Al ranges 823
Si/Al ratio
and butane cracking 368
determination by NMR 364, 397
effect on properties 18
$SiCl_4$-treatment 273, 274, 279
Si-O-T angles 399
Si-sites nonequivalent 397
Si-source 182
Sigma-2 213
Silanation of outer surface 843
Silane, in tuning accessibility 827
Silanol nests 264, 285
Silica coating of outer surface 990
Silicalite 18, 211, 829
Silicalite-2 211
Silicate clusters 192-196
Siliceous mordenite, ^{29}Si NMR 399
Siliceous offretite, ^{29}Si NMR 399
Siliceous omega, ^{29}Si NMR 399
Siliceous zeolites 399
Siliceous ZSM-5 398
Silver-exchanged zeolites in banana ripening 511
Silver-zeolites in carbohydrate coupling 832
Simulation of alkane adsorption 459
Single file diffusion 531
historical sequence 534
Six-ring sheets 73, 124
Skeletal isomerization of olefins 763
Slow release 832
Smectites 301, 302, 307, 310
Smith, J.V. 31
Sodalite 21, 43, 130, 206, 207, 395
Sodalite cage 76
Sodalite unit 39, 76
Sodium clusters in zeolites 975
Sodium-23 NMR 411
Solid state NMR 364, 387
Solvent effects in epoxidation 923
Sorbex separation processes 831
Sorption entropy 357
Sorption isotherms 353
Sorption of benzene 375
Sources, SiO_2, AlO_2 182
Spaciousness index 361, 1014
SPGK process 809-811
SSZ-33 151-155
STA-1, -2, -6, -7 237
Stability 267, 1017
Stacked-bed configuration 795
Stacking faults 57
Starch to sorbitol process 895
Steam stabilization 268, 269, 279, 282
Stefan Maxwell equation 530
Stereoselectivity, in MPV reduction 891
Structure Commission 2, 3, 14
Structured catalysts 689
by dipcoating 689
by in situ growth 690
Styrene oxide 839
Subunits 39
Supported zeolites 209
Supramolecular assemblies 957
Surface area, of clays 308
Swelling, of clays 308
Synthesis 175
$AlPO_4$-molecular sieves 28
chemical sources 181-183
organic 821
variables 205
zeolites 28
Synthesis gas to gasoline 807

T

T atom variation 677
T-atoms and acidity 825
T-O-T angles 351, 1015
t-plot 355
Talc 302, 332
TAPO (TiAPO) 231
TEM 350
Temperature programmed desorption 378
Templates
degradation 197
in synthesis 29, 185
Templates in synthesis Ti-Beta 919
Templates, zeolites
$AlPO_4$ 236
^{13}C NMR of 412
Terminal chlorination 833
Tetrasubstituted benzenes, separation 833
Theta-1 (TON) 210, 877
Theory, general 423
Theory on zeolite lattice stability 1017
Thiazole synthesis 877
Thiele modulus 552, 566
Thioethers, synthesis 889
Thiols, synthesis 889

Thiophene
    acylation 862
    methylation 855
    selective adsorption 830
THO 127, 128
Ti site in TS-1
    Bronsted acid activity 931
    Clerici-mechanism 934
    epoxidation transition states 933
    EXAFS and XANES 928
    infrared 928
    Lewis acidity 932
    mechanism of oxygenations 932
    reaction with H_2O_2 931
    reaction with protic substrates 930
    reactivity 929
    structure 926
    UV-Vis 928
    XRD data 927
Ti-Beta 839, 841, 915, 918, 919, 927
Ti-MCM-41 918, 919, 1019
Ti-UTD-1 919
Ti-ZSM-12 918
Ti-ZSM-48 917, 927
TIP process 759
Titanium-containing silicalites 914
Titanosilicates
    discovery 6
    lattice parameters 445
    TAM-5 3
Toluene bromination 865
Toluidine isomerization 845
Toyota 6, 1008
TPD 378
Tracer techniques
    in diffusion 534
    in ion exchange 487
Transacetalization 889
Transesterification 888
Transition state shape selectivity 666, 824, 913, 989
Transmission electron microscopy 350
Trends 812
Trimethylamine 880
Trisubstituted benzenes, separation 831, 847
TS-1 839, 914
    conversion catalysed 914
    synthesis 915
    use of fluoride 916
TS-2 917, 927
Tschernichite 25
Two- and multistep syntheses 892

U

UCSB-6, -8, -10 237, 244
Ultrastable Y 267
    characteristics of 269
    composition of 270
    effect of Na-content 273
    in catalytic cracking 779
    presence of mesopores 277
    Si/Al ratios 271
Union Carbide Corp. 5, 11, 14, 20, 230
Unit cell contraction 271, 273
UOP 14, 758, 808, 809, 831, 844, 1007
USY 263
UTD-1 13, 24, 25, 959
UV-VIS spectroscopy of coke 723

V

Van 't Hoff equation 542
Vanadium-containing silicalites
    catalysis by 941
    characterization 937
    ESR spectra 938
    in propane to propene conversion 941
    leaching 942
    overview of recipies 936
    structure V-site 940
    synthesis 935
    V-MEL 942
    V-NCL-1 937
    ^{51}V NMR chemical shifts 939
    V-ZSM-48 937
Vanadium-containing zeolites 934-943
Vanillin process of Rhodia 868
VAPO-5
    oxidation catalyst 945
    synthesis 944
VAPOs 944
Vermiculite 302, 332
Void fraction and structure 1018
VPI-5 21, 22, 233
    framework 51
    pore size 21
    synthesis 235, 241
    transition to AlPO-8 1006
VPI-5 (VFI) 51, 125, 235, 241, 959, 1005, 1010, 1012, 1013
VPI-9 (VNI) 53
VSV 159

W

Water adsorption 19, 829
Websites IZA 31
WEI 163
Window diameters of LTA 446
Window insulation 16

X

X zeolite single crystals 221
X-ray absorption spectroscopy 381
X-ray diffraction 347
X-ray fluorescence 61, 362
X-ray photoelectron spectroscopy 363
X-ray powder diffraction 58-64
XANES 381
Xe chemical shifts
    of KL 358
    of NaY 358
    of ZSM-5 358
Xe NMR 358
XPS 363, 380
XRD 347
XRD, of clays 303
XRD Powder patterns, collection of 62
XRF 362
Xylenes separation 827

Y

Y as host in ship-in-a-bottle systems 958-970
Y-zeolites
    in acylation 857
    in alkylation 853, 855
    in halogenation 864
YUG 118

Z

ZAPO (ZnAPO) 231
Zeolite A
    as drying agent 830
    in detergent formulations 4, 30, 516
    in nuclear waste cleanup 27
    Na/Ca/Mg ternary exchange 505
Zeolite L 25, 180
Zeolite NaA
    synthesis 188, 211
Zeolite omega 395, 399, 823
Zeolite P 63
Zeolite preparation, general 177
Zeolite-based technologies, maturity 812
Zeolites
    acidity 7
    in nuclear waste cleanup 27
    of commercial interest 3
    three letter code 14, 38
Zeolitic membranes 209
Zero-length column method 547
Zigzag chains 79
Zinc phosphates 24, 251-253
ZK-4 395
ZK-5 394
ZSM-5 38, 176, 179, 189, 206, 207, 219, 281, 335, 366, 371, 372, 379, 398, 410, 567, 701, 738, 7706, 837, 843, 849, 1010
    as synthesized 188
    crystal shapes 199, 221-223
    growth process 200
    single crystals 220
    IR spectroscopy 352
ZSM-5 as additive 781
ZSM-6 206
ZSM-11 212, 219
ZSM-11 (MEL) 48
ZSM-12 219
ZSM-22 212, 220, 664
ZSM-23 210, 219, 772
ZSM-35 189, 218
ZSM-39 189, 218
ZSM-47 206
ZSM-48 212, 218

STUDIES IN SURFACE SCIENCE AND CATALYSIS

Advisory Editors:
B. Delmon, Université Catholique de Louvain, Louvain-la-Neuve, Belgium
J.T. Yates, University of Pittsburgh, Pittsburgh, PA, U.S.A.

Volume 1 **Preparation of Catalysts I.** Scientific Bases for the Preparation of Heterogeneous Catalysts. Proceedings of the First International Symposium, Brussels, October 14–17, 1975
edited by **B. Delmon, P.A. Jacobs and G. Poncelet**

Volume 2 **The Control of the Reactivity of Solids.** A Critical Survey of the Factors that Influence the Reactivity of Solids, with Special Emphasis on the Control of the Chemical Processes in Relation to Practical Applications
by **V.V. Boldyrev, M. Bulens and B. Delmon**

Volume 3 **Preparation of Catalysts II.** Scientific Bases for the Preparation of Heterogeneous Catalysts. Proceedings of the Second International Symposium, Louvain-la-Neuve, September 4–7, 1978
edited by **B. Delmon, P. Grange, P. Jacobs and G. Poncelet**

Volume 4 **Growth and Properties of Metal Clusters.** Applications to Catalysis and the Photographic Process. Proceedings of the 32nd International Meeting of the Société de Chimie Physique, Villeurbanne, September 24–28, 1979
edited by **J. Bourdon**

Volume 5 **Catalysis by Zeolites.** Proceedings of an International Symposium, Ecully (Lyon), September 9–11, 1980
edited by **B. Imelik, C. Naccache, Y. Ben Taarit, J.C. Vedrine, G. Coudurier and H. Praliaud**

Volume 6 **Catalyst Deactivation.** Proceedings of an International Symposium, Antwerp, October 13–15, 1980
edited by **B. Delmon and G.F. Froment**

Volume 7 **New Horizons in Catalysis.** Proceedings of the 7th International Congress on Catalysis, Tokyo, June 30–July 4, 1980. Parts A and B
edited by **T. Seiyama and K. Tanabe**

Volume 8 **Catalysis by Supported Complexes**
by **Yu.I. Yermakov, B.N. Kuznetsov and V.A. Zakharov**

Volume 9 **Physics of Solid Surfaces.** Proceedings of a Symposium, Bechyňe, September 29–October 3, 1980
edited by **M. Láznička**

Volume 10 **Adsorption at the Gas–Solid and Liquid–Solid Interface.** Proceedings of an International Symposium, Aix-en-Provence, September 21–23, 1981
edited by **J. Rouquerol and K.S.W. Sing**

Volume 11 **Metal-Support and Metal-Additive Effects in Catalysis.** Proceedings of an International Symposium, Ecully (Lyon), September 14–16, 1982
edited by **B. Imelik, C. Naccache, G. Coudurier, H. Praliaud, P. Meriaudeau, P. Gallezot, G.A. Martin and J.C. Vedrine**

Volume 12 **Metal Microstructures in Zeolites.** Preparation - Properties – Applications. Proceedings of a Workshop, Bremen, September 22–24, 1982
edited by **P.A. Jacobs, N.I. Jaeger, P. Jirů and G. Schulz-Ekloff**

Volume 13 **Adsorption on Metal Surfaces.** An Integrated Approach
edited by **J. Bénard**

Volume 14 **Vibrations at Surfaces.** Proceedings of the Third International Conference, Asilomar, CA, September 1–4, 1982
edited by **C.R. Brundle and H. Morawitz**

Volume 15 **Heterogeneous Catalytic Reactions Involving Molecular Oxygen**
by **G.I. Golodets**

Volume 16 **Preparation of Catalysts III.** Scientific Bases for the Preparation of Heterogeneous Catalysts. Proceedings of the Third International Symposium, Louvain-la-Neuve, September 6–9, 1982
edited by **G. Poncelet, P. Grange and P.A. Jacobs**

Volume 17 **Spillover of Adsorbed Species.** Proceedings of an International Symposium, Lyon-Villeurbanne, September 12–16, 1983
edited by **G.M. Pajonk, S.J. Teichner and J.E. Germain**

Volume 18 **Structure and Reactivity of Modified Zeolites.** Proceedings of an International Conference, Prague, July 9–13, 1984
edited by **P.A. Jacobs, N.I. Jaeger, P. Jirů, V.B. Kazansky and G. Schulz-Ekloff**

Volume 19 **Catalysis on the Energy Scene.** Proceedings of the 9th Canadian Symposium on Catalysis, Quebec, P.Q., September 30–October 3, 1984
edited by **S. Kaliaguine and A. Mahay**

Volume 20 **Catalysis by Acids and Bases.** Proceedings of an International Symposium, Villeurbanne (Lyon), September 25–27, 1984
edited by **B. Imelik, C. Naccache, G. Coudurier, Y. Ben Taarit and J.C. Vedrine**

Volume 21 **Adsorption and Catalysis on Oxide Surfaces.** Proceedings of a Symposium, Uxbridge, June 28–29, 1984
edited by **M. Che and G.C. Bond**

Volume 22 **Unsteady Processes in Catalytic Reactors**
by **Yu.Sh. Matros**

Volume 23 **Physics of Solid Surfaces 1984**
edited by **J. Koukal**

Volume 24 **Zeolites: Synthesis, Structure, Technology and Application.** Proceedings of an International Symposium, Portorož-Portorose, September 3–8, 1984
edited by **B. Držaj, S. Hočevar and S. Pejovnik**

Volume 25 **Catalytic Polymerization of Olefins.** Proceedings of the International Symposium on Future Aspects of Olefin Polymerization, Tokyo, July 4–6, 1985
edited by **T. Keii and K. Soga**

Volume 26 **Vibrations at Surfaces 1985.** Proceedings of the Fourth International Conference, Bowness-on-Windermere, September 15–19, 1985
edited by **D.A. King, N.V. Richardson and S. Holloway**

Volume 27 **Catalytic Hydrogenation**
edited by **L. Cervený**

Volume 28 **New Developments in Zeolite Science and Technology.** Proceedings of the 7th International Zeolite Conference, Tokyo, August 17–22, 1986
edited by **Y. Murakami, A. Iijima and J.W. Ward**

Volume 29 **Metal Clusters in Catalysis**
edited by **B.C. Gates, L. Guczi and H. Knözinger**

Volume 30 **Catalysis and Automotive Pollution Control.** Proceedings of the First International Symposium, Brussels, September 8–11, 1986
edited by **A. Crucq and A. Frennet**

Volume 31 **Preparation of Catalysts IV.** Scientific Bases for the Preparation of Heterogeneous Catalysts. Proceedings of the Fourth International Symposium, Louvain-la-Neuve, September 1–4, 1986
edited by **B. Delmon, P. Grange, P.A. Jacobs and G. Poncelet**

Volume 32 **Thin Metal Films and Gas Chemisorption**
edited by **P. Wissmann**

Volume 33 **Synthesis of High-silica Aluminosilicate Zeolites**
edited by **P.A. Jacobs and J.A. Martens**

Volume 34 **Catalyst Deactivation 1987.** Proceedings of the 4th International Symposium, Antwerp, September 29–October 1, 1987
edited by **B. Delmon and G.F. Froment**

Volume 35 **Keynotes in Energy-Related Catalysis**
edited by **S. Kaliaguine**

Volume 36 **Methane Conversion.** Proceedings of a Symposium on the Production of Fuels and Chemicals from Natural Gas, Auckland, April 27–30, 1987
edited by **D.M. Bibby, C.D. Chang, R.F. Howe and S. Yurchak**

Volume 37 **Innovation in Zeolite Materials Science.** Proceedings of an International Symposium, Nieuwpoort, September 13–17, 1987
edited by **P.J. Grobet, W.J. Mortier, E.F. Vansant and G. Schulz-Ekloff**

Volume 38 **Catalysis 1987.** Proceedings of the 10th North American Meeting of the Catalysis Society, San Diego, CA, May 17–22, 1987
edited by **J.W. Ward**

Volume 39 **Characterization of Porous Solids.** Proceedings of the IUPAC Symposium (COPS I), Bad Soden a.Ts., April 26–29, 1987
edited by **K.K. Unger, J. Rouquerol, K.S.W. Sing and H. Kral**

Volume 40 **Physics of Solid Surfaces 1987.** Proceedings of the Fourth Symposium on Surface Physics, Bechyne Castle, September 7–11, 1987
edited by **J. Koukal**

Volume 41 **Heterogeneous Catalysis and Fine Chemicals.** Proceedings of an International Symposium, Poitiers, March 15–17, 1988
edited by **M. Guisnet, J. Barrault, C. Bouchoule, D. Duprez, C. Montassier and G. Pérot**

Volume 42 **Laboratory Studies of Heterogeneous Catalytic Processes**
by **E.G. Christoffel,** revised and edited by **Z. Paál**

Volume 43 **Catalytic Processes under Unsteady-State Conditions**
by **Yu. Sh. Matros**

Volume 44 **Successful Design of Catalysts.** Future Requirements and Development. Proceedings of the Worldwide Catalysis Seminars, July, 1988, on the Occasion of the 30th Anniversary of the Catalysis Society of Japan
edited by **T. Inui**

Volume 45 **Transition Metal Oxides.** Surface Chemistry and Catalysis
by **H.H. Kung**

Volume 46 **Zeolites as Catalysts, Sorbents and Detergent Builders.** Applications and Innovations. Proceedings of an International Symposium, Würzburg, September 4–8, 1988
edited by **H.G. Karge and J. Weitkamp**

Volume 47 **Photochemistry on Solid Surfaces**
edited by **M. Anpo and T. Matsuura**

Volume 48 **Structure and Reactivity of Surfaces.** Proceedings of a European Conference, Trieste, September 13–16, 1988
edited by **C. Morterra, A. Zecchina and G. Costa**

Volume 49 **Zeolites: Facts, Figures, Future.** Proceedings of the 8th International Zeolite Conference, Amsterdam, July 10–14, 1989. Parts A and B
edited by **P.A. Jacobs and R.A. van Santen**

Volume 50 **Hydrotreating Catalysts.** Preparation, Characterization and Performance. Proceedings of the Annual International AIChE Meeting, Washington, DC, November 27–December 2, 1988
edited by **M.L. Occelli and R.G. Anthony**

Volume 51 **New Solid Acids and Bases.** Their Catalytic Properties
by **K. Tanabe, M. Misono, Y. Ono and H. Hattori**

Volume 52 **Recent Advances in Zeolite Science.** Proceedings of the 1989 Meeting of the British Zeolite Association, Cambridge, April 17–19, 1989
edited by **J. Klinowsky and P.J. Barrie**

Volume 53 **Catalyst in Petroleum Refining 1989.** Proceedings of the First International Conference on Catalysts in Petroleum Refining, Kuwait, March 5–8, 1989
edited by **D.L. Trimm, S. Akashah, M. Absi-Halabi and A. Bishara**

Volume 54 **Future Opportunities in Catalytic and Separation Technology**
edited by **M. Misono, Y. Moro-oka and S. Kimura**

Volume 55	**New Developments in Selective Oxidation.** Proceedings of an International Symposium, Rimini, Italy, September 18–22, 1989 edited by **G. Centi and F. Trifiro**
Volume 56	**Olefin Polymerization Catalysts.** Proceedings of the International Symposium on Recent Developments in Olefin Polymerization Catalysts, Tokyo, October 23–25, 1989 edited by **T. Keii and K. Soga**
Volume 57A	**Spectroscopic Analysis of Heterogeneous Catalysts. Part A: Methods of Surface Analysis** edited by **J.L.G. Fierro**
Volume 57B	**Spectroscopic Analysis of Heterogeneous Catalysts. Part B: Chemisorption of Probe Molecules** edited by **J.L.G. Fierro**
Volume 58	**Introduction to Zeolite Science and Practice** edited by **H. van Bekkum, E.M. Flanigen and J.C. Jansen**
Volume 59	**Heterogeneous Catalysis and Fine Chemicals II.** Proceedings of the 2nd International Symposium, Poitiers, October 2–6, 1990 edited by **M. Guisnet, J. Barrault, C. Bouchoule, D. Duprez, G. Pérot, R. Maurel and C. Montassier**
Volume 60	**Chemistry of Microporous Crystals.** Proceedings of the International Symposium on Chemistry of Microporous Crystals, Tokyo, June 26–29, 1990 edited by **T. Inui, S. Namba and T. Tatsumi**
Volume 61	**Natural Gas Conversion.** Proceedings of the Symposium on Natural Gas Conversion, Oslo, August 12–17, 1990 edited by **A. Holmen, K.-J. Jens and S. Kolboe**
Volume 62	**Characterization of Porous Solids II.** Proceedings of the IUPAC Symposium (COPS II), Alicante, May 6–9, 1990 edited by **F. Rodríguez-Reinoso, J. Rouquerol, K.S.W. Sing and K.K. Unger**
Volume 63	**Preparation of Catalysts V.** Scientific Bases for the Preparation of Heterogeneous Catalysts. Proceedings of the Fifth International Symposium, Louvain-la-Neuve, September 3–6, 1990 edited by **G. Poncelet, P.A. Jacobs, P. Grange and B. Delmon**
Volume 64	**New Trends in CO Activation** edited by **L. Guczi**
Volume 65	**Catalysis and Adsorption by Zeolites.** Proceedings of ZEOCAT 90, Leipzig, August 20–23, 1990 edited by G. **Öhlmann, H. Pfeifer and R. Fricke**
Volume 66	**Dioxygen Activation and Homogeneous Catalytic Oxidation.** Proceedings of the Fourth International Symposium on Dioxygen Activation and Homogeneous Catalytic Oxidation, Balatonfüred, September 10–14, 1990 edited by **L.I. Simándi**
Volume 67	**Structure-Activity and Selectivity Relationships in Heterogeneous Catalysis.** Proceedings of the ACS Symposium on Structure-Activity Relationships in Heterogeneous Catalysis, Boston, MA, April 22–27, 1990 edited by **R.K. Grasselli and A.W. Sleight**
Volume 68	**Catalyst Deactivation 1991.** Proceedings of the Fifth International Symposium, Evanston, IL, June 24–26, 1991 edited by **C.H. Bartholomew and J.B. Butt**
Volume 69	**Zeolite Chemistry and Catalysis.** Proceedings of an International Symposium, Prague, Czechoslovakia, September 8–13, 1991 edited by **P.A. Jacobs, N.I. Jaeger, L. Kubelková and B. Wichterlová**
Volume 70	**Poisoning and Promotion in Catalysis based on Surface Science Concepts and Experiments** by **M. Kiskinova**

Volume 71 **Catalysis and Automotive Pollution Control II.** Proceedings of the 2nd International Symposium (CAPoC 2), Brussels, Belgium, September 10–13, 1990
edited by **A. Crucq**

Volume 72 **New Developments in Selective Oxidation by Heterogeneous Catalysis.** Proceedings of the 3rd European Workshop Meeting on New Developments in Selective Oxidation by Heterogeneous Catalysis, Louvain-la-Neuve, Belgium, April 8–10, 1991
edited by **P. Ruiz and B. Delmon**

Volume 73 **Progress in Catalysis.** Proceedings of the 12th Canadian Symposium on Catalysis, Banff, Alberta, Canada, May 25–28, 1992
edited by **K.J. Smith and E.C. Sanford**

Volume 74 **Angle-Resolved Photoemission. Theory and Current Applications**
edited by **S.D. Kevan**

Volume 75 **New Frontiers in Catalysis, Parts A-C.** Proceedings of the 10th International Congress on Catalysis, Budapest, Hungary, 19–24 July, 1992
edited by **L. Guczi, F. Solymosi and P. Tétényi**

Volume 76 **Fluid Catalytic Cracking: Science and Technology**
edited by **J.S. Magee and M.M. Mitchell, Jr.**

Volume 77 **New Aspects of Spillover Effect in Catalysis. For Development of Highly Active Catalysts.** Proceedings of the Third International Conference on Spillover, Kyoto, Japan, August 17–20, 1993
edited by **T. Inui, K. Fujimoto, T. Uchijima and M. Masai**

Volume 78 **Heterogeneous Catalysis and Fine Chemicals III.**
Proceedings of the 3rd International Symposium, Poitiers, April 5 - 8, 1993
edited by **M. Guisnet, J. Barbier, J. Barrault, C. Bouchoule, D. Duprez, G. Pérot and C. Montassier**

Volume 79 **Catalysis: An Integrated Approach to Homogeneous, Heterogeneous and Industrial Catalysis**
edited by **J.A. Moulijn, P.W.N.M. van Leeuwen and R.A. van Santen**

Volume 80 **Fundamentals of Adsorption.** Proceedings of the Fourth International Conference on Fundamentals of Adsorption, Kyoto, Japan, May 17–22, 1992
edited by **M. Suzuki**

Volume 81 **Natural Gas Conversion II.** Proceedings of the Third Natural Gas Conversion Symposium, Sydney, July 4–9, 1993
edited by **H.E. Curry-Hyde and R.F. Howe**

Volume 82 **New Developments in Selective Oxidation II.** Proceedings of the Second World Congress and Fourth European Workshop Meeting, Benalmádena, Spain, September 20–24, 1993
edited by **V. Cortés Corberán and S. Vic Bellón**

Volume 83 **Zeolites and Microporous Crystals.** Proceedings of the International Symposium on Zeolites and Microporous Crystals, Nagoya, Japan, August 22–25, 1993
edited by **T. Hattori and T. Yashima**

Volume 84 **Zeolites and Related Microporous Materials: State of the Art 1994.**
Proceedings of the 10th International Zeolite Conference, Garmisch-Partenkirchen, Germany, July 17–22, 1994
edited by **J. Weitkamp, H.G. Karge, H. Pfeifer and W. Hölderich**

Volume 85 **Advanced Zeolite Science and Applications**
edited by **J.C. Jansen, M. Stöcker, H.G. Karge and J. Weitkamp**

Volume 86 **Oscillating Heterogeneous Catalytic Systems**
by **M.M. Slin'ko and N.I. Jaeger**

Volume 87 **Characterization of Porous Solids III.** Proceedings of the IUPAC Symposium (COPS III), Marseille, France, May 9–12, 1993
edited by **J. Rouquerol, F. Rodriguez-Reinoso, K.S.W. Sing and K.K. Unger**

Volume 88 **Catalyst Deactivation 1994.** Proceedings of the 6th International Symposium, Ostend, Belgium, October 3–5, 1994
edited by **B. Delmon and G.F. Froment**

Volume 89 **Catalyst Design for Tailor-made Polyolefins.** Proceedings of the International Symposium on Catalyst Design for Tailor-made Polyolefins, Kanazawa, Japan, March 10–12, 1994
edited by **K. Soga and M. Terano**

Volume 90 **Acid-Base Catalysis II.** Proceedings of the International Symposium on Acid-Base Catalysis II, Sapporo, Japan, December 2–4, 1993
edited by **H. Hattori, M. Misono and Y. Ono**

Volume 91 **Preparation of Catalysts VI.** Scientific Bases for the Preparation of Heterogeneous Catalysts. Proceedings of the Sixth International Symposium, Louvain-La-Neuve, September 5–8, 1994
edited by **G. Poncelet, J. Martens, B. Delmon, P.A. Jacobs and P. Grange**

Volume 92 **Science and Technology in Catalysis 1994.** Proceedings of the Second Tokyo Conference on Advanced Catalytic Science and Technology, Tokyo, August 21–26, 1994
edited by **Y. Izumi, H. Arai and M. Iwamoto**

Volume 93 **Characterization and Chemical Modification of the Silica Surface**
by **E.F. Vansant, P. Van Der Voort and K.C. Vrancken**

Volume 94 **Catalysis by Microporous Materials.** Proceedings of ZEOCAT'95, Szombathely, Hungary, July 9–13, 1995
edited by **H.K. Beyer, H.G. Karge, I. Kiricsi and J.B. Nagy**

Volume 95 **Catalysis by Metals and Alloys**
by **V. Ponec and G.C. Bond**

Volume 96 **Catalysis and Automotive Pollution Control III.** Proceedings of the Third International Symposium (CAPoC3), Brussels, Belgium, April 20–22, 1994
edited by **A. Frennet and J.-M. Bastin**

Volume 97 **Zeolites: A Refined Tool for Designing Catalytic Sites.** Proceedings of the International Symposium, Québec, Canada, October 15–20, 1995
edited by **L. Bonneviot and S. Kaliaguine**

Volume 98 **Zeolite Science 1994: Recent Progress and Discussions.** Supplementary Materials to the 10th International Zeolite Conference, Garmisch-Partenkirchen, Germany, July 17–22, 1994
edited by **H.G. Karge and J. Weitkamp**

Volume 99 **Adsorption on New and Modified Inorganic Sorbents**
edited by **A. Dąbrowski and V.A. Tertykh**

Volume 100 **Catalysts in Petroleum Refining and Petrochemical Industries 1995.** Proceedings of the 2nd International Conference on Catalysts in Petroleum Refining and Petrochemical Industries, Kuwait, April 22-26, 1995
edited by **M. Absi-Halabi, J. Beshara, H. Qabazard and A. Stanislaus**

Volume 101 **11th International Congress on Catalysis - 40th Anniversary.** Proceedings of the 11th ICC, Baltimore, MD, USA, June 30-July 5, 1996
edited by **J.W. Hightower, W.N. Delgass, E. Iglesia and A.T. Bell**

Volume 102 **Recent Advances and New Horizons in Zeolite Science and Technology**
edited by **H. Chon, S.I. Woo and S. -E. Park**

Volume 103 **Semiconductor Nanoclusters - Physical, Chemical, and Catalytic Aspects**
edited by **P.V. Kamat and D. Meisel**

Volume 104 **Equilibria and Dynamics of Gas Adsorption on Heterogeneous Solid Surfaces**
edited by **W. Rudziński, W.A. Steele and G. Zgrablich**

Volume 105 **Progress in Zeolite and Microporous Materials**
Proceedings of the 11th International Zeolite Conference, Seoul, Korea, August 12-17, 1996
edited by **H. Chon, S.-K. Ihm and Y.S. Uh**

Volume 106 **Hydrotreatment and Hydrocracking of Oil Fractions**
Proceedings of the 1st International Symposium / 6th European Workshop, Oostende, Belgium, February 17-19, 1997
edited by **G.F. Froment, B. Delmon and P. Grange**

Volume 107 **Natural Gas Conversion IV**
Proceedings of the 4th International Natural Gas Conversion Symposium, Kruger Park, South Africa, November 19-23, 1995
edited by **M. de Pontes, R.L. Espinoza, C.P. Nicolaides, J.H. Scholtz and M.S. Scurrell**

Volume 108 **Heterogeneous Catalysis and Fine Chemicals IV**
Proceedings of the 4th International Symposium on Heterogeneous Catalysis and Fine Chemicals, Basel, Switzerland, September 8-12, 1996
edited by **H.U. Blaser, A. Baiker and R. Prins**

Volume 109 **Dynamics of Surfaces and Reaction Kinetics in Heterogeneous Catalysis.**
Proceedings of the International Symposium, Antwerp, Belgium, September 15-17, 1997
edited by **G.F. Froment and K.C. Waugh**

Volume 110 **Third World Congress on Oxidation Catalysis.**
Proceedings of the Third World Congress on Oxidation Catalysis, San Diego, CA, U.S.A., 21-26 September 1997
edited by **R.K. Grasselli, S.T. Oyama, A.M. Gaffney and J.E. Lyons**

Volume 111 **Catalyst Deactivation 1997.**
Proceedings of the 7th International Symposium, Cancun, Mexico, October 5-8, 1997
edited by **C.H. Bartholomew and G.A. Fuentes**

Volume 112 **Spillover and Migration of Surface Species on Catalysts.**
Proceedings of the 4th International Conference on Spillover, Dalian, China, September 15-18, 1997
edited by **Can Li and Qin Xin**

Volume 113 **Recent Advances in Basic and Applied Aspects of Industrial Catalysis.**
Proceedings of the 13th National Symposium and Silver Jubilee Symposium of Catalysis of India, Dehradun, India, April 2-4, 1997
edited by **T.S.R. Prasada Rao and G. Murali Dhar**

Volume 114 **Advances in Chemical Conversions for Mitigating Carbon Dioxide.**
Proceedings of the 4th International Conference on Carbon Dioxide Utilization, Kyoto, Japan, September 7-11, 1997
edited by **T. Inui, M. Anpo, K. Izui, S. Yanagida and T. Yamaguchi**

Volume 115 **Methods for Monitoring and Diagnosing the Efficiency of Catalytic Converters.**
A patent-oriented survey
by **M. Sideris**

Volume 116 **Catalysis and Automotive Pollution Control IV.**
Proceedings of the 4th International Symposium (CAPoC4), Brussels, Belgium, April 9-11, 1997
edited by **N. Kruse, A. Frennet and J.-M. Bastin**

Volume 117 **Mesoporous Molecular Sieves 1998**
Proceedings of the 1st International Symposium, Baltimore, MD, U.S.A., July 10-12, 1998
edited by **L. Bonneviot, F. Béland, C. Danumah, S. Giasson and S. Kaliaguine**

Volume 118 **Preparation of Catalysts VII**
Proceedings of the 7th International Symposium on Scientific Bases for the Preparation of Heterogeneous Catalysts, Louvain-la-Neuve, Belgium, September 1-4, 1998
edited by **B. Delmon, P.A. Jacobs, R. Maggi, J.A. Martens, P. Grange and G. Poncelet**

Volume 119 **Natural Gas Conversion V**
Proceedings of the 5th International Gas Conversion Symposium, Giardini-Naxos, Taormina, Italy, September 20-25, 1998
edited by **A. Parmaliana, D. Sanfilippo, F. Frusteri, A. Vaccari and F. Arena**

Volume 120A **Adsorption and its Applications in Industry and Environmental Protection.**
Vol I: Applications in Industry
edited by **A. Dąbrowski**

Volume 120B **Adsorption and its Applications in Industry and Environmental Protection.**
Vol II: Applications in Environmental Protection
edited by **A. Dąbrowski**

Volume 121 **Science and Technology in Catalysis 1998**
Proceedings of the Third Tokyo Conference in Advanced Catalytic Science and Technology, Tokyo, July 19-24, 1998
edited by **H. Hattori and K. Otsuka**

Volume 122 **Reaction Kinetics and the Development of Catalytic Processes**
Proceedings of the International Symposium, Brugge, Belgium, April 19-21, 1999
edited by **G.F. Froment and K.C. Waugh**

Volume 123 **Catalysis: An Integrated Approach**
Second, Revised and Enlarged Edition
edited by **R.A. van Santen, P.W.N.M. van Leeuwen, J.A. Moulijn and B.A. Averill**

Volume 124 **Experiments in Catalytic Reaction Engineering**
by **J.M. Berty**

Volume 125 **Porous Materials in Environmentally Friendly Processes**
Proceedings of the 1st International FEZA Conference, Eger, Hungary, September 1-4, 1999
edited by **I. Kiricsi, G. Pál-Borbély, J.B. Nagy and H.G. Karge**

Volume 126 **Catalyst Deactivation 1999**
Proceedings of the 8th International Symposium, Brugge, Belgium, October 10-13, 1999
edited by **B. Delmon and G.F. Froment**

Volume 127 **Hydrotreatment and Hydrocracking of Oil Fractions**
Proceedings of the 2nd International Symposium/7th European Workshop, Antwerpen, Belgium, November 14-17, 1999
edited by **B. Delmon, G.F. Froment and P. Grange**

Volume 128 **Characterisation of Porous Solids V**
Proceedings of the 5th International Symposium on the Characterisation of Porous Solids (COPS-V), Heidelberg, Germany, May 30- June 2, 1999
edited by **K.K. Unger, G. Kreysa and J.P. Baselt**

Volume 129 **Nanoporous Materials II**
Proceedings of the 2nd Conference on Access in Nanoporous Materials, Banff, Alberta, Canada, May 25-30, 2000
edited by **A. Sayari, M. Jaroniec and T.J. Pinnavaia**

Volume 130 **12th International Congress on Catalysis**
Proceedings of the 12th ICC, Granada, Spain, July 9-14, 2000
edited by **A. Corma, F.V. Melo, S. Mendioroz and J.L.G. Fierro**

Volume 131 **Catalytic Polymerization of Cycloolefins**
Ionic, Ziegler-Natta and Ring-Opening Metathesis Polymerization
by **V. Dragutan and R. Streck**

Volume 132 **Proceedings of the International Conference on Colloid and Surface Science, Tokyo, Japan, November 5-8, 2000**
25th Anniversary of the Division of Colloid and Surface Chemistry, The Chemical Society of Japan
edited by **Y. Iwasawa, N. Oyama and H. Kunieda**

Volume 133 **Reaction Kinetics and the Development and Operation of Catalytic Processes**
Proceedings of the 3rd International Symposium, Oostende, Belgium, April 22-25, 2001
edited by **G.F. Froment and K.C. Waugh**

Volume 134 **Fluid Catalytic Cracking V**
Materials and Technological Innovations
edited by **M.L. Occelli and P. O'Connor**

Volume 135 **Zeolites and Mesoporous Materials at the Dawn of the 21st Century.**
Proceedings of the 13th International Zeolite Conference, Montpellier, France, 8-13 July 2001
edited by **A. Galarneau, F. di Renzo, F. Fajula and J. Vedrine**

Volume 136 **Natural Gas Conversion VI.**
Proceedings of the 6th Natural Gas Conversion Symposium, June 17-22, 2001, Alaska, USA.
edited by **E. Iglesia, J.J. Spivey and T.H. Fleisch**

Volume 137 **Introduction to Zeolite Science and Practice**
2nd Completely Revised and Expanded Edition
edited by **H. van Bekkum, E.M. Flanigen, P.A. Jacobs and J.C. Jansen**